The Collected Papers of
Raymond D. Mindlin
Volume I

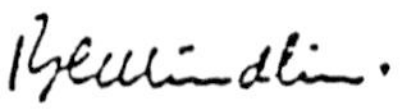

H. Deresiewicz M.P. Bieniek F.L. DiMaggio
Editors

The Collected Papers of Raymond D. Mindlin Volume I

The Late James Kip Finch Professor Emeritus
of Applied Science, Columbia University

Springer-Verlag
New York Berlin Heidelberg
London Paris Tokyo

H. Deresiewicz
Department of Mechanical Engineering
M.P. Bieniek
F.L. DiMaggio
Department of Civil Engineering
and Engineering Mechanics
Columbia University
New York, NY 10027
USA

Library of Congress Cataloging-in-Publication Data
Mindlin, Raymond D. (Raymond David), 1906–1987.
[Works. 1989]
The collected papers of Raymond D. Mindlin.
p. cm.
Bibliography: p.
1. Mechanics, Applied—Bibliography. 2. Mindlin, Raymond D.
(Raymond David), 1906–1987—Bibliography. I. Title.
Z7144.M4M56 1989
[TA350]
016.6201—dc19 88-38194

Printed on acid-free paper

Softcover reprint of the hardcover 1st edition 1989

9 8 7 6 5 4 3 2 1

ISBN-13: 978-1-4613-8867-8 e-ISBN-13: 978-1-4613-8865-4
DOI: 10.1007/978-1-4613-8865-4

Preface

The present volume makes available in one convenient place all the research papers published by Raymond D. Mindlin in the course of a half-century of professional activity. Included, in addition, is the full text of three of his extended review articles, because these remain timely as well as historically important.

Also to be found here is a brief sketch, written by one of us, his student and co-author, colleague and friend. It contains Mindlin's basic biographical data, although it dwells primarily on an analysis of his work and a chronicling of his accomplishments.

The papers reprinted herein are reproduced photographically as they appeared originally. We have excluded several short review articles; moreover, to our regret, because of constraints of space we have been unable to print three extended reviews. These are: "The photoelastic method of stress analysis," published in volume 10 of the *Journal of Applied Physics*; a short monograph, "Dynamics of package cushioning," which may be found in volume 24 of the *Bell System Technical Journal*; and the chapter entitled "Analogies," written for the *Handbook of Experimental Stress Analysis.*

In compiling the list of Publications (pp. xxi–xxviii) we have omitted abstracts and brief discussions; we have, however, included several early comprehensive discussions, deeming them, in substance and in length, each the equal of an independent paper. We offer this list in the firm belief that it represents the definitive bibliography of Mindlin's oeuvre, accurate and complete. The entries are arranged in the order of their appearance in print, rather than the order of submission (in many instances unavailable) or of composition (known only with unacceptable uncertainty). Happily, the sequences are known to differ in but a few particulars.

The reader seeking a thematic approach to the wealth of material presented here will find it useful to consult the topical arrangement given below. (The numbers in brackets refer to the corresponding items in the list of Publications.)

I. Photoelasticity and experimental mechanics
[1], [8], [10], [11], [16], [17], [18], [19], [20], [24], [26], [27], [31], [36], [113]

II. Classical elastostatics
Two-dimensional: [9], [12], [13], [14], [22], [24], [29]
Three-dimensional: [3], [4], [5], [32], [33], [42], [110], [112]

III. Vibrations and waves
[21], [25], [30], [49], [54], [94], [115], [128]
IV. Frictional contact and granular media
[28], [38], [41], [48], [56]
V. Rods, bars, beams, cylinders
[37], [60], [61], [62], [72], [114], [117], [118]
VI. Isotropic plates
[34], [46], [47], [51], [52], [55], [58], [59], [63], [64], [76], [123], [129]
VII. Crystal plates
[35], [43], [45], [50], [53], [54], [57], [65], [66], [69], [70], [71], [73], [77], [81], [84], [85], [86], [89], [90], [98], [119], [121], [122], [124], [125], [127]
VIII. Piezoelectric plates
[39], [44], [67], [68], [100], [103], [105], [107], [108], [109], [126]
IX. Generalized continua
[74], [75], [78], [79], [80], [82], [83], [87], [106]
X. Elastic dielectrics
[88], [95], [97], [99], [101], [102], [106], [111], [120]
XI. Lattice theory
[91], [92], [93], [95], [96], [105]
XII. Structural and geotechnical mechanics
[2], [6], [7], [15]
XIII. Other
[40]
XIV. Review articles
[16], [20], [21], [23], [31], [36], [48], [64], [97], [105], [116], [121]

Ray Mindlin's place in the annals of mechanics is made secure by his scientific output. But it seems to us that to omit mention of his influence as teacher, fount of inspiration, and example to emulate would leave a grossly incomplete picture of the man and the effect he has wrought on the field in which he labored with so much dedication and such signal distinction. He nurtured some thirty students to the doctoral degree, giving unstintingly of his time and energies, devoting himself to the task of showing the way. Many of those he trained occupy prestigious positions in universities and in industrial laboratories; not a few have gone on to direct their own research students and these, in turn, have produced others, in an ever more ramified "genealogy."

We are confident we speak for all his students when we say that they have always viewed him with warmth and genuine affection and, above all, taken great pride in acknowledging themselves his disciples.

H. Deresiewicz
M. P. Bieniek
F. L. DiMaggio

Volume I

Contents*

* Numbers in brackets refer to the Publications of Raymond D. Mindlin (see pages xxi–xxviii).

60th Birthday Celebration (Minneapolis, Minnesota, 1966)

Seated, l. to r.: L. E. Goodman, Mrs. Goodman, D. C. Drucker, Mrs. Mindlin, Mindlin, M. Hetényi, G. Herrmann, Mrs. Lubkin. *Standing, in the same order*: H. Deresiewicz, A. G. Lubowe, D. H. Cheng, J. J. McCoy, H. D. McNiven, M. A. Medick, H. F. Tiersten, J. Tasi, D. C. Gazis, Y.-H. Pao, L. Rongved, E. Sternberg, T. R. Kane, E. A. Fox, P. C. Y. Lee, J. Duffy, and J. L. Lubkin.

A Bio/Bibliographical Sketch

Raymond David Mindlin was born on September 17, 1906, in New York City, the second of three sons of Henry, a prosperous businessman, and Beatrice (née Levy). There are clear indications that the family, which subsequently came also to include a half-sister, was supportive and closely knit, and that Ray, as he came to be known to his friends and colleagues, enjoyed a tranquil childhood.

For his secondary education the young man was sent to the highly regarded Ethical Culture School, of which he had fond recollections in later years. Upon graduating in 1924 he enrolled at Columbia, beginning an association with the University that was to last for more than half a century. Four years later (during which time he distinguished himself as a sprinter on the varsity track team) he received the first of his four earned degrees, a B. A., followed by a B. S. in 1931, and in 1932 by a C. E. and the Illig medal for "proficiency in scholarship."

The economic depression then gripping the land was a major reason why he remained for graduate work, funded by the modest stipend that accompanied the research assistantship to which he was appointed. Love's *Treatise on the Mathematical Theory on Elasticity*, the fourth and last edition of which had appeared not long before, came to his hand, and from then on the direction of his professional work was firmly set.

Applied mechanics was not strongly represented on Columbia's faculty; in fact, there was little substantive research in it in the United States, and essentially no graduate instruction, until the advent on the scene of Stephen Timoshenko, exiled by the Bolshevik revolution in Russia. Some time after his appointment as professor at the University of Michigan in 1927, he organized a series of summer courses, in which each year one or another of the foremost names in this field, including L. Prandtl, R. V. Southwell, and H. M. Westergaard, participated. Mindlin attended in 1933, '34, and '35, and there is no doubt that the experience at Ann Arbor served to confirm him in his choice of his life's work.

His initial publication [1]* dates from the middle of this period, a solo, full-length paper describing a new type of polariscope for photoelastic analysis. April of 1935 saw the appearance of a lengthy discussion, written

* Numbers in brackets refer to corresponding entries in the list of publications (pages xxi–xxviii).

with Westergaard, of a paper on torsion of structural beams. During the year, Mindlin spent all available time in his basement office struggling to master Love's grand opus, a massive, difficult, demanding book, from time to time having to contend with complaints, from a faculty lacking in theoretical training and outlook, that the young man was wasting time. It was James Kip Finch, his Department chairman (and in later years a noted historian of engineering as well as Dean of the Engineering School), who served both as his protector and source of encouragement by issuing an edict to his colleagues, "Leave him alone, maybe something will come of it." Mindlin, in his turn, never forgot Finch's kindness and support, and it was at his persistent urging that the University, 35 years later, created the Finch chair (with Mindlin its first occupant).

For his doctoral research Mindlin set himself a fundamental problem in theoretical elasticity: determining the stresses in an elastic half-space subjected to a sub-surface point load. Working without any guidance at Columbia, he succeeded in finding the solution by employing the method of nuclei of strain. The results, nowadays referred to as "Mindlin's problem," represent a generalization of the two classical 19th century solutions respectively associated with the names of Kelvin and Boussinesq, and have become the basis for analytical formulations widely employed in geotechnical engineering. At Timoshenko's urging, in order to establish the author's priority of discovery, a summary of the results was dispatched to the *Comptes Rendus* of the *Académie des Sciences* in Paris, noted for almost instantaneous publication, where it appeared [3] in September of 1935. The full paper [4] was published in *Physics* (now the *Journal of Applied Physics*) in 1936, the year Mindlin received the Ph.D. degree. It was also the year in which he published his one and only paper to appear in a periodical devoted to pure mathematics, in essence a completeness proof of the Papkovich representation in classical elasticity [5]. It is interesting to note that nearly twenty years later he returned to the problem bearing his name and offered a solution by means of a systematic procedure grounded in potential theory [42].

Academic advancement, it appears, proceeded at a rather leisurely tempo in those days. Mindlin remained an assistant for another two years, at which point he was elevated to instructor in civil engineering, and only in 1940 did he receive promotion to assistant professor. His research in the six years following receipt of the doctorate was a mix of analytical and experimental endeavors, with emphasis on the photoelastic method of stress analysis. It included a work on three-dimensional photoelasticity [17], published jointly with D. C. Drucker, his first doctoral student, in which the method of oblique incidence was first suggested. An important part of current laboratory technique is based on this and a companion and two subsequent papers [18], [26], [27].

In 1942 Mindlin was co-opted by the Applied Physics Laboratory in Silver Spring (a Maryland suburb of the nation's capital), an institution engaged in naval ordnance work. There he played a significant role in the development of the proximity fuse, one of the major achievements in the scientific war effort. For his part in its success he was presented, shortly after the end of the war, with the Presidential Medal for Merit, the highest decoration awarded to civilians.

The swiftness of his climb up the academic ladder now made up for the earlier glacial pace: he came back to Columbia in 1945 as an associate professor, and two years later attained the rank of professor.

The year of his return to academia also marked the publication, in the *Bell System Technical Journal*, of a small monograph on the dynamics of package cushioning [21], which quickly became a vade mecum for designers of containers for protection of electronic equipment from mechanical shock and vibration. It was the first of several papers stimulated by his eight-year association as consultant to the Bell Telephone Laboratories.

A most important paper arising from his symbiosis with the Bell Laboratories stemmed from a perennial effort toward rational design of the carbon microphone. Published in 1949, it gave new direction to research on frictional phenomena at surfaces in contact [28]. The work, its basic consequences first tested and verified experimentally [38] and then generalized [41], led to an understanding of the process of wear due to local sliding and gave a mighty impetus to world-wide researches, analytical and experimental, in the industrially vital areas of sliding and rolling contacts, ball bearings, railway wheels, and the mechanics of granular media [56].

Shortly thereafter Mindlin began his trail-blazing studies, which he was to continue intermittently for more than two decades, on waves and vibrations in isotropic elastic plates [34] and, concommitantly, on high-frequency vibrations of crystal plates [35], soon extending his studies to account for the piezoelectric effect [39]. The method he had employed of reducing the exact, three-dimensional equations to describe the dynamics of plates, by expanding in series of the thickness coordinate (or in terms of selected functions thereof) and truncating the series, he also applied, suitably altered, to dynamical problems of bars [37], [60], [62], stimulating, as in the case of plates, an ever growing literature of the subject.

The work on isotropic bars and plates led to results in use world-wide in development and design of electromechanical filters and solid delay lines. But it was his studies on crystal plates that caused the greatest stir, for they were pioneering papers in a vexingly difficult mathematical field. In them, he elucidated a very complicated phenomenon of great technical importance and thereby pointed the way to major improvements in the design and performance of quartz crystals for filter circuits. The U.S. Army Signal Corps, long-time sponsor of the research on crystal plate vibrations, persuaded him to prepare a monograph on the subject by granting him a sole-supplier contract for the purpose. In 1955 the task was completed to the acclaim of the sponsor—it was in the form of a ca. 170-page-long report entitled "An Introduction to the Mathematical Theory of Vibrations of Elastic Plates." Yet even as its author was composing the text, better ways of doing the old were occurring to him and new ideas for what should, and could, be done were springing to his mind. Indeed, despite much subsequent work in other areas of mechanics, he kept returning to this subject to the very end of his life. The monograph seemed to haunt him, for he intermittently worked on a revision–more accurately, a complete reworking–of it. A few months before his death he spoke of plans to add yet another major part to what he had already written before he would consider the job done; ever the realist, he added, "But don't hold your breath!"*

It is strikingly clear in retrospect that the period of the late '40s and early '50s was a particularly fruitful one for him. In addition to the basic papers on elastic contact and on high-frequency vibrations of plates and bars, a pioneering article on the interpretation of optical birefringence in viscoelastic materials [27] and a seminal paper on the interaction of an elastic shell with a surrounding fluid [40], the second of these written in collaboration with his colleague H. H. Bleich, date to that time. Of evident importance to the Navy, the latter work triggered a great deal of subsequent research on the response of submarine hulls to shock loading. Also sandwiched into that period is the lengthy article on analogies [31], with another of his colleagues, M. G. Salvadori, as co-author, which appeared as a chapter in the *Handbook of Experimental Stress Analysis.*

Beginning in 1962, in response to what purported to be, but were seen by Mindlin as erroneously based, expanded theories of anisotropic materials, he was led to construct, first, a theory that took account of couple-stresses [74], characteristically applying it immediately thereafter to a two-dimensional case that seemed simple enough to be susceptible of experimental validation [75]. Two years later, in a paper on micro-structure [78], he treated a continuum endowed at each point with an internal displacement field, and showed that the lowest order of this formulation results in a theory that includes the optical lattice modes at the long-wavelength limit. In a paper published the following year [83], he formulated a linear theory of equilibrium of an elastic solid in which the stored energy depends on the strain and its first and second gradients. This proved to be the first continuum description capable of accounting for surface tension, without resorting to the artificial construct of a vanishingly thin surface membrane. The papers contributed to the start of lively activity in this field of generalized elastic continua, and stimulated applications to areas as diverse as the mechanics of laminated and of fiber-reinforced materials, and of framed structures.

In 1967 he was appointed James Kip Finch Professor of Applied Science, the first holder of a chair named in honor of the patron and protector of his days as a graduated student. He was to hold this post until his retirement in 1975.

A series of papers initiated with [88] introduced the contribution of the

*Shortly after making this observation, aware of the gravity of what proved to be his last illness, he entrusted the manuscript for completion (and, it is fervently hoped, timely publication) to P. C. Y. Lee, one of his last doctoral students.

polarization gradient (in addition to the strain and the polarization itself) to the stored energy of a solid. Mindlin was able to show that this augmented formulation of the elastic dielectric continuum accomodates the surface energy of deformation and polarization, predicts the experimentally observed anomalous capacitance of thin dielectric films [95], and accounts for acoustical and, when the magnetic field is included, optical activity [99].

More or less concurrently came one other group of highly innovative researches, in which his aim was to bridge the gap between continuum and lattice theories. Thus, he was able to show [91] that a model of a continuum, formed by two or more interpenetrating media, in which the stored energy depends not only on the strains of the individual constituents but also on their relative displacements, and simulating a (non-Bravais) lattice with a basis, contains optical as well as acoustical branches in the vibrational spectrum. In [95], a paper mentioned above, he demonstrated that his augmented theory of the elastic dielectric continuum yields the correct long-wave limit of a monatomic lattice theory with a shell model, and in [101], by endowing the model of [91] with individual polarizations, that the resulting continuum can yield the long-wave limit of a diatomic lattice with shell model.

The steady stream of his research was interrupted early in 1969 by illness, and he was obliged to submit to cardiac surgery. (This was before the coronary bypass operation became part of clinical procedure.) As he made amply clear at the time, he resented the lengthy convalescence because it was inhibiting his work. But by the end of that year he was back at his desk, and a glance at his list of publications after that date shows that he was able to resume his enviable scientific productivity.

His last hurrah, just a year before his death, came in the form of two papers, one dealing with free vibrations of a rectangular parallelepiped [128] and the other, with vibrations of a rectangular plate with all four edges free [129], classic problems against which he had butted his head, in vain, decades earlier. At long last he managed to construct solutions, apparently by inspired guesswork, reminiscent of the *tour de force* of his doctoral dissertation 50 years earlier. He was manifestly elated, particularly by what turned out to be his final publication, characterizing the accomplishment in the following words: "It's a pity there aren't many around, anymore, who would appreciate what an extraordinary feat that solution is." His letter, accompanying a photocopy of the work in holograph and dated January 9, 1986, continues, in evident allusion to the 19th century origin of the problem, "I'm reminded of Tony Biot [Maurice Anthony Biot] who was frequently a generation ahead of his time. I, on the other hand, seem to be some generations behind."

The preceeding narrative, though sketchy, has, I hope, given some indication of the breadth and depth of Mindlin's scholarship. Those who worked with him, and those others who knew him and his work, quickly came to appreciate his profound insights into the physics of any given situation, his surefootedness in discerning what is central and what is peripheral, and his multifaceted analytical and experimental skills. Nevertheless, a disservice would be done the subject of this account (as well as its reader) if nothing were said about him as colleague, mentor, and human being.

His professional colleagues treated him with deference at times bordering on awe. He was secure in the knowledge of his own worth, but wore the mantle of his eminence with genuine modesty. He was generous in giving or sharing credit, and unfailingly courteous to peers seeking his opinion or advice. Basically a shy, reserved person, he was, invariably and unexceptionally, the consummate gentleman.

His many disciples revered him and, one suspects, in his quiet way he enjoyed being the father figure. To celebrate his 60th birthday, his former doctoral students tendered him a dinner in Minneapolis, where most had journeyed ostensibly to attend a professional meeting. In 1974, in anticipation of his imminent retirement, he was presented with a published book, entitled *R. D. Mindlin and Applied Mechanics.* In eight substantial chapters, by fifteen of his erstwhile students, this novel kind of Festschrift contains summaries of those areas on which Mindlin's own endeavors have exercised a profound and lasting effect.

But it was by no means a one-way relationship. Whenever a former student of his was being feted, as such things tended to occur with increasing frequency as the years passed, Ray Mindlin would be inconspicuously present among the well-wishers. Perhaps the loyalty and affection made up to some extent for not having any children of his own. He married twice, each time losing a beloved mate by untimely death due to illness.

He served with devotion the profession which he made his life's work, through his research, his teaching, his advisory capacity to numerous government agencies, and his activities in various scientific and technical societies. Among the latter, mention is warranted of the following positions he held at various times: In the American Society of Mechanical Engineers (ASME), chairman, Applied Mechanics Division; member, Publications Committee, Engineering Societies Monographs Committee, Advisory Board of *Applied Mechanics Reviews.* In the American Society of Civil Engineers (ASCE), chairman, Committee on Applied Mechanics of the Structural Engineering Division (precursor of the Engineering Mechanics Division). In the Society for Experimental Stress Analysis (SESA), co-founder and president; member, executive committee. In the American Institute of Physics, associate editor, *Journal of Mathematical Physics.* Also, he was member of: the U.S. National Committee for Theoretical and Applied Mechanics; the General Assembly of the International Union of Theoretical and Applied Mechanics; the American Physical Society.

To be sure, his remarkable intellectual fecundity and gentle personality did not go unappreciated by his peers, and recognition came in the form of copious honors and awards. He was elected Fellow of the American Academy of Arts and Sciences (1958), of the ASME (1962), and of the Accoustical Society of America (1963); member of the National Academy of Engineering (1966) and of the National Academy of Sciences (1973); and Honorary Member of the ASME (1969).

He received the Research Prize (1958) and the von Karman Medal (1961) of the ASCE; the Timoshenko Medal (1964) and the ASME Medal (1976) from the ASME; the Trent-Crede Award of the Acoustical Society of America (1971); the Frocht Award of the SESA (1974); the Great Teacher Award (1960) and the Egleston Medal (1971) from Columbia University; and an Honorary D.Sc. degree from Northwestern University (1975); also, a Naval Ordnance Development Award (1945) and the C. B. Sawyer Award of the Army Electronics Command (1967). Remarkably, prestigious as these are, all but one were sandwiched between his most noteworthy honors, both bestowed by the U.S. Government: the Presidential Medal for Merit (1946), as noted earlier the highest civilian decoration of the second World War, and his ultimate accolade, the National Medal of Science, which he received in 1979.

Mindlin died on November 22, 1987, at age 81, in consequence of a debilitating illness that lingered for about a year. And yet, to the last, his mind remained acute and active, almost as if divorced from his frail body. His passing was mourned by all who knew him, but his name will live through his work.

Chaucer's lines, written about another scholar, of days long gone, might just as aptly have been indited for Raymond Mindlin:

Of study took he utmost care and heed
Not one word spoke he more than was his need
And that was said in fullest reverence
And short and quick and full of high good sense
Pregnant of moral virtue was his speech
And gladly would he learn and gladly teach.

H. Deresiewicz

Surrounded by well-wishers observing his 68th birthday and marking his forthcoming mandatory retirement from Columbia (Boulder, Colorado, 1974)

Seated, l. to r., front row: K. Robinson, F. L. DiMaggio, N. Perrone, R. Skalak, P. C. Y. Lee, H. F. Tiersten; *second row*: G. Herrmann, Mrs. Skalak, Mrs. Pao, Mindlin, Mrs. Mindlin, M. G. Salvadori, Mrs. Albrecht, J. H. Weiner.

Standing, in the same order: R. L. Schiffman, J. M. Garrelts, L. E. Goodman, J. J. McCoy, H. D. McNiven, M. L. Baron, E. H. Lee, B. A. Boley, J. Achenbach, B. Albrecht, J. Miklowitz (in rear), M. A. Medick, H. Liebowitz, D. C. Drucker, A. G. Lubowe, D. C. Gazis, J. Tasi, P. M. Naghdi, unidentified (in rear), N. Basdekas, J. Duffy, E. Sternberg, H. H. Bleich, and Y.-H. Pao.

Publications of Raymond D. Mindlin

[1] A reflection polariscope for photoelastic analysis, *Review of Scientific Instruments*, **5**, 224–228 (1934).

[2] Structural beams in torsion (with H. M. Westergaard), discussion of paper by I. Lyse and B. G. Johnston, *Proceedings of the American Society of Civil Engineers*, **61**, 509–512 (1935).

[3] Contribution au problème d'équilibre de l'élasticité d'un solide indéfinie limité par un plan, *Comptes Rendus de l'Académie des Sciences*, Paris, **201**, 536–537 (1935).

[4] Force at a point in the interior of a semi-infinite solid, *Physics*, **7**, 195–202 (1936).

[5] Note on the Galerkin and Papkovitch stress functions, *Bulletin of the American Mathematical Society*, **42**, 373–376 (1936).

[6] Stresses around circular holes in dams and buttresses, discussion of paper by I. K. Silverman, *Proceedings of the American Society of Civil Engineers*, **63**, 399–403 (1937).

[7] Pressure distributions on retaining walls, discussion of paper by M. G. Spangler, *Proceedings of the International Conference on Soil Mechanics and Foundation Engineering*, Cambridge, Mass.-1936, **III**, 155–156 (1937).

[8] Analysis of doubly refracting materials with circularly and elliptically polarized light, *Journal of the Optical Society of America*, **27**, 288–291 (1937).

[9] Stress systems in a circular disk under radial forces, *Journal of Applied Mechanics*, **4**, A-115–A-118 (1937).

[10] Distortion of the photoelastic fringe pattern in an optically unbalanced polariscope, *Journal of Applied Mechanics*, **4**, A-170–A-172 (1937).

[11] The equiangular strain-rosette, *Civil Engineering*, **8**, 546–547 (1938).

[12] Stresses in a heavy disk suspended from an eccentric peg, *Journal of Applied Physics*, **9**, 714–717 (1938).

[13] Stresses in an eccentrically rotating disk, *Philosophical Magazine*, Ser. 7, **xxvi**, 713–719 (1938).

[14] Gravitational stresses in bipolar coordinates, *Proceedings of the Fifth International Congress for Applied Mechanics*, Cambridge, Mass.-1938 112–116 (1939).

[15] Stress distribution around a tunnel, *Proceedings of the American Society of Civil Engineers*, **65**, 619–642 (1939); reprinted in *Transactions American Society of Civil Engineers*, **105**, 1117–1140 (1940).

*[16] A review of the photoelastic method of stress analysis, *Journal of Applied Physics*, **10**, 222–241, 273–294 (1939).

[17] Stress analysis by three-dimensional photoelastic methods (with D. C. Drucker), *Journal of Applied Physics*, **11**, 724–732 (1940).

[18] An extension of the photoelastic method of stress measurement to plates in transverse bending, discussion of paper by J. N. Goodier and G. H. Lee, *Journal of Applied Mechanics*, **8**, A-187–A-189 (1941).

[19] Optical aspects of three-dimensional photoelasticity, *Journal of the Franklin Institute*, **233**, 349–363 (1942).

*[20] The present status of three-dimensional photoelasticity, *Civil Engineering*, **12**, 255–258 (1942).

*[21] Dynamics of package cushioning, *Bell System Technical Journal*, **24**, 353–461 (1945).

[22] The analogy between multiply-connected slices and slabs, *Quarterly of Applied Mathematics*, **IV**, 279–290 (1946).

*[23] Activities of the Society for Experimental Stress Analysis, *Proceedings of the Society for Experimental Stress Analysis*, **IV**, no. 2, ix–xi (1947).

[24] Stress distribution around a hole near the edge of a plate under tension, *Proceedings of the Society for Experimental Stress Analysis*, **V**, no. 2, 56–68 (1948).

[25] Response of damped elastic systems to transient disturbances (with F. W. Stubner and H. L. Cooper), *Proceedings of the Society for Experimental Stress Analysis*, **V**, no. 2, 69–87 (1948).

[26] The optical equations of three-dimensional photoelasticity (with L. E. Goodman), *Journal of Applied Physics*, **20**, 89–95 (1949).

[27] A mathematical theory of photo-viscoelasticity, *Journal of Applied Physics*, **20**, 206–216 (1949).

[28] Compliance of elastic bodies in contact, *Journal of Applied Mechanics*, **16**, 259–268 (1949).

[29] Thermoelastic stress around a cylindrical inclusion of elliptic cross section (with H. L. Cooper), *Journal of Applied Mechanics*, **17**, 265–268 (1950).

[30] Beam vibrations with time-dependent boundary conditions (with L. E. Goodman), *Journal of Applied Mechanics*, **17**, 377–380 (1950).

*[31] Analogies (with M. G. Salvadori); in M. Hetényi(ed), *Handbook of Experimental Stress Analysis*, John Wiley & Sons, New York, 700–827 (1950).

[32] Nuclei of strain in the semi-infinite solid (with D. H. Cheng), *Journal of Applied Physics*, **21**, 926–930 (1950).

* Not part of this collection.

[33] Thermoelastic stress in the semi-infinite solid (with D. H. Cheng), *Journal of Applied Physics*, **21**, 931–933 (1950).

[34] Influence of rotatory inertia and shear on flexural motions of isotropic, elastic plates, *Journal of Applied Mechanics*, **18**, 31–38 (1951).

[35] Thickness-shear and flexural vibrations of crystal plates, *Journal of Applied Physics*, **22**, 316–323 (1951).

*[36] Development of the mathematical theory of three-dimensional photoelasticity, *Applied Mechanics Reviews*, **4**, 537–539 (1951).

[37] A one-dimensional theory of compressional waves in an elastic rod (with G. Herrmann), *Proceedings of the First U.S. National Congress of Applied Mechanics-1951*, 187–191 (1952).

[38] Effects of an oscillating tangential force on the contact surfaces of elastic spheres (with W. P. Mason, T. F. Osmer, and H. Deresiewicz), *Proceedings of the First U.S. National Congress of Applied Mechanics-*1951, 203–208 (1952).

[39] Forced thickness-shear and flexural vibrations of piezoelectric crystal plates, *Journal of Applied Physics*, **23**, 83–88 (1952).

[40] Response of an elastic cylindrical shell to a transverse, step shock wave (with H. H. Bleich), *Journal of Applied Mechanics*, **20**, 189–195 (1953).

[41] Elastic spheres in contact under varying oblique forces (with H. Deresiewicz), *Journal of Applied Mechanics*, **20**, 327–344 (1953).

[42] Force at a point in the interior of a semi-infinite solid, *Proceedings of the First Midwestern Conference on Solid Mechanics*, 56–59 (1953).

[43] Thickness-shear and flexural vibrations of contoured crystal plates (with M. Forray), *Journal of Applied Physics*, **25**, 12–20 (1954).

[44] Thickness-shear vibrations of piezoelectric crystal plates with incomplete electrodes (with H. Deresiewicz), *Journal of Applied Physics*, **25**, 21–24 (1954).

[45] Suppression of overtones of thickness-shear and flexural vibrations of crystal plates (with H. Deresiewicz), *Journal of Applied Physics*, **25**, 25–27 (1954).

[46] Thickness-shear and flexural vibrations of a circular disk (with H. Deresiewicz), *Journal of Applied Physics*, **25**, 1329–1332 (1954).

[47] Axially symmetric flexural vibrations of a circular disk (with H. Deresiewicz), *Journal of Applied Mechanics*, **22**, 86–88 (1955).

*[48] Mechanics of granular media, *Proceedings of the Second U.S. National Congress of Applied Mechanics-*1954, 13–20 (1955).

[49] Timoshenko's shear coefficient for flexural vibrations of beams (with H. Deresiewicz), *Proceedings of the Second U.S. National Congress of Applied Mechanics-*1954, 175–178 (1955).

[50] Thickness-shear and flexural vibrations of rectangular crystal plates (with H. Deresiewicz), *Journal of Applied Physics*, **26**, 1435–1442 (1955).

* Not part of this collection.

[51] High-frequency extensional vibrations of plates (with T. R. Kane), *Journal of Applied Mechanics*, **23**, 277–283 (1956).

[52] Flexural vibrations of rectangular plates (with A. Schacknow and H. Deresiewicz), *Journal of Applied Mechanics*, **23**, 430–436 (1956).

[53] Simple modes of vibration of crystals, *Journal of Applied Physics*, **27**, 1462–1466 (1956).

[54] Waves on the surface of a crystal (with H. Deresiewicz), *Journal of Applied Physics*, **28**, 669–671 (1957).

[55] Influence of width on velocities of long waves in plates (with D. C. Gazis), *Journal of Applied Mechanics*, **24**, 541–546 (1957).

[56] Stress-strain relations and vibrations of a granular medium (with J. Duffy), *Journal of Applied Mechanics*, **24**, 585–593 (1957).

[57] Vibrations of a monoclinic crystal plate (with E. G. Newman), *Journal of the Acoustical Society of America*, **29**, 1206–1218 (1957).

[58] Vibrations of an infinite, elastic plate at its cut-off frequencies, *Proceedings of the Third U.S. National Congress of Applied Mechanics*, 225–226 (1958).

[59] Extensional vibrations of elastic plates (with M. A. Medick), *Journal of Applied Mechanics*, **26**, 561–569 (1959).

[60] Axially symmetric waves in elastic rods (with H. D. McNiven), *Journal of Applied Mechanics*, **27**, 145–151 (1960).

[61] Vibrations and waves in elastic bars of rectangular cross section (with E. A. Fox), *Journal of Applied Mechanics*, **27**, 152–158 (1960).

[62] Dispersion of flexural waves in an elastic, circular cylinder (with Y.-H. Pao), *Journal of Applied Mechanics*, **27**, 513–520 (1960).

[63] Extensional vibrations and waves in a circular disk and a semi-infinite plate (with D. C. Gazis), *Journal of Applied Mechanics*, **27**, 541–547 (1960).

[64] Waves and vibrations in isotropic, elastic plates; in J. N. Goodier and N. J. Hoff (eds), *Structural Mechanics*, Pergamon Press, Oxford, 199–232 (1960).

[65] High frequency vibrations of crystal plates, *Quarterly of Applied Mathematics*, **XIX**, 51–61 (1961).

[66] On the equations of extensional motion of crystal plates (with H. L. Cooper), *Quarterly of Applied Mathematics*, **XIX**, 111–118 (1961).

[67] On the equations of motion of piezoelectric crystals; in *Problems of Continuum Mechanics* (N. I. Muskhelishvili 70th Birthday Volume), Society for Industrial and Applied Mathematics, Philadelphia, 282–290 (1961).

[68] Forced vibrations of piezoelectric crystal plates (with H. F. Tiersten), *Quarterly of Applied Mathematics*, **XX**, 107–119 (1962).

[69] Vibrations of an infinite, monoclinic crystal plate at high frequencies and long wavelengths (with R. K. Kaul), *Journal of the Acoustical Society of America*, **34**, 1895–1901 (1962).

[70] Frequency spectrum of a monoclinic crystal plate (with R. K. Kaul), *Journal of the Acoustical Society of America*, **34**, 1902–1910 (1962).

[71] Extensional vibrations of thin quartz disks (with A. G. Lubowe), *Journal of the Acoustical Society of America*, **34**, 1911–1918 (1962).

[72] Dispersion of axially symmetric waves in elastic rods (with M. Onoe and H. D. McNiven), *Journal of Applied Mechanics*, **29**, 729–734 (1962).

[73] Strong resonances of rectangular AT-cut quartz plates (with D. C. Gazis), *Proceedings of the Fourth U.S. National Congress of Applied Mechanics*, 305–310 (1962).

[74] Effects of couple-stresses in linear elasticity (with H. F. Tiersten), *Archive for Rational Mechanics and Analysis*, **11**, 415–448 (1962).

[75] Influence of couple-stresses on stress concentrations, *Experimental Mechanics*, **3**, 1–7 (1963).

[76] Extensional waves along the edge of an elastic plate (with J. J. McCoy), *Journal of Applied Mechanics*, **30**, 75–78 (1963).

[77] High frequency vibrations of plated, crystal plates; in *Progress in Applied Mechanics* (The Prager Anniversary Volume), Macmillan, New York, 73–84 (1963).

[78] Micro-structure in linear elasticity, *Archive for Rational Mechanics and Analysis*, **16**, 51–78 (1964).

[79] Complex representation of displacements and stresses in plane strain with couple-stresses; in *Applications of the Theory of Functions in Continuum Mechanics* (Proceedings of the International Symposium, Tbilisi, U.S.S.R., 1963), Nauka, Moscow, 256–259 (1965).

[80] On the equations of elastic materials with micro-structure, *International Journal of Solids and Structures*, **1**, 73–78 (1965).

[81] Thickness-twist vibrations of an infinite, monoclinic, crystal plate, *International Journal of Solids and Structures*, **1**, 141–145 (1965).

[82] Stress functions for a Cosserat continuum, *International Journal of Solids and Structures*, **1**, 265–271 (1965).

[83] Second gradient of strain and surface-tension in linear elasticity, *International Journal of Solids and Structures*, **1**, 417–438 (1965).

[84] Thickness-shear and flexural vibrations of partially plated, crystal plates (with P. C. Y. Lee), *International Journal of Solids and Structures*, **2**, 125–139 (1966).

[85] Bechmann's number for harmonic overtones of thickness/twist vibrations of rotated-Y-cut quartz plates, *Journal of the Acoustical Society of America*, **41**, 969–973 (1967).

[86] Anharmonic, thickness-twist overtones of thickness-shear and flexural vibrations of rectangular, *AT*-cut quartz plates (with W. J. Spencer), *Journal of the Acoustical Society of America*, **42**, 1268–1277 (1967).

[87] On first strain-gradient theories in linear elasticity (with N. N. Eshel), *International Journal of Solids and Structures*, **4**, 109–124 (1968).

[88] Polarization gradient in elastic dielectrics, *International Journal of Solids and Structures*, **4**, 637–642 (1968).

[89] Optimum sizes and shapes of electrodes for quartz resonators, *Journal of the Acoustical Society of America*, **43**, 1329–1331 (1968).

[90] Face-shear waves in rotated Y-cut quartz plates; in F. P. J. Rimrott and J. Schwaighofer (eds), *Mechanics of the Solid State* (Joseph Marin Memorial Volume), University of Toronto Press, Toronto, 143–145 (1968).

[91] Theories of elastic continua and crystal lattice theories; in E. Kröner (ed), *Mechanics of Generalized Continua*, (IUTAM Symposium, Freudenstadt-Stuttgart, 1967), Springer-Verlag, Berlin, 312–320 (1968).

[92] Lattice and continuum theories of simple modes of vibration in cubic crystal plates and bars; in *Dynamics of Structured Solids*, American Society of Mechanical Engineers, New York, 2–9 (1968).

[93] Crystal lattice theory of torsion of a rectangular bar of simple cubic structure, *Indian Journal of Mechanics and Mathematics*, special issue (B. Sen 70th Birthday Volume), part I, 35–40 (1968).

[94] Note on vibrations of prisms and polygonal plates; in *Problems of Hydrodynamics and Continuum Mechanics* (L. I. Sedov 60th Birthday Volume), *Society for Industrial and Applied Mathematics*, Philadelphia, 485–488 (1969).

[95] Continuum and lattice theories of influence of electromechanical coupling on capacitance of thin dielectric films, *International Journal of Solids and Structures*, **5**, 1197–1208 (1969).

[96] Lattice theory of shear modes of vibration and torsional equilibrium of simple-cubic crystal plates and bars, *International Journal of Solids and Structures*, **6**, 725–738 (1970).

[97] Polarization gradient in elastic dielectrics, *Courses and Lectures* No. 24, International Centre for Mechanical Sciences, Udine, Italy, 3–55 (1970).

[98] Thickness-twist vibrations of a quartz strip, *International Journal of Solids and Structures*, **7**, 1–4 (1971).

[99] Acoustical and optical activity in alpha quartz (with R. A. Toupin), *International Journal of Solids and Structures*, **7**, 1219–1227 (1971).

[100] Electromechanical vibrations of centrosymmetric cubic crystal plates, *PMM–Journal of Mechanics and Applied Mathematics*, **35**, 404–408 (1971).

[101] A continuum theory of a diatomic, elastic dielectric, *International Journal of Solids and Structures*, **8**, 369–383 (1972).

[102] Coupled elastic and electromagnetic fields in a diatomic, dielectric continuum, *International Journal of Solids and Structures*, **8**, 401–408 (1972).

[103] High frequency vibrations of piezoelectric crystal plates, *International Journal of Solids and Structures*, **8**, 895–906 (1972).

[104] Galerkin stress functions for non-local theories of elasticity, *International Journal of Solids and Structures*, **8**, 1407–1411 (1972).

[105] Elasticity, piezoelectricity and crystal lattice dynamics, *Journal of Elasticity*, **2**, 217–282 (1972).

[106] On the electrostatic potential of a point charge in a dielectric solid, *International Journal of Solids and Structures*, **9**, 233–235 (1973).

[107] Electromagnetic radiation from a vibrating quartz plate, *International Journal of Solids and Structures*, **9**, 697–702 (1973).

[108] Coupled piezoelectric vibrations of quartz plates, *International Journal of Solids and Structures*, **10**, 453–459 (1974).

[109] Equations of high frequency vibrations of thermopiezoelectric crystal plates, *International Journal of Solids and Structures*, **10**, 625–637 (1974).

[110] Note on torsional rigidity of elastic bars, *Mechanics Research Communications*, **1**, 57–59 (1974).

[111] Electromagnetic radiation from a vibrating, elastic sphere, *International Journal of Solids and Structures*, **10**, 1307–1314 (1974).

[112] Solution of St. Venant's torsion problem by power series, *International Journal of Solids and Structures*, **11**, 321–328 (1975).

[113] Rotation-dependence of the photoelastic effect; in J. Kestens (ed), *The Photoelastic Effect and its Applications*, Springer-Verlag, Berlin, 376–388 (1975).

[114] Low frequency vibrations of elastic bars, *International Journal of Solids and Structures*, **12**, 27–49 (1976).

[115] Note on axial-shear and contour vibrations of prisms, *International Journal of Solids and Structures*, **12**, 99–105 (1976).

*[116] Education and research in mechanics in the U.S.A., *Técnica* (Instituto Superior Técnico, Lisbon), **38**, 1–3 (1976).

[117] Torsion and flexure of piezoelectric crystal bars; in J. D. Achenbach, Y.-H. Pao and H. F. Tiersten (eds), *Application of Elastic Waves in Electrical Devices, Non-Destructive Testing and Seismology*, Northwestern University, Evanston, Illinois, 83–96 (1976).

[118] Frequencies of vibration of beams by Schelkunoff iteration with the HP-65 calculator, *Computers & Structures*, **7**, 639–650 (1977).

[119] On thickness-twist vibrations of a quartz strip; in *Some Aspects of Mechanics of Continua* (B. Sen Memorial Volume), Department of Mathematics, Jadavpur University, Calcutta, Part I, 29–31 (1977).

[120] A variational principle for the equations of piezoelectromagnetism in a compound continuum; in *Kompleksnii Analiz i ego Prilozhenia* (I. N. Vekua 70th Birthday Volume), Academy of Sciences U.S.S.R.; Nauka, Moscow, 397–400 (1978).

*[121] The sesquicentennial of the first crystal plate equations, *Proceedings*

* Not part of this collection.

of the 33rd Annual Symposium on Frequency Control, Electronic Industries Association, Washington, 1–3 (1979).

[122] Vibrations of quartz plates with the HP-67 pocket calculator, *Computers & Structures*, **10**, 751–759 (1979).

[123] On Reissner's equations for sandwich plates; in S. Nemat-Nasser (ed), *Mechanics Today*, **5** (E. Reissner 65th Birthday Volume), Pergamon Press, Oxford, 315–328 (1980).

[124] High frequency vibrations of quartz plates by expansion in series of Ekstein functions, *International Journal of Solids and Structures*, **16**, 785–791 (1980).

[125] Third overtone quartz resonator, *International Journal of Solids and Structures*, **18**, 809–817 (1982).

[126] Frequencies of piezoelectrically forced vibrations of electroded, doubly rotated, quartz plates, *International Journal of Solids and Structures*, **20**, 141–157 (1984).

[127] Vibrations of doubly-rotated-cut quartz plates with monoclinic symmetry, *International Journal of Solids and Structures*, **21**, 597–607 (1985).

[128] On vibrations of rectangular parallelepipeds, *International Journal of Solids and Structures*, **22**, 1423–1430 (1986).

[129] Flexural vibrations of rectangular plates with free edges, *Mechanics Research Communications*, **13**, 349–357 (1986).

The Papers

A Reflection Polariscope for Photoelastic Analysis

RAYMOND D. MINDLIN, *Research Assistant, Department of Civil Engineering, Columbia University*

(Received February 26, 1934)

The paper embodies the description of a polariscope which offers several advantages for photoelastic investigations and is applicable to the analysis of stress distributions in parallel plates and to Mesnager's modified three-dimensional analysis. The chief characteristic of the optical arrangement is that the light, after passing through the usual arrangement of polarizing prism, retardation plate and model, is reflected directly back through the same system.

INTRODUCTION

IN many cases encountered in photoelastic analysis it is necessary to determine the stress distribution in each member of a pair of parallel plates. A solution of this problem was recently presented by W. H. Haupt;[1] but, while it is a welcome contribution to the analysis of doubly refracting materials, Mr. Haupt's method involves a rather lengthy laboratory procedure which is likely to be viewed with apprehension by engineers who are interested in rapid and simple methods of analysis.

Another problem which has long evaded solution is the application of photoelastic analysis to three dimensional stress systems. Of the methods devised for this purpose that described by the late Professor Mesnager[2] seems most likely to be the first that will yield satisfactory results. The successful application of the scheme would lead to a determination of the extreme fiber stress distribution produced in a plate by forces applied at an angle to the plane of the plate. A polished metal slab is coated with a thin layer of transparent material which becomes doubly refracting when strained. Bending the plate out of its plane strains the coating and the

[1] Walter H. Haupt, *A Photoelastic Method of Stress Evaluation in Structures Involving Two Parallel Systems of Plane Stresses*, presented at the Cincinnati Meeting of the American Physical Society, December, 1933. Phys. Rev. **45**, 122A (1934).

[2] A. Mesnager, Comptes Rendus **190**, 1249 (1930), *Sur la détermination optique des tensions intérieures dans les solides à trois dimensions.*

resulting double refraction is measured by analyzing polarized light reflected from the polished metal surface. Mesnager suggested that Tardy's polariscope[3] be used for such an analysis, but an attempt at using Tardy's method proved unsatisfactory for photoelastic work because there is necessarily a small angle between the incident and reflected rays. Hence, the two rays follow different paths through the doubly refracting material with the result that the image is blurred and the analysis is inaccurate where the stress gradient is high.

The optical system and the method of analysis described below are believed to offer a simple and direct means for carrying out the solutions of both of the above problems. Tedious point by point explorations give way to a photographic record of the entire stress field; and even for ordinary photoelastic analysis the system has several important advantages over the older methods.

The Reflection Polariscope

The chief characteristic of the optical arrangement is that the light, after passing through the usual arrangement of polarizing prism, retardation plate and model, is reflected directly back through the same system.

A diagrammatic sketch of the reflection polariscope is given in Fig. 1. Monochromatic

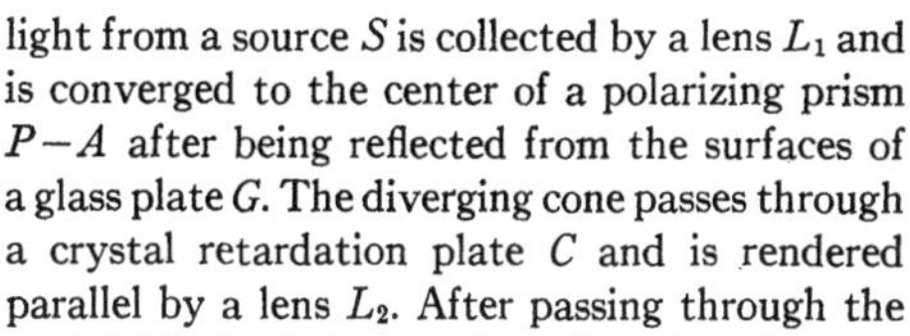

Fig. 1. Optical arrangement of the reflection polariscope.

light from a source S is collected by a lens L_1 and is converged to the center of a polarizing prism $P-A$ after being reflected from the surfaces of a glass plate G. The diverging cone passes through a crystal retardation plate C and is rendered parallel by a lens L_2. After passing through the model M, the light is reflected at R and passes

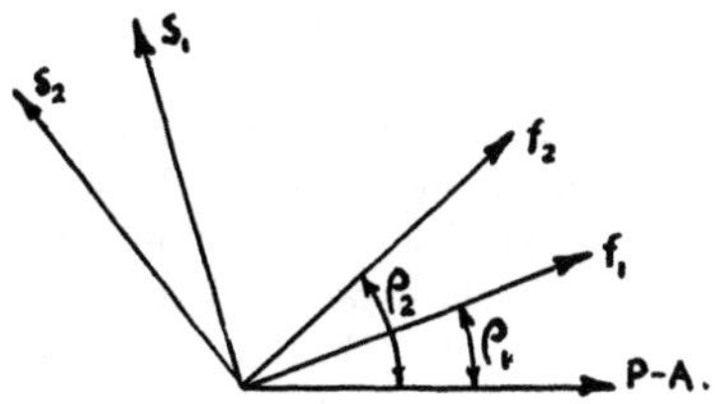

Fig. 2. Azimuth relations of optic axes.

back, in order, through the model, the lens L_2, the retardation plate, the prism and the glass plate G and then strikes the screen or photographic plate F.

The arrangement differs from the Norrenberg doubler[4] in two respects. First, a single prism is used as both polarizer and analyzer; and, second, a quarter-wave and a half-wave retardation plate at C play an important part in the operation of the instrument, as will be described below.

For Mesnager's method the application of the system is immediately evident, while in the case of the analysis of a pair of parallel plates, the reflector R is placed between the plates and each specimen is examined in turn.

The photoelastic method generally in use in this country, at the present time, involves the recording of two sets of interference patterns. A specimen examined in circularly polarized light exhibits a system of interference fringes known as $p-q$ fringes. These are independent of the orientation of the model and represent the loci of points along which the relative retardation is constant, corresponding to a constant difference between maximum and minimum principal stresses (p and q). When the specimen is viewed in a plane polarized field, there appears, superimposed on the $p-q$ fringes, another set of fringes known as isoclinics. The isoclinics occur whenever the optic axes in the specimen (which coincide with the axes of principal stress) are parallel to the direction of vibration of the light emerging from the polarizer. The second set of fringes is therefore dependent upon the relative orientations of the model and the plane of polarization. Between "crossed nicols" they

[3] M. H.-L. Tardy, *Méthode pratique d'examen de mesure de la biréfringence des verres d'optique*, Rev. d'Optique **8**, 59–69 (1929).

[4] An adaptation of the Norrenberg doubler to photoelastic analysis was described in a recent paper, Georges Mabboux, *Applications de la Photoélasticimetrie a l'etude des ouvrages en Beton*, Rev. d'Optique **12**, 501–507 (1932).

appear as dark areas; while between parallel nicols light and dark are reversed. Isoclinics are usually broad, ill-defined bands which obscure the $p-q$ fringes and it is for the purpose of clearing up the field that the model is also examined in circularly polarized light.

Circularly polarized light is produced when monochromatic light passes through a polarizing prism and a quarter-wave retardation plate with its optic axes at 45° to the direction of vibration of light transmitted through the polarizer. In the ordinary transmission polariscope the light, after passing through the circular polarizer and the model, is examined through a similar combination known as a circular analyzer. In the reflection polariscope there is, of course, only one prism and quarter-wave plate combination but, if it is remembered that a circular analyzer is independent of its orientation as long as the relative azimuths of the prism and quarter-wave plate are maintained constant, it becomes immediately evident that a circular polarizer may be used as its own analyzer; hence, it is possible to eliminate the isoclinics and obtain clear $p-q$ fringes in the reflection polariscope (see Figs. 3, 4 and 5a).

If the retardation plate C (Fig. 1) is removed, it may be seen that the reflection polariscope corresponds to a plane polariscope with parallel nicols. Isoclinics appear as bands of light which interrupt the $p-q$ fringes (Fig. 5b). In some cases it is easier to identify and locate the

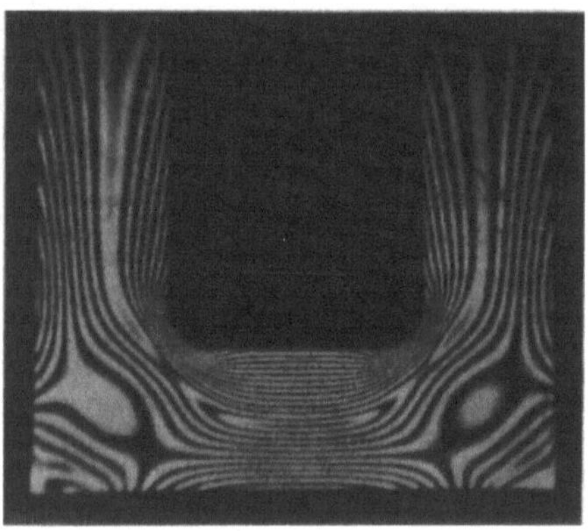

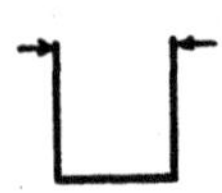

FIG. 3. $p-q$ fringes for a U-frame under horizontally applied forces at the free ends of the vertical members. Bakelite model, 0.300 in. thick.

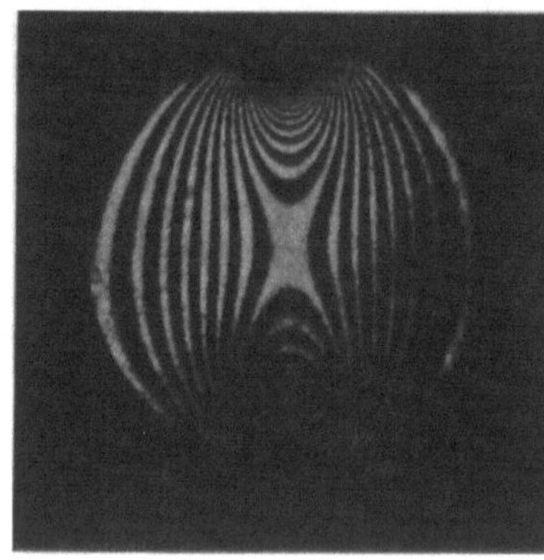

FIG. 4 $p-q$ fringes for a disk under a pair of diametral forces. Bakelite model, 0.169 in. thick.

(a) (b) (c)

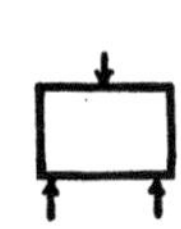

FIG. 5. Short, deep beam; simply supported, with centrally applied concentrated load. Glass model, 0.200 in. thick; edges slightly bevelled. (a) $p-q$ fringes; (b) 90° isoclinics; (c) 90° isoclinics reversed.

isoclinics if they are dark. This is readily brought about by reversing the negative (Fig. 5c).

In order to determine the directions of principal stress over the whole model, isoclinics are recorded for several relative orientations of the model and the plane of polarization. In the transmission polariscope this is accomplished by simultaneously rotating the two nicols, as it is usually inconvenient to rotate the model. Rotation of the nicols is usually accompanied by undesirably distortion and translation of the image. This may be avoided in the reflection polariscope by using a half-wave retardation plate at C (Fig. 1). Rotation of the half-wave plate through a given angle turns the direction of vibration through twice the angle. At points in the model where the optic axes are parallel to the vibration emerging from the half-wave plate, the vibration remains linear and is unchanged in azimuth; and, on returning through the half-wave plate, is rotated back again to its original position. Thus we still have the effect of parallel nicols and yet need rotate only one plate which, being plane-parallel, does not alter the shape or position of the image.

There is an additional reason for avoiding rotation of the polarizing prism in the reflection polariscope. The glass plate G acts as a partial polarizer for the light coming from L_1 and as a partial analyzer for the light on the return passage. There will thus be an optimum position of the polarizing prism for maximum intensity of the image. By using the half-wave plate the polarizing prism may be adjusted for maximum transmission through the glass plate and thereafter need not be moved.

Analytical Treatment

In Fig. 2, $P-A$ is the vibration direction of the polarizer-analyzer; f_1 and s_1 are, respectively, the directions of the fast and slow optic axes in the retardation plate; f_2 and s_2 are the fast and slow axes at a point in the model.

The linear vibration leaving the polarizer may be represented by

$$D_p = a\cos(\omega t-\alpha).$$

Emerging from the retardation plate there is, in general, an elliptical vibration with components

$$Df_1 = a\cos\rho_1\cos(\omega t-\alpha_1),$$

$$Ds_1 = -a\sin\rho_1\cos(\omega t-\alpha_1-\delta_1),$$

where δ_1 is the relative phase retardation produced in the crystal plate.

The wave leaving the model is given by

$$Df_2 = a\cos\rho_1\cos(\rho_2-\rho_1)\cos(\omega t-\alpha_2) - a\sin\rho_1\sin(\rho_2-\rho_1)\cos(\omega t-\alpha_2-\delta_1),$$

$$Ds_2 = -a\cos\rho_1\sin(\rho_2-\rho_1)\cos(\omega t-\alpha_2-\delta_2) - a\sin\rho_1\cos(\rho_2-\rho_1)\cos(\omega t-\alpha_2-\delta_1-\delta_2),$$

where δ_2 is the relative phase retardation introduced by the model.

On reflection at the surface R, both waves undergo a phase change of 180°. This has the effect of reversing the signs of Df_2 and Ds_2.

The wave leaving the model on the return passage is given by

$$D'f_2 = -a\cos\rho_1\cos(\rho_2-\rho_1)\cos(\omega t-\alpha_4) + a\sin\rho_1\sin(\rho_2-\rho_1)\cos(\omega t-\alpha_4-\delta_1),$$

$$D's_2 = +a\cos\rho_1\sin(\rho_2-\rho_1)\cos(\omega t-\alpha_4-2\delta_2) + a\sin\rho_1\cos(\rho_2-\rho_1)\cos(\omega t-\alpha_4-\delta_1-2\delta_2),$$

and, on emerging from the crystal plate,

$$\begin{aligned} D'f_1 = {} & -a\cos\rho_1\cos^2(\rho_2-\rho_1)\cos(\omega t-\alpha_5) \\ & +a\sin\rho_1\sin(\rho_2-\rho_1)\cos(\rho_2-\rho_1)\cos(\omega t-\alpha_5-\delta_1) \\ & -a\cos\rho_1\sin^2(\rho_2-\rho_1)\cos(\omega t-\alpha_5-2\delta_2) \\ & -a\sin\rho_1\cos(\rho_2-\rho_1)\sin(\rho_2-\rho_1)\cos(\omega t-\alpha_5-\delta_1-2\delta_2), \end{aligned}$$

$$\begin{aligned} D's_1 = {} & -a\cos\rho_1\cos(\rho_2-\rho_1)\sin(\rho_2-\rho_1)\cos(\omega t-\alpha_5-\delta_1) \\ & +a\sin\rho_1\sin^2(\rho_2-\rho_1)\cos(\omega t-\alpha_5-2\delta_1) \\ & +a\cos\rho_1\sin(\rho_2-\rho_1)\cos(\rho_2-\rho_1)\cos(\omega t-\alpha_5-\delta_1-2\delta_2) \\ & +a\sin\rho_1\cos^2(\rho_2-\rho_1)\cos(\omega t-\alpha_5-2\delta_1-2\delta_2). \end{aligned}$$

The final linear vibration emerging from the analyzer is given by

$$D_A = D'f_1 \cos \rho_1 + D's_1 \sin \rho_1$$

and its intensity is measured by the integral

$$I = (1/T)\int_0^T D_A^2 dt, \tag{1}$$

where T is the period of the vibration.

To obtain the $p-q$ fringes alone, let $\delta_1 = \pi/2$ and $\rho_1 = \pi/4$ and the evaluation of Eq. (1) results in

$$I = (a^2/2)\sin^2 \delta_2. \tag{2}$$

The fringe pattern accordingly depends solely upon the relative retardation in the model since ρ_2 does not appear in the equation.

In the ordinary transmission polariscope the corresponding equation is

$$I = (a^2/2)\sin^2(\delta_2/2). \tag{3}$$

Thus for a given load and thickness of model, we obtain double the number of fringes in the reflection polariscope; which is to be expected as the light passes through the model twice. This is an important point since the precision of a photoelastic analysis depends, to a large extent, upon the number of fringes which can be obtained without overstressing the model.

For the isoclinics let $\delta_1 = \pi$ and, from Eq. (1),

$$I = (a^2/2)[1 - \sin^2 2(2\rho_1 - \rho_2)\sin^2 \delta_2]. \tag{4}$$

Hence, we have maximum intensity when $\rho_2 = 2\rho_1, (2\rho_1 - \pi/2)$.

In the above analysis it is assumed that there is a first surface reflector at R, but in practice it is found that Eqs. (2) and (4) hold equally well when an ordinary second surface mirror is used.

Conclusions

That good results may be obtained by the reflection method is illustrated in Figs. 3, 4 and 5. The first two photographs were obtained from Bakelite models, while in the last example glass was used. The broad boundary line appearing around the specimen in the latter case is due to the bevelled edges of the glass plate.

Loss of intensity of light at reflecting surfaces would appear to be an obvious disadvantage inherent in the system; but this has proven to be of little practical consequence. Images of ample intensity for classroom demonstration were produced with a 500 watt projection lamp and, with a mercury vapor lamp, good photographs may be obtained with relatively short exposures. Exposures for the accompanying photographs varied between ten and fifteen seconds for Eastman Orthochromatic plates and a magnification factor of unity, with about an inch length of the tube of a General Electric Uviarc as a light source.

Advantages of the reflection method are:

1. For a given stress and thickness of model, the fringe order is doubled.
2. The fringes are sharp and clear since the light is transmitted and reflected through the model along the same path.
3. There is no displacement or distortion of the image while the isoclinics are being recorded as neither the model nor the polarizing prism is rotated.
4. In adjusting for circular polarization or in setting for isoclinics there is only a retardation plate to rotate.
5. Besides its use in ordinary photoelastic analysis, the optical arrangement finds its most important applications in the analysis of stress distributions in parallel plates and in Mesnager's modified three-dimensional analysis.
6. The apparatus is directly applicable to the interference method of determining isopachic lines.[5]
7. The length of the apparatus is approximately half that of the transmission polariscope and the total cost of the optical parts is considerably lower.

The author wishes to acknowledge his indebtedness to Professor G. B. Pegram and Professor H. W. Farwell of the Department of Physics at Columbia University for their helpful criticism in the preparation of this paper.

---

[5] V. Tesar, Rev. d'Optique **3**, 97–104 (1932); M. M. Frocht, *On the Application of Interference Fringes to Stress Analysis*, J. Frank. Inst. **216**, 73 (1933).

Reprinted from
Review of Scientific Instruments
5 (1934) 224-228

AMERICAN SOCIETY OF CIVIL ENGINEERS

Founded November 5, 1852

DISCUSSIONS

STRUCTURAL BEAMS IN TORSION

Discussion

BY MESSRS. H. M. WESTERGAARD AND R. D. MINDLIN, AND JOSEPH B. REYNOLDS

H. M. WESTERGAARD,[12] M. AM. SOC. C. E., and R. D. MINDLIN,[13] JUN. AM. SOC. C. E. (by letter).[13a]—In Fig. 9 the authors show curves which represent different formulas for the concentration of torsional shearing stresses at fillets. The curve which is identified by the names of the writers was derived as a result of correspondence with the authors, and for this reason its derivation is given herein.

The process of approximate analysis is closely related to those used by A. and L. Föppl[14] and by S. Timoshenko,[15] but differs in the following respects:

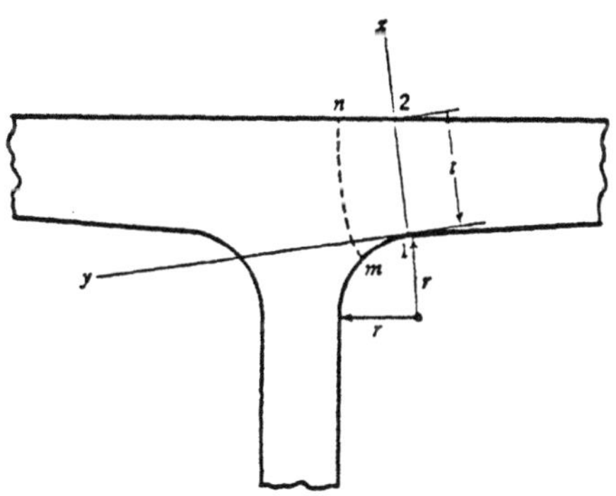

FIG. 27.—I-BEAM IN TORSION.

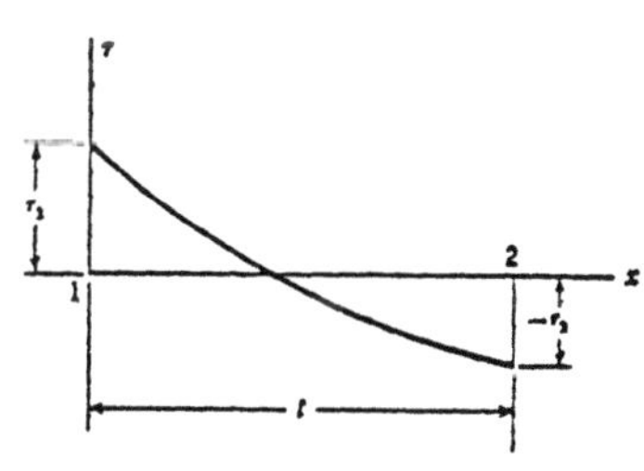

FIG. 28.—SHEARING STRESSES AT SECTION 1-2, IN FIG. 27.

First, consideration of the influence of the straight edge of the cross-section opposite the fillet led to a slightly greater concentration factor near the beginning of the fillet, at Point 1 in Fig. 27, than that obtained by Timoshenko's

NOTE.—The paper by Inge Lyse, M. Am. Soc. C. E., and Bruce G. Johnston, Jun. Am. Soc. C. E., is published in this number of *Proceedings*. This discussion is printed in *Proceedings* in order that the views expressed may be brought before all members for further discussion.

[12] Prof. of Theoretical and Applied Mechanics, Univ. of Illinois, Urbana, Illinois.

[13] Research Asst., Dept. of Civ. Eng., Columbia, Univ., New York, N. Y.

[13a] Received by the Secretary March 1, 1935.

[14] "Drang und Zwang," by A. and L. Föppl, Vol. 2, Second Edition, 1928, p. 73.

[15] "Theory of Elasticity," by S. Timoshenko, 1934, p. 259.

formula; and, second, the writers had before them the detailed data obtained by P. A. Cushman[16] in tests with soap films. These data show clearly that the maximum shearing stress occurs at a point such as m in Fig. 27, and that there is a notable increase of stress from Point 1 to Point m. Special consideration was given to this increase.

Let τ_0 denote the shearing stress that would exist at Point 1 in Fig. 27 if the edge of the cross-section at that point were straightened out by moving the beginning of the fillet toward the left, when, at the same time, the thickness of the flange, t, the torsion factor, K, and the total twisting moment, T, are left unchanged. The stress, τ_0 is defined by the formula:

$$\tau_0 = \frac{T\,t}{K} \qquad (57)$$

Let τ_1 and τ_m denote the actual shearing stresses at Points 1 and m, τ_m being the maximum stress at the fillet. Then $\frac{\tau_1}{\tau_0}$ is the concentration factor at Point 1 and $\frac{\tau_m}{\tau_0}$ is the desired concentration factor for the fillet.

Prandtl's soap film analogy, which the authors have used advantageously, furnishes the key to the solution. It is noted that the soap film is stretched over an opening shaped like the cross-section and is inflated a small amount by an excess air pressure on one side. The shearing stresses on the original cross-section follow the contour lines on the film, the edge being one of the contour lines. Furthermore, the shearing stresses are proportional to the slopes of the film, or proportional to the density of the contour lines.

The contour lines at Section 1–2 in Fig. 27 must be approximately perpendicular to that section. Accordingly, the shearing stresses, τ, at that section are approximately in the direction of y. Since the slope of the film is $c\tau$, c being a constant, the curvature of the film at Section 1–2 in the direction of x becomes $c\,\frac{d\tau}{dx}$. The curvature of the film in the direction of y at Section 1–2 is accounted for by the curving of the contour lines. If the radius of curvature of the contour lines at a particular point is R, then the curvature of the film in the direction of y at the same point becomes $c\,\frac{\tau}{R}$. The curvature of the surface is the sum of the curvatures in the directions of x and y; that is, $c\left(\frac{d\tau}{dx} + \frac{\tau}{R}\right)$.

The equilibrium of the film requires that this combined curvature be constant. Since $R = r$ at Point 1 and $R = \infty$ at Point 2, it follows that,

$$\left[\frac{d\tau}{dx}\right]_1 + \frac{\tau_1}{r} = \left[\frac{d\tau}{dx}\right]_2 \qquad (58)$$

[16] "Shearing Stresses in Torsion and Bending by Membrane Analogy," by P. A. Cushman, Doctoral Dissertation, Univ. of Michigan, 1932.

Since Points 1 and 2 are on the same contour line, it is also required that,

$$\int_0^t \tau \, dx = 0 \quad \text{.......................(59)}$$

The diagram of the shearing stresses at Section 1-2 must be shaped about as shown in Fig. 28. Equation (60) has been constructed so that it satisfies this general requirement of shape as well as the specific requirements in Equations (58) and (59), and, therefore, it may be assumed to represent the shearing stress approximately:

$$\tau = \tau_1 \left[1 - \left(2 + \frac{t}{3r} \right) \frac{x}{t} + \frac{x^2}{2\,tr} \right] \quad \text{..............(60)}$$

The constant curvature of the film may be computed as:

$$c \left[\frac{d\tau}{dx} \right]_2 = - \frac{2\,c\,\tau_1}{t} \left(1 - \frac{t}{3r} \right) \quad \text{................(61)}$$

If the fillet were some distance away, this curvature would remain the same, τ_1 would be replaced by τ_0, and the term containing r would disappear. Consequently,

$$\tau_0 = \tau_1 \left(1 - \frac{t}{3r} \right) \quad \text{....................(62)}$$

which gives the concentration factor at Point 1,

$$\frac{\tau_1}{\tau_0} = \frac{1}{1 - \dfrac{t}{3r}} \quad \text{........................(63)}$$

The important indications of Equation (63) are preserved when the following simpler formula is substituted:

$$\frac{\tau_1}{\tau_0} = 1 + \frac{t}{3r} \quad \text{.........................(64)}$$

As was mentioned, the maximum stress, τ_m, occurs at a point such as m in Fig. 27. Equation (64) can be used to obtain an estimate of this stress by making the following replacements: t is replaced by a distance, t_m, measured along a curved section as drawn from m to n; and τ_0 is replaced by a stress corresponding to straight edges and a thickness, t_m: that is, according to Equation (57), by $\tau_0 \dfrac{t_m}{t}$. Thus, the concentration factor becomes,

$$\frac{\tau_m}{\tau_0} = \frac{t_m}{t} \left(1 + \frac{t_m}{3r} \right) \quad \text{......................(65)}$$

A reasonable estimate of t_m is,

$$t_m = t + 0.3\,r \quad \text{.......................(66)}$$

This value, substituted in Equation (65), leads directly to the formula represented graphically by the authors,

$$\frac{\tau_m}{\tau_0} = 1.2 + \frac{1}{3}\left(\frac{t}{r} + \frac{r}{t}\right) \qquad (67)$$

Reprinted from
Proceedings of the American Society of Civil Engineers
61 (1935) 509-512

ÉLASTICITÉ. — *Contribution au problème d'équilibre d'élasticité d'un solide indéfini limité par un plan.* Note (¹) de M. **Raymond Mindlin**, présentée par M. Caquot.

Les tensions causées par des forces ayant leur point d'application à l'intérieur d'un solide homogène, limité par une surface plane mais indéfini d'un côté de ce plan, peuvent être exprimées au moyen de fonctions biharmoniques. (B. Galerkin, *Comptes rendus*, 190, 1930, p. 1047; H. M. Westergaard, *Bulletin of the American Mathematical Society*, 1935.)

Le solide est limité par le plan $z = 0$, la direction positive de l'axe z étant à l'intérieur de ce corps. Si une force P agit au point $(0, 0, +c)$ et a la

(¹) Séance du 16 septembre 1935.

direction de l'axe positif z, la fonction harmonique est alors

$$\Phi_3 = \frac{P}{8\pi(1-\mu)}\Big\{ R_1 + [8\mu(1-\mu)-1]R_2$$
$$+ 4(1-2\mu)[(1-\mu)z - \mu c]\log(R_2+z+c) - \frac{2cz}{R_2}\Big\},$$

où μ est le coefficient de Poisson,

$$R_1^2 = x^2+y^2+(z-c)^2 \quad \text{et} \quad R_2^2 = x^2+y^2+(z+c)^2.$$

Si la force est dirigée selon la direction positive de l'axe x, on a deux fonctions

$$\Phi_1 = \frac{P}{8\pi(1-\mu)}\Big\{ R_1 + R_2 - \frac{2c^2}{R_2} - 4(1-\mu)(1-2\mu)[R_2-(z+c)\log(R_2+z+c)]\Big\},$$
$$\Phi_3 = \frac{P}{8\pi(1-\mu)}\Big\{ \frac{2cx}{R_2} + 2(1-2\mu)x\log(R_2+z+c)\Big\}.$$

A l'aide de ces trois fonctions, les tensions causées par des forces quelconques à l'intérieur d'un corps limité par un plan, mais indéfini d'un côté de ce plan, peuvent être calculées au moyen des formules données par M. Galerkin.

Les solutions données permettent de résoudre, pour le solide considéré ici, le cas de deux forces égales et directement opposées dont les points d'application sont voisins et beaucoup d'autres cas particuliers comme dans le problème du solide indéfini.

Reprinted from
Comptes Rendus de l'Académie des Sciences, Paris
201 (1935) 536-537

Force at a Point in the Interior of a Semi-Infinite Solid

RAYMOND D. MINDLIN, *Department of Civil Engineering, Columbia University*

(Received March 13, 1936)

A solution of the three-dimensional elasticity equations for a homogeneous isotropic solid is given for the case of a concentrated force acting in the interior of a semi-infinite solid. This represents the fundamental solution having a singular point in a solid bounded by a plane. From it may be derived, by a known method of synthesis, the solutions for the semi-infinite solid which correspond to the solutions known as nuclei of strain in the solid of indefinite extent.

1. INTRODUCTION

ONE of the fundamental results in the theory of elasticity is the Kelvin solution[1] for a force applied at a point in a solid of indefinite extent. It is well known that stress distributions for a number of problems of practical importance may be obtained from the Kelvin solution by methods of synthesis and superposition.[2] For example, the problem, first solved by Lamé, of the stresses in a spherical container under uniform internal or external pressure may be treated by superposing stresses derived from Kelvin's results. First we form, from the Kelvin single force solution, the stresses for a double force, that is, a pair of equal and opposite forces acting at neighboring points. By superposing three mutually perpendicular double forces, we arrive at the solution known as the center of compression. Finally, the addition of a uniform tension or pressure to the center of compression leads to the solution of Lamé's problem.

The classical problem of Boussinesq dealing with a normal force applied at the plane boundary of a semi-infinite solid[3] has found practical application in the study of the distribution of foundation pressures, contact stresses, and in certain problems of soil mechanics. It is known that Boussinesq's problem may also be solved by superposing solutions derived from Kelvin's results. To do this we form a line of centers of compression by integrating the stresses for a single center of compression. By combining this result with the original Kelvin single-force solution, the Boussinesq formulas are obtained.

The Kelvin solution may be used in studying stresses due to a force applied at a great distance from a boundary while the Boussinesq solution is applicable in the case where the force acts at the surface. The solutions described in this paper fill in the gap between the two by giving the stresses for the case where the force is applied near the surface. Such a condition is approached in a number of practical problems such, for example, as in the case of a guy wire anchor. Again, by integrating these new solutions along a line, we may approximate the conditions produced by a friction pile or an anchor rod.

Solutions of the elasticity equations such as the single force, double force, center of compression, and line of centers of compression are referred to as nuclei of strain.[2] The number of nuclei of strain which may be obtained by synthesis from the Kelvin solution is unlimited and further combinations lead to the solutions for such important cases as Southwell's problem of the spherical cavity in an unlimited solid under simple tension[4] and Cerruti's problem of a force applied tangentially at the plane boundary of a semi-infinite solid.[5]

Each of the nuclei of strain obtained from Kelvin's results is a solution applicable to a solid of indefinite extent. With the solution for a force in the interior of a semi-infinite solid, as described in this paper, we may apply the method of synthesis to develop a new series of nuclei of strain for which the stresses vanish on the plane boundary of the semi-infinite solid. Such solutions may be termed half-space nuclei of strain.

[1] Sir W. Thomson, Cambridge and Dublin Math. J. (1848). See also A. E. H. Love, *Mathematical Theory of Elasticity*, fourth edition (Cambridge, 1927), p. 183; and S. Timoshenko, *Theory of Elasticity* (New York, 1934), p. 321.

[2] Love, p. 186, and Timoshenko, p. 323.

[3] J. Boussinesq, *Applications des Potentials* . . . (Paris, 1885). Also Love, p. 191, and Timoshenko, p. 331.

[4] R. V. Southwell and H. J. Gough, Phil. Mag. **1**, 71 (1926).

[5] V. Cerruti, Acc. Lincei. Mem. fis. mat., Roma **13**, 81 (1882). Also Love, p. 241.

The notion of such half-space nuclei of strain suggests some interesting possibilities. By employing these half-space nuclei as unit solutions, it may be possible to find solutions analogous to those obtained from the Kelvin nuclei but with the additional feature that a new plane boundary is introduced. For example, the solution analogous to Southwell's problem would pertain to a spherical cavity near the surface of a semi-infinite solid; the solution analogous to the problem of Boussinesq would be that for a force applied at the surface of a body bounded by two perpendicular planes (i.e., the transmission of force through "quarter-space"); the solution analogous to the hollow sphere under uniform pressure would apply to an eccentrically hollow sphere.

The present paper deals with the fundamental solution for the single force in the interior of a semi-infinite solid. The problem is divided into two parts: (1) force normal to the boundary; (2) force parallel to the boundary.

The discovery of the solutions followed a study of H. M. Westergaard's interpretation[6] of the problems of Kelvin, Boussinesq and Cerruti in terms of Galerkin vectors and the results obtained are shown to include these problems for certain limiting conditions.

2. The Galerkin Vector

The components of stress in an elastic isotropic solid were expressed by B. Galerkin[7] in terms of partial differential coefficients of three functions satisfying a fourth-order equation. P. F. Papkovitch[8] and H. M. Westergaard[9] have expressed the displacements and stresses in terms of a vector stress function whose scalar coefficients are the Galerkin functions. The displacements are given, in Westergaard's form of the Galerkin vector, by[9]

$$\begin{aligned} u&=(1/2G)[2(1-\mu)\Delta X-(\partial/\partial x)\operatorname{div}\mathbf{F}],\\ v&=(1/2G)[2(1-\mu)\Delta Y-(\partial/\partial y)\operatorname{div}\mathbf{F}],\\ w&=(1/2G)[2(1-\mu)\Delta Z-(\partial/\partial z)\operatorname{div}\mathbf{F}], \end{aligned}\tag{1}$$

where: X, Y, Z are the Galerkin functions,
$\mathbf{i}$, $\mathbf{j}$, $\mathbf{k}$ constitute an orthogonal system of unit vectors
$\mathbf{F}=\mathbf{i}X+\mathbf{j}Y+\mathbf{k}Z$ is the Galerkin vector,
Δ is Laplace's operator,
G is the modulus of rigidity,
μ is Poisson's ratio.

The stresses are given by[9]

$$\begin{aligned} \sigma_x&=2(1-\mu)\partial\Delta X/\partial x+(\mu\Delta-\partial^2/\partial x^2)\operatorname{div}\mathbf{F},\\ \sigma_y&=2(1-\mu)\partial\Delta Y/\partial y+(\mu\Delta-\partial^2/\partial y^2)\operatorname{div}\mathbf{F},\\ \sigma_z&=2(1-\mu)\partial\Delta Z/\partial z+(\mu\Delta-\partial^2/\partial z^2)\operatorname{div}\mathbf{F},\\ \tau_{yz}&=(1-\mu)(\partial\Delta Y/\partial z+\partial\Delta Z/\partial y)-(\partial^2/\partial y\partial z)\operatorname{div}\mathbf{F},\\ \tau_{zx}&=(1-\mu)(\partial\Delta Z/\partial x+\partial\Delta X/\partial z)-(\partial^2/\partial z\partial x)\operatorname{div}\mathbf{F},\\ \tau_{xy}&=(1-\mu)(\partial\Delta X/\partial y+\partial\Delta Y/\partial x)-(\partial^2/\partial x\partial y)\operatorname{div}\mathbf{F}. \end{aligned}\tag{2}$$

For the case of zero body force, the Galerkin vector must satisfy the biharmonic equation,

$$\Delta^2\mathbf{F}=0.\tag{3}$$

It was observed by Professor Westergaard[6] that the coefficient of a single axially symmetrical component of the Galerkin vector is identical with Love's function[10] for strain symmetrical about the corresponding axis.

3. Nuclei of Strain in Terms of Galerkin Vectors

The solutions given in sections 4 and 5 were obtained by superposition of nuclei of strain for an unlimited solid. It has been found convenient, in applying the method of superposition, to express the nuclei of strain in terms of Galerkin vectors.

In the following list of functions, A, B, C, $\cdots$ H are constants and $R=(x^2+y^2+z^2)^{\frac{1}{2}}$. It may be observed that each of the functions satisfies the biharmonic equation and that the stresses derived from them vanish at infinity.

[6] H. M. Westergaard, *Lectures on Elasticity while Visiting Professor at the University of Michigan* (Summer Session, 1934).

[7] B. Galerkin, Comptes rendus **190**, 1047 (1930).

[8] P. F. Papkovitch, Comptes rendus **195**, 513, 754 (1932).

[9] H. M. Westergaard, Bull. Am. Math. Soc. **41**, 695 (1935).

[10] Love, p. 274; Timoshenko, p. 309.

GALERKIN VECTORS FOR NUCLEI OF STRAIN IN AN UNLIMITED SOLID

A. Single force (Kelvin's problem).

(1) $\mathbf{i}AR$ (single force in x direction)
(2) $\mathbf{j}AR$ (single force in y direction)
(3) $\mathbf{k}AR$ (single force in z direction)

B. Double force

(1) $\mathbf{i}Bx/R$ (double force in x direction)
(2) $\mathbf{j}By/R$ (double force in y direction)
(3) $\mathbf{k}Bz/R$ (double force in z direction)

C. Double force with moment

(1) $\mathbf{i}Cz/R$ (double force in x direction with moment about y axis)
(2) $\mathbf{j}Cx/R$ (double force in y direction with moment about z axis)
(3) $\mathbf{k}Cy/R$ (double force in z direction with moment about x axis)
(4) $\mathbf{i}Cy/R$ (double force in x direction with moment about z axis)
(5) $\mathbf{j}Cz/R$ (double force in y direction with moment about x axis)
(6) $\mathbf{k}Cx/R$ (double force in z direction with moment about y axis)

D. Line of double forces with moment

(1) $\mathbf{i}Dz \log (R+x)$ (type $C(1)$ along x axis from $x=0$ to $x=-\infty$)
(2) $\mathbf{j}Dx \log (R+y)$ (type $C(2)$ along y axis from $y=0$ to $y=-\infty$)
(3) $\mathbf{k}Dy \log (R+z)$ (type $C(3)$ along z axis from $z=0$ to $z=-\infty$)
(4) $\mathbf{i}Dy \log (R+x)$ (type $C(4)$ along x axis from $x=0$ to $x=-\infty$)
(5) $\mathbf{j}Dz \log (R+y)$ (type $C(5)$ along y axis from $y=0$ to $y=-\infty$)
(6) $\mathbf{k}Dx \log (R+z)$ (type $C(6)$ along z axis from $z=0$ to $z=-\infty$)

E. Center of compression or dilatation

(1) $\mathbf{i}E \log (R+x)$
(2) $\mathbf{j}E \log (R+y)$ (All three yield the same stresses)
(3) $\mathbf{k}E \log (R+z)$

F. Line of compression or dilatation of constant strength

(1) $\mathbf{i}F[x \log (R+x)-R]$ (along x axis from $x=0$ to $x=-\infty$)
(2) $\mathbf{j}F[y \log (R+y)-R]$ (along y axis from $y=0$ to $y=-\infty$)
(3) $\mathbf{k}F[z \log (R+z)-R]$ (along z axis from $z=0$ to $z=-\infty$)

G. Doublet (double center of compression-dilatation)

(1) $\mathbf{i}G(1/R)$ (axis of doublet parallel to x axis)
(2) $\mathbf{j}G(1/R)$ (axis of doublet parallel to y axis)
(3) $\mathbf{k}G(1/R)$ (axis of doublet parallel to z axis)

H. Linearly varying line of doublets with strength proportional to distance from the origin

(1) $\mathbf{i}H[z \log (R+z)-R]$ (type $G(1)$ along z axis from $z=0$ to $z=-\infty$)
(2) $\mathbf{j}H[x \log (R+x)-R]$ (type $G(2)$ along x axis from $x=0$ to $x=-\infty$)
(3) $\mathbf{k}H[y \log (R+y)-R]$ (type $G(3)$ along y axis from $y=0$ to $y=-\infty$)
(4) $\mathbf{i}H[y \log (R+y)-R]$ (type $G(1)$ along y axis from $y=0$ to $y=-\infty$)
(5) $\mathbf{j}H[z \log (R+z)-R]$ (type $G(2)$ along z axis from $z=0$ to $z=-\infty$)
(6) $\mathbf{k}H[x \log (R+x)-R]$ (type $G(3)$ along x axis from $x=0$ to $x=-\infty$)

4. FORCE NORMAL TO THE BOUNDARY OF A SEMI-INFINITE SOLID

The semi-infinite solid is considered to be bounded by the plane $z=0$, the positive z axis penetrating into the body. A force P is applied at point $(0, 0, +c)$ and acts in the positive z direction (Fig. 1).

Since the stress will be symmetrical about the z axis, we need consider only the $\mathbf{k}$ component of $\mathbf{F}$. Transforming to cylindrical coordinates r, θ, z and noting that Z is independent of θ, we find, from Eqs. (1) and (2), the following expressions for the displacements and stresses in symmetrical cylindrical coordinates:[10]

$$\begin{aligned} U &= -(1/2G)(\partial^2 Z/\partial r\partial z), \\ w &= (1/2G)[2(1-\mu)\Delta Z-\partial^2 Z/\partial z^2], \\ \sigma_r &= (\partial/\partial z)[\mu\Delta Z-\partial^2 Z/\partial r^2], \\ \sigma_\theta &= (\partial/\partial z)[\mu\Delta Z-(1/r)(\partial Z/\partial r)], \\ \sigma_z &= (\partial/\partial z)[(2-\mu)\Delta Z-\partial^2 Z/\partial z^2], \\ \tau_{rz} &= (\partial/\partial r)[(1-\mu)\Delta Z-\partial^2 Z/\partial z^2]. \end{aligned} \tag{4}$$

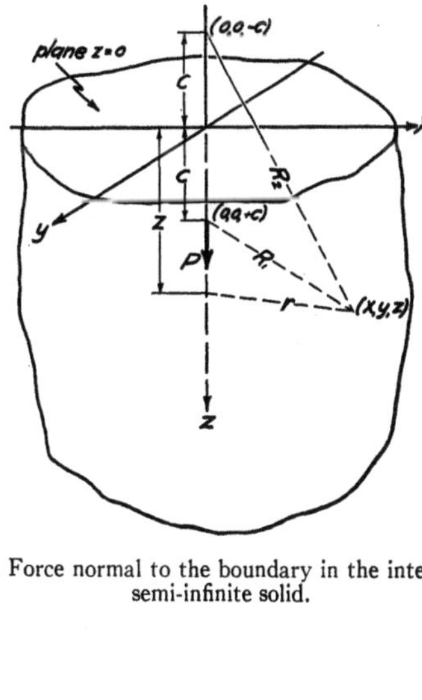

FIG. 1. Force normal to the boundary in the interior of a semi-infinite solid.

We first apply the Kelvin solution at point $(0, 0, +c)$ by replacing R with $R_1=(r^2+(z-c)^2)^{\frac{1}{2}}$ in solution $A(3)$ of section 3. We note that stresses σ_z and τ_{rz} are produced on the plane $z=0$. The boundary conditions for $z=0$ are

$$[\sigma_z]_{z=0}=[\tau_{rz}]_{z=0}=0. \tag{5}$$

It is therefore necessary to add further solutions to remove the boundary stresses and these solutions must introduce no new singularities in the region $z\geqq 0$. Nuclei of strain of types $A(3)$, $B(3)$, $E(3)$, $F(3)$ and $G(3)$, applied at $(0, 0, -c)$, are found to satisfy these conditions if the proper values are assigned to the constants A, B, E, F and G.

To transfer nuclei of strain from the origin to $(0, 0, -c)$, we replace z with $(z+c)$ and R with $R_2=(r^2+(z+c)^2)^{\frac{1}{2}}$ in the formulas of section 3.

The values of the constants are obtained from the boundary Eqs. (5) and the equilibrium condition

$$P=-\int_0^\infty 2\pi r\sigma_z dr, \quad (z>c). \tag{6}$$

This procedure leads us to the required Galerkin vector:

$$\mathbf{F}=[P\mathbf{k}/8\pi(1-\mu)]\{R_1+(3-4\mu)R_2-2c(z+c)/R_2-4(1-2\mu)c\log(R_2+z+c) \\ +4(1-\mu)(1-2\mu)[(z+c)\log(R_2+z+c)-R_2]+2c^2/R_2\}. \tag{7}$$

The solution for a force in the interior of a semi-infinite solid and normal to the boundary is seen to be compounded of six nuclei of strain which, in an unlimited solid, represent: (1) a single force at $(0, 0, +c)$; (2) a single force at $(0, 0, -c)$; (3) a double force at $(0, 0, -c)$; (4) a center of compression at $(0, 0, -c)$; (5) a line of centers of compression extending from $z=-c$ to $z=-\infty$; (6) a doublet at $(0, 0, -c)$.

Collecting terms in Eq. (7) we obtain:[11]

$$\mathbf{F}=[P\mathbf{k}/8\pi(1-\mu)]\{R_1+[8\mu(1-\mu)-1]R_2+4(1-2\mu)[(1-\mu)z-\mu c]\log(R_2+z+c)-2cz/R_2\}. \tag{8}$$

Eqs. (8) and (4) yield the displacements and stresses:

$$U=\frac{Pr}{16\pi G(1-\mu)}\left[\frac{z-c}{R_1^3}+\frac{(3-4\mu)(z-c)}{R_2^3}-\frac{4(1-\mu)(1-2\mu)}{R_2(R_2+z+c)}+\frac{6cz(z+c)}{R_2^5}\right],$$

$$w=\frac{P}{16\pi G(1-\mu)}\left[\frac{3-4\mu}{R_1}+\frac{8(1-\mu)^2-(3-4\mu)}{R_2}+\frac{(z-c)^2}{R_1^3}+\frac{(3-4\mu)(z+c)^2-2cz}{R_2^3}+\frac{6cz(z+c)^2}{R_2^5}\right],$$

$$\sigma_r=\frac{P}{8\pi(1-\mu)}\left[\frac{(1-2\mu)(z-c)}{R_1^3}-\frac{(1-2\mu)(z+7c)}{R_2^3}+\frac{4(1-\mu)(1-2\mu)}{R_2(R_2+z+c)}-\frac{3r^2(z-c)}{R_1^5}\right. \\ \left.+\frac{6c(1-2\mu)(z+c)^2-6c^2(z+c)-3(3-4\mu)r^2(z-c)}{R_2^5}-\frac{30cr^2z(z+c)}{R_2^7}\right], \tag{9}$$

$$\sigma_\theta=\frac{P(1-2\mu)}{8\pi(1-\mu)}\left[\frac{(z-c)}{R_1^3}+\frac{(3-4\mu)(z+c)-6c}{R_2^3}-\frac{4(1-\mu)}{R_2(R_2+z+c)}+\frac{6c(z+c)^2}{R_2^5}-\frac{6c^2(z+c)}{(1-2\mu)R_2^5}\right],$$

$$\sigma_z=\frac{P}{8\pi(1-\mu)}\left[-\frac{(1-2\mu)(z-c)}{R_1^3}+\frac{(1-2\mu)(z-c)}{R_2^3}-\frac{3(z-c)^3}{R_1^5}\right. \\ \left.-\frac{3(3-4\mu)z(z+c)^2-3c(z+c)(5z-c)}{R_2^5}-\frac{30cz(z+c)^3}{R_2^7}\right],$$

$$\tau_{rz}=\frac{Pr}{8\pi(1-\mu)}\left[-\frac{1-2\mu}{R_1^3}+\frac{1-2\mu}{R_2^3}-\frac{3(z-c)^2}{R_1^5}-\frac{3(3-4\mu)z(z+c)-3c(3z+c)}{R_2^5}-\frac{30cz(z+c)^2}{R_2^7}\right].$$

[11] R. D. Mindlin, Comptes rendus 201, 536 (1935).

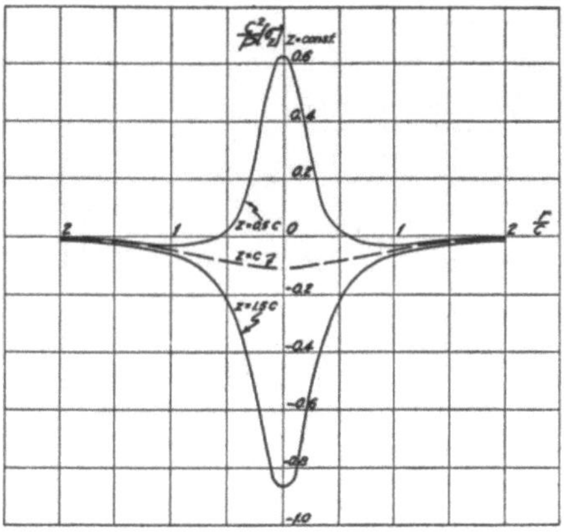

FIG. 2. Normal stress on planes parallel to boundary for the case of a force normal to the boundary ($\mu=0.3$).

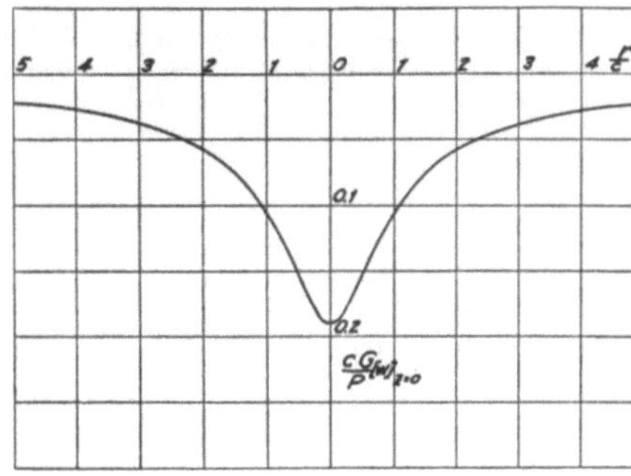

FIG. 3. Vertical deflection of the surface for the case of a force normal to the boundary ($\mu=0.3$).

Fig. 2 shows the distribution of normal stress on horizontal planes, and Fig. 3 shows the vertical deflection of the surface of the semi-infinite solid.

When $c\rightarrow\infty$ all terms in Eqs. (9) containing R_2 vanish and the solution becomes that for Kelvin's problem where the force is applied at $(0, 0, +c)$ in the positive z direction. The corresponding Galerkin vector is

$$\mathbf{F}=\mathbf{k}[PR_1/8\pi(1-\mu)]. \tag{10}$$

When $c\rightarrow 0$ Eqs. (9) give the stresses and displacements for Boussinesq's problem.[12] These stresses are represented by the Galerkin vector

$$\mathbf{F}=\mathbf{k}(P/2\pi)[2\mu R+(1-2\mu)z\log(R+z)]. \tag{11}$$

The solution for the problem of Boussinesq was given by Professor Westergaard[6] in the form of the Galerkin vector

$$\mathbf{F}=\mathbf{k}(PR/2\pi) \tag{12a}$$

combined with the potential function

$$\phi=-[(1-2\mu)/2\pi]P\log(R+z). \tag{12b}$$

The relation between ϕ and $\mathbf{F}$ is given by

$$\operatorname{grad}\phi=2(1-\mu)\Delta\mathbf{F}-\operatorname{grad}\operatorname{div}\mathbf{F}. \tag{13}$$

As a further verification of the solution, we may compare the stresses with those obtained for the corresponding two-dimensional case by E. Melan.[13] While the approach to the solution in three dimensions was simplified through the use of cylindrical coordinates, the comparison with Melan's formulas is facilitated by referring the components of stress to rectangular coordinates. Eqs. (9) then become:

---

[12] The stresses and displacements for the problems of Boussinesq and Cerruti are given by F. Vogt, *Det Norske Videnskaps Akademi i Oslo*, Math.-Naturv. klasse, No. 2, pp. 5, 20 (1925).

[13] E. Melan, Zeits. f. angew. Math. und Mech. **12**, 343 (1932). This comparison was suggested by Professor S. Timoshenko.

$$\sigma_x=\frac{P}{8\pi(1-\mu)}\left[\frac{(1-2\mu)(z-c)}{R_1^3}-\frac{3x^2(z-c)}{R_1^5}+\frac{(1-2\mu)[3(z-c)-4\mu(z+c)]}{R_2^3}\right.$$
$$-\frac{3(3-4\mu)x^2(z-c)-6c(z+c)[(1-2\mu)z-2\mu c]}{R_2^5}-\frac{30cx^2z(z+c)}{R_2^7}$$
$$\left.-\frac{4(1-\mu)(1-2\mu)}{R_2(R_2+z+c)}\left(1-\frac{x^2}{R_2(R_2+z+c)}-\frac{x^2}{R_2^2}\right)\right],$$

$$\sigma_y=\frac{P}{8\pi(1-\mu)}\left[\frac{(1-2\mu)(z-c)}{R_1^3}-\frac{3y^2(z-c)}{R_1^5}+\frac{(1-2\mu)[3(z-c)-4\mu(z+c)]}{R_2^3}\right.$$
$$-\frac{3(3-4\mu)y^2(z-c)-6c(z+c)[(1-2\mu)z-2\mu c]}{R_2^5}-\frac{30cy^2z(z+c)}{R_2^7}$$
$$\left.-\frac{4(1-\mu)(1-2\mu)}{R_2(R_2+z+c)}\left(1-\frac{y^2}{R_2(R_2+z+c)}-\frac{y^2}{R_2^2}\right)\right], \quad (14)$$

$$\sigma_z=\frac{P}{8\pi(1-\mu)}\left[-\frac{(1-2\mu)(z-c)}{R_1^3}+\frac{(1-2\mu)(z-c)}{R_2^3}-\frac{3(z-c)^3}{R_1^5}\right.$$
$$\left.-\frac{3(3-4\mu)z(z+c)^2-3c(z+c)(5z-c)}{R_2^5}-\frac{30cz(z+c)^3}{R_2^7}\right],$$

$$\tau_{yz}=\frac{Py}{8\pi(1-\mu)}\left[-\frac{(1-2\mu)}{R_1^3}+\frac{(1-2\mu)}{R_2^3}-\frac{3(z-c)^2}{R_1^5}-\frac{3(3-4\mu)z(z+c)-3c(3z+c)}{R_2^5}-\frac{30cz(z+c)^2}{R_2^7}\right],$$

$$\tau_{zx}=\frac{Px}{8\pi(1-\mu)}\left[-\frac{(1-2\mu)}{R_1^3}+\frac{(1-2\mu)}{R_2^3}-\frac{3(z-c)^2}{R_1^5}-\frac{3(3-4\mu)z(z+c)-3c(3z+c)}{R_2^5}-\frac{30cz(z+c)^2}{R_2^7}\right],$$

$$\tau_{xy}=\frac{Pxy}{8\pi(1-\mu)}\left[-\frac{3(z-c)}{R_1^5}-\frac{3(3-4\mu)(z-c)}{R_2^5}+\frac{4(1-\mu)(1-2\mu)}{R_2^2(R_2+z+c)}\left(\frac{1}{R_2+z+c}+\frac{1}{R_2}\right)-\frac{30cz(z+c)}{R_2^7}\right].$$

We consider a uniform distribution of forces of magnitude p per unit of length along a line through $(0, 0, +c)$ parallel to the y axis. If db is a small element of this line at a distance b from the z axis, the stresses due to the uniform pressure p acting on db are found by substituting pdb for P and $(y-b)$ for y in Eqs. (14). If, now, we integrate these formulas with respect to b between the limits $-\infty$ and $+\infty$, we obtain the corresponding solution for plane strain. By taking account of the relations between the elastic constants for plane strain and plane stress and also noting an interchange of axes, this procedure leads directly to the stresses given by Melan.

5. Force Parallel to the Boundary of a Semi-Infinite Solid

In this case there is no axial symmetry and we employ rectangular coordinates (x, y, z), and Eqs. (1) and (2). The force P is applied at $(0, 0, +c)$ and acts in the positive x direction (Fig. 4).

We again begin by applying the Kelvin solution at $(0, 0, +c)$, this time using type $A(1)$. To remove the stresses on plane $z=0$, introduced by the single force, we employ nuclei of strain of types $A(1)$, $C(6)$, $D(6)$, $G(1)$ and $H(1)$, again substituting $(z+c)$ for z and $R_2=(x^2+y^2+(z+c)^2)^{\frac{1}{2}}$ for R. The constants associated with these solutions are determined from the boundary conditions

$$[\sigma_z]_{z=0}=[\tau_{yz}]_{z=0}=[\tau_{zx}]_{z=0}=0 \quad (15)$$

and the equilibrium condition

$$P=-4\int_0^\infty\int_0^\infty \tau_{zx}dydx, \quad (z>c). \tag{16}$$

The result of these operations is the Galerkin vector[11]

$$\mathbf{F}=[P/8\pi(1-\mu)](\mathbf{i}\{R_1+R_2-2c^2/R_2 +4(1-\mu)(1-2\mu)[(z+c)\log(R_2+z+c)-R_2]\} +\mathbf{k}\{2cx/R_2+2(1-2\mu)x\log(R_2+z+c)\}). \tag{17}$$

Thus, for the case of a force in the interior of a semi-infinite solid, parallel to the boundary, the solution is obtained by superposition of six nuclei of strain for the unlimited solid: (1) a single force at $(0, 0, +c)$; (2) a single force at $(0, 0, -c)$; (3) a doublet at $(0, 0, -c)$; (4) a semi-infinite line of doublets extending from $z=-c$ to $z=-\infty$ with strength proportional to the distance from $z=-c$; (5) a double force with moment at $(0, 0, -c)$; (6) a semi-infinite line of double forces with moment extending from $z=-c$ to $z=-\infty$.

The displacements and stresses, obtained from Eqs. (1), (2) and (17), are found to be:

$$u=\frac{P}{16\pi G(1-\mu)}\left[\frac{3-4\mu}{R_1}+\frac{1}{R_2}+\frac{x^2}{R_1^3}+\frac{(3-4\mu)x^2}{R_2^3}+\frac{2cz}{R_2^3}\left(1-\frac{3x^2}{R_2^2}\right)+\frac{4(1-\mu)(1-2\mu)}{R_2+z+c}\left(1-\frac{x^2}{R_2(R_2+z+c)}\right)\right],$$

$$v=\frac{Pxy}{16\pi G(1-\mu)}\left[\frac{1}{R_1^3}+\frac{3-4\mu}{R_2^3}-\frac{6cz}{R_2^5}-\frac{4(1-\mu)(1-2\mu)}{R_2(R_2+z+c)^2}\right],$$

$$w=\frac{Px}{16\pi G(1-\mu)}\left[\frac{z-c}{R_1^3}+\frac{(3-4\mu)(z-c)}{R_2^3}-\frac{6cz(z+c)}{R_2^5}+\frac{4(1-\mu)(1-2\mu)}{R_2(R_2+z+c)}\right],$$

$$\sigma_x=\frac{Px}{8\pi(1-\mu)}\left[-\frac{(1-2\mu)}{R_1^3}+\frac{(1-2\mu)(5-4\mu)}{R_2^3}-\frac{3x^2}{R_1^5}-\frac{3(3-4\mu)x^2}{R_2^5}-\frac{4(1-\mu)(1-2\mu)}{R_2(R_2+z+c)^2}\left(3-\frac{x^2(3R_2+z+c)}{R_2^2(R_2+z+c)}\right)+\frac{6c}{R_2^5}\left(3c-(3-2\mu)(z+c)+\frac{5x^2z}{R_2^2}\right)\right],$$

$$\sigma_y=\frac{Px}{8\pi(1-\mu)}\left[\frac{(1-2\mu)}{R_1^3}+\frac{(1-2\mu)(3-4\mu)}{R_2^3}-\frac{3y^2}{R_1^5}-\frac{3(3-4\mu)y^2}{R_2^5}-\frac{4(1-\mu)(1-2\mu)}{R_2(R_2+z+c)^2}\left(1-\frac{y^2(3R_2+z+c)}{R_2^2(R_2+z+c)}\right)+\frac{6c}{R_2^5}\left(c-(1-2\mu)(z+c)+\frac{5y^2z}{R_2^2}\right)\right], \tag{18}$$

$$\sigma_z=\frac{Px}{8\pi(1-\mu)}\left[\frac{(1-2\mu)}{R_1^3}-\frac{(1-2\mu)}{R_2^3}-\frac{3(z-c)^2}{R_1^5}-\frac{3(3-4\mu)(z+c)^2}{R_2^5}+\frac{6c}{R_2^5}\left(c+(1-2\mu)(z+c)+\frac{5z(z+c)^2}{R_2^2}\right)\right],$$

$$\tau_{yz}=\frac{Pxy}{8\pi(1-\mu)}\left[-\frac{3(z-c)}{R_1^5}-\frac{3(3-4\mu)(z+c)}{R_2^5}+\frac{6c}{R_2^5}\left(1-2\mu+\frac{5z(z+c)}{R_2^2}\right)\right],$$

$$\tau_{zx}=\frac{P}{8\pi(1-\mu)}\left[-\frac{(1-2\mu)(z-c)}{R_1^3}+\frac{(1-2\mu)(z-c)}{R_2^3}-\frac{3x^2(z-c)}{R_1^5}-\frac{3(3-4\mu)x^2(z+c)}{R_2^5}-\frac{6c}{R_2^5}\left(z(z+c)-(1-2\mu)x^2-\frac{5x^2z(z+c)}{R_2^2}\right)\right],$$

$$\tau_{xy}=\frac{Py}{8\pi(1-\mu)}\left[-\frac{(1-2\mu)}{R_1^3}+\frac{(1-2\mu)}{R_2^3}-\frac{3x^2}{R_1^5}-\frac{3(3-4\mu)x^2}{R_2^5}-\frac{4(1-\mu)(1-2\mu)}{R_2(R_2+z+c)^2}\left(1-\frac{x^2(3R_2+z+c)}{R_2^2(R_2+z+c)}\right)-\frac{6cz}{R_2^5}\left(1-\frac{5x^2}{R_2^2}\right)\right].$$

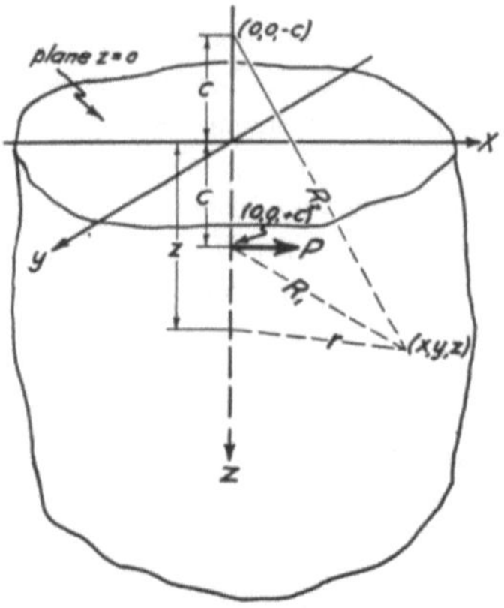

FIG. 4. Force parallel to the boundary in the interior of a semi-infinite solid.

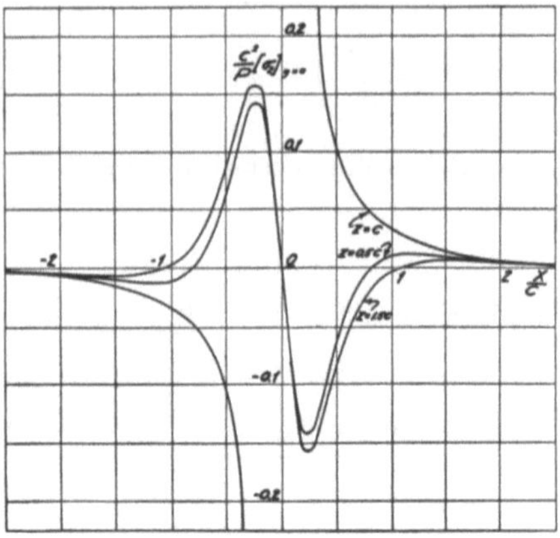

FIG. 5. Normal stress along lines parallel to the x axis on planes parallel to the boundary for the case of a force parallel to the boundary ($\mu=0.3$).

Fig. 5 shows the distribution of normal stress on horizontal planes, and Fig. 6 shows the vertical deflection of the surface of the solid along the x axis.

When $c\to\infty$ all terms in Eq. (18) containing R_2 vanish and the solution becomes that for Kelvin's problem with the force apolied at $(0, 0, +c)$ in the positive x direction. The corresponding Galerkin vector is

$$\mathbf{F}=\mathbf{i}[PR_1/8\pi(1-\mu)]. \tag{19}$$

When $c\to 0$ Eqs. (18) give the stresses and displacements for the problem of Cerruti.[12] These results are represented by the Galerkin vector

$$\begin{aligned}\mathbf{F}=[P/4\pi(1-\mu)](\mathbf{i}\{R+2(1-\mu)(1-2\mu)\\ \times[z\log(R+z)-R]\}\\ +\mathbf{k}\{(1-2\mu)x\log(R+z)\}). \quad (20)\end{aligned}$$

The solution for the problem of Cerruti was given by Professor Westergaard[6] in the form of the Galerkin vector

$$\begin{aligned}\mathbf{F}=[P/4\pi(1-\mu)][\mathbf{i}R\\ +\mathbf{k}(1-2\mu)x\log(R+z)] \quad (21a)\end{aligned}$$

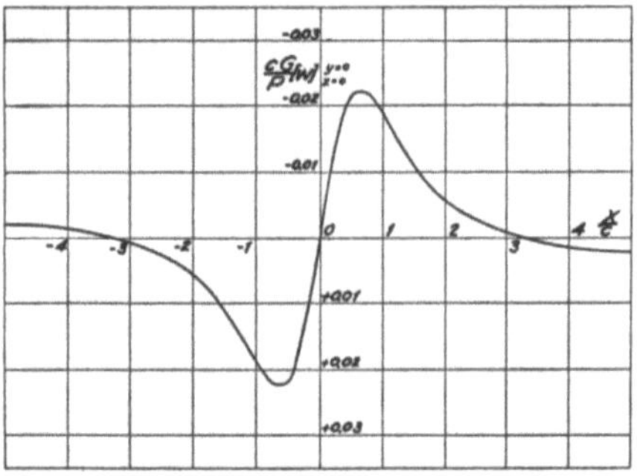

FIG. 6. Vertical deflection of the surface along the x axis for the case of a force parallel to the boundary ($\mu=0.3$).

combined with the potential function

$$\phi=[(1-2\mu)P/2\pi][x/(R+z)]. \tag{21b}$$

Following the same procedure as in the case of the normal force, the stresses given in Eqs. (18) may be reduced to Melan's formulas for the two-dimensional analog.[13]

The author is indebted to Professor H. M. Westergaard for valuable criticisms of the manuscript.

Reprinted from
Bulletin of the American Mathematical Society
42 (1936) 373-376

NOTE ON THE GALERKIN AND PAPKOVITCH STRESS FUNCTIONS

BY R. D. MINDLIN

1. *Introduction.* H. M. Westergaard* has given a useful interpretation of the Galerkin stress functions† as the components of a vector function satisfying a fourth order equation. From the Galerkin vector, P. F. Papkovitch‡ has developed a new solution of the three-dimensional elasticity equations for a homogeneous, isotropic solid in terms of harmonic functions. The same solution has been given by H. Neuber.§

Some interesting aspects of the Galerkin and Papkovitch functions may be observed when they are approached from a consideration of Helmholtz's theorem. In so doing, it is found that these functions may be reached by a direct analytical process and that they are connected through simple functional rela-

* H. M. Westergaard, this Bulletin, vol. 41 (1935), p. 695.

† B. Galerkin, Comptes Rendus, vol. 190 (1930), p. 1047. See also Todhunter and Pearson, *History of Elasticity*, vol. 2, part 2, pp. 268–270.

‡ P. F. Papkovitch, Comptes Rendus, vol. 195 (1932), pp. 513, 754. J. N. Goodler calls attention to Todhunter and Pearson, loc. cit., vol. 2, part 2, p. 373.

§ H. Neuber, Zeitschrift für angewandte Mathematik und Mechanik, vol. 14 (1934), p. 203.

tions with each other and with the potentials of the dilatation and rotation.

2. *The Helmholtz Transformation.* It is known that the displacement which gives rise to the dilatation is the gradient of a scalar potential function and the displacement which produces the rotation is the curl of a solenoidal vector potential function, so that the displacement may be expressed by

$$\boldsymbol{\rho} = \operatorname{grad} \phi + \operatorname{curl} \boldsymbol{S}, \tag{1}$$

where $\boldsymbol{\rho}$ is the vector displacement, ϕ is a scalar potential and $\boldsymbol{S}$ is a vector potential such that div $\boldsymbol{S}=0$. This transformation is commonly associated with the name of Helmholtz.

3. *Equilibrium Equation.* In vector form, the equilibrium equation in terms of the displacement is

$$\Delta\boldsymbol{\rho} + \frac{1}{1-2\mu} \operatorname{grad} \operatorname{div} \boldsymbol{\rho} = -\frac{\boldsymbol{K}}{G}, \tag{2}$$

where Δ is Laplace's operator, μ is Poisson's ratio, $\boldsymbol{K}$ is the vector body force per unit of volume, and G is the modulus of rigidity.

Combining equations (1) and (2) and writing

$$\alpha = \frac{2(1-\mu)}{1-2\mu},$$

we obtain the equilibrium equation in terms of the potential functions ϕ and $\boldsymbol{S}$:

$$\Delta[\alpha \operatorname{grad} \phi + \operatorname{curl} \boldsymbol{S}] = -\frac{\boldsymbol{K}}{G}. \tag{3}$$

4. *The Galerkin Vector.* Since $\boldsymbol{S}$ is solenoidal, it may be represented by the curl of a vector function. Let

$$\boldsymbol{S} = -\operatorname{curl} \boldsymbol{W}. \tag{4}$$

The functions ϕ and $\boldsymbol{S}$ are independent functions and we may therefore write

$$\phi = \frac{1}{\alpha} \operatorname{div} \boldsymbol{W}. \tag{5}$$

Substituting equations (4) and (5) in the equilibrium equation (3), we obtain

$$\Delta[\text{grad div } \boldsymbol{W} - \text{curl curl } \boldsymbol{W}] = -\frac{\boldsymbol{K}}{G},$$

and, making use of the identity

$$\text{curl curl} = \text{grad div} - \Delta, \tag{6}$$

we obtain

$$\Delta\Delta \boldsymbol{W} = -\frac{\boldsymbol{K}}{G}. \tag{7}$$

Substituting equations (4) and (5) in equation (1), and employing the identity given in equation (6), we obtain

$$\boldsymbol{\varrho} = \Delta \boldsymbol{W} - \frac{1}{2(1-\mu)} \text{grad div } \boldsymbol{W}. \tag{8}$$

The vector function $\boldsymbol{W}$, satisfying equation (7) and producing the displacement according to equation (8), is identical with the Galerkin vector in the form given by Papkovitch. The relation between $\boldsymbol{W}$ and the Galerkin vector is

$$\boldsymbol{W} = \frac{1-\mu}{G} \boldsymbol{F}.$$

5. *The Papkovitch Functions.* The quantity in parentheses in equation (3) represents a vector function, say $\boldsymbol{B}$. Hence

$$\boldsymbol{B} = \alpha \text{ grad } \phi + \text{curl } \boldsymbol{S}, \tag{9}$$

and

$$\text{div } \boldsymbol{B} = \alpha\Delta\phi. \tag{10}$$

If the body force is taken equal to zero, equations (3) and (9) specify that $\boldsymbol{B}$ is harmonic, so that the complete solution of equation (10) may be written as

$$\phi = \frac{1}{2\alpha}(\boldsymbol{R}\cdot\boldsymbol{B} + \beta), \tag{11}$$

where $\boldsymbol{R}$ is the position vector of a field point referred to the origin and

$$\Delta \boldsymbol{B} = 0, \qquad \Delta\beta = 0. \tag{12}$$

Equations (1), (9), and (11) yield the expression for the displacement:

$$\boldsymbol{\varrho} = \boldsymbol{B} - \frac{1}{4(1-\mu)} \operatorname{grad} (\boldsymbol{R}\cdot\boldsymbol{B} + \beta). \tag{13}$$

The functions $\boldsymbol{B}$ and β, satisfying equations (12) and producing the displacement according to equation (13), are the Papkovitch functions. They bear the following relations to the corresponding functions $\phi_0, \phi_1, \phi_2, \phi_3$ found by Neuber:

$$\boldsymbol{B} = \frac{2(1-\mu)}{G} (\boldsymbol{i}\phi_1 + \boldsymbol{j}\phi_2 + \boldsymbol{k}\phi_3),$$

$$\beta = \frac{2(1-\mu)}{G} \phi_0,$$

where $\boldsymbol{i}, \boldsymbol{j}, \boldsymbol{k}$ are unit orthogonal vectors.

6. *Relations Among the Stress Functions.* When the body force is zero, the Galerkin, Papkovitch, and potential functions are found to be connected by the following relations. From equations (4), (5), and (9), we have

$$\begin{aligned} \boldsymbol{B} &= \Delta \boldsymbol{W} \\ &= \alpha \operatorname{grad} \phi + \operatorname{curl} \boldsymbol{S}; \end{aligned} \tag{14}$$

from equations (5), (11), and (14), we find

$$\begin{aligned} \beta &= 2 \operatorname{div} \boldsymbol{W} - \boldsymbol{R}\cdot\Delta\boldsymbol{W} \\ &= \alpha\left(2\phi - \boldsymbol{R}\cdot\operatorname{grad}\phi - \frac{1}{\alpha}\boldsymbol{R}\cdot\operatorname{curl}\boldsymbol{S}\right), \end{aligned} \tag{15}$$

and we have already seen that

$$\boldsymbol{S} = -\operatorname{curl} \boldsymbol{W},$$

$$\phi = \frac{1}{\alpha}\operatorname{div}\boldsymbol{W} = \frac{1}{\alpha}(\boldsymbol{R}\cdot\boldsymbol{B} + \beta).$$

COLUMBIA UNIVERSITY

AMERICAN SOCIETY OF CIVIL ENGINEERS

Founded November 5, 1852

DISCUSSIONS

STRESSES AROUND CIRCULAR HOLES IN DAMS AND BUTTRESSES

Discussion

BY MESSRS. R. D. MINDLIN, AND CHESLEY J. POSEY

R. D. MINDLIN,[7] JUN. AM. SOC. C. E. (by letter).[7a]—In examining Equations (28), (29), (30), and (34) for the stresses around a circular hole resulting from the gravitational action of a dam, it is observed that the elastic constants of the material are not involved. This is surprising in the light of a recent extension by M. A. Biot[8] of a theorem by J. H. Michell[9].

Michell's theorem applies to a two-dimensional stress system in a region pierced by holes, no body forces being present. He showed that in such a system the stresses are independent of the elastic constants only if the resultant force over each hole vanishes. This conclusion is found necessary in order to assure single-valuedness of the displacements.

Biot has extended Michell's theorem to the case of gravity stresses. He has shown[8] that "in a solid two-dimensional homogeneous body the stresses due to gravity do not depend on the elasticity constants of the material; they do, however, in general for a body with holes." Therefore, a more detailed examination of the stresses due to mass is indicated.

The author has conveniently divided the stresses in the solid dam into two systems. In System I the stresses on any closed boundary produce zero resultant force. Hence, the additional stresses required to remove the normal and shearing stresses on the periphery of a circular hole also involve zero resultant force on the hole. No difficulty with many valued displacements is encountered, therefore, in dealing with System I, and the author has found the required stress system by a well known method.

NOTE.—The paper by I. K. Silverman, Jun. Am. Soc. C. E., was published in November, 1936, *Proceedings*. This discussion is printed in *Proceedings* in order that the views expressed may be brought before all members for further discussion of the paper.

[7] Research Asst., Dept. of Civ. Eng., Columbia Univ., New York, N. Y.

[7a] Received by the Secretary November 21, 1936.

[8] "Distributed Gravity and Temperature Loading in Two-Dimensional Elasticity Replaced by Boundary Pressures and Dislocations", by M. A. Biot, *Journal of Applied Mechanics*, Vol. 2, No. 2 (June, 1935), p. 41.

[9] "On the Direct Determination of Stress in an Elastic Solid, with Application to the Theory of Plates", by J. H. Michell, *Proceedings*, London Math. Soc. Vol. XXXI (1899), p. 100.

In dealing with System II, Mr. Silverman has observed that these stresses produce a resultant force on an internal circular boundary equal to the weight of the enclosed disk. In order to remove the stresses from the periphery of the hole, an additional stress system must be superposed which yields equal and opposite stresses on the circular boundary. The latter system, then, must involve a resultant stress on the hole equal and opposite to the weight of the enclosed disk of radius, r_h. Therefore, according to Michell's theorem, the superimposed system should involve the elastic constants. This was to be expected from Biot's theorem. In fact, the foregoing explanation is simply a derivation of Biot's theorem for the present case.

If the elastic constants do not appear in a case where the resultant force on an internal boundary does not vanish, then, by Michell's theorem, the displacements resulting from the stresses should be many valued. Furthermore, when the internal boundary is closed, as in this case, many-valued functions in the displacements are not permissible, and a solution containing them is not valid.

It is necessary, therefore, to proceed to examine the stress function (Equation (33)) which was used to annul the stresses on the boundary, $\rho = r_i$. Writing $c\, r^2_i = a_1$, Equation (33) becomes[9b]:

$$F_1 = \frac{a_1}{2}\, \rho\, \theta \sin \theta \quad \ldots\ldots\ldots\ldots\ldots\ldots\ldots (50)$$

The stress function given in Equation (50) is known to produce many-valued functions in the displacements of the form[10], $\frac{a_1}{2}\frac{(1-\nu)}{E}\, \theta \sin \theta$; and $\frac{a_1}{2}\frac{(1-\nu)}{E}\, \theta \cos \theta$, respectively, in which ν = Poisson's ratio. This was to be expected from the preceding analysis, and it may be concluded that the stress system represented by Equations (34) is not possible in a continuous elastic medium.

To obtain the correct system of stress required to remove the stresses on the periphery of the hole, a bi-harmonic stress function must be found which will satisfy the following conditions:

(1) The displacements introduced by the stress function must be single-valued;

[9b] Correction for *Transactions*: The author has written θ for $(\theta - \beta)$. This should be kept in mind when adding the stresses involved in System II to those of System I. The following corrections, also approved by the author, will be made before the paper is finally published in *Transactions*: In the line below Equation (8), change "Equation (1)" to "Equation (6)"; Equation (10) should read "$s_\rho = -\frac{\rho}{6 K^3}$ [........"; in Equation (25), "K_2" should be "k_2"; to Equation (27e), add "$= B_3 = C_3$"; in the line following Equation (32), change "θ" to θ'; change Line 2 following Equation (32), to read "$(s_\theta)_{II} = c\left(\frac{r^2_i}{r^2_o} - 1\right) \rho \cos \theta'$; $(s_s)_{II} = 0$; and, $\phi' = \omega - \beta$"; in Line 3 following Equation (34c), write "$s_\theta = -c r_h \cos \theta$"; in Equation (36), change "+0.00204" to read "+0.02204"; in Equation (37), insert a minus sign before 1.13; in Table 1, for $\theta = 180$, $s_\theta = -1.5877$; in Appendix I, delete the definition for K_2 and add the definition "$k_2 = \cot \alpha$" in the definition for k_n; and, in Equation (45a), insert a minus sign before "$G_n\, q_n$".

[10] See "Theory of Elasticity", by S. Timoshenko, p. 116; McGraw-Hill, 1934.

(2) The normal and shearing stresses on the circular boundary, $\rho = r_i$, must be equal and opposite to those of System II; and,

(3) All the stresses introduced by the new function must approach zero as ρ approaches infinity.

In order to satisfy Condition (1) one must combine with Equation (50) a stress function[10],

$$F_2 = b'_1 \rho \log \rho \cos \theta \quad \text{......(51)}$$

in which,

$$b'_1 = -\frac{a_1 (1 - \nu)}{4} \quad \text{......(52)}$$

The combined functions, F_1 and F_2, satisfy Conditions (1) and (3), but not Condition (2). The addition of F_2 to the system has served to introduce shearing stresses on the circular boundary (proportional to sin θ), which are not present in System II or in F_1. Therefore, another function, F_3, must be added, which will remove these shearing stresses without introducing normal stresses on the circular boundary other than those proportional to cos θ. Furthermore, F_3 must satisfy Conditions (1) and (3). Such a function is given by:

$$F_3 = a'_1 \rho^{-1} \cos \theta \quad \text{......(53)}$$

The complete stress function is then given by:

$$F = F_1 + F_2 + F_3 = \frac{a_1}{2} \rho \theta \sin \theta + b'_1 \rho \log \rho \cos \theta + a'_1 \rho^{-1} \cos \theta \quad \text{..(54)}$$

To determine the values of the constants it is necessary first to calculate the stresses. Using Equations (20) of the paper, these are found to be:

$$s_\rho = \left(\frac{a_1}{\rho} + \frac{b'_1}{\rho} - \frac{2\, a'_1}{\rho^3} - c\, \rho \right) \cos \theta \quad \text{......(55a)}$$

$$s_\theta = \left(\frac{b'_1}{\rho} + \frac{2\, a'_1}{\rho^3} - c\, \rho \right) \cos \theta \quad \text{......(55b)}$$

and,

$$s_s = \left(\frac{b'_1}{\rho} - \frac{2\, a'_1}{\rho^3} \right) \sin \theta \quad \text{......(55c)}$$

Condition (2) is expressed by the equation:

$$[s_\rho]_{\rho=r_i} = [s_s]_{\rho=r_i} = 0 \quad \text{......(56)}$$

and, hence:

$$\frac{a_1}{r_i} + \frac{b'_1}{r_i} - \frac{2\, a'_1}{r^3_i} - c\, r_i = 0 \quad \text{......(57a)}$$

and

$$\frac{b'_1}{r_i} - \frac{2\, a'_1}{r^3_i} = 0 \quad \text{......(57b)}$$

Condition (1) is expressed by Equation (52). Solving for the three unknowns in Equations (52) and (57):

$$a_1 = c\, r_i^2 \qquad \text{(58a)}$$

$$b'_1 = -\frac{1-\nu}{4}\, c\, r_i^2 \qquad \text{(58b)}$$

and,

$$a'_1 = -\frac{1-\nu}{8}\, c\, r_i^4 \qquad \text{(58c)}$$

Substituting these values in Equation (54) the final stress function becomes:

$$F = \frac{1}{2}\, c\, r_i^2\, \rho\, \theta \sin\theta - \frac{1-\nu}{4}\, c\, r_i^2\, \rho \log\rho \cos\theta - \frac{1-\nu}{8}\, c\, r_i^4\, \rho^{-1} \cos\theta \qquad \text{(59)}$$

The corresponding stresses are, from Equations (20):

$$s_\rho = c\left(\frac{3+\nu}{4}\,\frac{r_i^2}{\rho} + \frac{1-\nu}{4}\,\frac{r_i^4}{\rho^3} - \rho\right)\cos\theta \qquad \text{(60a)}$$

$$s_\theta = -\,c\left(\frac{1-\nu}{4}\,\frac{r_i^2}{\rho} + \frac{1-\nu}{4}\,\frac{r_i^4}{\rho^3} + \rho\right)\cos\theta \qquad \text{(60b)}$$

and,

$$s_s = c\left(-\frac{1-\nu}{4}\,\frac{r_i^2}{\rho} + \frac{1-\nu}{8}\,\frac{r_i^4}{\rho^3}\right)\sin\theta \qquad \text{(60c)}$$

Equations (60) express the correct stresses for System II. By virtue of Equation (52), Condition (1) is satisfied; by virtue of Equations (57*a*) and (57*b*). Condition (2) is satisfied; and an examination of Equations (60) shows that Condition (3) is satisfied. The hoop stresses at the circular boundary are found by substituting $\rho = r_i$ in Equation (60*b*)[9b]:

$$[s_\theta]_{\rho=r_i} = -\left(\frac{3-\nu}{2}\right) c\, r_i \cos\theta \qquad \text{(61)}$$

The foregoing analysis has been made for the case of plane stress. It is approximately applicable to a buttress if the diameter of the hole is small in comparison with its distance from the nearest edge of the buttress, and if the thickness of the buttress is small in comparison with the diameter of the hole. For the case of a longitudinal hole in a gravity section the stresses should be transformed to the corresponding system for plane strain.

Additional references should be given to the results of previous stress analyses for the wedge section. The stress systems represented by Equations (8) to (12) and (22) to (25), inclusive, are due to M. Levy[11] and may be obtained from equations given by Love[12]. This stress system applies only to

[11] "Sur l'équilibre elastic d'un barrage en maçonnerie à section triangulaire," by M. Lévy, *Comptes Rendus*, Paris, Vol. 127 (1898), p. 10.

[12] "Theory of Elasticity", by A. E. H. Love, Fourth Edition, 1927, Equations (34*A*) and (34*B*), pp. 212 and 213.

portions of a dam remote from the base. Hence, in addition to the requirement that the small hole be far from the faces of the dam, it must also be far from the base.

Equations (8) and (22) are seen to be of the same form, differing only in the constants associated with the terms of the third degree polynomial. Equation (22) represents the stress which the author has designated as System I. Hence, the additional stress functions, required to annul the stresses due to System I on a circular boundary, will be the same as the stress functions required to annul the stresses on the circular boundary resulting from the water loading, except for the constants. Therefore, a solution of the problem of the stresses due to mass really includes that of the problem of stresses due to water load.

The author is undoubtedly unaware of a previous solution of this problem by G. Nishimura and T. Takayama[13]. Of the two solutions, the writer prefers that of Mr. Silverman as being far less cumbersome. Very skilfully the author has applied what mathematicians would call an elegant method and, independently, has obtained the solution of an interesting problem.

Reprinted from
Proceedings of the American Society of Civil Engineers
63 (1937) 399-403

[13] "On the Stress Distribution in the Vicinity of a Horizontal Circular Hole in a Gravitating Wedge-Shaped Elastic Solid", by G. Nishimura and T. Takayama, *Bulletin*, Earthquake Research Inst., Tokyo, Vol. 10 (1932), p. 723.

Discussion
Pressure Distributions on Retaining Walls

DR. RAYMOND D. MINDLIN, INST. IN CIV. ENG.,
COLUMBIA UNIVERSITY, N.Y.

In Paper No. J-1, Vol. I, Mr. Spangler observes that the experimental pressure distribution curves for lateral pressure against a retaining wall due to concentrated surcharges are similar in shape to those obtained from the Boussinesq solution. He also notes that the Boussinesq solution gives values which are far too low and correctly attributes this discrepancy to the presence of the unyielding retaining wall for which no provision is made in the Boussinesq analysis. It was suggested by S. D. Carothers (Engineering, London, (1924) pp. 1, 156) that the action of a smooth rigid retaining wall may be taken into account by applying the method of images to the Boussinesq solution. It is the purpose of this discussion to compare Mr. Spangler's experimental data with the results obtained by the method of images and to point out how the method may be extended to include retaining walls of finite height and also walls with sloping back faces.

We consider the semi-infinite solid bounded by the plane $z = 0$ and apply to the free surface a concentrated load P at point $x = -a$, $y = 0$ and an equal load at $x = +a$, $y = 0$ (Fig. 1). Considerations of symmetry lead at once to the conclusion that, under this system of loading, there is no displacement in the x direction on plane $x = 0$. Also there is no shearing stress on this plane. The plane $x = 0$ may therefore be replaced by the back face of a smooth, rigid retaining wall. The lateral pressure against the wall is then simply double the Boussinesq pressures. If we assume incompressibility (Poisson's ratio = 1/2), the pressure against the wall, along the z axis, is given by

$$p = \frac{3P}{\pi} \cdot \frac{a^2 z}{(a^2 + z^2)^{5/2}}$$

The curves determined by this formula were calculated for all the cases of concentrated loading investigated by Mr. Spangler and were found to correspond remarkably well with the experimental results. Some typical cases are shown in Fig. 2. It is interesting to note that the maximum pressure always occurs at a distance below the surface equal to one-half the distance from the load point to the wall. It may be observed that the assumption of perfect rigidity results in higher pressures than would obtain if any degree of yielding were assigned to the wall.

The same procedure may be followed for the case of line loadings or for loadings distributed over an area.

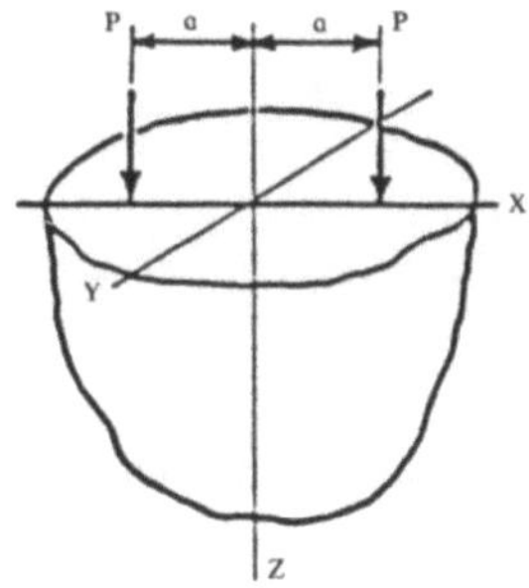

Fig. 1. Method of images for a smooth rigid retaining wall of infinite depth

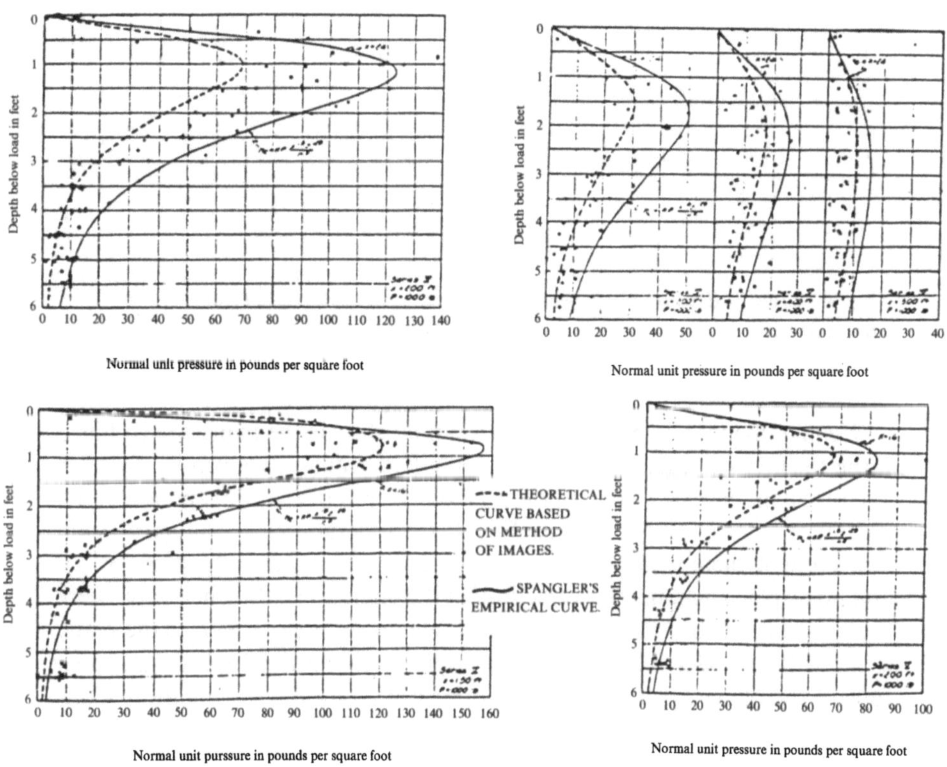

Fig. 2.

It is frequently desirable to consider as rigid not only the retaining wall, but also the foundation below the fill. In this case a similar expedient suggests itself. We avail ourselves of the known solutions for a body bounded by two parallel planes and loaded with a pair of equal and opposite forces acting normal to and at the boundary planes (Fig. 3). The two dimensional case is discussed by L. N. G. Filon (Phil. Trans. Roy. Soc., London, Vol. 201, A, 1903) and the three dimensional case by J. Dougall (Edinburgh Roy. Soc. Trans., Vol. 41, 1904). (See also, M. Biot, Physics, Vol. 6, No. 12, 1935).

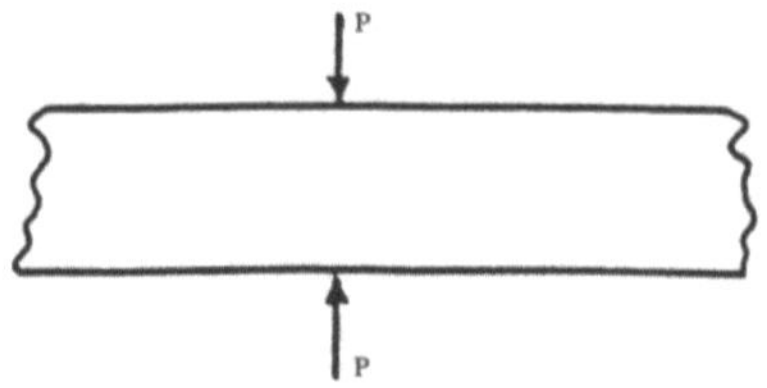

Fig. 3. Body bounded by two parallel planes under two equal and opposite forces

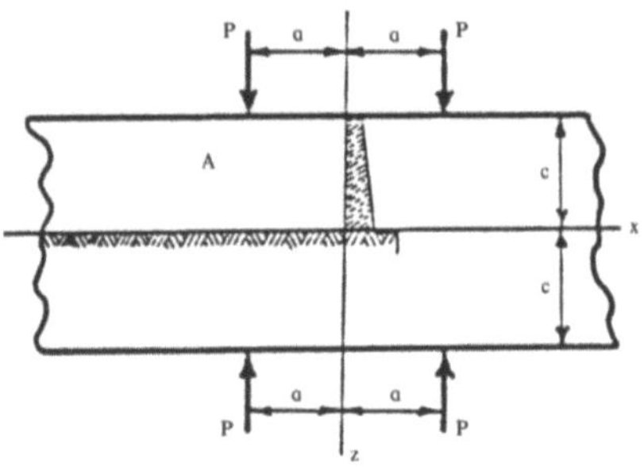

Fig. 4. Smooth rigid retaining wall of finite depth

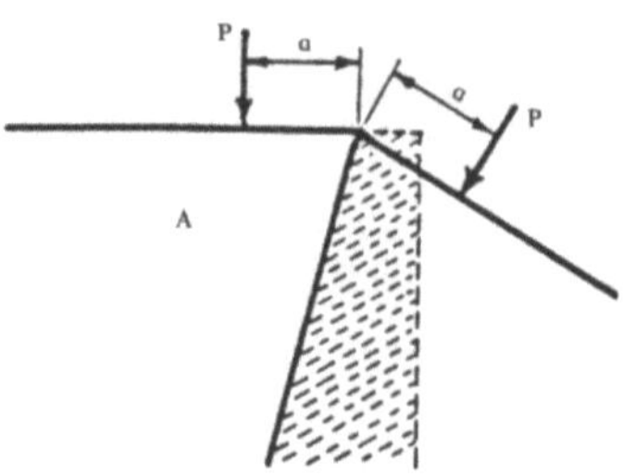

Fig. 5. Smooth rigid retaining wall with sloping back face.

We take the middle plane of the horizontal slice as the plane $z = 0$. Then, if the thickness of the slice is $2c$, the boundary planes are given by $z = \pm c$. We place one pair of loads at points ($x = -a$, $y = 0$, $z = \pm c$) and are equal pair at ($x = +a$, $y = 0$, $z = \pm c$), Fig. 4. Consideration of symmetry again indicate that there is no displacement in the x direction on the plane $x = 0$ and no displacement in the z direction on the plane $z = 0$. Furthermore, there are no shearing stresses on these planes. We may therefore replace the plane $x = 0$ with the back face of a smooth, rigid retaining wall and we may also replace the plane $z = 0$ with the smooth surface of a rigid foundation. Considering the quadrant marked A in Fig. 4, we see that it is bounded on top by a free surface, on bottom by a smooth rigid bed and on the right by a smooth rigid retaining wall. The normal pressures on both the wall and the foundation may then be calculated from the formulas given by Dougall or Filon, depending upon whether we wish to discuss a point loading or an infinite line load.

The cases of finite line loadings and areal loadings may be treated in a similar manner.

The pressures against retaining walls with inclined back faces may be calculated by utilizing a solution by W. M. Shepherd for stress systems in an infinite sector (Proc. Royal Soc., London, Series X, Vol. 148, 1935). See Fig. 5.

Finally, it is possible to introduce another smooth, rigid wall parallel to the plane $y = 0$, or a pair of rigid planes parallel to either $x = 0$ or $y = 0$ or both.

In fact we may, if we wish, calculate the state of stress in a rectangular parallelopiped which is loaded on a free surface and restrained on four sides by smooth rigid walls.

Reset from
Proceedings of the International Conference on Soil Mechanics and Foundation Engineering
Cambridge, Mass.-1936
III (1937) 155-156

Analysis of Doubly Refracting Materials with Circularly and Elliptically Polarized Light

RAYMOND D. MINDLIN
Department of Civil Engineering, Columbia University, New York, N. Y.
(Received May 6, 1937)

INTRODUCTION

When a doubly refracting crystal plate is examined at normal incidence between "crossed Nicols," two types of interference bands are observed. These are the fringes (isochromatics) and the brushes (isoclinics). The former depend upon the relative retardation of the ordinary and extraordinary rays and the latter depend upon the orientations of the polarizing axes of the crystal. It is often desirable to eliminate the isoclinics in order to better view the retardation fringes. This is usually accomplished by introducing, between the specimen and the polarizers, a pair of retardation plates with polarizing axes oriented at 45° with respect to the vibration directions of the polarizers. If the retardation plates are true quarter-wave plates for the monochromatic light source, it is well known that the isoclinics are completely removed, while the retardation fringes remain unaltered.

It is the purpose of this paper to show that, when the retardation plates are not accurately quarter-wave plates, not only are the isoclinics not completely removed but also the retardation fringes are not necessarily unchanged. These conclusions have been reached through an analysis of the equations governing the intensity

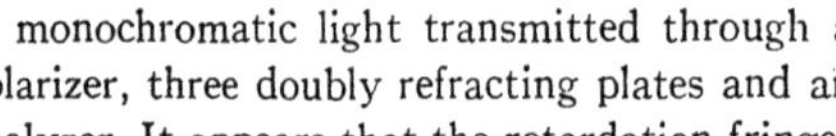

FIG. 1. Schematic diagram of polarizing apparatus. S=monochromatic light source; P=polarizer; 1 and 3 =retardation plates; 2=specimen; A=analyzer.

of monochromatic light transmitted through a polarizer, three doubly refracting plates and an analyzer. It appears that the retardation fringes are unaltered if the two retardation plates are identical, regardless of the retardation introduced by each one; but, if the two plates are not identical, the fringes are displaced by an amount which depends, in part, upon the difference in retardations of the two plates.

This phenomenon is important, for example, in photoelastic work in which the magnitude of the stress at a point is calculated from the order of the fringe appearing at that point. The displacement of a fringe, caused by inaccuracies in the retardation plates, introduces an error, in the stress, which may be calculated from the equations given below, provided that the retardations of the two retardation plates are known.

INTENSITY EQUATIONS

Figure 1 shows, schematically, the optical arrangement under consideration. Light from a monochromatic source, S, passes successively at normal incidence through a polarizer, P, a retardation plate, 1, the specimen, 2, a second retardation plate, 3, and an analyzer A. Fig. 2 depicts the orientations of the polarizing axes as seen by an observer viewing the oncoming light from the analyzer end of the system. P is the vibration direction of the light transmitted through the polarizer; ρ_1, ρ_2, ρ_3 are the orientations of the fast axes of the first retardation plate, the specimen, and the second retardation plate, respectively, measured positive counter clockwise from P; and θ is the orientation of the vibration direction of the analyzer, referred to P.

In the equations which follow, I is the intensity of light transmitted through the system; I_0 is the intensity transmitted through the system when $\rho_1=\rho_2=\rho_3=\theta=0$ and δ_1, δ_2, δ_3 are the relative phase retardations introduced by the first retardation plate, the specimen and the second retardation plate, respectively. For arbitrary values of all the orientations and retardations the equation which governs the intensity is[1]

[1] For a method of deriving formulas such as Eq. (1), see G. Szivessy, *Handbuch der Physik*, Vol. 19, p. 924.

$$\frac{2I}{I_0}=\sin 2(\theta-\rho_3)\left\{\begin{array}{l}-\cos 2\rho_1 \cos 2(\rho_2-\rho_1) \sin 2(\rho_3-\rho_2) \cos \delta_3 \\ +\sin 2\rho_1 \sin 2(\rho_2-\rho_1) \sin 2(\rho_3-\rho_2) \cos \delta_3 \cos \delta_1 \\ -\cos 2\rho_1 \sin 2(\rho_2-\rho_1) \cos 2(\rho_3-\rho_2) \cos \delta_3 \cos \delta_2 \\ +\cos 2\rho_1 \sin 2(\rho_2-\rho_1) \sin \delta_3 \sin \delta_2 \\ -\sin 2\rho_1 \cos 2(\rho_2-\rho_1) \cos 2(\rho_3-\rho_2) \cos \delta_1 \cos \delta_2 \cos \delta_3 \\ +\sin 2\rho_1 \cos 2(\rho_2-\rho_1) \sin \delta_2 \sin \delta_3 \cos \delta_1 \\ +\sin 2\rho_1 \cos 2(\rho_2-\rho_1) \cos \delta_3 \sin \delta_2 \sin \delta_1 \\ +\sin 2\rho_1 \sin \delta_3 \cos \delta_2 \sin \delta_1\end{array}\right\}$$

$$+\cos^2(\theta-\rho_3)\left\{\begin{array}{l}2\cos^2\rho_1 \cos^2(\rho_2-\rho_1) \cos^2(\rho_3-\rho_2) \\ +2\cos^2\rho_1 \sin^2(\rho_2-\rho_1) \sin^2(\rho_3-\rho_2) \\ +2\sin^2\rho_1 \sin^2(\rho_2-\rho_1) \cos^2(\rho_3-\rho_2) \\ +2\sin^2\rho_1 \cos^2(\rho_2-\rho_1) \sin^2(\rho_3-\rho_2) \\ -\sin 2\rho_1 \sin 2(\rho_2-\rho_1) \cos 2(\rho_3-\rho_2) \cos \delta_1 \\ -\cos 2\rho_1 \sin 2(\rho_2-\rho_1) \sin 2(\rho_3-\rho_2) \cos \delta_2 \\ -\sin 2\rho_1 \cos 2(\rho_2-\rho_1) \sin 2(\rho_3-\rho_2) \cos \delta_1 \cos \delta_2 \\ +\sin 2\rho_1 \sin 2(\rho_3-\rho_2) \sin \delta_1 \sin \delta_2\end{array}\right\}. \quad (1)$$

$$+\sin^2(\theta-\rho_3)\left\{\begin{array}{l}2\cos^2\rho_1 \cos^2(\rho_2-\rho_1) \sin^2(\rho_3-\rho_2) \\ +2\cos^2\rho_1 \sin^2(\rho_2-\rho_1) \cos^2(\rho_3-\rho_2) \\ +2\sin^2\rho_1 \sin^2(\rho_2-\rho_1) \sin^2(\rho_3-\rho_2) \\ +2\sin^2\rho_1 \cos^2(\rho_2-\rho_1) \cos^2(\rho_3-\rho_2) \\ +\sin 2\rho_1 \sin 2(\rho_2-\rho_1) \cos 2(\rho_3-\rho_2) \cos \delta_1 \\ +\cos 2\rho_1 \sin 2(\rho_2-\rho_1) \sin 2(\rho_3-\rho_2) \cos \delta_2 \\ +\sin 2\rho_1 \cos 2(\rho_2-\rho_1) \sin 2(\rho_3-\rho_2) \cos \delta_1 \cos \delta_2 \\ -\sin 2\rho_1 \sin 2(\rho_3-\rho_2) \sin \delta_1 \sin \delta_2\end{array}\right\}$$

Plane Polariscope

If the specimen is viewed simply between crossed Nicols, the optical system is known as a plane polariscope. The intensity of light transmitted through such a system is obtained by substituting, in Eq. (1), $\delta_1=\delta_3=0$ and $\theta=\pi/2$. Then

$$I=I_0 \sin^2(\delta_2/2) \sin^2 2\rho_2. \quad (2)$$

Extinction of light is observed when

$$\delta_2=2n\pi, \quad (n=\pm 1, \pm 2, \pm 3, \cdots) \quad (3)$$

and also when

$$\rho_2=n\pi/2. \quad (4)$$

Dark bands formed according to Eq. (3) are the retardation fringes and those governed by Eq. (4) are the isoclinics.

Circular Polariscope

To remove the isoclinics, quarter-wave retardations are assigned to plates 1 and 3 and the polarizing axes of the two plates are oriented at angles of 45° to the vibration directions of P and A. The optical arrangement is known as a circular polariscope. In Eq. (1), $\delta_1=\delta_3=\pi/2$, $\rho_1=\pi/4$, $\rho_3=3\pi/4$, $\theta=\pi/2$. The intensity of monochromatic light transmitted through a circular polariscope is, then, given by

$$I=I_0 \sin^2(\delta_2/2). \quad (5)$$

The intensity is accordingly independent of the orientation (ρ_2) of the polarizing axes in the specimen, so that the isoclinics do not appear; but the retardation fringes again occur according to Eq. (3).

Balanced Polariscope

Suppose, now, that the two retardation plates are not true quarter-wave plates for the monochromatic source, but that they are, nevertheless, identical. Substituting in Eq. (1) the values $\delta_1=\delta_3=\delta$ and $\rho_1=\pi/4$, $\rho_3=3\pi/4$, $\theta=\pi/2$, there results

$$I=I_0 \sin^2(\delta_2/2)(1-\cos^2 2\rho_2 \cos^2 \delta). \quad (6)$$

The quantity in parentheses in Eq. (6) shows that, in this case, the isoclinics are not completely removed. The proportion which remains depends upon the relative retardation of each of

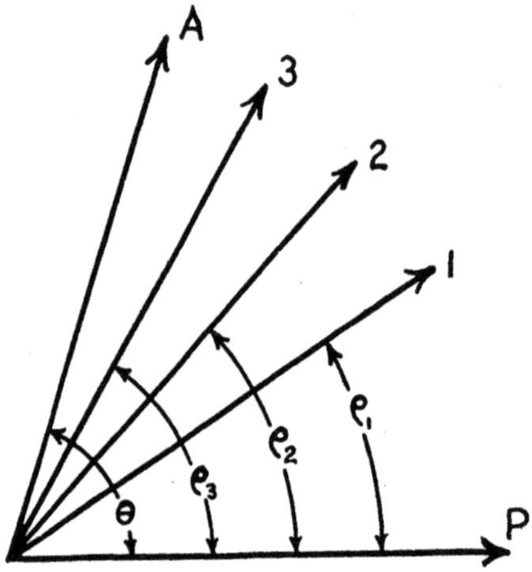

FIG. 2. Angular relations of vibration direction of analyzer A, fast polarizing axes of retardation plates (1, 3) and specimen (2); all referred to vibration direction of polarizer (P).

the two retardation plates and follows a cosine-squared law. When $\delta \to 0$ the isoclinics are completely restored and Eq. (6) becomes identical with Eq. (2). When $\delta \to \pi/2$ (quarter-wave plates) the isoclinics are completely removed and Eq. (6) approaches Eq. (5).

No matter what value is assigned to δ, the retardation fringes always occur according to Eq. (3) since δ_2 appears in Eq. (6) only in the form of the multiplying factor $\sin^2(\delta_2/2)$ just as in Eq. (2).

The relation expressed by Eq. (6) was used by R. V. Baud[2] to show that, with a *white* light surce, the isochromatic fringes vary with the orientation (ρ_2) of the polarizing axes in the specimen.

UNBALANCED POLARISCOPE

If the two retardation plates are not identical, it appears that even with monochromatic light, the fringe pattern is a function of the orientation of the polarizing axes in the specimen. Furthermore, for a given inclination of the polarizing axes the fringe pattern observed also depends upon the relative retardations of the two retardation plates.

The intensity equation governing this unbalanced case is obtained by substituting $\rho_1=\pi/4$, $\rho_3=3\pi/4$, $\theta=\pi/2$ in Eq. (1), and there results:

$$I=I_0\left\{\sin^2\frac{\delta_2}{2}[\cos(\delta_3-\delta_1)-\cos^2 2\rho_2\cos\delta_1\cos\delta_3]+\sin^2\frac{\delta_3-\delta_1}{2}-\tfrac{1}{2}\sin\delta_2\sin 2\rho_2\sin(\delta_3-\delta_1)\right\}. \quad (7)$$

The appearance of the term $\sin\delta_2$ in Eq. (7) indicates that, when $\delta_3\neq\delta_1$, the fringes no longer occur according to Eq. (3). There is then observed a displacement of the fringe pattern which is further complicated by the presence of the multiplying factor $\sin 2\rho_2$. Consequently, there is a complex distortion of the fringe pattern when both the relative retardation and the inclination of the polarizing axes are not uniform throughout the specimen, as is the case in a photoelastic model.

Instead of fringes appearing according to Eq. (3), maximum and minimum intensities will be observed in an unbalanced polariscope when

$$\tan\delta_2=\frac{\sin 2\rho_2\sin(\delta_3-\delta_1)}{\cos(\delta_3-\delta_1)-\cos^2 2\rho_2\cos\delta_1\cos\delta_3}. \quad (8)$$

For a given pair of retardation plates, the maximum error in calculating relative retardation from fringe order occurs when the polarizing axes in the specimen are at 45° to the vibration direction of the polarizer. Eq. (8) then reduces to

$$\tan\delta_2=\pm\tan(\delta_3-\delta_1)$$

and therefore a dark fringe appears when

$$\delta_2=2n\pi\pm(\delta_3-\delta_1). \quad (9)$$

The upper sign is associated with positive values of $\sin 2\rho_2$ and the lower sign with negative values of $\sin 2\rho_2$.

CONCLUSIONS

(1) Equation (7) shows that the fringe pattern is distorted in an optically unbalanced polari-

[2] R. V. Baud, J. O. S. A. 20, 386 (1930), and 21, 119 (1931).

scope. The distortion, for a given wave-length, is a function of the orientation of the specimen and the phase retardations of the two retardation plates.

(2) The magnitude of the error introduced by the optical unbalance (with monochromatic light) may be calculated from Eq. (8).

(3) Equation (9) shows that the *maximum* error (in calculating relative retardation from fringe order) that can occur in an optically unbalanced polariscope is given by the difference between the phase retardations of the two retardation plates.

(4) If the retardation plates in a polariscope are different from quarter-wave plates, but are nevertheless identical, it follows from Eq. (6) that the fringe pattern is unaffected. The only effect will be incomplete removal of isoclinics.

(5) Equation (6) also shows that if a monochromatic light source of wave-length other than that for which the instrument is designed is substituted in a balanced polariscope, the fringe pattern will remain unchanged.

Reprinted from
Journal of the Optical Society of America
27 (1937) 288-291

(*Reprinted from the* JOURNAL OF APPLIED MECHANICS *for September, 1937*)

Stress Systems in a Circular Disk Under Radial Forces

BY RAYMOND D. MINDLIN,[1] NEW YORK, N. Y.

The transformation of inversion is employed in this paper to generalize a particular solution obtained by Michell (1),[2] for plane stress in a circular disk. In Michell's solution the disk is loaded in its plane with a pair of equal and opposite forces; one at the center of the disk and the other on the periphery as shown in Fig. 1. In the transformed state, the force in the interior of the disk is no longer at the center, as shown in Fig. 3 and represented by Equation [9]. By simple superposition, the transformed solution is extended to include any self-equilibrating system of radial forces applied in the plane of the disk as shown in Fig. *6b*.

The transformation would involve only straightforward calculation were it not for the fact that the disk is multiply connected. This condition is caused by the presence of a singularity in the interior of the disk at the point of application of the force. In the inversion of plane stress in a multiply connected plate, the displacement functions for the transformed state are not necessarily single-valued even though they may have arisen from single-valued displacements in the original plate. This difficulty is overcome by the introduction of suitable functions which remove the many valued terms in the displacements, but do not disturb the stress conditions on the boundaries of the transformed disk.

The solution is shown to include, as particular cases two well-known problems: (*a*) A disk loaded on its periphery by forces acting along a diameter as shown in Fig. 4; and (*b*) a semi-infinite plate loaded near its edge by a force normal to the boundary as shown in Fig. 5.

MICHELL'S SOLUTION

CONSIDER first a solution of the problem of a circular disk loaded with a pair of equal and opposite forces in the plane of the disk, where one force is applied at the center of the disk and the other on the periphery as shown in Fig. 1. The stresses for this configuration were given by Michell (1).

The solution may be expressed in terms of a plane stress ϕ satisfying the biharmonic equation

$$\Delta^4\phi = 0 \qquad [1]$$

where $\nabla^4 = (\partial^4/\partial x^4) + 2(\partial^4/\partial x^2\partial y^2) + (\partial^4/\partial y^4)$. The stresses are expressed in terms of ϕ by the relations

$$\sigma_x = \frac{\partial^2\phi}{\partial y^2},\quad \sigma_y = \frac{\partial^2\phi}{\partial x^2},\quad \tau_{xy} = -\frac{\partial^2\phi}{\partial x\partial y} \qquad [2]$$

where σ_x and σ_y are the normal components of stress on planes perpendicular to the x- and y-axes, respectively, and τ_{xy} is the shearing stress on these planes.

For the problem indicated in Fig. 1, it may be verified that the stress function is given by

$$\phi_0 = -\frac{F}{2\pi}\left[y(\varphi + 2\theta) - \frac{1-\nu}{2}\,x\log\rho - \frac{1+\nu}{4a^2}\,x\rho^2 - \frac{\rho^2}{2a}\right] \qquad [3]$$

where x, y, ρ, φ, θ are coordinates as shown in Fig. 1; a is the radius of the disk; ν is Poisson's ratio; F is the magnitude of the applied forces. Terms in ϕ_0 involving the coordinate perpendicular to the x, y-plane are neglected.

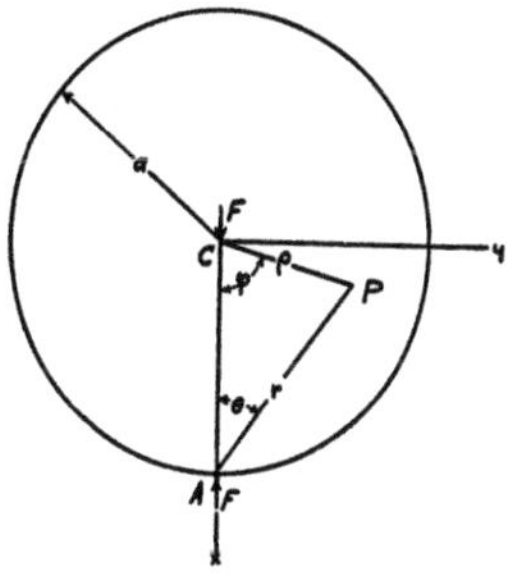

FIG. 1 DISK LOADED AT ITS CENTER AND PERIPHERY—MICHELL'S PROBLEM

We wish to transform ϕ_0 so that it will yield the stresses for the case where the force which acts in the interior of the disk is no longer at the center C.

TRANSFORMATION OF INVERSION

It is necessary at this point to indicate briefly some of the properties of the inversion transformation which will be used in succeeding paragraphs (2).

If O' is a point in a plane and P any other point, we may locate a third point P' on the line through $O'P$, such that

$$(O'P')\,(O'P) = k^2 \qquad [4]$$

where k is a constant length called the radius of inversion and O' is called the center of inversion. The circle whose center is O' and whose radius is k is known as the circle of inversion, and P and P' are termed mutual inverts with respect to the circle of inversion.

If, for each point P in the plane, we locate a new point P' according to Equation [4], the whole of the plane outside the circle of inversion is transferred to the interior of the circle and vice versa. In general, any circle in the plane is transformed into a circle; any straight line (except those passing through O') is transformed into a circle through O', and conversely. In particular, by properly choosing a center and radius of inversion, we may transform a circle into itself while points within the circle shift their relative positions. This property will be used in the sequel.

[1] Department of Civil Engineering, Columbia University.

[2] Numbers in parentheses refer to the Bibliography at the end of the paper.

Presented at the Joint Meeting of the Applied Mechanics and Hydraulics Divisions of THE AMERICAN SOCIETY OF MECHANICAL ENGINEERS, held at Cornell University, Ithaca, N. Y., June 25–26, 1937.

Discussion of this paper should be addressed to the Secretary, A.S.M.E., 29 West 39th Street, New York, N. Y., and will be accepted until November 10, 1937, for publication at a later date. Discussion received after the closing date will be returned.

In applying the transformation of inversion to plane stress problems, it is important to know:

1 A stress function ϕ satisfying Equation [1] is transformed to a function ϕ' such that $(O'P')^2\phi'$ (equals ϕ_1, say) satisfies Equation [1] when ∇^4 is expressed in terms of the transformed coordinates; hence ϕ_1 represents a stress function for a problem in plane stress.

2 A boundary free from stress is transformed into a boundary under uniform normal stress.

3 A concentrated force F is transformed into a concentrated force k^2F.

4 In a multiply connected plate, the stresses derived from the transformed stress function do not necessarily give rise to single-valued displacements in the transformed region.

Coordinate Relations

In Fig. 2 we wish to invert the disk of radius a into itself in such a manner that its center C goes to a point O distant c from C.

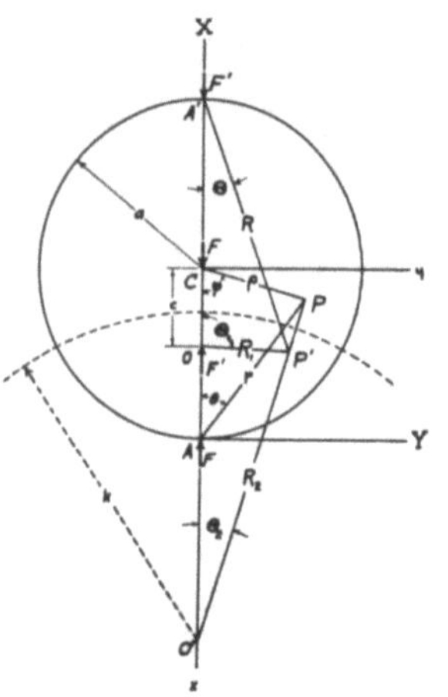

Fig. 2 Inversion of Michell's Problem

To do this we employ the procedure followed by Love in the analogous problem in the theory of bending of thin plates (3). The center of inversion O' is chosen so that O' and O are mutual inverts with respect to the periphery of the disk (hence $O'C = a^2/c$). The radius of inversion is taken as

$$k = [(a^4/c^2) - a^2]^{\frac{1}{2}} \qquad [5]$$

Then, in Fig. 2, the periphery of the disk inverts into itself; C goes to O; A goes to A'; and, in general, any point P goes to a point P' in such a manner as to satisfy Equations [4] and [5].

We require to know the coordinates of P in terms of the coordinates of P'. The coordinates of P are given by x, y, ρ, φ, r, θ as shown in Fig. 1. The coordinates of P' are given by R, Θ, R_1, Θ_1, R_2, Θ_2 as shown in Fig. 2. The latter are conveniently referred to a system of rectangular coordinates X, Y with origin at A as in Fig. 2. Then angle

$$AA'P = \Theta = \tan^{-1}\frac{Y}{2a - X}$$

angle

$$A'OP' = \Theta_1 = \tan^{-1}\frac{Y}{X - a + c} = \tan^{-1}\frac{Y}{X_1}$$

angle

$$A'O'P' = \Theta_2 = \tan^{-1}\frac{Y}{[X + (a^2/c) - a]} = \tan^{-1}\frac{Y}{X_2}$$

and

$$(A'P')^2 = R^2 = Y^2 + (2a - X)^2$$
$$(OP')^2 = R_1^2 = Y^2 + (X - a + c)^2 = Y^2 + X_1^2$$
$$(O'P')^2 = R_2^2 = Y^2 + [X + (a^2/c) - a]^2 = Y^2 + X_2^2$$

If, in Fig. 2, we note that triangle $O'PC$ is similar to triangle $O'OP'$, and that triangle $O'AP$ is similar to triangle $O'P'A'$, then the following relations between the coordinates in the original and transformed disks are found

$$\left.\begin{aligned} \varphi &= \Theta_1 - \Theta_2 \\ \theta &= \Theta + \Theta_2 \\ \rho &= \frac{a^2}{c}\frac{R_1}{R_2} \\ x &= \frac{a^2}{c} - \frac{X_2 k^2}{R_2^2} \\ y &= \frac{Y k^2}{R_2^2} \end{aligned}\right\} \qquad [6]$$

Inversion of Michell's Solution

To transform Michell's solution, we first substitute in Equation [3] the relations given in Equations [6] and so obtain a function ϕ' in terms of the transformed coordinates. Multiplying ϕ' by R_2^2 we obtain a function, say ϕ_1, which satisfies the transformed Equation [1] according to the properties of the inversion transformation stated previously in this paper. Further, the force F at C, shown in Fig. 2, is transformed into an oppositely directed force $F' = k^2F$ at O, and the force F at A is transformed into an oppositely directed force $F' = k^2F$ at A'; a uniform normal stress (easily removed) is produced on the circular boundary.

It remains only to examine the stress function for terms which may introduce many-valued functions in the displacements. This investigation is necessary because of the concentrated force at O. The stresses became infinite at O, and therefore this point must be excluded from the region for which the solution is applicable. Hence, the region under consideration is included between two closed curves, the circle of radius a and the closed boundary around the singularity at O. In such a multiply connected region, the inversion transformation may produce many-valued displacements, and these must be eliminated if the solution is to be valid.

Substituting Equations [6] into Equation [3], and multiplying by R_2^2, we obtain, tentatively, after some rearranging of terms

$$\phi_1 = -\frac{F'}{2\pi}\left[Y(2\Theta + \Theta_1 + \Theta_2) - \frac{1-\nu}{2}X_1\log\frac{R_1}{R_2} + \frac{1+\nu}{4}\frac{a^2}{c^2}X_2\frac{R_1^2}{R_2^2} + \alpha R_1^2 + \frac{1-\nu}{2}\frac{c}{a^2-c^2}R_1^2\log\frac{R_1}{R_2}\right] + (\text{terms linear in } X \text{ and } Y) \qquad [7]$$

where α is a constant to be adjusted so that the uniform normal stress on the circular boundary vanishes, and the terms linear in X and Y may be neglected as they yield no stress.

An examination of the function ϕ_1 reveals that the first two terms combined yield single-valued displacements within the disk as also do the third and fourth terms. However, the last term within the brackets introduces many-valued displacements in the region included in the disk. It is necessary to remove such terms without altering the stress conditions on the circular boundary of the disk. The reason for arranging the terms in the manner of Equation [7] appears when we notice that the function $R_1^2 \log (R_1/R_2)$ produces uniform normal stress on the circular boundary of the disk. Hence, we may drop this term and include its uniform normal boundary stress in the constant α.

Evaluating the normal stress on the boundary resulting from the first three terms, we find that to remove it, we must set

$$\alpha = -\frac{a+c}{4a^2c}\left[(c+a) - \nu(c-a)\right] \qquad [8]$$

The final stress function for the transformed problem as shown in Fig. 3, is then

$$\phi_1 = -\frac{F'}{2\pi}\left[Y(2\Theta + \Theta_1 + \Theta_2) - \frac{1-\nu}{2} X_1 \log\frac{R_1}{R_2} + \frac{1+\nu}{4}\frac{a^2}{c^2} X_2 \frac{R_1^2}{R_2^2} + \alpha R_1^2 \right] \qquad [9]$$

where α has the value given in Equation [8]. The stresses, obtained from Equations [2], (substituting X, Y for x, y) are found to be

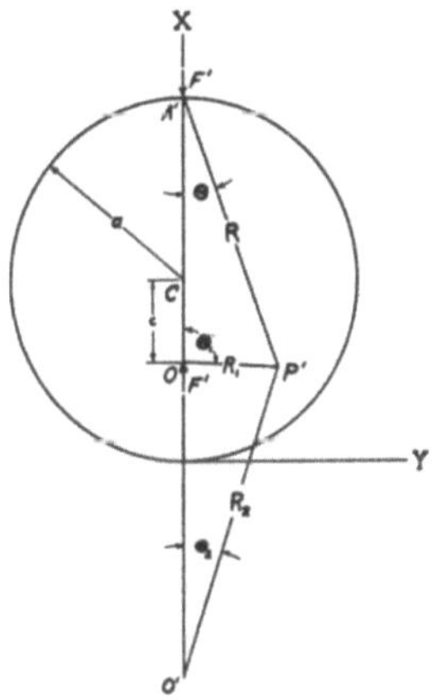

FIG. 3 THE TRANSFORMED PROBLEM

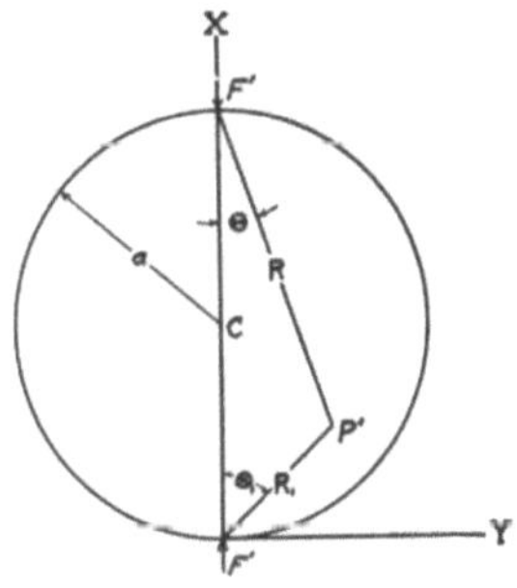

FIG. 4 THE ROLLER PROBLEM—PARTICULAR CASE WHEN $c = a$

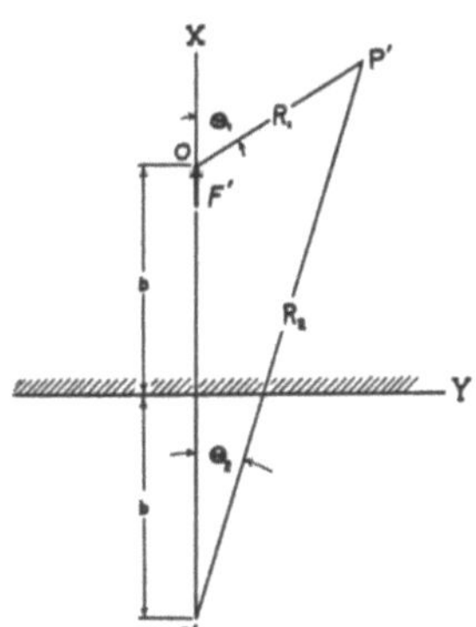

FIG. 5 MELAN'S PROBLEM—PARTICULAR CASE WHEN $a \to \infty$

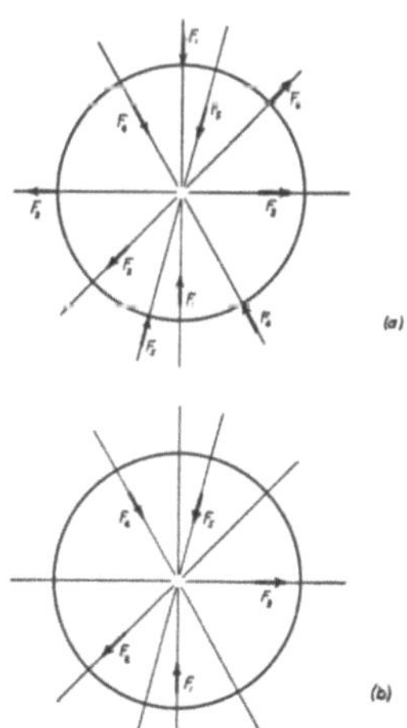

FIG. 6 GENERAL CASE—SELF-EQUILIBRATING SYSTEM OF RADIAL FORCES

$$\sigma_x = -\frac{F'}{2\pi}\left[\frac{4(2a-X)^3}{R^4} + \frac{2X_1^3}{R_1^4} + \frac{2X_2^3}{R_2^4} - \frac{1-\nu}{2}X_1\left(\frac{1}{R_1^2} - \frac{2Y^2}{R_1^4} - \frac{1}{R_2^2} + \frac{2Y^2}{R_2^4}\right) + \frac{(1+\nu)a^2}{2c^2}X_2(X_2^2 - X_1^2)\left(\frac{1}{R_2^4} - \frac{4Y^2}{R_2^6}\right) + 2\alpha\right]$$

$$\sigma_y = -\frac{F'}{2\pi}\left[2Y^2\left(\frac{2(2a-X)}{R^4} + \frac{X_1}{R_1^4} + \frac{X_2}{R_2^4}\right) - \frac{1-\nu}{2}\left(\frac{3X_1}{R_1^2} - \frac{2X_1^3}{R_2^4} - \frac{X_1+2X_2}{R_2^2} + \frac{2X_1X_2^2}{R_2^4}\right) + \frac{(1+\nu)a^2}{2c^2}\left(\frac{2X_1+X_2}{R_2^2} - \frac{4X_1X_2^2 + 3X_2R_1^2}{R_2^4} + \frac{4X_2^3R_1^2}{R_2^6}\right) + 2\alpha\right]$$

$$\tau_{xy} = -\frac{F'Y}{2\pi}\left[-\frac{4(2a-X)^2}{R^4} + \frac{2X_1^2}{R_1^4} + \frac{2X_2^2}{R_2^4} + \frac{1-\nu}{2}\left(\frac{1}{R_1^2} - \frac{1}{R_2^2} - \frac{2X_1^2}{R_1^4} + \frac{2X_1X_2}{R_2^4}\right) - \frac{(1+\nu)a^2}{2c^2}\left(\frac{3X_2+X_1}{R_2^4}\left\{\frac{a^2}{c} - c\right\} - \frac{4X_2^2(X_2^2 - X_1^2)}{R_2^6}\right)\right]$$

PARTICULAR CASES

Two well-known problems involving straight and circular boundaries may be shown to be special forms of the solution represented by Equation [9].

(a) *The Roller Problem.* Michell (1) gives the solution for the disk loaded normally on its periphery by a pair of forces at the ends of a diameter, as shown in Fig. 4. This may be obtained from Equation [9] and Fig. 3 by letting $c = a$. Then point O coincides with O' and in Equation [9] $\Theta_1 = \Theta_2$ and $R_1 = R_2$ so that, dropping terms linear in X and Y, the stress function for the case of Fig. 4 becomes

$$\phi_2 = -\frac{F'}{2\pi}\left[2Y(\Theta + \Theta_1) - \frac{R_1^2}{a}\right] \qquad [10]$$

(b) *Melan's Problem.* If, in Equation [9] and Fig. 3, we let $a \to \infty$ in such a way that $a - c = b$ (a constant), we obtain the solution for a semi-infinite plate loaded at a distance b from its edge by a force normal to the boundary, as shown in Fig. 5. In Equation [9] we have

$$\lim_{a \to \infty} \begin{cases} a/c = 1 \\ \Theta = 0 \\ R_1^2 = Y^2 + (X-b)^2 \\ R_2^2 = Y^2 + (X+b)^2 \end{cases}$$

Then, noting that $R_1^2 = R_2^2 - 4bX$, the stress function for Melan's problem becomes

$$\phi_3 = -\frac{F'}{2\pi}\left[Y(\Theta_1 + \Theta_2) - \frac{1-\nu}{2}(X-b)\log\frac{R_1}{R_2} - \frac{(1-\nu)bX(X+b)}{R_2^2}\right] \qquad [11]$$

If translated into Melan's notation, Equation [11]

will be found to be identical with his Equation [9] (4).

A final check on Equation [9] is obtained by evaluating its limit as $c \to 0$. Noting that $\Theta_2 \to 0$ and $c^2R_2^2 \to a^4$ as $c \to 0$, Equation [9] becomes

$$\phi = -\frac{F'}{2\pi}\left[Y(2\Theta + \Theta_1) - \frac{1-\nu}{2}X_1 \log R_1 - \frac{1+\nu}{4a^2}X_1R_1^2 - \frac{R_1^2}{2a}\right] \ldots [12]$$

In order to compare this equation with Equation [3] we observe that for the case $c = 0$, $Y = -y$, $\Theta = -\theta$, $\Theta = -\varphi$, $X_1 = x$, $R_1 = \rho$. Substituting these values in Equation [12], we verify that the latter is identical with Equation [3].

General Case

We can find the solution for any system of radial forces by applying solutions of type ϕ_1 (Equation [9]) at the required orientations as shown in Fig. 6*a*. If the system of forces in the interior of the disk is self-equilibrating, then the system on the boundary is also self-equilibrating and the latter may be removed by a well-known method (5). There remains, then, only the system shown in Fig. 6*b*.

BIBLIOGRAPHY

1 "Elementary Distribution of Plain Stress," by J. H. Michell, Proceedings of the London Mathematical Society, vol. 32, 1900, p. 44.

2 "Theory of Elasticity," by S. Timoshenko, McGraw-Hill Book Company, New York, N. Y., first edition, 1934, p. 177. Also: "A Treatise on the Mathematical Theory of Elasticity," by A. E. H. Love, Cambridge University Press, London, England, fourth edition, 1927, p. 215. Also: "A Treatise on Photo-Elasticity," by E. G. Coker and L. N. G. Filon, Cambridge University Press, London, England, 1931, p. 398.

3 "A Treatise on the Mathematical Theory of Elasticity," by A. E. H. Love, Cambridge University Press, London, England, fourth edition, 1927, p. 490.

4 "Der Spannungszustand der durch eine Einzelkraft im Innern beanspruchen Halbscheibe," by E. Melan, *Zeitschrift fur angewandte Mathematik und Mechanik*, vol. 12, December 6, 1932, p. 345.

5 "Theory of Elasticity," by S. Timoshenko, McGraw-Hill Book Company, New York, N. Y., first edition, 1934, p. 106.

(*Reprinted from the* JOURNAL OF APPLIED MECHANICS *for December, 1937, pp. A-170 to A-172*)

Distortion of the Photoelastic Fringe Pattern in an Optically Unbalanced Polariscope

BY RAYMOND D. MINDLIN,[1] NEW YORK, N. Y.

This paper describes the results of an investigation of the effect of inaccuracies in the two quarter-wave plates of a photoelastic polariscope on the configuration of the monochromatic fringe pattern. It is shown that the fringe pattern is distorted to an extent which depends upon the orientations of the axes of principal stress and the difference between the retardations of the two plates.

IN THE optical system ordinarily used in a photoelastic polariscope, three doubly-refracting plates are situated between a pair of polarizers. The arrangement is illustrated schematically in Fig. 1, in which the light passes from left to right from the source S, through the polarizer P, a retardation plate 1, the model 2, a second retardation plate 3, and the analyzer A. Plates 1 and 3 are usually crystal retardation plates which produce uniform relative retardation, whereas the phase retardation in the model varies from point to point in proportion to the principal stress difference.

For reasons stated later, it is desirable that the two crystal plates each introduce a relative retardation of one quarter-phase of the monochromatic vibration issuing from the light source. As a matter of practical construction, the production of a perfect quarter-wave plate is a difficult and costly operation. In most polariscopes now in use, it will very likely be found that not only are the retardation plates not true quarter-wave plates, but also that the retardation produced by one is different from that produced by the other.

In the following paragraphs, it will be shown that the isochromatic-fringe pattern, obtained with monochromatic light, is unaffected by the inaccuracies in the retardation plates only when the two plates are identical. When the retardation plates are not identical, the fringe pattern is distorted and no longer presents a true picture of the stress distribution. The relation which governs the intensity of light transmitted through such a polariscope is given by Eq [6]. The magnitude of the *maximum* error introduced depends upon the difference in phase between the retardations of the two plates and may be calculated from Eq [8]. In general, the error is also a function of the orientation of the axes of principal stress, and may be calculated from Eq [7].

The results of the investigation show that *small* errors in the retardation plates have a negligible effect on the precision of a photoelastic analysis in which the fringe method is used. The conclusion is reached that in designing a photoelastic polariscope it is unnecessary to specify, within narrow limits, the retardation characteristics of the crystal plates.

NOTATION

δ_1 = relative phase retardation of first retardation plate.
δ_2 = relative phase retardation of a point in the model.
δ_3 = relative phase retardation of second retardation plate.
ρ_1 = angular coordinate of fast axis of first retardation plate.
ρ_2 = angular coordinate of fast axis of a point in the model.
ρ_3 = angular coordinate of fast axis of second retardation plate.
θ = vibration direction of analyzer (all angles are referred to the vibration direction of the polarizer and are reckoned positive in the counterclockwise direction as seen by an observer viewing the oncoming light from the analyzer end of the polariscope).
I_0 = intensity of light transmitted through the polariscope when $\rho_1 = \rho_2 = \rho_3 = \theta = 0$.
I = relative intensity of light referred to I_0.
n = 0, ±1, ±2, ± . . . .

CIRCULAR POLARISCOPE

If the source S emits a homogeneous radiation for which plates 1 and 3 are quarter-wave plates ($\delta_1 = \delta_3 = \pi/2$) and if the axes

FIG. 1 SCHEMATIC DIAGRAM OF PHOTOELASTIC POLARISCOPE
(S = monochromatic light source, P = polarizer, 1 and 3 = retardation plates, 2 = model, A = analyzer.)

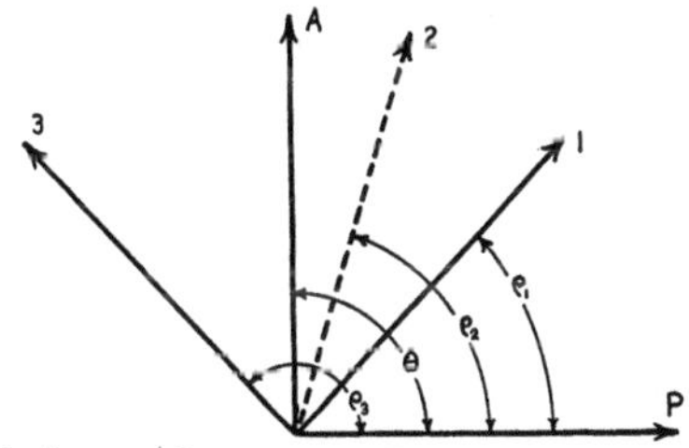

FIG. 2 ANGULAR RELATIONS OF FAST AXES OF RETARDATION PLATES AND MODEL AND VIBRATION DIRECTIONS OF POLARIZER AND ANALYZER, AS USED IN EQUATIONS [1] TO [8], INCLUSIVE
(ρ_1 = 45 deg, ρ_3 = 135 deg, θ = 90 deg.)

of the retardation plates are oriented in the directions indicated in Fig. 2, the system is included in the designation "circular polariscope."

The relative intensity of light transmitted through a circular polariscope is given by the well-known relation

$$I = I_0 \sin^2 \frac{\delta_2}{2} \qquad [1]$$

so that dark fringes are observed wherever, in the model,

$$\delta_2 = 2n\pi \text{ where } n = 0, \pm 1, \pm 2, \pm \ldots \qquad [2]$$

A fringe pattern for a circular-ring segment subjected to bending would appear, in a circular polariscope, as in Fig. 3.

[1] Department of Civil Engineering, Columbia University.

Presented at the Annual Meeting of THE AMERICAN SOCIETY OF MECHANICAL ENGINEERS, held in New York, N. Y., December 6–10, 1937.

Discussion of this paper should be addressed to the Secretary, A.S.M.E., 29 West 39th Street, New York, N. Y., and will be accepted until February 10, 1938, for publication at a later date. Discussion received after the closing date will be returned.

NOTE: Statements and opinions advanced in this paper are to be understood as individual expressions of the author, and not those of the Society.

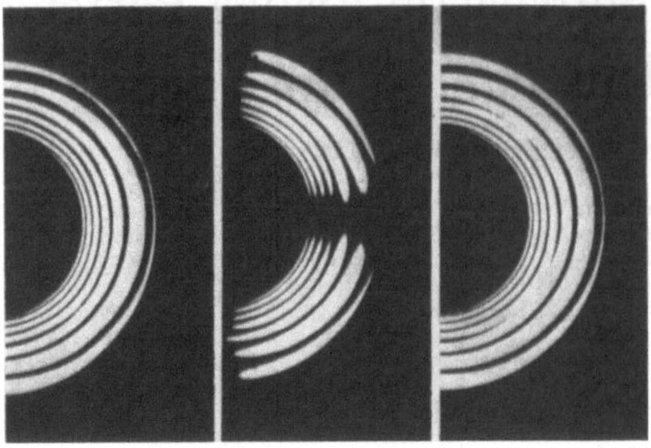

Fig. 3 Fig. 4 Fig. 5

FIG. 3 PHOTOELASTIC FRINGE PATTERN FOR THE BENDING OF A CIRCULAR-RING SEGMENT AS VIEWED IN A CIRCULAR POLARISCOPE ($\delta_3 = \delta_1 = \pi/2$.)

FIG. 4 FRINGE PATTERN AS OBSERVED IN A PLANE POLARISCOPE ($\delta_3 = \delta_1 = 0$.)

FIG. 5 FRINGE PATTERN AS OBSERVED IN A BALANCED POLARISCOPE ($\delta_3 = \delta_1 = 0.57\pi$.)

PLANE POLARISCOPE

When plates 1 and 3 are removed ($\delta_1 = \delta_3 = 0$) the resulting optical system is termed a plane polariscope. The light intensity transmitted through the instrument in this case is given by

$$I = I_0 \sin^2 \frac{\delta_2}{2} \sin^2 2\rho_2 \ldots\ldots\ldots\ldots\ldots [3]$$

Dark fringes are again observed according to Eq [2] and, in addition, extinction of light is also obtained when

$$\rho_2 = \frac{n\pi}{2} \ldots\ldots\ldots\ldots\ldots\ldots\ldots [4]$$

The loci of points in the model for which the orientations of the axes of principal stress satisfy Eq [4] are known as isoclinics.

Fig. 4 shows the appearance of the same model as in Fig. 3 when viewed through a plane polariscope. The isoclinics are the vertical and horizontal dark bands which cross the δ_2 fringes. It should be observed that the positions of the latter are unchanged, in accordance with Eq [3].

It is for the purpose of removing the isoclinics that the quarter-wave retardation plates are used.

BALANCED POLARISCOPE

Retardation plates are again introduced at 1 and 3, with phase retardations different from one quarter-wave but identical with each other ($\delta_3 = \delta_1 = \delta$). The intensity of light transmitted through the system is then given by

$$I = I_0 \sin^2 \frac{\delta_2}{2} (1 - \cos^2 2\rho_2 \cos^2 \delta) \ldots\ldots\ldots [5]$$

The relation expressed by Eq [5] was first introduced in photoelastic analysis by R. V. Baud[2] who used it to show that the isochromatic pattern obtained with a *white-light* source is in part a function of the orientation of the axes of principal stress.

With a monochromatic source, however, the isochromatics are unchanged because the intensity is again directly proportional to $\sin^2 (\delta_2/2)$. The quantity in parentheses in Eq [5] indicates simply that the isoclinics are not completely removed by the introduction of the retardation plates, but the δ_2 fringes still occur according to Eq [2].

Fig. 5 illustrates the effect obtained in a balanced polariscope. The variation of intensity with the orientation of the principal axes of stress is reduced, during photographic processing, to a variation of intensity of the zero-order fringe. The positions of the fringes, it may be observed, are unchanged.

UNBALANCED POLARISCOPE

When the relative retardations produced by the two retardation plates are different from each other ($\delta_3 \neq \delta_1$), the intensity of light transmitted through the polariscope is given by

$$I = I_0 \left\{ \sin^2 \frac{\delta_2}{2} [\cos (\delta_3 - \delta_1) - \cos^2 2\rho_2 \cos \delta_1 \cos \delta_3] + \sin^2 \frac{\delta_3 - \delta_1}{2} - \frac{1}{2} \sin \delta_2 \sin 2\rho_2 \sin (\delta_3 - \delta_1) \right\} \ldots [6]$$ [3]

The intensity is no longer directly proportional to $\sin^2 (\delta_2/2)$ on account of the appearance of the term involving $\sin \delta_2$ and dark fringes, in general, will not occur according to Eq [2].

Maximum and minimum intensities in an unbalanced polariscope occur when

$$\tan \delta_2 = \frac{\sin 2\rho_2 \sin (\delta_3 - \delta_1)}{\cos (\delta_3 - \delta_1) - \cos^2 2\rho_2 \cos \delta_1 \cos \delta_3} \ldots\ldots [7]$$

Hence, instead of a fringe appearing when $\delta_2 = 2n\pi$, there will be a maximum or minimum intensity when $\tan \delta_2$ is given by Eq [7]. This indicates a distortion of the fringe pattern as a function of the phase retardations of the crystal plates and the orientations of the axes of principal stress, since δ_1, δ_3, and ρ_2 are all involved in Eq [7].

The maximum displacement of the fringes occurs when the axes of principal stress are at 45 deg to the direction of polarization. In this case Eq [7] reduces to

$$\tan \delta_2 = \pm \tan (\delta_3 - \delta_1)$$

and therefore a dark fringe appears whenever

$$\delta_2 = 2n\pi \pm (\delta_3 - \delta_1) \ldots\ldots\ldots\ldots\ldots\ldots [8]$$

The upper sign is associated with positive values of $\sin 2\rho_2$ and the lower sign with negative values of $\sin 2\rho_2$. Eq [8] for the unbalanced polariscope is to be compared with Eq [2] for the balanced case.

Figs. 6(*a*), (*b*), (*c*), (*d*), and (*e*) were prepared to show that a distortion of the fringe pattern actually does occur when $\delta_3 \neq \delta_1$. In order to obtain these photographs a Babinet-Soleil compensator was substituted for the second retardation plate so that δ_3 could be varied at will, and any value of $\delta_3 - \delta_1$ obtained. The figures show, successively, the distortion of the fringe pattern as

[2] *Journal of the Optical Society of America*, vol. 20, no. 7, 1930, 386, and vol. 21, no. 2, 1931, p. 119.

[3] Equation [6] and all of the preceding equations may be derived from a more general equation given in a paper by the author on "Analysis of Doubly Refracting Materials With Circularly and Elliptically Polarized Light," *Journal of the Optical Society of America* vol. 27, 1937, p. 288.

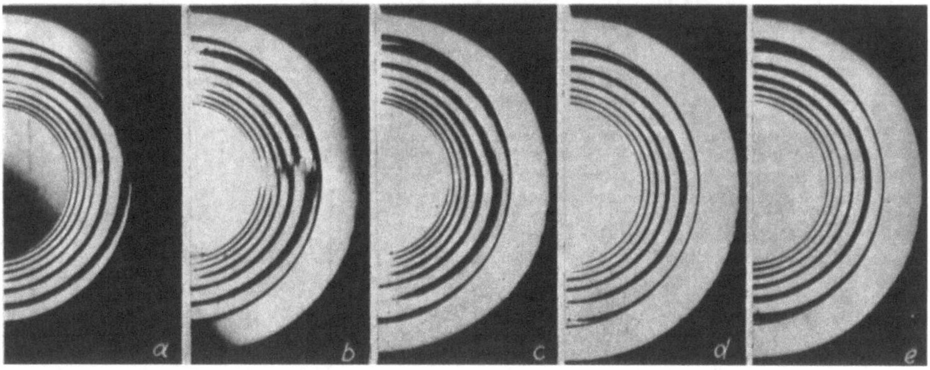

FIG. 6 FRINGE PATTERN AS OBSERVED IN AN UNBALANCED POLARISCOPE
[($\delta_3 - \delta_1 \neq 0$). (a) $\delta_3 - \delta_1 = \pi/4$; (b) $\delta_3 - \delta_1 = 0.375\ \pi$; (c) $\delta_3 - \delta_1 = \pi/2$; (d) $\delta_3 - \delta_1 = 3\ \pi/4$; (e) $\delta_3 - \delta_1 = \pi$.]

$\delta_3 - \delta_1$ is varied from 0 to π. It may be observed that the fringes move outward from the center of the ring in the upper quadrant and inward in the lower quadrant in accordance with the $\pm$ sign of Eq [8]. The amount of displacement is, from Eq [7], a function of δ_1, δ_3, and ρ_2 and may be calculated for any point in the model.

A detailed analysis of Eq [6] reveals that for certain values of δ_2 and $\delta_3 - \delta_1$ the intensity at $\rho_2 = 0$ is a maximum as $\rho_2 \rightarrow 0$ from the positive side and a minimum as $\rho_2 \rightarrow 0$ from negative values. It is therefore to be expected that there will be certain combinations for which the fringes are discontinuous at $\rho_2 = 0$. This phenomenon is illustrated in Fig. 6(b) for which $\delta_3 - \delta_1 = 0.37\ \pi$.

When the difference between the retardations of the two plates has passed through a half cycle ($\delta_3 - \delta_1 = \pi$) the light and dark areas have been completely reversed, Fig. 6(e), and the fringe pattern appears as it would in a polariscope with a right-circular polarizer and a left-circular analyzer.

CONCLUSIONS

(1) Eq [6] shows that the photoelastic fringe pattern is distorted in an optically unbalanced polariscope.

(2) Eq [8] permits the calculation of the allowable limits of error in the retardation plates for a specified precision of the fringe pattern. From the point of view of the stress analyst, the most important part of the fringe pattern is that in the region of highest stress. Assume, for example, that the tenth-order fringe appears in the highly stressed region and that a precision of 1 per cent is required in this region. Eq [8] shows that the permissible error in the difference between the relative retardations of the crystal plates is, in this case, 0.2π or 36 deg. Little skill or time is required for the production of such a pair of plates.

(3) It should be observed that, for a given pair of plates, the absolute magnitude of the error is fixed so that a pair of plates which produces a 1 per cent error in the stress calculated from the tenth fringe will produce a 10 per cent error in the stress calculated from the first fringe.

(4) In so far as the accuracy of the monochromatic fringe pattern is concerned, it is immaterial, for ordinary photoelastic work, whether or not the retardation plates produce one quarter-wave retardations, as long as the difference between their retardations is small.

(5) If the retardation plates in a polariscope are different from quarter-wave plates, but are nevertheless identical, it follows from Eq [6] that the fringe pattern is unaffected. The only effect will be incomplete removal of isoclinics.

(6) Eq [6] also shows that if a monochromatic-light source, of wave length other than that for which the instrument is designed is substituted in a balanced polariscope, the fringe pattern will remain unchanged.

(*Reprinted from* CIVIL ENGINEERING *for August 1938*)

The Equiangular Strain-Rosette

By RAYMOND D. MINDLIN, JUN. AM. SOC. C.E.
INSTRUCTOR, DEPARTMENT OF CIVIL ENGINEERING, COLUMBIA UNIVERSITY, NEW YORK, N.Y.

THERE is developing, today, an increasing interest in experimental studies of complex stress distributions in structural elements. Engineers recognize the need for an intimate knowledge of the detailed state of stress in the material at critical regions throughout the structure. One approach to this knowledge is by means of surface strain measurements of the type known as strain-rosette.

The state of stress at a point on the surface of a stressed material is completely determined only if the magnitudes of the two principal stresses and the orientation of the axes of principal stress are known. The specification of stress in such a two-dimensional system accordingly depends upon three quantities. In any experimental determination of two-dimensional stress, therefore, three independent quantities must be determined in some manner. That is, since it is not possible to measure stress directly, at least three quantities upon which stress depends must be measured.

In the strain-rosette method, the required quantities are obtained by measuring the relative displacements of three or more pairs of points in the neighborhood of a point, P, on the surface of the stressed specimen. The strain in the neighborhood of P is assumed to be homogeneous (that is, uniform) so that three or more unit elongations can be calculated directly from the displacements by dividing the relative displacements by the corresponding gage lengths. Strictly speaking, these strain data, combined with a stress-strain relation, describe the state of stress at P. But the description is in such a form as to be almost useless for purposes of analysis. The stress analyst wishes to know either the components of stress (or strain) acting on a certain plane, or the magnitudes and directions of the principal stresses (or strains). It is therefore necessary to have available a means for interpreting the rosette readings.

In the following paragraphs there are given the theoretical bases on which rosette interpretations are founded, and the theory is applied to a very convenient form of rosette in which the three gage lines make equal angles with each other. The equiangular arrangement is chosen as an example because it leads to the most symmetrical and simple formulas possible with the necessary minimum of three gage lines. The same procedure may be followed to obtain formulas for three gage lines inclined at any other angles with each other.

BASIC FORMULA

There is a fundamental relation in the analysis of strain which connects the unit elongation in any direction with the components of strain referred to a set of coordinate axes (see Case, *Strength of Materials*, 1925, page 79):

$$\epsilon_N = \epsilon_x \cos^2\theta + \epsilon_y \sin^2\theta + \epsilon_{xy} \sin\theta \cos\theta \ldots\ldots [1]$$

Here ϵ_x is the unit elongation in the x direction; ϵ_y is the unit elongation in the y direction, perpendicular to x; and ϵ_{xy} is the shearing strain, that is, the change in angle between the x and y directions during deformation. ϵ_N is the unit elongation in a direction N making an angle θ with the x direction. θ is measured as positive in the counterclockwise direction from the x axis.

In the derivation of Eq. 1, the assumption is made that ϵ_x, ϵ_y, ϵ_{xy} are small relative to unity.

Equation 1 shows that, in order to calculate the magnitude of the strain ϵ_N in any chosen direction specified by θ, three quantities (ϵ_x, ϵ_y, ϵ_{xy}) must be known. Conversely, in order to calculate the components of strain for any set of axes, three other independent quantities must be known. For example, there might be known the strain ϵ_N corresponding to three known values of θ. From these the three components ϵ_x, ϵ_y, ϵ_{xy} may be calculated. This is the basis of the strain-rosette procedure.

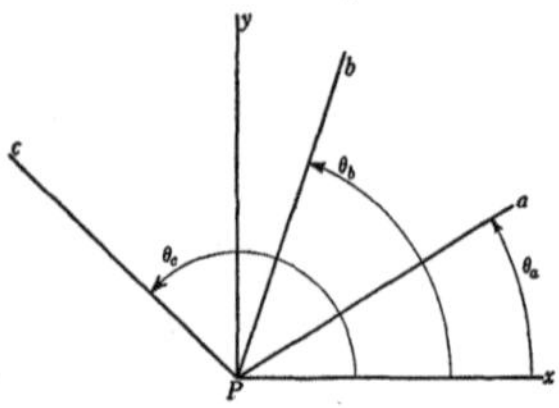

FIG. 1

The changes in length of three line segments Oa, Ob, Oc, oriented at three different angles θ_a, θ_b, θ_c with respect to an arbitrarily chosen x direction, are measured in the neighborhood of a point P on the surface of a stressed plate (Fig. 1). The three strains ϵ_a, ϵ_b, ϵ_c are the ratios of the measured changes in length to the original lengths. Three equations of the type of Eq. 1 may then be written:

$$\left.\begin{aligned}\epsilon_a &= \epsilon_x \cos^2\theta_a + \epsilon_y \sin^2\theta_a + \epsilon_{xy} \sin\theta_a \cos\theta_a\\ \epsilon_b &= \epsilon_x \cos^2\theta_b + \epsilon_y \sin^2\theta_b + \epsilon_{xy} \sin\theta_b \cos\theta_b\\ \epsilon_c &= \epsilon_x \cos^2\theta_c + \epsilon_y \sin^2\theta_c + \epsilon_{xy} \sin\theta_c \cos\theta_c\end{aligned}\right\} \cdots [2]$$

It is a simple process to solve for ϵ_x, ϵ_y, and ϵ_{xy}, in terms of ϵ_a, ϵ_b, ϵ_c for any given values of θ_a, θ_b, θ_c.

EQUIANGULAR STRAIN-ROSETTE

The expressions for ϵ_x, ϵ_y, and ϵ_{xy} in terms of ϵ_a, ϵ_b, and ϵ_c take very simple forms when the three gage lines make equal angles with each other and when the x axis is chosen to coincide with one of the gage lines. Figure 2

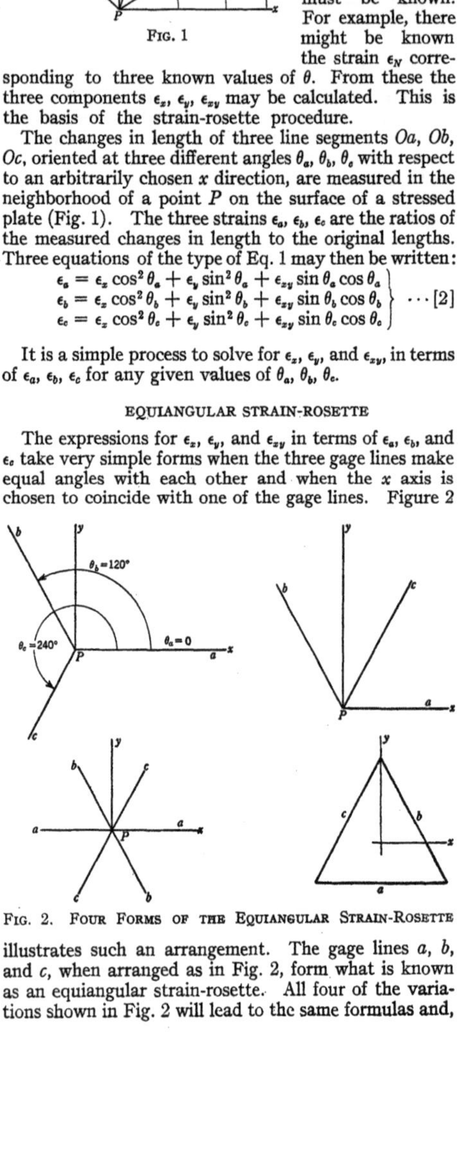

FIG. 2. FOUR FORMS OF THE EQUIANGULAR STRAIN-ROSETTE

illustrates such an arrangement. The gage lines a, b, and c, when arranged as in Fig. 2, form what is known as an equiangular strain-rosette. All four of the variations shown in Fig. 2 will lead to the same formulas and, on the basis of the assumption of homogeneous strain, will give equally good results.

Substituting the values $\theta_a = 0$, $\theta_b = 120°$, $\theta_c = 240°$ in Eqs. 2, there results

$$\left.\begin{aligned} \epsilon_x &= \epsilon_a \\ \epsilon_y &= {}^1/_3\,(2\epsilon_b + 2\epsilon_c - \epsilon_a) \\ \epsilon_{xy} &= \frac{2}{\sqrt{3}}\,(\epsilon_c - \epsilon_b) \end{aligned}\right\} \quad \ldots\ldots\ldots [3]$$

The expressions for the principal strains, ϵ_1, ϵ_2, in terms of the components of strain ϵ_x, ϵ_y, ϵ_{xy} are:

$$\epsilon_1 = {}^1/_2\,(\epsilon_x + \epsilon_y) + {}^1/_2\sqrt{(\epsilon_x - \epsilon_y)^2 + \epsilon_{xy}^2}$$

$$\epsilon_2 = {}^1/_2\,(\epsilon_x + \epsilon_y) - {}^1/_2\sqrt{(\epsilon_x - \epsilon_y)^2 + \epsilon_{xy}^2}$$

and the angle, ϕ, which the direction of the algebraically larger principal strain (ϵ_1) makes with the x axis is

$$\tan 2\phi = \frac{\epsilon_{xy}}{\epsilon_x - \epsilon_y}$$

(see Case, op cit, page 79).

To express these three quantities directly in terms of the equiangular strain readings, it is only necessary to

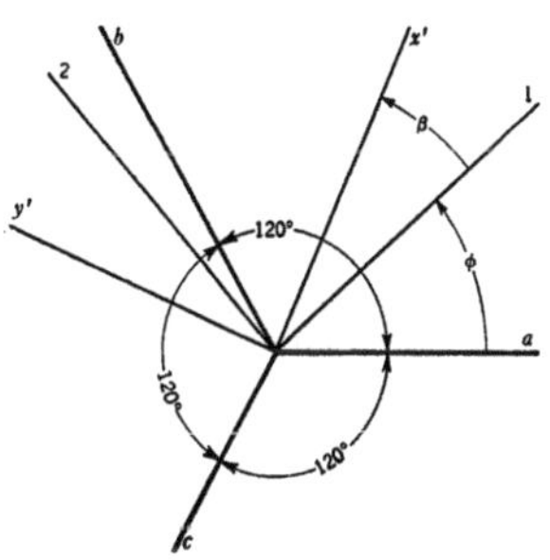

FIG. 3

substitute for ϵ_x, ϵ_y, and ϵ_{xy} their values given by Eqs. 3. There results

$$\left.\begin{aligned} \epsilon_1 &= {}^1/_3\,(\epsilon_a + \epsilon_b + \epsilon_c) + \\ &\quad \frac{\sqrt{2}}{3}\sqrt{(\epsilon_a - \epsilon_b)^2 + (\epsilon_b - \epsilon_c)^2 + (\epsilon_c - \epsilon_a)^2} = A + B \\ \epsilon_2 &= {}^1/_3(\epsilon_a + \epsilon_b + \epsilon_c) - \\ &\quad \frac{\sqrt{2}}{3}\sqrt{(\epsilon_a - \epsilon_b)^2 + (\epsilon_b - \epsilon_c)^2 + (\epsilon_c - \epsilon_a)^2} = A - B \\ \tan 2\phi &= \frac{\sqrt{3}\,(\epsilon_c - \epsilon_b)}{2\epsilon_a - \epsilon_b - \epsilon_c} \end{aligned}\right\} \quad [4]$$

The angle ϕ is measured as positive in the counterclockwise direction from the x-axis to the direction of ϵ_1. If $(\epsilon_c - \epsilon_b) > 0$, $0 < \phi < \frac{\pi}{2}$. If $(\epsilon_c - \epsilon_b) < 0$, $\frac{\pi}{2} < \phi < \pi$.

Equations 3 give, in terms of ϵ_a, ϵ_b, and ϵ_c, the components of strain referred to a set of coordinates for which one axis coincides with a gage line. It may be required to calculate the components of strain along some other set of rectangular axes, say x' and y'. It is convenient to designate the orientation of the new set of axes by reference to a principal stress direction.

Let the x' axis be located by an angle β measured positively in the counterclockwise sense from the axis of the algebraically larger principal strain (Fig. 3). Then

$$\left.\begin{aligned} \epsilon_{x'} &= A + B\cos 2\beta \\ \epsilon_{y'} &= A - B\cos 2\beta \\ \epsilon_{x'y'} &= -\,2B\sin 2\beta \end{aligned}\right\} \quad \ldots\ldots\ldots\ldots [5]$$

In order to calculate stresses from the strains, assumptions must be made as to the relations between stress and deformation in the material under test. It is customary to assume that the material follows Hooke's Law and is isotropic. These assumptions lead to the ordinary formulas for the stress-strain relations in terms of Young's modulus, E, and Poisson's ratio, ν. For the condition of two-dimensional stress which exists on a free surface, these are

$$\left.\begin{aligned} \sigma_x &= \frac{E}{1-\nu^2}\,(\epsilon_x - \nu\epsilon_y) \\ \sigma_y &= \frac{E}{1-\nu^2}\,(\epsilon_y - \nu\epsilon_x) \end{aligned}\right\} \quad \ldots\ldots\ldots\ldots [6]$$

(Case, op cit page 418).

Formulas for the principal stresses in terms of the strains ϵ_a, ϵ_b, ϵ_c are obtained by substituting the first two of Eqs. 4, in Eqs. 6, replacing x and y by 1 and 2 in the latter. The result is

$$\left.\begin{aligned} \sigma_1 &= E\left\{\frac{A}{1-\nu} + \frac{B}{1+\nu}\right\} \\ \sigma_2 &= E\left\{\frac{A}{1-\nu} - \frac{B}{1+\nu}\right\} \end{aligned}\right\} \quad \ldots\ldots [7]$$

The direction of the maximum principal stress is again given by the last of Eqs. 4.

The formulas for the components of stress referred to any set of axes x', y' are obtained by combining Eqs. 6 and Eqs. 5, with the result

$$\left.\begin{aligned} \sigma_{x'} &= E\left\{\frac{A}{1-\nu} + \frac{B}{1+\nu}\cos 2\beta\right\} \\ \sigma_{y'} &= E\left\{\frac{A}{1-\nu} - \frac{B}{1+\nu}\cos 2\beta\right\} \\ \sigma_{x'y'} &= -\frac{E}{1+\nu}B\sin 2\beta \end{aligned}\right\} \quad \ldots [8]$$

For those familiar with the Mohr circle diagram, Fig. 4 is given to show how simply Mohr's circle may be constructed directly from the strain-rosette data.

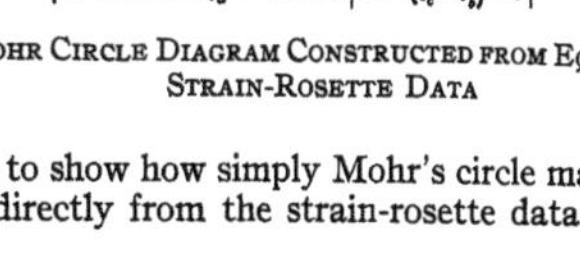

FIG. 4. MOHR CIRCLE DIAGRAM CONSTRUCTED FROM EQUIANGULAR STRAIN-ROSETTE DATA

Reprinted from JOURNAL OF APPLIED PHYSICS, Vol. 9, No. 11, November, 1938

Stresses in a Heavy Disk Suspended from an Eccentric Peg

RAYMOND D. MINDLIN
Department of Civil Engineering, Columbia University, New York, New York
(Received July 1, 1938)

An exact solution of the two-dimensional elasticity equations for small strain is obtained which satisfies the boundary conditions for a heavy disk suspended in a vertical plane from an eccentric peg of small diameter.

INTRODUCTION

A SOLUTION will be given for the stresses in a circular disk supported with its plane vertical by a peg located at any point, A, on a vertical diameter (Fig. 1). The diameter of the eccentric peg is assumed to be small in comparison with the diameter of the disk so that the peg may be replaced by a concentrated force F. The thickness of the disk is assumed to be small in comparison with its diameter and functions in the stresses involving the coordinate normal to the plane of the disk are neglected. The problem is therefore strictly two-dimensional.

On account of the eccentricity of the balancing force, the problem is most conveniently handled in bipolar coordinates. Such a coordinate system may be obtained from the conformal transformation

$$x+iy=ia\coth\tfrac{1}{2}(\alpha+i\beta), \tag{1}$$

whence

$$x=\frac{a\sin\beta}{\cosh\alpha-\cos\beta},\quad y=\frac{a\sinh\alpha}{\cosh\alpha-\cos\beta} \tag{2}$$

$$\alpha=\log(r_1/r_2),\qquad \beta=\theta_1-\theta_2. \tag{3}$$

The significance of these coordinates is illustrated in Fig. 1.

Important parts are taken in the solution by the stretch ratio, J, of the transformation and the angle ϕ which the tangent to the curve $\beta=$constant makes with the x-axis. These are given by

$$J=\left[\left(\frac{\partial x}{\partial\alpha}\right)^2+\left(\frac{\partial y}{\partial\alpha}\right)^2\right]^{\frac{1}{2}}=\frac{a}{\cosh\alpha-\cos\beta}, \tag{4}$$

$$\phi=\theta_1+\theta_2+\frac{\pi}{2}=\tan^{-1}\frac{\cosh\alpha\cos\beta-1}{\sinh\alpha\sin\beta} \tag{5}$$

The components of stress in plane bipolar coordinates are expressed in terms of a stress function χ and a body force potential Ω by[1]

$$\left.\begin{aligned}
a\widehat{\alpha\alpha}&=\left\{(\cosh\alpha-\cos\beta)\frac{\partial^2}{\partial\beta^2}-\sinh\alpha\frac{\partial}{\partial\alpha}-\sin\beta\frac{\partial}{\partial\beta}+\cosh\alpha\right\}\left(\frac{\chi}{J}\right)+a\Omega\\
a\widehat{\beta\beta}&=\left\{(\cosh\alpha-\cos\beta)\frac{\partial^2}{\partial\alpha^2}-\sinh\alpha\frac{\partial}{\partial\alpha}-\sin\beta\frac{\partial}{\partial\beta}+\cos\beta\right\}\left(\frac{\chi}{J}\right)+a\Omega\\
a\widehat{\alpha\beta}&=-(\cosh\alpha-\cos\beta)\frac{\partial^2(\chi/J)}{\partial\alpha\partial\beta}
\end{aligned}\right\}, \tag{6}$$

where Ω is related to the x and y components of the body force per unit of volume by

$$X=-\partial\Omega/\partial x,\quad Y=-\partial\Omega/\partial y \tag{7}$$

and χ satisfies the differential equation

$$\nabla^4\chi=-(1-\nu)\nabla^2\Omega, \tag{8}$$

in which ν is Poisson's ratio, ∇^2 is Laplace's operator, and $\nabla^4=\nabla^2\nabla^2$.

Since the disk is situated in a uniform gravitational field directed along the Y-axis, the body force system is given by $X=0$, $Y=w$ where w is the weight per unit of volume of the disk. Hence $\Omega=-wy=-wJ\sinh\alpha$ and $\nabla^2\Omega=0$, so that (8) reduces to

$$\nabla^4\chi=0. \tag{9}$$

[1] G. B. Jeffery, Phil. Trans. Roy. Soc. **A221**, 265–293 (1920).

A general solution of (9) in bipolar coordinates has been given by Jeffery:[1]

$$\frac{\chi}{J}=(E\cos\beta+F\sin\beta+G\cosh\alpha+H\sinh\alpha)\beta$$

$$+(A_0\cosh\alpha+B_0\alpha\cosh\alpha+C_0\sinh\alpha+D_0\alpha\sinh\alpha)$$

$$+(A_1\cosh 2\alpha+B_1+C_1\sinh 2\alpha+D_1\alpha)\cos\beta$$

$$+(A_1'\cosh 2\alpha+B_1'+C_1'\sinh 2\alpha+D_1'\alpha)\sin\beta$$

$$+\sum_{n=2}^{\infty}\left\{\begin{array}{l}[A_n\cosh(n+1)\alpha+B_n\cosh(n-1)\alpha\\ \quad+C_n\sinh(n+1)\alpha\\ \quad\quad+D_n\sinh(n-1)\alpha]\cos n\beta\\ +[A_n'\cosh(n+1)\alpha\\ \quad+B_n'\cosh(n-1)\alpha\\ \quad+C_n'\sinh(n+1)\alpha\\ \quad\quad+D_n'\sinh(n-1)\alpha]\sin n\beta\end{array}\right\}. \quad (10)$$

Boundary Conditions

The boundary of the disk is taken as the circle $\alpha=\text{constant}=\alpha_1$. The edge of the disk is free of stress, so that

$$\text{when } \alpha=\alpha_1:\left\{\begin{array}{l}\widehat{\alpha\beta}=0\\ \widehat{\alpha\alpha}=0\end{array}\right\}. \quad (11)$$

From the first of (11) and the third of (6) we have

$$(\partial/\partial\alpha)(\chi/J)=\xi\ (\text{a constant}). \quad (12)$$

From (12), the second of (11) and the first of (6) we have

$$\left\{(\cosh\alpha_1-\cos\beta)\frac{\partial^2}{\partial\beta^2}-\sin\beta\frac{\partial}{\partial\beta}+\cosh\alpha_1\right\}\left(\frac{\chi}{J}\right)$$
$$=waJ_1\sinh\alpha_1+\xi\sinh\alpha_1,$$

the solution of which is

$$\chi/J_1=\xi\tanh\alpha_1+\eta(\cosh\alpha_1\cos\beta-1)+\zeta\sin\beta+wa^2\operatorname{cosech}\alpha_1-\tfrac{1}{2}wa^2\operatorname{cosech}^2\alpha_1\phi_1\sin\beta, \quad (13)$$

where η and ζ are constants.

(13) may be written in the form

$$\chi=\zeta J_1\sin\beta+(\xi\operatorname{sech}\alpha_1+\eta\sinh\alpha_1)J_1\sinh\alpha_1-a\eta\cosh\alpha_1+wa^2J_1\operatorname{cosech}\alpha_1-\tfrac{1}{2}wa^2\operatorname{cosech}^2\alpha_1\phi_1J_1\sin\beta$$
$$=\zeta x_1+(\xi\operatorname{sech}\alpha_1+\eta\sinh\alpha_1)y_1-a\eta\cosh\alpha_1+wa^2J_1\operatorname{cosech}\alpha_1-\tfrac{1}{2}wa^2\operatorname{cosech}^2\alpha_1\phi_1J_1\sin\beta, \quad (14)$$

where x_1 and y_1 are the coordinates of a point on α_1. By adding a function linear in x and y to χ, the terms in (14) involving ξ, η, ζ may be removed for one boundary without affecting the stresses. Hence, there being only one boundary in this problem, ξ, η, and ζ may be taken equal to zero and the boundary conditions reduce to

$$\left.\begin{array}{l}(\partial/\partial\alpha)(\chi/J)=0\\ \chi/J=wa\operatorname{cosech}\alpha_1-\tfrac{1}{2}wa^2\operatorname{cosech}^2\alpha_1\phi_1\sin\beta\end{array}\right\}. \quad (15)$$

Construction of the Stress Function

The last term in the second of (15) warrants special consideration, for it is observed that no

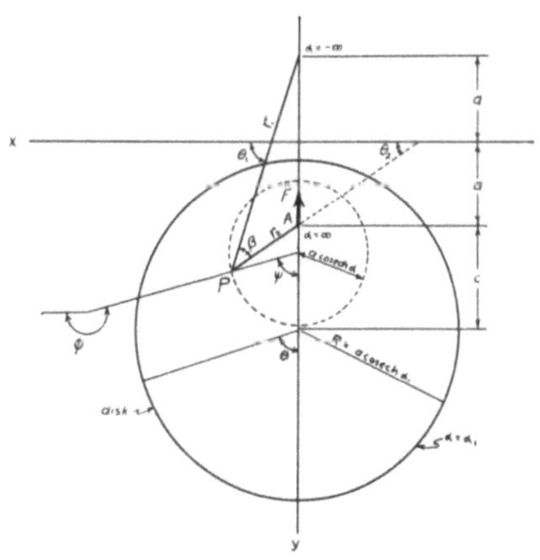

Fig. 1.

similar term appears in (10). It may be verified that the function

$$\chi_1=-\tfrac{1}{2}wa^2\operatorname{cosech}^2\alpha_1J\phi\sin\beta \quad (16)$$

satisfies (9), although it does not appear explicitly in (10). Observing that a cosech α_1 is the radius, R, of the disk; $\phi=\theta_1+\theta_2+\pi/2$; $J\sin\beta=x$; (16) may be written as

$$\chi_1=-\tfrac{1}{2}wR^2x(\theta_1+\theta_2+\pi/2). \qquad (17)$$

In (17) we recognize the stress function for a pair of single forces acting at the poles $(0, +a)$, $(0, -a)$ in the direction of the negative y-axis. The magnitudes of the forces are equal to $w\pi R^2$ which is the weight of the disk with the thickness taken as unity. Hence (16) is the stress function required to annul the *resultant* of the stresses produced on $\alpha=\alpha_1$ by the gravitational force. We shall therefore introduce it as part of the stress function for this problem.

It is known[2] that the stress function $x\theta_2$ produces many valued functions in the displacements for the region included in the disk. These may be removed by adding the stress function

$$\chi_2/J=\tfrac{1}{4}(1-\nu)wa^2\operatorname{cosech}^2\alpha_1 \times[\alpha\sinh\alpha-\alpha(\cosh\alpha-\cos\beta)], \qquad (18)$$

which is perhaps more readily recognized when it is written in our other notation:

$$\chi_2=\tfrac{1}{4}(1-\nu)wR^2 \times[(y-a)\log r_2-(y-a)\log r_1]. \qquad (19)$$

The term $(y-a)\log r_2$ in (19) serves to eliminate the many valued displacements in the disk which were introduced by the term $x\theta_2$ in (17).

Hence the awkward term $\phi\sin\beta$ of (15) may be properly eliminated by introducing as part of our stress function χ_1 and χ_2.

Although χ_1 and χ_2 serve to simplify the second boundary condition, they introduce a difficulty of another kind in the first boundary condition. If (16) is substituted in the first of (15), there results

$$\frac{\partial}{\partial\alpha}\left(\frac{\chi_1}{J}\right)=-\tfrac{1}{2}wa^2\operatorname{cosech}^2\alpha_1\frac{\sin^2\beta}{\cosh\alpha-\cos\beta}. \qquad (20)$$

Again, no term of this type appears explicitly in (10). It is true that (20) can be expanded into an infinite series of cosines. The same is true of $\phi\sin\beta$. But we shall again adopt the technique of introducing suitable finite functions to eliminate awkward terms.

[2] Timoshenko, *Theory of Elasticity* (New York, 1934), p. 116.

The function

$$\chi_3=Aa^{-2}y(x^2+y^2)+Ba^{-1}y^2 =AJ\sinh\alpha\frac{\cosh\alpha+\cos\beta}{\cosh\alpha-\cos\beta}+BJ\frac{\sinh^2\alpha}{\cosh\alpha-\cos\beta},$$

where A and B are constants, is a suitable biharmonic function, i.e., it satisfies (9). It may be verified that, if $A=\tfrac{1}{4}wa^2$ and $B=-\tfrac{1}{2}wa^2\coth\alpha_1$, the awkward term in (20) is eliminated on $\alpha=\alpha_1$ and there remain in the boundary conditions only terms of the form $C+D\cos\beta$.

We are now in a position to select suitable terms of this last type from Jeffery's general solution (10). These are

$$\chi_4/J=A_0\cosh\alpha +(A_1\cosh 2\alpha+C_1\sinh 2\alpha)\cos\beta. \qquad (21)$$

It is readily verified that the proper values for A_0, A_1 and C_1 are

$$A_0=\tfrac{1}{4}wa^2[2\coth\alpha_1-(1-\nu)(\coth\alpha_1-1) +(1+\nu)\operatorname{cosech}^2\alpha_1]$$

$$A_1=-C_1=-\tfrac{1}{8}(1+\nu)wa^2\frac{\operatorname{cosech}^4\alpha_1}{(\coth\alpha_1-1)^2}.$$

The complete stress function is therefore given by

$$\chi/J=\chi_1/J+\chi_2/J+\chi_3/J+\chi_4/J$$

$$=\tfrac{1}{4}wa^2\operatorname{cosech}^2\alpha_1\Big\{-2\phi\sin\beta +(1-\nu)[\alpha\sinh\alpha-\alpha(\cosh\alpha-\cos\beta)] +a^{-1}\sinh^2\alpha_1J\sinh\alpha(\cosh\alpha+\cos\beta) -a^{-1}\sinh 2\alpha_1J\sinh^2\alpha -\tfrac{1}{2}(1+\nu)\frac{\operatorname{cosech}^2\alpha_1}{(\coth\alpha_1-1)^2}(\cosh 2\alpha-\sinh 2\alpha)\cos\beta +[\sinh 2\alpha_1-(1-\nu)\sinh^2\alpha_1(\coth\alpha_1-1) +\tfrac{1}{2}(1+\nu)]\cosh\alpha\Big\}. \qquad (22)$$

STRESSES

The formulas for the bipolar components of stress in the disk are found by substituting (22) in (6). The results are

$$\widehat{\alpha\alpha}=-wJ\sinh\alpha-\tfrac{1}{2}w\,\mathrm{cosech}^2\,\alpha_1[a\sinh\alpha(\cosh\alpha-\cos\beta)-J\sinh^3\alpha]$$
$$+\tfrac{1}{4}(1-\nu)wa\,\mathrm{cosech}^2\,\alpha_1[\sinh\alpha(\cosh\alpha-\cos\beta)-\sinh^2\alpha]+\tfrac{1}{2}wJ[\sinh\alpha+(J/a)\sinh 2\alpha\sin^2\beta]$$
$$-wa^{-1}J^2\coth\alpha_1\sinh^2\alpha\sin^2\beta-(aw/8)(1+\nu)\,\mathrm{cosech}^2\,\alpha_1e^{2(\alpha_1-\alpha)}(1+2\sinh\alpha\cos\beta)$$
$$+\tfrac{1}{2}wa[\coth\alpha_1-\tfrac{1}{2}(1-\nu)(\coth\alpha_1-1)+\tfrac{1}{4}(1+\nu)\,\mathrm{cosech}^2\,\alpha_1], \quad (23)$$

$$\widehat{\beta\beta}=-wJ\sinh\alpha+\tfrac{1}{2}w\,\mathrm{cosech}^2\,\alpha_1J\sinh\alpha\sin^2\beta$$
$$+\tfrac{1}{4}(1-\nu)wa\,\mathrm{cosech}^2\,\alpha_1[(\cosh\alpha-\cos\beta)(2\cosh\alpha-\sinh\alpha)-\sinh^2\alpha]$$
$$+\tfrac{1}{2}wJ[3\sinh\alpha-(J/a)\sinh 2\alpha\sin^2\beta]-wa^{-1}J^2\coth\alpha_1(\cosh\alpha\cos\beta-1)^2$$
$$-(wa/8)(1+\nu)\,\mathrm{cosech}^2\,\alpha_1e^{2(\alpha_1-\alpha)}[4(\cosh\alpha-\cos\beta)+2\sinh\alpha\cos\beta+1]$$
$$+\tfrac{1}{2}wa[\coth\alpha_1-\tfrac{1}{2}(1-\nu)(\coth\alpha_1-1)+\tfrac{1}{4}(1+\nu)\,\mathrm{cosech}^2\,\alpha_1, \quad (24)$$

$$\widehat{\alpha\beta}=\tfrac{1}{2}w\,\mathrm{cosech}^2\,\alpha_1[J\sinh^2\alpha\sin\beta-a(\cosh\alpha-\cos\beta)\sin\beta]+\tfrac{1}{4}(1-\nu)wa\,\mathrm{cosech}^2\,\alpha_1(\cosh\alpha-\cos\beta)\sin\beta$$
$$+\tfrac{1}{2}wJ[\cosh 2\alpha\sin\beta-a^{-1}J\sinh 2\alpha\sinh\alpha\sin\beta]-\tfrac{1}{2}wJ\coth\alpha_1[\sinh 2\alpha\sin\beta-2a^{-1}J\sinh^3\alpha\sin\beta]$$
$$+\tfrac{1}{4}(1+\nu)wa\,\mathrm{cosech}^2\,\alpha_1e^{2(\alpha_1-\alpha)}(\cosh\alpha-\cos\beta)\sin\beta. \quad (25)$$

It is interesting to calculate the stresses $\widehat{\beta\beta}_1$ in the boundary of the disk. These are obtained by substituting $\alpha=\alpha_1$ in (24) with the result:

$$\widehat{\beta\beta}_1=\tfrac{1}{2}wa\,\mathrm{cosech}^2\,\alpha_1(\cosh\alpha_1-\cos\beta)\times[(1-\nu)e^{-\alpha_1}-(1+\nu)\cos\beta]. \quad (26)$$

This is more easily interpreted in terms of the physical dimensions of the disk. If d is the ratio of the distance of the peg from the center of the disk to the radius R of the disk, then $d=\cosh\alpha$. Also, let θ be the angle which a radius from the center of the disk makes with the y-axis as shown in Fig. 1. Then

$$\widehat{\beta\beta}_1=\frac{wR(1-d^2)}{2(1+d^2+2d\cos\theta)}\times\left\{(1-\nu)d-(1+\nu)\frac{(1+d^2)\cos\theta+2d}{1+d^2+2d\cos\theta}\right\}. \quad (27)$$

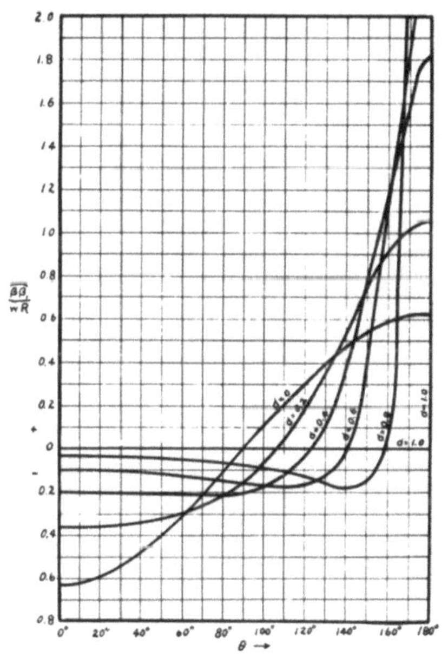

FIG. 2.

In Fig. 2, the ratio $\widehat{\beta\beta}_1/wR$ is plotted against θ for various values of $d=c/R$. It is interesting to observe that the stress in the boundary of the disk follows a simple cosine law when the peg is at the center ($d=0$). When the disk is supported at its outer edge ($d=1$) the stress in the boundary is zero except at the peg ($\theta=180°$) where the stress is infinite.

From the PHILOSOPHICAL MAGAZINE, Ser. 7, vol. xxvi. p. 713,
November 1938.

Stresses in an Eccentrically Rotating Disk. By

RAYMOND D. MINDLIN, *Ph.D., Department of Civil Engineering, Columbia University, New York City* *.

SYNOPSIS.

THE two-dimensoinal elasticity equations for small strain are solved in this paper to obtain the stresses in a circular disk which rotates about an axis normal to the plane of the disk. The axis of rotation may pass either inside or outside the disk. The unbalance introduced by the eccentricity of the axis of rotation is equilibrated by means of a concentrated force acting in the plane of the disk. The point of application of the force may be at any point in the disk along the straight line passing through the centre of rotation and the centre of the disk. The bipolar coordinate system is employed, and the solution is obtained in finite terms.

WE consider a circular disk of density ρ, radius R, and of small thickness taken as unity. The disk rotates with uniform angular velocity ω about an axis normal to its plane. The point of intersection, C, of the axis of rotation and the plane of the disk is located at point $(0, c)$ of an x, y system of rectangular coordinates as shown in the figure. Point C may be anywhere along the y-axis. The coordinates axes rotate with the disk, so that the centre, D, of the disk always lies on the y-axis at a distance d from the origin.

A centripetal force F of magnitude $\rho\omega^2\pi R^2(c-d)$ is required to maintain the circular motion. This force is considered concentrated at a point A with coordinates $(0, a)$. Whereas A must lie inside the disk, C may lie inside or outside the disk.

On account of the eccentricity of the balancing force the problem is most conveniently treated in bipolar coordinates. Such a coordinate system may be obtained from the conformal transformation.

$$x+iy=ia\coth\tfrac{1}{2}(\alpha+i\beta), \quad . \quad . \quad . \quad . \quad . \quad . \quad (1)$$

* Communicated by Prof. R. V. Southwell, M.A., F.R.S.

$$\text{whence}\quad x=\frac{a\sin\beta}{\cosh\alpha-\cos\beta},\quad y=\frac{a\sinh\alpha}{\cosh\alpha-\cos\beta}, \quad . \quad . \quad (2)$$

$$\alpha=\log\frac{r_1}{r_2},\quad \beta=\theta_1-\theta_2. \quad . \quad . \quad . \quad . \quad (3)$$

The significance of the coordinates appearing in (2) and (3) is illustrated in the figure. It is to be observed that the positive pole $\alpha=\infty$ coincides with the point of application $(0, a)$ of the balancing force. The boundary of the disk is taken as the circle $\alpha=\alpha_1$. The centre of this

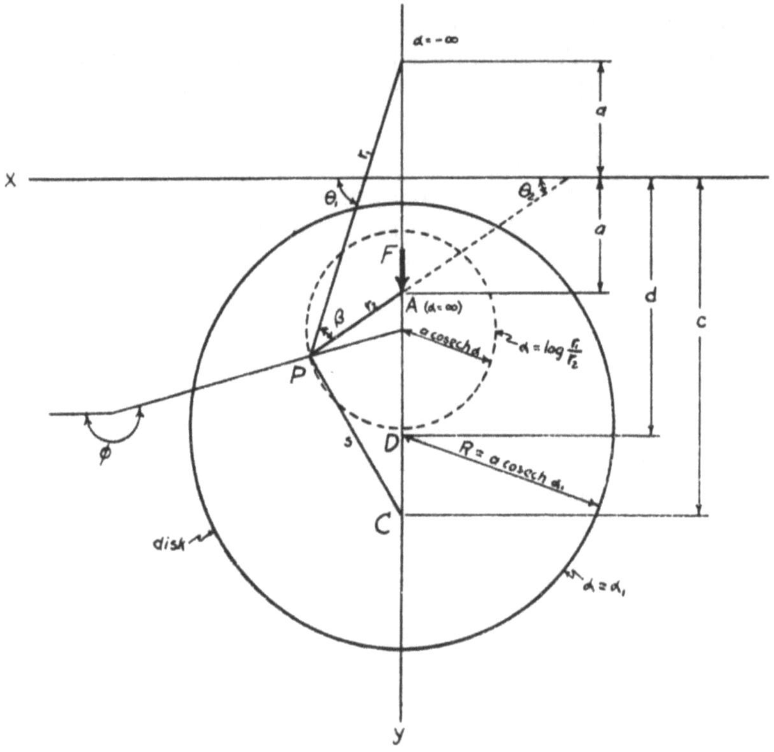

circle lies on the y-axis at a distance $d=a\coth\alpha_1$ from the origin, and its radius is $R=a\,\text{cosech}\,\alpha_1$.

Important parts are taken in the ensuing solution by the stretch ratio, J, of the transformation (1) and the angle ϕ which the tangent to the curve $\beta=$constant makes with the x-axis. These are given by

$$J=\sqrt{\left(\frac{\partial x}{\partial\alpha}\right)^2+\left(\frac{\partial y}{\partial\alpha}\right)^2}=\frac{a}{\cosh\alpha-\cos\beta}, \quad . \quad . \quad (4)$$

$$\phi=\theta_1+\theta_2+\frac{\pi}{2}=\tan^{-1}\frac{\cosh\alpha\cos\beta-1}{\sinh\alpha\sin\beta}. \quad . \quad . \quad (5)$$

The problem is treated as one of plane stress, and functions in the stresses involving the coordinate normal to the x, y-plane are neglected. The components of stress in plane curvilinear coordinates are expressed in terms of a stress function χ and a body force function Ω by [1]

$$\left.\begin{aligned} \widehat{\alpha\alpha} &= \frac{1}{J^2}\frac{\partial^2\chi}{\partial\beta^2}+\frac{1}{J^3}\left(\frac{\partial\chi}{\partial\alpha}\frac{\partial J}{\partial\alpha}-\frac{\partial\chi}{\partial\beta}\frac{\partial J}{\partial\beta}\right)+\Omega, \\ \widehat{\beta\beta} &= \frac{1}{J^2}\frac{\partial^2\chi}{\delta\alpha^2}+\frac{1}{J^3}\left(\frac{\partial\chi}{\partial\beta}\frac{\partial J}{\partial\beta}-\frac{\partial\chi}{\partial\alpha}\frac{\partial J}{\partial\alpha}\right)+\Omega, \\ \widehat{\alpha\beta} &= -\frac{1}{J^2}\frac{\partial^2\chi}{\partial\alpha\partial\beta}+\frac{1}{J^3}\left(\frac{\partial\chi}{\partial\alpha}\frac{\partial J}{\partial\beta}+\frac{\partial\chi}{\partial\beta}\frac{\partial J}{\partial\alpha}\right), \end{aligned}\right\} \quad . \quad . \quad (6)$$

where Ω is related to the x and y components of the body force per unit of volume by

$$X=-\frac{\partial\Omega}{\partial x},\quad Y=-\frac{\partial\Omega}{\partial y}, \quad . \quad . \quad . \quad . \quad . \quad (7)$$

and χ satisfies the differential equation

$$\nabla^4\chi=-(1-\nu)\nabla^2\Omega, \quad . \quad . \quad . \quad . \quad . \quad (8)$$

in which ν is Poisson's ratio, ∇^2 is Laplace's two-dimensional operator, and $\nabla^4=\nabla^2\nabla^2$.

The body force per unit of volume at a point P distant s from C is $\rho\omega^2 s$, and it acts in the direction $\overline{CP}$. Hence from (7),

$$\begin{aligned} \Omega &= -\tfrac{1}{2}\rho\omega^2 s^2+\text{constant}, \\ &= -\tfrac{1}{2}\rho\omega^2[x^2+(y-c)^2]+\text{constant}. \quad . \quad . \quad (9) \end{aligned}$$

Considering (2) and (4), and taking the constant equal to $\frac{1}{2}\rho\omega^2(c^2-a^2)$, (9) becomes

$$\Omega=\rho\omega^2 J\sinh\alpha(c-a\coth\alpha), \quad . \quad . \quad . \quad (10)$$

and $$\nabla^2\Omega=-2\rho\omega^2.$$

Hence (8) becomes . . . . (11)

$$\nabla^4\chi=2(1-\nu)\rho\omega^2. \quad . \quad . \quad . \quad . \quad . \quad . \quad (12)$$

The problem therefore consists in finding a solution of (12) which produces a concentrated force of magnitude $\rho\omega^2\pi R^2(c-d)$ at A, and which yields stresses according to (6) which satisfy the conditions

$$\widehat{\alpha\alpha}=\widehat{\alpha\beta}=0 \quad . \quad . \quad . \quad . \quad . \quad . \quad . \quad (13)$$

on the boundary of the disk.

It may be verified that

$$\chi_0=\tfrac{1}{8}(1-\nu)\rho\omega^2a^2J^2e^{-2\alpha} \quad . \quad . \quad . \quad . \quad (14)$$

is a particular solution of (12), so that we may take our stress function to be

$$\chi=\chi_0+\chi', \quad . \quad . \quad . \quad . \quad . \quad . \quad (15)$$

where χ' is governed by

$$\nabla^4\chi'=0. \quad . \quad . \quad . \quad . \quad . \quad . \quad (16)$$

We shall now construct a convenient stress function satisfying (16) and giving a concentrated force of magnitude F acting at A in the positive y-direction. Considering polar coordinates r_2, θ_2 with origin at A, the stress function for the required single force is known to be [2]

$$\chi_a=\frac{2\pi}{F}\left\{r_2\theta_2\cos\theta_2+\frac{1-\nu}{2}r_2\sin\theta_2\log r_2\right\}. \quad (17)$$

In order to express (17) in bipolar coordinates it is advantageous to first add to it the following stress function satisfying (16) and producing no singularities inside the disk :

$$\chi_b=\frac{F}{2\pi}\left\{r_1\left(\theta_1+\frac{\pi}{2}\right)\cos\theta_1-\frac{1-\nu}{2}r_2\sin\theta_2\log r_1\right\}. \quad (18)$$

Then, noting that

$$r_1\cos\theta_1=r_2\cos\theta_2=x=J\sin\beta,$$

$$\theta_1+\theta_2+\frac{\pi}{2}=\phi,$$

$$r_2\sin\theta_2=y-a=J(\sinh\alpha-\cosh\alpha+\cos\beta),$$

$$\log\frac{r_1}{r_2}=\alpha,$$

we find

$$\chi_1=\chi_a+\chi_b=\frac{FJ}{2\pi}\left\{\phi\sin\beta-\frac{1-\nu}{2}\alpha(\sinh\alpha-\cosh\alpha+\cos\beta)\right\}. \quad (19)$$

The magnitude of F is given by

$$F=\rho\omega^2\pi R^2(c-d),$$

$$=\rho\omega^2\pi a^2\operatorname{cosech}^2\alpha_1(c-a\coth\alpha_1). \quad . \quad . \quad (20)$$

The components of stress $\widehat{\alpha\alpha}$ and $\widehat{\alpha\beta}$ produced by Ω, χ_0 and χ_1, are found to be, from (6),

$$\left.\begin{aligned}
\widehat{\alpha\alpha}=&\rho\omega^2\mathrm{J}\sinh\alpha(c-a\coth\alpha)\\
&+\tfrac{1}{4}(1-\nu)\rho\omega^2(a\mathrm{J}e^{-\alpha}+\mathrm{J}^2e^{-2\alpha}\sin^2\beta)\\
&+\tfrac{1}{2}\rho\omega^2\operatorname{cosech}^2\alpha_1(c-a\coth\alpha_1)[a\sinh\alpha(\cosh\alpha\\
&\qquad-\cos\beta)-\mathrm{J}\sinh^3\alpha]\\
&+\tfrac{1}{4}(1-\nu)\rho\omega^2a\operatorname{cosech}^2\alpha_1(c-a\coth\alpha_1)[\sinh^2\alpha\\
&\qquad-\sinh\alpha(\cosh\alpha-\cos\beta)],\\
\widehat{\alpha\beta}=&-\tfrac{1}{4}(1-\nu)\rho\omega^2e^{-2\alpha}[a\mathrm{J}\sin\beta+\mathrm{J}^2\sinh\alpha\sin\beta]\\
&+\tfrac{1}{2}\rho\omega^2\operatorname{cosech}^2\alpha_1(c-a\coth\alpha_1)[a\sin\beta(\cosh\alpha\\
&\qquad-\cos\beta)-\mathrm{J}\sinh^2\alpha\sin\beta]\\
&-\tfrac{1}{4}(1-\nu)\rho\omega^2a\operatorname{cosech}^2\alpha_1(c-a\coth\alpha_1)\sin\beta\\
&\qquad\times(\cosh\alpha-\cos\beta).
\end{aligned}\right\}\quad(21)$$

These do not vanish when $\alpha=\alpha_1$, so that it is necessary to introduce further solutions of (16) in order to satisfy (13).

There exists a general method, due to Jeffery[3], for finding the required functions when the boundary stresses are expressed in the form

$$f(\alpha_1)\cos n\beta \quad\text{or}\quad f(\alpha_1)\sin n\beta.$$

In order to apply Jeffery's method in this case it would be necessary to expand in infinite series the terms in (21) which involve J. However, it is possible to avoid an infinite series treatment by introducing special functions which will remove the awkward J terms on $\alpha=\alpha_1$.

To this end we consider the simple biharmonic functions (*i. e.*, satisfying (16)):

$$\begin{aligned}
\chi_2&=\mathrm{G}a^{-1}y(x^2+y^2)+\mathrm{H}y^2\\
&=\mathrm{G}\mathrm{J}^2\sinh\alpha(\cosh\alpha+\cos\beta)+\mathrm{H}\mathrm{J}^2\sinh^2\alpha,
\end{aligned}\quad(22)$$

from which

$$\left.\begin{aligned}
\widehat{\alpha\alpha}=&2\mathrm{G}(\mathrm{J}a^{-1}\sinh\alpha+\mathrm{J}^2a^{-2}\sinh2\alpha\sin^2\beta)\\
&\qquad+2\mathrm{H}\mathrm{J}^2a^{-2}\sinh^2\alpha\sin^2\beta,\\
\widehat{\alpha\beta}=&2\mathrm{G}(\mathrm{J}a^{-1}\cosh2\alpha-\mathrm{J}^2a^{-2}\sinh2\alpha\sinh\alpha)\sin\beta\\
&\qquad+\mathrm{H}(\mathrm{J}a^{-1}\sinh2\alpha-2\mathrm{J}^2a^{-2}\sinh^3\alpha)\sin\beta.
\end{aligned}\right\}\quad(23)$$

The values of the constants G and H which will remove the terms in J and J^2 from $\widehat{\alpha\alpha}$ and $\widehat{\alpha\beta}$ when $\alpha=\alpha_1$ are found to be

$$\left.\begin{aligned}
\mathrm{G}&=\tfrac{1}{4}\rho\omega^2a[a\coth\alpha_1-c-\tfrac{1}{2}a(1-\nu)(\coth\alpha_1-1)],\\
\mathrm{H}&=-\tfrac{1}{8}\rho\omega^2a\operatorname{cosech}^2\alpha_1[2(a\coth\alpha_1-c)\sinh2\alpha_1\\
&\qquad-a(1-\nu)].
\end{aligned}\right\}\quad(24)$$

The combined functions Ω, χ_0, χ_1, and χ_2 produce only the following normal and shearing stresses on the boundary of the disk :

$$\left.\begin{aligned}\widehat{\alpha\alpha}_1&=\tfrac{1}{4}\rho\omega^2 a(c-a\coth\alpha_1)[(1+\nu)(\cosh\alpha_1\\&\qquad-\cos\beta)\operatorname{cosech}\alpha_1+1-\nu],\\ \widehat{\alpha\beta}_1&=\tfrac{1}{4}(1+\nu)\rho\omega^2 a(c-a\coth\alpha_1)(\cosh\alpha_1\\&\qquad-\cos\beta)\operatorname{cosech}^2\alpha_1\sin\beta.\end{aligned}\right\} . \quad (25)$$

To remove these last stresses from the boundary it is only necessary to consider a solution of (16) of the form $f_1(\alpha)+f_2(\alpha)\cos\beta$. This is known to be [(3)]

$$\chi_3=\mathrm{J}(\mathrm{A}_0\cosh\alpha+\mathrm{B}_0\alpha\cosh\alpha+\mathrm{C}_0\sinh\alpha+\mathrm{D}_0\,\alpha\sinh\alpha)$$
$$+\mathrm{J}(\mathrm{A}_1\cosh 2\alpha+\mathrm{B}_1+\mathrm{C}_1\sinh 2\alpha+\mathrm{D}_1\alpha)\cos\beta. \quad (26)$$

We require that there be no singularities inside the disk other than the single force at A as given by (17). Hence we must take

$$\left.\begin{aligned}\mathrm{B}_0=\mathrm{D}_0=\mathrm{D}_1=0,\\ \mathrm{A}_1+\mathrm{C}_1=0.\end{aligned}\right\} . \quad . \quad . \quad . \quad . \quad (27)$$

Furthermore, the term $\mathrm{C}_0\mathrm{J}\sinh\alpha$ produces no stress, and $\mathrm{A}_0\mathrm{J}\cosh\alpha$ produces the same stress as $\mathrm{B}_1\mathrm{J}\cos\beta$, so that we may drop the terms in A_0 and C_0. (26) therefore reduces to

$$\chi_3=\mathrm{J}[\mathrm{A}_1 e^{-2\alpha}+\mathrm{B}_1]\cos\beta, \quad . \quad . \quad . \quad . \quad (28)$$

from which

$$\left.\begin{aligned}\widehat{\alpha\alpha}&=\mathrm{B}_1 a^{-1}+\mathrm{A}_1 a^{-1}e^{-2\alpha}+2\mathrm{A}_1 a^{-1}e^{-2\alpha}\sinh\alpha\cos\beta,\\ \widehat{\alpha\beta}&=-2\mathrm{A}_1 a^{-1}e^{-2\alpha}(\cosh\alpha-\cos\beta)\sin\beta.\end{aligned}\right\} . \quad (29)$$

Setting $\alpha=\alpha_1$, and equating coefficients in (29) and (25) in the usual manner, we find

$$\left.\begin{aligned}\mathrm{A}_1&=\tfrac{1}{8}(1+\nu)\rho\omega^2 a^2 e^{2\alpha_1}\operatorname{cosech}^2\alpha_1(c-a\coth\alpha_1),\\ \mathrm{B}_1&=-\tfrac{1}{4}\rho\omega^2 a^2(c-a\coth\alpha_1)[(1+\nu)(\coth\alpha_1\\&\qquad+\tfrac{1}{2}\operatorname{cosech}^2\alpha_1)+1-\nu].\end{aligned}\right\} . \quad (30)$$

The complete solution is therefore given by

$$\Omega=\rho\omega^2\mathrm{J}\sinh\alpha(c-a\coth\alpha),$$

$$\begin{aligned}\chi&=\chi_0+\chi_1+\chi_2+\chi_3\\ &=\tfrac{1}{8}(1-\nu)\rho\omega^2 a^2\mathrm{J}^2 e^{-2\alpha}\\ &\quad+\tfrac{1}{2}\rho\omega^2 a^2\operatorname{cosech}^2\alpha_1(c-a\coth\alpha_1)\mathrm{J}[\phi\sin\beta-\tfrac{1}{2}(1-\nu)\alpha\\ &\qquad\times(\sinh\alpha-\cosh\alpha+\cos\beta)]\end{aligned}$$

$$+\tfrac{1}{4}\rho\omega^2 a[a \coth \alpha_1 - c - \frac{a}{2}(1-\nu)(\coth \alpha_1 - 1)]J^2 \sinh \alpha$$
$$\times(\cosh \alpha + \cos \beta)$$
$$-\tfrac{1}{8}\rho\omega^2 a \operatorname{cosech}^2 \alpha_1[2(a \coth \alpha_1 - c)\sinh 2\alpha_1 - a(1-\nu)]$$
$$\times J^2 \sinh^2 \alpha$$
$$+\tfrac{1}{8}(1+\nu)\rho\omega^2 a^2 e^{2\alpha_1} \operatorname{cosech}^2 \alpha_1 (c - a \coth \alpha_1) J e^{-2\alpha} \cos \beta$$
$$-\tfrac{1}{4}\rho\omega^2 a^2 (c - a \coth \alpha_1)[(1+\nu)(\coth \alpha_1 + \tfrac{1}{2}\operatorname{cosech}^2 \alpha_1)$$
$$+1-\nu]J \cos \beta.$$

References.

(1) Coker and Filon, ' Photo-Elasticity,' p. 163 (Cambridge, 1931).
(2) Coker and Filon, *ibid.* p. 327.
(3) G. B. Jeffery, Phil. Trans. Roy. Soc., Series A, ccxxi. pp. 265–293

Reprinted from the Proceedings of the
Fifth International Congress of Applied Mechanics, 1938

RAYMOND D. MINDLIN

GRAVITATIONAL STRESSES IN BIPOLAR COORDINATES

1. INTRODUCTION. A very general solution of the two-dimensional elasticity equations in bipolar coordinates has been given by G. B. JEFFERY[1] for the case of zero body force. In the present paper the influence of a constant body force is considered, simulating the action of a uniform gravitational field.

One family of curves in bipolar coordinates is a system of coaxial circles. It is shown how the boundary conditions on the circular boundaries can be expressed in a simple manner when these boundaries are free of stress and a body force is present. The method is then applied to a two-boundary problem and to a one-boundary problem. The latter is the problem of the heavy circular disk suspended in its plane from an eccentric peg. The former is the technically important case of the horizontal cylindrical hole of circular cross-section situated in a gravitating semi-infinite solid.

The two-boundary solution gives the disturbance of stress around an unlined tunnel and includes consideration of the fact that the plane surface of the gravitating solid is free of stress. When the hole is considered to be very close to the plane boundary, the solution may be used in the study of the dead load stresses in the arched structure included between the boundary of the hole and the upper surface of the solid.

2. STRESSES IN BIPOLAR COORDINATES. A plane bipolar coordinate system may be obtained from the conformal transformation

$$x + iy = i a \coth \tfrac{1}{2}(\alpha + i\beta) \qquad (1)$$

whence

$$x = \frac{a \sin\beta}{\cosh\alpha - \cos\beta}, \quad y = \frac{a \sinh\alpha}{\cosh\alpha - \cos\beta} \qquad (2)$$

$$\alpha = \log\frac{r_1}{r_2}, \quad \beta = \theta_1 - \theta_2 \qquad (3)$$

The significance of these coordinates is shown in Fig. 1.

Important parts are taken in the solution by the stretch ratio, J, of the transformation and the angle φ which the tangent to the curve β = const. makes with the x-axis. These are given by

$$J = \sqrt{\left(\frac{\partial x}{\partial\alpha}\right)^2 + \left(\frac{\partial y}{\partial\alpha}\right)^2} = \frac{a}{\cosh\alpha - \cos\beta} \qquad (4)$$

$$\varphi = \theta_1 + \theta_2 + \frac{\pi}{2} = \tan^{-1}\frac{\cosh\alpha\cos\beta - 1}{\sinh\alpha\sin\beta} \qquad (5)$$

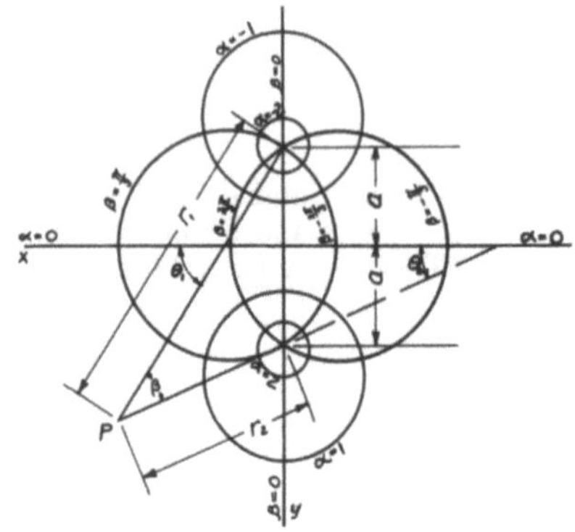

Fig.1. Bipolar Coordinates.

The two-dimensional components of stress in bipolar coordinates are expressed in terms of a stress function χ and a body force potential Ω by

$$\left.\begin{aligned} a\sigma_\alpha &= \left[(\cosh\alpha - \cos\beta)\frac{\partial^2}{\partial\beta^2} - \sinh\alpha\frac{\partial}{\partial\alpha} - \sin\beta\frac{\partial}{\partial\beta} + \cosh\alpha\right]\left(\frac{\chi}{J}\right) + a\Omega \\ a\sigma_\beta &= \left[(\cosh\alpha - \cos\beta)\frac{\partial^2}{\partial\alpha^2} - \sinh\alpha\frac{\partial}{\partial\alpha} - \sin\beta\frac{\partial}{\partial\beta} + \cos\beta\right]\left(\frac{\chi}{J}\right) + a\Omega \\ a\tau_{\alpha\beta} &= -(\cosh\alpha - \cos\beta)\frac{\partial^2}{\partial\alpha\partial\beta}\left(\frac{\chi}{J}\right) \end{aligned}\right\} \qquad (6)$$

where Ω is related to the x and y components of the body force per unit of volume by

$$X = -\frac{\partial\Omega}{\partial x}, \quad Y = -\frac{\partial\Omega}{\partial y} \qquad (7)$$

and χ satisfies the differential equation

$$\nabla^4\chi = -(1-\nu)\nabla^2\Omega \qquad (8)$$

in which ∇^2 is LAPLACE's operator and $\nabla^4 = \nabla^2\nabla^2$. ν is POISSON's ratio. (8) is suitable for plane stress but for plane strain ν' is to be substituted for ν where

1. JEFFERY, G.B., Plane Stress and Plane Strain in Bipolar Coordinates, Phil. Trans. Royal Soc., Series A, Vol. 221, pp. 265-293, November, 1920.

$$\nu' = \frac{\nu}{1-\nu} \tag{9}$$

We shall take as the body force system $X = 0$, $Y = w$ where w is the weight per unit of volume of the elastic solid. Hence $\Omega = -wy = -wJ\sinh\alpha$ and $\nabla^2\Omega = 0$, so that (8) reduces to

$$\nabla^4\chi = 0 \tag{10}$$

3. BOUNDARY CONDITIONS. The problems which we shall discuss will be those for which the curves α = const. are free of stress.

Since, on a free boundary α = const., $\tau_{\alpha\beta} = 0$, the third of (6) requires

$$\frac{\partial}{\partial\alpha}\left(\frac{\chi}{J}\right) = \xi \quad \text{a constant} \tag{11}$$

On the same boundary $\sigma_\alpha = 0$. Hence

$$\left[(\cosh\alpha - \cos\beta)\frac{\partial^2}{\partial\beta^2} - \sin\beta\frac{\partial}{\partial\beta} + \cosh\alpha\right]\left(\frac{\chi}{J}\right) = waJ\sinh\alpha + \xi\sinh\alpha$$

the solution of which is

$$\frac{\chi}{J} = \xi\tanh\alpha + \eta(\cosh\alpha\cos\beta - 1) + \zeta\sin\beta + wa^2\operatorname{cosech}\alpha - \tfrac{1}{2}wa^2\operatorname{cosech}^2\alpha\,\varphi\sin\beta \tag{12}$$

where η and ζ are constants.

(11) and (12) are the conditions which χ must satisfy on each boundary α = const.

The last term in (12) warrants special notice. If we examine the stress function

$$\chi = -\frac{1}{2}wa^2\operatorname{cosech}^2\alpha_1 J\varphi\sin\beta \tag{13}$$

and observe that: (1) $a\operatorname{cosech}\alpha_1$ is the radius, R, of the circle $\alpha = \alpha_1$; (2) $\varphi = \theta_1 + \theta_2 + \frac{\pi}{2}$; (3) $J\sin\beta = x$, so that (13) may be written as

$$\chi = -\frac{1}{2}wR^2x\left(\theta_1 + \theta_2 + \frac{\pi}{2}\right) \tag{14}$$

we notice that (13) or (14) is the stress function for a pair of single forces acting at the poles $(0, +a)$, $(0, -a)$ in the direction of the negative y axis. The magnitudes of the forces are equal to $w\pi R^2$ which is the weight per unit of thickness of the material included in the circle $\alpha = \alpha_1$. Hence (13) is the stress function required to annul the resultant of the stresses produced on $\alpha = \alpha_1$ by the body force.

4. EARLY TERMS OF THE STRESS FUNCTION. In order to satisfy the boundary conditions on $\alpha = \alpha_1$, we shall first annul the resultant of whatever stresses exist on the boundary and then apply a general solution, due to JEFFERY, to eliminate the remaining stresses.

As we have seen in the preceding section, the resultant of the stresses on a boundary $\alpha = \alpha_1$ may be removed by the stress function given in (13). This function introduces many-valued terms in the displacements. The problems which we shall consider require that the displacements be single-valued in the region $\alpha > 0$. This may be accomplished by adding the stress function

$$\frac{\chi}{J} = \frac{1}{4}(1-\nu)wa^2\operatorname{cosech}^2\alpha_1\left[\alpha\sinh\alpha - \alpha(\cosh\alpha - \cos\beta)\right] \tag{15}$$

(15) is perhaps more readily recognized when it is written in our other notation:

$$\chi = \frac{1}{4}(1-\nu)wR^2\left[(y-a)\log r_2 - (y-a)\log r_1\right] \tag{16}$$

The term $(y-a)\log r_2$ in (16) serves to eliminate the many-valued displacements in $\alpha > 0$ which were introduced by the term $x\theta_2$ in (14). We shall accordingly include in our stress function the early terms

$$\frac{\chi_0}{J} = -\frac{1}{2}wa^2\operatorname{cosech}^2\alpha_1\left\{\varphi\sin\beta - \frac{1}{2}(1-\nu)\left[\alpha\sinh\alpha - \alpha(\cosh\alpha - \cos\beta)\right]\right\} \tag{17}$$

This is a suitable function for a state of plane stress. For plane strain, we must substitute ν' for ν, where ν' is given by (9).

5. HORIZONTAL CYLINDRICAL HOLE IN A GRAVITATING SEMI-INFINITE SOLID. The problem is one of plane strain with the cylindrical surface $\alpha = \alpha_1$ and the plane $\alpha = 0$ as boundaries. The stress system is induced by the potential $\Omega = -wy$, so that the stress in the solid far from the hole is a hydrostatic pressure (Fig.2).

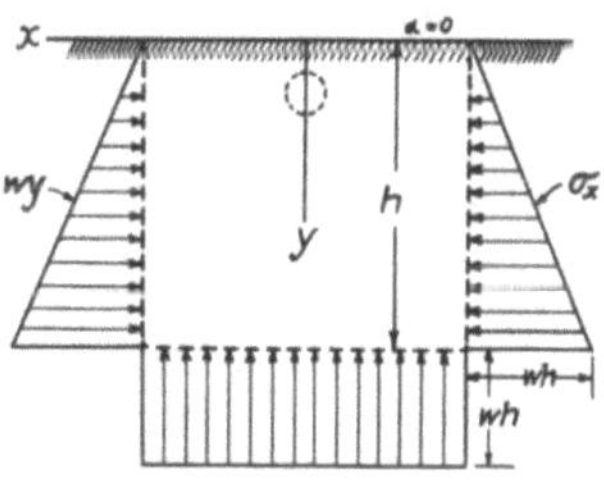

Fig.2. Horizontal Cylindrical Hole in a Gravitating Semi-Infinite Solid.

It is recognized that the stress far from the hole in a real solid might not be a hydrostatic pressure. Consideration will be given to this factor at another time.

At present we are concerned only with annulling on $\alpha = \alpha_1$ the stresses introduced there by the system

$$\Omega = \sigma_x = \sigma_y = -wy\ ; \quad \tau_{xy} = 0 \tag{18}$$

and at the same time maintaining $\alpha = 0$ free of stress.

We begin by introducing (17) (with ν replaced by ν') to annul the resultant of the stresses on $\alpha = \alpha_1$. This leaves only a self-equilibrating system of stress on $\alpha = \alpha_1$ but at the same time introduces stresses on the initially free boundary $\alpha = 0$.

It is advantageous to remove the latter before proceeding with the application of the boundary conditions (11) and (12). The stress function required to annul the stresses produced on $\alpha = 0$ by (17) is well known.[2] In terms of our bipolar notation it is

$$\frac{\chi_1}{J} = \frac{1}{8}(1+\nu')wa^2 \operatorname{cosech}^2\alpha_1 (\cosh 2\alpha - \sinh 2\alpha - 1)\cos\beta \tag{19}$$

It should be observed that (17) and (19) combined represent the two-dimensional solution for a force of magnitude $w\pi R^2$ applied along the y axis at point $(0, \mp a)$ in the interior of the semi-infinite solid $y > 0$.

It is now required to find an auxiliary stress function χ_2 satisfying (10) which will annul the stresses on $\alpha = \alpha_1$ introduced by (17), (19) and Ω and at the same time produce no stresses on $\alpha = 0$. For this purpose we make use of that part of JEFFERY's general solution of (10) which is even in β and which yields single-valued displacements:

$$\begin{aligned}\frac{\chi_2}{J} &= B_0\,\alpha(\cosh\alpha - \cos\beta) + C_0\sinh\alpha \\ &\quad + (A_1\cosh 2\alpha + B_1 + C_1\sinh 2\alpha)\cos\beta \\ &\quad + \sum_{n=2}^{\infty}\left\{\begin{matrix}A_n\cosh(n+1)\alpha + B_n\cosh(n-1)\alpha \\ +C_n\sinh(n+1)\alpha + D_n\sinh(n-1)\alpha\end{matrix}\right\}\cos n\beta\end{aligned} \tag{20}$$

Inspection of (11) and (12) (with $w = 0$) reveals that χ_2 will produce no stresses on $\alpha = 0$ if

$$\left[\frac{\chi_2}{J}\right]_{\alpha=0} = \left[\frac{\partial}{\partial\alpha}\left(\frac{\chi_2}{J}\right)\right]_{\alpha=0} = 0 \tag{21}$$

Substituting (20) in (21), we obtain the following relations among the constants which will assure the vanishing of σ_α and $\tau_{\alpha\beta}$ on $\alpha = 0$:

$$\left.\begin{aligned} A_n + B_n &= 0 \quad \text{for } n \geqq 1 \\ (n+1)C_n + (n-1)D_n &= 0 \quad \text{for } n \geqq 2 \\ 2C_1 + D_1 &= 0 \\ B_0 + C_0 &= 0 \end{aligned}\right\} \tag{22}$$

Substituting (22) in (20), and writing E_n for $C_n/(n-1)$ we find the following form for the auxiliary stress function:

$$\begin{aligned}\frac{\chi_2}{J} &= B_0\left[\alpha(\cosh\alpha - \cos\beta) + \sinh\alpha(\cosh\alpha\cos\beta - 1)\right] \\ &\quad + A_1(\cosh 2\alpha - 1)\cos\beta \\ &\quad + \sum_{n=2}^{\infty}\left\{\begin{matrix}A_n\left[\cosh(n+1)\alpha - \cosh(n-1)\alpha\right] \\ +E_n\left[(n-1)\sinh(n+1)\alpha - (n+1)\sinh(n-1)\alpha\right]\end{matrix}\right\}\cos n\beta\end{aligned} \tag{23}$$

The line of reasoning leading from (20) to (23) is essentially that followed by JEFFERY in applying his general solution to the problem of a semi-infinite plate containing an unstressed circular hole and under a uniform tension parallel to its straight edge. The process has been repeated here because JEFFERY considered correctly only the first of conditions (22), with the result that the term $B_0\sinh\alpha(\cosh\alpha\cos\beta - 1)$ is lacking in his Equation (66). Consequently, in his application, the boundary conditions on $\alpha = 0$ are not satisfied. A recalculation of JEFFERY's results shows that this error requires substantial changes in his factors of stress concentration, especially when the hole is close to the straight boundary.

Returning to the present problem, the complete stress function is given by

$$\frac{\chi}{J} = \frac{\chi_0}{J} + \frac{\chi_1}{J} + \frac{\chi_2}{J} \tag{24}$$

To determine the values of the constants B_0, A_1, A_n, E_n, we have to substitute (24) in (11) and (12) from which we obtain the required linear equations. The results of the solution of the latter are

$$\begin{aligned} B_0 &= \frac{1}{2}wa^2\operatorname{cosech}^2\alpha_1\left[2\coth\alpha_1 + \frac{1}{2}(1-\nu')(\coth\alpha_1 - 1)\right] \\ A_1 &= \frac{1}{2}wa^2\operatorname{cosech}^2\alpha_1 \\ A_n &= \frac{-1}{2}wa^2\operatorname{cosech}^2\alpha_1\frac{\epsilon^{-n\alpha_1}\sinh\alpha_1(\sinh n\alpha_1\cosh\alpha_1 - n\cosh\alpha_1\sinh\alpha_1)}{\sinh^2 n\alpha_1 - n^2\sinh^2\alpha_1} \\ E_n &= -\frac{1}{2}wa^2\operatorname{cosech}^2\alpha_1\frac{\epsilon^{-n\alpha_1}\sinh^2\alpha_1\sinh n\alpha_1}{\sinh^2 n\alpha_1 - n^2\sinh^2\alpha_1}\end{aligned} \tag{25}$$

Substituting (25) in (24), we find the final stress function to be

$$\begin{aligned}\frac{\chi}{J} = -\frac{1}{2}wa^2\operatorname{cosech}^2\alpha_1\Big\{&\varphi\sin\beta - \frac{1}{2}(1-\nu')\alpha\sinh\alpha \\ &+ \frac{1}{2}(5-\nu')\coth\alpha_1(\cosh\alpha - \cos\beta)\alpha \\ &- \frac{1}{4}(3-\nu')(\cosh 2\alpha - 1)\cos\beta \\ &+ \frac{1}{4}\left[(5-\nu')\coth\alpha_1 - 2\right]\sinh 2\alpha\cos\beta \\ &+ \sum_{n=2}^{\infty}\frac{2\sinh\alpha_1\epsilon^{n\alpha_1}}{\sinh^2 n\alpha_1 - n^2\sinh^2\alpha_1}\left[\begin{matrix}\sinh n\alpha_1\sinh n\alpha\sinh(\alpha-\alpha_1) \\ -n\sinh\alpha_1\sinh\alpha\sinh n(\alpha-\alpha_1)\end{matrix}\right]\cos n\beta\Big\}\end{aligned} \tag{26}$$

The stress $[\sigma_\beta]_{\alpha=\alpha_1}$ in the boundary of the hole is found to be, after some manipulations,

2. MELAN, E., Zeitschrift für angewandte Math. und Mech., Vol. 12, p. 345, Dec. 6, 1932. Also MINDLIN, R.D., Jour. of Applied Mechanics, Vol. 4, No. 3, p. 115, September, 1937, and Physics, Vol. 7, No. 5, p. 195, May, 1936.

$$[\sigma_\beta]_{\alpha=\alpha_1} = \frac{2wa(\cosh\alpha_1 - \cos\beta)}{\sinh\alpha_1}\left\{\frac{1-\cosh\alpha_1\cos\beta}{(\cosh\alpha_1-\cos\beta)^2} - \coth\alpha_1 - \frac{1}{4}(7-\nu')\operatorname{cosech}\alpha_1\cos\beta + 2\epsilon^{-\alpha_1}\cos\beta - \sum_{n=2}^{\infty} R_n \cos n\beta\right\} \quad (27)$$

where

$$R_n = \frac{n\epsilon^{-n\alpha_1}(\sinh n\alpha_1 \cosh n\alpha_1 - n\sinh\alpha_1\cosh\alpha_1)}{\sinh^2 n\alpha_1 - n^2\sinh^2\alpha_1} - n\epsilon^{-n\alpha_1}$$

When the hole is very far from the surface (27) is found to approach as a limit

$$[\sigma_\beta]_{\alpha=\alpha_1} = -2cw - \frac{3-4\nu}{2(1-\nu)} wR\cos\psi \quad \text{approximately} \quad (28)$$

where c is the distance of the center of the hole from the straight boundary, R is the radius of the hole, and ψ is the angle between the radius from the center of the hole and a normal to the straight boundary. The second term in (28) represents the effect of the weight of the material removed from the hole and is small in comparison with the first term, which represents the stress concentration at a hole in a uniform isotropic stress field of magnitude -cw.

In Fig. 3 there is plotted a curve showing the variation of the stress at the point on the periphery of the hole nearest to the straight boundary as the distance of the hole from the

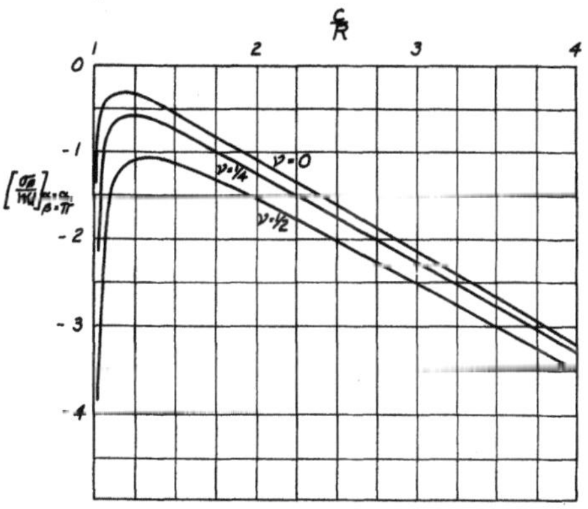

Fig.3. Stress at the top of a Horizontal Cylindrical Hole in a Gravitating Semi-Infinite Solid.

free surface is varied. The abscissa is the ratio c/R; the ordinate is the ratio $[\sigma_\beta]_{\alpha=\alpha_1}$ /2wR for $\beta = \pi$. The curve shows that the disturbing influence of the free straight boundary is not noticeable until the hole is very close to the upper surface of the solid. For very small values of α_1 however, there is a large increase in stress on this account. For larger values of c/R the stress approaches that given by (28).

Calculations of the stresses in both boundaries have been made for a complete range of values for α and β. It has also been found that the resultant force and couple on a section β = constant can be expressed in finite terms so that when α is small, direct comparisons may be made with the elementary arch theory for the dead load stresses in the arched structure included between the boundaries. Finally, consideration has been given to the fact that the stress in the solid far from the hole may not be a hydrostatic pressure. An additional stress function has been constructed to accommodate the present solution to other types of boundary conditions which are often postulated. Discussion of these factors is postponed on account of space limitations.

6. HEAVY DISK SUSPENDED FROM AN ECCENTRIC PEG. We consider the problem as one of plane stress. The disk is suspended from the pole (0, + a), (Fig. 4). Neglecting functions in-

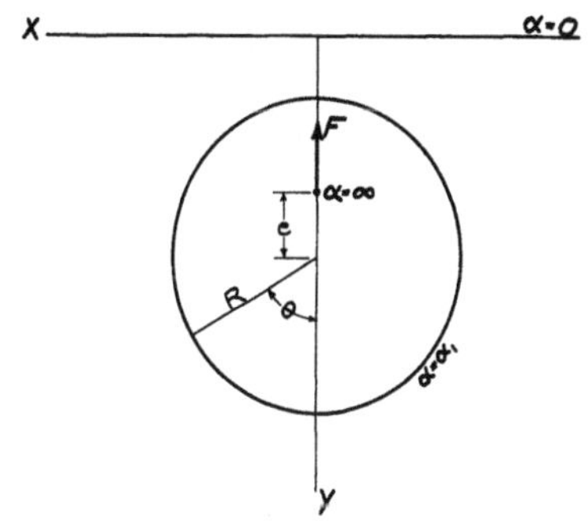

Fig.4. Heavy Disk Suspended from an Eccentric Peg.

volving the coordinate normal to the x, y-plane, the stress function is found to be

$$\mathfrak{X} = G_1\varphi\sin\beta + G_2[\alpha\sinh\alpha - \alpha(\cosh\alpha - \cos\beta)] + G_3 J\sinh\alpha(\cosh\alpha + \cos\beta) + G_4 J\sinh^2\alpha + G_5(\cosh 2\alpha - \sinh 2\alpha)\cos\beta + G_6\cosh\alpha$$

where

$$G_1 = -\frac{1}{2}wa^2\operatorname{cosech}^2\alpha_1$$

$$G_2 = +\frac{1}{4}(1-\nu)wa^2\operatorname{cosech}^2\alpha_1$$

$$G_3 = \frac{1}{4}wa$$

$$G_4 = -\frac{1}{2}wa\coth\alpha_1$$

$$G_5 = -\frac{1}{8}(1+\nu)wa^2\frac{\operatorname{cosech}^4\alpha_1}{(\cosh\alpha_1 - 1)^2}$$

$$G_6 = \frac{1}{2}wa^2\left[\coth\alpha_1 - \frac{1}{2}(1-\nu)(\coth\alpha_1 - 1) + \frac{1}{2}(1+\nu)\operatorname{cosech}^2\alpha_1\right]$$

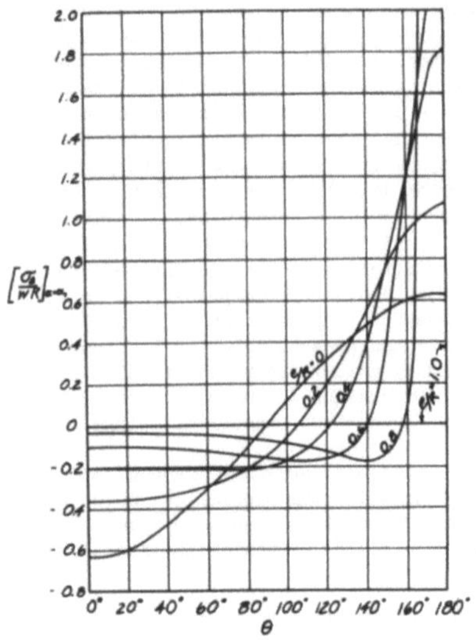

Fig.5. Stress Around the Edge of a Heavy D
Suspended from an Eccentric Peg.

In this case the stress function was found directly in finite terms through the introduction of the special functions $\varphi \sin\beta$, $J \sinh \alpha\ (\cosh\alpha + \cos\beta)$ and $J \sinh$ which do not appear as such in the series solution (20). A more detailed discussion of this problem is contemplated for a future publication.

Curves are given in Fig. 5 which show the variation of the stress around the edge of the disk for several values of the eccentricity (e/R) of the peg. It may be observed that the peripheral stress follows a simple cosine law when the peg is at the center. When the peg is at the edge the stress in the boundary is zero except at the point of support, where it is infinite.

AMERICAN SOCIETY OF CIVIL ENGINEERS

Founded November 5, 1852

TRANSACTIONS

Paper No. 2082

STRESS DISTRIBUTION AROUND A TUNNEL

By Raymond D. Mindlin,[1] Assoc. M. Am. Soc. C. E.

With Discussion by Messrs. W. O. Richmond, Jacob Feld, J. H. A. Brahtz, and Raymond D. Mindlin.

Synopsis

The problem of the disturbance of stress around a tunnel is treated in this paper from the point of view of the theory of elasticity. The tunnel is considered to be a horizontal cylindrical hole of circular cross-section situated in a semi-infinite elastic solid under the action of gravity. Through the introduction of the bipolar co-ordinate system an exact solution of the classical elasticity equations is obtained which satisfies the boundary conditions at both the upper free surface of the solid and the free surface of the hole.

Formulas for the stresses are given in terms of infinite series. The series have been evaluated and the results are given in the form of tables and curves which show that the stresses are important when the hole is either very close to the upper surface of the solid or very far from the surface.

Introduction

In the study of the equilibrium of elastic solids, problems are often encountered in which the shape of the structure and the distribution of the loading are so irregular that the ordinary methods of mechanics and strength of materials are not adequate to cope with the situation. The analysis of the disturbance of stress around a tunnel presents just these difficulties. This problem is concerned with an elastic solid of large dimensions, pierced by a hole (the tunnel) and bounded by an unloaded upper surface (Fig. 1). Before this mass is pierced by the tunnel, it is in a condition of elastic equilibrium. The problem considered here is that of the disturbance of this original condition of stress created when the mass is pierced by the tunnel.

The methods of the theory of elasticity have been adapted previously to studies of the tunnel problem.[2] The most nearly complete discussions were

Note.—Published in April, 1939, *Proceedings*.

[1] Instr., Dept. of Civ. Eng., Columbia Univ., New York, N. Y.

[2] Since this paper was submitted Prof. Z. Anzo has called the author's attention to a paper "On the Stresses in a Gravitating Elastic Body Having a Circular Hole near Its Horizontal Surface," by Z. Anzo (in Japanese), Technology Reports of the Kyushu Imperial University, Vol. XII, No. 3, June, 1937.

given by S. Yamaguti,[3] and H. Schmidt,[4] who investigated the state of stress around a horizontal hole of circular cross-section, situated in a gravitating solid of indefinite extent. In his analysis, Yamaguti correctly considered the effect on the stresses of (1) the weight of the material removed from the hole, and (2) the stress concentration induced by an assumed type of initial stress in the surrounding medium. However, the influence of conditions on the upper surface of the solid was neglected. Accordingly, by St. Venant's Principle, the solution is applicable only when the distance of the hole from the upper surface is large in comparison with the radius of the hole. When such a restriction is made, however, a complicated solution of the type used by Yamaguti is no longer necessary because, in this case, the effect of the weight of the material removed from the hole is negligible in comparison with the stress concentration induced by the initial stress in the surrounding material. When the hole is far from any other boundary the latter effect can be obtained quickly by a method[5] well known in the theory of elasticity. The hole, with its center at point A (Fig. 1), is considered simply as situated in a uniform stress field composed of the principal stresses which were initially present at point A.

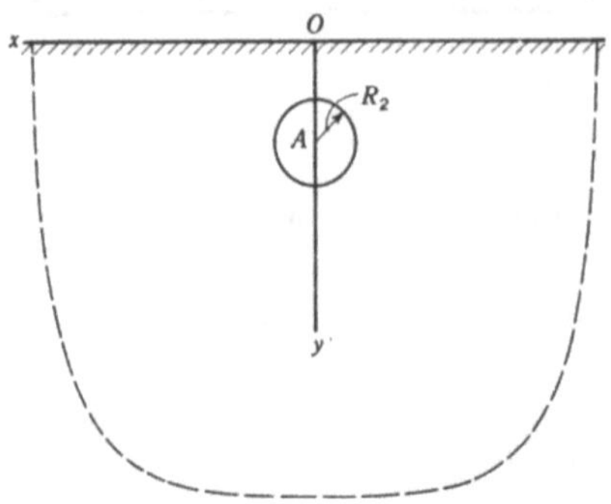

FIG. 1.—HORIZONTAL CLYINDRICAL HOLE OF CIRCULAR CROSS-SECTION IN A GRAVITATING, SEMI-INFINITE, ELASTIC SOLID

An exact solution of the elasticity equations is described here in which full consideration is given to the conditions at both the boundary of the hole and the free upper surface of the solid. The results are valid, therefore, regardless of the proximity of the two boundaries. In fact, the two boundaries may be taken so close together that the solution may be applied to the study of the dead load stresses in the thin-arched structure included between them. A simple method for treating this case is given in Appendix I. Finally, more detailed consideration is given to the assumed state of initial stress in the solid, which need not be the particular state assumed by Yamaguti.

NOTATION

The symbols used in this paper are defined where they first appear and are assembled alphabetically, for reference, in Appendix II.

STATEMENT OF THE PROBLEM

The tunnel is considered to be a horizontal cylindrical cavity of circular cross-section with its axis parallel to the z-axis of a rectangular co-ordinate system x, y, z. The surrounding material has, as its upper surface, the plane $y = 0$, but is otherwise unbounded. Technically such a body is known as a semi-infinite solid. The y-axis penetrates, in its positive direction, vertically

[3] "On the Stresses Around a Horizontal Circular Hole in a Gravitating Elastic Solid," by N. Yamaguti, *Journal* of the Civil Engineering Society of Japan, Vol. 15 (1929), pp. 291–303.

[4] "Statische Probleme des Tunnel- und Druckstollenbaues," by H. Schmidt, 1926.

[5] "Theory of Elasticity," by S. Timoshenko, New York, 1934, p. 80.

downward and bisects the cross-section of the tunnel (see Fig. 1). The semi-infinite solid has a uniform weight per unit of volume, w, so that the problem involves a constant body force, w, per unit of volume, acting in the positive y direction.

The length of the tunnel is assumed to be large in comparison with its diameter. This factor and the uniformity of the body force permit a treatment of the problem as one of plane strain.[6]

One method of studying problems in plain strain involves expressing the components of stress in terms of a body-force potential, Ω, and derivatives of a stress function χ. In rectangular co-ordinates these expressions take the form[7]

$$\sigma_x = \frac{\partial^2 \chi}{\partial y^2} + \Omega \quad \text{(1a)}$$

$$\sigma_y = \frac{\partial^2 \chi}{\partial x^2} + \Omega \quad \text{(1b)}$$

and,

$$\tau_{xy} = -\frac{\partial^2 \chi}{\partial x\, \partial y} \quad \text{(1c)}$$

The components (X, Y) of the body force per unit of volume are derived from the body-force potential by

$$X = -\frac{\partial \Omega}{\partial x} \quad \text{(2a)}; \qquad \text{and,} \qquad Y = -\frac{\partial \Omega}{\partial y} \quad \text{(2b)}$$

The stress function χ is a function of x and y only and is governed by the fourth-order differential equation

$$\nabla^4 \chi = -\frac{1 - 2\nu}{1 - \nu} \nabla^2 \Omega \quad \text{(3)}$$

in which ν is Poisson's ratio and $\nabla^4 = \nabla^2 \nabla^2$ where ∇^2 is Laplace's operator $\left(\frac{\partial^2}{\partial x^2} + \frac{\partial^2}{\partial y^2}\right)$.

In the present case, $X = 0$ and $Y = w$ so that, from Equations (2), $\Omega = - w\, y$; whence $\nabla^2 \Omega = 0$ and Equation (3) becomes

$$\nabla^4 \chi = 0 \quad \text{(4)}$$

The problem, then, is that of finding a solution of Equation (4) that satisfies a particular set of boundary conditions. The straight boundary and the circular boundary, in this case, are unloaded, and, therefore, there will be four boundary conditions to the effect that the normal stress and the shearing stress on each of these boundaries is zero.

On a horizontal plane, at a great distance below the tunnel, there will be zero shearing stress and a uniform normal stress equal to $- w\, y$. On vertical planes normal to the x-axis and situated at a great distance from the tunnel, the distribution of stress will depend upon the state of stress assumed to have existed in the semi-infinite solid before the boring of the tunnel. Three initial states will be assumed.

[6] "A Treatise on Photo-Elasticity," by E. G. Coker and L. N. G. Filon, Cambridge, University Press, 1931, p. 125.

[7] *Loc. cit.*, p. 128.

Case I.—It may be assumed, for example, that each element of the solid is initially under a state of isotropic pressure. The stress in the earth at great depths is probably in this state, approximately. Under this assumption the lateral boundary condition will be a linear distribution of normal pressure $\sigma_x = - w\,y$ as shown in Fig. 2(*a*). Such a stress distribution in the semi-

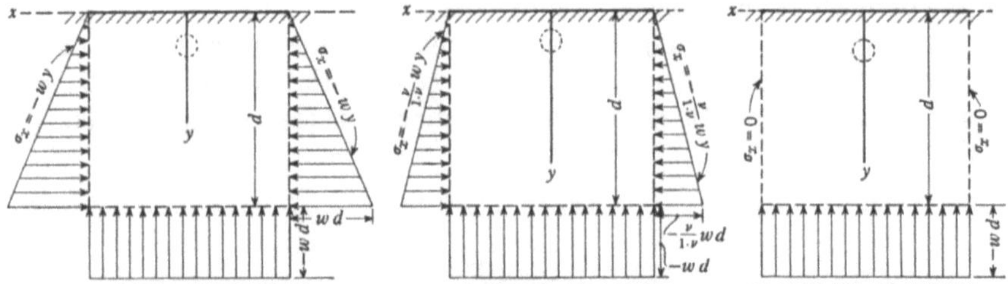

(*a*) CASE I, HYDROSTATIC PRESSURE (*b*) CASE II, NO LATERAL DEFORMATION (*c*) CASE III, NO LATERAL RESTRAINT

FIG. 2.—ASSUMED BOUNDARY CONDITIONS FOR THE SEMI-INFINITE SOLID

infinite solid, without the hole, may be expressed by means of the body-force potential

$$\Omega = - w\,y \quad \text{(5a)}$$

and the stress function

$$\chi_1 = 0 \quad \text{(5b)}$$

Case II.—It might be assumed, however, that initially (long before the boring of the tunnel) the material was restrained from lateral displacement during a supposed application of the gravitational field. In this case it may be shown that the lateral pressure is

$$\sigma_x = - \frac{\nu}{1 - \nu}\,w\,y$$

and that the initial state of stress may be expressed by

$$\Omega = - w\,y \quad \text{(6a)}$$

and

$$\chi_2 = \frac{1 - 2\,\nu}{6(1 - \nu)}\,w\,y^3 \quad \text{(6b)}$$

This is the state of initial stress (shown in Fig. 2(*b*)) which was assumed by Yamaguti.

Case III.—Another assumption which might be made is that there is no lateral stress at all (Fig. 2(*c*)) in which case the initial stress system may be represented by

$$\Omega = - w\,y \quad \text{(7a)}$$

and

$$\chi_3 = \tfrac{1}{6}\,w\,y^3 \quad \text{(7b)}$$

These three assumptions include a great range of possibilities for the initial state of stress.

CURVILINEAR CO-ORDINATES

Returning to a consideration of the four boundary conditions on the straight and circular boundaries, it is known that the mathematical form will be greatly simplified if the analysis is framed in a system of curvilinear co-ordinates which is particularly suited to the boundaries. The curvilinear co-ordinate system (α, β) which will be used is obtained through a conformal transformation of the type[8]

$$x + i\,y = f(\alpha + i\,\beta) \quad \text{(8)}$$

in which $i = \sqrt{-1}$. Important parts will be taken in the solution by the stretch ratio, J, of the transformation and the angle ϕ which the tangent to the curve β = *constant* makes with the x-axis. It is known[9] that these functions are given by

$$J\,e^{i\phi} = f'(\alpha + i\,\beta) \quad \text{(9)}$$

in which e = base of Naperian logarithms. The expressions for the radii of curvature of the co-ordinate curves will be used also. If ρ_1 and ρ_2 are the radii of curvature of β = *constant* and α = *constant*, respectively (both positive for counterclockwise rotation of the tangent moving in the positive direction along the curve), then[9]

$$\frac{1}{\rho_1} = \frac{1}{J}\frac{\partial \phi}{\partial \alpha} = -\frac{1}{J^2}\frac{\partial J}{\partial \beta} \quad \text{(10a)}$$

and

$$\frac{1}{\rho_2} = \frac{1}{J}\frac{\partial \phi}{\partial \beta} = \frac{1}{J^2}\frac{\partial J}{\partial \alpha} \quad \text{(10b)}$$

Finally, the components of stress in curvilinear co-ordinates are given by[10]

$$\sigma_\alpha = \frac{1}{J^2}\frac{\partial^2 \chi}{\partial \beta^2} + \frac{1}{J^3}\left(\frac{\partial \chi}{\partial \alpha}\frac{\partial J}{\partial \alpha} - \frac{\partial \chi}{\partial \beta}\frac{\partial J}{\partial \beta}\right) + \Omega \quad \text{(11a)}$$

$$\sigma_\beta = \frac{1}{J^2}\frac{\partial^2 \chi}{\partial \alpha^2} + \frac{1}{J^3}\left(\frac{\partial \chi}{\partial \beta}\frac{\partial J}{\partial \beta} - \frac{\partial \chi}{\partial \alpha}\frac{\partial J}{\partial \alpha}\right) + \Omega \quad \text{(11b)}$$

and

$$\tau_{\alpha\beta} = -\frac{1}{J^2}\frac{\partial^2 \chi}{\partial \alpha\,\partial \beta} + \frac{1}{J^3}\left(\frac{\partial \chi}{\partial \alpha}\frac{\partial J}{\partial \beta} + \frac{\partial \chi}{\partial \beta}\frac{\partial J}{\partial \alpha}\right) \quad \text{(11c)}$$

Bipolar Co-ordinates

A co-ordinate system that is ideally suited to the present requirements is the bipolar system. It was first applied to two-dimensional elasticity by G. B. Jeffery,[11] who gave a general solution of the elasticity equations in this co-ordinate system for the case of zero body force. His method will be adapted to the present requirements after account has been taken of the effect of the action of the gravitational force.

To obtain a system of bipolar co-ordinates, Equation (8) is particularized as

$$x + i\,y = i\,a \coth \tfrac{1}{2}(\alpha + i\,\beta) \quad \text{(12)}$$

[8] "A Treatise on Photo-Elasticity," by E. G. Coker and L. N. G. Filon, Cambridge, University Press, 1931, p. 152.

[9] *Loc. cit.*, p. 153.

[10] *Loc. cit.*, p. 163.

[11] "Plane Stress and Plane Strain in Bipolar Coordinates," by G. B. Jeffery, *Transactions* of the Royal Society, London, England, Series A, Vol. 221 (1920), pp. 265–293.

in which a = distance from the origin to a pole, in bipolar co-ordinates. Equating real and imaginary parts in Equation (12) there results

$$x = \frac{a \sin \beta}{\cosh \alpha - \cos \beta} \qquad (13a)$$

and

$$y = \frac{a \sinh \alpha}{\cosh \alpha - \cos \beta} \qquad (13b)$$

From Equations (9) and (12)

$$J\, e^{i\phi} = -\frac{i\,a}{2} \operatorname{csch}^2 \tfrac{1}{2}(\alpha + i\,\beta) \qquad (14)$$

whence

$$J = \frac{a}{\cosh \alpha - \cos \beta} \qquad (15a)$$

and,

$$\tan \phi = \frac{\cosh \alpha \cos \beta - 1}{\sinh \alpha \sin \beta} \qquad (15b)$$

Equation (12) may also be written in the form

$$\alpha + i\,\beta = \log \frac{x + i(y + a)}{x + i(y - a)} \qquad (16)$$

Writing $x + i(y + a) = r_1\, e^{i\theta_1}$ and $x + i(y - a) = r_2\, e^{i\theta_2}$, and equating real and imaginary parts in Equation (16), there results

$$\alpha = \log \frac{r_1}{r_2} \qquad (17a)$$

and

$$\beta = \theta_1 - \theta_2 \qquad (17b)$$

in which r_1 and r_2 are the radii to a point in the x, y-plane from the points $(0, -a)$, $(0, +a)$, respectively, and θ_1, θ_2 are the angles which these radii make with the x-axis, as shown in Fig. 3.

The curves α = *constant* are coaxial circles of radius $a \operatorname{csch} \alpha$ having the poles $(0, -a)$ and $(0, +a)$ as limiting points. The centers of these circles lie on the y-axis at distances $a \coth \alpha$ from the origin. The x-axis is the curve $\alpha = 0$. This and a circle α = *constant* (positive) represent the boundaries of the problem.

The curves β = *constant* are circular arcs passing through the poles $(0, \pm a)$; β is the angle included between the radii r_1 and r_2; it has a discontinuity of 2π on passing between the poles.

In terms of θ_1 and θ_2, the angle ϕ which the tangent to β = *constant* makes with the x-axis is given by

$$\phi = \theta_1 + \theta_2 + \frac{\pi}{2} \qquad (18)$$

From Equations (11) and (15a) the components of stress in bipolar co-ordinates are found to be

$$a\sigma_\alpha = \left\{(\cosh\alpha - \cos\beta)\frac{\partial^2}{\partial\beta^2} - \sinh\alpha\frac{\partial}{\partial\alpha} - \sin\beta\frac{\partial}{\partial\beta} + \cosh\alpha\right\}\left(\frac{\chi}{J}\right) + a\,\Omega \quad \text{(19a)}$$

$$a\sigma_\beta = \left\{(\cosh\alpha - \cos\beta)\frac{\partial^2}{\partial\alpha^2} - \sinh\alpha\frac{\partial}{\partial\alpha} - \sin\beta\frac{\partial}{\partial\beta} + \cos\beta\right\}\left(\frac{\chi}{J}\right) + a\,\Omega \quad \text{(19b)}$$

and

$$a\tau_{\alpha\beta} = -(\cosh\alpha - \cos\beta)\frac{\partial^2}{\partial\alpha\,\partial\beta}\left(\frac{\chi}{J}\right) \quad \text{(19c)}$$

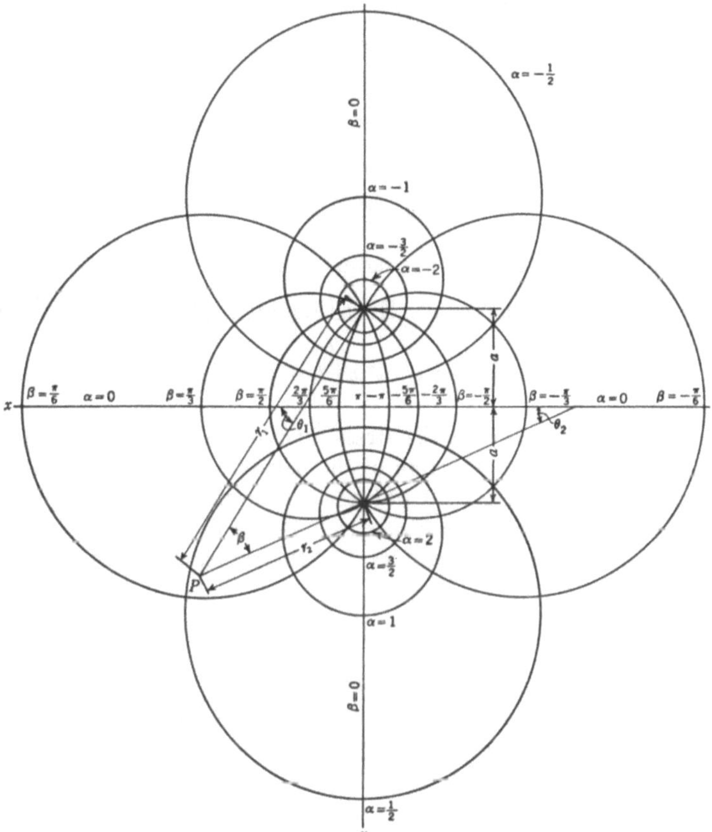

FIG. 3.—BIPOLAR CO-ORDINATES

BOUNDARY CONDITIONS

In each of the three cases, there are two curves, α = *constant*, which are free of stress. Since $\tau_{\alpha\beta} = 0$ on a free boundary α = *constant*, Equation (19c) requires that

$$\frac{\partial}{\partial\alpha}\left(\frac{\chi}{J}\right) = \xi \text{ (a constant)} \quad \text{(20)}$$

On the same boundary $\sigma_\alpha = 0$. Hence, noting that $-w\,y = -w\,J\sinh\alpha$: and substituting Equation (20) in Equation (19a) with $\sigma_\alpha = 0$, there results,

$$\left\{(\cosh\alpha - \cos\beta)\frac{\partial^2}{\partial\beta^2} - \sin\beta\frac{\partial}{\partial\beta} + \cosh\alpha\right\}\left(\frac{\chi}{J}\right) = w\,a\,J\sinh\alpha + \xi\sinh\alpha \quad (21)$$

Equation (21) is an ordinary differential equation the solution of which is found by the usual methods[12] to be

$$\frac{\chi}{J} = \xi\tanh\alpha + \eta\,(\cosh\alpha\cos\beta - 1) + \zeta\sin\beta + w\,a^2\operatorname{csch}\alpha - \tfrac{1}{2}w\,a^2\operatorname{csch}^2\alpha\,\phi\sin\beta \quad (22)$$

in which ξ, η, and ζ are constants.

Equations (20) and (22) are the conditions which $\frac{\chi}{J}$ must satisfy on each boundary α = *constant.*

The last term in Equation (22) warrants special notice. In the stress function

$$\chi = -\frac{w\,a^2}{2}\operatorname{csch}^2\alpha_1\,\phi\,J\sin\beta \quad (23)$$

it is observed that $a \operatorname{csch}\alpha_1$ is the radius, R_2, of the circle $\alpha = \alpha_1$; and that $\phi = \theta_1 + \theta_2 + \frac{\pi}{2}$ and $J\sin\beta = x$. Hence Equation (23) may be written as

$$\chi = -\tfrac{1}{2}w\,R_2^2\,x\left(\theta_1 + \theta_2 + \frac{\pi}{2}\right) \quad (24)$$

Equation (24) is recognized as the stress function for a pair of single forces[13] acting at the poles in the direction of the negative y-axis. The magnitudes of the forces are equal to $w\,\pi\,R_2^2$ which is the weight per unit of thickness of the material included in the circle $\alpha = \alpha_1$. Hence Equation (23) is the stress function required to annul the resultant of the stresses produced on α = *constant* by the body force.

Early Terms of the Stress Function

In order to satisfy the boundary conditions on a boundary α = *constant,* the following procedure suggested by Jeffery[11] will be followed. The resultant of whatever stresses exist on the boundary will be annulled first and then a general method, due to Jeffery, will be applied to eliminate the remaining stresses.

It was seen in the preceding section that the resultant of the stresses on a boundary $\alpha = \alpha_1$ may be removed by the stress function given in Equation (23). This function introduces many valued terms[14] in the displacements. In the present problem it is required that the displacements be single-valued in the region $\alpha > 0$. This may be accomplished by adding the stress function

$$\frac{\chi}{J} = \frac{1 - 2\nu}{4\,(1 - \nu)}\,a^2\operatorname{csch}^2\alpha_1\left[\alpha\sinh\alpha - \alpha\,(\cosh\alpha - \cos\beta)\right] \quad (25)$$

[12] "Differential Equations," by H. T. H. Piaggio, London (1929), Chapter VII.

[13] "A Treatise on Photo-Elasticity," by E. G. Coker and L. N. G. Filon, Cambridge, University Press, 1931, p. 327.

[14] "A Treatise on Photo-Elasticity," by E. G. Coker and L. N. G. Filon, Cambridge, University Press, 1931, p. 326.

Equation (25) is perhaps more readily recognized when it is written in terms of r_1, r_2, θ_1, θ_2:

$$\chi = \frac{1 - 2\nu}{4(1 - \nu)} w R_2^2 [(y - a) \log r_2 - (y - a) \log r_1] \ldots\ldots (26)$$

The term $(y - a) \log r_2$ in Equation (26) serves to eliminate the many-valued displacements in the region $\alpha > 0$ which were introduced by the term $x \theta_2$ in Equation (24).[13] Accordingly there should be included, in the stress function, the early terms

$$\frac{\chi_4}{J} = -\tfrac{1}{2} w a^2 \operatorname{csch}^2 \alpha_1$$

$$\times \left\{ \phi \sin \beta - \frac{1 - 2\nu}{2(1 - \nu)} [\alpha \sinh \alpha - \alpha (\cosh \alpha - \cos \beta)] \right\} \ldots\ldots (27)$$

First Auxiliary Stress Function

The stress function χ_4 leaves only a self equilibrating system of stress on $\alpha = \alpha_1$; but, at the same time, it introduces stresses on the initially free boundary $\alpha = 0$. It is advantageous to remove the stresses introduced on $\alpha = 0$ by Equation (27) before proceeding with the application of the boundary conditions, Equations (20) and (22). In this way advantage may be taken of a great simplification of the boundary equations suggested by Jeffery.

The stress function required to annul the stresses produced on $\alpha = 0$ by Equation (27) is well known.[15] In terms of the bipolar notation it is:

$$\frac{\chi_5}{J} = \frac{w a^2}{8(1 - \nu)} \operatorname{csch}^2 \alpha_1 (\cosh 2\alpha - \sinh 2\alpha - 1) \cos \beta \ldots\ldots (28)$$

It should be observed that Equations (27) and (28), combined, represent the two-dimensional solution for a force of magnitude $w \pi R_2^2$ applied along the y-axis at Point $(0, a)$ in the interior of the semi-infinite solid $y > 0$.

Second Auxiliary Stress Function

It is now required to find a second auxiliary stress function, satisfying Equation (4), which will annul the stresses on $\alpha = \alpha_1$ introduced by $\chi_{1,2,3}$, χ_4, χ_5 and Ω, and at the same time produce no stresses on $\alpha = 0$. In curvilinear co-ordinates Equation (4) becomes[16]

$$\left(\frac{\partial^2}{\partial \alpha^2} + \frac{\partial^2}{\partial \beta^2}\right)\left\{\frac{1}{J^2}\left(\frac{\partial^2 \chi}{\partial \alpha^2} + \frac{\partial^2 \chi}{\partial \beta^2}\right)\right\} = 0 \ldots\ldots (29a)$$

Jeffery showed that, for the case of bipolar co-ordinates, Equation (29*a*) can be reduced to a linear equation with constant coefficients by considering $\frac{\chi}{J}$

[15] "Der Spannungszustand der durch eine Einzelkraft im Innern beanspruchten Halbscheibe," by E. Melan, *Zeitschrift für Angewandte Mathematik und Mechanik*, Vol. 12 (1932), pp. 343–346; see also "Stress Systems in a Circular Disk under Radial Forces," by R. D. Mindlin, *Transactions*, A. S. M. E., Vol. 59 (1937), pp. A-115 to A-118; and "Force at a Point in the Interior of a Semi-Infinite Solid," by R. D. Mindlin, *Physics*, Vol. 7 (1936), pp. 195–202.

[16] "A Treatise on Photo-Elasticity," by E. G. Coker and L. N. G. Filon, Cambridge, University Press, 1931, p. 164.

instead of χ as the dependent variable. Equation (29a) then becomes

$$\left(\frac{\partial^4}{\partial\alpha^4} + 2\frac{\partial^4}{\partial\alpha^2\,\partial\beta^2} + \frac{\partial^4}{\partial\beta^4} - 2\frac{\partial^2}{\partial\alpha^2} + 2\frac{\partial^2}{\partial\beta^2} + 1\right)\left(\frac{\chi}{J}\right) = 0 \quad \ldots.(29b)$$

On account of the symmetry of the present problem, the solution must be even in β. Furthermore, since there is a discontinuity of $2\,\pi$ in β on passing between the poles, the solution must be periodic in β, of period $2\,\pi$. A solution of the form

$$\frac{\chi_6}{J} = f\,(\alpha)\cos n\,\beta \quad \ldots\ldots\ldots\ldots\ldots\ldots\ldots(30)$$

will therefore be considered. Substituting Equation (30) in Equation (29b) there results:

$$\left(\frac{d^4}{d\alpha^4} - 2\,(n^2 - 1)\frac{d^2}{d\alpha^2} + n^4 - 2\,n^2 + 1\right)f\,(\alpha) = 0 \quad \ldots\ldots(31)$$

Equation (31) is an ordinary differential equation whose solution is found to be, by the usual methods:[17a]

For $n \geqq 2$,

$$f_n\,(\alpha) = A_n \cosh\,(n+1)\,\alpha + B_n \cosh\,(n-1)\,\alpha + C_n \sinh\,(n+1)\,\alpha + D_n \sinh\,(n-1)\,\alpha \quad \ldots\ldots\ldots\ldots(32a)$$

for $n = 1$,

$$f_1\,(\alpha) = A_1 \cosh 2\,\alpha + B_1 + C_1 \sinh 2\,\alpha + D_1\,\alpha \quad \ldots\ldots\ldots(32b)$$

and, for $n = 0$,

$$f_0\,(\alpha) = A_0 \cosh\alpha + B_0\,\alpha\cosh\alpha + C_0 \sinh\alpha + D_0\,\alpha\sinh\alpha \quad \ldots..(32c)$$

Now, $A_0 \cosh\alpha$ and $B_1 \cos\beta$ give the same stresses; hence, A_0 may be taken equal to zero, which is equivalent to adding a constant term to the stress function. An inspection of Equations (20) and (22), with $w = 0$, reveals that χ_6 will produce no stresses on $\alpha = 0$, if

$$f\,(0) = 0 \quad \ldots\ldots\ldots\ldots\ldots\ldots\ldots(33a)$$

and

$$f'\,(0) = 0 \quad \ldots\ldots\ldots\ldots\ldots\ldots\ldots(33b)$$

Substituting Equations (32) in Equations (33), the following relations are obtained among the constants, which will assure the vanishing of σ_α and $\tau_{\alpha\beta}$ when $\alpha = 0$:

For $n \geqq 1$,

$$A_n + B_n = 0 \quad \ldots\ldots\ldots\ldots\ldots\ldots\ldots(34a)$$

for $n \geqq 2$,

$$(n+1)\,C_n + (n-1)\,D_n = 0 \quad \ldots\ldots\ldots\ldots\ldots(34b)$$

Also

$$2\,C_1 + D_1 = 0 \quad \ldots\ldots\ldots\ldots\ldots\ldots\ldots(34c)$$

and,

$$B_0 + C_0 = 0 \quad \ldots\ldots\ldots\ldots\ldots\ldots\ldots(34d)$$

Jeffery has shown that the condition that the displacements be single valued requires that

$$D_0 = 0 \quad \ldots\ldots\ldots\ldots\ldots\ldots\ldots(35a)$$

[17a] "Differential Equations," by H. T. H. Piaggio, London (1929), Chapter III.

and

$$D_1 + B_0 = 0 \qquad (35b)$$

Substituting Equations (34) and (35) in Equations (32) and writing E_n for $\frac{C_n}{(n-1)}$, the following form is found for the second auxiliary stress function:

$$\frac{\chi_6}{J} = B_0 [\alpha (\cosh \alpha - \cos \beta) + \sinh \alpha (\cosh \alpha \cos \beta - 1)]$$

$$+ A_1 (\cosh 2\alpha - 1) \cos \beta + \sum_{n=2}^{\infty} \{A_n [\cosh (n+1)\alpha - \cosh (n-1)\alpha]$$

$$+ E_n [(n-1) \sinh (n+1)\alpha - (n+1) \sinh (n-1)\alpha]\} \cos n\beta \ldots (36)$$

(The line of reasoning leading from Equations (33) to (36) is essentially that followed by Jeffery[11] in applying his general bipolar solution to the problem of a semi-infinite plate containing an unstressed circular hole, and under a uniform tension parallel to its straight edge. The process has been repeated here because Jeffery considered correctly only the first of the conditions given in Equations (34), with the result that the term $B_0 \sinh \alpha (\cosh \alpha \cos \beta - 1)$ is lacking in his Equation (66).[11] Consequently, in his application, the boundary conditions on $\alpha = 0$ are not satisfied.)

Complete Stress Function

The complete stress function (χ) is the sum of the initial stress function (χ_1, χ_2, or χ_3), the early terms (χ_4) and the two auxiliary stress functions (χ_5 and χ_6), as follows:

$$\frac{\chi}{J} = \frac{\chi_{1,2,3}}{J} - \frac{w\,a^2}{2} \operatorname{csch}^2 \alpha_1$$

$$\times \left\{ \phi \sin \beta - \frac{1 - 2\nu}{2(1-\nu)} [\alpha \sinh \alpha - \alpha (\cosh \alpha - \cos \beta)] \right\}$$

$$+ \frac{1}{8(1-\nu)} w\,a^2 \operatorname{csch}^2 \alpha_1 (\cosh 2\alpha - \sinh 2\alpha - 1) \cos \beta$$

$$+ B_0 [\alpha (\cosh \alpha - \cos \beta) + \sinh \alpha (\cosh \alpha \cos \beta - 1)]$$

$$+ A_1 (\cosh 2\alpha - 1) \cos \beta + \sum_{n=2}^{\infty} \Big\{ A_n [\cosh (n+1)\alpha - \cosh (n-1)\alpha]$$

$$+ E_n [(n-1) \sinh (n+1)\alpha - (n+1) \sinh (n-1)\alpha] \Big\} \cos n\beta \ldots (37)$$

There remains only the determination of the arbitrary constants. This is accomplished by substituting Equation (37) in Equations (20) and (22) with the proper value for $\chi_{1,2,3}$ for each of Cases I, II, and III.

Case I

In this case the first term on the right in Equation (37) is $\frac{\chi_1}{J} = 0$. The determination of the constants B_0, A_1, A_n, E_n by substitution of Equation (37) in the boundary conditions (Equations (20) and (22)) is perfectly straightforward except for one consideration. The difficulty arises from the appearance of a term $\frac{\sin^2 \beta}{(\cosh \alpha_1 - \cos \beta)}$ which results from the differentiation of the function $\phi \sin \beta$ with respect to α in Equation (20).

The values of the constants are determined by solving a set of simultaneous linear equations formed by equating coefficients of like terms of the type $\cos n\beta$. The term involving $(\cosh\alpha_1 - \cos\beta)$ in the denominator can only be adapted to this process by expanding it in infinite series as follows:

$$\frac{\sin^2\beta}{\cosh\alpha_1 - \cos\beta} = e^{-\alpha_1} + e^{-2\alpha_1} - 2\sinh\alpha_1 \sum_{n=2}^{\infty} e^{-n\alpha_1}\cos n\beta \ldots\ldots (38)$$

This is valid for $\alpha_1 > 0$. It is the appearance of this term in the first boundary condition which necessitates an infinite series solution. Were it not for this term the constants A_n and E_n would vanish and the final stress function would appear as the sum of a finite number of elementary functions.

Using Equation (38) after the substitution of Equation (37) in Equations (20) and (22), the following values for the constants result:

$$B_0 = -\tfrac{1}{2} w a^2 \operatorname{csch}^2\alpha_1 \left[2\coth\alpha_1 + \frac{1-2\nu}{2(1-\nu)}(\coth\alpha_1 - 1)\right] \ldots (39a)$$

$$A_1 = \tfrac{1}{2} w a^2 \operatorname{csch}^2\alpha_1 \ldots\ldots (39b)$$

$$A_n = -\tfrac{1}{2} w a^2 \operatorname{csch}^2\alpha_1 \times \frac{e^{-n\alpha_1}\sinh\alpha_1(\sinh n\alpha_1\cosh\alpha_1 - n\cosh n\alpha_1\sinh\alpha_1)}{\sinh^2 n\alpha_1 - n^2\sinh^2\alpha_1} \ldots\ldots (39c)$$

and

$$E_n = -\tfrac{1}{2} w a^2 \operatorname{csch}^2\alpha_1 \frac{e^{-n\alpha_1}\sinh^2\alpha_1\sinh n\alpha_1}{\sinh^2 n\alpha_1 - n^2\sinh^2\alpha_1} \ldots\ldots (39d)$$

Substituting Equations (39) in Equation (37), the stress function for Case I is found to be:

$$\begin{aligned}\frac{\chi}{J} = -\tfrac{1}{2} w a^2 \operatorname{csch}^2\alpha_1 \Bigg\{ \phi\sin\beta - \frac{1-2\nu}{2(1-\nu)}\alpha\sinh\alpha \\ + \frac{5-6\nu}{2(1-\nu)}\coth\alpha_1(\cosh\alpha - \cos\beta)\alpha - \frac{3-4\nu}{4(1-\nu)}(\cosh 2\alpha - 1)\cos\beta \\ + \frac{1}{2}\left(\frac{5-6\nu}{2(1-\nu)}\coth\alpha_1 - 1\right)\sinh 2\alpha\cos\beta + \sum_{n=2}^{\infty}\frac{2\sinh\alpha_1 e^{-n\alpha_1}}{\sinh^2 n\alpha_1 - n^2\sinh^2\alpha_1} \\ \times [\sinh n\alpha_1 \sinh n\alpha \sinh(\alpha - \alpha_1) \\ - n\sinh\alpha_1\sinh\alpha\sinh n(\alpha - \alpha_1)]\Bigg\}\cos n\beta \ldots\ldots (40)\end{aligned}$$

The most important stresses are those in the circular boundary. Using Equations (19b) and (40):

$$\begin{aligned}[\sigma_\beta]_{\alpha=\alpha_1} = \frac{w a(\cosh\alpha_1 - \cos\beta)}{2\sinh^2\alpha_1}\Bigg\{\frac{2(1-\cosh\alpha_1\cos\beta)\sinh\alpha_1}{(\cosh\alpha_1 - \cos\beta)^2} \\ - 4\cosh\alpha_1 - \left(\frac{5-6\nu}{1-\nu} + 2e^{-2\alpha_1}\right)\cos\beta \\ - 4\sinh\alpha_1\sum_{n=2}^{\infty} N_n\cos n\beta\Bigg\} \ldots\ldots (41)\end{aligned}$$

in which

$$N_n = \frac{n\, e^{-n\alpha_1} (\sinh n\,\alpha_1 \cosh n\,\alpha_1 - n \sinh \alpha_1 \cosh \alpha_1)}{\sinh^2 n\,\alpha_1 - n^2 \sinh^2 \alpha_1} \quad \text{.......(42)}$$

In order to obtain numerical values for $[\sigma_\beta]_{\alpha=\alpha_1}$ it is necessary to tabulate values for the series in Equation (41); but it is found that the series converges very slowly, especially for small values of α_1. To facilitate the computations, the more slowly converging part may be separated by letting

$$N_n = n\, e^{-n\alpha_1} + R_n \quad \text{.....................(43)}$$

Noting that

$$2 \sum_{n=1}^{\infty} n\, e^{-n\alpha_1} \cos n\,\beta = \frac{\cosh \alpha_1 \cos \beta - 1}{(\cosh \alpha_1 - \cos \beta)^2} \quad \text{.............(44)}$$

and substituting Equation (43) in Equation (41):

$$[\sigma_\beta]_{\alpha=\alpha_1} = \frac{2\, w\, a\, (\cosh \alpha_1 - \cos \beta)}{\sinh \alpha_1} \left\{ \frac{1 - \cosh \alpha_1 \cos \beta}{(\cosh \alpha_1 - \cos \beta)^2} - \coth \alpha_1 \right.$$
$$\left. - \frac{(7 - 8\,\nu)}{4\,(1 - \nu)} \frac{\cos \beta}{\sinh \alpha_1} + 2\, e^{-\alpha_1} \cos \beta - \sum_{n=2}^{\infty} R_n \cos n\,\beta \right\} \quad \text{........(45)}$$

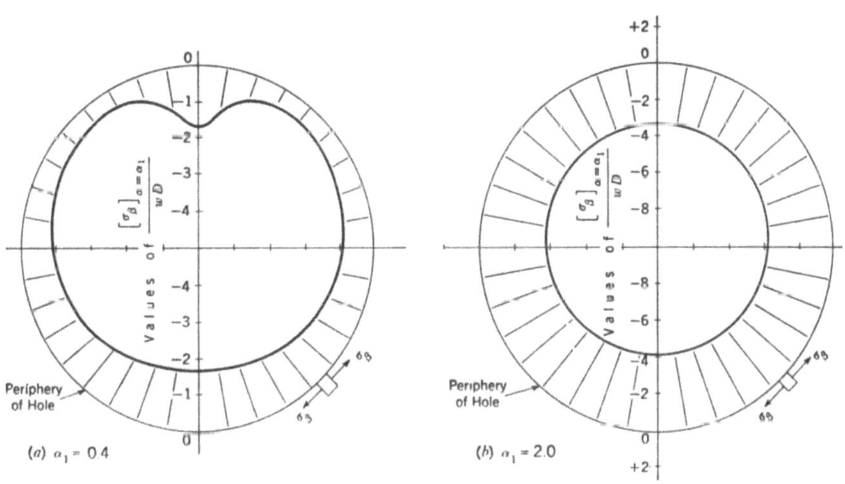

FIG. 4.—DIAGRAM OF STRESSES $\frac{[\sigma_\beta]_{\alpha=\alpha_1}}{w\,D}$ AROUND THE TUNNEL

Values of R_n are given in Table 1. Values of $[\sigma_\beta]_{\alpha=\alpha_1}$ around the hole have been calculated for a wide range of values of α_1 and for three values (0, $\frac{1}{4}$, and $\frac{1}{2}$) of ν (see Table 2). In the typical curves in Fig. 4 the dimensionless ratio $\frac{[\sigma_\beta]_{\alpha=\alpha_1}}{w\,D}$ (in which D is the diameter of the hole) is plotted radially from the periphery of the hole. Fig. 4(*a*) gives the variation of $[\sigma_\beta]_{\alpha=\alpha_1}$ around the hole for a small value (0.4) of α_1—that is, when the hole is close to the surface; and Fig. 4(*b*) is for a larger value (2.0) of α_1. Both are for $\nu = \frac{1}{2}$.

TABLE 1.—VALUES OF R_n

R_n	α_1 = 0.4	0.6	0.8	1.0	1.2	1.4	1.6	1.8	2.0	2.2	2.4
R_2	1.4663	0.5193	0.2043	0.0848	0.0364	0.0157	0.0069	0.0031	0.0014	0.0006	0.0003
R_3	0.7620	0.1852	0.0482	0.0127	0.0033	0.0034	0.0002	0.0001	....	....	....
R_4	0.3848	0.0615	0.0099	0.0016	0.0003	....	....	....	....	....	....
R_5	0.1884	0.0189	0.0018	0.002	....	....	....	....	....	....	....
R_6	0.0893	0.0054	0.0003	....	....	....	....	....	....	....	....
R_7	0.0409	0.0015	....	....	....	....	....	....	....	....	....
R_8	0.0181	0.0004	....	....	....	....	....	....	....	....	....
R_9	0.0078	0.0001	....	....	....	....	....	....	....	....	....
R_{10}	0.0033	....	....	....	....	....	....	....	....	....	....
R_{11}	0.0013	....	....	....	....	....	....	....	....	....	....
R_{12}	0.0005	....	....	....	....	....	....	....	....	....	....
R_{13}	0.0002	....	....	....	....	....	....	....	....	....	....
R_{14}	0.0001	....	....	....	....	....	....	....	....	....	....

Table 3 gives some idea of the relation between the radius (R_2) of the hole and the distance (c) of the center of the hole from the free surface for the range of α_1 for which calculations were made.

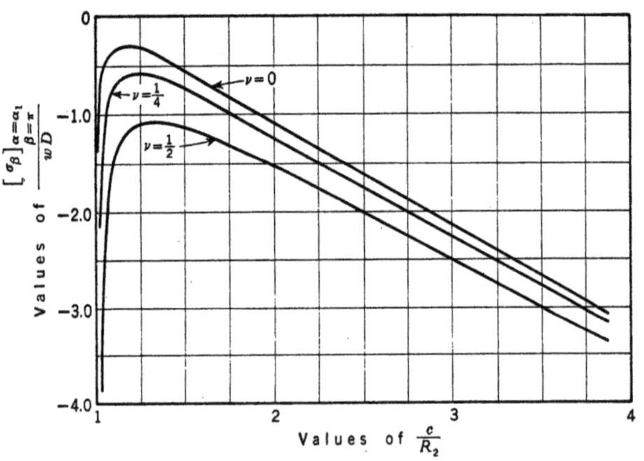

FIG. 5.—VARIATION OF THE QUANTITY $\frac{[\sigma_\beta]_{\alpha=\alpha_1, \beta=\pi}}{w D}$ AT THE HIGHEST POINT OF THE TUNNEL WALL *vs.* THE RATIO $\frac{c}{R_2}$ FOR CASE I

In Fig. 5 the quantity $\frac{[\sigma_\beta]_{\alpha=\alpha_1}}{w D}$ at the point of the hole nearest to the upper free surface is plotted against the ratio $\frac{c}{R_2}$ for three values of Poisson's ratio.

Figs. 4 and 5 and Table 2 illustrate the following significant facts for Case I:

(1) The disturbing influence of the upper free boundary is not felt until $\frac{c}{R_2}$ is less than about 1.5;

(2) For $\frac{c}{R_2} > 1.5$, the stresses increase almost linearly with the depth;

TABLE 2.—Values of $\frac{[\sigma_\beta]_{\alpha=\alpha_1}}{w\,D}$ (Case I)

α_1	β 0	20	40	60	80	100	120	140	160	180
	$\nu = 0$ (Case I)									
0.2	−1.63	....	....	....	....	....	....	....	....	−1.35
0.4	−1.69	−1.07	−0.77	−0.71	−0.69	−0.65	−0.57	−0.49	−0.43	−0.40
0.6	−1.80	−1.42	−1.13	−0.93	−0.86	−0.76	−0.62	−0.46	−0.34	−0.29
0.8	−1.96	−1.72	−1.37	−1.20	−1.07	−0.93	−0.74	−0.59	−0.45	−0.40
1.0	−2.18	−2.01	−1.71	−1.49	−1.33	−1.17	−0.98	−0.80	−0.66	−0.61
1.2	−2.46	−2.33	−2.07	−1.84	−1.64	−1.46	−1.27	−1.08	−0.95	−0.90
1.4	−2.81	−2.71	−2.48	−2.24	−2.03	−1.82	−1.64	−1.45	−1.31	−1.26
1.6	−3.25	−3.17	−2.97	−2.73	−2.50	−2.28	−2.07	−1.89	−1.76	−1.71
1.8	−3.79	−3.72	−3.54	−3.31	−3.07	−2.84	−2.62	−2.44	−2.31	−2.27
2.0	−4.45	−4.42	−4.22	−4.00	−3.75	−3.51	−3.29	−3.11	−2.98	−2.94
	$\nu = \frac{1}{4}$ (Case I)									
0.2	−1.63	....	....	....	....	....	....	....	....	−2.18
0.4	−1.67	−1.05	−0.73	−0.66	−0.66	−0.69	−0.74	−0.78	−0.81	−0.85
0.6	−1.77	−1.39	−1.02	−0.89	−0.84	−0.79	−0.73	−0.66	−0.60	−0.58
0.8	−1.93	−1.68	−1.33	−1.16	−1.05	−0.96	−0.83	−0.74	−0.65	−0.62
1.0	−2.14	−1.97	−1.67	−1.45	−1.31	−1.19	−1.06	−0.93	−0.83	−0.79
1.2	−2.41	−2.29	−2.03	−1.80	−1.63	−1.48	−1.34	−1.19	−1.09	−1.06
1.4	−2.76	−2.66	−2.44	−2.21	−2.01	−1.84	−1.69	−1.55	−1.43	−1.39
1.6	−3.19	−3.11	−2.92	−2.69	−2.49	−2.30	−2.13	−1.98	−1.88	−1.84
1.8	−3.73	−3.66	−3.49	−3.27	−3.04	−2.85	−2.67	−2.52	−2.42	−2.38
2.0	−4.39	−4.36	−4.17	−3.96	−3.74	−3.53	−3.34	−3.19	−3.08	−3.05
	$\nu = \frac{1}{2}$ (Cases I and II)									
0.2	−1.61	....	....	....	....	....	....	....	....	−3.85
0.4	−1.64	−0.99	−0.63	−0.54	−0.59	−0.78	−1.06	−1.35	−1.58	−1.67
0.6	−1.73	−1.33	−0.93	−0.80	−0.79	−0.85	−0.95	−1.05	−1.12	−1.15
0.8	−1.86	−1.61	−1.25	−1.08	−1.01	−1.01	−1.01	−1.04	−1.05	−1.06
1.0	−2.06	−1.89	−1.58	−1.37	−1.28	−1.23	−1.20	−1.18	−1.16	−1.15
1.2	−2.32	−2.20	−1.94	−1.73	−1.60	−1.52	−1.46	−1.41	−1.38	−1.37
1.4	−2.66	−2.56	−2.35	−2.13	−1.98	−1.87	−1.81	−1.74	−1.69	−1.67
1.6	−3.08	−3.00	−2.82	−2.62	−2.40	−2.33	−2.24	−2.16	−2.11	−2.00
1.8	−3.01	−3.55	−3.39	−3.20	−3.02	−2.88	−2.77	−2.69	−2.63	−2.61
2.0	−4.26	−4.24	−4.06	−3.88	−3.71	−3.56	−3.44	−3.35	−3.29	−3.27

Table 3.—Position of Hole with Respect to Horizontal Surface for Values of α_1 and $\frac{c}{R_2}$

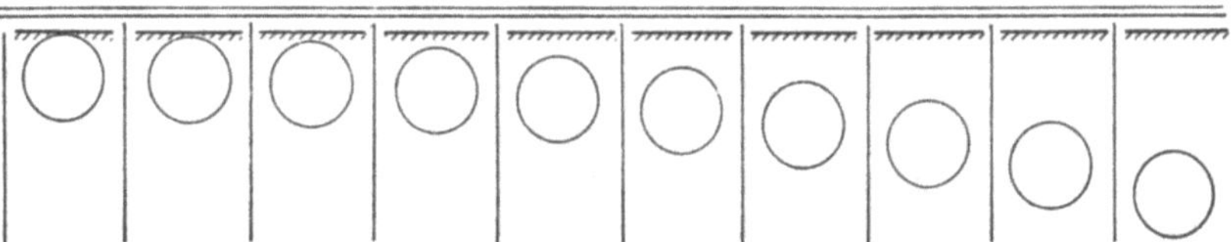

α_1	0.2	0.4	0.6	0.8	1.0	1.2	1.4	1.6	1.8	2.0
$\frac{c}{R_2}$	1.02	1.08	1.19	1.34	1.54	1.81	2.15	2.58	3.10	3.76

(3) For $\frac{c}{R_2} < 1.2$, the stresses increase with great rapidity as the hole approaches the upper surface of the solid;

(4) For materials of reasonable strength and density, the stresses are important only when the hole is very close to the upper surface of the solid or when it is very far from the surface;

(5) Variation of Poisson's ratio, ν, has little effect on the stresses except when the hole is very close to the upper surface; and,

(6) The stresses on the boundary of the tunnel are greatest at the lowest point of the tunnel cross-section except when α_1 is very small, in which case there are two equal maxima straddling the highest point.

In an analysis of this type, involving so much mathematical detail, it is important to determine whether or not the solution approaches, in a limiting case, a result that can be obtained directly by a simple process. When the hole is far from the upper surface, there are two factors that should be revealed in the solution: First, the stresses induced by the weight of the material removed from the hole should be very small in comparison with the stress concentration due to the initial stress in the surrounding material; and, second, by St. Venant's principle, the latter effect should be the same as for a hole in a uniform-stress field composed of the principal stresses present before the introduction of the hole. In Case I, the initial stress system is an isotropic pressure, and it is known[5] that in such a system the introduction of a hole doubles the stress.

To perform the limiting process in Equation (41), let $\alpha \to \infty$ while $a \to c$; $a \operatorname{csch} \alpha_1 \to R_2$; and $\beta \to \psi$ in which $\psi \left(= \frac{3\pi}{2} - \phi \right)$ is the angle between the radius from the center of the hole and the normal to the straight boundary, so that

$$\cos \psi = \frac{\cosh \alpha_1 \cos \beta - 1}{\cosh \alpha_1 - \cos \beta} \quad \text{(46a)}$$

and

$$\sin \psi = \frac{\sinh \alpha_1 \sin \beta}{\cosh \alpha_1 - \cos \beta} \quad \text{(46b)}$$

Making these substitutions in Equation (41):

$$[\sigma_\beta]_{\alpha=\alpha_1} = -2\,c\,w - R_2\,w \frac{3 - 4\,\nu}{2\,(1 - \nu)} \cos \psi \quad \text{(47)}$$

In Equation (47) the second term arises from the weight of the material removed from the hole and the first term is the stress concentration effect. The initial stress at a distance c below the surface is $-c\,w$, so that the term $-2\,c\,w$ reveals the predicted stress concentration factor of 2. If R_2 is small in comparison with c, the second term is small in comparison with the first. Therefore, if the hole is far from the upper surface, it is only necessary to invoke St. Venant's principle to arrive immediately at the resulting stresses.

Cases II and III

The same general stress function, Equation (37), which was used for Case I is applicable to Cases II and III. The latter two differ from the former only in the first term of Equation (37). In Case I the first term was obtained from Equation (5*b*). In Case II, the first term is obtained from Equation (6*b*); and for Case III, Equation (7*b*) is used. Thus Cases II and III differ from Case I only by a term

$$\chi_{2,3} = G\,w\,y^3 \qquad (48)$$

in which, for Case II,

$$G = \frac{1 - 2\nu}{6(1-\nu)} \qquad (49a)$$

and, for Case III,

$$G = \tfrac{1}{6} \qquad (49b)$$

To arrive at Cases II and III, therefore, it is only necessary to superpose on Case I the stress system arising from the initial stress represented by Equation (48). This stress system is obtained by combining the second auxiliary stress function (Equation (36)) with Equation (48) in such a manner that the stresses on the circle $\alpha = \alpha_1$ are annulled. Equation (48) can only be used with Equation (36) if it is expressed in the form $J\,f(\alpha)\cos n\beta$. To this end write

$$\chi_{2,3} = G\,w\,y^3 = \frac{G\,w\,a^3 \sinh^3\alpha}{(\cosh\alpha - \cos\beta)^3}$$

$$= G\,w\,a^2\,J\left\{\cosh\alpha + 2\sum_{n=1}^{\infty}(\cosh\alpha + n\sinh\alpha)\,e^{-n\alpha}\cos n\beta\right\} \qquad (50)$$

The complete stress function to be added to Case I to transform it to Cases II and III is, therefore,

$$\frac{\chi}{J} = G\,w\,a^2\left\{\cosh\alpha + 2\sum_{n=1}^{\infty}(\cosh\alpha + n\sinh\alpha)\,e^{-n\alpha}\cos n\beta\right.$$
$$+ B_0\left[\alpha(\cosh\alpha - \cos\beta) + \sinh\alpha(\cosh\alpha\cos\beta - 1)\right]$$
$$+ A_1(\cosh 2\alpha - 1)\cos\beta + \sum_{n=2}^{\infty}\Big(A_n\left[\cosh(n+1)\alpha - \cosh(n-1)\alpha\right]$$
$$\left. + E_n\left[(n-1)\sinh(n+1)\alpha - (n+1)\sinh(n-1)\alpha\right]\Big)\cos n\beta\right\} \qquad (51)$$

To determine the values of the constants, Equation (51) is substituted in Equations (20) and (22) with $w = 0$ in the latter; thus:

$$B_0 = 3\coth\alpha_1\,\mathrm{csch}^2\alpha_1 \qquad (52a)$$

$$A_1 = -\tfrac{3}{2}\,\mathrm{csch}^2\alpha_1 \qquad (52b)$$

$$A_n = -\frac{n(n^2-1)\sinh^2\alpha_1}{\sinh^2 n\alpha_1 - n^2\sinh^2\alpha_1} \qquad (52c)$$

and

$$E_n = \frac{n\sinh\alpha_1(n\sinh\alpha_1 + \cosh\alpha_1) + e^{-n\alpha_1}\sinh n\alpha_1}{\sinh^2 n\alpha_1 - n^2\sinh^2\alpha_1} \qquad (52d)$$

Again, the stresses $[\sigma_\beta]_{\alpha=\alpha_1}$ on the boundary of the hole are found by substituting Equations (52) and (51) in Equation (19*b*). The result is

$$[\sigma_\beta]_{\alpha=\alpha_1} = w\,G\,a\,(\cosh\alpha_1 - \cos\beta)\{6\coth\alpha_1\,\mathrm{csch}\,\alpha_1$$
$$+ 6\,\mathrm{csch}^2\alpha_1\cos\beta + 4\sinh\alpha_1\sum_{n=2}^{\infty}S_n\cos n\beta\} \qquad (53)$$

in which

$$S_n = \frac{n\,(n^2 - 1)\sinh^2 n\,\alpha_1}{\sinh^2 n\,\alpha_1 - n^2 \sinh^2 \alpha_1} \qquad (54)$$

The series in Equation (53) converges too slowly for calculation purposes. The slowly converging part may be separated out by letting

$$S_n = 2\,n\,(n^2 - 1)\,e^{-n\alpha_1} + T_n \qquad (55)$$

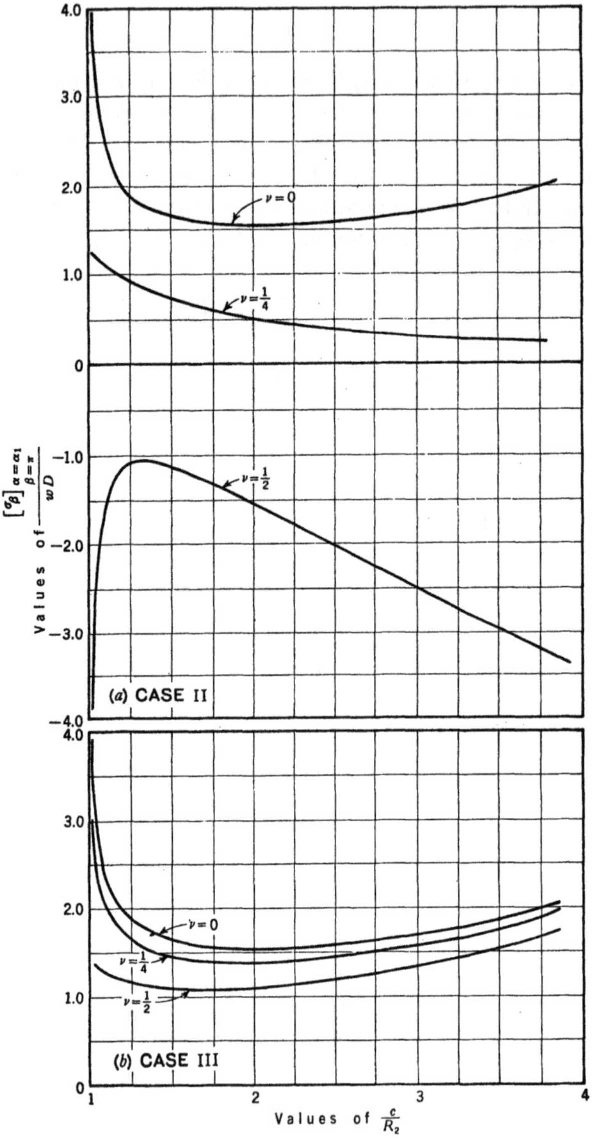

FIG. 6.—VARIATION OF THE QUANTITY $\frac{[\sigma_\beta]_{\alpha=\alpha_1,\ \beta=\pi}}{w\,D}$ AT THE HIGHEST POINT OF THE TUNNEL WALL *vs.* THE RATIO $\frac{c}{R_2}$

Then, noting that

$$8\,(\cosh\alpha_1 - \cos\beta)\sum_{n=1}^{\infty} n\,(n^2 - 1)\,e^{-n\alpha_1}\cos n\,\beta = 6\,\operatorname{csch}^2\alpha_1$$
$$\times\,(2\cosh\alpha_1\cos 2\,\psi + \cos\psi + \cos 3\,\psi)\ldots\ldots\ldots\ldots(56)$$

and substituting Equation (55) in Equation (53):

$$[\sigma_\beta]_{\alpha=\alpha_1} = w\,G\,a\,(\cosh\alpha_1 - \cos\beta)\{6\coth\alpha_1\operatorname{csch}\alpha_1 + 6\operatorname{csch}^2\alpha_1\cos\beta$$
$$+ 4\sinh\alpha_1\sum_{n=2}^{\infty} T_n\cos n\,\beta\} + 6\,w\,G\,a\operatorname{csch}\alpha_1$$
$$\times(\cos\psi + 2\cosh\alpha_1\cos 2\,\psi + \cos 3\,\psi)\ldots\ldots\ldots\ldots(57)$$

Values of T_n are given in Table 4. Values of $[\sigma_\beta]_{\alpha=\alpha_1}$ for Case II and Case III are obtained by adding Equations (45) and (57) with G given by Equations (49*a*) and (49*b*), respectively. The results are given in Table 5.

The curves in Fig. 6 show the variation of the stress in terms of $\frac{[\sigma_\beta]_{\alpha=\alpha_1}}{w\,D}$, at the highest point of the tunnel as $\frac{c}{R_2}$ is varied. The conclusions that were drawn for Case I also hold for Cases II and III with the following additional comments:

(1) The wide variation of the stresses for the three separate cases shows the necessity of knowing the initial state of stress in the solid; and,

(2) The apparent great change in the stresses of Case II with variation of Poisson's ratio is to be attributed to variation of the state of initial stress rather than to any marked change with Poisson's ratio.

TABLE 4.—VALUES OF T_n

T_n	α_1										
	0.4	0.6	0.8	1.0	1.2	1.4	1.6	1.8	2.0	2.2	2.4
T_2	41.4132	10.1670	3.3052	1.2281	0.4924	0.2047	0.0876	0.0382	0.0168	0.0074	0.0013
T_3	33.2080	6.1642	1.4073	0.3146	0.0867	0.0219	0.0055	0.0026	0.0003	0.0001	....
T_4	20.1894	3.1329	0.4583	0.0680	0.0099	0.0015	0.0001	....	....	....	....
T_5	16.2229	1.3750	0.1211	0.0103	0.0008	0.0001	....	....	....	....	....
T_6	10.1179	0.5320	0.0271	0.0012	0.0001	....	....	....	....	....	....
T_7	5.9144	0.1857	0.0053	0.0001	....	....	....	....	....	....	....
T_8	3.2620	0.0593	0.0005	....	....	....	....	....	....	....	....
T_9	1.7069	0.0176	0.0001	....	....	....	....	....	....	....	....
T_{10}	0.8523	0.0043	....	....	....	....	....	....	....	....	....
T_{11}	0.4100	0.0023	....	....	....	....	....	....	....	....	....
T_{12}	0.1882	0.0002	....	....	....	....	....	....	....	....	....
T_{13}	0.0828	....	....	....	....	....	....	....	....	....	....
T_{14}	0.0361	....	....	....	....	....	....	....	....	....	....
T_{15}	0.0140	....	....	....	....	....	....	....	....	....	....
T_{16}	0.0029	....	....	....	....	....	....	....	....	....	....
T_{17}	0.0005	....	....	....	....	....	....	....	....	....	....
T_{18}	0.0001	....	....	....	....	....	....	....	....	....	....

Once again it is desirable to check the results by applying a limiting process as in the preceding section. In this case there is available, for comparison, Yamaguti's approximate solution[3] for a hole far from the upper free surface with initial stresses corresponding to Case II.

TABLE 5.—VALUES OF $\dfrac{[\sigma_\beta]_{\alpha=\alpha_1}}{w\,D}$ FOR CASES II AND III

α_1	β = 0	20	40	60	80	100	120	140	160	180
	(*a*) CASE II AND CASE III, $\nu = 0$									
0.2	1.24	....	....	....	....	....	....	....	....	3.87
0.4	1.20	−1.62	−1.04	−1.03	−1.04	−0.53	0.73	1.89	2.16	2.05
0.6	1.25	−1.60	−1.69	−1.27	−1.00	−0.45	0.31	1.14	1.79	2.03
0.8	1.29	−2.43	−2.18	−1.68	−1.20	−0.63	0.15	0.93	1.54	1.77
1.0	1.36	−0.81	−2.53	−2.23	−1.47	−0.84	0.05	0.77	1.41	1.66
1.2	1.47	−0.48	−2.74	−2.89	−2.09	−1.07	−0.20	0.68	1.31	1.55
1.4	1.61	−0.22	−2.86	−3.53	−2.53	−1.44	−0.44	0.58	1.29	1.55
1.6	1.79	0.01	−2.96	−4.19	−3.59	−2.22	−0.75	0.48	1.31	1.60
1.8	2.04	0.24	−3.09	−4.89	−4.55	−3.00	−1.19	0.38	1.38	1.74
2.0	2.33	0.43	−3.24	−5.65	−5.65	−4.00	−1.79	0.19	1.52	1.98
	(*b*) CASE II, $\nu = \frac{1}{4}$									
0.2	0.29	....	....	....	....	....	....	....	....	1.29
0.4	0.26	−1.41	−0.90	−0.87	−0.89	−0.61	0.14	0.81	0.91	0.79
0.6	0.26	−1.51	−1.39	−1.11	−0.93	−0.58	−0.11	0.41	0.82	0.97
0.8	0.24	−2.15	−1.87	−1.48	−1.14	−0.75	−0.23	0.27	0.68	0.83
1.0	0.22	−1.17	−2.22	−1.95	−1.41	−0.97	−0.44	0.12	0.56	0.72
1.2	0.21	−1.05	−2.47	−2.50	−1.93	−1.22	−0.62	−0.02	0.42	0.68
1.4	0.19	−1.00	−2.69	−3.07	−2.35	−1.59	−0.89	−0.20	0.30	0.48
1.6	0.17	−0.99	−2.92	−3.67	−3.22	−2.26	−1.25	−0.40	0.17	0.37
1.8	0.15	−1.02	−3.19	−4.33	−4.03	−2.96	−1.72	−0.64	0.04	0.29
2.0	0.13	−1.12	−3.51	−5.06	−5.00	−3.85	−2.34	−0.99	−0.08	0.23
	(*c*) CASE III, $\nu = \frac{1}{4}$									
0.2	1.25	....	....	....	....	....	....	....	....	3.03
0.4	1.22	−1.60	−0.99	−0.97	−1.00	−0.57	0.57	1.61	1.78	1.61
0.6	1.27	−1.57	−1.57	−1.22	−0.97	−0.48	0.20	0.95	1.52	1.74
0.8	1.33	−2.39	−2.14	−1.64	−1.18	−0.65	0.07	0.78	1.34	1.55
1.0	1.40	−0.77	−2.49	−2.20	−1.45	−0.87	−0.13	0.64	1.25	1.48
1.2	1.52	−0.44	−2.70	−2.85	−2.08	−1.09	−0.27	0.57	1.17	1.40
1.4	1.66	−0.17	−2.81	−3.49	−2.52	−1.46	−0.50	0.48	1.16	1.41
1.6	1.85	0.07	−2.91	−4.15	−3.58	−2.23	−0.81	0.39	1.19	1.48
1.8	2.10	0.29	−3.04	−4.85	−4.53	−3.02	−1.24	0.30	1.27	1.62
2.0	2.39	0.49	−3.19	−5.61	−5.64	−4.01	−1.84	0.11	1.42	1.87
	(*d*) CASE III, $\nu = \frac{1}{2}$									
0.2	1.27	....	....	....	....	....	....	....	....	1.36
0.4	1.25	−1.54	−0.90	−0.86	−0.94	−0.66	0.25	1.03	1.01	0.78
0.6	1.32	−1.51	−1.49	−1.13	−0.93	−0.54	−0.02	0.56	1.00	1.17
0.8	1.39	−2.32	−2.05	−1.56	−1.14	−0.70	−0.11	0.48	0.94	1.11
1.0	1.48	−0.69	−2.41	−2.12	−1.42	−0.91	−0.27	0.39	0.92	1.12
1.2	1.61	−0.35	−2.61	−2.78	−2.04	−1.13	−0.39	0.35	0.89	1.09
1.4	1.76	−0.07	−2.72	−3.42	−2.49	−1.50	−0.61	0.28	0.90	1.13
1.6	1.96	0.18	−2.82	−4.08	−3.55	−2.27	−0.92	0.21	0.96	1.23
1.8	2.22	0.41	−2.94	−4.78	−4.51	−3.05	−1.34	0.13	1.06	1.39
2.0	2.52	0.61	−3.08	−5.54	−5.61	−4.04	−1.94	−0.05	1.21	1.65

Note: Case II, $\nu = \frac{1}{2}$ and Case I, $\nu = \frac{1}{2}$ are identical.

Applying the limiting conditions:

$$[\sigma_\beta]_{\alpha=\alpha_1} = \frac{w}{1-\nu}\{-c - \tfrac{1}{2}R_2\cos\psi + 2(1-2\nu)c\cos 2\psi + (1-2\nu)R_2\cos 3\psi\} \quad \ldots\ldots (58)$$

In order to compare $[\sigma_\beta]_{\alpha=\alpha_1}$ in Equation (58) with Yamaguti's σ_θ (written $\overline{\theta\theta_1}$) note that his symbol θ is given by $\theta = \pi - \psi$.

When this substitution is made, the two formulas check except for an error in sign[17b] in Yamaguti's work.

APPENDIX I

Dead Load Moment and Thrust in an Arch

The results of the preceding sections can be applied to the arch-like structure formed between $\alpha = 0$ and $\alpha = \alpha_1$ for small values of α_1; but, when α_1 is small, the infinite series in the stress formulas converge very slowly. This is due to the fact that, for $\alpha_1 = 0$, the series actually diverge. Consequently, the computations are exceedingly laborious. The difficulty may be avoided if it is observed that the resultant force and couple on a section $\beta =$ *constant* connecting the two boundaries may be expressed in finite terms. In what follows, this statement will be proved and the results applied to Case I as an illustrative example.

Let s_α be an arc of the curve $\beta =$ *constant*, measured positively in the direction of increasing α. In traveling along the curve in the positive direction, the right-hand side is called the outside and the left-hand side the inside. Let $\bar{X}$ and $\bar{Y}$ be the forces per unit of thickness parallel to the x-axis and the y-axis which the material outside of s_α exerts on the material inside of s_α per unit of length of s_α. Then, if l and m are the direction cosines of the outward drawn normal to s_α referred to the x-axis and the y-axis:[18]

$$\bar{X} = l\,\sigma_x + m\,\tau_{xy} \qquad \text{(59a)}$$

and

$$\bar{Y} = l\,\tau_{xy} + m\,\sigma_y \qquad \text{(59b)}$$

Now $l = \dfrac{dy}{ds_\alpha}$; $m = -\dfrac{dx}{ds_\alpha}$ and σ_x, σ_y and τ_{xy} are given, in terms of χ and Ω, by Equations (1). Hence

$$\bar{X} = \frac{dy}{ds_\alpha}\left(\frac{\partial^2\chi}{\partial y^2} + \Omega\right) + \frac{dx}{ds_\alpha}\frac{\partial^2\chi}{\partial x\,\partial y} = \frac{d}{ds_\alpha}\left(\frac{\partial\chi}{\partial y}\right) + \Omega\frac{dy}{ds_\alpha} \qquad \text{(60a)}$$

and

$$\bar{Y} = -\frac{dy}{ds_\alpha}\frac{\partial^2\chi}{\partial x\,\partial y} - \frac{dx}{ds_\alpha}\left(\frac{\partial^2\chi}{\partial x^2} + \Omega\right) = -\frac{d}{ds_\alpha}\left(\frac{\partial\chi}{\partial x}\right) - \Omega\frac{dx}{ds_\alpha} \qquad \text{(60b)}.$$

The x and y components of resultant force on the entire arc between the straight and circular boundaries are found by integrating Equations (60) along, s_α, from the point of intersection of s_α with $\alpha = 0$ to the point of intersection of s_α with $\alpha = \alpha_1$. Calling these total x and y components H and V, respectively:

$$H = \int_0^{\alpha_1} \bar{X}\,ds_\alpha = \left[\frac{\partial\chi}{\partial y}\right]_0^{\alpha_1} + \int_0^{\alpha_1}\Omega\,dy \qquad \text{(61a)}$$

and,

$$V = \int_0^{\alpha_1} \bar{Y}\,ds_\alpha = -\left[\frac{\partial\chi}{\partial x}\right]_0^{\alpha_1} - \int_0^{\alpha_1}\Omega\,dx \qquad \text{(61b)}$$

[17b] The error in Yamaguti's paper first appears in his Equations (16). The second term in the second of his Equations (16) should be preceded by a plus sign instead of a minus sign. The error is carried through the remainder of the paper so that his formula for $\overline{\theta\theta_1}$ is incorrect.

[18] "A Treatise on Photo-Elasticity," by E. G. Coker and L. N. G. Filon, Cambridge University Press, 1931, p. 507.

Now $\frac{\partial \chi}{\partial y} = \frac{\partial \alpha}{\partial y}\frac{\partial \chi}{\partial \alpha} + \frac{\partial \beta}{\partial y}\frac{\partial \chi}{\partial \beta}$; or,

$$\frac{\partial \chi}{\partial y} = \frac{\partial \alpha}{\partial y}\left(J\frac{\partial}{\partial \alpha} + \frac{\partial J}{\partial \alpha}\right)\left(\frac{\chi}{J}\right) + \frac{\partial \beta}{\partial y}\left(J\frac{\partial}{\partial \beta} + \frac{\partial J}{\partial \beta}\right)\left(\frac{\chi}{J}\right) \ldots\ldots (62a)$$

and, $\frac{\partial \chi}{\partial x} = \frac{\partial \alpha}{\partial x}\frac{\partial \chi}{\partial \alpha} + \frac{\partial \beta}{\partial x}\frac{\partial \chi}{\partial \beta}$; or,

$$\frac{\partial \chi}{\partial y} = \frac{\partial \alpha}{\partial x}\left(J\frac{\partial}{\partial \alpha} + \frac{\partial J}{\partial \alpha}\right)\left(\frac{\chi}{J}\right) + \frac{\partial \beta}{\partial x}\left(J\frac{\partial}{\partial \beta} + \frac{\partial J}{\partial \beta}\right)\left(\frac{\chi}{J}\right) \ldots\ldots (62b)$$

Therefore

$$H = \left[\frac{\partial \alpha}{\partial y}\left(J\frac{\partial}{\partial \alpha} + \frac{\partial J}{\partial \alpha}\right)\left(\frac{\chi}{J}\right) + \frac{\partial \beta}{\partial y}\left(J\frac{\partial}{\partial \beta} + \frac{\partial J}{\partial \beta}\right)\left(\frac{\chi}{J}\right) + \int \Omega\, dy\right]_0^{\alpha_1} \ldots (63a)$$

and,

$$V = -\left[\frac{\partial \alpha}{\partial x}\left(J\frac{\partial}{\partial \alpha} + \frac{\partial J}{\partial \alpha}\right)\left(\frac{\chi}{J}\right) + \frac{\partial \beta}{\partial x}\left(J\frac{\partial}{\partial \beta} + \frac{\partial J}{\partial \beta}\right)\left(\frac{\chi}{J}\right) + \int \Omega\, dx\right]_0^{\alpha_1} \ldots (63b)$$

To find the total moment on the section, take moments of $\tau_{\alpha\beta}$ on β = *constant* about the center of the circle β = *constant* as the moment center. Then

$$M = \rho_1 \int_0^{\alpha_1} \tau_{\alpha\beta}\, ds_\alpha = \frac{J^2}{\left(\frac{\partial J}{\partial \beta}\right)_{\alpha 1}} \left[\frac{\partial}{\partial \beta}\left(\frac{\chi}{J}\right)\right]_0^{\alpha_1} \ldots\ldots\ldots\ldots (64)$$

In Equations (63) and (64) $\frac{\chi}{J}$ occurs only in the form of its boundary values and the boundary values of its first derivatives with respect to α and β; but, by Equations (20) and (22), these quantities are given in finite terms. Hence, by using Equations (20) and (22) in Equations (63) and (64), formulas for the force and couple on a section β = *constant*, connecting the two boundaries, may be found readily in finite terms.

Applying Equations (63) and (64) to Case I:

$$H = \frac{w\, a^2}{2}\left(\frac{1 - 2\nu}{2(1-\nu)}\alpha_1 \operatorname{csch}^2 \alpha_1 - \coth \alpha_1\right) \ldots\ldots\ldots\ldots (65a)$$

$$V = -\frac{w\, a^2}{2}\left\{\left(\phi_1 - \frac{\pi}{2}\right)(\csc^2 \beta - \operatorname{csch}^2 \alpha_1) + \frac{2 \coth \alpha_1 \sin \beta}{\cosh \alpha_1 - \cos \beta} - \frac{(1 - \cosh \alpha_1 \cos \beta)(\cosh \alpha_1 + \cos \beta)}{\sin \beta \sinh \alpha_1 (\cosh \alpha_1 - \cos \beta)}\right\} \ldots\ldots\ldots\ldots (65b)$$

and,

$$M = \frac{w\, a^3}{2} \operatorname{csch}^2 \alpha_1 \left\{\left(\phi_1 - \frac{\pi}{2}\right)\cot \beta - \frac{\sinh \alpha_1}{\cosh \alpha_1 - \cos \beta} + \frac{5 - 6\nu}{2(1-\nu)}(\alpha_1 \coth \alpha_1 - 1) + \sinh \alpha_1\, e^{-\alpha_1}\right\} \ldots\ldots\ldots\ldots (65c)$$

In terms of the dimensions shown in Fig. 7 these become

$$H = \frac{w}{2}\left\{\frac{1 - 2\nu}{2(1-\nu)} R_2^2 \log \frac{c + a}{R_2} - a\, c\right\} \ldots\ldots\ldots\ldots (66a)$$

$$V = -\frac{w}{2}\left\{\Theta (R_1^2 - R_2^2) + y_1 f + x_1 c\right\} \ldots\ldots\ldots\ldots (66b)$$

and,

$$M = \frac{w}{2} R_2^2 \left\{ \Theta f - y_1 + \frac{5 - 6\nu}{2} \left(c \log \frac{c + a}{R_2} - a \right) + a^2 \frac{(c - a)}{R_2^2} \right\} \quad \text{..(66c)}$$

To determine the moment (M_0) at the crown of the arch it should be observed that

$$\lim_{\beta \to \pi} \Theta \cot \beta = - \frac{\sinh \alpha_1}{\cosh \alpha_1 + 1} = - \frac{y_0}{a} \quad \text{..............(67)}$$

Hence

$$M_0 = \frac{w}{2} R_2^2 \left\{ -2 y_0 + \frac{5 - 6\nu}{2} \left(c \log \frac{c + a}{R_2} - a \right) + \frac{a^2 (c - a)}{R_2^2} \right\} \quad \text{...(68)}$$

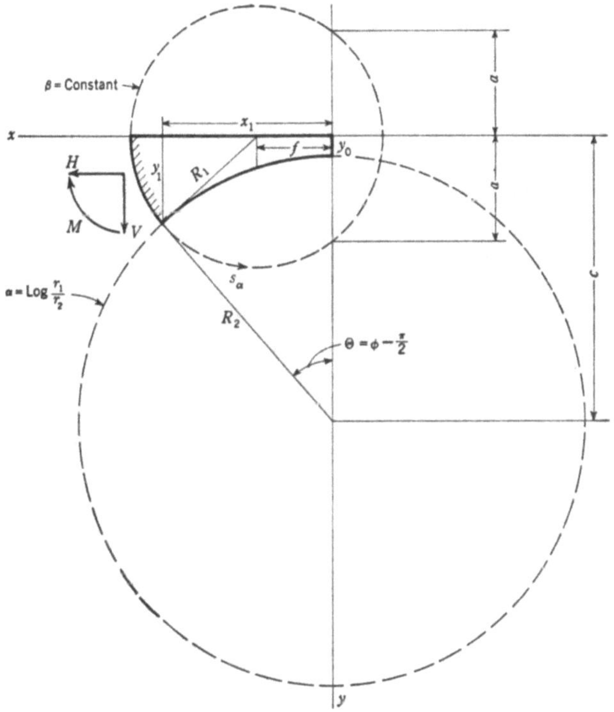

FIG. 7.—DIMENSIONS OF ARCH

APPENDIX II

NOTATION

The following notation conforms essentially with American Standard Symbols for Mechanics, Structural Engineering, and Testing Materials[19] compiled by a committee of the American Standards Association with Society representation, and approved by the Association in 1932. Where additional symbols are required, the notation used by Professor Timoshenko[20] is followed.

[19] ASA—Z10A—1932.

[20] "Theory of Elasticity," by S. Timoshenko, New York, 1934.

A = a constant ($A_0 \cdots A_n$; $B_0 \cdots B_n$; $C_0 \cdots C_n$; $D_0 \cdots D_n$);
a = a distance from the origin to a pole, in bipolar co-ordinates;
c = distance of the center of hole from the origin $= a \coth \alpha_1$
D = diameter of hole; (see also A)
d = depth below ground surface, to a horizontal plane (see Fig. 2);
e = Naperian base $= 2.718 \cdots$;
f = $a \cot \beta$ (see Fig. 7);
G = a constant defined in Equation (49);
H, V (see Fig. 7); H and V = horizontal and vertical components of a resultant force, on a section through the arch;
$i = \sqrt{-1}$;
J = stretch ratio of a conformal transformation of co-ordinates;
l, m = direction cosines (see Appendix I);
M = moment; total moment on a section through an arch; M_0 = moment at the crown of an arch (see Fig. 7);
N = a coefficient (N_n, R_n, S_n, T_n) = coefficients of Fourier series;
n = an integer ($1, 2, 3 \cdots n$);
R (see N) also R_2 = radius of hole $= a \operatorname{csch} \alpha_1$; R_1 = radius of circular section through an arch $= a \csc \beta$.
r = radius; r_1 and r_2 = radii to a point from the poles of a bipolar co-ordinate system (see Fig. 3);
s = arc of a curve; s_α = element of length along a curve β = *constant;*
w = weight per unit of volume of material;
X, Y Symbols X and Y = components of body force per unit of volume; $\bar{X}$ and $\bar{Y}$ = components of stress on a unit length of curve;
x, y Symbols x and y = rectangular co-ordinates; x_1 and y_1 = rectangular co-ordinates of the point of intersection of s_α and α_1 (see Fig. 7);
α Symbols α and β = curvilinear co-ordinates; α_1 = value of the co-ordinate α on the boundary of a hole;
ζ, η, ξ = boundary constants (see Equation (22));
$\Theta = \phi - \frac{\pi}{2}$;
θ_1, θ_2 = angles that r_1 and r_2 make with the x-axis;
ν = Poisson's ratio;
ρ = radius of curvature; ρ_1 and ρ_2 = radii of curvature of the curves β = *constant* and α = *constant,* respectively;
σ = unit normal stress; σ_x, σ_y = components of unit normal stress in rectangular co-ordinates; σ_α, σ_β = components of unit normal stress in curvilinear co-ordinates;
τ = unit shear stress; τ_{xy} = unit shear stress in rectangular co-ordinates; $\tau_{\alpha\beta}$ = unit shear stress in curvilinear co-ordinates;
ϕ = angle between the x-axis and the tangent to the curve β = *constant;*
χ = stress function;
$\psi = \frac{3\,\pi}{2} - \phi$;
Ω = body-force function; and,
∇^2 = Laplace's two-dimensional operator ($\nabla^4 = \nabla^2\nabla^2$).

DISCUSSION

W. O. RICHMOND,[21] ESQ. (by letter).—The stress distribution due to gravity in a semi-infinite elastic body pierced by a circular hole is an interesting problem and the author's treatment of it is a complete one The application of the results in practise depend, of course, upon the accuracy with which the assumptions are fulfilled. If the shape of the tunnel is other than circular (as it often is), a mathematical solution is practically impossible and recourse must be had to experimental methods.

An interesting experimental approach to this problem was developed by Mr. S. Yamaguti,[22] whose method consisted of casting a flat plate of agar-agar in a horizontal position. A network of lines was placed on the plate, and a hole was cut, of a shape appropriately representing the tunnel profile. When the model was placed in a vertical position, the strain in the material due to its own weight could be measured from the distortion of the originally parallel lines, and hence the stresses could be determined. For a round tunnel profile Mr. Yamaguti obtained good agreement between his mathematical solution and the results of his model test.

Mr. Yamaguti's method of test reproduces the author's Case II with the appropriate value of Poisson's ratio for agar-agar. The author's Case I could also be reproduced by this method. It would only be necessary to cast the agar-agar plate in a vertical position, so that the stress distribution in the unpierced plate would approximate a hydrostatic pressure.

A second experimental approach to the problem is by the method of photo-elasticity. If the tunnel profile is some distance below the surface, the stress distribution may be obtained approximately by the method of superposition, as stated by the author. Assuming that the original stress distribution is as represented by Case II, for instance, the boundary stress around the profile is just determined for a uniform compression in a vertical direction of magnitude.

$$\sigma_y = -w\,c \qquad (69)$$

Then the boundary stress is determined, for a stress at right angles, of magnitude

$$\sigma_x = -\frac{\nu}{1-2\nu}\,w\,c \qquad (70)$$

in which the symbols used are those of the author in Fig. 2, and c is the distance of the hole from the free surface. The stress on the boundary is then obtained by adding the two boundary stress distributions. For a non-circular profile the boundary stresses are most easily determined by photo-elastic tests.

This procedure is sufficiently accurate if the profile is some distance below the surface. For the case shown in Fig. 4(*b*) (that is, for a value of Poisson's ratio of ½ and $\alpha_1 = 2.0$), this approximate method gives a uniform compressive stress around the boundary of the tunnel of $-3.76\,w\,D$, in which w = density

[21] Asst. Prof. of Mech. Eng., Univ. of British Columbia, Vancouver, Canada.

[22] "On the Investigation of the Stress Distribution in a Tunnel with Agar-agar Model Experiments," by S. Yamaguti, *Proceedings*, Third International Cong. for Applied Mechanics, Vol. II, 1930, pp. 150–154.

of the material, and D = diameter of the circular profile. The maximum stress by the author's computation is $-4.26\,w\,D$. The error in the approximate method is thus slightly more than 10%, which is perhaps accurate enough for practical purposes. Thus, the conclusion may be drawn that if the tunnel is more than four radii, or two diameters from the surface, the approximate solution obtained by superposition is sufficiently accurate.

A few years ago the writer (working under the direction of A. Nadai at the Westinghouse Research Laboratories) made some photo-elastic tests on holes of tunnel profile at the suggestion of F. C. Carstarphen, M. Am. Soc. C. E. The profiles tested were squares with various fillet radii at the corners; the ratio of fillet radius, r, to side of square, D, had the values $\frac{r}{D} = \frac{1}{3}, \frac{1}{6}$, and $\frac{1}{12}$. The boundary stress distribution was obtained photo-elastically when the bakelite plate which contained the holes was subjected to a pure tension parallel to an axis of symmetry perpendicular to the sides of the holes. The stress concentration factors (that is, the ratio of the maximum stress on the boundary of the hole to the stress in the plate a long way from the hole) from these tests are

TABLE 6.—STRESS CONCENTRATION FACTORS

Item No.	Description	Ratio of Fillet Radius, r, to the Side of the Square, D			
		$\frac{1}{2}$	$\frac{1}{3}$	$\frac{1}{6}$	$\frac{1}{12}$
1	Stress concentration factor for tension or compression........	3	2.6	2.5	2.6
2	Maximum compressive stress on boundary of tunnel profile (Case II, $\nu = \frac{1}{4}$)........	2.7	2.4	2.4	2.5

shown in Table 6. For the sake of completeness, the case of the circular hole in a tensile stress field, for which $\frac{r}{D} = \frac{1}{2}$, is included, the results for which can be obtained by computation.[23] It is interesting to note that a minimum stress concentration factor seems to result for a value of $\frac{r}{D}$ of about $\frac{1}{6}$. The stress concentration factor for a square hole with sharp corners would be theoretically infinite.

The stress distribution around the boundary of tunnels of similar cross-section was then computed from the boundary stresses determined by the photo-elastic tests. The case solved by the writer was the author's Case II, in which the principal stresses before the tunnel existed were given by Equations (69) and (70). The results obtained by this procedure for a value of Poisson's ratio of $\frac{1}{4}$ are shown in Item 2, Table 6. These results should be sufficiently accurate, provided that the tunnel is at least two diameters below the surface, that the material in which the tunnel is driven behaves elastically, and that the original principal stresses are as assumed.

[23] "Theory of Elasticity," by S. Timoshenko, 1934, p. 75.

JACOB FELD,[24] M. AM. SOC. C. E. (by letter).—For the condition explained in this paper—namely, a circular tunnel in a uniform elastic material—Mr. Mindlin's solution is the most complete available. For use in practical cases, however, the following items must be taken into account:

I.—The internal stress in the rock in its natural condition, which stress may be relieved in the local vicinity of the tunnel by the removal of the tunnel rock.

II.—The percentage of error resulting from the use of the method and formulas in this paper when rock conditions cannot be assumed as isotropic. It is seldom that a tunnel is driven through a uniform granite, which is probably the nearest approach to the material assumed in this paper. The more usual cases are tunnels driven through laminar rock, such as shales, schists, or limestones. It is extremely doubtful whether the stress determinations resulting from the formulas in this paper can be applied to a case where a blocky limestone is encountered, or, to take a different example, where schist in thin layers, lying edgewise to the face of the tunnel, is encountered.

III.—The effect on stresses around the tunnel if temporary bracing is wedged against the exposed rock during driving, or the effect on the rock stresses as the tunnel is lined with concrete and grouted under pressure; also the effect of various grouting pressures in balancing or over-balancing rock stresses.

In taking the foregoing suggestions into consideration, it might be pertinent to consider the assumption made under the heading "Statement of the Problem" that, at the circular boundary, the normal stress and the shearing stress are equal to zero. Such an assumption is not correct when rock is encountered which requires bracing to prevent movement. It seems to the writer that in the three cases mentioned as covering the initial states of the rock, although they cover a great range of possibilities, such range is chiefly in the variation of the Poisson ratio in Case II. Case I is that of hydrostatic pressure which may be applicable at tremendous distances below the earth's surface in rocks that are under such pressure that they are practically liquid, and Case III would only apply where a tunnel is driven through blocks of rock with open vertical seams, which case could not be considered isotropic. With reference to Case II and its dependence upon the Poisson ratio of the material, it might well be emphasized that such ratio may vary for rocks, depending upon the amount of the stress. However, as stated in Mr. Mindlin's paper, such variation has little effect on the stresses themselves, except when the tunnel is unusually close to the open surface. It is for such a condition (namely, shallow tunnels) that a careful stress analysis is most necessary.

J. H. A. BRAHTZ,[25] ESQ. (by letter).—Congratulations are due the author for his excellent presentation of a very difficult theoretical subject. It is highly gratifying to have studies of this caliber brought to the attention of the engineering profession. In the mathematical theory of elasticity, relatively few exact solutions exist that may be applied directly to engineering problems. This paper is a definite contribution to this small group of exact solutions.

[24] Cons. Engr., New York, N. Y.

[25] Director, Photoelastic Laboratory, U. S. Dept. of the Interior, Bureau of Reclamation, Denver, Colo.

A characteristic behavior of additions to the exact-solution group seems to be that the underlying mathematics become more abstruse and complicated with each addition. This is particularly true in the cases treated by Mr. Mindlin, but in no way detracts from the value of the paper. However, it does limit its broad use and application to a relatively small group of engineers with a highly specialized mathematical background. Apparently, the author is well aware of this limitation and has minimized the attendant difficulties by presenting, in tabular and graphical form, the results of many computations. These simplify and reduce the labor greatly in applying the theoretical results to engineering problems.

In his study of the stresses due to body forces, Mr. Mindlin has considered three possible conditions that differ only in the ratio of horizontal stress to vertical stress along the vertical boundaries. To the mathematician the choice of this ratio represents no problem; it is merely called a boundary condition and is assumed to be known. To the practicing engineer the proper choice of this ratio is of utmost importance. For lack of definite information on the proper value of this ratio, the engineer is likely to employ a value that produces the most critical state of stress. This is on the side of safety but scarcely ever on the side of economy. Although the author has left this value "floating" (as he should, since a knowledge of it comes under the science of soil mechanics or geophysics, and not mathematics), it seems important to stress the need of experimental and statistical knowledge on this phase of foundation engineering.

It would have been interesting if the author had included one or more photoelastic checks and had compared the experimental maximum shear stresses with the computed values. It is realized, however, that body forces are difficult to apply to photoelastic models and that even when M. A. Biot's[26] method of employing boundary forces is used, formal and practical difficulties arise in the application of a continuously varying normal load on the closed boundary that corresponds to the tunnel surface.

Mr. Mindlin confined the application of the bipolar theory to body force or "dead load" stresses. For structures having shapes similar to those given in the paper, other common conditions of loading exist. These include the cases in which either the straight or circular boundaries are subjected to uniformly distributed loads. Although the general form for these cases has been given elsewhere,[27] it is felt that the particular cases in which one boundary is a straight line should be included in the paper, for completeness. Loadings of this type frequently arise in engineering practice. Usually they occur simultaneously with the case treated by the author. An example of this condition would be a circular tunnel in bed-rock below a river-bed or earth embankment. Sometimes they occur as the only loading; for example, a vertical shaft near the up-stream face of a masonry dam or a pressure pipe, or conduit, near a free surface. The latter is the case of uniform normal pressure on the inner boundary.

The writer is presenting the results obtained by auxiliary studies covering the conditions of uniform normal loads along the boundaries. The first part of

[26] *Journal of Applied Mechanics*, June, 1935.

[27] See, for example, "A Treatise on Photo-Elasticity," by E. G. Coker and L. N. G. Filon, Cambridge, University Press, 1931, p. 313, *et seq.*

the study is purely mathematical and is based upon Jeffery's general theory for stresses in a circular plate having an eccentric hole. A uniform normal pressure is applied to the outer boundary. If the radius of this boundary is assumed to increase without limit, the segment of the outer boundary in the vicinity of the hole approaches the limiting condition of a straight line. It should be noted that, in this limiting process, the outer boundary is still loaded on all sides, and the stress formulas obtained for the limiting case must be for the condition of the infinite boundaries, as well as the straight boundary, being loaded uniformly.

Although this is an important loading condition, a more common condition is one in which the straight boundary is loaded fully, and the vertical infinite boundaries are loaded only partly. The stresses for this condition can be obtained, however, by combining the results of a mathematical solution with those of photoelastic experiments, or by combining two mathematical solutions. The two mathematical solutions to be combined are the one given by the writer in this discussion and the solution by Mr. Jeffery referred to by the author.[11] However, Mr. Mindlin has called attention to an inconsistency in Mr. Jeffery's solution, and, rather than attempt to correct this fundamental error, the writer has used a combination of experiment and theory. The experimental problem treated by the writer is the problem treated by Mr. Jeffery. A table of numerical values for stresses at critical points has been developed by S. Timoshenko.[28] For the smallest value of $\frac{d}{r}$ given in the table, the boundary stress on the straight boundary opposite the hole is approximately twice the average stress in the plate, and of opposite sign. In his experimental study the writer was unable to note any change of sign in the stress along this boundary. An inspection of the isochromatic photographs included herewith bears out this point. Finally, the stresses are given for the cases in which a uniform pressure exists only on the internal boundary.

Another interesting application of bipolar co-ordinates arises in connection with the problem of percolation into a tunnel in pervious foundation or a shaft of circular cross-section situated near the up-stream face of a masonry dam.

Uniform Normal Loads on Boundaries.—Again making use of the bipolar transformation, shown in Fig. 1, the following boundary conditions are specified:

(I) The boundary $\alpha = \alpha_1$ is subjected to a normal load $\sigma_\alpha = 1$ and is free of shear—that is, $\tau_{\alpha\beta} = 0$.

(II) The boundary $\alpha = \alpha_2$ is free of load—that is, $\sigma_\alpha = \tau_{\alpha\beta} = 0$; and α_1 and α_2 are any of the positive α-circles shown in Fig. 1.

Now consider the function (see Equation (30))—

$$\frac{x_6}{J} = F = \sum_{n=0}^{\infty} f_n(\alpha) \cos n\beta \qquad (71)$$

The first two functions ($n = 0$ and $n = 1$) are

$$F_0 = A_0 \cosh\alpha + B_0\alpha\cosh\alpha + C_0\sinh\alpha + D_0\alpha\sinh\alpha \qquad (72a)$$

[28] "Theory of Elasticity," by S. Timoshenko, New York. 1934, p. 79.

and

$$F_1 = (A_1 \cosh 2\alpha + B_1 + C_1 \sinh 2\alpha + D_1 \alpha) \cos\beta \ldots\ldots (72b)$$

The relations, $D_0 = 0$ and $-D_1 = B_0$, have been shown by the author to exist in order that displacements may be single valued; and he has shown that A_0 may be placed equal to zero since $B_1 \cos\beta$ gives the same type of stress. Hence $F_0 = -D_1 \alpha \cosh\alpha$ and

$$F = F_0 - F_1 = (A \cosh 2\alpha + B + C \sinh 2\alpha) \cos\beta + D\alpha (\cos\beta - \cosh\alpha) \ldots\ldots (73)$$

Substitution of Equation (73) in the general stress equations (Equations (19)) gives

$$a\,\sigma_\alpha = A (\cosh 2\alpha - 2 \sinh 2\alpha \sinh\alpha \cos\beta) + B + C (\sinh 2\alpha - 2 \cosh 2\alpha \sinh\alpha \cos\beta) + D \sinh\alpha (\cosh\alpha - \cos\beta) \ldots\ldots (74a)$$

$$a\,\sigma_\beta = A [2 \cosh\alpha \cos\beta - \cosh 2\alpha (4\cos^2\beta - 1 - 2\cosh\alpha\cos\beta)] + B + C [2 \sinh\alpha\cos\beta - \sinh 2\alpha (4\cos^2\beta - 1 - 2\cosh\alpha\cos\beta)] - D \sinh\alpha (\cosh\alpha - \cos\beta) \ldots\ldots (74b)$$

and

$$a\,\tau_{\alpha\beta} = (2A \sinh 2\alpha + 2C \cosh 2\alpha + D)(\cosh\alpha - \cos\beta)\sin\beta \,.\, (74c)$$

The constants A, B, C, and D are now determined by the boundary conditions (I) and (II). Omitting mathematical details, the constants are found to be:

$$A = \frac{a}{2} \frac{\sinh(\alpha_2 + \alpha_1)}{\sinh(\alpha_2 - \alpha_1)[1 - \cosh(\alpha_2 + \alpha_1)\cosh(\alpha_2 - \alpha_1)]} \ldots (75a)$$

$$B = A \frac{\sinh(\alpha_2 - \alpha_1) - \sinh 2\alpha_2 \cosh(\alpha_2 - \alpha_1)}{\sinh(\alpha_2 + \alpha_1)} \ldots\ldots (75b)$$

$$C = -A \coth(\alpha_2 + \alpha_1) \ldots\ldots (75c)$$

and

$$D = 2A \frac{\cosh(\alpha_2 - \alpha_1)}{\sinh(\alpha_2 + \alpha_1)} \ldots\ldots (75d)$$

It has been stated that Equations (75) are applicable to any two circular boundaries, $\alpha = \alpha_1$ and $\alpha = \alpha_2$, as shown in Fig. 1. It gives the stresses for unit normal tension on the boundaries of a circular cylinder with an eccentric circular hole.[27] If the outer boundary is free of force, and if a unit compression force acts on the boundary of the hole, it is only necessary to superimpose a unit hydrostatic state of compression, $\sigma_\alpha' = \sigma_\beta' = -1$, $\tau_{\alpha\beta}' = 0$, on to the stresses of the first solution.

Application to a Circular Tunnel.—If the ground surface is assumed plane (that is, a cylinder of infinite radius), α_1 is placed equal to zero in the formulas; hence, Equations (75) become

$$A = \frac{a}{2(1 - \cosh^2\alpha_2)} \ldots\ldots (76a)$$

$$B = A(1 - 2\cosh^2\alpha_2) \ldots\ldots (76b)$$

$$C = -A \coth\alpha_2 \ldots\ldots (76c)$$

and

$$D = -2C \ldots\ldots (76d)$$

The values of the constants for this case are plotted in Fig. 8 for various ratios of $\frac{d}{r}$, in which d is the depth of the center below the ground surface and r is the radius of the tunnel. The next step is to plot the stresses for unit com-

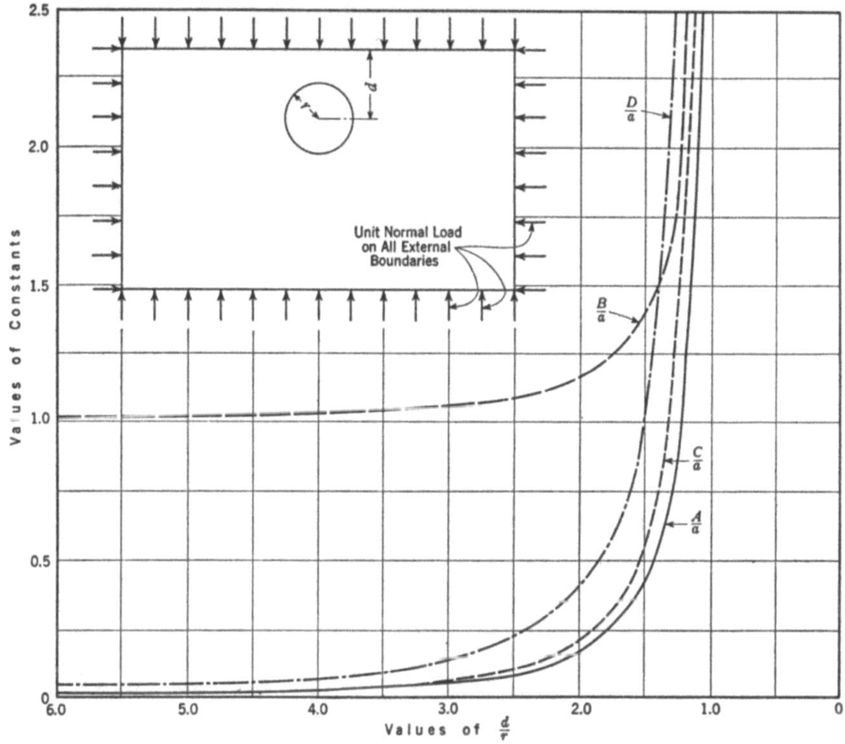

FIG. 8.—CONSTANTS A, B, C, AND D WHEN $\alpha_1 = 0$

pressive load, obtained by stress Equations (74), around the hole, along the ground surface, and along a vertical line through the center of the hole for $\frac{d}{r}$ = 6, 5, 4, 3, 2, 1, 1.5 and 1.25, respectively.[29] All stresses due to this loading are compression and can be plotted in terms of the unit load. Fig. 9 illustrates the form of curves for $\frac{d}{r} = 1.5$.

As stated previously, the computations made for a unit normal load acting on the outer boundary may also be used to determine the stresses in the foundation due to a unit normal load acting on the boundary of the hole. It is obvious that, if a unit normal load is applied simultaneously on both the outer and inner boundaries, all points in the body will have unit normal stress in all directions. Such a body is said to be in a hydrostatic state of stress. Principal stresses are equal to the same quantity at all points, and all shear stresses are zero.

[29] *Technical Memorandum No. 597*, U. S. Bureau of Reclamation.

Let a normal stress at any point due to a unit normal load acting on the boundary $\alpha = 0$ be denoted σ' and the shear by τ'. Furthermore, let a normal stress and shear at the same point due to a unit normal load acting on the

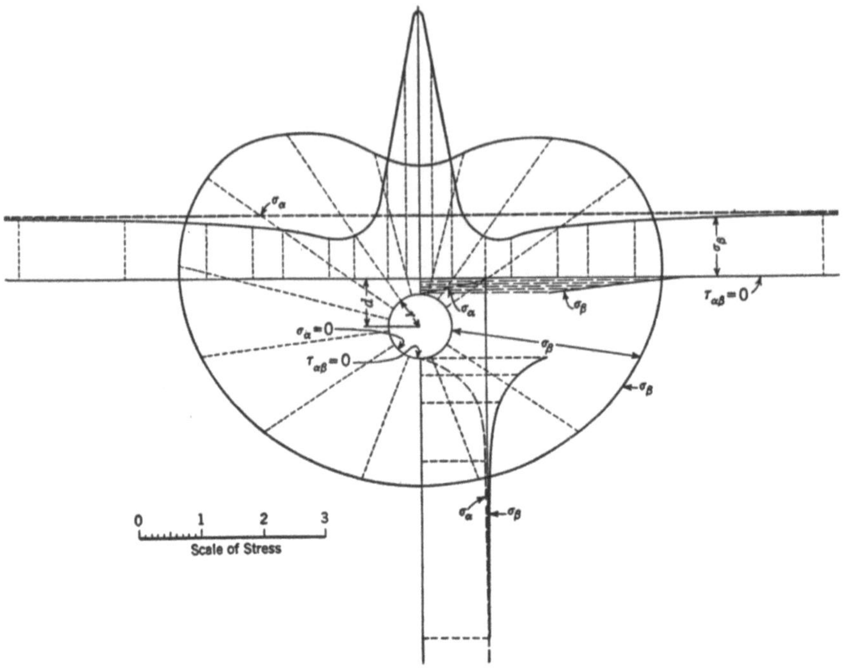

FIG. 9.—UNIT NORMAL COMPRESSION ON ALL EXTERNAL BOUNDARIES; $\frac{d}{r} = 1.5$

boundary $\alpha = \alpha_2$ be denoted by σ'' and τ'', respectively. Then, if unit pressures act simultaneously on both boundaries, the hydrostatic state gives:

$$\sigma' + \sigma'' = +1 \qquad (77a)$$

and

$$\tau' + \tau'' = 0 \qquad (77b)$$

if compression stress is considered positive.

By solving Equations (77), the normal stress σ'' and the shears τ'' can be found in terms of σ' and τ', respectively, as follows:

$$\sigma'' = 1 - \sigma' \qquad (78a)$$

and

$$\tau'' = -\tau' \qquad (78b)$$

A negative result will then denote tension. Hence, if the normal stresses found for a unit normal load acting on $\alpha = 0$ are deducted from unity, the stresses due to a unit normal load acting on $\alpha = \alpha_2$ will be obtained. The shears found at any point of the body for a unit load acting on $\alpha = 0$ will be of equal magnitude but opposite direction for a unit load acting on $\alpha = \alpha_2$.

Equations (78) are easily solved graphically by curves such as those in Fig. 9. It will be found that the stresses σ_α'' are still compressive and attain their maximum value at the hole, diminishing asymptotically to zero away from the hole. The stresses σ_β'' become tension, maximum at the hole, and also approach zero asymptotically.

In order to obtain the stresses for a boundary loading of any magnitude, it is only necessary to multiply those obtained for the unit loading by the magnitude of the actual boundary load. In other words, the stress diagrams in Fig. 9 are influence lines in the usual engineering sense.

Finally, if the external boundaries are loaded with a normal compressive force P_0 and the perimeter of the hole with a normal compressive force, P_1, the solution is obtained by multiplying the stresses σ' and τ' by P_0 and the stresses σ'' and τ'' by P_1, and then superimposing the results.

It is to be noted that the stresses σ' and τ' are due to a unit compression on the entire external boundary $\alpha = 0$. This includes both the plane surface of the foundation and the infinite boundaries. In many practical applications it is desirable to remove all or part of the load on the infinite vertical boundaries. For this purpose the theoretical solution by Mr. Jeffery [11] (mentioned parenthetically by Mr. Mindlin in the text following Equation (36)), is applicable. However, the author has raised a doubt about the strict correctness of Mr. Jeffery's application of his general bipolar solution to this case; therefore the writer has conducted a series of photoelastic experiments which will serve both a practical purpose and as an experimental check on Mr. Jeffery's theoretical results.

Photoelastic Study.—In the photoelastic study of Mr. Jeffery's problem, of course, it was impossible to use a bakelite sheet of infinite dimensions. It is believed that the results obtained by using a plate of finite dimensions give valid results when the diameter of the hole and the distance from the free edge are small in comparison with the over-all dimensions of the model, in accordance with Saint-Venant's principle.

The model was constructed in rectangular form of bakelite, 0.5 in. thick, with vertical boundaries 2.75 in. wide by 4.62 in. long. A quarter-inch hole was drilled near one free edge of the model, such that $\frac{d}{r}$ was 6 (d is the distance from the center of the hole to the straight boundary and r is the radius of the hole). When the experiment was completed for the case $\frac{d}{r} = 6.0$, the model was reduced in width successively so that the ratio $\frac{d}{r}$ was equal to 5.0, 4.0, 3.0, 2.0, 1.0 and 1.5. The normal load on the lateral boundaries was maintained at a constant intensity of 1 000 lb per sq in.

The photoelastic interferometer [30] was used to measure the individual principal stresses along the shortest straight line connecting the straight edge of the model with the hole. The boundary stresses around the hole and along the free edge were obtained directly from the isochromatic photographs.[29]

[30] "Direct Optical Measurement of Individual Principal Stresses," by J. H. A. Brahtz and John Soehrens, Jun. Am. Soc. C. E., *Journal of Applied Physics*, Vol. 10, No. 4, April, 1939.

For example, the stresses along these lines for $\frac{d}{r} = 1.5$ are shown in Fig. 10, and the isochromatic photographs for $\frac{d}{r} = 1.5$ and 6.0 are shown in Fig. 11. In these the calibration of principal stresses along a free boundary is 0.173 p per

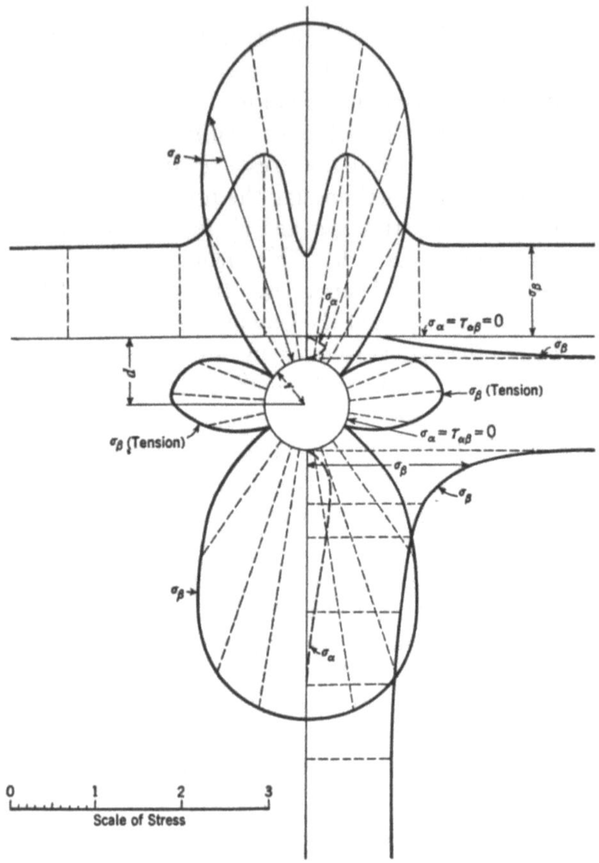

FIG. 10.—UNIT NORMAL COMPRESSION ON VERTICAL BOUNDARIES AT INFINITY; $\frac{d}{r} = 1.5$ (ALL STRESSES COMPRESSION EXCEPT AS NOTED)

fringe, in which p is the intensity of loading on the vertical boundaries and is assumed equal to unity. The probable error in the experimental stresses is ± 5 per cent. The calibration for maximum shears, of course, would be 0.0865 p. Note that the stresses along the straight boundary are compression throughout.

The stresses plotted as in Fig. 10 for unit end loading of the vertical infinite boundaries, in connection with the stresses plotted as in Fig. 9 and combined with the author's results, permit of the complete stress determination for various boundary loading and body forces.

The theoretical results of the paper may be used with sufficient accuracy for civil engineering purposes to estimate the stress conditions around vertical shafts and inspection galleries near the faces of the dam, and pressure pipes running through the dam. These problems have been treated elsewhere [31, 32] under the assumption that the openings were far removed from the faces, whereas the present theory permits the opening to come as close to the faces as practical considerations will permit. In this application of the theory, it is customary first to compute the principal stresses at the center of the opening, as if the opening did not exist, and then apply these principal stresses as normal boundary loading far removed from the opening and compute the stresses in the vicinity of the opening by the present theory.

(a) $\frac{d}{r} = 1.5$ (b) $\frac{d}{r} = 6.0$

FIG. 11.—ISOCHROMATIC PHOTOGRAPHS; MAXIMUM SHEAR CALIBRATION 0.0865 p PER FRINGE; VERTICAL LOAD

It is hoped that the author, in his closure, will comment further on the apparent discrepancies in the theoretical and photoelastic results for a plate with an eccentric circular opening and under a uniform traction. This problem is important as it occurs in many forms in practical structures.

Finally, it should be noted that the author's bipolar transformation can be used also for the determination of the liquid pore pressures which may exist in the material surrounding the tunnel and the seepage into the tunnel. The

[31] "Introduction to the Stresses in the Wall Rock of Tunnels," by Frederick C. Carstarphen, M. Am. Soc. C. E., *The Mines Magazine* (Colorado School of Mines, Golden, Colo.), May, 1939.

[32] "Stresses Around Circular Holes in Dams and Buttresses," by I. K. Silverman, Assoc. M. Am. Soc. C. E., *Transactions*, Am. Soc. C. E., Vol. 103 (1938), p. 133.

theory and formulas for these effects may be found in the memorandum [29] as well as the details of the various theoretical and photoelastic studies mentioned and exemplified by the writer.

Acknowledgments.—The writer wishes to express his appreciation to the staff of the photoelastic laboratory and others of the Bureau of Reclamation, Denver, Colo., for their valuable assistance in the preparation of this discussion. Mr. John E. Soehrens, Associate Engineer, Mr. H. B. Phillips, Junior Engineer, and Mr. H. J. Kahm, Engineer Aid, performed the photoelastic experiments. Mr. C. N. Zangar, Assistant Engineer, and Mr. J. R. Bruggeman, were responsible for the mathematical analyses.

RAYMOND D. MINDLIN,[33] ASSOC. M. AM. SOC. C. E. (by letter).—An analysis of the stresses around a horizontal tunnel of circular cross-section, situated in a gravitating elastic medium, is presented in the paper. It shows how the stresses are affected by (1) the proximity of the free upper surface of the medium, (2) a certain range of assumed initial stresses in the medium, and (3) variation of Poisson's ratio.

A result predicted from St. Venant's principle was verified—that is, when the tunnel is far from the surface the stresses may be calculated by considering the tunnel situated in a uniform stress field composed of the principal stresses initially present at the tunnel level. However, even in this comparatively simple case, there are available from the mathematical theory of elasticity only the solutions for openings of circular or elliptic section. Professor Richmond has shown how a very simple photo-elastic experiment, combined with a judicious use of elasticity theory, will solve the problem for any shape as long as the opening is far from the upper surface. In his application to the square section with circular fillets at the corners, Professor Richmond finds that there is an optimum fillet radius which produces a minimum stress concentration factor. This is a result of some importance which might be applied usefully in other fields of stress analysis and design.

Professor Richmond has listed correctly the limitations inherent in his suggestion. One of these in particular, which was also pointed out by Messrs. Brahtz and Feld, is of such importance in both the photo-elastic and mathematical treatments of the tunnel problem that it warrants another repetition. It is that the stresses in the vicinity of the tunnel cannot be predicted, even approximately, without advance information as to the state of initial stress in the medium. This was stated in the original paper when comparing the widely different results obtained for Cases I, II, and III.

Mr. Feld mentions the question of the effect of Poisson's ratio on the stress distribution. As stated in the paper, the reason for the great change in the stresses of Case II when Poisson's ratio is altered is that, in this case, the boundary conditions are functions of Poisson's ratio.

The effect on the stresses of bracing or lining in the tunnel, mentioned by Mr. Feld, can be calculated by the methods of elasticity theory. The effect of grouting pressure can be estimated from Jeffery's solution for the hole under uniform pressure in the semi-infinite plate. The results obtained from this

[33] Instr., Dept. of Civ. Eng., Columbia Univ., New York, N. Y.

solution are indicated in the first part of the discussion by Mr. Brahtz and are treated in greater detail in the source to which Mr. Brahtz refers.[29]

Mr. Brahtz reveals a very interesting point in his discussion of the problem of the semi-infinite plate with a circular hole near the straight boundary and under uniform stress parallel to the straight boundary. The point in question is whether or not there is a reversal of sign in the stress along the straight edge when the hole is close to this boundary. According to Jeffery's mathematical solution of the problem, there should be reversal of sign for $\frac{d}{r} < 2$. Mr. Brahtz made a photo-elastic experiment with $\frac{d}{r} = 1.5$ and found no reversal of sign, but he noted that the writer had called attention in his paper to an error in Jeffery's solution which prevents comparison with the experimental results.

The writer has corrected Jeffery's solution and finds that Mr. Brahtz's results are partly verified. The corrected solution indicates that there is no reversal of sign along the straight boundary. However, there are some discrepancies between the magnitudes of the stresses obtained in Mr. Brahtz's experiments and those found mathematically. These discrepancies result in percentage errors higher than the 5% estimated by Mr. Brahtz, especially in the regions of low stress. This is to be expected, since a small error in the measurement of phase retardation, while producing a negligible percentage error in regions of high stress, will produce a very large percentage error in regions of low stress.

The writer has calculated the corrected values for Jeffery's problem and has verified them photo-elastically for values of $\frac{d}{r}$ down to 1.1. The details of the results will be presented in another paper.

Reprinted from
Proceedings of the American Society of Civil Engineers
65 (1939) 619-642
reprinted in
Transactions American Society of Civil Engineers
105 (1940) 1117-1140

Reprinted from Journal of Applied Physics, Vol. 11, No. 11, 724–732, November, 1940

Stress Analysis by Three-Dimensional Photoelastic Methods*

Daniel C. Drucker and Raymond D. Mindlin
Department of Civil Engineering, Columbia University, New York, New York
(Received June 13, 1940)

This paper describes the results of an investigation and extension of the various proposed procedures and methods of analysis for the photoelastic determination of three-dimensional states of stress. Two important limitations of the previous developments are removed. A true extension to three dimensions is given by the determination of the effect on retardation of appreciable variation in the orientation of the secondary principal stresses in planes perpendicular to the wave normal (such variation being the rule rather than the exception for general states of stress). Also, methods for the analysis of whole planes are presented, thus avoiding the cumbersome and tedious point by point procedures that have been advanced. In the particular case of plane stress, these methods reduce to a purely optical technique for determining the principal stresses themselves, or more easily their sum, for the entire model.

Introduction

Two general practical methods have been proposed, and to some extent used, for the photoelastic determination of three-dimensional states of stress. In one, the stresses are "frozen" in the model and slices cut from it are analyzed; in the other, the phenomenon of scattering of light is employed. The technique of measurement required is different for the two cases and therefore they will be discussed separately. However, both methods depend upon the effects of stress on the propagation of light, and these are the same regardless of the method of observation. In the previous developments of the theory, only those effects were considered which are present in ordinary two-dimensional states of stress.

These effects in three-dimensional states of stress differ considerably from those present in the usual two-dimensional case. The rotation along the wave normal of the axes of secondary principal stress[1] in planes perpendicular to the wave normal is the principal cause of the difference and has not been considered previously.

Knowledge of what occurs in general states of stress permits broadening of the methods of observation to include entire planes rather than confining attention to a single point at a time.

Weller's Method

If one wishes to analyze a three-dimensional state of stress in a model under actual load, the only practical method of general application available at present is that proposed by Weller.[2] This method is based upon the fact that in a transparent medium, light scattered at right angles to the incident wave normal is plane polarized, the amplitude of the vibration reaching the eye of an observer being proportional to the projection of the light vector on the plane perpendicular to the direction of observation. The result is similar to that produced by the analyzer in a polariscope. (See Fig. 1.) If the light incident at a point is plane polarized, the intensity of scattered light observed at right angles to the wave normal will thus vary from zero when looking along the direction of vibration to a maximum when looking perpendicularly to this direction.

When photoelastic materials are stressed they become doubly refracting and, for a given wave normal, light can be transmitted only in vibrations along two mutually perpendicular axes lying in the planes perpendicular to the wave normal. These directions, which we may designate as the axes of secondary principal stress[1] are determined in the same manner as the principal directions for plane stress. (The normal and shearing stresses acting on the planes perpendicular to the wave normal need not be considered in measuring or calculating secondary principal stresses as the former may be varied

* Publication assisted by the Ernest Kempton Adams Fund for Physical Research of Columbia University.
[1] R. D. Mindlin, "A review of the photoelastic method of stress analysis," J. App. Phys. **10**, 229 (1939).

[2] R. Weller, "A new method for photoelasticity in three dimensions," J. App. Phys. **10**, 266 (1939). "Three-dimensional photoelastic analysis by scattered light," Papers presented at the 9th meeting of the Eastern Photoelasticity Conference, May 1939, p. 19.

at will without changing the latter, even though such variation changes the magnitudes and directions of the true principal stresses.) The light scattered from any point (O, Fig. 1) can therefore be considered as coming from two plane polarized vibrations (along p and q) differing in phase by an amount depending upon the integrated optical properties of the material along the path from the point of entrance (O') into the model to the point at which the scattering occurs. In general the two vibrations combine to form an elliptical vibration, the plane of the ellipse being perpendicular to the wave normal. The intensity of the light scattered at right angles to the wave normal is proportional to the square of the length of the projection (e.g., QQ or TT) of this ellipse on a plane perpendicular to the direction of observation.

If a whole plane of the model is illuminated by a sheet of parallel rays of polarized light formed by passage through a polarizer, collimator, and a slit, the light scattered in a given direction perpendicular to the wave normal will in general vary in intensity from point to point in the plane depending upon the projections of the ellipse of vibration associated with each point in the plane, the alternate lighter and darker regions forming a fringe pattern. This interference pattern will vary in distinctness and even in form as the direction of observation is changed, but in general it may be said that the distance between fringes in the direction of travel of the light is a function of the state of stress between the points of minimum intensity.

Weller's method for a general state of stress is to fix attention on any point desired and rotate the model until the minimum distance between fringes straddling the point is obtained. If the stress field is perfectly uniform in the vicinity, this minimum distance gives the maximum principal stress difference at the point according to the simple relation[3]

$$\sigma_3-\sigma_1=\frac{C}{d}, \tag{1}$$

where σ_3 and σ_1 are the algebraically largest and least principal stresses, respectively, C is the stress optical coefficient in stress units per unit thickness per fringe, and d is the distance between fringes measured along the wave normal, i.e., along the direction of the intermediate principal stress σ_2. The formula and reasoning are exactly the same as in two-dimensional work.

If σ_3 and σ_1 vary in magnitude along the wave normal, but remain fixed in direction, the stress

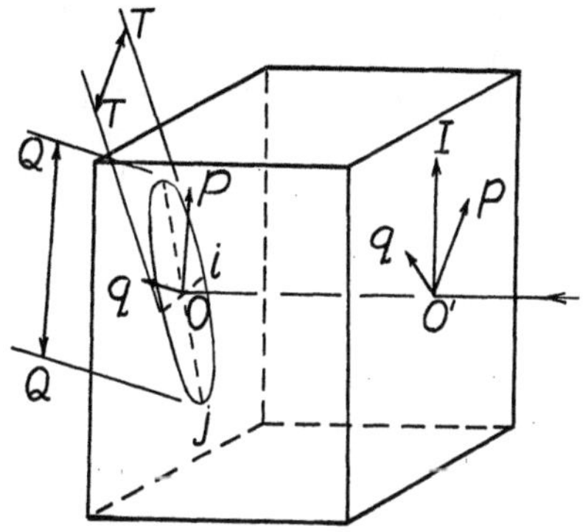

FIG. 1. Scattering of light. The entering vibration OI' breaks up into two components, one along p, the other along q. At any point O in the model the two components in general will combine to form an elliptic vibration with axes Oi and Oj in the plane perpendicular to the wave normal $O'O$. The intensity of scattered light seen by an observer is proportional to the square of the projection of the ellipse (e.g., QQ or TT) on a plane perpendicular to the direction of observation.

found from Eq. (1) will exist at some point intermediate between the two fringes. For small gradients of stress magnitude, the midpoint is close enough. If greater accuracy should be desired, plotting the fringe positions against their order will determine the point more closely.

If the orientations of the principal stresses do not vary, there is no theoretical objection to Weller's proposal. However, as will be shown in the following section, if there is rotation along the wave normal of the axes of primary or secondary principal stress in planes perpendicular to the wave normal, the distance between fringes is no longer dependent solely on the magnitudes of the stresses and Eq. (1) or its equivalent for secondary principal stress is no longer valid.

[3] R. Weller and J. K. Bussey, "Photoelastic analysis of three-dimensional stress systems using scattered light," Tech. Note No. 737 N. A. C. A.

Effect of Rotation of Axes on Retardation

As we are interested in determining the effect of rotation of the axes of primary or secondary principal stress, it is desirable to strip the problem of all stress variables except orientation. The fundamental case to be considered is thus one in which the magnitudes of the stresses remain fixed while their orientations, in planes perpendicular to the wave normal, change at a constant rate along the wave normal. In any case we have to solve a problem of wave propagation in a nonhomogeneous, nonisotropic medium. The mathematical formulation of such problems generally leads to a system of simultaneous, linear, partial differential equations of the second order with variable coefficients.

The differential equations may be formulated in terms of an elastic ether theory of light. From a general point of view, this theory is not the most satisfactory one available; but it is very convenient for photoelasticity as it is based upon the ideas of stress, strain, and displacement, and upon the equations of motion with which workers in elasticity and photoelasticity are familiar. The resulting differential equations are exactly the same as those obtained from the electromagnetic theory of light if the symbols representing displacement, density, and shear modulus are interpreted as magnetic force, magnetic permeability, and reciprocal of dielectric coefficient, respectively.

The equations of motion in terms of the stresses are

$$\left.\begin{aligned}\frac{\partial\sigma_x'}{\partial x}+\frac{\partial\tau_{xy}'}{\partial y}+\frac{\partial\tau_{xz}'}{\partial z}&=\rho\frac{\partial^2u'}{\partial t^2}\\ \frac{\partial\tau_{xy}'}{\partial x}+\frac{\partial\sigma_y'}{\partial y}+\frac{\partial\tau_{yz}'}{\partial z}&=\rho\frac{\partial^2v'}{\partial t^2}\\ \frac{\partial\tau_{xz}'}{\partial x}+\frac{\partial\tau_{yz}'}{\partial y}+\frac{\partial\sigma_z'}{\partial z}&=\rho\frac{\partial^2w'}{\partial t^2}\end{aligned}\right\}, \tag{2}$$

where σ_x', σ_y', σ_z', τ_{xy}', τ_{yz}', τ_{xz}' are the ether stresses, u', v', w' are the ether displacements, and ρ is the density.

The ether stresses may be expressed in terms of the ether displacements and the elastic coefficients of the ether by three equations of the type[4]

$$\sigma_x'=-2G_3\frac{\partial v'}{\partial y}-2G_2\frac{\partial w'}{\partial z}+2E_1\left(\frac{\partial v'}{\partial z}+\frac{\partial w'}{\partial y}\right) \tag{3}$$

and three of the type

$$\tau_{yz}'=2E_1\frac{\partial u'}{\partial x}+G_1\left(\frac{\partial v'}{\partial z}+\frac{\partial w'}{\partial y}\right)-E_3\left(\frac{\partial w'}{\partial x}+\frac{\partial u'}{\partial z}\right)-E_2\left(\frac{\partial u'}{\partial y}+\frac{\partial v'}{\partial x}\right). \tag{4}$$

The elastic coefficients (the E_i and G_i) are variables for a nonhomogeneous medium. They are related to the principal wave velocities a, b, c by equations of the type

$$a^2=\frac{G_1'}{\rho}, \tag{5}$$

in which G_1' is a stationary value of the G_i.

Finally, the principal velocities a, b, c are determined from the stress-optical laws[5] by three equations of the type

$$a^2=v_0^2+\alpha(\sigma_2+\sigma_3)+\beta\sigma_1, \tag{6}$$

in which v_0 is the velocity in the unstressed medium and α and β are stress-optical coefficients, constant for a given material and wavelength, λ, of light.

Considering a Cartesian coordinate system x, y, z with the origin at entrance to the medium, the simplest state of stress that will satisfy the conditions of the present problem is a simple tension, $\sigma=\sigma_3$, in the yz-plane inclined to the z axis at a variable angle $\varphi=(A/2)x$, where A is a constant. If the wave normal is in the direction of the x axis, $\Delta=(2\pi\sigma/C)x=2kx$ is the relative retardation in radians caused by the magnitude of σ alone. The ratio of rotation to retardation is φ/Δ a constant which we shall call $R/2$.

Using these specifications for the state of stress, and combining Eqs. (2), (3), (4), (5), and (6) we find the following simultaneous differential equations for v' and w':

---

[4] M. G. Lamé, *Leçons sur la Théorie Mathématique de l'Élasticité* (Paris, 1866), p. 231.

[5] E. G. Coker and L. N. G. Filon, *Treatise on Photoelasticity* (Cambridge University Press, 1931), p. 198.

$$\left.\begin{aligned}\frac{\partial^2 v'}{\partial t^2}&=H\frac{\partial^2 v'}{\partial x^2}-B\cos Ax\frac{\partial^2 v'}{\partial x^2}+BA\sin Ax\frac{\partial v'}{\partial x}\\&\quad+B\sin Ax\frac{\partial^2 w'}{\partial x^2}+BA\cos Ax\frac{\partial w'}{\partial x}\\\frac{\partial^2 w'}{\partial t^2}&=H\frac{\partial^2 w'}{\partial x^2}+B\cos Ax\frac{\partial^2 w'}{\partial x^2}-BA\sin Ax\frac{\partial w'}{\partial x}\\&\quad+B\sin Ax\frac{\partial^2 v'}{\partial x^2}+BA\cos Ax\frac{\partial v'}{\partial x}\end{aligned}\right\}\quad(7)$$

in which

$$B=\left(\frac{\alpha-\beta}{2}\right)\sigma,\quad H=v_0^2+\left(\frac{\alpha+\beta}{2}\right)\sigma.$$

Equations (7) may be reduced to equations with constant coefficients by referring the dependent variables to the axes of principal stress. Calling the displacements along the 2 and 3 axes (perpendicular and parallel to the direction of σ) v and w, we find

$$\left.\begin{aligned}F\frac{\partial^2 v}{\partial x^2}-\frac{A^2D}{4}v+HA\frac{\partial w}{\partial x}=\frac{\partial^2 v}{\partial t^2}\\D\frac{\partial^2 w}{\partial x^2}-\frac{A^2F}{4}w-HA\frac{\partial v}{\partial x}=\frac{\partial^2 w}{\partial t^2}\end{aligned}\right\}\quad(8)$$

in which

$$F=v_0^2+\beta\sigma=H-B;\quad D=v_0^2+\alpha\sigma=H+B.$$

The independent variables may be separated in the usual manner by setting

$$\left.\begin{aligned}v&=V(x)e^{\pm i\omega t}\\w&=W(x)e^{\pm i\omega t}\end{aligned}\right\}\quad(9)$$

and, finally, the dependent variables V and W may be separated by differentiation and addition of the two resulting equations. This results in the same fourth-order linear differential equation with constant coefficients for both V and W:

$$DF\frac{\partial^4 V}{\partial x^4}+\left[(D+F)\omega^2-\frac{A^2}{4}(D^2+F^2)+H^2A^2\right]\frac{\partial^2 V}{\partial x^2}+\left[\omega^4-\frac{A^2\omega^2}{4}(D+F)+\frac{A^4}{16}DF\right]=0.\quad(10)$$

The exact solution of this equation is so cumbersome that it is quite unusable. It can however be simplified by taking into account the orders of magnitude of the quantities involved. If the product $2k\lambda$ is much less than unity, most of the terms in the solution disappear. Even in highly stressed Bakelite, $2k\lambda$ will be less than 0.01 so that the simplification is justified, especially so as the square of the term appears and not the term itself.

Considering only the forward traveling waves and taking $v=M\cos\omega t$ at $x=0$, we then find

$$v=[K\sin kSx+M\cos kSx]\cos\left(\omega t-\frac{\omega}{\sqrt{H}}x\right)+(L\sin kSx)\sin\left(\omega t-\frac{\omega}{\sqrt{H}}x\right),\quad(11)$$

where $S=(1+R^2)^{\frac{1}{2}}$ and K and L are constants. Taking $w=N\cos\omega t$ at $x=0$, the same equation will hold for w except for constants. Thus there are four constants to be determined. They can be obtained in terms of M and N by substituting in Eqs. (8). When this is done, apparently eight relations exist among the four constants. Making the assumption above as to the magnitude of the retardation and also a similar one for the magnitude of $A=2\varphi/x$ and also for $S=(1+R^2)^{\frac{1}{2}}$, the eight relations reduce to four which determine the constants uniquely. Even if a rotation of axes of 360° takes place within a distance as small as a millimeter, and if $2k$ is as small as one-tenth of a fringe per inch, λA and λS are small compared to unity and can be neglected, so that the assumptions are entirely justified.

v and w can then be expressed in terms of their initial amplitudes M and N as follows:

$$\left.\begin{aligned}v&=\left[-\frac{R}{S}N\sin kSx+M\cos kSx\right]\times\cos\left(\omega t-\frac{\omega}{\sqrt{H}}x\right)-\left(\frac{M}{S}\sin kSx\right)\sin\left(\omega t-\frac{\omega}{\sqrt{H}}x\right)\\w&=\left[+\frac{R}{S}M\sin kSx+N\cos kSx\right]\times\cos\left(\omega t-\frac{\omega}{\sqrt{H}}x\right)+\left(\frac{N}{S}\sin kSx\right)\sin\left(\omega t-\frac{\omega}{\sqrt{H}}x\right)\end{aligned}\right\}\quad(12)$$

assuming them to be in phase on entrance. If φ is set equal to zero these reduce to the expressions which would be obtained for simple tension along z. If v and w have an arbitrary phase relation on entrance, it is only necessary to add a phase constant ϵ to the terms multiplied by M and another phase constant ϵ' to the terms multiplied by N. For example, if circularly polarized light enters the model, $\pi/2$ is added to the phase of either the M or N terms depending upon whether the polarization is right or left, and M is then set equal to N. The term $(\omega/\sqrt{H})x$ appears in all of the phase angles with the same sign and therefore has no influence on the relative retardation.

From Eqs. (12) for the components of vibration, or more clearly from the combined form

$$\left.\begin{aligned} v=&-\frac{R}{S}N\sin kSx\cos\left(\omega t-\frac{\omega}{\sqrt{H}}x\right)\\ &+\left[M\left(1-\frac{R^2}{S^2}\sin^2 kSx\right)^{\frac{1}{2}}\right]\\ &\times\cos\left[\omega t-\frac{\omega}{\sqrt{H}}x+\arctan\left(\frac{\tan kSx}{S}\right)\right]\\ w=&+\frac{R}{S}M\sin kSx\cos\left(\omega t-\frac{\omega}{\sqrt{H}}x\right)\\ &+\left[N\left(1-\frac{R^2}{S^2}\sin^2 kSx\right)^{\frac{1}{2}}\right]\\ &\times\cos\left[\omega t-\frac{\omega}{\sqrt{H}}x-\arctan\left(\frac{\tan kSx}{S}\right)\right]\end{aligned}\right\}\quad(13)$$

it may be seen that, wherever $\tan kSx=0$, v and w are in phase. This requires $kSx=n\pi$, or $x=n\pi/kS$. The distance between points in phase is thus $d=\pi/kS$. Now π/k is the spacing one would find if only the magnitude of the stresses affected the retardation, so that the factor $1/S$ is the effect of rotation. Hence the distance between points of equal phase is decreased, or conversely the retardation is increased by rotation. Intensity expressions for either plane or circularly polarized light on entrance can be set up. From these it is found that minima of intensity occur at intervals of π/kS for observation in any direction fixed in space or for a direction rotating with the axes of principal stress.

The same results were obtained by a quite different method of approach which is of value because of its closer physical connection with the problem. A series of parallel identical crystal plates (corresponding to constant magnitude of principal stresses) each with its polarizing axes inclined at a constant angle to those of the previous plate (uniform rotation of the axes of principal stress) was considered. If the number of plates is made infinite while the total relative retardation due to the magnitude of the stresses and the total rotation remain finite, the conditions of the problem are satisfied.

There are two essential parts to the solution. First, general expressions are obtained for the vibrations along each of the two crystal axes in any plate by a deductive-inductive process; second, the limits of these expressions, in usable form, are found as the number of plates becomes infinite. Both of these steps are rather long and involved. A doubly infinite series must be set up and then summed. Series of Bessel functions of the zero and first order appear, which reduce eventually to the simple trigonometric terms of Eq. (12).

It is of interest to note that the justifiable assumptions as to the magnitudes of the rotation and retardation found necessary in the previous method do not appear explicitly in this one though obviously they must be contained in it.

From Eq. (12) we see that if the entering vibration coincides with an initial direction of principal stress, say the p direction, then the vibration along q at any point is given by

$$w=\left(\frac{R}{S}M\sin kSx\right)\cos\left(\omega t-\frac{\omega}{\sqrt{H}}x\right).\quad(14)$$

If the ratio, $R/2$, of the rotation to the retardation (due to the magnitude of stress) is small, even though the magnitude of the rotation itself be large, the intensity along q will be very small, the light will be approximately plane polarized at all points. The light vector may be said to follow along the direction of principal stress, i.e., practically no light will be scattered in this direction. The eye of an observer moving along a helix (the intersection of a line along the

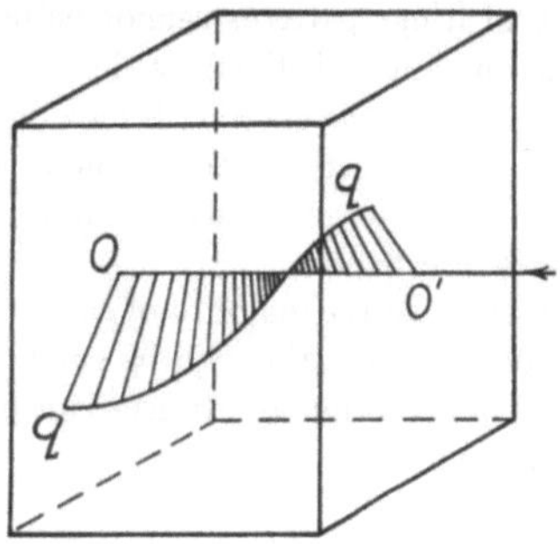

FIG. 2. The light vector follows the axis of principal stress. $O'q$—entering direction of vibration of light coincides with a principal axis of stress.

principal axis of stress with a cylinder having the points observed as its axis) will perceive a continuous dark band. (See Fig. 2.)

Equation (14) shows that even if the ratio of rotation to retardation is high, the intensity along q is zero wherever the vibration is plane polarized. There is a following around of the light vector, but between the points of plane polarization the intensity along q may be high. Fringes will therefore appear when there is rotation even though the entering vibration coincides with an initial direction of principal stress, while if there is no rotation, uniform intensity will prevail no matter what the direction of observation. Furthermore, the spacing of the fringes will be the same as the spacing when the entering light has components along both p and q.

The distance between fringes is decreased due to the rotation of the axes of principal stress. As can be seen from Eqs. (12) and (13), plane polarization or circular polarization will reoccur at intervals given by:

$$d=\frac{\pi}{k}\cdot\frac{1}{S}. \qquad (15)$$

The retardation is thus seen to be increased by the factor $S=(1+R^2)^{\frac{1}{2}}$ where $R=2\varphi/\Delta$, or the distance between fringes decreased by $1/S$. The difference between the secondary principal stresses in the general case is thus

$$\sigma_p-\sigma_q=\frac{C}{d}\cdot\frac{1}{S}, \qquad (16)$$

which is to be compared with Eq. (1). As S can be quite large, neglect of it may, in many cases, produce appreciable error, though on the "safe" side as the apparent stress will be higher than the true. If the total rotation is small, the formulas show that in practical cases it may be neglected as far as its effect on the distance between fringes is concerned. Usually one is interested in regions of high stress where the fringes will be closely spaced. In such cases, the total rotation probably will not be very large and the ratio of rotation to retardation probably small, though often appreciable in effect.

While true extinction will occur only if the eye follows the directions of principal stress in a helix, it is nevertheless true that for a fixed direction of observation, minima will occur at the same spacing as given in (15) and although the fringe pattern will not be as distinct it will yield the same results. The amount of rotation between two fringes may be obtained by finding the difference in the orientations for which each appears darkest.

Analysis of General States of Stress

Having the interpretation of fringe spacing in terms of stress magnitude and rotation of axes, it is possible to consider the procedures proposed for making the necessary optical measurements. Weller's suggestion of universal rotation of the model while under load would require a very complicated loading device. Also, the stress at only one point at a time would be determined. The suggestion is therefore advanced that whole planes be investigated.

The state of stress at a point is specified by six quantities, three normal stresses and three shearing stresses across the planes on which the normal stresses act. As the effect of an isotropic stress is only to change the initial index of refraction, it cannot be detected by a photoelastic pattern and therefore only the differences between the normal stresses plus the three shearing stresses, or the difference between the principal stresses and the planes on which they act, can be found. Without *a priori* knowledge as to the state of stress, five independent measurements must be made to determine the unknown quantities.

Let a plane be chosen. Consider some line in the plane as a reference direction. If a narrow beam of circularly polarized light is passed in the plane so that the wave normal is first in the direction of the reference line, then at one angle to it and next at another, and photographs are taken of the light scattered at right angles to the wave normals, three independent measurements will have been made for each point in the plane. Call the reference line the x axis, the plane the xy plane. The fringe spacings obtained for three positions of the wave normal do not completely determine the state of stress. Two more independent measurements are required to get $\sigma_x-\sigma_y$, $\sigma_y-\sigma_z$, τ_{xy}, τ_{yz}, τ_{xz}. Two positions of the wave normal in the xz plane other than along x will supply the additional information required. If, however, the orientations of the secondary principal stresses can be determined or are known, one need work only in the xy plane, the three fringe spacings and the orientations are sufficient.

The general problem of finding the orientation of the secondary or primary principal stresses is rather complicated. If the rotation ratio is small at all points, then plane polarized light entering along a direction of principal stress will indicate the directions of principal stress all along the wave normal. Only when the entering light is vibrating in such a direction will a continuous dark band appear. If the ratio is large, no such reliable way is available. A continuous dark band will not exist, even the entering direction of the principal stresses cannot be determined. Distinctness of fringe patterns cannot be used as a criterion; minimum distinctness does not necessarily occur when the observer is looking along a principal stress direction, and in general it would be difficult to determine such a minimum with any degree of exactness.

Fortunately, in the majority of cases the ratio of rotation to retardation will be small and the directions of the principal stresses can probably be determined.

Analysis of Plane Stress

In plane stress, the magnitude and direction of a principal stress is known at every point, ($\sigma_z=0$ acts perpendicular to the plane of the model). There are thus only three quantities to be determined, the principal stresses in the plane, σ_1, σ_2, and their orientation.

Consider rectangular axes x, y in the plate (Fig. 3). σ_y, the normal stress in the y direction, is determined from the fringe spacing when light is passed in the x direction, σ_ξ and σ_η when the wave normal is at two other known orientations with respect to the x axis. Taking these positions symmetrically about x, at some angle γ to it, and calling the distance between fringes obtained along these directions d_x, d_ξ, d_η, we have $\sigma_y=C/d_x$, $\sigma_\xi=C/d_\xi$, $\sigma_\eta=C/d_\eta$. The three normal stresses determine the principal stresses and their orientation in the plane. If γ is taken as 60°, the formulas become rather simple. They are entirely analogous to those developed for the equiangular strain rosette.[6]

$$\sigma_1=\frac{C}{3}\left(\frac{1}{d_x}+\frac{1}{d_\xi}+\frac{1}{d_\eta}\right)+\frac{\sqrt{2}}{3}C\left[\left(\frac{1}{d_x}-\frac{1}{d_\xi}\right)^2+\left(\frac{1}{d_\xi}-\frac{1}{d_\eta}\right)^2+\left(\frac{1}{d_\eta}-\frac{1}{d_x}\right)^2\right]^{\frac{1}{2}},$$

$$\sigma_2=\frac{C}{3}\left(\frac{1}{d_x}+\frac{1}{d_\xi}+\frac{1}{d_\eta}\right)-\frac{\sqrt{2}}{3}C\left[\left(\frac{1}{d_x}-\frac{1}{d_\xi}\right)^2+\left(\frac{1}{d_\xi}-\frac{1}{d_\eta}\right)^2+\left(\frac{1}{d_\eta}-\frac{1}{d_x}\right)^2\right]^{\frac{1}{2}},$$

$$\tan 2\theta=\sqrt{3}\left(\frac{1}{d_\xi}-\frac{1}{d_\eta}\right)\Big/\left(\frac{2}{d_x}-\frac{1}{d_\xi}-\frac{1}{d_\eta}\right), \tag{17}$$

$$\sigma_1-\sigma_2=\frac{2\sqrt{2}}{3}C\left[\left(\frac{1}{d_x}-\frac{1}{d_\xi}\right)^2+\left(\frac{1}{d_\xi}-\frac{1}{d_\eta}\right)^2+\left(\frac{1}{d_\eta}-\frac{1}{d_x}\right)^2\right]^{\frac{1}{2}},$$

$$\sigma_1+\sigma_2=\frac{2}{3}C\left(\frac{1}{d_x}+\frac{1}{d_\xi}+\frac{1}{d_\eta}\right).$$

[6] R. D. Mindlin, "The equiangular strain-rosette," Civ. Eng. 8, 546 (1938).

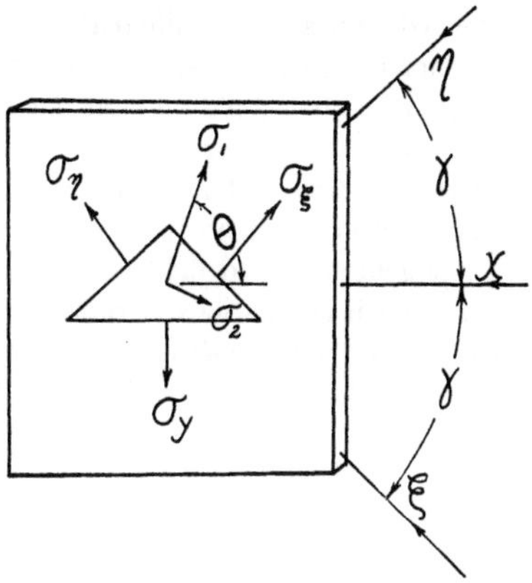

FIG. 3. Analysis of plane stress. Fringe spacings obtained from three positions of the wave normal in the plane determine the principal stresses and their orientation.

Similar formulas result for $\gamma=45°$.

As $\sigma_x+\sigma_y=\sigma_1+\sigma_2$, two measurements at right angles are sufficient to determine the sum of the principal stresses. It will be more accurate for two-dimensional work to take an isochromatic picture in the usual manner and then find the sum of the principal stresses as above, using the additional result as a check on the accuracy of measurement. For the principal stresses themselves, or for their sum and difference, care will have to be taken with the sign of the distance between fringes. If called positive when due to tension, it is negative when due to compression. The sign will often be apparent, if it is not, white light may be employed, the color sequence will show whether the stress is of the same or opposite sign as the stress at the previous point, and following back to a known point, as for example one on the boundary, the sign can be determined. Due to dispersion, it may be necessary to reduce the load for this determination. The isoclinics can also be used for determination of the sign. In the case of plane stress, the variation of stress through the thickness is small and a wide beam of light can be employed. The wider beam will give a much greater intensity of scattered light and may therefore be desirable.

It should be noted that a sheet of photoelastic material may appear free from initial birefringence when observed in a circular polariscope and yet have a fairly high equivalent stress perpendicular to the plane, which may make it unsuitable for this proposed method.

A two-dimensional state of stress also exists at an unloaded plane boundary. An isochromatic picture can be obtained to a certain extent but it would have little accuracy. The principal stresses and their orientation can be found by the method and formulas, (17), given for plane stress.

A plane of symmetry or a principal plane in a three-dimensional case is very similar to the plane stress case. The quantities found from the three measurements at 60° will be the differences between the principal stresses and their orientations, which is all that one can expect to obtain. Calling the normal stress on the plane σ_z, just substitute $\sigma_1-\sigma_z$ for σ_1 and $\sigma_2-\sigma_z$ for σ_2 in (17).

THE FREEZING METHOD

The preceding development has all been applied specifically to the determination of the state of stress from the patterns produced by the light scattered from a plane. It applies also with but few modifications to the "freezing" method. In this method[7,8] the model is heated, loaded, then cooled under load. Most of the birefringence present when the model is under load at the elevated temperature remains after cooling, and it is not disturbed by careful sawing. Thin slices are cut from the model and analyzed. When the slice is placed in a polariscope, an integrated effect through the thickness is obtained. If the stress in the neighborhood of a point is constant, the fringe order appearing there when the slice is in a circular polariscope is

$$n=\frac{(\sigma_p-\sigma_q)t}{C}, \tag{18}$$

where, as before, σ_p and σ_q are the secondary principal stresses in the plane perpendicular to the wave normal (in general not the plane of the slice), t is the thickness of the slice in the direction of the wave normal, and C is the stress-optical coefficient. The difference between the secondary

[7] G. Oppel, "Polarisationsoptische Untersuchung räumlicher Spannungs- und Dehnungszustände," Forschung auf dem Gebiete des Ingenieurwesens **7**, 240 (1936).

[8] M. Hetényi, "The application of hardening resins in three-dimensional photoelastic studies," J. App. Phys. **10**, 295 (1939).

principal stresses is therefore determined directly from the fringe order, their orientation can be obtained in the same way one obtains the isoclinics for a model under plane stress. If the fringe orders and the orientations are obtained for three different directions of the wave normal, all information possible as to the state of stress is determined. The method is entirely analogous to that indicated in the section on "Analysis of General States of Stress" and applies with possibly even more accuracy because of the greater ease of determination of the directions of the secondary principal stresses. Little difficulty is caused by a small stress gradient provided the directions of σ_p and σ_q remain unchanged along the wave normal.

Rotation of the slice about only one axis (can be taken in the slice) is required, and only three positions need be investigated in this rotation. Immersion in a liquid of approximately the same refractive index will eliminate reflections.

If the p and q axes do rotate, it should be possible to find the amount of this rotation in a plane polariscope if the stress is sufficiently high, and generally one is interested only in the highly stressed portions. As shown in the preceding work, plane polarized light entering along a principal stress direction will rotate with the principal stress axes and emerge practically plane polarized. Independent rotation of the polarizer and analyzer will find the orientations and the amount of rotation. The correction factor to the fringe order to find the stress is, as before, $1/S$. The difference between the secondary principal stresses is given by

$$\sigma_p - \sigma_q = \frac{nC}{t} \cdot \frac{1}{S}. \tag{19}$$

The corrected values for the differences between the secondary principal stresses and the orientations found as above should be used referred to the central plane of the slice.

Hiltscher[9] proposed analyzing the slice cut out from the model in a microscope point by point using convergent polarized light as one does in crystallographic investigations. If rotation of the principal axes through the slice is at all appreciable, the proposal does not appear practicable.

Conclusion

The investigation of the effect of rotation of the principal axes in planes perpendicular to the wave normal led to two important results:

(1) Rotation increases the relative phase retardation by a factor $S=(1+(2\varphi/\Delta)^2)^{\frac{1}{2}}$ which at times may be quite large. If the ratio $2\varphi/\Delta$ is not constant, the ratio of the increment in 2φ to the increment in Δ between fringes can be used in regions of high stress.

(2) At high stress, the light vector rotates with the principal stress directions, plane polarized light entering along a direction of principal stress will remain practically plane polarized in the direction of this stress through the entire model.

The method of working in planes rather than point by point should be of value for general states of stress when a large portion of the model is to be analyzed. In the particular case of plane stress it yields a purely optical method of determining the sum of the principal stresses, or the principal stresses themselves, over the entire model.

---

[9] R. Hiltscher, "Polarisationsoptische Untersuchung des raumlichen Spannungszustandes im konvergenten Licht;" Forschung auf dem Gebiete des Ingenieurwesens **9**, 91 (1938).

An Extension of the Photoelastic Method of Stress Measurement to Plates in Transverse Bending[1]

R. D. Mindlin.[2] The writer would like to discuss the following three aspects of the photoelastic study of plate-bending problems presented in this paper: (1) The exact theory of the propagation of light through the plate; (2) the geometric optics of the polariscope; (3) alternative photoelastic methods which might be used in plate-bending problems.

Propagation of Light Through the Plate

The relation of direct proportionality, connecting relative phase retardation with principal stress difference, is deduced from the laws of propagation of light in a homogeneous crystal. When the crystal is nonhomogeneous along the direction of propagation of light, it is essential to re-examine the optical situation to determine whether or not the elementary relation still applies, either exactly or approximately. The authors did this by means of a preliminary experiment and found an affirmative answer for their special case. A similar investigation performed mathematically is useful because it brings out the order of magnitude of an approximation which is involved. It may be accomplished by a study of the solution of the differential equations governing the propagation of light for the particular case at hand. The general differential equations are three, of the type

$$b^2 \frac{\partial \omega_\eta}{\partial \zeta} - c^2 \frac{\partial \omega_\zeta}{\partial \eta} = \frac{\partial^2 u}{\partial t^2} \qquad [1]$$

(the remaining two being obtained by cyclical permutation of u-v-w, ξ-η-ζ, and a-b-c) in which u, v, w are the components of the light vector referred to the principal axes of optical symmetry (principal axes of stress); ξ, η, ζ are rectangular coordinates, measured along the principal axes; ω_ξ, ω_η, ω_ζ are the components of the curl of the light vector; a, b, c are the principal velocities of light and t is the time. The principal velocities are determined from the state of stress according to the stress-optical laws given by three equations of the type

$$a^2 = V_0^2 + \alpha(\sigma_2 + \sigma_3) + \beta\sigma_1 \qquad [2]$$

in which σ_1, σ_2, σ_3 are the principal stresses, α and β are stress-optical coefficients for frequency ω, and V_0 is the velocity in the unstressed medium.

For the present case, it is sufficient to take $\sigma_1 = -\sigma_3 = \sigma_m z/h$ and $\sigma_2 = 0$, where σ_1 and σ_3 are stresses parallel to, and z the coordinate perpendicular to, the central plane of the plate; $2h$ is the thickness of the plate and σ_m is the maximum stress at the surfaces of the plate. As optical boundary conditions, we take a plane-polarized light wave entering the plate at normal incidence (in the positive z direction) with the plane of polarization bisecting the angle between σ_1 and σ_3, and analyze the resulting waves after they penetrate into the plate to a point located by the coordinate z. Then, omitting details, the solution of Equations [1] lead to the following expression for light intensity I

$$\frac{\pi I}{I_0} = \frac{1}{2}\left\{[J_0(\varphi)]^2 + [Y_0(\varphi)]^2 + [J_0(\psi)]^2 + [Y_0(\psi)]^2\right\}$$
$$+ \frac{[J_0(\varphi)J_0(\psi) - Y_0(\varphi)Y_0(\psi)][J_0(\varphi_{-h})J_0(\psi_{-h}) - Y_0(\varphi_{-h})Y_0(\psi_{-h})]}{[J_0(\psi_{-h})]^2 + [Y_0(\psi_{-h})]^2}$$
$$+ \frac{[J_0(\varphi)Y_0(\psi) + J_0(\psi)Y_0(\varphi)][J_0(\varphi_{-h})Y_0(\psi_{-h}) + J_0(\psi_{-h})Y_0(\varphi_{-h})]}{[J_0(\psi_{-h})]^2 + [Y_0(\psi_{-h})]^2} \qquad [3]$$

in which I_0 is the intensity observed for the unstressed plate, J_0 and Y_0 are the zero-order Bessel functions of the first and second kinds, the arguments being

$$\left.\begin{aligned} \varphi &= \frac{2\omega V_0 h}{(\alpha - \beta)\sigma_m}\sqrt{1 + \frac{(\alpha - \beta)\sigma_m z}{V_0^2 h}} \\ \psi &= \frac{2\omega V_0 h}{(\alpha - \beta)\sigma_m}\sqrt{1 - \frac{(\alpha - \beta)\sigma_m z}{V_0^2 h}} \end{aligned}\right\} \qquad [4]$$

The value of the argument at $z = -h$ is to be taken where the subscript $-h$ appears in Equation [3].

For bakelite and visible light, the order of magnitude of the quantity $2\omega V_0 h/(\alpha - \beta)\sigma_m$ is about 10^{8} so that the zeros of the Bessel functions are for all practical purposes equally spaced over their range of present use. It is therefore permissible to substitute damped trigonometric functions for the Bessel functions, with the result that Equation [3] becomes

$$\frac{\pi I}{I_0} = \frac{1}{\varphi} + \frac{1}{\psi} + \frac{2\psi_{-h}\cos(\varphi + \psi - \varphi_{-h} - \psi_{-h})}{\sqrt{\varphi_{-h}\psi_{-h}\varphi\psi}} \qquad [5]$$

We do not yet have a relation of simple proportionality between stress and retardation, but, noting that the quantity $(\alpha - \beta)\sigma_m/V_0^2 h$ in Equations [4] is of order of magnitude 10^{-3} for bakelite, we are justified in neglecting powers higher than

[1] By J. N. Goodier and G. H. Lee, published in the March, 1941, issue of the Journal of Applied Mechanics, Trans. A.S.M.E., vol. 63, p. A-27.

[2] Department of Civil Engineering, Columbia University, New York, N. Y. Mem. A.S.M.E.

the second in the binomial expansions of the radicals. Performing this operation we find that

$$I = I_0 \sin^2 \frac{\pi C \sigma_m (h^2 - z^2)}{\lambda h} \quad [6]$$

where C is the relative stress-optical coefficient and λ is the wave length in vacuo. The relative phase retardation observed in the authors' experiment is twice that obtained from Equation [6] for $z = 0$ on account of reflection. Hence, for their experiment

$$I = I_0 \sin^2 \frac{2\pi C \sigma_m h}{\lambda} \quad [7]$$

This is exactly the expression which would be obtained for the intensity if the assumption were made at the start that the relative phase retardation is proportional to the principal stress difference.

It should be observed that, if the stress-optical sensitivity and the stress were each multiplied by a factor of 30, the simple relation would not hold at all, even for rough work.

Another fact which has been shown by a similar study[3] is that even for present materials and stresses, the simple relation does not hold when the orientations of the principal stresses in the x, y plane vary through the thickness. Although this does not pertain to the experiments performed by the authors, it does apply to their suggestion that their method can be used where a mean stress is developed in bending. In this case, the principal stress directions, in general, do vary through the thickness, and optical observations of a different type from those made by the authors are required.[3]

Reflection Polariscope

The authors state that, in their polariscope, reflected light from numerous optical surfaces hinders photography and that image intensity is low. In its physical optics, their apparatus is identical in design with one devised by the writer,[4] the distinguishing feature of which is that a single polarizing element is used as both polarizer and analyzer. The two instruments differ, however, in their geometrical optics and to this may be attributed the fact that the one has intensity and multiple-reflection difficulties while the other has not. It is better to place the axis of the telescope along the collimation axis of the instrument and not at right angles to it, as the authors have done, i.e., light source and telescope should be interchanged. Their half-silvered mirror may then be replaced by a clear-glass plate, and the polarizer analyzer should be set so that the plane of polarization is perpendicular to the plane of the glass plate. The light intensity is markedly increased by this change and no difficulty is encountered with reflections at the illumination and observation end of the apparatus. To eliminate undesirable reflections from the other end of the apparatus, the polaroid plate should be replaced by a Nicol or Halle prism, as in these the end faces are not perpendicular to the collimation axis. If one of these prisms is not available, a long-focal-length collimating lens should be used and the polaroid plate tilted slightly to cast its reflection out of the field. The long-focal-length collimating lens is desirable in any case as then it and the retardation plate and compensator may also be tilted and thus eliminate all of the undesirable reflections except that from the front surface of the model. To improve the geometric quality of the image, it is also desirable to place a diaphragm stop at the focal point of the collimating lens.

Alternative Methods

(*a*) The first suggested method requires a "sandwich" as in the authors' method, but does not require a reflecting surface between the two elements of the model.

The plastic "lucite" has about the same "elastic" properties as bakelite but its stress-optical sensitivity is much less, the ratio being about 1:70. The suggestion is to make a sandwich of lucite and bakelite, without an intervening reflecting surface, and transmit the light through the model. The double refraction in the lucite being negligible, the average principal stress difference in the bakelite half is obtained directly.

(*b*). The second suggested method requires neither a sandwich nor even measurement of relative retardation in the bent plate.

A bakelite plate is first prepared for use by "freezing" in it a known stress distribution, say a simple tension, following the procedure used in the freezing method for three-dimensional photoelasticity. After cooling, the relative retardation is measured and this retardation corresponds to a simple tension, say, σ_x' in the cold plate. The plate is then cut to the required shape and subjected to bending, resulting in a superposed stress system in which the stresses parallel to the central plane of the plate will be called σ_x'', σ_y'', τ_{xy}''.

The latter vary linearly through the thickness; say, $\sigma_x'' = Az$, $\sigma_y'' = Bz$, $\tau_{xy}'' = Cz$ where A, B, C are constants. The total stress system parallel to the x, y plane is then $\sigma_x = \sigma_x' + Az$, $\sigma_y = Bz$, $\tau_{xy} = Cz$ and the angle θ, which the algebraically larger principal stress in the x, y plane makes with the x axis, is given by

$$\tan 2\theta = \frac{2Cz}{(A - B)z + \sigma_x'} \quad [8]$$

Hence, the directions of the principal stresses in the x, y plane vary through the thickness of the plate.

Let the values of θ at the surfaces, $z = \pm h$, of the plate be θ_{+h} and θ_{-h}, respectively. Then

$$\left.\begin{aligned} \tan 2\theta_{-h} &= \frac{2Ch}{(A - B)h - \sigma_x'} \\ \tan 2\theta_{+h} &= \frac{2Ch}{(A - B)h + \sigma_x'} \end{aligned}\right\} \quad [9]$$

hence

$$\left.\begin{aligned} A - B &= \frac{\sigma_x' \sin 2(\theta_{-h} + \theta_{+h})}{h \sin 2(\theta_{-h} - \theta_{+h})} \\ C &= \frac{\sigma_x' \sin 2\theta_{-h} \sin 2\theta_{+h}}{h \sin 2(\theta_{-h} - \theta_{+h})} \end{aligned}\right\} \quad [10]$$

If the angles θ_{-h} and θ_{+h} can be measured, the required quantities $(A - B)$ and C are determined. The procedure for measuring θ_{-h} and θ_{+h} at each point of the plate is known,[3] the method being based upon a peculiar phenomenon[3] which occurs when plane-polarized light is incident on the plate. If the plane of polarization is parallel to a principal stress direction on entry, it will rotate with the principal stress through the plate and the light will emerge plane-polarized parallel to the principal stress direction at the second surface of the plate. Hence, the angles at each point may be determined by independent rotation of polarizer and analyzer[3] in the usual photoelastic apparatus. It remains to be seen if θ_{-h} and θ_{+h} can be measured accurately enough.

The method fails at points in the plate where the direction of the frozen stress σ_x' happens to coincide with the direction of a

[3] "Stress Analysis by Three-Dimensional Photoelastic Methods," by D. C. Drucker and R. D. Mindlin, *Journal of Applied Physics*, vol. 11, 1940, pp. 724–732.

[4] "A Reflection Polariscope for Photoelastic Analysis," by R. D. Mindlin, *Review of Scientific Instruments*, vol. 5, 1934, pp. 224–228.

principal bending stress, but this may be avoided at points of interest by judicious cutting of the frozen plate.

S. P. Timoshenko.[5] This paper gives a very interesting application of the photoelastic method in studying the bending of plates. It seems that the method can be useful in many cases of bending of plates which cannot be easily treated in an analytical way. There is, however, some limitation in the use of the method. The method assumes that the stresses have a linear distribution over the thickness of the plate, hence it cannot be applied, for instance, in studying local stresses at a point of application of a concentrated force. The difficulty arises also in the application of the method to the investigation of stress concentration around a small circular hole. The investigation shows that a high stress concentration takes place around the hole and that the width of the region of high stresses is small in comparison with the thickness of the plate. Under such conditions we cannot assume that the stresses have a linear distribution.

We think, also, that the elementary-plate theory cannot be applied at the edge of the hole without some additional consideration. Under such circumstances, it seems that the measurements of strain around the hole in a bent plate can give us more reliable information regarding stress concentration than the photoelastic method does, and that the applicability of the photoelastic method in this case can be established only if it will be shown that the results obtained photoelastically coincide with those obtained by strain measurements.

Authors' Closure

Dr. Mindlin's proposals for alternative methods are a valuable extension of the paper. His method (2) is particularly interesting, since it abolishes the sandwich and the difficulty of making it of anything but the rather unsatisfactory celluloid. The inapplicability of the method of the paper where there is mean stress as well as bending stress is clear, in the general case, but in those numerous problems which have axes of symmetry, the difficulty does not arise for measurements on these axes, and frequently such measurements are sufficient. Several modifications of the optical system, some as suggested in the discussion, were considered and tried. The optical arrangement was described in the paper as a matter of recording what was done in these particular circumstances, and is certainly susceptible of improvement.

Professor Timoshenko's discussion raises two points—that there will not be a linear distribution of stress (*a*) near a concentrated load, or (*b*) at boundaries. These are real difficulties in the subject and were debated at some length while this work was in progress. There is little to be said about (*a*) except that such regions are outside the province of this method, unless a multiple sandwich can be made, giving reflections from various depths, and so the actual distribution of stress over the thickness. The ordinary photoelastic method for plane stress suffers from a similar restriction. Near a concentrated load much variation through the thickness may be expected. In (*b*) the bent plate again is on much the same footing as the plane-stress models, if we confine attention only to the smallness of the region of concentration and its comparability with the thickness. In each case, the standard distribution is departed from in favor of a more complicated régime. However, the bent plate has of course an additional complication at the boundary. The ordinary theory only satisfies a condition of the type $Q + \frac{\partial H}{\partial S} = 0$ at a free boundary, instead of the strict conditions $Q = 0$, $H = 0$, Q being shearing force and H twisting couple per unit length. A theoretical solution may then permit an appreciable $H \left(\text{or } \frac{\partial H}{\partial S}\right)$ provided it is counterbalanced by a suitable Q. It is known that this departure from the strict conditions can only disturb an edge zone comparable with the thickness, but it does mean of course that there is no necessity for agreement of photoelastic and theoretical stresses within such a zone. However, the tests of Dumont[6] were made with strain gages on metal plates, and agreed well with the theory by Goodier.[7] As far as the evidence goes, from the present tests and those of Dumont, it suggests that these boundary complications are not serious. But it may well be mentioned here that Dr. D. C. Drucker has recently obtained photoelastic results on the hole in the bent plate, using a modification of Dr. Mindlin's second method, which do not conform. No explanation of this discrepancy has as yet been established.

[5] Professor of Theoretical and Applied Mechanics, Stanford University, Stanford University, Calif. Fellow A.S.M.E.

[6] Refer to Bibliography (3) of paper.

[7] Refer to Bibliography (1) of paper.

Reprinted from
Journal of Applied Mechanics
8 (1941) A-187 - A-189

Reprinted from the JOURNAL OF THE FRANKLIN INSTITUTE,
Vol. 233, No. 4, April, 1942

OPTICAL ASPECTS OF THREE-DIMENSIONAL PHOTOELASTICITY.*

BY

RAYMOND D. MINDLIN,

Department of Civil Engineering, Columbia University.

1. TWO-DIMENSIONAL PHOTOELASTICITY.

In two-dimensional photoelasticity a plate of transparent isotropic material is stressed by a uniplanar force system coplanar with the middle plane of the plate (Fig. 1). The material of the plate thereby becomes optically doubly refracting and certain of the resulting optical properties are determined by examining their effects on polarized light passed through the plate at normal incidence. The state of stress in the plate is then calculated on the basis of experimentally determined relations between stress and optical effect carried out under similar conditions.

The state of stress, at a point in a plate loaded only by forces in its plane, is a particular case of a more general stress condition; the most general being characterized by six independent quantities, namely, the magnitudes and the directions of the three mutually perpendicular principal stresses. The three principal stresses will be denoted in this paper by the symbols σ_3, σ_2, σ_1, and their relative magnitudes will be taken as $\sigma_3 > \sigma_2 > \sigma_1$.

In a plate loaded in its plane, one of the principal stresses is always normal to the plane of the plate and its magnitude is zero, so that three of the six quantities are known *a priori*. The remaining three quantities are the magnitudes of the two principal stresses, lying in the plane of the plate, and their orientation. An important characteristic of such a stress distribution is that, at each point in the plate, one principal plane of stress is parallel to the middle plane of the

* From a paper presented at the Eighth Semi-Annual Meeting of the Eastern Photoelasticity Conference, December 10, 1938, Columbia University, New York City.

plate. It should be noticed that the two principal stresses in the plane of the plate may be σ_1 and σ_2 or σ_2 and σ_3 or σ_3 and σ_1.

When polarized light is passed normally through the plate, the relative phase difference of the two waves, into

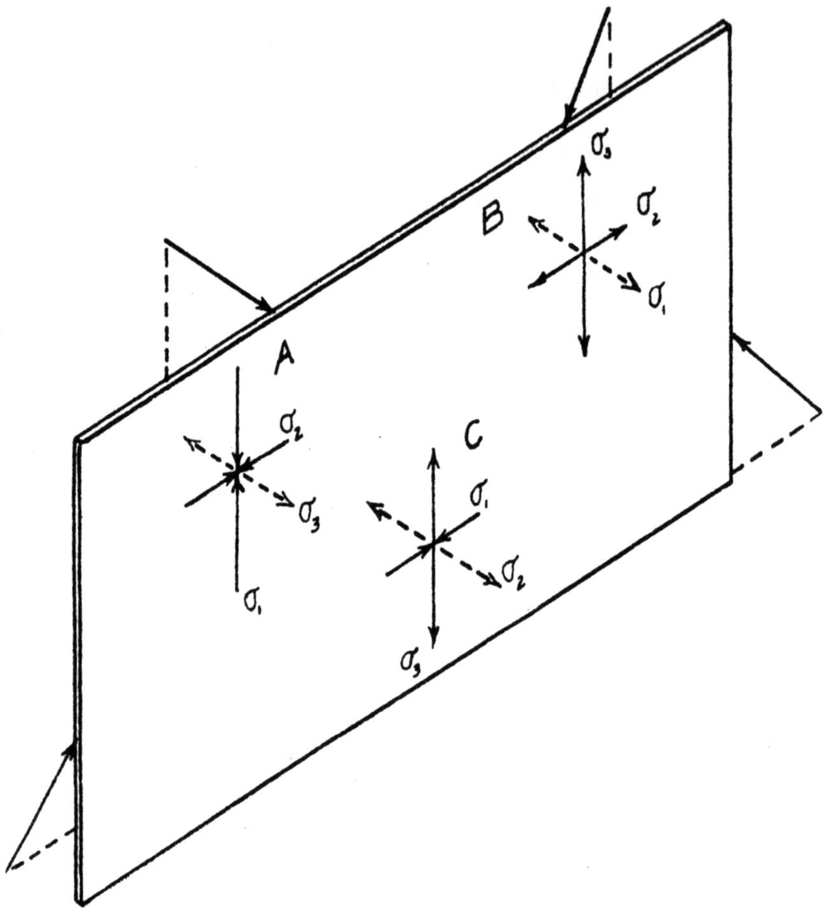

FIG. 1.

which the light is physically resolved, is found to be directly proportional to the difference between the two principal stresses in the plane of the plate. Since two of the principal stresses vanish on a free boundary of the plate, the relative phase difference there is directly a measure of the third principal stress which is, of course, parallel to the edge of the plate.

The maximum shearing stress in the plate is a quantity of prime engineering importance. It is given by half the difference between the algebraically largest and smallest principal stresses. If, at any point in the plate, the two principal stresses in the plane of the plate are of opposite sign (one tension and the other compression as at point *C* in Fig. 1), the relative phase difference is there a direct measure of the maximum shearing stress. In general, however, it cannot be said that the relative phase difference measures the maximum shearing stress in the plate. It measures the maximum shearing stress on planes normal to the plane of the plate and this will be the true maximum only when $\sigma_2 = 0$.

Experiments also show that the polarizing axes of the two-dimensionally stressed plate coincide with the axes of principal stress in the plane of the plate. Hence it is a simple matter to determine the directions of the principal stresses by observing extinction positions as the plate is rotated in its plane between crossed nicols.

Regarding the independent determination of the magnitudes of the principal stresses in a two-dimensional stress system, there is an extensive literature, to which reference is not essential for the purposes of this paper.

2. THREE-DIMENSIONAL PHOTOELASTICITY.

There is now in the course of development a tecnhique for optically analyzing three-dimensional stress systems.[1,2,3,4,5] In brief, the process consists in heating a model made from

[1] "A New Photoelastic Material," by A. G. Solakian. *Mechanical Engineering*, **57**, 767–771 (1935).

[2] "Polarisationsoptische Untersuchung raümlicher Spannungs- und Dehnungszustände," by G. Oppel. *Forschung auf dem Gebiete des Ingenieurwesens*, **7**, No. 5, 240–248 (1936).

[3] "Polarisationsoptische Untersuchung des räumlichen Spannungszustándes im konvergenten Licht," by R. Hiltscher. *Forschung auf dem Gebiete des Ingenieurwesens*, **9**, No. 2, 91–108 (1938).

[4] "Das Kunstharz Phenolformaldeheyd in der Spannungsoptik," by A. Kuske. *Forschung auf dem Gebiete des Ingenieurwesens*, **9**, No. 3, 139–149 (1938).

[5] "Photoelastic Studies of Three-Dimensional Stress Problems," by M. Hetényi, presented at the Fifth International Congress for Applied Mechanics, Cambridge, Mass., September, 1938; also "The Fundamentals of Three-Dimensional Photoelasticity," by M. Hetényi. *Journal of Applied Mechanics*, **5**, No. 4, -1A49–155 (1938).

a hardening resin (for example, a phenolformaldehyde resin or a glycerine phthallic anhydride resin) above a certain critical temperature, applying loads, cooling under load, and then removing the loads. The resulting material is optically doubly refracting and its optical properties are closely connected with the elastic state of stress present in the material when it is in the loaded condition at the elevated

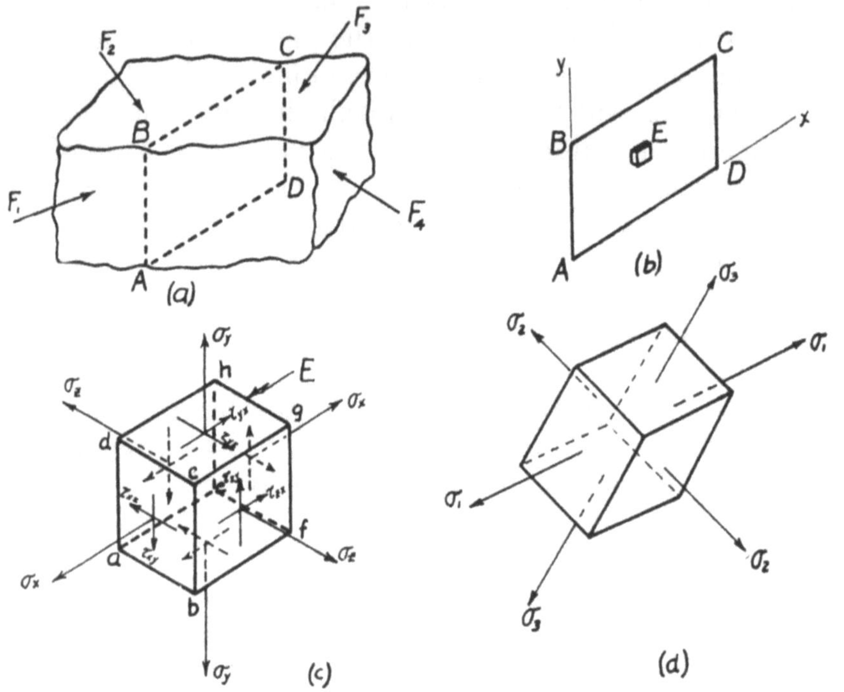

FIG. 2.

temperature. One of the most striking features of the process is that the optical properties of the material are apparently unchanged by cutting the specimen into thin slices. The photoelastic properties of hardening resins at elevated temperatures have been well understood as a result of the work of Hetényi (see footnote No. 5) and reference to his papers should be made for the details of the thermal process.

It is therefore possible to determine the stress distribution in an entire specimen by determining the optical properties of the slices. The most important characteristic of the slices, which distinguishes them from the plates employed in two-dimensional work, is that the optical properties in general represent a stress distribution involving both normal and shearing stresses on the plane of the slice. Fig. 2(*a*) represents a three-dimensional model, acted upon by a general system of forces F_1, $F_2 \cdots$, from which a slice *ABCD* is cut after the model has been carried through the thermal and loading process perfected by Hetényi. Fig. 2(*b*) represents the slice which is taken as the x, y-plane and shows a small element, E, in the shape of a rectangular parallelepiped whose faces are parallel or perpendicular to the plane of the slice. The state of stress on the element is depicted in Fig. 2(*c*). Plane *bcgf*, which is parallel to the plane of the slice, is not a principal plane of stress, since it has shearing components τ_{zx} and τ_{zy} acting on it. To obtain the orientation of the element so that only normal (principal) stresses act on it, it would be necessary to rotate it until it reaches a position such as is indicated in Fig. 2(*d*).

The relative retardation for light passed normally through the slice in the usual manner (the z-direction in Fig. 2(*c*)) is a measure of the difference between the *secondary principal stresses*, p and q, in the x, y-plane where

$$p, q = \frac{\sigma_x + \sigma_y}{2} \pm \sqrt{(\sigma_x - \sigma_y)^2 + 4\tau_{xy}{}^2}.$$

Since p and q are restricted to lie in the x, y-plane, they do not, in general, correspond to σ_3 and σ_2 or σ_2 and σ_1 or σ_3 and σ_1.

Hence, if there are shearing stresses on the plane of the slice, the latter is not a plane of principal stress and the relative phase difference, measured in the usual manner, is proportional, not to a difference between principal stresses but to the difference between the *secondary principal stresses* lying in the plane of the slice. The relative phase difference for the slice will be related to the principal stress difference through certain angles which fix the orientation of the slice with respect to the principal axes.

Auxiliary measurements (that is, measurements supplementing those made in two-dimensional photoelasticity) must therefore be made to obtain the data requisite for converting the phase difference of the slice to maximum phase difference. The principles involved in this process are well known to students of crystal optics and there are several methods widely used by petrologists for making the measurements. It is only necessary for workers in photoelasticity to adapt one of the petrological techniques to the requirements of photoelasticity and to ascertain the relations between the observed effects and stress.

3. SOME PRINCIPLES OF CRYSTAL OPTICS.

The optical phenomena encountered in photoelasticity are adequately described by a wave theory of light. Of the wave theories which have been devised, the electromagnetic theory is, in many respects, the most complete. Chronologically, the electromagnetic theory displaced an elastic ether theory which, before it was abandoned, had been developed to such a stage that it accurately explained all of the phenomena of crystal optics which are now observed in photoelasticity. Workers in photoelasticity are mostly civil and mechanical engineers who are, on the whole, more familiar with the stress equations of motion and the fundamental ideas of stress and strain than with the principles of electricity and magnetism. Since the former are the bases of the elastic ether theories, it seems likely that engineers will more easily master the principles of crystal optics via an elastic ether theory [6] than via the electromagnetic theory.

Regardless of which theory is used, the conclusion is reached that light is propagated in a transparent, non-isotropic medium only in the form of plane polarized waves and that, for each wave normal, there are in general two wave velocities corresponding to two waves whose polarization directions are perpendicular to each other and to the wave normal.

The wave velocities and directions of polarization may be conveniently described in terms of the geometry of *Fresnel's ellipsoid*. The semi-axes of a central section of the ellipsoid, perpendicular to a wave normal, give the reciprocals of the

[6] See, for example, Lamé, "Théorie de l'Élasticité," Paris, 1866. Chapters XVII to XXIV.

velocities for that wave normal; and the direction of a semi-axis of the section is the direction of polarization of the wave having the velocity given by the reciprocal of the other

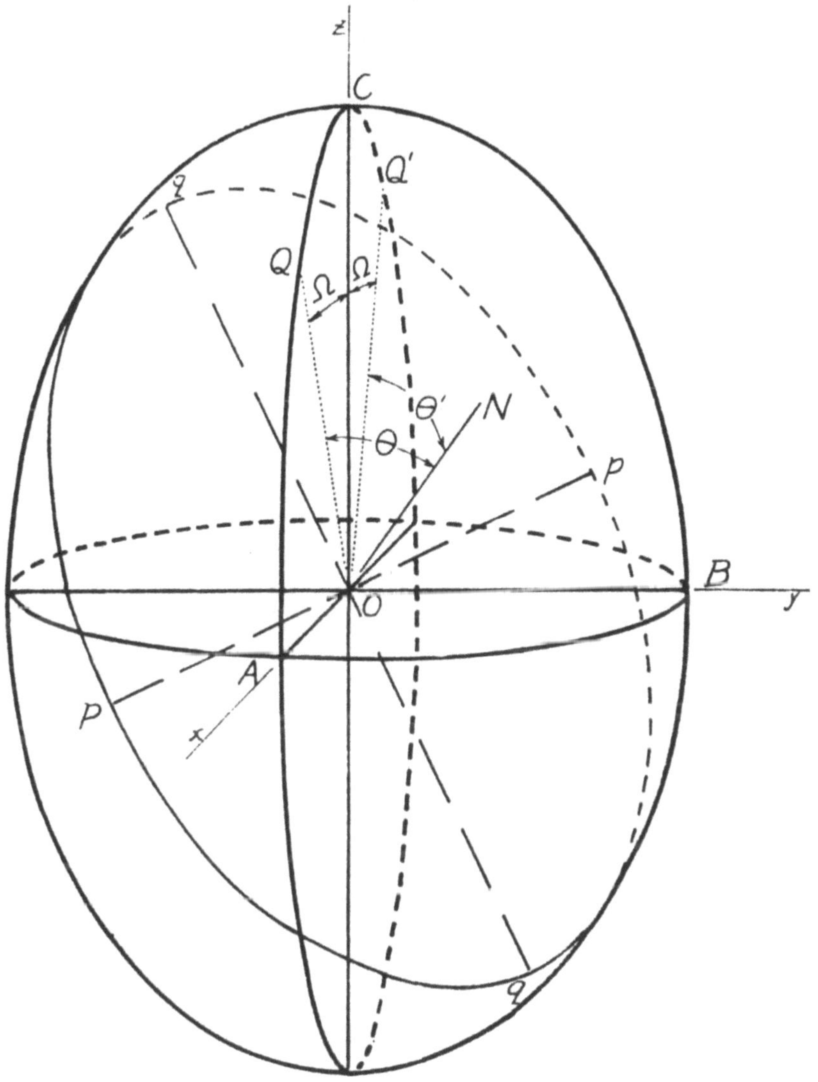

FIG. 3.

semi-axis. The three principal semi-axes of the ellipsoid, as distinguished from the two semi-axes of an arbitrary central section, thus represent the stationary values (maximum, minimum and minimax) of wave velocity. These three axes are termed the principal axes of optical symmetry of the medium.

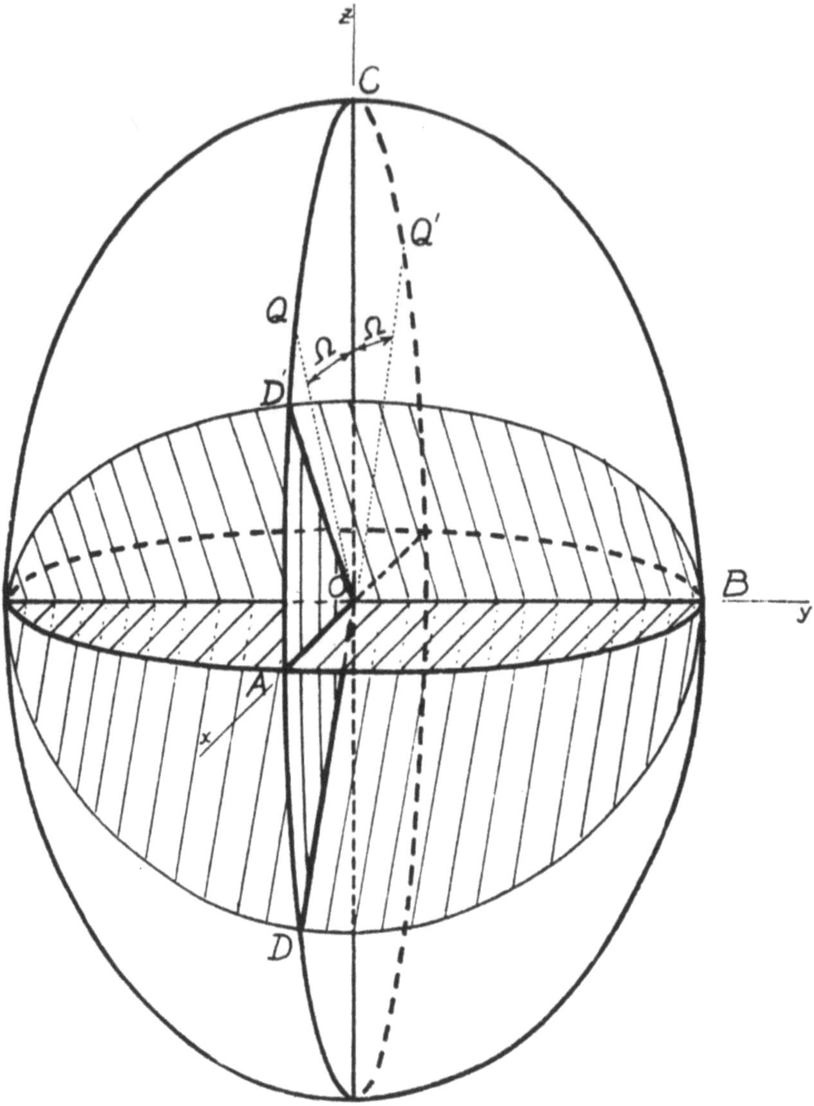

FIG. 4.

In Fig. 3, OA, OB, OC are the principal semi-axes of the ellipsoid. If a, b, and c are the magnitudes of the principal wave-velocities, then

$$OA = 1/a, \qquad OB = 1/b, \qquad OC = 1/c.$$

The ellipse $pqpq$ is the intersection with the ellipsoid of a plane passing through the center O, the normal to the plane being ON. The seim-axes Op and Oq of the ellipse are the reciprocals of the wave velocities k_1 and k_2, respectively, for a wave normal ON. Op is the polarization direction of k_2, and Oq is the polarization direction of k_1. We also have

$$a \geqq k_1 \geqq b \geqq k_2 \geqq c.$$

There are two planes (DOB and $D'OB$ of Fig. 4) through the intermediate principal axis (OB) of the ellipsoid which intersect the ellipsoid in circles. Since the semi-axes of each of these sections are equal, a wave front parallel to either one has only one velocity. The normals (OQ and OQ' in Fig. 4) to these planes are called the optic axes; they lie in the x, z-plane and make equal angles Ω with the z-axis. The angle 2Ω is termed the *optic axial angle*. From the known geometrical properties of the ellipsoid we have

$$\sin \Omega = \sqrt{\frac{a^2 - b^2}{a^2 - c^2}}, \qquad \cos \Omega = \sqrt{\frac{b^2 - c^2}{a^2 - c^2}}. \tag{1}$$

Furthermore, if θ (Fig. 3) is the angle between the wave normal ON and the optic axis OQ, and if θ' is the angle between the wave normal and the optic axis OQ', we have, from the geometry of the ellipsoid

$$\left.\begin{aligned} k_1^2 &= \tfrac{1}{2}[a^2 + c^2 + (a^2 - c^2)\cos(\theta - \theta')], \\ k_2^2 &= \tfrac{1}{2}[a^2 + c^2 + (a^2 - c^2)\cos(\theta + \theta')], \end{aligned}\right\} \tag{2}$$

so that

$$k_1^2 - k_2^2 = (a^2 - c^2)\sin\theta\sin\theta'. \tag{3}$$

In Fig. 3, plane $pqpq$ is a central section of Fresnel's ellipsoid drawn parallel to the wave front. For light traversing the slice at normal incidence, this plane is parallel to the plane of the slice, and its normal, ON, is both the wave normal and the normal to the slice.

The planes of polarization are NOp and NOq. These planes bisect, internally and externally, the dihedral angles between the planes containing the wave normal ON and the optic axes OQ and OQ'. That is, planes NOp and NOq bisect the angles between planes NOQ and NOQ'.

The most general crystal is *biaxial*, that is, there are two optic axes; but, if $b = c$ or $b = a$, the crystal becomes *uniaxial*. Considering Fresnel's ellipsoid (Fig. 3), with its greatest axis $2/c$, mean axis $2/b$ and least axis $2/a$, if we decrease the mean axis until it equals the least axis, the ellipsoid degenerates to a prolate spheroid for which the z-axis is both the axis of revolution and the optic axis. The optic axial angle (2Ω) decreases to zero so that θ and θ' are equal. This is the ellipsoid for a *positive uniaxial* crystal.

To obtain a *negative uniaxial* crystal, increase the mean axis of Fresnel's ellipsoid until it equals the greatest axis $2/c$. The ellipsoid degenerates to an oblate spheroid with the x-axis as both the axis of revolution and the optic axis. The optic axial angle increases to π and $\theta + \theta' = \pi$.

The optic axial angle, 2Ω, is defined as that angle between the optic axes which is bisected by the axis of least velocity (z). When $2\Omega < \pi/2$, the biaxial crystal resembles, somewhat, a positive uniaxial crystal and the more acute the angle, the closer is the similarity. Such a crystal is called *positive biaxial*. When $2\Omega > \pi/2$, the crystal resembles a negative uniaxial crystal and, the closer 2Ω approaches π, the closer is the similarity. Such a crystal is *negative biaxial*. When $2\Omega = \pi/2$, we have, from Eq. (1)

$$a^2 - b^2 = b^2 - c^2. \tag{4}$$

The relative phase difference, δ, of the two waves, traveling a distance ρ through the crystalline medium with velocities k_1 and k_2 is given by

$$\delta = \frac{2\pi\rho}{\lambda}(n_2 - n_1), \tag{5}$$

where n_1 and n_2 are the indices of refraction corresponding to k_1 and k_2, respectively; i.e., if k is the velocity of light *in vacuo*,

$$n_1 = \frac{k}{k_1}; \qquad n_2 = \frac{k}{k_2}. \tag{6}$$

Since the velocity differences are small in comparison with the velocities themselves, we may write, for Equations

(1), (3) and (4)

$$\sin \Omega = \sqrt{\frac{n_a - n_b}{n_a - n_c}}, \qquad \cos \Omega = \sqrt{\frac{n_b - n_c}{n_a - n_c}}, \tag{7}$$

$$n_2 - n_1 = (n_c - n_a) \sin \theta \sin \theta', \tag{8}$$

$$n_a - n_b = n_b - n_c \quad \text{when} \quad 2\Omega = \pi/2. \tag{9}$$

Hence

$$\delta = \frac{2\pi\rho}{\lambda} (n_c - n_a) \sin \theta \sin \theta'. \tag{10}$$

4. STRESS-OPTICAL RELATIONS.

If the scale of Fresnel's ellipsoid is adjusted so that a unit of length represents the velocity of light *in vacuo*, the semi-axes of any section represent indices of refraction. Experiments indicate that the same type of relation exists between stress and change of index of refraction as between stress and strain.[7] Hence, the optical properties which an originally isotropic material assumes on the application of stress may be visualized by imagining that Fresnel's ellipsoid is the result of the deformation of a sphere of radius n_0 subjected to principal stresses applied parallel to the principal axes of the ellipsoid. The mathematical statement of this principle is

$$\left.\begin{aligned} n_a - n_0 &= C_1\sigma_1 + C_2(\sigma_2 + \sigma_3), \\ n_b - n_0 &= C_1\sigma_2 + C_2(\sigma_3 + \sigma_1), \\ n_c - n_0 &= C_1\sigma_3 + C_2(\sigma_1 + \sigma_2), \end{aligned}\right\} \tag{11}$$

in which C_1 and C_2 are constants, $n_c > n_b > n_a$ and $\sigma_3 > \sigma_2 > \sigma_1$. The axes a, b, c, and 1, 2, 3 and x, y, z are coincident in the order named, i.e., a, 1 and x are coincident, etc. If C_1 is positive, a unidirectional tension ($\sigma_2 = \sigma_1 = 0$) will elongate the sphere to a prolate spheroid and so produce a positive uniaxial crystal. If, combined with this tension, a compression is applied at right angles, a positive biaxial crystal is produced when the absolute magnitude of the compression is less than the absolute magnitude of the tension. When the magnitudes are reversed, a negative biaxial crystal results.

[7] "On the Equilibrium of Elastic Solids," by J. C. Maxwell. *Transactions of the Royal Society of Edinburgh*, **20**, 87–120 (1853).

When we have a simple shear, i.e., when absolute magnitude of the tension and compression are equal, we find, substituting $\sigma_3 = -\sigma_1$ and $\sigma_2 = 0$ in Eq. (11) and using Eq. (9), that the optic axes intersect at right angles.

It is apparent that a knowledge of the three principal indices of refraction is necessary for the calculation of the three principal stresses from Equations (11). The differences between refractive indices are generally easier to measure than the indices themselves. These differences are proportional to the principal stress differences. Thus from Equations (11)

$$\left.\begin{aligned} n_c - n_a &= C(\sigma_3 - \sigma_1), \\ n_a - n_b &= C(\sigma_1 - \sigma_2), \\ n_b - n_c &= C(\sigma_2 - \sigma_3), \end{aligned}\right\} \tag{12}$$

where $C = C_1 - C_2$.

From Equations (12), (10) and (7),

$$\left.\begin{aligned} \sigma_3 - \sigma_1 &= \frac{\lambda\delta}{2\pi\rho C} \cdot \frac{1}{\sin\theta \sin\theta'}, \\ \sigma_2 - \sigma_1 &= \frac{\lambda\delta}{2\pi\rho C} \cdot \frac{\sin^2\Omega}{\sin\theta \sin\theta'}, \\ \sigma_3 - \sigma_2 &= \frac{\lambda\delta}{2\pi\rho C} \cdot \frac{\cos^2\Omega}{\sin\theta \sin\theta'}. \end{aligned}\right\} \tag{13}$$

Of the quantities on the right in Equations (13), δ is a phase retardation which may be measured with a compensator. The wave length, λ, length, ρ, and stress optical coefficient C are determined in the usual manner. Hence, if we can measure the angles θ and θ' between the wave normal and the optic axes, we can determine the maximum shearing stress $\frac{1}{2}(\sigma_3 - \sigma_1)$. If, in addition, we can measure the optic axial angle 2Ω, we can determine all three principal stress differences.

5. SPECIAL CASES.

There are certain special cases for which it is not necessary to measure all of the three angles θ, θ' and Ω; and others for which the measurement of these angles will yield more information than simply the differences between principal stresses.

Most important is the case when the stress distribution has a plane of symmetry. This will occur when both the model and the loading system are geometrically symmetrical about a plane. The plane of symmetry is a principal plane of stress so that a slice which contains this plane will have a relative retardation proportional to one of the three principal stress differences. The plane which contains the optic axes will be either parallel or normal to the plane of symmetry. In the latter case, the bisector of the optic axial angle is either parallel or perpendicular to the plane of the slice. Hence we have $\theta = \theta' = \pi/2$ or $\theta = \theta' = \Omega$ or $\theta = \theta' = \pi/2 - \Omega$, depending on which principal plane of optical symmetry coincides with the plane of the slice. Hence, if we measure θ and Ω in addition to δ, we may calculate the three principal stress differences from Equations (13).

The two-dimensional stress system ordinarily obtained in two-dimensional photoelasticity is a special form of the case of symmetry about a plane; for, in a plate in a state of plane stress, the plane of the plate is a principal plane of stress and, in addition, the principal stress normal to the plate has zero magnitude. As in the more general case of symmetry about a plane, we measure θ and Ω (θ and θ' being equal) but now, since one of the principal stresses is zero, the other two may be determined separately from Equations (13). This method for the independent determination of the two principal stresses in two-dimensional photoelasticity was first described by Hiltscher.[3]

Hiltscher[3] also observed that, since one of the principal stresses is zero on a free boundary in a three-dimensional stress field, the other two principal stresses may be determined if θ and Ω are known.

In a prismatic or cylindrical bar subjected to St. Venant torsion,[8] one of the principal stresses at any point in the bar is zero and the other two are equal in magnitude and opposite in sign. The zero principal stress direction is perpendicular to a generator of the bar and the other two principal stress directions make angles of 45° with the generator. This state of stress corresponds to a biaxial crystal with an optic axial angle of $\pi/2$, the optic axes being parallel and perpendicular to the generator. A slice of the bar normal to a generator

[8] "Theory of Elasticity," by S. Timoshenko. McGraw-Hill Book Co., New York, 1934. P. 228.

therefore gives zero phase difference for light at normal incidence; but a slice at any other angle to a generator will give a phase difference from which we may calculate the principal stresses. To effect this calculation, it is necessary to to know the angles (θ and θ') which the wave normal makes with the optic axes. One of these will be constant and equal to the angle between the generator and the normal to the slice, but the other will vary over the section. For each point in the slice, we need therefore measure only the latter angle (θ') and the phase difference δ. Then Equations (13) give the principal stresses directly, provided the wave-length λ, the slice thickness ρ, and the stress optical coefficient C are known. It was observed by Hetényi [5] that a 45° slice exhibits the maximum number of fringes, but it should be noted that the fringes do not represent loci of constant principal stress difference. It is necessary to know the angle θ' for each point in the 45° slice. This angle may be determined in the following manner: Let O be a point in the 45° slice, at which θ' is to be found. The normal to the slice at O and the line through O parallel to a generator of the bar determine a plane which intersects the plane of the slice in a line OA. Let OB be a polarizing axis of the slice at O (determined from the isoclinics in the usual manner) and let angle $AOB = \alpha$. Then it may be shown that

$$\sin \theta' = (1 + \cos^2 2\alpha)^{1/2}.$$

Similar formulas hold for slices cut at other angles or viewed obliquely.

6. OPTICAL TECHNIQUES.

There are three standard methods for measuring n_a n_b and n_c but none of them appears to be suitable for the current development in three-dimensional photoelasticity.

In Stokes' method,[9] a plane parallel plate of the crystal is placed over a rectangular grid of ruled lines and the latter are observed through a microscope. On account of the double refraction in the plate, each family of parallel lines in the grid will appear in focus at two positions of the microscope objective. The three indices can be calculated from

[9] "Analytical Theory of Light," by J. Walker, Cambridge University Press, 1904. Pp. 225–230.

the relative focus positions. However, the photoelastic double refraction is so weak that this method does not appear feasible.

The prism method [10] would require the cutting of a prism from the model for each point at which the stress is desired.

The total reflection methods [11] are very accurate but they measure the optical properties of a surface layer of the specimen. In a photoelastic slice the surface properties are not a reliable indication of the internal optical characteristics.

For determining the differences between principal refractive indices in a doubly refracting plate, we have seen that it is necessary to measure the relative phase retardation of the plate, the optic axial angle (2Ω) and the angles (θ and θ') between the optic axes and the plate normal. The measurement of the three angles is ordinarily effected by one of two methods; one involving the use of the universal rotating stage [12] and the other involving observations with converging polarized light.[13] The latter is the method used by Hiltscher.[3] On account of the very weak double refraction of photoelastic materials, and the necessity of using thin slices, the converging light method is likely to be inaccurate if not impossible in many cases. A combination of the universal rotating stage for measuring the three angles θ, θ' and Ω with a wedge compensator for measuring δ seems to offer good possibilities for the basis of an optical technique for three-dimensional photoelasticity.

Note added in proof:

Since this paper was written, another photoelastic method for analyzing three-dimensional systems has been developed by R. Weller. Reports of recent work in this field may be found in the following papers:

R. Weller, *Journal of Applied Physics*, Vol. 10, p. 266 (1939).

R. Weller and J. K. Bussey, *Tech. Note No. 737, Nat. Adv. Comm. for Aeronautics* (1939).

R. Weller and B. Fried, *Bull. No. 106, Ohio State Univ. Eng. Exp. Sta.* (1940).

D. C. Drucker and R. D. Mindlin, *Journal of Applied Physics*, Vol. 11, p. 724 (1940).

P. R. Rosenberg, *Proc. 13th Semi-annual Eastern Photoelasticity Conference*, Mass. Inst. of Technology, Cambridge, Mass., June, 1941.

---

[10] J. Walker, op. cit., pp. 230–242.

[11] J. Walker, op. cit., pp. 230–242.

[12] "Manual of Petrographic Methods," by A. Johannsen. McGraw-Hill Book Co., New York, 1918. Pp. 487–507.

[13] A. Johannsen, op. cit., pp. 449–486. See also reference No. 3.

Reprinted from
Quarterly of Applied Mathematics
IV (1946) 279-290

THE ANALOGY BETWEEN MULTIPLY-CONNECTED SLICES AND SLABS*

BY

RAYMOND D. MINDLIN

Department of Civil Engineering, Columbia University

1. Introduction. The analogy between the two-dimensional field of stress and the transverse flexure of a thin plate was first applied by K. Wieghardt[1] to the solution of a problem involving boundary loading of a simply-connected body. As is well known, the analogy establishes the proportionality of the curvatures of the surface of the plate to the components of stress in the two-dimensional field of stress. H. M. Westergaard[2] introduced the useful terminology of slab and slice, free slice and constrained slice, and gave the boundary conditions for the slab when the slice is multiply-connected and is stressed by boundary loads having no resultant force on an internal boundary. Westergaard also proposed the use of the analogy in the investigation of the stresses in the Boulder Canyon Dam,[3] a problem involving gravity and boundary loading of a simply connected body. An improvement in experimental technique was contributed by H. Cranz[4] in introducing an optical spherometer[5] for measuring the components of surface curvature. Cranz's application was to boundary load problems in simply connected bodies.

It is the purpose of this paper to give the general boundary conditions for the slab when the slice is multiply-connected and is stressed by any combination of boundary loading, body forces, dislocations and thermal dilatations. The analogy has, in fact, its most useful applications in the last three cases as they are either difficult to reproduce, or the resulting stresses are difficult to measure, in an experimental model of the slice itself, while the analogous conditions for the slab, developed below, are easy to handle.

In order to proceed, it is necessary, first, to set down the general boundary value problem for the slice. It is convenient to do this along the lines established by Michell,[6] with the additional consideration of dislocations and thermal dilatations.

2. Airy's stress function and its differential equations. In a state of plane strain defined by setting

---

* Received April 9, 1946.

[1] K. Wieghardt, *Über ein neues Verfahren, verwickelte Spannungsverteilungen in elastischen Körpern auf experimentellem Wege zu finden*, Mitteilungen über Forschungsarbeiten a. d. Gebiete d. Ingenieurwesens, **49**, 15–30 (1908).

[2] H. M. Westergaard, *Graphostatics of stress functions*, Transactions, Amer. Soc. Mech. Eng., **56**, 141–150 (1934).

[3] United States Department of the Interior, Bureau of Reclamation, Boulder Canyon Project, Final Reports (1938), Part V, Technical Investigations: Bulletin 2, *Slab analogy experiments*; Bulletin 4, *Stress studies for Boulder Dam.*

[4] H. Cranz, *Die experimentelle Bestimmung der Airyschen Spannungsfunktion mit Hilfe des Plattengleichnisses*, Ingenieur-Archiv, **10**, 159–166 (1939).

[5] E. Einsporn, *Ebenheit*, Zeitschrift für Instrumentenkunde, **57**, 265–285 (1937).

[6] J. H. Michell, *On the di. ct determination of stress in an elastic solid, with application to the theory of plates*, Proc. London Math. Soc., **31**, 100–124 (1899).

$$\gamma_{yz} = \gamma_{zx} = \epsilon_z = 0$$

and restricting the displacements u and v to be functions of x and y only, the relations between strain, displacement, stress and temperature in an isotropic elastic body are

$$\epsilon_x = \frac{\partial u}{\partial x} = \frac{1}{E_1}\left[(1 - \nu_1^2)\sigma_x - \nu_1(1 + \nu_1)\sigma_y\right] + (1 + \nu_1)\alpha_1 T, \tag{2.1a}$$

$$\epsilon_y = \frac{\partial v}{\partial y} = \frac{1}{E_1}\left[(1 - \nu_1^2)\sigma_y - \nu_1(1 + \nu_1)\sigma_x\right] + (1 + \nu_1)\alpha_1 T, \tag{2.1b}$$

$$\gamma_{xy} = \frac{\partial v}{\partial x} + \frac{\partial u}{\partial y} = \frac{2(1 + \nu_1)}{E_1}\tau_{xy}. \tag{2.1c}$$

These are the relations for a constrained slice. The notations for stress, strain and displacement are the usual ones and E_1, ν_1 are Young's Modulus and Poisson's ratio for the material of the slice, α_1 is the coefficient of linear thermal expansion, and T is the temperature in excess of a uniform initial temperature.

When the stresses are expressed in terms of Airy's stress function (ϕ) and a body force potential (V) by

$$\sigma_x = \frac{\partial^2 \phi}{\partial y^2} + V, \qquad \sigma_y = \frac{\partial^2 \phi}{\partial x^2} + V, \qquad \tau_{xy} = -\frac{\partial^2 \phi}{\partial x \partial y}, \tag{2.2}$$

the equations of equilibrium are satisfied and the strain relation

$$\frac{\partial^2 \epsilon_x}{\partial y^2} + \frac{\partial^2 \epsilon_y}{\partial x^2} = \frac{\partial^2 \gamma_{xy}}{\partial x \partial y} \tag{2.3}$$

yields the differential equation governing ϕ:

$$\nabla^4 \phi = -\frac{1 - 2\nu_1}{1 - \nu_1}\nabla^2 V - \frac{1 + \nu_1}{1 - \nu_1}\alpha_1 \nabla^2 T. \tag{2.4}$$

In a state of plane stress, defined by

$$\sigma_z = \tau_{yz} = \tau_{zx} = 0,$$

the strain-displacement-stress-temperature relations become

$$\epsilon_x = \frac{\partial u}{\partial x} = \frac{1}{E_1}(\sigma_x - \nu_1 \sigma_y) + \alpha_1 T, \tag{2.5a}$$

$$\epsilon_y = \frac{\partial v}{\partial y} = \frac{1}{E_1}(\sigma_y - \nu_1 \sigma_x) + \alpha_1 T, \tag{2.5b}$$

$$\epsilon_z = \frac{\partial w}{\partial z} = -\frac{\nu_1}{E}(\sigma_x + \sigma_y) + \alpha_1 T, \tag{2.5c}$$

$$\gamma_{xy} = \frac{\partial v}{\partial x} + \frac{\partial u}{\partial y} = \frac{2(1 + \nu_1)}{E_1}\tau_{xy}. \tag{2.5d}$$

These are the relations for a free slice. If the components of stress are again expressed in terms of an Airy function and a body force potential by (2.2), the equilibrium equations are identically satisfied and the strain relations reduce to

$$\nabla^4\phi = -(1-\nu_1)\nabla^2 V - (1+\nu_1)\alpha\nabla^2 T \tag{2.6}$$

if terms associated with the coordinate z are neglected.

In what follows, the case of plane strain (constrained slice) will be treated, but the results are directly applicable to plane stress (neglecting z-dependent terms) if Young's modulus E_1, Poisson's ratio ν_1 and the linear thermal expansion coefficient α_1 are replaced by E_1', ν_1' and α_1' where

$$E_1' = \frac{E_1(1+2\nu_1)}{(1+\nu_1)^2}, \qquad \nu_1' = \frac{\nu_1}{1+\nu_1}, \qquad \alpha_1' = \frac{\alpha_1(1+\nu_1)}{1+2\nu_1}. \tag{2.7}$$

3. Conditions on ϕ at a point on a boundary of the slice. Michell[6] gave the conditions to be satisfied, at each point of each boundary, by ϕ and its derivative normal to the boundary:

$$\phi = \int_0^s (Bl - Am)ds + \alpha x + \beta y + \gamma, \tag{3.1}$$

$$\frac{d\phi}{dn} = Al + Bm + \alpha l + \beta m, \tag{3.2}$$

where α, β, γ are constants, in general different for each boundary, ds is an element of arc of a boundary, dn an element of normal to that boundary, and

$$l = \frac{dy}{ds}, \qquad m = \frac{-dx}{ds}, \tag{3.3}$$

$$A = -\int_0^s \overline{Y}ds + \int_0^s Vmds, \qquad B = \int_0^s \overline{X}ds - \int_0^s Vlds, \tag{3.4}$$

$$\overline{X} = \sigma_x l + \tau_{xy}m, \qquad \overline{Y} = \tau_{xy}l + \sigma_y m. \tag{3.5}$$

In a simply connected body, α, β, γ may be assigned arbitrary (including zero) values as the addition of a linear function of x and y to ϕ does not affect the stresses. In a multiply-connected body, three additional conditions on ϕ are required for determining α, β, γ, on each additional boundary. Equations (3.1) to (3.5) are not altered by introducing thermal dilatations and dislocations of the type considered here.

4. Conditions on ϕ for each boundary of the slice. The additional conditions on ϕ are obtained by assuming the strains (and hence the stresses) to be continuous and requiring the rotations and displacements (a) to be single-valued or (b) to have prescribed discontinuities (dislocations). Michell[6] gave the conditions for case (a). The conditions for case (b), including, also, thermal dilatations, are derived by following Michell's procedure with modifications along the lines indicated by Volterra.[7]

(*i*) *Rotation condition.* Considering the rotation

$$\omega_z = \frac{1}{2}\left(\frac{\partial v}{\partial x} - \frac{\partial u}{\partial y}\right), \tag{4.1}$$

[7] Love, *Theory of elasticity*, 4th ed., Cambridge Univ. Press, Cambridge, 1927, pp. 221–228.

we require that the line integral of its differential have a value, say c, after one complete circuit around (and along) a boundary. Thus,

$$c = \oint d\omega_z. \tag{4.2}$$

Now,

$$\oint d\omega_z = \oint \frac{\partial \omega_z}{\partial x} dx + \frac{\partial \omega_z}{\partial y} dy$$

$$= \oint \left(\frac{1}{2} \frac{\partial \gamma_{xy}}{\partial x} - \frac{\partial \epsilon_x}{\partial y}\right) dx + \left(\frac{\partial \epsilon_y}{\partial x} - \frac{1}{2} \frac{\partial \gamma_{xy}}{\partial y}\right) dy. \tag{4.3}$$

Replacing the strain components by their expressions in terms of ϕ, V and T, we find

$$\frac{E_1 c}{1+\nu_1} = (1-\nu_1) \oint \left(\frac{\partial}{\partial x}(\nabla^2\phi) dy - \frac{\partial}{\partial y}(\nabla^2\phi) dx\right) + (1-2\nu_1) \oint \left(\frac{\partial V}{\partial x} dy - \frac{\partial V}{\partial y} dx\right)$$

$$+ E_1\alpha_1 \oint \left(\frac{\partial T}{\partial x} dy - \frac{\partial T}{\partial y} dx\right).$$

Then

$$\oint \frac{d(\nabla^2\phi)}{dn} ds = \frac{E_1 c}{1-\nu_1^2} - \frac{1-2\nu_1}{1-\nu_1} \oint \frac{dV}{dn} ds - \frac{E_1\alpha_1}{1-\nu_1} \oint \frac{dT}{dn} ds. \tag{4.4}$$

This is the first of Michell's three conditions on ϕ for each boundary of the slice. It may be observed that, if the circuit of the line integral in (4.3) were reducible, the integral would vanish because, by Green's theorem,

$$\oint \left(\frac{1}{2} \frac{\partial \gamma_{xy}}{\partial x} - \frac{\partial \epsilon_x}{\partial y}\right) dx + \left(\frac{\partial \epsilon_y}{\partial x} - \frac{1}{2} \frac{\partial \gamma_{xy}}{\partial y}\right) dy$$

$$= \iint \left(\frac{\partial^2 \epsilon_y}{\partial x^2} + \frac{\partial^2 \epsilon_x}{\partial y^2} - \frac{\partial^2 \gamma_{xy}}{\partial x \partial y}\right) dx dy; \tag{4.5}$$

and the surface integral vanishes by virtue of (2.3).

(*ii*) *Displacement conditions.* We admit a translational dislocation a parallel to x and set

$$a = \oint du = \oint \frac{\partial u}{\partial x} dx + \frac{\partial u}{\partial y} dy = \oint \left(\epsilon_x dx + \frac{1}{2} \gamma_{xy} dy\right) - \oint \omega_z dy.$$

Now

$$\oint \omega_z dy = y_0 \oint d\omega_z - \oint y d\omega_z = y_0 c - \oint y\, d\omega_z,$$

where y_0 is the y-coordinate of the starting point of integration. Also

$$\oint y d\omega_z = \oint y \left(\frac{\partial \omega_z}{\partial x} dx + \frac{\partial \omega_z}{\partial y} dy\right)$$

$$= \oint y \left(\frac{1}{2} \frac{\partial \gamma_{xy}}{\partial x} - \frac{\partial \epsilon_x}{\partial y}\right) dx + \oint y \left(\frac{\partial \epsilon_y}{\partial x} - \frac{1}{2} \frac{\partial \gamma_{xy}}{\partial y}\right) dy.$$

Hence

$$a + y_0c = \oint\left[\epsilon_x + y\left(\frac{1}{2}\frac{\partial\gamma_{xy}}{\partial x} - \frac{\partial\epsilon_x}{\partial y}\right)\right]dx$$
$$+ \oint\left[\frac{1}{2}\gamma_{xy} + y\left(\frac{\partial\epsilon_y}{\partial x} - \frac{1}{2}\frac{\partial\gamma_{xy}}{\partial y}\right)\right]dy. \tag{4.6}$$

We now note that

$$\oint \epsilon_x dx + \frac{1}{2}\gamma_{xy}dy = [x\epsilon_x]_0^0 + \frac{1}{2}[y\gamma_{xy}]_0^0 - \oint\left(x\frac{\partial\epsilon_x}{\partial x}dx + \frac{1}{2}y\frac{\partial\gamma_{xy}}{\partial y}dy\right)$$
$$= -\oint\left(x\frac{\partial\epsilon_x}{\partial x}dx + \frac{1}{2}y\frac{\partial\gamma_{xy}}{\partial y}dy\right),$$

the terms outside the integrals vanishing because of the assumption of continuous strains. Equation (4.6) then becomes

$$a + y_0c = \oint\left[y\left(\frac{1}{2}\frac{\partial\gamma_{xy}}{\partial x} - \frac{\partial\epsilon_x}{\partial y}\right) - x\frac{\partial\epsilon_x}{\partial x}\right]dx + \oint y\left(\frac{\partial\epsilon_y}{\partial x} - \frac{\partial\gamma_{xy}}{\partial y}\right)dy. \tag{4.7}$$

When the strain components in (4.7) are replaced by their expressions in terms of ϕ, V and T, we find

$$\frac{E_1(a + y_0c)}{1 + \nu_1} = (1 - \nu_1)\oint y\left[\frac{\partial}{\partial x}(\nabla^2\phi)dy - \frac{\partial}{\partial y}(\nabla^2\phi)dx\right]$$
$$+ (1 - 2\nu_1)\oint y\left(\frac{\partial V}{\partial x}dy - \frac{\partial V}{\partial y}dx\right)$$
$$+ E_1\alpha_1\oint y\left(\frac{\partial T}{\partial x}dy - \frac{\partial T}{\partial y}dx\right)$$
$$- \oint x\left[(1 - \nu_1)\frac{\partial}{\partial x}(\nabla^2\phi) + (1 - 2\nu_1)\frac{\partial V}{\partial x} + E_1\alpha_1\frac{\partial T}{\partial x}\right]dx$$
$$+ \oint\left(x\frac{\partial^3\phi}{\partial x^3}dx + y\frac{\partial^3\phi}{\partial x\partial y^2}dy\right). \tag{4.8}$$

Now,

$$\oint\left(x\frac{\partial^3\phi}{\partial x^3}dx + y\frac{\partial^3\phi}{\partial x\partial y^2}dy\right) = \left[x\frac{\partial^2\phi}{\partial x^2} + y\frac{\partial^2\phi}{\partial x\partial y}\right]_0^0 - \oint\left(\frac{\partial^2\phi}{\partial x^2}dx + \frac{\partial^2\phi}{\partial x\partial y}dy\right)$$
$$= -\oint\frac{d}{ds}\left(\frac{\partial\phi}{\partial x}\right)ds,$$

the term outside the integral vanishing because the stresses are continuous. But, from (2.2), (3.3) and (3.5),

$$\frac{d}{ds}\left(\frac{\partial\phi}{\partial x}\right) = Vm - \overline{Y}.$$

Hence (4.8) may be written:

$$\oint \left(y \frac{d(\nabla^2\phi)}{dn} - x \frac{d(\nabla^2\phi)}{ds} \right) ds = \frac{E_1(a + y_0 c)}{1 - \nu_1^2} - \frac{1 - 2\nu_1}{1 - \nu_1} \oint \left(y \frac{dV}{dn} - x \frac{dV}{ds} \right) ds$$
$$- \frac{E_1\alpha_1}{1 - \nu_1} \oint \left(y \frac{dT}{dn} - x \frac{dT}{ds} \right) ds$$
$$- \frac{1}{1 - \nu_1} \oint (\overline{Y} - Vm) ds. \tag{4.9}$$

This is Michell's second condition on ϕ for each boundary of the slice.

Similarly, admitting a translational dislocation b in the y-component of displacement, we set

$$b = \oint dv$$

and we find

$$\oint \left(y \frac{d(\nabla^2\phi)}{ds} + x \frac{d(\nabla^2\phi)}{dn} \right) ds = - \frac{E_1(b - x_0 c)}{1 - \nu_1^2} - \frac{1 - 2\nu_1}{1 - \nu_1} \oint \left(y \frac{dV}{ds} + x \frac{dV}{dn} \right) ds$$
$$- \frac{E_1\alpha_1}{1 - \nu_1} \oint \left(y \frac{dT}{ds} + x \frac{dT}{dn} \right) ds$$
$$- \frac{1}{1 - \nu_1} \oint (\overline{X} - Vl) ds, \tag{4.10}$$

which is the last of Michell's three conditions.

Corresponding to (4.5), a similar application of Green's theorem to (4.6) reveals that the right hand side of the latter vanishes for reducible circuits and the same result is found for the corresponding step in the development of Michell's third condition.

The differential equation (2.6), the boundary conditions (3.1) and (3.2), and the three conditions (4.4), (4.9) and (4.10) constitute a statement of the boundary value problem of plane elasticity for stresses induced by boundary loading, body forces, dislocations, and thermal dilatations. The general formulation of the problem reveals the analogies, discovered by M. A. Biot,[8] between gravity loading and boundary pressures, and between thermal loading and boundary pressures and dislocations.

5. The slab equations. In the approximate theory of the bending of thin plates[9] (slabs), the deflection (w) is governed by the differential equation

$$D\nabla^4 w = Z, \tag{5.1}$$

where D is the flexural rigidity of the plate and Z is the surface load, normal to the middle plane.

The components of curvature in the y, z and x, z planes are given by

[8] M. A. Biot, *Distributed gravity and temperature loading in two-dimensional elasticity replaced by boundary pressures and dislocations*, J. Appl. Mech., **2**, A 41–A 95 (1935).

[9] Love, loc. cit., p. 487.

$$\kappa_x = \frac{\partial^2 w}{\partial y^2}, \qquad \kappa_y = \frac{\partial^2 w}{\partial x^2}. \tag{5.2}$$

On a boundary of the slab, the shearing force (N) normal to the middle plane, the flexural couple (G), and the torsional couple (H) (all per unit of arc length s) are

$$N = -D\frac{\partial}{\partial n}(\nabla^2 w), \tag{5.3a}$$

$$G = -D\left[\frac{\partial^2 w}{\partial n^2} + \nu_2\left(\frac{\partial^2 w}{\partial s^2} + \frac{1}{\rho'}\frac{\partial w}{\partial n}\right)\right], \tag{5.3b}$$

$$H = (1-\nu_2)D\frac{\partial}{\partial n}\left(\frac{\partial w}{\partial s}\right), \tag{5.3c}$$

where ρ' is the radius of curvature of the boundary of the unflexed slab and ν_2 is Poisson's ratio for the slab material.

The resultant force and the components, parallel to the x and y axes, of the resultant couple on a complete boundary are[10]

$$F_z = \oint\left(N - \frac{\partial H}{\partial s}\right)ds, \tag{5.4a}$$

$$M_x = \oint\left[y\left(N - \frac{\partial H}{\partial s}\right) + G\frac{dx}{ds}\right]ds, \tag{5.4b}$$

$$M_y = \oint\left[G\frac{dy}{ds} - x\left(N - \frac{\partial H}{\partial s}\right)\right]ds. \tag{5.4c}$$

Substituting (5.3) in (5.4) we find

$$F_z = -D\oint\left[\frac{\partial}{\partial n}(\nabla^2 w) + (1-\nu_2)\frac{\partial}{\partial s}\frac{\partial}{\partial n}\left(\frac{\partial w}{\partial s}\right)\right]ds, \tag{5.5a}$$

$$M_x = -D\oint\left\{y\left[\frac{\partial}{\partial n}(\nabla^2 w) + (1-\nu_2)\frac{\partial}{\partial s}\frac{\partial}{\partial n}\left(\frac{\partial w}{\partial s}\right)\right]\right.$$
$$\left. + \frac{dx}{ds}\left[\frac{\partial^2 w}{\partial n^2} + \nu_2\left(\frac{\partial^2 w}{\partial s^2} + \frac{1}{\rho'}\frac{\partial w}{\partial n}\right)\right]\right\}ds, \tag{5.5b}$$

$$M_y = -D\oint\left\{\frac{dy}{ds}\left[\frac{\partial^2 w}{\partial n^2} + \nu_2\left(\frac{\partial^2 w}{\partial s^2} + \frac{1}{\rho'}\frac{\partial w}{\partial n}\right)\right]\right.$$
$$\left. - x\left[\frac{\partial}{\partial n}(\nabla^2 w) + (1-\nu_2)\frac{\partial}{\partial s}\frac{\partial}{\partial n}\left(\frac{\partial w}{\partial s}\right)\right]\right\}ds. \tag{5.5c}$$

6. The analogy for singly-connected bodies. Noting the similarity between the differential equations (2.6) and (5.1) for ϕ and w, we set

$$w = K\phi, \tag{6.1}$$

[10] Love, loc. cit., p. 460.

where K is a conversion constant having the dimensions of length/force.

Then, from (6.1) and (2.6),

$$\nabla^4 w = -\frac{1-2\nu_1}{1-\nu_1} K\nabla^2 V - \frac{1+\nu_1}{1-\nu_1} K\alpha_1 \nabla^2 T \tag{6.2}$$

becomes the differential equation for the deflection of the analogous slab. Hence

$$Z = -\frac{1-2\nu_1}{1-\nu_1} KD\nabla^2 V - \frac{1+\nu_1}{1-\nu_1} KD\alpha_1 \nabla^2 T \tag{6.3}$$

is the normal surface loading to be applied to the face of the slab. In the case of a steady state temperature distribution,

$$\nabla^2 T = 0. \tag{6.4}$$

If, in addition, the body force potential is harmonic, the slab is subjected to edge loading only. If either V or T is not harmonic, transverse loading is required on the surface of the slab, and the load may vary slowly with time.

The edge conditions (i.e., the elevation and slope at each point of a boundary) of the slab are specified by substituting $w=K\phi$ in (3.1) and (3.2). Thus

$$\frac{w}{K} = \int_0^s (Bl - Am)ds + \alpha x + \beta y + \gamma, \tag{6.5}$$

$$\frac{1}{K}\frac{dw}{dn} = Al + Bm + \alpha l + \beta m. \tag{6.6}$$

The normal components of stress in the slice are obtained by combining (2.2), (5.2) and (6.1), with the result

$$\sigma_x = \frac{\kappa_x}{K} + V, \qquad \sigma_y = \frac{\kappa_y}{K} + V. \tag{6.7}$$

The principal stresses and their directions may be calculated from two sets of curvature measurements at each point.[4] If the boundary of the slab is a scale model of the boundary of the slice, e.g., if the ratio of a linear dimension of the slab to the corresponding linear dimension of the slice is k, the stress components in the slice are given by

$$\sigma_x = \frac{k^2\kappa_x}{K} + V, \qquad \sigma_y = \frac{k^2\kappa_y}{K} + V. \tag{6.8}$$

For a singly-connected body, (6.1) to (6.8) completely specify the analogy, since the unknown constants α, β, γ may be given arbitrary values.

7. Additional conditions on the slab for multiply-connected bodies. For a multiply-connected body, α, β, γ must be prescribed for each boundary. Now, it will be observed, from (6.5) and (6.6), that α, β, γ specify a rigid body translation and rotation of each complete boundary of the slab. Such rigid body movements may be effected by applying, on each boundary, a resultant force, normal to the middle plane of the slab, and a couple about an axis properly oriented in the plane of the slab. The magnitudes of the force and the x and y components of the couple on each boundary are determined by expressing F_z, M_x, M_y (see (5.5)) in terms of the specified boundary loadings, body forces, dislocations, and temperature distribution of the slice.

i. Resultant force on a boundary of the slab. Replacing w by $K\phi$ in (5.5a), we have

$$F_z = -KD \oint \left[\frac{\partial}{\partial n}(\nabla^2\phi) + (1 - \nu_2)\frac{\partial}{\partial s}\frac{\partial}{\partial n}\left(\frac{\partial\phi}{\partial s}\right)\right] ds. \tag{7.1}$$

Now

$$\oint \frac{\partial}{\partial s}\frac{\partial}{\partial n}\left(\frac{\partial\phi}{\partial s}\right) ds = 0$$

because of the assumption of continuity of the components of stress in the slice. Hence

$$F_z = -KD \oint \frac{\partial}{\partial n}(\nabla^2\phi) ds. \tag{7.2}$$

Therefore, from (4.4),

$$\frac{(1 - \nu_1)F_z}{KD} = -\frac{E_1 c}{1 + \nu_1} + (1 - 2\nu_1)\oint \frac{dV}{dn} ds + E_1\alpha_1 \oint \frac{dT}{dn} ds, \tag{7.3}$$

whereby F_z is expressed in terms of known quantities.

ii. x-component of couple on a boundary of the slab. Substituting $K\phi$ for w in (5.5b):

$$M_x = -KD \oint \left\{ y\left[\frac{\partial}{\partial n}(\nabla^2\phi) + (1 - \nu_2)\frac{\partial}{\partial s}\frac{\partial}{\partial n}\left(\frac{\partial\phi}{\partial s}\right)\right] + \frac{dx}{ds}\left[\frac{\partial^2\phi}{\partial n^2} + \nu_2\left(\frac{\partial^2\phi}{\partial s^2} + \frac{1}{\rho'}\frac{\partial\phi}{\partial n}\right)\right]\right\} ds. \tag{7.4}$$

Eliminating

$$\oint y \frac{\partial}{\partial n}(\nabla^2\phi) ds$$

between (7.4) and (4.9), we find

$$\begin{aligned}\frac{M_x}{KD} = &-\oint \left\{ x\frac{d(\nabla^2\phi)}{ds} + (1 - \nu_2)y\frac{\partial}{\partial s}\frac{\partial}{\partial n}\left(\frac{\partial\phi}{\partial s}\right) + \frac{dx}{ds}\left[\frac{\partial^2\phi}{\partial n^2} + \nu_2\left(\frac{\partial^2\phi}{\partial s^2} + \frac{1}{\rho'}\frac{\partial\phi}{\partial n}\right)\right]\right\} ds \\ &- \frac{E_1(a + y_0 c)}{1 - \nu_1^2} + \frac{1 - 2\nu_1}{1 - \nu_1}\oint\left(y\frac{dV}{dn} - x\frac{dV}{ds}\right) ds \\ &+ \frac{E_1\alpha_1}{1 - \nu_1}\oint\left(y\frac{dT}{dn} - x\frac{dT}{ds}\right) ds + \frac{1}{1 - \nu_1}\oint(\overline{Y} - Vm) ds.\end{aligned} \tag{7.5}$$

Now,

$$\oint \left[x \frac{\partial(\nabla^2\phi)}{\partial s} + (1 - \nu_2) y \frac{\partial}{\partial s} \frac{\partial}{\partial n} \left(\frac{\partial \phi}{\partial s} \right) \right] ds = \left[x \nabla^2 \phi \right]_0^0 + (1 - \nu_2) \left[y \frac{\partial}{\partial n} \left(\frac{\partial \phi}{\partial s} \right) \right]_0^0$$

$$- \oint \left[\frac{dx}{ds} \nabla^2 \phi + (1 - \nu_2) \frac{dy}{ds} \frac{\partial}{\partial n} \left(\frac{\partial \phi}{\partial s} \right) \right] ds. \tag{7.6}$$

The terms outside the integral vanish on account of the assumption of continuity of the stress components. Therefore the first integral on the right hand side of (7.5) becomes

$$\oint \left\{ (1 - \nu_2) \frac{dy}{ds} \frac{\partial}{\partial n} \left(\frac{\partial \phi}{\partial s} \right) + \frac{dx}{ds} \left[\nabla^2 \phi - \frac{\partial^2 \phi}{\partial n^2} - \nu_2 \left(\frac{\partial^2 \phi}{\partial s^2} + \frac{1}{\rho'} \frac{\partial \phi}{\partial n} \right) \right] \right\} ds. \tag{7.7}$$

On a boundary

$$\nabla^2 \phi = \frac{\partial^2 \phi}{\partial n^2} + \frac{1}{\rho'} \frac{\partial \phi}{\partial n} + \frac{\partial^2 \phi}{\partial s^2}, \tag{7.8}$$

so that (7.7) becomes

$$(1 - \nu_2) \oint \left[\frac{dy}{ds} \frac{\partial}{\partial n} \left(\frac{\partial \phi}{\partial s} \right) + \frac{dx}{ds} \left(\frac{\partial^2 \phi}{\partial s^2} + \frac{1}{\rho'} \frac{\partial \phi}{\partial n} \right) \right] ds. \tag{7.9}$$

However, along a boundary,

$$\frac{\partial}{\partial n} \left(\frac{\partial \phi}{\partial s} \right) = - \tau_{ns}, \tag{7.10}$$

$$\frac{\partial^2 \phi}{\partial s^2} + \frac{1}{\rho'} \frac{\partial \phi}{\partial n} = \sigma_n - V, \tag{7.11}$$

$$\frac{dy}{ds} \tau_{ns} - \frac{dx}{ds} \sigma_n = \overline{Y}. \tag{7.12}$$

Hence, (7.9) becomes

$$- (1 - \nu_2) \oint \left(\overline{Y} + V \frac{dx}{ds} \right) ds.$$

Substituting back in (7.5), we have, finally,

$$\frac{(1 - \nu_1) M_x}{KD} = - \frac{E_1 (a + y_0 c)}{1 + \nu_1} + (1 - 2\nu_1) \oint \left(y \frac{dV}{dn} - x \frac{dV}{ds} \right) ds$$

$$+ E_1 \alpha_1 \oint \left(y \frac{dT}{dn} - x \frac{dT}{ds} \right) ds$$

$$- \left[(1 - \nu_2)(1 - \nu_1) + 1 \right] \oint (\overline{Y} - Vm) ds. \tag{7.13}$$

This gives M_x in terms of known quantities.

iii. y-component of couple on a boundary of the slab. Substituting $K\phi$ for w in (5.5c),

$$M_y = -KD\oint\left\{\frac{dy}{ds}\left[\frac{\partial^2\phi}{\partial n^2}+\nu_2\left(\frac{\partial^2\phi}{\partial s^2}+\frac{1}{\rho'}\frac{\partial\phi}{\partial n}\right)\right]\right.$$
$$\left.-x\left[\frac{\partial(\nabla^2\phi)}{\partial n}+(1-\nu_2)\frac{\partial}{\partial s}\frac{\partial}{\partial n}\left(\frac{\partial\phi}{\partial s}\right)\right]\right\}ds. \tag{7.14}$$

Eliminating

$$\oint x\frac{\partial}{\partial n}(\nabla^2\phi)ds$$

between (7.14) and (4.10), we have

$$\frac{M_y}{KD} = -\oint\left\{y\frac{d}{ds}(\nabla^2\phi)-(1-\nu_2)x\frac{\partial}{\partial s}\frac{\partial}{\partial n}\left(\frac{\partial\phi}{\partial s}\right)\right.$$
$$\left.+\frac{dy}{ds}\left[\frac{\partial^2\phi}{\partial n^2}+\nu_2\left(\frac{\partial^2\phi}{\partial s^2}+\frac{1}{\rho'}\frac{\partial\phi}{\partial n}\right)\right]\right\}ds$$
$$-\frac{E_1(b-x_0c)}{1-\nu_1^2}-\frac{1-2\nu_1}{1-\nu_1}\oint\left(y\frac{dV}{ds}+x\frac{dV}{dn}\right)ds$$
$$-\frac{E_1\alpha_1}{1-\nu_1}\oint\left(y\frac{dT}{ds}+x\frac{dT}{dn}\right)ds-\frac{1}{1-\nu_1}\oint(X-Vl)ds. \tag{7.15}$$

Now,

$$\oint\left[y\frac{d(\nabla^2\phi)}{ds}-(1-\nu_2)x\frac{\partial}{\partial s}\frac{\partial}{\partial n}\left(\frac{\partial\phi}{\partial s}\right)\right]ds = [y\nabla^2\phi]_0^0-(1-\nu_2)\left[x\frac{\partial}{\partial n}\left(\frac{\partial\phi}{\partial s}\right)\right]_0^0$$
$$-\oint\left[\frac{dy}{ds}\nabla^2\phi-(1-\nu_2)\frac{dx}{ds}\frac{\partial}{\partial n}\left(\frac{\partial\phi}{\partial s}\right)\right]ds. \tag{7.16}$$

The terms outside the integrals in (7.16) vanish on account of the assumption of continuity of stresses. Therefore the first integral on the right hand side of (7.15) becomes

$$\oint\left\{\frac{dy}{ds}\left[\nabla^2\phi-\frac{\partial^2\phi}{\partial n^2}-\nu_2\left(\frac{\partial^2\phi}{\partial s^2}+\frac{1}{\rho'}\frac{\partial\phi}{\partial n}\right)\right]-(1-\nu_2)\frac{dx}{ds}\frac{\partial}{\partial n}\left(\frac{\partial\phi}{\partial s}\right)\right\}ds. \tag{7.17}$$

Then, using (7.8), (7.10) and (7.11) and noting that

$$\tau_{ns}\frac{dx}{ds}+\sigma_n\frac{dy}{ds}=\overline{X}, \tag{7.18}$$

(7.17) may be written in the form

$$(1-\nu_2)\oint(\overline{X}-Vl)ds. \tag{7.19}$$

Substituting back in (7.15), we have

$$\frac{(1-\nu_1)M_y}{KD} = -\frac{E_1(b - x_0c)}{1+\nu_1} - (1-2\nu_1)\oint\left(y\frac{dV}{ds} + x\frac{dV}{dn}\right)ds$$
$$- E_1\alpha_1\oint\left(y\frac{dT}{ds} + x\frac{dT}{dn}\right)ds$$
$$+ [(1-\nu_2)(1-\nu_1) - 1]\oint(\overline{X} - Vl)ds. \tag{7.20}$$

8. Recapitulation. The stresses, in a multiply-connected slice, resulting from boundary loadings $\overline{X}$, $\overline{Y}$, a body force potential V, dislocations a, b, c and temperatures T, are related to the curvatures of a slab according to

$$\sigma_x = \frac{\kappa_x}{K} + V, \qquad \sigma_y = \frac{\kappa_y}{K} + V$$

if the following conditions are satisfied on the slab:
(i) The surface loading on the slab is

$$Z = -\frac{1-2\nu_1}{1-\nu_1}KD\nabla^2 V - \frac{1+\nu_1}{1-\nu_1}KD\alpha_1\nabla^2 T; \tag{6.3}$$

(ii) The boundaries of the slab are geometrically identical with those of the slice, with elevations and normal slopes given by

$$\frac{w}{K} = \int_0^s (Bl - Am)ds + \alpha x + \beta y + \gamma, \quad \frac{1}{K}\frac{dw}{dn} = Al + Bm + \alpha l + \beta m, \tag{6.5}$$

at each point of each boundary;
(iii) There are a resultant force (F_z) and resultant couples (M_x) and (M_y), on each boundary, with magnitudes given by

$$\frac{(1-\nu_1)F_z}{KD} = -\frac{E_1c}{1+\nu_1} + (1-2\nu_1)\oint\frac{dV}{dn}ds + E_1\alpha_1\oint\frac{dT}{dn}ds, \tag{7.2}$$

$$\frac{(1-\nu_1)M_x}{KD} = -\frac{E_1(a + y_0c)}{1+\nu_1} + (1-2\nu_1)\oint\left(y\frac{dV}{dn} - x\frac{dV}{ds}\right)ds$$
$$+ E_1\alpha_1\oint\left(y\frac{dT}{dn} - x\frac{dT}{ds}\right)ds$$
$$- [(1-\nu_1)(1-\nu_2) - 1]\oint(\overline{Y} - Vm)ds, \tag{7.13}$$

$$\frac{(1-\nu_1)M_y}{KD} = -\frac{E_1(b - x_0c)}{1+\nu_1} - (1-2\nu_1)\oint\left(y\frac{dV}{ds} + x\frac{dV}{dn}\right)ds$$
$$- E_1\alpha_1\oint\left(y\frac{dT}{ds} + x\frac{dT}{dn}\right)ds$$
$$+ [(1-\nu_1)(1-\nu_2) - 1]\oint(\overline{X} - Vl)ds. \tag{7.20}$$

Reprinted from QUARTERLY OF APPLIED MATHEMATICS
Vol. V, No. 2, July, 1947

CORRECTIONS TO MY PAPER

THE ANALOGY BETWEEN MULTIPLY-CONNECTED SLICES AND SLABS*

QUARTERLY OF APPLIED MATHEMATICS, **3**, 279–290 (1946)

BY RAYMOND D. MINDLIN (*Columbia University*)

Equations (2.4), (2.6), (6.2), (6.3): replace $(1+\nu_1)$ by E_1 in the coefficient of $\nabla^2 T$.

Equation (2.6): replace α by α_1.

Equation (6.2) is obtained from (6.1) and (2.4) and not from (6.1) and (2.6).

The reference to Biot's analogies at the end of Sec. 4 should read "analogies . . . between gravity loading and boundary pressures and dislocations and between thermal loading and dislocations."

The integration by parts following Eq. (4.6) is incomplete with the result that the following corrections should be made:

Equations (4.9), (4.10), (7.5), (7.15): multiply the last integral by 2.

Equations (7.13), (7.20): replace $[(1-\nu_1)(1-\nu_2)-1]$ by $[(1-\nu_1)(1-\nu_2)-2]$.

* Received April 22, 1947.

Reprinted from
Proceedings of the Society for Experimental Stress Analysis
V, no. 2 (1948) 56-68

STRESS DISTRIBUTION AROUND A HOLE NEAR THE EDGE OF A PLATE UNDER TENSION

RAYMOND D. MINDLIN, Columbia University

INTRODUCTION

This paper deals with the stress in a wide plate under simple tension parallel to a straight edge and containing a small circular hole located near the edge (see Fig. 1). The problem was first studied by G. B. Jeffery (1)* as an application of his general solution of the two-dimensional equations of elasticity in bipolar coordinates. Jeffery's results have been widely quoted (2) and have led to some discussion, especially because he predicted that the component of stress parallel to the straight edge becomes compressive in the neighborhood of point m (Fig. 1) when the hole is very close to the edge of the plate.

In the course of the solution of another problem, (3) (hereinafter called the "Tunnel" problem) in which Jeffery's general method was employed, it was noticed that Jeffery had made an error in his application to the problem under present discussion. As a result of the error, the boundary conditions on the straight edge were not satisfied and this led to incorrect stresses in the neighborhood of points m and n (Fig. 1), the defect becoming more pronounced as the hole is located closer to the straight edge. The corrected mathematical solution, reported in this paper, shows that the stress at point m always is tensile, but approaches zero as the hole approaches the edge of the plate. The stress at point n (the largest stress) is found to be higher than that reported by Jeffery.

In a discussion of the Tunnel paper,† J. H. A. Brahtz gave some of the results of a photoelastic study conducted at the U. S. Bureau of Reclamation for the purpose of comparison with Jeffery's solution. (4) In the study at the Bureau of Reclamation, six values were used for the ratio of the distance (c) of the center of the hole from the straight edge to the radius (R_2) of the hole as follows: $c/R_2 = 6.0, 5.0, 4.0, 3.0, 2.0, 1.5$. According to Jeffery's results, the stress at m (Fig. 1) for $c/R_2 = 2.0$ should have been zero and for $c/R_2 = 1.5$, should have been negative. Brahtz found the stresses along the straight edge to be tensile throughout.

*Numbers in parentheses refer to bibliography at end of paper.
†Mindlin, op. cit., p. 1143.

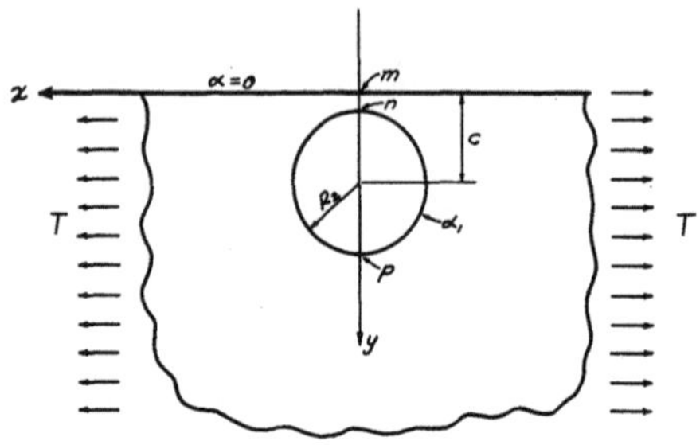

FIG. 1. Semi-infinite plate with a circular hole near the straight edge and under tension parallel to the straight edge.

The corrected mathematical solution revealed such pronounced peculiarities in the stress distribution for $c/R_2 < 1.5$ that it was considered desirable to extend the photoelastic studies to the smaller values. Accordingly, experiments were made for $c/R_2 = 1.5, 1.2$, and 1.1. In the last case the shortest distance between the edge of the hole and the edge of the plate was only 5% of the diameter of the hole, so that unusual care was required in the experimental procedure.

Statement of the Mathematical Problem

We consider a semi-infinite plate of unit thickness, with the x, y-plane as its middle surface. The plate is bounded by the plane $y = 0$ (the straight edge) and contains a cylindrical hole of radius R_2, with its center on the y-axis distant c from the origin. Far from the hole, a uniform tensile stress T acts parallel to the x axis.

The problem is treated as one of plane stress, ignoring terms depending upon the z-coordinate. The components of stress can then be expressed in terms of a plane biharmonic function χ. In rectangular coordinates these expressions take the form (5)

$$\widehat{xx} = \frac{\partial^2\chi}{\partial y^2}, \tag{1a}$$

$$\widehat{yy} = \frac{\partial^2\chi}{\partial x^2}, \tag{1b}$$

$$\widehat{xy} = -\frac{\partial^2\chi}{\partial x\partial y}, \tag{1c}$$

$$\widehat{zx} = \widehat{zy} = \widehat{zz} = 0,$$

and χ is a function of x and y satisfying the equation

$$\nabla^4\chi = 0 \tag{2}$$

in which $\nabla^4 = \nabla^2\nabla^2$ and ∇^2 is Laplace's operator $\left(\frac{\partial^2}{\partial x^2} + \frac{\partial^2}{\partial y^2}\right)$.

It is required to find a solution of equation (2) for which the normal and shearing components of stress vanish on the straight and circular boundaries and all the stresses reduce to a simple tension $\widehat{xx} = T$ far from the hole.

Curvilinear Coordinates

The curvilinear coordinate system (α,β) used in the solution is obtained through a conformal transformation*

$$x + iy = f(\alpha + i\beta). \tag{3}$$

The stretch ratio, J, of the transformation and the angle φ which the tangent to the curve β = *constant* makes with the x-axis are given by †

$$Je^{i\varphi} = f'(\alpha + i\beta) \tag{4}$$

If ρ_1 and ρ_2 are the radii of curvature of the curves β = constant and α = constant, respectively, (both positive for counterclockwise rotation of the tangent moving in the positive direction along the curve) then†

$$\frac{1}{\rho_1} = \frac{1}{J}\frac{\partial\varphi}{\partial\alpha} = -\frac{1}{J^2}\frac{\partial J}{\partial\beta}, \tag{5a}$$

$$\frac{1}{\rho_2} = \frac{1}{J}\frac{\partial\phi}{\partial\beta} = \frac{1}{J^2}\frac{\partial J}{\partial\alpha}. \tag{5b}$$

Finally, the components of stress in curvilinea coordinates are given by‡

$$\widehat{\alpha\alpha} = \frac{1}{J^2}\frac{\partial^2\chi}{\partial\beta^2} + \frac{1}{J^3}\left(\frac{\partial\chi}{\partial\alpha}\frac{\partial J}{\partial\alpha} - \frac{\partial\chi}{\partial\beta}\frac{\partial J}{\partial\beta}\right), \tag{6a}$$

$$\widehat{\beta\beta} = \frac{1}{J^2}\frac{\partial^2\chi}{\partial\alpha^2} + \frac{1}{J^3}\left(\frac{\partial\chi}{\partial\beta}\frac{\partial J}{\partial\beta} - \frac{\partial\chi}{\partial\alpha}\frac{\partial J}{\partial\alpha}\right), \tag{6b}$$

$$\widehat{\alpha\beta} = -\frac{1}{J^2}\frac{\partial^2\chi}{\partial\alpha\partial\beta} + \frac{1}{J^3}\left(\frac{\partial\chi}{\partial\alpha}\frac{\partial J}{\partial\beta} + \frac{\partial\chi}{\partial\beta}\frac{\partial J}{\partial\alpha}\right). \tag{6c}$$

*Coker and Filon, op. cit., p. 152.
†Coker and Filon, op. cit., p. 153.
‡Coker and Filon, op. cit., p. 163.

Bipolar Coordinates

Following Jeffery, the bipolar coordinate system is used. This system is obtained by particularizing equation (3) as

$$x + iy = ia \coth \tfrac{1}{2}(\alpha + i\beta) \tag{7}$$

where a is the distance from the origin to a pole.

Equating real and imaginary parts in equation (7), there results

$$x = \frac{a \sin\beta}{\cosh\alpha - \cos\beta}, \tag{8a}$$

$$y = \frac{a \sinh\alpha}{\cosh\alpha - \cos\beta}. \tag{8b}$$

From equations (4) and (7),

$$Je^{i\varphi} = -\frac{ia}{2}\operatorname{csch}^2 \tfrac{1}{2}(\alpha + i\beta), \tag{9}$$

whence

$$J = \frac{a}{\cosh\alpha - \cos\beta}, \tag{10a}$$

$$\tan\varphi = \frac{\cosh\alpha\cos\beta - 1}{\sinh\alpha\sin\beta}. \tag{10b}$$

Equation (7) may also be written in the form

$$\alpha + i\beta = \log\frac{x + i(y + a)}{x + i(y - a)}. \tag{11}$$

Writing

$$x + i(y + a) = r_1e^{i\theta_1},$$

$$x + i(y - a) = r_2e^{i\theta_2},$$

and equating real and imaginary parts in equation (11) we have

$$\alpha = \log\frac{r_1}{r_2} \tag{12a}$$

$$\beta = \theta_1 - \theta_2 \tag{12b}$$

in which r_1 and r_2 are the radii to a point in the xy plane from the points $(0, -a)$, $(0, a)$, respectively, and θ_1, θ_2 are the angles which these radii make with the x-axis, as shown in Fig. 2.

Referring to Fig. 2, the curves α = *constant* are coaxial circles of radius a *csch* α, in accordance with equations (5b) and (10). The poles $(0, a)$, $(0, -a)$ are the limiting points of these circles as $\alpha \to \pm\infty$. The centers of the circles α = constant lie on the y-axis at distances

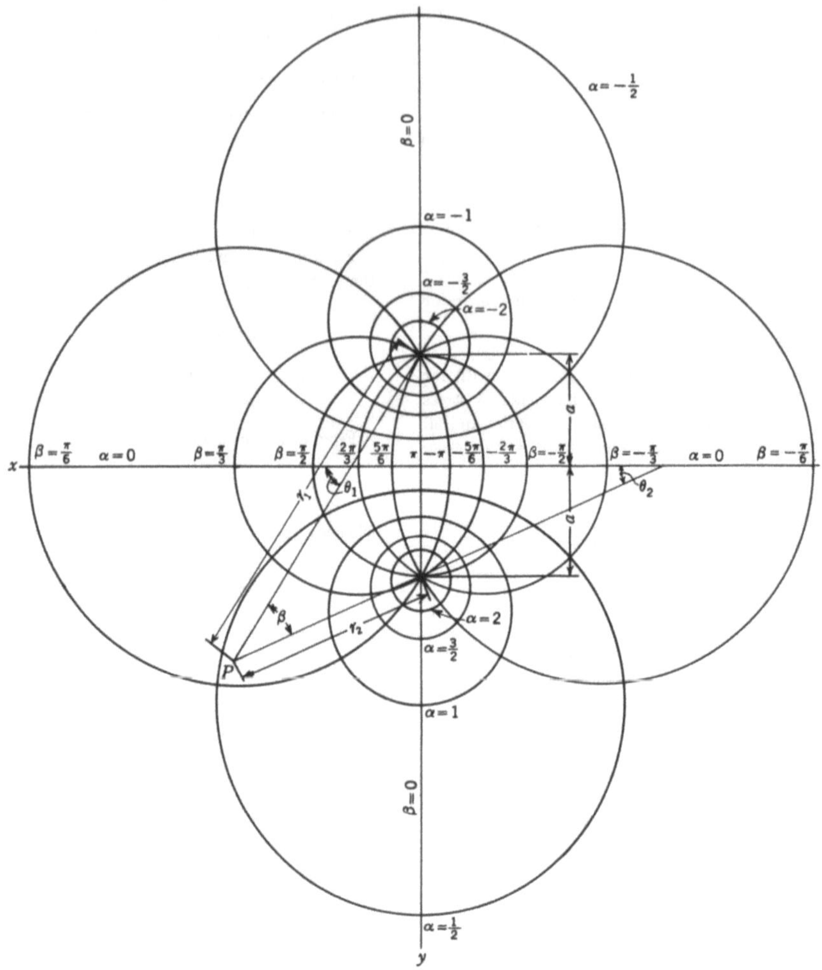

FIG. 2. Bipolar coordinates.

$a \coth \alpha$ from the origin. The x-axis is the curve $\alpha = 0$. This and a circle α = constant (positive) are the boundaries.

The curves β = constant are coaxial circular arcs passing through the poles $(0, \pm a)$. β is the angle included between the radii r_1 and r_2; it has a discontinuity of 2π on passing between the poles.

In terms of θ_1 and θ_2, the angle φ which the tangent to β = constant makes with the x-axis is

$$\varphi = \theta_1 + \theta_2 + \frac{\pi}{2} \tag{13}$$

From equations (6) and (10a) the components of stress in bipolar coordinates are found to be

$$\widehat{\alpha\alpha} = \left[(\cosh\alpha - \cos\beta)\frac{\partial^2}{\partial\beta^2} - \sinh\alpha\frac{\partial}{\partial\alpha} - \sin\beta\frac{\partial}{\partial\beta} + \cosh\alpha\right]\left(\frac{\chi}{J}\right), \quad (14a)$$

$$\widehat{\beta\beta} = \left[(\cosh\alpha - \cos\beta)\frac{\partial^2}{\partial\alpha^2} - \sinh\alpha\frac{\partial}{\partial\alpha} - \sin\beta\frac{\partial}{\partial\beta} + \cos\beta\right]\left(\frac{\chi}{J}\right), \quad (14b)$$

$$\widehat{\alpha\beta} = -(\cosh\alpha - \cos\beta)\frac{\partial^2}{\partial\alpha\partial\beta}\left(\frac{\chi}{J}\right). \quad (14c)$$

Boundary Conditions

Identifying the straight boundary by the subscript 0 and the circular boundary by the subscript 1, the conditions that the normal and shearing components of stress must vanish on these two boundaries may be written as

$$\widehat{\alpha\beta}_i = 0 \quad (i = 0, 1) \quad (15a)$$

$$\widehat{\alpha\alpha}_i = 0 \quad (i = 0, 1). \quad (15b)$$

Substituting (14c) in (15a) and integrating with respect to β, we find that, on each boundary

$$\frac{\partial}{\partial\alpha}\left(\frac{\chi}{J}\right) = \xi_i, \text{ (a constant)}. \quad (16)$$

Substituting (14a) in (15b) and using (16) we find

$$\left[(\cosh\alpha_i - \cos\beta)\frac{d^2}{d\beta^2} - \sin\beta\frac{d}{d\beta} + \cosh\alpha_i\right]\left(\frac{\chi}{J}\right) = \xi_i \sinh\alpha_i. \quad (17)$$

This is an ordinary differential equation whose solution is

$$\left(\frac{\chi}{J}\right)_i = \xi_i \tanh\alpha_i + \eta_i(\cosh\alpha_i \cos\beta - 1) + \zeta_i \sin\beta, \ (i = 0, 1)$$

in which ξ_i, η_i, and ζ_i are constants. They are, in fact, linear combinations of Michell's (6) three boundary constants which, it is known, can be taken equal to zero on one boundary, as this is equivalent to adding a function linear in x and y to the stress function and accordingly does not affect the stress.

Taking the boundary constants on the straight boundary to be zero, the boundary conditions on the straight and circular boundaries become:

on $\alpha = \alpha_0 = 0$:

$$\frac{\partial}{\partial\alpha}\left(\frac{\chi}{J}\right) = 0, \quad (18a)$$

$$\frac{\chi}{J} = 0 \quad (18b)$$

on $\alpha = \alpha_1$:

$$\frac{\partial}{\partial\alpha}\left(\frac{\chi}{J}\right) = \zeta_1, \quad (19a)$$

$$\frac{\chi}{J} = \zeta_1 \tanh\alpha_1 + \eta_1(\cosh\alpha_1, \cos\beta - 1) + \zeta_1 \sin\beta. \quad (19b)$$

At a great distance from the hole ($\alpha \to 0$, $\beta \to 0$), the state of stress must reduce to the system

$$\widehat{xx} = T, \quad (20a)$$

$$\widehat{yy} = 0, \quad (20b)$$

$$\widehat{xy} = 0. \quad (20c)$$

The stress function for this state of stress is

$$\chi_0 = \tfrac{1}{2} T y^2 \quad (21a)$$

or

$$\frac{\chi_0}{J} = \tfrac{1}{2} aT \frac{\sinh^2\alpha}{\cosh\alpha - \cos\beta} \quad (21b)$$

$$= \tfrac{1}{2} aT \sinh\alpha\left(1 + 2\sum_{n=1}^{\infty} e^{-n\alpha}\cos n\beta\right) \quad (21c)$$

where the series expansion is valid for $\alpha > 0$, which is the region of our plate.

The stress function χ_0 satisfies the boundary conditions at infinity (equations (20)) and on $\alpha = 0$ (equations (18)) but it does not satisfy the boundary conditions on $\alpha = \alpha_1$, (equations (19)). It is necessary to add an auxiliary stress function (χ_1) which satisfies equations (18) and gives no stress at infinity and which, when added to χ_0, gives a stress function

$$\chi = \chi_0 + \chi_1 \quad (22)$$

which satisfies equations (19).

Auxiliary Stress Function χ_1.

The auxiliary stress function χ_1 must, of course, satisfy equation (2) which becomes,* in curvilinear coordinates,

$$\left(\frac{\partial^2}{\partial\alpha^2} + \frac{\partial^2}{\partial\beta^2}\right)\left[\frac{1}{J^2}\left(\frac{\partial^2\chi_1}{\partial\alpha^2} + \frac{\partial^2\chi_1}{\partial\beta^2}\right)\right] = 0. \quad (23)$$

Jeffery showed that, for the case of bipolar coordinates, equation (23) can be reduced to a linear equation with constant coefficients by considering χ/J, instead of χ, as the dependent variable. Then equation (23) becomes

*Coker and Filon, op. cit., p. 164.

$$\left(\frac{\partial^4}{\partial\alpha^4} + 2\frac{\partial^4}{\partial\alpha^2\partial\beta^2} + \frac{\partial^4}{\partial\beta^4} - 2\frac{\partial^2}{\partial\alpha^2} + 2\frac{\partial^2}{\partial\beta^2} + 1\right)\left(\frac{\chi_1}{J}\right) = 0. \tag{24}$$

On account of the symmetry of the present problem, the solution must be even in β. Furthermore, since there is a discontinuity of 2π in β on passing between the poles, the stresses must be periodic in β of period 2π.

A solution of the form

$$\frac{\chi_1}{J} = f_n(\alpha) \cos n\beta \tag{25}$$

is therefore considered. Substituting equation (25) in equation (24), there results

$$\left[\frac{d^4}{d\alpha^4} - 2(n^2-1)\frac{d^2}{d\alpha^2} + n^4 - 2n^2 + 1\right] f_n(\alpha) = 0. \tag{26}$$

The solution of this ordinary differential equation is, for $n \geqq 2$:

$$f_n(\alpha) = A_n \cosh (n+1)\alpha + B_n \cosh (n-1)\alpha + C_n \sinh (n+1)\alpha + D_n \sinh (n-1)\alpha, \tag{27a}$$

for $n = 1$:

$$f_1(\alpha) = A_1 \cosh 2\alpha + B_1 + C_1 \sinh 2\alpha + D_1\alpha \tag{27b}$$

and, for $n = 0$:

$$f_0(\alpha) = A_0 \cosh\alpha + B_0\alpha \cosh\alpha + C_0 \sinh\alpha + D_0 \alpha \sinh \alpha. \tag{27c}$$

Now, $A_0 \cosh\alpha$ and $B_1 \cos\beta$ are stress functions yielding the same stresses, hence A_0 may be taken equal to zero. Also, Jeffery has shown that the condition of single valuedness of displacements requires that

$$D_0 = 0, \tag{28a}$$

$$D_1 + B_0 = 0. \tag{28b}$$

Substituting equation (25) in the boundary conditions, equations (18), for the straight boundary we find the conditions

$$f_n(0) = 0, \tag{29a}$$

$$f_n'(0) = 0. \tag{29b}$$

Equations (29) specify relations among the constants A_n, B_n, C_n, D_n which must be satisfied if the straight boundary is to be free of stress.

Up to this point, the solution has followed Jeffery's work without change. Jeffery appears to have made an error in substituting equations (27) in equations (29). When this substitution is made, we find

$$A_n + B_n = 0, \; (n \geqq 1) \tag{30a}$$

$$(n+1)C_n + (n-1)D_n = 0, \; (n \geqq 2) \tag{30b}$$

$$2C_1 + D_1 = 0, \tag{30c}$$

$$B_0 + C_0 = 0. \tag{30d}$$

Jeffery considered correctly only the first of equations (30), with the result that his stress function does not leave the straight boundary free of traction.

Taking account of equations (30) and writing E_n for $C_n/(n-1)$ the auxiliary stress function becomes

$$\begin{aligned}\frac{\chi_1}{J} &= B_0\Big[\alpha(\cosh\alpha - \cos\beta) + \sinh\alpha(\cosh\alpha\cos\beta - 1)\Big] \\ &+ A_1(\cosh 2\alpha - 1)\cos\beta \\ &+ \sum_{n=2}^{\infty}\{A_n[\cosh(n+1)\alpha - \cosh(n-1)\alpha] \\ &+ E_n[(n-1)\sinh(n+1)\alpha \\ &- (n+1)\sinh(n-1)\alpha]\}\cos n\beta.\end{aligned} \tag{31}$$

Complete Stress Function

The complete stress function may now be written in the following form

$$\begin{aligned}\chi = aT\Big\{&\tfrac{1}{2}\sinh\alpha\,(1 + 2\sum_{n=1}^{\infty} e^{-n\alpha}\cos n\beta) \\ &+ B_0[\alpha(\cosh\alpha - \cos\beta) + \sinh\alpha(\cosh\alpha\cos\beta - 1)] \\ &+ A_1(\cosh 2\alpha - 1)\cos\beta \\ &+ \sum_{n=2}^{\infty}\{A_n[\cosh(n+1)\alpha - \cosh(n-1)\alpha] \\ &+ E_n[(n-1)\sinh(n+1)\alpha \\ &- (n+1)\sinh(n-1)\alpha]\}\cos n\beta\Big\}.\end{aligned} \tag{32}$$

All that remains is to assign values to the constants B_0, A_1, A_n, E_n. This is done by substituting (32) in the remaining boundary conditions (equations (19)), with the results:

$$\xi_1 = \frac{aT}{2}(\cosh\alpha_1 + \alpha_1 \operatorname{csch}\alpha_1)$$

$$\eta_1 = \frac{aT}{2}(\operatorname{csch}\alpha_1 - \alpha_1 \operatorname{csch}^3\alpha_1 \tanh\alpha_1)$$

$$\zeta_1 = 0$$

$$B_0 = \tfrac{1}{2}\operatorname{csch}^2\alpha_1$$

$$A_1 = -\tfrac{1}{2}e^{-\alpha_1}\operatorname{csch}\alpha_1$$

$$A_n = -\frac{n^2\sinh^2\alpha_1 - n\sinh\alpha_1\cosh\alpha_1 + e^{-n\alpha_1}\sinh n\alpha_1}{2(\sin\mathrm{h}^2\, n\alpha_1 - n^2 \sin\mathrm{h}^2\,\alpha_1)}$$

TABLE I: $\widehat{\beta\beta_0}/T$

β \ α_1	0.2	0.4	0.6	0.8	1.0	1.2	1.4	1.6	1.8	2.0
180°	0.081	0.212	0.314	0.412	0.506	0.596	0.681	0.758	0.822	0.873
171°	0.326	0.332	0.392	0.468	0.549	0.630	0.708	0.778	0.837	0.884
162°	1.034	0.679	0.617	0.631	0.674	0.738	0.785	0.838	0.881	0.916
153°	2.138	1.221	0.968	0.884	0.867	0.879	0.903	0.927	0.948	0.964
144°	3.529	1.902	1.409	1.202	1.109	1.067	1.048	1.036	1.027	1.020
135°	5.072	2.655	1.896	1.553	1.374	1.271	1.201	1.149	1.108	1.077
126°	6.614	3.406	2.381	1.900	1.634	1.466	1.344	1.251	1.180	1.126
117°	8.003	4.081	2.814	2.207	1.860	1.628	1.457	1.338	1.231	1.161
108°	9.103	4.613	3.151	2.443	2.025	1.737	1.526	1.370	1.256	1.176
99°	9.807	4.948	3.359	2.580	2.107	1.778	1.540	1.371	1.253	1.171
90°	10.043	5.053	3.415	2.601	2.093	1.743	1.500	1.334	1.223	1.149
81°	9.789	4.915	3.310	2.496	1.981	1.637	1.412	1.267	1.174	1.114
72°	9.070	4.548	3.051	2.270	1.782	1.478	1.293	1.182	1.115	1.074
63°	7.956	3.986	2.655	1.943	1.527	1.294	1.167	1.097	1.058	1.036
54°	6.555	3.282	2.149	1.558	1.260	1.120	1.055	1.025	1.012	1.006
45°	4.887	2.495	1.589	1.186	1.035	0.987	0.976	0.978	0.982	0.987
36°	3.456	1.681	1.074	0.911	0.897	0.917	0.940	0.958	0.971	0.980
27°	2.061	0.942	0.748	0.796	0.861	0.909	0.941	0.961	0.974	0.983
18°	0.905	0.551	0.716	0.837	0.904	0.942	0.964	0.977	0.985	0.990
9°	0.326	0.742	0.889	0.945	0.970	0.983	0.989	0.993	0.996	0.997
0°	1	1	1	1	1	1	1	1	1	1

TABLE II: $\widehat{\beta\beta_1}/T$

β \ α_1	0.2	0.4	0.6	0.8	1.0	1.2	1.4	1.6	1.8	2.0
180°	19.960	10.190	6.952	5.387	4.494	3.944	3.597	3.376	3.238	3.151
171°	19.747	9.999	6.826	5.290	4.414	3.874	3.532	3.314	3.175	3.086
162°	18.667	9.452	6.456	5.005	4.178	3.669	3.344	3.132	2.995	2.895
153°	16.947	8.593	5.868	4.553	3.803	3.341	3.040	2.840	2.093	2.587
144°	14.723	7.468	5.104	3.963	3.313	2.909	2.639	2.444	2.295	2.177
135°	12.141	6.160	4.215	3.274	2.738	2.399	2.159	1.972	1.817	1.686
126°	9.411	4.768	3.261	2.393	2.114	1.837	1.624	1.443	1.283	1.141
117°	6.630	3.372	2.300	1.781	1.475	1.252	1.060	0.884	0.721	0.576
108°	4.049	2.052	1.391	1.064	0.853	0.671	0.494	0.323	0.167	0.030
99°	1.788	0.894	0.583	0.416	0.275	0.119	−0.046	−0.203	−0.340	−0.455
90°	−0.045	−0.053	−0.088	−0.138	−0.239	−0.380	−0.528	−0.656	−0.756	−0.830
81°	−1.386	−0.763	−0.601	−0.586	−0.676	−0.802	−0.913	−0.989	−1.032	−1.050
72°	−2.223	−1.221	−0.952	−0.928	−1.018	−1.108	−1.154	−1.155	−1.123	−1.074
63°	−2.588	−1.439	−1.159	−1.168	−1.236	−1.250	−1.198	−1.103	−0.987	−0.874
54°	−2.551	−1.459	−1.257	−1.292	−1.278	−1.164	−0.990	−0.798	−0.613	−1.447
45°	−2.218	−1.340	−1.270	−1.241	−1.056	−0.782	−0.495	−0.232	−0.007	−0.181
36°	−1.708	−1.171	−1.144	−0.878	−0.470	−0.066	0.277	0.553	0.770	0.941
27°	−1.149	−0.987	−0.643	−0.030	0.572	0.935	1.235	1.453	1.612	1.731
18°	−0.724	−0.360	0.616	1.332	1.769	2.032	2.197	2.303	2.376	2.427
9°	−0.045	1.761	2.497	2.770	2.874	2.911	2.919	2.917	2.911	2.905
0°	3.812	3.655	3.525	3.414	3.323	3.249	3.189	3.141	3.104	3.076

$$E_n = \frac{n \sinh^2 \alpha_1}{2(\sinh^2 n\alpha_1 - n^2 \sinh^2 \alpha_1)}.$$

With these values of the constants, χ satisfies the boundary conditions (18) and (19) and reduces to χ_0 at infinity.

Stress along the Boundaries

The most important components of stress are $\widehat{\beta\beta}_0$ and $\widehat{\beta\beta}_1$, *i.e.*, those along the straight and circular boundaries respectively.

To calculate these, we note from (14) that, when $\widehat{\alpha\alpha} = 0$, $\widehat{\beta\beta}$ becomes

$$a\widehat{\beta\beta} = (\cos\beta - \cosh\alpha)\left(\frac{\partial^2}{\partial\beta^2} - \frac{\partial^2}{\partial\alpha^2} + 1\right)\left(\frac{\chi}{J}\right). \quad (33)$$

Substituting (32) in (33) and letting $\alpha = 0$ for $\widehat{\beta\beta}_0$ and $\alpha = \alpha_1$ for $\widehat{\beta\beta}_1$, we find

$$\widehat{\beta\beta}_0 = T\left\{1 - 2(1-\cos\beta)\left[\operatorname{csch}\alpha_1 e^{-\alpha_1}\cos\beta - 2\sum_{n=2}^{\infty} n A_n \cos n\beta\right]\right\}$$

$$\widehat{\beta\beta}_1 = 2T\left[1 - \frac{2\sinh^2\alpha \sin^2\beta}{(\cosh\alpha_1 - \cos\beta)^2}\right] + 2T(\cosh\alpha - \cos\beta)\left[\tfrac{1}{2}\operatorname{csch}\alpha_1 + 2e^{-2\alpha_1}\cos\beta + \sum_{n=2}^{\infty} N_n \cos n\beta\right]$$

where

$$N_n = \frac{n^2 \sinh\alpha_1 \cosh n\alpha_1 - n \sinh n\alpha_1 \cosh\alpha_1}{\sinh^2 n\alpha_1 - n^2 \sinh^2\alpha_1} - 2n(n\sinh\alpha_1 - \cosh\alpha_1)e^{-n\alpha_1}.$$

$\widehat{\beta\beta}_0$ and $\widehat{\beta\beta}_1$ were computed at 9° intervals of β for ten values of α_1, at intervals of 0.2 from 0.2 to 2.0. The values of c/R_2 corresponding to the values of α_1 used in the computations are given in Fig. 3, along with diagrams showing the proximity of the hole to the straight boundary. The results of the computations are given in Tables I and II and some of these are shown graphically in Figs. 4. and 5.

The extraordinary appearance of $\widehat{\beta\beta}_0$ for small values of α_1, shown in Fig. 4, was what prompted the photoelastic experiments.

Resultant Force Transmitted between the Hole and the Straight Edge

An interesting remark can be made about the total load passing between the hole and the edge of the plate.

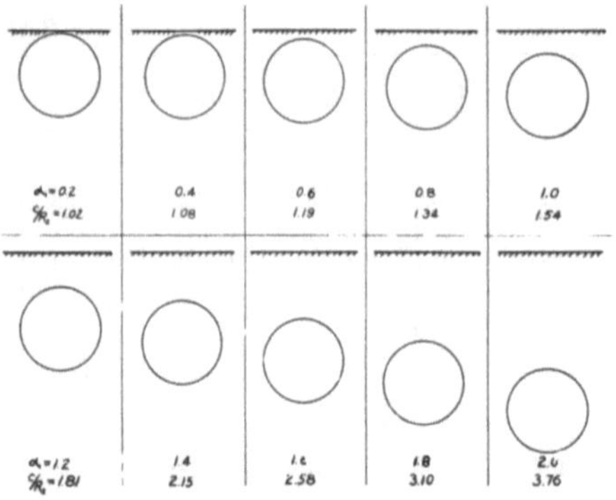

FIG. 3. Proximity of hole to edge of plate for values of α_1 (and the corresponding values of c/R_2) selected for computation.

The resultant force H acting on the section $m-n$ is given by*

$$H = \left\{\left[\frac{\partial\alpha}{\partial y}\left(J\frac{\partial}{\partial\alpha} + \frac{\partial J}{\partial\alpha}\right)\left(\frac{\chi}{J}\right) + \frac{\partial\beta}{\partial y}\left(J\frac{\partial}{\partial\beta} + \frac{\partial J}{\partial\beta}\right)\left(\frac{\chi}{J}\right)\right]_0^{\alpha_1}\right\}_{\beta=\pi} \quad (34)$$

In (34) χ/J occurs only in the form of its boundary value and the boundary values of its first derivatives with respect to α and β. Referring to (18) and (19) it may be seen that these quantities are given in closed form, so that a closed expression is found for H. In the present instance, this expression is particularly simple. It is

$$H = Ta = T\sqrt{c^2 - R_2^2}. \quad (35)$$

This result can be represented by a simple geometrical construction. Through point m draw a tangent to the circle. The distance between m and the point of tangency is a. With m as center and a as radius, swing an arc intersecting mp at q. Draw qr parallel to the straight edge. That portion of the load which, far from the hole, is carried by the material between qr and the straight edge, passes between the hole and the straight edge. The remainder passes around the other side of the hole.

Photoelastic Experiments

The design of the photoelastic model is shown in Fig. 6. The model material was Bakelite BT-61-893 and the grips were of brass. The latter were clamped to the bakelite with No. 10-32 screws through holes, drilled and reamed to push-fit, in the positions shown.

*Mindlin, op. cit., p. 1138.

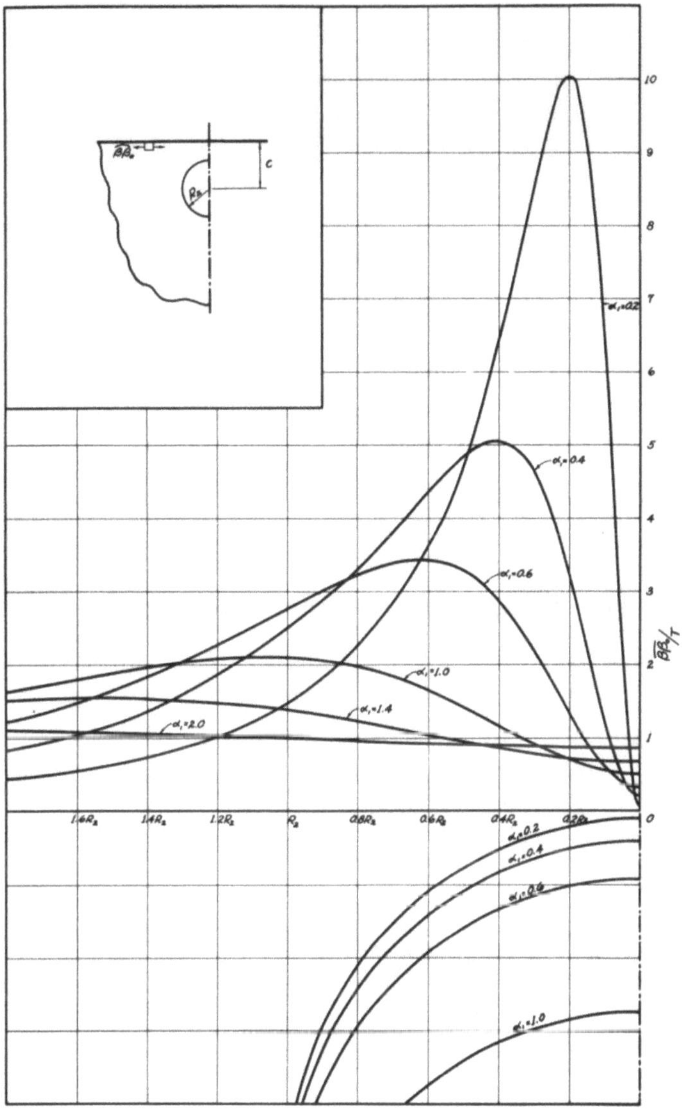

FIG. 4. Variation of stress along the straight edge (from mathematical solution).

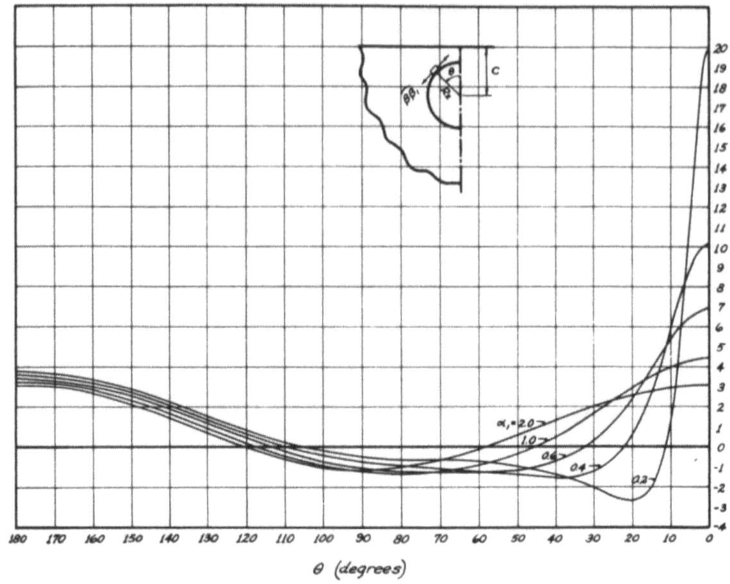

FIG. 5. Variation of stress along the edge of the hole (from mathematical solution).

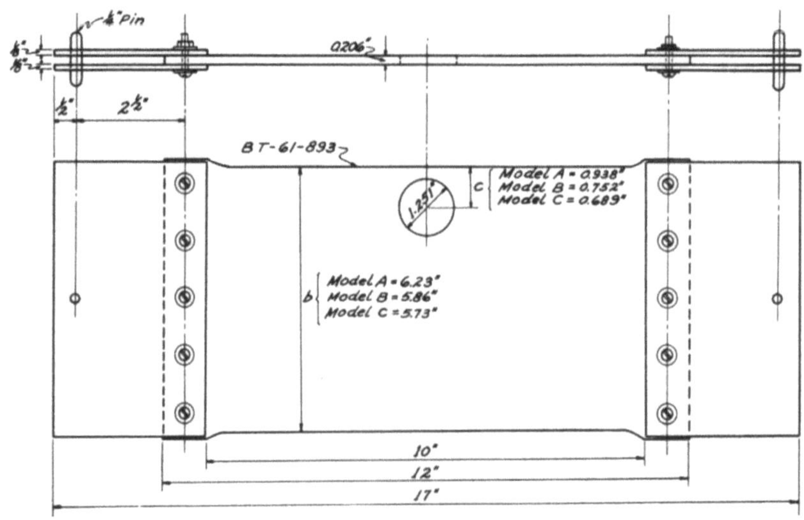

FIG. 6. Model used in photoelastic experiments.

FIG. 7. Fringe photograph: $c/R_2 = 1.5$.

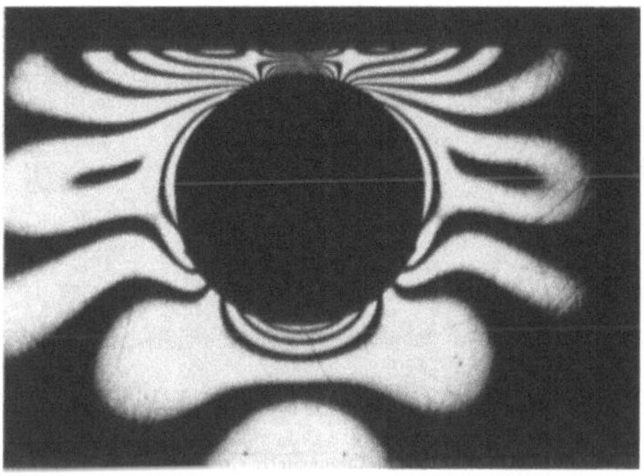

FIG. 8. Fringe photograph: $c/R_2 = 1.2$

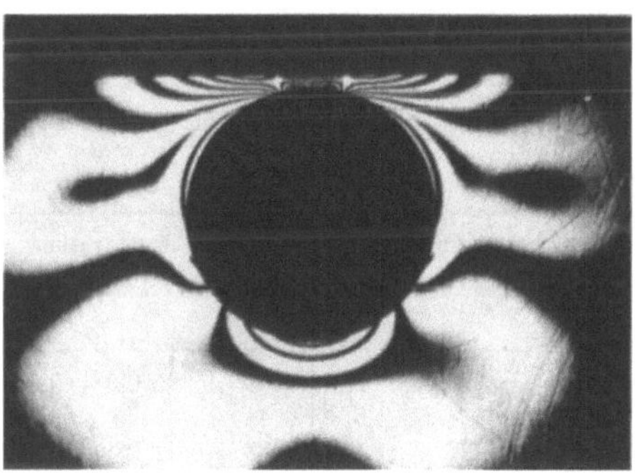

FIG. 9. Fringe photograph: $c/R_2 = 1.1$.

Three tests were planned, using the same plate, at ratios of $c/R_2 = 1.5$, 1.2, and 1.1. It was planned that the model would be loaded, in each case, to obtain about ten fringes between the edge of the hole and the near edge of the plate and it was deemed advisable to make the minimum distance, between the hole and the straight edge, not less than 1/16″. This fixed the minimum allowable diameter of the hole at 1¼″.

The procedure for cutting the hole and the straight edges was as follows. A 5/8″ diameter hole was drilled, in approximately the correct location, with a twist drill at low speed and intermittent feed, so as not to heat the bakelite. The brass plates were attached and the model was supported, in a vertical plane, on its loading pins on a milling machine table. First the straight edge farthest from the hole was milled with a side milling cutter. Then the hole was bored to size with a fly cutter and finally the remaining straight edge was milled. This was Model A, Fig. 6.

The model was then placed in the loading machine and load was applied until about ten fringes appeared between the hole and the near straight edge. Photographs were then taken through a circularly polarizing polariscope, using the 5461 Å radiation from a G.E. Uviarc tube and Eastman D.C. Orthochromatic plates. Following this the fringes were observed through the polariscope using the mercury vapor lamp without a filter. A sketch was made of the fringes and the orders identified.

The model was then replaced on the milling machine and the straight edges were milled down to the dimensions shown for Model B, Fig. 6, the final cut being made on the edge closest to the hole. The hole diameter was not increased. After loading, sketching and identifying fringes, and photographing, the procedure was repeated for Model C. The total elapsed time was about four hours.

The plates were developed in Eastman D-11 developer for 5 minutes at 65°F and fixed in Kodak Acid Fixing Solution for 15 minutes at 65°F. The fringe patterns are shown in Figs. 7, 8, and 9. There is no detectable "edge effect" on the three straight edges as a result of making the final, careful cuts on these edges immediately before photographing. This schedule was followed because it was the stress along the straight edge that was of greatest interest. The edge effect on the circular boundary can be observed to increase in the three successive photographs. The effect has penetrated a distance of several thousandths of an inch in the photograph of Model C.

The fringe orders at points m, n, and p were determined by extrapolation in the following manner. Using

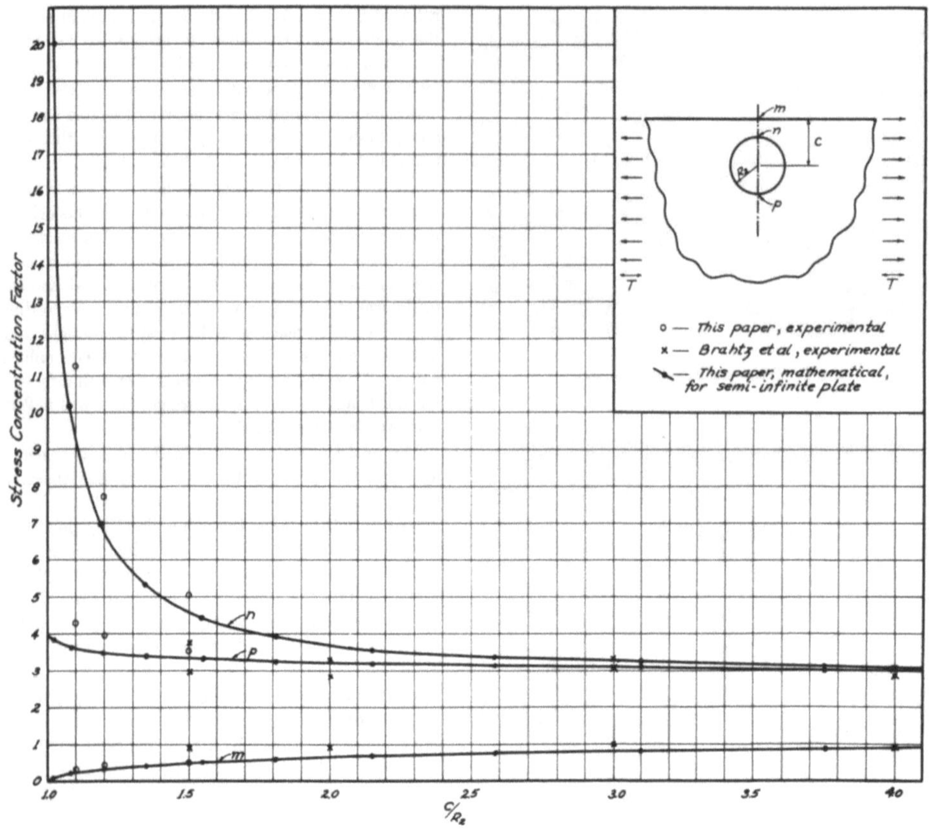

FIG. 10. Stress concentration factors.

TABLE III

Model	$\frac{c}{R_2}$	Width (b, inches)	Load (L, lbs)	Av. Fringe Order ($L/88.5b$)	Fringe Order at m	n	p	Stress Concentration Factor m	n	p
A	1.5	6.23	1150	2.09	1.1	10.6	7.4	0.53	5.08	3.54
B	1.2	5.86	760	1.47	0.65	11.4	5.8	0.44	7.76	3.95
C	1.1	5.73	530	1.05	0.4	11.8	4.5	0.38	11.28	4.30

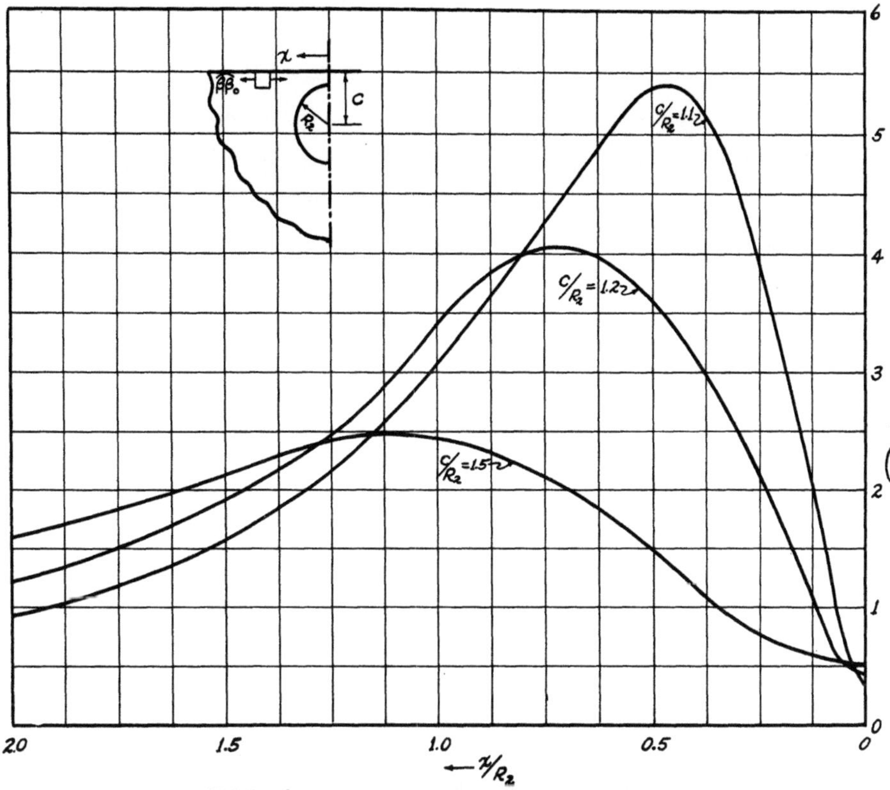

FIG. 11. Variation of stress along the straight edge (from photoelastic experiments).

a travelling micrometer microscope, the relative positions of the edges and the centers of the fringes were determined along the line $\overline{mnp}$ on the photographic plates. The ratios of $\overline{mn}/\overline{np}$ were then compared with the corresponding ratios previously measured on the model with micrometer calipers. The ratios checked to within 1 in 300, 1 in 150, and 1 in 30 for Models A, B, and C respectively. This step was necessary in order to be assured that the observed edge on the photographic plate was really the true edge. A rectangular coordinate plot was then made of fringe order versus fringe location for each model. A curve was drawn through the plotted points and extrapolated to the edge locations to determine the fringe orders at *m*, *n*, and *p*.

Stress concentration factors were determined by ratio to the average fringe order over the gross section. The latter was determined, for each model, by dividing the recorded load by the product of the gross width and the stress optical coefficient. The stress optical coefficient, as determined from a tensile test on a ½″ wide specimen cut from the model, was 88.5 pounds per inch per fringe. The results are shown in Table III and are plotted in Fig. 10 along with the results of Brahtz *et al* and the mathematical solution.

It should be noted that the experimental and mathematical results cannot be compared directly, since the former are for a plate of finite width while the latter apply to a plate of infinite width. The stress concentration factors, reported here, lie above the mathematical curve, but this is only because the factors are referred to the gross section. If the stress concentration factors had been referred to the net section they would have fallen below the mathematical curve. The comparable value must lie between the two, and thus we can at least say there is qualitative agreement.

Curves of the edge stresses all around the hole and

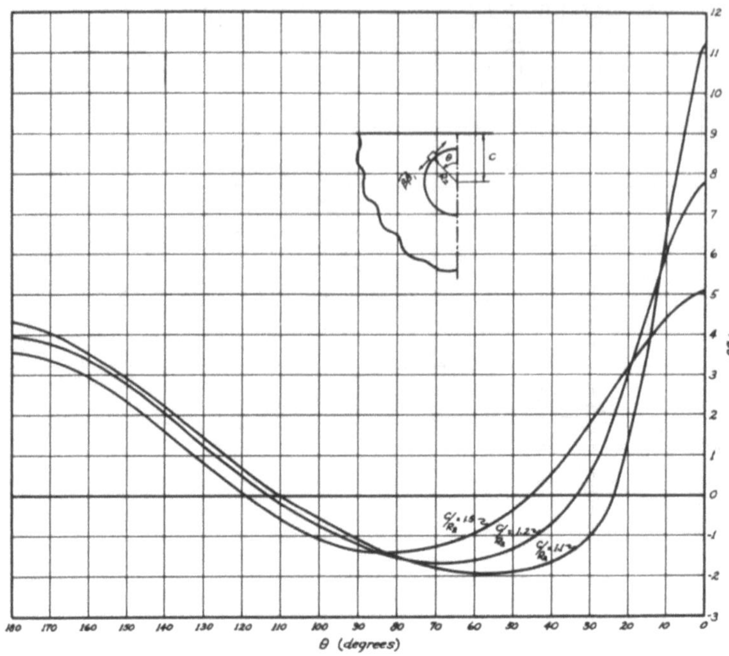

FIG. 12. Variation of stress along the edge of the hole (from photoelastic experiments).

along the straight boundary were obtained by erecting ordinates proportional to fringe order at the intersections of the fringes with the edges. The results are plotted in Figs. 11 and 12. These are to be compared with Figs. 4 and 5. Again, a direct comparison cannot be made because of the finite width of the experimental plate, but the qualitative agreement can be observed. The shapes of the curves are almost the same and the algebraic signs of the ordinates are the same; in particular, the stress along the straight edge is tensile at all points.

ACKNOWLEDGEMENT

I wish to thank Dr. W. H. Hoppmann and Dr. Y. Nubar, who assisted in the experiments, and Mr. R. Contini, who assisted in the computations, while they were graduate students in the Department of Civil Engineering at Columbia University in 1939-1940.

BIBLIOGRAPHY

1. G. B. Jeffery, "Plane Stress and Plane Strain in Bipolar Coordinates," Transactions of the Royal Society, London, England, Series A, Vol. 221, (1920), pp. 265–293.
2. S. Timoshenko, "Theory of Elasticity", New York, (1934), p. 79.
3. R. D. Mindlin, "Stress Distribution around a Tunnel", Proceedings of the American Society of Civil Engineers, April, (1939) and Transactions of the American Society of Civil Engineers, Vol. 105 (1940), pp. 1117–1153.
4. Additional results are given in "Stresses and Pore Pressures around Circular Openings near a Boundary", by J. H. A. Brahtz, J. E. Soehrens, C. M. Zangar and J. A. Bruggeman, Technical Memorandum No. 597, Bureau of Reclamation, U. S. Department of the Interior, Denver, Colorado, January 12, 1940.
5. E. G. Coker and L. N. G. Filon, "A Treatise on Photo-Elasticity", Cambridge, University Press, 1931, p. 128.
6. J. H. Michell, "On the Direct Determination of Stress in an Elastic Solid, with Application to the Theory of Plates", Proceedings of the London Mathematical Society, Vol. 31 (1899), pp. 100–124.

Reprinted from
Proceedings of the Society for Experimental Stress Analysis
V, no. 2 (1948) 69-87

RESPONSE OF DAMPED ELASTIC SYSTEMS TO TRANSIENT DISTURBANCES

R. D. MINDLIN, Columbia University, and F. W. STUBNER and H. L. COOPER,
Bell Telephone Laboratories

1. Introduction and Summary

When an elastic system is subjected to a transient disturbance, the response of the system is influenced strongly by the ratios of its natural frequencies of vibration to the frequency components of the disturbance.

If the ratio (p_1/ω) of the lowest natural frequency of the system to the dominant frequency component of the disturbance is large, vibrations of the system at its natural frequencies are not excited appreciably and the response may be computed as if the loads were applied statically.

As the ratio p_1/ω is reduced, the fundamental mode of vibration of the system is excited and this must be taken into account in calculating maximum displacements, strains, and accelerations. If p_1/ω is not so small that modes of the system higher than the first are excited, it is sufficient to treat the system as one with a single degree of freedom having the same frequency as that of the fundamental mode and velocity damping yielding the same energy loss per cycle. In the special case of a cantilever whose clamped end is subjected to a transverse pulse acceleration in the form of a half-sine wave, this treatment gives results which check experimental measurements within a few percent if $p_1/\omega > 0.3$. Furthermore, in this frequency range and with the small amount of damping present in most structural materials, omission of the damping of the system does not result in serious discrepancies in the case of the single pulse disturbance.

For very small ratios p_1/ω, many of the higher modes of the elastic system may be excited appreciably. It then becomes important to be able to predict what the contributions of the higher modes will be, under what circumstances they may be significant, and how to take them into account. To these ends, elastic systems with an infinite number of degrees of freedom were studied.

Elastic systems may be divided into two classes: dispersive and nondispersive. A dispersive system is one in which velocity of wave propagation depends upon wave length while, in a nondispersive system, velocity is independent of wave length. In the latter, the frequencies of the higher modes are integral multiples of the fundamental, while this is not true for dispersive systems. Flexural motions of a bar are dispersive; longitudinal motions of a bar are essentially nondispersive, as long as the wave length is not too samll in comparison with the radius of gyration of the cross section of the bar. The effects of the higher modes in response to a pulse disturbance are quite different for the two classes of elastic systems.

In the case of a bar, one end of which experiences a half-sine-wave pulse acceleration *parallel* to the axis of the bar, the mathematical solution indicates that it is not necessary to consider the higher modes in calculating maximum strain. The one-degree-of-freedom system gives adequate results, the error being on the safe side.

On the other hand, the clamped-free bar, subjected to a half-sine-wave pulse acceleration of its clamped end *transverse* to the axis of the bar, may require consideration of the higher modes. If the latter are ignored, appreciable error is made on the unsafe side when the pulse duration is short in comparison with the period of the fundamental mode of the system. For example, when p_1/ω is 0.1, the measured maximum strain at the clamped end is 30% greater than that predicted by considering only the first mode.

As a first step in taking into account higher mode flexural motions, it is natural to study the predictions of the Bernoulli-Euler equation. This was done by Symonds and Vigness (1)* for the case of the transverse pulse acceleration of the clamped end of a cantilever. They found that, as the pulse duration decreases, this theory predicts that the maximum strain at the clamped end increases *without limit.* At $p_1/\omega = 0.1$ the predicted strain is 15% higher than the measured strain reported below and at $p_1/\omega = 0.01$ the error is probably about 35%.

*Numbers in parentheses refer to bibliography at end of paper

Thus, the one-degree-of-freedom theory predicts strains that are too low, while the Bernoulli-Euler theory predicts strains that are too high. The defect in the former is due to omission of the higher mode contributions. A detailed comparison of the mathematical results given below with the experimental records indicates that the excess in the latter case is due mainly to neglect of the rapid dissipation of energy in the higher modes. This was also one of the sources of error suggested by Symonds and Vigness as contributing to the absurd predictions of the Bernoulli-Euler theory for very short duration pulse disturbances.

We have now arrived at the main purpose of this paper, namely, to discuss a procedure, suitable for engineering applications, for taking into account energy dissipation in the various modes of flexural motions of a bar.

In 1927 Sezawa (2) proposed a modification of the Bernoulli-Euler equation to accommodate the effect of internal damping. Sezawa's modification is based on the assumption that the material of the bar follows the Stokes' law of viscosity, *i.e.*, stress is a linear function of strain and strain-rate. As a result of this assumption, it is predicted that the fraction of critical damping (or the logarithmic decrement or the percent energy loss per cycle or the reciprocal of Q, etc.) of each mode of vibration is directly proportional to the frequency of the mode. Now, it is well known (3), (4) that most structural materials do not follow Stokes' law but that, at least to a first approximation, Stokes' viscosity coefficient is inversely proportional to frequency, *i.e.*, the fraction of critical damping is found to be nearly independent of the frequency. Consequently, it may be expected that the use of the Sezawa equation will result in excessive suppression of the higher modes and, accordingly, there will be predicted strains and accelerations that are too small. This is confirmed by the calculations and experimental results described below.

It does not appear to be useful, for engineering design purposes, to derive a beam equation based on such laws as are known for internal friction in solids. Solutions of the equation would be very difficult to obtain.

As an alternative, the following procedure is proposed.(5) Assume, at first, that the Sezawa equation is applicable and proceed with its solution, which is no more difficult, although possibly more tedious, than the solution of the Bernoulli-Euler equation for the same initial and boundary conditions. In the resulting solution the Stokes viscosity coefficient will be found to appear always as a product with the ratio of a frequency to Young's modulus. Replace this product by a constant, the same for all frequencies. The constant is, in fact, twice the fraction of critical damping or, for small damping, the reciprocal of Q; i.e., $c\omega/E = Q^{-1}$.

This technique was tested by comparison of mathematical with experimental results for two types of transient disturbances and was found to be satisfactory. In one experiment the strain was measured at the clamped end of a cantilever subjected to a transverse half-sine-wave pulse acceleration and in the second experiment the acceleration was measured at the free end of a plucked cantilever. In each case the value of the constant Q (or constant fraction of critical damping) was obtained by measuring the decay of strains or displacements during free vibration in the fundamental mode at amplitudes approximating those used in the transient tests.

A possible explanation for the close correspondence of the mathematical and experimental results may be the following. In the first place, it appears that, for short duration disturbances, the *initial* distribution of strain and acceleration among the various modes is insensitive to the law of damping if the damping is small. Hence no great error is introduced in initial distributions by using an erroneous equation of motion (Sezawa's). Subsequently, the strains and accelerations are very sensitive to the law of damping, but only to the extent that the energy loss per cycle in each mode must be about the same in the mathematical solution as in the actual structure. This criterion is apparently satisfied by the assumption of constant Q.

In contemplating the possible application of the technique proposed here to a variety of beam problems, two important questions arise:

(1) Is it permissible to assign all energy losses to internal damping alone? In many cases it is likely that much energy is dissipated through supports or other surrounding media. However, the method is valid if the percent energy loss per cycle, due to all causes lumped together, is independent of frequency and if the initial energy distribution is insensitive to the law of damping. This appears to be so for the two experiments described below.

(2) What are the consequences of the assumption that damping is independent of amplitude? One consequence is, of course, that the differential equation becomes linear and, therefore, solutions can be obtained without encountering extraordinary difficulties. Since *maximum* strains are usually of most importance, part of the defect in the linear equation may be overcome by choosing, for the constant value of Q, one in the range of the maximum strain that is considered permissible for the beam. This procedure was satisfactory for the two experiments described below. It would probably

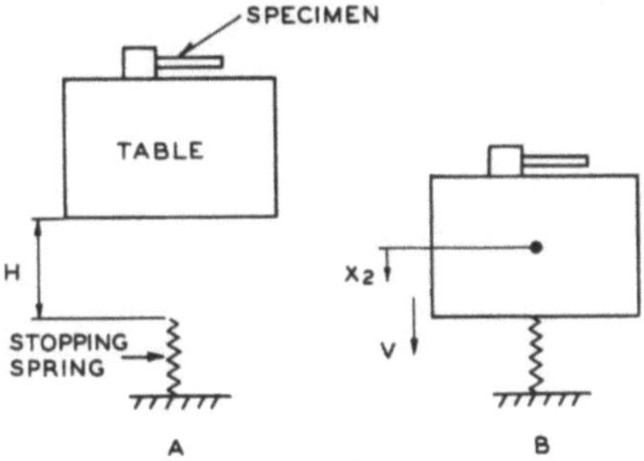

Fig. 2.1. Schematic diagram of a drop test.

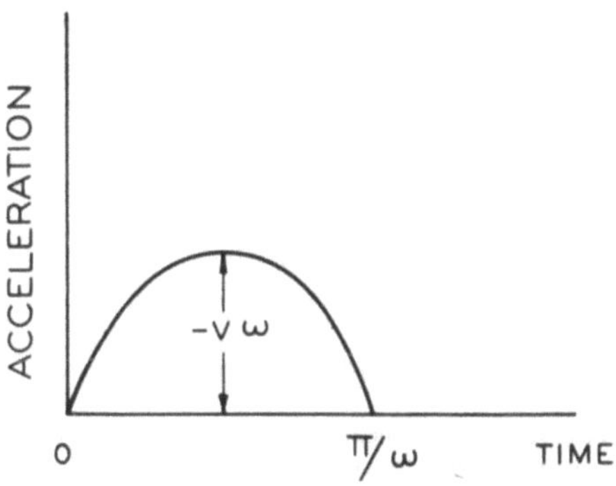

Fig. 2.2. Ideal half-since-wave pulse acceleration.

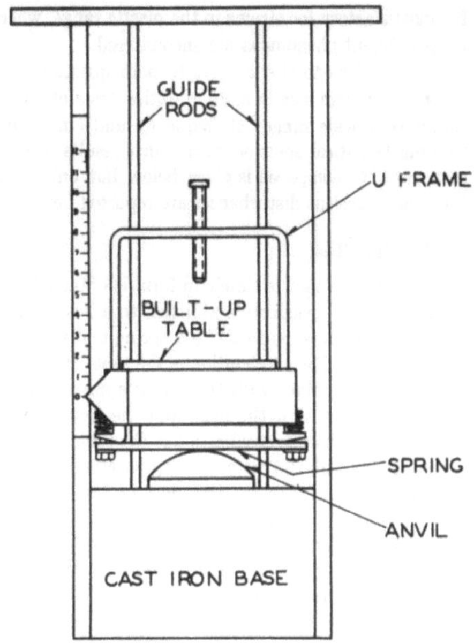

Fig. 2.3. American Standards Association drop tester.

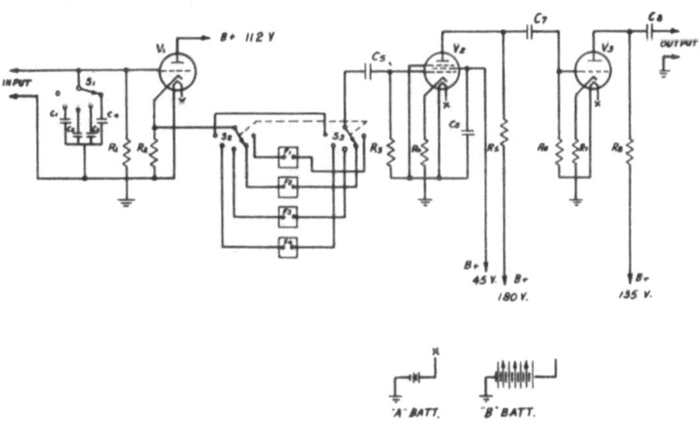

DESIG.	APPARATUS	
C 1	.01	COND.
C 2	.025	
C 3	.05	
C 4	.075	
C 5	50 MFD 25 V	
C 6	.25 MFD 400V	
C 7	.5 MFD 400V	
C 8	.5 MFD 400V	
F 1	W-76492	FILTER
F 2	W-76498	
F 3	W-76672	
F 4	W-76576	
R 1	10 MEG.	RESISTOR
R 2	1500 ω	
R 3	1 MEG.	
R 4	820 ω	
R 5	50,000 ω	
R 6	1 MEG.	
R 7	2700 ω	
R 8	50000 ω	
S 1		NO. 2505 CENTRALAS SW.
S 2		
S 3		
V 1	6 SF5	VACUUM TUBE
V 2	6 SJ7	
V 3	6 G5	

Fig. 2.4. Filter and pre-amplifier circuit.

be unsatisfactory for strains in the plastic range, where quite different phenomena are encountered.

Contributions to the answers to both questions can be obtained from steady state vibration tests of a cantilever over wide ranges of frequency and amplitude. The mathematical solution with which results of such tests are to be compared is given below but only tests involving transient disturbances are reported here.

2. The Drop Test

A disturbance that is simple in form is obtained in a drop test. The specimen is attached to a heavy table which is dropped so as to strike a spring and rebound. (See Fig. 2.1.) The acceleration of the table during the interval of its contact with the spring approximates a half-sine-wave pulse if the mass and the frequency of the fundamental mode of vibration of the table are large in comparison with the corresponding parameters of the specimen and the spring. The part of the specimen that is clamped to the table then experiences a half-sine-wave pulse acceleration (Fig. 2.2) whose duration is the half-period of vibration (π/ω) of the table on the spring and whose acceleration amplitude is $-v\omega$, where v is the striking velocity of the table; ω is equal to $(k/m)^{\frac{1}{2}}$, where k is the spring rate of the stopping spring; and m is the mass of the table and attachments. $v\omega$ is made large in comparison with the acceleration of gravity so that the latter may be neglected.

Calling x_2 the displacement of the table, measured positively downward from the position of the table at first contact with the spring (Fig. 2.1(b)), we have

$$x_2 = \frac{v}{\omega} \sin \omega t, \qquad 0 \leq t \leq \pi/\omega. \qquad (2.1)$$

The table acceleration is then

$$\ddot{x}_2 = -v\omega \sin \omega t, \qquad 0 \leq t \leq \pi/\omega, \qquad (2.2)$$

$$\ddot{x}_2 = O, \qquad 0 \geq t \geq \pi/\omega. \qquad (2.3)$$

This acceleration was produced reasonably closely on an adaptation of the Shock-testing Mechanism for Electrical Indicating Instruments as specified by the American Standards Association. (6) (See Fig. 2.3.) The changes that were made in the machine came as a result of measurements of the acceleration of the table. A Westinghouse quartz crystal accelerometer was bolted to the table and the output of the accelerometer was fed, through the circuit shown in Fig. 2.4, to the vertical amplifier of a cathode-ray oscilloscope. A microswitch, actuated by the table just before contact of the spring and anvil, triggered a single sweep circuit on the horizontal amplifier of the oscilloscope. The resulting trace was photographed with a still camera mounted in front of the cathode-ray tube.

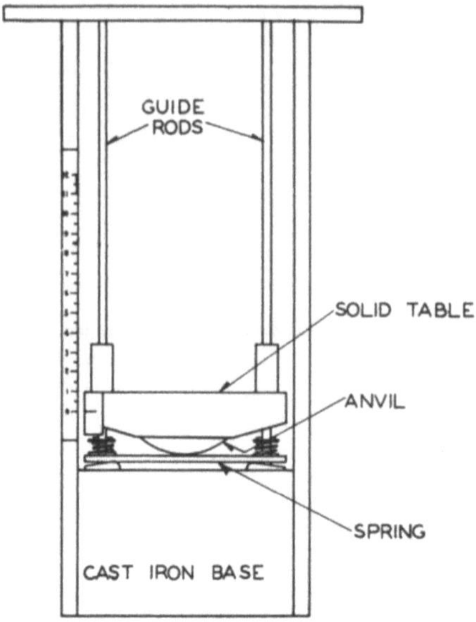

FIG. 2.5. Modified drop tester.

As a result of high-frequency vibration of the carriage platform, U-frame, and spring assembly attached to the carriage, the trace was not recognizable as a half-sine-wave before modifications were made in the machine. To reduce these vibrations, the carriage was replaced by a block of dural and the U-frame was eliminated. The positions of the spring and anvil were interchanged, *i.e.*, the anvil (a hardened steel spherical segment instead of the original cylindrical anvil) was bolted to the underside of the table and the stopping spring was supported on the base of the machine (Fig. 2.5). Four springs were supplied, to give nominal shock durations of 0.002, 0.005, 0.007, and 0.012 seconds. A typical accelerogram, recorded after these changes were made, is shown in Fig. 2.6. The remaining high-frequency vibration is that of the fundamental mode of the table at 5500 cps. This frequency is high enough to have no influence on the strain measurements which were made later. In recording Fig. 2.6 a "10,000 cps" low-pass filter was used to cut out the 23,000 cps natural frequency of the accelerometer. The characteristics of the filter are shown in Fig. 2.7.

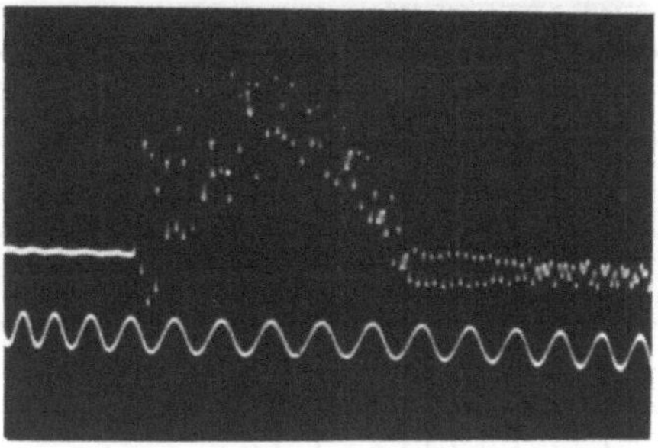

FIG. 2.6. Typical record of table acceleration of modified drop tester. Timing wave is 1000 cps, 10,000 cps filter was used. Trace shows 5,500 cps table vibration superposed on pulse.

The maximum ordinate of the average line drawn through the recorded acceleration trace was converted to volts by comparison with the ordinate of a steady audio signal of frequency approximately equal to the "stopping" frequency, fed through the same system and measured with a vacuum-tube voltmeter. The voltage was then converted to acceleration, using the manufacturer's calibration of the accelerometer. The accelerations checked within 5% those calculated from the weight of the table, the spring rate of the stopping spring and the height of drop. The duration of the acceleration pulse was obtained by comparison with a 1000 cps signal recorded on the oscillogram.

After the experiments were completed, it was found that the table vibration could be greatly reduced by placing a drop of castor oil on the spring, covering the surface of contact with the anvil. The resulting acceleration record is shown in Fig. 2.8.

3. Theory of Drop Test of an Undamped System with One Degree of Freedom

In Fig. 3.1 is shown an idealization of the drop test in which the specimen is represented by a lumped mass (m_1), a massless spring with spring rate k_1, and damping coefficient c_1. The displacement of m_1, measured positively downward from its position when the table first touches the spring, is x_1.

Neglecting damping, at first, the equation of motion of m_1 is

$$m_1\ddot{x}_1 + k_1(x_1 - x_2) = 0. \qquad (3.1)$$

Let x be the relative displacement of m_1 with respect to m, *i.e.*,

$$x = x_1 - x_2. \qquad (3.2)$$

Now, x is proportional to the force (k_1x) in the spring

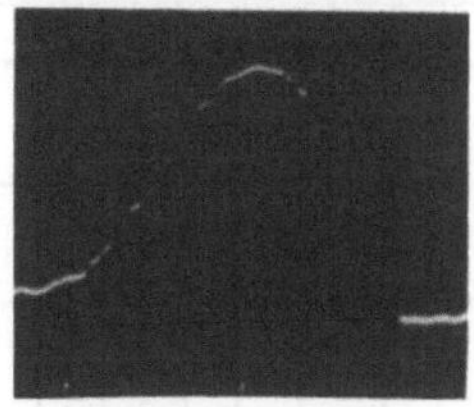

FIG. 2.8. Record of table acceleration with oil drop between anvil and spring of modified tester.

and to the acceleration of m_1, and hence is proportional to the deflection, strain, and stress in the beam provided only the first mode of the beam is excited.

Substituting (3.2) into (3.1) and dividing by m_1 we have

$$\ddot{x} + p_1^2 x = -\ddot{x}_2, \qquad (3.3)$$

where

$$p_1 = \sqrt{\frac{k_1}{m_1}}. \qquad (3.4)$$

Since m_1 is assumed to be small in comparison with m, the acceleration of the latter is not altered and (2.2) and (2.3) are used for $\ddot{x}_2$ in (3.3). With the initial conditions

$$[x]_{t=0} = [\dot{x}]_{t=0} = 0, \qquad (3.5)$$

the solution of (3.3) is (7)

$$\left.\begin{aligned} x &= \frac{v\omega}{\omega^2 - p_1^2}\left[\frac{\omega}{p_1}\sin p_1 t - \sin \omega t\right], \quad 0 \le t \le \pi/\omega, \\ x &= \frac{2v\omega^2 \cos(\pi p_1/2\omega)}{p_1(\omega^2 - p_1^2)} \sin p_1\left(t - \frac{\pi}{2\omega}\right), \quad t \ge \pi/\omega. \end{aligned}\right\} \qquad (3.6)$$

It may be seen that, during the time of impact, x is composed of a forced displacement at the acceleration frequency (ω) on which is superposed a free vibration at the natural frequency (p_1) of the beam. After impact the beam executes only free vibrations.

From (3.6) the maximum value of x is found to be

$$\left.\begin{aligned} x_{\max} &= \frac{v}{p_1\left(\dfrac{p_1}{\omega} - 1\right)} \sin \frac{2n\pi}{\dfrac{p_1}{\omega} + 1}, \quad 0 \le t \le \frac{\pi}{\omega}, \\ x_{\max} &= \frac{2v\omega^2 \cos \dfrac{\pi p_1}{2\omega}}{p_1(\omega^2 - p_1^2)}, \quad t \ge \frac{\pi}{\omega}, \end{aligned}\right\} \qquad (3.7)$$

where n is a positive integer so chosen as to make the sine term as large as possible while the argument remains less than π.

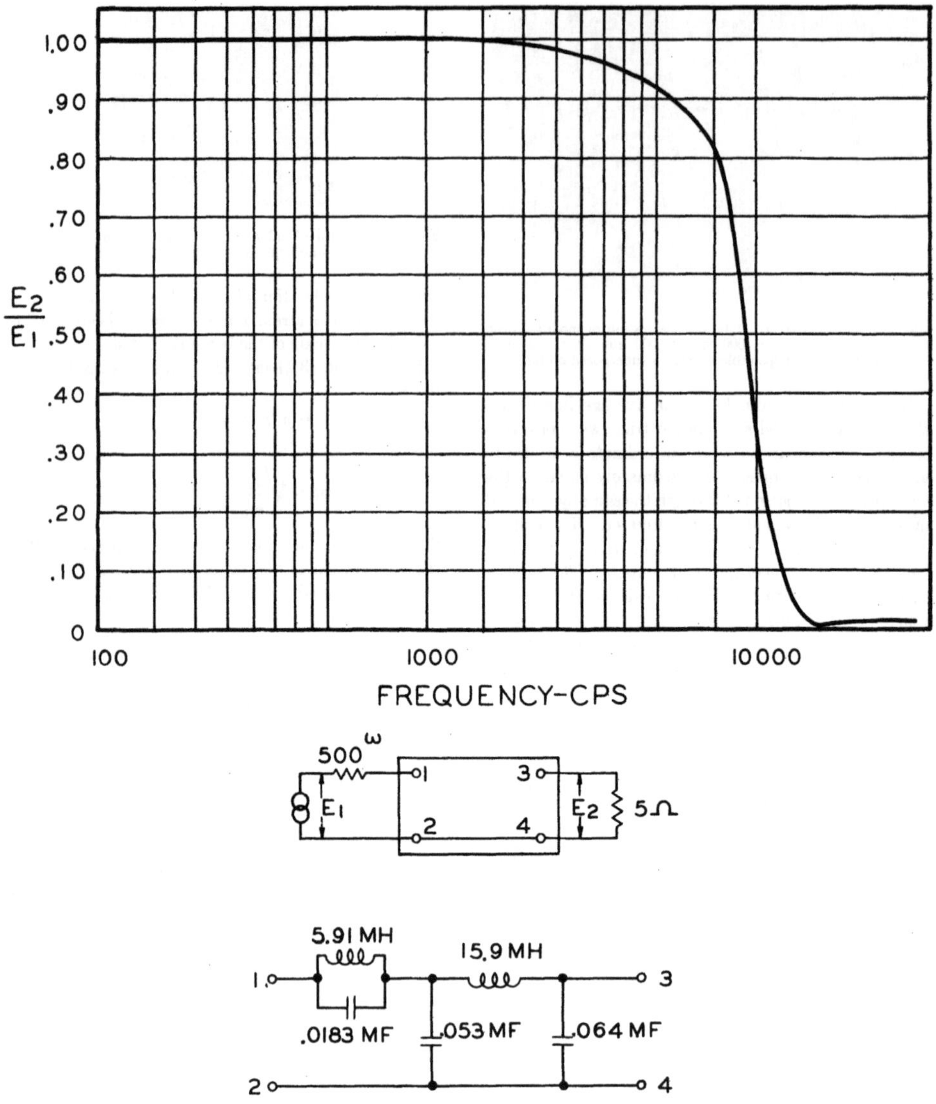

FIG. 2.7. Design and voltage characteristic of "10,000 cps" filter.

6

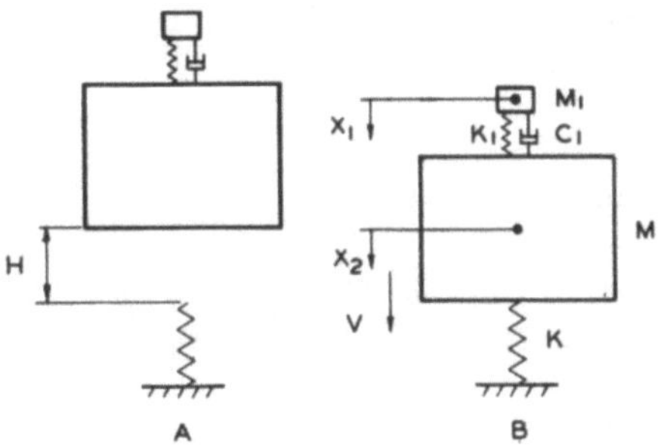

FIG. 3.1. Idealization of drop test.

4. Amplification Factor

It is convenient to express the results of Section 3 in the form of an *amplification factor* defined as the ratio of the absolute value of x_{max} to the absolute value (x_{st}) that x would have if the acceleration $\ddot{x}_2$ reached its peak value ($v\omega$) in a very long time so as not to excite vibrations of the system. x_{st} may be found from (3.3) by omitting the time-dependent terms, thus

$$x_{st} = \frac{v\omega}{p_1^2}. \tag{4.1}$$

Hence the amplification factor (A) is given by

$$\left.\begin{aligned} A &= \frac{x_{max}}{x_{st}} = \frac{\frac{p_1}{\omega}}{\frac{p_1}{\omega}-1}\sin\frac{2n\pi}{\frac{p_1}{\omega}+1}, \quad 0 \le t \le \pi/\omega, \\ A &= \frac{x_{max}}{x_{st}} = \frac{2\frac{p_1}{\omega}\cos\frac{\pi p_1}{2\omega}}{1-\frac{p_1^2}{\omega^2}}, \quad t \ge \pi/\omega, \end{aligned}\right\} \tag{4.2}$$

and we observe that A depends only on p_1/ω.

We find that the second of (4.2) gives the larger value when $p_1 < \omega$ while the first of (4.2) gives the larger value when $p_1 > \omega$. That is, when (and only when) the duration of impact is shorter than the half-period of vibration of the beam, the maximum strain occurs after the impact is over.

The curve marked $\beta_1 = 0$ in Fig. 4.1 is a plot of the largest value of A from (4.2). The second of (4.2) was used for $p_1/\omega \le 1$ and the first of (4.2) was used for $p_1/\omega \ge 1$.

To calculate the maximum dynamic stress in a structure, the amplification factor curve is used in accordance with the following procedure:

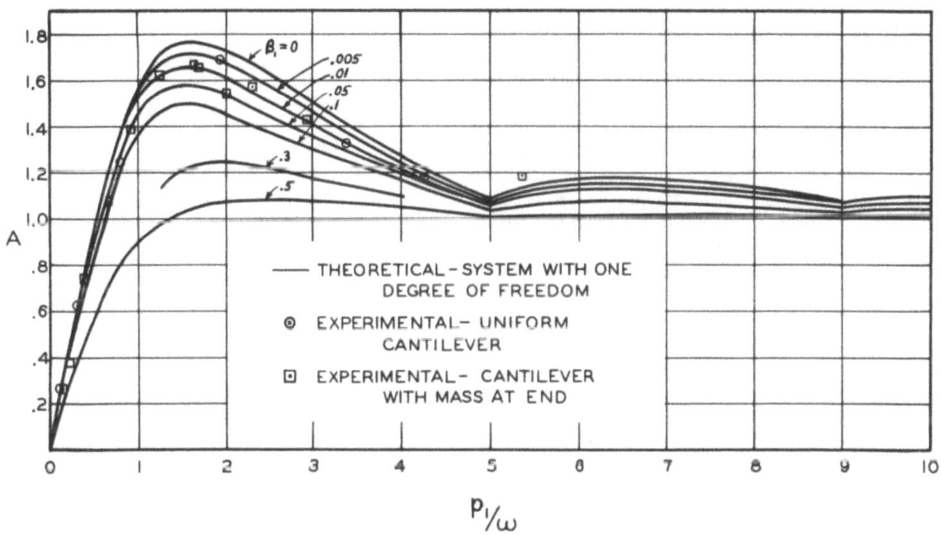

FIG. 4.1. Amplification factor curves for a damped system with one degree of freedom subjected to a half-sine-wave pulse acceleration.

7

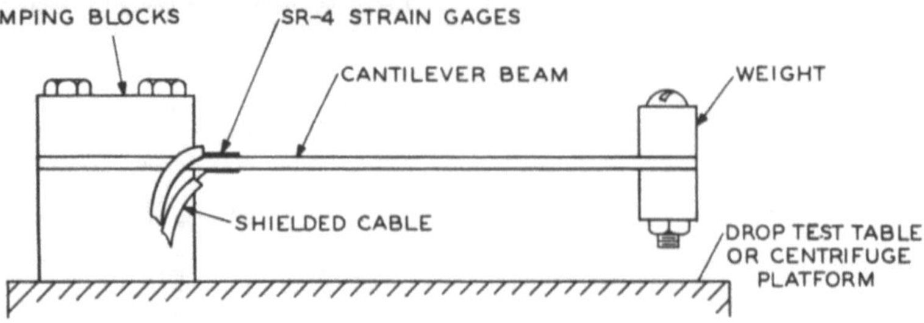

FIG. 5.1. Cantilever specimen used in drop tests.

(1) Determine the frequency of vibration (p_1) of the fundamental mode of the structure, divide by the stopping frequency (ω), and enter the curve at abscissa p_1/ω, obtaining the ordinate A.

(2) Multiply the weight of the structure by the maximum acceleration (in units of "number of times gravity") and calculate the stress as a "dead load" problem.

(3) The product of the results obtained in (1) and (2) is the maximum dynamic stress.

5. Drop Tests of Cantilever Bars

Twenty tests were made in which the surface strain was measured close to the clamped end of cold rolled steel cantilever bars. Each bar was clamped between steel blocks and the blocks were bolted to the drop test table as shown in Fig. 5.1. Five of the bars were of uniform rectangular cross section while the remainder had ead weights attached to the free end. The strain was detected by *SR*-4, type *C*-14, bonded electric wire gages cemented to the top and bottom faces of the cantilevers close to the clamped end. The gages formed two arms of a 2000 cps A.C. bridge as shown in Fig. 5.2. The output of the bridge was fed through the circuit shown in Fig. 2.4 (without filters), to a cathode-ray oscilloscope. Typical oscillograms are illustrated in Figs. 5.3 and 5.4. These show the modulation of the amplitude of the 2000 cps carrier wave resulting from the transient strain. The maximum ordinate of each oscillogram was converted to volts (V_{dyn}) by comparison with the ordinate of a steady A.C. signal whose voltage was measured with a vacuum-tube voltmeter.

The cantilevers in their clamping blocks were then transferred to a centrifuge and a "static" run was made in order to obtain data for the denominator of the amplification factor. In the centrifuge test the voltage output of the strain gage and amplifier circuit was measured for a range of accelerations and extrapolated (or interpolated) to the voltage output (V_{st}) for the acceleration equal to the maximum acceleration of the drop test. The ratio V_{dyn}/V_{st} is the amplification factor. (It will be noticed that it is not necessary to know the calibration factor for the gages.)

The numerator of the ratio p_1/ω was obtained by observing the Lissajou figure on a cathode-ray oscilloscope with the horizontal amplifier fed by a Hewlett-Packard oscillator and the vertical amplifier fed by one of the strain gages, while the beam was excited in its fundamental mode on a voice-coil vibrator. Finally, measurements of damping were made on a few cantilevers by

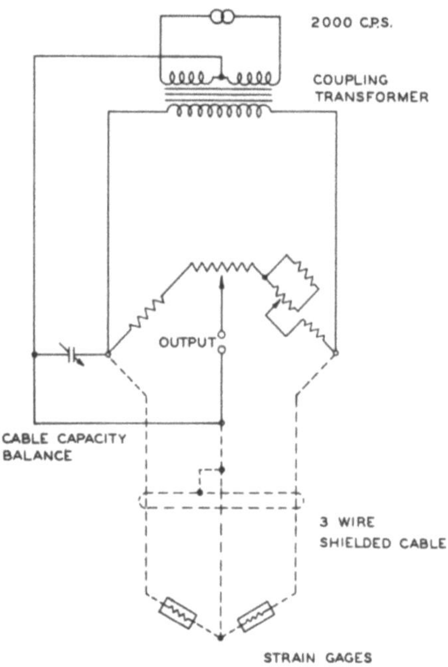

FIG. 5.2. Strain gage bridge.

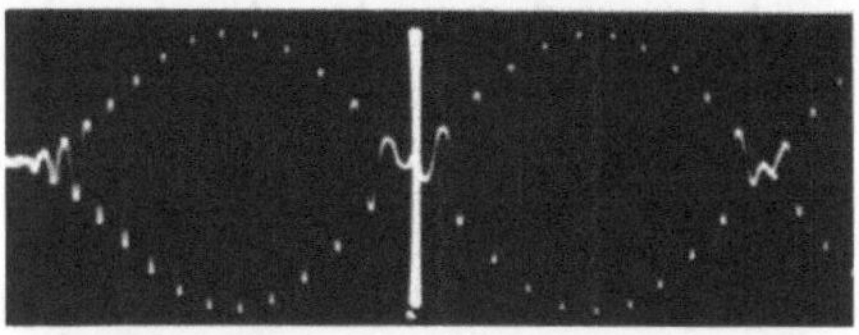

Fig. 5.3. Strain at clamped end of cantilever in drop test. Carrier wave frequency is 2000 cps, $p_1/\omega = 0.9$. Specimen No. 6, Table I. (Vertical line is trace of calibration voltage.)

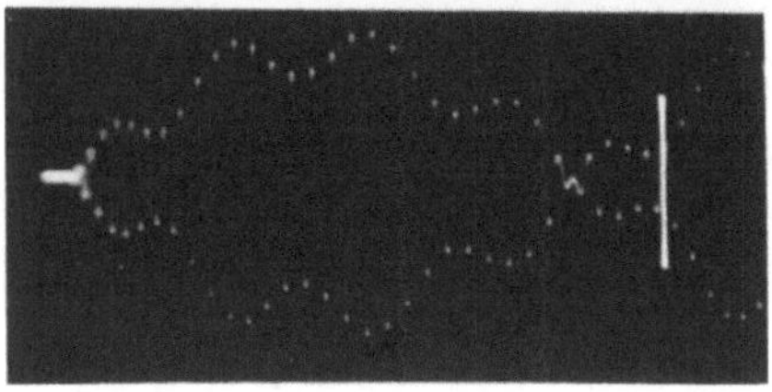

Fig. 5.4. Strain at clamped end of cantilever in drop test. Carrier wave frequency is 2000 cps, $p_1/\omega = 0.1$. Specimen No. 18, Table I.

exciting the fundamental mode and recording the decay of the signal from the strain gages.

An outline of the data is given in Table I and the amplification factors are plotted in Fig. 4.1. It may be seen that the curve marked $\beta_1 = 0$ follows the general trend of the experimental points, so that there is qualitative agreement. However, the theoretical curve lies above the experimental data in the range $.5 < p_1/\omega < 5$, while it falls below the experimental data for very small values of p_1/ω. It is useful, though fruitless, to explore the possibility that these discrepancies may be due to the neglect of damping in the system with one degree of freedom. This is done in the following section.

6. Theory of Drop Test of a Damped System with One Degree of Freedom

If viscous damping is considered in the system with one degree of freedom, equation (3.1) is altered to

$$m_1\ddot{x}_1 + c_1(\dot{x}_1 - \dot{x}_2) + k_1(x_1 - x_2) = 0. \quad (6.1)$$

TABLE I

Cantilever Drop-test Data

Cantilevers

No.	Length (in)	Width (in)	Depth (in)	End Weight (lbs)	Frequency 1st Mode cps	β_1	G	p_1/ω	Amplification Factor
1	6	½	¼	0	193		195	0.80	1.25
2	6	½	¼	0	193	.006	92	1.93	1.69
3	6	½	¼	¼	92		187	0.38	0.74
4	6	½	¼	¼	92	.005	142	0.38	0.75
5	6	½	¼	¼	92		67	1.24	1.62
6	6	½	¼	¼	92		76	0.92	1.39
7	6	½	¼	½	66		64	0.66	1.07
8	6	½	¼	½	66	.008	48	1.65	1.67
9	6	½	¼	1	50		120	0.21	0.38
10	5	1	¼	0	250		68	3.37	1.32
11	5	1	¼	¼	147		67	2.00	1.54
12	5	1	¼	½	120		66	1.67	1.66
13	4	1	¼	¼	215		95	2.92	1.42
14	4	1	¼	¼	215		48	5.35	1.18
15	4	1	¼	½	170		78	2.30	1.57
16	4	1	¼	½	170		48	4.25	1.18
17	17	½	¼	0	24		65	0.32	0.63
18	17	½	¼	0	24	.005	122	0.10	0.27
19	8	⅝	3/32	¼	23		55	0.31	0.63
20	8	⅝	3/32	¾	14		60	0.14	0.27

9

It is convenient to express the amount of damping as the fraction of critical damping:

$$\beta_1 = \frac{c_1}{2\sqrt{m_1 k_1}}. \tag{6.2}$$

Then (6.1) becomes

$$\ddot{x} + 2\beta_1 p_1 \dot{x} + p_1^2 x = -\ddot{x}_2, \tag{6.3}$$

where x, p_1 and x_2 have the same significance as before (see equations (2.2), (2.3), (3.2), (3.4)).

The amplification factors for equation (6.3) with initial conditions (3.5) were obtained on the Westinghouse Mechanical Transients Analyzer (8) for $\beta_1 =$ 0.005, 0.01, 0.05, 0.10, 0.30, 0.50, and 1.00. The resulting curves are shown in Fig. 4.1.

The test specimens had fractions of critical damping in the neighborhood of 0.005. The curve marked $\beta_1 = 0.005$ falls closer to the experimental points in the range $0.5 < p_1/\omega < 5$ than does the curve $\beta_1 = 0$, but the discrepancy is made greater in the range $p_1/\omega < 0.5$.

What is required is a theory which, in comparison with the system with one degree of freedom, will give lower amplification factors in the range $0.5 < p_1/\omega < 5$ and higher amplification factors in the range $p_1/\omega < 0.5$.

7. Bernoulli-Euler System

It is reasonable to suppose that the system with one degree of freedom yields amplification factors that are too low for short duration pulses, as a result of neglecting the contributions of the higher modes of vibrations. All of the flexural modes of vibration of a beam are considered in the Bernoulli-Euler equation:

$$a^2 \frac{\partial^4 w}{\partial x^4} + \frac{\partial^2 w}{\partial t^2} = 0, \tag{7.1}$$

where

x = coordinate parallel to the axis of the beam
w = transverse displacement
$a^2 = Ek^2/\rho$
E = Young's modulus
ρ = density
k = radius of gyration of the cross section with respect to a principal axis (assumed to be normal to the plane of bending).

The initial and boundary conditions on (7.1) for a cantilever of length l in a drop test are

$$\Big[w\Big]_{t=0} = 0, \tag{7.2}$$

$$\left[\frac{\partial w}{\partial t}\right]_{t=0} = -v, \tag{7.3}$$

$$\Big[w\Big]_{x=l} = -\frac{v}{\omega} \sin \omega t \, M\left(\frac{\pi}{\omega} - t\right) + v\left(t - \frac{\pi}{\omega}\right) \times M\left(t - \frac{\pi}{\omega}\right), \tag{7.4}$$

$$\left[\frac{\partial w}{\partial x}\right]_{x=l} = 0, \tag{7.5}$$

$$\left[\frac{\partial^2 w}{\partial x^2}\right]_{x=0} = 0, \tag{7.6}$$

$$\left[\frac{\partial^3 w}{\partial x^3}\right]_{x=0} = 0, \tag{7.7}$$

where

$$\left.\begin{aligned} M(y) &= 1 \text{ for } y \geq 0 \\ M(y) &= 0 \text{ for } y \leq 0 \end{aligned}\right\} \tag{7.8}$$

This problem was solved by Symonds and Vigness (1) with the following results:

for $0 \leq t \leq \frac{\pi}{\omega}$,

$$w = -\frac{v\psi(x)\sin\omega t}{2\omega} - \frac{2v\omega^2}{l}\sum_{n=1}^{\infty}\frac{\varphi_n(x)\sin p_n t}{\mu_n p_n(\omega^2 - p_n^2)}, \tag{7.9}$$

for $t \geq \frac{\pi}{\omega}$,

$$w = v\left(t - \frac{\pi}{\omega}\right) - \frac{4v\omega^2}{l}\sum_{n=1}^{\infty}\frac{\varphi_n(x)\cos\frac{\pi p_n}{2\omega}\sin p_n\left(t - \frac{\pi}{2\omega}\right)}{\mu_n p_n(\omega^2 - p_n^2)}, \tag{7.10}$$

where

$$K\psi(x) = (\cosh bx + \cos bx)(\cosh bl + \cos bl) - (\sinh bx + \sin bx)(\sinh bl - \sin bl),$$

$$K = 1 + \cosh bl \cos bl,$$

$$L\varphi_n(x) = (\cosh \mu_n x + \cos \mu_n x)(\cosh \mu_n l + \cos \mu_n l) - (\sinh \mu_n x + \sin \mu_n x)(\sinh \mu_n l - \sin \mu_n l),$$

$$L = \cosh \mu_n l \sin \mu_n l - \sinh \mu_n l \cos \mu_n l,$$

$\mu_n l$ = nth real positive root of $1 + \cosh \mu_n l \cos \mu_n l = 0$,

$$b = \sqrt{\frac{\omega}{a}},$$

$p_n = \mu_n^2 a$ (frequency of the nth mode of vibration).

The extreme fiber strain at the clamped end of the bar is given by

10

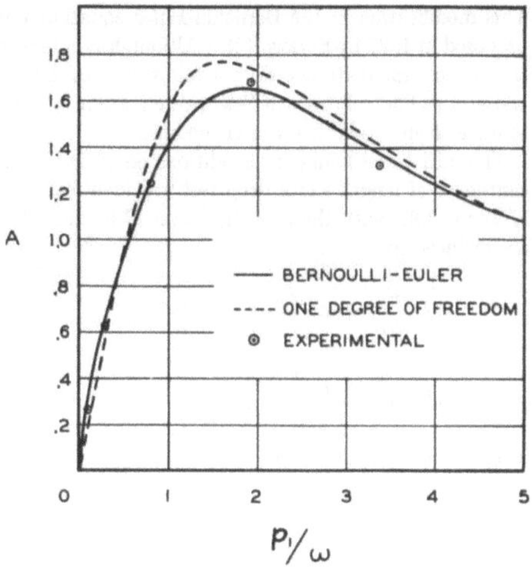

FIG. 7.1. Amplification factor for half-sine-wave pulse acceleration.

$$\epsilon_m = \frac{d}{2}\left[\frac{\partial^2 w}{\partial x^2}\right]_{x=l}, \tag{7.11}$$

where d is the depth of the beam. The extreme fiber strain, at the clamped end, that would result if the maximum acceleration ($v\omega$) were reached very slowly, as in the centrifuge tests, is

$$\epsilon_{st} = \frac{v\omega l^2 d}{4a^2}. \tag{7.12}$$

The amplification factor for strain at the clamped end is the maximum absolute value of ϵ_m/ϵ_{st}:

$$A = \left|\frac{2a^2\left(\dfrac{\partial^2 w}{\partial x^2}\right)_{x=l}}{v\omega l^2}\right|_{max} \tag{7.13}$$

The amplification factor (7.13) is plotted against p_1/ω in Fig. 7.1. On the same figure are plotted the amplification factors for the undamped system with one degree of freedom and the experimental data for the uniform cantilevers. It may be seen that the Bernoulli-Euler theory is very close to the experimental data over most of the range of frequency ratios. However, the appearance is deceiving in the range of very low frequency ratios. It should be remembered that the ordinate of the curve is a number by which ϵ_{st} is to be multiplied to obtain the maximum dynamic strain. Hence, the *percent* discrepancy between theory and experiment is directly the percent error in dynamic strain. Since the ordinate is very small for small values of p_1/ω, very small differences in ordinates may mean large percent errors.

Although amplification factor curves are useful in design work, it is better to use another set of coordinates in making comparisons between theory and experiment. For example, in Fig. 7.2 the ordinate is $A\omega/p_1$ and the abscissa is ω/p_1 (the reciprocal of the abscissa in Fig. 7.1). The ordinate is proportional to the maximum strain (ϵ_m) at the clamped end:

$$A\omega/p_1 = 1.138\,\frac{\epsilon_m a}{vd}. \tag{7.14}$$

In moving to the right in the figure it is as if, in the drop test, the height of drop is maintained constant while the stopping spring is made stiffer. The ordinate at $\omega/p_1 = \infty$ in Fig. 7.2 is equal to the slope of the curve at $p_1/\omega = 0$ in Fig. 7.1. These coordinates are similar to those used by Symonds and Vigness.

Curves I and II in Fig. 7.2 are for the system with one degree of freedom and the Bernoulli-Euler system respectively. For frequency ratios $\omega/p_1 < 3$ the Bernoulli-Euler system fits the experimental data remarkably well, even though damping is neglected in the theory, whereas the beam specimens had damping of approximately one-half percent of critical. For frequency ratios $\omega/p_1 > 3$ the system with one degree of freedom gives strains that are too low. That this discrepancy is due to the fact that higher mode contributions are neglected, is apparent from the fact that the Bernoulli-Euler theory gives much larger strains for short duration shocks in which higher modes can be excited. However, the Bernoulli-Euler system predicts strains that are too large. For example, at $\omega/p_1 = 0.5$ the experiment gave $A\omega/p_1 = 2.57$, while the theory gives $A\omega/p_1 = 2.93$, an error of 14%. Although experiments at shorter duration shocks were not made, the discrepancy would undoubtedly be greater because the experimental strains would probably level off, while the Bernoulli-Euler strain becomes larger without limit. The latter can be ascertained by examining (7.10) and (7.11) for $\omega \to \infty$ (instantaneous velocity change, of magnitude $2v$, of the clamped end of the beam):

$$\lim_{\omega\to\infty} \epsilon_m = \frac{4vd}{a}\sum_{n=1}^{\infty}\frac{\sin p_n t}{\mu_n l\,(\coth\mu_n l - \cot\mu_n l)}.$$

After the first few terms, the amplitudes of the $\sin p_n t$ terms decrease like $1/n$, so that the sum of the amplitudes is divergent. The frequencies p_n are not integral multiples of each other and hence there will be a time, t_m, when all the modes are in phase and the strain will be infinite. However, it is clear that, before t_m is reached, a large portion of the energy in the higher modes will have been dissipated in a real beam so that the strain amplitudes in these modes will be negligible.

8. Sezawa Equation

In the derivation of the Bernoulli-Euler equation, it is assumed that the axial stress(σ) is directly proportional to the axial strain (ϵ), the constant of proportionality being Young's modulus (E). Thus

$$\sigma = E\epsilon. \tag{8.1}$$

The simplest law, which introduces internal damping in an elastic solid, is Stokes' law, which assumes that stress is a linear function of strain and strain-rate. In the case of the beam

$$\sigma = E\epsilon + c\dot{\epsilon}, \tag{8.2}$$

where c is the coefficient of viscosity and $\dot{\epsilon}$ is the strain-rate. To pass from (8.1) to (8.2) it is only necessary to replace E in (8.1) by $\left(E + c\frac{\partial}{\partial t}\right)$. If this is done in the Bernoulli-Euler equation (7.1), we obtain

$$a^2 \frac{\partial^4 w}{\partial x^4} + \frac{ck^2}{\rho}\frac{\partial^5 w}{\partial x^4 \partial t} + \frac{\partial^2 w}{\partial t^2} = 0. \tag{8.3}$$

This modification of the Bernoulli-Euler equation was proposed in 1927 by Sezawa (2). Although it is known that structural materials do not follow Stokes' law, it is useful to find solutions of Sezawa's equation in order to have them available for later modification.

The initial and boundary conditions on (8.3) for the cantilever of length l in a drop test are identical with (7.2) to (7.8), with the exception that (7.6) and (7.7) are replaced by

$$\left[E\frac{\partial^2 w}{\partial x^2} + c\frac{\partial^3 w}{\partial x^2 \partial t}\right]_{x=0} = 0, \tag{8.4}$$

$$\left[E\frac{\partial^3 w}{\partial x^3} + c\frac{\partial^4 w}{\partial x^3 \partial t}\right]_{x=0} = 0. \tag{8.5}$$

The solution is found to be:

for $0 \le t \le \pi/\omega$: (8.6)

$$w = -\left[\frac{v}{2\omega}\xi(x)\cos\omega t + \eta(x)\sin\omega t\right] - \frac{2v\omega^2}{l}\sum_{n=1}^{\infty}\frac{\varphi_n(x)\,e^{-\beta_n p_n t}\sin(p_n' t + \gamma_n)}{\mu_n p_n'\sqrt{(\omega^2 - p_n^2)^2 + 4\beta_n^2 p_n^2\omega^2}},$$

and, for $t \ge \pi/\omega$:

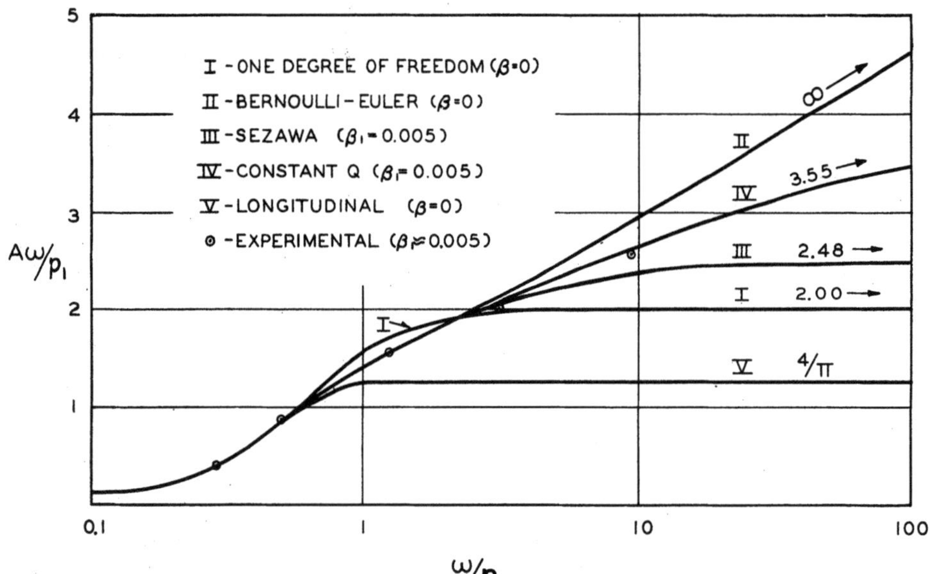

FIG. 7.2. Strain versus ratio of stopping frequency to frequency of fundamental mode.

12

$$w = v\left(t - \frac{\pi}{\omega}\right) - \frac{4v\omega^2}{l}\sum_{n=1}^{\infty}\varphi_n(x)e^{-\beta_n p_n(t-\pi/2\omega)}$$

$$\times \frac{\sqrt{\sinh^2\left(\frac{\beta_n p_n \pi}{2\omega}\right) + \cos^2\left(\frac{p_n'\pi}{2\omega}\right)}}{\mu_n p_n'\sqrt{(\omega^2 - p_n^2)^2 + 4\beta_n^2 p_n^2\omega^2}} \sin(p_n't + \epsilon_n), \quad (8.7)$$

where p_n, μ_n, and $\varphi_n(x)$ have the same meanings as before and

$$\xi(x) = \frac{1}{M}\left[c_1\psi_1(x) + c_2\psi_2(x) + c_3\psi_3(x) + c_4\psi_4(x)\right],$$

$$\eta(x) = \frac{1}{M}\left[c_2\psi_1(x) - c_1\psi_2(x) + c_4\psi_3(x) - c_3\psi_4(x)\right],$$

$$\psi_1(x) = \cos fx \cosh gx + \cosh fx \cos gx,$$

$$\psi_2(x) = \sinh fx \sin gx - \sin fx \sinh gx,$$

$$\psi_3(x) = \sin fx \cosh gx + \sinh fx \cos gx,$$

$$\psi_4(x) = \cos fx \sinh gx + \cosh fx \sin gx,$$

$$M = (\cos fl \cosh fl + \cos gl \cosh gl)^2 + (\sin fl \sin gl + \sinh fl \sinh gl)^2,$$

$$c_1 = \sinh fl \sin gl (\sin^2 fl - \sinh^2 gl) - \sin fl \sinh gl (\sin^2 gl - \sinh^2 fl),$$

$$c_2 = (\cos gl \cosh gl + \cos fl \cosh fl) \times (\cos fl \cos gl + \cosh fl \cosh gl),$$

$$c_3 = \cos fl \sinh gl - \cosh fl \sin gl + \sinh gl \cosh fl (\cos gl \cosh gl - \sin fl \sinh fl) - \sin gl \cos fl (\cos gl \cosh gl + \sin fl \sinh fl),$$

$$c_4 = \sin fl \cosh gl - \sinh fl \cos gl + \cosh gl \sinh fl (\sin gl \sinh gl - \cos fl \cosh fl) + \cos gl \sin fl (\sin gl \sinh gl + \cos fl \cosh fl),$$

$$f = \sqrt{\frac{\omega}{a}}\cos^{\frac{1}{2}}\theta \cos(\theta/4),$$

$$g = -\sqrt{\frac{\omega}{a}}\cos^{\frac{1}{2}}\theta \sin(\theta/4),$$

$$\tan\theta = \frac{c\omega}{E},$$

$$\tan\gamma_n = \frac{2\beta_n p_n^2\sqrt{1-\beta_n^2}}{2\beta_n^2 p_n^2 - p_n^2 + \omega^2},$$

$$\tan\epsilon_n = \frac{e^{-\beta_n p_n \pi/\omega}\sin\gamma_n - \sin(p_n'\pi/\omega - \gamma_n)}{e^{-\beta_n p_n \pi/\omega}\cos\gamma_n + \cos(p_n'\pi/\omega - \gamma_n)},$$

$$p_n' = p_n\sqrt{1-\beta_n^2},$$

$$\beta_n = \frac{p_n c}{2E}.$$

β_n is the fraction of critical damping in the nth mode. Thus, if β_1 is the fraction of critical damping in the first mode, then

$$\beta_n = \beta_1 \frac{p_n}{p_1}.$$

As a result of the assumption of Stokes' law of damping, the Sezawa equation makes the fraction of critical damping in any mode directly proportional to the frequency of that mode. It may be verified that (8.6) and (8.7) reduce to (7.9) and (7.10) when c is set equal to zero.

It was pointed out in Section 1 that tests of many materials show that, at least to a first approximation, damping is independent of frequency. Hence, we may expect that the higher modes in (8.6) and (8.7) will be excessively damped so that predicted strains will be too small for short duration shocks. This is confirmed in the experiments, as may be seen in Fig. 7.2, in which Curve III is drawn for equations (8.6) and (8.7) and $\beta_1 = 0.005$ (the approximate value of first mode damping of the specimens). At a frequency ratio $\omega/p_1 = 10$, the predicted strain is about 10% less than the measured strain.

The Sezawa equation does give the required result that the strain remain finite as the shock duration approaches zero. This may be verified by substituting (8.7) in (7.11) and letting $\omega \to \infty$:

$$\lim_{\omega\to\infty}\epsilon_m = \frac{4vd}{a}\sum_{n=1}^{\infty}\frac{e^{-\beta_n p_n t}\sin p_n't}{\mu_n l\sqrt{1-\beta_n^2}(\coth\mu_n l - \cot\mu_n l)}. \quad (8.8)$$

Equation (8.8) pertains to the solution of the Sezawa equation for an instantaneous velocity change, of magnitude $2v$, of the clamped end of the beam, a problem which was solved by Stowell, Schwartz and Houbolt. (9)

Using equation (8.8) the quantity

$$\lim_{\omega\to\infty} A\omega/p_1 = \frac{1.138a}{vd}\lim_{\omega\to\infty}\epsilon_m \quad (8.9)$$

is plotted against β_1 as the lower curve in Fig. 8.1. It should be remembered that the ordinate of Fig. 8.1 at any β_1 is the ordinate of Fig. 7.2 for $\omega/p_1 \to \infty$ and the same β_1. It is seen that, for $\beta_1 > 0$, the maximum strain is finite and it requires only a very small amount of damping to reduce the strain from the infinity of the Bernoulli-Euler theory to reasonable values. However, as noted above, the reduction is excessive.

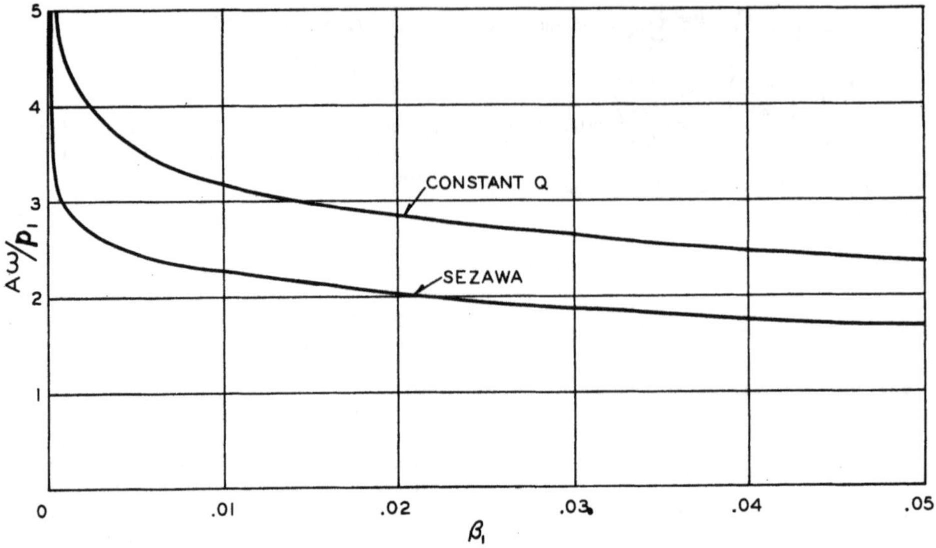

FIG. 8.1. Influence of internal damping on the maximum strain at the clamped end of a cantilever subjected to a sudden transverse velocity change.

9. Constant Q Hypothesis

The fraction of critical damping, β_n, which appears in free vibration solutions of the Sezawa equation (the summations in (8.6) and (8.7)), is given by

$$\beta_n = \frac{p_n c}{2E}.$$

Since c and E were assumed to be constant, β_n is directly proportional to the frequency of the corresponding mode. This is contrary to the evidence (3), (4) sighted in Section 1. Much closer correlation with previous experiments would be obtained if β_n were independent of frequency. This would be the case if Stokes' viscosity coefficient (c) were inversely proportional to p_n.

Similarly, in the forced vibration part of (8.6) (*i.e.*, the term outside the summation) c appears only in the combination $c\omega/E$. In this case c should be taken inversely proportional to ω. Accordingly, we shall arbitrarily set

$$\frac{cp_n}{E} = \frac{c\omega}{E} = 2\beta = \frac{1}{Q}, \tag{9.1}$$

where β and Q are constants.

The magnitude of β or Q may be determined experimentally by observations on the decay of free vibrations of the first mode or by a forced vibration experiment. The former method was used on the cantilever specimens and the fraction of critical damping was found to be approximately .005, *i.e.*, $Q = 100$. With this value of Q the quantity $A\omega/p_1$ was calculated, from (7.14), (7.11), (8.6), and (8.7), and the results are given as Curve IV of Fig. 7.2. It may be seen that the curve matches the experimental results exceedingly well.

The constant Q hypothesis also satisfies the requirement that the strain remain finite as the shock duration approaches zero. The limiting values for various values of Q ($= 1/2\beta$) were calculated from (8.8) and (8.9) and are plotted as the upper curve in Fig. 8.1.

It is interesting to examine the percent contribution of each mode to the maximum strain according to the Sezawa, Constant Q, and Bernoulli-Euler theories. The data are given in Table II for $\beta_1 = 0.005$ and frequency ratios $\omega/p_1 = 1$, 10, 100, and ∞. It should be remembered that the maximum strain occurs at a time, t_m, considerably later than the time of initiation of the disturbance. During that interval the strain amplitude decays, in the various modes, at a rate governed by the assumed law of damping. As might be expected, the contributions to maximum strain come from a greater number of modes as the shock duration decreases and as the law of damping passes from Sezawa to Constant Q to Bernoulli-Euler. The percentages in the last

TABLE II

Per Cent Contribution of Each Mode to Maximum Strain

ω/p_1 / n	Sezawa				Constant Q				Bernoulli-Euler			
	1	10	100	∞	1	10	100	∞	1	10	100	∞
1	100.0	70.3	67.4	67.5	100.0	61.9	46.2	44.9	98.5	59.8	38.6	0.0
2	0.0	29.7	30.7	31.0	0.0	30.7	19.7	17.3	1.4	30.3	21.3	0.0
3		0.0	1.8	1.5		7.3	12.5	11.5	0.0	8.7	12.5	0.0
4			0.0	0.0		0.0	8.9	8.0		0.8	8.7	0.0
5							6.2	6.0		0.3	6.5	0.0
6							3.9	4.4		0.1	4.8	0.0
7							2.0	3.2		0.1	3.4	0.0
8							0.6	2.2		0.0	2.2	0.0
9							0.0	1.4			1.2	0.0
10								0.8			0.4	0.0
11								0.3			0.1	0.0
12								0.1			0.2	0.0
13								0.0			0.1	0.0
14											0.0	0.0

TABLE III

Ratios of Initial Strain Amplitudes

ω/p_1 / n	Sezawa				Constant Q			
	1	10	100	∞	1	10	100	∞
1	1.00	1.00	1.00	1.00	1.00	1.00	1.00	1.00
2		.97	1.00	1.00		.99	1.00	1.00
3			.98	1.00		.99	1.00	1.00
4							1.00	1.00
5							1.00	1.00
6							.99	1.00
7							.99	1.00
8							.99	1.00
9								1.00
10								1.00
11								1.00
12								1.00
13								1.00

column of Table II are all zero because the maximum strain is infinite when $\omega/p_1 \to \infty$ in the absence of damping.

The strain amplitude which *enters* each mode is affected very little by the choice of a law of damping as long as the damping is small. This may be verified by calculating the ratio of the initial strain amplitude for each mode of the Sezawa or Constant Q solutions to the initial strain amplitude for the corresponding mode of the Bernoulli-Euler solution. Results are shown in Table III for $\beta_1 = 0.005$. Only those modes are considered which contribute more than 0.1% to the final maximum strain. As can be seen from the table, the initial strain amplitudes are almost identical for the corresponding contributing modes in all three solutions. This may be considered as justification for switching from one law of damping to another *after* solving the differential equation. The subsequent motion is, of course, influenced by damping, but it is reasonable to suppose that the energy dissipation per cycle per mode is the controlling factor. The choice of the Constant Q law apparently yields the proper rates of dissipation for the various modes.

It should be noted that the term in brackets preceding the summation in equation (8.6) is the solution for the case of steady state sinusoidal oscillation of the clamped end of the cantilever. This is the solution, mentioned at the end of Section 1, with which steady-state vibration experiments should be compared.

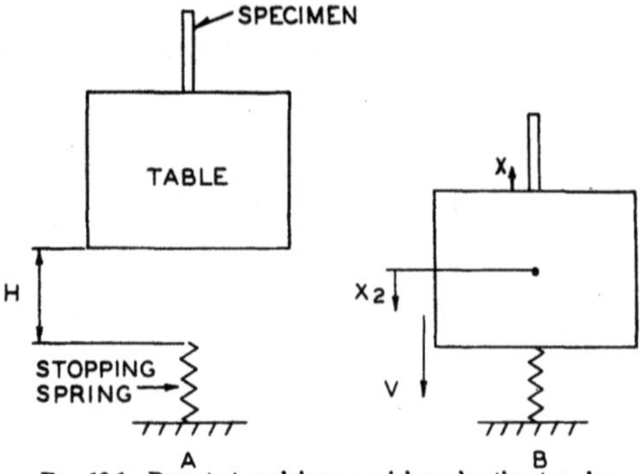

FIG. 10.1 Drop test applying an axial acceleration to a bar.

10. A Nondispersive System

The necessity for considering damping in the transient flexural motions of a beam arises from the fact that there is dispersion in the propagation of flexural waves.

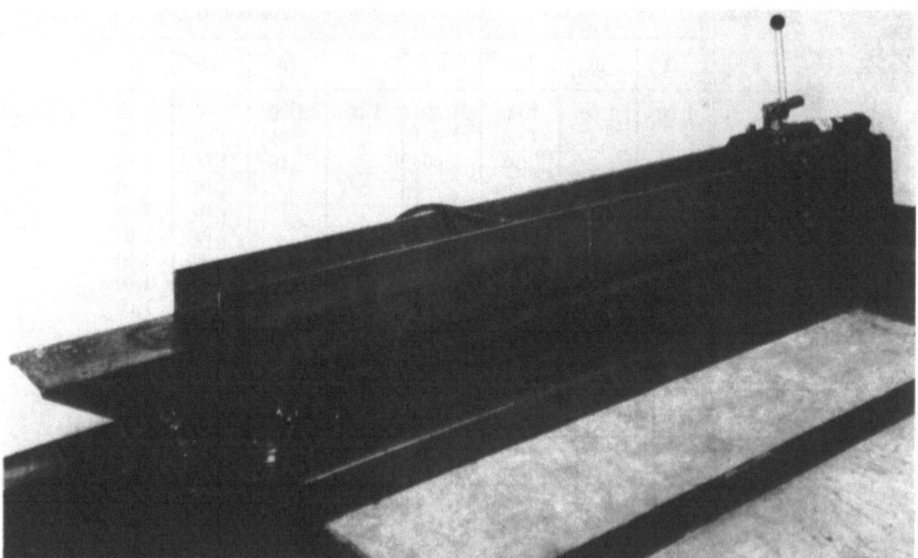

16

After a transverse disturbance is initiated, flexural waves travel back and forth along the beam and the shape of the disturbance changes continuously because its component waves travel with different velocities. There will be a time when large numbers of wave peaks are in phase at some section of the beam. This time may be after many cycles of the higher modes but, if there is no damping, the amplitudes of the higher modes will not have been diminished and will sum to large strains and accelerations. In a real beam, damping will have reduced the amplitudes of the higher modes so that, when coincidence of peaks occurs, the sum will not be large.

In a nondispersive system, all waves travel with the same velocity, so that a disturbance maintains its shape in travelling back and forth. Hence no delayed coincidence of wave peaks can occur to produce unusually high strains.

The nondispersive analogue of the problem of flexure treated above is obtained by up-ending the beam in the drop test so as to apply the pulse acceleration parallel to the axis of the bar (Fig. 10.1). The motion of the bar is then assumed to be governed by the one-dimensional wave equation:

$$\frac{E}{\rho}\frac{\partial^2 u}{\partial x^2} = \frac{\partial^2 u}{\partial t^2}, \tag{10.1}$$

where u is the axial displacement of a section of the bar distant x from the end attached to the table. The solution of (10.1) for the appropriate initial and boundary conditions was given in an earlier paper* and the product $(A\omega/p_1)$ of frequency ratio and amplification factor for strain at the end of the bar attached to the table is plotted as Curve V in Fig. 7.2. It may be seen that $A\omega/p_1$ remains finite as $\omega/p_1 \to \infty$, even though damping has been omitted. In fact $A\omega/p_1 = 4/\pi$ for $\omega/p_1 \geq 1$. Hence the difficulty of infinite strains that appears in the analogous solution of the Bernoulli-Euler equation is not encountered.

*Reference 7(a), pp. 451–457; or reference 7(b), pp. 99–105.

It may also be seen that Curve V lies below Curve I for all values of ω/p_1. Hence, in design work, the substitution of a system with one degree of freedom for a distributed nondispersive system will give strains on the safe side when the disturbance is a single pulse.

11. Plucked Reed

A very sensitive measure of occurrences in the higher modes of vibration is acceleration, since this quantity is proportional to the square of the frequency. Furthermore, acceleration is easy to record and it is also easy to excite large acceleration amplitudes in the higher modes. The latter was accomplished by fixing the clamped end of a cantilever, deflecting the free end, and suddenly releasing the applied force. The cantilever was, in this case, a large H-beam (see Fig. 11.1) originally designed for the shock-testing of vacuum tubes. The beam was 6½ feet long and its section was a 6½″ x 8″ H. The natural frequency of the cantilever was 30 cps with $\beta_1 = 0.016$. The fraction of critical damping was measured by fastening a pencil to the free end of the beam, plucking the latter, and recording the pencil trace on a large card drawn quickly, by hand, across the moving pencil point. The high damping ($\beta_1 = 0.016$) is attributed mostly to energy losses through the sup-

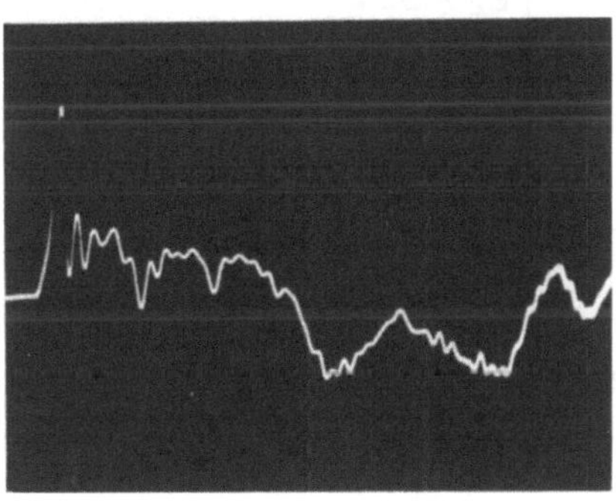

FIG. 11.2. Acceleration of free end of plucked reed.

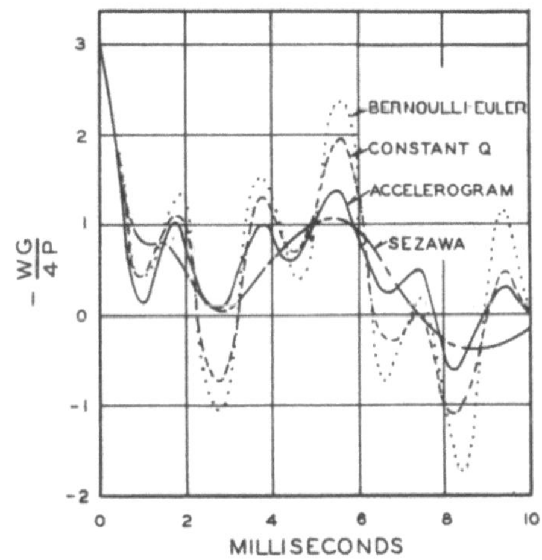

FIG. 11.4. Comparison of recorded acceleration with sum of first three modes for three theories.

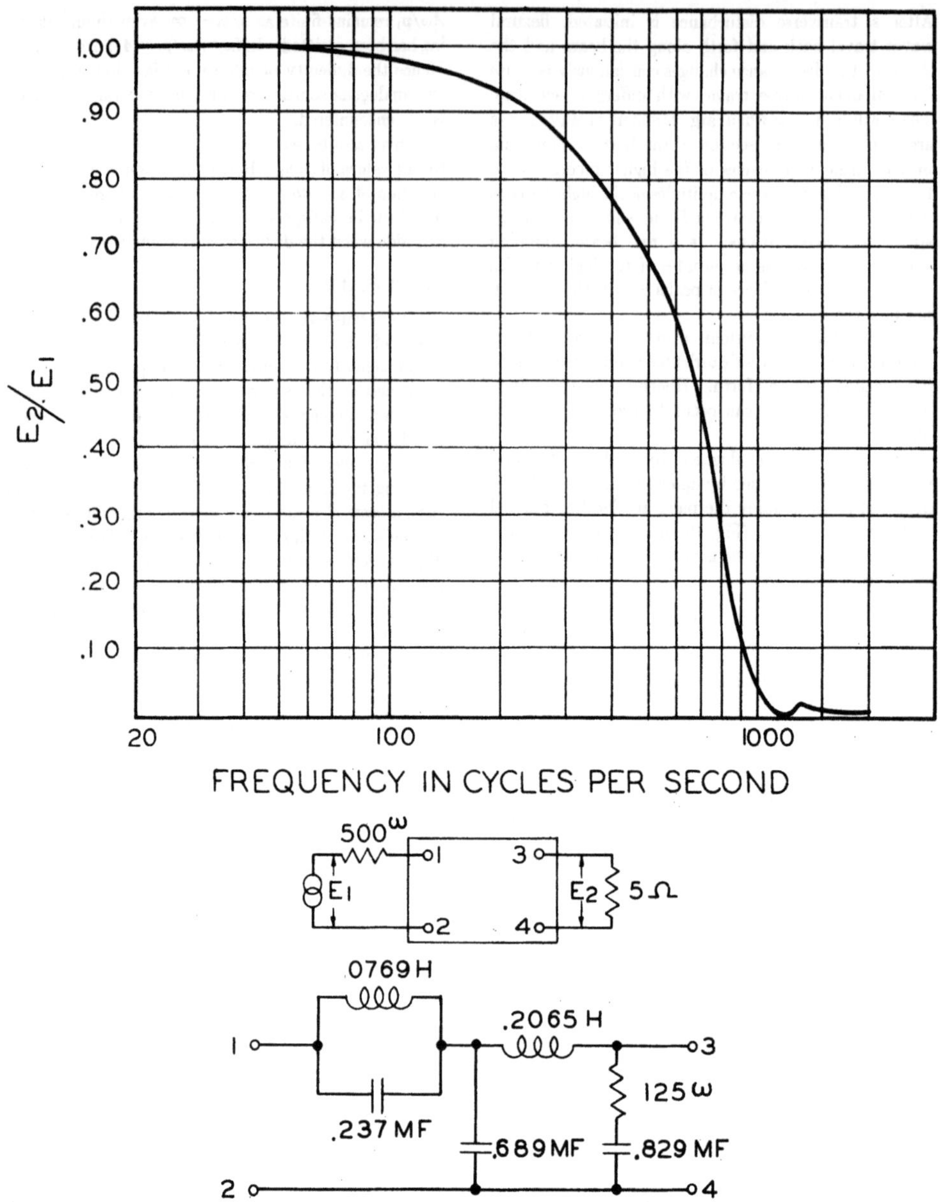

FIG. 11.3. Design and voltage characteristic of "1000 cps" filter.

port. The higher modes do not appear in the pencil trace, since more than 97% of the end deflection is contained in the first mode. The accelerations, however, are distributed more uniformly and this makes available the opportunity to study the decays of the higher mode vibrations and to compare them with the various theories.

To record acceleration, a Westinghouse quartz crystal accelerometer was bolted to the free end of the cantilever and its output was fed through the circuit, shown in Fig. 2.4, to a cathode-ray oscilloscope. In Fig. 11.2 is shown an accelerogram taken with a "1000 cps" filter in the circuit. The time span of Fig. 11.2 is approximately one cycle of the fundamental mode of the beam. The transmission characteristics of the filter are shown in Fig. 11.3. This low pass filter was used so as to cut out all modes higher than the third for convenience in making comparisons with theory.

Mathematically, the problem was treated as a solution of the Sezawa equation (equation (8.3)) with initial conditions

$$\left[w\right]_{t=0} = \frac{P}{EI}\left(\frac{x^2 l}{2} - \frac{x^3}{6}\right), \tag{11.1}$$

$$\left[\frac{\partial w}{\partial t}\right]_{t=0} = 0 \tag{11.2}$$

and the boundary conditions of a cantilever beam. In (11.1), P is the initial deflecting force applied to the free end ($x = l$) of the beam. From the solution, the acceleration of the free end, in units of "number of times gravity" is found to be

$$G = -\frac{4P}{W}\sum_{n=1}^{\infty} e^{-\beta_n p_n t}\cos(p_n' t + \alpha), \tag{11.3}$$

where β_n, p_n, and p_n' are the same as defined above, W is the weight of the cantilever and

$$\tan\alpha = \frac{\beta_n}{\sqrt{1-\beta_n^2}}.$$

In Fig. 11.4 are plotted the sum of the first three modes of (11.3) for each of the three theories, along with a sketch of the trace from Fig. 11.2. The dotted curve represents the Bernoulli-Euler theory, obtained from (11.3) by setting $\beta_n = 0$. The dashed curve represents the Constant Q hypothesis, obtained from (11.3) by setting $\beta_n = 0.02$. The dot-dashed curve represents the Sezawa theory, obtained from (11.3) by setting $\beta_n = 0.02\, p_n/p_1$. The time span of Fig. 11.4 is about one-third of the period of the fundamental mode.

It may be seen that the Sezawa theory gives a very poor prediction. In it the third mode has completely disappeared after two milliseconds, whereas the oscillogram shows that at 10 milliseconds the third mode is still present to the extent of one-quarter of its initial amplitude. The Bernoulli-Euler theory,of course, gives accelerations that are much too large and the discrepancy between it and the oscillogram becomes greater as time progresses, since there is no damping at all in that theory. The Constant Q theory may be seen to give results that are closest to the oscillogram.

Usually accelerograms are so complicated in appearance that it seems futile to expect that the curve conforms with any theory. A remarkable feature of Fig. 11.4 is that all the crests and troughs of the oscillograph trace are present in the Constant Q theory in the proper order and in approximately the correct shape.

BIBLIOGRAPHY

1. P. Symonds and I. Vigness, "Influence of Impact on Damage to Some Elastic Systems", Report No. 0–2293, May 15, 1944, Naval Research Laboratory, Bellevue, D. C.

2. K. Sezawa, "On the Decay of Waves in Visco-elastic Solid Bodies", Bulletin of the Earthquake Research Institute, Tokyo Imperial University, Vol. III, September, 1927, p. 50.

3. R. L. Wegel and H. Walther, "Internal Dissipation in Solids for Small Cyclic Strains", Physics, Vol. 6, p. 141 (1935).

4. A. L. Kimball, "Vibration Problems, Part V-Friction and Internal Damping", A.S.M.E., Journal of Applied Mechanics Reprint Series, Book 2, Design Data on Mechanics, May, 1944, p. A–22.

5. R. D. Mindlin, "Response of Damped Elastic Systems to a Pulse Acceleration", Proceedings of the 6th International Congress for Applied Mechanics, Paris, France, September, 1946.

6. Shock-testing Mechanism for Electrical Indicating Instruments, American War Standard C39.3-1943, American Standards Association, 29 West 39th Street, New York 18, New York.

7. (a) R. D. Mindlin, "Dynamics of Package Cushioning", Bell System Technical Journal, Vol. 24, July-October, 1945, p. 424; or (b) Bell Telephone System Monograph B–1369, p. 72.

8. H. E. Criner, G. D. McCann and C. E. Warren, "A New Device for the Solution of Transient-vibration Problems by the Method of Electrical-mechanical Analogy". Journal of Applied Mechanics, Vol. 12, No. 3 (1945) pp. A–135 to A–141.

9. E. Z. Stowell, E. B. Schwartz and J. C. Houbolt, "Bending and Shear Stresses Developed by the Instantaneous Arrest of the Root of a Moving Cantilever Beam", National Advisory Committee for Aeronautics, Wartime Report, originally issued November 1944 as Advance Restricted Report No. L4127.

Reprinted from JOURNAL OF APPLIED PHYSICS, Vol. 20, No. 1, 89–95, January, 1949

The Optical Equations of Three-Dimensional Photoelasticity

RAYMOND D. MINDLIN AND LAWRENCE E. GOODMAN
Department of Civil Engineering, Columbia University, New York, New York
(Received July 12, 1948)

The fundamental optical problem encountered in three-dimensional photoelasticity is that of the propagation of light in a heterogeneous, aeolotropic medium. Although there is an extensive literature on the homogeneous, aeolotropic medium and the heterogeneous, isotropic medium, only a few special cases of the combined problem have been studied from the point of view of Maxwell's electromagnetic equations. With regard to approximate treatments, the most general is that of F. Neumann (1841), derived from purely kinematical considerations. In the present paper the passage from Maxwell's to Neumann's equations is executed and the nature and order of magnitude of the necessary approximations are disclosed. During the passage, there are discovered a pair of equations, of relatively simple form, which open the way to a variety of new solutions in closed form and reveal the essential parameter upon which the optical phenomena depend.

I. INTRODUCTION

IN applying the photoelastic method of stress analysis to states of stress which vary in three dimensions, the experimenter is confronted with the problem of interpreting the effects observed when light is passed through a "crystal" whose properties vary from point to point in a very general manner. The dynamical equations governing the propagation of light in homogeneous-aeolotropic media and in heterogeneous-isotropic media have been explored extensively but there appear to be only two exact solutions of the equations for cases in which the medium is both heterogeneous and aeolotropic. One of these[1] is for the case in which the principal axes of the section of the index ellipsoid by a plane perpendicular to the wave normal vary linearly along the wave normal. Another[2] is for the case in which the principal axes of the section of the index ellipsoid rotate as a linear function of the distance along the wave normal. Both of these solutions differ little from the analogous solution of the approximate equations whose derivation is based upon simple kinematical considerations.

As more complex states of stress are contemplated, solutions of the dynamical equations become increasingly difficult and it is very likely that, even if they were found, they would differ only slightly from the corresponding solutions of the kinematical equations. However, since the two types of equations are derived quite independently, the nature and order of magnitude of the approximations involved in the kinematical equations are not apparent at the outset. One purpose of this paper is to reveal these approximations by executing the passage from the dynamical to the kinematical equations. This is accomplished in two steps. First it is shown that an approximation, usually ignored, is involved in the type of medium produced by a two-dimensional state of stress. This approximation is adopted at the outset of the three-dimensional considerations and the additional approximations are then disclosed.

II. ELECTROMAGNETIC AND STRESS-OPTICAL EQUATIONS

For the dynamical system we choose Maxwell's electromagnetic equations. These become, for

[1] R. D. Mindlin, J. App. Mech. 8, A187 (1941).
[2] D. C. Drucker and R. D. Mindlin, J. App. Phys. 11, 724 (1940).

our case,[3]

$$\dot{\mathbf{D}}=\nabla\times\mathbf{H},\quad \nabla\cdot\mathbf{D}=0,\tag{II.1}$$

$$\dot{\mathbf{B}}=-\nabla\times\mathbf{E},\quad \nabla\cdot\mathbf{B}=0,\tag{II.2}$$

$$\mathbf{B}=\mu\mathbf{H},\tag{II.3}$$

$$\mathbf{E}=\mathbf{\Pi}\cdot\mathbf{D},\tag{II.4}$$

where $\mathbf{E}$ is the electric vector, $\mathbf{H}$ is the magnetic vector, $\mathbf{D}$ is the electric induction, $\mathbf{B}$ is the magnetic induction and μ is the (constant) magnetic permeability. In the electrostatic system of units $\mu=v_0^{-2}$, where v_0 is the velocity of light in vacuum. The heterogeneity and aeolotropy of the medium are characterized by the self-conjugate index dyadic $\mathbf{\Pi}$ whose elements are the reciprocals of the squares of the indices of refraction n_{rs} $(r, s=1, 2, 3)$. Thus, if $\mathbf{r}$ is the position vector, the index (Fresnel) ellipsoid is given by the equation

$$\mathbf{r}\cdot\mathbf{\Pi}\cdot\mathbf{r}=1.\tag{II.5}$$

In a medium possessing isotropic elasticity, the stress-optical relation may be written as

$$\mathbf{\Pi}-n_0^{-2}\mathbf{I}=C\mathbf{\Psi}+C'\Theta\mathbf{I},\tag{II.6}$$

where $\mathbf{I}$ is the idemfactor, C and C' are stress-optical coefficients, $\mathbf{\Psi}$ is the stress dyadic and Θ its first invariant. The symbol n_0 represents, at the least, the index of refraction of the unstressed material, but may include a symmetric function of the principal stresses. Equation (II.6) is equivalent to the law proposed by Coker and Filon.[4]

It is convenient, for our purposes, to write the equations of motion in terms of the magnetic vector. From (II.2) and (II.3),

$$\mu\dot{\mathbf{H}}=-\nabla\times\mathbf{E}.\tag{II.7}$$

From (II.7) and (II.4)

$$\mu\dot{\mathbf{H}}=-\nabla\times(\mathbf{\Pi}\cdot\mathbf{D}).\tag{II.8}$$

Finally, operating on (II.8) with $\partial/\partial t$ and using (II.1),

$$\mu\ddot{\mathbf{H}}=-\nabla\times(\mathbf{\Pi}\cdot\nabla\times\mathbf{H}),\quad \nabla\cdot\mathbf{H}=0.\tag{II.9}$$

Equations (II.9), with $\mathbf{\Pi}$ given by (II.6), are the dynamical equations governing the propagation of light in our medium.

III. TWO-DIMENSIONAL CONSIDERATIONS

Let us suppose, first, that the stress dyadic is a function of x and y only and has one principal axis everywhere parallel to z. Identifying the axes 1, 2, 3 of the index dyadic with the rectangular coordinate axes x, y, z respectively, we have, from (II.6),

$$n_{13}^{-1}=n_{31}^{-1}=n_{23}^{-1}=n_{32}^{-1}=0,\tag{III.1}$$

$$n_{rs}=n_{rs}(x, y).\tag{III.2}$$

Using (III.1) and (III.2), Eqs. (II.9) become the following four equations governing u, v, w (the x, y, z components of $\mathbf{H}$):

$$\mu\frac{\partial^2 u}{\partial t^2}=n_{22}^{-2}\frac{\partial^2 u}{\partial z^2}-n_{21}^{-2}\frac{\partial^2 v}{\partial z^2}-n_{22}^{-2}\frac{\partial^2 w}{\partial x\partial z}+n_{21}^{-2}\frac{\partial^2 w}{\partial y\partial z}-\frac{\partial}{\partial y}\left[n_{33}^{-2}\left(\frac{\partial v}{\partial x}-\frac{\partial u}{\partial y}\right)\right],\tag{III.3}$$

$$\mu\frac{\partial^2 v}{\partial t^2}=n_{11}^{-2}\frac{\partial^2 v}{\partial z^2}-n_{12}^{-2}\frac{\partial^2 u}{\partial z^2}-n_{11}^{-2}\frac{\partial^2 w}{\partial y\partial z}+n_{12}^{-2}\frac{\partial^2 w}{\partial x\partial z}+\frac{\partial}{\partial x}\left[n_{33}^{-2}\left(\frac{\partial v}{\partial x}-\frac{\partial u}{\partial y}\right)\right],\tag{III.4}$$

$$\mu\frac{\partial^2 w}{\partial t^2}=-\frac{\partial}{\partial x}\left[n_{21}^{-2}\left(\frac{\partial w}{\partial y}-\frac{\partial v}{\partial z}\right)+n_{22}^{-2}\left(\frac{\partial u}{\partial z}-\frac{\partial w}{\partial x}\right)\right]+\frac{\partial}{\partial y}\left[n_{11}^{-2}\left(\frac{\partial w}{\partial y}-\frac{\partial v}{\partial z}\right)+n_{12}^{-2}\left(\frac{\partial u}{\partial z}-\frac{\partial w}{\partial x}\right)\right],\tag{III.5}$$

$$\frac{\partial u}{\partial x}+\frac{\partial v}{\partial y}+\frac{\partial w}{\partial z}=0.\tag{III.6}$$

Anticipating solutions in the form of waves travelling in the z direction it is awkward to have the derivatives, with respect to z, of both u and v in each of (III.3) and (III.4). This complication may be eliminated by using the stress trajectories as a system of orthogonal curvilinear coordinates.

The central section of the ellipsoid (II.5) by

[3] L. N. G. Coker and E. G. Filon, *A Treatise on Photo-Elasticity* (Cambridge University Press, Teddington, 1931), Chap. I.
[4] See reference 3, p. 198, Eq. (3.102).

a plane $z=\text{const.}$ is the ellipse

$$n_{11}^{-2}x^2+2n_{12}^{-2}xy+n_{22}^{-2}y^2=1. \quad \text{(III.7)}$$

The principal semi-axes of (III.7) have lengths given by the stationary values of

$$n_N^{-2}=n_{11}^{-2}\cos^2\theta+n_{22}^{-2}\sin^2\theta +2n_{12}^{-2}\sin\theta\cos\theta, \quad \text{(III.8)}$$

where θ is the angle between the x-axis and a direction N in the x, y plane. The stationary values occur at $\theta=\varphi$, $\varphi+\pi/2$, where

$$\tan 2\varphi=2n_{12}^{-2}/(n_{11}^{-2}-n_{22}^{-2}), \quad \text{(III.9)}$$

and have magnitudes

$$n_a^{-2}=[n_N^{-2}]_{\theta=\varphi}=n_{11}^{-2}\cos^2\varphi+n_{22}^{-2}\sin^2\varphi +n_{12}^{-2}\sin 2\varphi, \quad \text{(III.10)}$$

$$n_b^{-2}=[n_N^{-2}]_{\theta=\varphi+\pi/2}=n_{11}^{-2}\sin^2\varphi +n_{22}^{-2}\cos^2\varphi-n_{12}^{-2}\sin 2\varphi, \quad \text{(III.11)}$$

where n_a and n_b are principal indices of refraction, in view of the present restrictions on $\mathbf{\Psi}$. The principal velocities a, b, c are given by

$$a^2=\mu^{-1}n_a^{-2}, \quad b^2=\mu^{-1}n_b^{-2}, \quad c^2=\mu^{-1}n_{33}^{-2}. \quad \text{(III.12)}$$

Our curvilinear coordinates s_1, s_2 increase along curves whose tangents make angles φ and $\varphi+\pi/2$, respectively, with the x axis. Hence

$$\frac{\partial}{\partial x}=\cos\varphi\frac{\partial}{\partial s_1}-\sin\varphi\frac{\partial}{\partial s_2}, \qquad \frac{\partial}{\partial y}=\sin\varphi\frac{\partial}{\partial s_1}+\cos\varphi\frac{\partial}{\partial s_2}. \quad \text{(III.13)}$$

Finally, let u' and v' be the components of $\mathbf{H}$ in the directions s_1 and s_2 (increasing) respectively. Then

$$u=u'\cos\varphi-v'\sin\varphi, \quad v=v'\cos\varphi+u'\sin\varphi. \quad \text{(III.14)}$$

Using (III.9) to (III.14), Eqs. (III.3) and (III.4) become

$$\frac{\partial^2u'}{\partial t^2}-b^2\frac{\partial^2u'}{\partial z^2}=b^2\frac{\partial\Delta_1}{\partial s_1}-\frac{\partial(c^2\omega_1)}{\partial s_2}, \quad \text{(III.15)}$$

$$\frac{\partial^2v'}{\partial t^2}-a^2\frac{\partial^2v'}{\partial z^2}=a^2\frac{\partial\Delta_1}{\partial s_2}+\frac{\partial(c^2\omega_1)}{\partial s_1}, \quad \text{(III.16)}$$

where

$$\Delta_1=-\frac{\partial w}{\partial z}=\frac{\partial u}{\partial x}+\frac{\partial v}{\partial y} =\frac{\partial u'}{\partial s_1}+\frac{\partial v'}{\partial s_2}+u'\frac{\partial\varphi}{\partial s_2}-v'\frac{\partial\varphi}{\partial s_1}, \quad \text{(III.17)}$$

$$\omega_1=\frac{\partial v}{\partial x}-\frac{\partial u}{\partial y}=\frac{\partial v'}{\partial s_1}-\frac{\partial u'}{\partial s_2}+v'\frac{\partial\varphi}{\partial s_2}+u'\frac{\partial\varphi}{\partial s_1}. \quad \text{(III.18)}$$

It is customary, in two-dimensional photoelasticity, to employ solutions of the type

$$u'=A\cos(\psi-\varphi)\cos\omega[t-(z/b)], \quad \text{(III.19)}$$

$$v'=A\sin(\psi-\varphi)\cos\omega[t-(z/a)], \quad \text{(III.20)}$$

where ω is the frequency and A^2 is proportional to the intensity of the incident light and ψ is the (constant) angle between the x axis and the direction of polarization at incidence. It may be seen that (III.19) and (III.20) are solutions of (III.15) and (III.16) only for homogeneous stress (which transforms the material into a homogeneous crystal) whereupon the right-hand sides of (III.15) and (III.16) vanish.

To find the nature and order of magnitude of the terms that are ordinarily neglected, substitute (III.19) and (III.20) in (III.15) and (III.16) and consider the ratios of the amplitudes of the terms on the right to those on the left. These ratios are of the following six types:

$$\text{(i)}\quad \frac{b^2}{\omega^2}\frac{\partial^2\varphi}{\partial s^2} = O\left[\lambda^2\frac{\partial^2\varphi}{\partial s^2}\right],$$

$$\text{(ii)}\quad \left(\frac{b}{\omega}\frac{\partial\varphi}{\partial s}\right)^2 = O\left[\lambda\frac{\partial\varphi}{\partial s}\right]^2,$$

$$\text{(iii)}\quad \left(\frac{b}{\omega}\frac{\partial\varphi}{\partial s}\right)\left(bz\frac{\partial b^{-1}}{\partial s}\right) = O\left[\lambda z\frac{\partial\varphi}{\partial s}\frac{\partial n}{\partial s}\right],$$

$$\text{(iv)}\quad -bz\frac{b}{\omega}\frac{\partial^2 b^{-1}}{\partial s^2} = O\left[\lambda z\frac{\partial^2 n}{\partial s^2}\right],$$

$$\text{(v)}\quad \frac{1}{\omega^2}\frac{\partial\varphi}{\partial s}\frac{\partial c^2}{\partial s} = O\left[\lambda^2\frac{\partial n}{\partial s}\frac{\partial\varphi}{\partial s}\right],$$

$$\text{(vi)}\quad \frac{z}{\omega}\frac{\partial c^2}{\partial s}\frac{\partial b^{-1}}{\partial s} = O\left[\lambda z\left(\frac{\partial n}{\partial s}\right)^2\right],$$

where s, n and b are representative lengths in the x, y plane, index of refraction, and wave velocity, respectively, λ is the wave-length of light in vacuum and $O[\ ]$ stands for "order of magnitude of." To estimate the order of magnitude of these ratios, consider a wave-length of 6×10^{-4} mm, a stress gradient of 10^8 dynes/cm^2 ($\cong$1500 lb./in.^{2}) per mm and a stress-optical coefficient of 100 Brewsters. Under these extreme conditions, $\lambda\partial n/\partial s=O[10^{-7}]$ and, for a model 5 mm thick, $z\partial n/\partial s=O[10^{-3}]$. With regard to terms involving φ, if the rotation of the principal axes were 2π radians in one millimeter, $\lambda\partial\varphi/\partial s=O[10^{-3}]$. Since the ratios involve squares or products of terms not exceeding $O[10^{-3}]$ the ratios themselves do not exceed $O[10^{-6}]$.

It is true that, in the neighborhood of an isotropic point, $\partial\varphi/\partial s$ becomes very large. However, $\partial\varphi/\partial s$ may be expressed in terms of $\partial n/\partial s$ and the principal-stress difference through the use of the stress equations of equilibrium and the stress-optical law, following which it may be shown, by a limiting process, that the ratios involving $\partial\varphi/\partial s$ are, again, small.

Hence, all of the terms on the right-hand sides of (III.15) and (III.16) are negligible and these equations may be replaced, to a good approximation, by

$$b^2\frac{\partial^2 u'}{\partial z^2}=\frac{\partial^2 u'}{\partial t^2}, \qquad \text{(III.21)}$$

$$a^2\frac{\partial^2 v'}{\partial z^2}=\frac{\partial^2 v'}{\partial t^2}. \qquad \text{(III.22)}$$

Equations (III.19) and (III.20) are solutions of (III.21) and (III.22) and lead directly to the usual laws of photoelasticity, namely, the directions of polarization coincide with the directions of principal stress in the plane perpendicular to the wave normal and the relative phase retardation is

$$\delta=\omega z\left(\frac{1}{b}-\frac{1}{a}\right)=\frac{2\pi z}{\lambda}(n_b-n_a). \qquad \text{(III.23)}$$

It is useful to observe that the order of Eqs. (III.21) and (III.22) may be reduced if we are interested only in waves moving in one direction. In this case we may write

$$\frac{\partial u'}{\partial z}=\mp\frac{1}{b}\frac{\partial u'}{\partial t}, \qquad \text{(III.24)}$$

$$\frac{\partial v'}{\partial z}=\mp\frac{1}{a}\frac{\partial v'}{\partial t}. \qquad \text{(III.25)}$$

The minus signs apply to forward moving waves, such as (III.19) and (III.20) while the plus signs refer to backward moving waves which we may ignore for most purposes. It may be verified, by substitution, that (III.24) and (III.25) are indeed solutions of (III.21) and (III.22).

The restriction of z independence placed on the stress dyadic at the beginning of this section is a rather severe one if we remember that this is possible in relatively few states of stress even in "two-dimensional" photoelasticity. However, the purpose of this section is to dispose of x and y dependence so that the more difficult case of z dependence may be treated separately.

IV. THE OPTICAL WAVE EQUATIONS OF THREE-DIMENSIONAL PHOTOELASTICITY

In general the index dyadic is a function of all three coordinates. However, if we are interested in plane waves with wave normal in, say, the z direction, the results of the preceding section show that, for variations of the index dyadic that might be encountered in photoelasticity, the dependence on x and y may be neglected. In fact, a detailed study of the general equations shows that all of the x and y dependent terms are of the order of magnitude of those already considered in the preceding section. Accordingly we may assume that the n_{rs} are functions of z only and, hence, u, v and w are functions of z and t only. If these restrictions are introduced into (II.9) the differential equations reduce to $w=0$ and

$$\mu\frac{\partial^2 u}{\partial t^2}=n_{22}^{-2}\frac{\partial^2 u}{\partial z^2}-n_{21}^{-2}\frac{\partial^2 v}{\partial z^2}+\frac{dn_{22}^{-2}}{dz}\frac{\partial u}{\partial z}-\frac{dn_{21}^{-2}}{dz}\frac{\partial v}{\partial z}, \qquad \text{(IV.1)}$$

$$\mu\frac{\partial^2 v}{\partial t^2}=n_{11}^{-2}\frac{\partial^2 v}{\partial z^2}-n_{12}^{-2}\frac{\partial^2 u}{\partial z^2}+\frac{dn_{11}^{-2}}{dz}\frac{\partial v}{\partial z}-\frac{dn_{12}^{-2}}{dz}\frac{\partial u}{\partial z}. \qquad \text{(IV.2)}$$

The equation of the central section of the index ellipsoid by a plane perpendicular to the z axis is again given by (III.7). The principal axes of this section are oriented at angles φ and $\varphi+\pi/2$ with respect to the x axis, where φ is given by (III.9). It is to be noted that these axes are not in general principal axes of the index ellipsoid, as they were in Section III, but are principal axes of the central section perpendicular to the z axis. These have been called secondary principal axes. The lengths of the secondary principal semi-axes of the section are given by the stationary values, say n_1 and n_2 of (III.8) where

$$n_1^{-2}=[n_N^{-2}]_{\theta=\varphi}=n_{11}^{-2}\cos^2\varphi+n_{22}^{-2}\sin^2\varphi+n_{12}^{-2}\sin 2\varphi, \quad \text{(IV.3)}$$

$$n_2^{-2}=[n_N^{-2}]_{\theta=\varphi+\pi/2}=n_{11}^{-2}\sin^2\varphi+n_{22}^{-2}\cos^2\varphi-n_{12}^{-2}\sin 2\varphi. \quad \text{(IV.4)}$$

The velocities corresponding to n_1 and n_2 will be designated by k_1 and k_2, where

$$k_1^2=\mu^{-1}n_1^{-2}, \quad k_2^2=\mu^{-1}n_2^{-2}. \quad \text{(IV.5)}$$

It is convenient to refer (IV.1) and (IV.2) to secondary principal axes. Using (IV.3) to (IV.5) and (III.14), we find

$$\frac{\partial^2 u'}{\partial t^2}=\frac{\partial}{\partial z}\left[k_2^2\left(\frac{\partial u'}{\partial z}-v'\frac{d\varphi}{dz}\right)\right]-k_1^2\frac{d\varphi}{dz}\left(\frac{\partial v'}{\partial z}+u'\frac{d\varphi}{dz}\right), \quad \text{(IV.6)}$$

$$\frac{\partial^2 v'}{\partial t^2}=\frac{\partial}{\partial z}\left[k_1^2\left(\frac{\partial v'}{\partial z}+u'\frac{d\varphi}{dz}\right)\right]+k_2^2\frac{d\varphi}{dz}\left(\frac{\partial u'}{\partial z}-v'\frac{d\varphi}{dz}\right). \quad \text{(IV.7)}$$

Equations (IV.6) and (IV.7) are the simplest form of Maxwell's equations when dependence on x and y is neglected. If k_1, k_2 and φ are independent of z, (IV.6) and (IV.7) reduce to (III.21) and (III.22).

We now seek a simplification, by reduction of order, analogous to (III.24) and (III.25). To this end we consider the possibility that

$$\frac{\partial u'}{\partial z}-v'\frac{d\varphi}{dz}=\mp\frac{1}{k_2}\frac{\partial u'}{\partial t}, \quad \text{(IV.8)}$$

$$\frac{\partial v'}{\partial z}+u'\frac{d\varphi}{dz}=\mp\frac{1}{k_1}\frac{\partial v'}{\partial t} \quad \text{(IV.9)}$$

may be solutions of (IV.6) and (IV.7). If we substitute (IV.8) and (IV.9) into (IV.6) and (IV.7) we find

$$\frac{\partial^2 u'}{\partial t^2}=\frac{\partial^2 u'}{\partial t^2}\pm(k_1-k_2)\frac{d\varphi}{dz}\frac{\partial v'}{\partial t}\mp\frac{dk_2}{dz}\frac{\partial u'}{\partial t}, \quad \text{(IV.10)}$$

$$\frac{\partial^2 v'}{\partial t^2}=\frac{\partial^2 v'}{\partial t^2}\pm(k_1-k_2)\frac{d\varphi}{dz}\frac{\partial u'}{\partial t}\mp\frac{dk_1}{dz}\frac{\partial v'}{\partial t}. \quad \text{(IV.11)}$$

Equations (IV.10) and (IV.11) fail to be identities by terms whose coefficients have ratios, to the coefficients of the leading terms, of order of magnitude

$$\lambda\frac{dn}{dz} \quad \text{and} \quad [(n_1-n_2)/n]\lambda\frac{d\varphi}{dz}. \quad \text{(IV.12)}$$

The first of these has been shown, in Section III, to be of order of magnitude 10^{-7}. Hence, if the product of $(n_1-n_2)/n$ and $\lambda d\varphi/dz$ is negligible, (IV.6) and (IV.7) may be replaced by (IV.8) and (IV.9). In a material with a stress-optical coefficient of 100 Brewsters under a stress of 10^8 dynes/cm², the order of magnitude of $(n_1-n_2)/n$ is 10^{-3} while a rotation of the secondary principal axes in the amount 2π radians per millimeter makes $\lambda d\varphi/dz$ of order of magnitude 10^{-3}.

It is not impossible for $d\varphi/dz$ to become very large. An example is the case of combined tension and torsion of a cylindrical bar. For a wave-normal intersecting the axis of the bar, $d\varphi/dz$ approaches infinity as the tension approaches zero. However, at the same time $(n_1-n_2)/n$ approaches zero and the product of the two has a zero limit, in the neighborhood of the axis of the bar.

As a result of the considerations in this and the preceding section, we conclude that the propagation of a forward moving wave in a photoelastic body under a three-dimensional state of stress is governed, to a close approximation, by the differential equations

$$\frac{\partial u'}{\partial z} - v'\frac{d\varphi}{dz} = -\frac{1}{k_2}\frac{\partial u'}{\partial t}, \qquad \text{(IV.13)}$$

$$\frac{\partial v'}{\partial z} + u'\frac{d\varphi}{dz} = -\frac{1}{k_1}\frac{\partial v'}{\partial t}. \qquad \text{(IV.14)}$$

V. PASSAGE TO NEUMANN'S EQUATIONS

Consider solutions of (IV.13) and (IV.14) of the form

$$u' = U\cos(\omega t - \alpha), \qquad \text{(V.1)}$$

$$v' = V\cos(\omega t - \beta), \qquad \text{(V.2)}$$

where the amplitudes U and V and the phase angles α and β are functions of z. Substituting these expressions for u' and v' in (IV.13) and (IV.14) and equating coefficients of like trigonometric terms of argument ωt, we find

$$U\frac{d\alpha}{dz}\cos\alpha + \frac{dU}{dz}\sin\alpha - V\frac{d\varphi}{dz}\sin\beta = \frac{\omega U}{k_2}\cos\alpha, \qquad \text{(V.3)}$$

$$-U\frac{d\alpha}{dz}\sin\alpha + \frac{dU}{dz}\cos\alpha - V\frac{d\varphi}{dz}\cos\beta = -\frac{\omega U}{k_2}\sin\alpha, \qquad \text{(V.4)}$$

$$V\frac{d\beta}{dz}\cos\beta + \frac{dV}{dz}\sin\beta + U\frac{d\varphi}{dz}\sin\alpha = \frac{\omega V}{k_1}\cos\beta, \qquad \text{(V.5)}$$

$$-V\frac{d\beta}{dz}\sin\beta + \frac{dV}{dz}\cos\beta + U\frac{d\varphi}{dz}\cos\alpha = -\frac{\omega V}{k_1}\sin\beta. \qquad \text{(V.6)}$$

Now, multiply (V.3) by $\sin\alpha$ and (V.4) by $\cos\alpha$ and add; multiply (V.5) by $\sin\beta$ and (V.6) by $\cos\beta$ and add; multiply (V.3) by $\cos\alpha$ and (V.4) by $\sin\alpha$ and subtract; multiply (V.5) by $\cos\beta$ and (V.6) by $\sin\beta$ and subtract. The following four equations result:

$$\frac{dU}{dz} = V\frac{d\varphi}{dz}\cos\delta, \qquad \text{(V.7)}$$

$$\frac{dV}{dz} = -U\frac{d\varphi}{dz}\cos\delta, \qquad \text{(V.8)}$$

$$U\frac{d\alpha}{dz} + V\frac{d\varphi}{dz}\sin\delta = \frac{\omega U}{k_2}, \qquad \text{(V.9)}$$

$$V\frac{d\beta}{dz} + U\frac{d\varphi}{dz}\sin\delta = \frac{\omega V}{k_1}, \qquad \text{(V.10)}$$

where δ is the relative phase retardation:

$$\delta = \alpha - \beta. \qquad \text{(V.11)}$$

Subtracting (V.10) from (V.9) we have

$$\frac{d\delta}{dz} + \left(\frac{V}{U} - \frac{U}{V}\right)\frac{d\varphi}{dz}\sin\delta = -\omega\left(\frac{1}{k_1} - \frac{1}{k_2}\right) = \frac{-2\pi}{\lambda}(n_1 - n_2). \qquad \text{(V.12)}$$

Now, let the amplitude ratio be given by

$$V/U = \tan\gamma, \qquad \text{(V.13)}$$

so that

$$(V/U) - (U/V) = -2\cot 2\gamma, \qquad \text{(V.14)}$$

and

$$\frac{d\gamma}{dz} = \left(U\frac{dV}{dz} - V\frac{dU}{dz}\right) \Big/ (U^2 + V^2). \qquad \text{(V.15)}$$

Substituting into (V.15) the values of dU/dz and dV/dz given in (V.7) and (V.8) we find

$$d\gamma/dz = -(d\varphi/dz)\cos\delta \qquad \text{(V.16)}$$

and, substituting (V.14) into (V.12), we find

$$\frac{d\delta}{dz} + \frac{2\pi}{\lambda}(n_1 - n_2) = 2\frac{d\varphi}{dz}\cot 2\gamma \sin\delta. \qquad \text{(V.17)}$$

Equations (V.16) and (V.17) are Neumann's equations,[5] derived in 1841 from purely kinematical considerations. Alternative derivations of Neumann's equations, again from kinematical considerations, are given by Coker and Filon[6]

[5] F. E. Neumann, Abh. d. Kön. Akad. d. Wissenschaften zu Berlin (1841) Part II, p. 1.

[6] See reference 3, p. 253.

and Kammerer.[7] These equations appear to have been used but little. Coker and Filon[6] employed them to estimate the effect on δ and γ of the quadratic variation of optical properties through the thickness of a plate under a state of plane stress. Menges[8] devised a graphical method of solving for (n_1-n_2) and φ from data obtained by measurement of δ and γ.

VI. REDUCTION TO A CANONICAL FORM

The contributions of the various optical properties of the medium are revealed by discovering the characteristic curves of (IV.13) and (IV.14) and then reducing the equations to a canonical form.

If we solve (IV.13) for v', differentiate and substitute in (IV.14), the result is a second order partial differential equation in one dependent variable, u', and two independent variables, z and t:

$$\frac{\partial^2 u'}{\partial z^2}+(k_2^{-1}+k_1^{-1})\frac{\partial^2 u'}{\partial z\partial t}+(k_2^{-1}k_1^{-1})\frac{\partial^2 u'}{\partial t^2}+\left(\frac{d^2\varphi/dz^2}{d\varphi/dz}\right)\frac{\partial u'}{\partial z}+\left(\frac{dk_2^{-1}}{dz}+k_2^{-1}\frac{d^2\varphi/dz^2}{d\varphi/dz}\right)\frac{\partial u'}{\partial t}+u'\left(\frac{d\varphi}{dz}\right)^2=0. \quad \text{(VI.1)}$$

The quadratic form associated with this equation is

$$Q(p)=ap^2+2bp+c=p^2+(k_2^{-1}+k_1^{-1})p+k_2^{-1}k_1^{-1}=(p+k_2^{-1})(p+k_1^{-1}). \quad \text{(VI.2)}$$

Since the discriminant, $ac-b^2$, of this quadratic form is always negative, Eq. (VI.1) is of the hyperbolic type. The differential equations which determine the characteristic curves are

$$\frac{\partial\sigma}{\partial z}=-k_1^{-1}\frac{\partial\sigma}{\partial t} \quad\text{and}\quad \frac{\partial\tau}{\partial z}=-k_2^{-1}\frac{\partial\tau}{\partial t}, \quad \text{(VI.3)}$$

whence the characteristic curves for Eq. (VI.1), and also for its counterpart obtained by eliminating u', are given by the constant values of

$$\sigma=\omega\left(t-\int k_1^{-1}dz\right), \quad \text{(VI.4)}$$

$$\tau=\omega\left(t-\int k_2^{-1}dz\right). \quad \text{(VI.5)}$$

These reveal the manner in which the wave solutions of (IV.13) and (IV.14) are propagated along the positive z-direction.

Transforming from variables z and t to σ and τ and defining

$$R=\frac{d\varphi/dz}{\omega[(1/k_2)-(1/k_1)]}=\frac{d\varphi/dz}{(2\pi/\lambda)(n_2-n_1)}, \quad \text{(VI.6)}$$

Eqs. (IV.13) and (IV.14) become, finally,

$$Ru'=\partial v'/\partial\tau, \quad \text{(VI.7)}$$

$$Rv'=\partial u'/\partial\sigma. \quad \text{(VI.8)}$$

The quantity R is thus shown to be the essential parameter controlling optical wave propagation in three-dimensional photoelasticity. R is a ratio whose numerator is the rate of rotation of the secondary principal axes and whose denominator is the rate of relative phase retardation that would obtain in the absence of rotation. This ratio is a generalization of a similar quantity discovered in the solution of Maxwell's equations for the case in which both $d\varphi/dz$ and n_2-n_1 are constants.[2] In that solution it was found that the influence, on observed relative phase retardation, of the uniform rotation of the secondary principal axes could be expressed entirely in terms of R. It now appears that this must be true regardless of the manner in which both $d\varphi/dz$ and n_2-n_1 vary. The specific functional dependence of relative phase retardation on R will, of course, depend on the form of R itself. It would be useful to discover the influence of R in a number of special cases, a process which now appears feasible upon noticing the relative simplicity of Eqs. (VI.7) and (VI.8) in comparison with Maxwell's or Neumann's equations. In particular it is to be noted that, if u' and v' are separated between (VI.7) and (VI.8), the equation governing each of them is a special form of the general one-dimensional wave equation.

[7] A. Kammerer, *Recherches sur la photoélasticimétrie* (Hermann et Cie., Paris, 1944), p. 37.
[8] H. J. Menges, Zeits. f. Angewandte Math. u. Mech. **20**, 210 (1940).

Reprinted from JOURNAL OF APPLIED PHYSICS, Vol. 20, No. 2, 206–216, February, 1949

A Mathematical Theory of Photo-Viscoelasticity

RAYMOND D. MINDLIN
Department of Civil Engineering, Columbia University, New York, New York and *Bell Telephone Laboratories, Murray Hill, New Jersey*

(Received August 23, 1948)

A phenomenological law of optical birefringence, induced by small strain in viscoelastic materials, is formulated and applied to an idealized medium. The resulting optical-stress-strain-time-temperature relations are used in finding conditions under which models made of viscoelastic materials may be employed in the photoelastic method of solving boundary value problems in the linear theory of elasticity.

I. INTRODUCTION

A DESCRIPTION of photoelastic phenomena in viscoelastic media must include time-dependent relations between three second-rank tensors: birefringence, stress, and strain. From the viewpoint of mechanics of continua it is desirable to include, also, space-dependence (non-homogeneous strain) and certain aspects of temperature dependence. All of these variables can be incorporated in a relatively simple theory if the system is assumed to be linear. The attribute of linearity is not attained, of course, without some loss of generality. The chief sacrifices are finite strain and non-linear stress-strain and stress-strain-rate relations. In addition, thermodynamic effects have been neglected. What remains, however, is of considerable interest and wide applicability.

In Section II, a strain-optical law is formulated for small strain in viscoelastic media. The stress-strain properties of the medium to which the law is applied are described in Section III. These stress-strain properties are both more and less general than those usually employed for similar purposes. They are more general in including instantaneous elastic compressibility and time-dependent viscosity and elasticity, both of which are considered to be of interest in photo-visco-elastic phenomena. The medium is less general in including only one element, each, of instantaneous elasticity, retarded elasticity and flow. However, as indicated at appropriate places in the text, most of the results take the same form regardless of the number of shear moduli, viscosity coefficients, and time derivatives entering into the stress-strain relations. The general strain-optical law is applied to the idealized medium in Section IV, and conclusions are drawn regarding the relations between birefringence and stress and strain in various types of viscoelastic media.

The next two sections deal with the relations between stress, strain, and the optical properties that are most easily measured, and it is shown that, in general, the relations are not simple. This is followed by a study of the conditions under which these relations reduce to forms similar to those for an elastic medium.

Sections VIII to XII contain an investigation of the conditions under which the birefringence in a viscoelastic body may be translated directly to the stress in an elastic body of the same shape. The results are summarized in Section XIII. The essence of the conclusions is that viscoelastic materials may be used for models in photo-elasticity except if there are loads which move over the surface of the model.

II. GENERAL STRAIN-OPTICAL EQUATION

Consider a continuous, viscoelastic medium of which the mechanical model[1] is an assemblage of m elastic elements and an equal, greater, or lesser number of viscous elements. Let $\mathbf{\Phi}_k'$, $k=1, 2, \cdots m$, be the small, pure-strain dyadic[2] of the kth elastic element and let θ_k' be its mean stretch (i.e., one-third of its cubical dilatation). The strain deviation of the kth elastic element

[1] See T. Alfrey, *Mechanical Behavior of High Polymers* (Interscience Publishers, Inc., New York, 1948) as a general reference on the use of mechanical models of visco-elastic materials and especially Appendices II and III in which are summarized Alfrey's investigations of homogeneous and non-homogeneous stress and strain in a very general viscoelastic medium.

[2] For the dyadic notation used in this paper, see C. E. Weatherburn, *Advanced Vector Analysis* (G. Bell and Sons, London, 1928).

will be represented by $\mathbf{\Phi}_{dk}'=\mathbf{\Phi}_k'-\theta_k'\mathbf{I}$ where $\mathbf{I}$ is the idemfactor.

Let $\mathbf{\Pi}$ be the index dyadic, so that

$$\mathbf{r}\cdot\mathbf{\Pi}\cdot\mathbf{r}=1 \tag{2.01}$$

is the equation of the index (Fresnel) ellipsoid[3] if $\mathbf{r}$ is the radius vector drawn from the center of the ellipsoid to its surface.

If it is assumed that the birefringence arises solely from the deformation of the elastic elements, the most general linear, isotropic relation that can exist between birefringence and strain is

$$\mathbf{\Pi}-n_0^{-2}\mathbf{I}=\sum_{k=1}^{m}(c_k\mathbf{\Phi}_{dk}'+d_k\theta_k'\mathbf{I}), \tag{2.02}$$

where c_k and d_k are strain-optical coefficients and n_0 is the index of refraction of the unstrained medium.

Before the birefringence can be expressed in terms of the macroscopic strain and stress, it is necessary to specify the mechanical model of the viscoelastic medium and its stress-strain properties.

III. STRESS-STRAIN RELATIONS FOR AN IDEALIZED VISCOELASTIC MATERIAL

Aside from failure, the important mechanical phenomena exhibited by viscoelastic materials are *instantaneous elasticity*, *retarded elasticity*, and *flow*. All three of these are exhibited, in turn, by a simple mechanical model, illustrated in Fig. 1, which has been used frequently to represent an idealized viscoelastic medium.[1]

In Fig. 1, the spring marked (1)′ represents the instantaneous elasticity. This element is assumed to possess linear, isotropic elasticity, with shear modulus G_1 and Poisson's ratio ν_1. It is customary to assume that viscoelastic materials are incompressible and this is, indeed, very nearly true especially for retarded elasticity and flow. However, if complete incompressibility is assumed, the mean strain-optical coefficient d_k vanishes. Since there is not sufficient experimental evidence to warrant the omission of d_k, the attribute of instantaneous elastic compressibility will be retained until non-homogeneous stress is considered, at which time (Section IX) the reason for the restriction to $\nu_1=\frac{1}{2}$ will become apparent.

The spring marked (2)′ and the dash-pot marked (2)″ in Fig. 1, together represent the retarded elasticity. Incompressibility is assumed for this phenomenon and, hence, the linear, isotropic, elastic element (2)′ has only a shear modulus G_2 and the linear, isotropic viscous element (2)″ has only a viscosity coefficient η_2. Finally, the dash-pot marked (3)″ represents the incompressible flow with viscosity coefficient η_3.

In some materials G_1 and G_2 are quite insensitive to temperature over a limited range, whereas η_2 and η_3 depend strongly on temperature. Use of this attribute is made in the freezing method of three-dimensional photoelasticity. In order to include a wide variety of materials, G_1 and G_2 will also be considered temperature dependent. It will be assumed that the temperature is a known function of time and independent of the space coordinates and hence we shall consider that *the viscosity coefficients and shear moduli are functions of time only.*

The following notation will be used:

$\mathbf{\Phi}_1'$, $\mathbf{\Phi}_2'$, $\mathbf{\Phi}_2''$, $\mathbf{\Phi}_3''$ = pure strain dyadic of elements (1)′, (2)′, (2)″, (3)″, respectively,

θ_1', θ_2', θ_2'', θ_3'' = mean stretch of elements (1)′, (2)′, (2)″, (3)″, respectively,

$\mathbf{\Psi}_1'$, $\mathbf{\Psi}_2'$, $\mathbf{\Psi}_2''$, $\mathbf{\Psi}_3''$ = stress dyadic of elements (1)′, (2)′, (2)″, (3)″, respectively.

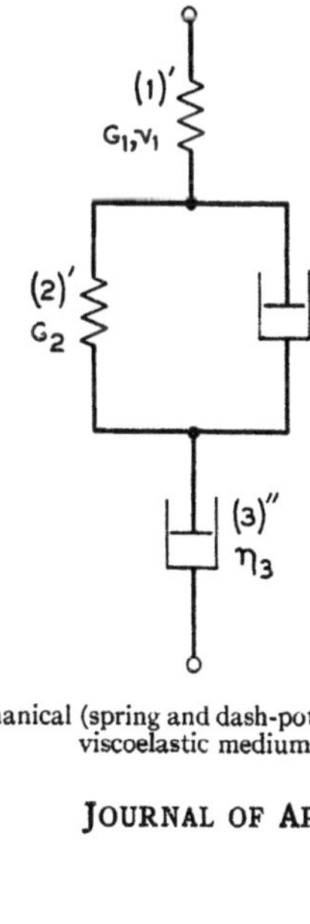

FIG. 1. Mechanical (spring and dash-pot) model of idealized viscoelastic medium.

[3] E. G. Coker and L. N. G. Filon, *A Treatise on Photo-Elasticity* (Cambridge University Press, London, 1932), p. 17.

$\Theta_1', \Theta_2', \Theta_2'', \Theta_3''$ = mean tension of elements (1)′, (2)′, (2)″, (3)″, respectively,

$\mathbf{\Phi}$ = macroscopic pure strain dyadic,
θ = macroscopic mean stretch,
$\mathbf{\Phi}_d=\mathbf{\Phi}-\theta\mathbf{I}$ = macroscopic strain deviation,
$\mathbf{\Psi}$ = macroscopic stress dyadic,
Θ = macroscopic mean tension,
$\mathbf{\Psi}_d=\mathbf{\Psi}-\Theta\mathbf{I}$ = macroscopic stress deviation.

The assumption of incompressibility of retarded elasticity and flow requires

$$\theta_2'=\theta_2''=\theta_3''=0, \tag{3.01}$$

i.e., the pure strain dyadics $\mathbf{\Phi}_2', \mathbf{\Phi}_2'', \mathbf{\Phi}_3''$ are equal to their respective deviations. The geometry of the model requires

$$\mathbf{\Phi}_2'=\mathbf{\Phi}_2'', \tag{3.02}$$

$$\mathbf{\Phi}_1'+\mathbf{\Phi}_2'+\mathbf{\Phi}_3''=\mathbf{\Phi}. \tag{3.03}$$

The geometry and equilibrium of the model require

$$\mathbf{\Psi}_1'=\mathbf{\Psi}_2'+\mathbf{\Psi}_2''=\mathbf{\Psi}_3''=\mathbf{\Psi}, \tag{3.04}$$

$$\Theta_1'=\Theta_2'+\Theta_2''=\Theta_3''=\Theta. \tag{3.05}$$

Finally, (3.01) and (3.03) require

$$\theta_1'=\theta. \tag{3.06}$$

Then, neglecting thermodynamic effects, the stress-strain relations are, for element (1)′,

$$\mathbf{\Psi}_d=2G_1(\mathbf{\Phi}_1'-\theta\mathbf{I}), \quad =2G_1\mathbf{\Phi}_{d1}', \tag{3.07}$$

$$(1-2\nu_1)\Theta=2G_1(1+\nu_1)\theta; \tag{3.08}$$

for elements (2)′ and (2)″,

$$\mathbf{\Psi}_d=2G_2\mathbf{\Phi}_2'+2\eta_2\frac{\partial\mathbf{\Phi}_2'}{\partial t}; \tag{3.09}$$

and, for element (3)″,

$$\mathbf{\Psi}_d=2\eta_3\frac{\partial\mathbf{\Phi}_3''}{\partial t}, \tag{3.10}$$

where t is the time.

Equations (3.07), (3.09), and (3.10) may be solved for $\mathbf{\Phi}_{d1}'$ and $\mathbf{\Phi}_{d2}'$ (which are required for the strain-optical equation) by using (3.03) to eliminate $\mathbf{\Phi}_3''$ and obtaining the additional equation, that is required, by differentiating (3.07) with respect to t. The results are

$$2G_1\mathbf{\Phi}_{d1}'=\mathbf{\Psi}_d, \tag{3.11}$$

$$2G_2\mathbf{\Phi}_{d2}'=\left(1+\frac{\eta_2}{\eta_3}+\eta_2\frac{dG_1^{-1}}{dt}+\frac{\eta_2}{G_1}\frac{\partial}{\partial t}\right)\mathbf{\Psi}_d-2\eta_2\frac{\partial\mathbf{\Phi}_d}{\partial t}, \tag{3.12}$$

in which $d/dt=(dT/dt)(d/dT)$, where T is the temperature.

The relation between $\mathbf{\Psi}_d$ and $\mathbf{\Phi}_d$, independent of $\mathbf{\Phi}_1', \mathbf{\Phi}_2', \mathbf{\Phi}_3''$, is obtained by substituting (3.12) into (3.09), with the result

$$\left[\frac{G_2}{\eta_3}+F(t)+\left(1+\frac{G_2}{G_1}+\frac{2\eta_2}{G_1}\frac{dG_1^{-1}}{dt}+\frac{G_2}{G_1}\frac{d(\eta_2/G_2)}{dt}+\frac{\eta_2}{\eta_3}\right)\frac{\partial}{\partial t}+\frac{\eta_2}{G_1}\frac{\partial^2}{\partial t^2}\right]\mathbf{\Psi}_d$$
$$=2\left[G_2\left(1+\frac{d(\eta_2/G_2)}{dt}\right)\frac{\partial}{\partial t}+\eta_2\frac{\partial^2}{\partial t^2}\right]\mathbf{\Phi}_d, \tag{3.13}$$

where

$$F(t)=\frac{d(\eta_2/\eta_3)}{dt}+\frac{G_2(\eta_3+\eta_2)}{G_1\eta_3}\frac{d(G_1/G_2)}{dt}+\left(G_1+G_2+\frac{G_1\eta_2}{\eta_3}+G_2\frac{d(\eta_2/G_2)}{dt}\right)\frac{dG_1^{-1}}{dt}+\eta_2\frac{d^2G_1^{-1}}{dt^2}.$$

Equations (3.08) and (3.13) describe the relation between stress and strain in the idealized visco-elastic medium represented by Fig. 1.

Let

$$\mathbf{P}[\ \ ]=\left[\frac{G_2}{\eta_3}+F(t)+\left(1+\frac{G_2}{G_1}+\frac{2\eta_2}{G_1}\frac{dG_1^{-1}}{dt}+\frac{G_2}{G_1}\frac{d(\eta_2/G_2)}{dt}+\frac{\eta_2}{\eta_3}\right)\frac{\partial}{\partial t}+\frac{\eta_2}{G_1}\frac{\partial^2}{\partial t^2}\right], \tag{3.14}$$

$$\mathbf{Q}[\ \ ]=\left[G_2\left(1+\frac{d(\eta_2/G_2)}{dt}\right)\frac{\partial}{\partial t}+\eta_2\frac{\partial^2}{\partial t^2}\right]. \tag{3.15}$$

Then (3.13) becomes[4]

$$\mathbf{P}[\mathbf{\Psi}_d]=2\mathbf{Q}[\mathbf{\Phi}_d]. \tag{3.16}$$

One of the materials used very frequently in photoelasticity is a cross-linked polymer (Bakelite BT-61-893) which exhibits instantaneous and retarded elasticity, but no flow. For materials of this type $\eta_3=\infty$, and G_1 and G_2 are constants.

[4] See reference 1, p. 561, for comparison.

Hence, (3.13) reduces to

$$\left[1+\frac{G_2}{G_1}+\frac{1}{G_1}\frac{d\eta_2}{dt}+\frac{\eta_2}{G_1}\frac{\partial}{\partial t}\right]\mathbf{\Psi}_d = 2\left[G_2+\frac{d\eta_2}{dt}+\eta_2\frac{\partial}{\partial t}\right]\mathbf{\Phi}_d. \quad (3.17)$$

Setting $\eta_3=\infty$, $G_1=$constant, in (3.12), and using (3.17), the expression for $\mathbf{\Phi}_{d2}'$ reduces to

$$2G_1\mathbf{\Phi}_{d2}'=\left(1+\frac{1}{G_2}\frac{d\eta_2}{dt}\right)(2G_1\mathbf{\Phi}_d-\mathbf{\Psi}_d). \quad (3.18)$$

IV. STRESS-STRAIN-OPTICAL EQUATION FOR THE IDEALIZED MEDIUM

The general strain-optical equation (2.02) reduces, for the idealized medium, to

$$\mathbf{\Pi}-n_0^{-2}\mathbf{I}=c_1\mathbf{\Phi}_{d1}'+c_2\mathbf{\Phi}_{d2}'+d_1\theta\mathbf{I}. \quad (4.01)$$

Thus three strain-optical coefficients, c_1, c_2, d_1, are required to describe the optical properties of the idealized, isotropic, viscoelastic medium as opposed to two coefficients for an isotropic, elastic medium. Two of the coefficients in (4.01), namely, c_1 and c_2, are *relative* strain-optical coefficients and contribute only to relative phase retardation, as will be shown below. The third coefficient, d_1, is a *mean* strain-optical coefficient whose magnitude can be determined only by measurement of absolute phase retardation. All of the coefficients may be temperature-dependent.

The relation between birefringence and macroscopic stress and strain is obtained by substituting, into (4.01), the expressions for $\mathbf{\Phi}_{d1}'$ and $\mathbf{\Phi}_{d2}'$ given in (3.11) and (3.12), with the result*

$$\mathbf{\Pi}-n_0^{-2}\mathbf{I}=\frac{1}{2}\left[\frac{c_1}{G_1}+\frac{c_2}{G_2}\left(1+\frac{\eta_2}{\eta_3}+\eta_2\frac{dG_1^{-1}}{dt}+\frac{\eta_2}{G_1}\frac{\partial}{\partial t}\right)\right]\mathbf{\Psi}_d-\frac{c_2\eta_2}{G_2}\frac{\partial\mathbf{\Phi}_d}{\partial t}+d_1\theta\mathbf{I}. \quad (4.02)$$

Let

$$\mathbf{R}[\;]=\frac{1}{2}\left[\frac{c_1}{G_1}+\frac{c_2}{G_2}\left(1+\frac{\eta_2}{\eta_3}+\eta_2\frac{dG_1^{-1}}{dt}+\frac{\eta_2}{G_1}\frac{\partial}{\partial t}\right)\right], \quad (4.03)$$

$$-\mathbf{S}[\;]=\frac{c_2\eta_2}{2G_2}\frac{\partial}{\partial t}. \quad (4.04)$$

Then the stress-strain-optical equation becomes

$$\mathbf{\Pi}-n_0^{-2}\mathbf{I}=\mathbf{R}[\mathbf{\Psi}_d]+2\mathbf{S}[\mathbf{\Phi}_d]+d_1\theta\mathbf{I}. \quad (4.05)$$

In the case of the medium exhibiting no flow ($\eta_3=\infty$, G_1 and $G_2=$constants), the relation between birefringence and stress and strain is obtained by substituting (3.11) and (3.18) into (4.01), with the result

$$\mathbf{\Pi}-n_0^{-2}\mathbf{I}=\frac{1}{2G_1}\left[c_1-c_2\left(1+\frac{1}{G_2}\frac{d\eta_2}{dt}\right)\right]\mathbf{\Psi}_d+c_2\left(1+\frac{1}{G_2}\frac{d\eta_2}{dt}\right)\mathbf{\Phi}_d+d_1\theta\mathbf{I}. \quad (4.06)$$

It is interesting to observe that, in the medium exhibiting flow, the relative birefringence depends upon the stress, stress-rate, and strain-rate and is independent of the strain itself. In the medium exhibiting no flow, the relative birefringence is simply a linear function of the stress and strain and is independent of both the stress- and strain-rate. In the case of a medium exhibiting flow but no retarded elasticity, $\mathbf{\Phi}_2'$ is absent from (4.01), and the expression for birefringence is obtained by setting $c_2=0$ in (4.02), whence the relative birefringence depends only on the stress.

Stress-strain-optical equations analogous to (4.02) can be obtained for a viscoelastic medium represented by a model having any number and arrangement of elastic and viscous elements since the system of stress-strain equations for the elements can always be solved for the $\mathbf{\Phi}_{dk}'$ to obtain expressions corresponding to (3.11) and (3.12). A similar situation obtains in regard to stress-strain relations analogous to (3.13). Thus, wherever the linear operators **P[ ]**, **Q[ ]**, **R[ ]**, **S[ ]** appear, they may be replaced by another set of linear operators appropriate to a different material.

V. OPTICAL MEASUREMENTS

The optical quantities which are most easily, and therefore most commonly, measured are relative phase retardation (ρ) and amplitude ratio (γ).

In the case of homogeneous birefringence, the

* It may be observed that, inasmuch as the elastic elements are assumed to be linearly elastic, the general strain-optical relation, Eq. (2.02), might just as well have been expressed in terms of the stresses in the elastic elements, with similar results.

relative phase retardation is given by

$$\rho=2\pi L(n_1-n_2)/\lambda, \qquad (5.01)$$

where L is the path length, λ is the wave-length in vacuum, and n_1 and n_2 are the principal indices of refraction in the plane of the wave front (secondary principal indices). The amplitude ratio gives, directly, the angle φ which a polarizing axis makes with a reference direction in the plane of the wave front.

In the case of non-homogeneous birefringence, n_1-n_2 and φ may be obtained from measurements of ρ and γ by solving Neumann's equations:[5]

$$\frac{d\gamma}{dz}=-\frac{d\varphi}{dz}\cos\rho, \qquad (5.02)$$

$$\frac{d\rho}{dz}-\frac{2\pi}{\lambda}(n_1-n_2)=2\frac{d\varphi}{dz}\cot 2\gamma \sin\rho. \qquad (5.03)$$

We require formulas relating n_1-n_2 and φ to the stress and strain in a viscoelastic medium. Since n_1 and n_2 differ little from n_0 and from each other, n_1-n_2 is nearly equal to $n_0^3(n_2^{-2}-n_1^{-2})/2$, and hence it will be sufficient to find a relation involving $n_2^{-2}-n_1^{-2}$.

VI. RELATIVE STRESS-STRAIN-OPTICAL RELATIONS IN THE PLANE OF THE WAVE FRONT

In terms of the orthogonal unit vectors **i**, **j**, **k**, parallel to the axes of x, y, z, respectively, the nonian forms of the dyadics $\mathbf{\Pi}$, $\mathbf{\Phi}$, $\mathbf{\Psi}$ are

$$\mathbf{\Pi}=n_{xx}^{-2}\mathbf{ii}+n_{yy}^{-2}\mathbf{jj}+n_{zz}^{-2}\mathbf{kk}+n_{yz}^{-2}(\mathbf{jk}+\mathbf{kj})+n_{zx}^{-2}(\mathbf{ki}+\mathbf{ik})+n_{xy}^{-2}(\mathbf{ij}+\mathbf{ji}), \qquad (6.01)$$

$$\mathbf{\Phi}=\epsilon_{xx}\mathbf{ii}+\epsilon_{yy}\mathbf{jj}+\epsilon_{zz}\mathbf{kk}+\epsilon_{yz}(\mathbf{jk}+\mathbf{kj})+\epsilon_{zx}(\mathbf{ki}+\mathbf{ik})+\epsilon_{xy}(\mathbf{ij}+\mathbf{ji}), \qquad (6.02)$$

$$\mathbf{\Psi}=\sigma_{xx}\mathbf{ii}+\sigma_{yy}\mathbf{jj}+\sigma_{zz}\mathbf{kk}+\sigma_{yz}(\mathbf{jk}+\mathbf{kj})+\sigma_{zx}(\mathbf{ki}+\mathbf{ik})+\sigma_{xy}(\mathbf{ij}+\mathbf{ji}). \qquad (6.03)$$

We suppose the wave normal to be parallel to the z axis. Then the Cartesian equation of the central section of the index ellipsoid perpendicular to the wave normal is obtained by setting $\mathbf{r}=\mathbf{i}x+\mathbf{j}y$ in (2.01) and expanding, with the result

$$n_{xx}^{-2}x^2+2n_{xy}^{-2}xy+n_{yy}^{-2}y^2=1, \qquad (6.04)$$

[5] See reference 3, p. 256.

or in polar coordinates $r^2=x^2+y^2$, $\vartheta=\tan^{-1}(y/x)$,

$$n_{xx}^{-2}\cos^2\vartheta+2n_{xy}^{-2}\sin\vartheta\cos\vartheta+n_{yy}^{-2}\sin^2\vartheta=r^{-2}. \qquad (6.05)$$

Setting the derivative of (6.05) with respect to ϑ equal to zero, we find the angles, $\vartheta=\varphi$, $\varphi+\pi/2$, which the stationary positions (polarizing axes) of the radius vector make with the x axis:

$$\tan 2\varphi=2n_{xy}^{-2}/(n_{xx}^{-2}-n_{yy}^{-2}). \qquad (6.06)$$

Substituting $\vartheta=\varphi$, $\varphi+\pi/2$ in (6.05) we obtain the magnitudes of the stationary values of the radius vector, i.e., the indices of refraction n_1, n_2 (where $n_1>n_2$) in the plane of the wave front:

$$n_2^{-2}=n_{xx}^{-2}\cos^2\varphi+n_{yy}^{-2}\sin^2\varphi+n_{xy}^{-2}\sin 2\varphi, \qquad (6.07)$$

$$n_1^{-2}=n_{xx}^{-2}\sin^2\varphi+n_{yy}^{-2}\cos^2\varphi-n_{xy}^{-2}\sin 2\varphi. \qquad (6.08)$$

Similarly, for the strain dyadic

$$\epsilon_1=\epsilon_{xx}\cos^2\varphi'+\epsilon_{yy}\sin^2\varphi'+\epsilon_{xy}\sin 2\varphi', \qquad (6.09)$$

$$\epsilon_2=\epsilon_{xx}\sin^2\varphi'+\epsilon_{yy}\cos^2\varphi'-\epsilon_{xy}\sin 2\varphi', \qquad (6.10)$$

$$\tan 2\varphi'=2\epsilon_{xy}/(\epsilon_{xx}-\epsilon_{yy}), \qquad (6.11)$$

and, for the stress dyadic,

$$\sigma_1=\sigma_{xx}\cos^2\varphi''+\sigma_{yy}\sin^2\varphi''+\sigma_{xy}\sin 2\varphi'', \qquad (6.12)$$

$$\sigma_2=\sigma_{xx}\sin^2\varphi''+\sigma_{yy}\cos^2\varphi''-\sigma_{xy}\sin 2\varphi'', \qquad (6.13)$$

$$\tan 2\varphi''=2\sigma_{xy}/(\sigma_{xx}-\sigma_{yy}), \qquad (6.14)$$

where ϵ_1, ϵ_2 and σ_1, σ_2 are secondary principal strains and stresses ($\epsilon_1>\epsilon_2$, $\sigma_1>\sigma_2$) in the plane of the wave front, and φ' and φ'' are the angles which the directions of ϵ_1 and σ_1, respectively, make with the x axis. In an isotropic elastic medium $\varphi=\varphi'=\varphi''$, but this is not necessarily true in a viscoelastic medium.

Returning to the stress-strain-optical equation (4.05) and equating, in turn, the coefficients of the dyads **ii**, **jj**, **ij**, we obtain:

$$n_{xx}^{-2}-n_0^{-2}=\mathbf{R}[\sigma_{xx}-\Theta]+2\mathbf{S}[\epsilon_{xx}-\theta]+d_1\theta, \qquad (6.15)$$

$$n_{yy}^{-2}-n_0^{-2}=\mathbf{R}[\sigma_{yy}-\Theta]+2\mathbf{S}[\epsilon_{yy}-\theta]+d_1\theta, \qquad (6.16)$$

$$n_{xy}^{-2}=\mathbf{R}[\sigma_{xy}]+2\mathbf{S}[\epsilon_{xy}]. \qquad (6.17)$$

Subtracting (6.16) from (6.15),

$$n_{xx}^{-2}-n_{yy}^{-2}=\mathbf{R}[\sigma_{xx}-\sigma_{yy}]+2\mathbf{S}[\epsilon_{xx}-\epsilon_{yy}]. \qquad (6.18)$$

Now, from (6.07), (6.08), (6.17), and (6.18),

$$n_2^{-2}-n_1^{-2}=(n_{xx}^{-2}-n_{yy}^{-2})\cos 2\varphi+2n_{xy}^{-2}\sin 2\varphi,$$
$$=\{\mathbf{R}[\sigma_{xx}-\sigma_{yy}]+2\mathbf{S}[\epsilon_{xx}-\epsilon_{yy}]\}\cos 2\varphi$$
$$+2\{\mathbf{R}[\sigma_{xy}]+2\mathbf{S}[\epsilon_{xy}]\}\sin 2\varphi. \quad (6.19)$$

From (6.09) to (6.14),

$$\epsilon_{xx}-\epsilon_{yy}=(\epsilon_1-\epsilon_2)\cos 2\varphi', \quad (6.20)$$
$$2\epsilon_{xy}=(\epsilon_1-\epsilon_2)\sin 2\varphi', \quad (6.21)$$
$$\sigma_{xx}-\sigma_{yy}=(\sigma_1-\sigma_2)\cos 2\varphi'', \quad (6.22)$$
$$2\sigma_{xy}=(\sigma_1-\sigma_2)\sin 2\varphi'', \quad (6.23)$$

and, hence,

$$n_2^{-2}-n_1^{-2}=\{(\mathbf{R}[(\sigma_1-\sigma_2)\cos 2\varphi'']$$
$$+2\mathbf{S}[(\epsilon_1-\epsilon_2)\cos 2\varphi']\}\cos 2\varphi$$
$$+\{\mathbf{R}[(\sigma_1-\sigma_2)\sin 2\varphi'']$$
$$+2\mathbf{S}[(\epsilon_1-\epsilon_2)\sin 2\varphi']\}\sin 2\varphi. \quad (6.24)$$

Equation (6.24) is the stress-strain-optical relation in the plane of the wave front.

In the case of the medium exhibiting instantaneous and retarded elasticity, but no flow ($\eta_3=\infty$, G_1 and G_2=const.), the partial differential coefficients are absent from the operators **R** and **S**, and (6.24) reduces to

$$2G_1(n_2^{-2}-n_1^{-2})=\left(c_1-c_2-\frac{c_2}{G_2}\frac{d\eta_2}{dt}\right)$$
$$\times(\sigma_1-\sigma_2)\cos 2(\varphi-\varphi'')$$
$$+2G_1c_2\left(1+\frac{1}{G_2}\frac{d\eta_2}{dt}\right)(\epsilon_1-\epsilon_2)\cos 2(\varphi-\varphi'). \quad (6.25)$$

The corresponding equation for a medium exhibiting instantaneous elasticity and flow, but no retarded elasticity, is obtained by setting $c_2=0$ in (6.24), with the result

$$2G_1(n_2^{-2}-n_1^{-2})=c_1(\sigma_1-\sigma_2)\cos 2(\varphi-\varphi''). \quad (6.26)$$

Finally, for an elastic medium, $\varphi=\varphi''$ and (6.26) reduces to

$$2G_1(n_2^{-2}-n_1^{-2})=c_1(\sigma_1-\sigma_2). \quad (6.27)$$

When account is taken of the fact that

$$n_2^{-2}-n_1^{-2}\approx 2(n_1-n_2)/n_0^3, \quad (6.28)$$

Eq. (6.27), and the relation $\varphi=\varphi''$ comprise the usual stress-optical laws for isotropic, elastic media, in which case the determination of n_1-n_2, and φ gives directly, $\sigma_1-\sigma_2$ and φ''. For a visco-elastic medium, however, a knowledge of n_1-n_2 and φ is not sufficient, in general, to determine $\sigma_1-\sigma_2$ and φ'', as may be seen by inspecting (6.24) or (6.25) or (6.26).

VII. SIMPLIFICATION WHEN THE STRESS DEVIATION IS A PRODUCT FUNCTION

The complexity of the stress-strain-optical equations is due to the fact that, in general, the secondary principal axes of stress and strain, in the plane of the wave front, do not coincide with each other or with the polarizing axes and also the relative orientations of these axes may vary with time.

Consider a state of stress in which the stress deviation is the product of two functions, one of which depends only on x, y, and z and the other only on t; thus

$$\mathbf{\Psi}_d=\bar{\mathbf{\Psi}}_d(x,y,z)f(t). \quad (7.01)$$

(In what follows, a bar over a symbol indicates that the quantity is a function of x, y, z only.) As a result of the stress-strain relation (3.16), $\mathbf{\Phi}_d$ will then be a product function also; thus

$$\mathbf{\Phi}_d=\bar{\mathbf{\Phi}}_d(x,y,z)g(t), \quad (7.02)$$

and, accordingly,

$$\bar{\mathbf{\Psi}}_d/2\bar{\mathbf{\Phi}}_d=\mathbf{Q}[g]/\mathbf{P}[f]=G_0, \quad (7.03)$$

where G_0 is an arbitrary constant of the nature and dimensions of a shear modulus. Hence, the assumption (7.01) implies[6]

$$\bar{\mathbf{\Psi}}_d=2G_0\bar{\mathbf{\Phi}}_d, \quad (7.04)$$
$$\mathbf{Q}[g]=G_0\mathbf{P}[f]. \quad (7.05)$$

Now, from (7.01), (7.02), (6.02), and (6.03),

$$\sigma_{xy}=\overline{\sigma_{xy}}(x,y,z)f(t),$$
$$\sigma_{xx}-\sigma_{yy}=\overline{\sigma_{xx}-\sigma_{yy}}(x,y,z)f(t), \quad (7.06)$$
$$\epsilon_{xy}=\overline{\epsilon_{xy}}(x,y,z)g(t),$$
$$\epsilon_{xx}-\epsilon_{yy}=\overline{\epsilon_{xx}-\epsilon_{yy}}(x,y,z)g(t), \quad (7.07)$$

whence, from (6.11) and (6.14), φ' and φ'' are functions of x, y, z only. Also, from (7.04),

$$\overline{\sigma_{xy}}=2G_0\overline{\epsilon_{xy}}, \quad \overline{\sigma_{xx}-\sigma_{yy}}=2G_0\overline{\epsilon_{xx}-\epsilon_{yy}}, \quad (7.08)$$

so that $\varphi'=\varphi''$. Finally, dividing (6.17) by (6.18)

[6] See reference 1, p. 563, for comparisons.

and using (7.06), (7.07), (7.08), and (6.06), we find

$$\varphi = \varphi' = \varphi'' = \varphi(x, y, z). \quad (7.09)$$

Thus the condition (7.01) makes the secondary principal axes of birefringence, strain and stress coincide.

Noting that

$$\sigma_1 - \sigma_2 = \overline{\sigma_1 - \sigma_2} f(t) \quad \text{and} \quad \epsilon_1 - \epsilon_2 = \overline{\epsilon_1 - \epsilon_2} g(t),$$

(6.24) becomes

$$n_2^{-2} - n_1^{-2} = \overline{\sigma_1 - \sigma_2} \mathbf{R}[f] + 2\overline{\epsilon_1 - \epsilon_2} \mathbf{S}[g]. \quad (7.10)$$

Finally, since $\overline{\sigma_1 - \sigma_2} = 2G_0 \overline{\epsilon_1 - \epsilon_2}$,

$$n_2^{-2} - n_1^{-2} = \overline{\sigma_1 - \sigma_2} \{\mathbf{R}[f] + \mathbf{S}[g]/G_0\}, \quad (7.11)$$

or, assuming that (6.28) is an equality,

$$n_1 - n_2 = \overline{\sigma_1 - \sigma_2} C(t), \quad (7.12)$$

where

$$C(t) = n_0^3 \{\mathbf{R}[f] + \mathbf{S}[g]/G_0\}/2. \quad (7.13)$$

Thus, the condition $\Psi_d = \overline{\Psi}_d f(t)$ leads to a simple stress-optical relation for viscoelastic media entirely analogous to that for an elastic medium. The time- and temperature-dependent relative stress-optical coefficient $C(t)$ can be determined in a polariscope by an experiment in simple tension in which the tensile stress σ is applied according to the desired law $\bar{\sigma} f(t)$ and the temperature is varied with time in the desired manner.

Similarly a strain-optical relation

$$n_1 - n_2 = \overline{\epsilon_1 - \epsilon_2} c(t) \quad (7.14)$$

may be written, where

$$c(t) = n_0^3 \{G_0 \mathbf{R}[f] + \mathbf{S}[g]\}. \quad (7.15)$$

The time- and temperature-dependent, relative strain-optical coefficient $c(t)$ may be obtained in a polariscope by an elongation experiment in which the stretch ϵ follows a prescribed law $\bar{\epsilon} g(t)$ and the temperature is varied with time in the desired manner.

These experiments may be used, also, to solve Eq. (7.05).

VIII. SOME THEOREMS FROM ELASTICITY THEORY

In the remainder of this paper, photoelasticity will be considered as an experimental method for finding the state of stress in an elastic body, i.e., for solving the equations of elasticity, and we shall study the conditions under which viscoelastic materials may be used as models for this purpose. We have already seen that the condition $\Psi_d = \overline{\Psi}_d f(t)$ is sufficient to assure the simple relations (7.09) and (7.12) between birefringence and stress. We require, then,

A. The necessary and sufficient conditions on the viscoelastic model for the stress in it to be of this form.

In addition, we require,

B. The necessary and sufficient conditions for the stress in the viscoelastic model to correspond to the stress in an elastic body of the same shape under analogous boundary conditions.

The problem of finding these conditions is simplified by the fact that, from the viewpoint of mechanics of continua, elastic and viscoelastic media differ only in the relation between stress and strain. Hence, all of the theorems of elasticity theory which are independent of the stress-strain relation apply as well to viscoelastic bodies. The theorems which we shall need are:[7]

(1) In a body in equilibrium the stress dyadic, in the absence of body forces, satisfies the stress equation of equilibrium:

$$\nabla \cdot \Psi = 0. \quad (8.01)$$

(In a viscoelastic body, a state of "equilibrium" is one in which the inertia term, in the equation of motion, is negligible.)

(2) The pure small-strain dyadic satisfies the compatibility equation:

$$\nabla \times \Phi \times \nabla = 0. \quad (8.02)$$

(3) In a simply connected body, continuity and single valuedness of strain imply single valuedness of displacement.

(4) In a simply connected body in which the strain is continuous and single valued, the compatibility condition ((8.02)) is a sufficient condition for single valuedness of displacement.

(5) In an $(s+1)$-connected body, in which the strain is continuous and single valued, the displacement may be many valued to the extent of possessing s definite cyclic functions, which may be interpreted as s discontinuities across s sur-

[7] A. E. H. Love, *Theory of Elasticity* (Cambridge University Press, London, 1927), fourth edition, pp. 85, 221–225.

faces σ_i. The surfaces σ_i are those which reduce the multiply-connected body to a simply connected one, and the discontinuities must be of the form

$$\delta \mathbf{u}_i = \mathbf{a}_i + \mathbf{r}\cdot(\mathbf{I}\times\mathbf{b}_i), \quad i=1, 2, \cdots s, \quad (8.03)$$

where $\mathbf{a}_i$ and $\mathbf{b}_i$ are independent of x, y, z and $\mathbf{r}=\mathbf{i}x+\mathbf{j}y+\mathbf{k}z$.

In addition, a uniqueness theorem will be established along lines paralleling closely the Kirchhoff uniqueness theorem of elasticity theory.[8]

IX. COMPATIBILITY CONDITION

The stress-strain relation for an elastic medium with shear modulus G and Poisson's ratio ν may be written as

$$2G\mathbf{\Phi} = \mathbf{\Psi} - \alpha\Theta\mathbf{I}, \quad (9.01)$$

where

$$\alpha = 3\nu/(1+\nu),$$

and ν is Poisson's ratio. Hence the compatibility condition (8.02) requires that the stress in an elastic medium satisfy the equation

$$\nabla\times(\mathbf{\Psi} - \alpha\Theta\mathbf{I})\times\nabla = 0. \quad (9.02)$$

The stress-strain relations (3.08) and (3.16) for the idealized viscoelastic medium may be written in the form

$$2\mathbf{Q}[\mathbf{\Phi}] = \mathbf{P}[\mathbf{\Psi} - \alpha_1\Theta\mathbf{I}] - (1-\alpha_1)(\mathbf{P} - \mathbf{Q}/G_1)[\Theta\mathbf{I}], \quad (9.03)$$

where

$$\alpha_1 = 3\nu_1/(1+\nu_1).$$

In any physical experiment the functions $\mathbf{\Phi}$, $\mathbf{\Psi}$, and Θ will be such that the order of differentiation with respect to t and x, y, z may be reversed.[9] Hence, the compatibility condition requires that the stress in the viscoelastic medium satisfy the equation

$$\mathbf{P}[\nabla\times(\mathbf{\Psi} - \alpha_1\Theta\mathbf{I})\times\nabla] - (1-\alpha_1)(\mathbf{P} - \mathbf{Q}/G_1)[\nabla\times\Theta\mathbf{I}\times\nabla] = 0. \quad (9.04)$$

Conditions B, Section VIII, require the stress in the viscoelastic medium to satisfy differential equations of the same form as those for an elastic medium. Comparing (9.04) with (9.02), we see that this is true only if $\alpha_1 = 1$ or if Θ is independent of x, y, and z. The form of Θ is usually not known in advance, is difficult to measure, and, moreover, is not likely to be found to be independent of the space coordinates even if measured. Hence, we shall have to require $\alpha_1 = 1$, i.e., the viscoelastic medium must be incompressible. Fortunately, many viscoelastic materials suitable for photoelasticity have Poisson's ratios nearly equal to one-half.

It should be noted that the stress in an incompressible viscoelastic medium satisfies a differential equation ((9.04) with $\alpha_1=1$) of the same form as the differential equation ((9.02) with $\alpha=1$) governing the stress in an *incompressible* elastic medium.[10]

X. NECESSARY AND SUFFICIENT CONDITIONS FOR $\mathbf{\Psi}_d = \overline{\mathbf{\Psi}}_d f(t)$ IN A SIMPLY CONNECTED BODY

We note, first of all, that in an incompressible medium $\mathbf{\Phi} = \mathbf{\Phi}_d$ and, since $\mathbf{I}:\mathbf{\Phi}_d = 0$,

$$\mathbf{\Psi}:\mathbf{\Phi} = (\mathbf{\Psi}_d + \Theta\mathbf{I}):\mathbf{\Phi}_d = \mathbf{\Psi}_d:\mathbf{\Phi}_d. \quad (10.01)$$

Also, the pure strain dyadic $\mathbf{\Phi}$ is the self-conjugate part of the strain dyadic $\nabla\mathbf{u}$, where $\mathbf{u}$ is the displacement vector. Thus

$$2\mathbf{\Phi} = 2\mathbf{\Phi}_d = \nabla\mathbf{u} + \mathbf{u}\nabla. \quad (10.02)$$

Then, since $\mathbf{\Psi}$ is also self-conjugate,

$$\mathbf{\Psi}:\mathbf{\Phi} = \tfrac{1}{2}\mathbf{\Psi}:(\nabla\mathbf{u} + \mathbf{u}\nabla) = \mathbf{\Psi}:\nabla\mathbf{u} = \nabla\cdot(\mathbf{\Psi}\cdot\mathbf{u}) - (\nabla\cdot\mathbf{\Psi})\cdot\mathbf{u}. \quad (10.03)$$

But, in a body in equilibrium, $\nabla\cdot\mathbf{\Psi} = 0$. Hence, in an incompressible body in equilibrium,

$$\mathbf{\Psi}_d:\mathbf{\Phi}_d = \nabla\cdot(\mathbf{\Psi}\cdot\mathbf{u}), \quad (10.04)$$

and we may write

$$\iiint \mathbf{\Psi}_d:\mathbf{\Phi}_d d\tau = \iiint \nabla\cdot(\mathbf{\Psi}\cdot\mathbf{u})d\tau, \quad (10.05)$$

where $d\tau$ is an element of volume and the integrations are extended throughout the volume of the viscoelastic body under consideration.

As noted in Section VIII, in a *simply connected* body no displacement discontinuities are possible if the strain is continuous and single valued.

[8] See reference 7, p. 170.

[9] It is assumed that the temperature distribution throughout the body is homogeneous, so that η_i and G_i depend only on t and not on x, y, or z.

[10] See reference 1, p. 561

In such a body the divergence theorem yields

$$\iiint \nabla\cdot(\Psi\cdot \mathbf{u})d\tau=\iint \mathbf{n}\cdot\Psi\cdot \mathbf{u}d\sigma, \quad (10.06)$$

where $d\sigma$ is an element of the surface, $\mathbf{n}$ is its unit, outward-drawn, normal vector and the surface integral is extended over the whole surface of the body. (In the surface integral, $\mathbf{u}$ is the displacement at the surface of the body and the traction

$$\mathbf{T}=\mathbf{n}\cdot\Psi \quad (10.07)$$

is the surface traction.) Hence,

$$\iiint \Psi_d:\Phi_d d\tau=\iint \mathbf{T}\cdot\mathbf{u}d\sigma. \quad (10.08)$$

We have seen (Section VII), that, if

$$\Psi_d=\overline{\Psi}_d f(t),$$

Φ_d becomes $\bar{\Phi}_d g(t)$. Hence, from (10.08), a necessary condition for Ψ_d to be of the form $\overline{\Psi}_d f(t)$ is

$$\mathbf{T}\cdot\mathbf{u}=\bar{\mathbf{T}}\cdot\bar{\mathbf{u}}f(t)g(t), \quad (10.09)$$

and further, in view of (10.02),

$$\mathbf{T}=\bar{\mathbf{T}}(x, y, z)f(t), \quad (10.10)$$

$$\mathbf{u}=\bar{\mathbf{u}}(x, y, z)g(t) \quad (10.11)$$

on the boundary. This excludes, for example, *moving loads*.

Equation (10.09) is not a sufficient condition for Ψ_d to be of the required form; for there may be present another stress system Ψ^0 (an "initial stress") for which $\mathbf{n}\cdot\Psi^0\cdot\mathbf{u}^0=0$ on the boundary. The conditions are completed by requiring Ψ_d^0 to be either zero or of the form

$$\Psi_d^0=\overline{\Psi}_d^0(x, y, z)f(t). \quad (10.12)$$

The presence of initial stress is easily detected optically but is usually difficult to measure. Generally it will not be of the form (10.12) unless it is a *residual* stress remaining from a previous experiment with boundary conditions (10.09), in which case the calibration test (Section XIII) must include the complete history of $f(t)$, $g(t)$, $\eta_i(t)$, and $G_i(t)$ from the unstressed state.

XI. UNIQUENESS THEOREM FOR A SIMPLY CONNECTED BODY

From (10.08), (10.10), (10.11), (7.01), (7.02), and (7.04),

$$\iiint \overline{\Psi}_d:\overline{\Psi}_d d\tau=2G_0\iint \bar{\mathbf{T}}\cdot\bar{\mathbf{u}}d\sigma. \quad (11.01)$$

Let Ψ_d^I and Ψ_d^{II} be the stress deviations, vanishing at $t=0$, of two equilibrium states of stress Ψ^I and Ψ^{II} in an incompressible body. We suppose the tractions $\mathbf{T}^I$, $\mathbf{T}^{II}$ and the displacements $\mathbf{u}^I$, $\mathbf{u}^{II}$ to be of such form that, on the boundary,

$$\mathbf{T}^I\cdot\mathbf{u}^I=\bar{\mathbf{T}}^I\cdot\bar{\mathbf{u}}^I f(t)g(t)$$

and

$$\mathbf{T}^{II}\cdot\mathbf{u}^{II}=\bar{\mathbf{T}}^{II}\cdot\bar{\mathbf{u}}^{II} f(t)g(t).$$

Then, by Section X,

$$\Psi_d^I=\overline{\Psi}_d^I f(t), \quad \Psi_d^{II}=\overline{\Psi}_d^{II} f(t).$$

Let

$$\Psi^*=\Psi^I-\Psi^{II},$$

$$\Psi_d^*=\overline{\Psi}_d^I-\overline{\Psi}_d^{II},$$

$$\bar{\mathbf{T}}^*\cdot\bar{\mathbf{u}}^*=(\bar{\mathbf{T}}^I-\bar{\mathbf{T}}^{II})\cdot(\bar{\mathbf{u}}^I-\bar{\mathbf{u}}^{II}).$$

Then, by (11.01),

$$\iiint \overline{\Psi}_d^*:\overline{\Psi}_d^* d\tau=2G_0\iint \bar{\mathbf{T}}^*\cdot\bar{\mathbf{u}}^* d\sigma. \quad (11.02)$$

Now, if

$$\bar{\mathbf{T}}^*\cdot\bar{\mathbf{u}}^*=0 \quad (11.03)$$

on the boundary, the surface integral in (11.02) vanishes and hence the volume integral also vanishes. But $\overline{\Psi}_d^*:\overline{\Psi}_d^*$ is equal to the sum of the squares of the differences between the principal stresses of Ψ^* and its mean tension. Hence, $\overline{\Psi}_d^*:\overline{\Psi}_d^*$ is always positive or zero. Thus, if the volume integral in (11.02) vanishes, $\Psi_d^*:\Psi_d^*=0$ and, consequently, Ψ^I and Ψ^{II} can differ only by a mean tension.

A mean tension produces no relative birefringence and cannot, in general, be determined from measurements of relative phase retardation and orientation of polarizing axes. An exception is when one or more of the principal stresses is known, *a priori*, to vanish, as at a point on a free boundary or in a state of plane stress. Then, however, the condition $\overline{\Psi}_d^*:\overline{\Psi}_d^*=0$ makes $\Psi^I=\Psi^{II}$.

Let

$$\mathbf{T}=T_\lambda\boldsymbol{\lambda}+T_\mu\boldsymbol{\mu}+T_\nu\boldsymbol{\nu},\quad \mathbf{u}=u_\lambda\boldsymbol{\lambda}+u_\mu\boldsymbol{\mu}+u_\nu\boldsymbol{\nu},$$

where $\boldsymbol{\lambda}$, $\boldsymbol{\mu}$, $\boldsymbol{\nu}$ constitute a set of mutually perpendicular unit vectors at each point of the boundary. Then (11.03) may be written as

$$(\bar{T}_\lambda{}^I-\bar{T}_\lambda{}^{II})(\bar{u}_\lambda{}^I-\bar{u}_\lambda{}^{II})+(\bar{T}_\mu{}^I-\bar{T}_\mu{}^{II})(\bar{u}_\mu{}^I-\bar{u}_\mu{}^{II})$$
$$+(\bar{T}_\nu{}^I-\bar{T}_\nu{}^{II})(\bar{u}_\nu{}^I-\bar{u}_\nu{}^{II})=0,\quad (11.04)$$

where the superscripts and bars have the same significance as before. Hence, the specification of any one of eight combinations of boundary conditions (one traction, one displacement and six mixed conditions) will make the two states of stress identical except, possibly, for a mean tension.

XII. EXTENSION TO MULTIPLY CONNECTED BODIES

As stated in Section VIII, in an $(s+1)$-connected body there may be s displacement discontinuities $\delta\mathbf{u}_i$, of the form (8.03), across s surfaces σ_i which reduce the multiply connected body to a simply connected one. In this case the divergence theorem, instead of having the form (10.06), becomes

$$\iiint\nabla\cdot(\boldsymbol{\Psi}\cdot\mathbf{u})d\tau=\iint\mathbf{n}\cdot\boldsymbol{\Psi}\cdot\mathbf{u}d\sigma$$
$$+\sum_{i=1}^{s}\iint\mathbf{n}_i\cdot\boldsymbol{\Psi}\cdot\delta\mathbf{u}_i d\sigma_i,\quad (12.01)$$

where the surface integrals under the summation sign are taken over the surfaces of discontinuity.

Since the additional terms in (12.01) are of the same form as the integral over the external surface, all that has been said above for the quantity $\mathbf{n}\cdot\boldsymbol{\Psi}\cdot\mathbf{u}$ applies also to each of the quantities $\mathbf{n}_i\cdot\boldsymbol{\Psi}\cdot\delta\mathbf{u}_i$.

The additional stress introduced by the "dislocations" $\delta\mathbf{u}_i$ may be looked upon from two points of view. First, the additional stress may be regarded as an initial stress and treated in the same manner as $\boldsymbol{\Psi}^0$ above. Second, it may be required to introduce dislocations deliberately, as in experiments to determine the effect of a difference in Poisson's ratio between model and prototype or in applications of Biot's analogy between thermal stress and dislocations. In such cases the only additional condition that must be satisfied is that the $\delta\mathbf{u}_i$ be of the form $\delta\bar{\mathbf{u}}_i g(t)$.

XIII. RECAPITULATION AND REMARKS

For an $(s+1)$-connected, viscoelastic body in a state of equilibrium, the necessary and sufficient conditions, for the stress deviation to be of the form $\bar{\boldsymbol{\Psi}}_d f(t)$ and for $\bar{\boldsymbol{\Psi}}_d$ to correspond to the stress in an elastic body of the same shape under analogous boundary conditions, are

(1) The viscoelastic material must be incompressible.**

(2) As in elasticity theory, over each and every portion of the surface of the body and the s surfaces of discontinuity of displacement, one of the following eight combinations of boundary conditions must be prescribed:

$$T_\lambda,\ T_\mu,\ T_\nu;\ T_\lambda,\ T_\mu,\ u_\nu;\ T_\lambda,\ u_\mu,\ T_\nu;\ u_\lambda,\ T_\mu,\ T_\nu;$$
$$u_\lambda,\ u_\mu,\ u_\nu;\ u_\lambda,\ u_\mu,\ T_\nu;\ u_\lambda,\ T_\mu,\ u_\nu;\ T_\lambda,\ u_\mu,\ u_\nu.$$

(Note that different "mixtures" may be prescribed over different portions of the surfaces. The first three mixed conditions consist of a component of displacement and the resultant traction in the plane perpendicular to it, while the second three mixed conditions consist of a component of traction and the resultant displacement in the plane perpendicular to it.)

(3) Components of boundary traction, where prescribed, must be of the form

$$\bar{T}_j(x,y,z)f(t),\quad j=\lambda,\mu,\nu$$

and components of boundary displacement, where prescribed, must be of the form

$$\bar{u}_j(x,y,z)g(t),\quad j=\lambda,\mu,\nu,$$

where $\mathbf{Q}[g]=G_0\mathbf{P}[f]$.

(4) The initial stress must be either zero or of the form $\bar{\boldsymbol{\Psi}}_d{}^0 f(t)$, in which case $f(t)$ must be reckoned from a zero state of stress.

Under these conditions the stress and strain in the viscoelastic body are related to the birefringence through the simple formulas

$$\varphi=\varphi'=\varphi'',$$
$$n_1-n_2=\overline{\sigma_1-\sigma_2}C(t),$$
$$n_1-n_2=\overline{\epsilon_1-\epsilon_2}c(t),$$

and the barred components of stress and strain

** In practice, Poisson's ratio should differ little from ½.

are those for an incompressible elastic body with shear modulus G_0.

The time- and temperature-dependent stress- and strain-optical coefficients C and c may be determined by experiments on a simple-tension specimen of the same material as the model. For example, the viscoelastic model, held at constant temperature, may be loaded by several forces whose directions and points of application on the boundary are fixed, but whose magnitudes, while maintaining constant ratios, vary with time according to a law $f(t)$. A photographic record of the fringe pattern is made at a time t_1 This is the most common type of photoelastic test. (The time-dependence may be introduced simply by the process of reaching and maintaining constant loads.) The calibration bar must be loaded according to the same time-law, at the same temperature, and the ratio $(n_1-n_2)/\overline{\sigma_1-\sigma_2}$ calculated at time t_1. This gives the stress-optical coefficient $C(t_1)$ for the photograph.

It is interesting to notice that if the temperature is varied during the model test, as in the freezing method of three-dimensional photoelasticity, the only *additional* requirement is that, when finding the calibration factor $C(t_1)$, the calibration specimen must be subjected to the same temperature-time cycle as the model. Thus, the freezing method is not restricted to models made of cross-linked polymers.

In the case of mixed, non-zero, boundary conditions a preliminary experiment must be performed, for example, in simple tension, to determine the $g(t)$ corresponding to a given $f(t)$ or vice versa.

In the determination of stress-concentration factors it is not necessary to know the stress-optical coefficient when using elastic materials or viscoelastic materials in which η_2 and η_3 are so large that "optical creep" is negligible. Similarly, it is not necessary to find $C(t_1)$ in stress-concentration factor tests with viscoelastic materials that *do* exhibit optical creep because, if the above conditions are met, the ratio of the stress at any two points of the model is a constant, equal to the corresponding ratio for the analogous elastic body. This is true regardless of flow or of temperature variation with time.

In the transfer, from model to prototype, of results obtained with a model having a Poisson's ratio different fron that of the prototype, the same difficulties are encountered with viscoelastic models as with elastic models. The few theorems governing such transfers that are available for elastic models are applicable to viscoelastic models.

(*Reprinted from the Journal of Applied Mechanics for September, 1949*)

Compliance of Elastic Bodies in Contact

By R. D. MINDLIN,[1] NEW YORK, N. Y.

A small tangential force and a small torsional couple are applied across the elliptic contact surface of a pair of elastic bodies which have been pressed together. If there is no slip at the contact surface, considerations of symmetry and continuity lead to the conclusion that there is no change in the normal component of traction across the surface and, aside from warping of the surface, there is no relative displacement of points on the contact surface. The problem is thus reduced to a "problem of the plane" in which the tangential displacements and normal component of traction are given over part of the boundary and the three components of traction are given over the remainder. In the case of the tangential force it is observed that, when Poisson's ratio is zero, the problem is a simple one, in potential theory, which is then generalized by means of a special device. An expression for tangential compliance is found as a linear combination of complete elliptic integrals. In general, the compliance is greater in the direction of the major axis of the elliptic contact surface than in the direction of the minor axis. Both components of tangential compliance increase as Poisson's ratio decreases and become equal when Poisson's ratio is zero. Over the practical range of Poisson's ratio, the tangential compliance is greater than the normal compliance, but never more than twice as great as long as there is no slip. The tangential traction on the contact surface is everywhere parallel to the applied force. Contours of constant traction are ellipses homothetic with the elliptic boundary. The magnitude of the traction rises from one half the average at the center of the contact surface to infinity at the edge. Due to this infinity, there will be slip, the effect of which is studied for the circular contact surface. In the case of the torsional couple, the solution is obtained by generalizing a solution by H. Neuber pertaining to a hyperbolic groove in a twisted shaft. The torsional compliance is expressed in terms of complete elliptic integrals and, for the circular contact area, reduces to that found by E. Reissner and H. F. Sagoci. The resultant traction at a point rises from zero at the center to infinity at the edge of the contact surface, but is constant along and parallel to homothetic ellipses only in the case of the circular contact area.

Notation

Love's notation is used for elastic constants and components of displacement and traction:

μ = modulus of rigidity
σ = Poisson's ratio
$\lambda = 2\mu\sigma/(1-2\sigma)$, Lamé's constant
u, v, w = rectangular components of displacement
X_ν, Y_ν, Z_ν = rectangular components of traction at a point on plane surface $z = 0$ of body $z > 0$.

Additional symbols are defined where they are introduced.

Introduction

Consider two homogeneous, isotropic, elastic bodies in contact at a point, O, in an unstressed state, as shown in Fig. 1(*a*). The surfaces of the bodies have a common tangent plane and a common normal at O. Let the tangent plane be the plane of x and y, and the normal the z-axis of a rectangular co-ordinate system. If the bodies are pressed together with a force P_z, parallel to Oz, they will come into contact over a small surface in the neighborhood of O, as illustrated in Fig. 1(*b*). According

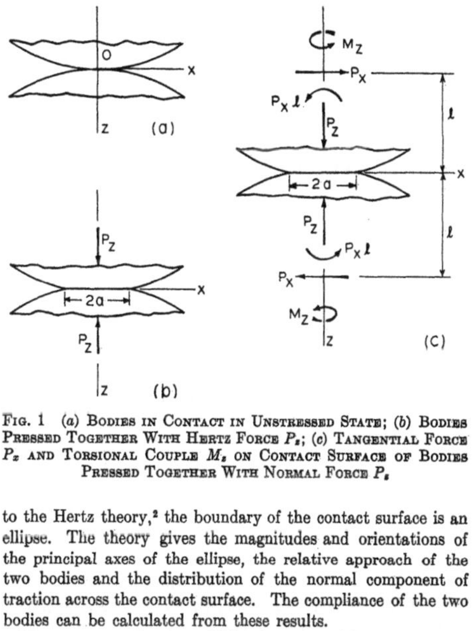

Fig. 1 (*a*) Bodies in Contact in Unstressed State; (*b*) Bodies Pressed Together With Hertz Force P_z; (*c*) Tangential Force P_x and Torsional Couple M_z on Contact Surface of Bodies Pressed Together With Normal Force P_z

to the Hertz theory,[2] the boundary of the contact surface is an ellipse. The theory gives the magnitudes and orientations of the principal axes of the ellipse, the relative approach of the two bodies and the distribution of the normal component of traction across the contact surface. The compliance of the two bodies can be calculated from these results.

Suppose, now, that an additional system of forces is applied to the bodies such that, across the contact surface, one body exerts on the other a small force perpendicular to Oz and a small couple with axis Oz, leaving P_z unchanged, as shown in Fig. 1(*c*). It is required to find the tangential and torsional compliances of the two bodies.

Hertz Solution

The results of the Hertz solution are[2]

$$p = \frac{3P_z}{2\pi ab}\left(1 - \frac{x^2}{a^2} - \frac{y^2}{b^2}\right)^{1/2} \qquad [1]$$

[1] Professor of Civil Engineering, Columbia University, and Consultant, Bell Telephone Laboratories, Inc. Mem. ASME.

Presented at the National Meeting of the Applied Mechanics Division, Chicago, Ill., June 17–19, 1948, of The American Society of Mechanical Engineers.

Discussion of this paper should be addressed to the Secretary, ASME, 29 West 39th Street, New York, N. Y., and will be accepted until October 10, 1949, for publication at a later date. Discussion received after the closing date will be returned.

[2] "Treatise on the Mathematical Theory of Elasticity," by A. E. H. Love, Cambridge University Press, fourth edition, 1927, pp. 193–198.

$$\alpha = \frac{3}{4} P_z(\vartheta_1 + \vartheta_2) \int_0^\infty [(a^2 + \tau)(b^2 + \tau)\tau]^{-1/2} d\tau \ldots [2]$$

$$A = \frac{3}{4} P_z (\vartheta_1 + \vartheta_2) \int_0^\infty (a^2 + \tau)^{-3/2} [(b^2 + \tau)\tau]^{-1/2} d\tau .. [3]$$

$$B = \frac{3}{4} P_z (\vartheta_1 + \vartheta_2) \int_0^\infty (b^2 + \tau)^{-3/2} [(a^2 + \tau)\tau]^{-1/2} d\tau .. [4]$$

where p is the normal pressure on the contact surface, a and b are the principal semiaxes of the elliptic boundary of the contact surface, α is the relative approach of the two bodies, A and B depend only the shapes and relative orientation of the two bodies in the unstressed state and ϑ_1 and ϑ_2 are elastic constants:

$$\vartheta_1 = \frac{\lambda_1 + 2\mu_1}{4\pi\mu_1(\lambda_1 + \mu_1)} = \frac{1 - \sigma_1}{2\pi\mu_1} \ldots\ldots\ldots\ldots [5]$$

$$\vartheta_2 = \frac{\lambda_2 + 2\mu_2}{4\pi\mu_2(\lambda_2 + \mu_2)} = \frac{1 - \sigma_2}{2\pi\mu_2} \ldots\ldots\ldots\ldots [6]$$

where λ and μ are Lamé's constants and σ is Poisson's ratio, the subscripts referring to the two bodies. Equations [3] and [4] determine a and b, while Equation [2] determines α when a and b are known.

To obtain the normal compliance

$$C_z = \frac{d\alpha}{dP_z} \ldots\ldots\ldots\ldots\ldots\ldots\ldots [7]$$

in a form suitable for quantitative comparison with the tangential compliance, let

$$\tau = a^2 \tan^2 \varphi$$

whereby Equations [2] to [4] are transformed to

$$2b\alpha = 3P_z(\vartheta_1 + \vartheta_2) \int_0^{\frac{\pi}{2}} a^{-1}\Phi^{-1/2} d\varphi \ldots\ldots\ldots [8]$$

$$2a^2bA = 3P_z(\vartheta_1 + \vartheta_2) \int_0^{\frac{\pi}{2}} a^{-1}\Phi^{-1/2} \cos^2\varphi \, d\varphi \ldots\ldots [9]$$

$$2b^3B = 3P_z(\vartheta_1 + \vartheta_2) \int_0^{\frac{\pi}{2}} a^{-3}\Phi^{-3/2} \cos^2\varphi \, d\varphi \ldots. [10]$$

where

$$\Phi = a^{-2}\cos^2\varphi + b^{-2}\sin^2\varphi \ldots\ldots\ldots\ldots [11]$$

The integrals in Equations [8] to [10] are complete elliptic integrals and are functions of a/b only. But a/b is a function of A/B only, as may be seen by dividing Equation [9] by Equation [10]. Since A/B is independent of P_z, each of the integrals is independent of P_z. The calculation of the compliance is now straightforward (remembering that b in Equation [8] is, according to Equation [10], proportional to the cube root of P_z) with the result

$$C_z = (\vartheta_1 + \vartheta_2)\, a^{-1}b^{-1} \int_0^{\frac{\pi}{2}} \Phi^{-1/2} d\varphi \ldots\ldots\ldots [12]$$

Tangential Force: Boundary Conditions

We shall consider, first, the effect of the tangential force. Taking the axes of x and y to coincide with the principal axes of the elliptic boundary of the contact surface, we resolve the tangential force into components P_x and P_y parallel to x and y, respectively. Without loss of generality, we may consider the effect of P_x alone.

To simplify the establishment of the boundary conditions we impose, temporarily, the following restrictions on the shape, orientation and elastic properties of the unstressed bodies:

1 If R_1, R_1' are, respectively, the maximum and minimum radii of curvature of the unstressed surface, at O, of one body and R_2, R_2' the corresponding radii of the other, then $R_1 = R_2$, $R_1' = R_2'$. (The radii are positive when the centers of curvature are inside the respective bodies.)

2 The normal sections of the two surfaces at O, containing the radii R_1 and R_2, coincide.

3 The two bodies have the same elastic properties.

With these restrictions the contact surface, under the action of P_z alone, is the portion of a plane bounded by an ellipse. We shall designate the contact surface by the symbol S_1 when its elliptic boundary curve is not included and by $\bar{S}_1$, when the boundary curve is included.

The system composed of the two bodies is geometrically and elastically symmetrical with respect to each of the three coordinate planes. To this system is applied the force system P_x, which is symmetrical with respect to the x, z-plane and antisymmetrical with respect to the x, y, and y, z-planes.

If u, v, w are the x, y, z components of displacement due to P_x and if we set (as we may, by Saint Venant's principle)

$$\lim_{x,y,z\to\infty} (u, v, w) = 0 \ldots\ldots\ldots\ldots\ldots [13]$$

then, in consequence of the foregoing symmetries, we must have

$$u(x, y, z) = -u(x, y, -z) + \delta_x \ldots\ldots\ldots [14]$$

$$v(x, y, z) = -v(x, y, -z) \ldots\ldots\ldots\ldots\ldots [15]$$

$$w(x, y, z) = w(x, y, -z) \ldots\ldots\ldots\ldots\ldots [16]$$

where δ_x is a component of rigid-body displacement, the only one of the six components permissible in view of Equation [13], the x, z-symmetry and the y, z-antisymmetry.

We now assume that there is no slip between the two bodies; i.e., u and v are continuous across $\bar{S}_1$. Setting $z = 0$ in Equations [14] and [15], we find

$$u = \delta_x/2,\ v = 0 \text{ on } \bar{S}_1 \ldots\ldots\ldots\ldots\ldots [17]$$

The third boundary condition on $\bar{S}_1$ is found by examining the expression, in terms of displacements, of the normal component of traction, due to P_x, exerted on the plane surface of the body lying in the region $z \geq 0$:

$$Z_\nu = -\lambda\left(\frac{\partial u}{\partial x} + \frac{\partial v}{\partial y}\right) - (\lambda + 2\mu)\frac{\partial w}{\partial z} \ldots\ldots [18]$$

Using Equation [17], Equation [18] reduces to

$$Z_\nu = -(\lambda + 2\mu)\frac{\partial w}{\partial z} \text{ on } \bar{S}_1 \ldots\ldots\ldots\ldots [19]$$

We now suppose that the two bodies remain in contact and in equilibrium across the contact surface. Then w and Z_ν are continuous across $\bar{S}_1$. The latter requires $\partial w/\partial z$ to be continuous across $\bar{S}_1$, by Equation [19]. Since w and $\partial w/\partial z$ are continuous across $\bar{S}_1$ and, by Equation [16], w is an even function of z, it is necessary that $\partial w/\partial z = 0$ on $\bar{S}_1$. Hence, by Equation [19],

$$Z_\nu = 0 \text{ on } \bar{S}_1 \ldots\ldots\ldots\ldots\ldots\ldots [20]$$

Considering the body lying in the region $z \geq 0$, the part of its boundary outside and in the neighborhood of $\bar{S}_1$ is approximated, as in the Hertz theory, by the plane $z = 0$. This part of the boundary, designated as S_2, is required to be free of traction. We can now assemble the complete boundary conditions for the body $z \geqslant 0$:

$$u = \delta_x/2, \; v = Z_\nu = 0 \text{ on } \bar{S}_1 \quad [21]$$

$$X_\nu = Y_\nu = Z_\nu = 0 \text{ on } S_2 \quad [22]$$

$$\lim_{x,y,z\to\infty} (u, v, w) = 0 \quad [23]$$

where

$$X_\nu = -\mu\left(\frac{\partial u}{\partial z} + \frac{\partial w}{\partial x}\right)_{z=0}, \quad Y_\nu = -\mu\left(\frac{\partial v}{\partial z} + \frac{\partial w}{\partial y}\right)_{z=0} \quad [24]$$

It may be seen that the contact surface, as regards displacements in the x, y-plane, shifts uniformly in the x-direction, without change of size or shape. This shift is proportional to P_x since the differential equations of the system are linear. Hence, if there is no slip on the contact surface, the tangential compliance of the two bodies is simply

$$C_x = \delta_x/P_x \quad [25]$$

and the problem of finding it is the problem of finding the value of δ_x corresponding to a given P_x. In the case of slip (see section entitled "Influence of Slip"), the displacement is not simply proportional to P_x because, although the differential equations are linear, the boundary conditions vary with P_x.

We now remove the restrictions imposed on the shapes, relative orientation and elastic properties of the bodies. Elimination of restrictions 1 and 2 introduces a slight warping of the contact surface under P_x alone. Within the limits of the small strain, small rotation theory, this warping may be ignored, as in the Hertz theory. Removal of restriction 3 makes the conditions of continuity of w and vanishing of Z_ν across $\bar{S}_1$ incompatible. In general, to insure continuity of w across the contact surface, there is required a component Z_ν on $\bar{S}_1$. This corresponds to the components X_ν and Y_ν which are ignored in the Hertz theory in the case of bodies of unlike elasticities. We shall adopt the analogous expedient here by retaining the condition $Z_\nu = 0$ on $\bar{S}_1$.

When the elastic properties of the two bodies are different, the relative displacement of $\bar{S}_1$ with respect to a distant point in the body $z > 0$ is not the same as the relative displacement with respect to a distant point in the body $z < 0$. Designating the former displacement by $\delta_x'/2$ and the latter by $\delta_x''/2$, the tangential compliances of the two bodies individually are

$$C_x' = \frac{\delta_x'}{2P_x}, \quad C_x'' = \frac{\delta_x''}{2P_x} \quad [26]$$

and the tangential compliance of the two bodies together is

$$C_x = \frac{\delta_x' + \delta_x''}{2P_x} \quad [27]$$

Method of Solution

In the case of the tangential force, we require the solution of the three equations of equilibrium

$$\mu\nabla^2(u, v, w) + (\lambda + \mu)\left(\frac{\partial}{\partial x}, \frac{\partial}{\partial y}, \frac{\partial}{\partial z}\right)\left(\frac{\partial u}{\partial x} + \frac{\partial v}{\partial y} + \frac{\partial w}{\partial z}\right) = 0 \quad [28]$$

satisfying the boundary conditions given in Equations [21] to [23].

It is convenient to express the components of displacement in terms of the potential functions of Boussinesq and Cerruti in the following two forms:[3]

$$u = \frac{1}{2\pi\mu}\frac{\partial^2 F_1}{\partial z^2} + \frac{\lambda}{4\pi\mu(\lambda+\mu)}\frac{\partial}{\partial x}\left(\frac{\partial F_1}{\partial x} + \frac{\partial G_1}{\partial y}\right) - \frac{1}{4\pi\mu} z \frac{\partial^2}{\partial x \partial z}\left(\frac{\partial F_1}{\partial x} + \frac{\partial G_1}{\partial y}\right) \quad [29]$$

$$v = \frac{1}{2\pi\mu}\frac{\partial^2 G_1}{\partial z^2} + \frac{\lambda}{4\pi\mu(\lambda+\mu)}\frac{\partial}{\partial y}\left(\frac{\partial F_1}{\partial x} + \frac{\partial G_1}{\partial y}\right) - \frac{1}{4\pi\mu} z \frac{\partial^2}{\partial y \partial z}\left(\frac{\partial F_1}{\partial x} + \frac{\partial G_1}{\partial y}\right) \quad [30]$$

$$w = \frac{1}{4\pi(\lambda+\mu)}\frac{\partial}{\partial z}\left(\frac{\partial F_1}{\partial x} + \frac{\partial G_1}{\partial y}\right) - \frac{1}{4\pi\mu} z \frac{\partial^2}{\partial z^2}\left(\frac{\partial F_1}{\partial x} + \frac{\partial G_1}{\partial y}\right) \quad [31]$$

$$F_1 = \iint X_\nu(\xi, \eta)\Omega d\xi d\eta, \quad G_1 = \iint Y_\nu(\xi, \eta)\Omega d\xi d\eta \quad [32]$$

$$\Omega = z \log(r + z) - r \quad [33]$$

$$r^2 = (\xi - x)^2 + (\eta - y)^2 + z^2 \quad [34]$$

$$\nabla^2 F_1 = 0, \; \nabla^2 G_1 = 0 \quad [35]$$

$$u = -\frac{1}{2\pi}\frac{\partial L}{\partial z} + \frac{\lambda+\mu}{2\pi(\lambda+2\mu)} z \frac{\partial}{\partial x}\left(\frac{\partial L}{\partial x} + \frac{\partial M}{\partial y}\right) \quad [36]$$

$$v = -\frac{1}{2\pi}\frac{\partial M}{\partial z} + \frac{\lambda+\mu}{2\pi(\lambda+2\mu)} z \frac{\partial}{\partial y}\left(\frac{\partial L}{\partial x} + \frac{\partial M}{\partial y}\right) \quad [37]$$

$$w = -\frac{\mu}{2\pi(\lambda+2\mu)}\left(\frac{\partial L}{\partial x} + \frac{\partial M}{\partial y}\right) + \frac{\lambda+\mu}{2\pi(\lambda+2\mu)} z \frac{\partial}{\partial z}\left(\frac{\partial L}{\partial x} + \frac{\partial M}{\partial y}\right) \quad [38]$$

$$L = \iint \frac{u(\xi, \eta)}{r} d\xi d\eta, \quad M = \iint \frac{v(\xi, \eta)}{r} d\xi d\eta \quad [39]$$

$$\nabla^2 L = 0, \; \nabla^2 M = 0 \quad [40]$$

In either form, the equations of equilibrium are satisfied and five of the nine boundary conditions are satisfied, namely, those expressed by Equation [23] and the last of Equations [21] and [22]. If either X_ν, Y_ν or u, v were specified over the whole boundary $z = 0$, the solution of the problem would reduce to the evaluation of the integrals in Equations [32] or [39]. However, in our case, u, v are specified over part of the boundary and X_ν, Y_ν over the remainder.

The four boundary conditions which are yet to be satisfied are expressed in terms of the functions L and M as follows:

$$u = -\frac{1}{2\pi}\frac{\partial L}{\partial z} = \frac{\delta_x'}{2} \text{ on } \bar{S}_1 \quad [41]$$

$$v = -\frac{1}{2\pi}\frac{\partial M}{\partial z} = 0 \text{ on } \bar{S}_1 \quad [42]$$

$$X_\nu = \frac{\mu}{2\pi}\frac{\partial^2 L}{\partial z^2} - \frac{\mu\lambda}{2\pi(\lambda+2\mu)}\frac{\partial}{\partial x}\left(\frac{\partial L}{\partial x} + \frac{\partial M}{\partial y}\right) = 0 \text{ on } S_2 \quad [43]$$

$$Y_\nu = \frac{\mu}{2\pi}\frac{\partial^2 M}{\partial z^2} - \frac{\mu\lambda}{2\pi(\lambda+2\mu)}\frac{\partial}{\partial y}\left(\frac{\partial L}{\partial x} + \frac{\partial M}{\partial y}\right) = 0 \text{ on } S_2 \quad [44]$$

We note that, if Poisson's ratio is zero (so that $\lambda = 0$), the boundary conditions reduce to

[3] Reference 2, pp. 242 and 244.

$$\frac{\partial L_0}{\partial z} = -\pi\delta_x', \quad \frac{\partial M_0}{\partial z} = 0 \text{ on } \bar{S}_1 \qquad [45]$$

$$\frac{\partial^2 L_0}{\partial z^2} = 0, \quad \frac{\partial^2 M_0}{\partial z^2} = 0 \text{ on } S_2 \qquad [46]$$

where L_0 and M_0 are the values of L and M for $\lambda = 0$. Hence, $M_0 = 0$ and the problem of finding $\partial L_0/\partial z$ is the same as that of finding the velocity potential of the irrotational flow of a perfect fluid through a circular or elliptic aperture in an infinite plane; or, it is same as the problem of finding the distribution of electric charge on a circular or elliptic disk whose potential is held constant. The solutions of these problems are known, so that we can find L_0.

To find L and M for an arbitrary value of Poisson's ratio, we assume, tentatively,

$$\frac{\partial L}{\partial x} + \frac{\partial M}{\partial y} = C_1 \frac{\partial L_0}{\partial x} \qquad [47]$$

$$\frac{\partial M}{\partial x} - \frac{\partial L}{\partial y} = -C_2 \frac{\partial L_0}{\partial y} \qquad [48]$$

where C_1 and C_2 are constants which are equal when λ and M are zero, i.e., the tentative solution reduces to $L = L_0$ when Poisson's ratio is zero.

From Equations [47] and [48] we find, using $\nabla^2 L = 0$,

$$-\frac{\partial^2 L}{\partial z^2} = C_1 \frac{\partial^2 L_0}{\partial x^2} + C_2 \frac{\partial^2 L_0}{\partial y^2} \qquad [49]$$

$$-\frac{\partial^2 M}{\partial z^2} = (C_1 - C_2) \frac{\partial^2 L_0}{\partial x \partial y} \qquad [50]$$

Substituting Equations [47] and [49] in Equation [43], there results, on $z = 0$,

$$X_\nu = -\frac{C_1\mu(\lambda + \mu)}{\pi(\lambda + 2\mu)} \frac{\partial^2 L_0}{\partial x^2} - \frac{C_2\mu}{2\pi} \frac{\partial^2 L_0}{\partial y^2} \qquad [51]$$

If we set

$$C_2 = \frac{2(\lambda + \mu)}{(\lambda + 2\mu)} C_1 \qquad [52]$$

Equation [51] reduces (using $\nabla^2 L_0 = 0$) to

$$X_\nu = \frac{C_1\mu(\lambda + \mu)}{\pi(\lambda + 2\mu)} \frac{\partial^2 L_0}{\partial z^2} \qquad [53]$$

whence, by Equation [46], the boundary condition expressed by Equation [43] is satisfied.

Again, substituting Equations [47] and [50] in Equation [44] and using Equation [52], we find

$$Y_\nu = 0 \qquad [54]$$

i.e., Y_ν vanishes over the entire boundary $z = 0$.

To verify the assumed form of solution, it remains to ascertain that the boundary conditions expressed by Equations [41] and [42] are satisfied. These become, using Equations [49], [50], [52], [40], [45],

$$\frac{C_1(3\lambda + 4\mu)}{4\pi(\lambda + 2\mu)} \frac{\partial L_0}{\partial z} + \frac{C_1\lambda}{4\pi(\lambda + 2\mu)} \int \left(\frac{\partial^2 L_0}{\partial x^2} - \frac{\partial^2 L_0}{\partial y^2}\right) dz = -\frac{\delta_x'}{2} \qquad [55]$$

$$\frac{C_1\lambda}{2\pi(\lambda + 2\mu)} \int \frac{\partial^2 L_0}{\partial x \partial y} dz = 0 \qquad [56]$$

Since $\partial L_0/\partial z$ is a constant on $\bar{S}_1$, and since the solution must hold for $0 \leqslant \lambda \leqslant \infty$, we require

$$\int \left(\frac{\partial^2 L_0}{\partial x^2} - \frac{\partial^2 L_0}{\partial y^2}\right) dz = \text{const on } \bar{S}_1 \qquad [57]$$

$$\int \frac{\partial^2 L_0}{\partial x \partial y} dz = 0 \text{ on } \bar{S}_1 \qquad [58]$$

There is an alternative method of testing the displacement boundary conditions. In view of the result $Y_\nu = 0$ over the entire plane $z = 0$, we conclude that, if the assumptions embodied in Equations [47] and [48] are indeed valid, the displacements must be expressible in terms of the function F_1 alone, i.e., we should set $G_1 = 0$ and F_1 proportional to

$$\int\int \frac{\partial^2 L_0}{\partial z^2} \Omega d\xi d\eta \qquad [59]$$

Then, using Equations [29, 30, 35], the displacement boundary conditions become

$$\frac{3\lambda + 4\mu}{8\pi\mu(\lambda + \mu)} \frac{\partial^2 F_1}{\partial z^2} + \frac{\lambda}{8\pi\mu(\lambda + \mu)} \left(\frac{\partial^2 F_1}{\partial x^2} - \frac{\partial^2 F_1}{\partial y^2}\right) = \frac{\delta_x'}{2} \text{ on } \bar{S}_1 \qquad [60]$$

$$\frac{\lambda}{4\pi\mu(\lambda + \mu)} \frac{\partial^2 F_1}{\partial x \partial y} = 0 \text{ on } \bar{S}_1 \qquad [61]$$

Since the boundary conditions must hold for $0 \leqslant \lambda \leqslant \infty$, we require

$$\frac{\partial^2 F_1}{\partial x^2} - \frac{\partial^2 F_1}{\partial y^2} = \text{const on } \bar{S}_1 \qquad [62]$$

$$\frac{\partial^2 F_1}{\partial x \partial y} = 0 \text{ on } \bar{S}_1 \qquad [63]$$

$$\frac{\partial^2 F_1}{\partial z^2} = \text{const on } \bar{S}_1 \qquad [64]$$

The method outlined in this section will be applied, first, to the special case of the circular contact area because of its simplicity.

Tangential Force, Circular Contact Area, $\lambda = 0$

Let the boundary of the contact surface be a circle of radius a and let $\rho^2 = x^2 + y^2$. For the case $\lambda = 0$, we must find the harmonic function L_0 satisfying the boundary conditions (see Equations [45, 46])

$$\frac{\partial L_0}{\partial z} = -\pi\delta_x', \ \rho \leqslant a \qquad [65]$$

$$\frac{\partial^2 L_0}{\partial z^2} = 0, \ \rho > a \qquad [66]$$

From the analogous problem in hydrodynamics, we have[4]

$$\frac{\partial L_0}{\partial z} = -2\delta_x' \int_0^\infty e^{-kz} J_0(k\rho)k^{-1} \sin ka \, dk \qquad [67]$$

where J_0 is the zero order Bessel function of the first kind. Hence

$$L_0 = 2\delta_x' \int_0^\infty e^{-kz} J_0(k\rho)k^{-2} \sin ka \, dk \qquad [68]$$

Tangential Force, Circular Contact Area, $\lambda \neq 0$

To remove the restriction $\lambda = 0$, we apply the first procedure

[4] "Hydrodynamics," by H. Lamb, sixth edition, Cambridge University Press, 1932, p. 137.

of the section entitled "Method of Solution." We find, from Equation [68],

$$\int\left(\frac{\partial^2 L_0}{\partial x^2}-\frac{\partial^2 L_0}{\partial y^2}\right)dz = -2\delta_x' \cos 2\theta \int_0^\infty e^{-kz} J_2(k\rho)k^{-1}\sin ka\, dk \quad [69]$$

$$\int \frac{\partial^2 L_0}{\partial x \partial y} dz = -\delta_x' \sin 2\theta \int_0^\infty e^{-kz} J_2(k\rho)k^{-1}\sin ka\, dk \quad [70]$$

where $\theta = \tan^{-1}(y/x)$. Now[5]

$$\int_0^\infty J_2(k\rho)k^{-1}\sin ka\, dk = \begin{Bmatrix} 0, \ \rho \leqslant a, \\ a(\rho^2-a^2)^{1/2}/\rho^2, \ \rho \geqslant a \end{Bmatrix} \quad [71]$$

Hence Equations [57] and [58] are satisfied.

To evaluate the constant C_1, we substitute Equations [65, 69, 71] in Equation [55], with the result

$$C_1 = \frac{2(\lambda+2\mu)}{(3\lambda+4\mu)} \quad [72]$$

The tangential traction over the contact surface is found by inserting the values of C_1 and L_0 from Equations [72] and [68] into Equation [53]. Noting that

$$\left.\frac{\partial^2 L_0}{\partial z^2}\right]_{z=0} = 2\delta_x' \int_0^\infty J_0(k\rho)\sin ka\, dk = \begin{Bmatrix} 2\delta_x'(a^2-\rho^2)^{-1/2}, \ \rho < a \\ \infty, \ \rho = a \\ 0, \ \rho > a \end{Bmatrix} \quad [73]$$

we find

$$X_\nu = \frac{4\mu(\lambda+\mu)\delta_x'}{\pi(3\lambda+4\mu)}(a^2-\rho^2)^{-1/2}, \ \rho < a \quad [74]$$

The relation between the tangential force and the displacement is found from

$$P_x = \int_0^a \int_0^{2\pi} X_\nu \rho\, d\rho\, d\theta = \frac{8\mu(\lambda+\mu)a\delta_x'}{3\lambda+4\mu} \quad [75]$$

so that

$$X_\nu = \frac{P_x}{2\pi a}(a^2-\rho^2)^{-1/2}, \ \rho < a \quad [76]$$

We see that, in the case of a tangential force across a circular contact area, the tangential traction is everywhere parallel to the direction of the applied force. Contours of constant tangential traction are concentric circles. The magnitude of the traction rises from one-half the average at the center to infinity at the edge.

The tangential compliance of one body is

$$C_x' = \frac{\delta_x'}{2P_x} = \frac{3\lambda+4\mu}{16\mu(\lambda+\mu)a} = \frac{2-\sigma}{8\mu a} \quad [77]$$

The tangential compliance of two bodies, with unlike elastic constants, is

$$C_x = \frac{1}{8a}\left(\frac{2-\sigma_1}{\mu_1}+\frac{2-\sigma_2}{\mu_2}\right) \quad [78]$$

It is interesting to notice that all of the results of this section can be extracted from a solution by H. Neuber[6] for the stress concentration at a hyperbolic notch.

Influence of Slip

A practical feature of the foregoing solution which requires additional study is the infinite traction at the edge of the contact surface. Presumably, the tangential component of traction cannot exceed the product of the coefficient of friction f and the normal component of traction p. Thus, for the circular contact area, we should have, from Equation [1],

$$X_\nu = \frac{3fP_z}{2\pi a^2}\left(1-\frac{\rho^2}{a^2}\right)^{1/2}, \ a' \leqslant \rho \leqslant a \quad [79]$$

$$X_\nu < \frac{3fP_z}{2\pi a^2}\left(1-\frac{\rho^2}{a^2}\right)^{1/2}, \ \rho \leqslant a' \quad [80]$$

$$Y_\nu = 0 \quad [81]$$

where we have assumed that slip has penetrated from the outer radius a to an inner radius a' and that, in the included annulus, the tangential traction remains at its greatest possible value fp.

If slip has just progressed over the entire contact surface, so that Equation [79] holds throughout, we may calculate the corresponding tangential displacements by substituting Equation [79] into Equation [32], and the combination into Equations [29] and [30], with the results

$$8\pi^2\mu a^2 u = 3fP_z \int_0^{\omega_0}\int_0^{2\pi}(2-\sigma+\sigma\cos 2\varphi)\Lambda_0\, d\omega\, d\varphi \quad [82]$$

$$8\pi^2\mu a^2 v = 3\sigma f P_z \int_0^{\omega_0}\int_0^{2\pi}\Lambda_0 \sin 2\varphi\, d\omega\, d\varphi \quad [83]$$

where $\Lambda_0 = (-\Phi_0\omega^2 - 2\Psi_0\omega - \Gamma_0)^{1/2}$ and

$$\omega = [(\xi-x)^2+(\eta-y)^2]^{1/2} = -\Omega_{z=0} \quad [84]$$

$$\varphi = \tan^{-1}\frac{\eta-y}{\xi-x} \quad [85]$$

$$\Phi_0 = a^{-2} \quad [86]$$

$$\Psi_0 = (x\cos\varphi + y\sin\varphi)a^{-2} \quad [87]$$

$$\Gamma_0 = (\rho^2-a^2)a^{-2} \quad [88]$$

and ω_0 is the distance from the point (x, y) to the circle, i.e.,

$$\omega_0 = \left(-\Psi_0+\sqrt{\Psi_0^2-\Phi_0\Gamma_0}\right)\Phi_0^{-1} \quad [89]$$

Considering the surface integral in Equation [83], and integrating with respect to ω, we arrive at

$$\int^{2\pi}\frac{(\Psi_0^2-\Phi_0\Gamma_0)\pi}{4\Phi_0^{3/2}}\sin 2\varphi\, d\varphi - \int_0^{2\pi}\left[\frac{\Psi_0(-\Gamma_0)^{1/2}}{2\Phi_0}\right]\sin 2\varphi\, d\varphi - \int_0^{2\pi}\left[\frac{\Psi_0^2-\Phi_0\Gamma_0}{2\Phi_0^{3/2}}\sin^{-1}\frac{\Psi_0}{\sqrt{\Psi_0^2-\Phi_0\Gamma_0}}\right]\sin 2\varphi\, d\varphi$$

We note that the algebraic signs of the quantities in brackets change if we substitute φ-π for φ. Hence, the integrals of the bracketed terms, over the interval 2π, vanish. The same terms occur in Equation [82]. The remaining integrations in Equations [82] and [83] are straightforward, and we find

[5] "Treatise on the Theory of Bessel Functions," by G. N. Watson, Cambridge University Press, 1922, p. 405.

[6] "Kerbspannungslehre," by H. Neuber, Julius Springer, Berlin, Germany, 1937, pp. 85–90.

$$64\,\mu a^3 u = 3fP_z[2(2-\sigma)(2a^2-\rho^2) + \sigma\rho^2 \cos 2\theta] \quad [90]$$

$$64\,\mu a^3 v = 3fP_z\,\sigma\rho^2 \sin 2\theta \quad [91]$$

If we now consider a distribution of traction

$$X_\nu = -\frac{3fP_zQ}{2\pi a'^2}\left(1-\frac{\rho^2}{a'^2}\right)^{1/2}, \quad \rho < a' \quad [92]$$

$$Y_\nu = 0 \quad [93]$$

where Q is a constant, we find corresponding displacements u' and v':

$$64\mu a'^3 u' = -3QfP_z[2(2-\sigma)(2a'^2-\rho^2) + \sigma\rho^2\cos 2\theta] \quad [94]$$

$$64\,\mu a'^3 v' = -3QfP_z\,\sigma\rho^2 \sin 2\theta \quad [95]$$

If, now, we set $Q = a'^3/a^3$ and add the two sets of displacements we have

$$u + u' = \frac{3fP_z(2-\sigma)}{16\,\mu a}\left(1-\frac{a'^2}{a^2}\right), \quad \rho < a' \quad [96]$$

$$v + v' = 0, \ \rho < a' \quad [97]$$

Hence, the system of surface tractions[7]

$$X_\nu = \frac{3fP_z}{2\pi a^2}\left(1-\frac{\rho^2}{a^2}\right)^{1/2}, \quad a' < \rho < a \quad [98]$$

$$X_\nu = \frac{3fP_z}{2\pi a^2}\left(1-\frac{\rho^2}{a^2}\right)^{1/2} - \frac{3fP_z a'}{2\pi a^3}\left(1-\frac{\rho^2}{a'^2}\right)^{1/2}, \quad \rho < a' \quad [99]$$

produces displacements which satisfy the conditions of no slip within the circle of radius a'. To find a' we calculate

$$P_x = \int_0^a \int_0^{2\pi} X_\nu \rho \, d\rho \, d\theta$$

$$= fP_z\left(1-\frac{a'^3}{a^3}\right) \quad [100]$$

whence

$$a' = a\left(1-\frac{P_x}{fP_z}\right)^{1/3} \quad [101]$$

The distribution of traction, represented by Equations [98] and [99], is illustrated in Fig. 2 for the case $P_x/fP_z = 0.3$.

To find the compliance, first eliminate a' between Equations [101] and [96] and obtain

$$u + u' = \frac{3fP_z(2-\sigma)}{16\mu a}\left[1-\left(1-\frac{P_x}{fP_z}\right)^{2/3}\right] \quad [102]$$

Then the tangential compliance of one body is

$$C_s' = \frac{d(u+u')}{dP_x} = \frac{2-\sigma}{8\mu a}\left(1-\frac{P_x}{fP_z}\right)^{-1/3} \quad [103]$$

This is to be compared the with compliance without slip given in Equation [77]. The ratio

$$\frac{C_x'}{C_s'} = \left(1-\frac{P_x}{fP_z}\right)^{-1/3} \quad [104]$$

is plotted in Fig. 3, which shows the range over which the compliance calculated on the assumption of no slip is useful. For small ratios of P_x to fP_z the error is small. The remainder of this paper is confined to the case of no slip.

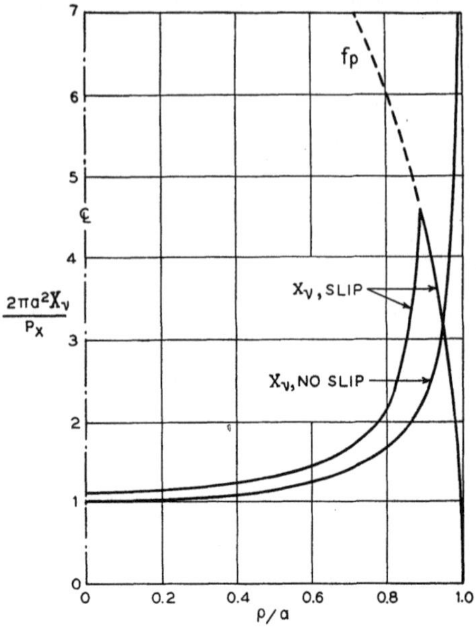

FIG. 2 DISTRIBUTION OF TANGENTIAL TRACTION X_ν DUE TO A TANGENTIAL FORCE P_x ACROSS A CIRCULAR CONTACT AREA (Curve marked "X_ν, SLIP" is for $P_x/fP_z = 0.3$.)

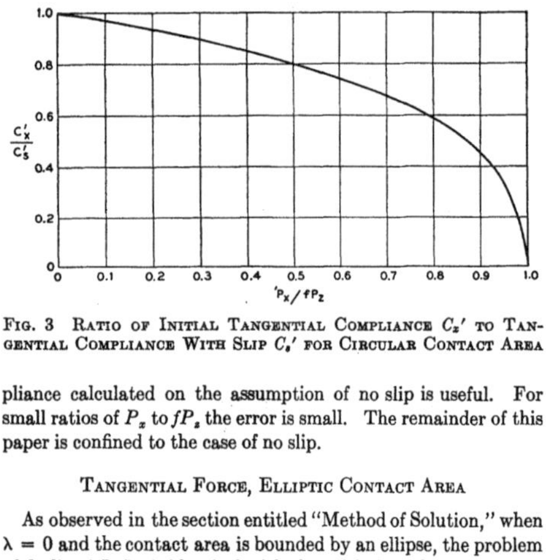

FIG. 3 RATIO OF INITIAL TANGENTIAL COMPLIANCE C_x' TO TANGENTIAL COMPLIANCE WITH SLIP C_s' FOR CIRCULAR CONTACT AREA

TANGENTIAL FORCE, ELLIPTIC CONTACT AREA

As observed in the section entitled "Method of Solution," when $\lambda = 0$ and the contact area is bounded by an ellipse, the problem of finding $\partial L_0/\partial z$ is identical with that of finding the velocity potential of the irrotational flow of a perfect fluid through an elliptic aperture in an infinite plane. From the solution[8] of that problem we need only the result that $\partial^2 L_0/\partial z^2$ is proportional to

$$(1 - a^{-2}x^2 - b^{-2}y^2)^{-1/2} \quad [105]$$

on $\bar{S}_1$ and vanishes on S_2.

Applying the alternative procedure for $\lambda \neq 0$, described at the

[7] Dr. Stewart Way has called the author's attention to a paper by C. Cattaneo, "Sul contatto di due corpi elastici: distribuzione locale degli sforzi," Accademia dei Lincei, Rendiconti, series 6, vol. XXVII, 1938 pp. 342–348, 434–436, 474–478, in which formulas, corresponding to Equations [98] and [99], are derived for the elliptic contact surface.

[8] Reference 4, p. 151.

end of the section entitled "Method of Solution," we take

$$X_\nu = D(1 - a^{-2}x^2 - b^{-2}y^2)^{-1/2}, \quad Y_\nu = 0, \text{ on } \bar{S}_1 \ldots [106]$$

$$X_\nu = Y_\nu = 0, \text{ on } S_2 \ldots\ldots [107]$$

where D is a constant, determined by the condition

$$P_x = \int\int_{\bar{S}_1} X_\nu \, dx \, dy \ldots\ldots [108]$$

To evaluate the integral in Equation [108] let

$$x = \rho \cos\theta, \quad y = \rho \sin\theta \ldots\ldots [109]$$

Then

$$P_x = D \int_0^{\rho_1} \int_0^{2\pi} \frac{\rho \, d\rho \, d\theta}{(1 - \Theta \rho^2)^{1/2}} \ldots\ldots [110]$$

where

$$\Theta = a^{-2}\cos^2\theta + b^{-2}\sin^2\theta \ldots\ldots [111]$$

and ρ_1 is the distance from the origin of x and y to a point on the ellipse, i.e.,

$$\rho_1 = \Theta^{-1/2} \ldots\ldots [112]$$

The integration then proceeds without difficulty, with the result

$$P_x = 2\pi abD \ldots\ldots [113]$$

Subject to verification of the displacement boundary conditions, we conclude that the tangential traction on the contact surface is everywhere parallel to the applied tangential force. Contours of constant tangential traction are homothetic ellipses, i.e., geometrically similar, and with principal axes parallel to the elliptic boundary. The magnitude of the traction rises from one half the average at the center to infinity at the edge.

It remains to test the displacement boundary conditions, as given in Equations [62] to [64], and to determine δ_x'. To do this we have to find the values of the second derivatives of

$$F_1 = \int\int_{\bar{S}_1} X_\nu(\xi, \eta)\Omega \, d\xi \, d\eta \ldots\ldots [114]$$

on $\bar{S}_1$.

As before, let

$$\omega = [(\xi - x)^2 + (\eta - y)^2]^{1/2} = \Omega_{z=0} \ldots\ldots [115]$$

$$\xi - x = \omega \cos\varphi \ldots\ldots [116]$$

$$\eta - y = \omega \sin\varphi \ldots\ldots [117]$$

Then

$$X_\nu = D(-\Phi\omega^2 - 2\Psi\omega - \Gamma)^{-1/2} \ldots\ldots [118]$$

where

$$\Phi = a^{-2}\cos^2\varphi + b^{-2}\sin^2\varphi \ldots\ldots [119]$$

$$\Psi = a^{-2}x\cos\varphi + b^{-2}y\sin\varphi \ldots\ldots [120]$$

$$\Gamma = a^{-2}x^2 + b^{-2}y^2 - 1 \ldots\ldots [121]$$

Accordingly, on $\bar{S}_1$,

$$\frac{\partial^2 F_1}{\partial x^2} - \frac{\partial^2 F_1}{\partial y^2} = D \int_0^{\omega_1} \int_0^{2\pi} \frac{\cos 2\varphi \, d\omega d\varphi}{(-\Phi\omega^2 - 2\Psi\omega - \Gamma)^{1/2}} \ldots [122]$$

$$\frac{\partial^2 F_1}{\partial x \partial y} = \frac{D}{2} \int_0^{\omega_1} \int_0^{2\pi} \frac{\sin 2\varphi \, d\omega d\varphi}{(-\Phi\omega^2 - 2\Psi\omega - \Gamma)^{1/2}} \ldots [123]$$

$$\frac{\partial^2 F_1}{\partial z^2} = D \int_0^{\omega_1} \int_0^{2\pi} \frac{d\omega d\varphi}{(-\Phi\omega^2 - 2\Psi\omega - \Gamma)^{1/2}} \ldots [124]$$

where ω_1 is the distance from the point (x, y) to a point on the ellipse. This distance is expressed in terms of φ by noting that, on the ellipse,

$$a^{-2}\xi^2 + b^{-2}\eta^2 - 1 = 0 \ldots\ldots [125]$$

or

$$\Phi\omega^2 + 2\Psi\omega + \Gamma = 0 \ldots\ldots [126]$$

The significant root of Equation [126] is

$$\omega_1 = (-\Psi + \sqrt{\Psi^2 - \Phi\Gamma})\Phi^{-1} \ldots\ldots [127]$$

Returning to Equation [122] and performing the integration with respect to ω, we find

$$\begin{aligned} \frac{\partial^2 F_1}{\partial x^2} - \frac{\partial^2 F_1}{\partial y^2} &= D \int_0^{2\pi} \left[\Phi^{-1/2} \sin^{-1} \frac{\Phi\omega + \Psi}{\sqrt{\Psi^2 - \Phi\Gamma}} \right]_0^{\omega_1} \cos 2\varphi \, d\varphi \\ &= \frac{\pi D}{2} \int_0^{2\pi} \Phi^{-1/2} \cos 2\varphi \, d\varphi \\ &\quad - D \int_0^{2\pi} \Phi^{-1/2} \cos 2\varphi \sin^{-1} \frac{\Psi}{\sqrt{\Psi^2 - \Phi\Gamma}} \, d\varphi. \quad [128] \end{aligned}$$

In the second integral in Equation [128], we note that

$$\Phi(\varphi) = \Phi(\varphi + \pi)$$

$$\Psi(\varphi) = -\Psi(\varphi + \pi)$$

$$\cos 2\varphi = \cos 2(\varphi + \pi)$$

and Γ is independent of φ. Hence, the integrand reverses sign on substituting $\varphi + \pi$ for φ. Accordingly, the second integral of Equation [128] vanishes and we have

$$\left.\begin{aligned} \frac{\partial^2 F_1}{\partial x^2} - \frac{\partial^2 F_1}{\partial y^2} &= \frac{\pi D}{2} \int_0^{2\pi} \Phi^{-1/2} \cos 2\varphi \, d\varphi \\ &= 2\pi D \int_0^{\frac{\pi}{2}} \Phi^{-1/2} \cos 2\varphi \, d\varphi \\ &= \text{const} \end{aligned}\right\} \ldots [129]$$

so that Equation [62] is satisfied.

Similarly, from Equation [123],

$$\frac{\partial^2 F_1}{\partial x \partial y} = \frac{\pi D}{4} \int_0^{2\pi} \Phi^{-1/2} \sin 2\varphi \, d\varphi - \frac{D}{2} \int_0^{2\pi} \Phi^{-1/2} \sin 2\varphi \sin^{-1} \frac{\Psi}{\sqrt{\Psi^2 - \Phi\Gamma}} \, d\varphi \ldots [130]$$

Since $\sin 2\varphi = \sin 2(\varphi + \pi)$ the second integral in Equation [130] may be shown to vanish for the same reason that the second integral in Equation [128] vanished. Regarding the first integral in Equation [130], calling the integrand $g(\varphi)$, we note

$$g(\varphi) = g(\varphi + \pi)$$

and

$$\begin{aligned} \int_0^{2\pi} g(\varphi) d\varphi &= 2 \int_0^{\pi} g(\varphi) \, d\varphi \\ &= 2 \int_0^{\frac{\pi}{2}} g(\varphi) \, d\varphi + 2 \int_{\frac{\pi}{2}}^{\pi} g(\varphi) \, d\varphi \end{aligned}$$

But

$$\sin 2\varphi = -\sin 2(\pi - \varphi)$$

$$\Phi(\varphi) = \Phi(\pi - \varphi)$$

Hence

$$\int_{\frac{\pi}{2}}^{\pi} g(\varphi)d\varphi = -\int_0^{\frac{\pi}{2}} g(\varphi)d\varphi$$

and, therefore,

$$\int_0^{2\pi} g(\varphi)d\varphi = 0$$

Thus

$$\frac{\partial^2 F_1}{\partial x \partial y} = 0$$

on $\bar{S}_1$ and Equation [63] is satisfied.

Finally

$$\left.\begin{aligned}\frac{\partial^2 F_1}{\partial z^2} &= \frac{\pi D}{2}\int_0^{2\pi}\Phi^{-1/2}\,d\varphi \\ &\quad - D\int_0^{2\pi}\Phi^{-1/2}\sin^{-1}\frac{\Psi}{\sqrt{\Psi^2-\Phi\Gamma}}\,d\varphi \\ &= 2\pi D\int_0^{\frac{\pi}{2}}\Phi^{-1/2}\,d\varphi \\ &= \text{const}\end{aligned}\right\} \quad [131]$$

so that the last displacement boundary condition, given in Equation [64], is satisfied.

To find the displacement δ_x' due to the force P_x we substitute the results expressed by Equations [129] and [131] into Equation [60] and use Equation [113] to obtain

$$2\pi ab\mu_1\delta_x' = P_x(2-\sigma_1)\int_0^{\frac{\pi}{2}}\Phi^{-1/2}\,d\varphi + P_x\sigma_1\int_0^{\frac{\pi}{2}}\Phi^{-1/2}\cos 2\varphi\;d\varphi \quad [132]$$

in which μ has been replaced by μ_1 and λ by $2\mu_1\sigma_1/(1-2\sigma_1)$.

Formulas for Tangential Compliance

To find numerical values of the tangential compliance from Equation [132], it is advisable to transform the integrals in that equation to forms that are tabulated. For the first integral we write

$$\int_0^{\frac{\pi}{2}}\Phi^{-1/2}\,d\varphi = \int_0^{\frac{\pi}{2}}\frac{d\varphi}{\sqrt{a^{-2}\cos^2\varphi + b^{-2}\sin^2\varphi}}$$

$$= \left\{\begin{array}{l} a\displaystyle\int_0^{\frac{\pi}{2}}\frac{d\varphi}{\sqrt{1-k^2\sin^2\varphi}},\ a<b \\ a\pi/2,\ a=b \\ b\displaystyle\int_0^{\frac{\pi}{2}}\frac{d\varphi}{\sqrt{1-k_1^2\sin^2\varphi}},\ a>b\end{array}\right\} \quad [133]$$

where

$$k^2 = 1-\frac{a^2}{b^2},\quad k_1^2 = 1-\frac{b^2}{a^2} \quad [134]$$

Hence

$$\int_0^{\frac{\pi}{2}}\Phi^{-1/2}\,d\varphi = \left\{\begin{array}{l} a\mathbf{K},\ a<b \\ a\,\pi/2,\ a=b \\ b\mathbf{K}_1,\ a>b\end{array}\right\} \quad [135]$$

where $\mathbf{K}$ and $\mathbf{K}_1$ are the complete elliptic integrals of the first kind of arguments $(\pi/2, k)$ and $(\pi/2, k_1)$ respectively.

For the second integral, considering first the case $a < b$, we write

$$\int_0^{\frac{\pi}{2}}\Phi^{-1/2}\cos 2\varphi\,d\varphi = a\int_0^{\frac{\pi}{2}}\frac{(1-2\sin^2\varphi)d\varphi}{\sqrt{1-k^2\sin^2\varphi}} = a\mathbf{K} - 2a\int_0^{\frac{\pi}{2}}\frac{\sin^2\varphi d\varphi}{\sqrt{1-k^2\sin^2\varphi}}$$

Now

$$\int_0^{\frac{\pi}{2}}\frac{\sin^2\varphi\,d\varphi}{\sqrt{1-k^2\sin^2\varphi}} = \frac{1}{k^2}\int_0^{\frac{\pi}{2}}\frac{d\varphi}{\sqrt{1-k^2\sin^2\varphi}} - \frac{1}{k^2}\int_0^{\frac{\pi}{2}}\sqrt{1-k^2\sin^2\varphi}\,d\varphi = k^{-2}(\mathbf{K}-\mathbf{E})$$

where $\mathbf{E}$ is the complete elliptic integral of the second kind, of argument $(\pi/2, k)$. Hence

$$\int_0^{\frac{\pi}{2}}\Phi^{-1/2}\cos 2\varphi\,d\varphi = -\frac{a}{k}\left[\left(\frac{2}{k}-k\right)\mathbf{K} - \frac{2}{k}\mathbf{E}\right],\ a<b \quad [136]$$

The expression in brackets in Equation [136] appears in electromagnetic theory as a quantity proportional to the mutual inductance of a pair of coaxial circular currents. An elaborate tabulation of the quantity

$$\mathbf{N} = 4\pi\left[\left(\frac{2}{k}-k\right)\mathbf{K} - \frac{2}{k}\mathbf{E}\right] \quad [137]$$

has been made by Nagaoka and Sakurai.[9]

For the case $a > b$, we replace φ by $\pi/2 - \varphi$ in the second integral of Equation [132] and find, by a similar procedure

$$\int_0^{\frac{\pi}{2}}\Phi^{-1/2}\cos 2\varphi\,d\varphi = \frac{b}{k_1}\left[\left(\frac{2}{k_1}-k_1\right)\mathbf{K}_1 - \frac{2}{k_1}\mathbf{E}_1\right],\ a>b \quad [138]$$

where $\mathbf{E}_1$ is the complete elliptic integral of the second kind of argument $(\pi/2, k_1)$.

The difference in algebraic sign between Equation [138] and Equation [136] arises from the fact that

$$\cos 2\varphi = -\cos 2\left(\frac{\pi}{2}-\varphi\right)$$

We find, then,

$$\int_0^{\frac{\pi}{2}}\Phi^{-1/2}\cos 2\varphi\,d\varphi = \left\{\begin{array}{l} -a\mathbf{N}/4\pi k,\ a<b \\ 0,\ a=b \\ b\,\mathbf{N}_1/4\pi k_1,\ a>b\end{array}\right\} \quad [139]$$

[9] "Table No. 2," by H. Nagaoka and S. Sakurai, Scientific Papers of the Institute of Physical and Chemical Research, Tokyo, Sept., 1927. See also Table No. 1, Dec., 1922, for $\mathbf{K}$ and $\mathbf{E}$.

Inserting the expressions given in Equations [136] and [139] into Equation [132], we find for $a < b$:

$$\frac{2\mu_1 b\delta_x'}{P_x} = \frac{2-\sigma_1}{2}\left[\frac{2\mathsf{K}}{\pi} - \frac{\sigma_1}{2\pi^2(2-\sigma_1)}\frac{\mathsf{N}}{k}\right] \ldots\ldots [140]$$

for $a = b$:

$$\frac{2\mu_1 a\delta_x'}{P_x} = \frac{2-\sigma_1}{2} \ldots\ldots\ldots\ldots\ldots\ldots [141]$$

and for $a > b$:

$$\frac{2\mu_1 a\delta_x'}{P_x} = \frac{2-\sigma_1}{2}\left[\frac{2\mathsf{K}_1}{\pi} + \frac{\sigma_1}{2\pi^2(2-\sigma_1)}\frac{\mathsf{N}_1}{k_1}\right] \ldots\ldots [142]$$

The last three equations express the ratio of the tangential compliance to the tangential compliance of a body with Poisson's ratio zero and a circular contact area of diameter equal to the maximum diameter of the elliptic contact area. This ratio is plotted as the ordinate in Fig. 4, using Nagaoka and Sakurai's tables.

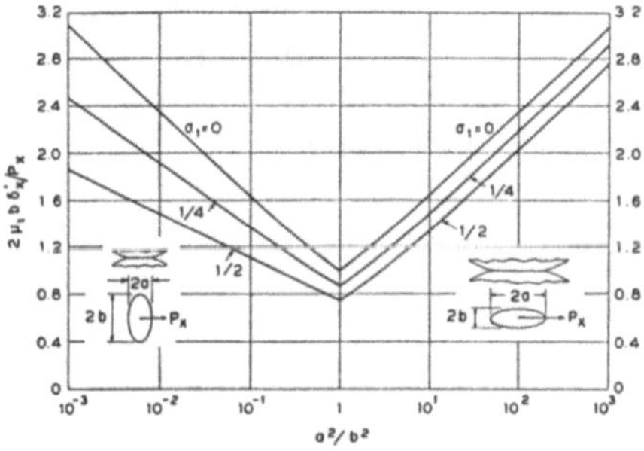

a, b = principal semi-axes of elliptic contact surface
P_x = force in direction of axis a
δ_x' = relative displacement in direction a
μ_1 = modulus of rigidity
σ_1 = Poisson's ratio

FIG. 4 INITIAL TANGENTIAL COMPLIANCE OF ELASTIC BODIES IN CONTACT

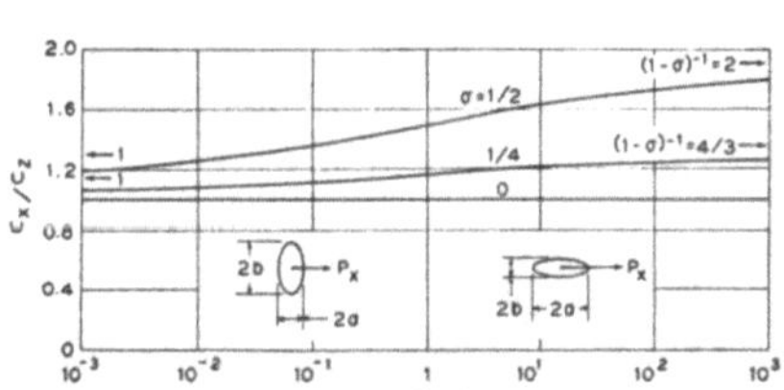

FIG. 5 RATIO OF INITIAL TANGENTIAL COMPLIANCE C_x TO NORMAL COMPLIANCE C_z OF BODIES WITH LIKE ELASTIC CONSTANTS

It may be seen that the tangential compliance decreases as Poisson's ratio increases. When $\sigma = 0$, the tangential compliance is isotropic, but when $0 < \sigma \leqslant {}^1/_2$ the compliance in the direction of the major axis is greater than the compliance in the direction of the minor axis. Hence, if the tangential force is not parallel to a principal axis of the ellipse, the displacement is parallel to the force only if $\sigma = 0$.

RATIO OF TANGENTIAL TO NORMAL COMPLIANCE

The interesting features of the ratio of the tangential to the normal compliance are illustrated adequately by examining the case in which the two bodies have the same elastic properties. Then, from Equation [12],

$$C_z = \begin{Bmatrix} (1-\sigma)\,\mathsf{K}/\pi\mu b, & a < b \\ (1-\sigma)/2\mu a, & a = b \\ (1-\sigma)\,\mathsf{K}_1/\pi\mu a, & a > b \end{Bmatrix} \ldots\ldots\ldots [143]$$

Hence, from Equations [143] and [140] to [142],

$$\frac{C_x}{C_z} = \begin{cases} \dfrac{\pi(2-\sigma)}{4(1-\sigma)}\cdot\dfrac{1}{\mathsf{K}}\left[\dfrac{2\mathsf{K}}{\pi} - \dfrac{\sigma}{2\pi^2(2-\sigma)}\dfrac{\mathsf{N}}{k}\right], & a < b \\ \dfrac{2-\sigma}{2(1-\sigma)}, & a = b \\ \dfrac{\pi(2-\sigma)}{4(1-\sigma)}\cdot\dfrac{1}{\mathsf{K}_1}\left[\dfrac{2\mathsf{K}_1}{\pi} + \dfrac{\sigma}{2\pi^2(2-\sigma)}\dfrac{\mathsf{N}_1}{k_1}\right], & a > b \end{cases}$$

To plot this ratio it is only necessary to divide the ordinates in Fig. 4 by $2(1-\sigma)\mathsf{K}/\pi$, for $a < b$ and by $2(1-\sigma)\,\mathsf{K}_1/\pi$ for $a > b$. The results are plotted in Fig. 5. Several limiting cases of interest are as follows:

$$\text{For } \sigma = 0,\ \frac{C_x}{C_z} = 1$$

$$\text{For } \frac{a}{b} \to 0,\ \frac{C_x}{C_z} = 1$$

$$\text{For } \frac{a}{b} \to \infty,\ \frac{C_x}{C_z} = \frac{1}{1-\sigma}$$

$$\text{For } \frac{a}{b} = 1,\ \frac{C_x}{C_z} = \frac{2-\sigma}{2(1-\sigma)}$$

It may be seen that the tangential compliance, over the practical range $0 < \sigma < {}^1/_2$, is always greater than the normal compliance, but never more than twice as great as long as there is no slip.

TWISTING COUPLE: BOUNDARY CONDITIONS

By considerations of symmetry and continuity similar to those in the section entitled "Tangential Force: Boundary Conditions," it may be shown that, in the case of a moment about the z-axis, no normal component of traction is induced on the contact surface and the surface rotates about the z-axis without distortion in the x, y-plane. The boundary conditions for the case of twist are, therefore,

$$u = \beta y/2,\ v = -\beta x/2,\ Z_\nu = 0 \text{ on } \bar{S}_1 \ldots\ldots [144]$$

$$X_\nu = Y_\nu = Z_\nu = 0 \text{ on } S_2 \ldots\ldots\ldots\ldots [145]$$

$$\lim_{x,y,z\to\infty} (u, v, w) = 0 \ldots\ldots\ldots\ldots\ldots\ldots [146]$$

where $\beta/2$ is the angle of rotation of the contact surface with respect to a distant point in the body $z \gg 0$. For the circular contact area, these boundary conditions are satisfied by a limiting case of a solution by Neuber[10] for the stress concentration at a hyperbolic groove in a twisted shaft. In that solution the traction across the contact surface is found to be

$$X_\nu = \frac{3M_z y}{4\pi a^4}\left(1-\frac{\rho^2}{a^2}\right)^{-1/2},\ Y_\nu = -\frac{3M_z x}{4\pi a^4}\left(1-\frac{\rho^2}{a^2}\right)^{-1/2} \ldots [147]$$

and the compliance of a single body is

[10] See ref. 6, pp. 90–92. See also "Forced Torsional Oscillations of an Elastic Half-Space," by E. Reissner and H. F. Sagoci, *Journal of Applied Physics*, vol. 15, 1944, pp. 652–654.

$$\frac{\beta}{2M_z} = \frac{3}{16\mu a^3} \quad \text{[148]}$$

By analogy with our previous results, we consider the possibility that, in the case of the elliptic contact area, the traction may be

$$\left.\begin{aligned} X_\nu &= T'y(1 - a^{-2}x^2 - b^{-2}y^2)^{-1/2}, \\ Y_\nu &= -T''x(1 - a^{-2}x^2 - b^{-2}y^2)^{-1/2}, \\ Z_\nu &= 0 \end{aligned}\right\} \text{on } S_1 \quad \text{[149]}$$

$$X_\nu = Y_\nu = Z_\nu = 0 \text{ on } S_2 \quad \text{[150]}$$

where T' and T'' are constants. With these surface tractions, all of the boundary conditions, except possibly the first two of Equations [144], are satisfied. If the assumed tractions give the required values of u and v on $\bar{S}_1$, the assumptions expressed in Equations [149] are verified as correct.

Torsional Compliance

To calculate u and v, we substitute Equations [149] and [150] in Equations [29], [30], and [32] and perform the indicated operations. We find, on $\bar{S}_1$,

$$\mu u = y\left[T'a^{-2}\int_0^{\frac{\pi}{2}} \Phi^{-3/2}\cos^2\varphi d\varphi + \sigma(T''b^{-2} - T'a^{-2})\int_0^{\frac{\pi}{2}} \Phi^{-3/2}\sin^2\varphi\cos^2\varphi\, d\varphi\right] \quad \text{[151]}$$

$$\mu v = x\left[-T''b^{-2}\int_0^{\frac{\pi}{2}} \Phi^{-3/2}\sin^2\varphi\, d\varphi + \sigma(T''b^{-2} - T'a^{-2})\int_0^{\frac{\pi}{2}} \Phi^{-3/2}\sin^2\varphi\cos^2\varphi\, d\varphi\right] \quad \text{[152]}$$

These expressions have the correct form.

We note that

$$\int_0^{\frac{\pi}{2}} \Phi^{-3/2}\sin^2\varphi\, d\varphi = a^3[\mathsf{E} - (1 - k^2)\mathsf{K}]/[k^2(1 - k^2)]$$

$$\int_0^{\frac{\pi}{2}} \Phi^{-3/2}\cos^2\varphi\, d\varphi = a^3[\mathsf{K} - \mathsf{E}]/k^2$$

$$\int_0^{\frac{\pi}{2}} \Phi^{-3/2}\sin^2\varphi\cos^2\varphi\, d\varphi = a^3[(2 - k^2)\mathsf{K} - 2\mathsf{E}]/k^4$$

The functions

$$\mathsf{D} = (\mathsf{K} - \mathsf{E})/k^2$$
$$\mathsf{B} = [\mathsf{E} - (1 - k^2)\mathsf{K}]/k^2$$
$$\mathsf{C} = [(2 - k^2)\mathsf{K} - 2\mathsf{E}]/k^4$$

are tabulated by Jahnke and Emde.[11] Using the values of u and v from Equations [144], we solve Equations [151] and [152] for T' and T'' with the following results:

$$T' = \frac{\mu\beta}{2a}\frac{\mathsf{B} - 2\sigma(1 - k^2)\mathsf{C}}{\mathsf{BD} - \sigma\mathsf{CE}} \quad \text{[153]}$$

$$T'' = \frac{\mu\beta}{2a}\frac{\mathsf{D} - 2\sigma\mathsf{C}}{\mathsf{BD} - \sigma\mathsf{CE}} \quad \text{[154]}$$

To find the relation between β and M_z, we calculate

$$M_z = \int\int_{S_1} (X_\nu y - Y_\nu x)dx\,dy$$
$$= \frac{2}{3}\pi ab(T'b^2 + T''a^2)$$
$$= \frac{1}{3}\pi b^3\mu\beta\,\frac{\mathsf{E} - 4\sigma(1 - k^2)\mathsf{C}}{\mathsf{BD} - \sigma\mathsf{CE}} \quad \text{[155]}$$

Hence, the torsional compliance of one body is

$$\frac{\beta}{2M_z} = \frac{3}{16b^3\mu}\frac{8\{\mathsf{BD} - \sigma\mathsf{CE}\}}{\pi\{\mathsf{E} - 4\sigma(1 - k^2)\mathsf{C}\}} \quad \text{[156]}$$

Equation [156] is plotted in Fig. 6.

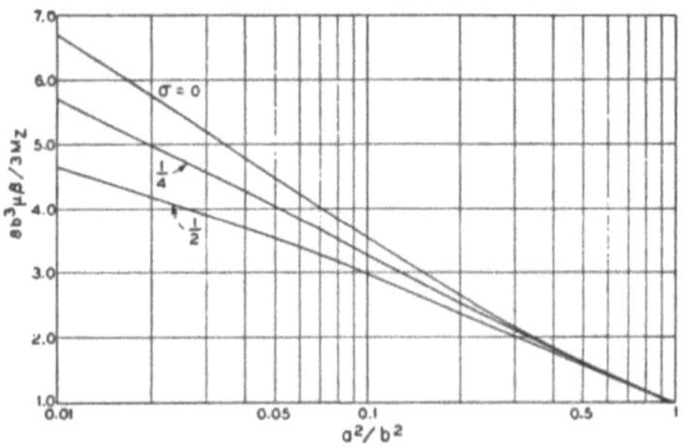

Fig. 6 Initial Torsional Compliance

Noting that

$$\mathsf{B}\left(\frac{\pi}{2}, 0\right) = \mathsf{D}\left(\frac{\pi}{2}, 0\right) = \pi/4, \quad \mathsf{C}\left(\frac{\pi}{2}, 0\right) = \pi/16$$

it may be verified that T', T'' and the compliance reduce to the appropriate values for the limiting case of the circular contact area.

[11] "Tables of Functions," by E. Jahnke and F. Emde, Dover Publications, New York, N. Y., 1943, pp. 82–83.

Paper No. 49—A-20

Thermoelastic Stress Around a Cylindrical Inclusion of Elliptic Cross Section

By RAYMOND D. MINDLIN[1] and HILDA L. COOPER,[2] MURRAY HILL, N. J.

The two-dimensional equations of thermoelasticity are solved for the case of a uniform temperature change of an infinite medium containing a cylindrical inclusion of elliptic cross section. The problem is treated as one of plane strain in elliptic co-ordinates, and the solution is given in closed form. Formulas and curves are given for the maximum values of various components of stress at the interface between the inclusion and the surrounding medium.

Introduction

If an elastic body containing an elastic inclusion is heated, uniformly or not, an elastic state of stress is induced if the inclusion has a coefficient of linear thermal expansion different from that of the surrounding material. The problem considered here is one of uniform heating, or cooling, of a body composed of an infinite elastic medium in which is embedded an infinitely long cylinder of elliptic cross section. The cylinder and the surrounding medium have not only different coefficients of thermal expansion, but also different constants of elasticity. The displacement and traction, across the interface between the two materials, are assumed to be continuous. The problem is treated as one of plane strain in elliptic co-ordinates.

Elliptic Co-Ordinates

We shall use elliptic co-ordinates ξ, η defined, in terms of rectangular co-ordinates x, y, by[3]

$$x + iy = c \cosh (\xi + i\eta) = c \cosh \zeta$$

so that c is one half the distance between the foci of the family of confocal ellipses ξ = const and the orthogonal family of confocal hyperbolas η = const, Fig. 1. The co-ordinate ξ is restricted to be positive with the same value assigned to the whole of one ellipse.

The elliptic boundary between the inner and outer bodies is taken as the curve $\xi = \xi_0$. Hence, if b and a are, respectively, the minor and major semiaxes of the boundary, their ratio is given by

$$\gamma = b/a = \tanh \xi_0$$

The stretch ratio J, and the angle, φ, which the tangent to η = const makes with the x axis, are given by $Je^{i\varphi} = d(x + iy)/d\zeta$,

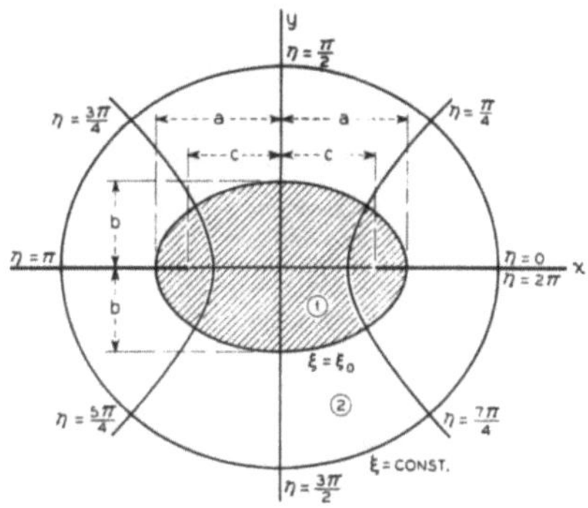

Fig. 1 Elliptic Co-Ordinates

so that

$$2J^2 = c^2 (\cosh 2\xi - \cos 2\eta)$$

$$\tan \varphi = \coth \xi \tan \eta$$

Formulas for Stress, Strain, and Displacement

The components of stress are expressed in terms of Airy's stress function χ, by the usual formulas[4]

$$\sigma_\xi = \frac{1}{J^2}\frac{\partial^2\chi}{\partial\eta^2} + \frac{1}{J^3}\left(\frac{\partial\chi}{\partial\xi}\frac{\partial J}{\partial\xi} - \frac{\partial\chi}{\partial\eta}\frac{\partial J}{\partial\eta}\right)$$

$$\sigma_\eta = \frac{1}{J^2}\frac{\partial^2\chi}{\partial\xi^2} + \frac{1}{J^3}\left(\frac{\partial\chi}{\partial\eta}\frac{\partial J}{\partial\eta} - \frac{\partial\chi}{\partial\xi}\frac{\partial J}{\partial\xi}\right)$$

$$\tau_{\xi\eta} = -\frac{1}{J^2}\frac{\partial^2\chi}{\partial\xi\partial\eta} + \frac{1}{J^3}\left(\frac{\partial\chi}{\partial\xi}\frac{\partial J}{\partial\eta} + \frac{\partial\chi}{\partial\eta}\frac{\partial J}{\partial\xi}\right)$$

so that

$$\sigma_\xi + \sigma_\eta = \frac{1}{J^2}\left(\frac{\partial^2\chi}{\partial\xi^2} + \frac{\partial^2\chi}{\partial\eta^2}\right) = \nabla^2\chi$$

and

$$\nabla^4\chi = 0 \qquad [1]$$

The stress-strain relation is expressed, in rectangular co-ordinates, by three equations of the type

$$E\epsilon_x = \sigma_x - \nu(\sigma_y + \sigma_z) + E\alpha T$$

and three of the type

$$G\gamma_{xy} = \tau_{xy}$$

where E is Young's modulus, G is the shear modulus, ν is Poisson's ratio, α is the coefficient of linear thermal expansion, and T

---

[1] Columbia University, New York, N. Y., and Consultant, Bell Telephone Laboratories. Mem. ASME.

[2] Bell Telephone Laboratories.

[3] "A Treatise on Photo-Elasticity," by E. G. Coker and L. N. G. Filon, Cambridge University Press, London, England, 1931, p. 159.

Contributed by the Applied Mechanics Division and presented at the Annual Meeting, New York, N. Y., November 27–December 2, 1949, of The American Society of Mechanical Engineers.

Discussion of this paper should be addressed to the Secretary, ASME, 29 West 39th Street, New York, N. Y., and will be accepted until one month after final publication of the paper itself in the Journal of Applied Mechanics.

[4] Ibid., p. 163.

is the temperature above a uniform initial temperature. For plane strain

$$\gamma_{yz} = \gamma_{xz} = 0$$

and we may take

$$\epsilon_z = k = \text{const}$$

Then

$$\sigma_z = \nu(\sigma_x + \sigma_y) - E\alpha T + Ek$$

and the stress-strain relation becomes, in curvilinear co-ordinates

$$2G\,\epsilon_\xi = (1-\nu)(\sigma_\xi + \sigma_\eta) - \sigma_\eta + E\alpha T - 2\nu Gk$$

$$2G\,\epsilon_\eta = (1-\nu)(\sigma_\xi + \sigma_\eta) - \sigma_\xi + E\alpha T - 2\nu Gk$$

$$\epsilon_z = k, \quad G\gamma_{\xi\eta} = \tau_{\xi\eta}$$

$$\gamma_{\eta z} = \gamma_{z\xi} = \tau_{\eta z} = \tau_{z\xi} = 0$$

The components of displacement, U_ξ, U_η, in the directions ξ increasing and η increasing, respectively, are given by

$$\left.\begin{aligned} 4GJU_\xi &= 2(1-\nu)Ju_\xi - 2\partial\chi/\partial\xi + c^2(E\alpha T - 2\nu Gk)\sinh 2\xi \\ 4GJU_\eta &= 2(1-\nu)Ju_\eta - 2\partial\chi/\partial t - c^2(E\alpha T - 2\nu Gk)\sin 2\eta \end{aligned}\right\} \quad \ldots\ldots\ldots [2]$$

where

$$u_\xi + iu_\eta = e^{i\varphi}\int Je^{i\varphi}(\nabla^2\chi + iR)d\zeta$$

and R is the conjugate of $\nabla^2\chi$. Equations [2] are the usual forms[5] with the addition of the terms containing $E\alpha T - 2\nu Gk$, which account for the thermal dilatation and axial strain.

We shall designate the portion within the elliptic boundary as Region 1 and the surrounding medium as Region 2. Quantities associated with Region 1 will be identified by primes, and those of Region 2 by double primes.

Boundary Conditions

In view of the assumptions of continuity of traction and displacement across the common boundary of Regions 1 and 2, we require

$$\sigma_\xi' = \sigma_\xi'',\ \tau_{\xi\eta}' = \tau_{\xi\eta}'' \text{ on } \xi = \xi_0 \quad \ldots\ldots\ldots\ldots [\text{I}]$$

$$U_\xi' = U_\xi'',\ U_\eta' = U_\eta'',\ U_z' = U_z'' \text{ on } \xi = \xi_0 \quad \ldots . [\text{II}]$$

At points in Region 2 far from its inner boundary, the stress must vanish, that is

$$\lim_{\xi\to\infty}(\sigma_\xi'', \sigma_\eta'', \sigma_z'', \tau_{\xi\eta}'') = 0 \quad \ldots\ldots\ldots\ldots [\text{III}]$$

If, as is likely in practice, the composite body is bounded by two traction-free planes z = const, the solution to be given should not be used if $E' >> E''$ and, in any case, not within a distance a from the intersections of the cylindrical surface $\xi = \xi_0$ with the two planes z = const. These restrictions arise from the assumption of plane strain in which we can require, at the most, on z = const

$$\int_0^{2\pi}\int_0^{\xi_0}\sigma_z' J^2 d\xi d\eta + \lim_{\xi_1\to\infty}\int_0^{2\pi}\int_{\xi_0}^{\xi_1}\sigma_z'' J^2 d\xi d\eta = 0 \quad ..[\text{IV}]$$

[5] Ibid., p. 166.

The third of Conditions [II] requires

$$\epsilon_z' = k' = \epsilon_z'' = k'' = k \text{ (say)}$$

and Condition [III] requires

$$k = \alpha'' T$$

If the components of stress and displacement are derived from stress functions which are solutions of Equation [1], and if Boundary Conditions [I, II, III] are satisfied, Condition [IV] will be satisfied identically.

Stress Function

The appropriate Airy functions, χ' for Region 1 and χ'' for Region 2, were selected from a list[6] of biharmonic functions in elliptic co-ordinates, leading to single-valued displacements. The possibility that the solution might be expressed in closed form was indicated by Donnell's solution[7] of a similar problem. The functions which generate the solution of the present problem are

$$\left.\begin{aligned} \chi' &= A\cosh 2\xi\cos 2\eta + B(\cosh 2\xi + \cos 2\eta) \\ \chi'' &= De^{-2\xi}\cos 2\eta + F\xi + H(e^{-2\xi} + \cos 2\eta) \end{aligned}\right\} \ldots [3]$$

where A, B, D, F, H are constants whose values are found, from the boundary conditions, to be

$$\left.\begin{aligned} A &= c^2\delta(1-\gamma^2)\gamma^{-1}K/4 = A_0c^2/4 \\ B &= [n\gamma - (1+\gamma)^2](1-\gamma^2)^{-1}A = B_0c^2/4 \\ D &= -2\gamma(1-\gamma)^{-2}A = D_0c^2/4 \\ F &= 2[n(1+\gamma^2) - 2(1+\gamma)^2]\gamma(1-\gamma^2)^{-2}A \\ H &= n\gamma(1-\gamma^2)^{-1}A = H_0c^2/4 \end{aligned}\right\} .. [4]$$

in which

$$K = \frac{(2m'+1)(m''+1)^2(2m''+1)^{-1}}{n[n(m'-1) - 4(m'+1) + 2(m''+1)] - 2(m''+1-n)\gamma^{-1}(1+\gamma)^2}$$

$$m' = \nu'/(1-\nu'), \quad m'' = \nu''/(1-\nu'')$$

$$n = (1 - G''/G')/(1-\nu''), \quad \delta = E''(\alpha'' - \alpha')T$$

This completes the formal solution of the problem as stated, and we proceed to the computation of the important components of stress.

Boundary Stress

On the boundary $\xi = \xi_0$, the components of stress (identified by overbars) are found to be

$$\left.\begin{aligned} \bar\sigma_\xi' &= \bar\sigma_\xi'' = B_0 + A_0\Phi \\ \bar\sigma_\eta' &= B_0 - A_0\Phi \\ \bar\sigma_\eta'' &= H_0(1-2\Psi) - D_0(\Phi + \Psi\cos 2\eta) \\ \bar\tau_{\xi\eta}' &= \tau_{\xi\eta}'' = A_0\Psi\sin 2\eta \\ \sigma_z' &= (E'/E'')\delta + 2\nu' B_0 \\ \bar\sigma_z'' &= 2\nu'' H_0(1-\Psi) \end{aligned}\right\} \ldots\ldots [5]$$

where

$$\Phi = \frac{1-\gamma^2-(1+\gamma^2)\cos 2\eta}{1+\gamma^2-(1-\gamma^2)\cos 2\eta}$$

$$\Psi = \frac{2\gamma}{1+\gamma^2-(1-\gamma^2)\cos 2\eta}$$

and A_0, B_0, D_0, H_0 are defined in Equations [4].

[6] Ibid., p. 536.

[7] "Stress Concentrations Due to Elliptical Discontinuities in Plates Under Edge Forces," by L. H. Donnell, Th. von Kármán Anniversary Volume, California Institute of Technology, 1941.

The principal stresses on $\xi = \xi_0$ are given by the formulas

$$p, q = \frac{1}{2}(\bar{\sigma}_\xi + \bar{\sigma}_\eta) \pm \frac{1}{2}[(\bar{\sigma}_\xi - \bar{\sigma}_\eta)^2 + 4\bar{\tau}_{\xi\eta}^2]^{1/2} \ldots\ldots [6]$$

where p is associated with the + sign and is, hence, always the algebraically larger of the two principal stresses. Both p and q are found to be independent of η within the boundary (Region 1) and dependent on η outside the boundary (Region 2).

Substituting the expressions for $\bar{\sigma}_\xi''$, $\bar{\sigma}_\eta''$, $\tau_{\xi\eta}''$ from Equations [5] into Equations [6] and setting the derivatives, with respect to η, of the resulting expressions for p'' and q'' equal to zero, we find that both p'' and q'' have stationary values at[8] $\eta = 0, \pi/2, \eta_0$ where

$$\cos 2\eta_0 = \frac{1}{1-\gamma^2}\left[1 + \gamma^2 - \frac{2n\gamma^2(2 \pm n^{1/2})}{(1+\gamma)^2 - n\gamma}\right] \ldots [7]$$

When using Equation [7] the following must be noted:

(a) It applies only if $n \geqslant 0$.

(b) It applies only if n and γ have such values that the right-hand side lies between −1 and +1.

(c) If $\delta > 0$, the plus sign applies to q and the minus sign to p.

(d) If $\delta < 0$, the plus sign applies to p and the minus sign to q.

The magnitudes of the stationary values of p'' and q'' are found by substituting $\eta = 0, \pi/2, \eta_0$ into Equations [6], with the following results

$$p''_{\eta=0,\,\delta>0} = -q''_{\eta=0,\,\delta<0} = K|\delta|[n - 2(1 + \gamma^{-1})]$$

$$q''_{\eta=0,\,\delta>0} = -p''_{\eta=0,\,\delta<0} = K|\delta|[n(1 - 2\gamma^{-1}) + 2(1 + \gamma^{-1})]$$

$$p''_{\eta=\pi/2,\,\delta>0} = -q''_{\eta=\pi/2,\,\delta<0} = K|\delta|\,[n - 2(1 + \gamma)]$$

$$q''_{\eta=\pi/2,\,\delta>0} = -p''_{\eta=\pi/2,\,\delta<0} = K|\delta|\,[n(1 - 2\gamma) + 2(1 + \gamma)]$$

$$p''_{\eta=\eta_0,\,\delta>0} = -q''_{\eta=\eta_0,\,\delta<0} = K|\delta|\left[n - \frac{2[(1+\gamma)^2 - n\gamma]}{(2 - n^{1/2})\gamma n^{1/2}}\right]$$

$$q''_{\eta=\eta_0,\,\delta>0} = -p''_{\eta=\eta_0,\,\delta<0} = K|\delta|\left[n + \frac{2[(1+\gamma)^2 - n\gamma]}{(2 + n^{1/2})\gamma n^{1/2}}\right]$$

where $|\delta|$ denotes the absolute value of δ.

Maximum Tension in Outer Region

Of special interest is the algebraically largest of the values of p'' given by the preceding formulas. This value, denoted by p_{max}'' is plotted in Fig. 2 for $\delta > 0$, and in Fig. 3 for $\delta < 0$, with $\nu' = \nu'' = 0.3$. It is found that p_{max}'' is always a tension, whereas the other stationary values of p'' may be positive or negative. When Equation [7] is applicable, p_{max}'' occurs at $\eta = \eta_0$, otherwise it occurs at $\eta = 0$. The dashed curve in each of Figs. 2 and 3 is the locus of combinations of b/a and E'/E'' for which p_{max}'' shifts from $\eta = 0$ to $\eta = \eta_0$. Below and to the left of the dashed curves, p_{max}'' occurs at $\eta = 0$; above and to the right of the dashed curves, p_{max}'' occurs at $\eta = \eta_0$. For $b/a = 0$, p_{max}'' occurs at $\eta = 0$ for all values of E'/E''.

For $\delta > 0$, the largest deviation of η_0 from zero is only about 0.03 radians and the largest deviation of p_{max}'' from $p''_{\eta=0,\,\delta>0}$ is less than 1 per cent. However, when $\delta < 0$, p_{max}'' may occur at any value of η, and may have values considerably larger (algebraically) than $p''_{\eta=0,\,\delta<0}$. In fact, when $\delta < 0$, both principal stresses at $\eta = 0$ may be compressive over a wide range of b/a and E'/E''; but, in such cases, p_{max}'' occurs at $\eta = \eta_0$ and is tensile. This is contrary to the usual rule that the most dangerous stress occurs where the boundary has the sharpest curvature.

[8] Because of symmetry it is necessary to consider only one quadrant of the x, y-plane.

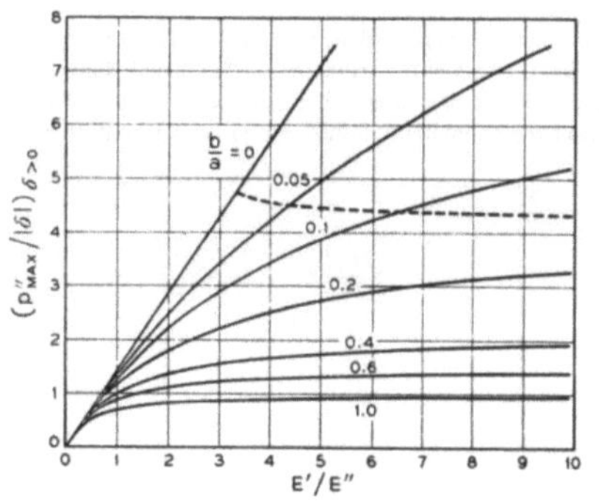

FIG. 2 Maximum Tension in Outer Medium at Interface When Inclusion Tends to Expand Less, or Contract More, Than Surrounding-Medium ($\delta > 0$)

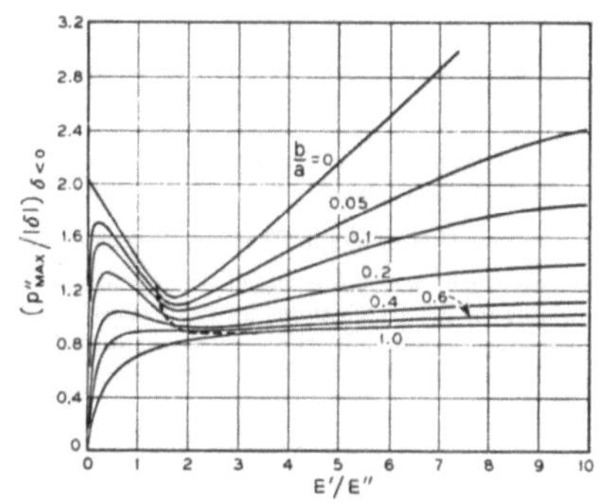

FIG. 3 Maximum Tension in Outer Medium at Interface When Inclusion Tends to Expand More, or Contract Less, Than Surrounding Medium ($\delta < 0$)

Principal Stress Difference in Outer Region

The difference between the principal stresses is of interest as it is proportional to one of the simplest measurements that can be made with the photoelastic technique. Stationary values of $p'' - q''$ on $\xi = \xi_0$ in Region 2 occur at $\eta = 0, \pi/2, \eta_1$, where

$$\cos 2\eta_1 = \frac{1}{1-\gamma^2}\left[1 + \gamma^2 - \frac{n\gamma^2(4 - n)}{(1+\gamma)^2 - n\gamma}\right]$$

The magnitudes of the stationary values of $p'' - q''$ are

$$(p'' - q'')_{\eta=0} = 2K\,|\delta|\,\gamma^{-1}\,[n - 2(1 + \gamma)]$$

$$(p'' - q'')_{\eta=\pi/2} = 2K\,|\delta|\,\gamma[n - 2(1 + \gamma^{-1})]$$

$$(p'' - q'')_{\eta=\eta_1} = 4K\,|\delta|\,\gamma^{-1}\,[(1+\gamma)^2 - n\gamma]\,[n(4 - n)]^{-1/2}$$

The algebraically largest value of these quantities, denoted by $(p'' - q'')_{max}$, is plotted in Fig. 4 for the case $\nu' = \nu'' = 0.3$. Here, again, the dashed curve is the locus of combinations of the parameters b/a and E'/E'' for which $(p'' - q'')_{max}$ shifts from $\eta = 0$ to $\eta = \eta_1$. To the left and below the dashed curve, $(p'' - q'')_{max}$ occurs at $\eta = 0$; to the right and above the dashed curve, $(p'' - q'')_{max}$ occurs at $\eta = \eta_1$.

It should be noticed that, for corresponding values of b/a and E'/E'', the sum of the ordinates in Figs. 2 and 3 gives the ordinate in Fig. 4 only when all three points fall below and to the left of the appropriate dashed curves, as it is only in such cases that the maximum values of the three stresses occur at the same point on the boundary (i.e., $\eta = 0$).

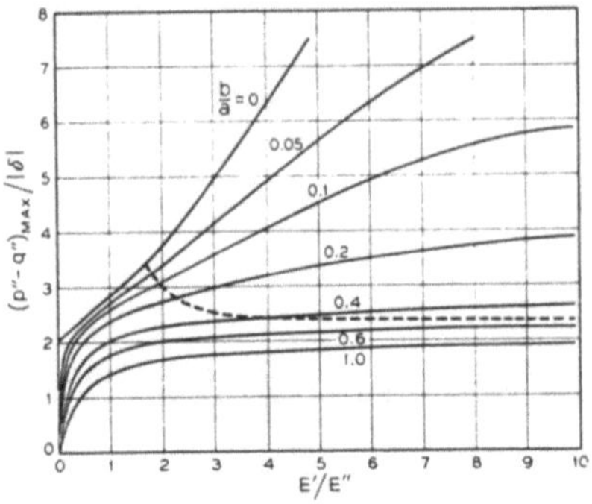

Fig. 4 Maximum Difference Between Principal Stresses in Outer Medium at Interface

Tangential Component of Traction Across Interface

The maximum value of the tangential component of the traction across the interface between the two regions is of interest in considering the possibility of slip. This quantity, denoted by $(\bar{\tau}_{\xi\eta})_{\max}$ and obtained from the fourth of Equations [5], is found to be

$$(\bar{\tau}_{\xi\eta})_{\max} = K\delta\gamma^{-1}(1 - \gamma)^2 \qquad [8]$$

and occurs at $\eta = \eta_2$ where

$$\cos 2\eta_2 = (1 - \gamma^2)(1 + \gamma^2)^{-1}$$

Equation [8] is plotted in Fig. 5 for $\nu' = \nu'' = 0.3$.

Stress in Inner Region

The state of stress in Region 1 is very simple. This may be recognized by transforming the first of Equations [3] to rectangular co-ordinates, obtaining

$$\chi' = A[2c^{-2}(x^2 - y^2) - 1] + 2Bc^{-2}(x^2 + y^2)$$

It is readily seen that the stress is homogeneous throughout Region 1, the principal stresses being parallel to the principal axes of the ellipse. When $\delta > 0$, the algebraically larger principal stress is a tension of magnitude

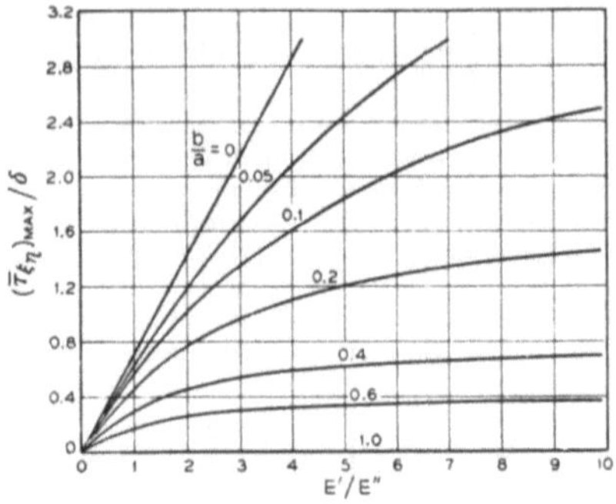

Fig. 5 Maximum Value of Tangential Component of Traction Across Interface

$$p'_{\delta>0} = K|\delta|[n - 2(1 + \gamma^{-1})] = p'_{\eta=0,\,\delta>0}$$

Hence for $\nu' = \nu'' = 0.3$, Fig. 2 gives $p'_{\delta>0}$ exactly to the left and below the dashed curve and to within 1 per cent over the remainder. When $\delta < 0$, the algebraically larger principal stress in Region 1 is a compression.

The principal stress difference throughout Region 1 is of course constant and equal to the maximum value of the tangential component of traction across the interface. Thus

$$p' - q' = 2(\bar{\tau}_{\xi\eta})_{\max}$$

where $(\bar{\tau}_{\xi\eta})_{\max}$ is given by Equation [8] and is plotted in Fig. 5 for $\nu' = \nu'' = 0.3$.

Axial Stress

The components of stress parallel to the axis of the cylinder are given by the last two of Equations [5]. The formula given for σ_z' is valid throughout Region 1 and, as may be seen, σ_z' is a constant. The formula given for $\bar{\sigma}_z''$ pertains only to the boundary $\xi = \xi_0$. The stationary values of $\bar{\sigma}_z''$ occur at $\eta = 0, \pi/2$, with magnitudes

$$(\bar{\sigma}_z'')_{\eta=0} = 2\nu''K\delta n(1 - \gamma^{-1})$$

$$(\bar{\sigma}_z'')_{\eta=\pi/2} = 2\nu''K\delta n(1 - \gamma)$$

Whereas, for the cylinder of circular cross section ($\gamma = 1$), σ_z is zero throughout, it may become very large at the ends of the major axis of an elongated elliptic cross section.

Reprinted from
Journal of Applied Mechanics
17 (1950) 265-268

Paper No. 49—A-19

Beam Vibrations With Time-Dependent Boundary Conditions

By R. D. MINDLIN[1] and L. E. GOODMAN,[2] NEW YORK, N. Y.

A procedure is described for extending the method of separation of variables to the solution of beam-vibration problems with time-dependent boundary conditions. The procedure is applicable to a wide variety of time-dependent boundary-value problems in systems governed by linear partial differential equations.

Introduction

It happens frequently that vibration problems have to deal with continuous systems of which one or more boundaries are constrained to undergo displacements or tractions which vary with time. Simple examples include a cantilevered beam whose clamped end experiences a lateral displacement pulse, a rod subjected at its ends to varying longitudinal pressure, or a plate whose edge is twisted by an oscillating edge moment. In other cases the time-dependence of the boundary condition is itself the object of investigation as when one structural element interacts with another, or when we wish to infer the state of strain at one end of a structure from continuous measurements made at another end. All of these problems are characterized by the fact that the boundary conditions are not "stationary" and on this account, solutions are not, in general, obtainable by the classical method of separation of variables, although solutions do appear in the literature for a number of special problems of this type.[3]

The present paper includes these solutions as special cases. In it is developed a general method by means of which the time-dependence is removed from the boundary conditions. The remaining problem then can be solved by any one of the classical or numerical methods available for handling the free- or forced-vibration problem. The method is here developed for and applied to the problem of the flexural vibrations of beams. It is equally applicable to time-dependent boundary-value problems in the flexural vibrations of plates, the torsional vibrations of shafts and, in fact, in a wide variety of systems governed by linear partial differential equations.

Problems of this type have been solved by the method of Laplace transform, but the present method employs more elementary mathematical techniques and is applicable in certain cases where the transforms do not exist or their inverses are not known.

Statement of Problem

Consider a structure whose displacement is governed by the equation

$$a^2 \frac{\partial^4 w}{\partial x^4} + \frac{\partial^2 w}{\partial t^2} = \frac{q(x)p(t)}{\rho A} \qquad [1]$$

This is the case if the structure is a prismatic beam executing flexural vibrations according to the simple Bernoulli-Euler or classical theory of flexure. Then the symbols have the meanings

- w = deflection of beam
- x = position along beam; $x = 0$ is one end of beam, $x = l$ is the other end
- ρ = density
- A = cross-sectional area of beam
- a^2 = $EI/\rho A$, where E and I are Young's modulus and the second moment of area of the cross section of the beam, respectively
- $q(x)p(t)$ = external load per unit length of beam. If the load does not vary with time, $p(t) = 1$

The method to be described applies equally well if the rotatory inertia, shear, damping, and elastic foundation terms are included in the differential equation, but they are omitted for simplicity.

In the following development, the symbol D_i is used to represent a linear differential operator of order 0, 1, 2, or 3, as the boundary conditions of the problem dictate, that is, $D_i[w]$ stands for w, or $\partial w/\partial x$, or $\partial^2 w/\partial x^2$, or $\partial^3 w/\partial x^3$, or a linear combination of these operations. With this notation the boundary conditions can be written

$$\left.\begin{aligned} D_i[w(0, t)] &= f_i(t), \quad i = 1,2 \\ D_i[w(l, t)] &= f_i(t), \quad i = 3,4 \end{aligned}\right\} \qquad [2]$$

For example, if the beam in question is a cantilever clamped at $x = 0$

$$D_1 = 1,\ D_2 = \partial/\partial x,\ D_3 = \partial^2/\partial x^2,\ D_4 = \partial^3/\partial x^3$$

while the displacement and slope at $x = 0$, and the moment and shear at $x = l$ may be required to vary with time according to $f_i(t)$. Linear combinations of the operators occur, for example, when the end of the beam is spring-restrained against displacement or rotation.

The initial conditions of the motion are specified by two arbitrary functions

$$\left.\begin{aligned} w(x, 0) &= w_0(x) \\ \left.\frac{\partial w}{\partial t}\right]_{t=0} &= \dot{w}_0(x) \end{aligned}\right\} \qquad [3]$$

A difficulty in solving problems of this type by the method of separation of variables arises when not all of the functions $f_i(t)$ vanish. The method of separation of variables breaks down when applied directly because it is not possible to satisfy Equations [2] by adjustment of the x-dependent function.

Method of Solution

The foregoing difficulty may be resolved by separating the solution into two parts, one of which is later adjusted so as to simplify the boundary conditions on the other. We take

[1] Professor of Civil Engineering, Columbia University, and Consultant, Bell Telephone Laboratories, Inc., Murray Hill, N. J. Mem. ASME.

[2] Instructor in Civil Engineering, Columbia University.

[3] See references (1) to (4). Numbers in parentheses refer to the Bibliography at the end of the paper.

Contributed by the Applied Mechanics Division and presented at the Annual Meeting, New York, N. Y., November 27–December 2, 1949, of The American Society of Mechanical Engineers.

Discussion of this paper should be addressed to the Secretary, ASME, 29 West 39th Street, New York, N. Y., and will be accepted until one month after final publication of the paper itself in the Journal of Applied Mechanics.

$$w = \zeta(x, t) + \sum_{i=1}^{4} f_i(t)g_i(x) \qquad [4]$$

Substituting Equation [4] into Equation [1] we find that ζ must satisfy the differential equation

$$a^2 \frac{\partial^4\zeta}{\partial x^4} + \frac{\partial^2\zeta}{\partial t^2} = \frac{q(x)p(t)}{\rho A} - \sum_{i=1}^{4} (a^2 f_i g_i^{\mathrm{IV}} + \ddot{f}_i g_i) \qquad [5]$$

where Roman superscripts indicate differentiations with respect to x and dots indicate differentiations with respect to time.

In addition, the assumed expression for w, Equation [4], must satisfy the boundary conditions, Equations [2]. Hence

$$\left.\begin{aligned} D_i[\zeta(0, t)] &= f_i(t) - \sum_{j=1}^{4} D_i[g_j(0)]f_j(t), \; i = 1,2 \\ D_i[\zeta(l, t)] &= f_i(t) - \sum_{j=1}^{4} D_i[g_j(l)]f_j(t), \; i = 3,4 \end{aligned}\right\} \qquad [6]$$

Finally, the initial conditions, Equations [3], become

$$\left.\begin{aligned} \zeta(x, 0) &= w_0 - \sum_{i=1}^{4} f_i(0)g_i(x) \\ \left.\frac{\partial\zeta}{\partial t}\right]_{t=0} &= \dot{w}_0 - \sum_{i=1}^{4} \dot{f}_i(0)g_i(x) \end{aligned}\right\} \qquad [7]$$

The functions $g_i(x)$ are now chosen so as to reduce to zero the right-hand sides of Equations [6]. To assure this, it is sufficient to satisfy the 16 conditions

$$\left.\begin{aligned} &D_1[g_1(0)] = 1 \qquad D_1[g_2(0)] = 0 \\ &D_2[g_1(0)] = 0 \qquad D_2[g_2(0)] = 1 \\ &D_3[g_1(l)] = 0 \qquad D_3[g_2(l)] = 0 \\ &D_4[g_1(l)] = 0 \qquad D_4[g_2(l)] = 0 \\ &D_1[g_3(0)] = 0 \qquad D_1[g_4(0)] = 0 \\ &D_2[g_3(0)] = 0 \qquad D_2[g_4(0)] = 0 \\ &D_3[g_3(l)] = 1 \qquad D_3[g_4(l)] = 0 \\ &D_4[g_3(l)] = 0 \qquad D_4[g_4(l)] = 1 \end{aligned}\right\} \qquad [8]$$

or, in more abbreviated notation

$$\left.\begin{aligned} D_j[g_i(0)] &= \delta_{ij}, \; j = 1,2, \; i = 1,2,3,4 \\ D_j[g_i(l)] &= \delta_{ij}, \; j = 3,4, \; i = 1,2,3,4 \end{aligned}\right\} \qquad [9]$$

where $\delta_{ij} = 0$ for $i \neq j$ and $\delta_{ij} = 1$ for $i = j$. Each column of Equations [8] provides four conditions on one of the four functions g_i. In order to be certain of being able to satisfy these conditions in all cases the g_i are taken to be polynomials of the fifth degree in x

$$g_i = a_i + b_i x + c_i x^2 + d_i x^3 + e_i x^4 + f_i x^5, \; i = 1,2,3,4 \qquad [10]$$

choosing the coefficients according to the following rule:

"Substitute each of the g_i in the appropriate (ith) column of Equations [8]. In each case, there will result a set of four linear algebraic equations governing the coefficients of the particular function g_i. If more than four of the constants $a_i \ldots f_i$ appear in these expressions, reduce to four the number which do appear by setting equal to zero the coefficient of the term of highest degree in x and also, if necessary, the coefficient of the term of second highest degree. If any of the constants $a_i \ldots f_i$ does not appear, set it equal to zero." In this way the four conditions on the coefficients of each of the g_i will always be expressed by means of four linear, independent, algebraic equations in four unknowns. These equations determine the constants of Equations [10] so as to satisfy Equations [8]. With this choice of the functions g_i, the boundary conditions expressed by Equations [6] become those of a "stationary" problem

$$\left.\begin{aligned} D_i[\zeta(0, t)] &= 0 \qquad i = 1,2 \\ D_i[\zeta(l, t)] &= 0 \qquad i = 3,4 \end{aligned}\right\} \qquad [11]$$

Usually only a third-degree polynomial is required for the functions g_i. The additional terms are included to accommodate exceptional cases as, for example, when time-dependent moment and shear are prescribed at both ends of the beam. It should be noticed, also, that it is necessary only to compute those of the g_i for which the corresponding $f_i(t)$ do not vanish.

It remains to find the function $\zeta(x, t)$ satisfying the differential equation, Equation [5], the boundary conditions, Equations [11], and the initial conditions, Equations [7]. This can be done in the classical manner. We seek a solution in the form

$$\zeta = \sum_{n=1}^{\infty} X_n T_n \qquad [12]$$

where

$$X_n = X_n(x), \quad T_n = T_n(t)$$

and assume that the functions X_n will be orthogonal[4] with respect to the interval 0,l so that $q(x)$, $g_i(x)$ and $g_i^{\mathrm{IV}}(x)$ can be expanded in series of functions X_n by means of the expansion formulas

$$\left.\begin{aligned} q(x) &= \sum_{n=1}^{\infty} Q_n X_n \\ g_i(x) &= \sum_{n=1}^{\infty} G_{in} X_n \\ g_i^{\mathrm{IV}}(x) &= \sum_{n=1}^{\infty} G_{in}^{*} X_n \end{aligned}\right\} \qquad [13]$$

where the constants Q_n, G_{in}, and G_{in}^* are given by the expressions

$$\left.\begin{aligned} Q_n &= \frac{\int_0^l q(x)X_n dx}{\int_0^l X_n^2 dx} \\ G_{in} &= \frac{\int_0^l g_i(x)X_n dx}{\int_0^l X_n^2 dx}, \quad i = 1, 2, 3, 4 \\ G_{in}^* &= \frac{\int_0^l g_i^{\mathrm{IV}}(x)X_n dx}{\int_0^l X_n^2 dx}, \quad i = 1, 2, 3, 4 \end{aligned}\right\} \qquad [14]$$

After substituting Equations [12] and [13] in Equation [5] and separating variables, the equations governing X_n and T_n are readily solved

$$X_n = C_n \cos\frac{m_n x}{l} + D_n \sin\frac{m_n x}{l} + E_n \cosh\frac{m_n x}{l} + F_n \sinh\frac{m_n x}{l} \qquad [15]$$

$$T_n = A_n \cos\omega_n t + B_n \sin\omega_n t + \frac{1}{\omega_n}\int_0^t P_n(\tau)\sin\omega_n(t - \tau)d\tau \qquad [16]$$

---

[4] The conditions under which the functions X_n are orthogonal with respect to the interval 0,l are well known (5). It is sufficient to note that this is the case if the ends of the beam are fixed or free or simply supported or restrained against translation or rotation by linear springs. These include all the types of end conditions which can arise from Equation [11]. In a private communication, Dr. R. W. Hamming has shown how to extend the method to nonorthogonal functions by introducing the adjoint equation.

where

$$P_n(\tau) = Q_n\frac{p(\tau)}{\rho A} - \sum_{i=1}^{4}[G_{in}\ddot{f}_i(\tau) + a^2G_{in}^*f_i(\tau)], \quad m_n^2 = \frac{\omega_n l^2}{a}$$

Equations [15], [16], and [12] determine the form of ζ. The constants of integration C_n, D_n, E_n, and F_n are adjusted to satisfy the simple boundary conditions, Equations [11], which also determine the transcendental equation governing m_n and hence the natural frequencies ω_n. This part of the solution is identical with the free-vibration case. The initial conditions given in Equations [7] serve to determine the constants of integration A_n and B_n. Since

$$\zeta(x, 0) = \sum_{n=1}^{\infty} A_nX_n$$

$$\left.\frac{\partial\zeta}{\partial t}\right]_{t=0} = \sum_{n=1}^{\infty} B_nX_n\omega_n$$

we must have

$$\left.\begin{aligned} A_n &= \frac{\int_0^l\left[w_0 - \sum_{i=1}^{4} f_i(0)g_i(x)\right]X_n dx}{\int_0^l X_n^2 dx} \\ B_n &= \frac{\int_0^l\left[\dot{w}_0 - \sum_{i=1}^{4} \dot{f}_i(0)g_i(x)\right]X_n dx}{\omega_n\int_0^l X_n^2 dx}\end{aligned}\right\} \quad \text{.....[17]}$$

This completes the formal solution of the problem.

Applications of the Method

Example 1. As a first illustration, consider the vibrations which arise when the "free" end of a cantilever beam is actuated by a cam. Taking the fixed end of the beam at $x = l$, the boundary conditions are

$$w(0, t) = f_1(t), \quad \left.\frac{\partial^2 w}{\partial x^2}\right]_{x=0} = \left.\frac{\partial w}{\partial x}\right]_{x=l} = w(l, t) = 0$$

where $f_1(t)$ is the transverse displacement enforced by the cam. For convenience take $w_0 = \dot{w}_0 = 0$. Applying the rule of the previous section we have

$$g_1 = 1 - \frac{3x}{2l} + \frac{x^3}{2l^3} \quad \text{................[18]}$$

Application of the boundary conditions to Equation [15] yields

$$X_n = \sinh m_n \sin\frac{m_n x}{l} - \sin m_n \sinh\frac{m_n x}{l} \quad \text{.....[19]}$$

where

$$\tan m_n = \tanh m_n$$

From Equation [14]

$$G_{in} = 2/[m_n(\sinh m_n - \sin m_n)]$$

and from Equations [17]

$$A_n = -f_1(0)G_{1n}$$

$$B_n = -\dot{f}_1(0)G_{1n}/\omega_n$$

so that a general solution of this problem is

$$w = f_1(t)\left(1 - \frac{3x}{2l} + \frac{x^3}{2l^3}\right) - \sum_{n=1}^{\infty}\frac{2X_n\left[f_1(t) - \omega_n\int_0^t f_1(\tau)\sin\omega_n(t-\tau)d\tau\right]}{m_n(\sinh m_n - \sin m_n)} \quad \text{.....[20]}$$

Special cases of this general solution may be obtained by substituting the corresponding values of f_1 in Equation [20] (6).

Example 2. The clamped end of a cantilever beam undergoes a damped sine-wave displacement transient, starting from rest. As before, we take the clamped end of the beam at $x = l$; then we are given

$$D_1 = \frac{\partial^2}{\partial x^2}, \quad D_2 = \frac{\partial^3}{\partial x^3}, \quad D_3 = 1, \quad D_4 = \frac{\partial}{\partial x}$$

$$f_1 = f_2 = f_4 = 0, \quad f_3 = Ae^{-\beta t}\sin\omega_f t$$

From Equations [15] and [16]

$$X_n = \left(\sin\frac{m_n x}{l} + \sinh\frac{m_n x}{l}\right)(\cos m_n + \cosh m_n) - \left(\cos\frac{m_n x}{l} + \cosh\frac{m_n x}{l}\right)(\sin m_n + \sinh m_n)$$

$$\cos m_n \cosh m_n = -1$$

$$g_3 = 1$$

$$G_{3n} = [(\coth m_n - \cot m_n)m_n]^{-1}$$

$$A_n = 0$$

$$B_n = -AG_{3n}\frac{\omega_f}{\omega_n}$$

$$T_n = -A\frac{\omega_f}{\omega_n}G_{3n}\sin\omega_n t - \frac{1}{\omega_n}G_{3n}\int_0^t \ddot{f}_3(\tau)\sin\omega_n(t-\tau)d\tau$$

and a complete solution is

$$w = Ae^{-\beta t}\sin\omega_f t + \sum_{n=1}^{\infty}\frac{AX_n}{m_n(\coth m_n - \cot m_n)}[\omega_f L\sin(\omega_n t - \varphi_1) + \omega_n Le^{-\beta t}\sin(\omega_f t + \varphi_2) - e^{-\beta t}\sin\omega_f t] \quad \text{.... [21]}$$

$$L = \omega_n\left\{[\beta^2 + (\omega_f + \omega_n)^2][\beta^2 + (\omega_f - \omega_n)^2]\right\}^{-1/2}$$

$$\varphi_1 = \tan^{-1} 2\beta\omega_n/(\beta^2 + \omega_f^2 - \omega_n^2)$$

$$\varphi_2 = \tan^{-1} 2\beta\omega_f/(\beta^2 + \omega_n^2 - \omega_f^2)$$

A Special Case

When the time-dependent functions in the boundary conditions are simple circular or hyperbolic or exponential functions, it is sometimes advantageous to choose the $g_i(x)$ in other than polynomial forms. Although these cases can be handled by the general method described in the preceding sections, the solution may be put in a form more suitable for computation by means of the following device:

If, for example, the $f_i(t)$ are simple trigonometric functions, then

$$\ddot{f}_i(t) = -k^2f_i(t)$$

If we choose

$$g_i(x) = a_i \sin \sqrt{\frac{k}{a}}\,x + b_i \cos \sqrt{\frac{k}{a}}\,x + c_i \cosh \sqrt{\frac{k}{a}}\,x + d_i \sinh \sqrt{\frac{k}{a}}\,x \quad \ldots\ldots [22]$$

then the right-hand side of Equation [5] will reduce to zero (except for the distributed-load term, if any), and at the same time the constants $a_i \ldots d_i$ of Equations [22] may be adjusted so as to satisfy the 16 conditions given in Equations [8]. Having removed the time-dependence in this manner, the solution proceeds exactly as before except that the summations with respect to i do not appear in Equation [16].

This variant of the general method takes advantage of a special form of the time-dependence of the boundary conditions and therefore, cannot be used in general. Where it can be employed, it serves to remove from the summation sign the forced part of the solution. This part of the solution is expressed in closed form.

Example 3. As a simple example of the special method, consider a simply supported beam, initially at rest, one of whose ends $x = l$, is subjected to an oscillating bending moment of amplitude M and frequency ω_f. In this case the given quantities are

$$D_1 = 1,\; D_2 = \frac{\partial^2}{\partial x^2},\; D_3 = 1,\; D_4 = EI\frac{\partial^2}{\partial x^2}$$

$$f_1 = f_2 = f_3 = 0,\; f_4 = -M \sin \omega_f t$$

$$w_0 = \quad {}_0 = 0$$

The constants of Equation [22] are to be adjusted so as to satisfy the conditions expressed by Equations [8]. When this is done

$$Mg_4(x) = \frac{M}{EI}\frac{a}{2\omega_f}\left[\sinh \sqrt{\frac{\omega_f}{a}}\,x \operatorname{cosech} \sqrt{\frac{\omega_f}{a}}\,l - \sin \sqrt{\frac{\omega_f}{a}}\,x \operatorname{cosec} \sqrt{\frac{\omega_f}{a}}\,l\right] \quad \ldots\ldots [23]$$

Equation [23] gives directly, in closed form, the amplitude of the steady state or forced vibration. It remains to complete the solution by writing out the transient. From Equation [15], together with the boundary conditions of Equations [11]

$$X_n = \sin\frac{n\pi x}{l},\quad \omega_n = \frac{n^2\pi^2 a}{l^2}$$

From Equation [16] and [17]

$$A_n = 0$$

$$B_n = \frac{M}{EI}\frac{2\omega_f a}{n\pi}\frac{(-1)^n}{\omega_n^2 - \omega_f^2}$$

$$T_n = B_n \sin \omega_n t$$

The complete solution of the problem is therefore

$$w = -\frac{M}{EI}\frac{a}{2\omega_f}\left[\sinh \sqrt{\frac{\omega_f}{a}}\,x \operatorname{cosech} \sqrt{\frac{\omega_f}{a}}\,l - \sin \sqrt{\frac{\omega_f}{a}}\,x \operatorname{cosec} \sqrt{\frac{\omega_f}{a}}\,l\right]\sin \omega_f t + \frac{M}{EI}\frac{a}{2\omega_f}\frac{4}{\pi}\sum_{n=1}^{\infty}\frac{(-1)^n}{n\left(\frac{\omega_n^2}{\omega_f^2} - 1\right)}\sin\frac{n\pi x}{l}\sin\frac{n^2\pi^2 a t}{l^2} \quad \ldots [24]$$

BIBLIOGRAPHY

1 "Sur les vibrations d'un système quelconque de points matériels," by J. M. C. Duhamel, *Journal de l'École Polytechnique*, Cah. 23, Paris, France, 1834, pp. 1–36.

2 "Solution de divers problèmes de Mécanique, etc.," by E. Phillips, *Journal de Math.*, Paris, France, vol. 9, 1864, pp. 25–83.

3 "L'Élasticité des Corps Solides," by A. Clebsch, with notes by B. de St.Venant, Paris, France, 1883, pp. 538–547.

4 "History of the Theory of Elasticity," by I. Todhunter and K. Pearson, Cambridge, England, 1886, vol. 1, p. 486; vol. 2, part 1, pp. 240–241, 470–472.

5 "Theory of Sound," by Lord Rayleigh, The Macmillan Company, London, England, second edition, vol. 1, 1894, p. 262.

6 "Vibration of a Cantilever Beam With Prescribed End Motion," by G. A. Nothmann, Trans. ASME, vol. 70, 1948, p. 327.

Reprinted from
Journal of Applied Mechanics
17 (1950) 377-380

Reprinted from JOURNAL OF APPLIED PHYSICS, Vol. 21, No. 9, 926–930, September, 1950

Nuclei of Strain in the Semi-Infinite Solid

RAYMOND D. MINDLIN AND DAVID H. CHENG
Columbia University, New York, New York
(Received April 28, 1950)

Stress functions are given for forty nuclei of strain in the semi-infinite elastic solid. The corresponding stresses have the property that the traction vanishes across the plane boundary of the solid. The nuclei are derived, by processes of superposition, differentiation and integration, from the solution of the elasticity equations for the single force in the interior of the semi-infinite solid.

I. INTRODUCTION

THERE is an interesting group of solutions of the equations of the linear theory of elasticity known as nuclei of strain. For the isotropic solid of indefinite extent the fundamental solution of this type is that of Kelvin[1] for a single force applied at a point. By differentiation of this solution a family of additional nuclei can be obtained, examples of which are[1] the double force and the double force with moment. These, in turn, can be combined to form nuclei such as the center of dilatation and the center of rotation. By further differentiation, various types of doublets and multiplets are obtained. Alternatively, the preceding solutions may be integrated to form line distributions, such as[1] semi-infinite lines of centers of dilatation, or doublets, or double forces with moment, etc.

The solutions of many interesting problems may be expressed in terms of combinations of these nuclei or have been discovered by a process of superposition of nuclei. Examples are Lamé's problem[2] of the spherical container under internal and external pressure, Boussinesq's and Cerruti's problems[3] of the single force acting at the surface of a semi-infinite solid, Southwell's problem[2] of the spherical cavity in an infinite solid under simple tension, the analogous solution by Goodier[4] for the spherical inclusion and Mindlin's solution[5] for the single force in the semi-infinite solid.

The last named solution occupies the same position for the semi-infinite solid as Kelvin's solution does for the solid of indefinite extent. The nuclei derived from it, by differentiation and integration, all satisfy the condition of vanishing traction on a plane boundary so that, in contemplating the solution of new problems with at least one plane, traction-free surface, a portion of the boundary conditions is already satisfied.

Since the solution for the single force in the semi-infinite solid is itself formed from the superposition of eighteen nuclei derived from Kelvin's solution (six for each of the three components of the force), the calculation of nuclei for the semi-infinite solid is a formidable one. The possible usefulness of such nuclei is so great, however, that this task was undertaken and the results for forty nuclei are reported below.[6] The methods employed are the same as those described by Love[1] in connection with the calculation of nuclei in the solid of indefinite extent. The results are given in very condensed form by expressing them in terms of the Galerkin vector stress function. The displacement vector $\mathbf{u}$ is given in terms of the Galerkin vector $\mathbf{F}$ by

$$2G\mathbf{u}=2(1-\nu)\nabla^2\mathbf{F}-\nabla\nabla\cdot\mathbf{F}, \tag{1}$$

where G is the shear modulus, ν is Poisson's ratio and ∇, $\nabla\cdot$ and ∇^2 are the gradient, divergence and Laplacian operators, respectively.

II. REDUCTION FROM THREE COMPONENTS TO ONE

Certain nuclei are obtained by the superposition of three others and the immediate result is a three-component Galerkin vector. This may be simplified, in some cases, by reduction to a one-component vector, say $\mathbf{F}'=\mathbf{k}Z'$, so that

$$2G\mathbf{u}=2(1-\nu)\mathbf{k}\nabla^2 Z'-\nabla(\partial Z'/\partial z). \tag{2}$$

To find the function $\mathbf{F}'$ corresponding to a given $\mathbf{F}$, (1) and (2) are equated to obtain the differential equation

$$2(1-\nu)\mathbf{k}\nabla^2 Z'-\nabla(\partial Z'/\partial z)=2(1-\nu)\nabla^2\mathbf{F}-\nabla\nabla\cdot\mathbf{F}, \tag{3}$$

which is to be solved for Z' in terms of $\mathbf{F}$.

Case 1: $\nabla^2\mathbf{F}=0$

Operating on (3) with $\nabla\cdot$ and $\nabla\times$ and using the condition $\nabla^2\mathbf{F}=0$, there results

$$\nabla\nabla^2 Z'=0$$

from which $\nabla^2 Z'$ is a constant which may be taken as zero since, by (2), it contributes only a rigid body displacement. Hence (3) reduces to

$$\nabla(\partial Z'/\partial z)=\nabla\nabla\cdot\mathbf{F} \tag{4}$$

the first integral of which is

$$\partial Z'/\partial z=\nabla\cdot\mathbf{F}$$

[1] A. E. H. Love, *Mathematical Theory of Elasticity* (Cambridge University Press, London, 1927), fourth edition, Chapter VIII.
[2] S. Timoshenko, *Theory of Elasticity* (McGraw-Hill Book Company, Inc., New York, 1934), Chapter 11.
[3] See reference 1, Chapter X.
[4] J. N. Goodier, "Concentration of stress around spherical and cylindrical inclusions and flaws," J. App. Mech. 1, 39–44 (1933).
[5] R. D. Mindlin, "Force at a point in the interior of a semi-infinite solid," Physics 7, 195–202 (1936).
[6] An application of one of them is given in J. App. Phys. 21, 931 (1950).

plus a constant which may be omitted since it does not contribute to the displacement. The second integral of (4) is

$$Z'=\int \nabla\cdot\mathbf{F}dz$$

plus a function of x and y which may be omitted since it does not contribute to the displacement.

Case 2: $\nabla\times\nabla\times\mathbf{F}=\nabla^2\nabla\cdot\mathbf{F}=0$

Since

$$\nabla\times\nabla\times\mathbf{F}=\nabla\nabla\cdot\mathbf{F}-\nabla^2\mathbf{F},$$

Eq. (3) reduces, in this case, to

$$2(1-\nu)\mathbf{k}\nabla^2Z'-\nabla(\partial Z'/\partial z)=(1-2\nu)\nabla\nabla\cdot\mathbf{F}. \quad (5)$$

Operating on (5) with $\nabla\cdot$ and $\nabla\times$ and using the condition $\nabla^2\nabla\cdot\mathbf{F}=0$, we find, as before, that $\nabla^2Z'=0$. Hence (5) reduces to

$$\nabla(\partial Z'/\partial z)=-(1-2\nu)\nabla\nabla\cdot\mathbf{F}$$

from which, by the procedure used in Case 1,

$$Z'=-(1-2\nu)\int\nabla\cdot\mathbf{F}dz.$$

III. HALF-SPACE NUCLEI IN TERMS OF GALERKIN VECTORS

The half-space considered is defined by $Z\geqslant 0$. Point nuclei are at $(0, 0, c)$ and one end of each line nucleus is at the same point. $\mathbf{i}$, $\mathbf{j}$, $\mathbf{k}$ are unit vectors in the x, y, z directions, respectively, and

$$R_1^2=x^2+y^2+(z-c)^2,\quad R_2^2=x^2+y^2+(z+c)^2.$$

A multiplying constant is omitted in each stress function.

A. Single Force

(1) Single force in x direction

$$\mathbf{i}\left\{R_1+R_2-\frac{2c^2}{R_2}+4(1-\nu)(1-2\nu)\times[(z+c)\log(R_2+z+c)-R_2]\right\}+\mathbf{k}\left[\frac{2cx}{R_2}+2(1-2\nu)x\log(R_2+z+c)\right].$$

(2) Single force in y direction

$$\mathbf{j}\left\{R_1+R_2-\frac{2c^2}{R_2}+4(1-\nu)(1-2\nu)\times[(z+c)\log(R_2+z+c)-R_2]\right\}+\mathbf{k}\left[\frac{2cy}{R_2}+2(1-2\nu)y\log(R_2+z+c)\right]$$

(3) Single force in z direction

$$\mathbf{k}\left\{R_1+[8\nu(1-\nu)-1]R_2+4(1-2\nu)[(1-\nu)z-\nu c]\log(R_2+z+c)-\frac{2cz}{R_2}\right\}.$$

B. Double Force

(1) Double force in x direction

$$\mathbf{i}x\left[\frac{1}{R_1}+\frac{1}{R_2}-\frac{4(1-\nu)(1-2\nu)}{R_2+z+c}+\frac{2c^2}{R_2^3}\right]+\mathbf{k}\left[\frac{2c}{R_2}-\frac{2cx^2}{R_2^3}+\frac{2(1-2\nu)x^2}{R_2(R_2+z+c)}+2(1-2\nu)\log(R_2+z+c)\right].$$

(2) Double force in y direction

$$\mathbf{j}y\left[\frac{1}{R_1}+\frac{1}{R_2}-\frac{4(1-\nu)(1-2\nu)}{R_2+z+c}+\frac{2c^2}{R_2^3}\right]+\mathbf{k}\left[\frac{2c}{R_2}+2(1-2\nu)\log(R_2+z+c)+\frac{2(1-2\nu)y^2}{R_2(R_2+z+c)}-\frac{2cy^2}{R_2^3}\right].$$

(3) Double force in z direction

$$\mathbf{k}\left[\frac{z-c}{R_1}-(1-4\nu)\frac{z-c}{R_2}-\frac{2cz(z+c)}{R_2^3}+4\nu(1-2\nu)\log(R_2+z+c)\right].$$

C. Double Force with Moment

(1) Double force in x direction with moment about y axis

$$\mathbf{i}\left[\frac{z-c}{R_1}-\frac{z-3c}{R_2}-\frac{2c^2(z+c)}{R_2^3}-4(1-\nu)(1-2\nu)\log(R_2+z+c)\right]+\mathbf{k}2x\left[\frac{c(z+c)}{R_2^3}-\frac{2(1-\nu)}{R_2}\right].$$

(2) Double force in x direction with moment about z axis

$$\mathbf{i}y\left[\frac{1}{R_1}+\frac{1}{R_2}+\frac{2c^2}{R_2^3}-\frac{4(1-\nu)(1-2\nu)}{R_2+z+c}\right]+\mathbf{k}2xy\left[\frac{1-2\nu}{R_2(R_2+z+c)}-\frac{c}{R_2^3}\right].$$

(3) Double force in y direction with moment about x axis

$$\mathbf{j}\left[\frac{z-c}{R_1}-\frac{z-3c}{R_2}-\frac{2c^2(z+c)}{R_2^3}-4(1-\nu)(1-2\nu)\log(R_2+z+c)\right]+\mathbf{k}2y\left[\frac{c(z+c)}{R_2^3}-\frac{2(1-\nu)}{R_2}\right].$$

(4) Double force in y direction with moment about z axis

$$\mathbf{j}x\left[\frac{1}{R_1}+\frac{1}{R_2}+\frac{2c^2}{R_2^3}-\frac{4(1-\nu)(1-2\nu)}{R_2+z+c}\right]+\mathbf{k}2xy\left[\frac{1-2\nu}{R_2(R_2+z+c)}-\frac{c}{R_2^3}\right].$$

(5) Double force in z direction with moment about x axis

$$\mathbf{k}y\left\{\frac{1}{R_1}+\frac{8\nu(1-\nu)-1}{R_2}+\frac{4(1-2\nu)[(1-\nu)z-\nu c]}{R_2(R_2+z+c)}+\frac{2cz}{R_2^3}\right\}.$$

(6) Double force in z direction with moment about y axis

$$\mathbf{k}x\left\{\frac{1}{R_1}+\frac{8\nu(1-\nu)-1}{R_2}+\frac{4(1-2\nu)[(1-\nu)z-\nu c]}{R_2(R_2+z+c)}+\frac{2cz}{R_2^3}\right\}.$$

D. Line of Double Forces with Moment

(1) Type C(1) along x axis from $x=0$ to $x=-\infty$

$$\mathbf{i}\left[(z-c)\log(R_1+x)-(z-3c)\log(R_2+x)-\frac{2c^2(z+c)x}{R_2(R_2^2-x^2)}\right]-\mathbf{k}\left[\frac{2c(z+c)}{R_2}+8\nu(1-\nu)R_2+4(1-\nu)(1-2\nu)(z+c)\log(R_2+z+c)\right].$$

(2) Type C(2) along x axis from $x=0$ to $x=-\infty$

$$\mathbf{i}y\left[\log(R_1+x)+\log(R_2+x)+\frac{2c^2x}{R_2(R_2^2-x^2)}\right]+\mathbf{k}y\left[2\nu(1-2\nu)\log(R_2+z+c)-2(1-\nu)(1-2\nu)\frac{z+c}{R_2+z+c}+\frac{2c}{R_2}\right].$$

(3) Type C(3) along y axis from $y=0$ to $y=-\infty$

$$\mathbf{j}\left[(z-c)\log(R_1+y)-(z-3c)\log(R_2+y)-\frac{2c^2(z+c)y}{R_2(R_2^2-y^2)}\right]-\mathbf{k}\left[\frac{2c(z+c)}{R_2}+8\nu(1-\nu)R_2+4(1-\nu)(1-2\nu)(z+c)\log(R_2+z+c)\right].$$

(4) Type C(4) along y axis from $y=0$ to $y=-\infty$

$$\mathbf{j}x\left[\log(R_1+y)+\log(R_2+y)+\frac{2c^2y}{R_2(R_2^2-y^2)}\right]+\mathbf{k}x\left[2\nu(1-2\nu)\log(R_2+z+c)-2(1-\nu)(1-2\nu)\frac{z+c}{R_2+z+c}+\frac{2c}{R_2}\right].$$

(5) Type C(5) along z axis from $z=c$ to $z=\infty$

$$\mathbf{k}y\left\{\log(R_1-z+c)+[8\nu(1-\nu)-1]\log(R_2+z+c)-\frac{4(1-2\nu)z}{R_2+z+c}-2\nu(1-2\nu)\left[\log(R_2+z+c)+\frac{R_2}{R_2+z+c}\right]-2z\left[\frac{1}{R_2}+\frac{z(z+c)}{R_2(x^2+y^2)}-\frac{z}{(x^2+y^2)}\right]\right\}.$$

(6) Type C(6) along z axis from $z=c$ to $z=\infty$

$$\mathbf{k}x\left\{\log(R_1-z+c)+[8\nu(1-\nu)-1]\log(R_2+z+c)-\frac{4(1-2\nu)z}{R_2+z+c}-2\nu(1-2\nu)\left[\log(R_2+z+c)+\frac{R_2}{R_2+z+c}\right]-2z\left[\frac{1}{R_2}+\frac{z(z+c)}{R_2(x^2+y^2)}-\frac{z}{(x^2+y^2)}\right]\right\}.$$

E. Center of Dilatation

(1)

$$\mathbf{i}[\log(R_1+x)+(1-4\nu)\log(R_2+x)]+\mathbf{k}\frac{2z}{R_2}.$$

(2)

$$\mathbf{j}[\log(R_1+y)+(1-4\nu)\log(R_2+y)]+\mathbf{k}\frac{2z}{R_2}.$$

(3)

$$\mathbf{k}\left[\log(R_1+z-c)+(1-4\nu)\log(R_2+z+c)+\frac{2z}{R_2}\right].$$

All three yield the same displacements and stresses.

F. Line of Centers of Dilatation

(1) Along x axis from $x=0$ to $x=-\infty$

$$\mathbf{i}\{x\log(R_1+x)-R_1+(1-4\nu)[x\log(R_2+x)-R_2]\}+\mathbf{k}2z\log(R_2+x).$$

(2) Along y axis from $y=0$ to $y=-\infty$

$$\mathbf{j}\{y\log(R_1+y)-R_1+(1-4\nu)[y\log(R_2+y)-R_2]\}+\mathbf{k}2z\log(R_2+y).$$

(3) Along z axis from $z=c$ to $z=\infty$

$$\mathbf{k}\{R_1-(z-c)\log(R_1+z-c)+(1-4\nu)\times[(z+c)\log(R_2+z+c)-R_2]+2z\log(R_2+z+c)\}.$$

G. Doublet

(1) Axis of doublet parallel to x axis

$$\mathbf{i}\left(\frac{1}{R_1}+\frac{1-4\nu}{R_2}\right)-\mathbf{k}\frac{2xz}{R_2^3}.$$

(2) Axis of doublet parallel to y axis

$$\mathbf{j}\left(\frac{1}{R_1}+\frac{1-4\nu}{R_2}\right)-\mathbf{k}\frac{2yz}{R_2^3}.$$

(3) Axis of doublet parallel to z axis

$$\mathbf{k}\left[\frac{1}{R_1}-\frac{1-4\nu}{R_2}+\frac{2z(z+c)}{R_2^3}\right].$$

H. Linearly Varying Line of Doublets with Strength Proportional to Distance from the Origin

(1) Type G(1) along y axis from $y=0$ to $y=-\infty$

$$\mathbf{i}\{R_1-y\log(R_1+y)+(1-4\nu)[R_2-y\log(R_2+y)]\}+\mathbf{k}\frac{2xz}{R_2+y}.$$

(2) Type G(1) along z axis from $z=c$ to $z=\infty$

$$\mathbf{i}\{R_1+z\log(R_1-z+c)+(1-4\nu)[R_2-z\log(R_2+z+c)]\}+\mathbf{k}2xz\left[\frac{1}{R_2}+\frac{z(z+c)}{R_2(x^2+y^2)}-\frac{z}{x^2+y^2}\right].$$

(3) Type G(2) along x axis from $x=0$ to $x=-\infty$

$$\mathbf{j}\{R_1-x\log(R_1+x)+(1-4\nu)[R_2-x\log(R_2+x)]\}+\mathbf{k}\frac{2yz}{R_2+x}.$$

(4) Type G(2) along z axis from $z=c$ to $z=\infty$

$$\mathbf{j}\{R_1+z\log(R_1-z+c)+(1-4\nu)[R_2-z\log(R_2+z+c)]\}+\mathbf{k}2yz\left[\frac{1}{R_2}+\frac{z(z+c)}{R_2(x^2+y^2)}-\frac{z}{x^2+y^2}\right].$$

(5) Type G(3) along x axis from $x=0$ to $x=-\infty$

$$\mathbf{k}\left\{R_1-(1-4\nu)R_2-x[\log(R_1+x)-(1-4\nu)\log(R_2+x)]-\frac{2z(z+c)}{R_2+x}\right\}.$$

(6) Type G(3) along y axis from $y=0$ to $y=-\infty$

$$\mathbf{k}\left\{R_1-(1-4\nu)R_2-y[\log(R_1+y)-(1-4\nu)\log(R_2+y)]-\frac{2z(z+c)}{R_2+y}\right\}.$$

I. Center of Rotation

(1) Center of rotation about x axis (Type C(5)—Type C(3))

$$\mathbf{j}\left[-\frac{z-c}{R_1}+\frac{z-3c}{R_2}+4(1-\nu)(1-2\nu)\log(R_2+z+c)+\frac{2c^2(z+c)}{R_2^3}\right]+\mathbf{k}y\left\{\frac{1}{R_1}+\frac{4\nu(1-2\nu)+3}{R_2}-\frac{2c^2}{R_2^3}+\frac{4(1-2\nu)[(1-\nu)z-\nu c]}{R_2(R_2+z+c)}\right\}.$$

(2) Center of rotation about y axis (Type C(6)—Type C(1))

$$\mathbf{i}\left[-\frac{z-c}{R_1}+\frac{z-3c}{R_2}+4(1-\nu)(1-2\nu)\log(R_2+z+c)+\frac{2c^2(z+c)}{R_2^3}\right]+\mathbf{k}x\left\{\frac{1}{R_1}+\frac{4\nu(1-2\nu)+3}{R_2}-\frac{2c^2}{R_2^3}+\frac{4(1-2\nu)[(1-\nu)z-\nu c]}{R_2(R_2+z+c)}\right\}.$$

(3) Center of rotation about z axis (Type C(4)—Type C(2))

$$(\mathbf{j}x-\mathbf{i}y)\left[\frac{1}{R_1}+\frac{1}{R_2}+\frac{2c^2}{R_2^3}-\frac{4(1-\nu)(1-2\nu)}{R_2+z+c}\right].$$

J. Line of Centers of Rotation

(1) Type I(1) along y axis from $y=0$ to $y=-\infty$

$$\mathbf{j}\left[-(z-c)\log(R_1+y)+(z-3c)\log(R_2+y)+\frac{2c^2(z+c)y}{R_2(R_2^2-y^2)}\right]+\mathbf{k}\left\{R_1-[4(1-2\nu)^2-3]R_2+4(1-2\nu)[z+(1-2\nu)(z+c)]\log(R_2+z+c)+\frac{2c^2}{R_2}\right\}.$$

(2) Type I(1) along z axis from $z=c$ to $z=\infty$

$$\mathbf{j}\left\{R_1+R_2-\frac{2c^2}{R_2}+4(1-\nu)(1-2\nu)\times[(z+c)\log(R_2+z+c)-R_2]\right\}+\mathbf{k}y\left\{\log(R_1-z+c)+[2\nu(1-2\nu)+1]\log(R_2+z+c)+\frac{2c^2}{R_2(R_2+z+c)}-\frac{2[R_2(z-c)-z^2]}{x^2+y^2}+\frac{2(1-2\nu)}{R_2+z+c}[\nu R_2+2\nu(z+c)-2z]\right\}.$$

(3) Type I(2) along x axis from $x=0$ to $x=-\infty$

$$\mathbf{i}\left[-(z-c)\log(R_1+x)+(z-3c)\log(R_2+x)+\frac{2c^2(z+c)x}{R_2(R_2^2-x^2)}\right]+\mathbf{k}\left\{R_1-[4(1-2\nu)^2-3]R_2+4(1-2\nu)[z+(1-2\nu)(z+c)]\log(R_2+z+c)+\frac{2c^2}{R_2}\right\}.$$

(4) Type I(2) along z axis from $z=c$ to $z=\infty$

$$\mathbf{i}\left\{R_1+R_2-\frac{2c^2}{R_2}+4(1-\nu)(1-2\nu)\times[(z+c)\log(R_2+z+c)-R_2]\right\}+\mathbf{k}x\left\{\log(R_1-z+c)+[2\nu(1-2\nu)+1]\log(R_2+z+c)+\frac{2c^2}{R_2(R_2+z+c)}-\frac{2[R_2(z-c)-z^2]}{x^2+y^2}+\frac{2(1-2\nu)}{R_2+z+c}[\nu R_2+2\nu(z+c)-2z]\right\}.$$

(5) Type I(3) along x axis from $x=0$ to $x=-\infty$

$$-\mathbf{i}y\left[\log(R_1+x)+\log(R_2+x)+\frac{2c^2x}{R_2(R_2^2-x^2)}\right]+\mathbf{j}\left[R_1+R_2-\frac{2c^2}{R_2}\right].$$

(6) Type I(3) along y axis from $y=0$ to $y=-\infty$

$$\mathbf{i}\left[R_1+R_2-\frac{2c^2}{R_2}\right]-\mathbf{j}x\left[\log(R_1+y)+\log(R_2+y)+\frac{2c^2y}{R_2(R_2^2-y^2)}\right].$$

IV. COMPONENTS OF DISPLACEMENT AND STRESS

The formulas for the components of displacement and stress corresponding to each of the nuclei listed in the preceding section are listed in Appendices I and II.[7]

[7] On file at the Department of Civil Engineering, Columbia University, New York 27, New York.

Reprinted from JOURNAL OF APPLIED PHYSICS, Vol. 21, No. 9, 931–933, September, 1950

Thermoelastic Stress in the Semi-Infinite Solid

RAYMOND D. MINDLIN AND DAVID H. CHENG
Columbia University, New York, New York
(Received April 28, 1950)

The solution of the elasticity equations for the center of dilatation in the semi-infinite solid is introduced into Goodier's theory of thermoelastic stress. It is shown that the problem of potential to be solved is identical with that for the solid of indefinite extent. The results are applied to the case of an expanding (or contracting) spherical inclusion embedded in an elastic body near its surface.

I. INTRODUCTION

GOODIER[1] has shown that the elastic effect of a non-uniform distribution of temperature (T) in an infinite body is the same as that of a distribution of centers of dilatation of strength $-\beta/4\pi$ where

$$\beta=\alpha T(1+\nu)/(1-\nu) \qquad (1)$$

and α and ν are the coefficient of linear thermal expansion and Poisson's ratio, respectively. Since α and T always appear together, what is said of a non-uniform distribution of temperature applies also to a space variation of the coefficient of linear thermal expansion. Goodier's theory has been applied to two problems of thermal stress in the infinite elastic body by Myklestad.[2]

If the body is not infinite the singularities within it give rise to tractions and displacements at its surface and the removal of these is a boundary-value problem in elasticity theory. However, Goodier noted that, if the solution of the elasticity equations for the center of dilatation is known for the bounded body, the thermoelastic problem is reduced, again, to an integration if the method of singularities is employed.

In this paper it is shown that, for the semi-infinite body with traction-free surface, the additional integral has the same form as that for the infinite body so that, if the latter is known, the calculation of displacement and stress reduces to a process of differentiation only. The result is then applied to a case of a spherical inclusion in the semi-infinite body. Use is made of the solution, given in an accompanying paper,[3] for the center of dilatation in the semi-infinite body.

II. THERMOELASTIC DISPLACEMENT

The Galerkin vector stress function for a center of dilatation of strength A at (ξ, η, ζ) in the infinite, isotropic, elastic solid is[4]

$$\mathbf{F}=-2AG\mathbf{k}\log(R_1'+z-\zeta),$$

where G is the modulus of rigidity, $\mathbf{k}$ is a unit vector in the z direction and

$$R_1'^2=(x-\xi)^2+(y-\eta)^2+(z-\zeta)^2.$$

Hence Goodier's nucleus of thermoelastic strain is given by

$$\mathbf{F}=(G\beta/2\pi)\mathbf{k}\log(R_1'+z-\zeta).$$

In an accompanying paper[3] it is shown that the corresponding nucleus at (ξ, η, ζ) in the semi-infinite body $z\geqslant 0$ (with the plane $z=0$ free of traction) is represented by

$$\mathbf{F}_0=(G\beta/2\pi)\mathbf{k}[\log(R_1'+z-\zeta) +(1-4\nu)\log(R_2'+z+\zeta)+2(z+\zeta)/R_2'-2\zeta/R_2'],$$

where

$$R_2'^2=(x-\xi)^2+(y-\eta)^2+(z+\zeta)^2.$$

Thus, for each center of dilatation at (ξ, η, ζ) in $z>0$, the plane $z=0$ is freed of traction by applying, at the image point $(\xi, \eta, -\zeta)$, a center of dilatation, a double force and a doublet, each of the proper strength.

The displacement $\mathbf{u}_0$ associated with $\mathbf{F}_0$ is calculated from

$$2G\mathbf{u}_0=2(1-\nu)\nabla^2\mathbf{F}_0-\nabla\nabla\cdot\mathbf{F}_0$$

and may be written in terms of operations on $1/R_1'$ and $1/R_2'$ as follows:

$$4\pi\mathbf{u}_0=-\beta\nabla(1/R_1')-\beta\nabla_2(1/R_2'),$$

where

$$\nabla_2=(3-4\nu)\nabla+2\nabla z(\partial/\partial z)-4(1-\nu)\mathbf{k}\nabla^2 z$$

is a vector operator in which, for example, the scalar operator $\nabla^2 z$ signifies that the operand is first to be multiplied by z and then operated upon by ∇^2.

For a distribution $\beta(\xi, \eta, \zeta)$ in a region V_1 in $z>0$, the displacement is given by

$$\begin{aligned}4\pi\mathbf{u}&=4\pi\int_{V_1}\mathbf{u}_0 d\tau\\ &=-\nabla\int_{V_1}\frac{\beta}{R_1'}d\tau-\nabla_2\int_{V_1}\frac{\beta}{R_2'}d\tau\\ &=-\nabla\int_{V_1}\frac{\beta}{R_1'}d\tau-\nabla_2\int_{V_2}\frac{\beta}{R_1'}d\tau\\ &=-\nabla\varphi_1-\nabla_2\varphi_2,\end{aligned} \qquad (2)$$

[1] J. N. Goodier, "On the integration of the thermoelastic equations," Phil. Mag. 7 (No. 23), 1017–1032 (1937).
[2] N. O. Myklestad, "Two problems of thermal stress in the infinite solid," J. App. Mech. 9, A-136-143 (1942).
[3] R. D. Mindlin and D. H. Cheng, "Nuclei of strain in the semi-infinite solid," J. App. Phys. 21, 926 (1950).
[4] R. D. Mindlin, "Force at a point in the interior of a semi-infinite solid," Physics 7, 195–202 (1936).

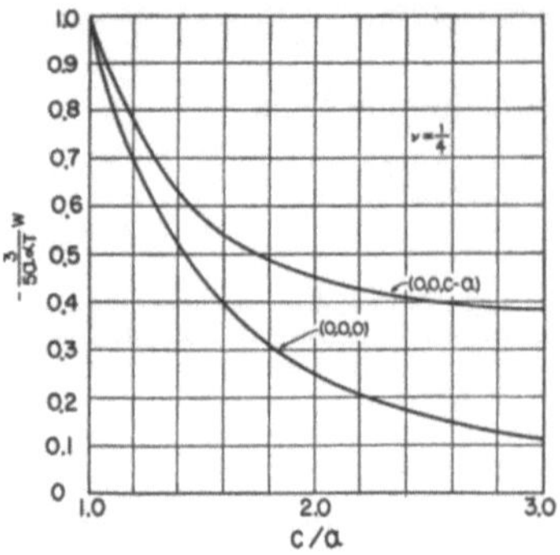

FIG. 1. Displacement normal to surface at points (0, 0, 0) and (0, 0, $c-a$).

where $d\tau = d\xi d\eta d\zeta$ and V_2 is the image of V_1 in the plane $z=0$. The volume integrals pass through ∇ and ∇_2 since the latter are x, y, z operators. The first integral is the potential (φ_1) of a distribution of matter of density β in V_1 as in Goodier's theory. The second integral (φ_2) is simply the reflection transformation of φ_1 in the plane $z=0$, so that the integration need not be repeated. For any distribution of β whose potential is known, the calculation of displacement in the semi-infinite body thus reduces to the simple process of applying the operators ∇ and ∇_2. Although φ_1 and φ_2 are mutual images, the corresponding displacements are not. This is due to the difference between ∇ and ∇_2.

III. APPLICATION TO SPHERICAL INCLUSION

Let V_1 be a sphere of radius a with center at $(0, 0, c)$ where $a \leqslant c$. The radii to a field point (x, y, z) from $(0, 0, c)$ and from its image $(0, 0, -c)$ are denoted by $R_1 = [x^2+y^2+(z-c)^2]^{\frac{1}{2}}$ and $R_2 = [x^2+y^2+(z+c)^2]^{\frac{1}{2}}$ respectively. Let α be the excess of the coefficient of thermal expansion of the material within V_1 over that of the surrounding medium and let T be a uniform

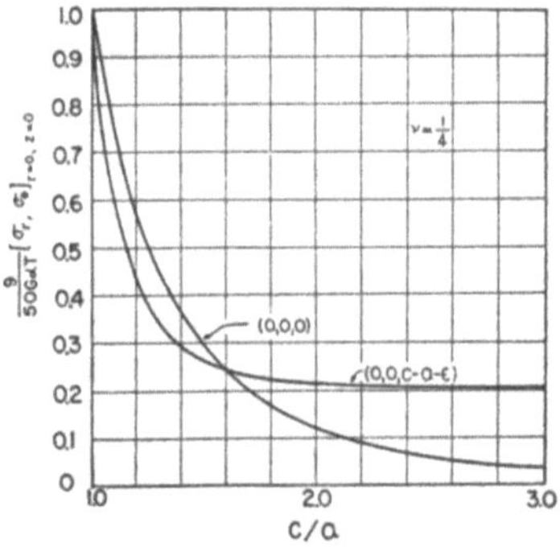

FIG. 2. σ_r and σ_θ at points (0, 0, 0), (0, 0, $c-a-\epsilon$).

temperature rise. Then β is given by (1) in $R_1 < a$ and is zero in $z \geqslant 0$, $R_1 > a$. The elastic constants of the two regions are assumed to be the same.

From the well-known results for the potential of a homogeneous sphere[5] we have, for exterior points ($z \geqslant 0$, $R_1 > a$)

$$\varphi_1 = 4\pi a^3\beta/3R_1, \quad \varphi_2 = 4\pi a^3\beta/3R_2$$

and, for interior points ($R_1 < a$)

$$\varphi_1 = 2\pi\beta(3a^2 - R_1^2)/3, \quad \varphi_2 = 4\pi a^3\beta/3R_2.$$

Substituting in (2) and performing the operations ∇ and ∇_2, we have, for exterior points,

$$\mathbf{u}_e = \frac{a^3\beta}{3}\left\{\frac{\mathbf{R}_1}{R_1^3} + \frac{(3-4\nu)\mathbf{R}_2}{R_2^3} - \frac{6z(z+c)\mathbf{R}_2}{R_2^5} - \frac{2\mathbf{k}}{R_2^3}[(3-4\nu)(z+c)-z]\right\},$$

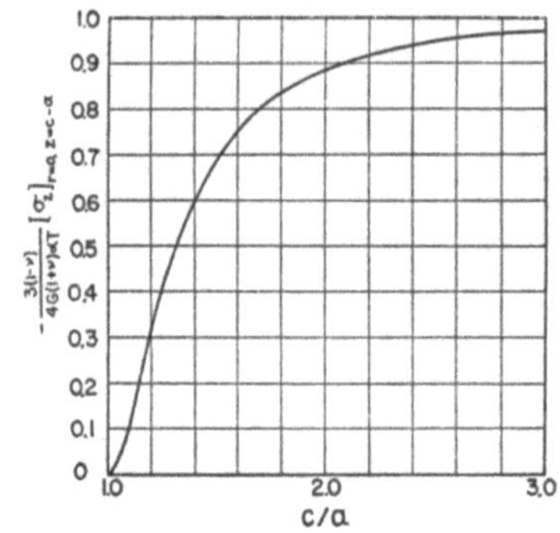

FIG. 3. σ_z at point (0, 0, $c-a$).

and, for interior points,

$$\mathbf{u}_i = \mathbf{u}_e + \frac{a^3\beta\mathbf{R}_1}{3}\left(\frac{1}{a^3} - \frac{1}{R_1^3}\right),$$

where $\mathbf{R}_1$ and $\mathbf{R}_2$ are the position vectors from $(0, 0, c)$ and $(0, 0, -c)$, respectively. It may be observed that, for exterior points, the displacement is the same as that for a single center of dilatation at the center of the sphere.

The stress is calculated from the displacement by means of the usual thermoelastic relation. In dyadic notation[6] this is

$$\Psi = \lambda \mathbf{I}[\nabla\cdot\mathbf{u} - \beta(1-\nu)/\nu] + G(\nabla\mathbf{u} + \mathbf{u}\nabla)$$

where Ψ is the stress dyadic, $\mathbf{I}$ is the idemfactor and

[5] W. D. MacMillan, *Theory of the Potential* (McGraw-Hill Book Company, Inc., New York, 1930), p. 40.
[6] C. E. Weatherburn, *Advanced Vector Analysis* (G. Bell and Sons, London, 1928).

$\lambda=2G\nu/(1-2\nu)$. In performing the calculations for interior points, β is given by (1) while for exterior points $\beta=0$. The cylindrical components are, for exterior points,

$$\sigma_r=\sigma_\theta-2G\beta a^3r^2\left[\frac{1}{R_1^5}+\frac{3-4\nu}{R_2^5}-\frac{10z(z+c)}{R_2^7}\right],$$

$$\sigma_\theta=\frac{2G\beta a^3}{3}\left[\frac{1}{R_1^3}+\frac{3-8\nu}{R_2^3}-\frac{6z(z+c)}{R_2^5}+\frac{12\nu(z+c)^2}{R_2^5}\right],$$

$$\sigma_z=\frac{2G\beta a^3}{3}\left[\frac{1}{R_1^3}-\frac{1}{R_2^3}-\frac{18z(z+c)}{R_2^5}\right.$$
$$\left.+\frac{3(z+c)^2}{R_2^5}-\frac{3(z-c)^2}{R_1^5}+\frac{30z(z+c)^3}{R_2^7}\right],$$

$$\tau_{rz}=-2G\beta a^3r\left[\frac{z-c}{R_1^5}+\frac{3z+c}{R_2^5}-\frac{10z(z+c)^2}{R_2^7}\right],$$

$$\tau_{\theta r}=\tau_{z\theta}=0.$$

For interior points the formulas are the same except that in each of the normal components all the terms containing R_1 are replaced by $-2/a^3$, while the term containing R_1 in the shear component vanishes.

The magnitudes of the components of stress and displacement at the point on the sphere nearest the free surface and at the point on the free surface nearest the sphere are plotted in the Figs. 1, 2, and 3 as functions of the depth of the center of the sphere below the surface. It is interesting to observe that for an expanding sphere close to the surface the tensile stress in the surrounding medium can become $4(1+\nu)$ times as great as for the sphere in an infinite medium.

Paper No.
50—APM-22

Influence of Rotatory Inertia and Shear on Flexural Motions of Isotropic, Elastic Plates

By R. D. MINDLIN,[1] NEW YORK, N. Y.

A two-dimensional theory of flexural motions of isotropic, elastic plates is deduced from the three-dimensional equations of elasticity. The theory includes the effects of rotatory inertia and shear in the same manner as Timoshenko's one-dimensional theory of bars. Velocities of straight-crested waves are computed and found to agree with those obtained from the three-dimensional theory. A uniqueness theorem reveals that three edge conditions are required.

Introduction

According to the classical two-dimensional theory of flexural motions of elastic plates, leading to Lagrange's equation

$$D\nabla^4\bar{w} + \rho h \frac{\partial^2 \bar{w}}{\partial t^2} = q(x, y, t) \quad [1]$$

the wave velocity of straight-crested waves is inversely proportional to the wave length. The exact solution, by Rayleigh (1),[2] Lamb (2), and Timoshenko (3), of the general equations of the linear theory of elasticity, for the case of straight-crested flexural waves, confirms this result only for waves which are long in comparison with the thickness of the plate. As the wave length diminishes, the velocity in the three-dimensional theory has as its upper limit the velocity of Rayleigh surface waves. Hence the classical plate theory cannot be expected to give good results for sharp transients or for the frequencies of modes of vibration of high order. A similar situation exists in regard to the classical one-dimensional theory of flexural motions of elastic bars, leading to the Bernoulli-Euler equation.

In the case of bars, correction terms have been supplied by Rayleigh and Timoshenko. Rayleigh (4) introduced the effect of rotatory inertia but, although the upper limit of the velocity is thereby made finite, its magnitude is too large. It was not until Timoshenko (5, 6) included the effect of transverse shear deformation that a one-dimensional theory of flexural motions of bars was obtained which gives satisfactory results for short waves and high modes of vibration.

In the case of plates, equations of motion analogous to Timoshenko's bar equations have been given by Uflyand (7) and a corresponding theory of plate equilibrium has been given by Hencky (8). One purpose of the present paper is to show how a more comprehensive two-dimensional theory of flexural motions of plates, analogous to Timoshenko's one-dimensional theory for bars, may be deduced directly from the three-dimensional equations of elasticity. Uflyand's equations of motion are reached, but are later modified in the light of a comparison of the velocity of straight-crested waves with the corresponding velocity obtained from the three-dimensional theory. The modification consists of the selection of the value of a constant, appearing in the relation between average transverse shear stress and strain, variously chosen as 2/3 and 8/9 by Timoshenko, and 2/3 by Uflyand. A formula for this constant, involving Poisson's ratio, is suggested (see Equation [47]). By its use, the limiting velocity for very short waves is made identical with the velocity of Rayleigh surface waves, and velocities intermediate between very long and very short waves are brought into close agreement with the three-dimensional theory.

In addition to the derivation of equations of motion, there are derived theorems dealing with kinetic and potential energies, uniqueness, initial and boundary conditions, and single-valuedness of displacements corresponding to similar theorems in the three-dimensional theory.

At various stages of the development, attention is directed to the very close similarities between the theory under discussion and E. Reissner's theory (9, 10) of flexural equilibrium of plates. Of special interest is the conclusion that, as in Reissner's theory, three boundary conditions are to be satisfied rather than the two of classical plate theory.

Plate-Stress Components

The plate is referred to an x, y, z-system of rectangular co-ordinates. The faces of the plate are the planes $z = \pm h/2$ and its cylindrical surfaces are defined by plane curves or polygons C_i parallel to the x, y-plane. The faces of the plate are taken to be free from tangential traction but under normal pressures q_1 and q_2. Thus in the usual notation

$$\tau_{zx}]_{z=\pm h/2} = \tau_{yz}]_{z=\pm h/2} = 0 \quad [2]$$

$$\sigma_z]_{z=h/2} = -q_1(x, y, t),\ \sigma_z]_{z=-h/2} = -q_2(x, y, t) \quad [3]$$

For convenience the notation $q = q_2 - q_1$ will be used. Normal pressures are retained on both faces in order that one of them may, if desired, be made proportional to the transverse displacement to simulate the effect of an elastic foundation.

The bending and twisting moments and the transverse shearing forces, all per unit of length, are defined in the customary manner

$$\left.\begin{aligned}(M_x, M_y, M_{yx}) &= \int_{-h/2}^{h/2} (\sigma_x, \sigma_y, \tau_{yx})\, z dz \\ (Q_x, Q_y) &= \int_{-h/2}^{h/2} (\tau_{zx}, \tau_{yz})\, dz\end{aligned}\right\} \quad [4]$$

[1] Professor of Civil Engineering, Columbia University; and Consultant, Bell Telephone Laboratories, Murray Hill, N. J. Mem. ASME.

[2] Numbers in parentheses refer to the Bibliography at the end of the paper.

For presentation at the Annual Conference of the Applied Mechanics Division, Purdue University, Lafayette, Ind., June 22–24, 1950, of The American Society of Mechanical Engineers.

Discussion of this paper should be addressed to the Secretary, ASME, 29 West 39th Street, New York, N. Y., and will be accepted until one month after final publication of the paper itself in the Journal of Applied Mechanics.

Note: Statements and opinions advanced in papers are to be understood as individual expressions of their authors and not those of the Society. Manuscript received at ASME Headquarters, March 29, 1950. Paper No. 50—APM-22.

For brevity these will be designated as the "plate-stress components."

Stress-Strain Relations

In three-dimensional elasticity theory there are six components of stress which are expressed in terms of six components of strain through Hooke's law. The stress-strain relations and the expressions for the six components of strain in terms of three components of displacement are then used to reduce to three the number of unknowns in the three equations of motion. In the present theory there are only five plate-stress components and these will be expressed in terms of the same number of "plate-strain components." The latter will then be expressed in terms of three "plate-displacement components."

Consider the general Hooke's law in the form wherein each of the six components of strain is expressed in terms of the six components of stress. Of the six equations, one is ignored, namely, the one containing the unit elongation (ϵ_z) normal to the faces of the plate. The remaining five equations are then solved for $\sigma_x, \sigma_y, \tau_{yz}, \tau_{zx}$ and τ_{yz} in terms of $\epsilon_x, \epsilon_y, \gamma_{yz}, \gamma_{zx}, \gamma_{yz}$ and σ_z. Alternatively, one may start with the expressions for the components of stress in terms of the components of strain, solve the one containing σ_z for ϵ_z and use this to eliminate ϵ_z from the remaining five equations. Integrations are now performed on these equations so as to convert the stress components into plate-stress components in accordance with Equations [4]. The results are then altered in two respects: (*a*) The integrals containing σ_z are dropped; (*b*) the coefficients of the integrals containing γ_{zx} and γ_{yz} are replaced by constants whose magnitudes are to be determined later. This procedure is applicable to anisotropic materials. In the case of an isotropic material, we find

$$\left.\begin{aligned} M_x &= D(\Gamma_x + \mu\Gamma_y), \quad M_y = D(\Gamma_y + \mu\Gamma_x), \\ M_{yx} &= (1-\mu)\, D\Gamma_{yx}/2 \\ Q_x &= G'h\Gamma_{zx}, \quad Q_y = G'h\Gamma_{yz} \end{aligned}\right\} \quad \ldots [5]$$

where the plate modulus D is expressed in terms of Poisson's ratio μ and Young's modulus E, or the modulus of rigidity G, by

$$D = \frac{Eh^3}{12(1-\mu^2)} = \frac{Gh^3}{6(1-\mu)} \quad \ldots\ldots\ldots\ldots [6]$$

and

$$G' = \kappa^2 G \quad \ldots\ldots\ldots\ldots [7]$$

so that G' is the constant which replaces the coefficient, G, of the integrals containing γ_{zx} and γ_{yz} in the integrated stress-strain relations and κ is the constant, referred to in the introduction, for which a formula involving Poisson's ratio will be given presently. The Γ's in Equations [5] are the plate-strain components defined by

$$\left.\begin{aligned} (\Gamma_x, \Gamma_y, \Gamma_{yx}) &= 12h^{-3}\int_{-h/2}^{h/2} (\epsilon_x, \epsilon_y, \gamma_{yx})\, z\,dz \\ (\Gamma_{zx}, \Gamma_{yz}) &= h^{-1}\int_{-h/2}^{h/2} (\gamma_{zx}, \gamma_{yz})\, dz \end{aligned}\right\} \quad \ldots [8]$$

Equations [5] would have been obtained if it had been assumed that $\sigma_z = 0$ at the start. However, the procedure adopted reveals that only a linearly weighted, average effect of σ_z is neglected, rather than σ_z itself.

Plate-Displacement Components

To proceed toward the establishment of plate-strain-displacement relations, the strain-displacement relations of the three-dimensional theory are integrated in accordance with the right-hand sides of Equations [8]. Thus

$$\left.\begin{aligned} \int_{-h/2}^{h/2} (\epsilon_x, \epsilon_y, \gamma_{yx})\, z\,dz &= \int_{-h/2}^{h/2} \left(\frac{\partial u}{\partial x}, \frac{\partial v}{\partial y}, \frac{\partial v}{\partial x} + \frac{\partial u}{\partial y}\right) z\,dz \\ \int_{-h/2}^{h/2} (\gamma_{zx}, \gamma_{yz})\,dz &= \int_{-h/2}^{h/2} \left(\frac{\partial u}{\partial z} + \frac{\partial w}{\partial x}, \frac{\partial v}{\partial z} + \frac{\partial w}{\partial y}\right) dz \end{aligned}\right\} \quad \ldots\ldots\ldots [9]$$

It is now assumed that u and v are proportional to z and w is independent of z

$$u = z\psi_x(x, y, t),\ v = z\psi_y(x, y, t),\ w = \bar{w}(x, y, t) \ldots [10]$$

The functions ψ_x and ψ_y correspond to the negative of the function ψ introduced by Timoshenko (6) in his bar theory and to the negatives of the functions ψ/h and χ/h used by Hencky (8).

Questions raised by Goodier (11), in connection with the appearance of Equations [10] in the first version (9) of Reissner's theory of plates, were answered in Reissner's second version (10) by replacing Equations [10] with averages while retaining his assumptions regarding the z-dependence of the components of stress. It appears that the reverse procedure leads to the consideration of shear deformation in the manner adopted by Timoshenko.

From Equations [8], [9], and [10], we have the relations between the plate-strain components and the plate-displacement components $\psi_x, \psi_y, \bar{w}$

$$\left.\begin{aligned} \Gamma_x &= \frac{\partial\psi_x}{\partial x},\ \Gamma_y = \frac{\partial\psi_y}{\partial y},\ \Gamma_{yx} = \frac{\partial\psi_y}{\partial x} + \frac{\partial\psi_x}{\partial y} \\ \Gamma_{zx} &= \psi_x + \frac{\partial\bar{w}}{\partial x},\ \Gamma_{yz} = \psi_y + \frac{\partial\bar{w}}{\partial y} \end{aligned}\right\} \quad \ldots\ldots [11]$$

The analogous relations for classical plate theory are obtained by setting $\Gamma_{zx} = \Gamma_{yz} = 0$.

Substituting Equations [11] into Equations [5], the following relations are obtained between the plate-stress and plate-displacement components

$$\left.\begin{aligned} M_x &= D\left(\frac{\partial\psi_x}{\partial x} + \mu\frac{\partial\psi_y}{\partial y}\right),\ M_y = D\left(\frac{\partial\psi_y}{\partial y} + \mu\frac{\partial\psi_x}{\partial x}\right) \\ M_{yx} &= \frac{1-\mu}{2} D\left(\frac{\partial\psi_y}{\partial x} + \frac{\partial\psi_x}{\partial y}\right) \\ Q_x &= \kappa^2 Gh\left(\frac{\partial\bar{w}}{\partial x} + \psi_x\right),\ Q_y = \kappa^2 Gh\left(\frac{\partial\bar{w}}{\partial y} + \psi_y\right) \end{aligned}\right\} \quad \ldots [12]$$

Equations [12] are similar to the corresponding formulas in E. Reissner's theory.[3] His "equivalent changes of slope," β_x and β_y, are here replaced by ψ_x and ψ_y; his "weighted average displacement" is here replaced by $\bar{w}$. The transverse load q does not appear in the present expressions for M_x and M_y due to the omission of the integrals containing σ_z in the integrated stress-strain relations. The constant $\kappa^2 Gh$ plays a role similar to Reissner's C_s, which has the value $(5/6)Gh$ for isotropic materials and a quadratic variation of transverse shear stress through the thickness of the plate. Hence κ^2 in Reissner's equilibrium theory has the value 5/6. In the present theory the z-dependence of the components of stress is not specified and κ is, as yet, not determined.

Equations of Motion

The stress-equations of motion of three-dimensional elasticity theory

[3] See reference (10), Equations [8], [23], and [28].

$$\left.\begin{aligned}\frac{\partial \sigma_x}{\partial x}+\frac{\partial \tau_{yx}}{\partial y}+\frac{\partial \tau_{zx}}{\partial z}&=\rho\frac{\partial^2 u}{\partial t^2}\\ \frac{\partial \tau_{yx}}{\partial x}+\frac{\partial \sigma_y}{\partial y}+\frac{\partial \tau_{yz}}{\partial z}&=\rho\frac{\partial^2 v}{\partial t^2}\\ \frac{\partial \tau_{zx}}{\partial x}+\frac{\partial \tau_{yz}}{\partial y}+\frac{\partial \sigma_z}{\partial z}&=\rho\frac{\partial^2 w}{\partial t^2}\end{aligned}\right\} \quad [13]$$

are converted to plate-stress equations of motion in the following manner.[4] The first two of Equations [13] are multiplied by z and integrated over the plate thickness. These become, after making use of Equations [2], [4], and [10]

$$\left.\begin{aligned}\frac{\partial M_x}{\partial x}+\frac{\partial M_{yx}}{\partial y}-Q_x&=\frac{\rho h^3}{12}\frac{\partial^2\psi_x}{\partial t^2}\\ \frac{\partial M_{yx}}{\partial x}+\frac{\partial M_y}{\partial y}-Q_y&=\frac{\rho h^3}{12}\frac{\partial^2\psi_y}{\partial t^2}\end{aligned}\right\} \quad [14]$$

(The right-hand sides of Equations [14] represent the influence of rotatory inertia.) The third of Equations [13] is integrated over the plate thickness and, after using Equations [3], [4], and [10], becomes

$$\frac{\partial Q_x}{\partial x}+\frac{\partial Q_y}{\partial y}+q=\rho h\frac{\partial^2 \bar{w}}{\partial t^2} \quad [15]$$

The three plate-stress equations of motion, Equations [14] and [15], may be expressed in terms of the plate-displacements ψ_x, ψ_y, $\bar{w}$ by using Equations [12], with the result

$$\left.\begin{aligned}&\frac{D}{2}\left[(1-\mu)\nabla^2\psi_x+(1+\mu)\frac{\partial\Phi}{\partial x}\right]-\kappa^2Gh\left(\psi_x+\frac{\partial\bar{w}}{\partial x}\right)=\frac{\rho h^3}{12}\frac{\partial^2\psi_x}{\partial t^2}\\ &\frac{D}{2}\left[(1-\mu)\nabla^2\psi_y+(1+\mu)\frac{\partial\Phi}{\partial y}\right]-\kappa^2Gh\left(\psi_y+\frac{\partial\bar{w}}{\partial y}\right)=\frac{\rho h^3}{12}\frac{\partial^2\psi_y}{\partial t^2}\\ &\kappa^2Gh(\nabla^2\bar{w}+\Phi)+q=\rho h\frac{\partial^2\bar{w}}{\partial t^2}\end{aligned}\right\} \quad [16]$$

where

$$\Phi=\frac{\partial\psi_x}{\partial x}+\frac{\partial\psi_y}{\partial y}$$

and ∇^2 is Laplace's two-dimensional operator.

Equations [16] are the same as Uflyand's equations of motion (7) except for the value of κ.

Energy Functions

The strain-energy-function W in the three-dimensional theory is given by

$$\left.\begin{aligned}2W&=\sigma_x\epsilon_x+\sigma_y\epsilon_y+\sigma_z\epsilon_z+\tau_{xy}\gamma_{xy}+\tau_{yz}\gamma_{yz}+\tau_{zx}\gamma_{zx}\\ &=\sigma_x\frac{\partial u}{\partial x}+\sigma_y\frac{\partial v}{\partial y}+\sigma_z\frac{\partial w}{\partial z}+\tau_{xy}\left(\frac{\partial v}{\partial x}+\frac{\partial u}{\partial y}\right)\\ &\quad+\tau_{yz}\left(\frac{\partial w}{\partial y}+\frac{\partial v}{\partial z}\right)+\tau_{zx}\left(\frac{\partial u}{\partial z}+\frac{\partial w}{\partial x}\right)\end{aligned}\right\} \quad [17]$$

[4] This method was employed by Boussinesq (12) to derive plate equations without using the calculus of variations. A. E. Green (13) has shown, recently, how the equations of E. Reissner's plate theory (10) may be obtained in a similar manner.

The analogous plate-strain-energy-function $\bar{W}$ is obtained by substituting Equations [10] into Equations [17] and integrating over the plate thickness, with the result

$$\left.\begin{aligned}2\bar{W}&=M_x\frac{\partial\psi_x}{\partial x}+M_y\frac{\partial\psi_y}{\partial y}+M_{yx}\left(\frac{\partial\psi_y}{\partial x}+\frac{\partial\psi_x}{\partial y}\right)\\ &\quad+Q_x\left(\frac{\partial\bar{w}}{\partial x}+\psi_x\right)+Q_y\left(\frac{\partial\bar{w}}{\partial y}+\psi_y\right)\\ &=M_x\Gamma_x+M_y\Gamma_y+M_{yx}\Gamma_{yx}+Q_x\Gamma_{zx}+Q_y\Gamma_{yz}\end{aligned}\right\} \quad [18]$$

and this becomes, through Equations [5]

$$4\bar{W}=D(1+\mu)(\Gamma_x+\Gamma_y)^2+D(1-\mu)\left[(\Gamma_x-\Gamma_y)^2+\Gamma_{yx}^2\right]+2\kappa^2Gh(\Gamma_{yz}^2+\Gamma_{zx}^2) \quad [19]$$

If $E/(1+\mu)$ and $E/(1-\mu)$ are required to be positive, $\bar{W}$ is positive. Then, if $\bar{W}=0$, all the plate-strain components vanish and, through Equations [5], so do the plate-stress components.

From Equations [19] and [5], we find

$$\left.\begin{aligned}&\frac{\partial\bar{W}}{\partial\Gamma_x}=M_x,\quad\frac{\partial\bar{W}}{\partial\Gamma_y}=M_y,\quad\frac{\partial\bar{W}}{\partial\Gamma_{yx}}=M_{yx}\\ &\frac{\partial\bar{W}}{\partial\Gamma_{zx}}=Q_x,\quad\frac{\partial\bar{W}}{\partial\Gamma_{yz}}=Q_y\end{aligned}\right\} \quad [20]$$

The kinetic energy per unit of volume, according to the general linear theory, is

$$\frac{\rho}{2}\left[\left(\frac{\partial u}{\partial t}\right)^2+\left(\frac{\partial v}{\partial t}\right)^2+\left(\frac{\partial w}{\partial t}\right)^2\right] \quad [21]$$

By using Equations [10] and integrating over the thickness, this becomes, for the plate

$$\frac{\rho h^3}{24}\left[\left(\frac{\partial\psi_x}{\partial t}\right)^2+\left(\frac{\partial\psi_y}{\partial t}\right)^2\right]+\frac{\rho h}{2}\left(\frac{\partial\bar{w}}{\partial t}\right)^2 \quad [22]$$

Total Energy and External Work

The kinetic energy in the plate at time t is given by

$$\bar{T}=\int\int\left\{\frac{\rho h^3}{24}\left[\left(\frac{\partial\psi_x}{\partial t}\right)^2+\left(\frac{\partial\psi_y}{\partial t}\right)^2\right]+\frac{\rho h}{2}\left(\frac{\partial\bar{w}}{\partial t}\right)^2\right\}dx\,dy \quad [23]$$

and the potential energy is

$$\bar{V}=\int\int\bar{W}\,dx\,dy \quad [24]$$

where the integrations are extended over the surface of the plate.

The total energy at time t is the sum of $\bar{T}$ and $\bar{V}$, which may be written as

$$\begin{aligned}\bar{T}+\bar{V}&=\int_{t_0}^{t}dt\int\int\frac{\partial}{\partial t}\left\{\frac{\rho h^3}{24}\left[\left(\frac{\partial\psi_x}{\partial t}\right)^2+\left(\frac{\partial\psi_y}{\partial t}\right)^2\right]+\frac{\rho h}{2}\left(\frac{\partial\bar{w}}{\partial t}\right)^2\right\}dx\,dy\\ &\quad+\int_{t_0}^{t}dt\int\int\frac{\partial\bar{W}}{\partial t}dx\,dy+\bar{T}_0+\bar{V}_0\end{aligned} \quad [25]$$

where $\bar{T}_0$ and $\bar{V}_0$ are the values of $\bar{T}$ and $\bar{V}$ at an initial time t_0.

Performing the operations $\partial/\partial t$ as indicated in Equation [25], the first integrand becomes

$$\frac{\rho h^3}{12}\left[\frac{\partial\psi_x}{\partial t}\frac{\partial^2\psi_x}{\partial t^2}+\frac{\partial\psi_y}{\partial t}\frac{\partial^2\psi_y}{\partial t^2}\right]+\rho h\frac{\partial\bar{w}}{\partial t}\frac{\partial^2\bar{w}}{\partial t^2} \quad [26]$$

and the second integrand is

$$\frac{\partial \overline{W}}{\partial t} = \frac{\partial \overline{W}}{\partial \Gamma_x}\frac{\partial \Gamma_x}{\partial t} + \frac{\partial \overline{W}}{\partial \Gamma_y}\frac{\partial \Gamma_y}{\partial t} + \frac{\partial \overline{W}}{\partial \Gamma_{yz}}\frac{\partial \Gamma_{yz}}{\partial t} + \frac{\partial \overline{W}}{\partial \Gamma_{zx}}\frac{\partial \Gamma_{zx}}{\partial t}$$

$$+ \frac{\partial \overline{W}}{\partial \Gamma_{yx}}\frac{\partial \Gamma_{yx}}{\partial t} = \left(M_x \frac{\partial}{\partial x} + M_{yx}\frac{\partial}{\partial y} + Q_x\right)\frac{\partial \psi_x}{\partial t}$$

$$+ \left(M_{yx}\frac{\partial}{\partial x} + M_y \frac{\partial}{\partial y} + Q_y\right)\frac{\partial \psi_y}{\partial t} + \left(Q_x \frac{\partial}{\partial x} + Q_y \frac{\partial}{\partial y}\right)\frac{\partial \bar{w}}{\partial t} \quad \ldots\ldots\ldots [27]$$

by Equations [11] and [20].

In the surface integral of $\partial\overline{W}/\partial t$, Equation [25], the terms containing space derivatives may be integrated by parts to obtain

$$\iint \frac{\partial \overline{W}}{\partial t}\,dx\,dy = \oint \left[\frac{\partial \psi_x}{\partial t}(M_x l + M_{yx} m) + \frac{\partial \psi_y}{\partial t}(M_{yx} l + M_y m) + \frac{\partial \bar{w}}{\partial t}(Q_x l + Q_y m)\right] ds$$

$$- \iint \left[\frac{\partial \psi_x}{\partial t}\left(\frac{\partial M_x}{\partial x} + \frac{\partial M_{yx}}{\partial y} - Q_x\right) + \frac{\partial \psi_y}{\partial t}\left(\frac{\partial M_{yx}}{\partial x} + \frac{\partial M_y}{\partial y} - Q_y\right) + \frac{\partial \bar{w}}{\partial t}\left(\frac{\partial Q_x}{\partial x} + \frac{\partial Q_y}{\partial y}\right)\right] dx\,dy \ldots\ldots [28]$$

in which l and m are the direction cosines of the normal to the boundary and the line integral is taken around all (external and internal) boundaries. In the integration by parts it is assumed that ψ_x, ψ_y, and $\bar{w}$ are single-valued, otherwise additional line integrals would be required.

In terms of components referred to co-ordinates ν and s measured normal to and along the boundary, respectively, the line integral in Equation [28] may be written as

$$\oint \left(\frac{\partial \psi_\nu}{\partial t} M_\nu + \frac{\partial \psi_s}{\partial t} M_{\nu s} + \frac{\partial \bar{w}}{\partial t} Q_\nu\right) ds$$

Combining these results, we have

$$\overline{T} + \overline{V} = \int_{t_0}^{t} dt \oint \left(\frac{\partial \psi_\nu}{\partial t} M_\nu + \frac{\partial \psi_s}{\partial t} M_{\nu s} + \frac{\partial \bar{w}}{\partial t} Q_\nu\right) ds$$

$$+ \int_{t_0}^{t} dt \iint \left[\frac{\partial \psi_x}{\partial t}\left(\frac{\rho h^3}{12}\frac{\partial^2 \psi_x}{\partial t^2} - \frac{\partial M_x}{\partial x} - \frac{\partial M_{yx}}{\partial y} + Q_x\right) + \frac{\partial \psi_y}{\partial t}\left(\frac{\rho h^3}{12}\frac{\partial^2 \psi_y}{\partial t^2} - \frac{\partial M_{yx}}{\partial x} - \frac{\partial M_y}{\partial y} + Q_y\right) + \frac{\partial \bar{w}}{\partial t}\left(\rho h \frac{\partial^2 \bar{w}}{\partial t^2} - \frac{\partial Q_x}{\partial x} - \frac{\partial Q_y}{\partial y}\right)\right] dx\,dy + \overline{T}_0 + \overline{V}_0$$

If Equations [14] and [15] (the plate equations of motion) are satisfied, this reduces to

$$\overline{T} + \overline{V} = \int_{t_0}^{t} dt \oint \left(\frac{\partial \psi_\nu}{\partial t} M_\nu + \frac{\partial \psi_s}{\partial t} M_{\nu s} + \frac{\partial \bar{w}}{\partial t} Q_\nu\right) ds + \int_{t_0}^{t} dt \iint q \frac{\partial \bar{w}}{\partial t}\,dx\,dy + \overline{T}_0 + \overline{V}_0 \ldots\ldots [29]$$

Equation [29] is simply the form, in the present theory, of the statement that the total energy in the plate at time t is equal to the sum of the energy at time t_0 and the work done by the external forces along the edge and over the surface of the plate during the interval $t - t_0$.

Initial and Boundary Conditions

Appropriate initial and boundary conditions for a system of differential equations are those which are sufficient to assure a unique solution. In the preceding two sections the groundwork has been laid for the application of a plate analog of F. Neumann's uniqueness theorem (14).

Consider two sets of plate displacements and surface loads. If the components of each set satisfy the equations of motion, so will their differences. Then the energies and the plate-stress components calculated from the differences will satisfy an equation of the same form as Equation [29]. If the right-hand side of Equation [29] vanishes, $\overline{T}$ and $\overline{V}$ vanish separately, since they are both positive. If $\overline{T}$ vanishes, the kinetic energy per unit of volume vanishes, since it is positive, and hence the plate velocities vanish. If $\overline{V}$ vanishes, $\overline{W}$ vanishes, since it is positive, and if $\overline{W}$ vanishes, so do the plate-strain and plate-stress components, as long as $E/(1 + \mu)$ and $E/(1 - \mu)$ are required to be positive. Hence, if the right-hand side of Equation [29] (calculated from the differences) vanishes, the two systems must be identical, except possibly for a rigid-body displacement (independent of time, since the velocities vanish) and the latter can be eliminated by requiring the initial displacements to be the same.

Returning, now, to a single set of solutions, appropriate initial and boundary conditions to be specified for the present theory are as follows:

1 On the edges of plate: any combination which contains one member of each of the three pairs of terms in the line integral of Equation [29].

2 Throughout the plate: (*a*) either q or $\bar{w}$, and (*b*) the initial values of ψ_x, ψ_y, $\bar{w}$ and their time derivatives.

The relation (see Equations [12])

$$\psi_s = \frac{Q_s}{\kappa^2 G h} - \frac{\partial \bar{w}}{\partial s}$$

is replaced, in the classical theory of plates, by

$$\psi_s = -\frac{\partial \bar{w}}{\partial s}$$

since the transverse shear deformation $Q_s/\kappa^2 Gh$ vanishes. The term $M_{\nu s}\,\partial\psi_s/\partial t$ in Equation [29] would then be replaced by $-M_{\nu s}\partial^2\bar{w}/\partial s\,\partial t$. But

$$-\oint M_{\nu s}\frac{\partial}{\partial s}\left(\frac{\partial \bar{w}}{\partial t}\right) ds = \oint \frac{\partial \bar{w}}{\partial t}\frac{\partial M_{\nu s}}{\partial s}\,ds$$

Hence the second and third terms in the integrand of the line integral of Equation [29] would combine into the single term

$$\frac{\partial \bar{w}}{\partial t}\left(\frac{\partial M_{\nu s}}{\partial s} + Q_\nu\right)$$

leaving only the two edge conditions of Kirchoff. In the present theory, however, three edge conditions are required, as in E. Reissner's theory.

Compatibility Equations

In the integration by parts leading to Equation [29] it was assumed that ψ_x, ψ_y, and $\bar{w}$ are single-valued. If this restriction were relaxed, additional line integrals would appear in Equation [29], and hence there would be additional requirements for uniqueness. If the restriction to single-valued displacements is retained, it is necessary to specify that

$$\oint d\psi_x = 0, \quad \oint d\psi_y = 0, \quad \oint d\bar{w} = 0 \ldots\ldots [30]$$

where the line integrals are taken around any closed path in the plate.

Usually, single-valuedness may be ascertained by inspection. However, if the differential equations are expressed entirely in terms of plate-stress components, as they may be in equilibrium problems, additional conditions are introduced by Equations [30]. These may be found as follows

$$\oint d\psi_x = \oint \frac{\partial \psi_x}{\partial x} dx + \frac{\partial \psi_x}{\partial y} dy$$

$$= \oint \left(\Gamma_x \, dx + \frac{1}{2} \Gamma_{yx} \, dy \right) - \oint \Omega_z dy$$

where

$$\Omega_z = \frac{1}{2}\left(\frac{\partial \psi_y}{\partial x} - \frac{\partial \psi_x}{\partial y} \right)$$

Now

$$-\oint \Omega_z dy = \oint (y - y_0) d\Omega_z$$

where y_0 is the y-co-ordinate of the starting point of integration. Also

$$d\Omega_z = \frac{\partial \Omega_z}{\partial x} dx + \frac{\partial \Omega_z}{\partial y} dy$$

and

$$2\frac{\partial \Omega_z}{\partial x} = \frac{\partial \Gamma_{yx}}{\partial x} - 2\frac{\partial \Gamma_x}{\partial y}, \quad 2\frac{\partial \Omega_z}{\partial y} = 2\frac{\partial \Gamma_y}{\partial x} - \frac{\partial \Gamma_{yx}}{\partial y}$$

Combining these results, we have

$$\oint d\psi_x = \oint (\xi_x dx + \eta_x \, dy) \ldots\ldots\ldots\ldots [31]$$

where

$$\xi_x = \Gamma_x + (y - y_0)\frac{1}{2}\left(\frac{\partial \Gamma_{yx}}{\partial x} - \frac{\partial \Gamma_x}{\partial y} \right)$$

$$\eta_x = \frac{1}{2}\Gamma_{yx} + (y - y_0)\left(\frac{\partial \Gamma_y}{\partial x} - \frac{1}{2}\frac{\partial \Gamma_{yx}}{\partial y} \right)$$

Similarly

$$\oint d\psi_y = \oint (\xi_y \, dx + \eta_y \, dy) \ldots\ldots\ldots\ldots [32]$$

where

$$\xi_y = \frac{1}{2}\Gamma_{yx} - (x - x_0)\left(\frac{1}{2}\frac{\partial \Gamma_{yx}}{\partial x} - \frac{\partial \Gamma_x}{\partial y} \right)$$

$$\eta_y = \Gamma_y - (x - x_0)\left(\frac{\partial \Gamma_y}{\partial x} - \frac{1}{2}\frac{\partial \Gamma_{yx}}{\partial y} \right)$$

For the condition on $\bar{w}$, we have

$$\oint d\bar{w} = \left(\frac{\partial \bar{w}}{\partial x} dx + \frac{\partial \bar{w}}{\partial y} dy \right)$$

$$= \frac{1}{2}\oint (\Gamma_{zx} \, dx + \Gamma_{yz} \, dy) - \oint (\Omega_y \, dx - \Omega_x \, dy)$$

where

$$\Omega_x = \frac{1}{2}\left(\frac{\partial \bar{w}}{\partial y} - \psi_y \right), \quad \Omega_y = \frac{1}{2}\left(\psi_x - \frac{\partial \bar{w}}{\partial x} \right)$$

(Note that Ω_x, Ω_y, Ω_z are the plate-rotation components.)

Now

$$-\oint (\Omega_y \, dx - \Omega_x dy) = \oint [(x - x_0) d\Omega_y - (y - y_0) d\Omega_x]$$

where (x_0, y_0) is the starting point of integration. Also

$$d\Omega_x = \frac{\partial \Omega_x}{\partial x} dx + \frac{\partial \Omega_x}{\partial y} dy, \quad d\Omega_y = \frac{\partial \Omega_y}{\partial x} dx + \frac{\partial \Omega_y}{\partial y} dy$$

and

$$2\frac{\partial \Omega_x}{\partial x} = \frac{\partial \Gamma_{zx}}{\partial y} - \Gamma_{yz}, \; 2\frac{\partial \Omega_x}{\partial y} = \frac{\partial \Gamma_{yz}}{\partial y} - 2\Gamma_y$$

$$2\frac{\partial \Omega_y}{\partial y} = \Gamma_{yz} - \frac{\partial \Gamma_{yz}}{\partial x}, \; 2\frac{\partial \Omega_y}{\partial x} = 2\Gamma_x - \frac{\partial \Gamma_{zx}}{\partial x}$$

Combining these results, we have

$$\oint d\bar{w} = \oint (\xi_z \, dx + \eta_z \, dy) \ldots\ldots\ldots\ldots [33]$$

where

$$\xi_z = \frac{1}{2}\Gamma_{xz} + (x - x_0)\left(\Gamma_x - \frac{1}{2}\frac{\partial \Gamma_{zx}}{\partial x} \right) + \frac{1}{2}(y - y_0)\left(\Gamma_{yz} - \frac{\partial \Gamma_{zx}}{\partial y} \right)$$

$$\eta_z = \frac{1}{2}\Gamma_{yz} + \frac{1}{2}(x - x_0)\left(\Gamma_{yz} - \frac{\partial \Gamma_{yz}}{\partial x} \right) + (y - y_0)\left(\Gamma_y - \frac{1}{2}\frac{\partial \Gamma_{yz}}{\partial y} \right)$$

Equations [31], [32], and [33] constitute the plate analog of Cesàro's theorem[5] for closed paths. Applying Green's theorem to these equations, we find, for reducible paths

$$\left.\begin{aligned}
\oint d\psi_x &= \iint (y - y_0, \left(\frac{\partial^2 \Gamma_y}{\partial x^2} + \frac{\partial^2 \Gamma_x}{\partial y^2} - \frac{\partial^2 \Gamma_{yx}}{\partial x \partial y} \right) dx \, dy \\
\oint d\psi_y &= -\iint (x - x_0)\left(\frac{\partial^2 \Gamma_y}{\partial x^2} + \frac{\partial^2 \Gamma_x}{\partial y^2} - \frac{\partial^2 \Gamma_{yx}}{\partial x \partial y} \right) dx \, dy \\
\oint d\bar{w} &= \frac{1}{2}\iint (y - y_0)\left(2\frac{\partial \Gamma_y}{\partial x} - \frac{\partial \Gamma_{yx}}{\partial y} \frac{\partial^2 \Gamma_{yz}}{\partial x \partial y} + \frac{\partial^2 \Gamma_{zx}}{\partial y^2} \right) dx \, dy \\
&\quad - \frac{1}{2}\iint (x - x_0)\left(2\frac{\partial \Gamma_x}{\partial y} - \frac{\partial \Gamma_{yx}}{\partial x} - \frac{\partial^2 \Gamma_{zx}}{\partial x \partial y} + \frac{\partial^2 \Gamma_{yz}}{\partial x^2} \right) dx \, dy
\end{aligned}\right\} \ldots [34]$$

Hence, three necessary conditions on the plate strains for single-valuedness of ψ_x, ψ_y, $\bar{w}$ are

[5] See reference (14), p. 223.

$$\left.\begin{aligned}&\frac{\partial^2\Gamma_y}{\partial x^2}+\frac{\partial^2\Gamma_x}{\partial y^2}=\frac{\partial^2\Gamma_{yx}}{\partial x\,\partial y}\\&2\frac{\partial\Gamma_x}{\partial y}=\frac{\partial}{\partial x}\left(-\frac{\partial\Gamma_{yz}}{\partial x}+\frac{\partial\Gamma_{zx}}{\partial y}+\Gamma_{yx}\right)\\&2\frac{\partial\Gamma_y}{\partial x}=\frac{\partial}{\partial y}\left(-\frac{\partial\Gamma_{zx}}{\partial y}+\Gamma_{yx}+\frac{\partial\Gamma_{yz}}{\partial x}\right)\end{aligned}\right\}\quad\ldots\ldots[35]$$

and these are also sufficient conditions for a simply connected plate. In an $(n + 1)$ connected plate, each application of Green's theorem introduces n integrals around the n internal boundaries. Hence there are $3n$ additional conditions

$$\oint_{Ci} d\psi_x = 0,\quad \oint_{Ci} d\psi_y = 0,\quad \oint_{Ci} d\bar{w} = 0,\ i = 1, 2 \ldots n$$

where the paths C_i are the internal boundaries.

Equations [35] are the analogs, in the present theory, of the six compatibility equations in 3-dimensional linear elasticity theory. That they are necessary conditions for single-valuedness of ψ_x, ψ_y, and $\bar{w}$ may be determined directly by eliminating these quantities from Equations [11]. The plate-compatibility equations may be expressed in terms of the plate-stress components by using Equations [5].

Equations Governing $\bar{w}$

A single differential equation for $\bar{w}$ may be obtained by eliminating ψ_x and ψ_y from Equations [16] in the following manner: Differentiate the first and second of Equations [16] with respect to x and y, respectively, and add to obtain

$$\left(D\nabla^2 - G'h - \frac{\rho h^3}{12}\frac{\partial^2}{\partial t^2}\right)\Phi = G'h\nabla^2\bar{w}\ldots\ldots\ldots[36]$$

Then eliminate Φ between Equation [36] and the third of Equations [16] with the result

$$\left(\nabla^2 - \frac{\rho}{G'}\frac{\partial^2}{\partial t^2}\right)\left(D\nabla^2 - \frac{\rho h^3}{12}\frac{\partial^2}{\partial t^2}\right)\bar{w} + \rho h\frac{\partial^2\bar{w}}{\partial t^2} = \left(1 - \frac{D\nabla^2}{G'h} + \frac{\rho h^2}{12G'}\frac{\partial^2}{\partial t^2}\right)q\ldots\ldots[37]$$

Equation [37] is the two-dimensional analog of Timoshenko's beam equation (6).

If the rotatory inertia terms are omitted from Equation [37], it reduces to

$$D\left(\nabla^2 - \frac{\rho}{G'}\frac{\partial^2}{\partial t^2}\right)\nabla^2\bar{w} + \rho h\frac{\partial^2\bar{w}}{\partial t^2} = \left(1 - \frac{D\nabla^2}{G'h}\right)q\ldots[38]$$

If the transverse shear deformation is neglected, but the rotatory inertia terms are retained, Equation [37] reduces to

$$\left(D\nabla^2 - \frac{\rho h^3}{12}\frac{\partial^2}{\partial t^2}\right)\nabla^2\bar{w} + \rho h\frac{\partial^2\bar{w}}{\partial t^2} = q\ldots\ldots\ldots[39]$$

Finally, if both the transverse shear deformation and rotatory inertia terms are omitted, Equation [37] reduces to Equation [1] in which $\nabla^4 = \nabla^2\nabla^2$.

Straight-Crested Waves

The exact solution of the three-dimensional equations of elasticity is known for the case of straight-crested flexural waves in an infinite plate (1, 2, 3). This solution will be used to test the present theory and to complete the equations of motion and the stress-strain-displacement relations by choosing the value of κ.

Three-Dimensional Theory. In the three-dimensional theory, the wave velocity c of wave length λ is given in the form of the trancendental equation (1, 2, 3)

$$\frac{4c_s^2\sqrt{(c_s^2 - \alpha c^2)(c_s^2 - c^2)}}{(2c_s^2 - c^2)^2} = \frac{\tanh\frac{\pi h}{\lambda c_s}\sqrt{c_s^2 - \alpha c^2}}{\tanh\frac{\pi h}{\lambda c_s}\sqrt{c_s^2 - c^2}},\ 0 < \frac{c}{c_s} < 1\ldots\ldots\ldots[40]$$

where

$$c_s = \sqrt{G/\rho}$$

i.e., the velocity of shear waves and

$$\alpha = \frac{1 - 2\mu}{2(1 - \mu)}$$

Lamb's values (2) for c/c_s versus h/λ (with $\mu = 1/2$) are plotted

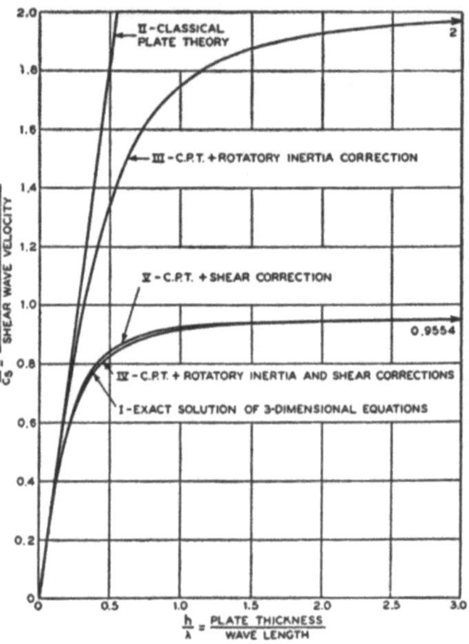

Fig. 1

as curve I in Fig. 1. For long waves ($\lambda >> h$) Equation [40] reduces to

$$\frac{c^2}{c_s^2} = \frac{2\pi^2}{3(1 - \mu)}\left(\frac{h}{\lambda}\right)^2\ldots\ldots\ldots\ldots\ldots[41]$$

while the limit for short waves ($\lambda \to 0$) is

$$4c_s^2\sqrt{(c_s^2 - \alpha c^2)(c_s^2 - c^2)} = (2c_s^2 - c^2)^2,\ 0 < \frac{c}{c_s} < 1\ldots[42]$$

i.e., the equation for the velocity of Rayleigh surface waves. For $\mu = 1/2$, Equation [42] gives $c = 0.9554\,c_s$.

Classical Plate Theory. If the solution for a straight-crested wave

$$\bar{w} = \cos\frac{2\pi}{\lambda}(x - ct) \qquad [43]$$

is substituted into Equation [1], with $q = 0$, we find

$$c^2 = \frac{D}{\rho h}\left(\frac{2\pi}{\lambda}\right)^2 = \frac{2\pi^2 c_s^2}{3(1-\mu)}\left(\frac{h}{\lambda}\right)^2 \qquad [44]$$

which is identical with Equation [41]. Thus, as observed by Rayleigh (1), classical plate theory gives good results for long waves. Equation [44] is plotted (for $\mu = 1/2$) as curve II in Fig. 1. It may be seen that when the wave length becomes less than 5 or 10 times the plate thickness the classical plate theory departs markedly from the three-dimensional theory.

Rotatory Inertia Correction. We consider, next, Equation [39] (with $q = 0$), which contains only the rotatory inertia correction. Insertion of $\bar{w}$ from Equation [43] results in

$$\frac{c^2}{c_s^2} = \frac{2\pi^2}{3(1-\mu)}\left(\frac{h}{\lambda}\right)^2\left[1 + \frac{\pi^2}{3}\left(\frac{h}{\lambda}\right)^2\right]^{-1} \qquad [45]$$

Equation [45], with $\mu = 1/2$, is plotted in Fig. 1 as curve III. For $\lambda \gg h$, Equation [45] reduces to [41] as it should; but the limit of c^2/c_s^2 as $\lambda \to 0$ is $2/(1 - \mu)$, a value that is much too large in comparison with the solution of Equation [42].

Rotatory Inertia and Shear Corrections. In this case the complete left-hand side of Equation [37] is used, with the result

$$\frac{\pi^2}{3}\left(\frac{h}{\lambda}\right)^2\left(1 - \frac{c^2}{\kappa^2 c_s^2}\right)\left(\frac{c_p^2}{c^2} - 1\right) = 1 \qquad [46]$$

where

$$c_p = \sqrt{\frac{E}{\rho(1-\mu^2)}}$$

Accordingly, there are two roots for c^2. Confining attention to the smaller root, Equation [46] reduces to Equation [41] for long waves, as it should. The limit of Equation [46] as $\lambda \to 0$ is

$$c^2 = \kappa^2 c_s^2$$

According to the three-dimensional theory, this should be the velocity of Rayleigh surface waves. Hence, by Equation [42] the appropriate value of κ^2 is the root of

$$4\sqrt{(1-\alpha\kappa^2)(1-\kappa^2)} = (2-\kappa^2)^2,\quad 0 < \kappa < 1 \qquad [47]$$

For anisotropic plates the analogous constants would be obtained from the algebraic equations whose roots give the velocities of surface waves appropriate to the medium. The dependence of κ^2 on Poisson's ratio is illustrated by noting that it ranges almost linearly from 0.76 for $\mu = 0$ to 0.91 for $\mu = 1/2$.

The smaller root of Equation [46], with $\mu = 1/2$ and $\kappa = 0.9554$, is plotted as curve IV in Fig. 1. The correspondence with the exact solution of the three-dimensional equations is so close that no difference between the two can be detected in computations carried to three figures. Curves IV and I are therefore the same in Fig. 1.

Shear Correction Only. It is interesting to notice that the transverse shear deformation accounts almostly entirely for the discrepancy between classical plate theory and the three-dimensional theory over the whole wave-length spectrum. Equation [38], which contains the shear-correction term only, yields

$$\frac{c^2}{c_s^2} = \frac{2\pi^2}{3(1-\mu)}\left(\frac{h}{\lambda}\right)^2\left[1 + \frac{2\pi^2}{3\kappa^2(1-\mu)}\left(\frac{h}{\lambda}\right)^2\right]^{-1} \qquad [48]$$

This is plotted, for $\mu = 1/2$, as curve V in Fig. 1.

Thickness-Shear Motion

The circular frequency of the first antisymmetric mode of thickness-shear vibration, according to the exact theory, is[6]

$$p = \pi c_s/h \qquad [49]$$

The corresponding solution in the present theory is obtained by setting

$$\psi_y = \bar{w} = 0,\quad \psi_x = e^{ipt}$$

in Equations [16], with the result

$$p = \kappa c_s\sqrt{12}/h \qquad [50]$$

To make Equations [49] and [50] identical requires

$$\kappa^2 = \pi^2/12 \qquad [51]$$

a value very close to Reissner's 5/6.

Substituting $\kappa^2 = \pi^2/12$ into Equation [47] we find

$$\mu = 0.176$$

Thus, in a material with a Poisson's ratio of 0.176, there is no conflict between Equations [51] and [47]. For other values of μ, one must choose or compromise between Equations [51] and [47] in accordance with the relative importance of the two modes of motion.

Reduction to Wave Equations

In the absence of surface loading, the equations of motion may be transformed to an exceptionally simple form with the loss of only trivial solutions.

If two of the plate-displacement components are expressed in terms of the potentials which give rise to the areal dilatation and rotation

$$\psi_x = \partial\varphi/\partial x + \partial H/\partial y,\quad \psi_y = \partial\varphi/\partial y - \partial H/\partial x \qquad [52]$$

and a factor e^{ipt} (hereafter omitted) is assumed, Equations [16] become

$$\left.\begin{aligned}
&\frac{\partial}{\partial x}[\nabla^2\varphi + (R\delta_0^4 - S^{-1})\varphi S - S^{-1}\bar{w}] \\
&\qquad + \frac{1-\mu}{2}\frac{\partial}{\partial y}(\nabla^2 + \omega^2)H = 0 \\
&\frac{\partial}{\partial y}[\nabla^2\varphi + (R\delta_0^4 - S^{-1})\varphi - S^{-1}\bar{w}] \\
&\qquad - \frac{1-\mu}{2}\frac{\partial}{\partial x}(\nabla^2 + \omega^2)H = 0 \\
&\nabla^2(\varphi + \bar{w}) + S\delta_0^4\bar{w} = 0
\end{aligned}\right\} \qquad [53]$$

where

$$R = h^2/12,\quad S = D/\kappa^2 Gh,\quad \delta_0^4 = \rho p^2 h/D,$$
$$\omega^2 = 2(R\delta_0^4 - S^{-1})/(1-\mu) \qquad [54]$$

The constants R and S represent the effects of rotary inertia and transverse shear deformation, respectively.

H may be separated from φ and $\bar{w}$ by differentiation, addition, and subtraction of the first two of Equations [53]. These equations become

$$\left.\begin{aligned}
&\nabla^2(\nabla^2 + \omega^2)H = 0 \\
&\nabla^2[\nabla^2\varphi + (R\delta_0^4 - S^{-1})\varphi - S^{-1}\bar{w}] = 0
\end{aligned}\right\} \qquad [55]$$

[6] Reference (2), p. 122.

It now remains to separate φ and $\bar{w}$ between the second of Equations [55] and the third of Equations [53]. It will be observed that these equations are satisfied by

$$\varphi = (\sigma - 1)\bar{w} \qquad [56]$$

where σ is a constant, provided

$$\sigma^{-1}S\delta_0^4 = R\delta_0^4 - S^{-1} - [S(\sigma - 1)]^{-1} = \delta^2 \text{ (say)} \qquad [57]$$

and

$$\nabla^2\bar{w} + \delta^2\bar{w} = 0 \qquad [58]$$

Solution of Equations [57] yields

$$\delta_1^2, \delta_2^2 = \frac{1}{2}\delta_0^4\left[R + S \pm \sqrt{(R - S)^2 + 4\delta_0^{-4}}\right] \qquad [59]$$

$$\sigma_1, \sigma_2 = (\delta_2^2, \delta_1^2)(R\delta_0^4 - S^{-1})^{-1} \qquad [60]$$

Hence, two deflection functions ($\bar{w}_1$ and $\bar{w}_2$) are obtained by setting σ, in Equation [56], equal to σ_1 or σ_2 and these functions are governed by separate wave equations obtained by setting δ equal to δ_1 or δ_2 in Equation [58]. Equation [56] is now shown to be nonrestrictive by noting that Equation [37] may be written as

$$(\nabla^2 + \delta_1^2)(\nabla^2 + \delta_2^2)\bar{w} = 0 \qquad [61]$$

when $q = 0$. Finally, in view of Equations [56] and [58], the bracketed terms in Equations [53] vanish, so that the equation governing H reduces to

$$(\nabla^2 + \omega^2)H = 0$$

To sum up, we may write

$$\psi_x = (\sigma_1 - 1)\partial\bar{w}_1/\partial x + (\sigma_2 - 1)\partial\bar{w}_2/\partial x + \partial H/\partial y$$
$$\psi_y = (\sigma_1 - 1)\partial\bar{w}_1/\partial y + (\sigma_2 - 1)\partial\bar{w}_2/\partial y - \partial H/\partial x \qquad [62]$$
$$\bar{w} = \bar{w}_1 + \bar{w}_2$$

and

$$(\nabla^2 + \omega^2)H = 0 \qquad (\nabla^2 + \delta_1^2)\bar{w}_1 = 0 \qquad (\nabla^2 + \delta_2^2)\bar{w}_2 = 0 \qquad [63]$$

It may be observed that, if $R = S = 0$, H and σ vanish and $\delta^2 = \pm\delta_0^2$, i.e., the present equations all reduce to those of classical plate theory when the effects of rotatory inertia and shear are removed.

BIBLIOGRAPHY

1 "On the Free Vibrations of an Infinite Plate of Homogeneous Isotropic Elastic Matter," by Lord Rayleigh, Proceedings of the London Mathematical Society, London, England, vol. 10, 1889, pp. 225–234.

2 "On Waves in an Elastic Plate," by H. Lamb, Proceedings of the Royal Society of London, England, series A, vol. 93, 1917, pp. 114–128.

3 "On the Transverse Vibrations of Bars of Uniform Cross Section," by S. Timoshenko, *Philosophical Magazine*, series 6, vol. 43, 1922, pp. 125–131.

4 "Theory of Sound," by Lord Rayleigh, second edition, The Macmillan Company, New York, N. Y., vol. 1, p. 258.

5 "On the Correction for Shear of the Differential Equation for Transverse Vibrations of Prismatic Bars," by S. Timoshenko, Philosophical Magazine, series 6, vol. 41, 1921, pp. 744–746.

6 "Vibration Problems in Engineering," by S. Timoshenko, second edition, D. Van Nostrand Company, Inc., New York, N. Y., 1937, p. 337.

7 "The Propagation of Waves in the Transverse Vibrations of Bars and Plates," by Ya. S. Uflyand, Akad. Nauk SSSR. Prikl. Mat. Meh., vol. 12, 1948, pp. 287–300 (Russian).

8 "Über die Berücksichtigung der Schubverzerrung in ebenen Platten," by H. Hencky, Ingenieur Archiv., vol. 16, 1947, pp. 72–76.

9 "The Effect of Transverse Shear Deformation on the Bending of Elastic Plates," by E. Reissner, Journal of Applied Mechanics, Trans. ASME, vol. 67, 1945, p. A-69.

10 "On Bending of Elastic Plates," by E. Reissner, *Quarterly o Applied Mathematics*, vol. 5, 1947, pp. 55–68.

11 Discussion of Ref. (9), by J. N. Goodier, Journal of Applied Mechanics, Trans. ASME, vol. 68, 1946, p. A-251.

12 "Étude nouvelle sur l'équilibre et le mouvement des corps solides élastiques dont certaines dimensions sont très-petites par rapport à d'autres," by J. Boussinesq, Journ. de Math., Paris, France, series 2, vol. 16, 1871, pp. 125–274, and series 3, vol. 5, 1879, pp. 329–344

13 "On Reissner's Theory of Bending of Elastic Plates," by A. E. Green, *Quarterly of Applied Mathematics*, vol. 7, 1949, pp. 223–228.

14 "Theory of Elasticity," by A. E. H. Love, fourth edition, The Macmillan Company, New York, N. Y., p. 176.

Reprinted from
Journal of Applied Mechanics
18 (1951) 31-38

Reprinted from JOURNAL OF APPLIED PHYSICS, Vol. 22, No. 3, 316–323, March, 1951

Thickness-Shear and Flexural Vibrations of Crystal Plates

RAYMOND D. MINDLIN
Columbia University, New York, and Bell Telephone Laboratories, Murray Hill, New Jersey
(Received September 18, 1950)

The theory of flexural motions of elastic plates, including the effects of rotatory inertia and shear, is extended to crystal plates. The equations are solved approximately for the case of rectangular plates excited by thickness-shear deformation parallel to one edge. Results of computations of resonant frequencies of rectangular, *AT*-cut, quartz plates are shown and compared with experimental data. Simple algebraic formulas are obtained relating frequency, dimensions, and crystal properties for resonances of special interest in design.

INTRODUCTION

EXACT solutions of the three-dimensional equations governing the vibrations of crystals have been obtained for only a few body shapes and modes of motion. Special interest centers around a few shapes and modes which have been found useful as frequency control and filter elements in electric circuits. Among these are the modes excited by forced thickness-shear deformation of plates.

The resonant frequencies of thickness-shear modes in an infinite crystal plate were found by Koga[1] in closed form, but correspondingly simple solutions of the thickness-shear type are not obtainable for a bounded plate. Using Koga's solution as a zero-order approximation, Ekstein[2] applied a perturbation method to obtain approximate frequencies of a plate with a pair of parallel bounding planes normal to the faces of the plate. He arrived at a formula for the resonant frequencies which he showed to have the same form as an empirical formula devised by Sykes.[3] This formula, however, fits Sykes' experimental data only at discrete points whose locations cannot be found from the formula. The curves of frequency *versus* length to thickness ratio obtained from Sykes' and Ekstein's formulas depart markedly from the experimental data in the neighborhood of flexural resonances and become progressively worse both at higher thickness-shear overtones and lower length-to-thickness ratios. Ekstein advanced a qualitative explanation for the discrepancies in the neighborhood of flexural resonances by pointing out that

[1] I. Koga, Physics 3, 70 (1932).
[2] H. Ekstein, Phys. Rev. 68, 11 (1945).
[3] R. A. Sykes, Chapter VI, *Quartz Crystals for Electrical Circuits*, edited by R. A. Heising (D. Van Nostrand, New York, 1946), p. 218, Eqs. 6.9.

flexural modes in an infinite plate constitute another zero-order solution, and he had shown previously[4] that "coupling" effects become important when the frequencies of two zero-order modes are nearly equal.

In Ekstein's method of solving plate vibration problems, a start is made with one or more exact solutions (zero-order approximations) of the three-dimensional equations, satisfying some of the boundary conditions (e.g., those on the faces of the plate), and a perturbation theory is applied to satisfy, approximately, the remaining boundary conditions. On the other hand, for bodies having at least one small dimension, the classical procedure in elasticity theory is to reduce the number of variables in the differential equations of the system at the start by omitting certain variables deemed to be unimportant and employing average values or restricted regions of others. In this way an approximate theory is formed (two- or one-dimensional) in which some of the boundary conditions are satisfied identically, and some or all of the remaining ones may be satisfied exactly. Familiar examples are the approximate theories of flexural motions of bars and plates leading to the bernoulli-euler and lagrange equations of motion. In their classical form, these theories are not suitable for the present problem as they contain no shear deformation and, as Rayleigh[5] pointed out, they give good results for flexural wave velocities only when the wavelength is long in comparison with the thickness.

In the case of bars, Rayleigh[6] added a correction term for rotatory inertia to the classical theory with some, but not much, improvement in wave velocities. It was not until Timoshenko[7] introduced a correction for transverse shear deformation that the one-dimensional beam theory was made suitable for short wavelengths and high frequencies.

In the case of isotropic plates it has been shown[8] that the theory corresponding to Timoshenko's beam theory contains the appropriate approximations to the first thickness-shear mode and all the flexure modes. This theory is extended to crystal plates in the present paper. The additional boundaries considered by Ekstein are then introduced and a transcendental frequency equation of simple form is obtained. Notwithstanding its simplicity, the frequency equation contains, in great detail, the elaborate spectrum of frequencies obtained experimentally. The zeros and infinities of the frequency equation correspond to points in the frequency spectrum of special interest in design. For these points explicit solutions are obtained in the form of a pair of algebraic formulas. One of the formulas is similar to that of Sykes and Ekstein but contains an additional term which accommodates the higher thickness-shear overtones and smaller length to thickness ratios. The second formula locates the points at which the curves, computed from the first one, cross the curves computed from the transcendental equation. Thus, both frequency and length to thickness ratio are obtained easily for the points of interest without solving the transcendental equation.

The frequency equation that is obtained is simply the one for flexural resonances according to the improved theory of plates and is quite independent of the notion of thickness shear modes which, of course, do not exist in pure form in a bounded plate. From the point of view of plate theory, shear deformation is always present in flexural motion so that flexural resonance may be excited by forcing shear deformation (piezoelectrically or electrostrictively) at the resonant frequencies of flexure. From this point of view, the resonances in a bounded plate commonly designated as thickness-shear, or thickness-shear overtones, are simply local regions in the spectrum of flexural resonances over which the frequency does not change, as rapidly as elsewhere, with change of plate dimensions.

DERIVATION OF PLATE EQUATIONS

The plate is referred to an x, y, z system of rectangular coordinates. The faces of the plate are the planes $y=\pm h/2$, and its cylindrical surfaces (edges) have their generators normal to the x, z plane. The elastic properties of the material are assumed to have monoclinic symmetry. This includes the crystal classes and cuts commonly employed as thickness-shear resonators. With this symmetry and the x-axis as the principal axis, the stress-displacement relations are (in the I.R.E. Standard notation)

$$\begin{aligned}
T_1&=c_{11}(\partial u_1/\partial x)+c_{12}(\partial u_2/\partial y)+c_{13}(\partial u_3/\partial z)\\
&\qquad+c_{14}[(\partial u_3/\partial y)+(\partial u_2/\partial z)],\\
T_2&=c_{12}(\partial u_1/\partial x)+c_{22}(\partial u_2/\partial y)+c_{23}(\partial u_3/\partial z)\\
&\qquad+c_{24}[(\partial u_3/\partial y)+(\partial u_2/\partial z)],\\
T_3&=c_{13}(\partial u_1/\partial x)+c_{23}(\partial u_2/\partial y)+c_{33}(\partial u_3/\partial z)\\
&\qquad+c_{34}[(\partial u_3/\partial y)+(\partial u_2/\partial z)],\\
T_4&=c_{14}(\partial u_1/\partial x)+c_{24}(\partial u_2/\partial y)+c_{34}(\partial u_3/\partial z)\\
&\qquad+c_{44}[(\partial u_3/\partial y)(\partial u_2/\partial z)],\\
T_5&=c_{55}[(\partial u_1/\partial z)+(\partial u_3/\partial x)]\\
&\qquad+c_{56}[(\partial u_2/\partial x)+(\partial u_1/\partial y)],
\end{aligned}\tag{1}$$

and

$$\begin{aligned}
T_6&=c_{56}[(\partial u_1/\partial z)+(\partial u_3/\partial x)]\\
&\qquad+c_{66}[(\partial u_2/\partial x)+(\partial u_1/\partial y)].
\end{aligned}$$

In the three-dimensional theory, substitution of these expressions for the components of stress into the stress equations of motion

$$\begin{aligned}
&(\partial T_1/\partial x)+(\partial T_6/\partial y)+(\partial T_5/\partial z)=\rho(\partial^2 u_1/\partial t^2),\\
&(\partial T_6/\partial x)+(\partial T_2/\partial y)+(\partial T_4/\partial z)=\rho(\partial^2 u_2/\partial t^2),\\
\text{and}\\
&(\partial T_5/\partial x)+(\partial T_4/\partial y)+(\partial T_3/\partial z)=\rho(\partial^2 u_3/\partial t^2)
\end{aligned}\tag{2}$$

yields the differential equations governing the displacements.

[4] H. Ekstein, Phys. Rev. **66**, 108 (1944).
[5] Lord Rayleigh, Proc. London Math. Soc. **10**, 225 (1889).
[6] Lord Rayleigh, *Theory of Sound* (Macmillan and Company, London, 1926), second edition, Vol. 1, p. 258.
[7] S. P. Timoshenko, Phil. Mag. **41**, 744 (1921).
[8] R. D. Mindlin, Trans. Am. Soc. Mech. Engr., Paper No. 50-APM-22 (1950).

The corresponding two-dimensional relations of plate theory may be obtained as follows. The normal components of the tractions on planes parallel to the surfaces of the plate are assumed to contribute negligible amounts to the flexural and shear deformations of the plate. This is equivalent to assuming that

$$T_2=0 \tag{3}$$

throughout the plate. The plane faces of the plate are completely traction-free, so that, in addition,

$$T_6=T_4=0 \quad \text{on} \quad y=\pm h/2. \tag{4}$$

The displacements are assumed to have the form

$$u_1=y\psi_x(x,z,t),\quad u_2=\eta(x,z,t),\quad \text{and} \quad u_3=y\psi_z(x,z,t). \tag{5}$$

Equations (3) and (5) serve to eliminate all compressional and face-shear modes and all thickness-shear modes except the first and its overtones. There remain only the possibilities of approximations to plate flexural, torsional, and first thickness-shear modes and their overtones.

Three stress couples and two stress resultants are defined, in the usual manner, by

$$\begin{aligned}(M_x, M_z, M_{xz})&=\int_{-h/2}^{h/2}(T_1,T_3,T_5)y\,dy\\ (Q_x,Q_z)&=\int_{-h/2}^{h/2}(T_6,T_4)dy.\end{aligned} \tag{6}$$

By means of (3), the second of Eq. (1) is solved for $\partial u_2/\partial y$, and this expression is used to eliminate $\partial u_2/\partial y$ from the remaining five stress-displacement relations. Equations (5) are now substituted in the remaining five of (1) and integrations are performed so as to convert the left-hand sides of (1) to stress couples and stress resultants in accordance with (6). The formulas for the two stress resultants (Q_x, Q_z) are then altered by introducing a constant multiplying factor whose magnitude is to be determined later by matching an exact solution of the three-dimensional equations. This procedure converts (1) to

$$\begin{aligned}M_x&=D_1(\partial\psi_x/\partial x)+D_2\partial\psi_z/\partial z,\\ M_z&=D_2(\partial\psi_x/\partial x)+D_3\partial\psi_z/\partial z,\\ M_{xz}&=D_5[(\partial\psi_z/\partial x)+\partial\psi_x/\partial z],\\ Q_z&=D_4(\psi_z+\partial\eta/\partial z),\quad \text{and}\quad Q_x=D_6(\psi_x+\partial\eta/\partial x),\end{aligned} \tag{7}$$

where

$$\begin{aligned}D_1&=(c_{11}-c_{12}^2/c_{22})h^3/12,\quad D_2=(c_{13}-c_{12}c_{23}/c_{22})h^3/12,\\ D_3&=(c_{33}-c_{23}^2/c_{22})h^3/12,\quad D_4=\kappa^2(c_{44}-c_{24}^2/c_{22})h,\\ D_5&=c_{55}h^3/12,\quad \text{and}\quad D_6=\kappa^2c_{66}h,\end{aligned} \tag{8}$$

in which κ^2 is the multiplying factor mentioned above. Of the 13 elastic constants, three (c_{14}, c_{34}, c_{56}) do not contribute, and the remaining ten enter in only six combinations. It may be seen that κ controls the transverse shear correction. For example, if κ is infinite the last two of (7) require

$$\psi_z=-\partial\eta/\partial z,\quad \psi_x=-\partial\eta/\partial x, \tag{9}$$

which are the relations of classical plate theory in which shear deformation is neglected.

The equations of motion (2) are converted, similarly, by integrating over the thickness of the plate after multiplying the first and third by y. If Eqs. (3)–(6) are used, the results are

$$\begin{aligned}&(\partial M_x/\partial x)-Q_x+(\partial M_{xz}/\partial z)=(\rho h^3/12)(\partial^2\psi_x/\partial t^2),\\ &(\partial Q_x/\partial x)+(\partial Q_z/\partial z)=\rho h(\partial^2\eta/\partial t^2),\\ &\text{and}\\ &(\partial M_{xz}/\partial x)-Q_z+(\partial M_z/\partial z)=(\rho h^3/12)(\partial^2\psi_z/\partial t^2).\end{aligned} \tag{10}$$

Finally, inserting (7) into (10), the plate equations of motion are

$$\begin{aligned}&D_1(\partial^2\psi_x/\partial x^2)+D_5(\partial^2\psi_x/\partial z^2)\\ &\quad+(D_2+D_5)(\partial^2\psi_z/\partial x\partial z)-D_6(\psi_x+\partial\eta/\partial x)\\ &\qquad=(\rho h^3/12)(\partial^2\psi_x/\partial t^2),\\ &D_6[(\partial\psi_x/\partial x)+(\partial^2\eta/\partial x^2)]\\ &\quad+D_4[(\partial\psi_z/\partial z)+(\partial^2\eta/\partial z^2)]=\rho h(\partial^2\eta/\partial t^2),\\ &\text{and}\\ &D_3(\partial^2\psi_z/\partial z^2)+D_5(\partial^2\psi_z/\partial x^2)\\ &\quad+(D_2+D_5)(\partial^2\psi_x/\partial x\partial z)-D_4(\psi_z+\partial\eta/\partial z)\\ &\qquad=(\rho h^3/12)(\partial^2\psi_z/\partial t^2).\end{aligned} \tag{11}$$

The right-hand sides of the first and third of (11) represent the effects of rotatory inertia. If these are omitted, and if the shear deformation terms (i.e., those containing κ) are eliminated, Eqs. (11), with the help of (9), reduce to the classical theory of orthotropic plates. The retention of the shear and rotatory inertia terms permits the application of the theory of plates to high frequency flexural modes and the accompanying thickness-shear motions.

Equations (7) and (11) replace Hooke's law and the equations of motion of the three-dimensional theory. The associated boundary conditions may be found by converting Neumann's uniqueness theorem[9] to apply to plates. This may be done by using (5) and performing the integrations with respect to y that appear in the volume and surface integrals in Neumann's theorem.[10] In terms of components referred to coordinates ν and s measured, respectively, normal to and along the boundary of the plate, the theorem reveals that, at each point of the edge of the plate, one member of each of the following three products must be specified:

$$M_\nu(\partial\psi_\nu/\partial t),\quad M_{\nu s}(\partial\psi_s/\partial t),\quad Q_\nu(\partial\eta/\partial t). \tag{12}$$

On a free edge, for example,

$$M_\nu=M_{\nu s}=Q_\nu=0. \tag{13}$$

Thus, three boundary conditions are required at each

[9] A. E. H. Love, *Theory of Elasticity* (Cambridge University Press, London, 1927), fourth edition, p. 176.

[10] The development of a plate-analog of Neumann's uniqueness theorem is described in detail in reference 8.

point on the edge, as opposed to two in the classical theory of plates. This is in agreement with E. Reissner's theory of equilibrium of plates[11] in which shear deformation is accounted for in a different manner.

THICKNESS-SHEAR VIBRATION OF AN INFINITE PLATE

The solution of the three-dimensional equations for thickness-shear vibration parallel to the x axis is obtained by setting

$$u_1=u(y)e^{i\omega t}, \quad u_2=u_3=0. \tag{14}$$

The equations of motion (2) then reduce to

$$c_{66}(\partial^2u/\partial y^2)+\rho\omega^2u=0.$$

Applying the boundary conditions (4), the usual result[1] is obtained for the circular frequency $\bar{\omega}$ of the first mode:

$$\bar{\omega}^2=\pi^2c_{66}/\rho h^2. \tag{15}$$

The corresponding solution of the plate, Eqs. (11), is obtained by setting

$$\psi_x=B\exp(i\bar{\omega}t), \quad \eta=\psi_z=0, \tag{16}$$

where B is a constant. The equations of motion (11) are satisfied provided

$$\bar{\omega}^2=12\kappa^2c_{66}/\rho h^2. \tag{17}$$

The value of the coefficient κ is obtained by equating (15) and (17), with the result

$$\kappa=\pi/(12)^{\frac{1}{2}}. \tag{18}$$

It may be seen that the first antisymmetric mode of thickness-shear motion is contained in plate theory when rotatory inertia and shear terms are included.

THICKNESS-SHEAR EXCITATION OF A RECTANGULAR PLATE

Suppose, now, that the thickness shear vibration

$$\psi_x=Be^{i\omega t}, \quad \eta=\psi_z=0 \tag{19}$$

is forced (piezoelectrically, for example) in a rectangular plate with free edges $x=\pm l/2$, $z=\pm$const. The boundary conditions to be satisfied are, from (13),

$$M_x=M_{xz}=Q_x=0 \quad \text{on} \quad x=\pm l/2 \tag{20}$$

$$M_z=M_{xz}=Q_z=0 \quad \text{on} \quad z=\pm\text{const.} \tag{21}$$

Substitution of (19) into (7) shows that all the boundary conditions are satisfied with the exception that

$$Q_x=D_6Be^{i\omega t} \quad \text{on} \quad x=\pm l/2. \tag{22}$$

Additional modes are required such that, combined with (19), the equations of motion (11) and the boundary conditions (20) and (21) will be satisfied. Since (11) are essentially the equations of flexural motion, it is apparent that the presence of boundaries results, in this case, in the excitation of flexural modes.

[11] E. Reissner, J. Appl. Mech. 12, A-69 (1945); Quart. Appl. Math. 5, 55 (1947).

SOLUTION NEGLECTING WIDTH

If the additional motions introduced to remove (22) are taken as independent of z, the boundary conditions (20) on $x=\pm l/2$ can be satisfied, but small stress couples and resultants will be introduced on $z=\pm$const. Extensive experiments by Sykes on rotated Y-cut quartz plates show that, over a considerable range, the resonant frequencies are practically independent of the width. With the assumption of z-independence, ψ_z disappears from the first two equations of motion (11). These become, after introducing a factor $e^{i\omega t}$, omitted hereafter,

$$\begin{aligned}(\partial^2\psi_x/\partial x^2)+(R\delta_0^4-S^{-1})\psi_x=S^{-1}\partial\eta/\partial x\\ (\partial\psi_x/\partial x)+[(\partial^2/\partial x^2)+S\delta_0^4]\eta=0,\end{aligned} \tag{23}$$

where

$$R=h^2/12, \quad S=D_1/D_6,$$

and

$$\delta_0^4=\rho h\omega^2/D_1=\omega^2/(\bar{\omega}^2RS). \tag{24}$$

The constants R and S represent the effects of rotatory inertia and transverse shear deformation, respectively. The ratio

$$g=R/S=\pi^2c_{66}/12(c_{11}-c_{12}^2/c_{22}) \tag{25}$$

plays an important role. It is the ratio of the thickness-shear modulus (κ^2c_{66}) to the plate flexure modulus. In the case of an isotropic material, c_{66} becomes the shear modulus, and $c_{11}-c_{12}^2/c_{22}$ reduces to $E/(1-\nu^2)$, where E is Young's modulus and ν is the poisson ratio.

Equations (23) are satisfied by

$$\psi_x=(\sigma-1)\partial\eta/\partial x, \tag{26}$$

where σ is a constant, provided

$$\sigma^{-1}S\delta_0^4=R\delta_0^4-S^{-1}-[S(\sigma-1)]^{-1}=\delta^2, \text{ (say)} \tag{27}$$

and

$$(\partial^2\eta/\partial x^2)+\delta^2\eta=0. \tag{28}$$

Solution of (27) yields

$$2(\delta_1^2, \delta_2^2)=\delta_0^4\{R+S\pm[(R-S)^2+4\delta_0^{-4}]^{\frac{1}{2}}\}, \tag{29}$$

and

$$(\sigma_1, \sigma_2)=(\delta_1^{-2}, \delta_2^{-2})S\delta_0^4. \tag{30}$$

Hence, two deflection functions (η_1 and η_2) are obtained by setting δ equal to δ_1 or δ_2 in (28). They are the solutions of the fourth-order equation

$$[(\partial^2/\partial x^2)+\delta_1^2][(\partial^2/\partial x^2)+\delta_2^2]\eta=0, \tag{31}$$

which can be obtained directly by eliminating ψ_x from (23). Equation (31) is, in fact, no more than an anisotropic plate version of Timoshenko's beam equation.[7]

To sum up, we may write

$$\eta=\eta_1+\eta_2, \tag{32}$$

$$\psi_x=(\sigma_1-1)(\partial\eta_1/\partial x)+(\sigma_2-1)\partial\eta_2/\partial x, \tag{33}$$

$$(\partial^2\eta_1/\partial x^2)+\delta_1^2\eta_1=0, \quad (\partial^2\eta_2/\partial x^2)+\delta_2^2\eta_2=0, \tag{34}$$

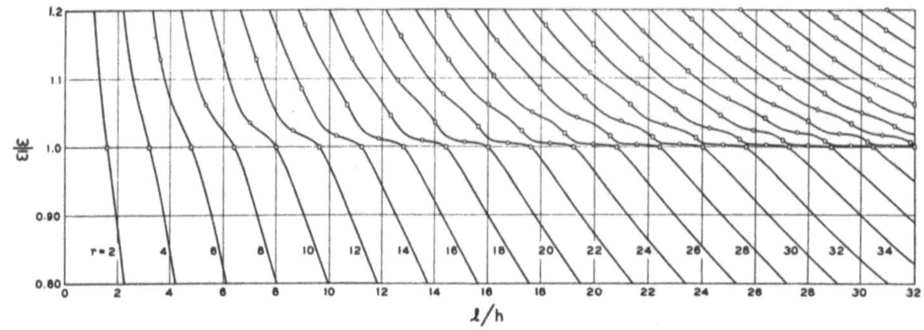

FIG. 1. Resonant frequencies of the even modes of vibration of an *AT*-cut quartz plate as computed from Eqs. (42), (43), and (44). The ordinate is the ratio of the resonant frequency (ω) to the frequency ($\bar{\omega}$) of the first thickness-shear mode of an infinite plate. The abscissa is the ratio of the length (l) to the thickness (h). The circles and squares locate points for which the transcendental frequency equation can be solved explicitly. They are also the points at which the corresponding curves for the odd modes would cross the curves shown.

and, from (7), $M_{xz}=0$,

$$M_x=-D_1[(\sigma_1-1)\delta_1^2\eta_1+(\sigma_2-1)\delta_2^2\eta_2],$$

and (35)

$$Q_x=D_6[(\sigma_1\partial\eta_1/\partial x)+\sigma_2\partial\eta_2/\partial x].$$

The solutions of (34), which are even in x (i.e., the symmetric, or odd order, flexure modes), make Q_x odd in x. Hence, they cannot annul (22), which is even in x, and play no part in satisfying the boundary conditions on $x=\pm l/2$. This is the counterpart of the experimental observation that there is no coupling between the fundamental thickness-shear mode and the odd flexure modes. The remaining solutions of (34) are,

$$\eta_1=A_1\sin\delta_1 x, \quad \text{and} \quad \eta_2=A_2\sin\delta_2 x. \tag{36}$$

The constants A_1 and A_2 are evaluated from the boundary conditions, i.e., (36) are substituted into (35) and then, at $x=\pm l/2$, $M_x=0$, $Q_x=-D_6B$ to annul (22). This results in the equations

$$A_1(\sigma_1-1)\delta_1^2\sin(\delta_1 l/2)+A_2(\sigma_2-1)\delta_2^2\sin(\delta_2 l/2)=0,$$

and (37)

$$A_1\sigma_1\delta_1\cos(\delta_1 l/2)+A_2\sigma_2\delta_2\cos(\delta_2 l/2)=-B.$$

These equations give A_1 and A_2 in terms of the forcing amplitude B. Resonance occurs when the determinant of (37) vanishes, i.e., when

$$\sigma_2\delta_1(\sigma_1-1)\tan(\delta_1 l/2)=\sigma_1\delta_2(\sigma_2-1)\tan(\delta_2 l/2). \tag{38}$$

The roots of this equation give the resonant frequencies in terms of the dimensions and elastic constants of the plate.

FREQUENCY EQUATION

To put the frequency Eq. (38) into a form suitable for computing the roots, let

$$\gamma=\delta_1 l/2, \tag{39}$$

and

$$a=\delta_2/\delta_1\doteq[(1-b)/(1+b)]^{\frac{1}{2}}, \tag{40}$$

where b is the positive root of

$$b^2=1+4g(\bar{\omega}^2/\omega^2-1)(1+g)^{-2}. \tag{41}$$

It may be seen, from (29), that δ_1 is always real, while δ_2 is real if $\omega>\bar{\omega}$ and imaginary if $\omega<\bar{\omega}$. From (40), the same applies to a. (The case $\delta_2=0$, $a=0$, and $\omega=\bar{\omega}$ will be considered later along with other special roots.) When a is imaginary, let $a=ia_1$, where a_1 is real. Then (38) may be written as

$$(a^2-g)\tan\gamma=a(1-ga^2)\tan a\gamma, \qquad \omega>\bar{\omega},$$

and (42)

$$(a_1^2+g)\tan\gamma=a_1(1+ga_1^2)\tanh a_1\gamma, \quad \omega<\bar{\omega}.$$

From (40) and (41),

$$\begin{aligned} \omega/\bar{\omega}&=[1-a^2(1+g)^2/g(1+a^2)^2]^{-\frac{1}{2}}, & \omega>\bar{\omega} \\ \omega/\bar{\omega}&=[1+a_1^2(1+g)^2/g(1-a_1^2)^2]^{-\frac{1}{2}}, & \omega<\bar{\omega}, \end{aligned} \tag{43}$$

and, from (39),

$$\begin{aligned} l/h&=\gamma(\bar{\omega}/\omega)[(1+a^2)/3(1+g)]^{\frac{1}{2}}, & \omega>\bar{\omega} \\ l/h&=\gamma(\bar{\omega}/\omega)[(1-a_1^2)/3(1+g)]^{\frac{1}{2}}, & \omega<\bar{\omega}. \end{aligned} \tag{44}$$

The elastic constants affect the frequencies only through the constants g and c_{66} (the latter is contained in $\bar{\omega}$). Thus, there is a family of roots of (42) for each value of g, i.e., for each crystal class and orientation of cut. Selection of g and a or a_1 fixes $\omega/\bar{\omega}$ according to (43) and defines roots γ of (42). The values of γ are computed and inserted in (44) to give the values of l/h corresponding to the chosen g and a or a_1, i.e., to the chosen $\omega/\bar{\omega}$. For a given g, this procedure is repeated over the desired range of $\omega/\bar{\omega}$ (or a and a_1).

The results of a set of computations (for $g=0.283$) are shown in Fig. 1. The curves will be recognized as being characteristic of the frequency spectrum of resonances excited by thickness-shear vibration of rectangular quartz plates.[12] The circled and squared points in Fig. 1 represent special roots of (42) for which both sides of the equation are infinity and zero, respectively. These roots will be discussed in detail later.

COMPARISON WITH FREQUENCIES OF *AT* QUARTZ PLATES

Mason's values[13] for the elastic constants of quartz, referred to the principal axis of crystal symmetry, are, in units of 10^{10} dynes/cm^2, $c_{11}^0=86.05$, $c_{12}^0=5.05$, $c_{13}^0=10.45$, $c_{14}^0=18.25$, $c_{33}^0=107.1$, $c_{44}^0=58.65$, and $c_{66}^0=(c_{11}^0-c_{12}^0)/2=40.5$.

The constants which are required, referred to rotated axes, are given by [14]

$$c_{66}=c_{44}^0s^2+(c_{11}^0-c_{12}^0)c^2/2-2c_{14}^0sc,$$
$$c_{22}=c_{11}^0c^4+c_{33}^0s^4+2(2c_{44}^0+c_{13}^0)s^2c^2+4c_{14}^0sc^3,$$
$$c_{12}=c_{12}^0c^2+c_{13}^0s^2-2c_{14}^0sc,\quad c_{11}=c_{11}^0,$$

where $s=\sin\theta$ and $c=\cos\theta$. For the *AT* cut, $\theta=35°15'$. Hence, in units of 10^{10} dynes/cm^2, $c_{66}=29.34$, $c_{22}=129.9$, $c_{12}=-10.49$, $c_{11}=86.05$, and, from (25) and (18), $g=0.283$. Thus, the curves in Fig. 1 are for the *AT* cut.

Portions of these curves are drawn to a larger scale in Figs. 2 and 3.

Data obtained by Sykes with *AT* plates are plotted as circles in Figs. 2 and 3. The dimensions of the plates were initially 25×25×0.807 mm for Fig. 2 and 22×14 ×1.000 mm for Fig. 3. In successive experiments the x-dimensions (l) were cut down in decrements of 0.1 mm to 21.8 mm and 8 mm, respectively. Hence, l/h varies from 28.38 to 30.98 in Fig. 2 and 8 to 14 in Fig. 3. The recorded resonant frequencies (cps) were converted to $\omega/\bar{\omega}$ by multiplying by $2h/(c_{66}/\rho)^{\frac{1}{2}}=0.485\times10^{-6}$, in which $h=0.807$ and 1.000, $\rho=2.65$, and $c_{66}=29.34\times10^{-10}$. A better fit between experiment and theory would have been obtained in the neighborhood of $\omega/\bar{\omega}=1$ if c_{66} had been calculated from Sykes' data on the *AT* plate rather than from Mason's primary data on the nonrotated constants.

It may be seen that there is a relative displacement, between theory and experiment, of about 0.1 to 0.3 percent along the l/h axis. This could be due to approximations made in the theory or to errors in the elastic constants. In the latter case, for example, a 1 percent error in c_{11} would produce a 0.1 percent displacement of the type observed.

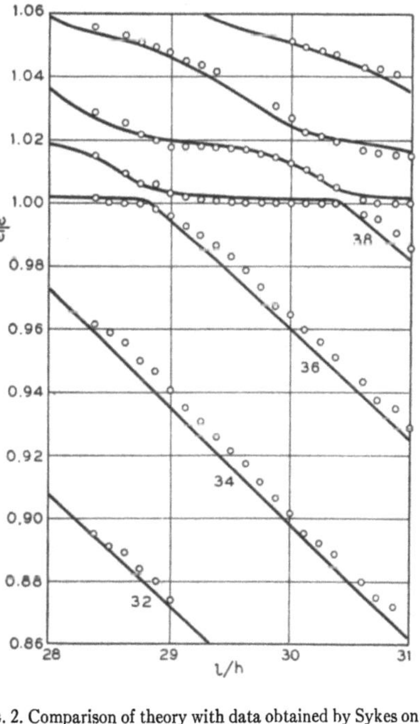

FIG. 2. Comparison of theory with data obtained by Sykes on an *AT* quartz plate, initially 25×25×0.807 mm, cut down in decrements of 0.1 mm to 25×21.8×0.807 mm.

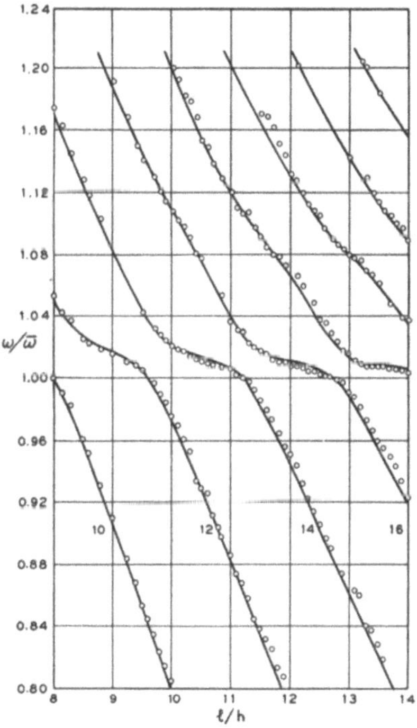

FIG. 3. Comparison of theory with data obtained by Sykes on an *AT* quartz plate, initially 22×14×1.000 mm cut down in decrements of 0.1 mm to 22×8×1.000 mm.

[12] See reference 3, p. 233, Fig. 6.15.
[13] W. P. Mason, *Piezoelectric Crystals and Their Applications to Ultrasonics* (D. Van Nostrand Company, Inc., New York, 1950), p. 84.
[14] See reference 3, p. 247.

SOLUTION OF FREQUENCY EQUATION BY SUCCESSIVE APPROXIMATIONS

Special interest centers around the frequencies of the higher modes in the neighborhood of the thickness-shear frequency $\bar{\omega}$. In this range, the following solution of the frequency equation by successive approximations (suggested by S. A. Schelkunoff) converges very rapidly. Writing the first of (42) in the form

$$\tan\gamma = \varphi(g, a)\tan a\gamma, \quad \omega > \bar{\omega}, \tag{45}$$

let

$$\gamma = r\pi + \beta, \quad -\pi/2 < \beta < \pi/2, \quad r = 1, 2, 3 \ldots.$$

Since $\tan(r\pi+\beta) = \tan\beta$, (45) may be written

$$\beta = \tan^{-1}[\varphi(g, a)\tan a(r\pi+\beta)], \quad \omega > \bar{\omega}.$$

As a first approximation take

$$\beta_1 = \tan^{-1}[\varphi(g, a)\tan ar\pi], \quad \omega > \bar{\omega},$$

and, as successive approximations,

$$\beta_{i+1} = \tan^{-1}[\varphi(g, a)\tan a(r\pi+\beta_i)], \quad \omega > \bar{\omega}.$$

It is usually necessary to take only one or two steps.

Similarly, for $\omega < \bar{\omega}$,

$$\tan\gamma = \varphi_1(g, a_1)\tanh a_1\gamma, \quad \omega < \bar{\omega}, \tag{46}$$

and

$$\beta_{i+1} = \tan^{-1}[\varphi_1(g, a_1)\tanh a_1(r\pi+\beta_i)], \quad \omega < \bar{\omega}.$$

The product $a_1 r\pi$ does not have to become very large before this can be written as

$$\beta = \tan^{-1}\varphi_1(g, a_1),$$

which accounts for the monotonic, regularly spaced curves for the low frequency portions of the flexural modes in Fig. 1.

ODD MODES

While only the even flexural modes are excited by the forcing function (19), the odd modes may be excited by forcing a thickness-shear deformation antisymmetrical with respect to x. In this case, the residual stress resultants corresponding to (22) will be odd in x and the even solutions

$$\eta_1 = B_1\cos\delta_1 x, \quad \eta_2 = B_2\cos\delta_2 x \tag{47}$$

of (34) will be required to annul them. The only change in the results, therefore, is the interchange of sine and cosine throughout. The transcendental equations are the same as those for the even modes except for the replacement of $\varphi(g, a)$ and $\varphi_1(g, a_1)$ by their reciprocals and a reversal of sign for the case $\omega/\bar{\omega} < 1$. The treatment of these equations is then very similar. In particular, the circled and squared points are also roots and hence the odd-mode curves of $\omega/\bar{\omega}$ *vs* l/h, corresponding to Fig. 1, cross the curves of Fig. 1 at these points. Whereas, for the even modes, the circled points are on the flat portions of the curves (slow variation of $\omega/\bar{\omega}$ with l/h) and the squared points are on the steep portions (rapid variation of $\omega/\bar{\omega}$ with l/h), the converse is true for the odd flexural modes.

ALGEBRAIC FORMULAS

The circled points in Fig. 1 are those for which γ and $a\gamma$ in the first of (42) are odd multiples of $\pi/2$. In the case of excitation by thickness-shear, even in x, these points are of special interest, as they are far from the frequencies of neighboring modes and center in regions where the frequency changes slowly with change of dimensions. The squared points, for which γ and $a\gamma$ in the first of (42) are even multiples of $\pi/2$, have the same properties for thickness-shear excitation odd in x. Because of the degeneracy of the transcendental equation for both sets of points, algebraic formulas may be obtained for them.

In the first of (42), the roots under discussion are

$$\begin{aligned} \gamma &= r\pi/2, \quad r = 1, 2, 3 \ldots \\ a\gamma &= n\pi/2, \quad n = 0, 1, 2 \ldots, \end{aligned} \tag{48}$$

from which

$$a = n/r. \tag{49}$$

Substituting (48) and (49) in the first of (43) and (44), the latter become

$$\omega/\bar{\omega} = g^{\frac{1}{2}}(r^2+n^2)[(gr^2-n^2)(r^2-gn^2)]^{-\frac{1}{2}} \tag{50}$$

$$l/h = (\pi/2)[(gr^2-n^2)(r^2-gn^2)/3g(1+g)(r^2+n^2)]^{\frac{1}{2}}. \tag{51}$$

Equations (50) and (51) give the coordinates ($\omega/\bar{\omega}$, l/h) of the circled points (odd r and n) and squared points (even r and n). These points are plotted in Fig. 4 for the AT plate ($g = 0.283$) over the ranges 0 to 1.2 for $\omega/\bar{\omega}$ and 0 to 32 for l/h.

The equations of the two families of curves passing through the points $n = \text{const}$ and $r = \text{const}$ are obtained by eliminating r and n, respectively, between (50) and (51) with the results

$$2(\omega/\bar{\omega})^2 = 1 + (kn^2h^2/l^2) + [(1+kn^2h^2/l^2)^2 - 4(\kappa nh/l)^4/g]^{\frac{1}{2}}, \tag{52}$$

and

$$2(\omega/\bar{\omega})^2 = 1 + (kr^2h^2/l^2) - [(1+kr^2h^2/l^2)^2 - 4(\kappa rh/l)^4/g]^{\frac{1}{2}}, \tag{53}$$

where

$$k = \kappa^2(1+g)/g = (\pi^2/12) + (c_{11}c_{22} - c_{12}^2)/c_{22}c_{66}. \tag{54}$$

These curves are illustrated in Fig. 4 for the AT cut.

Equation (52) is to be compared with the empirical equation given by Sykes which, for the present case and in the present notation, is

$$(\omega/\bar{\omega})^2 = 1 + kn^2h^2/l^2, \tag{55}$$

where k is an experimental constant. Sykes' equation was derived by Ekstein, from his theory, with a formula for k more elaborate than (54) but yielding approximately the same numerical values. Equations (55) and (52) differ, in form, only by the term $4(\kappa nh/l)^4/g$ in the

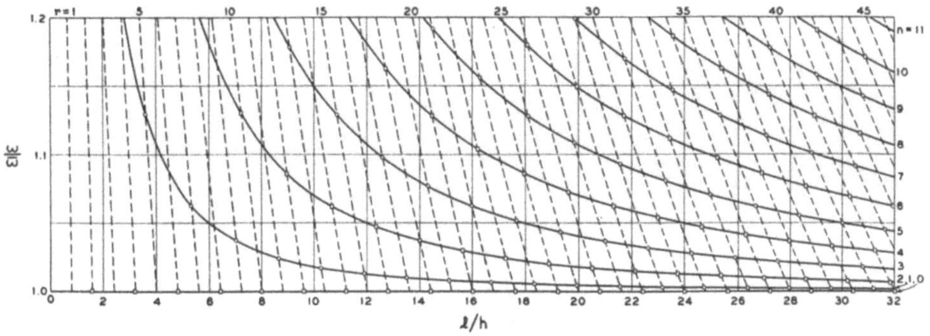

FIG. 4. The circles and squares are those from the upper half of Fig. 1. Their coordinates are given by the algebraic formulas (50) and (51). They are connected by full and dashed curves given by Eqs. (52) and (53), respectively. Points on these curves do not correspond to resonant frequencies except at the intersections marked with a circle or square. The full curves $n=2, 3, 4$ . . . are those commonly designated as the first, second, third . . . overtones of the fundamental thickness-shear mode.

latter. The additional term accommodates the higher overtones of thickness-shear vibration (large n) and small values of l/h.

It should be observed, however, that equations like (52) or (55) have meaning only at discrete points. These may be located by (50) and (51) or by the intersections of the families of curves (52) and (53) as illustrated in Fig. 4.

Regarding Eq. (53), it may be observed that the curves $r=$constant are continuations, into the region $\omega/\bar{\omega}>1$, of the flexural mode curves (Fig. 1) from the region $\omega/\bar{\omega}<1$. Alternatively, curves $n+r=$constant are close approximations to the curves in the upper region of Fig. 1.

SHAPES OF MODES

The shapes of the modes of vibration may be determined very easily for the frequencies discussed in the preceding section. Noting that $\delta_1=r\pi/l$ and $\delta_2=n\pi/l$, we find, from (32), (33), and (36),

$$\eta=A_1\sin(r\pi x/l)+A_2\sin(n\pi x/l), \tag{56}$$

and

$$h\psi_z=U_1\sin(r\pi x/l)+U_2\sin(n\pi x/l), \tag{57}$$

where

$$U_1=A_1h(\sigma_1-1)\delta_1,\quad U_2=A_2h(\sigma_2-1)\delta_2.$$

These give, respectively, the displacements that would be observed normal and tangential to the surface of the plate. Each contains a short wavelength part, $\sin(r\pi x/l)$, and a long wavelength part, $\sin(n\pi x/l)$. The former is commonly associated with the notion of flexural modes and the latter with thickness-shear modes. Alternatively, each type of mode contains both normal and tangential components of displacement.

The relative amplitudes of the four parts of the surface displacement may be found by using the first of (37), from which

$$\left|\frac{A_2}{A_1}\right|=\left|\frac{(\sigma_1-1)\delta_1^2\sin(\delta_1 l/2)}{(\sigma_2-1)\delta_2^2\sin(\delta_2 l/2)}\right|. \tag{58}$$

Noting that $\sigma_1/\sigma_2=a^2=n^2/r^2$ and $\sigma_1=(1+a^2)/(1+g)$, this becomes, for r and n odd,

$$|A_2/A_1|=(gr^2-n^2)/(r^2-gn^2). \tag{59}$$

For r and n even, the even solutions (47) are used, and the resulting amplitude ratio is the same as (59). Usually r/n is large, and hence $|A_2/A_1|$ is approximately equal to g. Thus, regarding displacements normal to the plate, the part associated with thickness-shear is less than that associated with flexure in the ratio $g:1$ (approximately). The corresponding ratio for the tangential displacements is

$$U_2/U_1=r/n;$$

i.e., the tangential displacement associated with thickness-shear is usually much larger than that associated with the accompanying flexural motion. Similarly, it can be shown that $|U_2/A_2|$ is approximately r/ng times $|U_1/A_1|$; also, $|U_2/A_1|$ is r/n times $|U_1/A_1|$. Thus, in general, for the circled points the predominant motion is tangential to the surface of the plate.

Reprinted from
Proceedings of the First U.S. National Congress of Applied Mechanics-1951
(1952) 187-191

A ONE-DIMENSIONAL THEORY OF COMPRESSIONAL WAVES IN AN ELASTIC ROD

by R. D. Mindlin
G. Herrmann

Department of Civil Engineering, Columbia University

THE mathematical theory of the longitudinal motions of an elastic bar had its beginnings in the investigations by Daniel Bernoulli, in 1741, on the vibrations of a column of air. Both the bar and air column theories, in their classical form, lead to the one-dimensional wave equation

$$c_e^2 \frac{\partial^2 u_z}{\partial z^2} = \frac{\partial^2 u_z}{\partial t^2} \qquad [1]$$

with constant phase velocity c_e, governing the displacement, u_z parallel to the axis, z, of the bar. Later, an exact solution, by Pochhammer[1], of the equations of the linear theory of elasticity, for waves in an infinite bar of uniform circular section, disclosed, in detail, the dispersive nature of the medium. The constant

$$c_e = \left(\frac{E}{\rho}\right)^{1/2} \qquad [2]$$

where E is Young's modulus and ρ is the density, appears as a first approximation to the velocity of the first mode of motion, valid for waves which are long in comparison with the radius of the bar. For shorter waves, the velocity diminishes and, in the limit, approaches the velocity, c_R, of Rayleigh surface waves (see Curve 1 of Fig. 1).

The complexity of the Pochhammer solution reveals the impracticability of attempting to employ the general equations of elasticity in attacking problems of vibrations of short bars, high frequencies in long bars, or transients with steep wave-fronts. All of these problems involve the effects accompanying short wave-lengths so that, on the other hand, the classical first approximation is inadequate. There is need, therefore, for an intermediate theory which contains the essential physics of short-wave phenomena but retains the one-dimensional simplicity of the first ap-

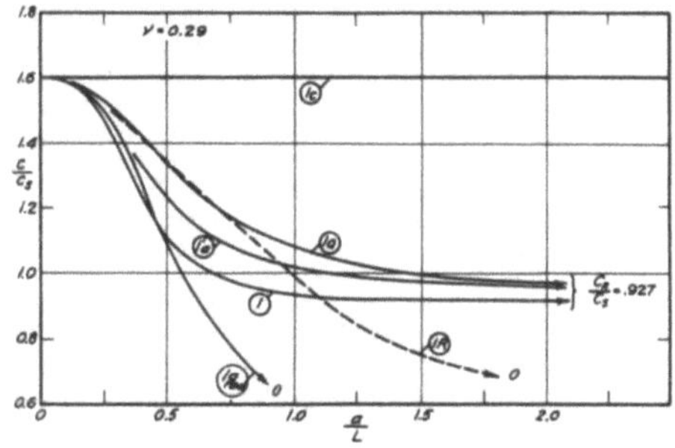

Fig. 1. LONGITUDINAL WAVE VELOCITIES OF FIRST MODE OF MOTION ACCORDING TO:

1. Pochhammer solution (after Davies).
1a. Present theory with K correction.
1'a. Present theory with K and K_1 corrections.
1a (rad). Present theory with radial inertia term, but without radial shear term.
1R. Classical theory with Rayleigh radial inertia term added (after Davies).
1c. Classical first approximation.

proximation. This led to a second approximation, by Rayleigh[2], which takes into account the effects of radial inertia and yields a phase velocity

$$c^2 = \frac{c_e^2}{\left(1 + \frac{2\pi^2 \nu^2 a^2}{L^2}\right)} \qquad [3]$$

where ν is Poisson's ratio, a is the radius of the bar and L is the wave-length. The equation of motion

$$c_e^2 \frac{\partial^2 u_z}{\partial z^2} = \frac{\partial^2 u_z}{\partial t^2} - \frac{\nu^2 a^2}{2} \frac{\partial^4 u_z}{\partial z^2 \partial t^2} \qquad [4]$$

which incorporates the Rayleigh radial inertia correction, appears to have been derived, first, by Love.[3]

As may be seen, (Curve 1R, Fig. 1) the Rayleigh correction gives improved velocities for moderately short

[1] A. E. H. Love, "Theory of Elasticity," (Cambridge University Press, London, 1927), 4th Ed., p. 287.

[2] Lord Rayleigh, "Theory of Sound," (Dover Pub., N. Y., 1945), p. 251.

[3] Love, p. 428.

This work was assisted, in part, by the Office of Naval Research, under Contract Nonr 266 (09).

wave-lengths; but, as the wave-length continues to diminish, the velocity drops to zero (as $a/L \to \infty$) instead of approaching the velocity of Rayleigh surface waves. Curves 1 and 1R are taken from a paper by R. M. Davies.[4]

It is apparent, from physical considerations, that, for short waves, radial inertia effects are no more important than those of radial shear. Both contributions are taken into account in the present paper, leading to a theory which appears to be satisfactory, at least for the first mode of motion, over the full range of wave-lengths. As might be expected, for example from the Pochhammer solution, coupling through Poisson's ratio is exhibited between the first (longitudinal) mode and a second (radial) mode. The importance of such coupling has been recognized previously; especially by Giebe and Blechschmidt[5] who evolved a qualitative theory based on an assumed frequency equation for a coupled system of two degrees of freedom.

The theory derived in the sequel is deduced from the general equations of elasticity in a manner similar to that employed recently[6] for considering transverse shear deformation in flexural motions of plates. Initial assumptions are made as to the form of the displacements, following which energy considerations lead to forms of the equations of motion, weighted average stresses and boundary conditions appropriate to the assumed displacements. Any degree of approximation may be chosen, but it is expedient to explore only the third. Rayleigh's second approximation raises the order of the equation of motion from the second to the fourth and a third approximation necessitates no further increase in order. However, higher approximations increase the order above the fourth and do not appear to be practical. The fourth order equations found here are very similar in form to the equations of Timoshenko's[7] theory of flexural motions of beams and may be treated in a similar manner.

ENERGY CONSIDERATIONS

In the general theory of small motions of an elastic continuum, the increment in energy of a volume V, in an interval of time, $t - t_o$, is

$$\Delta U = \int_{t_o}^{t} dt \int_V \frac{\partial}{\partial t}(T + W)\, d\tau \qquad [5]$$

where $d\tau$ is the element of volume and T and W are the kinetic energy and elastic strain energy, respectively, per unit of volume. In cylindrical coordinates r, θ, z, we have

$$2T = \rho\,(\dot{u}_r^2 + \dot{u}_\theta^2 + \dot{u}_z^2) \qquad [6]$$

where u_r, u_θ, u_z are the components of displacement and the dot indicates partial differentiation with respect to time. Also

$$\frac{\partial W}{\partial t} = \frac{\partial W}{\partial \epsilon_{rr}}\dot{\epsilon}_{rr} + \dots \frac{\partial W}{\partial \epsilon_{\theta z}}\dot{\epsilon}_{\theta z} + \dots$$

$$= \sigma_{rr}\frac{\partial \dot{u}_r}{\partial r} + \dots\ \sigma_{\theta z}\left(\frac{1}{r}\frac{\partial \dot{u}_z}{\partial \theta} + \frac{\partial \dot{u}_\theta}{\partial z}\right) + \dots \qquad [7]$$

where $\epsilon_{rr}, \dots, \epsilon_{\theta z}, \dots$ and $\sigma_{rr}, \dots, \sigma_{\theta z}, \dots$ are the components of strain and stress, respectively.

We now consider a bar of circular cross-section, $r = a$, and assume, as an approximation, axially symmetric displacements

$$u_r = r a^{-1} u(z, t), \quad u_\theta = 0, \quad u_z = w(z, t) \qquad [8]$$

The volume integral in [5] then reduces to

$$\int_V f(r, \theta, z, t)\, d\tau = 2\pi \int_o^a \int_b^c f(r, z, t)\, r\, dr\, dz \qquad [9]$$

where $c - b$ is the length of the bar.

Substituting [8] into [6] and [7], the latter into [5] and using [9], we are led to the definition of four "bar-stress components"

$$P_r = \int_o^a \sigma_{rr}\, r\, dr, \qquad P_\theta = \int_o^a \sigma_{\theta\theta}\, r\, dr$$

$$P_z = \int_o^a \sigma_{zz}\, r\, dr, \qquad Q = \int_o^a \sigma_{zr}\, a^{-1} r^2\, dr \qquad [10]$$

and the increment of energy reduces to

$$\frac{2\Delta U}{\pi} = \rho a^2 \int_{t_o}^{t} dt \int_b^c (\dot{u}\ddot{u} + 2\dot{w}\ddot{w})\, dz$$
$$+ 4\int_{t_o}^{t} dt \int_b^c [\dot{u} a^{-1}(P_r + P_\theta) + \dot{w}' P_z$$
$$+ \dot{u}' Q]\, dz \qquad [11]$$

where prime indicates partial differentiation with respect to z.

Finally, integrating the last two terms in the bracket of [11] and rearranging,

$$\frac{\Delta U}{2\pi} = \int_{t_o}^{t} dt \int_b^c \dot{u}\left(\frac{\rho a^2}{4}\ddot{u} + \frac{P_r + P_\theta}{a} - Q'\right) dz$$
$$+ \int_{t_o}^{t} dt \int_b^c \dot{w}\left(\frac{\rho a^2}{2}\ddot{w} - P_z'\right) dz$$
$$+ \int_{t_o}^{t} [\dot{w} P_z + \dot{u} Q]_b^c\, dt \qquad [12]$$

---

[4] R. M. Davies, *Phil. Trans. Roy. Soc.*, A, *240*, 1948, p. 375.

[5] E. Giebe and E. Blechschmidt, *Ann. d. Physik*, *18*, 1933, p. 417.

[6] R. D. Mindlin, *Jour. Appl. Mechanics*, *18*, 1951, p. 31 and *Jour. Appl. Phys.*, *22*, 1951, p. 316.

[7] S. Timoshenko, *Phil. Mag.*, *41*, 1921, p. 744 and *43*, 1922, p. 125.

STRESS EQUATIONS OF MOTION

The quantities in parentheses in [12] may be replaced by surface tractions through the use of the stress equations of motion. In the axially symmetric case under consideration, the equations of motion, from the general theory, reduce to

$$\left.\begin{aligned} \frac{\partial\sigma_{rr}}{\partial r} + \sigma_{zr}' + (\sigma_{rr} - \sigma_{\theta\theta})\, r^{-1} &= \rho \ddot{u}_r \\ \frac{\partial\sigma_{zr}}{\partial r} + \sigma_{zz}' + \sigma_{zr}\, r^{-1} &= \rho\, \ddot{u}_z \end{aligned}\right\} \qquad [13]$$

To convert to bar equations, the first of [13] is multiplied by r^2, and the second by r and the equations are then integrated over the radius of the section. The motivation for this process is, of course, the form of the quantities in parentheses in [12]. Using [8] and [10] and performing integrations by parts on the terms containing σ_{rr}' and σ_{zr}', we obtain

$$\left.\begin{aligned} Q' - a^{-1}(P_r + P_\theta) + aR &= \frac{\rho a^2 \ddot{u}}{4} \\ P_z' + aZ &= \frac{\rho a^2 \ddot{w}}{2} \end{aligned}\right\} \qquad [14]$$

where

$$R = \sigma_{rr}\Big]_{r=a}, \qquad Z = \sigma_{zr}\Big]_{r=a} \qquad [15]$$

are the radial and axial components, respectively, of the traction on the cylindrical surface of the bar. Equations [14] are the stress equations of motion of the bar consisttent with the assumptions [8].

BOUNDARY AND INITIAL CONDITIONS

Substituting [14] in [12], we find

$$\Delta U/2\pi = a \int_{t_0}^{t} dt \int_{b}^{c} (\dot{u}\, R + \dot{w}\, Z)\, dz + \int_{t_0}^{t} [\dot{u}\, Q + \dot{w}\, P_z]_b^c\, dt \qquad [16]$$

which is simply the form, in the present theory, of the theorem that the rate of increase of energy in the bar is equal to the rate at which work is done on the surfaces of the bar. By an argument of uniqueness, analogous to that derived by Neumann[8], the boundary and initial conditions appropriate to the present theory are:

1. At each point along the length of the bar, one member of each of the products $\dot{u}R$ and $\dot{w}Z$ must be specified.
2. At each end of the bar one member of each of the products $\dot{u}Q$ and $\dot{w}P_z$ must be specified.
3. The initial displacements and velocities must be specified.

Equations [14] and [16] would complete the theory were it not for the assumptions of continuity and single-valuedness of displacements (u and w) implicit in the integrations performed in arriving at [16]. In the absence of conditions assuring continuity or specifying discontinuities, many solutions of [14] may be found for the same boundary and initial conditions. The difficulty may be avoided by converting [14] to displacement equations of motion, following which the suitability of the displacements may be determined by inspection. The conversion, however, requires the establishment of stress-displacement relations.

DISPLACEMENT EQUATIONS OF MOTION

Hooke's Law of the general theory specializes in the present isotropic, axisymmetric case to

$$\left.\begin{aligned} \sigma_{rr} &= \lambda\Delta + 2\,\mu\epsilon_{rr} \\ \sigma_{\theta\theta} &= \lambda\Delta + 2\,\mu\epsilon_{\theta\theta} \\ \sigma_{zz} &= \lambda\Delta + 2\,\mu\epsilon_{zz} \\ \sigma_{zr} &= \mu\epsilon_{zr} \end{aligned}\right\} \qquad [17]$$

where λ and μ are Lamé's constants and Δ is the dilatation. In [17] the strains are first replaced by their usual expressions in terms of displacements. The latter, in turn, are replaced by the approximations [8]. Then integrations are performed on [17] so as to convert the left-hand members to bar-stresses in accordance with [10], with the result

$$\left.\begin{aligned} 2\,P_r = 2\,P_\theta &= 2\,a\,(\lambda + \mu)\,u + a^2\lambda\, w' \\ 2\,P_z &= 2\,a\,\lambda\, u + a^2\,(\lambda + 2\mu)\, w' \\ 4\,Q &= a^2\mu\, u' \end{aligned}\right\} \qquad [18]$$

Finally, in order to make the wave-velocities of the approximate theory fit those of the Pochhammer solution more closely, constants κ and κ_1 are introduced into the stress displacement relations in the following manner:

$$\left.\begin{aligned} 2\,P_r = 2\,P_\theta &= \kappa_1^2\,[2\,a\,(\lambda + \mu)\,u + a^2\lambda w'] \\ 2\,P_z &= 2\,a\lambda u + a^2\,(\lambda + 2\mu) w' \\ 4\,Q &= \kappa^2 a^2 \mu\, u' \end{aligned}\right\} \qquad [19]$$

The displacement-equations of motion are now obtained by combining [14] and [19], with the result

$$\left.\begin{aligned} a^2\,\kappa^2\mu u'' - 8\kappa_1^2(\lambda + \mu)\,u - 4\,a\kappa_1^2\,\lambda w' + 4aR &= \rho a^2 \ddot{u} \\ 2\,a\lambda u' + a^2(\lambda + 2\mu)\,w'' + 2aZ &= \rho a^2 \ddot{w} \end{aligned}\right\} \qquad [20]$$

The right member, $\rho a^2\ddot{u}$, of the first of [20] is the radial inertia correction and the first term, $a^2\kappa^2\mu u''$, of the left member is the radial shear correction.

---

[8] Love, p. 176.

189

The similarity between equations [20] for radial and longitudinal motions of a bar and Timoshenko's equations for rotational and transverse motions of a beam is striking. An interesting distinction is that Timoshenko's theory reduces to the next lower approximation by neglecting transverse shear *strain*, whereas the present theory reduces to the next lower approximation by neglecting radial shear *stress.*

WAVES IN AN INFINITE BAR

To obtain the solution of [20] for waves in an infinite bar, let

$$u = Ae^{i\gamma (z-ct)}, \quad w = B e^{i\gamma (z-ct)}, \quad R = Z = 0 \qquad [21]$$

where $\gamma = 2\pi/L$ and c is the wave-velocity.

Then [20] reduces to a pair of homogeneous, linear, algebraic equations, in A and B, whose determinant, set equal to zero, yields the frequency, or velocity, equation

$$\left[\frac{\kappa^2 \mu}{\rho} + \frac{8\kappa_1^2(\lambda+\mu)}{a^2\gamma^2\rho} - c^2\right]\left[\frac{\lambda+2\mu}{\rho} - c^2\right] - \frac{8\kappa_1^2\lambda^2}{a^2\gamma^2\rho^2} = 0 \qquad [22]$$

This is the frequency equation of a two-mode, coupled system.

For long waves $(a/L \to 0)$, equation [22] yields

$$c^2 = \frac{(3\lambda+2\mu)\mu}{(\lambda+\mu)\rho} = \frac{E}{\rho} = c_e^2, \quad c^2 = \infty \qquad [23]$$

which are identical with the corresponding limiting velocities of the first two modes in the exact (Pochhammer) solution.

For short waves $(a/L \to \infty)$, equation [22] yields

$$c^2 = \frac{\kappa^2\mu}{\rho} = \kappa^2 c_s^2, \quad c^2 = \frac{(\lambda+2\mu)}{\rho} = c_c^2 \qquad [24]$$

In the exact theory, the first of these is the velocity of Rayleigh surface waves. Hence, to make the present approximation match in this respect, κ^2 should be taken as the root of[9]

$$4\sqrt{(1-\alpha\kappa^2)(1-\kappa^2)} = (2-\kappa^2)^2, \quad 0<\kappa<1 \qquad [25]$$

where

$$\alpha = \frac{(1-2\nu)}{2(1-\nu)} \qquad [26]$$

The second of [24] is $c^2 = \mu/\rho$, in the exact solution, instead of $c^2 = (\lambda+2\mu)/\rho$. This appears to be a defect in the present theory, at first sight, but perhaps judgement should be withheld until solutions for finite bars of very short length are studied in detail.

Curves of c/c_e versus a/L are shown in Fig. 2 for several values of Poisson's ratio. These curves exhibit a typical coupling phenomenon between two modes of motion. The analogous curves for the first two modes of the Pochhammer solution are given by R. M. Davies[10] for $\nu = 0.29$. When $\nu = 0$ (and, hence, $\lambda = 0$) the two modes are uncoupled and have velocities

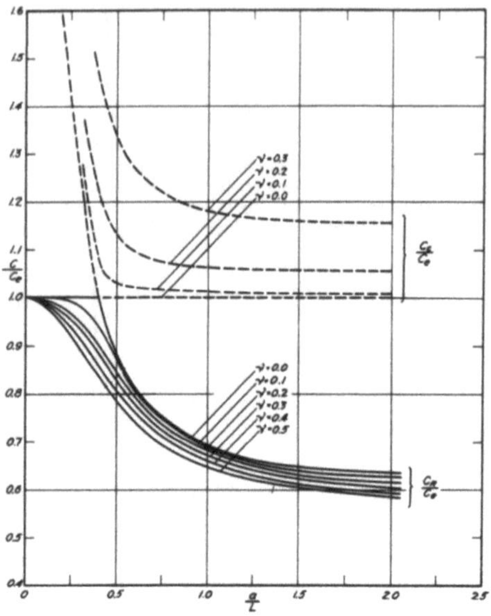

Fig. 2. EFFECT OF COUPLING, THROUGH POISSON'S RATIO, ON WAVE VELOCITIES OF AXIAL AND RADIAL MODES ACCORDING TO THE THEORY OF THIS PAPER.

$$c^2 = \frac{2\mu}{\rho} = \frac{E}{\rho}, \quad c^2 = \frac{\kappa^2\mu}{\rho} + \frac{8\kappa_1^2\mu}{a^2\gamma^2\rho} \qquad [27]$$

Referring to [20], the first of these is associated with purely axial motion. The uncoupled axial motion has a constant velocity, the same as in the exact solution. The uncoupled radial motion varies hyperbolically with a/L and has the correct limiting values. The intermediate values depend upon κ_1 which is taken as unity in Fig. 2.

The value of κ_1 is obtained from a comparison of first mode curves, in the exact and present solutions, of a/L versus c/c_s where $c_s = \mu/\rho$. These are shown, for the exact solution, in Fig. 3, using Bancroft's[11] data. It was observed by Hudson[12] that the curves for various values of Poisson's ratio are tangent at a common point whose coordinates are: $a/L = 0.293$, $c/c_s = \sqrt{2}$. The corresponding curves for the present approximate theory, with $\kappa_1 = 1$ and κ computed from [25], are shown in Fig. 4. These curves may be adjusted so as to be tangent at the

[9] Love, p. 308.

[10] R. M. Davies, *loc. cit.*

[11] D. Bancroft, *Phys. Rev.*, 59, 1941, p. 588.

[12] G. E. Hudson, *Phys. Rev.*, 63, 1943, p. 46.

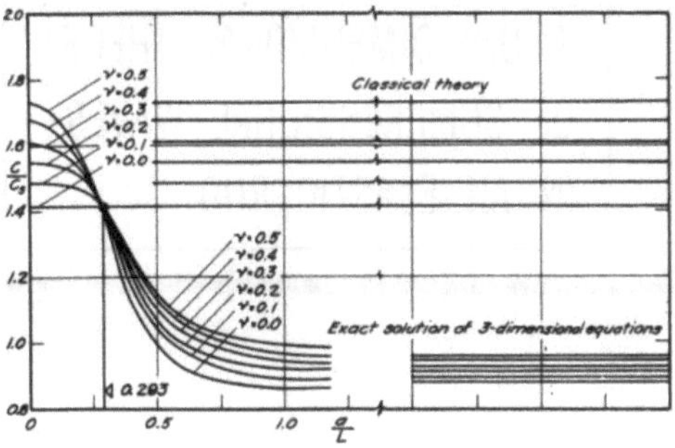

Fig. 3. FIRST MODE VELOCITIES FROM POCHHAMMER SOLUTION (AFTER BANCROFT AND HUDSON).

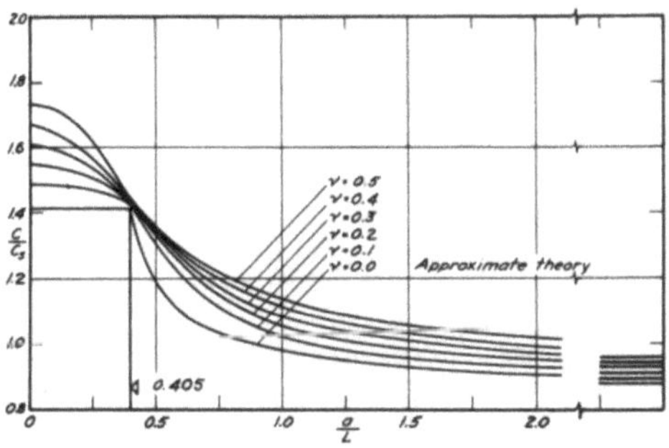

Fig. 4. FIRST MODE VELOCITIES FROM PRESENT SOLUTION WITH $\kappa_1 = 1$ and κ computed from [25].

correct common point by appropriate choice of κ_1. From [22] and the Bancroft-Hudson coordinates of the common point, we find

$$\kappa_1^2 = 0.422\,(2 - \kappa^2) \qquad [28]$$

The resulting improvement in the first mode velocity is illustrated in Fig. 1, by the shift of Curve 1a to 1'a.

EQUATION OF AXIAL MOTION

A single equation on w may be obtained by eliminating u from [20] with the result

$$\frac{E}{\rho}\frac{\partial^2 w}{\partial z^2} - \frac{\partial^2 w}{\partial t^2} + \frac{2Z}{\rho a}$$
$$+ \frac{a^2\rho}{8\kappa_1^2(\lambda+\mu)}\frac{\partial^2}{\partial t^2}\left[\frac{\lambda+2\mu}{\rho}\frac{\partial^2 w}{\partial z^2} - \frac{\partial^2 w}{\partial t^2} + \frac{2Z}{\rho a}\right]$$
$$- \frac{a^2\kappa^2\mu}{8\kappa_1^2(\lambda+\mu)}\frac{\partial^2}{\partial z^2}\left[\frac{\lambda+2\mu}{\rho}\frac{\partial^2 w}{\partial z^2} - \frac{\partial^2 w}{\partial t^2} + \frac{2Z}{\rho a}\right] \qquad [29]$$
$$+ \frac{\lambda}{\rho(\lambda+\mu)\kappa_1^2}\frac{\partial R}{\partial z} = 0$$

The first line of [29] contains the classical terms while the second and third lines are the radial inertia and radial shear corrections, respectively. It is interesting to note that the radial inertia correction in the present theory is not the same as Rayleigh's. This is revealed by comparing the second line in [29] with the last term in [4] and is illustrated in Fig. 1 where Curve 1a (rad) represents the present theory with the shear correction omitted and only the correction for radial inertia retained. It may be seen that the radial inertia correction that is found by the method of the present paper is better where these corrections are applicable. The reason for the difference between the two corrections is that Rayleigh, and also Love, made the additional assumption $u = -\nu a w'$. If this is introduced into [29] and if the shear and surface traction terms are dropped, [29] reduces to [4].

CONCLUDING REMARKS

The approximate theory of bars derived in this paper is completely contained in the equations of motion [20], the stress-displacement relations [18] and the boundary and initial conditions implied by [16]. The solution of the special problem of waves in an infinite bar has served two purposes. One purpose was to show the significance of the factors κ and κ_1 and to illustrate means for choosing their values. The other purpose was to test the theory by comparing solutions obtainable in both the approximate and exact theories. Since this comparison is favorable, further solutions of the approximate theory, not available in the exact theory, may be essayed. Work is now in progress on free vibrations of finite rods and an impact problem exhibiting dispersion.

R. D. Mindlin
G. Herrmann

A ONE-DIMENSIONAL THEORY OF COMPRESSIONAL WAVES IN AN ELASTIC ROD

Columbia University

It is the purpose of this brief note to correct an error in a paper, under the same title, which appeared in the *Proceedings of the First U. S. National Congress of Applied Mechanics.*

After the one-dimensional stress-displacement relations for the rod were derived, the elastic constants were altered so as to obtain a better fit with the three-dimensional theory. In so doing, the essential symmetry of the relations was inadvertently destroyed. The first two of Equations [19] should have been

$$2P_r = 2P_\theta = 2\kappa_1^2 a(\lambda + \mu)\, u + \kappa_1 a^2 \lambda w'$$

$$2P_z = 2\kappa_1 a\lambda u + a^2(\lambda + 2\mu)\, w'$$

and the differential equations [20] are correspondingly changed.

Fortunately, this change does not affect the wave velocities which were calculated. However, the change would affect solutions of problems involving boundary conditions at the ends of the rod.

Reprinted from
Proceedings of the First U.S. National Congress of Applied Mechanics-1951
(1952) 203-208

by R. D. Mindlin
W. P. Mason
T. F. Osmer
H. Deresiewicz

EFFECTS OF AN OSCILLATING TANGENTIAL FORCE ON THE CONTACT SURFACES OF ELASTIC SPHERES

Columbia University; Bell Telephone Laboratories, Murray Hill, N. J.; Bell Telephone Laboratories, Murray Hill, N. J.; Applied Physics Laboratory, The John Hopkins University

INTRODUCTION

If two elastic spheres are pressed together with a force, N, normal to their common initial tangent plane, they will come into contact on a circular area whose radius, a, is given by the Hertz theory[1]. The theory also predicts a purely normal traction, Z_z, on the contact surface, with distribution

$$Z_z = p\,(1-\rho^2/a^2)^{1/2}, \qquad [1]$$

where ρ is the distance from the center of the contact surface and $p = 3N/2\pi a^2$ is the maximum contact pressure.

The addition of a system of forces resulting in a monotonically increasing force, T, tangent to the contact surface, with N unchanged, has been studied previously in some detail,[2,3] with the following main results.

An assumption of no slip on the contact surface is untenable as it results in a distribution of tangential traction X_z, in the direction of T, which rises to infinity at the edge of the contact area. Hence, slip may be expected to begin at the edge and progress inward with increasing T. Since X_z, in the absence of slip, is radially symmetric, it is reasonable to suppose that slip occurs on an annulus. Further, as a first approximation, it is assumed that $X_z = fZ_z$ on the annulus, where f is the coefficient of friction. That is, wherever slip occurs, the magnitude of the tangential component of traction is equal to the product of the coefficient of friction (assumed constant) and the normal component of traction; the sense of X_z is that of the slip.

The radius of the inner boundary of the annulus is

$$c = a(1-T/fN)^{1/3}, \qquad [2]$$

while the tangential component of traction on the contact surface is

$$X_z = fp(1-\rho^2/a^2)^{1/2} \qquad c \le \rho \le a$$
$$X_z = fp(1-\rho^2/a^2)^{1/2} - fp(c/a)(1-\rho^2/c^2)^{1/2}, \quad \rho \le c \qquad [3]$$

This distribution is illustrated by curve aAB in Fig. 1.

Finally, the displacement, in the direction of T, of points far from the contact surface with respect to the *uniform* displacement of the portion of the contact surface over which no slip has occurred, is

$$\delta = \frac{3(2-\sigma)fN}{16\mu a}(1-c^2/a^2)$$
$$= \frac{3(2-\sigma)fN}{16\mu a}\left[1-\left(1-\frac{T}{fN}\right)^{2/3}\right]. \qquad [4]$$

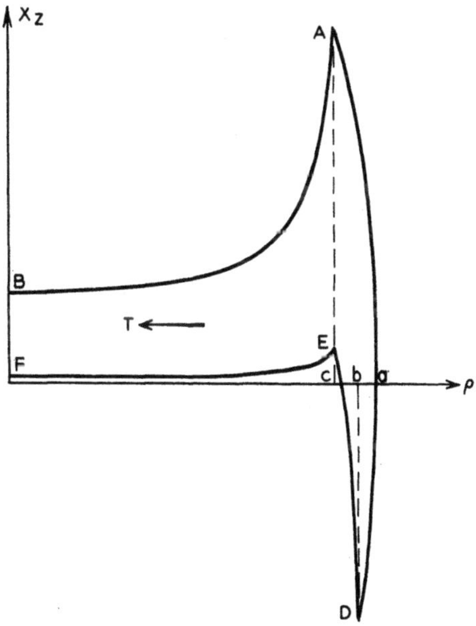

Fig. 1. THEORETICAL DISTRIBUTIONS OF TANGENTIAL COMPONENT OF TRACTION ON CONTACT SURFACE. CURVE aAB IS FOR P OF PATH OP IN FIG. 2. CURVE aDEF IS FOR R OF PATH OPR IN FIG. 2.

where σ is Poisson's ratio and μ is the shear modulus. This load-displacement relation is illustrated by curve OPQ in Fig. 2.

DIMINISHING TANGENTIAL FORCE

Suppose that the tangential force T, having reached a value T^* such that $0 < T^* < fN$, is reduced. If slip were prevented, X_z would again be infinite at $\rho = a$. Hence, slip must commence at $\rho = a$ and is tentatively presumed to penetrate to a radius b where $c \leq b \leq a$. Since this slip is opposite in sense to the initial slip, the sense of X_z on the annulus $b \leq \rho \leq a$ must be reversed, but its absolute magnitude is not altered. Thus, we must add

$$X_z = -2fp(1-\rho^2/a^2)^{1/2}, \quad b \leq \rho \leq a, \qquad [5]$$

to the initial traction [3]. Further, in order to maintain the required condition of uniform tangential displacement, over the portion of the contact surface on which no slip occurs, we must add also, according to Cattaneo's theory,

$$X_z = -2fp(1-\rho^2/a^2)^{1/2} + 2fp(b/a)(1-\rho^2/b^2)^{1/2}, \quad \rho \leq b, \qquad [6]$$

to the initial traction [3].

By employing the procedure leading to [4], the displacement on $\rho \leq b$, accompanying [5] and [6], is found to be

$$\delta = -\frac{3(2-\sigma)fN}{8\mu a}(1-b^2/a^2), \rho \leq b. \qquad [7]$$

Adding [5] and [6] to [3], we have, for the traction on the contact surface during diminution of T,

$$X_z = -fp(1-\rho^2/a^2)^{1/2}, \quad b \leq \rho \leq a,$$

$$X_z = -fp(1-\rho^2/a^2)^{1/2} + 2fp(b/a)(1-\rho^2/b^2)^{1/2}, \quad c \leq \rho \leq b$$

$$X_z = -fp(1-\rho^2/a^2)^{1/2} + 2fp(b/a)(1-\rho^2/b^2)^{1/2} - fp(c/a)(1-\rho^2/c^2)^{1/2}, \quad \rho \leq c. \qquad [8]$$

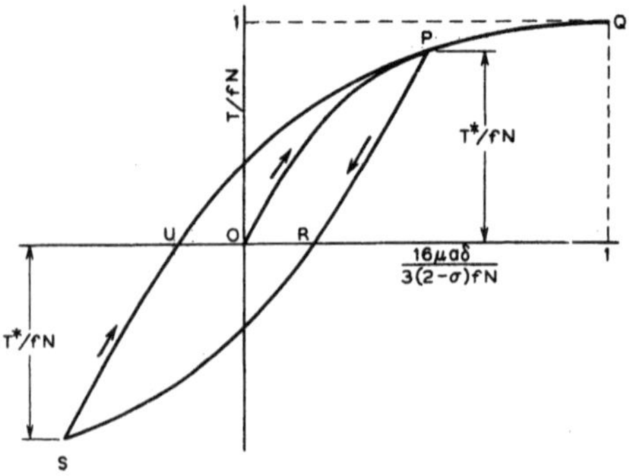

Fig. 2. THEORETICAL TANGENTIAL FORCE-DISPLACEMENT CURVES.

This distribution is illustrated by curve $aDEF$ in Fig. 1.

Again, adding [7] and [4], we find

$$\delta_d = \frac{3(2-\sigma)fN}{16\mu a}\left(\frac{2b^2}{a^2} - \frac{c^2}{a^2} - 1\right) \qquad [9]$$

where the subscript on δ_d designates decreasing T.

The magnitude of b is found from the equilibrium condition

$$T = 2\pi \int_0^a \rho X_z \, d\rho$$

$$= fN(1-c^3/a^3) - 2fN(1-b^3/a^3). \qquad [10]$$

But, from [2],

$$c = a(1 - T^*/fN)^{1/3}; \qquad [11]$$

hence, from [10] and [11],

$$b = a\left(1 - \frac{T^* - T}{2fN}\right)^{1/3} \qquad [12]$$

Thus, if $|T| \leq T^*$, $c \leq b \leq a$.

Substituting [11] and [12] in [9], we have

$$\delta_d = \frac{3(2-\sigma)fN}{16\mu a}\left[2\left(1 - \frac{T^*-T}{2fN}\right)^{2/3} - \left(1 - \frac{T^*}{fN}\right)^{2/3} - 1\right] \qquad [13]$$

This load-displacement relation for diminishing tangential force is illustrated by curve PRS in Fig. 2.

Comparing [11] and [12] we observe that $b = c$ when $T = -T^*$. That is, counter slip penetrates to the inner radius of the initial annulus of slip when the tangential load is fully reversed. The tangential traction, at this stage, is obtained by placing $b = c$ in [8]. The result is identical with [3] except for a reversal of sign of X_z. Thus, the traction is distributed as shown by aAB in Fig. 1 but with opposite sign. Similarly, the displacement, at this stage, is obtained by letting $T = -T^*$ in [13], so that the latter reduces to [4] with $T = T^*$. Hence, the displacement at S in Fig. 2 is equal and opposite to that at P in the same figure.

OSCILLATING TANGENTIAL FORCE

Since the entire situation at $T = -T^*$ is identical with that at $T = T^*$, except for sign reversal, a subsequent increase of T from $-T^*$ to T^* will be accompanied by the same events as occurred during the decrease from T^* to $-T^*$, except for reversal of sign. Thus, in starting along SU, slip starts at $\rho = a$ in the same sense as occurred along path OP. At an intermediate point of SU the traction will be like $aDEF$ with reversed sign. When the traction again reaches T^*, b will again have penetrated to c and the traction will be exactly aAB. The displacement along the path SU is

$$\delta_i = -\delta_d(-T)$$

$$= -\frac{3(2-\sigma)fN}{16\mu a}\left[2\left(1 - \frac{T^*+T}{2fN}\right)^{2/3} - \left(1 - \frac{T^*}{fN}\right) - 1\right] \qquad [14]$$

Hence, when $T = T^*$, the terminal points P of SUP and P of OP are identical, as may be seen by comparing [14] and [4]. Subsequent diminution of T will then produce a repetition of the events accompanying the first diminution. It is now evident that, during oscillation of T between $\pm T^*$, the load-displacement curve traverses the loop $PRSUP$. It is interesting that the slopes at O, P and S of the segments OP, PR and SU are identical.

The area enclosed in the loop (Fig. 2) gives the frictional energy loss per cycle:

$$W = \int_{-T^*}^{T^*} (\delta_d - \delta_i)\, dT$$

$$= \frac{9(2-\sigma) f^2 N^2}{10\, \mu\, a} \left\{ 1 - \left(1 - \frac{T^*}{fN}\right)^{5/3} - \frac{5T^*}{6fN}\left[1 + \left(1 - \frac{T^*}{fN}\right)^{2/3}\right] \right\} \qquad [15]$$

For small T^*/fN, [15] reduces to

$$W = \frac{(2-\sigma) T^{*3}}{36\, \mu\, a f N} \qquad [16]$$

i.e., the energy loss per cycle varies as the cube of the tangential force.

It may be useful to collect those aspects of the foregoing theory which constitute assumptions and those which are results of deductive reasoning.

The assumptions are:

1. Slip is initiated when and where, in the absence of slip, infinite tangential traction is found in the solution of the appropriate boundary value problem in the theory of elasticity.

2. Slip occurs only on annular surfaces.

3. At any point on a surface along which slip has just taken place, the tangential component of traction has the same sense as that of the slip and its magnitude is equal to the product of a constant coefficient of friction and the normal component of traction at that point.

On the other hand, the constancy of the tangential component of the displacement, over a portion of the contact surface where no slip occurs, is a result of considerations of symmetry. The tractions on and displacements of such portions are obtained from the solutions of boundary value problems in the theory of elasticity. The same holds for the dimensions of the annuli of slip.

EXPERIMENTAL TESTS

If experiments are made for various ratios of T^*/fN, and a plot is made of measured values of $1 - (c/a)^3$ versus T^*/N, the data, according to [11], should fall on a straight line whose slope is the coefficient of friction.

Such experiments were performed employing polished crown glass lenses. To obtain symmetry in the driving system, a central doubly-convex lens, pressed on from either side by the convex portions of plano-convex lenses, was driven by a barium titanate transducer. The complete arrangement is shown in Fig. 3 and a close-up of the lenses is shown in Fig. 4. All lenses were 1 cm in diameter, 2 mm thick and had 4.85 inch radii of curvature. The lower lens was cemented into a circular recess in a brass anvil screwed to the base; the upper, into a recess in a hinged lever, placed parallel to the transducer in order to increase its stiffness in the direction of motion. Normal forces, on the lenses, of the order of 10 pounds were used, producing contact surface diameters of about 0.030 inches. A conical hole in the lever, above the lenses, permitted rays of light to be passed through the lenses and reflected back, at the various surfaces, into a microscope. By observing the relative positions of the Newton's rings formed around the upper and lower contact surfaces, proper alignment of the lenses could be obtained.

Since the coefficient of friction of glass is about 0.25 and there were two sets of surfaces in contact, tangential forces of the order of 5 pounds were required to bring the central lens to the point of sliding. Previous experiments[4] with hollow cylinders of barium titanate had shown that forces larger than this were possible with r.m.s. voltages

Fig. 3. GENERAL VIEW OF APPARATUS.

Fig. 4. CLOSE-UP VIEW OF ANVIL, LENSES AND ENDS OF TRANSDUCER AND LEVER.

of the order of 3000 volts. In order to avoid resonance effects, a low-frequency (60 c.p.s.) excitation was used. This was obtained from a 115-volt power main, stepped up by means of a neon transformer and regulated by a variac in the primary.

The equation[5] for one-dimensional stress in polarized barium titanate can be written in the form

$$X_x = Y_o x_x - d_{31} Y_o E_3 \,, \qquad [17]$$

where X_x is the stress, x_x the strain, Y_o Young's modulus, d_{31} the effective piezoelectric constant for the degree of poling used, and E_3 the electric field. If E_3 is taken to be the voltage gradient applied through the thickness of the hollow cylinder between the inner and outer plating, and the stress is expressed in pounds per square inch, the constants Y_o and d_{31} for the 4% lead titanate-barium titanate ceramic used are: $Y_o = 16.4 \times 10^6$ pounds per square inch; $d_{31} = 14.5 \times 10^{-10}$ inches per volt.

The cross-sectional area of the barium titanate cylinder (1 inch O.D., ½ inch I.D.) was 0.59 square inches; the wall-thickness in the direction of the applied voltage, ¼ inch. Hence, if the device were clamped so that the strain $x_x = 0$, the force exerted by an applied voltage of 3000 volts would be

$$F = \frac{0.59 \times 14.5 \times 10^{-10} \times 3000 \times 16.4 \times 10^6}{0.25} = 167 \textit{ pounds}$$

On the other hand, if the stress were zero, the expansion in the plated length of 4 inches would be

$$4\, x_x = \frac{4 \times 14.5 \times 10^{-10} \times 3000}{0.25} = 0.7 \times 10^{-4} \textit{ inches.}$$

The actual tangential force applied depends on the relative displacement of the glass lenses due to deformation and slip. To measure the force, a lead titanate-barium titanate disk, poled in the thickness direction, was placed between the driver and the bracket which clamped the center lens as shown in Figs. 3 and 4. Thus, the entire force applied to the center lens was exerted through the disk so that the voltage generated by the disk, and measured with a high-impedance voltmeter, gave a direct indication of the tangential force on the lens. The calibration, obtained by replacing the lenses with a spring of known stiffness (1.97×10^4 pounds per inch), and measuring the resulting displacement by means of a velocity pickup, was found to be 2.65 pounds/volt. Due to symmetry, the tangential load was assumed to be divided equally between each of the contact areas. The validity of this assumption was borne out by examination of the wear patterns on the lenses. Thus, the tangential load was measured with a calibrated barium titanate disk; the normal load, by a fixed-weight lever system.

Enlarged photographs of typical patterns of wear (diameter approximately 0.03 inch) are shown in Fig. 5(a) and (b). The outer edge of the contact surface was usually well defined, but the inner edge of the annulus was sometimes so irregular that an estimate of its diameter was difficult. In such cases measurements were made by independent observers. This accounts for the double and triple points in Fig. 6.

The fact that annuli of surface wear were observed and that the plotted data of T^*/N versus $1-(c/a)^3$ fall closely on a straight line through the origin (Fig. 6) are considered to be good evidence of the validity of the theory derived in the first part of this paper. Furthermore, the coefficient of friction, obtained in this very indirect manner, agrees quite well with direct measurements made by others.[6]

An attempt was also made to check the theoretical force-displacement curves (Fig. 2) and the energy loss per cycle (Equation [15]). The applied force was measured directly, as before, by the barium titanate disk; the displacement, by a velocity pickup inserted between the anvil and the lens end of the transducer. The voltages were transmitted to the vertical and horizontal plates, respectively, of a cathode ray oscillograph. A phase shifter, placed in the velocity pickup circuit, was adjusted until the force-displacement relation was a straight line with the calibrating spring in place, thus assuring that relative phase shifts in the circuit were balanced out. With the lens system in the setup, curves for six values of T^*/N, shown in Fig. 7, were obtained. These are roughly in agreement with the loop in Fig. 2. Also a curve drawn tangent to the long axes of the loops in Fig. 7 is similar to OP in Fig. 2. Since the voltages had previously been calibrated in terms of force and displacement, the area of each loop in pound inches could be evaluated. The results are shown by the circles joined by the full line in Fig. 8. The dashed line in the same figure is drawn according to Equation [15]. At the larger displacements the agreement between experiment and theory is good; but, for small displacements, the energy dissipation for cycle varies as the square of the displacement whereas the theory predicts a cube law.

Several interesting side phenomena were observed in the experiments. When wear patterns appeared, it was

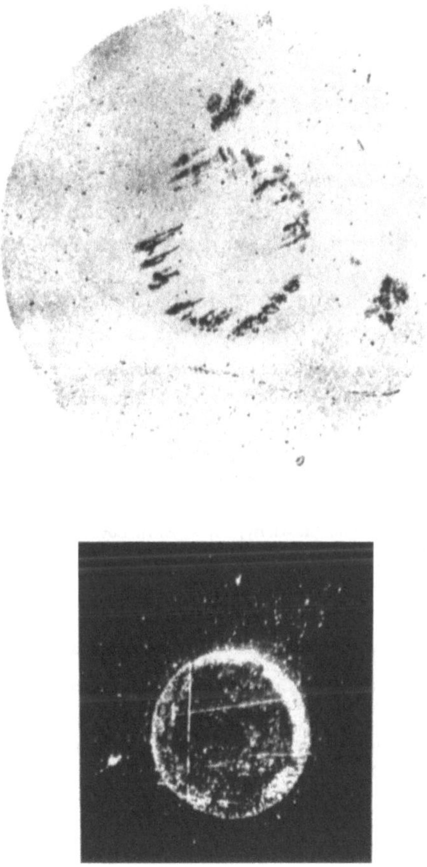

Fig. 5. TWO EXAMPLES OF ANNULI OF WEAR.

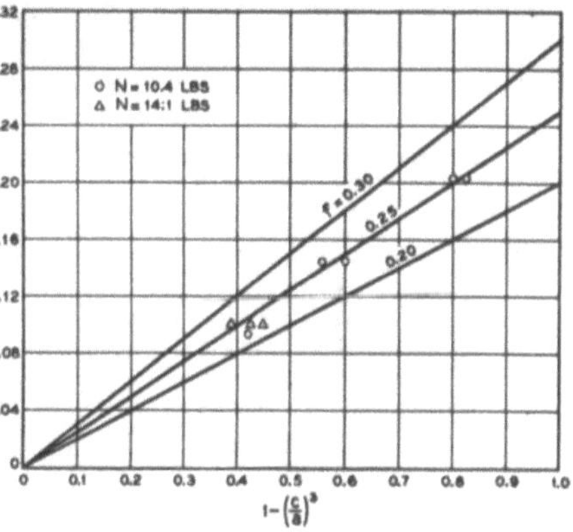

Fig. 6. COMPARISON OF THEORY AND EXPERIMENT ON BASIS OF VARIATION OF ANNULUS DIMENSIONS WITH APPLIED LOADS. THE STRAIGHT LINES THROUGH THE ORIGIN ARE THEORETICAL CURVES FOR VARIOUS VALUES OF COEFFICIENT OF FRICTION.

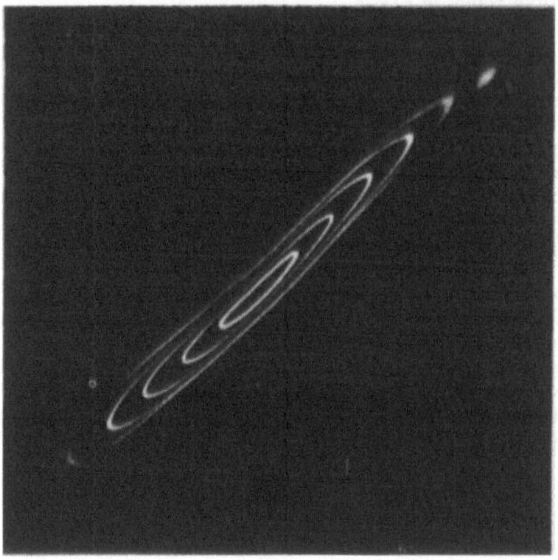

Fig. 7. SUPERPOSED OSCILLOGRAMS OF FORCE-DISPLACEMENT LOOPS. (COMPARE WITH FIG. 2).

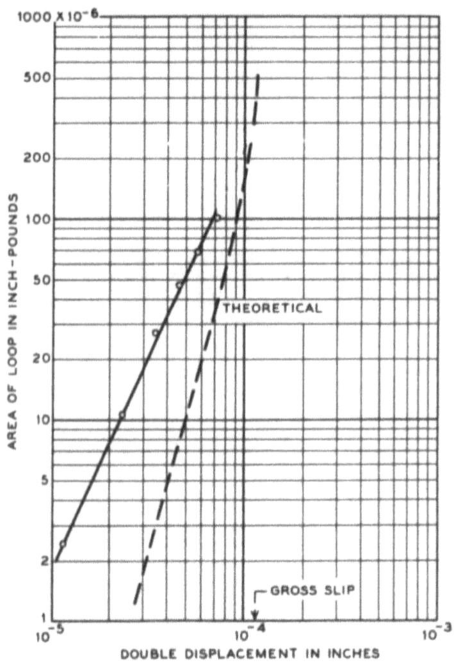

Fig. 8. COMPARISON OF THEORETICAL AND EXPERIMENTAL ENERGY DISSIPATION PER CYCLE.

usually after a period of one-half to three hours. However, in about half the runs no surface markings were observed at all, even after a full day's run.

Fig. 9 illustrates two other side effects. One is the eccentricity of the ring of wear, which may have been due to misalignment of the lenses. The other is the development of cracks in the glass. These invariably started at the edge of the contact surface, on the diameter parallel to the direction of the tangential force, and spread laterally, roughly at right angles to this diameter. This is probably a good indication of the location and direction of the maximum tensile stress in the glass.

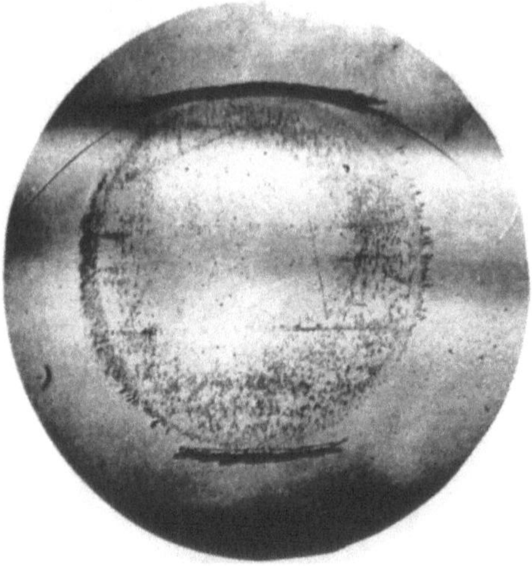

Fig. 9. EXAMPLE OF CRACKS AND ECCENTRIC RING OF WEAR.

BIBLIOGRAPHY

1. "Theory of Elasticity," by A. E. H. Love, Cambridge, 4th Ed., 1927, pp. 193-198.

2. C. Cattaneo, *Accad. dei Lincei, Rendiconti,* Ser. 6, Vol. 27, 1938, pp. 342-348, 434-436, 474-478.

3. R. D. Mindlin, *Jour. Appl. Mech.*, Vol. 16, 1949, pp. 259-268.

4. W. P. Mason and R. F. Wick, *Jour. Acous. Soc. Amer.*, Vol. 23, 1951, pp. 209-214.

5. "Piezoelectric Crystals and Their Application to Ultrasonics," by W. P. Mason, Chap. XII, equation (12.27) D. Van Nostrand Co., 1950.

6. I. Simon, O. McMahon and R. J. Bowen, *Jour. App. Phys.*, Vol. 22, 1951, pp. 177-184.

Reprinted from JOURNAL OF APPLIED PHYSICS, Vol. 23, No. 1, 83–88, January, 1952

Forced Thickness-Shear and Flexural Vibrations of Piezoelectric Crystal Plates

RAYMOND D. MINDLIN
Department of Civil Engineering, Columbia University, New York
(Received July 10, 1951)

An approximate theory is given for the vibrations of piezoelectric crystal plates in thickness-shear and flexural modes, including the interaction between the elastic and electric fields. The problem of the rectangular, AT, quartz plate is solved and computed frequencies are compared with measurements by Sykes. Algebraic formulas are obtained relating resonant frequencies with dimensions, elastic and electric constants, and orientation of cut.

I. INTRODUCTION

IN two previous papers procedures were described for studying various aspects of shear and flexural motions of elastic plates. In the first paper[1] (referred to in the sequel as I) the three-dimensional equations of isotropic elasticity were reduced to two-dimensional, plate-flexure equations and, in the course of the reduction, the terms representing rotatory inertia and shear were retained. It was shown that these terms account for the discrepancy between the velocities of short waves, predicted by the classical theory of plates and the complete three-dimensional theory. As a consequence, it was concluded that the improved plate theory must contain high frequency vibration phenomena beyond the range of applicability of classical plate theory. In the second paper[2] (referred to in the sequel as II) the improved plate theory was extended to include crystal plates. A problem of vibrations of a rectangular plate was solved and computations of frequencies were made for AT-cut quartz plates having finite length to thickness ratios. The theory revealed, both qualitatively and quantitatively, the intricate high frequency spectrum observed in the laboratory.

In the present paper the improved plate theory is further extended to accommodate the piezoelectric relations and the electric field equations. The theory includes, then, the interaction between the elastic and electric fields in a vibrating piezoelectric crystal.

The problem of the rectangular plate, analogous to that considered in II, is reconsidered as one of piezoelectrically forced vibrations. The electric-elastic equations are solved simultaneously and the boundary conditions yield a transcendental frequency equation of the same form as that found in II. The effect of the interaction of the two fields on the frequencies is contained in two correction terms: one on the thickness-shear modulus and one on the plate-flexure modulus. One of the results is a formula for the surface charge density, in terms of the shear strain and the surface potential, from which the current through the vibrating plate may be calculated.

The main advantage gained through the use of an adequate, two-dimensional theory of plates is that the mathematical solution for a plate of finite length is no more difficult than the solution for an infinite plate in

[1] R. D. Mindlin, J. Appl. Mech. **18**, No. 1, 31–38 (March, 1951).
[2] R. D. Mindlin, J. Appl. Phys. **22**, 316–323 (1951).

the three-dimensional theory. This is done at the sacrifice of higher modes through the thickness; but the crystalline nature of the medium, the length modes, and the coupling between thickness-shear and flexural modes are retained, and these appear to contain the essence of the observed phenomena.

THREE-DIMENSIONAL EQUATIONS

It is expedient to record, first, the three-dimensional equations of the elastic and electric fields and the piezoelectric relations connecting them. The elastic and electric properties of the material are assumed to have monoclinic, sphenoidal symmetry, with the x-axis an axis of binary symmetry. Then the piezoelectric relations are

$$\begin{aligned}
T_1&=c_{11}S_1+c_{12}S_2+c_{13}S_3+c_{14}S_4-e_{11}E_1,\\
T_2&=c_{12}S_1+c_{22}S_2+c_{23}S_3+c_{24}S_4-e_{12}E_1,\\
T_3&=c_{13}S_1+c_{23}S_2+c_{33}S_3+c_{34}S_4-e_{13}E_1,\\
T_4&=c_{14}S_1+c_{24}S_2+c_{34}S_3+c_{44}S_4-e_{14}E_1,\\
T_5&=c_{55}S_5+c_{56}S_6-e_{25}E_2-e_{35}E_3,\\
T_6&=c_{56}S_5+c_{66}S_6-e_{26}E_2-e_{36}E_3.
\end{aligned}\tag{1}$$

$$\begin{aligned}
P_1&=e_{11}S_1+e_{12}S_2+e_{13}S_3+e_{14}S_4+\kappa_{11}E_1,\\
P_2&=e_{25}S_5+e_{26}S_6+\kappa_{22}E_2+\kappa_{23}E_3,\\
P_3&=e_{35}S_5+e_{36}S_6+\kappa_{23}E_2+\kappa_{33}E_3,
\end{aligned}\tag{2}$$

where T_i, S_i, E_i, P_i are the components of stress, strain, electric force, and polarization, respectively, and c_{ij}, e_{ij}, κ_{ij} are the elastic, piezoelectric, and susceptibility coefficients, respectively.

The strain-displacement relations are

$$\begin{aligned}
S_1&=\partial u_1/\partial x, & S_4&=\partial u_3/\partial y+\partial u_2/\partial z,\\
S_2&=\partial u_2/\partial y, & S_5&=\partial u_1/\partial z+\partial u_3/\partial x,\\
S_3&=\partial u_3/\partial z, & S_6&=\partial u_2/\partial x+\partial u_1/\partial y.
\end{aligned}\tag{3}$$

The stress equations of motion are

$$\begin{aligned}
\frac{\partial T_1}{\partial x}+\frac{\partial T_6}{\partial y}+\frac{\partial T_5}{\partial z}&=\rho\frac{\partial^2 u_1}{\partial t^2},\\
\frac{\partial T_6}{\partial x}+\frac{\partial T_2}{\partial y}+\frac{\partial T_4}{\partial z}&=\rho\frac{\partial^2 u_2}{\partial t^2},\\
\frac{\partial T_5}{\partial x}+\frac{\partial T_4}{\partial y}+\frac{\partial T_3}{\partial z}&=\rho\frac{\partial^2 u_3}{\partial t^2},
\end{aligned}\tag{4}$$

and the pertinent electric field equations are

$$(E_1, E_2, E_3)=-\left(\frac{\partial V}{\partial x}, \frac{\partial V}{\partial y}, \frac{\partial V}{\partial z}\right),\tag{5}$$

$$\frac{\partial^2 V}{\partial x^2}+\frac{\partial^2 V}{\partial y^2}+\frac{\partial^2 V}{\partial z^2}=4\pi\left(\frac{\partial P_1}{\partial x}+\frac{\partial P_2}{\partial y}+\frac{\partial P_3}{\partial z}\right),\tag{6}$$

where V is the electrostatic potential. The influence of the magnetic field is omitted.

Through the use of Eqs. (1), (2), (3), and (5), Eqs. (4) and (6) may be expressed in terms of the dependent variables u_1, u_2, u_3, and V. Thus, consideration of the electric field adds one differential equation and one dependent variable to the usual equations of elasticity.

DERIVATION OF PLATE EQUATIONS

As in II, the faces of the plate are the planes $y=\pm h/2$ of an x, y, z system of rectangular coordinates. The normal components of the tractions on planes parallel to the faces are neglected, i.e.,

$$T_2=0\tag{7}$$

throughout the plate. In addition the faces are assumed to be traction-free, so that

$$T_6=T_4=0 \quad \text{on} \quad y=\pm h/2.\tag{8}$$

Five plate-stress components (three stress couples and two stress resultants) are defined, in the usual manner, by

$$\begin{aligned}
(M_x, M_z, M_{xz})&=\int_{-h/2}^{h/2}(T_1, T_3, T_5)y\,dy,\\
(Q_x, Q_z)&=\int_{-h/2}^{h/2}(T_6, T_4)dy.
\end{aligned}\tag{9}$$

In addition, for the purposes of this paper, three plate-polarization components are defined by

$$\begin{aligned}
(P_x, P_z)&=\int_{-h/2}^{h/2}y(P_1, P_3)dy,\\
P_y&=\int_{-h/2}^{h/2}P_2dy.
\end{aligned}\tag{10}$$

(In order to avoid a new symbol, true polarization components are designated by the letter P with number subscripts, while plate-polarization components are designated by the letter P with letter subscripts.)

The four dependent variables of the three-dimensional theory are now assumed to have the form

$$hu_1=y\xi,\quad u_2=\eta,\quad hu_3=y\zeta,\quad hV=y\varphi,\tag{11}$$

where ξ, η, ζ and φ are functions of x, z and t only.

The conversion of the three-dimensional piezoelectric relations (1) and (2) to two dimensions proceeds as follows. Setting $T_2=0$, the second of (1) is solved for S_2 and this is used to eliminate S_2 from the remaining five of (1) and from (2). In the resulting eight equations, the strain and electric force components are expressed in terms of the displacements and electric potential in accordance with (3) and (5). The latter, in turn, are expressed in terms of ξ, η, ζ, and φ through (11). Integrations are now performed on the eight equations so as to convert the left hand sides to plate-stress and plate-polarization components, in accordance with (9) and (10), with the result:

$$
\begin{aligned}
M_x&=D_1\partial\xi/\partial x+D_2\partial\zeta/\partial z+F_1\partial\varphi/\partial x,\\
M_z&=D_2\partial\xi/\partial x+D_3\partial\zeta/\partial z+F_2\partial\varphi/\partial x,\\
M_{xz}&=D_5(\partial\xi/\partial z+\partial\zeta/\partial x)+F_5\partial\varphi/\partial z,\\
Q_z&=D_4(\zeta+h\partial\eta/\partial z),\\
Q_x&=D_6(\xi+h\partial\eta/\partial x)+F_6\varphi.
\end{aligned}
\tag{12}
$$

$$
\begin{aligned}
P_x&=F_1\partial\xi/\partial x+F_2\partial\zeta/\partial z-C_1\partial\varphi/\partial x,\\
P_y&=F_6(\xi+h\partial\eta/\partial x)-C_2\varphi,\\
P_z&=F_5(\partial\xi/\partial z+\partial\zeta/\partial x)-C_3\partial\varphi/\partial z,
\end{aligned}
\tag{13}
$$

where

$$
\begin{aligned}
&D_1=(c_{11}-c_{12}{}^2/c_{22})R, && D_2=(c_{13}-c_{12}c_{23}/c_{22})R,\\
&D_3=(c_{33}-c_{23}{}^2/c_{22})R, && D_4=\kappa^2(c_{44}-c_{24}{}^2/c_{22}),\\
&D_5=c_{55}R, && D_6=\kappa^2c_{66},\\
&F_1=(e_{11}-c_{12}e_{12}/c_{22})R, && F_2=(e_{13}-c_{23}e_{12}/c_{22})R,\\
&F_5=e_{35}R, && F_6=e_{26},\\
&C_1=(\kappa_{11}+e_{12}{}^2/c_{22})R, && C_2=\kappa_{22},\quad C_3=\kappa_{33}R,
\end{aligned}
\tag{14}
$$

and

$$R=h^2/12.$$

As in I and II, the factor κ^2 is introduced in the constants D_4 and D_6 in order to accommodate, as well as possible, the assumed simple y-dependence to the true y-dependence which varies markedly with wavelength and frequency. As before, the value of κ^2 will be determined by comparison of an exact solution of the three-dimensional equations with the corresponding solution of the plate equations.

The three-dimensional differential equations are converted to plate equations by integrating over the thickness of the plate after multiplying (6) and the first and third of (4) by y. Using Eqs. (7) to (11), the result is

$$
\begin{aligned}
&\frac{\partial M_x}{\partial x}-Q_x+\frac{\partial M_{xz}}{\partial z}=\rho R\frac{\partial^2\xi}{\partial t^2},\\
&\frac{\partial Q_x}{\partial x}+\frac{\partial Q_z}{\partial z}=\rho h\frac{\partial^2\eta}{\partial t^2},\\
&\frac{\partial M_{xz}}{\partial x}-Q_z+\frac{\partial M_z}{\partial z}=\rho R\frac{\partial^2\zeta}{\partial t^2},\\
&R\left(\frac{\partial^2\varphi}{\partial x^2}+\frac{\partial^2\varphi}{\partial z^2}\right)=4\pi\left(\frac{\partial P_x}{\partial x}+\frac{\partial P_z}{\partial z}\right).
\end{aligned}
\tag{15}
$$

Finally, the plate-stress and plate-polarization components may be eliminated from (15), through the use of the plate-piezoelectric relations (12) and (13), leaving four differential equations governing the four dependent variables ξ, η, ζ, φ.

BOUNDARY CONDITIONS

If the electric terms are omitted in (12), (13), and (15), the equations reduce to those found in II. It was shown in I how the boundary conditions appropriate to the latter equations may be obtained from a plate-analogue of Neumann's uniqueness theorem. In terms of components referred to coordinates ν and s, measured, respectively, normal to and along the boundary of the plate, it was found that, at each point on the edge of the plate, there must be specified one member of each of the three products

$$M_\nu\frac{\partial\xi_\nu}{\partial t},\quad M_{\nu s}\frac{\partial\xi_s}{\partial t},\quad Q_\nu\frac{\partial\eta}{\partial t},\tag{16}$$

where ξ_ν and ξ_s are the components of the vector (ξ, ζ) in the ν and s directions. On a free edge, for example,

$$M_\nu=M_{\nu s}=Q_\nu=0.\tag{17}$$

To account for the electric terms it is only necessary to add the electrostatic energy, V_e, and proceed as before. Thus,

$$
\begin{aligned}
8\pi V_e&=\iiint(E_1D_1+E_2D_2+E_3D_3)dxdydz\\
&=\iiint(E_1{}^2+E_2{}^2+E_3{}^2)dxdydz\\
&\quad+4\pi\iiint(E_1P_1+E_2P_2+E_3P_3)dxdydz.
\end{aligned}
\tag{18}
$$

From (5) and the last of (11), this becomes

$$
\begin{aligned}
8\pi V_e=h^{-2}&\iiint\left[\left(y\frac{\partial\varphi}{\partial x}\right)^2+\varphi^2+\left(y\frac{\partial\varphi}{\partial z}\right)^2\right]dxdydz-4\pi h^{-1}\\
&\times\iiint\left(P_1y\frac{\partial\varphi}{\partial x}+P_2\varphi+P_3y\frac{\partial\varphi}{\partial z}\right)dxdydz.
\end{aligned}
\tag{19}
$$

Integrating over the thickness of the plate and using (10), we have

$$
\begin{aligned}
8\pi V_e=h^{-1}&\iint\left[R\left(\frac{\partial\varphi}{\partial x}\right)^2+\varphi^2+R\left(\frac{\partial\varphi}{\partial z}\right)^2\right]dxdz\\
&-4\pi h^{-1}\iint\left(P_x\frac{\partial\varphi}{\partial x}+P_y\varphi+P_z\frac{\partial\varphi}{\partial z}\right)dxdz.
\end{aligned}
\tag{20}
$$

The four terms containing derivatives are integrated by parts to obtain

$$
\begin{aligned}
8\pi hV_e=&\iint\left[4\pi\left(\frac{\partial P_x}{\partial x}+\frac{\partial P_z}{\partial z}\right)-R\left(\frac{\partial^2\varphi}{\partial x^2}+\frac{\partial^2\varphi}{\partial z^2}\right)\right]\varphi dxdz\\
&-\iint(4\pi P_y-\varphi)\varphi dxdz\\
&-\oint\left(4\pi P_\nu-R\frac{\partial\varphi}{\partial\nu}\right)\varphi ds.
\end{aligned}
\tag{21}
$$

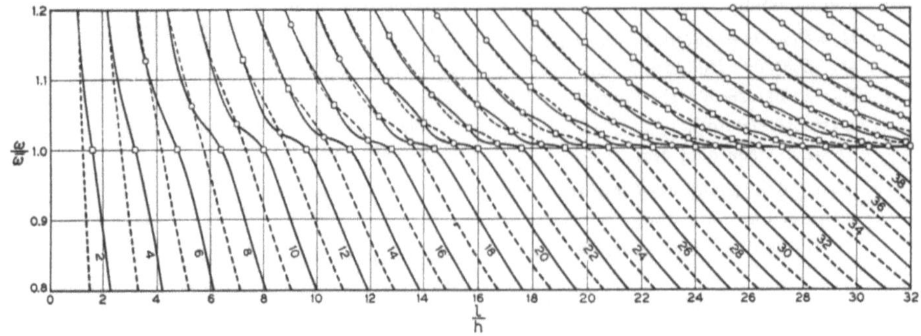

FIG. 1. Resonant frequencies of an AT-cut, rectangular, quartz plate as calculated from Eqs. (42), (43), and (44). For the circled and squared points, at which resonant frequencies of even and odd modes coincide, the transcendental frequency equation degenerates and the coordinates are given by simple algebraic formulas.

Noting that the first integral vanishes because of the last of (15) and writing

$$4\pi h\sigma_s = 4\pi P_y - \varphi, \qquad 4\pi h^2\sigma_e = 4\pi P_\nu - R\partial\varphi/\partial\nu, \tag{22}$$

we have

$$2V_e = -\int\int \sigma_s \varphi dx dz - h \oint \sigma_e \varphi ds. \tag{23}$$

It will be observed that σ_s is the surface charge density, σ_e is an "edge charge density," and φ is the potential difference between the two surfaces of the plate. Thus, by the argument used in I, the additional boundary conditions are:

1. At each point on the surface of the plate, either φ or σ_s may be specified.
2. At each point on the edge of the plate, either φ or σ_e may be specified.

A useful formula for the surface charge density may be obtained from (22), (13), and (14):

$$h\sigma_s = F_6(\xi + h\partial\eta/\partial x) - (1+4\pi\kappa_{22})\varphi/4\pi.$$

It is convenient to introduce the dielectric constants

$$K_i = (1+4\pi\kappa_{ii})/4\pi, \quad i=1,2 \tag{24}$$

so that

$$h\sigma_s = F_6(\xi + h\partial\eta/\partial x) - K_2\varphi. \tag{25}$$

INFINITE PLATE: DETERMINATION OF κ^2

The solution of the three-dimensional equations for the circular frequency, ω, of thickness-shear vibrations of an infinite plate with infinite plated electrodes is given by[3]

$$\omega^2 = \frac{4\alpha_n^2 c_{66}}{\rho h^2}\left(1+\frac{e_{26}^2}{c_{66}K_2}\right), \tag{26}$$

where α_n are the roots of

$$\tan\alpha_n = \alpha_n\left(1+\frac{c_{66}K_2}{e_{26}^2}\right). \tag{27}$$

The corresponding solution of the plate equations is obtained by setting

$$\varphi = \varphi_0 \exp(i\bar{\omega}t), \quad \xi = \xi_0 \exp(i\bar{\omega}t), \quad \eta = \zeta = 0,$$

where φ_0 and ξ_0 are constants. Then, from (12) and (13),

$$M_x = M_z = M_{xz} = Q_z = P_x = P_z = 0,$$
$$Q_x = (D_6\xi_0 + F_6\varphi_0)\exp(i\bar{\omega}t),$$
$$P_y = (F_6\xi_0 - C_2\varphi_0)\exp(i\bar{\omega}t).$$

Hence the last three of (15) are satisfied identically and the first gives

$$\xi_0 = F_6\varphi_0/(\rho\bar{\omega}^2 R - D_6),$$

which is the displacement as a function of impressed voltage and frequency and the elastic and piezoelectric constants. Resonance occurs when

$$\bar{\omega}^2 = D_6/\rho R = 12\kappa^2 c_{66}/\rho h^2. \tag{28}$$

Comparing (26) and (28), we find

$$3\kappa^2 = \alpha_1^2(1+e_{26}^2/c_{66}K_2), \tag{29}$$

where α_1 is the smallest root (>0) of (27).

In the absence of piezoelectric effect ($e_{26}=0$), $\alpha_1 = \pi/2$ and $\kappa^2 = \pi^2/12$ as in I and II.

PLATE OF FINITE LENGTH

As observed in II, extensive experimental evidence obtained by Sykes on rotated Y-cut quartz plates shows that, over a considerable range, the frequencies are practically independent of the z-dimension. This suggests taking $\zeta=0$ and ξ, η, and φ independent of z.

[3] A. W. Lawson, Phys. Rev. 62, 71 (1942). Equation (27), is a special case of Lawson's Eq. (57).

Equations (12) and (13) then become

$$\begin{aligned} M_x &= D_1\xi' + F_1\varphi', \\ M_z &= D_2\xi' + F_2\varphi', \\ Q_x &= D_6(\xi + h\eta') + F_6\varphi, \\ P_x &= F_1\xi' - C_1\varphi', \\ P_y &= F_6(\xi + h\eta') - C_2\varphi, \\ M_{xz} &= Q_z = P_z = 0, \end{aligned} \tag{30}$$

where prime indicates differentiation with respect to x. Inserting (30) into (15) (the equations of motion and the electric field equation) the latter become, after introducing a factor $\exp(i\omega t)$,

$$\begin{aligned} D_1\xi'' - D_6(\xi + h\eta') + F_1\varphi'' - F_6\varphi + \rho\omega^2 R\xi &= 0, \\ D_6(\xi' + h\eta'') + F_6\varphi' + \rho\omega^2 h\eta &= 0, \\ (R + 4\pi C_1)\varphi'' - 4\pi F_1\xi'' &= 0. \end{aligned} \tag{31}$$

From the last of (31),

$$\varphi = 4\pi F_1\xi/(R + 4\pi C_1) + Cx + \varphi_0, \tag{32}$$

where C and φ_0 are constants. C may be taken equal to zero, in the present case. The resulting expression for φ is then introduced into the first two of (31), yielding a pair of equations for ξ and η. Eliminating ξ, the equation governing η is found to be

$$(\partial^2/\partial x^2 + \delta_1^2)(\partial^2/\partial x^2 + \delta_2^2)\eta = 0, \tag{33}$$

where

$$\delta_1^2, \delta_2^2 = \frac{S\delta_0^4}{2}\{g^* + 1 \pm [(g^* - 1)^2 + 4g^*\bar{\omega}^{*2}/\omega^2]^{\frac{1}{2}}\},$$

and

$$S = D_1/D_6,$$

$$\delta_0^4 = \rho\omega^2/D_1 = \omega^2/(RS\bar{\omega}^2),$$

$$\bar{\omega}^* = \bar{\omega}(1 + N)^{\frac{1}{2}},$$

$$g^* = R/S(1 + J),$$

$$J = \frac{(e_{11} - e_{12}c_{12}/c_{22})^2}{(c_{11} - c_{12}^2/c_{22})(K_1 + e_{12}^2/c_{22})},$$

$$N = \frac{12e_{26}(e_{11} - e_{12}c_{12}/c_{22})}{\pi^2 c_{66}(K_1 + e_{12}^2/c_{22})}.$$

If the electric field is neglected, $J = N = 0$ and δ_0, δ_1 and δ_2 become identical with the corresponding terms in II.

The solution of (33) may be written as

$$\begin{aligned} \eta &= \eta_1 + \eta_2, \\ \eta_1 &= A_1 \sin\delta_1 x + B_1 \cos\delta_1 x, \\ \eta_2 &= A_2 \sin\delta_2 x + B_2 \cos\delta_2 x. \end{aligned} \tag{34}$$

Then, from (31),

$$(1 + N)\xi = (\sigma_1 - 1)h\eta_1' + (\sigma_2 - 1)h\eta_2' + \frac{F_6\varphi_0\bar{\omega}^{*2}}{D_6(\omega^2 - \bar{\omega}^{*2})}, \tag{35}$$

where

$$\sigma_1, \sigma_2 = (\delta_1^{-2}, \delta_2^{-2})S\delta_0^4. \tag{36}$$

The boundary conditions are $M_x = Q_x = M_{xz} = 0$ on $x = \pm l/2$. These may be applied separately to the even and odd modes. For the former, $B_1 = B_2 = 0$ in (34). Then, inserting (34) and (35) in (30), the boundary conditions become

$$\begin{aligned} &A_1(\sigma_1 - 1)\delta_1^2 \sin(\delta_1 l/2) \\ &\qquad + A_2(\sigma_2 - 1)\delta_2^2 \sin(\delta_2 l/2) = 0, \\ &A_1\sigma_1\delta_1 \cos(\delta_1 l/2) \\ &\qquad + A_2\sigma_2\delta_2 \cos(\delta_2 l/2) = \frac{F_6\varphi_0\omega^2}{D_6 h(\bar{\omega}^{*2} - \omega^2)}. \end{aligned} \tag{37}$$

Resonance occurs when the determinant of (37) vanishes, i.e., when

$$\sigma_2\delta_1(\sigma_1 - 1)\tan(\delta_1 l/2) = \sigma_1\delta_2(\sigma_2 - 1)\tan\delta_2 l/2. \tag{38}$$

This frequency equation has the same form as that in II and differs only in the definitions of σ_1, σ_2, δ_1 and δ_2. Proceeding as in II, let

$$\gamma = \delta_1 l/2 \tag{39}$$

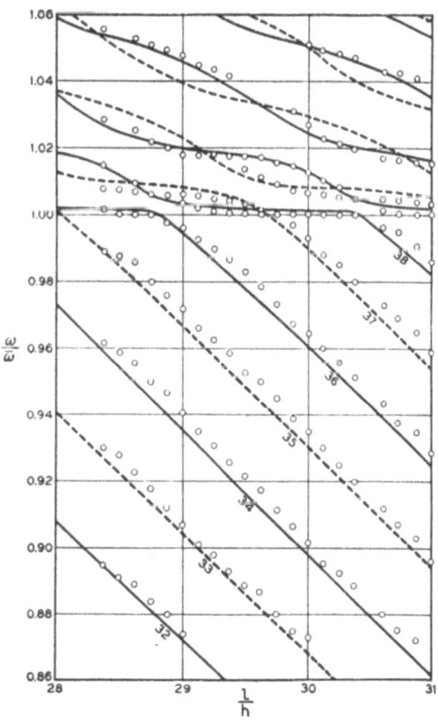

FIG. 2. Comparison of theory with data obtained by Sykes on an AT quartz plate, initially 25×25×0.807 mm, cut down in decrements of 0.1 mm to 25×21.8×0.807 mm.

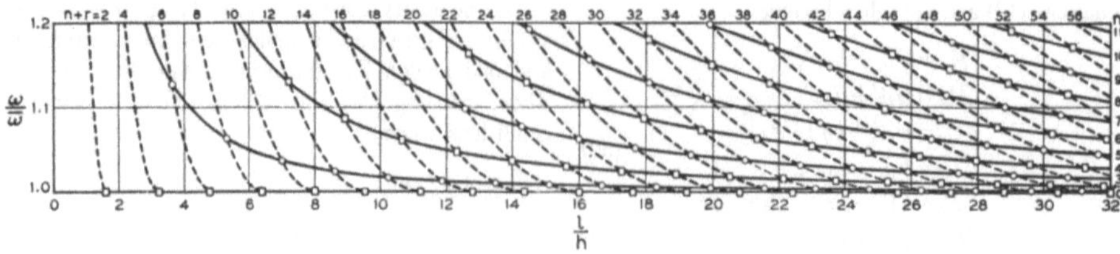

FIG. 3. The circled and squared points are those from the upper half of Fig. 1. The full curves $n=2, 3, 4\cdots$ are those commonly designated as the first, second, third$\cdots$ overtones of the fundamental thickness-shear mode. Only the circled and squared points on these curves correspond to resonant frequencies. The dashed curves, whose equation is (46), are exact at the circled and squared points and, at intermediate points, are approximations to the exact curves in the upper half of Fig. 1.

and

$$a=\delta_2/\delta_1=[(1-b)/(1+b)]^{\frac{1}{2}}, \tag{40}$$

where b is the positive root of

$$b^2=1+4g^*(\bar{\omega}^{*2}/\omega^2-1)(1+g^*)^{-2}. \tag{41}$$

Since δ_1 is always real while δ_2 and a are real if $\omega>\bar{\omega}^*$ and imaginary if $\omega<\bar{\omega}^*$, (38) may be written as

$$\begin{aligned}(a^2-g^*)\tan\gamma&=a(1-g^*a^2)\tan a\gamma, & \omega>\bar{\omega}^*,\\ (a_1^2+g^*)\tan\gamma&=a_1(1+g^*a_1^2)\tanh a_1\gamma, & \omega<\bar{\omega}^*,\end{aligned} \tag{42}$$

where $a_1=-ia$. From (40) and (41)

$$\begin{aligned}\omega/\bar{\omega}^*&=[1-a^2(1+g^*)^2/g^*(1+a^2)^2]^{-\frac{1}{2}}, & \omega>\bar{\omega}^*,\\ \omega/\bar{\omega}^*&=[1+a_1^2(1+g^*)^2/g^*(1-a_1^2)^2]^{-\frac{1}{2}}, & \omega<\bar{\omega}^*;\end{aligned} \tag{43}$$

and, from (39),

$$\begin{aligned}l/h&=\gamma(\bar{\omega}^*/\omega)[(1+a^2)/3(1+g^*)]^{\frac{1}{2}}, & \omega>\bar{\omega}^*,\\ l/h&=\gamma(\bar{\omega}^*/\omega)[(1-a_1^2)/3(1+g^*)]^{\frac{1}{2}}, & \omega<\bar{\omega}^*.\end{aligned} \tag{44}$$

Equations (42), (43), and (44) are identical in form with the correspondingly numbered equations in II and the computation of data for constructing curves of frequency *versus* plate dimensions, for a given crystal and orientation of cut, is carried out in the same manner as before. The only difference is that

$$g=\pi^2c_{66}/12(c_{11}-c_{12}^2/c_{22})$$

of II is replaced here by

$$g^*=g/(1+J),$$

and

$$\bar{\omega}^2=\pi^2c_{66}/\rho h^2$$

of II is replaced here by

$$\bar{\omega}^{*2}=4c_{66}\alpha_1^2(1+e_{26}^2/c_{66}K_2)(1+N)/\rho h^2.$$

For the AT cut of quartz, for example, we find

$$\begin{aligned}g/g^*&=1.000558,\\ \bar{\omega}/\bar{\omega}^*&=1.000552.\end{aligned} \tag{45}$$

It is apparent that it is unnecessary to recompute the curves of II for the new values of g and $\bar{\omega}$. Instead, the computations for the old values were extended to include the odd modes and the results are shown in Figs. 1 and 2 along with additional data by Sykes in the latter figure.

The algebraic formulas for the coordinates of the circled and squared points of Fig. 1 are the same as those given in Eqs. (50) and (51) of II except for the replacement of g and $\bar{\omega}$ by g^* and $\bar{\omega}^*$. Similarly the equations of curves passing through points $n=$constant, $r=$constant and $n+r=$constant are unchanged in form, but the new values for g, $\bar{\omega}$, k, and κ should be used. Figure 4 and Eqs. (52) and (53) of II illustrate the families of curves passing through $n=$constant, and $r=$constant. Figure 3 of the present paper illustrates the curves passing through $n=$constant and $n+r=m$ $=$constant. The equation of the latter is

$$\frac{(1-g)^2}{g}\frac{\omega^2}{\bar{\omega}^2}=\frac{m^2kh^2}{l^2}-2-\left[\left(\frac{m^2kh^2}{l^2}-2\right)^2-\left(\frac{m^2kh^2}{l^2}\frac{(1-g)}{(1+g)}\right)^2\right]^{\frac{1}{2}}. \tag{46}$$

An interesting feature of (46) is that it gives the frequency spectrum not only exactly at the circled and squared points, but also approximately at intermediate points. This may be seen by comparing the dashed curves in Fig. 3 with the curves in Fig. 1 for $\omega/\bar{\omega}>1$.

Response of an Elastic Cylindrical Shell to a Transverse, Step Shock Wave

BY R. D. MINDLIN[1] AND H. H. BLEICH,[2] NEW YORK, N. Y.

A plane shock wave in an acoustic medium encounters a long cylindrical shell whose axis is parallel to the wave front. An approximate mathematical solution is obtained for the elastic response of the shell.

INTRODUCTION

IN this paper there is described a mathematical study of the response of a cylindrical shell to a shock wave. The shell is elastic, circular in cross section, homogeneous, and infinitely long. Its motions are restricted to dilatational, translational, and inextensional-flexural modes. The shell is submerged in an acoustic fluid, i. e., one whose motions conform to the linear theory of waves of expansion. The shock wave is assumed to be plane, with its wave front parallel to the axis of the shell.

By means of an approximation, which is valid for the early stages, the potential of the diffraction and radiation is eliminated between the simultaneous equations which couple the generalized co-ordinates of the shell and the fluid. As a result, a simple differential equation is obtained for each generalized co-ordinate describing the motion of the shell. The solution is reduced to a definite integral during the time of transit of the shock-wave front across the shell and, for the case of the step shock wave, is expressed in terms of tabulated functions thereafter. Explicit formulas are given for the displacement, velocity, acceleration, and pressure associated with each mode of motion of the shell.

EQUATIONS GOVERNING INTERACTION BETWEEN SHELL AND FLUID

Let w be the radial displacement of the shell and Z the pressure on it, both positive inward, Fig. 1. Let

$$w = \sum_{n=0}^{\infty} q_n(t) \cos n\theta \qquad [1]$$

$$Z = \sum_{n=0}^{\infty} Q_n(t) \cos n\theta \qquad [2]$$

Then, one form of the equation of cylindrical motion of a shell may be written

$$\ddot{q}_n + \omega_n^2 q_n = Q_n/m_n \qquad [3]$$

where ω_n are the natural frequencies of cylindrical vibration of the shell under a uniform pressure (but not submerged) and

$$m_n = m(n^2 + 1)/n^2, \; n > 0 \qquad [4]$$

$$m_0 = m, \; n = 0 \qquad [5]$$

where m is the mass per unit of surface area of the shell.

The mode $n = 0$ is purely dilatational; $n = 1$ is translational, parallel to $\theta = 0$; $n > 1$ are inextensional, flexural modes. Extensional deformations in $n > 0$ are neglected.

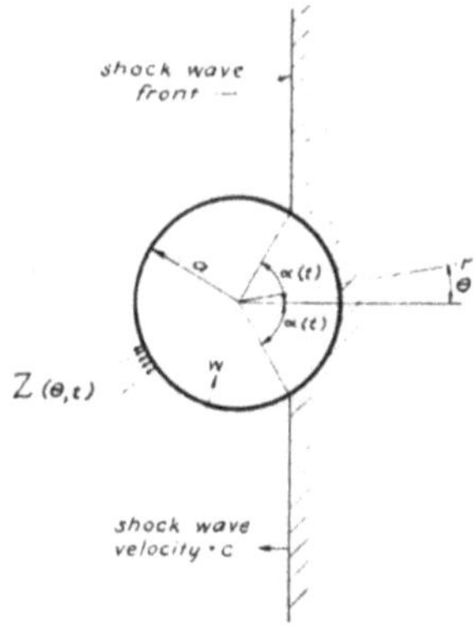

FIG. 1 GEOMETRY OF PROBLEM CONSIDERED

The pressure on the shell is due to the shock wave and to the waves reflected and radiated from the vibrating shell:

$$Z = p(\theta, t) + \rho\dot{\varphi}(a, \theta, t) \qquad [6]$$

where $p(\theta, t)$ is the shock-wave pressure at the surface of the shell ($r = a$), ρ is the density of the fluid, and $\varphi(r, \theta, t)$ is the velocity potential of the reflected and radiated waves, governed by the equation

$$c^2\nabla^2\varphi = \ddot{\varphi} \qquad [7]$$

where c is the velocity of sound in the fluid.

Let

$$p(\theta, t) = \sum_{n=0}^{\infty} P_n(t) \cos n\theta \qquad [8]$$

$$\varphi(r, \theta, t) = \sum_{n=0}^{\infty} \psi_n(r, t) \cos n\theta \qquad [9]$$

Then

$$Q_n(t) = P_n(t) + \rho\dot{\psi}_n(a, t) \qquad [10]$$

so that, from Equation [3], on $r = a$

$$\ddot{q}_n + \omega_n^2 q_n = (P_n + \rho\dot{\psi}_n)/m_n \qquad [11]$$

This is the first of the two equations controlling the interaction between the shell and the fluid.

[1] Professor of Civil Engineering, Columbia University. Member ASME.

[2] Associate Professor of Civil Engineering, Columbia University.

Contributed by the Applied Mechanics Division and presented at the Annual Meeting, New York, N. Y., November 30–December 5, 1952, of THE AMERICAN SOCIETY OF MECHANICAL ENGINEERS.

Discussion of this paper should be addressed to the Secretary, ASME, 29 West 39th Street, New York, N. Y., and will be accepted until one month after final publication of the paper itself in the JOURNAL OF APPLIED MECHANICS.

 Manuscript received by ASME Applied Mechanics Division, July 14, 1952. Paper No. 52—A-19.

The radial velocity of a point on the shell must equal the sum of the radial components of the particle velocities of the shock wave and the diffracted wave. Thus

$$w = v(\theta, t) + \frac{\partial\varphi(a, \theta, t)}{\partial a} \qquad [12]$$

where $v(\theta, t)$ is the radial component of the particle velocity in the shock wave at $r = a$. Let

$$\rho c v(\theta, t) = \sum_{n=0}^{\infty} V_n(t) \cos n\theta \qquad [13]$$

Then, on $r = a$

$$\dot{q}_n = \frac{V_n}{\rho c} + \frac{\partial\psi_n}{\partial a} \qquad [14]$$

This is the second of the two equations governing the interaction between the shell and the fluid.

The essential difficulty, in finding the response of the shell, is the awkwardness involved in the process and result of eliminating ψ_n between Equations [11] and [14].

Approximate Equation of Motion During Early Stages

The process of eliminating ψ_n is begun by expanding it in Fourier series over a time interval as yet unspecified

$$\psi_n(r, t) = \sum_{s=1}^{\infty} \chi_{ns}(r)e^{i\xi_s ct/a} \qquad [15]$$

where

$$\xi_s = ks\pi/2, \quad s = 1,2,\ldots, \qquad [16]$$

and k is a number which fixes the interval of expansion. Then, from Equations [7], [9] and [15]

$$\frac{d^2\chi_{ns}}{dr^2} + \frac{1}{r}\frac{d\chi_{ns}}{dr} + \left(\frac{\xi_s^2}{a^2} - \frac{n^2}{r^2}\right)\chi_{ns} = 0 \qquad [17]$$

The solution of Equation [17] for diverging waves is

$$\chi_{ns} = A_{ns}H_n(\xi_s r/a) \qquad [18]$$

where H_n is the Hankel function of the second kind (ordinarily written $H_n^{(2)}$). Hence, during $0 \leq t \leq 2a/kc$

$$\psi_n(a, t) = \sum_{s=1}^{\infty} A_{ns}H_n(\xi_s)e^{i\xi_s ct/a} \qquad [19]$$

$$\dot{\psi}_n(a, t) = i(c/a)\sum_{s=1}^{\infty} A_{ns}\xi_s H_n(\xi_s)e^{i\xi_s ct/a} \qquad [20]$$

$$\frac{\partial\psi_n(a, t)}{\partial a} = \frac{1}{a}\sum_{s=1}^{\infty} A_{ns}\xi_s H_n'(\xi_s)e^{i\xi_s ct/a} \qquad [21]$$

where $H_n'(\xi_s) = \partial H_n(\xi_s)/\partial\xi_s$.

The approximation is now made that ξ_s is large enough so that

$$H_n(\xi_s) \approx iH_n'(\xi_s), \ (\xi_s \gg 1) \qquad [22]$$

For the later terms ($s \gg 1$) of the expansion, Equation [15], the approximation is very good. For the early terms, the value of k, which determines the interval of expansion, plays an important role. The larger the value of k the shorter is the interval of expansion and the better is the approximation. Thus the solution to be obtained may be expected to be good at the early stages of envelopment and to become progressively poorer at later times

To this approximation, then, from Equations [20] and [21]

$$\frac{\partial\psi_n}{\partial a} = -\frac{1}{c}\frac{\partial\psi_n}{\partial t} \qquad [23]$$

Thus the main effect of the approximation is to give the reflection and radiation, associated with each mode of the shell, the propagation characteristics of plane waves rather than cylindrical waves.

From Equations [10], [14], and [23]

$$Q_n = P_n + V_n - \rho c\dot{q}_n \qquad [24]$$

and hence from Equation [3]

$$\ddot{q}_n + \frac{\rho c}{m_n}\dot{q}_n + \omega_n^2 q_n = \frac{P_n + V_n}{m_n} \qquad [25]$$

Thus the response of the shell to a given shock wave is obtained simply by integrating Equation [25] with the appropriate $P_n + V_n$. The fluid pressure on the shell is then obtained from Equation [24].

There is no difficulty in solving Equation [25] for quite general shock-wave properties, since only one integration is involved. However, a minimum of computations is required for a step wave.

Form of $P_n + V_n$ for a Step Shock Wave

The pressure in a step shock wave, at the periphery of the shell, is

$$p(\theta, t) = \begin{Bmatrix} p, & -\alpha \leqq \theta \leqq \alpha \\ 0, & \alpha \leqq \theta \leqq 2\pi - \alpha \end{Bmatrix}, \ 0 \leqq t \leqq 2a/c \qquad [26]$$

$$p(\theta, t) = p, \quad -\pi \leqq \theta \leqq \pi, \quad t \geqq 2a/c$$

where p is the constant peak pressure and

$$\alpha = \alpha(t) = \cos^{-1}(1 - ct/a) \qquad [27]$$

The particle velocity at $r = a$ is

$$v(\theta, t) = p(\theta, t)\cos\theta/\rho c \qquad [28]$$

Inserting Equations [26] and [28] in [8] and [13], the usual methods of expansion in series lead to the following:

(a) During $0 \leqq t \leqq 2a/c$

$$P_n = \begin{Bmatrix} \dfrac{p\alpha}{\pi}, & n = 0 \\ \dfrac{2p}{n\pi}\sin n\alpha, & n > 0 \end{Bmatrix} \qquad [29]$$

$$V_n = \begin{Bmatrix} \dfrac{p}{\pi}\sin\alpha, & n = 0 \\ \dfrac{p}{\pi}(\alpha + \sin\alpha\cos\alpha), & n = 1 \\ \dfrac{p}{\pi}\left[\dfrac{\sin(n-1)\alpha}{n-1} + \dfrac{\sin(n+1)\alpha}{n+1}\right], & n > 1 \end{Bmatrix} \qquad [30]$$

(b) During $t \geq 2a/c$

$$P_n = \begin{Bmatrix} p, \ n = 0 \\ 0, \ n > 0 \end{Bmatrix} \qquad [31]$$

$$V_n = \begin{Bmatrix} 0, \ n = 0 \\ p, \ n = 1 \\ 0, \ n > 1 \end{Bmatrix} \qquad [32]$$

It is interesting to observe that the impulse delivered to each mode contains a factor

$$\int_0^{2a/c} (P_n + V_n)dt = \begin{cases} 3pa/2c, & n = 0 \\ 2pa/c, & n = 1 \\ pa/2c, & n = 2 \\ 0, & n \geqq 3 \end{cases} \quad \ldots\ldots [33]$$

DILATATIONAL MODE ($n = 0$)

In Equation [25] let $n = 0$ and

$$\beta_0 = \rho c/2m\omega_0 \quad \ldots\ldots [34]$$

Then the equation of motion of the dilatational mode becomes

$$\ddot{q}_0 + 2\beta_0\omega_0\dot{q}_0 + \omega_0^2 q_0 = (P_0 + V_0)/m \quad \ldots\ldots [35]$$

Now, β_0 is the fraction of critical damping. This is greater than unity in the applications contemplated. Hence an appropriate form of the solution of Equation [35] is

$$m\omega_0\sqrt{\beta_0^2 - 1}\, q_0 = \int_0^t [P_0(\lambda) + V_0(\lambda)]e^{-\beta_0\omega_0(t-\lambda)} \sinh(\omega_0\sqrt{\beta_0^2 - 1}\,(t - \lambda))d\lambda \quad \ldots [36]$$

Let

$$\gamma = \cos^{-1}(1 - c\lambda/a) \quad \ldots\ldots [37]$$

$$S_0(\gamma) = P_0(\lambda) + V_0(\lambda) = \begin{cases} \frac{p}{\pi}(\gamma + \sin\gamma), & 0 \leqq \gamma \leqq \pi \\ p, & \gamma \geqq \pi \end{cases} \quad \ldots [38]$$

Then

$$c\,d\lambda = a \sin\gamma d\gamma \quad \ldots\ldots [39]$$

$$c(t - \lambda) = a(\cos\gamma - \cos\alpha) \quad \ldots\ldots [40]$$

and Equation [36] takes the form

$$2cm\omega_0\sqrt{\beta_0^2 - 1}\, q_0 = ae^{\bar{z}_0 \cos\alpha}\int_0^\alpha S_0(\gamma)\sin\gamma e^{-\bar{z}_0\cos\gamma}d\gamma - ae^{z_0\cos\alpha}\int_0^\alpha S_0(\gamma)\sin\gamma e^{-z_0\cos\gamma}d\gamma \quad \ldots [41]$$

where

$$\left.\begin{aligned} z_0 &= \frac{\omega_0 a}{c}(\beta_0 + \sqrt{\beta_0^2 - 1}), \\ \bar{z}_0 &= \frac{\omega_0 a}{c}(\beta_0 - \sqrt{\beta_0^2 - 1}), \end{aligned}\right\} \quad \ldots\ldots [42]$$

During the time of transit ($0 \leqq \alpha \leqq \pi$, i.e., $0 \leqq t \leqq 2a/c$) of the shock-wave front across the shell, the integrals in Equation [41] are incomplete modified Bessel functions of the first kind. These functions do not appear to be tabulated, but the integrations are performed numerically without difficulty. Upon and after full envelopment ($\alpha \geqq \pi$, i. e., $t \geqq 2a/c$) the integrals have the forms

$$\left.\begin{aligned} \int_0^\pi \gamma \sin\gamma e^{-z\cos\gamma}d\gamma &= \pi z^{-1}[e^z - I_0(z)] \\ \int_0^\pi \sin^2\gamma e^{-z\cos\gamma}d\gamma &= \pi z^{-1}I_1(z) \\ \int_\pi^\alpha \sin\gamma e^{-z\cos\gamma}d\gamma &= z^{-1}(e^{-z\cos\alpha} - e^z) \end{aligned}\right\} \quad \ldots [43]$$

where $I_n(z)$ is the modified Bessel function of the first kind.

The solution of Equation [35] after envelopment is, then

$$\frac{2cm\omega_0\sqrt{\beta_0^2 - 1}}{ap}\, q_0 = \frac{1}{\bar{z}_0} - \frac{1}{z_0} - \frac{e^{\bar{z}_0\cos\alpha}}{\bar{z}_0}[I_0(\bar{z}_0) - I_1(\bar{z}_0)] + \frac{e^{z_0\cos\alpha}}{z_0}[I_0(z_0) - I_1(z_0)] \quad \ldots\ldots [44]$$

where

$$\cos\alpha = 1 - ct/a \quad \ldots\ldots [45]$$

Noting that

$$\left.\begin{aligned} z_0 - \bar{z}_0 &= \frac{2\omega_0 a}{c}\sqrt{\beta_0^2 - 1} \\ z_0\bar{z}_0 &= \left(\frac{\omega_0 a}{c}\right)^2 \end{aligned}\right\} \quad \ldots\ldots [46]$$

and, also, setting

$$q_0^* = p/m\omega_0^2 \quad \ldots\ldots [47]$$

i.e., q_0^* is the static displacement of the shell under a uniform pressure p, the solution after envelopment may be written

$$\frac{q_0}{q_0^*} = 1 - \frac{z_0 e^{\bar{z}_0\cos\alpha}}{z_0 - \bar{z}_0}[I_0(\bar{z}_0) - I_1(\bar{z}_0)] + \frac{\bar{z}_0 e^{z_0\cos\alpha}}{z_0 - \bar{z}_0}[I_0(z_0) - I_1(z_0)] \quad \ldots [48]$$

In typical applications $z_0 \gg \bar{z}_0$ and $z_0 \gg 1$. Then $I_0(z_0) - I_1(z_0) \ll 1$ and $z_0 \cos\alpha$ is large and negative. Hence the second line of Equation [48] is small and vanishes rapidly. Since $I_0(\bar{z}_0) > I_1(\bar{z}_0)$, the displacement (and therefore the strain) at the instant of complete envelopment is less than the static displacement under the same pressure, but is increasing; and the difference $q_0 - q_0^*$ diminishes essentially exponentially with time constant $a/c\bar{z}_0$.

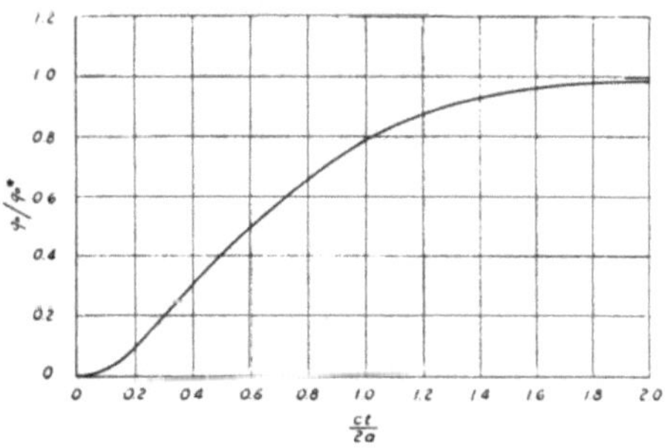

FIG. 2 DISPLACEMENT OF DILATATIONAL MODE
(q_0^* = displacement under static pressure p.)

A typical displacement response (q_0) is illustrated in Fig. 2 for $z_0 = 8.40$, $\bar{z}_0 = 1.33$. The displacement increases almost linearly with time up to 80 per cent of the static displacement at the time of full envelopment. Thereafter the displacement approaches the static displacement with a time constant of about 40 per cent of the transit time.

The ratio of the velocity ($\dot{q}_0$) of the dilatational mode to the particle velocity ($p/\rho c$) in the shock wave is readily found to be, during the transit time

$$\frac{\rho c\dot{q}_0}{p} = \frac{z_0 + \bar{z}_0}{p(z_0 - \bar{z}_0)}\left[-\bar{z}_0 e^{\bar{z}_0\cos\alpha}\int_0^\alpha S_0(\gamma)e^{-\bar{z}_0\cos\gamma}\sin\gamma d\gamma + z_0 e^{z_0\cos\alpha}\int_0^\alpha S_0(\gamma)e^{-z_0\cos\gamma}\sin\gamma d\gamma\right] \quad \ldots [49]$$

and, after the transit time

$$\frac{\rho c\dot{q}_0}{p} = \frac{z_0 + \bar{z}_0}{z_0 - \bar{z}_0}\left\{e^{\bar{z}_0\cos\alpha}[I_0(\bar{z}_0) - I_1(\bar{z}_0)] - e^{\bar{z}_0\cos\alpha}[I_0(z_0) - I_1(z_0)]\right\} \quad \ldots [50]$$

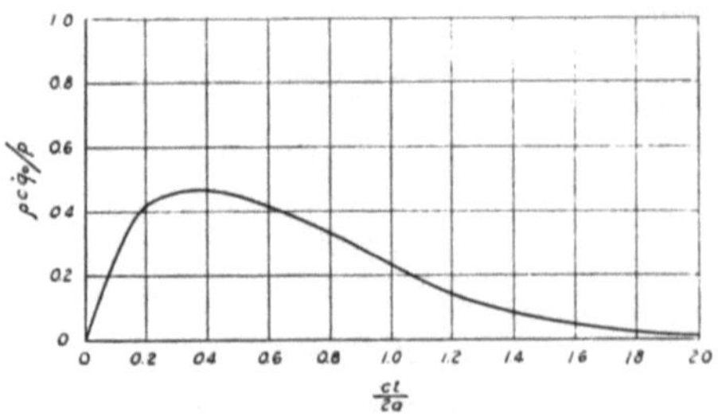

FIG. 3 VELOCITY OF DILATATIONAL MODE
($p/\rho c$ = particle velocity in shock wave.)

This is illustrated in Fig. 3.

The pressure in the dilatational mode is simply

$$Q_0 = P_0 + V_0 - \rho c\dot{q}_0 \quad [51]$$

and the acceleration is given by

$$\frac{2a\rho\ddot{q}_0}{p} = 2(z_0 + \bar{z}_0)\left(\frac{Q_0}{p} - \frac{q_0}{q_0^*}\right) \quad [52]$$

These are illustrated in Figs. 4 and 5.

It may be seen that, despite the apparent limitation of the validity of the solution to short times, all the results for the dilatational mode approach the correct limits for large t.

TRANSLATIONAL MODE ($n = 1$)

In Equation [25] set $n = 1$ and note that $m_1 = 2m$ and $\omega_1 = 0$. Thus the equation of motion of the translational mode is

$$\ddot{q}_1 + \frac{\rho c}{2m}\dot{q}_1 = \frac{P_1 + V_1}{2m} \quad [53]$$

Hence the velocity is given by

$$\rho c\dot{q}_1 = z_1 e^{z_1 \cos\alpha} \int_0^\alpha S_1(\gamma) \sin\gamma\, e^{-z_1\cos\gamma}\, d\gamma \quad [54]$$

where

$$z_1 = \rho a/2m \quad [55]$$

$$S_1(\gamma) = \begin{cases} \dfrac{p}{\pi}(\gamma + 2\sin\gamma + \sin\gamma\cos\gamma), & 0 \leq \gamma \leq \pi \\ p, \ \gamma \geq \pi \end{cases} \quad [56]$$

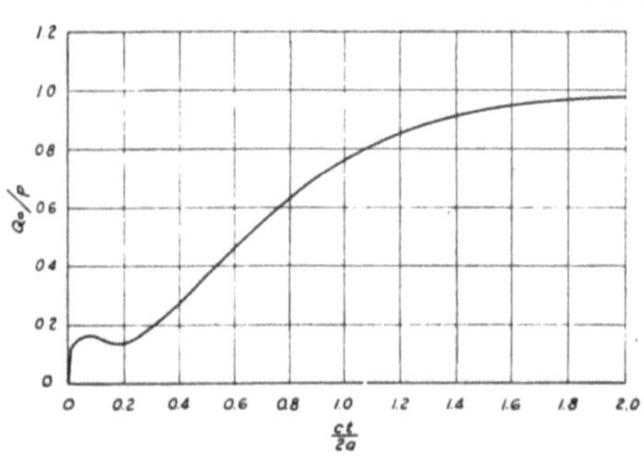

FIG. 4 PRESSURE IN DILATATIONAL MODE

During the transit time the integral in Equation [54] is again composed of incomplete Bessel functions, but it can easily be integrated numerically, especially since $z_1 = 1$ for a body in neutral buoyancy; i.e., only one integration has to be performed for all floating shells.

For the period $t \geq 2a/c$ there are four integrals in Equation [54]. Three of them are identical with Equation [43] and the fourth is

$$\int_0^\pi \sin 2\gamma \sin\gamma e^{-z_1\cos\gamma}\, d\gamma = -2\pi z_1^{-1} I_2(z_1) \quad [57]$$

Hence, for $t \geq 2a/c$

$$\rho c\dot{q}_1/p = 1 - e^{z_1\cos\alpha}[I_0(z_1) - 2I_1(z_1) + I_2(z_1)] \quad [58]$$

For the neutrally buoyant body ($z_1 = 1$) the quantity in brackets in Equation [58] is 0.1. Thus the shell as a whole reaches $^9/_{10}$ of the particle velocity in the shock wave at the instant of complete envelopment and thereafter approaches the particle velocity exponentially, with time constant a/c, i.e., half the transit time. (The translational velocity during and after envelopment is shown in Fig. 6.)

The displacement of the shell at the end of the transit time is found to be

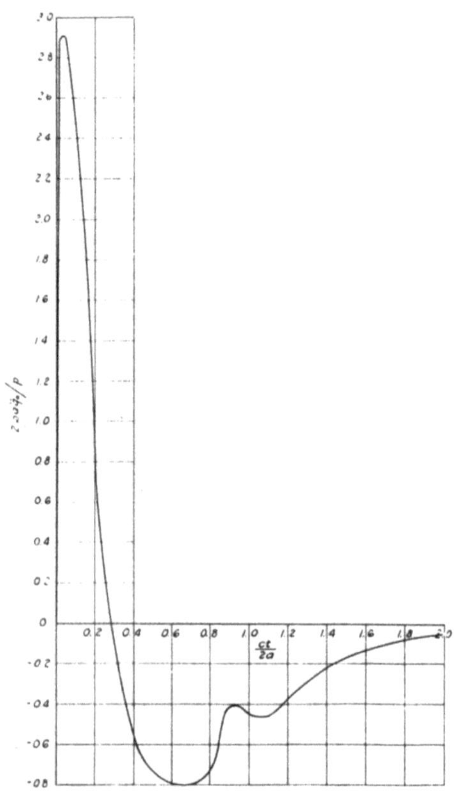

FIG. 5 ACCELERATION IN DILATATIONAL MODE
($p/2\rho a$ = uniform acceleration which would produce particle velocity in time $2a/c$.)

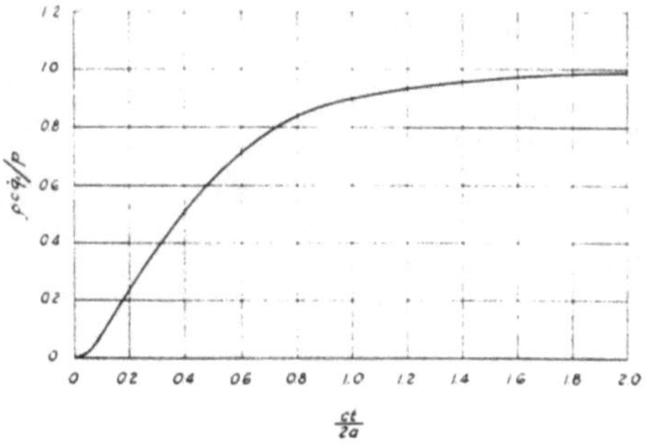

FIG. 6 VELOCITY IN TRANSLATIONAL MODE
($p/\rho c$ = particle velocity in shock wave.)

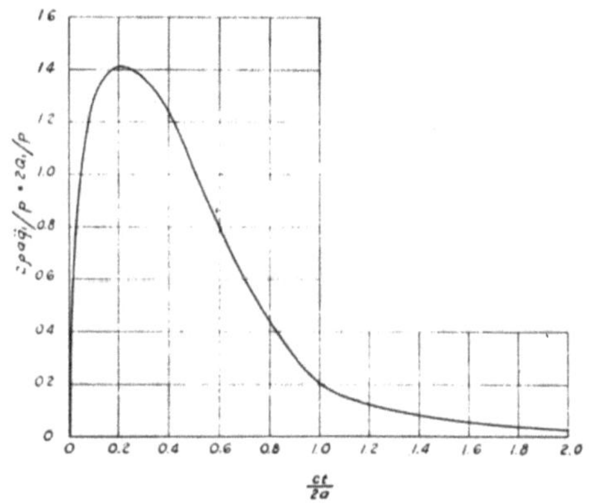

FIG. 7 ACCELERATION AND PRESSURE IN TRANSLATIONAL MODE

$$q_1 = 1.1 \frac{pa}{\rho c^2}, \quad t = 2a/c \qquad [59]$$

The translational acceleration is found by substituting Equations [54] and [58] in [53] and is plotted in Fig. 7 as a ratio to

$$\ddot{q}_1^* = p/2\rho a \qquad [60]$$

i.e., to the constant acceleration, during the transit time, that would produce a shell velocity equal to the particle velocity in the shock wave.

The generalized pressure in the translational mode is given by

$$Q_1 = m_1\ddot{q}_1 = 2m\ddot{q}_1 \qquad [61]$$

corresponding to a real pressure distribution

$$Z_1 = Q_1 \cos\theta \qquad [62]$$

Thus for a shell in neutral buoyancy

$$\frac{Q_1}{p} = \frac{1}{2}\frac{\ddot{q}_1}{\ddot{q}_1^*} \qquad [63]$$

i.e., the ordinates in Fig. 7 are also equal to $2Q_1/p$.

As in the case of the dilatational mode, the solution for the translational mode also approaches the correct limit for large t, despite the approximation.

FLEXURAL MODES ($n \geqq 2$)

The solution of Equation [25] for $n \geqq 2$ may be written

$$q_n = \frac{ap}{2\pi c m_n \omega_n \sqrt{\beta_n^2 - 1}} (\bar{B}_n e^{\bar{z}_n \cos\alpha} - B_n e^{z_n \cos\alpha}) \qquad [64]$$

where

$$\beta_n = \rho c/2m_n\omega_n \qquad [65]$$

$$\left.\begin{aligned} z_n &= \frac{\omega_n a}{c}(\beta_n + \sqrt{\beta_n^2 - 1}) \\ \bar{z}_n &= \frac{\omega_n a}{c}(\beta_n - \sqrt{\beta_n^2 - 1}) \end{aligned}\right\} \qquad [66]$$

$$B_n = \left\{\begin{aligned} &\frac{\pi}{p}\int_0^\alpha (P_n + V_n)\sin\gamma\, e^{-z_n\cos\gamma}\, d\gamma, \quad 0 \leqq \alpha \leqq \pi \\ &\frac{2\pi(-1)^n}{z_n}[I_n'(z_n) - I_n(z_n)], \quad \alpha \geqq \pi \end{aligned}\right\} \qquad [67]$$

and $\bar{B}_n$ is the same as B_n except that z_n is replaced by $\bar{z}_n$.

A comparison of the displacement in a flexural mode with that in the dilatational mode has little significance, but it is interesting to compare the maximum stress σ_n, in the flexural mode with the stress σ_0^* in the dilatational mode under constant pressure p. The former is

$$\sigma = Ed/R_n \qquad [68]$$

where

$$\frac{1}{R_n} = \frac{n^2 - 1}{a^2} q_n \qquad [69]$$

is the change of curvature, E is the appropriate modulus of elasticity, and d is the distance from the neutral axis to the extreme fiber. Also

$$\sigma_0^* = pa/h \qquad [70]$$

where h is the shell thickness. Hence, after some rearrangement

$$\frac{\sigma_n}{\sigma_0^*} = \frac{dhE(n^2-1)(z_n + \bar{z}_n)}{\pi a^2 \kappa (z_n - \bar{z}_n)} (\bar{B}_n e^{\bar{z}_n\cos\alpha} - B_n e^{z_n\cos\alpha}) \qquad [71]$$

where $\kappa = \rho c^2$ is the bulk modulus of the fluid. This ratio is plotted in Fig. 8 for the mode $n = 2$ and $\bar{z}_2 = 0.01$, $z_2 = 7.0$.

The later portion of the curve in Fig. 8 is dashed to indicate that it is unreliable as a result of the approximation.

The velocity in $n \geqq 2$ is given by

$$\frac{\rho c \dot{q}_n}{p} = \frac{z_n + \bar{z}_n}{\pi(z_n - \bar{z}_n)} (z_n B_n e^{z_n\cos\alpha} - \bar{z}_n \bar{B}_n e^{\bar{z}_n\cos\alpha}) \qquad [72]$$

This is plotted in Fig. 9 for $n = 2$.

The pressure is computed from

$$\frac{Q_n}{p} = \frac{P_n + V_n}{p} - \frac{\rho c \dot{q}_n}{p} \qquad [73]$$

and, finally, the acceleration, calculated from the preceding results is

$$\frac{\ddot{q}_n}{\ddot{q}_1^*} = 2(z_n + \bar{z}_n)\frac{Q_n}{p} + \frac{2z_n\bar{z}_n(z_n + \bar{z}_n)}{\pi(z_n - \bar{z}_n)} (B_n e^{z_n\cos\alpha} - \bar{B}_n e^{\bar{z}_n\cos\alpha}) \qquad [74]$$

The pressure and acceleration are plotted in Figs. 10 and 11 for $n = 2$.

Flexural Instability

After the diffraction effects are over, the flexural modes are

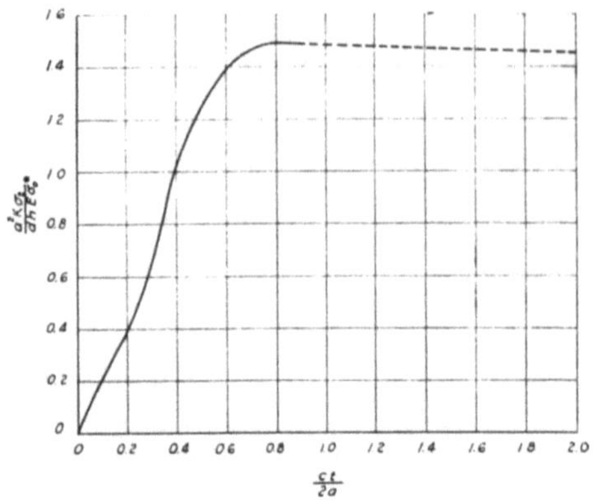

FIG. 8 STRESS IN FIRST FLEXURAL MODE

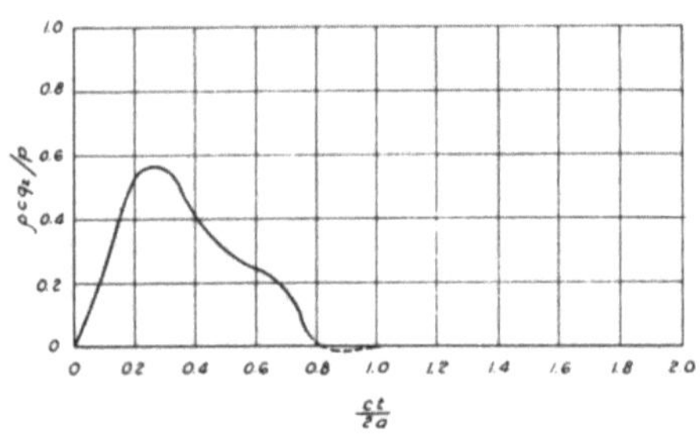

FIG. 9 VELOCITY IN FIRST FLEXURAL MODE

essentially in a state of free vibration. In this case the solution takes a different form. Returning to Equation [15], let

$$\psi_n = \chi_n(r)e^{i\Omega_n t} \qquad [75]$$

where Ω_n are the natural frequencies of the shell immersed in the fluid. Then the equation on χ_n becomes

$$\frac{d^2\chi_n}{dr^2} + \frac{1}{r}\frac{d\chi_n}{dr} + \left(\frac{\Omega_n^2}{c^2} - \frac{n^2}{r^2}\right)\chi_n = 0 \qquad [76]$$

whence

$$\left.\begin{aligned} \psi_n(a, t) &= A_n H_n(\eta_n)e^{i\Omega_n t} \\ \dot{\psi}_n(a, t) &= i(c/a)A_n\eta_n H_n(\eta_n)e^{i\Omega_n t} \\ \frac{\partial\psi_n(a, t)}{\partial a} &= \frac{1}{a}A_n\eta_n H_n'(\eta_n)e^{i\Omega_n t} \end{aligned}\right\} \qquad [77]$$

where $\eta_n = \Omega_n a/c$.

Noting that, for $n \geqq 2$, $P_n = V_n = 0$ in this interval, Equation [11] becomes

$$\ddot{q}_n + \omega_n^2 q_n = \rho\dot{\psi}_n/m_n \qquad [78]$$

and Equation [14] may be written as

$$\ddot{q}_n = \partial\dot{\psi}_n/\partial a \qquad [79]$$

But, from Equation [77]

$$\dot{\psi}_n = \frac{aH_n(\eta_n)}{\eta_n H_n'(\eta_n)}\frac{\partial\dot{\psi}_n}{\partial a} \qquad [80]$$

Hence

$$\left[1 - \frac{\rho a H_n(\eta_n)}{m_n\eta_n H_n'(\eta_n)}\right]\ddot{q}_n + \omega_n^2 q_n = 0 \qquad [81]$$

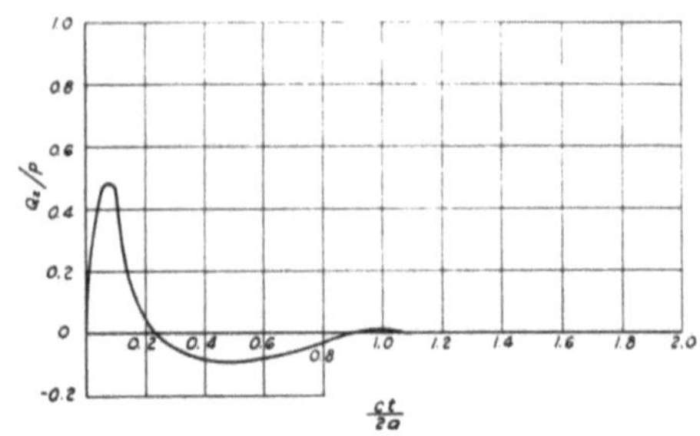

FIG. 10 PRESSURE IN FIRST FLEXURAL MODE

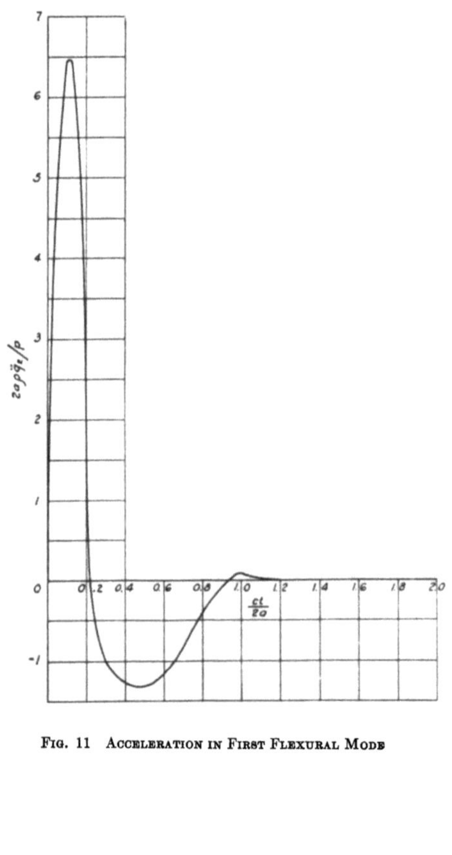

FIG. 11 ACCELERATION IN FIRST FLEXURAL MODE

is the equation governing the free vibrations. This also gives the transcendental equation

$$\Omega_n^2 \left[1 - \frac{\rho a H_n(\eta_n)}{m_n \eta_n H_n'(\eta_n)} \right] = \omega_n^2 \qquad [82]$$

whose roots determine the natural frequencies Ω_n of the submerged shell when the frequencies ω_n under a uniform pressure are known.

The flexural modes of greatest interest are the lower ones and for these η_n is small. But

$$\frac{H_n(\eta_n)}{\eta_n H_n'(\eta_n)} \approx -\frac{1}{n}, \qquad \eta_n \ll 1 \qquad [83]$$

Hence

$$\Omega_n^2 \approx \omega_n^2(1 + \rho a / n m_n) \qquad [84]$$

Thus Ω_n is essentially real as long as the ratio of the pressure to the stiffness is small enough for ω_n to be real. In this case the solution of Equation [81] is oscillatory. If the pressure increases sufficiently ω_n, and hence Ω_n can become imaginary. Then the solution of Equation [81] becomes nonoscillatory and buckling ensues.

Reprinted from
Journal of Applied Mechanics
20 (1953) 189-195

(Reprinted from the Journal of Applied Mechanics for September 1953, with permission of Copyright Owner)

Elastic Spheres in Contact Under Varying Oblique Forces[1]

By R. D. MINDLIN[2] and H. DERESIEWICZ,[3] NEW YORK, N. Y.

An investigation is made of the phenomena occurring at the contact surfaces of like elastic spheres subjected to a variety of applied forces. Owing to the presence of slip, with its associated energy dissipation and permanent set, the changes in tractions and displacements depend not only upon the initial state of loading, but upon the entire past history of loading and the instantaneous relative rates of change of the normal and tangential forces. On the basis of three of the simplest cases of varying oblique forces, a set of rules of procedure is assembled and then applied to two types of problems. In the first, the initial tangential compliances are calculated for a variety of past histories and instantaneous rates of loading; in the other, a detailed history of a varying oblique force is investigated.

1 Introduction

The mathematical study of the effects produced by mutual compression of elastic bodies was initiated by Hertz[4] who considered the case in which the forces between bodies are normal to the contact surfaces. Recent studies[5, 6, 7] of the effects of forces tangential to the contact surfaces have revealed the necessity for taking slip into account. This phenomenon, with its accompanying energy dissipation and permanent set, introduces nonlinearities of a different character than the Hertz nonlinearity. Not only do the changes in stresses and displacements depend upon the initial state of loading, but also upon the entire past history of loading and the instantaneous relative rates of change of the normal and tangential forces. For example, different phenomena are involved and different results obtained depending upon whether the normal or tangential force is held constant, while the other varies; whether they both vary, and whether the sense of the variation is such that one increases while the other decreases, both increase, or both decrease; whether the relative rate of change is greater or less than the coefficient of friction; whether the immediate past history of loading was in the same or opposite sense as the current loading. This situation makes it necessary to study a variety of special cases. The ones selected are those which seem likely to arise in applications and, at the same time, reveal new phenomena or require new techniques. In every case, the elastic bodies considered are identical spheres.

After a brief statement, in Section 2, of the results of the Hertz theory, three of the simplest cases of varying oblique forces are considered in Sections 3 to 5. In these, the normal force is held constant while the tangential force increases, decreases, and oscillates. On the basis of the considerations employed and the results obtained in the preceding cases, a set of rules for further procedure is stated in Section 6. The rules are then applied to two types of problems. In the first (Sections 7 to 13), the initial tangential compliances are calculated for a variety of past histories and instantaneous rates of loading. In the second (Sections 14 to 19), a detailed history of a varying oblique force is investigated. An integral is encountered which necessitates a definite choice of loading history in order to proceed further. The one selected is the case of an initial normal load, followed by an oblique force whose inclination remains constant while its magnitude varies. The force first increases, then decreases, and, finally, oscillates.

2 Normal Force Varying, Tangential Force Zero

This is the case to which the Hertz theory[4] is applicable. The theory predicts a plane, circular contact surface of radius

$$a = (KNR)^{1/3} \qquad [1]$$

where N is the normal force, R is the radius of the spheres, and $K = 3(1 - \nu^2)/4E$, in which ν and E are Poisson's ratio and Young's modulus, respectively, of the material.

The distribution of normal traction on the contact surface is given by

$$\sigma = \frac{3N}{2\pi a^3}(a^2 - \rho^2)^{1/2} \qquad [2]$$

where ρ is the radial distance from the center of the contact surface.

The theory also gives the relative approach of the spheres

$$\alpha = 2(KN/R^{1/2})^{2/3} \qquad [3]$$

Hence the normal compliance of one sphere is

$$c_n = \frac{1}{2}\frac{d\alpha}{dN} = \frac{2}{3}\left(\frac{K^2}{RN}\right)^{1/3} = \frac{1-\nu}{4\mu a} \qquad [4]$$

where μ is the shear modulus of the material.

3 Normal Force Constant, Tangential Force Increasing

This is the case investigated by Cattaneo[5] and Mindlin.[6] The normal force is applied first, following which the tangential force

---

[1] This investigation was supported by the Office of Naval Research under Contract Nonr-266(09).

[2] Professor of Civil Engineering, Columbia University. Mem. ASME.

[3] Assistant Professor of Civil Engineering, Columbia University. Jun. ASME.

[4] "Theory of Elasticity," by S. Timoshenko and J. N. Goodier, McGraw-Hill Book Company, Inc., New York, N. Y., 1951, p. 372.

[5] "Sul contatto di due Corpi Elastici," by C. Cattaneo, Accademia dei Lincei, Rendiconti, Series 6, vol. 27, 1938, pp. 342–348, 434–436, and 474–478.

[6] "Compliance of Elastic Bodies in Contact," by R. D. Mindlin, Journal of Applied Mechanics, Trans. ASME, vol. 71, 1949, pp. A-259–268.

[7] "Effects of an Oscillating Tangential Force on the Contact Surfaces of Elastic Spheres," by R. D. Mindlin, W. P. Mason, T. F. Osmer, and H. Deresiewicz, Proceedings of the First U. S. National Congress of Applied Mechanics, 1951, pp. 203–208.

Presented at the National Conference of the Applied Mechanics Division, Minneapolis, Minn., June 18–20, 1953, of The American Society of Mechanical Engineers.

Discussion of this paper should be addressed to the Secretary, ASME, 29 West 39th Street, New York, N. Y., and will be accepted until October 10, 1953, for publication at a later date. Discussion received after the closing date will be returned.

 Manuscript received by ASME Applied Mechanics Division, November 5, 1952. Paper No. 53—APM-14.

T increases monotonically from zero. Considerations of symmetry show that the distribution of normal traction is not altered during application of T, as long as the two spheres are alike.[3]

If it is assumed that no slip occurs, symmetry considerations lead to the conclusion that the tangential component of displacement at the contact surface is that of a rigid body. The situation is thus reduced to a mixed boundary-value problem in elasticity. On the contact surface, the normal component of traction (zero) and the tangential component of displacement (constant) are given. On the remainder of the surface of the sphere, approximated as plane, the three components of traction are given (all zero). Solution of this problem gives the tangential component of traction τ on the contact surface. This is everywhere parallel to the displacement (and the force T), is axially symmetric in magnitude and increases to infinity at the edge of the contact surface. In addition to yielding the initial tangential compliance, this solution reveals that slip must accompany tangential forces, as infinite tractions are required in the absence of slip.

Slip may be expected to start at the edge of the contact surface, where infinite tangential tractions would occur without slip, and progress radially inward. Since τ, in the absence of slip, is radially symmetric, it is reasonable to suppose that slip occurs on an annulus. Further, as a first approximation, it is assumed that the tangential traction on the annulus of slip is in the direction of the tangential force and has the magnitude $\tau = f\sigma$, where f is a constant coefficient of friction. On account of symmetry, the tangential component of displacement is constant on the portion of the contact surface on which no slip occurs (the "adhered portion"). Thus the situation is reduced to another mixed boundary-value problem in elasticity. The tangential displacement (constant) and the normal traction (zero) are given on the adhered portion and the traction is given over the remainder of the boundary (tangential traction $= f\sigma$, normal traction zero on the annulus of slip and all three components of traction zero outside). Solution of this problem[5,6] yields the inner radius of the annulus of slip

$$c = a\,(1 - T/fN)^{1/3} \qquad [5]$$

the distribution of tangential traction on the contact surface

$$\left.\begin{aligned} \tau &= \frac{3fN}{2\pi a^3}(a^2 - \rho^2)^{1/2}, \quad c \leqslant \rho \leqslant a \\ \tau &= \frac{3fN}{2\pi a^3}\left[(a^2 - \rho^2)^{1/2} - (c^2 - \rho^2)^{1/2}\right], \quad \rho \leqslant c \end{aligned}\right\} \qquad [6]$$

and the displacement of distant points with respect to the uniform displacement of the adhered portion

$$\left.\begin{aligned} \delta &= \frac{3(2-\nu)fN}{16\mu a}\left(1 - \frac{c^2}{a^2}\right) \\ &= \frac{3(2-\nu)fN}{16\mu a}\left[1 - \left(1 - \frac{T}{fN}\right)^{2/3}\right] \end{aligned}\right\} \qquad [7]$$

The distribution of tangential traction on the contact surface and the relation between tangential load and displacement are illustrated in Figs. 1 and 2, respectively. As T approaches fN, point c in Fig. 1 approaches the center of the contact surface (point O) in accordance with Equation [5]. For any $T < fN$ there corresponds a point on the curve in Fig. 2, in accordance with Equation [7]. When $T = fN$, the adhered portion of the contact surface has shrunk to zero and the displacement has reached the abscissa of point F in Fig. 2, at which point the displacement becomes indeterminate, as rigid-body sliding takes place over the whole contact surface. A distinction is thus made between slip and slide. The term slip is used here when there is relative tangential displacement over only a portion of contiguous surfaces. When T reaches the value for which the inner boundary of the slip surface shrinks to zero, there is relative displacement over the whole contact surface; this situation is termed slide.

[3] A case in which the normal traction is changed as a result of application of a tangential force is considered by H. Poritzky, Journal of Applied Mechanics, Trans. ASME, vol. 72, 1950, pp. 191–201.

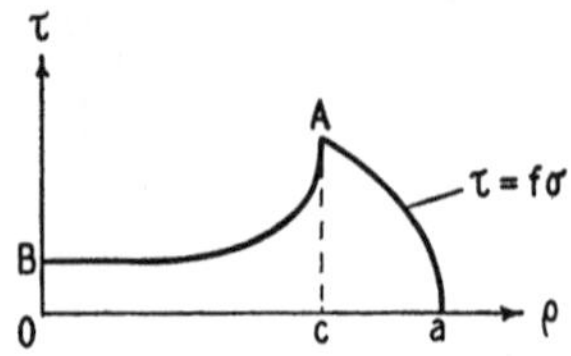

Fig. 1

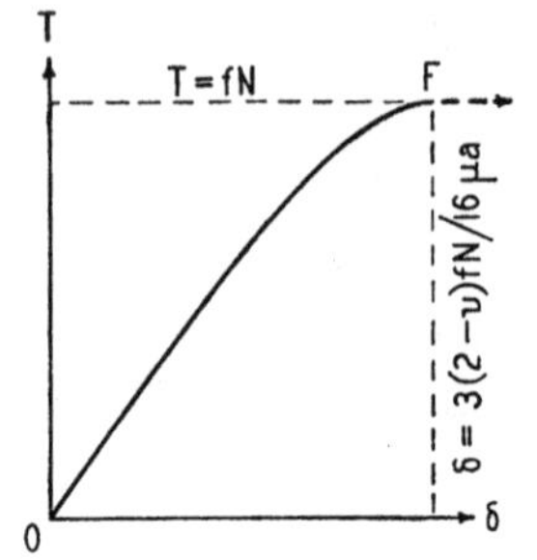

Fig. 2

The tangential compliance for the case of constant normal force and monotonically increasing tangential force is the reciprocal of the slope of the curve in Fig. 2 and is given by the formula

$$c_t = \frac{d\delta}{dT} = \frac{2-\nu}{8\mu a}\left(1 - \frac{T}{fN}\right)^{-1/3} \qquad [8]$$

Thus the tangential compliance is $(2 - \nu)/8\mu a$ at $T = 0$ and is infinite at $T = fN$; i.e., free sliding takes place at the latter load.

The ratio of the initial tangential compliance, $(2 - \nu)/8\mu a$, to the normal compliance, Equation [4], is $(2-\nu)/2(1-\nu)$. Hence the ratio ranges from unity, for $\nu = 0$, to 3/2, for $\nu = 1/2$.

4 Normal Force Constant, Tangential Force Decreasing

Suppose, now, that after having reached a value T^*, where $0 < T^* < fN$, the tangential force T is reduced. If slip were prevented, the tangential traction τ would tend to negative infinity on the boundary $\rho = a$. This conclusion is reached from the solution of the appropriate boundary-value problem in elasticity, as described in the preceding section. Hence slip, opposite in sense to the initial slip, is presumed to start at $\rho = a$ and penetrate to a radius b, assumed, temporarily, to lie in the interval $c \leqslant b \leqslant a$. As in the preceding section, the tangential traction on the annulus $b \leqslant \rho \leqslant a$ is equal to $f\sigma$ but its sense is now opposite to that of the initial traction. Hence the change of traction over the annulus $b \leqslant \rho \leqslant a$ is $-2f\sigma$. Since no additional slip occurs on the surface $\rho \leqslant b$, the change of tangential displacement in that region must, from symmetry, be that of a rigid body. Thus the change due to reduction of T presents a boundary-value problem in elasticity identical in form with that encountered in the preceding section. Hence the change in tangential traction, by analogy with Equations [6], is

$$\left.\begin{aligned}\tau_c &= -\frac{3fN}{\pi a^3}(a^2-\rho^2)^{1/2}, \quad b \leqslant \rho \leqslant a \\ \tau_c &= -\frac{3fN}{\pi a^3}[(a^2-\rho^2)^{1/2}-(b^2-\rho^2)^{1/2}], \quad \rho \leqslant b\end{aligned}\right\} \quad [9]$$

This distribution is illustrated by curve a-A'-B' in Fig. 3.

The resultant traction accompanying a reduction in T is obtained by adding the initial traction, Equations [6], and the change, Equations [9], with the result

$$\left.\begin{aligned}\tau &= -\frac{3fN}{2\pi a^3}(a^2-\rho^2)^{1/2}, \quad b \leqslant \rho \leqslant a \\ \tau &= -\frac{3fN}{2\pi a^3}[(a^2-\rho^2)^{1/2}-2(b^2-\rho^2)^{1/2}], \quad c \leqslant \rho \leqslant b \\ \tau &= -\frac{3fN}{2\pi a^3}[(a^2-\rho^2)^{1/2}-2(b^2-\rho^2)^{1/2} \\ &\qquad + (c^2-\rho^2)^{1/2}], \quad \rho \leqslant c\end{aligned}\right\} \quad [10]$$

This is illustrated, in Fig. 3, by curve a-D-E-F, which is the sum of a-A-B and a-A'-B'.

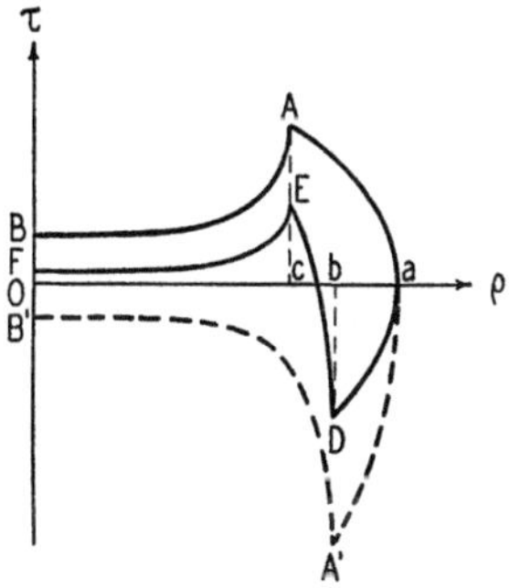

FIG. 3

The equilibrium condition

$$T = \int_0^{2\pi}\int_0^a \tau\, \rho d\rho d\theta \quad [11]$$

yields, upon insertion of τ from Equations [10]

$$T = \frac{fN}{a^3}(a^3-c^3) - \frac{2fN}{a^3}(a^3-b^3) \quad [12]$$

Noting, from Equation [5], that the first term on the right of Equation [12] is the initial tangential force T^*, the depth of penetration of counterslip is

$$b = a\left(1-\frac{T^*-T}{2fN}\right)^{1/3} \quad [13]$$

Thus the assumption $c \leqslant b \leqslant a$ is valid as long as $-T^* \leqslant T \leqslant T^*$. When $T = -T^*$, i.e., when the tangential load is fully reversed, $b = c$, i.e., counterslip has penetrated to the depth of the initial slip. At the same time, Equations [10] reduce to Equations [6] with signs reversed. The traction is then distributed just as the initial traction at $T = T^*$ was, but with opposite sense. The situation is the same as if no positive T had ever been applied, but only a negative T of magnitude $-T^*$. Upon further decrease of T to $-fN$ the phenomenon is again the same as the one described in the preceding section for a monotonic T.

The associated displacement of the adhered portion is found by a similar process of superposition. The change in displacement is obtained by multiplying the first of Equations [7] by -2 and replacing c by b as given in Equation [13]. The initial displacement is given by the second of Equations [7] with T replaced by T^*. The sum of the two is

$$\begin{aligned}\delta_d &= \frac{3(2-\nu)fN}{16\mu a}\left(2\frac{b^2}{a^2}-\frac{c^2}{a^2}-1\right) \\ &= \frac{3(2-\nu)fN}{16\mu a}\left[2\left(1-\frac{T^*-T}{2fN}\right)^{2/3}-\left(1-\frac{T^*}{fN}\right)^{2/3}-1\right]\end{aligned} \quad [14]$$

The tangential load-displacement curve is illustrated by the full lines in Fig. 4.

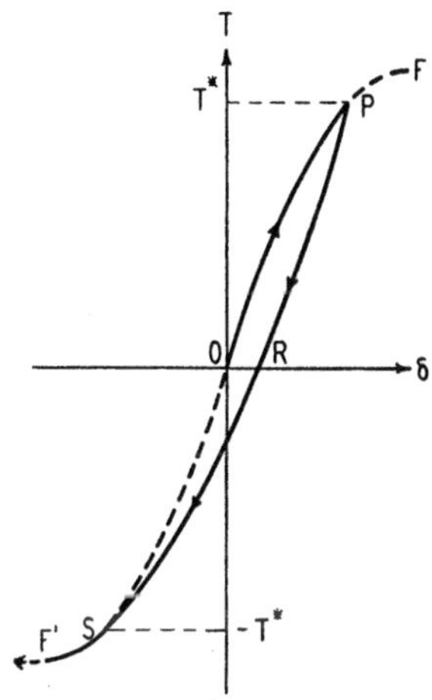

FIG. 4

The compliance for unloading is

$$\frac{d\delta_d}{dT} = \frac{2-\nu}{8\mu a}\left(1-\frac{T^*-T}{2fN}\right)^{-1/3} \quad [15]$$

It is interesting to note that the initial compliance on unloading ($T = T^*$ in Equation [15]) is the same as the initial compliance on loading ($T = 0$ in Equation [8]); that is, in Fig. 4, the slope of PR at P is the same as the slope of OP at O.

When T has been reduced from T^* to zero, there is a permanent set given by OR, in Fig. 4, the magnitude of which is obtained by setting $T = 0$ in Equation [14]. The accompanying traction is not zero, but is a self-equilibrating distribution obtained by setting $b = a(1-T^*/2fN)^{1/3}$ in Equations [10].

When T has been reduced to $-T^*$, the displacement has reached the negative of the displacement at $T = T^*$, i.e., the abscissa of S in Fig. 4 is the negative of the abscissa of P, and the compliance is identical with that of curve OP at $T = T^*$. Hence the unloading curve P-R-S is tangent, at S, to the central perversion OS, of the loading curve OP. Upon further decrease of T^* to $-fN$, the unloading curve follows the reflected loading curve SF', in accordance with the results for monotonic T given in Section 3.

5 Normal Force Constant, Tangential Force Oscillating

In the preceding section the entire situation at $T = -T^*$ is identical with that at $T = T^*$, except for reversal of sign. Hence a subsequent increase of T from $-T^*$ to T^* will be accompanied by the same events as occurred during the decrease from T^* to $-T^*$, except for reversal of sign. Thus, in starting along SU, Fig. 5, the compliance at S is the same as the compliance at P of PR. Slip again starts at $\rho = a$ in the same sense as occurred

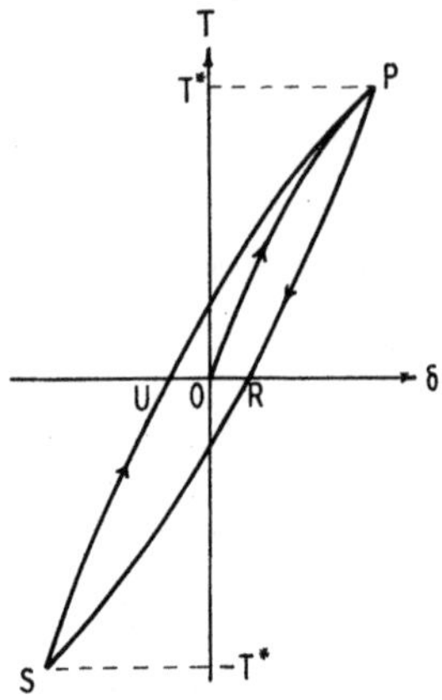

Fig. 5

along path OP. At an intermediate point of SU, the traction will be like a-D-E-F, Fig. 3, with sign reversed. When the tangential force once more reaches T^*, b will again have penetrated to c and the traction will be exactly a-A-B.

The displacement along path S-U-P is

$$\delta_i = -\delta_d(-T) = -\frac{3(2-\nu)fN}{16\mu a}\left[2\left(1-\frac{T^*+T}{2fN}\right)^{2/3} - \left(1-\frac{T^*}{fN}\right)^{2/3} - 1\right] \quad [16]$$

Hence, when $T = T^*$, the terminal points P of S-U-P and P of OP, Fig. 5, are identical, as may be seen by comparing Equation [16] with Equation [7] for that value of T.

Subsequent diminution of T will then produce a repetition of the events accompanying the first diminution.

It is now evident that during oscillation of T between $\pm T^*$, with N maintained constant, the load-displacement curve traverses the loop P-R-S-U-P, Fig. 5.

The area enclosed in the loop gives the frictional energy loss per cycle

$$F = \int_{-T^*}^{T^*} (\delta_d - \delta_i)dT = \frac{9(2-\nu)f^2N^2}{10\mu a}\left\{1 - \left(1-\frac{T^*}{fN}\right)^{5/3} - \frac{5T^*}{6fN}\left[1 + \left(1-\frac{T^*}{fN}\right)^{2/3}\right]\right\} \quad [17]$$

For small T^*/fN, Equation [17] reduces to

$$F = \frac{(2-\nu)T^{*3}}{36\mu a f N} \quad [18]$$

i.e., the energy loss per cycle varies as the cube of the maximum tangential force.

Some of the conclusions reached in this section have been subjected to experimental test.[7] The occurrence of the annulus of slip has been verified, as well as the constancy of the coefficient of friction and the relation between the size of the inner radius of the annulus and the magnitude of T^*. In the tests, the energy dissipation, for magnitudes of T^*/fN close to unity, agreed with the values predicted by the theory; for T^*/fN in the vicinity of zero, however, it appeared to vary more nearly as the square than the cube of the maximum tangential force.

6 Rules of Procedure

The results of the two preceding sections have been obtained by means of extensions of Cattaneo's and Mindlin's solutions for the case of constant N and monotonically increasing T. Before proceeding to cases of greater complexity, it may be helpful to set down, explicitly, the rules upon which the two preceding and the subsequent developments are based:

Rule 1. The radius of the contact surface and the normal component of traction on it are given by the Hertz formulas, Equations [1] and [2].

Rule 2. With every application or change of tangential force, slip will be initiated whenever, in the absence of slip, the solution of the appropriate boundary-value problem in the theory of elasticity yields a tangential traction, at any point, in excess of the product of a constant coefficient of friction and the normal component of traction at that point.

Rule 3. Slip, in the direction of the force causing it,[9] progresses concentrically, radially inward from the boundary of the contact surface, forming an "annulus of slip."

Rule 4. At any point on a contact surface, the magnitude of the tangential component of traction is at most equal to the product of a constant coefficient of friction and the normal component of traction at that point. The equality necessarily holds at a point at which slip has just occurred, in which case the traction has the same sense as the slip.[9]

Rule 5. The adhered portion of the contact surface, i.e., the portion encircled by an annulus on which slip occurs, is subjected to a change of tangential traction and undergoes a rigid-body tangential displacement. The radius of the adhered portion, the distribution of the traction and the magnitude of the displacement are obtained from Cattaneo's and Mindlin's formulas, Equations [5], [6], [7].

Rule 6. Beginning with an equilibrium position, for which the displacement and the distribution of traction have been established in accordance with the preceding rules, the effects of a change in the state of loading are obtained by advancing to the desired state through a sequence of equilibrium positions, each of which is obtained from its predecessor by applying Rules 1 to 5. In particular, the condition of equilibrium, Equation [11], is useful.

In the following sections these rules are applied to the calculation of tangential compliances for a number of past histories, initial states, and variations of both N and T.

7 N Increasing, T Increasing

Suppose that, after an initial state has been reached by first applying a normal force N and then a tangential force T, both N and T are to increase at an arbitrary relative rate.

The initial state is given by Equations [1], [2], [5], [6], and [7]. The distribution of tangential traction is illustrated by

[9] There is a small lateral component of relative tangential displacement which accompanies the major slip in the direction of the applied force. The lateral tangential traction, which accompanies the lateral slip, is neglected.

curve a-A-B in Fig. 6, the tangential load-displacement history by curve OA in Fig. 7.

Now, increase N by an increment ΔN. The radius of the contact surface increases to a value a_1 given by Equation [1] with N replaced by $N + \Delta N$

$$a_1 = [K(N + \Delta N)R]^{1/3} \quad \ldots\ldots\ldots\ldots [19]$$

The tangential traction remains unchanged.

With the normal load held constant at $N + \Delta N$, increase the tangential force by an increment $\Delta T < f\Delta N$. In accordance with the rules of Section 6, slip begins at $\rho = a_1$ and progresses inward to a radius c_1 given by Equation [5] with c, a, T, and N replaced by c_1, a_1, ΔT_1, and $N + \Delta N$, respectively

$$c_1 = a_1\,[1 - \Delta T/f(N + \Delta N)]^{1/3} \quad \ldots\ldots\ldots [20]$$

From Equations [1] and [19]

$$a^3(N + \Delta N) = a_1^3 N \quad \ldots\ldots\ldots [21]$$

Hence

$$c_1 = a(fN)^{-1/3}\,(fN + f\Delta N - \Delta T)^{1/3} \quad \ldots\ldots\ldots [22]$$

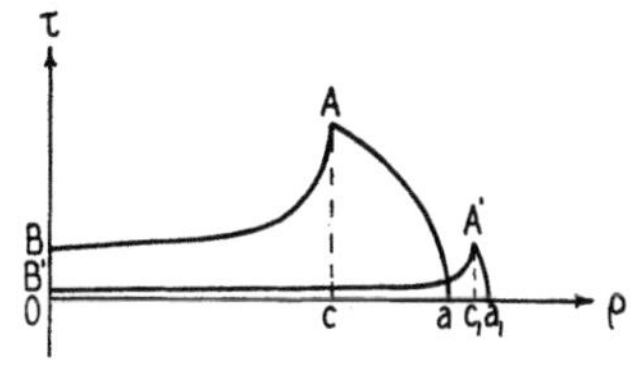

FIG. 6

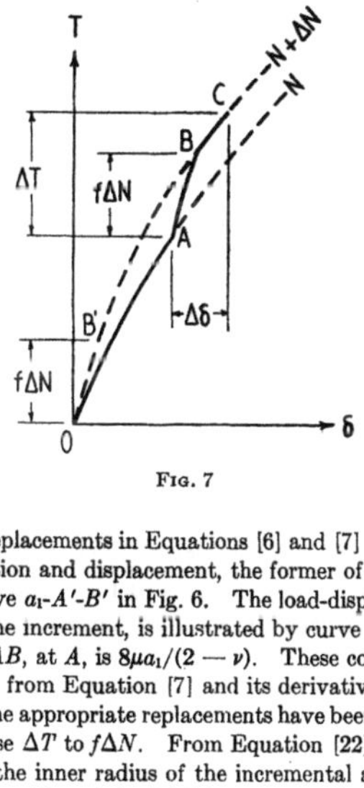

FIG. 7

The same replacements in Equations [6] and [7] give the increments in traction and displacement, the former of which is illustrated by curve a_1-A'-B' in Fig. 6. The load-displacement relation, during the increment, is illustrated by curve AB, in Fig. 7. The slope of AB, at A, is $8\mu a_1/(2 - \nu)$. These conclusions may be ascertained from Equation [7] and its derivative with respect to ΔT, after the appropriate replacements have been made.

Now increase ΔT to $f\Delta N$. From Equation [22], c_1 progresses in to a; i.e., the inner radius of the incremental annulus of slip shrinks to the radius of the initial contact surface. The increment in tangential traction is now given by

$$\left.\begin{aligned}\Delta\tau &= \frac{3f(N + \Delta N)}{2\pi a_1^3}\,(a_1^2 - \rho^2)^{1/2}, \quad a \leqslant \rho \leqslant a_1\\ \Delta\tau &= \frac{3f(N + \Delta N)}{2\pi a_1^3}\,[(a_1^2 - \rho^2)^{1/2} - (a^2 - \rho^2)^{1/2}], \; \rho \leqslant a\end{aligned}\right\} \ldots [23]$$

and is illustrated by curve a_1-A'-B' in Fig. 8. The total tangential traction is the sum of Equations [6] and [23]. Using Equation [21], this becomes

$$\left.\begin{aligned}\tau_R &= \tau + \Delta\tau = \frac{3f(N + \Delta N)}{2\pi a_1^3}\,(a_1^2 - \rho^2)^{1/2}, \; c \leqslant \rho \leqslant a_1\\ \tau_R &= \frac{3f(N + \Delta N)}{2\pi a_1^3}\,[(a_1^2 - \rho^2)^{1/2} - (c^2 - \rho^2)^{1/2}], \; \rho \leqslant c\end{aligned}\right\} \ldots [24]$$

But this is the distribution that would have been obtained if the entire history had been an initial application of normal force $N + \Delta N$ followed by a single application of tangential force $T + f\Delta N$. Thus the sum of curves a-A-B and a_1-A'-B', in Fig. 8,

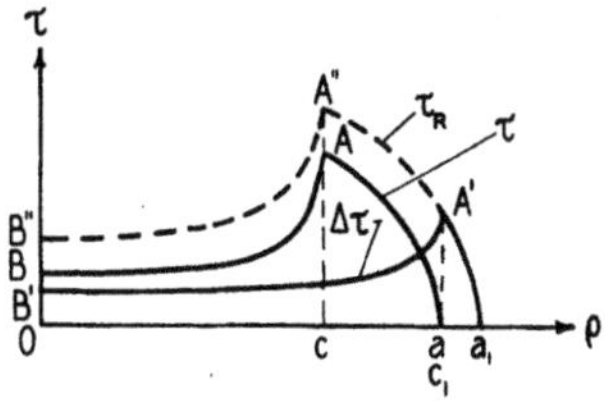

FIG. 8

is a_1-A''-B'', which is the curve corresponding to Equations [24]. Further increase in ΔT will then produce tractions in accordance with the findings in Section 3 for monotonic increase in T.

Further, when $\Delta T = f\Delta N$, the total displacement is the same as would have been obtained if an initial normal force $N + \Delta N$ had been followed by a tangential force $T + f\Delta N$; i.e., point B of AB, in Fig. 7, coincides with point B of the load-displacement curve, O-B'-B, for initial normal force $N + \Delta N$. Upon further increase in T, the load-displacement curve follows path BC.

The compliance at A of curve AB, in Fig. 7, is

$$\frac{2 - \nu}{8\mu a_1}$$

The compliance at B of curve O-B-C is

$$\frac{2 - \nu}{8\mu a_1}\left(1 - \frac{T}{f(N + \Delta N)}\right)^{-1/3}$$

Hence, for small $\Delta T \geqslant f\Delta N$, the increment in displacement is

$$\Delta\delta = \frac{2 - \nu}{8\mu a}\left[f\Delta N + \left(1 - \frac{T}{f(N + \Delta N)}\right)^{-1/3}(\Delta T - f\Delta N)\right]$$

The compliance at (N, T), for N and T increasing, is then

$$\begin{aligned}c_t &= \lim_{\substack{\Delta N \to 0\\ \Delta T \to 0}} \frac{\Delta\delta}{\Delta T}\\ &= \frac{2 - \nu}{8\mu a}\left[f\frac{dN}{dT} + \left(1 - f\frac{dN}{dT}\right)\left(1 - \frac{T}{fN}\right)^{-1/3}\right], 0 \leqslant \frac{dN}{dT} \leqslant \frac{1}{f}\end{aligned} \quad \ldots\ldots\ldots [25]$$

If $\Delta T < f\Delta N$, the load-displacement curve does not extend beyond B in Fig. 7. Hence

$$\Delta\delta = \frac{2 - \nu}{8\mu a}\,\Delta T$$

and the compliance is

$$c_t = \frac{2 - \nu}{8\mu a}, \quad \frac{dN}{dT} \geqslant \frac{1}{f} \quad \ldots\ldots\ldots\ldots [26]$$

8 N Decreasing, T Increasing

Starting with the same initial state as in Section 7, it is desired first to decrease N by an amount ΔN, with the tangential load held constant. Such a decrease would reduce the contact radius to a value a_1, given by Equation [1] with N replaced by $N - \Delta N$. However, the portions of the spheres in annulus $a_1 \leqslant \rho \leqslant a$ would no longer be in contact, and would be unable to sustain tangential traction. Consequently, before a reduction of normal load may be effected, it is necessary to remove the existing tangential traction from this region. This removal may be accomplished by means of the following procedure:

(*a*) The contact area $\rho \leqslant a_1$ is "frozen"; i.e., no slip is permitted to take place.

(*b*) A distribution of traction is added, such that it will cause the annulus $a_1 \leqslant \rho \leqslant a$ to become free of traction, and, at the same time, obey the injunction stated in (*a*).

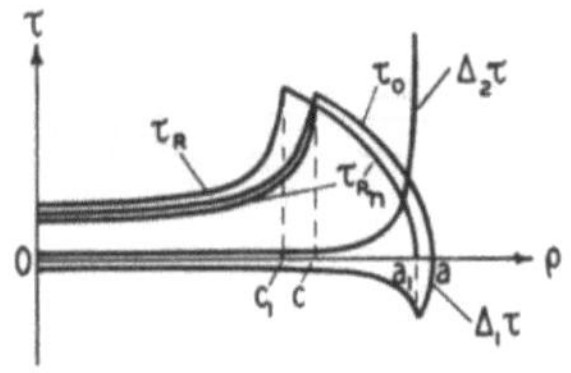

Fig. 9

Such a distribution (see Fig. 9) is given by

$$\left.\begin{aligned} \Delta_1\tau &= -\frac{3fN}{2\pi a^3}(a^2-\rho^2)^{1/2}, \quad a_1 \leqslant \rho \leqslant a \\ \Delta_1\tau &= -\frac{3fN}{2\pi a^3}\left[(a^2-\rho^2)^{1/2}-(a_1{}^2-\rho^2)^{1/2}\right], \quad \rho \leqslant a_1 \end{aligned}\right\} \quad ..[27]$$

The resultant traction (which, it should be noted, corresponds to a condition of equilibrium with the force $T - f\Delta N$, but not with T) will be

$$\left.\begin{aligned} \tau_{R_n} &= 0, \quad a_1 \leqslant \rho \leqslant a \\ \tau_{R_n} &= \frac{3f(N-\Delta N)}{2\pi a_1{}^3}(a_1{}^2-\rho^2)^{1/2}, \quad c \leqslant \rho \leqslant a_1 \\ \tau_{R_n} &= \frac{3f(N-\Delta N)}{2\pi a_1{}^3}\left[(a_1{}^2-\rho^2)^{1/2}-(c^2-\rho^2)^{1/2}\right], \quad \rho \leqslant c \end{aligned}\right\} \quad ..[28]$$

Since annulus $a_1 \leqslant \rho \leqslant a$ is now free of traction, it is permissible to decrease N by ΔN. The distribution of traction, Equations [28], is the same as if the order of loading had been: (1) $N - \Delta N$, (2) $T_1 = f(N - \Delta N)\,(1 - c^3/a_1{}^3) = T - f\Delta N$, where the value of T_1 is obtained by applying the integral condition, Equation [11]. The displacement may be thought to follow path AB, Fig. 10.

A portion of the tangential force, represented by the volume under the traction distribution, Equations [27], was removed in the course of the reduction of the normal load. But the spheres must be in equilibrium with T; hence the force $f\Delta N$ is transferred to the remaining contact surface (i.e., $\rho \leqslant a_1$). In consequence of the assumption stated in (*a*) of this section, the additional tangential traction[10] resulting from this transfer is given by (see Fig. 9)

$$\Delta_2\tau = \frac{f\Delta N}{2\pi a_1}(a_1{}^2-\rho^2)^{-1/2}, \quad \rho \leqslant a_1 \quad ....[29]$$

[10] See Equation [76], reference (6).

This is the distribution, referred to in Section 3, which increases without limit as ρ approaches a_1.

Now unfreeze the contact surface. Slip will progress in the direction of the applied tangential force in accordance with the fundamental rules. The resultant tangential traction will be distributed as given by Equations [28] with c replaced by c_1, where

$$c_1 = a_1\left(1 - \frac{T}{f(N-\Delta N)}\right)^{1/3}$$

The displacement may be thought to take path BC in Fig. 10.

Now, holding the normal force constant (at $N - \Delta N$), increase

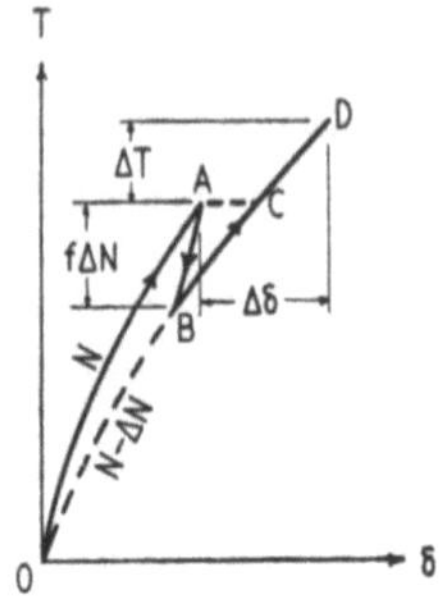

Fig. 10

the tangential force by an amount ΔT. At this point the tangential traction will be given by Equations [28] with c replaced by c_2, where

$$c_2 = a_1\left(1 - \frac{T+\Delta T}{f(N-\Delta N)}\right)^{1/3} < c_1$$

It is interesting that the resultant traction has, qualitatively, the same form as the initial traction, Equations [6].

The displacement will follow path CD on curve O-B-D, Fig. 10. Hence the compliance will be

$$\begin{aligned} c_t &= \lim_{\substack{\Delta N\to 0\\ \Delta T\to 0}}\left\{\frac{1}{\Delta T}\,\frac{2-\nu}{8\mu a}\left[-f\Delta N + \left(1-\frac{T}{fN}\right)^{-1/3}(f\Delta N+\Delta T)\right]\right\} \\ &= \frac{2-\nu}{8\mu a}\left[-f\frac{dN}{dT} + \left(1+f\frac{dN}{dT}\right)\left(1-\frac{T}{fN}\right)^{-1/3}\right] \quad ..[30] \end{aligned}$$

Since, inherently, $dN/dT < 0$, its absolute value must be used in this formula.

9 N Increasing, T Decreasing

1 Apply N, increase the tangential force monotonically from zero to T^*, then reduce it to T. The tangential traction is given by Equations [10]; its distribution is shown by τ_0 in Fig. 11.

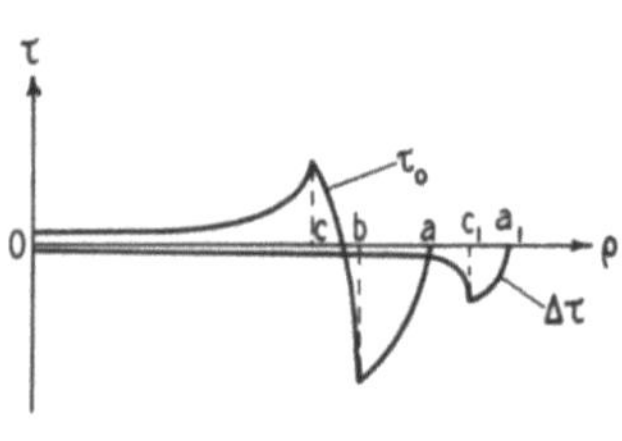

Fig. 11

2 Keeping the tangential load constant, increase the normal load to $N + \Delta N$. The radius a_1 of the new contact area is given by Equation [19]. The tangential traction remains unchanged.

3 Keeping the normal force constant (at $N + \Delta N$), reduce the tangential force by ΔT. The additional tangential traction (see $\Delta\tau$, Fig. 11) will be given by Equations [23] with sign reversed.

Again, if $\Delta T = f\Delta N$, so that $c_1 = a$ (Fig. 11), the resultant traction will be

$$\left.\begin{aligned}
\tau_R &= -\frac{3f(N+\Delta N)}{2\pi a_1^3}(a_1^2-\rho^2)^{1/2}, \quad b \leqslant \rho \leqslant a_1 \\
\tau_R &= -\frac{3f(N+\Delta N)}{2\pi a_1^3}[(a_1^2-\rho^2)^{1/2}-2(b^2-\rho^2)^{1/2}], \\
&\qquad\qquad c \leqslant \rho \leqslant b \\
\tau_R &= -\frac{3f(N+\Delta N)}{2\pi a_1^3}[(a_1^2-\rho^2)^{1/2}-2(b^2-\rho^2)^{1/2} \\
&\qquad + (c^2-\rho^2)^{1/2}], \ \rho \leqslant c
\end{aligned}\right\} \quad \text{..[31]}$$

Hence the situation is the same as that obtained by (1) imposing a normal load $N + \Delta N$, (2) applying a tangential load T_1^*, (3) reducing the tangential load to $T_1 = T - f\Delta N$. T_1^* may be found by noting that $T_1^* = f(N + \Delta N)(1 - c^3/a_1^3) = T^* + f\Delta N$.

4 The displacement (see Equation [14]) at the conclusion of step 1 will have traversed the path O-A-B (Fig. 12); following step 3 it will have reached point C, at which stage path O-A-B-C meets path O-A'-C.

If $\Delta T > f\Delta N$, splitting the decrement into $f\Delta N$ and $\Delta T - f\Delta N$ and applying these parts consecutively will yield a result entirely similar to that found in part 3 of this section. The additional displacement will follow path CD along curve $A'D$ (Fig. 12).

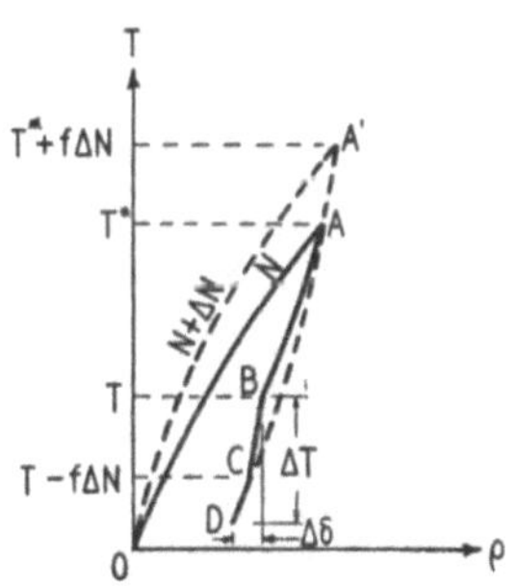

Fig. 12

Remembering that, again, inherently, $dN/dT < 0$, so that its absolute value must be employed, the compliance is

$$\left.\begin{aligned}
c_t &= \lim_{\substack{\Delta N\to 0\\ \Delta T\to 0}} \left\{\frac{1}{\Delta T}\frac{2-\nu}{8\mu a}\left[f\Delta N + \left(1-\frac{T^*-T}{2fN}\right)^{-1/2}(\Delta T - f\Delta N)\right]\right\} \\
&= \frac{2-\nu}{8\mu a}\left[f\frac{dN}{dT} + \left(1 - f\frac{dN}{dT}\right)\left(1-\frac{T^*-T}{2fN}\right)^{-1/2}\right], \ 0 > \frac{dN}{dT} \geqslant -\frac{1}{f} \\
c_t &= \frac{2-\nu}{8\mu a}, \quad \frac{dN}{dT} \leqslant -\frac{1}{f}
\end{aligned}\right\} \quad \text{..[32]}$$

10 N Decreasing, T Decreasing

1 Repeat step 1 of Section 9.

2 A decrease of normal load by an amount ΔN must be preceded by the expedients discussed in (*a*) and (*b*) of Section 8, i.e., a freezing of the contact area, and freeing of the annulus $a_1 \leqslant \rho \leqslant a$ of tangential traction by adding, with sign reversed, $\Delta\tau$ of Equations [27]. The resultant traction, which is not in equilibrium with the force T, will be (Fig. 13)

$$\left.\begin{aligned}
\tau_{R_n} &= 0, \quad a_1 \leqslant \rho \leqslant a \\
\tau_{R_n} &= -\frac{3f(N-\Delta N)}{2\pi a_1^3}(a_1^2-\rho^2)^{1/2}, \ b \leqslant \rho \leqslant a_1 \\
\tau_{R_n} &= -\frac{3f(N-\Delta N)}{2\pi a_1^3}[(a_1^2-\rho^2)^{1/2}-2(b^2-\rho^2)^{1/2}], \\
&\qquad\qquad c \leqslant \rho \leqslant b \\
\tau_{R_n} &= -\frac{3f(N-\Delta N)}{2\pi a_1^3}[(a_1^2-\rho^2)^{1/2}-2(b^2-\rho^2)^{1/2} \\
&\qquad + (c^2-\rho^2)^{1/2}], \ \rho \leqslant c
\end{aligned}\right\} \quad \text{..[33]}$$

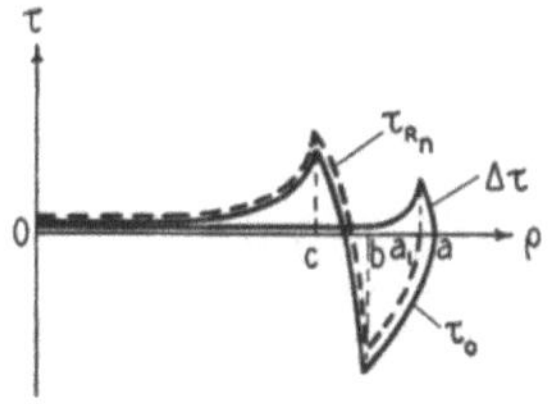

Fig. 13

The decrease in normal load is effected concurrently with the removal of the traction. This distribution may be reached by: (1) applying $N - \Delta N$, (2) applying $T_1^* = T^* - f\Delta N$, (3) reducing from T_1^* to $T + f\Delta N$. The transferred load (see Section 8) gives rise to a distribution of traction given by Equation [29] with sign reversed.

3 Unfreeze. Slip will progress in the direction opposite to that of the initial tangential force. The displacement may be thought to take path B-B'-C (Fig. 14). The traction will be given by Equations [33] with b replaced by b_1, where

$$b_1 = a_1\left(1 - \frac{T^* - T}{2f(N-\Delta N)} + \frac{\Delta N}{2(N-\Delta N)}\right)^{1/2}$$

4 Further displacement, due to a decrease of tangential load at constant normal load ($N - \Delta N$), will proceed along curve A'-B'-C to point D. Thus the compliance will be

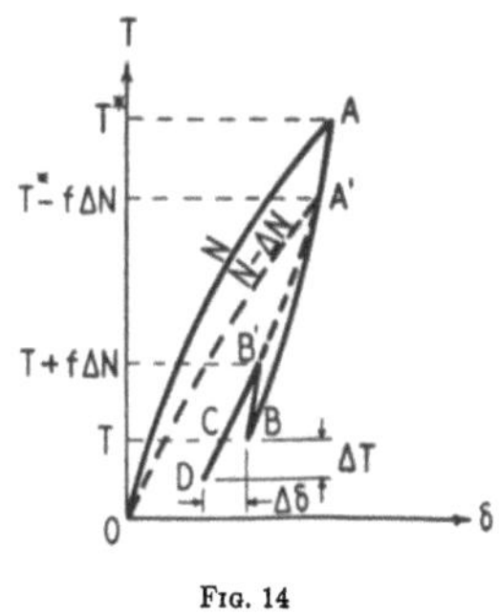

Fig. 14

$$c_t = \lim_{\substack{\Delta N \to 0 \\ \Delta T \to 0}} \left\{ \frac{1}{\Delta T} \frac{2-\nu}{8\mu a} \left[-f\Delta N + \left(1 - \frac{T^* - T}{2fN}\right)^{-1/3} (f\Delta N + \Delta T) \right] \right\}$$

$$= \frac{2-\nu}{8\mu a} \left[-f \frac{dN}{dT} + \left(1 + f\frac{dN}{dT}\right) \left(1 - \frac{T^* - T}{2fN}\right)^{-1/3} \right], \quad \frac{dN}{dT} \geqslant 0 \ldots . [34]$$

In the preceding four sections the discussion concerned tangential compliances due to simultaneously varying normal and tangential forces applied at nonzero levels of increasing or decreasing tangential load, in particular, at levels having a given previous history in the direction of the additional tangential force. In each case, the magnitude of the initial tangential force at these levels was reached at constant initial normal load.

In addition, it is of interest to find the values of compliances resulting from oblique loading at levels of tangential load which have no previous loading history in the direction of the tangential component of the system of oblique forces. The cases

(a) N increasing, T increasing, at $T^* = 0$
(b) N decreasing, T increasing, at $T^* = 0$
(c) N increasing, T decreasing, at $T = T^*$

are no more than special instances of loading considered in Sections 7, 8, and 9, respectively. The compliances are obtained directly from the general expressions given in the corresponding sections. On the other hand, there are several cases of interest for which the results cannot be obtained from previously considered cases. These are considered in Sections 11 to 13.

11 N Decreasing, T Decreasing, at $T = T^*$

1 Repeat step 1 of Section 9, with the proviso $T^* - T = \Delta T$ is an infinitesimal.

2 The reduction of the normal load by ΔN again must be preceded by freezing the contact surface and removal of traction, as described in Sections 8 and 10. Here, however, it is necessary to treat separately the cases $\Delta T \geqslant 2f\Delta N$ and $\Delta T < 2f\Delta N$.

(a) $\Delta T \geqslant 2f\Delta N$. Reference to Equations [1] and [13] reveals that

$$\left(\frac{b}{a}\right)^3 = 1 - \frac{\Delta T}{2fN} \leqslant 1 - \frac{\Delta N}{N} = \left(\frac{a_1}{a}\right)^3$$

i.e., $a_1 \geqslant b$. Hence the situation is the same as in the general case (Section 10); the compliance, a special case of Equation [34], is given by

$$c_t = \frac{2-\nu}{8\mu a}$$

The connection between this loading and the one discussed in Section 10 may be established by noting that, in Section 10, the load is decreased from T^* with $\Delta N = 0$, i.e., $dT/dN = \infty > 2f$.

(b) $\Delta T < 2f\Delta N$. In this case

$$\left(\frac{b}{a}\right)^3 = 1 - \frac{\Delta T}{2fN} > 1 - \frac{\Delta N}{N} = \left(\frac{a_1}{a}\right)^3$$

i.e., $a_1 < b$.

Now freeze the contact area. In order to free annulus $a_1 \leqslant \rho \leqslant a$ of tangential traction, add the following distributions of traction (Fig. 15) neither of which causes slip over circle $\rho \leqslant a_1$

$$\left.\begin{aligned} \Delta_1\tau &= \frac{3fN}{\pi a^3}(a^2-\rho^2)^{1/2}, \quad b \leqslant \rho \leqslant a \\ \Delta_1\tau &= \frac{3fN}{\pi a^3}[(a^2-\rho^2)^{1/2} - (b^2-\rho^2)^{1/2}], \quad \rho \leqslant b \end{aligned}\right\} \ldots [35]$$

$$\left.\begin{aligned} \Delta_2\tau &= -\frac{3fN}{2\pi a^3}(a^2-\rho^2)^{1/2}, \quad a_1 \leqslant \rho \leqslant a \\ \Delta_2\tau &= -\frac{3fN}{2\pi a^3}[(a^2-\rho^2)^{1/2} - (a_1^2-\rho^2)^{1/2}], \quad \rho \leqslant a_1 \end{aligned}\right\} \ldots [36]$$

The resultant traction (not in equilibrium with T), found by adding tractions of Equations [10], [35], and [36], is

$$\left.\begin{aligned} \tau_{R_n} &= 0, \quad a_1 \leqslant \rho \leqslant a \\ \tau_{R_n} &= \frac{3f(N-\Delta N)}{2\pi a_1^3}(a_1^2-\rho^2)^{1/2}, \quad c \leqslant \rho \leqslant a_1 \\ \tau_{R_n} &= \frac{3f(N-\Delta N)}{2\pi a_1^3}[(a_1^2-\rho^2)^{1/2} - (c^2-\rho^2)^{1/2}], \quad \rho \leqslant c \end{aligned}\right\} \ldots [37]$$

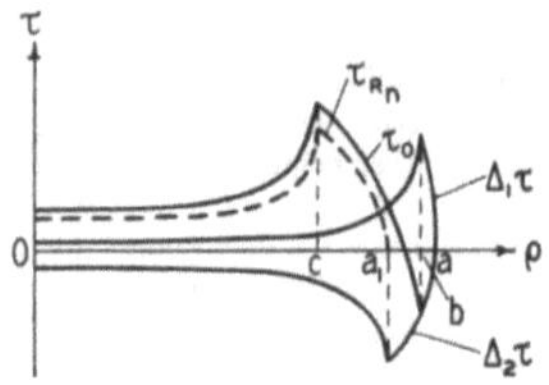

Fig. 15

Again, the normal load may be reduced by ΔN, provided that, concurrently, a traction distribution, similar to Equation [29], is applied over the area $\rho < a_1$. Its magnitude will be such as to provide the tangential force contained in the difference of volumes under the traction surfaces given by Equations [35] and [36]. Hence

$$\Delta\tau = -\frac{\Delta T - f\Delta N}{2\pi a_1}(a_1^2-\rho^2)^{-1/2}, \quad \rho < a_1 \ldots . [38]$$

The traction of Equations [37] is the same as if the order of loading had been: (1) $N - \Delta N$, (2) $T_1^* = T^* - f\Delta N$.

Now unfreeze; slip will progress in the direction indicated by the sign of the traction of Equation [38], i.e., for $\Delta T < f\Delta N$, in the direction of initial tangential force, for $f\Delta N < \Delta T < 2f\Delta N$, in the opposite direction. In the former case, the path of the displacement may be thought to be O-A-B-A-C-D (Fig. 16); the expression for the compliance is

$$c_t = \lim_{\substack{\Delta N \to 0 \\ \Delta T \to 0}} \left\{ \frac{1}{\Delta T} \left[\frac{2-\nu}{8\mu a} f\Delta N - \frac{2-\nu}{8\mu a_1}\left(1 - \frac{T_1^*}{f(N-\Delta N)}\right)^{-1/3} (f\Delta N - \Delta T) \right] \right\}$$

$$= \frac{2-\nu}{8\mu a}\left[f\frac{dN}{dT} + \left(1 - f\frac{dN}{dT}\right)\left(1 - \frac{T^*}{fN}\right)^{-1/3} \right],$$

$$\frac{dN}{dT} > \frac{1}{f} \ldots . . [39]$$

In the latter case, the displacement may be thought to traverse path O-A-B-A-C-D in Fig. 17, where CD is part of an unloading path such as $A'B'$ in Fig. 14. The compliance is $c_t = (2-\nu)/8\mu a$; thus this expression is valid for all $0 \leqslant dN/dT \leqslant 1/f$.

The traction of Equation [38] vanishes for $\Delta T = f\Delta N$, indicating that, in this instance, Equations [37] correspond to a condition of equilibrium with the force T; hence no slip accompanies unfreezing.

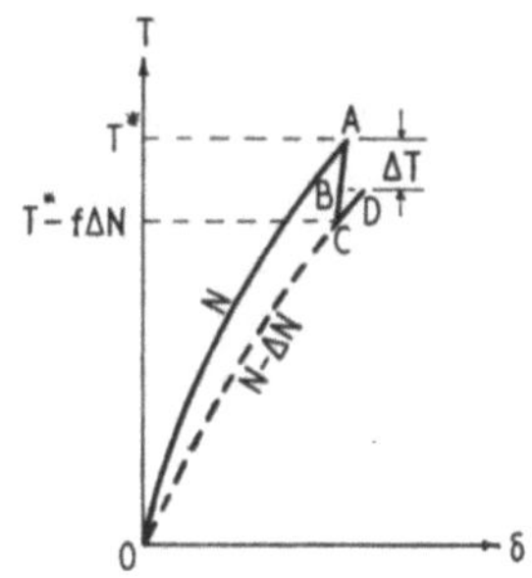

FIG. 16

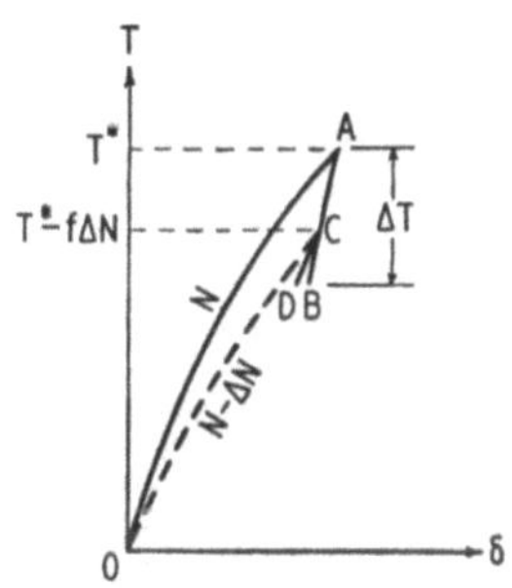

FIG. 17

12 N Increasing, T Increasing, at Point on Unloading Curve

1 Repeat steps 1 and 2 of Section 9.

2 Keeping the normal force constant (at $N + \Delta N$), increase the tangential force by ΔT.

Three cases arise, depending on whether $\Delta T \gtreqless f\Delta N$.

(a) $\Delta T = f\Delta N$. The additional traction will be such as to cause slip, in the direction of the newly applied force, in an annulus $c_1 \leqslant \rho \leqslant a_1$ and no further slip on circle $\rho \leqslant c_1$. Hence

$$\left.\begin{aligned}\Delta\tau &= \frac{3f(N+\Delta N)}{2\pi a_1^3}(a_1^2-\rho^2)^{1/2}, \quad c_1 \leqslant \rho \leqslant a_1\\ \Delta\tau &= \frac{3f(N+\Delta N)}{2\pi a_1^3}[(a_1^2-\rho^2)^{1/2}-(c_1^2-\rho^2)^{1/2}],\\ &\qquad \rho \leqslant c_1\end{aligned}\right\} \quad [40]$$

where

$$c_1 = a_1\left(1 - \frac{\Delta T}{f(N+\Delta N)}\right)^{1/3}$$

Since $\Delta T = f\Delta N$, $c_1 = a$; hence the resultant traction (the sum of Equations [10] and [40]) becomes (see Fig. 18)

$$\left.\begin{aligned}\tau_R &= \frac{3f(N+\Delta N)}{2\pi a_1^3}(a_1^2-\rho^2)^{1/2}, \quad a \leqslant \rho \leqslant a_1\\ \tau_R &= \frac{3f(N+\Delta N)}{2\pi a_1^3}[(a_1^2-\rho^2)^{1/2}-2(a^2-\rho^2)^{1/2}],\\ &\qquad b \leqslant \rho \leqslant a\\ \tau_R &= \frac{3f(N+\Delta N)}{2\pi a_1^3}[(a_1^2-\rho^2)^{1/2}-2(a^2-\rho^2)^{1/2}\\ &\qquad + 2(b^2-\rho^2)^{1/2}], \ c \leqslant \rho \leqslant b\\ \tau_R &= \frac{3f(N+\Delta N)}{2\pi a_1^3}[(a_1^2-\rho^2)^{1/2}-2(a^2-\rho^2)^{1/2}\\ &\qquad + 2(b^2-\rho^2)^{1/2}-(c^2-\rho^2)^{1/2}], \ \rho \leqslant c\end{aligned}\right\} \quad [41]$$

Hence the situation is the same as if the order of loading had been: (1) $N + \Delta N$, (2) T_1^*, (3) reduction of tangential load to T_1, (4) its increase by $\overline{\Delta T}$. It may be noted that

$$A \text{ — } \quad T_1^* = f(N+\Delta N)\left(1-\frac{c^3}{a_1^3}\right) = T^* + f\Delta N$$

$$B \text{ — } \quad T_1^* - T_1 = 2f(N+\Delta N)\left(1-\frac{b^3}{a_1^3}\right) = T^* - T + 2f\Delta N$$

$$\therefore T_1 = T - f\Delta N$$

$$C \text{ — } \quad \overline{\Delta T} = 2f(N+\Delta N)\left(1-\frac{a^3}{a_1^3}\right) = 2f\Delta N$$

Thus, on the load-displacement curve, Fig. 19, path O-A-B-C-C' may be replaced by path O-A'-B'-C'. The compliance is given by

$$c_t = \frac{2-\nu}{8\mu a}$$

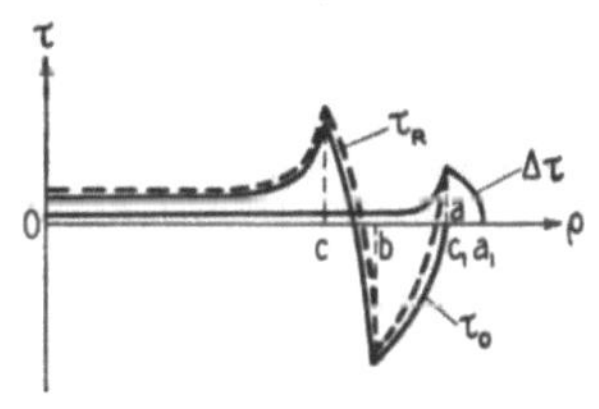

FIG. 18

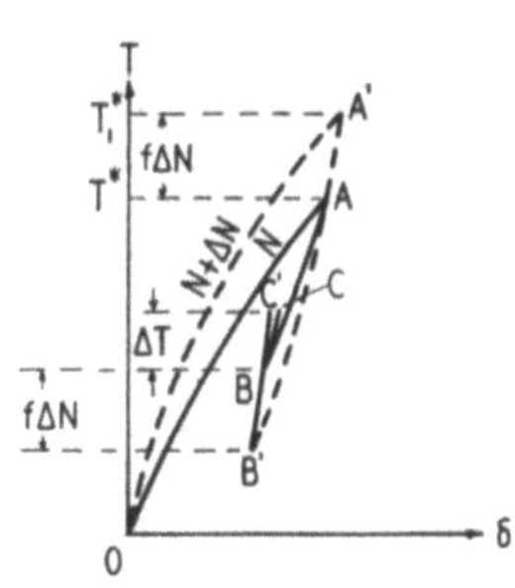

FIG. 19

(*b*) $\Delta T > f\Delta N$ ($c_1 < a$). If the increment ΔT is applied in two steps: $\Delta_1 T = f\Delta N$, $\Delta_2 T = \Delta T - f\Delta N$, in the usual manner, the course of events will be found the same as in (*a*) of this section, with the exception that $\overline{\Delta T} = \Delta T + f\Delta N$. Hence, again

$$c_t = \frac{2-\nu}{8\mu a}$$

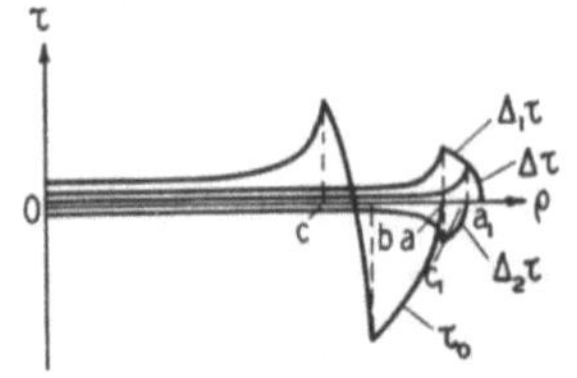

Fig. 20

(*c*) $\Delta T < f\Delta N$ ($c_1 > a$). The additional traction $\Delta\tau$, Equations [40], may be replaced by an equivalent distribution in the form (see Fig. 20)

$$\left.\begin{aligned}\Delta_1\tau &= \frac{3f(N+\Delta N)}{2\pi a_1^3}(a_1^2-\rho^2)^{1/2}, \quad a \leqslant \rho \leqslant a_1\\ \Delta_1\tau &= \frac{3f(N+\Delta N)}{2\pi a_1^3}[(a_1^2-\rho^2)^{1/2}-(a^2-\rho^2)^{1/2}],\\ &\qquad \rho \leqslant a\end{aligned}\right\} \quad \ldots [42]$$

and

$$\left.\begin{aligned}\Delta_2\tau &= 0, \quad c_1 \leqslant \rho \leqslant a_1\\ \Delta_2\tau &= \frac{3fN_{c_1}}{2\pi c_1^3}(c_1^2-\rho^2)^{1/2}, \quad a \leqslant \rho \leqslant c_1\\ \Delta_2\tau &= \frac{3fN_{c_1}}{2\pi c_1^3}[(c_1^2-\rho^2)^{1/2}-(a^2-\rho^2)^{1/2}], \quad \rho \leqslant a\end{aligned}\right\} \quad \ldots [43]$$

where $N_{c_1}/c_1^3 = N/a^3$.

It is seen that $\tau_0 + \Delta_1\tau$ is the resultant traction τ_R, Equations [41], obtained in case (*a*). Hence the load-displacement curve will be as shown in Fig. 21. There, path *O-A-B-C-D* may be replaced by path *O-A'-B'-C'-D*. The compliance becomes

$$c_t = \lim_{\substack{\Delta N \to 0\\ \Delta T \to 0}} \left\{\frac{1}{\Delta T}\left[\frac{2-\nu}{8\mu a_1}f\Delta N - \frac{2-\nu}{8\mu c_1}(f\Delta N - \Delta T)\right]\right\}$$

$$= \frac{2-\nu}{8\mu a} \quad \ldots\ldots\ldots [44]$$

Thus, Equation [44] is valid for all values of $dN/dT > 0$.

13 *N* Decreasing, *T* Increasing, at Point on Unloading Curve

1 Repeat step 1 of Section 9.

2 Keeping the normal force constant, increase the tangential load by an amount ΔT. In view of the discussion leading to Equations [10], the increment in traction will be (Fig. 22)

$$\left.\begin{aligned}\Delta_1\tau &= \frac{3fN}{\pi a^3}(a^2-\rho^2)^{1/2}, \quad b_1 \leqslant \rho \leqslant a\\ \Delta_1\tau &= \frac{3fN}{\pi a^3}[(a^2-\rho^2)^{1/2}-(b_1^2-\rho^2)^{1/2}], \quad \rho \leqslant b_1\end{aligned}\right\} \quad \ldots [45]$$

where

$$b_1 = a\left(1-\frac{\Delta T}{2fN}\right)^{1/2}$$

Hence the resultant traction τ_{R0} is given by Equations [41] in which $N + \Delta N$, a_1 and a are replaced by N, a and b_1, respectively.

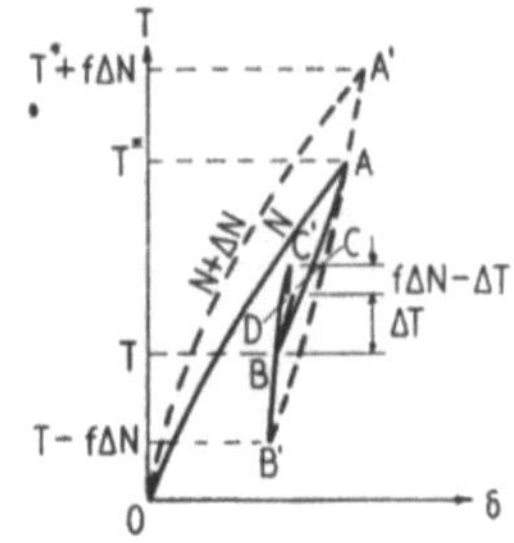

Fig. 21

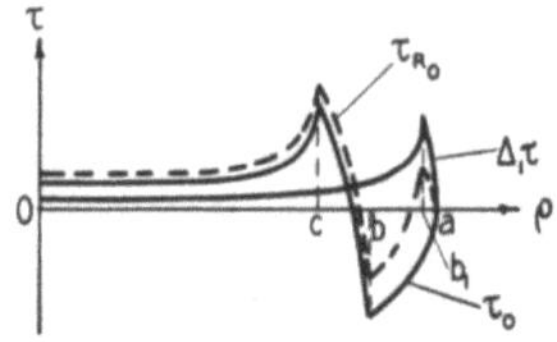

Fig. 22

3 Now, keeping the tangential force constant (at $T + \Delta T$), decrease the normal force by ΔN. The new contact radius is a_1. Several cases arise depending on the relative magnitudes of ΔT and $f\Delta N$. The solution follows the pattern established in Section 8; i.e., (1) freeze the contact area, (2) remove the traction in annulus $a_1 \leqslant \rho \leqslant a$ in such a way that surface $\rho \leqslant a_1$ remains frozen, (3) decrease N by ΔN, (4) unfreeze and permit slip to occur until equilibrium with force $T + \Delta T$ is re-established.

(*a*) $\Delta T \geqslant 2f\Delta N$ ($a_1 \geqslant b_1$). The additional traction necessary to free annulus $a_1 \leqslant \rho \leqslant a$ is $\Delta_2\tau$ (see Fig. 23) given by Equations [36]. Hence the resultant traction (which is not in equilibrium with $T + \Delta T$) becomes

$$\left.\begin{aligned}\tau_{R_n} &= 0, \quad a_1 \leqslant \rho \leqslant a\\ \tau_{R_n} &= \frac{3f(N-\Delta N)}{2\pi a_1^3}(a_1^2-\rho^2)^{1/2}, \quad b_1 \leqslant \rho \leqslant a_1\\ \tau_{R_n} &= \frac{3f(N-\Delta N)}{2\pi a_1^3}[(a_1^2-\rho^2)^{1/2}-2(b_1^2-\rho^2)^{1/2}],\\ &\qquad b \leqslant \rho \leqslant b_1\\ \tau_{R_n} &= \frac{3f(N-\Delta N)}{2\pi a_1^3}[(a_1^2-\rho^2)^{1/2}-2(b_1^2-\rho^2)^{1/2}\\ &\qquad + 2(b^2-\rho^2)^{1/2}], \quad c \leqslant \rho \leqslant b\\ \tau_{R_n} &= \frac{3f(N-\Delta N)}{2\pi a_1^3}[(a_1^2-\rho^2)^{1/2}-2(b_1^2-\rho^2)^{1/2}\\ &\qquad + 2(b^2-\rho^2)^{1/2}-(c^2-\rho^2)^{1/2}], \quad \rho \leqslant c\end{aligned}\right\} \quad \ldots [46]$$

This distribution of traction may be attained by applying (1) $N - \Delta N$, (2) T_1^*, then (3) reducing tangential force to T_1, and

(4) again increasing it by $\overline{\Delta T}$, where the values of these load levels are

$$A— \quad T_1^* = f(N - \Delta N)\left(1 - \frac{c^3}{a_1^3}\right) = T^* - f\Delta N$$

$$B— \quad T_1^* - T_1 = 2f(N - \Delta N)\left(1 - \frac{b^3}{a_1^3}\right) = T^* - T - 2f\Delta N$$

hence

$$T_1 = T + f\Delta N$$

$$C— \quad \overline{\Delta T} = 2f(N - \Delta N)\left(1 - \frac{b^3}{a_1^3}\right) = \Delta T - 2f\Delta N$$

$$\left.\begin{aligned}
\tau_{R_n} &= 0, \quad a_1 \leqslant \rho \leqslant a \\
\tau_{R_n} &= -\frac{3f(N-\Delta N)}{2\pi a_1^3}(a_1^2-\rho^2)^{1/2}, \quad b \leqslant \rho \leqslant a_1 \\
\tau_{R_n} &= -\frac{3f(N-\Delta N)}{2\pi a_1^3}[(a_1^2-\rho^2)^{1/2} - 2(b^2-\rho^2)^{1/2}], \\
&\qquad c \leqslant \rho \leqslant b \\
\tau_{R_n} &= -\frac{3f(N-\Delta N)}{2\pi a_1^3}[(a_1^2-\rho^2)^{1/2} - 2(b^2-\rho^2)^{1/2} \\
&\qquad + (c^2-\rho^2)^{1/2}], \quad \rho \leqslant c
\end{aligned}\right\} \quad [47]$$

Again, the situation is the same as if the order of loading had been: (1) $N - \Delta N$, (2) $T_1^* = T^* - f\Delta N$, then (3) decrease of tangential load to $T_1 = T + f\Delta N$.

If $f\Delta N \leqslant \Delta T < 2f\Delta N$, path O-A-B-C-B-B' in load-displacement plane, Fig. 26, may be replaced by path O-A'-B'; unfreezing

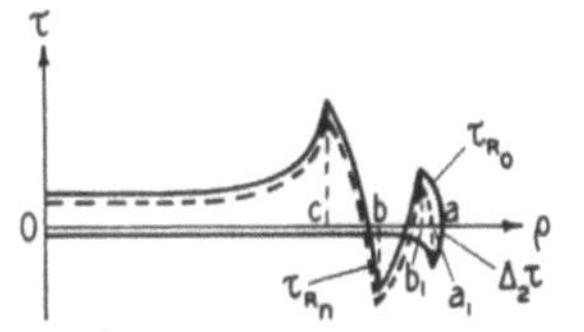

FIG. 23

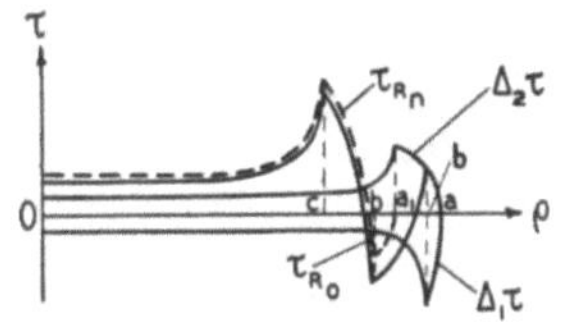

FIG. 25

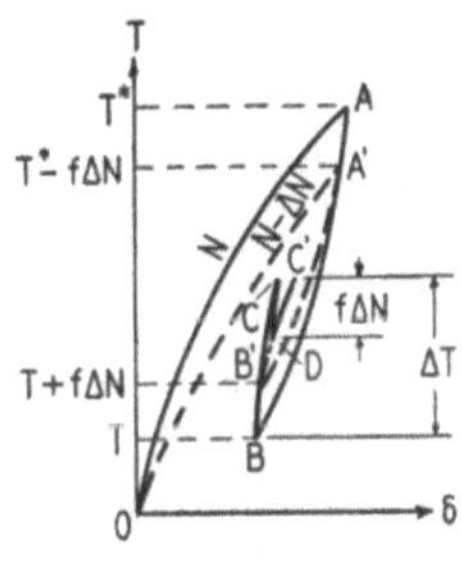

FIG. 24

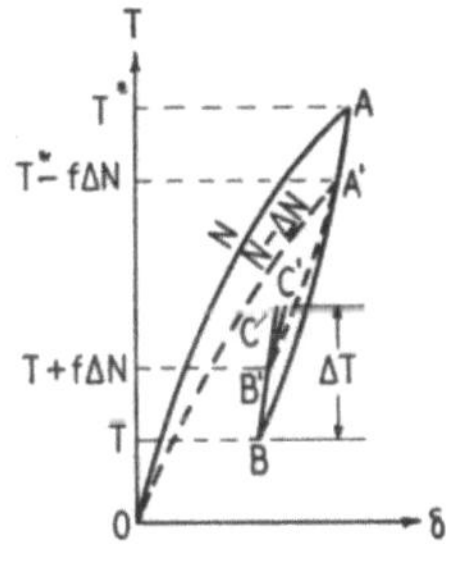

FIG. 26

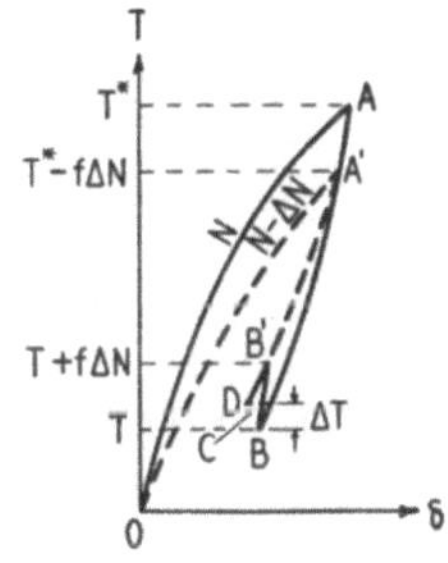

FIG. 27

Thus, in the load-displacement plane, Fig. 24, path O-A-B-C-D may be thought of as replaced by path O-A'-B'-D. Unfreezing will be followed by displacement along DC'.

The resulting compliance is

$$c_t = \lim_{\substack{\Delta N \to 0 \\ \Delta T \to 0}} \left\{ \frac{1}{\Delta T}\left[\frac{2-\nu}{8\mu a}\left(1 - \frac{\Delta T}{2fN}\right)^{-1/3}\Delta T - \frac{2-\nu}{8\mu a}\left(1 - \frac{\Delta T - f\Delta N}{2fN}\right)^{-1/3} f\Delta N + \frac{2-\nu}{8\mu a_1}\left(1 - \frac{f\Delta N}{2f(N-\Delta N)}\right)^{-1/3} f\Delta N\right]\right\} = \frac{2-\nu}{8\mu a}$$

(*b*) $\Delta T < 2f\Delta N$ ($a_1 < b_1$). The addition to the traction in step 1 of $\Delta_1\tau$, Equations [35], and $\Delta_2\tau$, Equations [36], both with signs reversed, and with b_1 replacing b in the expression for $\Delta_1\tau$, will serve to remove the traction in annulus $a_1 \leqslant \rho \leqslant a$. Hence the resultant traction (not in equilibrium with $T + \Delta T$) will be (see Fig. 25)

follows path $B'C'$. The compliance may be seen to be $c_t = (2-\nu)/8\mu a$.

If $\Delta T < f\Delta N$, unfreezing follows path $B'D$, Fig. 27. The compliance is given by

$$c_t = \lim_{\substack{\Delta N \to 0 \\ \Delta T \to 0}} \left\{ \frac{1}{\Delta T}\left[\frac{2-\nu}{8\mu a} f\Delta N - \frac{2-\nu}{8\mu a_1}\left(1 - \frac{T_1^* - T_1}{2f(N-\Delta N)}\right)^{-1/3}(f\Delta N - \Delta T)\right]\right\}$$
$$= \frac{2-\nu}{8\mu a}\left[f\frac{dN}{dT} + \left(1 - f\frac{dN}{dT}\right)\left(1 - \frac{T^*-T}{2fN}\right)^{-1/3}\right], \quad \frac{dN}{dT} \leqslant -\frac{1}{f}$$

and

$$c_t = \frac{2-\nu}{8\mu a}, \quad 0 > \frac{dN}{dT} \geqslant -\frac{1}{f} \quad \ldots\ldots\ldots [48]$$

Here, once again, the absolute value of dN/dT is to be used in the formula.

In examining the cases of Sections 11 and 13, the reader will discover an apparent paradox to the effect that the expressions for the tangential compliances depend on the order of application of ΔN and ΔT. However, an imposition of a large number of infinitesimal increments, followed by a limiting process, leads to the results obtained regardless of the initial order of application.

14 Tangential Tractions Due to Oblique Loading

In this section a calculation is made of the distribution of tangential traction on the contact area of two spheres, pressed together by an initial force N_0 normal to their common tangent plane, and acted upon by a monotonically increasing oblique force. It is assumed that the resultant force remains planar. A study is made of the traction distribution because, by means of it, it will be possible to calculate the relative displacement of the two spheres.

Let a_0 denote the radius of the initial contact circle. An increase, by an amount $\Delta_i N$, of the normal force increases the contact radius to a value a_i; an application, at constant normal force, of an incremental tangential force $\Delta_i T$ gives rise to a tangential traction $\Delta_i \tau$, expressed by Equations [6] in which the appropriate values of N and T have been inserted. The three cases $\Delta_i T \lesseqgtr f\Delta_i N$ are considered separately.

(a) $\Delta_i T < f\Delta_i N$ $(c_i > a_{i-1})$. After n increments, by amounts $\Delta_i N$, $\Delta_i T$ $(i = 1, 2, \ldots, n)$ of the normal and tangential forces, respectively, the resultant traction on circle $\rho < a_0$ will be, from Equations [6] and [1]

$$\tau_n = \frac{3f}{2\pi KR}\sum_{i=1}^{n}[(a_i^2-\rho^2)^{1/2}-(c_i^2-\rho^2)^{1/2}]$$

$$= \frac{3f}{2\pi KR}\sum_{i=1}^{n} a_i(1-\rho^2/a_i^2)^{1/2}\left[1-\left(1-\frac{1-c_i^2/a_i^2}{1-\rho^2/a_i^2}\right)^{1/2}\right] \ldots\ldots [49]$$

At the ith stage the value of the normal load is

$$N_i = N_0 + \sum_{\mu=1}^{i}\Delta_\mu N$$

Now, let

$$\sum_{\mu=1}^{i}\Delta_\mu N = \psi_i(T) \ldots\ldots [50]$$

so that

$$N_i = N_0 + \psi_i(T) \ldots\ldots [51]$$

Hence

$$a_i = [KR(N_0+\psi_i)]^{1/3} \ldots\ldots [52]$$

and

$$c_i = a_i\left[1-\frac{\Delta_i T}{f(N_0+\psi_i)}\right]^{1/3} \ldots\ldots [53]$$

Now, substitution of Equations [52] and [53] in Equation [49], and passage to the limit as $\Delta_i N$ and $\Delta_i T$ approach zero and n becomes infinite, yield

$$\tau = \frac{S^2}{2\pi\rho^2}\int_0^T\left[1-\frac{S^2}{(N_0+\psi)^{2/3}}\right]^{-1/2}(N_0+\psi)^{-2/3}dT \ldots [54]$$

where

$$S = \rho(KR)^{-1/3}$$

To proceed further it is necessary to specify $\psi(T)$, i.e., the relation between N and T. The simplest case is

$$\frac{dT}{dN} = \text{const, say, } \beta \ldots\ldots [55]$$

Using Equations [50] and [55], the integration in Equation [54] can be performed, with the result

$$\tau = \frac{3\beta(N_0+N^*)}{2\pi a_*^3}[(a_*^2-\rho^2)^{1/2}-(a_0^2-\rho^2)^{1/2}],\quad \rho < a_0 \ldots [56]$$

where $N_0 + N^*$ is the final normal load and a_* the corresponding contact radius.

The resultant traction in annulus $a_{r-1} \leqslant \rho \leqslant c_r$ at the ith stage of loading will be

$$\tau_n = \frac{3f}{2\pi KR}\sum_{i=r}^{n}[(a_i^2-\rho^2)^{1/2}-(c_i^2-\rho^2)^{1/2}]$$

$$= \frac{3f}{2\pi KR}\sum_{i=r}^{n} a_i(1-\rho^2/a_i^2)^{1/2}\left[1-\left(1-\frac{1-c_i^2/a_i^2}{1-\rho^2/a_i^2}\right)^{1/2}\right]$$

which may be rewritten

$$\tau_n = \frac{3f}{2\pi KR}a_r(1-\rho^2/a_r^2)^{1/2}\left[1-\left(1-\frac{1-c_r^2/a_r^2}{1-\rho^2/a_r^2}\right)^{1/2}\right]$$

$$+\frac{3f}{2\pi KR}\sum_{i=r+1}^{n} a_i(1-\rho^2/a_i^2)^{1/2}\left[1-\left(1-\frac{1-c_i^2/a_i^2}{1-\rho^2/a_i^2}\right)^{1/2}\right] \ldots\ldots [57]$$

Again, employment of Equations [52] and [53] and passage to the limit as vanishing increments of force are applied infinitely many times, followed by imposition of the condition in Equation [55] in conjunction with Equation [50], yield for the value of the resultant traction in Equation [57]

$$\tau = \frac{3\beta(N_0+N^*)}{2\pi a_*^3}(a_*^2-\rho^2)^{1/2},\quad a_0 \leqslant \rho \leqslant a_* \ldots [58]$$

Equations [56] and [58] constitute the expressions for the resultant traction. It may be noted they are exactly like Equations [6], with β playing the role of the coefficient of friction. Since $\tau < f\sigma$, no slip, in the form heretofore understood, has taken place. Instead, there appears to be an overlapping action, proceeding outward from a_0 to a_*, in the direction of the tangential force.

(b) $\Delta_i T = f\Delta_i N$ $(c_i = a_{i-1})$, i.e., $\beta = f$. The resultant traction in annulus $a_{r-1} \leqslant \rho \leqslant a_r$ is given by

$$\tau_n = \frac{3f}{2\pi KR}\left\{(a_r^2-\rho^2)^{1/2} + \sum_{i=r+1}^{n}[(a_i^2-\rho^2)^{1/2}-(a_{i-1}^2-\rho^2)^{1/2}]\right\}$$

$$= \frac{3f}{2\pi KR}(a_n^2-\rho^2)^{1/2}$$

$$= \frac{3fN_n}{2\pi a_n^3}(a_n^2-\rho^2)^{1/2}$$

Thus

$$\tau = \lim_{\substack{\Delta N,\Delta T\to 0\\ n\to\infty}}\{\tau_n\} = \frac{3f(N_0+N^*)}{2\pi a_*^3}(a_*^2-\rho^2)^{1/2},$$

$$a_0 \leqslant \rho \leqslant a_* \ldots\ldots [59]$$

On circle $\rho \leqslant a_0$

$$\tau_n = \frac{3f}{2\pi KR}\sum_{i=1}^{n}[(a_i^2-\rho^2)^{1/2}-(a_{i-1}^2-\rho^2)^{1/2}]$$

$$= \frac{3fN_n}{2\pi a_n^3}[(a_n^2-\rho^2)^{1/2}-(a_0^2-\rho^2)^{1/2}]$$

so that

$$\tau = \lim_{\substack{\Delta N,\Delta T\to 0\\ n\to\infty}}\{\tau_n\} = \frac{3f(N_0+N^*)}{2\pi a_*^3}[(a_*^2-\rho^2)^{1/2} - (a_0^2-\rho^2)^{1/2}],\quad \rho\leqslant a_0 \ldots\ldots [60]$$

Equations [59] and [60], which are essentially the same as Equations [6], completely specify the distribution of traction in this case.

(c) $\Delta_i T > f\Delta_i N$, i.e., $\beta > f$. The usual division of $\Delta_i T$ into $\Delta_i N$ and $\Delta_i T - f\Delta_i N$ is made and the two parts applied consecutively. After n applications the expression for the resultant traction will be

$$\tau_n = \frac{3f}{2\pi KR}(a_n^2-\rho^2)^{1/2},\quad c_n\leqslant\rho\leqslant a_n$$

$$\tau_n = \frac{3f}{2\pi KR}[(a_n^2-\rho^2)^{1/2}-(c_n^2-\rho^2)^{1/2}],\quad \rho\leqslant c_n$$

where

$$c_n = a_n\left(1-\frac{\sum_{i=1}^{n}\Delta_i T}{fN_n}\right)^{1/3} < a_0$$

Passage to the limit results in

$$\left.\begin{aligned}\tau &= \frac{3f(N_0+N^*)}{2\pi a_*^3}(a_*^2-\rho^2)^{1/2},\quad c_*\leqslant\rho\leqslant a_*\\ \tau &= \frac{3f(N_0+N^*)}{2\pi a_*^3}[(a_*^2-\rho^2)^{1/2}-(c_*^2-\rho^2)^{1/2}],\\ &\qquad\qquad \rho\leqslant c_*\end{aligned}\right\}\ldots[61]$$

where

$$c_* = a_*\left(1-\frac{T^*}{f(N_0+N^*)}\right)^{1/3}$$

15 Tangential Tractions Due to Oblique Unloading

Let it now be required to find the tractions when the loading described in the preceding section is followed by a monotonically decreasing oblique force of constant inclination.

(a) $\beta \lessgtr f$. The traction at the start of unloading is given by Equations [56] and [58]. Keeping the normal load constant, decrease the tangential force by $\Delta_1 T$. Slip will take place in the direction opposite to that of the initial tangential force. By the fundamental rules, the traction over the region on which new slip has occurred will be in the direction of slip and of magnitude $f\sigma$. Let b_1 denote the inner radius of the new slip annulus. Then, in order that no new slip occur in circle $\rho\leqslant b_1$, the additional traction must be the sum, with sign reversed, of the tractions of Equations [56] and [58], in which a_0 is replaced by b_1, and of Equations [6], in which b_1 replaces c. Hence the resultant traction will be

$$\tau_{R0} = -\frac{3f(N_0+N^*)}{2\pi a_*^3}(a_*^2-\rho^2)^{1/2},\quad b_1\leqslant\rho\leqslant a_*$$

$$\tau_{R0} = -\frac{3f(N_0+N^*)}{2\pi a_*^3}\left[(a_*^2-\rho^2)^{1/2} - \left(1+\frac{\beta}{f}\right)(b_1^2-\rho^2)^{1/2}\right],\quad a_0\leqslant\rho\leqslant b_1$$

$$\tau_{R0} = -\frac{3f(N_0+N^*)}{2\pi a_*^3}\left[(a_*^2-\rho^2)^{1/2} - \left(1+\frac{\beta}{f}\right)(b_1^2-\rho^2)^{1/2} + \frac{\beta}{f}(a_0^2-\rho^2)^{1/2}\right],\quad \rho\leqslant a_0$$

The equilibrium condition, Equation [11], yields

$$b_1 = a_*\left(1-\frac{\Delta T}{\left(1+\frac{\beta}{f}\right)f(N_0+N^*)}\right)^{1/3}$$

Reduction of the normal load by $\Delta_1 N$ to N_1, resulting in a contact surface of radius a_1 ($a_1 < b_1$, since $\Delta_1 N > \Delta_1 T/f$), must be preceded by a removal of the traction from annulus $a_1\leqslant\rho\leqslant a$ in such a manner that no slip occur on $\rho\leqslant a_1$ (see, for example, Section 11, 2b). This may be accomplished by: (1) subtracting from the foregoing τ_{R0} the same tractions which, when added to the traction of Equations [56] and [58], yielded τ_{R0}, (2) adding, with sign reversed, tractions of Equations [56] and [58], in which a_0 is replaced by a_1. This will result in a distribution of traction

$$\tau_{R1} = \frac{3\beta N_1}{2\pi a_1^3}(a_1^2-\rho^2)^{1/2},\quad a_0\leqslant\rho\leqslant a_1$$

$$\tau_{R1} = \frac{3\beta N_1}{2\pi a_1^3}[(a_1^2-\rho^2)^{1/2}-(a_0^2-\rho^2)^{1/2}],\quad \rho\leqslant a_1$$

where

$$N_1 = N_0 + N^* - \Delta_1 N$$

and

$$\frac{a_0}{a_1} = \left(1-\frac{T^*-\Delta_1 T}{\beta N_1}\right)^{1/3}$$

which, because it corresponds to a condition of equilibrium (with force $T^* - \Delta_1 T$), remains unaltered by the reduction of normal load and subsequent unfreezing. But this distribution of traction is, qualitatively, the same as at the start of unloading; hence the unloading process may be repeated with similar results.

After the nth reduction the traction will be

$$\tau_{Rn} = \frac{3\beta N_n}{2\pi a_n^3}(a_n^2-\rho^2)^{1/2},\quad a_0\leqslant\rho\leqslant a_n$$

$$\tau_{Rn} = \frac{3\beta N_n}{2\pi a_n^3}[(a_n^2-\rho^2)^{1/2}-(a_0^2-\rho^2)^{1/2}],\quad \rho\leqslant a_0$$

where

$$N_n = N_0 + N^* - \sum_{i=1}^{n}\Delta_i N$$

a_n is the contact radius at normal load N_n and

$$\frac{a_0}{a_n} = \left(1-\frac{T^*-\sum_{i=1}^{n}\Delta_i T}{\beta N_n}\right)^{1/3}$$

Finally

$$
\left.\begin{aligned}
\tau &= \lim_{\substack{\Delta N, \Delta T \to 0 \\ n \to \infty}} \{\tau_{B_n}\} \\
&= \frac{3\beta(N_0 + N)}{2\pi a^3}(a^2 - \rho^2)^{1/2}, \quad a_0 \leqslant \rho \leqslant a \\
\tau &= \frac{3\beta(N_0 + N)}{2\pi a^3}[(a^2 - \rho^2)^{1/2} - (a_0^2 - \rho^2)^{1/2}], \\
&\qquad \rho \leqslant a_0
\end{aligned}\right\} \quad \text{[62]}
$$

where

$$\frac{a_0}{a} = \left(1 - \frac{T}{\beta(N_0 + N)}\right)^{1/2} = \left(\frac{N_0}{N_0 + N}\right)^{1/2}$$

and $N_0 + N$, a, $T(> 0)$ denote final normal load, contact radius, and tangential load, respectively.

It may be observed that Equations [62] have the same form as their counterpart in loading, Equations [56] and [58].

(*b*) $\beta \geqslant f$. The traction at the start of unloading is given by Equations [61]. Reduction, at constant normal load, of the tangential load by $\Delta_1 T$ changes the traction to that given by Equations [10], in which N, a and $T^* - T$ are replaced by $N_0 + N^*$, a_* and $\Delta_1 T$, respectively.

In the process of reducing the normal force it is necessary to distinguish between the two cases $f \leqslant \beta < 2f$ and $\beta \geqslant 2f$; however, both follow closely the pattern of events discussed in Section 11. Thus the resultant equilibrium traction after one cycle of unloading will be

$$\tau_{B_1} = -\frac{3fN_1}{2\pi a_1^3}(a_1^2 - \rho^2)^{1/2}, \quad b_1 \leqslant \rho \leqslant a_1$$

$$\tau_{B_1} = -\frac{3fN_1}{2\pi a_1^3}[(a_1^2 - \rho^2)^{1/2} - 2(b_1^2 - \rho^2)^{1/2}], \quad c \leqslant \rho \leqslant b_1$$

$$\tau_{B_1} = -\frac{3fN_1}{2\pi a_1^3}[(a_1^2 - \rho^2)^{1/2} - 2(b_1^2 - \rho^2)^{1/2} + (c^2 - \rho^2)^{1/2}], \quad \rho \leqslant c$$

where the expressions

$$c = a_1\left(1 - \frac{T^*}{f(N_0 + N^*)}\right)^{1/2}\left(\frac{N_0 + N^*}{N_1}\right)^{1/2}$$

and

$$b_1 = a_1\left(1 + \frac{\Delta_1 N}{2N_1} - \frac{\Delta_1 T}{2fN_1}\right)^{1/2}$$

are obtained by using Equation [11].

After the nth reduction the traction will be given by the foregoing equations in which the subscript 1 is replaced by n and the symbol $\Delta_n \equiv \sum_{i=1}^{n} \Delta_i$. Hence

$$
\left.\begin{aligned}
\tau_B &= \lim_{\substack{\Delta N, \Delta T \to 0 \\ n \to \infty}} \{\tau_{B_n}\} \\
&= -\frac{3f(N_0 + N)}{2\pi a^3}(a^2 - \rho^2)^{1/2}, \quad b \leqslant \rho \leqslant a \\
\tau_B &= -\frac{3f(N_0 + N)}{2\pi a^3}[(a^2 - \rho^2)^{1/2} - 2(b^2 - \rho^2)^{1/2}], \\
&\qquad c_* \leqslant \rho \leqslant b \\
\tau_B &= -\frac{3f(N_0 + N)}{2\pi a^3}[(a^2 - \rho^2)^{1/2} - 2(b^2 - \rho^2)^{1/2} \\
&\qquad + (c_*^2 - \rho^2)^{1/2}], \quad \rho \leqslant c_*
\end{aligned}\right\} \quad \text{[63]}
$$

where $N_0 + N$, a, $T > 0$ are defined in Equations [62]

$$\frac{c_*}{a} = \left(1 - \frac{T^*}{f(N_0 + N^*)}\right)^{1/2}\left(\frac{N_0 + N^*}{N_0 + N}\right)^{1/2}$$

and

$$\frac{b}{a} = \left(1 - \frac{T^* - T}{2f(N_0 + N)} + \frac{N^* - N}{2(N_0 + N)}\right)^{1/2}$$

16 Tangential Tractions Due to Oscillating Oblique Force, $dT/dN \geqslant f$

It is now possible to calculate the distribution of traction on the contact surface of a pair of elastic spheres when the spheres are subjected to an initial normal compression, followed by an oscillating oblique force. This is the same problem as that of Section 5 except that the restriction to constant normal force is now removed. As will be seen, the additional variable introduces the complication that a steady state is not reached until after an additional half cycle.

Again, it is necessary to distinguish between cases $dT/dN \geqslant f$ and $dT/dN \leqslant f$. In this section the discussion is limited to the former.

1 Starting with an initial normal load N_0, apply a monotonically increasing oblique load until the contact surface of the two spheres is acted upon by normal and tangential forces $N_0 + N^*$ and T^*, respectively. It should be noted that $N^* = T^*/\beta$. The distribution of tangential traction is given by Equations [61] in which the expression for c_* may now be rewritten

$$\frac{c_*}{a_*} = \left(1 - \frac{T^*/fN_0}{1 + T^*/\beta N_0}\right)^{1/2} \quad \text{[64]}$$

Since the expression for a_0/a_* may be put in the form

$$\frac{a_0}{a_*} = \left(\frac{N_0}{N_0 + N^*}\right)^{1/2} = \left(\frac{1}{1 + T^*/\beta N_0}\right)^{1/2} = \left(1 - \frac{T^*/\beta N_0}{1 + T^*/\beta N_0}\right)^{1/2} \quad \text{[65]}$$

it is evident from a comparison of Equations [64] and [65] that $c_* \leqslant a_0$.

2 Now unload from $\{N_0 + N^*, T^*\}$. The traction at level $\{N_0 + N, T\}$ will be given by Equations [63] in which the expressions for b/a and c_*/a may be written

$$\frac{b}{a} = \left[1 - \frac{(T^* - T)/2fN_0}{1 + T/\beta N_0} + \frac{(T^* - T)/2\beta N_0}{1 + T/\beta N_0}\right]^{1/2} \quad \text{[66]}$$

and

$$\frac{c_*}{a} = \left[1 - \frac{(T^* - T)/2fN_0}{1 + T/\beta N_0} - \frac{(T^* + T)/2fN_0}{1 + T/\beta N_0} + \frac{(T^* - T)/\beta N_0}{1 + T/\beta N_0}\right]^{1/2} \quad \text{[67]}$$

It is seen from Equations [66] and [67] that $b = c_*$ for that value of T which satisfies the relation

$$\frac{T}{T^*} = \frac{1/\beta - 1/f}{1/\beta + 1/f} = \lambda \quad \text{[68]}$$

where $-1 \leqslant \lambda \leqslant 0$.

At $T = \lambda T^*$ the traction of Equations [63] degenerates to

$$\tau = -\frac{3f(N_0 + N_\lambda)}{2\pi a_\lambda^3}(a_\lambda^2 - \rho^2)^{1/2}, \quad c_* \leqslant \rho \leqslant a_\lambda$$

$$\tau = -\frac{3f(N_0 + N_\lambda)}{2\pi a_\lambda^3}[(a_\lambda^2 - \rho^2)^{1/2} - (c_*^2 - \rho^2)^{1/2}], \quad \rho \leqslant c_* \quad \text{......[69]}$$

where

$$N_\lambda = \frac{T_\lambda}{\beta} = \frac{\lambda T^*}{\beta}$$

and

$$\frac{a_\lambda}{a_0} = \left(1 + \frac{\lambda T^*}{\beta N_0}\right)^{1/3} \leqslant 1$$

Further unloading to $\{N_0 - N^*, -T^*\}$ causes the traction to become

$$\left.\begin{aligned}\tau &= -\frac{3f(N_0 - N^*)}{2\pi a_{-*}^3}(a_{-*}^2 - \rho^2)^{1/2}, \quad c_{-*} \leqslant \rho \leqslant a_{-*} \\ \tau &= -\frac{3f(N_0 - N^*)}{2\pi a_{-*}^3}[(a_{-*}^2 - \rho^2)^{1/2} - (c_{-*}^2 - \rho^2)^{1/2}], \\ &\qquad \rho \leqslant c_{-*}\end{aligned}\right\} \quad \text{..[70]}$$

where a_{-*}, the contact radius corresponding to a normal force $N_0 - N^*$, satisfies the relation

$$\frac{a_{-*}}{a_0} = \left(\frac{N_0 - N^*}{N_0}\right)^{1/3} = \left(1 - \frac{T^*}{\beta N_0}\right)^{1/3} \quad \text{.......[71]}$$

and c_{-*} is given by

$$\frac{c_{-*}}{a_{-*}} = \left(1 - \frac{T^*}{f(N_0 - N^*)}\right)^{1/3} = \left(1 - \frac{T^*/fN_0}{1 - T^*/\beta N_0}\right)^{1/3} \quad \text{..[72]}$$

Comparison of Equation [67] at $T = 0$ and Equation [71] reveals that $c_* \gtrless a_{-*}$, according as $\beta \lessgtr 2f$.

3 Reload; the traction at stage $\{N_0 + N, T\}$ will be

$$\left.\begin{aligned}\tau &= \frac{3f(N_0 + N)}{2\pi a^3}(a^2 - \rho^2)^{1/2}, \quad b_r \leqslant \rho \leqslant a \\ \tau &= \frac{3f(N_0 + N)}{2\pi a^3}[(a^2 - \rho^2)^{1/2} - 2(b_r^2 - \rho^2)^{1/2}], \\ &\qquad c_{-*} \leqslant \rho \leqslant b_r \\ \tau &= \frac{3f(N_0 + N)}{2\pi a^3}[(a^2 - \rho^2)^{1/2} - 2(b_r^2 - \rho^2)^{1/2} \\ &\qquad + (c_{-*}^2 - \rho^2)^{1/2}], \quad \rho \leqslant c_{-*}\end{aligned}\right\} \quad \text{..[73]}$$

where

$$|N| \leqslant N^*, \qquad |T| \leqslant T^*$$

Using the relation

$$\frac{c_{-*}}{a} = \frac{c_{-*}}{a_{-*}}\,\frac{a_{-*}}{a}$$

and the equilibrium condition, Equation [11], one obtains for the value of b_r/a

$$\frac{b_r}{a} = \left[\frac{1}{2}\left(1 + \frac{N_0 - N^*}{N_0 + N}\right) - \frac{T^* + T}{2f(N_0 + N)}\right]^{1/3} \quad \text{.... [74]}$$

At $\{N_0 + N^*, T^*\}$

$$a = a_*$$

$$\begin{aligned}\frac{c_{-*}}{a_*} &= \left(1 - \frac{T^*}{f(N_0 - N^*)}\right)^{1/3}\left(\frac{N_0 - N^*}{N_0 + N^*}\right)^{1/3} \\ &= \left[1 - \frac{(1 + 2f/\beta)T^*/fN_0}{1 + T^*/\beta N_0}\right]^{1/3} \quad \text{..............[75]}\end{aligned}$$

and

$$\begin{aligned}\frac{b_*}{a_*} &= \left(\frac{N_0}{N_0 + N^*} - \frac{T^*}{f(N_0 + N^*)}\right)^{1/3} \\ &= \left[1 - \frac{(1 + f/\beta)T^*/fN_0}{1 + T^*/\beta N_0}\right]^{1/3} \quad \text{...........[76]}\end{aligned}$$

Hence, comparison of Equations [64], [75], and [76] shows that

$$c_* > b_* > c_{-*}$$

4 Unload once again; the traction at load level $\{N_0 + N, T\}$ will be

$$\left.\begin{aligned}\tau &= -\frac{3f(N_0 + N)}{2\pi a^3}(a^2 - \rho^2)^{1/2}, \quad b_u \leqslant \rho \leqslant a \\ \tau &= -\frac{3f(N_0 + N)}{2\pi a^3}[(a^2 - \rho^2)^{1/2} - 2(b_u^2 - \rho^2)^{1/2}], \\ &\qquad b_* \leqslant \rho \leqslant b_u \\ \tau &= -\frac{3f(N_0 + N)}{2\pi a^3}[(a^2 - \rho^2)^{1/2} - 2(b_u^2 - \rho^2)^{1/2} \\ &\qquad + 2(b_*^2 - \rho^2)^{1/2}], \quad c_{-*} \leqslant \rho \leqslant b_* \\ \tau &= -\frac{3f(N_0 + N)}{2\pi a^3}[(a^2 - \rho^2)^{1/2} - 2(b_u^2 - \rho^2)^{1/2} \\ &\qquad + 2(b_*^2 - \rho^2)^{1/2} - (c_{-*}^2 - \rho^2)^{1/2}], \quad \rho \leqslant c_{-*}\end{aligned}\right\} \quad \text{..[77]}$$

In order for equilibrium with the force T to obtain, the relation

$$\frac{b_u}{a} = \left[\frac{1}{2}\left(1 + \frac{N_0 + N^*}{N_0 + N}\right) - \frac{T^* - T}{2f(N_0 + N)}\right]^{1/3} \quad \text{....[78]}$$

must hold (see Equation [11]).

When $T = -T^*$, the contact radius is once again a_{-*}, and $b_u = b_{-*}$. Furthermore

$$\frac{b_{-*}}{a_{-*}} = \left[\frac{N_0}{N_0 - N^*} - \frac{T^*}{f(N_0 - N^*)}\right]^{1/3} = \frac{b_*}{a_{-*}}$$

so that $b_{-*} = b_*$. Hence the distribution of traction, Equations [77], is the same as that obtained on first unloading to $\{N_0 - N^*, -T^*\}$ (see Equations [70]). Henceforth the cycle may be repeated with identical results, i.e., a steady state is reached after one and three-quarter cycles.

17 Load-Displacement Relation for Oscillating Oblique Force, $dT/dN \geqslant f$

The load-displacement relations corresponding to the tractions of the preceding section may be obtained by superposition of equations in the generic form of Equation [7]. By the aid of Equation [1] and the relation $E = 2(1 + \nu)\mu$, the first factor of the right-hand side of Equation [7] may be transformed to

$$\frac{3f(2 - \nu)N}{16\mu a} = \frac{f(2 - \nu)}{2(1 - \nu)}\,\frac{a^2}{R} \quad \text{............[79]}$$

Hence, during the first loading from $\{N_0, 0\}$ to $\{N_0 + N^*, T^*\}$ the displacement will be given by (see curve OP, Fig. 28)

$$\frac{2(1-\nu)}{f(2-\nu)}\frac{R\delta_l}{a_0^2} = \frac{a^2}{a_0^2}\left(1-\frac{c^2}{a^2}\right)$$

$$= (1+\theta L)^{2/3} - [1-(1-\theta)L]^{2/3}, \quad 0 \leqslant L \leqslant L^* \ldots\ldots [80]$$

where

$$L = \frac{T}{fN_0}, \qquad L^* = \frac{T^*}{fN_0}, \qquad \theta = \frac{f}{\beta}$$

and the subscript on δ_l stands for "loading." The range of θ is $0 \leqslant \theta \leqslant 1$.

From Equation [80], and the fact that

$$a = a_0(1+\theta L)^{1/3} \ldots\ldots\ldots\ldots\ldots\ldots [81]$$

the expression for the tangential compliance in this load interval is obtained. Thus

$$c_l = \frac{d\delta_l}{dT} = \frac{2-\nu}{8\mu a}\left[\theta + (1-\theta)\left(1-\frac{L}{1+\theta L}\right)^{-1/3}\right] \ldots [82]$$

If, now, Equation [55] and the definition of θ (see Equation [80]) are substituted into Equation [25] for the compliance due to the corresponding incremental process, the resulting relation becomes identical with Equation [82]. This result is to be expected, since the tractions in both cases are identical.

During the unloading process following initial loading, the load-displacement relation takes two different forms, according as $L \gtrless \lambda L^*$. With the aid of the expressions defined in Equations [66] and [67] one obtains (curves P-R-Y and YS, Fig. 28)

$$\frac{2(1-\nu)}{f(2-\nu)}\frac{R\delta_{u11}}{a_0^2} = \frac{a^2}{a_0^2}\left(2\frac{b^2}{a^2} - \frac{c_*^2}{a^2} - 1\right)$$

$$= 2\left[1+\frac{\theta}{2}(L^*+L) - \frac{1}{2}(L^*-L)\right]^{2/3}$$

$$- [1-(1-\theta)L^*]^{2/3} - (1+\theta L)^{2/3},$$

$$L^* \geqslant L \geqslant \lambda L^* \ldots . [83]$$

where λ is defined in Equation [68] and

$$\frac{2(1-\nu)}{f(2-\nu)}\frac{R\delta_{u12}}{a_0^2} = -(1+\theta L)^{2/3}$$

$$+ [1+(1+\theta)L]^{2/3}, \quad \lambda L^* \geqslant L \geqslant -L^* \ldots\ldots [84]$$

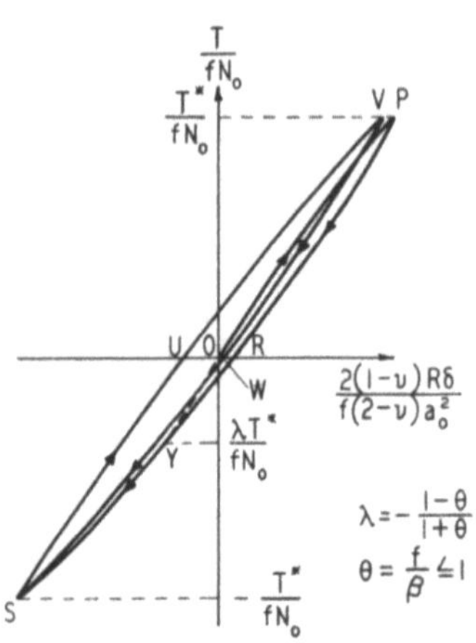

FIG. 28

The subscripts on δ_{u11} and δ_{u12} stand for "first portion of first unloading" and "second portion of first unloading," respectively. At point Y (Fig. 28), curves P-R-Y and YS have a common tangent.

In the load interval from $\{N_0 + N^*, T^*\}$ to $\{N_0 + N_\lambda, \lambda T^*\}$ the compliance is

$$c_{u11} = \frac{d\delta_{u11}}{dT}$$

$$= \frac{2-\nu}{8\mu a}\left\{-\theta + (1+\theta)\left[1-(1-\theta)\frac{L^*-L}{2(1+\theta L)}\right]^{-1/3}\right\}$$

$$\ldots\ldots [85]$$

It will be recalled that the compliance obtained by the corresponding incremental process is given by Equation [34] which, however, does not yield Equation [85] when fdN/dT is replaced by θ. This is because the respective traction distributions, although qualitatively alike, differ quantitatively; that is, in both cases

$$\left.\begin{aligned}
\tau &= -\frac{3f\mathfrak{N}}{2\pi a^3}(a^2-\rho^2)^{1/2}, \quad b \leqslant \rho \leqslant a \\
\tau &= -\frac{3f\mathfrak{N}}{2\pi a^3}[(a^2-\rho^2)^{1/2} - 2(b^2-\rho^2)^{1/2}], \\
&\qquad\qquad c_* \leqslant \rho \leqslant b \\
\tau &= -\frac{3f\mathfrak{N}}{2\pi a^3}[(a^2-\rho^2)^{1/2} - 2(b^2-\rho^2)^{1/2} \\
&\qquad + (c_*^2-\rho^2)^{1/2}], \quad \rho \leqslant c_*
\end{aligned}\right\} \ldots [86]$$

where $\mathfrak{N}$ denotes the total normal load. However, while in the "incremental" case

$c_*/a = (1 - T^*/f\mathfrak{N})^{1/3}$ and $b/a = [1-(T^*-T)/2f\mathfrak{N}]^{1/3}$

in the "continuous" case

$$\frac{c_*}{a} = \left(1-\frac{T^*}{f(N_0+N^*)}\right)^{1/3}\left(\frac{N_0+N^*}{N_0+N}\right)^{1/3}$$

$$= \left(1-\frac{T^*}{f\mathfrak{N}} + \frac{N^*-N}{\mathfrak{N}}\right)^{1/3}$$

and

$$\frac{b}{a} = \left(1-\frac{T^*-T}{2f\mathfrak{N}} + \frac{N^*-N}{\mathfrak{N}}\right)^{1/3}$$

where $|N| \leqslant N^*$.

In the load interval from $\{N_0 + N_\lambda, \lambda T^*\}$ to $\{N_0 - N^*, -T^*\}$ the compliance is

$$c_{u12} = \frac{d\delta_{u12}}{dT} = \frac{2-\nu}{8\mu a}\left[-\theta + (1+\theta)\left(1+\frac{1}{1+\theta L}\right)^{-1/3}\right]$$

$$\ldots\ldots [87]$$

The constant λ, incidentally, expressed in terms of θ, becomes

$$\lambda = -\frac{1-\theta}{1+\theta} \leqslant 0 \ldots\ldots\ldots\ldots\ldots [88]$$

The displacement during reloading from $\{N_0 - N^*, -T^*\}$ to $\{N_0 + N^*, T^*\}$, obtained by aid of Equations [72] and [74], is (curve S-U-V, Fig. 28)

$$\frac{2(1-\nu)}{f(2-\nu)}\frac{R\delta_r}{a_0^2} = \frac{a^2}{a_0^2}\left(1+\frac{c_*^2}{a^2} - 2\frac{b_r^2}{a^2}\right)$$

$$= (1+\theta L)^{2/3} + [1-(1+\theta)L^*]^{2/3}$$

$$-2\left[1-\frac{\theta}{2}(L^*-L) - \frac{1}{2}(L^*+L)\right]^{2/3}, \quad -L^* \leqslant L \leqslant L^*$$

$$\ldots . [89]$$

while the displacement during the second unloading process becomes, with reference to Equations [75], [76], and [78] (curve *V-W-S*, Fig. 28)

$$\frac{2(1-\nu)}{f(2-\nu)}\frac{R\delta_{u2}}{a_0^2} = \frac{a^2}{a_0^2}\left(2\frac{b_u^2}{a^2} - 2\frac{b_*^2}{a^2} + \frac{c_{-*}^2}{a^2} - 1\right)$$

$$= 2\left[1 + \frac{\theta}{2}(L^* + L) - \frac{1}{2}(L^* - L)\right]^{2/3} - 2(1 - L^*)^{2/3}$$

$$+ [1 - (1+\theta)L^*]^{2/3} - (1 + \theta L)^{2/3}, \quad L^* \geqslant L \geqslant -L^* \qquad \text{[90]}$$

The subscripts on δ_r and δ_{u2} stand for "reloading" and "second unloading," respectively.

Further repetition of the loading cycle will, of course, yield the displacements of Equations [89] and [90]. These displacements have the same meaning for the case of an oblique oscillating force as displacements of Equations [14] and [16] have for the case of pure tangential oscillating force; the two pairs of equations become identical when $\theta = 0$, so that the loop *S-U-V-W-S*, Fig. 28, degenerates to *S-U-P-R-S*, Fig. 5.

It should be noted that the character of the load-displacement curves, Fig. 28, varies with θ. Thus, the curvature (with respect to the abscissa) of *OP* at *O* is negative, as in Fig. 30, for values of θ in the interval $1/2 < \theta \leqslant 1$ and becomes positive, as shown in Fig. 28, for $\theta < 1/2$. The curvature of *S-U-V* at *S* is positive, as shown in Fig. 28, for values of $\theta < \sqrt{2} - 1$.

For the sake of completeness, the tangential compliances are given for the load intervals corresponding to which the displacements are δ_r and δ_{u2}, respectively. Thus

$$C_r = \frac{d\delta_r}{dT}$$

$$= \frac{2-\nu}{8\mu a}\left\{\theta + (1-\theta)\left[1 - (1+\theta)\frac{L^* + L}{2(1+\theta L)}\right]^{-1/3}\right\} \qquad \text{[91]}$$

and

$$C_{u2} = \frac{d\delta_{u2}}{dT}$$

$$= \frac{2-\nu}{8\mu a}\left\{-\theta + (1+\theta)\left[1 - (1-\theta)\frac{L^* - L}{2(1+\theta L)}\right]^{-1/3}\right\} \qquad \text{[92]}$$

In all equations of this section there exists a limitation on the value of T^*, namely

$$0 \geqslant \frac{-T^*}{f(N_0 - N^*)} \geqslant -1$$

which may be put in the form

$$0 \leqslant L^* \leqslant \frac{1}{1+\theta} \qquad \text{[93]}$$

18 Oscillating Oblique Force, $dT/dN \leqslant f$

1 Repeat step 1 of Section 16; the traction is given by Equations [56] and [58]. It will be recalled (see Section 14) that no slip occurs during loading.

2 During unloading from $\{N_0 + N^*, T^*\}$ to $\{N_0, 0\}$ the traction at any level of loading is given by Equations [62], and vanishes at $T = 0$. Again, no slip has taken place.

Further unloading constitutes, in effect, the situation treated in Section 8, except for change in sign of the tangential force and, hence, of the tangential traction. Consequently, the present process yields a distribution of traction expressed by Equations [6] with sign reversed, in which, of course, N stands for the total normal load. At $\{N_0 - N^*, -T^*\}$, $a = a_{-*}$ and $c = c_{-*}$. The expression for the resultant traction is given by Equations [70]; for c_{-*}/a_{-*}, by Equation [72].

It may be noted that, since, in the load interval $0 > T \geqslant -T^*$, the absolute value of the traction in annulus $c_{-*} \leqslant \rho \leqslant a_{-*}$ is $f\sigma$, slip does take place during this portion of the loading cycle.

3 During reloading to $\{N_0 + N^*, T^*\}$, the additional traction to be superposed on the traction of Equations [70] is obtained by the procedure of Section 14, leading, at the terminal load, to Equations [56] and [58] with a_0 replaced by a_{-*}. The distribution of the initial and the additional tractions is illus-

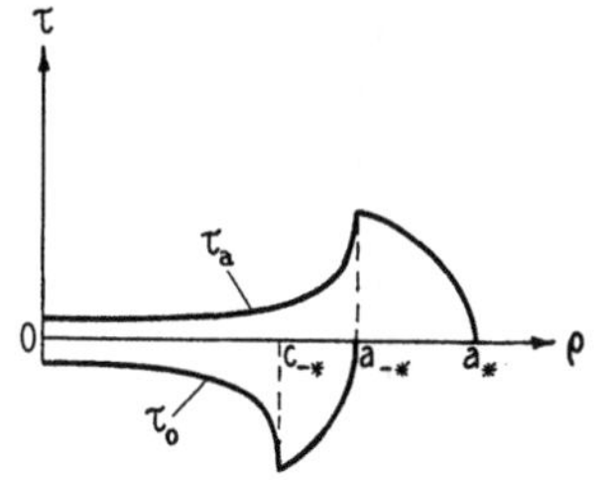

Fig. 29

trated by curves τ_0 and τ_a, respectively, in Fig. 29. Their algebraic sum yields the following resultant traction

$$\left.\begin{aligned}
\tau &= \frac{3\beta(N_0 + N^*)}{2\pi a_*^3}(a_*^2 - \rho^2)^{1/2}, \quad a_{-*} \leqslant \rho \leqslant a_* \\
\tau &= \frac{3\beta(N_0 + N^*)}{2\pi a_*^3}\left[(a_*^2 - \rho^2)^{1/2} - \left(1 + \frac{f}{\beta}\right)(a_{-*}^2 - \rho^2)^{1/2}\right], \quad c_{-*} \leqslant \rho \leqslant a_{-*} \\
\tau &= \frac{3\beta(N_0 + N^*)}{2\pi a_*^3}\left[(a_*^2 - \rho^2)^{1/2} - \left(1 + \frac{f}{\beta}\right)(a_{-*}^2 - \rho^2)^{1/2} + \frac{f}{\beta}(c_{-*}^2 - \rho^2)^{1/2}\right], \quad \rho \leqslant c_{-*}
\end{aligned}\right\} \qquad \text{[94]}$$

Here

$$\frac{a_{-*}}{a_*} = \left(\frac{N_0 - N^*}{N_0 + N^*}\right)^{1/3} = \left(1 - \frac{2T^*/\beta N_0}{1 + T^*/\beta N_0}\right)^{1/3} \qquad \text{[95]}$$

and

$$\frac{c_{-*}}{a_*} = \frac{c_{-*}}{a_{-*}}\frac{a_{-*}}{a_*} = \left(1 - \frac{(2 + \beta/f)T^*/\beta N_0}{1 + T^*/\beta N_0}\right)^{1/3} \qquad \text{[96]}$$

4 In part 1 of Section 15 it was shown that, for the loading under consideration, the distribution of traction during unloading is the same as that during loading. Hence, during unloading from $\{N_0 + N^*, T^*\}$ to $\{N_0 - N^*, -T^*\}$ the additional traction (τ_a in Fig. 29), discussed in part 3 of this section, is gradually removed, until, at $\{N_0 - N^*, -T^*\}$, the remaining traction is, once again, the same as at the end of the first unloading (see part 2 of this section and τ_0 in Fig. 29). A stable cycle is thus established, during which no slip takes place (for further discussion see Section 19).

The load-displacement relations are obtained in a manner

similar to those in Section 17. Thus, during initial loading and unloading

$$\frac{2(1-\nu)}{f(2-\nu)}\frac{R\delta}{a_0^2} = \frac{1}{\theta}\left[(1+\theta L)^{2/3} - 1\right], \quad L^* \geqslant L \geqslant 0 \quad\text{..[97]}$$

$$\frac{2(1-\nu)}{f(2-\nu)}\frac{R\delta}{a_0^2} = -(1+\theta L)^{2/3} + [1+(1+\theta)L]^{2/3}, \quad 0 \geqslant L \geqslant -L^* \quad\text{..[98]}$$

The relations of Equation [97] are illustrated by curves *OP* and *PO* in Fig. 30; of Equation [98], by curve *OS* in the same figure.

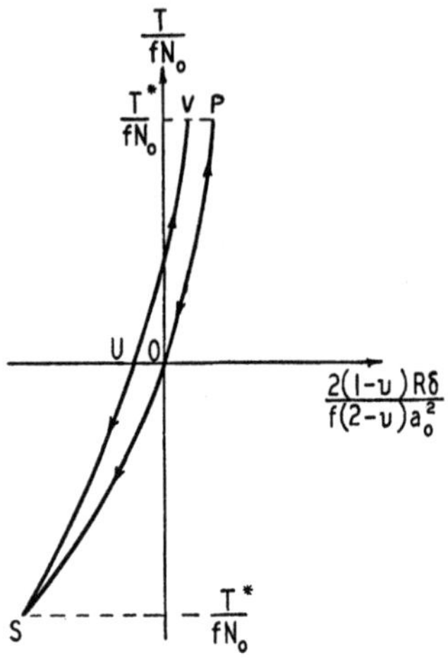

Fig. 30

The tangential compliance corresponding to the cycle in Equation [97] is

$$c_t = \frac{2-\nu}{8\mu a} \quad\text{....................[99]}$$

that corresponding to the load interval of Equation [98]

$$c_t = \frac{2-\nu}{8\mu a}\left[-\theta + (1+\theta)\left(1+\frac{L}{1+\theta L}\right)^{-1/3}\right] \quad\text{.[100]}$$

During each subsequent loading and unloading the displacement will be given by (curves *S-U-V* and *V-U-S*, Fig. 30)

$$\frac{2(1-\nu)}{f(2-\nu)}\frac{R\delta}{a_0^2} = \frac{1}{\theta}(1+\theta L)^{2/3} - \left(1+\frac{1}{\theta}\right)(1-\theta L^*)^{2/3} + [1-(1+\theta)L^*]^{2/3}, \quad |L| \leqslant L^* \quad\text{..[101]}$$

The compliance during the stable load cycle $|L| \leqslant L^*$ is the same as in Equation [99].

In Equations [97], [98] and [101] the range of θ is $1 \leqslant \theta \leqslant \infty$.

19 Frictional Energy Loss per Cycle

As in Section 5, the frictional energy loss per cycle is given by the area of the load-displacement loop (*S-U-V-W-S*, Fig. 28)

$$F = \int_{-T^*}^{T^*} (\delta_{u2} - \delta_r)\, dT$$
$$= \frac{9(2-\nu)(fN_0)^2}{10\mu a_0}\left\{\frac{1}{4\theta}\left[\frac{1+\theta}{1-\theta}(1-\theta L^*)^{5/3} - \frac{1-\theta}{1+\theta}(1+\theta L^*)^{5/3}\right] - \frac{1}{1-\theta^2}\left(1-\frac{1+5\theta^2}{6}L^*\right)(1-L^*)^{2/3}\right\} \quad\text{....[102]}$$

in which $0 \leqslant \theta \leqslant 1$, i.e., $\beta \geqslant f$. The expressions for δ_{u2} and δ_r were obtained from Equations [90] and [89], respectively.

For $\theta = 0$, i.e., N = const, Equation [102] reduces to Equation [17].

For small values of L^*, series expansion of Equation [102] yields

$$F = \frac{(2-\nu)T^{*3}}{36\,\mu a_0 f N_0}(1-\theta^2) \quad\text{............[103]}$$

which expression should be compared with Equation [18].

It should be noted that, in the case for which $1 \leqslant \theta \leqslant \infty$, i.e., $\beta \leqslant f$, the displacement retraces its path in loading and unloading (see Equation [101], and curves *S-U-V* and *V-U-S*, Fig. 30), so that no loop is formed; i.e., there is no frictional energy loss involved. This result is to be expected since, by one of the fundamental rules, slip does not occur if $\tau < f\sigma$ (see discussion following Equation [58]). It is interesting that, for $\beta = f$, i.e., $\theta = 1$, the energy, Equation [102], vanishes, indicating that dT/dN must exceed the coefficient of friction if energy is to be dissipated.

Force at a Point in the Interior of a Semi-Infinite Solid

R. D. MINDLIN, Columbia University, New York, N.Y.

Introduction

In a paper[1] under the same title, the solution of the linear equations of equilibrium of an elastic body was given for the case of a force acting at a point within an isotropic body bounded by a plane. The result was obtained by starting with Kelvin's solution for a force in an infinite body and guessing the nuclei of strain to add outside of the semi-infinite body so as to annul the tractions on the plane boundary. In the present paper it is shown how these results may be obtained, directly, by means of an application of potential theory.

Papkovitch Functions

In an isotropic elastic body in equilibrium, the displacement $\mathbf{u}$ is governed by the equation[2]

$$\mu \nabla^2 \mathbf{u} + \frac{\mu}{1-2\nu} \nabla\nabla\cdot\mathbf{u} + \mathbf{F} = 0 \tag{1}$$

where μ is the shear modulus, ν is Poisson's ratio and $\mathbf{F}$ is the body force per unit of volume. The stress, $\boldsymbol{\sigma}$, is related to the displacement by

$$\boldsymbol{\sigma} = \lambda \nabla\cdot\mathbf{u}\mathbf{I} + \mu(\nabla\mathbf{u} + \mathbf{u}\nabla) \tag{2}$$

where $\lambda = 2\mu\nu/(1-2\nu)$.

By Helmholtz's theorem[3], $\mathbf{u}$ may be resolved into lamellar and solenoidal components:

$$\mathbf{u} = \nabla\varphi + \nabla\times\mathbf{H}, \quad \nabla\cdot\mathbf{H} = 0, \tag{3}$$

so that (1) may be written

$$\mu \nabla^2(\alpha\nabla\varphi + \nabla\times\mathbf{H}) + \mathbf{F} = 0 \tag{4}$$

where $\alpha = 2(1-\nu)/(1-2\nu)$.

The quantity in parenthesis in (4) is a vector, say, $\mathbf{B} = \mathbf{i}B_x + \mathbf{j}B_y + \mathbf{k}B_z$ i.e.,

$$\alpha\nabla\varphi + \nabla\times\mathbf{H} = \mathbf{B}, \tag{5}$$

$$\mu \nabla^2 \mathbf{B} = -\mathbf{F} \tag{6}$$

Operating on (5) with $\nabla\cdot$, we find

$$\alpha \nabla^2\varphi = \nabla\cdot\mathbf{B} \tag{7}$$

the complete solution of which is

$$2\alpha\varphi = \mathbf{r}\cdot\mathbf{B} + \beta, \tag{8}$$

where β is a scalar function, which satisfies

$$\mu \nabla^2\beta = \mathbf{r}\cdot\mathbf{F}, \tag{9}$$

and $\mathbf{r} = \mathbf{i}x + \mathbf{j}y + \mathbf{k}z$ is the position vector.

Substituting (8) in (3) and eliminating $\nabla\times\mathbf{H}$ by means of (5), there results

$$\mathbf{u} = \mathbf{B} - \frac{1}{4(1-\nu)} \nabla(\mathbf{r}\cdot\mathbf{B} + \beta) \tag{10}$$

$$\mu \nabla^2 \mathbf{B} = -\mathbf{F} \tag{11}$$

$$\mu \nabla^2\beta = \mathbf{r}\cdot\mathbf{F} \tag{12}$$

Thus the displacement is expressed in terms of the Papkovitch functions, $\mathbf{B}$ and β, whose Laplacians are known if the body force $\mathbf{F}$ is known.

The proof of completeness of the Papkovitch functions, given above, is an extension, to include the body force, of one given in a previous paper[4], where $\mathbf{B}$ and β were called Papkovitch[5] functions after the originator of the solution (10) of the elasticity equations. Recent writers associate these functions with the name of Boussinesq[6], who introduced B_z and β, but employed functions of a different type where B_x and B_y could have been used.

Green's Formula

The value of a function V, at any point in a region, may be expressed in terms of its values at the boundary, its Laplacian and Green's function, G, for the region, by means of Green's formula[7]

$$-4\pi V = \int V\mathbf{n}\cdot\nabla G \, dS + \int G\nabla^2 V \, dv. \tag{13}$$

For the region $z \geqslant 0$, Green's function is

$$G = r_1^{-1} - r_2^{-1} \tag{14}$$

where

$$r_1^2 = (x-\xi)^2 + (y-\eta)^2 + (z-\zeta)^2 \tag{15}$$

$$r_2^2 = (x-\xi)^2 + (y-\eta)^2 + (z+\zeta)^2 \tag{16}$$

in which (x, y, z) are the coordinates of a point $P(x, y, z)$ in the region and (ξ, η, ζ), $(\xi, \eta, -\zeta)$ are the coordinates of a source point $Q(\xi, \eta, \zeta)$ and its image $Q'(\xi, \eta, -\zeta)$ respectively.

---

[1] R. D. Mindlin, Physics, Vol.7 (1936), pp. 195-202.

[2] For the vector notation used in this paper, see C. E. Weatherburn, Advanced Vector Analysis, G.Bell nnd Sons, Ltd. , London, 1928.

[3] Weatherburn, p.44.

[4] R. D. Mindlin, Bull. Am. Math. Soc., Vol. 42 (1936), pp. 373-376.

[5] P. F. Papkovitch, Comptes Rendus, Acad. des Sciences, Paris, Vol. 195 (1922), pp 513-515 and 754 - 756.

[6] J. Boussinesq, "Application des Potentiels à l'Étude de l'Équilibre et du Mouvement des Solides Élastiques," Gauthier-Villars, Paris, 1885, pp. 63 and 72.

[7] Weatherburn, p. 34.

Force at a Point

Kelvin's definition of a force at a point takes the following form, in the present case. Consider a distribution of body forces $\mathbf{F}$ in a closed region T within $z>0$, with $\mathbf{F}=0$ outside T but within $z \geqslant 0$. Diminish T indefinitely, always enclosing the point $C(0,0,c)$, but let

$$\lim_{T\to 0}\int_T \mathbf{F}\, dv = \mathbf{P} \tag{17}$$

where $\mathbf{P}$ is a constant force at C.

For later use, we note that the limit, as T approaches zero, of

$$Q(\xi,\eta,\zeta) = C(0,0,c), \tag{18}$$

$$Q'(\xi,\eta,-\zeta) = C'(0,0,-c), \tag{19}$$

$$r_1^2 = x^2+y^2+(z-c)^2 = R_1^2, \tag{20}$$

$$r_2^2 = x^2+y^2+(z+c)^2 = R_2^2, \tag{21}$$

$$G = \frac{1}{R_1} - \frac{1}{R_2}, \tag{22}$$

$$\frac{\partial G}{\partial \zeta} = -\frac{\partial}{\partial z}\left(\frac{1}{R_1}+\frac{1}{R_2}\right), \tag{23}$$

$$\frac{\partial^2 G}{\partial \zeta^2} = \frac{\partial^2}{\partial z^2}\left(\frac{1}{R_1}-\frac{1}{R_2}\right), \tag{24}$$

$$\frac{\partial G}{\partial \xi} = -\frac{\partial}{\partial x}\left(\frac{1}{R_1}-\frac{1}{R_2}\right). \tag{25}$$

Force Normal to the Plane Boundary

In this case we take $F_x = F_y = 0$ and $B_x = B_y = 0$. The remaining Papkovitch (in this case Boussinesq) functions, B_z and β, must satisfy the condition of vanishing traction on $z=0$. Thus, we have, from (2) and (10), on $z=0$,

$$\sigma_{zz} = \frac{\mu}{2(1-\nu)}\left[2(1-\nu)\frac{\partial B_z}{\partial z} - \frac{\partial^2\beta}{\partial z^2}\right] = 0, \tag{26}$$

$$\sigma_{zx} = \frac{\mu}{2(1-\nu)}\left[(1-2\nu)\frac{\partial B_z}{\partial x} - \frac{\partial^2\beta}{\partial x\,\partial z}\right] = 0, \tag{27}$$

$$\sigma_{zy} = \frac{\mu}{2(1-\nu)}\left[(1-2\nu)\frac{\partial B_z}{\partial y} - \frac{\partial^2\beta}{\partial y\,\partial z}\right] = 0. \tag{28}$$

The function in brackets in (26) is one whose Laplacian

$$\nabla^2\left[2(1-\nu)\frac{\partial B_z}{\partial z} - \frac{\partial^2\beta}{\partial z^2}\right] = \frac{-2(1-\nu)}{\mu}\frac{\partial F_z}{\partial z} - \frac{1}{\mu}\frac{\partial^2 (zF_z)}{\partial z^2} \tag{29}$$

is known throughout $z \geqslant 0$ and whose boundary value is zero. Hence, by (13),

$$2(1-\nu)\frac{\partial B_z}{\partial z} - \frac{\partial^2\beta}{\partial z^2} = \frac{1}{4\pi\mu}\int G\left[2(1-\nu)\frac{\partial F_z}{\partial \zeta} + \frac{\partial^2(\zeta F_z)}{\partial \zeta^2}\right] dv. \tag{30}$$

Now, integrating the first term in the volume integral by parts,

$$\iiint G\frac{\partial F_z}{\partial \zeta}\, d\xi\, d\eta\, d\zeta = \iint G F_z\, d\xi\, d\eta - \iiint F_z \frac{\partial G}{\partial \zeta}\, d\xi\, d\eta\, d\zeta. \tag{31}$$

The surface integral in (31) vanishes because $F_z = G = 0$ on the boundary of the body. Then, by (17) and (23),

$$\lim_{T\to 0}\int G\frac{\partial F_z}{\partial \zeta}\, dv = P_z\frac{\partial}{\partial z}\left(\frac{1}{R_1}+\frac{1}{R_2}\right). \tag{32}$$

Similarly, integrating by parts twice and using (17), (18) and (24),

$$\lim_{T\to 0}\int G\frac{\partial^2 F_z}{\partial \zeta^2}\, dv = cP_z\frac{\partial^2}{\partial z^2}\left(\frac{1}{R_1}-\frac{1}{R_2}\right). \tag{33}$$

Hence

$$2(1-\nu)B_z - \frac{\partial\beta}{\partial z} = \frac{P_z}{4\pi\mu}\left[2(1-\nu)\left(\frac{1}{R_1}+\frac{1}{R_2}\right) + c\frac{\partial}{\partial z}\left(\frac{1}{R_1}-\frac{1}{R_2}\right)\right], \tag{34}$$

where one integration with respect to z has been performed. (The arbitrary function of x and y, thereby introduced, must vanish since $2(1-\nu)B_z - \partial\beta/\partial z$ must vanish as $z\to\infty$).

Returning to the boundary conditions, we note that (27) and (28) can be integrated with respect to x and y, respectively, so that, on $z=0$,

$$(1-2\nu)B_z - \frac{\partial\beta}{\partial z} = 0. \tag{35}$$

The Laplacian

$$\nabla^2\left[(1-2\nu)B_z - \frac{\partial\beta}{\partial z}\right] = \frac{-(1-2\nu)}{\mu}F_z - \frac{1}{\mu}\frac{\partial(zF_z)}{\partial z} \tag{36}$$

is known throughout $z \geqslant 0$ and hence, by (13),

$$(1-2\nu)B_z - \frac{\partial\beta}{\partial z} = \frac{1}{4\pi\mu}\int G\left[(1-2\nu)F_z + \frac{\partial}{\partial\zeta}(\zeta F_z)\right] dv. \tag{37}$$

By the same process as before, using (17), (18), (22) and (23), we find

$$(1-2\nu)B_z - \frac{\partial\beta}{\partial z} = \frac{P_z}{4\pi\mu}\left[(1-2\nu)\left(\frac{1}{R_1}-\frac{1}{R_2}\right) + c\frac{\partial}{\partial z}\left(\frac{1}{R_1}+\frac{1}{R_2}\right)\right]. \tag{38}$$

Finally, from (34) and (38),

$$B_z = \frac{P_z}{4\pi\mu}\left[\frac{1}{R_1} + \frac{3-4\nu}{R_2} + \frac{2c(z+c)}{R_2^3}\right] \quad (39)$$

$$\beta = \frac{P_z}{4\pi\mu}\left[4(1-\nu)(1-2\nu)\log(R_2+z+c) - \frac{c}{R_1} - \frac{(3-4\nu)c}{R_2}\right] \quad (40)$$

These two functions constitute the solution for the case of the force at ($0, 0, c$) normal to the plane boundary.

Force Parallel to the Plane Boundary

In this case we take $F_y = F_z = 0$ and $B_y = 0$. The boundary conditions then become, on $z = 0$,

$$\sigma_{zz} = \frac{\mu}{2(1-\nu)}\left[2(1-\nu)\frac{\partial B_z}{\partial z} + 2\nu\frac{\partial B_x}{\partial x} - x\frac{\partial^2 B_x}{\partial z^2} - \frac{\partial^2 \beta}{\partial z^2}\right] = 0 \quad (41)$$

$$\sigma_{zx} = \frac{\mu}{2(1-\nu)}\left[(1-2\nu)\left(\frac{\partial B_z}{\partial x} + \frac{\partial B_x}{\partial z}\right) - x\frac{\partial^2 B_x}{\partial x \partial z} - \frac{\partial^2 \beta}{\partial x \partial z}\right] = 0 \quad (42)$$

$$\sigma_{zy} = \frac{\mu}{2(1-\nu)}\left[(1-2\nu)\frac{\partial B_z}{\partial y} - x\frac{\partial^2 B_x}{\partial y \partial z} - \frac{\partial^2 \beta}{\partial y \partial z}\right] = 0. \quad (43)$$

Differentiating (42) with respect to y and (43) with respect to x and subtracting, we find, on $z = 0$,

$$\frac{\partial^2 B_x}{\partial y \partial z} = 0 .$$

Hence, on $z = 0$,

$$\frac{\partial B_x}{\partial z} = 0 \quad (44)$$

Also, in $z \geq 0$

$$\nabla^2 \frac{\partial B_x}{\partial z} = -\frac{1}{\mu}\frac{\partial F_x}{\partial z}$$

Hence, from (13),

$$\frac{\partial B_x}{\partial z} = \frac{1}{4\pi\mu}\int G \frac{\partial F_x}{\partial \zeta}\, dv$$

$$= \frac{P_x}{4\pi\mu}\frac{\partial}{\partial z}\left(\frac{1}{R_1} + \frac{1}{R_2}\right)$$

by (32). Thus

$$B_x = \frac{P_x}{4\pi\mu}\left(\frac{1}{R_1} + \frac{1}{R_2}\right). \quad (45)$$

From (43), on $z = 0$,

$$(1-2\nu)B_z - \frac{\partial \beta}{\partial z} = x\frac{\partial B_x}{\partial z} = 0$$

by (44). Also, in $z \geq 0$

$$\nabla^2\left[(1-2\nu)B_z - \frac{\partial \beta}{\partial z}\right] = -\frac{x}{\mu}\frac{\partial F_x}{\partial z} .$$

Hence, by (13)

$$(1-2\nu)B_z - \frac{\partial \beta}{\partial z} = \frac{1}{4\pi\mu}\int G\,\xi\,\frac{\partial F_x}{\partial \zeta}\, dv. \quad (46)$$

But the right hand side of (46) vanishes since $\xi \to 0$ as $T \to 0$. Hence

$$(1-2\nu)B_z - \frac{\partial \beta}{\partial z} = 0 \quad (47)$$

throughout the region $z \geq 0$.

From (41), on $z = 0$,

$$2(1-\nu)\frac{\partial B_z}{\partial z} - \frac{\partial^2 \beta}{\partial z^2} + 2\nu\frac{\partial B_x}{\partial x} - x\frac{\partial^2 B_x}{\partial z^2} = 0 \quad (48)$$

Now, on $z = 0$, we have, from (45),

$$2\nu\frac{\partial B_x}{\partial x} - x\frac{\partial^2 B_x}{\partial z^2} = \frac{(1-2\nu)xP_x}{2\pi\mu R_0^3} - \frac{3P_x c^2 x}{2\pi\mu R_0^5}$$

where $R_0^2 = x^2 + y^2 + c^2$. But, on $z = 0$

$$-(1-2\nu)\frac{\partial B_x}{\partial x} = \frac{(1-2\nu)xP_x}{2\pi\mu R_0^3}$$

and

$$\frac{P_x c}{2\pi\mu}\frac{\partial^2}{\partial x \partial z}\left(\frac{1}{R_2}\right) = \frac{3P_x c^2 x}{2\pi\mu R_0^5} .$$

Hence, (48) may be rewritten as

$$\chi = 2(1-\nu)\frac{\partial B_z}{\partial z} - \frac{\partial^2 \beta}{\partial z^2} - (1-2\nu)\frac{\partial B_x}{\partial x} - \frac{P_x c}{2\pi\mu}\frac{\partial^2 R_2^{-1}}{\partial x \partial z} = 0 \quad (49)$$

on $z = 0$. The Laplacian of the left side of (49) is, in the region $z \geq 0$,

$$\nabla^2 \chi = -\frac{x}{\mu}\frac{\partial^2 F_x}{\partial z^2} + \frac{1-2\nu}{\mu}\frac{\partial F_x}{\partial x} .$$

Hence, from (13),

$$\chi = \frac{1}{4\pi\mu}\int G\left[\xi\frac{\partial^2 F_x}{\partial \zeta^2} - (1-2\nu)\frac{\partial F_x}{\partial \xi}\right] dv. \quad (50)$$

The first term in the integrand vanishes since $\xi \to 0$ as $T \to 0$. The second term is integrated by parts and the surface integral vanishes, leaving

$$\chi = \frac{1-2\nu}{4\pi\mu}\int F_x \frac{\partial G}{\partial \xi}\, dv$$

which, by (17) and (25), is

$$\chi = -\frac{(1-2\nu)P_x}{4\pi\mu}\frac{\partial}{\partial x}\left(\frac{1}{R_1} - \frac{1}{R_2}\right) . \quad (51)$$

Then, from (51), (49) and (45),

$$2(1-\nu)\frac{\partial B_z}{\partial z} - \frac{\partial^2 \beta}{\partial z^2} = -\frac{(1-2\nu)P_x x}{2\pi\mu R_2^3} + \frac{3P_x c x(z+c)}{2\pi\mu R_2^5} \quad (52)$$

or

$$2(1-\nu)B_z - \frac{\partial \beta}{\partial z} = \frac{(1-2\nu)P_x x}{2\pi\mu R_2(R_2+z+c)} - \frac{P_x c x}{2\pi\mu R_2^3} . \quad (53)$$

Finally, from (53) and (47), we have

$$B_z = \frac{(1-2\nu)\, P_x\, x}{2\pi\mu\, R_2\, (R_2+z+c)} - \frac{P_x\, c\, x}{2\pi\mu\, R_2^{\,3}} \tag{54}$$

$$\beta = -\frac{(1-2\nu)^2\, P_x\, x}{2\pi\mu\, (R_2+z+c)} - \frac{(1-2\nu)\, P_x\, c\, x}{2\pi\mu\, R_2\, (R_2+z+c)}\,. \tag{55}$$

These two functions, in addition to (45), comprise the solution for the case of the force at $(0, 0, c)$ parallel to the plane boundary.

Comparison with Previous Results

The previous solution, mentioned in the Introduction, was given in terms of the Galerkin vector $\mathbf{F}$ (not to be confused with the body force $\mathbf{F}$ in the present paper).

For the case of a force normal to the plane boundary the solution obtained was

$$\mathbf{F} = \frac{P_z\, \mathbf{k}}{8\pi(1-\nu)} \Big\{ R_1 + [8\nu(1-\nu)-1]\, R_2 - 2cz\, R_2^{-1} \tag{56}$$

$$+ 4(1-2\nu)[(1-\nu)z - \nu c] \log (R_2+z+c) \Big\}$$

and, for the case of a force parallel to the plane boundary,

$$\mathbf{F} = \frac{P_x\, \mathbf{i}}{8\pi(1-\nu)} \Big\{ R_1 + R_2 - 2c^2 R_2^{-1}$$

$$+ 4(1-\nu)(1-2\nu)\big[(z+c) \log (R_2+z+c) - R_2\big] \Big\} \tag{57}$$

$$+ \frac{P_x\, \mathbf{k}}{8\pi(1-\nu)} \Big\{ 2cx\, R_2^{-1} + 2(1-2\nu)x \log (R_2+z+c) \Big\}.$$

The relation between the Galerkin and Papkovitch functions has been shown to be [4]

$$\mu\, \mathbf{B} = (1-\nu)\, \nabla^2 \mathbf{F} \tag{58}$$

$$\mu\beta = (1-\nu)(2\, \nabla\cdot\mathbf{F} - \mathbf{r}\cdot \nabla^2 \mathbf{F}) \tag{59}$$

By inserting (56) and (57) in (58) and (59), it may be verified that the previous and present solutions are identical.

Reprinted from
Proceedings of the First Midwestern Conference on Solid Mechanics (1953) 56-59

Reprinted from JOURNAL OF APPLIED PHYSICS, Vol. 25, No. 1, 12–20, January, 1954

Thickness-Shear and Flexural Vibrations of Contoured Crystal Plates*

R. D. MINDLIN AND M. FORRAY
Department of Civil Engineering, Columbia University, New York, New York
(Received May 7, 1953)

An approximate method is devised for calculating frequencies of thickness-shear and flexural vibrations of crystal plates of varying thickness. Applications are made to plates having the shape of a double wedge and to plates with beveled edges.

INTRODUCTION

IN previous papers,[1] equations of motion were derived and illustrative problems were solved for the thickness-shear and flexural vibrations of isotropic, crystalline, and piezoelectric plates of uniform thickness. The investigations are now extended to include contoured, crystal plates, i.e., plates of varying thickness.

As might be expected, the equations of motion of contoured plates are not easy to solve. Even when they are specialized to contours and motions depending on but one spatial coordinate, as in this paper, they reduce only to equations of the fourth order with variable coefficients; and these are of a type for which there do not appear to be solutions in terms of tabulated functions.

There are two major physical factors which contribute to the complexity of the equations. One is the variation in thickness, which we wish to retain. The other is the coupling between the flexural and thickness-shear modes, each of which is an infinite system. This coupling occurs also, of course, in uniform plates, so that it is appropriate to examine solutions for uniform plates in an effort to devise simplifying approximations for contoured plates.

In II and III there were given the frequencies of thickness-shear and flexural vibrations of AT-cut, quartz plates of uniform thickness. The spectrum obtained is shown in Fig. 1. The data for these curves were calculated from the roots of a transcendental equation: the secular equation obtained from applying boundary conditions to an appropriate solution of the equations of motion. At the points of intersection of the curves for the even and odd modes (identified by circles and squares in Fig. 1) the transcendental equation degenerates to an algebraic formula. As a result, it was possible to find simple algebraic equations for two families of curves whose alternate intersections are at the circled and squared points. The algebraic curves are shown in Fig. 2. Now, as the coupling is diminished, the loops in Fig. 1 spread and approach as limits, for zero coupling, the curves in Fig. 2. This is illustrated, for a portion of the spectrum, in Fig. 3.

It was shown, in II, that the portions of the curves in Fig. 1, which are nearly parallel to the curves n = constant in Fig. 2, correspond to motions which are predominately thickness-shear. Also, the portions of the curves in Fig. 1, which are nearly parallel to the curves r = constant in Fig. 2, correspond to motions which are essentially flexural. Thus, in the limit, for zero coupling, the curves of Fig. 2 give the true frequencies not only at the alternate intersection points but also throughout; i.e., the curves n = constant and r = constant give the frequencies of the uncoupled thickness-shear and flexural modes, respectively. This situation is typical of the coupling of two infinite sets of modes.

It would be useful to be able to write separate equa-

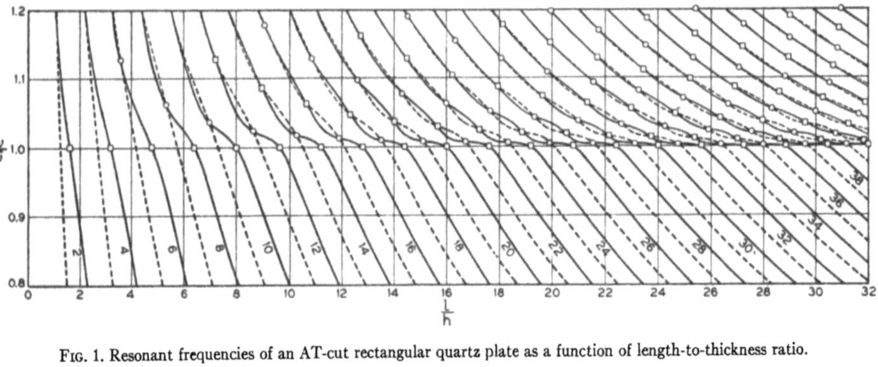

FIG. 1. Resonant frequencies of an AT-cut rectangular quartz plate as a function of length-to-thickness ratio.

* This investigation was supported by the U. S. Army Signal Corps Engineering Laboratories under a contract with Columbia University.

[1] R. D. Mindlin, J. Appl. Mech. 18, 31–38 (1951); J. Appl. Phys. 22, 316–323 (1951); 23, 83–88 (1952). These papers are referred to in the text as I, II and III, respectively.

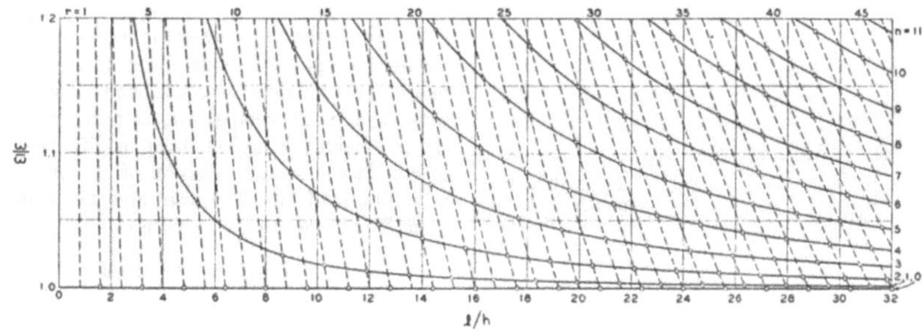

FIG. 2. Frequencies of uncoupled thickness-shear and flexural modes of an AT-cut rectangular quartz plate. The circled and squared points are those from the upper half of Fig. 1.

tions of motion, for the thickness-shear and flexural modes, which would lead directly to the curves in Fig. 2; but this does not appear to be possible. However, based on a physical interpretation of the coupling phenomenon, it is possible to write separate sets of equations of motion which lead approximately to the desired, uncoupled frequencies. These equations are, of course, much simpler than the equations which include the coupling. The thickness-shear equations are obtained by dropping the flexural-rigidity terms from the full equations and the flexural equations are obtained by dropping the thickness-shear deformation and rotatory inertia terms. In each case an arbitrary parameter is inserted in the transverse inertia term of the resulting equations. The values of these parameters are obtained by comparing the solutions of the separated equations, for the case of a uniform plate, with the known solution of the coupled equations. Calculations of the coordinates of the intersection points of Fig. 2 from the solutions of the coupled and uncoupled equations are thus made to agree very closely.

Consequently, we have two sets of uncoupled equations of motion which, for the uniform plate, give the curves of Fig. 2 with adequate precision. These equations are then solved for the case of a contoured plate having the shape of a double wedge. This is the simplest case which shows the effects of a greater thickness at the center of the plate than at the edge. It is found, for example, that the frequency of the first thickness-shear mode of a wedge is not much higher than the corresponding frequency of a uniform plate whose thickness is the same as the maximum thickness of the wedge. This result is closely associated with the fact that the thickness-shear motion is localized near the center of the wedge. In the thickness-shear overtones, however, a larger proportion of the wedge participates and, consequently, there is a greater spread in the frequencies of the overtones in a wedge than in a uniform plate. On the other hand, the flexural motion is predominant at the thin edges of the wedge and, with this, is associated the result that, in a given frequency range, there are approximately twice as many flexural modes in a wedge as in a uniform plate.

The thickness-shear equations are next solved for the case of a beveled plate, i.e., one with a portion of uniform thickness between the two wings of the double wedge. It is shown that a very small uniform portion serves to bring the frequency of the first thickness-shear mode very close to that of a uniform plate. This is in conformity with the finding that the thickness-shear motion is localized near the center.

PLATE EQUATIONS

The plate is referred to an x, y, z system of rectangular coordinates, with $y=0$ as its middle plane and the x axis as the axis of symmetry of a monoclinic crystal. The varying thickness, h, is then a function of x and z. For thickness-shear and flexural motions, the plate-stress equations take the same form as in II. These become, after introducing a factor $\exp(i\omega t)$:

$$\begin{aligned} &\partial M_x/\partial x - Q_x + \partial M_{xz}/\partial z + \rho h^3\omega^2\psi_x/12 = 0,\\ &\partial Q_x/\partial x + \partial Q_z/\partial z + \rho h\omega^2\eta = 0,\\ &\partial M_{xz}/\partial x - Q_z + \partial M_z/\partial z + \rho h^3\omega^2\psi_z/12 = 0, \end{aligned} \qquad (1)$$

where

$$\begin{aligned} M_x &= D_1\partial\psi_x/\partial x + D_2\partial\psi_z/\partial z,\\ M_z &= D_2\partial\psi_x/\partial x + D_3\partial\psi_z/\partial z,\\ M_{xz} &= D_5(\partial\psi_z/\partial x + \partial\psi_x/\partial z),\\ Q_z &= D_4(\psi_z + \partial\eta/\partial z),\\ Q_x &= D_6(\psi_x + \partial\eta/\partial x), \end{aligned} \qquad (2)$$

and

$$\begin{aligned} D_1 &= (C_{11} - C_{12}^2/C_{22})h^3/12,\\ D_2 &= (C_{13} - C_{12}C_{23}/C_{22})h^3/12,\\ D_3 &= (C_{33} - C_{23}^2/C_{22})h^3/12,\\ D_4 &= \pi^2(C_{44} - C_{24}^2/C_{22})h/12,\\ D_5 &= C_{55}h^3/12,\\ D_6 &= \pi^2 C_{66}h/12. \end{aligned} \qquad (3)$$

The boundary conditions are prescribed by specifying one member of each of the three products

$$M_\nu\psi_\nu, \quad M_{\nu s}\psi_s, \quad Q_\nu\eta, \qquad (4)$$

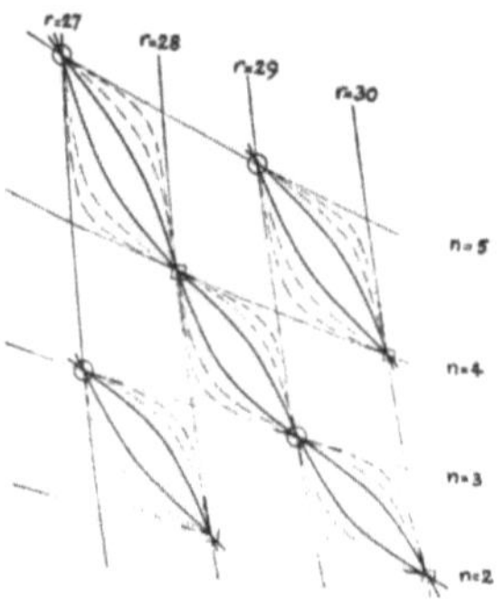

FIG. 3. Illustrating variation of coupling between thickness-shear and flexural modes.

at each point of the edges of the plate, where the normal is ν and the tangent is s.

MOTION INDEPENDENT OF z

In this paper, the problems considered involve contouring and motion independent of z, i.e.,

$$h=h(x), \quad \psi_x=\psi(x,t), \quad \psi_z=0, \quad \eta=\eta(x,t). \tag{5}$$

As a result, upon elimination of the moments and shears through the use of Eq. (2), the equations of motion (1) reduce to

$$\begin{aligned}(D_1\psi')'-D_6(\psi+\eta')+\rho h^3\omega^2\psi/12=0,\\ [D_6(\psi+\eta')]'+\rho h\omega^2\eta=0,\end{aligned} \tag{6}$$

where $'\equiv d/dx$. The boundary conditions are similarly reduced to the specification, on edges $x=$constant, of one member of each of the two products

$$M_x\psi \quad \text{and} \quad Q_x\eta, \tag{7}$$

where

$$M_x=D_1\psi' \quad \text{and} \quad Q_x=D_6(\psi+\eta'). \tag{8}$$

Equations (6)–(8) have the same form as those for a Timoshenko beam of varying section. There is no difficulty in eliminating one of ψ and η from Eq. (6) to obtain a fourth order equation on the other. Because of the x dependence of h, D_1, and D_6, however, the resulting equation defines, in general, functions which are not tabulated. This would be no great obstacle for the calculation of frequencies of the lower modes of vibration; but, for frequencies in the neighborhood of the lowest thickness-shear frequency, the thickness-shear motion is strongly coupled with high modes of flexure in a plate which is long in comparison with its thickness. Hence, energetic or perturbation methods are very lengthy.

RESOLUTION OF EQUATIONS

Equations (6)–(8) may be resolved into two sets of equations which govern, approximately, the thickness-shear and flexural motions separately.

To obtain the thickness-shear equations, we first set the flexural rigidity D_1 equal to zero in the first equation of (6). This restricts the equations to modes involving little flexure. To compensate, in part, for what little flexural rigidity is neglected, we modify the transverse inertia in the second equation of (6) by introducing a multiplying factor K_2 whose value will be determined later. Equations (6) then reduce to

$$\begin{aligned}12D_6(\psi_2+\eta_2')-\rho h^3\omega^2\psi_2=0,\\ [D_6(\psi_2+\eta_2')]'+K_2\rho h\omega^2\eta_2=0,\end{aligned} \tag{9}$$

where ψ_2 and η_2 are the displacements in the uncoupled shear modes. The boundary conditions reduce to the specification of one member of the single product

$$Q_x\eta_2, \tag{10}$$

on each of two boundaries $x=$constant. Q_x is still given by the second equation of (8).

The equations of flexure are obtained from Eqs. (6)–(8) by dropping the rotatory inertia and shear deformation terms and again modifying the transverse inertia. The result is, of course, the equation of classical beam theory,

$$(D_1\eta_1'')''-K_1\rho h\omega^2\eta_1=0, \tag{11}$$

where η_1 is uncoupled flexural deflection and K_1 is a multiplying factor to be determined later. The boundary conditions remain as in Eq. (7), but now

$$\psi_1=-\eta_1', \quad M_x=-D_1\eta_1'', \quad Q_x=-(D_1\eta_1'')'. \tag{12}$$

Equations (9) and (11) must now be completed by determining K_2 and K_1. This is done by solving (9) and (11) for the case of uniform thickness, for which the solution of the full equations (6) is known, and choosing expressions for K_2 and K_1 which will make the frequencies obtained from the resolved and combined equations match as well as possible.

DETERMINATION OF K_2

For a plate of uniform thickness, h and D_6, in Eqs. (9), are constants. Eliminating ψ_2 from (9) leaves η_2 governed by

$$\eta_2''+\delta_2^2\eta_2=0, \tag{13}$$

where

$$\delta_2^2=\frac{12K_2}{h^2}\left(\frac{\omega^2}{\bar{\omega}^2}-1\right) \tag{14}$$

and

$$\bar{\omega}^2=12D_6/\rho h^3, \tag{15}$$

i.e., $\bar{\omega}$ is the frequency of the first thickness-shear mode in an infinite plate of thickness h.

Upon inserting the solution of Eq. (13) in the boundary conditions for a free edge (i.e., $Q_x=0$ on $x=\pm l/2$), the secular equation is found to be

$$\delta_2 l=n\pi, \quad n=1,2,3,\cdots, \tag{16}$$

and hence from Eq. (14),

$$\frac{\omega^2}{\bar{\omega}^2}=1+\frac{\pi^2 n^2 h^2}{12K_2 l^2}. \quad (17)$$

This should be, approximately, the equation of the family of curves represented by full lines in Fig. 2. From II, it is known that this family of curves is given by

$$2(\omega/\bar{\omega})^2=1+kn^2h^2/l^2 +[(1+kn^2h^2/l^2)^2-4(Knh/l)^4/g]^{\frac{1}{2}} \quad (18)$$

and is represented approximately, for the small n and large l/h in which we are interested, by

$$(\omega/\bar{\omega})^2=1+kn^2h^2/l^2, \quad (19)$$

where $K^2=\pi^2/12$ and

$$k=\pi^2(1+g)/12g, \quad (20)$$

$$g=\frac{\pi^2 C_{66}}{12(C_{11}-C_{12}^2/C_{22})}. \quad (21)$$

Since $k>\pi^2/12$, it is apparent that Eq. (17) gives frequencies which are too low. It will be recalled, from II, that g is the ratio of the thickness-shear modulus to the plate-flexure modulus. Now, in deriving Eq. (17), the plate flexure modulus was set equal to zero, which is the same as making g infinite. But, with g infinite, Eq. (19) reduces to Eq. (17) if $K_2=1$. Thus, the discrepancy between Eqs. (17) and (19) is due to the omission of the effect of flexural rigidity. This effect may now be replaced, approximately, by setting

$$K_2=g/(1+g), \quad (22)$$

as a result of which Eqs. (17) and (19) become identical. Thus the omission of flexural rigidity is balanced by a decrease in transverse inertia, the same for all modes. It should be observed that Eqs. (17) and (22) still do not give the thickness-shear curves of Fig. 2 exactly, but only as well as Eq. (19) does. The discrepancies which remain are discussed below in the section entitled, "Test of Procedure".

DETERMINATION OF K_1

For a uniform plate, (11) reduces to

$$\eta_1^{\text{iv}}-\delta_1^4\eta_1=0, \quad (23)$$

where

$$\delta_1^4=K_1\rho h\omega^2/D_1. \quad (24)$$

The solution of Eq. (23) with the boundary conditions for free edges leads to the well-known secular equation

$$\cosh\delta_1 l \cos\delta_1 l=1, \quad (25)$$

whose roots, for the high modes of present interest, are given very closely by

$$\delta_1 l=(2r+1)\pi/2, \quad r=1,2,3,\cdots. \quad (26)$$

Hence,

$$(\omega/\bar{\omega})^2=[(2r+1)Kh/l]^4/16gK_1. \quad (27)$$

This is intended to represent the dashed curves in Fig. 2, whose exact equation, from II, is

$$2(\omega/\bar{\omega})^2=1+kr^2h^2/l^2 -[(1+kr^2h^2/l^2)^2-4(Krh/l)^4/g]^{\frac{1}{2}}. \quad (28)$$

The frequencies given by Eq. (27) are higher than those given by Eq. (28). This is essentially because the approximation involves making the shear modulus infinite; and the additional stiffness raises the frequencies.

Now, the main interest is in intersections of the dashed and full curves, of Fig. 2, in the neighborhood of $\omega/\bar{\omega}=1$. Therefore, it is desirable to have the curves of Eq. (27) pass through $\omega/\bar{\omega}=1$ at the same values of l/h that are obtained from Eq. (28) by setting $\omega/\bar{\omega}=1$, i.e., at

$$l/h=\pi r[12(1+g)]^{-\frac{1}{2}}. \quad (29)$$

This is accomplished by setting $\omega/\bar{\omega}=1$ in Eqs. (27) and (28) and then eliminating l/h between the two equations, with the result

$$K_1=\frac{(1+g)^2}{g}\left(\frac{2r+1}{2r}\right)^4. \quad (30)$$

Thus, the increase in shear stiffness is balanced by an increase in transverse inertia, slightly greater for the lower modes than for the higher.

TEST OF PROCEDURE

The coordinates of the intersection points of Eqs. (17) and (27), for r and n even and r and n odd, are to be compared with the corresponding coordinates obtained from Eqs. (18) and (28). The formulas for the coordinates are found by eliminating first $\omega/\bar{\omega}$ and then l/h from the two pairs of equations. Thus, the coordinates from the uncoupled equations (9) and (11) are given by

$$2(l/h)^2=kn^2\{-1+[1+(2K^4r^2/k^2gn^2)^2]^{\frac{1}{2}}\},$$
$$\omega/\bar{\omega}=(k^2gn^2/2K^4r^2)\{1+[1+(2K^4r^2/k^2gn^2)^2]^{\frac{1}{2}}\}, \quad (31)$$

and the coordinates from the full equations (6), as obtained in II, are given by

$$(l/h)^2=K^2(gr^2-n^2)(r^2-gn^2)/g(1+g)(r^2+n^2),$$
$$(\omega/\bar{\omega})^2=g(r^2+n^2)^2/(gr^2-n^2)(r^2-gn^2). \quad (32)$$

Table I contains the coordinates, calculated from Eqs. (31) and (32), for the values of r and n at which circled and squared intersection points occur in Fig. 2. In each group of four coordinates, the lower figures are those for l/h and $\omega/\bar{\omega}$, respectively, as calculated from the approximation (31) and the upper figures are those from the "exact" formulas (32). It may be seen that correspondence is very good, especially for

TABLE I. Comparison of intersection points of frequency spectrum obtained from coupled and uncoupled equations of motion of uniform plate.

n	0		1		2		3		4		5		6		7		8		9		10	
r	l/h	ω/ω	l/h	$\omega/\bar{\omega}$	l/h	$\omega/\bar{\omega}$	l/h	$\omega/\bar{\omega}$	l/h	$\omega/\bar{\omega}$	l/h	$\omega/\bar{\omega}$	l/h	$\omega/\bar{\omega}$	l/h	$\omega/\bar{\omega}$	l/h	$\omega/\bar{\omega}$	l/h	$\omega/\bar{\omega}$	l/h	$\omega/\bar{\omega}$
1																						
2	1.601	1.000																				
	1.601	1.000																				
3																						
4	3.203	1.000																				
	3.203	1.000																				
5			3.617	1.129																		
			3.778	1.123																		
6	4.804	1.000																				
	4.804	1.000																				
7			5.329	1.062																		
			5.441	1.061																		
8	6.405	1.000																				
	6.405	1.000																				
9			6.992	1.037																		
			7.078	1.037																		
10	8.007	1.000			7.234	1.129																
	8.007	1.000			7.555	1.123																
11			8.632	1.025																		
			8.702	1.024																		
12	9.608	1.000			8.965	1.087																
	9.608	1.000			9.228	1.084																
13			10.260	1.017			9.069	1.178														
			10.319	1.017			9.636	1.167														
14	11.209	1.000			10.658	1.062																
	11.209	1.000			10.882	1.061																
15			11.881	1.013			10.851	1.129														
			11.932	1.013			11.333	1.123														
16	12.810	1.000			12.328	1.047																
	12.810	1.000			12.523	1.046																
17			13.498	1.010			12.589	1.098														
			13.543	1.010			13.009	1.095														
18	14.412	1.000			13.983	1.037			12.693	1.163												
	14.412	1.000			14.155	1.037			13.416	1.154												
19			15.111	1.008			14.298	1.077														
			15.151	1.008			14.671	1.075														
20	16.013	1.000			15.627	1.030			14.468	1.129												
	16.013	1.000			15.782	1.030			15.110	1.123												
21			16.722	1.007			15.987	1.062			14.509	1.191										
			16.758	1.007			16.322	1.061			15.489	1.178										
22	17.614	1.000			17.264	1.025			16.210	1.104												
	17.614	1.000			17.404	1.024			16.789	1.101												
23			18.331	1.006			17.660	1.052			16.314	1.155										
			18.365	1.005			17.965	1.051			17.196	1.147										
24	19.216	1.000			18.894	1.021			17.929	1.087												
	19.216	1.000			19.023	1.020			18.456	1.084												
25			19.939	1.005			19.322	1.043			18.085	1.129										
			19.970	1.005			19.602	1.043			18.888	1.123										
26	20.817	1.000			20.520	1.017			19.630	1.073			18.138	1.178								
	20.817	1.000			20.639	1.017			20.113	1.071			19.272	1.167								
27			21.546	1.004			20.975	1.037			19.830	1.109										
			21.575	1.004			21.233	1.037			20.568	1.105										
28	22.418	1.000			22.143	1.015			21.316	1.062			19.933	1.150								
	22.418	1.000			22.253	1.015			21.763	1.061			20.974	1.142								
29			23.153	1.003			22.621	1.032			21.555	1.093			19.947	1.197						
			23.179	1.003			22.861	1.032			22.238	1.090			21.341	1.184						
30	24.020	1.000			23.763	1.013			22.991	1.054			21.701	1.129								
	24.020	1.000			23.865	1.013			23.407	1.053			22.665	1.123								
31			24.758	1.003			24.260	1.028			23.264	1.081			21.763	1.169						
			24.783	1.004			24.485	1.028			23.900	1.079			23.053	1.159						
32	25.621	1.000			25.380	1.011			24.657	1.047			23.449	1.112								
	25.621	1.000			25.476	1.011			25.046	1.046			24.346	1.107								
33			26.363	1.003			25.896	1.025			24.960	1.071			23.552	1.147						
			26.386	1.002			26.106	1.024			25.555	1.069			24.753	1.139						
34	27.222	1.000			26.995	1.010			26.315	1.042			25.178	1.098			23.579	1.186				
	27.222	1.000			27.086	1.010			26.680	1.041			26.019	1.095			25.125	1.174				
35			27.968	1.002			27.527	1.022			26.645	1.062			25.318	1.129						
			27.990	1.002			27.725	1.022			27.204	1.061			26.443	1.123						
36	28.824	1.000			28.609	1.009			27.967	1.037			26.894	1.087			25.385	1.163				
	28.824	1.000			28.695	1.009			28.311	1.037			27.684	1.084			26.833	1.154				
37			29.572	1.002			29.155	1.019			28.321	1.056			27.067	1.114						
			29.593	1.001			29.342	1.019			28.848	1.055			28.124	1.110						
38	30.425	1.000			30.222	1.008			29.613	1.033			28.597	1.077			27.170	1.144				
	30.425	1.000			30.303	1.008			29.939	1.033			29.343	1.075			28.531	1.137				
39			31.176	1.002			30.781	1.017			29.989	1.050			28.800	1.102			27.207	1.178		
			31.196	1.001			30.958	1.017			30.488	1.049			29.799	1.098			28.908	1.167		
40	32.026	1.000			31.833	1.007			31.255	1.030			30.290	1.069			28.935	1.129				
	32.026	1.000			31.910	1.007			31.564	1.030			30.996	1.068			30.220	1.123				
41			32.780	1.002			32.404	1.016			31.651	1.045			30.520	1.091			29.007	1.159		
			32.798	1.002			32.572	1.016			32.125	1.044			31.466	1.088			30.612	1.150		
42	33.628	1.000			33.444	1.007			32.893	1.027			31.974	1.062			30.685	1.116			29.018	1.191
	33.628	1.000			33.517	1.007			33.187	1.027			32.645	1.061			31.902	1.111			30.978	1.178

small n and large r; i.e., for the low thickness-shear overtones of a thin plate.

The shapes of the uncoupled and coupled modes may also be compared. For example, the deflections for the uncoupled, antisymmetric thickness-shear modes are, from Eq. (13) and Eq. (16),

$$\eta_2 = A_2 \sin(n\pi x/l),$$

and, the corresponding uncoupled flexural deflections are, from Eq. (23),

$$\eta_1 = A_1[\sin\delta_1 x + \sin(\delta_1 l/2)\,\mathrm{cosech}(\delta_1 l/2)\,\sinh\delta_1 x],$$

where δ_1 is given by Eq. (26). These are to be compared with the corresponding shapes obtained from the coupled equations [see II, Eq. (56)],

$$\eta=\eta_1+\eta_2$$
$$=A_1\sin(r\pi x/l)+A_2\sin(n\pi x/l).$$

Thus, the thickness-shear, or long wavelength, components have exactly the same shape in the two cases; but the flexural, or short wavelength, components differ.

THICKNESS-SHEAR VIBRATIONS OF A DOUBLE WEDGE

The thickness-shear equations (9) will be solved, in this section, for the case of a double wedge with free edges. The plate is of length l and its thickness (see Fig. 4) is

$$h=ax,\quad 0\leqslant x\leqslant l/2,$$
$$h=a(l-x),\quad l/2\leqslant x\leqslant l, \tag{33}$$

where $a=2h_0/l$ and h_0 is the maximum thickness of the wedge. By treating the symmetric and antisymmetric modes separately, only the first equation (33) is required.

Equations (9) are conveniently resolved into

$$12K_2h\eta_2=-(h^3\psi_2)'$$
$$D_6[h^{-1}(h^3\psi_2)']'+K_2(\rho h^3\omega^2-12D_6)\psi_2=0. \tag{34}$$

Then, with h given by the first of (33), the second of (34) becomes

$$\psi_2''+5x^{-1}\psi_2'+[\beta^2+(4-p^2)x^{-2}]\psi_2=0, \tag{35}$$

where

$$\beta^2=K_2\rho\omega^2/K^2C_{66}, \tag{36}$$

$$p^2=1+12K_2/a^2. \tag{37}$$

The solution of Eq. (35) is

$$\psi_2=Ax^{-2}J_p(\beta x)+Bx^{-2}Y_p(\beta x), \tag{38}$$

where J_p and Y_p are the Bessel functions of the first and second kinds and A and B are constants.

For the antisymmetric modes (n odd) the boundary conditions (10) are

$$Q_x(0)=0,\quad \eta(l/2)=0. \tag{39}$$

Hence, $B=0$ and

$$(1-p)J_p(\gamma)+\gamma J_{p-1}(\gamma)=0,\quad (n \text{ odd}), \tag{40}$$

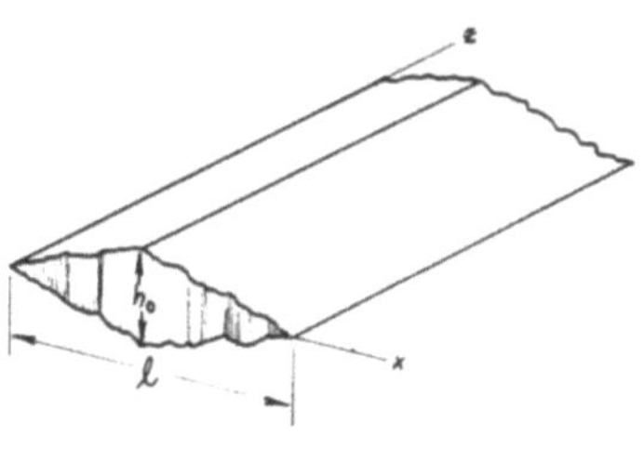

FIG. 4. Double wedge.

TABLE II. Thickness-shear frequencies of double wedge.

p	l/h_0	γ_n $n=1$	$n=2$	$n=3$	ω_n/ω^*, ($g=0.283$) $n=1$	$n=2$	$n=3$
8	9.76	9.95	12.23	14.22	1.25	1.54	1.79
16	19.64	18.31	21.09	23.35	1.15	1.32	1.46
24	29.50	26.57	29.71	32.18	1.11	1.24	1.34
32	39.29	34.78	38.22	40.86	1.09	1.20	1.28

where

$$\gamma=\beta l/2. \tag{41}$$

For the symmetric modes (n even) the boundary conditions are

$$Q_x(0)=0,\quad Q_x(l/2)=0. \tag{42}$$

Hence, $B=0$ and

$$J_p(\gamma)=0,\quad (n \text{ even}). \tag{43}$$

The frequencies are now obtained, in terms of the roots of Eqs. (40) and (43), from Eqs. (41) and (36). Introducing a reference frequency

$$\omega^*=(\pi/h_0)(C_{66}/\rho)^{\frac{1}{2}}, \tag{44}$$

i.e., the frequency of the first thickness-shear mode of an infinite plate of uniform thickness h_0, and using the expression for K_2 from Eq. (22), we have, for the frequency ω_n of the nth mode,

$$\omega_n/\omega^*=(\gamma_nh_0/l)[(1+g)/3g]^{\frac{1}{2}}. \tag{45}$$

The numbers γ_n, in Eq. (45), are the roots of Eqs. (40) and (43) taken alternately, i.e., when n is odd, γ is the $(n+1)/2$th root of Eq. (40); when n is even, γ is the $n/2$th root of Eq. (43). Table II contains the first three roots for four values of p, i.e., for four values of wedge angle. Also included in Table II are the corresponding frequencies calculated from Eq. (45) for the AT-cut of quartz, for which $g=0.283$ (see II). The thickness-shear frequencies of the first three modes are plotted as full lines in Fig. 5.

Equation (45) is the equation of a family of curves, for the wedge, which corresponds to the family of full lines, in Fig. 2, for the uniform plate. It is interesting to compare the influence of the length-thickness ratio on the frequencies in the two cases. In the case of the wedge, the frequency cannot be expressed explicitly in terms of l/h_0 because the latter is contained in the order, p, of the Bessel functions appearing in the secular equations. However, the leading terms can be found in the following manner. For the odd modes, the secular equation (40) can be written as

$$[\gamma J_p(\gamma)]'=0, \tag{46}$$

so that it is seen that the zeros of Eq. (40) lie near the stationary values of $J_p(\gamma)$. Hence, approximately, the first root of Eq. (40) is at the first maximum of $J_p(\gamma)$ which, by Watson's formula is

$$\gamma_1=p+0.8086^{\frac{1}{3}}p+O(p^{-\frac{1}{3}}) \tag{47}$$

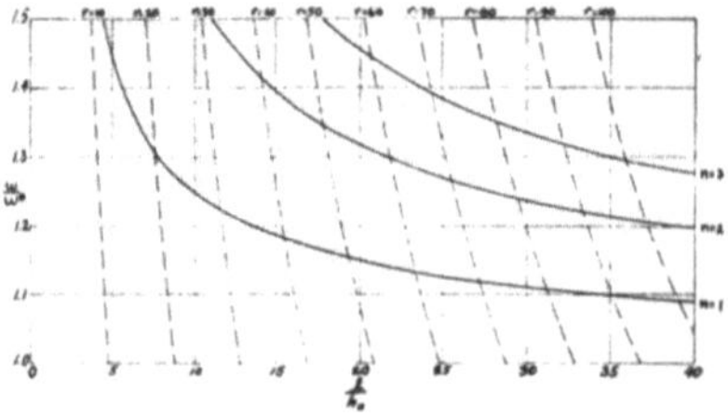

FIG. 5. Frequencies of uncoupled thickness-shear and flexural modes of a double wedge.

for large p, i.e., for large l/h_0. Inserting the expression for p from Eq. (37) and expanding, we find, approximately,

$$\omega_1=\omega^*+0.66\omega^*(kh_0^2/l^2)^{1/3}. \tag{48}$$

On the other hand, the frequency of the first thickness-shear mode of a uniform plate, for large l/h, is, from Eq. (18) or (19),

$$\omega_1=\bar{\omega}+(1/2)\bar{\omega}(kh^2/l^2). \tag{49}$$

In Eq. (49) the first term on the right is the thickness-shear frequency of an infinite plate and the second term gives the first order influence of the finite length of the plate. In Eq. (48) the first term on the right is the thickness-shear frequency of an infinite plate of uniform thickness equal to the maximum thickness of the wedge and the second term gives the first order influence of the ratio of length to maximum thickness. In both cases the second term is much smaller than the first but the correction term for the wedge is greater than that for the uniform plate. Hence, the thickness-shear frequency of a double wedge is only slightly greater than that of a uniform plate of thickness equal to the maximum thickness of the wedge. This is despite the fact that the average thickness of the wedge is only half that of the uniform plate, from which it might be conjectured that the wedge frequency would be double that of the uniform plate. The reason this is not so is that only the central portion of the wedge contributes to the shear motion, as will be shown below. Further, it may be seen, from the presence of the power $\frac{1}{3}$ in Eq. (48), that changes in length of the wedge have much less influence on the thickness-shear frequency than corresponding changes in length of a uniform plate. This also is due to the fact that the outer portions of the wedge do not participate in the shear motion.

Another influence of the wedge shape is on the spread of the thickness-shear frequency spectrum. To study this, we first find, approximately, the frequency of the second mode, which is obtained from the first root of Eq. (43). For large p, the first zero of $J_p(\gamma)$ is, by Watson's formula,

$$\gamma_2=p+1.8557p^{1/3}+O(p^{-1/3}), \tag{50}$$

so that, by the same procedure as before,

$$\omega_2=\omega^*+1.37\omega^*(kh_0^2/l^2)^{1/3}. \tag{51}$$

The corresponding frequency for the uniform plate is

$$\omega_2=\bar{\omega}+2\bar{\omega}(kh^2/l^2). \tag{52}$$

Hence, approximately, for plate and wedge of the same maximum thickness,

$$\frac{(\omega_2-\omega_1)_{\text{wedge}}}{(\omega_2-\omega_1)_{\text{plate}}}=0.47(l^2/kh^2)^{2/3}. \tag{53}$$

For example, for an AT quartz plate, $g=0.283$ and, hence, $k=3.73$. If the plate is of uniform thickness, the difference in frequencies, when $l/h=30$, is about 10 kilocycles per second for a plate one millimeter thick. By Eq. (53), the difference in frequency between the first two modes of the wedge would be about 180 kilocycles per second.

FLEXURAL VIBRATIONS OF A DOUBLE WEDGE

When $h=ax$, Eq. (11) becomes

$$16(x^3\eta_1'')''-\lambda^4x\eta_1=0, \tag{54}$$

where

$$\frac{\lambda^4}{16}=\frac{12K_1\rho\omega^2}{a^2(C_{11}-C_{12}^2/C_{22})}. \tag{55}$$

The solution of Eq. (54) is

$$x^{1/2}\eta_1=AJ_1(\lambda x^{1/2})+BJ_1(i\lambda x^{1/2})+CY_1(\lambda x^{1/2})+DY_1(i\lambda x^{1/2}). \tag{56}$$

For the symmetric modes, the boundary conditions are

$$M_x(0)=Q_x(0)=\eta_1'(l/2)=Q_x(l/2)=0. \tag{57}$$

Hence, B, C, and D vanish and

$$\delta J_0(\delta)-2J_1(\delta)=0, \tag{58}$$

where

$$\delta=\lambda(l/2)^{1/2}. \tag{59}$$

For the antisymmetric modes, the boundary conditions are

$$M_x(0)=Q_x(0)=\eta_1(l/2)=M_x(l/2)=0. \tag{60}$$

Hence, $C=D=0$, $B=AJ_1(\delta)/I_1(\delta)$ and

$$\delta I_1(\delta)J_1(\delta)-2[I_0(\delta)J_1(\delta)-I_1(\delta)J_0(\delta)]=0, \tag{61}$$

where I_0 and I_1 are modified Bessel functions.

From Eqs. (55) and (59), the frequency of the rth

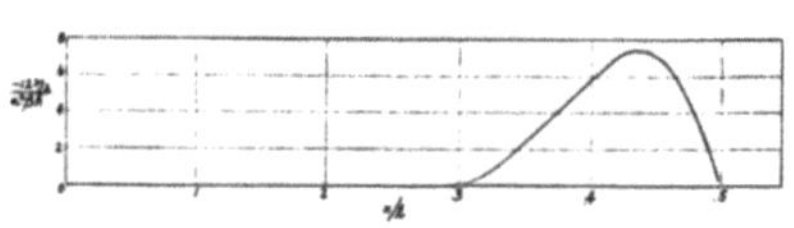

FIG. 6. Shape of transverse deflection of the first antisymmetric thickness-shear mode of a double wedge ($l/h_0=19.64$).

mode is given by

$$\frac{\omega_r}{\omega^*}=\frac{\delta_r^2 h_0^2}{12l^2(1+g)}\left(\frac{2r}{2r+1}\right)^2, \tag{62}$$

where δ_r is the $(r+1)/2$th root of Eq. (58) when r is odd; and, when r is even, δ_r is the $r/2$th root of Eq. (61).

Equation (62) is the equation of a family of curves, for the wedge, which corresponds to the family of dashed curves, in Fig. 2, for the uniform plate. The curves showing the uncoupled flexural frequencies for the wedge are the dashed lines in Fig. 5. Only every tenth mode is drawn in the figure.

For high modes, equations (58) and (61) give, approximately,

$$\delta_r=(2r+5)\pi/4. \tag{63}$$

Then, from Eqs. (30), (62), and (63), the frequencies of the high, uncoupled, flexural modes of a wedge are

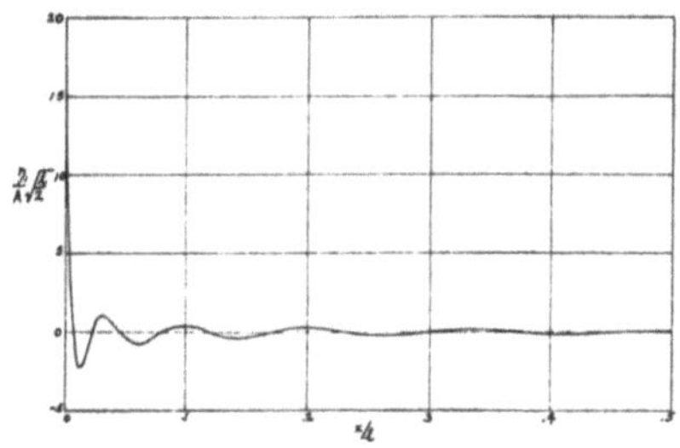

FIG. 7. Shape of transverse deflection of the tenth symmetric flexural mode of a double wedge.

given by

$$(\omega/\omega^*)^2=[(2r+5)Kh_0/2l]^4/16K_1 g. \tag{64}$$

This is to be compared with Eq. (27) which gives the frequencies of the high, uncoupled, flexural modes of a uniform plate. It may be seen that the rth mode of the wedge has the same frequency as the $(r+2)$th mode of a uniform plate of (1) twice the length and of thickness equal to the maximum thickness of the wedge or (2) the same length and one-quarter the maximum thickness of the wedge. Alternatively, for a uniform plate and a wedge of the same maximum dimensions, an uncoupled flexural mode of the wedge has a frequency one-quarter that of the next-but-one higher mode of the plate. This is in contradistinction to the relation between the frequencies of the uncoupled, thickness-shear modes of plate and wedge. These were found to be about the same.

It may also be seen that, for a given l/h, the order of the flexural mode which couples with the first thickness-shear mode is about twice as high for the wedge as for the uniform plate.

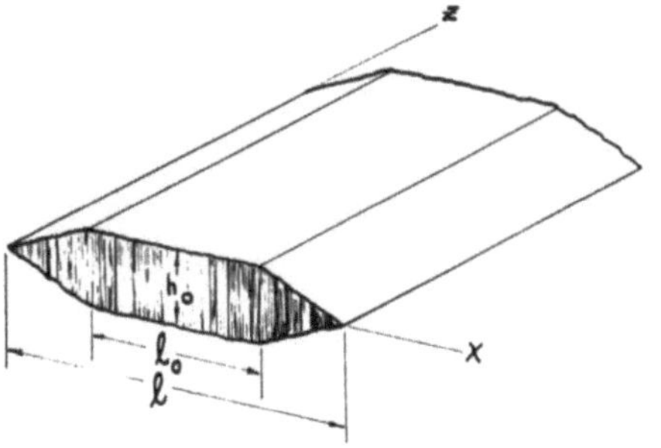

FIG. 8. Plate with beveled edges.

SHAPES OF MODES

For the uncoupled thickness-shear motion, the rotation is given by

$$\psi_2=Ax^{-2}J_p(\beta x),$$

and the deflection is given, in terms of ψ_2, by the first equation of (34). Thus, the mode shapes are described essentially by Bessel functions of the first kind of very large order (since p is very large for small wedge angles) and of small argument (since βx is small for the lower modes). A characteristic of such functions is that they are very small for an appreciable range in the neighborhood of zero argument. As illustrated in Fig. 6, for $l/h_0=19.64$, the deflection in the first antisymmetric mode is zero at the end of the wedge and is practically zero into about 60 percent of the distance to the center. Thus, only about 40 percent of the wedge (the central portion) participates in the thickness-shear motion.

On the other hand, the uncoupled flexural modes are described by Bessel functions of low order. For example, the deflection for the symmetric modes is

$$\eta_1=Ax^{-\frac{1}{2}}J_1(\lambda x^{\frac{1}{2}}).$$

The Bessel function of order unity has a diminishing amplitude with increasing argument. For example (see Fig. 7), for the tenth symmetric mode (which has twenty nodes) the amplitude at the edge of the wedge is about one hundred times as large as the amplitude at the center.

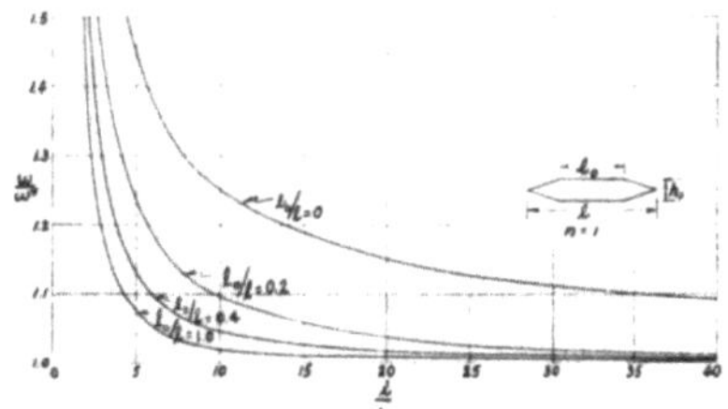

FIG. 9. Influence of length, of portion of uniform thickness, on frequency of first thickness-shear mode of a beveled plate.

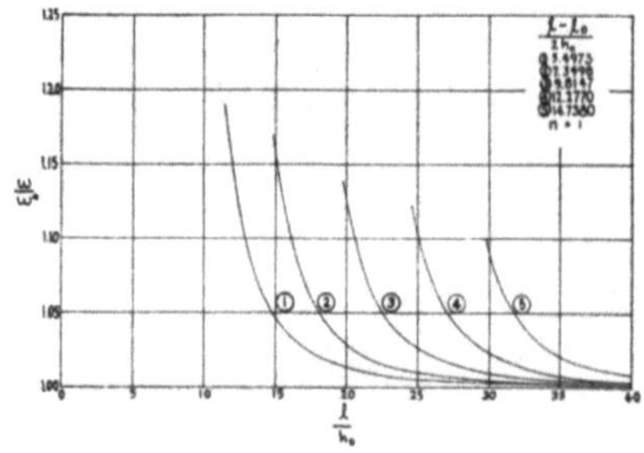

FIG. 10. Influence of wedge angle on frequency of first thickness-shear mode of a beveled plate.

Thus, while the thickness-shear motion is concentrated near the center of the wedge, the flexural motion is concentrated near the edges.

BEVELED PLATE

In this section we investigate the effect, on the thickness-shear frequencies, of a uniform portion of plate between the two wings of the wedge. The cross section of the plate is illustrated in Fig. 8, i.e., we have

$$h=ax,\quad 0\leqslant x\leqslant (l-l_0)/2;$$
$$h=h_0,\quad (l-l_0)/2\leqslant x\leqslant l/2. \qquad (65)$$

For the wedge portion we have, from Eqs. (34) and (38),

$$\psi=Ax^{-2}J_p(\beta x),$$
$$12K_2h\eta=-(h^3\psi)'. \qquad (66)$$

For the uniform portion, we have from Eqs. (9) and (13), for the antisymmetric modes,

$$\eta=B\sin\delta(l/2-x);$$
$$(1-\omega^2/\omega^{*2})\psi=B\delta\cos\delta(l/2-x). \qquad (67)$$

The boundary conditions require that ψ and η be continuous at $x=(l-l_0)/2$, i.e., at the interface between the wedge and uniform portions. Equating the appropriate members of Eqs. (66) and (67), and eliminating A and B, we find the secular equation

$$\alpha\gamma_0J_p(\gamma_0)\tan(C\alpha\gamma_0)=[\gamma_0J_p(\gamma_0)]', \qquad (68)$$

where

$$C^{-1}=(l/l_0-1),$$
$$\gamma_0=\beta l_0/2C, \qquad (69)$$
$$\alpha=\delta/\beta=(1-\omega^{*2}/\omega^2)^{1/2}.$$

The solution of Eqs. (68) and (69) for the frequency of the first thickness-shear mode as a function of l/h_0, for various values of l_0/l, is shown in Fig. 9. The uppermost curve is that for the double wedge and the lowest curve is that for the uniform plate. It may be seen that a very small uniform portion causes the frequency to be much closer to that of the uniform plate than to the frequency of a double wedge. This is in conformity with the previous result that the thickness-shear motion is concentrated near the center of the plate. Since, in this case, the central portion has a uniform thickness, it follows that the frequency is very nearly that of a uniform plate.

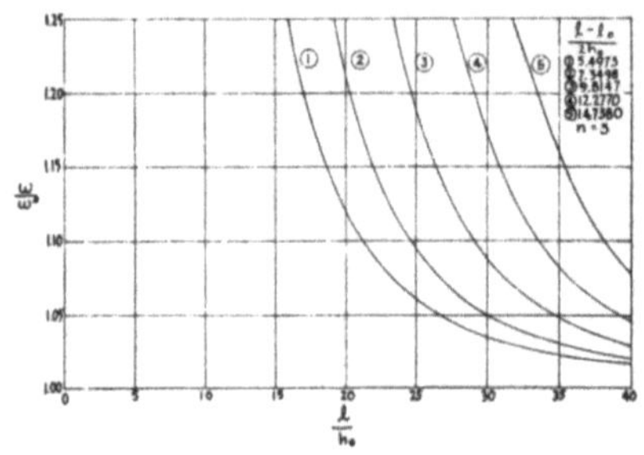

FIG. 11. Influence of wedge angle on frequency of third thickness-shear mode of a beveled plate.

Additional computations were made, from Eqs. (68) and (69), of ω/ω^* *vs* l/h_0 for 5 values of wedge angle. The results are given in Figs. 10 and 11 for the first ($n=1$) and second ($n=3$) antisymmetric modes, respectively.

Reprinted from JOURNAL OF APPLIED PHYSICS, Vol. 25, No. 1. 21–24, January, 1954

Thickness-Shear Vibrations of Piezoelectric Crystal Plates with Incomplete Electrodes*

R. D. MINDLIN AND H. DERESIEWICZ
Department of Civil Engineering, Columbia University, New York, New York
(Received May 7, 1953)

In this paper the influence of length of electrodes on thickness-shear frequencies of piezoelectric crystal plates is investigated.

IN the third of a series of papers[1] on thickness-shear and flexural vibrations of plates, solutions were obtained for the forced vibrations of rectangular piezoelectric crystal plates whose surfaces are completely coated by the electrodes. The present paper deals with crystal plates having rectangular electrodes which cover only a portion of their surfaces.

The presence or absence of electrode is equivalent to a change in the thickness-shear modulus of elasticity and piezoelectric constant. Thus, coated and uncoated portions of the plate, if separated, would have different natural frequencies of vibration even if they had the same dimensions. The full plate, then, exhibits coupling between the modes of the coated and uncoated portions. Since each of the latter contains coupling between the thickness-shear and flexural modes, the resulting frequency spectrum is very complicated. Although the secular equation is easy to obtain, the calculation of its roots is very laborious. In order to study the effect of coupling between the coated and uncoated portions, use is made of a device, developed previously (IV), whereby approximations to the thickness-shear and flexural modes may be investigated separately. The coupling phenomena between the thickness-shear modes of the coated and uncoated portions are then explored in detail. The thickness-shear frequencies of the partially coated plate are shown to be the result of coupling between two systems. One of these is a fully coated plate, in thickness-shear motion, whose length is equal to the length of the coated portion of the partially coated plate; the other is an uncoated plate, in similar motion, of length equal to that of the uncoated portion of the partially coated plate.

RESUMÉ OF EQUATIONS

Referred to a rectangular coordinate system, the faces of the plate are at $y=\pm h/2$ and only motion independent of z is considered. Then, from III, Equation (31), the equations of motion, at frequency ω, are

$$\begin{aligned} D_1\xi''-D_6(\xi+h\eta')+F_1\varphi''-F_6\varphi_0+\rho\omega^2R\xi&=0,\\ D_6(\xi'+h\eta'')+F_6\varphi_0'+\rho\omega^2h\eta&=0, \quad (1)\\ (R+4\pi C_1)\varphi''-4\pi F_1\xi''&=0, \end{aligned}$$

where primes denote differentiation with respect to x, while ξ and η are the x and y components of displacement, and φ is the electric potential. The coefficients are given by

$$\begin{aligned} D_1&=(c_{11}-c_{12}^2/c_{22})R; & F_1&=(e_{11}-c_{12}e_{12}/c_{22})R,\\ D_6&=\kappa^2c_{66}; & F_6&=\kappa^2e_{26}, \quad (2)\\ C_1&=(\kappa_{11}+e_{12}^2/c_{22})R; & R&=h^2/12, \end{aligned}$$

where c_{ij}, e_{ij}, κ_{ij} are the elastic, piezoelectric, and susceptibility coefficients, respectively, and where κ is defined later. Note that F_6 differs in definition from that in III by the factor κ^2.

The moment and shear in the plate are, respectively,

$$M_x=D_1\xi'+F_1\varphi'; \quad (3)$$

$$Q_x=D_6(\xi+h\eta')+F_6\varphi_0. \quad (4)$$

In Equations (1) and (4), taken from III, wherever F_6 is found as a coefficient, φ has been replaced by φ_0. The necessity for this has been brought to the authors' attention by Dr. J. A. Lewis, and arises from considering the sequence of operations in computing the relation

$$Q_x=\int_{-h/2}^{h/2}T_6dy. \quad (5)$$

In III, the expression for T_6 in terms of ξ, η, and φ was inserted in the integrand before integration. Consequently, it was not recognized that the term φ in the resulting expression should have appeared as φ_0, i.e., the applied potential, as would have been apparent if the integration had been performed first.

A similar change is required in the expression for

$$P_y=\int_{-h/2}^{h/2}P_2dy, \quad (6)$$

but this does not affect the differential equations for the modes of motion considered here.

INFINITE PLATE: DETERMINATION OF κ^2

The pure, thickness-shear motion of an infinite plate is obtained by setting $\eta=0$ and $\xi=\xi_0$ (a constant) in (1). Hence,

$$(\rho R\omega^2-\kappa^2c_{66})\xi_0=F_6\varphi_0. \quad (7)$$

From (7), resonance in an infinite plate occurs when

$$\omega^2=\kappa^2c_{66}/\rho R. \quad (8)$$

* This investigation was supported by the U. S. Army Signal Corps Engineering Laboratories under a contract with Columbia University.

[1] R. D. Mindlin, J. Appl. Mech. 18, 31–38 (1951); J. Appl. Phys. 22, 316–323 (1951), 23, 83–88 (1952); R. D. Mindlin and M. Forray, J. Appl. Phys. 24, 12 (1953). These papers are referred to as I, II, III and IV, respectively.

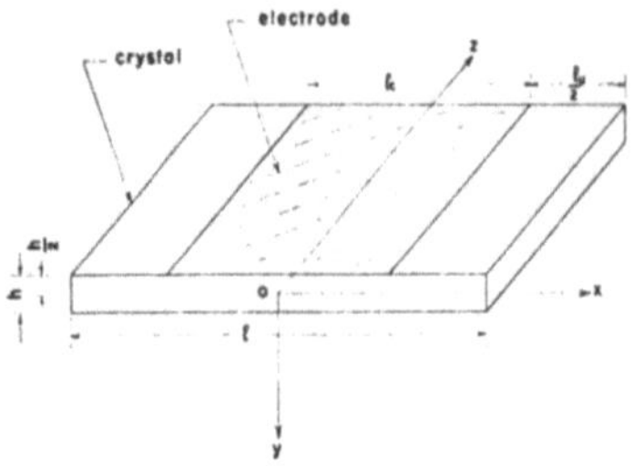

FIG. 1. Plate with electrode coating on a portion of the surface.

This is to be compared with Lawson's[2] solution of the three-dimensional equations:

$$\omega^2=\frac{4\alpha_n^2c_{66}}{\rho h^2}\left(1+\frac{e_{26}^2}{c_{66}K_2}\right), \tag{9}$$

where

$$K_2=(1+4\pi\kappa_{22})/4\pi$$

and α_n are the roots of

$$\tan\alpha_n=\alpha_n(1+c_{66}K_2/e_{26}^2) \tag{10}$$

for a plate with coated electrodes, and

$$\alpha_n=n\pi/2 \tag{11}$$

for an uncoated plate.

Comparing (8) and (9), we find

$$\kappa_i^2=\alpha_i^2(1+e_{26}^2/c_{66}K_2)/3,\quad i=c,\,u \tag{12}$$

where α_c is the first root of (10) and $\alpha_u=\pi/2$. Thus, κ has different values, κ_c and κ_u, according as the plate has or has not electrode coating. Since the moduli D_6 and F_6 are proportional to κ^2, the effect of electrode coating appears in this theory as an alteration of an elastic and piezoelectric property. From (8) and (12), infinite coated and uncoated plates have natural frequencies ω_c and ω_u, respectively, given by

$$\omega_i^2=\kappa_i^2c_{66}/\rho R,\quad i=c,\,u. \tag{13}$$

PARTIALLY COATED PLATE OF FINITE LENGTH

We consider, now, a plate, bounded by edges $x=\pm l/2$, with electrodes covering a portion $-l_c/2\leq x\leq l_c/2$ (see Fig. 1). The motions of the coated and uncoated portions are each governed by equations of the form (1) but with different values of the constants D_6 and F_6. In addition, $\varphi_0=0$ for the uncoated portion. The simultaneous solution of the two sets of equations must be found with boundary conditions $M_x=Q_x=0$ at $x=\pm l/2$ and M_x, Q_x, ξ and η continuous at $x=\pm l_c/2$. There is no difficulty in finding this solution and the accompanying secular equation. However, the frequency spectrum is very complicated because the motion involves four-fold coupling; i.e., between thickness-shear and flexure as well as between coated and uncoated portions. Since we are interested mainly in the effects of partial coating on the thickness-shear overtones, it is expedient to introduce an approximation whereby the thickness-shear modes may be studied independently of the flexural modes. This is a device employed previously (IV) to study plates of varying thickness. It consists, simply, of omitting the flexural resistance terms, which play only a minor role in the thickness-shear modes, and modifying the transverse inertia to compensate for the omission.

THICKNESS-SHEAR EQUATIONS

If the flexural resistance $D_1\xi'+F_1\varphi'$ is omitted from (1), these become

$$D_6(\xi+h\eta')+F_6\varphi_0-\rho\omega^2R\xi=0;$$
$$D_6(\xi'+h\eta'')+F_6\varphi_0'+\lambda\rho\omega^2h\eta=0, \tag{14}$$

where λ is the correction factor, for transverse inertia, which we now proceed to determine by comparing solutions of (14), for fully coated and uncoated plates of finite length, with the corresponding solutions of (1).

Eliminating ξ from (14) we find, for the equation governing η,

$$\eta''+\delta_i^2\eta=0 \tag{15}$$

where

$$\delta_i^2=\frac{\lambda_i}{R}\left(\frac{\omega^2}{\omega_i^2}-1\right),\quad i=c,\,u. \tag{16}$$

The resonant frequencies of the antisymmetric modes of a fully coated or a fully uncoated plate, of length l, are obtained by applying the boundary conditions $Q_x=0$ at $x=\pm l/2$ to the solution, $\eta=B\sin\delta_ix$, of (15), with the result $\delta_il=n\pi$, $n=1, 3, 5, \cdots$, or,

$$\frac{\omega^2}{\omega_i^2}=1+\frac{Rn^2\pi^2}{\lambda_il^2}. \tag{17}$$

The corresponding solution of (1) was shown in II and III to yield, as a first approximation,

$$\frac{\omega^2}{\omega_i^2}=1+\frac{k_in^2h^2}{l^2}, \tag{18}$$

where

$$k_i=\kappa_i^2(1+1/g_i),\quad g_i=R\kappa_i^2c_{66}/D_1(1+J),$$

$$J=\frac{(e_{11}-e_{12}c_{12}/c_{22})^2}{(c_{11}-c_{12}^2/c_{22})(K_1+e_{12}^2/c_{22})},$$

$$K_1=(1+4\pi\kappa_{11})/4\pi.$$

From (17) and (18), we find

$$\lambda_i=\pi^2/12k_i,\quad i=c,\,u. \tag{19}$$

Hence,

$$\delta_i^2=\frac{\pi^2}{k_ih^2}\left(\frac{\omega^2}{\omega_i^2}-1\right). \tag{20}$$

[2] A. W. Lawson, Phys. Rev. 62, 71 (1942).

In the absence of piezoelectric effect, $J=0$ and $\kappa^2=\pi^2/12$, so that (19) reduces to $\lambda=(1+g)/g$, as in IV.

THICKNESS-SHEAR VIBRATIONS OF A PARTIALLY COATED PLATE OF FINITE LENGTH

Equation (15), with δ_i given by (20), is now to be solved for the case illustrated in Fig. 1. Considering first the antisymmetric modes, the solution of (15) is taken as

$$\eta_c=B\sin\delta_c x,\qquad \eta_u=C\cos\delta_u x+D\sin\delta_u x \tag{21}$$

for the coated and uncoated portions, respectively.

The boundary conditions require the evanescence of the shear at $x=l/2$ and the continuity of shear and deflection at $x=l_c/2$. These become, respectively,

$$\eta_u'(l/2)=0;$$

$$\frac{\kappa_c^2c_{66}h}{1-\omega_c^2/\omega^2}\eta_c'(l_c/2)+\frac{\kappa_c^2e_{26}\varphi_0}{1-\omega_c^2/\omega^2}=\frac{\kappa_u^2c_{66}h}{1-\omega_u^2/\omega^2}\eta_u'(l_c/2); \tag{22}$$

$$\eta_c(l_c/2)=\eta_u(l_c/2).$$

Substituting (21) into (22), the secular equation is found to be

$$\kappa_u^2k_c\delta_c\tan(\delta_cl_c/2)=\kappa_c^2k_u\delta_u\cot(\delta_ul_u/2), \tag{23}$$

where $l_u=l-l_c$. To put this equation into a form suitable for computing its roots, let

$$\delta_u/\delta_c=a,\quad \gamma=\delta_cl_c/2,\quad r=l_c/l. \tag{24}$$

Referring to (20), and noting that $\omega_u>\omega_c$, we see that a is real when ω is outside the range of frequencies between ω_c and ω_u and that a is imaginary when ω lies between ω_c and ω_u. In the latter case, let $a=ia_1$. Then (23) becomes

$$\begin{aligned}&\tan\gamma\tan(r^{-1}-1)a\gamma=ak_u\kappa_c^2/k_c\kappa_u^2,\quad \omega<\omega_c,\,\omega>\omega_u;\\&\tan\gamma\tanh(r^{-1}-1)a_1\gamma=a_1k_u\kappa_c^2/k_c\kappa_u^2,\quad \omega_c<\omega<\omega_u.\end{aligned} \tag{25}$$

From the first of (24)

$$\omega/\omega_c=[g_u(k_c-k_ua^2)/(g_ck_c-g_uk_ua^2)]^{\frac{1}{2}},\qquad \omega<\omega_c,\,\omega>\omega_u; \tag{26}$$

$$\omega/\omega_c=[g_u(k_c+k_ua_1^2)/(g_ck_c+g_uk_ua_1^2)]^{\frac{1}{2}},\quad \omega_c<\omega<\omega_u;$$

and, from the second of (24),

$$l/h=(2\gamma/\pi r)[(g_ck_c-g_uk_ua^2)/(g_u-g_c)]^{\frac{1}{2}},\qquad \omega<\omega_c,\,\omega>\omega_u; \tag{27}$$

$$l/h=(2\gamma/\pi r)[(g_ck_c+g_uk_ua_1^2)/(g_u-g_c)]^{\frac{1}{2}},\quad \omega_c<\omega<\omega_u.$$

Equations (25), (26), and (27) constitute the solution of the problem of finding the frequencies for the case of antisymmetric modes. For the symmetric modes, (26) and (27) remain unaltered, while in (25) tan and tanh are replaced by cot and coth, respectively, and the algebraic sign of the right-hand side of the second of (25) is changed.

In calculating the relation between ω/ω_c and l/h, in (25), (26), and (27), the elastic constants enter only in the form κ_c, g_u/g_c, and k_u/k_c. For a given plate g_i is determined by the material constants in accordance with (12) and (18). The calculations are performed by selecting a value of a or a_1, which fixes the ratio ω/ω_c by (26) and the roots, γ, of (25), for a particular ratio of electrode coating r. The values of γ are then inserted in (27) to obtain the series of plate dimensions, l/h, which have the selected frequency ratio ω/ω_c. Such computations were performed for the AT cut of quartz and $r=\frac{1}{2}$, i.e., with electrodes covering half the surface. The resulting frequency spectrum is shown in Fig. 2. In the sets of three curves, in Fig. 2, the lowest is for the fully coated, the central for half coated, and the upper for the fully uncoated plate.

VARIATION OF FREQUENCY WITH LENGTH OF COATING

The influence of the fraction of coating (r) on the frequencies of a plate of given length may be obtained as follows. Replace γ, in (25), by its expression in terms of r and l/h, according to (27). In the revised equation (25), choose a value of l/h and the corresponding set of values of ω/ω_c (or a) determined previously. For l/h and ω/ω_c, calculate the roots, r, of the

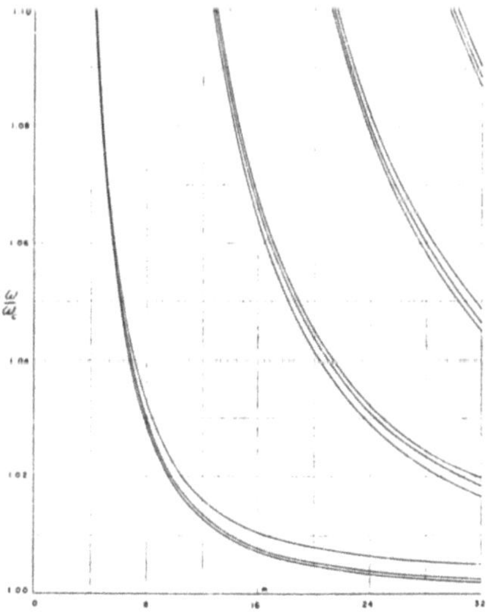

FIG. 2. Frequencies of thickness-shear vibration of a plate with various fractions of electrode coating on the surface.

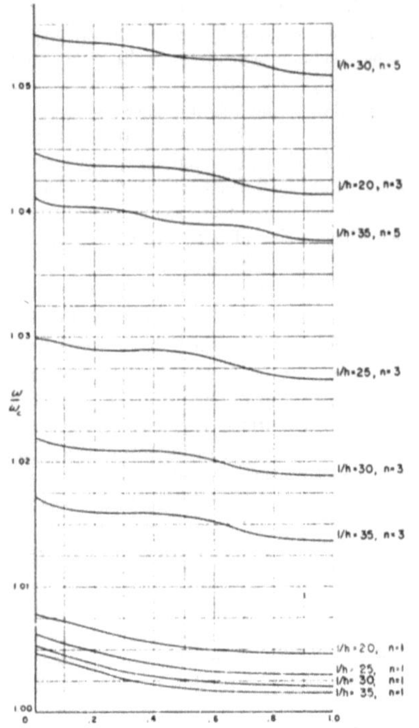

FIG. 3. Variation of thickness-shear frequencies with fraction of surface which is coated with electrode.

revised (25). This procedure yields data from which may be plotted curves of frequency (ω/ω_c) *versus* fraction of coating (r) for each value of l/h. Such curves are illustrated in Fig. 3.

UNCOUPLED THICKNESS-SHEAR MODES

Additional insight into the coupling phenomenon involved is obtained by studying the special roots of the first of (25) for which the transcendental equation degenerates to an algebraic one. We note that the first of (25) is satisfied by

$$\gamma = n\pi/2, \quad n=1,3,5,\cdots; \qquad (r^{-1}-1)a\gamma = m\pi/2, \quad m=0,2,4,\cdots. \tag{28}$$

Hence, the transcendental equation (25) for the frequencies reduces to

$$a = mr/n(1-r). \tag{29}$$

These expressions for a and γ are now inserted in (26) and (27) to obtain frequencies ω/ω_c and corresponding lengths l/h in terms of m and n; by eliminating m and n, in turn, between the formulas for ω/ω_c and l/h, there results

$$\omega^2/\omega_c^2 = 1 + k_c n^2 h^2/l_c^2, \qquad \omega^2/\omega_u^2 = 1 + k_u m^2 h^2/l_u^2. \tag{30}$$

These are the equations of the uncoupled thickness-shear modes. Comparing (30) with (18), we see that the thickness-shear overtones of a partially coated plate act as though they were produced by the coupling of two simple systems. One of these is comprised of the thickness-shear overtones of a fully coated plate whose length is that of the coated portion of the partially coated plate. The other system is comprised of the thickness-shear overtones of a completely uncoated plate whose length is the total uncoated length of the partially coated plate.

Reprinted from JOURNAL OF APPLIED PHYSICS, Vol. 25, No. 1, 25–27, January, 1954

Suppression of Overtones of Thickness-Shear and Flexural Vibrations of Crystal Plates*

R. D. MINDLIN AND H. DERESIEWICZ
Department of Civil Engineering, Columbia University, New York, New York
(Received May 7, 1953)

It is shown how electrodes, coated on the surfaces of thickness-shear plates, may be shaped so that, when the driving frequency coincides with a particular resonant frequency, only that resonant mode is excited.

INTRODUCTION

ELECTRODES, coated on the surface of a piezoelectric crystal, may be shaped with either of two purposes in view. It may be desired, in the case of filters, to suppress resonances of all modes, except one, regardless of the frequency of the driving voltage. This problem has been considered by J. J. Vormer,[1] for longitudinal vibrations of a rod, and by J. A. Lewis[2] for coupled thickness-shear and flexural vibrations. On the other hand, in the case of oscillators, it is desirable that, when the driving frequency coincides with the appropriate resonant frequency, no other modes of motion be excited. In the present paper it is shown how this may be done in the case of coupled thickness-shear and flexural vibrations of plates, within the framework of Mindlin's[3] plate equations.

PLATE EQUATIONS

The thickness (h) of the plate is taken parallel to the y direction of a rectangular coordinate system, and only vibratory motion, independent of the z coordinate, at frequency ω, is considered. As shown in III and in an accompanying paper,[4] the equations of motion are then

$$\begin{aligned} D_1\xi''-D_6(\xi+h\eta')+F_1\varphi''-F_6\varphi_0+\rho\omega^2R\xi&=0;\\ D_6(\xi'+h\eta'')+F_6\varphi_0'+\rho\omega^2h\eta&=0;\\ (R+4\pi C_1)\varphi''-4\pi F_1\xi''&=0, \end{aligned}\tag{1}$$

where primes indicate differentiation with respect to x, ξ and η are the tangential and normal components of displacement, φ is the electric potential, and φ_0 is the potential applied at the surface electrodes. The coefficients in Eqs. (1) are given by

$$D_1=(c_{11}-c_{12}^2/c_{22})R;\qquad F_1=(e_{11}-c_{12}e_{12}/c_{22})R;$$
$$D_6=\kappa_c^2c_{66};\qquad F_6=\kappa_c^2e_{26};$$
$$C_1=(\kappa_{11}+e_{12}^2/c_{22})R;\qquad R=h^2/12;$$
$$\kappa_c^2=\alpha_c^2(1+e_{26}^2/c_{66}K_2)/3;$$
$$K_i=(1+4\pi\kappa_{ii})/4\pi,\ i=1,2;$$
$$\alpha_c=\tan^{-1}[\alpha_c(1+K_2c_{66}/e_{26}^2)],\ (1^{st}\text{ root})$$

in which c_{ij}, e_{ij}, κ_{ij} are the elastic, piezoelectric, and susceptibility coefficients, respectively.

Eliminating φ from Eqs. (1), we have

$$\begin{aligned} \omega_c^2Rg_c^{-1}\xi''+(\omega^2-\omega_c^2)\xi-\omega_c^2h\eta'&=F_6\varphi_0/\rho R;\\ \omega_c^2h\eta''+12\omega^2h^{-1}\eta+\omega_c^2\xi'&=0, \end{aligned}\tag{2}$$

where

$$\omega_c^2=D_6/\rho R,$$
$$g_c=R/S(1+J),\quad S=D_1/D_6,$$
$$J=\frac{(e_{11}-e_{12}c_{12}/c_{22})^2}{(c_{11}-c_{12}^2/c_{22})(K_1+e_{12}^2/c_{22})}.$$

If η is eliminated from Eqs. (2), the equation governing ξ is found to be

$$(d^2/dx^2+\delta_1^2)(d^2/dx^2+\delta_2^2)\xi=C\varphi_0,\tag{3}$$

where

$$C=F_6\omega^2g_c/\rho R^3\omega_c^4,$$
$$2(\delta_1^2,\delta_2^2)=S\delta_0^4\{1+g_c\pm[(1-g_c)^2+4g_c\omega_c^2/\omega^2]^{\frac{1}{2}}\},$$
$$\delta_0^4=\omega^2/RS\omega_c^2.$$

The boundary conditions for free edges $x=\pm l/2$ are

$$\left.\begin{aligned} M_x&=D_1\xi'+F_1\varphi'=0;\\ Q_x&=D_6(\xi+h\eta')+F_6\varphi_0=0, \end{aligned}\right\}\text{on } x=\pm l/2$$

and these reduce to

$$\left.\begin{aligned} \xi'&=0;\\ \xi''+S\delta_0^4g_c\xi&=0, \end{aligned}\right\}\text{on } x=\pm l/2.\tag{4}$$

Thus, the differential equation to be solved is Eq. (3) with boundary conditions (4).

ELECTRODE SHAPE

It may be seen, from Eqs. (2), that the applied potential, φ_0, plays the mechanical role of a tangential surface traction. Suppose, now, that the electrodes do not cover the surfaces of the plate completely, but have varying widths (in the z direction). Then, strictly speaking, some z-dependent motion will be produced which will be coupled with the x-dependent motion governed by Eqs. (3) and (4). However, if we neglect this coupling, the x-dependent motion will be as if forced by an electrode potential which varies with the x coordinate in proportion to the electrode width. To

* This investigation was supported by the U. S. Army Signal Corps Engineering Laboratories under a contract with Columbia University.
[1] J. J. Vormer, Proc. Inst. Radio Engrs. 39, 1086–1087 (1951).
[2] J. A. Lewis, Bell Telephone Laboratories, Tech. Mem. MM 52-140-22, June 25, 1952.
[3] R. D. Mindlin, J. Appl. Mech. 18, 31–38 (1951); J. Appl. Phys. 22, 316–323 (1951), 23, 83–88 (1952). These papers are referred to as I, II, and III.
[4] R. D. Mindlin and H. Deresiewicz, J. Appl. Phys. 24, 21 (1953).

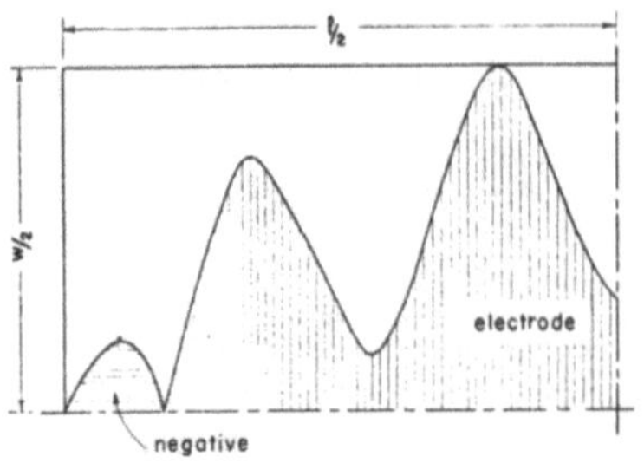

FIG. 1. Electrode shape for exciting[2] the coupled mode comprised of the first thickness-shear mode and the ninth flexural mode ($n=1$, $r=9$; $l/h=6.992$, $\omega/\omega_c=1.037$).

this approximation, then, $\varphi_0=\varphi_0(x)$, where φ_0 is now proportional to the electrode width in the z direction.

Consider the function

$$\varphi_0=\varphi_1\cos\beta_1 x+\varphi_2\cos\beta_2 x, \tag{5}$$

where φ_1, φ_2, β_1, β_2 are constants. Substituting Eq. (5) in Eq. (3), we find that the differential equation has the particular integral

$$\xi=U_1\cos\beta_1 x+U_2\cos\beta_2 x, \tag{6}$$

provided

$$U_1=\frac{C\varphi_1}{(\beta_1^2-\delta_1^2)(\beta_1^2-\delta_2^2)},\qquad U_2=\frac{C\varphi_2}{(\beta_2^2-\delta_1^2)(\beta_2^2-\delta_2^2)}. \tag{7}$$

Now, δ_1 and δ_2 are functions of ω. Let β_1 be the value which δ_1 approaches as ω approaches a frequency ω'. Also, let β_2 be that value which δ_2 approaches as δ_1 approaches β_1. Then U_1 and U_2 approach infinity, but

$$\lim_{\omega\to\omega'}\left\{\frac{U_1}{U_2}\right\}=-\frac{\varphi_1}{\varphi_2}\lim_{\omega\to\omega'}\left\{\frac{(\beta_2^2-\delta_2^2)}{(\beta_1^2-\delta_1^2)}\right\}$$

$$=-\frac{\varphi_1}{\varphi_2}\lim_{\omega\to\omega'}\left\{\frac{d\delta_2^2/d\omega^2}{d\delta_1^2/d\omega^2}\right\}$$

$$=-\frac{\varphi_1}{\varphi_2}\,\frac{a^4(1+g_c)^2-g_c(1+a^2)^2}{(1+g_c)^2-g_c(1+a^2)^2}, \tag{8}$$

where $a=\delta_2/\delta_1=\beta_2/\beta_1$, and φ_1 and φ_2 are arbitrary for an infinite plate.

Considering, now, a finite plate of length l, the boundary conditions (4) give

$$U_1/U_2=a\sin a\gamma/\sin\gamma; \tag{9}$$

$$(a^2-g_c)\tan\gamma=a(1-a^2g_c)\tan a\gamma, \tag{10}$$

where $\gamma=\delta_1 l/2$. Hence, for a plate of finite length, it is necessary, from Eqs. (8) and (9), that

$$\frac{\varphi_1}{\varphi_2}=-\frac{a\sin a\gamma}{\sin\gamma}\,\frac{(1+g_c)^2-g_c(1+a^2)^2}{a^4(1+g_c)^2-g_c(1+a^2)^2}. \tag{11}$$

As shown in II and III, Eq. (10) is the secular equation for the free vibrations of the plate. The frequencies are obtained by selecting a value of a, which fixes the frequency at

$$\omega=\omega_c[1-a^2(1+g_c)^2/g_c(1+a^2)^2]^{-\frac{1}{2}}, \tag{12}$$

following which the roots, γ, of Eq. (10) are computed. Then the lengths corresponding to the chosen ω are found from

$$l/h=\gamma(\omega_c/\omega)[(1+a^2)/3(1+g_c)]^{\frac{1}{2}}. \tag{13}$$

Hence, the electrode shape

$$\varphi_0=\varphi_1\cos(2\gamma x/l)+\varphi_2\cos(2a\gamma x/l), \tag{14}$$

where φ_1/φ_2 is given by Eq. (11), excites only one normal mode, of frequency given by Eq. (12), of a plate whose length is given by Eq. (13).

The solution has been carried out for the antisymmetric modes. The same results are obtained for the symmetric modes (excited by split electrodes) except that sine and cosine are interchanged.

ALGEBRAIC FORMULAS

The secular equation (10) is satisfied by (see II and III)

$$\gamma=r\pi/2,\qquad r=1,2,3,\cdots,$$
$$a\gamma=n\pi/2,\qquad n=0,1,2,\cdots, \tag{15}$$

where r and n are odd for the antisymmetric modes and even for the symmetric modes.

For the modes defined by Eqs. (15), the solution (14) takes the simple form

$$\varphi_0=\varphi_1\cos(r\pi x/l)+\varphi_2\cos(n\pi x/l) \tag{16}$$

for the antisymmetric modes and

$$\varphi_0=\varphi_1\sin(r\pi x/l)+\varphi_2\sin(n\pi x/l) \tag{17}$$

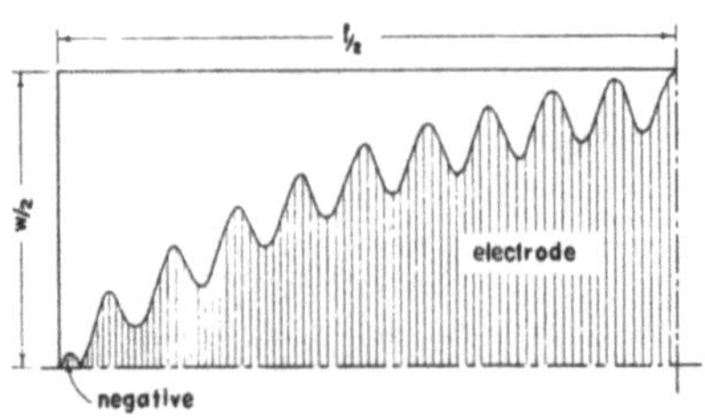

FIG. 2. Electrode shape for exciting the coupled mode comprised of the first thickness-shear mode and the thirty-ninth flexural mode ($n=1$, $r=39$; $l/h=31.176$, $\omega/\omega_c=1.002$).

for the symmetric modes, where

$$\frac{\varphi_1}{\varphi_2}=-\frac{n}{r}(-1)^{(3n+r)/2}\frac{r^4(1+g_c)^2-g_c(r^2+n^2)^2}{n^4(1+g_c)^2-g_c(r^2+n^2)^2}. \tag{18}$$

Examples of these shapes are shown in Figs. 1 and 2 for the AT cut of quartz, for which $g_c=0.283$.

The corresponding frequencies and lengths are given by

$$\omega/\omega_c=g_c^{\frac{1}{2}}(r^2+n^2)[(g_cr^2-n^2)(r^2-g_cn^2)]^{-\frac{1}{2}};$$

$$l/h=(\pi/2)[(g_cr^2-n^2)(r^2-g_cn^2)/3g_c(1+g_c)(r^2+n^2)]^{\frac{1}{2}}.$$

It may be seen that the first term in φ_0 has the mode shape of the flexural overtone and the second term in φ_0 has the mode shape of the thickness-shear overtone. That is, the first term has r half-waves and the second term n half-waves along the length of the plate. Generally, a low order thickness-shear mode is used in a thin plate. Then the coupling takes place with a high-order flexural mode, i.e., n/r is small. In this case

$$\frac{\varphi_1}{\varphi_2}\approx-\frac{n}{r}(-1)^{(3n+r)/2}\frac{(1+g_c)^2-g_c}{g_c}\ll 1,$$

so that the electrode has essentially the mode shape of the thickness-shear overtone, with its edge scalloped according to the mode shape of the flexural overtone, as illustrated in Fig. 2. For a given thickness-shear overtone (n) the amplitude of the scallop diminishes with increasing order of flexural overtone, i.e., with increasing length of the plate, as shown in Fig. 3.

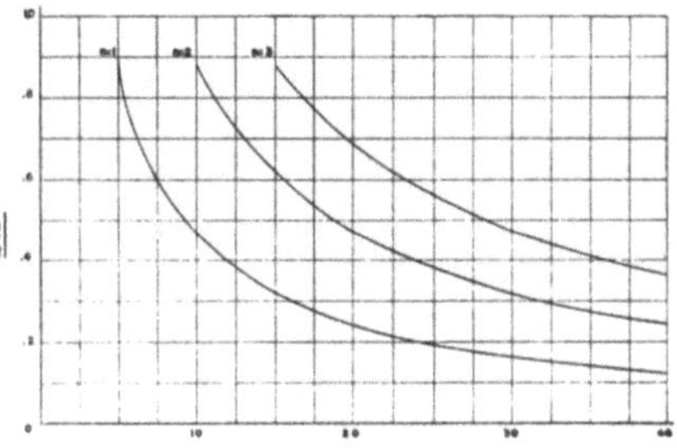

FIG. 3. Ratio of amplitudes of short and long wavelength portions of electrode shape as a function of the order (r) of flexural overtone, for various values of the order (n) of thickness-shear overtone.

ACKNOWLEDGMENT

We wish to acknowledge an illuminating discussion with Dr. J. A. Lewis.

Reprinted from JOURNAL OF APPLIED PHYSICS, Vol. 25, No. 10, 1329–1332, October, 1954

Thickness-Shear and Flexural Vibrations of a Circular Disk*

R. D. MINDLIN AND H. DERESIEWICZ
Department of Civil Engineering, Columbia University, New York, New York
(Received February 4, 1954)

Equations governing the coupled thickness-shear and flexural vibrations of isotropic, elastic plates are solved for the case of a circular disk with free edges. The results of computations are given for the resonant frequencies analogous to those which would be excited in an AT-cut quartz disk with full electrode coating. A marked difference is observed between the frequency spectrum of the disk and the spectrum of cylindrical vibrations of a rectangular plate. The difference is due to the presence of thickness-twist modes of motion, in the disk, which, along with the thickness-shear modes, are coupled to the flexural modes.

INTRODUCTION

THIS is the seventh in a series of papers on vibrations of plates. The first paper[1] contained a derivation of isotropic plate equations which accommodate the phenomenon of coupling between flexural and thickness-shear modes of motion. These equations were later extended to take into account crystal plates,[2] piezoelectric crystal plates, plates of varying thickness, plates with incomplete electrodes, and plates with specially shaped electrodes. (The first[1] and second[2] papers are referred to as I and II in the sequel.) The solutions of the various equations, given as illustrative examples in these papers, were all confined to cylindrical motions, that is, to one-dimensional problems.

We return, now, to the simplest (isotropic) form of the equations to solve a two-dimensional problem. We have in mind the antisymmetric modes of motion that are excited in an AT-cut quartz disk. However, the additional mathematical complexities introduced by the circular shape of the plate make it advisable to consider, first, the corresponding motions of an isotropic disk. It seems likely that the modes of motion and the frequency spectrum of the isotropic disk will be found to differ insignificantly from those of the quartz disk. This is because the two controlling parameters are the shape of the plate and the ratio of the plate-flexure modulus to the thickness-shear modulus. The former is the same for the two cases, and the latter has been made identical, in the computations, simply by choosing Poisson's ratio, for the isotropic material, to give the modulus ratio appropriate to the AT-cut of quartz. The lateral elastic constants, it is true, are not quite the same, but the discrepancy is small, and its influence on the motion is probably slight.

We give, here, the exact solutions of the equations for the isotropic, free disk and the results of detailed computations of the frequencies in the neighborhood of the thickness-shear frequency of an infinite plate. The frequency spectrum of the disk is found to have a different character from that obtained in II for the cylindrical vibrations of a rectangular plate. At first glance, it appears that the thickness-shear fundamental and its overtones are multivalued, i.e., corresponding to the fundamental and to each overtone of thickness-shear in the rectangular plate, there appears to be more than one resonant frequency in the disk. Upon closer examination, it develops that the additional resonances are due to the presence of a thickness-twist mode and its overtones, in which the displacement is a twist about an axis normal to the plate. The complete frequency spectrum is the result of coupling between thickness-shear and flexure and between thickness-twist and flexure.

PLATE EQUATIONS

In I and II, the plate equations were deduced from the three-dimensional equations of elasticity, referred to a rectangular coordinate system. The resulting equations may be transformed to a polar coordinate system, appropriate to the circular disk, but it is simpler to rederive the plate equations in the new system of coordinates. Thus, the assumed form of the displacements, in place of I(10) or II(5), is

$$u_r \approx y\psi_r(r,\theta,t), \quad u_\theta \approx y\psi_\theta(r,\theta,t), \quad u_y \approx \eta(r,\theta,t), \tag{1}$$

where r and θ are polar coordinates defined by $x=r\cos\theta$, $z=r\sin\theta$, i.e., the y axis is normal to the plane of the plate.

The derivation of the plate equations proceeds as before. Thus, corresponding to I(14,15) or II(10), we have plate-stress equations of motion

$$\begin{aligned} &\frac{\partial M_r}{\partial r}+\frac{1}{r}\frac{\partial M_{r\theta}}{\partial\theta}+\frac{M_r-M_\theta}{r}-Q_r=\frac{\rho h^3}{12}\frac{\partial^2\psi_r}{\partial t^2} \\ &\frac{\partial M_{r\theta}}{\partial r}+\frac{1}{r}\frac{\partial M_\theta}{\partial\theta}+\frac{2M_{r\theta}}{r}-Q_\theta=\frac{\rho h^3}{12}\frac{\partial^2\psi_\theta}{\partial t^2} \\ &\frac{\partial Q_r}{\partial r}+\frac{1}{r}\frac{\partial Q_\theta}{\partial\theta}+\frac{Q_r}{r}=\rho h\frac{\partial^2\eta}{\partial t^2}, \end{aligned} \tag{2}$$

where ρ is the density, h is the thickness of the plate and the plate-stress components are given in terms of the

* This investigation was supported by the U. S. Army Signal Corps and the U. S. Office of Naval Research.
[1] R. D. Mindlin, J. Appl. Mech. 18, 31–38 (1951).
[2] R. D. Mindlin, J. Appl. Phys. 22, 316–323 (1951).

plate-displacements by

$$M_r=D\left[\frac{\partial\psi_r}{\partial r}+\frac{\nu}{r}\left(\psi_r+\frac{\partial\psi_\theta}{\partial\theta}\right)\right]$$

$$M_\theta=D\left[\frac{1}{r}\left(\psi_r+\frac{\partial\psi_\theta}{\partial\theta}\right)+\nu\frac{\partial\psi_r}{\partial r}\right]$$

$$M_{r\theta}=\frac{D}{2}(1-\nu)\left[\frac{1}{r}\left(\frac{\partial\psi_r}{\partial\theta}-\psi_\theta\right)+\frac{\partial\psi_\theta}{\partial r}\right] \tag{3}$$

$$Q_r=\kappa^2\mu h\left(\psi_r+\frac{\partial\eta}{\partial r}\right)$$

$$Q_\theta=\kappa^2\mu h\left(\psi_\theta+\frac{1}{r}\frac{\partial\eta}{\partial\theta}\right),$$

where $D=Eh^3/12(1-\nu^2)$, E, μ, ν are Young's modulus, the shear modulus and Poisson's ratio, respectively, and $\kappa^2=\pi^2/12$.

If (3) were substituted in (2) the resulting plate-displacement equations of motion would be those corresponding to I(16) or II(11). However, in the present case it is more convenient to work with the alternative formulation of the equations given by I(62,63). In polar coordinates, again omitting a factor e^{ipt}, these become

$$\psi_r=(\sigma_1-1)\frac{\partial\eta_1}{\partial r}+(\sigma_2-1)\frac{\partial\eta_2}{\partial r}+\frac{1}{r}\frac{\partial H}{\partial\theta}$$

$$\psi_\theta=(\sigma_1-1)\frac{\partial\eta_1}{r\partial\theta}+(\sigma_2-1)\frac{\partial\eta_2}{r\partial\theta}-\frac{\partial H}{\partial r} \tag{4}$$

$$\eta=\eta_1+\eta_2,$$

$$(\nabla^2+\delta_1^2)\eta_1=0$$

$$(\nabla^2+\delta_2^2)\eta_2=0 \tag{5}$$

$$(\nabla^2+\omega^2)H=0,$$

where

$$\delta_1^2,\ \delta_2^2=\tfrac{1}{2}\delta_0^4\{R+S\pm[(R-S)^2+4\delta_0^{-4}]^{\frac{1}{2}}\}, \tag{6}$$

$$\sigma_1,\ \sigma_2=(\delta_2^2,\delta_1^2)(R\delta_0^4-S^{-1})^{-1}, \tag{7}$$

$$\omega^2=2(R\delta_0^4-S^{-1})/(1-\nu), \tag{8}$$

$$R=h^2/12,\quad S=D/\kappa^2\mu h,\quad \delta_0^4=\rho p^2h/D. \tag{9}$$

SOLUTION OF EQUATIONS OF MOTION

In view of the modes of motion of interest in the present case, the appropriate solutions of (5) are

$$\eta_1=A_1J_1(\delta_1r)\cos\theta$$

$$\eta_2=A_2J_1(\delta_2r)\cos\theta \tag{10}$$

$$H=A_3J_1(\omega r)\sin\theta,$$

where $J_1(x)$ is Bessel's function of the first kind, of order unity, and A_1, A_2, A_3 are arbitrary constants. These are the modes of circumferential order unity. There are also modes of higher and lower circumferential order in which, for example, $\eta_1=A_1J_n(\delta_1r)\cos n\theta$, $n=0, 1, 2, \cdots$, but we are not concerned with them here.

For the vibrations of a plate with free edges, the boundary conditions are

$$M_r=M_{r\theta}=Q_r=0\quad\text{at}\quad r=a, \tag{11}$$

where a is the radius of the disk. By using (3), (4), and (10), Eqs. (11) become

$$\left[\left(2-\frac{\delta_1^2a^2}{1-\nu}\right)J_1(\delta_1a)-\delta_1aJ_0(\delta_1a)\right](\sigma_1-1)A_1$$

$$+\left[\left(2-\frac{\delta_2^2a^2}{1-\nu}\right)J_1(\delta_2a)-\delta_2aJ_0(\delta_2a)\right](\sigma_2-1)A_2$$

$$+[\omega aJ_0(\omega a)-2J_1(\omega a)]A_3=0,$$

$$[\delta_1aJ_0(\delta_1a)-2J_1(\delta_1a)](\sigma_1-1)A_1$$

$$+[\delta_2aJ_0(\delta_2a)-2J_1(\delta_2a)](\sigma_2-1)A_2$$

$$+\left[\left(2-\frac{\omega^2a^2}{2}\right)J_1(\omega a)-\omega aJ_0(\omega a)\right]A_3=0, \tag{12}$$

$$[\delta_1aJ_0(\delta_1a)-J_1(\delta_1a)]\sigma_1A_1$$

$$+[\delta_2aJ_0(\delta_2a)-J_1(\delta_2a)]\sigma_2A_2+J_1(\omega a)A_3=0.$$

FREQUENCY EQUATION

Equations (12) comprise a system of three homogeneous, algebraic equations in the unknowns A_1, A_2, A_3. The vanishing of the determinant of their coefficients yields an equation for the determination of the frequency p through δ_1, δ_2, and ω. This secular equation, expressed in a form convenient for desk-calculator computation, reads

$$(1-\beta^2)\gamma^2\Gamma_1\Gamma_2+\frac{1-g\beta^2}{1+g}\left[\frac{2(1+\beta^2)}{(1-\nu)(1+g)}\right]^{\frac{1}{2}}\gamma^2\Gamma_1\Gamma_3$$

$$-\frac{1}{\beta}\frac{1-g\beta^2}{1+g}\left(3+\frac{1+\beta^2}{1+g}+\frac{g-\beta^2}{1-g\beta^2}-\frac{\beta^2}{1-\nu}\gamma^2\right)\gamma\Gamma_1$$

$$+\frac{1}{\beta}\frac{g-\beta^2}{1+g}\left[\frac{2(1+\beta^2)}{(1-\nu)(1+g)}\right]^{\frac{1}{2}}\gamma^2\Gamma_2\Gamma_3$$

$$-\frac{g-\beta^2}{1+g}\left(3+\frac{1}{\beta^2}\frac{1+\beta^2}{1+g}+\frac{1-g\beta^2}{g-\beta^2}-\frac{\gamma^2}{1-\nu}\right)\gamma\Gamma_2$$

$$-\frac{1-g\beta^2}{1+g}\left(1+\frac{1}{\beta^2}\frac{g-\beta^2}{1-g\beta^2}\right)\left[\frac{2(1+\beta^2)}{(1-\nu)(1+g)}\right]^{\frac{1}{2}}\gamma\Gamma_3$$

$$+\frac{1-\beta^2}{\beta}\left[4-\frac{g(1+\beta^2)}{(1-\nu)(1+g)}\gamma^2\right]=0, \tag{13a}$$

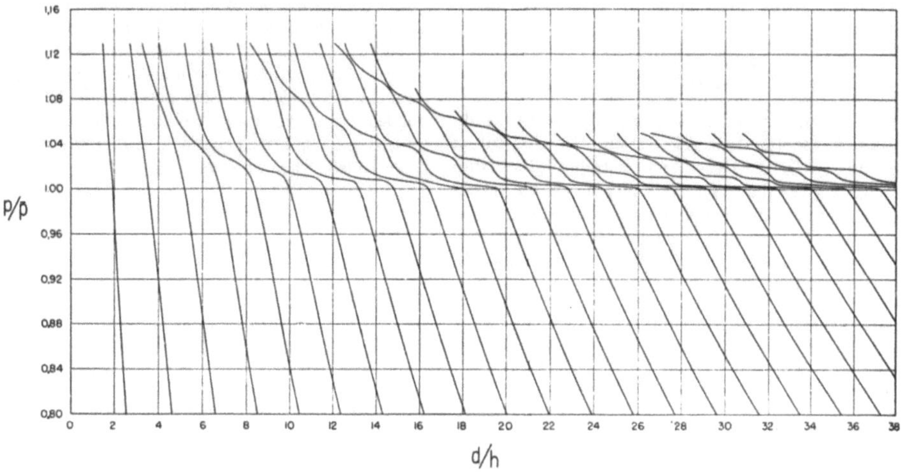

FIG. 1. Resonant frequencies of modes of vibration of circumferential order unity of a circular disk ($\nu=0.312$) as computed from Eqs. (13), (14), and (15). The ordinate is the ratio of the resonant frequency (p) to the frequency ($\bar{p}$) of the first thickness-shear mode of an infinite plate. The abscissa is the ratio of the diameter (d) to the thickness (h) of the disk.

where

$$\beta=\delta_2/\delta_1, \quad \gamma=\delta_1 a, \quad g=R/S$$

$$\Gamma_i=J_0(c_i\gamma)/J_1(c_i\gamma), \quad i=1, 2, 3$$

$$c_1=1, \quad c_2=\beta, \quad c_3=\beta\left[\frac{2(1+g)}{(1-\nu)(1+\beta^2)}\right]^{\frac{1}{2}}.$$

Now, $\delta_1^2>0$ for all values of $p>0$ while $\delta_2^2\gtreqless 0$ according as $p\gtreqless\bar{p}$, where $\bar{p}=\pi(\mu/\rho)^{\frac{1}{2}}/h$ is the frequency of the first thickness-shear mode of an infinite plate. Hence, β will be real or imaginary according as $p\gtrless\bar{p}$. Equation (13a) is a useful form for $p\geqslant\bar{p}$. In the range $p<\bar{p}$, let $\beta=i\beta_1$, so that the frequency equation reads

$$(1+\beta_1^2)\gamma^2\Gamma_1G_2+\frac{1+g\beta_1^2}{1+g}\left[\frac{2(1-\beta_1^2)}{(1-\nu)(1+g)}\right]^{\frac{1}{2}}\gamma^2\Gamma_1G_3$$
$$-\frac{1}{\beta_1}\frac{1+g\beta_1^2}{1+g}\left(3+\frac{1-\beta_1^2}{1+g}+\frac{g+\beta_1^2}{1+g\beta_1^2}+\frac{\beta_1^2}{1-\nu}\gamma^2\right)\gamma\Gamma_1$$
$$-\frac{1}{\beta_1}\frac{\beta_1^2+g}{1+g}\left[\frac{2(1-\beta_1^2)}{(1-\nu)(1+g)}\right]^{\frac{1}{2}}\gamma^2G_2G_3$$
$$-\frac{\beta_1^2+g}{1+g}\left(3-\frac{1}{\beta_1^2}\frac{1-\beta_1^2}{1+g}+\frac{1+g\beta_1^2}{g+\beta_1^2}-\frac{\gamma^2}{1-\nu}\right)\gamma G_2$$
$$+\frac{1+g\beta_1^2}{1+}\left(-1+\frac{1}{\beta_1^2}\frac{\beta_1^2+g}{1+g\beta_1^2}\right)\left[\frac{2(1-\beta_1^2)}{(1-\nu)(1+g)}\right]^{\frac{1}{2}}\gamma G_3$$
$$+\frac{1+\beta_1^2}{\beta_1}\left[4-\frac{g(1-\beta_1^2)}{(1-\nu)(1+g)}\gamma^2\right]=0, \quad (13b)$$

where

$$G_2=I_0(\beta_1\gamma)/I_1(\beta_1\gamma),$$

$$G_3=I_0(c'\gamma)/I_1(c'\gamma),$$

$$c'=\beta_1\left[\frac{2(1+g)}{(1-\nu)(1-\beta_1^2)}\right]^{\frac{1}{2}},$$

and $I_0(x)$, $I_1(x)$ are modified Bessel's functions of the first kind.

From the relation $\beta=\delta_2/\delta_1$ and Eq. (6), explicit expressions for the frequency may be obtained. Thus,

$$p/\bar{p}=[1-\beta^2(1+g)^2/g(1+\beta^2)^2]^{-\frac{1}{2}}, \quad p>\bar{p}$$
$$p/\bar{p}=[1+\beta_1^2(1+g)^2/g(1-\beta_1^2)^2]^{-\frac{1}{2}}, \quad p<\bar{p}. \quad (14)$$

Further, from the relation $\gamma=\delta_1 a$,

$$d/h=\gamma(\bar{p}/p)[(1+\beta^2)/3(1+g)]^{\frac{1}{2}}, \quad p>\bar{p}$$
$$d/h=\gamma(\bar{p}/p)[(1-\beta_1^2)/3(1+g)]^{\frac{1}{2}}, \quad p<\bar{p}, \quad (15)$$

where d is the plate diameter.

COMPUTATION OF FREQUENCIES

Equations (13), (14), and (15) constitute the solution of the problem. For a given material the values of ν and g are fixed. Choosing a value of β or β_1 fixes $p/\bar{p}$ by (14) and determines an infinite set of roots γ of (13). For the particular β or β_1 each of these roots determines a ratio d/h, thus giving an infinite set of d/h corresponding to every value of $p/\bar{p}$.

For the isotropic plate, the ratio of the thickness-shear modulus to the plate flexure modulus is

$$g=\pi^2(1-\nu)/24,$$

while, for the AT-cut of quartz,

$$g=\frac{\pi^2 c_{66}}{12(c_{11}-c_{12}^2/c_{22})},$$

where the elastic constants are referred to the coordinates of the (rotated) AT section. For this cut of quartz, $g=0.283$. Hence, an isotropic plate with $\nu=0.312$ gives the same value of g as an AT quartz plate. The computations were made for this value of ν.

To facilitate the task of finding the roots γ of Eqs. (13), for a series of values of $p/\bar{p}$, tables were computed of the functions:

$J_0(x)/J_1(x)$ 0(0.01)100

$I_0(x)/I_1(x)$ 0(0.01)10, 10(0.02)16, 16(0.1)20, 20(1)37

to six decimals.[3] The subsequent numerical work, carried out on a desk calculator, yielded the resonant frequencies of vibration depicted in Fig. 1.

The flexural modes, in the lower part of Fig. 1, are those of even order $r=2,4,6,\cdots$, and are similar to those in the lower part of Fig. 1 of II. However, the upper parts ($p/\bar{p}>1$) of the two figures are different. For example, in the case of the rectangular plate the thickness-shear fundamental couples with every even flexural mode in turn, but, in the disk, the thickness-shear fundamental appears to couple only with alternate, even, flexural modes. As a result, the thickness-shear fundamental appears to be double-valued.

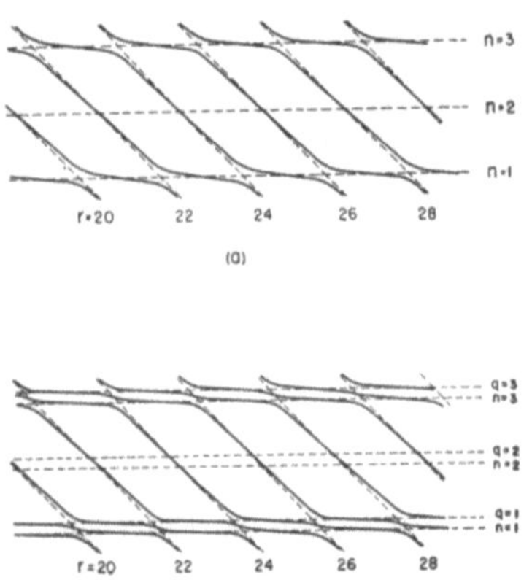

FIG. 2. (a) Frequency spectrum for coupling between thickness-shear and flexural modes. (b) Frequency spectrum for coupling between thickness-shear and flexural modes and between thickness-twist and flexural modes, with no coupling between shear and twist modes.

[3] The computation of the tables of $J_0(x)/J_1(x)$ was performed by S. Poley on equipment made available by the Watson Scientific Computing Laboratory of the International Business Machines Corporation. Both tables are stored in the files of the Department of Civil Engineering, Columbia University.

The apparent anomaly is resolved when it is recognized that there is present a set of thickness-twist modes in addition to thickness-shear and flexural modes. Mathematically, the additional modes derive from the solenoidal, retarded potential H. Their uncoupled frequencies may be obtained, approximately, by setting $\eta=0$, in the equations, so that H is the only potential left. Then we have

$$(\nabla^2+\omega^2)H=0$$

$$\psi_r=-\frac{1}{r}\frac{\partial H}{\partial\theta},\quad \psi_\theta=-\frac{\partial H}{\partial r}$$

$$M_{r\theta}=-\frac{D(1-\nu)}{2}\left(2\frac{\partial^2}{\partial r^2}+\omega^2\right)H$$

$$=-D(1-\nu)\omega^2 A_3 f(\omega r)\sin\theta,$$

where

$$f(\omega r)=\left(\frac{2}{\omega^2 r^2}-\frac{1}{2}\right)J_1(\omega r)-\frac{1}{\omega r}J_0(\omega r).$$

Applying only the boundary condition $M_{r\theta}=0$ on $r=a$, we obtain the secular equation $f(\omega a)=0$, or

$$2c_3\gamma\Gamma_3=4-c_3^2\gamma^2. \tag{16}$$

The roots of (16), in conjunction with (14) and (15), give the approximate frequency spectrum ($p/\bar{p}$ *vs* d/h) of the thickness-twist modes. The curves are asymptotic to $p/\bar{p}=1$ and are very nearly tangent to the almost horizontal portions of the curves in Fig. 1. That is, the thickness-twist modes have frequencies very close to those of the corresponding thickness-shear modes. The effect, of two such neighboring families, on the coupling with the flexural modes is illustrated in Fig. 2. In Fig. 2a is shown the usual spectrum resulting from the coupling of a family of odd thickness-shear modes ($n=1, 3, \cdots$) and even flexural modes ($r=20, 22, \cdots$). In Fig. 2b a family of thickness-twist modes ($q=1, 3, \cdots$), closely paralleling the thickness-shear modes, is added. If the result were to be interpreted as coupling between flexure and thickness-shear alone, it would appear that the thickness-shear modes are double-valued and couple only with alternate flexure modes. This would be especially so if the coupling were strong enough to smooth out the kinks in the almost horizontal portions of the curves, as is the case in Fig. 1. However, when the presence of the thickness-twist modes is taken into account, there are no anomalies.

It should be observed that the case described here is a degenerate one of coupling of three families of modes, because, although there is coupling between thickness-shear and flexure and between thickness-twist and flexure, there is no coupling between thickness-shear and thickness-twist. An example in which there is full coupling of the three families will be described in a subsequent paper.

Paper No.
54—A-15

Axially Symmetric Flexural Vibrations of a Circular Disk[1]

By H. DERESIEWICZ[2] and R. D. MINDLIN,[3] NEW YORK, N. Y.

At high frequencies, the flexural vibrations of a plate are described very poorly by the classical (Lagrange) theory because of neglect of the influence of coupling with thickness-shear vibrations. The latter may be taken into account by inclusion of rotatory inertia and shear-deformation terms in the equations. The resulting frequency spectrum is given, in this paper, for the case of axially symmetric vibrations of a circular disk with free edges and is compared with the spectrum predicted by the classical theory.

Introduction

The problem of free vibrations of a thin, isotropic, elastic, circular disk, of constant thickness, was first attacked by Poisson (1)[4] over a century ago. Basing his investigation on the classical (Lagrange) theory of plates, he obtained the lower frequencies of the axially symmetric flexural motions of such a disk with a traction-free boundary.

It is well known that the classical theory satisfactorily predicts actual behavior only for the first few flexural modes of motion of a plate whose thickness is small in comparison with its other dimensions. For the higher flexural modes the influence of coupling with the thickness-shear mode of motion becomes increasingly important. Hence, the classical theory, in which this effect is not taken into account, ceases to yield reliable information (2). In particular, at the frequency of pure thickness-shear vibration of an infinite plate (i.e., displacement constant in direction and parallel to the plane of the plate) a drastic change occurs in the frequency spectrum of a finite plate (3, 4).

The influence of coupling between flexure and shear is taken into account by inclusion of rotatory inertia and shear-deformation terms in the equations. The resulting change in the frequency spectrum has been given previously for the case of a free-free beam (3) and a class of antisymmetric modes of motion of a circular disk (4). The present paper contains a discussion of the axially symmetric, flexural vibrations of a free disk, with emphasis on behavior in the neighborhood of the thickness-shear frequency. The spectrum obtained with consideration of the effects of rotatory inertia and shear deformation is compared with that predicted by the classical theory.

---

[1] This investigation was supported by the Office of Naval Research under Contract Nonr-266(09).

[2] Assistant Professor of Civil Engineering, Columbia University. Assoc. Mem. ASME.

[3] Professor of Civil Engineering, Columbia University. Mem. ASME.

[4] Numbers in parentheses refer to the Bibliography at the end of the paper.

Contributed by the Applied Mechanics Division for presentation at the Annual Meeting, New York, N. Y., November 28–December 3, 1954, of The American Society of Mechanical Engineers.

Discussion of this paper should be addressed to the Secretary, ASME, 29 West 39th Street, New York, N. Y., and will be accepted until one month after final publication of the paper itself in the Journal of Applied Mechanics.

Note: Statements and opinions advanced in papers are to be understood as individual expressions of their authors and not those of the Society. Manuscript received by ASME Applied Mechanics Division, February 5, 1954. Paper No. 54—A-15.

Plate Equations

If account is taken of rotatory inertia and shear deformation, the plate stress-displacement relations in polar co-ordinates, for the axially symmetric case,[5] become

$$\left.\begin{aligned} M_r &= D\left(\frac{\partial\psi}{\partial r} + \frac{\nu}{r}\psi\right) \\ M_\theta &= D\left(\nu\frac{\partial\psi}{\partial r} + \frac{\psi}{r}\right) \\ Q_r &= \kappa^2\mu h\left(\psi + \frac{\partial w}{\partial r}\right) \\ M_{r\theta} &= Q_\theta = 0 \end{aligned}\right\} \qquad [1]$$

where $D = Eh^3/12(1-\nu^2)$, E, μ, ν, and h are Young's modulus, the shear modulus, Poisson's ratio, and the thickness, respectively, and $\kappa^2 = \pi^2/12$. The functions ψ and w are related to the radial and axial components of the displacement according to the approximations

$$\left.\begin{aligned} u_r &\approx z\psi(r,t) \\ u_z &\approx w(r,t) \end{aligned}\right\} \qquad [2]$$

u_θ, the circumferential component of the displacement, is zero.

The plate equations of motion,[6] for the present problem, are

$$\left.\begin{aligned} \frac{\partial M_r}{\partial r} + \frac{M_r - M_\theta}{r} - Q_r &= \frac{\rho h^3}{12}\frac{\partial^2\psi}{\partial t^2} \\ \frac{\partial Q_r}{\partial r} + \frac{Q_r}{r} &= \rho h\frac{\partial^2 w}{\partial t^2} \end{aligned}\right\} \qquad [3]$$

where ρ is the density of the plate.

If we now insert Equations [1] in [3] we obtain the plate displacement equations of motion. Omitting the time factor e^{ipt}, these become

$$\left.\begin{aligned} &\left(\frac{d^2}{dr^2} + \frac{1}{r}\frac{d}{dr} - \frac{1}{r^2} + \frac{\rho p^2 h^3}{12D}\right)\psi - \frac{\kappa^2\mu h}{D}\left(\psi + \frac{dw}{dr}\right) = 0 \\ &\left(\frac{d}{dr} + \frac{1}{r}\right)\psi + \left(\frac{d^2}{dr^2} + \frac{1}{r}\frac{d}{dr} + \frac{\rho p^2}{\kappa^2\mu}\right)w = 0 \end{aligned}\right\} \qquad [4]$$

Equations [4] may be uncoupled by differentiating the second equation once and subtracting the result from the first one. This procedure yields an expression for ψ in terms of w, which, when inserted into the second of Equations [4], gives a single equation on w only

$$\left(\frac{d^2}{dr^2} + \frac{1}{r}\frac{d}{dr} + \delta_1^2\right)\left(\frac{d^2}{dr^2} + \frac{1}{r}\frac{d}{dr} + \delta_2^2\right)w = 0 \qquad [5]$$

---

[5] These are given, in the general case, by equations [3] of reference (4).

[6] For the general case, see equations [2] of reference (4).

1

where

$$\delta_1^2, \delta_2^2 = \frac{\delta_0^4}{2}\left[R + S \pm \sqrt{(R-S)^2 + 4\delta_0^{-4}}\right]$$

$$S = D/\kappa^2\mu h, \delta_0^4 = \rho p^2 h/D, R = h^2/12$$

The expression for ψ in terms of w is given by

$$\psi = (R\delta_0^4 - S^{-1})^{-1}\frac{d}{dr}\left[\frac{d^2}{dr^2} + \frac{1}{r}\frac{d}{dr} + (S\delta_0^4 + S^{-1})\right]w \quad [6]$$

Solution of Equations of Motion

Equation [5] may be solved for w by noting that

$$w = w_1 + w_2 \quad [7]$$

where w_1 and w_2 satisfy, respectively, the equations

$$\left(\frac{d^2}{dr^2} + \frac{1}{r}\frac{d}{dr} + \delta_i^2\right)w_i = 0, \quad i = 1, 2 \quad [8]$$

Hence, the shear displacement ψ is found to be

$$\psi = (\sigma_1 - 1)\frac{dw_1}{dr} + (\sigma_2 - 1)\frac{dw_2}{dr} \quad [9]$$

where

$$\sigma_1, \sigma_2 = (\delta_2^2, \delta_1^2)(R\,\delta_0^4 - S^{-1})^{-1}$$

Both of Equations [8] are Bessel equations of zero order. In dealing with a solid disk, their appropriate solutions are

$$\left.\begin{aligned} w_1 &= A_1 J_0(\delta_1 r) \\ w_2 &= A_2 J_0(\delta_2 r) \end{aligned}\right\} \quad [10]$$

where A_1, A_2 are arbitrary constants.

For a plate with a traction-free edge, the boundary conditions are

$$M_r = Q_r = 0 \quad \text{at } r = a \quad [11]$$

where a is the radius of the plate.

Inserting Equations [1], [9], and [10] in Equations [11], the boundary conditions become

$$\left.\begin{aligned} &(\sigma_1 - 1)[\delta_1^2 a^2 J_0''(\delta_1 a) + \nu\delta_1 a J_0'(\delta_1 a)]A_1 \\ &\quad + (\sigma_2 - 1)[\delta_2^2 a^2 J_0''(\delta_2 a) + \nu\delta_2 a J_0'(\delta_2 a)]A_2 = 0 \\ &A_1\sigma_1\delta_1 J_0'(\delta_1 a) + A_2\sigma_2\delta_2 J_0'(\delta_2 a) = 0 \end{aligned}\right\} \quad [12]$$

where primes indicate differentiation with respect to the argument.

Frequency Equation

The secular equation, governing the frequency, is obtained by setting the determinant of Equations [12] equal to zero. The resulting equation may be written in the form

$$\frac{g - \beta^2}{1 + g}\gamma\Gamma_1 + \frac{1 - g\beta^2}{1 + g}\beta\gamma\Gamma_2 - (1 - \nu)(1 - \beta^2) = 0 \quad [13a]$$

where

$$\beta = \delta_2/\delta_1, \quad \gamma = \delta_1 a, \quad g = R/S$$

$$\Gamma_1 = J_0(\gamma)/J_1(\gamma), \quad \Gamma_2 = J_0(\beta\gamma)/J_1(\beta\gamma)$$

It may be observed that g is a material constant, depending only on Poisson's ratio, while the remaining functions in Equation [13a] depend on the frequency.

Since δ_1 is real for all positive values of the frequency p, while δ_2 is real or imaginary, depending on whether p is greater or less than $\bar{p} = \pi(\mu/\rho)^{1/2}/h$, β will be real or imaginary according as $p \gtrless \bar{p}$.[7] For the range $p < \bar{p}$, if we let $\beta = i\beta_1$, the frequency equation may be transformed to the more convenient form

$$\frac{g + \beta_1^2}{1 + g}\gamma\Gamma_1 + \frac{1 + g\beta_1^2}{1 + g}\beta_1\gamma G_2 - (1 - \nu)(1 + \beta_1^2) = 0 \quad [13b]$$

where $G_2 = I_0(\beta_1\gamma)/I_1(\beta_1\gamma)$, and $I_0(x)$, $I_1(x)$ are modified Bessel functions of the first kind.

We may find an explicit formula for the frequency by means of the relation $\beta = \delta_2/\delta_1$ and the expressions for δ_1 and δ_2 which immediately follow Equation [5]. Thus

$$\left.\begin{aligned} p/\bar{p} &= [1 - \beta^2(1 + g)^2/g\,(1 + \beta^2)^2]^{-1/2}, \; p > \bar{p} \\ p/\bar{p} &= [1 + \beta_1^2(1 + g)^2/g(1 - \beta_1^2)^2]^{-1/2}, \; p < \bar{p} \end{aligned}\right\} \quad [14]$$

In addition, from the relation $\gamma = \delta_1 a$, we find

$$\left.\begin{aligned} d/h &= \gamma(\bar{p}/p)[(1 + \beta^2)/3(1 + g)]^{1/2}, \; p > \bar{p} \\ d/h &= \gamma(\bar{p}/p)[(1 - \beta_1^2)/3(1 + g)]^{1/2}, \; p < \bar{p} \end{aligned}\right\} \quad [15]$$

where d is the diameter of the disk.

The complete solution of the problem is contained in Equations [13], [14], and [15]. With the value of ν specified for the material of the plate, a choice of β or β_1 determines $p/\bar{p}$ by Equations [14] and yields an infinite set of roots γ of Equations [13]. For the chosen β or β_1 each of these roots furnishes a ratio d/h through Equations [15], so that there results an infinite set of values of d/h corresponding to each value of $p/\bar{p}$.

Frequency Spectrum

The resonant frequencies of axially symmetric flexural vibrations of a circular plate with free boundaries, as predicted by the present theory, are shown, for the range $0.80 \leq p/\bar{p} \leq 1.12$, in Fig. 1. The computations were made for $g = 0.283$, corresponding to $\nu = 0.312$. These curves are characteristic of a frequency spectrum arising from the coupling of two infinite systems.[8] The first system consists of the flexural modes, whose frequency spectrum may be visualized as an extension, into the region $p > \bar{p}$, of the curves in the region $p < \bar{p}$. The second system is comprised of the fundamental thickness-shear mode and its overtones. Their spectrum may be discerned as formed by the loci of the flat, nearly horizontal portions of the resonance curves in the region above the thickness-shear frequency. They are, clearly, asymptotic to this frequency.

Comparison With Classical Theory

In order to observe the effect, on the resonant frequencies, of the coupling with thickness-shear motion, it is necessary to make a comparison with the corresponding frequency spectrum obtained from the classical theory, in which this effect is absent. Here, the relation between frequency and diameter to thickness ratio is,[9] in the present notation

$$d/h = \gamma(\bar{p}/p)^{1/2}[8/3\pi^2(1 - \nu)]^{1/4} \quad [16]$$

where the values of γ are given by the roots of the secular equation[10]

$$\gamma(\Gamma_1 + G_1) - 2(1 - \nu) = 0 \quad [17]$$

[7] $\bar{p}$ is the frequency of the first thickness-shear mode of an infinite plate.

[8] See, for example, Fig. 1 of reference (3).

[9] Reference (5), article 217, equations [1] and [4].

[10] Reference (5), article 219, equation [2].

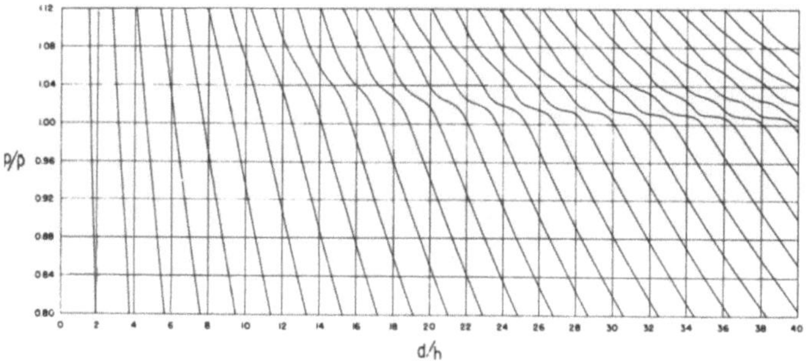

Fig. 1 Frequency Spectrum of Axially Symmetric Vibrations of a Free, Circular Disk, Illustrating Results of Coupling of Flexure and Thickness Shear
(d/h = ratio of diameter to thickness; $p/\bar{p}$ = ratio of resonant frequency to frequency of pure thickness-shear vibration of an infinite plate of thickness h; ν = 0.312.)

in which

$$G_1 = I_0(\gamma)/I_1(\gamma)$$

The resulting curves, for $\nu = 0.312$ (corresponding to $g = 0.283$), are shown in Fig. 2.

The presence of the thickness-shear mode is represented, in the present theory, by the rotatory-inertia coefficient R and the shear-deformation coefficient S. (In the sequel, this theory is referred to as RS and the classical theory as C.) It may be verified that the secular equation of RS (Equation [13]) degenerates to that of C (Equation (17]) for the limiting case $R = S = 0$ (so that $\delta_1^2, \delta_2^2 = \pm\delta_0^2$). Hence the curves in Fig. 1, when extended down to $p/\bar{p} = 0$, approach those of Fig. 2 asymptotically. The suppression of shear deformation in C serves as a constraint which raises the frequencies, in that theory, above those of RS. As a result, for a given plate, RS reveals that there are many more resonances, in a given frequency range, than are predicted by C. For example, for $d/h = 40$, there are 25 resonances in the range $0 < p/\bar{p} < 1$, whereas C predicts only 16.

Furthermore, in designing a plate to resonate at a certain frequency, in its fundamental mode, one would be led to choose too large a plate on the basis of C. For example, for $p/\bar{p} = 0.05$, the diameter would be in excess by 2.61 per cent. This discrepancy increases with the ratio of plate thickness to wave length of mode and, hence, with the order of the mode. For example, if the second mode of the plate is to vibrate at $p/\bar{p} = 0.05$, the error in diameter chosen on the basis of C would be 2.86 per cent. The discrepancies are more pronounced at higher frequencies. Thus, the error in diameter of a plate designed, according to C, to resonate at $p/\bar{p} = 0.1$, in the first mode, is 5.33 per cent and, in the second mode, 5.79 per cent.

The most striking difference between the two theories occurs at frequencies above that of the fundamental thickness-shear mode, as may be seen from Figs. 1 and 2.

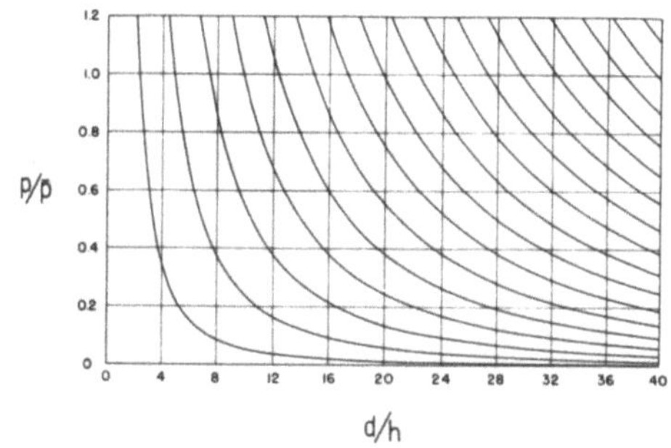

Fig. 2 Frequency Spectrum of Axially Symmetric Vibrations of a Free, Circular Disk According to the Classical (Lagrange) Theory of Flexure of Plates (ν = 0.312)

BIBLIOGRAPHY

1 "Mémoire sur l'équilibre et le mouvement des corps élastiques," by S. D. Poisson, Paris, Mémoires de l'Académie, vol. 8, 1829; see also "A History of the Theory of Elasticity," by I. Todhunter and K. Pearson, Cambridge University Press, England, 1886, vol. 1, p. 262f.

2 "Influence of Rotatory Inertia and Shear on Flexural Motions of Isotropic, Elastic Plates," by R. D. Mindlin, Journal of Applied Mechanics, Trans. ASME, vol. 73, 1951, pp. 31–38.

3 "Thickness-Shear and Flexural Vibrations of Crystal Plates," by R. D. Mindlin, *Journal of Applied Physics*, vol. 22, 1951, pp. 316–323.

4 "Thickness-Shear and Flexural Vibrations of an Isotropic, Elastic, Circular Disk," by R. D. Mindlin and H. Deresiewicz, Technical Report No. 11, Office of Naval Research Contract Nonr-200(09), December, 1953.

5 "Theory of Sound," by Lord Rayleigh, Dover Publications, New York, N. Y., 1944.

Reprinted from
Journal of Applied Mechanics
22 (1955) 86-88

R. D. Mindlin
H. Deresiewicz

TIMOSHENKO'S SHEAR COEFFICIENT FOR FLEXURAL VIBRATIONS OF BEAMS

Professor of Civil Engineering,
Columbia University, New York, N. Y.
Assistant Professor of Civil Engineering,
Columbia University, New York, N. Y.

It is pointed out that Timoshenko's shear coefficient, in his equations of flexural vibrations of beams, depends on both the shape of the cross-section and the mode of motion. It is shown how the latter may be taken into account by matching solutions, of Timoshenko's equations and the three-dimensional equations, for simple thickness-shear motions of infinite beams of various cross-sectional shapes.

Introduction

The range of applicability of the one-dimensional theory of flexural vibrations of beams was extended to higher frequencies by Timoshenko when he took into account the effect of transverse shear deformation. He arrived at the free-vibration equations [1]*

$$EI\,\frac{\partial^2\psi}{\partial x^2} + k\left(\frac{\partial y}{\partial x} - \psi\right)AG - \frac{I\gamma}{g}\frac{\partial^2\psi}{\partial t^2} = 0$$
$$\frac{\gamma A}{g}\frac{\partial^2 y}{\partial t^2} - k\left(\frac{\partial^2 y}{\partial x^2} - \frac{\partial\psi}{\partial x}\right)AG = 0 \qquad (1)$$

governing the transverse deflection, y, and the slope, ψ, of the deflection curve when the shear is neglected. The coefficient k is defined as the ratio of the average shear stress on a section to the product of the shear modulus and the angle of shear at the neutral axis. This ratio depends upon the distribution of shear stress on the section and, hence, k depends upon the shape of the section, as Timoshenko observed. However, the distribution of shear stress on a section depends also on the mode of motion of the beam. For example, for low modes of motion of a slender beam, the shear stress has a maximum at the neutral axis, while, for very high modes, the shear stress has a minimum at the same place. Thus, k depends both on the shape of the section and the frequency of vibration.

In the solution of Equations (1), the simplest interpretation of k is that it is a constant. For a beam of given cross-sectional shape, then, k can have the correct value for only one frequency. In the past, several calculations of k have been made, for various cross-sectional shapes, on the basis of statical considerations, that is, at zero frequency. These values are satisfactory for the low modes of motion of slender beams, where, in fact, the influence of transverse shear deformation is small in comparison with its influence at high frequencies.

As the frequency is increased, a point is reached at which a drastic change occurs in the spectrum. This is the frequency, ω', of the first thickness-shear mode of a beam of infinite length, i.e., the lowest frequency at which the infinite beam can vibrate with no transverse deflection, the displacement being entirely parallel to the axis of the beam.

The reason for the change in character of the frequency spectrum is that, at the frequencies of the first thickness-shear mode and its overtones, there is strong coupling between the flexural and thickness-shear modes of motion. Examples of this phenomenon have been described in detail in previous papers [2, 3] for special cases of crystal plates for which the equations of motion are the same as those for a Timoshenko beam.

It is desirable, then, that the thickness-shear frequency calculated from Equations (1) should be the same as that calculated from the equations to which (1) are an approximation, namely, the three-dimensional equations of small vibrations of an elastic body. In this connection it must be recognized that, whereas Timoshenko interpreted ψ as the slope of the deflection curve when the shear deformation is neglected, the product of ψ and the depth of the beam may be interpreted as the maximum axial displacement of a transverse section. Thus, ω' is calculated from (1) by setting $y = 0$ and ψ proportional to $\exp(i\omega' t)$, following which ω' is set equal to the corresponding frequency calculated from the three-dimensional equations. Since ω' depends upon k, the result is a formula for k.

It has been shown [4] that, for a rectangular section, this procedure leads to $k = \pi^2/12 \approx 0.822$. Comparison with experiments [2,3] shows that, when k is calculated in this manner, Timoshenko's equations give good results for both low and high frequencies of beams with free ends. This good agreement may be expected to hold, for slender beams, regardless of the end conditions, even for high modes, since the frequencies of high modes are

*Numbers in brackets refer to Bibliography at the end of the paper.

This investigation was supported by the Office of Naval Research under Contract Nonr-266(09) with Columbia University.

insensitive to the boundary conditions.

In the present paper, k is calculated in the same manner for a variety of cross-sectional shapes. To do this, it is necessary to solve the three-dimensional equations of elasticity for the appropriate frequency and equate it with the value of ω' obtained from Equations (1). In the case of motion parallel to the axis of the bar, the three-dimensional equations and boundary conditions reduce to equations governing a familiar hydrodynamical problem for which many solutions are known. The determination of k is thus reduced to an interpretation of these solutions and some additional computations. Results are given for the following sections: circle, ellipse, orthogonal parabolas and a variety of ovaloids. The values of k computed for these sections all lie within about 10 per cent of that for the rectangular section.

Thickness-Shear Motion: Timoshenko Theory

If γ is set equal to zero in Equations (1), the second of the equations gives $\partial\psi/\partial x = 0$, so that the first of (1) reduces to

$$\frac{d^2\psi}{dt^2} + k\left(\frac{c}{r}\right)^2 \psi = 0 \qquad (2)$$

where $c = (Gg/\gamma)^{1/2}$, i.e., the velocity of shear waves in an infinite, isotropic, elastic medium, and $r = (I/A)^{1/2}$, i.e., the radius of gyration of the cross section. Hence, the frequency of pure thickness-shear vibration of a Timoshenko beam is

$$\omega' = ck^{1/2}/r \qquad (3)$$

This is the frequency which is to be equated to the one calculated from the solution of the three-dimensional equations for each section.

Thickness-Shear Motion: Three-Dimensional Theory

Let the neutral axis of the beam be the z-axis of a rectangular coordinate system and let u, v and w be components of displacement in the x, y and z directions, respectively. For pure thickness-shear motion,

$$u = v = 0$$

$$w = \zeta(x, y)\, e^{i\omega t} \qquad (4)$$

Then the three-dimensional equations of motion [5] reduce to

$$\nabla^2\zeta + \delta^2\zeta \quad 0 \qquad (5)$$

where $\delta = \omega/c$ and $\nabla^2 = \partial^2/\partial x^2 + \partial^2/\partial y^2$. The condition that the traction vanish on the cylindrical surface of the beam reduces to

$$\partial\zeta/\partial n = 0 \qquad (6)$$

on the boundary, where n is the normal to the boundary.

The differential equation (5) and the boundary condition (6) are the same as those governing the small oscillations of a fluid in a basin of uniform depth [6] and the small vibrations of a gas in a rigid cylindrical container [7]. Solutions are available for a variety of boundaries.

Rectangle

In the case of a rectangular section, the frequency is independent of the width. If the depth is $2a$, the frequency is [6]

$$\omega = \pi c/2a \qquad (7)$$

Equating (7) and (3), and noting that $r^2 = a^2/3$, the shear constant, k, is found to be $\pi^2/12$, or, approximately, 0.822.

Circle

For a circular section of radius a, the lowest antisymmetric mode has a frequency [6]

$$\omega = 1.841\, c/a \qquad (8)$$

The radius of gyration of a circle about a diameter is $a/2$. Hence, equating (8) and (3), $k = 0.847$.

Ellipse

The elliptic section was treated by Jeffreys [8], who computed frequencies for two values of the eccentricity

$$e = (1 - b^2/a^2)^{1/2} \qquad (9)$$

where a and b are the semi-major and semi-minor axes, respectively. Additional values were computed and are listed in the columns headed $\omega a/c$ and $\omega b/c$ in Table I. The corresponding values of ω were then equated to that in Equation (3) to obtain the values of k listed in Table I.

It is interesting to notice that the frequency ($1.886\, c/a$) of the first antisymmetric mode of a very narrow ellipse, about its minor axis, is greater than that ($1.571\, c/a$) of a

TABLE I

	Motion antisymmetric about minor axis		Motion antisymmetric about major axis	
	$\omega a/c$	k	$\omega b/c$	k
Rectangle	1.571	.822	1.571	.822
Ellipse				
e = 1.0	1.886*	.889	**	
e = 0.9	1.878	.882	**	
e = 0.8	1.87 *	.87	1.78 *	.79
e = 0.7	1.858	.863	1.814	.823
e = 0.6	1.856	.861	1.823	.831
e = 0.3	1.845	.851	1.837	.844
e = 0 (circle)	1.841	.847	1.841	.847
Orthogonal Parabolas	2.117	.896		

* Jeffreys (8)
** Convergence slow

rectangle of depth equal to the major diameter of the ellipse. Jeffreys observed that this is due to the concentration of motion near the center of the ellipse. On the other hand, the corresponding frequency for a very wide ellipse appears to approach that of the rectangle. However, the corresponding values of k are not the same because of the difference in radii of gyration.

Parabolas

The hydrodynamic problem for a symmetric section bounded by a pair of orthogonal parabolas was studied by Hidaka [9]. Such a section has the property $a/b = 2$, where a and b are the semi-major and semi-minor axes, respectively. For motion antisymmetric with respect to the minor axis, Hidaka found a secular equation which can be reduced to

$$J_{-1/4}\,(\delta\, a/2) = 0 \qquad (10)$$

where $J(x)$ is the Bessel function of the first kind. The lowest root of Equation (10) is [10]

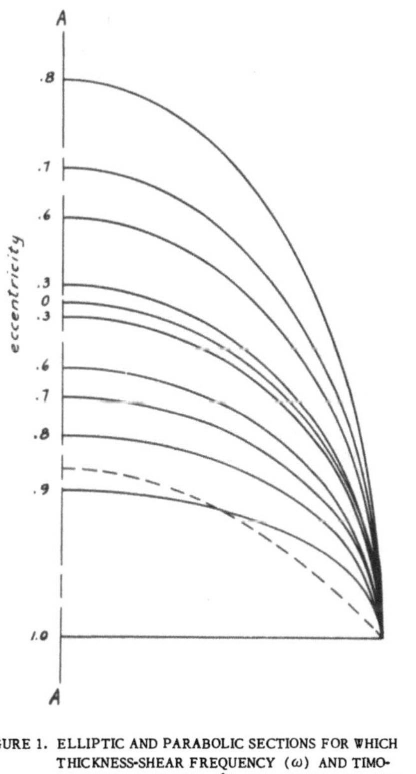

FIGURE 1. ELLIPTIC AND PARABOLIC SECTIONS FOR WHICH THICKNESS-SHEAR FREQUENCY (ω) AND TIMOSHENKO'S CONSTANT (k) ARE GIVEN IN TABLE I. DASHED LINE IS PARABOLIC SECTION. THICKNESS-SHEAR MOTION IS ANTISYMMETRIC ABOUT AXIS AA.

$$\delta\, a/2 = 1.0585 \qquad (11)$$

so that

$$\omega = 2.117\, c/a \qquad (12)$$

Equating (12) and (3) and noting that $r^2 = a^2/5$, the result is $k = 0.896$.

This completes the calculations for sections for which exact solutions are available. The results are assembled in Table I and the sections are illustrated in Fig. 1.

Ovaloids

Approximate frequencies, good for narrow sections, may be obtained by neglecting the variation of displacement across the width. Hidaka [9] did this for sections bounded by the curves

$$\frac{x^2}{a^2} + \left(\frac{y^2}{b^2}\right)^{1/2m} = 1$$
$$m = 2^{-\mu},\quad \mu = 0, 1, 2, \ldots \qquad (13)$$

A set of these curves, for the case $a = 2b$, is shown in Fig. 2.

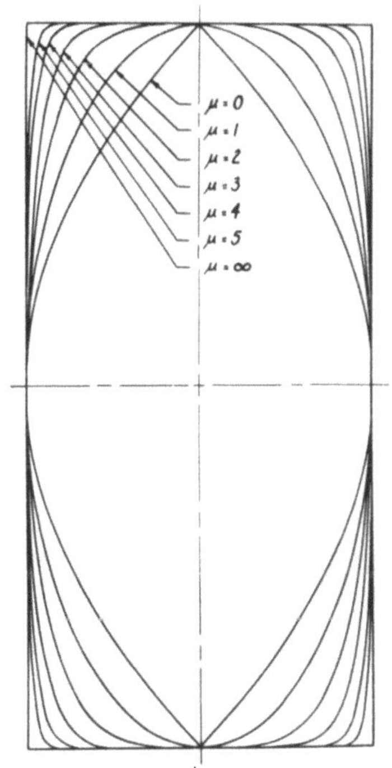

FIGURE 2. OVALOID SECTIONS FOR WHICH APPROXIMATE VALUES OF ω and k ARE GIVEN IN TABLE II.

TABLE II

μ	m	$\omega a/c$	r^2/a^2	k
0	1 *	2.150	1/5	.924
1	1/2 **	1.886	1/4	.889
2	1/4	1.737	2/7	.862
3	1/8	1.655	4/13	.843
4	1/16	1.616	8/25	.836
5	1/32	1.591	16/49	.826
∞	0 ***	1.571	1/3	.822

*parabolas; **ellipse; ***rectangle.

Hidaka's results for the lowest mode, antisymmetric about the horizontal axis of Fig. 2, are given in Table II, along with the corresponding values of k. In finding k, the radius of gyration (r), defined by

$$r^2 = \frac{\int_0^a yx^2\,dx}{\int_0^a y\,dx} \tag{14}$$

must be computed. Inserting y from Equation (13) into Equation (14), and making the substitution $x = a \sin\theta$ in the integrals, we find

$$\frac{r^2}{a^2} = 1 - \frac{\int_0^{\pi/2} \cos^{(3+2m)}\theta\,d\theta}{\int_0^{\pi/2} \cos^{(1+2m)}\theta\,d\theta} \tag{15}$$

By aid of the formula [11]

$$\int_0^{\pi/2} \cos^n x\,dx = \frac{\sqrt{\pi}}{2}\,\frac{\Gamma\left[\frac{n+1}{2}\right]}{\Gamma\left[\frac{n}{2}+1\right]}\,, \quad n > -1$$

and the recursion formula for the Gamma function, Equation (15) reduces to

$$r^2 = a^2/(3+2m) \tag{16}$$

The results in Table II are the same as in Table I for the rectangle ($\mu = \infty$) for any b/a since the assumption of uniform displacement across the width is exact in pure thickness-shear motion. For the ellipse ($\mu = 1$), Table II gives the same value as Table I only for $b/a = 0$ (i.e., $e = 1$), since here, too, there is no variation of displacement across the width. As the section becomes wider, there is some discrepancy, but it is not great. Even for $a = b$ (the circle), Table I gives $k = 0.847$, while the approximation yields 0.889. The approximation is not intended, of course, to apply to $b > a$. Another comparison may be made for the parabola with $a = 2b$. The value of k in Table I is 0.896, while the approximation gives 0.924. Thus it may be expected that the approximate solution of Equation (5) will give results good to within a few per cent for sections at least twice as deep as they are wide.

Bibliography

1. S. Timoshenko, "Vibration Problems in Engineering," D. Van Nostrand Company, New York, 2nd Edition, 1937, p. 338.

It is of some historical interest that both the rotatory inertia correction, usually attributed to Rayleigh ("Theory of Sound," Cambridge, England, first edition, 1877; current edition, reference [7], art. 162), and the transverse shear correction, usually attributed to Timoshenko (Philosophical Magazine, Ser. 6, Vol. 41, 1921, pp. 744–746, and Vol. 43, 1922, pp. 125–131), are given by M. Bresse in his "Cours de Mécanique Appliquée," Mallet-Bachelier, Paris, 1859, p. 126.

2. R. D. Mindlin, "Thickness-Shear and Flexural Vibrations of Crystal Plates," J. Appl. Phys., Vol. 22, 1951, pp. 316–323.
3. R. D. Mindlin, "Forced Thickness-Shear and Flexural Vibrations of Piezoelectric Crystal Plates," J. Appl. Phys., Vol. 23, 1952, pp. 83–88.
4. R. D. Mindlin, "Influence of Rotatory Inertia and Shear on Flexural Motions of Isotropic, Elastic Plates," J. Appl. Mech., Vol. 18, 1951, pp. 31–38.
5. S. Timoshenko and J. N. Goodier, "Theory of Elasticity," McGraw-Hill Book Co., New York, 2nd Edition, 1951, p. 452.
6. H. Lamb, "Hydrodynamics," Dover Publications, New York, 6th Edition, 1945, pp. 282–290.
7. Lord Rayleigh, "Theory of Sound," Dover Publications, New York, 2nd Edition, 1945, art. 339.
8. H. Jeffreys, "The Free Oscillations of Water in an Elliptic Lake," Proc. London Math. Soc., Ser. 2, Vol. 23, 1924, pp. 455–476.
9. K. Hidaka, "The Oscillations of Water in Spindle-Shaped and Elliptic Basins as well as the Associated Problems," Mem. Imp. Marine Observ., Kobe, Japan, Vol. 4, 1931, pp. 99–219.
10. "Tables of Bessel Functions of Fractional Order," Columbia University Press, New York, 1948, Vol. 1, p. 384.
11. B. O. Peirce, "A Short Table of Integrals," Ginn and Company, Boston, 3rd revised edition, 1929, No. 483.

Reprinted from
Proceedings of the Second U.S. National Congress of Applied Mechanics -1954
(1955) 175-178

Reprinted from JOURNAL OF APPLIED PHYSICS, Vol. 26, No. 12, 1435–1442, December, 1955

Thickness-Shear and Flexural Vibrations of Rectangular Crystal Plates*

R. D. MINDLIN AND H. DERESIEWICZ
Department of Civil Engineering and Engineering Mechanics, Columbia University, New York, New York
(Received July 1, 1955)

Equations governing thickness-shear and flexural vibrations of crystal plates are solved for the infinite plate, the simply-supported rectangular plate and the rectangular plate with one pair of parallel edges free and the other pair simply-supported. The equations permit three types of sinusoidal waves, with sinusoidal crests, in an infinite plate. Each of these undergoes a simple reflection upon normal incidence at a simply-supported straight edge, so that the frequency spectrum of a simply-supported rectangular plate has a relatively simple character. The results of a typical computation are given for the AT-cut of quartz. At a free edge each type of incident wave gives rise, in general, to all three types of reflected wave. Consequently, the frequency spectrum of a plate with a pair of parallel, free edges exhibits an intricate coupling of three infinite systems of modes. The development of the coupling is traced continuously by means of a solution involving elastically supported edges.

1. PLATE EQUATIONS

In a previous paper[1] (referred to in the sequel as I) there are derived approximate equations governing the flexural and first thickness-shear modes of motion of a crystal plate having an axis of monoclinic symmetry in the plane of the plate. The essence of the approximation is the restriction of the components of displacement to the forms

$$u_1=y\psi_x(x,z,t),\quad u_2=\eta(x,z,t),\quad u_3=y\psi_z(x,z,t),\tag{1}$$

where the y-axis is normal to the plane of the plate and u_1, u_2, u_3 are the components of displacement parallel to x,y,z, respectively. Then ψ_x,η,ψ_z are governed by

$$\begin{aligned}
&D_1\partial^2\psi_x/\partial x^2+D_5\partial^2\psi_x/\partial z^2+D_7\partial^2\psi_z/\partial x\partial z\\
&\qquad-D_6(\psi_x+\partial\eta/\partial x)=(\rho h^3/12)\partial^2\psi_x/\partial t^2,\\
&D_6(\partial\psi_x/\partial x+\partial^2\eta/\partial x^2)+D_4(\partial\psi_z/\partial z+\partial^2\eta/\partial z^2)\\
&\qquad=\rho h\partial^2\eta/\partial t^2,\\
&D_3\partial^2\psi_z/\partial z^2+D_5\partial^2\psi_z/\partial x^2+D_7\partial^2\psi_x/\partial x\partial z\\
&\qquad-D_4(\psi_z+\partial\eta/\partial z)=(\rho h^3/12)\partial^2\psi_z/\partial t^2,
\end{aligned}\tag{2}$$

where ρ is the density and h is the thickness of the plate. Further,

$$\begin{aligned}
&D_1=(c_{11}-c_{12}^2/c_{22})h^3/12,\quad D_2=(c_{13}-c_{12}c_{23}/c_{22})h^3/12,\\
&D_3=(c_{33}-c_{23}^2/c_{22})h^3/12,\quad D_4=\kappa^2c_{44}h,\\
&D_5=c_{55}h^3/12,\quad D_6=\kappa^2c_{66}h,\quad D_7=D_2+D_5,
\end{aligned}\tag{3}$$

in which $\kappa^2=\pi^2/12$ and the c_{ij} are elastic constants referred to x,y,z axes fixed in the plate. (Note that the definition of D_4 has been corrected.)

Boundary conditions sufficient to assure unique solutions of Eq. (2) are obtained by specifying, at each point of the edge of the plate, one member of each of the three products

$$M_\nu\partial\psi_\nu/\partial t,\quad M_{\nu s}\partial\psi_s/\partial t,\quad Q_\nu\partial\eta/\partial t,\tag{4}$$

where ν and s are coordinates normal and tangential, respectively, to the edge. In rectangular coordinates,

* This investigation was supported by the U. S. Army Signal Corps and the Office of Naval Research.
[1] R. D. Mindlin, J. Appl. Phys. 22, 316 (1951).

the couples $M_\nu, M_{\nu s}$ and the force Q_ν are given by

$$\begin{aligned} M_x &= D_1\partial\psi_x/\partial x + D_2\partial\psi_z/\partial z, \\ M_z &= D_3\partial\psi_z/\partial z + D_2\partial\psi_x/\partial x, \\ M_{xz} &= D_5(\partial\psi_z/\partial x + \partial\psi_x/\partial z), \\ Q_x &= D_6(\psi_x + \partial\eta/\partial x), \quad Q_z = D_4(\psi_z + \partial\eta/\partial z). \end{aligned} \tag{5}$$

In I a solution was given for the case $\psi_x = \psi_x(x,t)$, $\eta = \eta(x,t)$, $\psi_z = 0$ and a pair of free edges parallel to z. Results of computations were compared with experimental data obtained by Sykes for modes in which η is odd in x. In a subsequent paper[2] (II) the computations were extended to include η even in x and were compared with additional measurements by Sykes. Dependence of ψ_x, η and ψ_z on both x and z is considered in the following sections.

2. INFINITE PLATE

Equations (2) have a solution

$$\begin{aligned} \psi_x &= A \cos\xi x \sin\zeta z \cos\omega t \\ \eta &= Bh \sin\xi x \sin\zeta z \cos\omega t \\ \psi_z &= C \sin\xi x \cos\zeta z \cos\omega t \end{aligned} \tag{6}$$

provided

$$\begin{aligned} AP + B\pi f\alpha \quad + Cd_{71}\alpha\gamma &= 0 \\ A\alpha + B\pi fQ \quad + Cd_{46}\gamma &= 0 \\ Ad_{71}\alpha\gamma + B\pi fd_{46}\gamma + CR &= 0 \end{aligned} \tag{7}$$

where

$$\begin{aligned} &\alpha = \xi h/\pi, \quad \gamma = \zeta h/\pi, \quad d_{k1} = D_k/D_1 \\ &P = \alpha^2 + d_{51}\gamma^2 + f(1-\omega^2/\omega_x^2) \\ &Q = (\alpha^2 + d_{46}\gamma^2 - \omega^2/\kappa^2\omega_x^2)/f \\ &R = d_{51}\alpha^2 + d_{31}\gamma^2 + f(d_{46} - \omega^2/\omega_x^2) \\ &\omega_x = (\pi/h)(c_{66}/\rho)^{\frac{1}{2}}, \quad f = c_{66}/(c_{11} - c_{12}^2/c_{22}). \end{aligned} \tag{8}$$

Note that α and γ are the ratios of the thickness of the plate to the distances between nodes in the x- and z-directions, respectively; f is the ratio of the xy thickness-shear modulus to the xy flexure modulus; ω_x is the frequency of the first xy thickness-shear mode of an infinite plate.

Equations (7) determine amplitude ratios

$$\begin{aligned} B/A &= (d_{46}P - d_{71}\alpha^2)/\alpha\pi f(d_{71}Q - d_{46}) \\ C/A &= \gamma(d_{46}P - d_{71}\alpha^2)/\alpha(R - d_{46}d_{71}\gamma^2) \end{aligned} \tag{9}$$

and a relation between the wave numbers ξ, ζ and the frequency ω, according to the determinantal equation

$$\begin{vmatrix} P & \alpha & d_{71}\alpha\gamma \\ \alpha & Q & d_{46}\gamma \\ d_{71}\alpha\gamma & d_{46}\gamma & R \end{vmatrix} = 0. \tag{10}$$

This is a bicubic equation in ω as a function of ξ and ζ. For given real, positive ξ and ζ, there are three real, positive roots $\omega_1, \omega_2, \omega_3$ to each of which there corresponds a set of amplitude ratios, in accordance with Eq. (9). Thus, for a given pair of wavelengths, there are three modes of vibration of an infinite plate. In general, each mode has its own frequency and requires the coupling of all three components of displacement ψ_x, η, ψ_z. An understanding of the physical significance of the modes is gained by examining several special cases, which, moreover, are required for the subsequent analysis of rectangular plates.

For the limit $\xi h = 0$, $\zeta h = 0$ (long wavelengths), the positive roots of Eq. (10) and the corresponding amplitudes are

$$\begin{aligned} \omega_1 &= 0 = \omega_y, & A &= C = 0, \\ \omega_2 &= (\pi/h)(c_{66}/\rho)^{\frac{1}{2}} = \omega_x, & B &= C = 0, \\ \omega_3 &= (\pi/h)(c_{44}/\rho)^{\frac{1}{2}} = \omega_z, & A &= B = 0. \end{aligned} \tag{11}$$

The second and third of these are, respectively, the first xy and the first zy thickness-shear modes of an infinite plate. The first mode in Eq. (11) is flexural ($B \neq 0$). Its frequency is more recognizable, in its next higher order approximation,

$$\omega_y = \{[D_1\xi^4 + D_3\zeta^4 + 2(D_2 + 2D_5)\xi^2\zeta^2]/\rho h\}^{\frac{1}{2}}, \tag{12}$$

as the frequency of flexural vibration of an orthotropic plate according to the classical theory of thin plates. In the isotropic case,

$$\begin{aligned} D_1 &= D_3 = D = Eh^3/12(1-\nu^2), \\ D_2 &= \nu D, \quad D_5 = (1-\nu)D/2, \end{aligned} \tag{13}$$

where ν is Poisson's ratio and E is Young's modulus. Equation (12) then reduces to the usual form

$$\omega_y = (\xi^2 + \zeta^2)(D/\rho h)^{\frac{1}{2}}. \tag{14}$$

Thus, when the wavelengths in both the x- and z-directions are very long, in comparison with the thickness of the plate, the three components of displacement ψ_x, η, ψ_z essentially uncouple and produce two simple thickness-shear modes and one flexural mode. In general, $\omega_y < \omega_x, \omega_z$ and, for the AT-cut of quartz, $\omega_y < \omega_x < \omega_z$. The latter inequality is adopted in what follows.

Consider, now, the case in which only the wavelength in the z-direction becomes large ($\zeta = 0$). Then the bicubic Eq. (10) has three real roots $\omega_1, \omega_2, \omega_3$ given by

$$\begin{aligned} 2(\omega_{1,2}/\omega_x)^2 &= 1 + k\alpha^2 \pm [(1+k\alpha^2)^2 - 4\kappa^4\alpha^4/g]^{\frac{1}{2}} \\ (\omega_3/\omega_x)^2 &= (c_{44}/c_{66})[1 + (c_{55}/c_{44})\alpha^2] \end{aligned} \tag{15}$$

where

$$\begin{aligned} g &= \kappa^2 c_{66}/(c_{11} - c_{12}^2/c_{22}) \\ k &= \kappa^2(1+g)/g. \end{aligned} \tag{16}$$

The corresponding amplitudes are

$$\begin{aligned} \omega_{1,2}:& \quad C = 0, \quad A/B = [(\omega_{1,2}/\kappa\alpha\omega_x)^2 - 1]\pi\alpha \\ \omega_3:& \quad A = B = 0. \end{aligned} \tag{17}$$

Hence, when $\zeta = 0$, the displacement ψ_z is not coupled with ψ_x or η and forms a simple yz thickness-shear mode which varies sinusoidally with x. In each of the other

[2] R. D. Mindlin, J. Appl. Phys. 23, 83 (1952).

two modes the displacements ψ_x and η are coupled. For the mode with the lower frequency $|A/B|<1$ whereas the mode with the higher frequency has $|A/B|>1$. Thus, the mode with frequency ω_1 is essentially flexural while the mode with frequency ω_2 is essentially thickness-shear. These are the two types of motion that were considered in I.

Coming, now, to the case of arbitrary ξ and ζ, we observe that the three roots $\omega_1,\omega_2,\omega_3$ of Eq. (10) are associated with modes which are essentially flexural, xy thickness-shear and zy thickness-shear, respectively, although each mode contains all three components of displacement to some degree. For a fixed ζ, all three frequencies approach infinity as the wavelength in the x-direction approaches zero (i.e., as $\xi\to\infty$). On the other hand, as $\xi\to 0$, all three frequencies diminish monotonically and approach, asymptotically, cut-off values $\omega_1^*,\omega_2^*,\omega_3^*$ obtained as the roots of Eq. (10) with $\xi=0$. Thus

$$2(\omega_{1,3}^*/\omega_x)^2=(c_{44}/c_{66})\{1+k'\gamma^2 \mp[(1+k'\gamma^2)^2-4\kappa^4\gamma^4/g']^{\frac{1}{2}}\} \quad (18)$$

$$(\omega_2^*/\omega_x)^2=1+(c_{55}/c_{66})\gamma^2$$

where

$$g'=\kappa^2c_{44}/(c_{33}-c_{23}^2/c_{22})$$
$$k'=\kappa^2(1+g')/g'. \quad (19)$$

The behavior of the three roots of Eq. (10) is shown in Fig. 1 for the AT-cut of quartz[3] with $\zeta h/\pi=0.2$. Note that, if d is the distance between nodes in the z-direction, $\zeta h/\pi=h/d$.

It may be observed that, when the displacement components have phase reversals in the z-direction, the cut-off frequencies are higher than otherwise. Thus, $\omega_1^*>\omega_y$, $\omega_2^*>\omega_x$, $\omega_3^*>\omega_z$ when $h/d>0$. Cut-off frequencies for a range of values of h/d are given in Table I for the AT-cut of quartz.

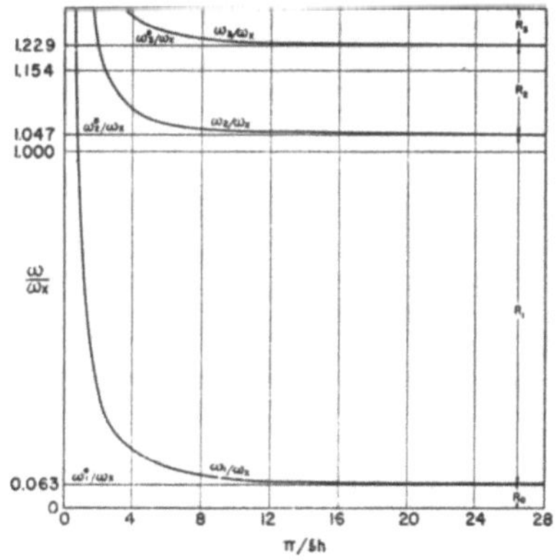

FIG. 1. Frequencies of vibration of an infinite plate. Variation with wavelength in x-direction for fixed wavelength in z-direction. AT-cut of quartz; $\zeta h/\pi=0.2$.

[3] The elastic constants used in the computations are those given by W. P. Mason, *Piezoelectric Crystals and Their Applications to Ultrasonics* (D. Van Nostrand Company, Inc., New York, 1950), p. 84.

TABLE I. Cut-off frequencies as functions of the ratio (h/d) of plate thickness to nodal spacing along z for the AT-cut of quartz.

h/d	Flexure ω_1^*/ω_x	Thickness-shear xy ω_2^*/ω_x	Thickness-shear zy ω_3^*/ω_x
0	0	1	1.154
0.1	0.016	1.012	1.173
0.2	0.063	1.047	1.229
0.3	0.134	1.102	1.313
0.4	0.220	1.175	1.419
0.5	0.316	1.263	1.540
0.6	0.420	1.363	1.673
0.7	0.527	1.472	1.814
0.8	0.636	1.589	1.962
0.9	0.747	1.711	2.115
1.0	0.858	1.839	2.272

3. SIMPLY-SUPPORTED RECTANGULAR PLATE

In the classical theory of plates, a simply-supported edge is characterized by vanishing bending moment (M_ν) and displacement (η) normal to the plane of the plate. With these boundary conditions, the modes of vibration of a rectangular plate are described by simple product functions such as those in Eq. (6). Equations (2), however, require an additional boundary condition. Referring to Eq. (4), this may be a condition on the twisting couple $M_{\nu s}$ or the associated rotation ψ_s. It is when the latter is set equal to zero that the solutions of Eq. (2), for the modes of vibration of a rectangular plate, are analogous to those of classical plate theory, i.e., ψ_x,η, and ψ_z are expressible as simple product functions. Accordingly, for a plate bounded by $x=\pm a$, $z=\pm c$, the boundary conditions for simply-supported edges are taken to be

$$M_x=0,\quad \partial\eta/\partial t=0,\quad \partial\psi_z/\partial t=0,\quad \text{on}\quad x=\pm a;$$
$$M_z=0,\quad \partial\eta/\partial t=0,\quad \partial\psi_x/\partial t=0,\quad \text{on}\quad z=\pm c. \quad (20)$$

We consider three solutions (one for each of $i=1,2,3$)

$$\psi_x^i=A_i\cos\xi_ix\sin\zeta_iz\cos\omega_it,$$
$$\eta^i=B_ih\sin\xi_ix\sin\zeta_iz\cos\omega_it, \quad (21)$$
$$\psi_z^i=C_i\sin\xi_ix\cos\zeta_iz\cos\omega_it.$$

To satisfy the equations of motion, Eq. (2), the wave numbers and frequency must again satisfy Eq. (10) with ξ,ζ,ω replaced by ξ_i,ζ_i,ω_i. Then, inserting Eq. (21) into Eq. (5), where necessary, and then into Eq. (20), the boundary conditions reduce to

$$\sin\xi_ia=0,\quad \sin\zeta_ic=0,\quad i=1,2,3. \quad (22)$$

It may be seen that the boundary conditions do not induce coupling between the three types of motion $i=1,2,3$, since each satisfies Eq. (22) separately. Hence, any nodal line of the vibrating infinite plate may be

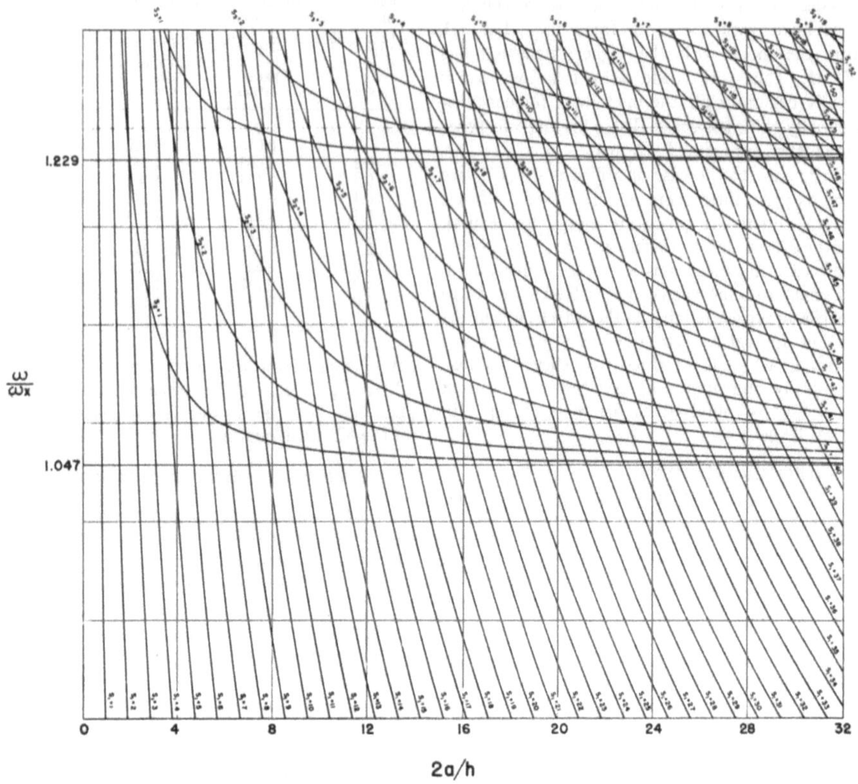

FIG. 2. Frequencies of vibration of a simply-supported rectangular plate. Variation with ratio $(2a/h)$ of length-to-thickness for fixed ratio $(h/d=0.2)$ of thickness-to-distance between nodes along width. Curves s_1, s_2, s_3 = constant give frequencies of flexural, xy thickness-shear and zy thickness-shear, respectively. Curves s_i odd, even, are for modes even and odd, respectively, in x and either even or odd in z (AT-cut of quartz).

taken as a simply-supported edge and the resulting vibrations may be regarded as superposed sinusoidal waves, with sinusoidal crests, undergoing simple reflections at these edges.

The symmetry of the modes (i.e., oddness or evenness in x and z) may be identified with the symmetry of η. In Eq. (21), η is odd in both x and z. Modes even in x and odd in z are obtained by interchanging $\sin\xi_i x$ and $\cos\xi_i x$ in all three of Eq. (21). The boundary conditions then reduce to

$$\cos\xi_i a=0, \quad \sin\zeta_i c=0, \quad i=1,2,3. \tag{23}$$

Similarly, for modes odd in x and even in z,

$$\sin\xi_i a=0, \quad \cos\zeta_i c=0, \quad i=1,2,3 \tag{24}$$

and, finally, for modes even in both x and z,

$$\cos\xi_i a=0, \quad \cos\zeta_i c=0, \quad i=1,2,3. \tag{25}$$

The solutions of all of Eqs. (22) to (25) may be written as

$$\xi_i=s_i\pi/2a, \quad \zeta_i=p_i\pi/2a, \quad i=1,2,3, \tag{26}$$

where the s_i are even integers for modes odd in x and odd integers for modes even in x. Similarly, the p_i are even integers for modes odd in z and odd integers for modes even in z.

Hence, the frequencies of vibration of a simply-supported plate are given by the roots of

$$\begin{vmatrix} P_i & \alpha_i & d_{71}\alpha_i\gamma_i \\ \alpha_i & Q_i & d_{46}\gamma_i \\ d_{71}\alpha_i\gamma_i & d_{46}\gamma_i & R_i \end{vmatrix}=0, \tag{27}$$

where

$$\left.\begin{aligned} &\alpha_i=s_i h/2a, \quad \gamma_i=p_i h/2c \\ &P_i=\alpha_i^2+d_{51}\gamma_i^2+f(1-\omega_i^2/\omega_x^2) \\ &Q_i=(\alpha_i^2+d_{46}\gamma_i^2-\omega_i^2/\kappa^2\omega_x^2)/f \\ &R_i=d_{51}\alpha_i^2+d_{31}\gamma_i^2+f(d_{46}-\omega_i^2/\omega_x^2). \end{aligned}\right\} \tag{28}$$

The associated amplitude ratios are

$$B_i/A_i=(d_{46}P_i-d_{71}\alpha_i^2)/\alpha_i\pi f(d_{71}Q_i-d_{46})$$
$$C_i/A_i=\gamma_i(d_{46}P_i-d_{71}\alpha_i^2)/\alpha_i(R_i-d_{46}d_{71}\gamma_i^2). \tag{29}$$

The frequency spectrum determined by Eq. (27) is plotted in Fig. 2 for the AT-cut of quartz and $\gamma_i=0.2$. It may be observed that the three families of curves in Fig. 2 may be obtained from the three curves in Fig. 1 by multiplying the abscissas of the latter by 1,2,3,4...etc. The curves $s_1,s_2,s_3=$constant give the frequencies of flexural, xy thickness-shear and zy thickness-shear modes, respectively. s_i odd, even are for modes even and odd, respectively, in x. All the curves apply to modes either odd or even in z.

4. COMPARISON WITH SYKES'S FORMULA

Sykes[4] has given a formula for the frequencies of thickness-shear vibrations:

$$f=\tfrac{1}{2}\left[\frac{c_{jj}}{\rho}\left(\frac{m^2}{t^2}+k\frac{n^2}{l^2}+k_1\frac{(p-1)^2}{w^2}\right)\right]^{\frac{1}{2}} \tag{30}$$

where k and k_1 are empirical coefficients. To convert to the notation of the present paper, note that

$$t=h,\quad l=2a,\quad w=2c$$
$$m=1,\quad n=s_2,\quad p-1=p_2 \tag{31}$$
$$(\pi/t)(c_{jj}/\rho)^{\frac{1}{2}}=\omega_x,\quad f=\omega_2/2\pi.$$

Then Eq. (30) becomes

$$(\omega_2/\omega_x)^2=1+ks_2^2h^2/4a^2+k_1p_2^2h^2/4c^2. \tag{32}$$

This is to be compared with the root of Eq. (27) corresponding to $i=2$.

Now, a frequency of the same form as Eq. (32) can be obtained from the equations of motion, Eq. (2), if we assume $\eta=\psi_z=0$ and if the presence of ψ_x in the second and third of Eq. (2) is ignored. As a result, Eq. (27) reduces to $P_2=0$, which has the same form as Eq. (32). Hence, what is contained in Sykes's coefficients k and k_1 is the effect of coupling of η and ψ_z with ψ_x. This coupling arises from two sources, namely, through the elastic constants and as a result of reflections of waves at the plane surfaces $y=\pm h/2$. The source of the second type of coupling is not apparent in the equations of motion, Eq. (2), since it has been masked as a result of the integration through the thickness. The effects of both types of coupling are contained in the frequencies and amplitude ratios given by Eqs. (27) and (29).

5. ONE PAIR OF PARALLEL EDGES FREE, THE OTHER PAIR SIMPLY-SUPPORTED

In the presence of a free edge, a third type of coupling occurs. If we consider waves of the type that generate the three solutions ($i=1,2,3$) in Eq. (21), then, on incidence at a free edge, each wave gives rise, in general, to reflected waves of all three types.[5] Hence, in a vibrating plate with free edges, the modes are composed of coupled modes of a plate with simply-supported edges.

TABLE II. Real and imaginary character of wave numbers.

Frequency range	Real	Imaginary
$R_3(\omega>\omega_3^*)$	ξ_1,ξ_2,ξ_3	
$R_2(\omega_2^*<\omega<\omega_3^*)$	ξ_1,ξ_2	ξ_3
$R_1(\omega_1^*<\omega<\omega_2^*)$	ξ_1	ξ_2,ξ_3
$R_0(\omega<\omega_1^*)$		ξ_1,ξ_2,ξ_3

For a plate free along $x=\pm a$ and simply-supported along $z=\pm c$, the boundary conditions are

$$M_x=0,\quad Q_x=0,\quad M_{xz}=0\quad \text{on}\quad x=\pm a$$
$$M_z=0,\quad \partial\eta/\partial t=0,\quad \partial\psi_x/\partial t=0\quad \text{on}\quad z=\pm c. \tag{33}$$

We consider displacements

$$\psi_x=\sum_{i=1}^{3} A_i\cos\xi_ix\sin\zeta z\cos\omega t$$
$$\eta=\sum_{i=1}^{3} B_ih\sin\xi_ix\sin\zeta z\cos\omega t \tag{34}$$
$$\psi_z=\sum_{i=1}^{3} C_i\sin\xi_ix\cos\zeta z\cos\omega t.$$

This is a solution of Eq. (2) provided that

$$\begin{vmatrix} P_i & \alpha_i & d_{71}\alpha_i\gamma \\ \alpha_i & Q_i & d_{46}\gamma \\ d_{71}\alpha_i\gamma & d_{46}\gamma & R_i \end{vmatrix}=0 \tag{35}$$

and

$$B_i/A_i=(d_{46}P_i-d_{71}\alpha_i^2)/\alpha_i\pi f(d_{71}Q_i-d_{46})$$
$$C_i/A_i=\gamma(d_{46}P_i-d_{71}\alpha_i^2)/\alpha_i(R_i-d_{46}d_{71}\gamma^2), \tag{36}$$

where

$$\alpha_i=\xi_ih/\pi,\quad \gamma=\zeta h/\pi$$
$$P_i=\alpha_i^2+d_{51}\gamma^2+f(1-\omega^2/\omega_x^2)$$
$$Q_i=(\alpha_i^2+d_{46}\gamma^2-\omega^2/\kappa^2\omega_x^2)/f \tag{37}$$
$$R_i=d_{51}\alpha_i^2+d_{31}\gamma^2+f(d_{46}-\omega^2/\omega_x^2).$$

Equation (35) is to be regarded as a bicubic in ξ with ζ and ω as parameters. For positive, real ζ and ω there are three roots ξ_i that are positive real or imaginary depending on the frequency range, as shown in Table II. These are the ξ_i which appear in Eqs. (34) to (37). The frequencies ω_1^*, ω_2^*, ω_3^* at which ξ_1, ξ_2, ξ_3, respectively, change from real to imaginary are those given in Eq. (18).

The boundary conditions, Eq. (33), on $z=\pm c$ are satisfied if

$$\zeta=p\pi/2c, \tag{38}$$

[4] R. A. Sykes in *Quartz Crystals for Electric Circuits*, edited by R. A. Heising (D. Van Nostrand Company, Inc., New York, 1946), chap. VI, p. 218, Eq. 6.9.

[5] This is shown for the isotropic plate by T. R. Kane, J. Appl. Mech. 21, 213 (1954).

where p is an even integer. For modes even in z, $\sin\zeta z$ and $\cos\zeta z$ are interchanged in Eq. (34). Then the boundary conditions on $z=\pm c$ are satisfied if ζ is given by Eq. (38) and p is an odd integer.

To find the requirements imposed by the boundary conditions on $x=\pm a$ substitute Eq. (34) first in Eq. (5) and then in Eq. (33), with the result

$$\begin{aligned} A_1\beta_{11}\sin\xi_1a+A_2\beta_{12}\sin\xi_2a+A_3\beta_{13}\sin\xi_3a&=0,\\ A_1\beta_{21}\cos\xi_1a+A_2\beta_{22}\cos\xi_2a+A_3\beta_{23}\cos\xi_3a&=0,\\ A_1\beta_{31}\cos\xi_1a+A_2\beta_{32}\cos\xi_2a+A_3\beta_{33}\cos\xi_3a&=0, \end{aligned} \tag{39}$$

where

$$\begin{aligned} \beta_{1i}&=d_{12}\alpha_i+\gamma C_i/A_i,\\ \beta_{2i}&=\gamma+\alpha_iC_i/A_i,\\ \beta_{3i}&=1+\pi\alpha_iB_i/A_i. \end{aligned} \tag{40}$$

Equations (39) determine amplitude ratios $A_1:A_2:A_3$. Thus, except for special cases to be considered later, all three types of the modes of a simply-supported plate are coupled as a result of introducing free edges. For a nontrivial solution, the determinant of Eq. (39) must vanish. This condition yields the transcendental equation

$$L_1\tan\varphi+L_2\tan M\varphi+L_3\tan N\varphi=0, \tag{41}$$

where

$$\begin{aligned} L_1&=\beta_{11}(\beta_{22}\beta_{33}-\beta_{32}\beta_{23}),\\ L_2&=\beta_{12}(\beta_{23}\beta_{31}-\beta_{33}\beta_{21}),\\ L_3&=\beta_{13}(\beta_{21}\beta_{32}-\beta_{31}\beta_{22}),\\ \varphi&=\xi_1a,\quad M=\xi_2/\xi_1,\quad N=\xi_3/\xi_1. \end{aligned} \tag{42}$$

For modes even in x, tangent is replaced by cotangent in Eq. (41).

The frequency spectrum may be computed in the following manner. The ratios of elastic constants d_{kl} are computed for the material of the plate and orientation of cut. Then choose a value of γ, the ratio of plate thickness to distance between nodes in the z-direction. The further choice of ω/ω_x permits the calculation of the α_i from Eq. (35) and the B_i/A_i and C_i/A_i from Eq. (36), so that the β_{ki} may be found from Eq. (40). With these results, the L_i and M and N may be computed from Eq. (42). Hence, all the coefficients in Eq. (41) are known and the equation becomes a transcendental equation in the single variable φ. A sequence of roots of Eq. (41) may then be computed by successive approximations. To each root there corresponds a length-thickness ratio according to the relation

$$2a/h=2\varphi/\pi\alpha_1. \tag{43}$$

Thus, a set of length-thickness ratios is found for the chosen value of ω/ω_x. The computation may be repeated for successive values of ω/ω_x until sufficient points can be plotted (ω/ω_x *vs* $2a/h$) to enable the curves of the frequency spectrum to be drawn for the chosen value of γ. The process may then be repeated for a new value of γ.

Before proceeding to an analysis of the frequency spectrum, it is advisable to study the relation between the uncoupled spectrum for simply-supported edges and the coupled spectrum for free edges in a simpler case.

6. CYLINDRICAL VIBRATIONS OF A STRIP

Returning to the problem of the rectangular plate simply-supported on all edges, we consider the limiting case ($\zeta h=0$) of motion independent of z. The asymptotic frequencies, Eq. (18), degenerate to $\omega_1^*=0$, $\omega_2^*=\omega_x$, $\omega_3^*=\omega_z$ and the frequencies themselves [from Eq. (27) with $\gamma_i=0$] become

$$\begin{aligned} 2(\omega_{1,2}/\omega_x)^2&=1+k\alpha_{1,2}^2\mp[(1+k\alpha_{1,2}^2)^2-4\kappa^4\alpha_{1,2}^4/g]^{\frac{1}{2}}\\ (\omega_3/\omega_x)^2&=(c_{44}/c_{66})[1+(c_{55}/c_{44})\alpha_3^2]. \end{aligned} \tag{44}$$

The expressions for ω_1 and ω_2 are the same as those given by Eqs. (53) and (52), respectively, of I. (Note that r,n,l and $\bar{\omega}$ of I are s_1, s_2, $2a$ and ω_x, respectively, here.) Hence, Fig. 4 of I is a part of the frequency spectrum of a simply-supported strip, of width $l=2a$, vibrating in modes whose shapes do not vary along the length. A typical portion of the spectrum of the xy thickness-shear and flexural modes, in the range $\omega>\omega_x$, is shown by the curves $s_i=$constant in Fig. 3. The curves s_i even are for modes odd in x and derive from the boundary conditions

$$\sin\xi_1a=0,\quad \sin\xi_2a=0. \tag{45}$$

The curves s_i odd are for modes even in x and derive from

$$\cos\xi_1a=0,\quad \cos\xi_2a=0. \tag{46}$$

Next, consider the limiting case $\zeta h=0$ in the problem solved in Sec. 5. The boundary conditions, Eq. (39), reduce to

$$\begin{aligned} A_1(\sigma_1-1)\xi_1^2\sin\xi_1a+A_2(\sigma_2-1)\xi_2^2\sin\xi_2a&=0\\ A_1\sigma_1\xi_1\cos\xi_1a+A_2\sigma_2\xi_2\cos\xi_2a&=0\\ A_3\xi_3\cos\xi_3a&=0, \end{aligned} \tag{47}$$

where

$$\sigma_i=\omega^2/(\kappa^2\alpha_i^2\omega_x^2), \tag{48}$$

and Eq. (41) becomes

$$(M^2-g)\tan\varphi+M(gM^2-1)\tan M\varphi=0. \tag{49}$$

Since A_3 occurs only in the last boundary condition, the zy thickness-shear modes are not coupled with the xy thickness-shear and flexural modes. However, the latter two are coupled with each other, since A_1 and A_2 appear in both of the first two conditions. This is the case which was treated in I. The frequency spectrum for modes odd in x appears as Fig. 1 of I and a typical portion of it, in the range $\omega>\omega_x$, is shown here as the curve marked K=0 in Fig. 3. (The spectrum for modes even in x is given in Fig. 1 of II).

We note that the first two of Eqs. (47) have the solutions Eqs. (45) and (46). Hence, the curve K=0, in Fig. 3, passes through the intersections of the curves s_1 even and s_2 even, and also through the intersections of the curves s_1 odd and s_2 odd. Between these intersections, however, the condition $\omega_1=\omega_2$, required by the boundary conditions for free edges, is not satisfied and hence the spectrum of the strip with free edges (K=0) does not coincide with the spectrum of the strip with simply-supported edges (s_i=constant). The development of the coupling, which produces this result, may be traced by considering the elastically supported strip, which contains the free and simply-supported strips as limiting cases.[6]

For elastically supported edges, the boundary conditions on $x=\pm a$ are

$$M_x=0, \quad Q_x=\mp K\eta, \tag{50}$$

where K is the spring constant of the elastic support. Accordingly, for xy thickness-shear and flexural modes odd in x, the boundary conditions become

$$A_1(\sigma_1-1)\xi_1^2 \sin\xi_1 a + A_2(\sigma_2-1)\xi_2^2 \sin\xi_2 a = 0,$$

$$D_6(A_1\sigma_1\xi_1 \cos\xi_1 a + A_2\sigma_2\xi_2 \cos\xi_2 a) = -K(A_1 \sin\xi_1 a + A_2 \sin\xi_2 a). \tag{51}$$

When $K=\infty$, Eqs. (51) reduce to Eqs. (45) and when K=0, Eqs. (51) reduce to the first two of Eqs. (47). When $0<K<\infty$, Eqs. (51) still have the roots determined by Eqs. (45), so that the curves giving the frequencies of modes odd in x of the elastically supported strip pass through the intersections of the curves s_1 even, s_2 even of the spectrum of modes odd in x of the simply-supported strip. The portions of the curves (for $0<K<\infty$) between these intersections do not, however, pass through the intersections of s_1 odd, s_2 odd because Eqs. (51) are not satisfied by Eqs. (46). In fact, these portions of the curves may be expected to fall above the intersections s_1 odd, s_2 odd, because the introduction of the elastic support is an additional constraint which serves to raise the frequencies above those of the strip with free edges. Calculation of the roots and corresponding frequencies shows that this is indeed the result. A typical portion of the spectrum for modes odd in x is shown by the dashed lines in Fig. 3 for two values of $0<K<\infty$.

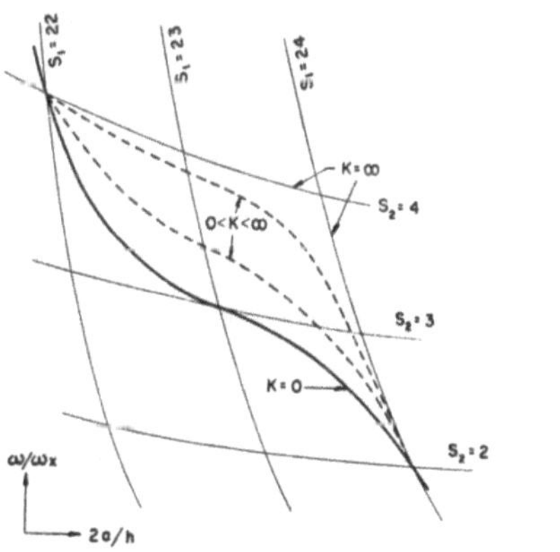

FIG. 3. Typical portion of spectrum of frequencies of cylindrical vibrations of a strip in range $\omega>\omega_x$, illustrating development of coupling. Curves s_i=constant are for simply supported edges; s_1 flexure, s_2 thickness-shear; s_i odd, even, are for modes even and odd, respectively, in x. Curve K=0 is for a mode, odd in x, of a strip with free edges. Curves $0<K<\infty$ are for modes, odd in x, of a strip with elastically supported edges. (K is the spring constant of the elastic support.)

[6] This was suggested by E. Sternberg.

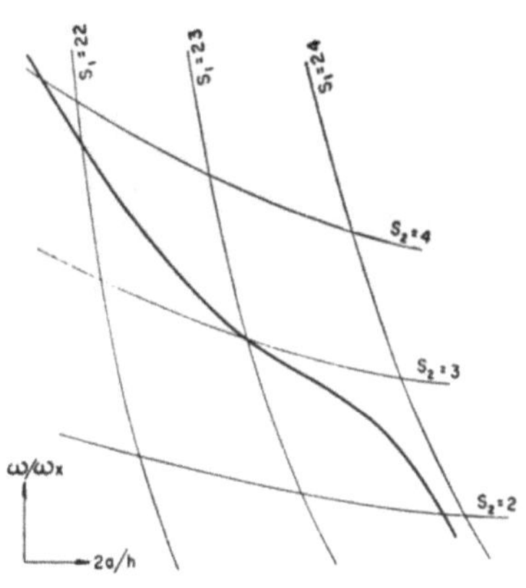

FIG. 4. Typical portion of spectrum, in range R_2, for a mode odd in x, of a plate free along one pair of parallel edges and simply supported along the other pair.

The transition from the simple frequency spectrum of uncoupled modes of the simply-supported strip to the frequency spectrum of the coupled modes of the free strip may now be traced continuously in Fig. 3, for modes odd in x. For modes even in x, the situation is similar. The only alteration required in the explanation in the preceding paragraphs is the interchange of "odd" and "even."

7. FREQUENCY SPECTRUM OF A PLATE WITH A PAIR OF PARALLEL EDGES FREE AND THE OTHER PAIR SIMPLY-SUPPORTED

The frequency spectrum governed by Eqs. (41) and (43) has a different character in each of the four ranges listed in Table II.

Range R_0, $\omega<\omega_1^*$.—In this range all three ξ_i are imaginary so that all three transcendental functions in Eq. (41) are hyperbolic. Accordingly, Eq. (41) has no roots in this range, i.e., there are no frequencies of vibration $\omega<\omega_1^*$.

Range R_1, $\omega_1^*<\omega<\omega_2^*$.—Here ξ_1 is real and ξ_2 and ξ_3 are imaginary. Only one of the three transcendental functions in Eq. (41) is oscillatory and the remaining two are almost unity. Hence, the spectrum is very much like that in the same range for the simply-sup-

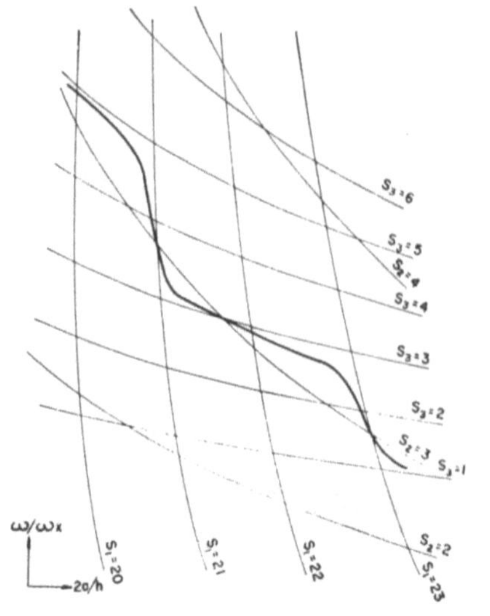

FIG. 5. Typical portion of spectrum, in range R_3, for a mode odd in x, of a plate free along one pair of parallel edges and simply supported along the other pair.

ported plate, as shown in Fig. 2. Near $\omega=\omega_2^*$ the curves are slightly lower than those in Fig. 2, but approach the latter asymptotically as $\omega\to\omega_1^*$.

Range R_2, $\omega_2^*<\omega<\omega_3^*$.—Both ξ_1 and ξ_2 are real while ξ_3 is imaginary in this range. Thus there are two trigonometric functions in Eq. (41) just as there are in Eq. (49) for the free strip. However, the presence of the third term causes the spectrum to differ from that illustrated by the heavy line in Fig. 3. Referring to Eq. (39), from which Eq. (41) was derived, it may be seen that the equations have a solution

$$\cos\xi_1 a=0, \quad \cos\xi_2 a=0 \tag{52}$$

$$A_3=0, \quad A_1/A_2=\pm\beta_{12}/\beta_{11} \tag{53}$$

but not a solution $\sin\xi_1 a=0$, $\sin\xi_2 a=0$. Hence, the frequency curve for a mode odd in x passes through the intersections of curves s_1 odd, s_2 odd but not through the intersections of s_1 even, s_2 even. The curve may be expected to fall below, rather than above, the latter, because the relaxation of the condition $\zeta h=0$ is a relaxation of a constraint. Hence, instead of the curve K=0 in Fig. 3, we have the curve shown by the heavy line in Fig. 4. It is to be noted that the curve for a mode odd in x, of a plate with a pair of free edges, passes through the intersections s_i odd, which are points on the spectrum for modes even in x of a simply-supported plate. It is also evident, from Eq. (53), that at these intersection points the zy thickness-shear mode is absent.

Range R_3, $\omega>\omega_3^*$.—Here all three ξ_i are real. Hence Eqs. (39) and (41) again have the solutions Eqs. (52), (53) and, in addition, two sets of solutions obtained by cyclical permutation of the subscripts in Eqs. (52) and (53). (Only the second subscript in β_{ki} is to be permuted.) Accordingly, all that was said of the intersections of curves s_1, s_2 constant in range R_2 applies to the intersections of s_1, s_2 constant, s_2, s_3 constant, and s_3, s_1 constant in range R_3. Note that, in each case, it is the zy thickness-shear mode, the flexure mode, and the xy thickness-shear mode, respectively, that is absent at the intersection points. A typical curve of the spectrum is illustrated by the heavy line in Fig. 5.

On the basis of these results, a set of rules may be formulated for sketching the frequency spectrum of the plate, with one pair of parallel edges free and the other pair simply-supported, on the network of curves s_i=constant of the simply-supported plate (Fig. 2). For modes odd in x the rules are:

(1) Starting at the low-frequency end of the spectrum, proceed to sketch a line upward. (The line should tend to the left, since, for a given mode, increasing frequencies correspond to decreasing length-thickness ratios.)

(2) The line cannot cross a curve s_i odd except at an intersection with another curve s_i odd.

(3) On approaching a curve s_i odd, the line must bear upward and to the left until it crosses the curve at the nearest intersection with another curve s_i odd.

(4) The line may cross curves s_i even, but must not pass through intersections of curves s_i even.

For modes even in x, it is only necessary to interchange odd and even in the rules.

High-Frequency Extensional Vibrations of Plates[1]

By T. R. KANE[2] and R. D. MINDLIN[3]

The equations of generalized plane stress are applicable to the extensional vibrations of plates only for the low modes of motion of thin plates. In this paper equations are derived which are applicable to both low and high modes of thin plates and to low modes of thick plates. As an example, the equations are solved for the case of axially symmetric vibrations of a circular disk and comparisons are made with the theory of generalized plane stress and with experiments.

1 Introduction

Extensional vibrations of thin plates are described satisfactorily by the equations of generalized plane stress provided the wave lengths, or distances between nodes, are large in comparison with the thickness of the plate. In terms of frequencies, that is to say that the theory of generalized plane stress gives good results for frequencies which are much lower than the frequency of the first mode of thickness vibration of the plate. The essential reason for the limitation of the equations of generalized plane stress is the absence, in them, of the phenomenon of coupling between extensional and thickness modes of motion.

Consider, for example, the radial vibrations of a circular disk. As the radius of the disk diminishes, the frequency of the fundamental mode of vibration increases without limit, according to the equations of generalized plane stress. This is illustrated by Curve I in Fig. 1. However, it is apparent that, in a disk whose diameter is considerably smaller than its thickness (i.e., a rod) there must be a mode of, essentially, thickness vibration whose frequency is much lower than that of the radial vibration. In a real disk, the coupling between the thickness and radial vibrations affects both the mode shapes and the frequencies to a very marked degree even at frequencies well below the fundamental frequency of thickness vibration.

It is the purpose of this paper to describe the derivation of a set of differential equations and boundary conditions which take into account coupling between extensional motions and the first mode of thickness vibration and to give the solution for the axially symmetric vibrations of a circular disk as an illustrative example, followed by a comparison with experimental measurements. The scheme adopted for the derivation of the equations is intended to exhibit the nature of the approximations by which the three-dimensional theory is reduced to a two-dimensional one without carrying the reduction so far as to arrive at the theory of generalized plane stress. An alternative course is to start with the equations of generalized plane stress and add correction terms, as was done by E. Reissner (1)[4] in a related investigation aimed at obtaining improved expressions for the variation of stress through the thickness of the plate. The equations derived in the present paper are not applicable to that problem.

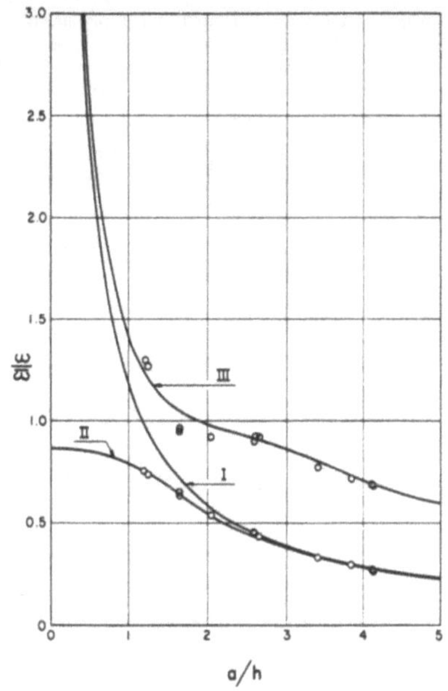

Fig. 1 Curves II and III Show Frequencies of First Two Modes of Axially Symmetric, Extensional Vibrations of Circular Disk According to Equations of This Paper
(Circled points are from experimental data obtained with barium-titanate disks at Bell Telephone Laboratories. Curve I shows frequencies of first mode according to theory of generalized plane stress.)

2 Three-Dimensional Equations

It is useful to review a derivation of a few of the classical equations of the linear theory of elasticity, arranged in such a manner as to serve as a guide in the derivation of the plate equations.

The law of conservation of energy (considering only mechanical effects) states that the rate of increase of energy in a body is equal to the rate at which work is done by the external forces. Thus, omitting body forces

[1] This investigation was supported by the Office of Naval Research, the Signal Corps Engineering Laboratories, and Bell Telephone Laboratories, Inc.

[2] Assistant Professor of Mechanical Engineering, University of Pennsylvania, Philadelphia, Pa. Assoc. Mem. ASME.

[3] Professor of Civil Engineering, Columbia University, New York, N. Y. Mem. ASME.

Contributed by the Applied Mechanics Division and presented at the Diamond Jubilee Annual Meeting, Chicago, Ill., November 13–18, 1955, of The American Society of Mechanical Engineers.

Discussion of this paper should be addressed to the Secretary, ASME, 29 West 39th Street, New York, N. Y., and will be accepted until July 10, 1956, for publication at a later date. Discussion received after the closing date will be returned.

Note: Statements and opinions advanced in papers are to be understood as individual expressions of their authors and not those of the Society. Manuscript received by ASME Applied Mechanics Division, April 8, 1955. Paper No. 55—A-50.

[4] Numbers in parentheses refer to the Bibliography at the end of the paper.

$$\iiint\left(\frac{\partial T}{\partial t}+\frac{\partial V}{\partial t}\right)dxdydz = \iint\left(\bar{X}\frac{\partial u_x}{\partial t}+\bar{Y}\frac{\partial u_y}{\partial t}+\bar{Z}\frac{\partial u_z}{\partial t}\right)dS \quad [1]$$

where T and V are the kinetic and potential energy-densities, respectively, and $\bar{X}$, $\bar{Y}$, $\bar{Z}$ (2) and u_x, u_y, u_z are the rectangular components of the surface traction and displacement, respectively. The triple integral is taken throughout the volume of the body and the double integral over its surface. Note that if the time derivatives are replaced by variations, Equation [1] is the variational equation of motion.[5]

Now
$$T = \frac{\rho}{2}\left[\left(\frac{\partial u_x}{\partial t}\right)^2+\left(\frac{\partial u_y}{\partial t}\right)^2+\left(\frac{\partial u_z}{\partial t}\right)^2\right] \quad [2]$$

so that
$$\frac{\partial T}{\partial t} = \rho\left(\frac{\partial^2 u_x}{\partial t^2}\frac{\partial u_x}{\partial t}+\frac{\partial^2 u_y}{\partial t^2}\frac{\partial u_y}{\partial t}+\frac{\partial^2 u_z}{\partial t^2}\frac{\partial u_z}{\partial t}\right) \quad [3]$$

Also, since the potential energy is a function of the components of strain ($\epsilon_x, \ldots, \gamma_{yz}, \ldots$) which are, in turn, functions of time, the rate of increase of V may be expressed as

$$\frac{\partial V}{\partial t} = \frac{\partial V}{\partial \epsilon_x}\frac{\partial \epsilon_x}{\partial t}+\ldots+\frac{\partial V}{\partial \gamma_{yz}}\frac{\partial \gamma_{yz}}{\partial t}+\ldots \quad [4]$$

Now
$$\epsilon_x = \frac{\partial u_x}{\partial x}, \ldots, \quad \gamma_{yz} = \frac{\partial u_z}{\partial y}+\frac{\partial u_y}{\partial z}, \ldots \quad [5]$$

and (3)
$$\sigma_x = \frac{\partial V}{\partial \epsilon_x}, \ldots, \quad \tau_{yz} = \frac{\partial V}{\partial \gamma_{yz}}, \ldots \quad [6]$$

Hence the rate of increase of potential energy-density may be written in the form

$$\left.\begin{aligned}\frac{\partial V}{\partial t} &= \left(\sigma_x\frac{\partial}{\partial x}+\tau_{yx}\frac{\partial}{\partial y}+\tau_{zx}\frac{\partial}{\partial z}\right)\frac{\partial u_x}{\partial t}+\ldots+\ldots\\ &= \frac{\partial}{\partial x}\left(\sigma_x\frac{\partial u_x}{\partial t}+\tau_{yx}\frac{\partial u_y}{\partial t}+\tau_{zx}\frac{\partial u_z}{\partial t}\right)+\ldots+\ldots\\ &\quad -\left(\frac{\partial \sigma_x}{\partial x}+\frac{\partial \tau_{yx}}{\partial y}+\frac{\partial \tau_{zx}}{\partial z}\right)\frac{\partial u_x}{\partial t}-\ldots-\ldots\end{aligned}\right\} \quad [7]$$

By integration by parts

$$\begin{aligned}\iiint&\left[\frac{\partial}{\partial x}\left(\sigma_x\frac{\partial u_x}{\partial t}+\tau_{yx}\frac{\partial u_y}{\partial t}+\tau_{zx}\frac{\partial u_z}{\partial t}\right)+\ldots\right]dxdydz\\ &= \iint\left[(\sigma_x l+\tau_{xy}m+\tau_{xz}n)\frac{\partial u_x}{\partial t}+\ldots+\ldots\right]dS\\ &= \iint\left(\bar{X}\frac{\partial u_x}{\partial t}+\bar{Y}\frac{\partial u_y}{\partial t}+\bar{Z}\frac{\partial u_z}{\partial t}\right)dS\end{aligned}$$

where l, m, n are the direction cosines of the outward drawn normal to S. Hence, substituting Equations [3] and [7] in [1], the law of conservation of energy reduces to

$$\iiint\left[\left(\frac{\partial \sigma_x}{\partial x}+\frac{\partial \tau_{yx}}{\partial y}+\frac{\partial \tau_{zx}}{\partial z}-\rho\frac{\partial^2 u_x}{\partial t^2}\right)\frac{\partial u_x}{\partial t}+\ldots+\ldots\right]dxdydz = 0 \quad [8]$$

Since this must hold for all volumes, the integrand must vanish, and, since the integrand must vanish for all velocities, the quantities in parentheses must vanish. This leads to three stress equations of motion of the type

$$\frac{\partial \sigma_x}{\partial x}+\frac{\partial \tau_{yx}}{\partial y}+\frac{\partial \tau_{zx}}{\partial z} = \rho\frac{\partial^2 u_x}{\partial t^2} \quad [9]$$

In an isotropic elastic body (3)

$$2V = \lambda(\epsilon_x+\epsilon_y+\epsilon_z)^2+2G(\epsilon_x^2+\epsilon_y^2+\epsilon_z^2) + G(\gamma_{yz}^2+\gamma_{zx}^2+\gamma_{xy}^2) \quad [10]$$

where λ and G are Lamé's constants of elasticity. Hence the stress-strain relation, from Equations [6] and [10], is

$$\left.\begin{aligned}\sigma_x &= (\lambda+2G)\epsilon_x+\lambda(\epsilon_y+\epsilon_z)\\ \sigma_y &= (\lambda+2G)\epsilon_y+\lambda(\epsilon_z+\epsilon_x)\\ \sigma_z &= (\lambda+2G)\epsilon_z+\lambda(\epsilon_x+\epsilon_y)\\ (\tau_{yz}, \tau_{zx}, \tau_{xy}) &= G(\gamma_{yz}, \gamma_{zx}, \gamma_{xy})\end{aligned}\right\} \quad [11]$$

Substituting these in Equation [9] and using Equations [5], there result the displacement equations of motion

$$\left.\begin{aligned}G\nabla^2 u_x+(\lambda+G)\frac{\partial e}{\partial x} &= \rho\frac{\partial^2 u_x}{\partial t^2}\\ G\nabla^2 u_y+(\lambda+G)\frac{\partial e}{\partial y} &= \rho\frac{\partial^2 u_y}{\partial t^2}\\ G\nabla^2 u_z+(\lambda+G)\frac{\partial e}{\partial z} &= \rho\frac{\partial^2 u_z}{\partial t^2}\end{aligned}\right\} \quad [12]$$

where
$$e = \frac{\partial u_x}{\partial x}+\frac{\partial u_y}{\partial y}+\frac{\partial u_z}{\partial z}$$

Boundary conditions appropriate to these equations of motion are obtained from Neumann's uniqueness theorem (4). By this theorem it is shown that, subject to the usual restrictions as to continuity, single valuedness, singularities and behavior at infinity, it is sufficient to specify, at each point of the boundary, one member of each of the three products appearing in the integrand of

$$\iint\left(\bar{X}\frac{\partial u_x}{\partial t}+\bar{Y}\frac{\partial u_y}{\partial t}+\bar{Z}\frac{\partial u_z}{\partial t}\right)dS \quad [13]$$

or their equivalents referred to co-ordinates normal and tangential to the surface.

3 Plate-Stress Equations of Motion

Consider a plate bounded by planes $z = \pm h$ and a right cylindrical surface intersecting the middle plane in a closed curve C. The components of displacement in the plate are assumed to be given to sufficient approximation by

$$\left.\begin{aligned}u_x &= v_x(x, y, t)\\ u_y &= v_y(x, y, t)\\ u_z &= zh^{-1}v_z(x, y, t)\end{aligned}\right\} \quad [14]$$

These expressions are substituted into Equations [8], which are integrated with respect to z between limits $\pm h$. The integrations are carried out explicitly wherever possible. Then, following the procedure leading from Equations [8] to [9], there result the plate-stress equations of motion

$$\left.\begin{aligned}\frac{\partial N_x}{\partial x}+\frac{\partial N_{xy}}{\partial y}+F_x &= 2\rho h\frac{\partial^2 v_x}{\partial t^2}\\ \frac{\partial N_{xy}}{\partial x}+\frac{\partial N_y}{\partial y}+F_y &= 2\rho h\frac{\partial^2 v_y}{\partial t^2}\\ \frac{\partial R_x}{\partial x}+\frac{\partial R_y}{\partial y}-N_z+hF_z &= \frac{2}{3}\rho h^2\frac{\partial^2 v_z}{\partial t^2}\end{aligned}\right\} \quad [15]$$

[5] See reference (4), p. 167.

where

$$(N_x, N_y, N_z, N_{xy}) = \int_{-h}^{h} (\sigma_x, \sigma_y, \sigma_z, \tau_{xy})dz \ldots\ldots [16]$$

$$(R_x, R_y) = \int_{-h}^{h} (\tau_{xz}, \tau_{zy})zdz \ldots\ldots\ldots\ldots [17]$$

$$(F_x, F_y, F_z) = [\tau_{xz}, \tau_{zy}, \sigma_z]_{z=h} - [\tau_{xz}, \tau_{zy}, -\sigma_z]_{z=-h} \ldots [18]$$

N_x, N_y, and N_{xy} are the usual forces per unit of length which appear in the elementary theory of plates acted upon by forces in the middle plane (5). N_z is $2h$ times the average transverse stress. R_x and R_y are components of "pinching" shear. They play a role, in the last of Equations [15], analogous to that of the components of transverse shear force, Q_x and Q_y, in the corresponding equation of the theory of flexure of plates (6). F_x, F_y, and F_z are the components of surface traction. They play the role in Equations [15] that is played by body forces in the three-dimensional theory.

The first two of Equations [15] are identical with those of the theory of generalized plane stress, but none of the terms of the third equation appears in that theory.

4 Plate-Displacement Equations of Motion

The relations between the plate-stresses (N_x, N_y, N_z, N_{xy}, R_x, R_y) and the plate-displacements (v_x, v_y, v_z) are obtained, as in the three-dimensional theory, from the potential energy-density V. In order to improve the results, the thickness strain, ϵ_z, in Equation [10] is replaced by $\kappa\epsilon_z$ where κ is a constant which plays a role analogous to that of the square root of the constant k' in Timoshenko's theory of beams (7). Thus

$$2V = \lambda(\epsilon_x + \epsilon_y + \kappa\epsilon_z)^2 + 2G(\epsilon_x^2 + \epsilon_y^2 + \kappa^2\epsilon_z^2) + G(\gamma_{yz}^2 + \gamma_{zx}^2 + \gamma_{xy}^2) \ldots [19]$$

The motivation for the introduction of κ, with the concomitant, fictitious anisotropy, is discussed in Section 7.

Using Equations [19] and [6], there results

$$\left.\begin{aligned} \sigma_x &= (\lambda + 2G)\epsilon_x + \lambda(\epsilon_y + \kappa\epsilon_z) \\ \sigma_y &= (\lambda + 2G)\epsilon_y + \lambda(\epsilon_x + \kappa\epsilon_z) \\ \sigma_z &= (\lambda + 2G)\kappa^2\epsilon_z + \lambda\kappa(\epsilon_x + \epsilon_y) \\ (\tau_{yz}, \tau_{zx}, \tau_{xy}) &= G(\gamma_{yz}, \gamma_{zx}, \gamma_{xy}) \end{aligned}\right\} \ldots\ldots [20]$$

The Expressions [5] for the strains in terms of the displacements are now inserted in Equations [20], following which the Approximations [14] are used to replace the displacements u_x, u_y, u_z. The resulting expressions for the stresses in terms of v_x, v_y, v_z are inserted in the right-hand sides of Equations [16] and [17] and the integrations are performed, with the results

$$\left.\begin{aligned} N_x &= 2h\left[(\lambda + 2G)\frac{\partial v_x}{\partial x} + \lambda\frac{\partial v_y}{\partial y} + \kappa\lambda\frac{v_z}{h}\right] \\ N_y &= 2h\left[(\lambda + 2G)\frac{\partial v_y}{\partial y} + \lambda\frac{\partial v_x}{\partial x} + \kappa\lambda\frac{v_z}{h}\right] \\ N_z &= 2h\left[\kappa^2(\lambda + 2G)\frac{v_z}{h} + \kappa\lambda\left(\frac{\partial v_x}{\partial x} + \frac{\partial v_y}{\partial y}\right)\right] \\ N_{xy} &= 2hG\left(\frac{\partial v_y}{\partial x} + \frac{\partial v_x}{\partial y}\right) \\ R_x &= \frac{2h^2G}{3}\frac{\partial v_z}{\partial x}, \quad R_y = \frac{2h^2G}{3}\frac{\partial v_z}{\partial y} \end{aligned}\right\} \ldots [21]$$

The displacement equations of motion are now obtained by substituting Equations [21] into [15], with the result

$$\left.\begin{aligned} G\nabla_1^2 v_x + (\lambda + G)\frac{\partial e_1}{\partial x} + \frac{\kappa\lambda}{h}\frac{\partial v_z}{\partial x} + \frac{F_x}{2h} &= \rho\frac{\partial^2 v_x}{\partial t^2} \\ G\nabla_1^2 v_y + (\lambda + G)\frac{\partial e_1}{\partial y} + \frac{\kappa\lambda}{h}\frac{\partial v_z}{\partial y} + \frac{F_y}{2h} &= \rho\frac{\partial^2 v_y}{\partial t^2} \\ G\nabla_1^2 v_z - \frac{3\kappa\lambda e_1}{h} - \frac{3\kappa^2}{h^2}(\lambda + 2G)v_z + \frac{3F_z}{2h} &= \rho\frac{\partial^2 v_z}{\partial t^2} \end{aligned}\right\} \ldots [22]$$

where

$$e_1 = \frac{\partial v_x}{\partial x} + \frac{\partial v_y}{\partial y}, \qquad \nabla_1^2 = \frac{\partial^2}{\partial x^2} + \frac{\partial^2}{\partial y^2}$$

It may be observed that Equations [22] degenerate to those of generalized plane stress if the third equation is ignored and if v_z, in the first two, is replaced by its value, in terms of v_x and v_y, obtained from the third of Equations [21] by setting $N_z = 0$, and $\kappa = 1$. See p. 497, Bibliography (4).

5 Boundary Conditions

Boundary conditions, on the edge of the plate, sufficient to assure a unique solution of Equations [22], are obtained by substituting Equation [14] into [13] and performing the integration with respect to z, with the result

$$\oint_C \left(\bar{X}_1\frac{\partial v_x}{\partial t} + \bar{Y}_1\frac{\partial v_y}{\partial t} + \frac{\bar{Z}_1}{h}\frac{\partial v_z}{\partial t}\right)ds \ldots\ldots [23]$$

where s is measured along the edge of the plate

$$\left.\begin{aligned} \bar{X}_1 &= N_xl + N_{xy}m \\ \bar{Y}_1 &= N_{xy}l + N_ym \\ \bar{Z}_1 &= R_xl + R_ym \end{aligned}\right\} \ldots\ldots\ldots\ldots [24]$$

and l and m are the direction cosines of the outward drawn normal to the edge.

The integrand of Expression [23] may be referred to co-ordinates n, s, normal and tangential to the edge, by means of the transformations

$$\left.\begin{aligned} v_n &= v_xl + v_ym \\ v_s &= v_yl - v_xm \\ N_n &= N_xl^2 + N_ym^2 + 2N_{xy}lm \\ N_{ns} &= (N_y - N_x)lm + N_{xy}(l^2 - m^2) \\ R_n &= R_xl + R_ym \end{aligned}\right\} \ldots\ldots [25]$$

with the result that Equation [23] becomes

$$\oint_C \left(N_n\frac{\partial v_n}{\partial t} + N_{ns}\frac{\partial v_s}{\partial t} + \frac{R_n}{h}\frac{\partial v_z}{\partial t}\right)ds \ldots\ldots [26]$$

The first two products in Expression [26] are identical with those occurring in the theory of generalized plane stress but it is seen that an additional boundary condition is required for Equations [22]; namely, either the shear R_n or the velocity $\partial v_z/\partial t$ must be specified.

6 Alternative Procedure

The results in Sections 3, 4, and 5 were obtained by inserting the Approximations [14] into equations at several stages of the development outlined in Section 2. This type of exposition was adopted in order to emphasize the relations between corresponding equations in the three-dimensional and plate theories. Alternatively, the approximations may be inserted directly into Equation [1] and the integration with respect to z performed once and for all. The plate equations then follow logically (except for the arbitrary introduction of κ) by a procedure analogous to that in the remainder of Section 2. One advantage of this method is the appearance of explicit expressions for the kinetic-energy density of the plate

$$T^* = \rho h\left[\left(\frac{\partial v_x}{\partial t}\right)^2 + \left(\frac{\partial v_y}{\partial t}\right)^2 + \frac{1}{3}\left(\frac{\partial v_z}{\partial t}\right)^2\right]$$

and the potential energy-density of the plate

$$2V^* = \lambda h(e_x + e_y + \kappa e_z)^2 + 2Gh(e_x^2 + e_y^2 + \kappa^2 e_z^2) + Gh\left(\frac{h^2}{3} g_{yz}^2 + \frac{h^2}{3} g_{zx}^2 + g_{xy}^2\right)$$

where

$$e_x = \frac{\partial v_x}{\partial x} \qquad g_{yz} = \frac{1}{h}\frac{\partial v_z}{\partial y}$$

$$e_y = \frac{\partial v_y}{\partial y} \qquad g_{zx} = \frac{1}{h}\frac{\partial v_z}{\partial x}$$

$$e_z = \frac{v_z}{h} \qquad g_{xy} = \frac{\partial v_y}{\partial x} + \frac{\partial v_x}{\partial y}$$

are the components of plate strain. Also, the relations

$$N_x = \frac{\partial V^*}{\partial e_x}, \ldots, \qquad R_y = \frac{\partial V^*}{\partial g_{yz}}$$

are found.

In addition, a rigorous proof of the theorem on uniqueness of solutions may be developed along the lines described in a previous paper (8) on flexural motions of plates.

7 Determination of κ

The constant κ was inserted in the expression for the strain energy in order to overcome a difficulty inherent in approximate equations of the type described here. In the present case the source of the difficulty is the assumption of a particular (linear) functional dependence, on z, of the thickness displacement u_z. This can be correct only at one frequency. At low frequencies the thickness stresses, τ_{xz}, τ_{zy}, σ_z, and the thickness inertia $\rho\partial^2 u_z/\partial t^2$ are small and, accordingly, the thickness displacement follows the extensional stresses, σ_x, σ_y, as in statical deformation. Since the extensional stresses are essentially independent of z, at low frequencies, the best assumption for u_z, in this range, is proportionality to z. This is the assumption used in Equation [14]. However, as the frequency increases, a range is reached where strong coupling exists between extensional motions in the plane of the plate and the first mode of pure thickness vibration of the plate, in which u_z varies as $\sin(\pi z/2h)$. For this range it would be preferable to start with the assumption

$$u_z = v_z(x, y, t)\sin(\pi z/2h)$$

but then the low-frequency results would be in error. A two-term approximation could be used, at the expense of increased order of the differential equations. Alternatively, the factor κ can be introduced and its value fixed so that the frequency of the first thickness mode, at which the important coupling occurs, is correct despite the incorrect distribution of displacement. At low frequencies, all the terms containing κ are negligible so that the equations reduce to the appropriate form of generalized plane stress.

The value of κ is chosen, then, by equating the frequency of pure thickness vibration, obtained from the plate equations of motion, Equations [22], with the corresponding frequency obtained from the three-dimensional equations.

In Equations [22] let

$$\left.\begin{aligned} v_x = v_y = F_x = F_y = F_z = 0 \\ v_z = C\exp(i\bar{\omega}t)\end{aligned}\right\} \qquad [27]$$

Then the first two of Equations [22] are satisfied identically and the third yields

$$\bar{\omega}^2 = \frac{3\kappa^2}{h^2}\frac{\lambda + 2G}{\rho} \qquad [28]$$

It may be observed that no such frequency is contained in the equations of generalized plane stress. It is, in fact, the presence of such a mode of vibration in Equations [22] that makes them applicable to much higher frequencies than those accommodated by the equations of generalized plane stress.

The frequency, corresponding to Equation [28], obtained from the three-dimensional equations is

$$\bar{\omega}^2 = \frac{\pi^2}{4h^2}\frac{\lambda + 2G}{\rho} \qquad [29]$$

Equating [28] and [29], we find

$$\kappa^2 = \pi^2/12 \qquad [30]$$

8 Free Vibrations

In the case of free vibrations, the equations of motion, Equations [22], take a simple and revealing form when expressed in terms of the potentials of the dilatation and rotation in the plane of the plate.

Let

$$\left.\begin{aligned} v_x &= \frac{\partial \varphi}{\partial x} + \frac{\partial H}{\partial y} \\ v_y &= \frac{\partial \varphi}{\partial y} - \frac{\partial H}{\partial x}\end{aligned}\right\} \qquad [31]$$

Then, introducing a time factor $\exp(i\omega t)$ (omitted in the sequel), Equations [22] are satisfied, in the absence of surface tractions, if

$$\left.\begin{aligned} c_2^2\nabla_1^2 H + \omega^2 H &= 0 \\ c_1^2\nabla_1^2\varphi + \omega^2\varphi + (\kappa\lambda/\rho h)v_z &= 0 \\ c_2^2\nabla_1^2 v_z + (\omega^2 - 3\kappa^2 c_1^2/h^2)v_z - (3\kappa\lambda/\rho h)\nabla_1^2\varphi &= 0\end{aligned}\right\} \qquad [32]$$

where

$$c_1^2 = (\lambda + 2G)/\rho, \quad c_2^2 = G/\rho \qquad [33]$$

Eliminating v_z between the second and third of Equations [32], the equation governing φ is found to be

$$(\nabla_1^2 + \delta_1^2)(\nabla_1^2 + \delta_2^2)\varphi = 0 \qquad [34]$$

where

$$\left.\begin{aligned} \delta_i^2 &= \frac{3\kappa^2}{2\beta h^2}\left[(\alpha + \beta)\frac{\omega^2}{\bar{\omega}^2} - 1 - (-1)^i\psi\right], \quad i = 1, 2 \\ \psi &= \left\{\left[(\alpha + \beta)\frac{\omega^2}{\bar{\omega}^2} - 1\right]^2 + 4\alpha\beta\frac{\omega^2}{\bar{\omega}^2}\left(1 - \frac{\omega^2}{\bar{\omega}^2}\right)\right\}^{1/2} \\ \alpha &= \frac{c_1^2}{c_3^2}, \quad \beta = \frac{c_2^2}{c_3^2}, \quad c_3^2 = \frac{4G(\lambda + G)}{\rho(\lambda + 2G)} = \frac{E}{\rho(1 - \nu^2)}\end{aligned}\right\} \qquad [35]$$

in which E is Young's modulus and ν is Poisson's ratio.

Hence there are three independent displacement potentials: φ_1, φ_2, and H (where $\varphi_1 + \varphi_2 = \varphi$). They are governed by the three equations

$$\left.\begin{aligned} (\nabla_1^2 + \delta_1^2)\varphi_1 &= 0 \\ (\nabla_1^2 + \delta_2^2)\varphi_2 &= 0 \\ (\nabla_1^2 + \delta_3^2)H &= 0\end{aligned}\right\} \qquad [36]$$

where $\delta_3^2 = \omega^2/c_2^2$. It may be seen that there are three phase velocities in this theory, given by ω/δ_i, $i = 1, 2, 3$. The velocity, $\omega/\delta_3 = c_2$, of the rotation, is the same as in the theory of generalized plane stress; but there are two dilatational velocities (ω/δ_1 and ω/δ_2) as opposed to only one (c_3) in the theory of generalized plane stress. Furthermore, the two dilatational velocities are dispersive, whereas c_3 is not. Examination of the expressions for δ_1 and δ_2 shows that δ_1 is always real but δ_2 is real for $\omega/\bar{\omega} > 1$ and imaginary for $\omega/\bar{\omega} < 1$. Hence, an important change in the vibration spectrum of the plate occurs at the frequency $\bar{\omega}$. This is why the appropriate choice of the value of κ is essential.

The plate displacements are expressed in terms of the three potentials by

$$\left.\begin{aligned} v_x &= \frac{\partial\varphi_1}{\partial x} + \frac{\partial\varphi_2}{\partial x} + \frac{\partial H}{\partial y} \\ v_y &= \frac{\partial\varphi_1}{\partial y} + \frac{\partial\varphi_2}{\partial y} - \frac{\partial H}{\partial x} \\ v_z &= \sigma_1\varphi_1 + \sigma_2\varphi_2 \end{aligned}\right\} \qquad [37]$$

where

$$\sigma_i = \frac{h(\lambda + 2G)}{\kappa\lambda}\left(\delta_i^2 - \frac{\omega^2}{c_1^2}\right), \quad i = 1, 2 \qquad [38]$$

The expression for v_z in terms of φ_1 and φ_2 comes from the second of Equations [32] and the first two of Equations [36].

9 Symmetrical Vibrations of a Circular Disk

In this case the variables are expressed in terms of polar coordinates r, θ, by means of the relations $x = r\cos\theta$, $y = r\sin\theta$ and the transformation Equations [25]. For symmetrical vibrations, $H = 0$ and $\varphi = \varphi(r)$, so that Equations [36] reduce to

$$\left(\frac{d^2}{dr^2} + \frac{1}{r}\frac{d}{dr} + \delta_i^2\right)\varphi_i = 0, \quad i = 1, 2 \qquad [39]$$

The solutions of Equation [39], for a solid disk, are

$$\varphi_1 = A_1 J_0(\delta_1 r), \qquad \varphi_2 = A_2 J_0(\delta_2 r) \qquad [40]$$

where J is the Bessel function of the first kind and A_1 and A_2 are constants. Hence

$$\left.\begin{aligned} v_r &= -A_1\delta_1 J_1(\delta_1 r) - A_2\delta_2 J_1(\delta_2 r) \\ v_\theta &= 0 \\ v_z &= A_1\sigma_1 J_0(\delta_1 r) + A_2\sigma_2 J_0(\delta_2 r) \end{aligned}\right\} \qquad [41]$$

We consider the case of a free edge $r = a$. The appropriate components of edge traction are, from Equations [25], [26], and [21]

$$\left.\begin{aligned} N_r &= 2h\left[(\lambda + 2G)\frac{dv_r}{dr} + \lambda\frac{v_r}{r} + \kappa\lambda\frac{v_z}{h}\right] \\ N_{r\theta} &= 0 \\ R_r &= \frac{2Gh^2}{3}\frac{dv_z}{dr} \end{aligned}\right\} \qquad [42]$$

A unique solution is obtained by setting these equal to zero at $r = a$. This results in a pair of simultaneous,[6] homogeneous equations in A_1 and A_2. The determinant of the coefficients, when equated to zero, yields the frequency equation

$$\frac{\beta}{\alpha} - \frac{\eta\gamma}{\xi}\frac{J_0(\xi\gamma)}{J_1(\xi\gamma)} = \frac{\xi^2 - 2\eta}{1 - 2\eta}\left[\frac{\beta}{\alpha} - \eta\gamma\frac{J_0(\gamma)}{J_1(\gamma)}\right] \qquad [43]$$

where

$$\xi = \delta_2/\delta_1, \qquad \gamma = \delta_1 a$$

$$\eta = \frac{\omega^2}{2c_1^2\delta_1^2} = \beta\frac{\omega^2}{\bar{\omega}^2}\left[(\alpha + \beta)\frac{\omega^2}{\bar{\omega}^2} - 1 + \psi\right]^{-1} \qquad [44]$$

From the last two of Equations [44], noting that $\bar{\omega} = \pi c_1/2h$

$$\frac{a}{h} = \frac{\gamma\bar{\omega}}{\pi\omega}\sqrt{8\eta} \qquad [45]$$

The frequency spectrum may be computed from Equations [43], [44], and [45] in the following manner: First, it is noted that α and β depend only on Poisson's ratio ν, while ξ and η depend only on ν and $\omega/\bar{\omega}$. Hence a choice of ν and $\omega/\bar{\omega}$ reduces Equation [43] to a transcendental equation in γ only. For each root, γ, of this equation, there corresponds a value of a/h in accordance with Equation [45]. Thus, for a given material and thickness, a sequence of disk diameters is found that will vibrate at a given frequency. Then the frequency is changed and a new set of diameters is calculated in the same manner. The process is repeated until sufficient data are obtained to enable curves to be drawn of $\omega/\bar{\omega}$ versus a/h. Results of such computations for the first two modes ($\nu = 0.3$) are shown in Table 1 and illustrated by Curves II and III in Fig. 1.

[6] It may be observed that for boundary conditions $v_r = 0$, $R_r = 0$ (analogous to a simply supported edge in the theory of flexure) the two modes of motion represented by φ_1 and φ_2 are not coupled and Equation [43] is replaced by the two simpler equations $J_1(\delta_1 a) = 0$ and $J_1(\delta_2 a) = 0$.

TABLE 1 COMPUTED FREQUENCIES OF FIRST TWO MODES OF VIBRATION OF CIRCULAR DISK

First Mode									
a/h	0	0.496	0.98	1.43	1.81	2.24	2.88	3.88	5
$\omega_1/\bar{\omega}$	0.863	0.85	0.8	0.7	0.6	0.5	0.4	0.3	0.235

Second Mode											
a/h	0.472	0.67	0.81	0.925	1.1	1.47	2.30	3.09	3.43	4.11	4.94
$\omega_2/\bar{\omega}$	3.0	2.0	1.7	1.5	1.3	1.1	0.95	0.85	0.80	0.70	0.60

TABLE 2 EXPERIMENTAL DATA

	1 diam, cm	2 thickness, cm	3 a/h	4 $\omega_1/2\pi$, kc	5 $\omega_2/2\pi$, kc	6 $\omega_1/\bar{\omega}$	7 $\omega_2/\bar{\omega}$
1	2.68	0.648	4.14	121.1	301.5	0.276	0.686
2	2.63	0.640	4.11	123.4	306.5	0.278	0.690
3	2.16	0.564	3.84	150.4	363.6	0.298	0.720
4	2.16	0.637	3.39	150.1	345.4	0.336	0.773
5	2.17	0.638	3.39	149.7	345.8	0.336	0.774
6	2.16	0.729	2.97	...	...	...	...
7	1.58	0.596	2.66	208.3	438.0	0.436	0.929
8	1.59	0.612	2.60	208.5	422.0	0.449	0.910
9	1.58	0.604	2.62	208.7	436.6	0.443	0.926
10	1.58	0.580	2.72	...	...	...	...
11	1.31	0.639	2.05	237.6	410.5	0.534	0.922
12	1.05	0.636	1.65	285.4	430.0	0.638	0.962
13	1.05	0.637	1.64	289.0	427.0	0.648	0.956
14	0.783	0.629	1.25	334.0	572.0	0.739	1.265
15	0.777	0.636	1.22	338.6	579.0	0.757	1.295

10 Discussion of Disk Frequencies

The interpretation of the results for the disk is facilitated by a study of several limiting cases.

Low Frequency, Large Diameter. For $\omega/\bar{\omega} \ll 1$, the first and third of Equations [44] reduce, approximately, to

$$\xi^2 = -\bar{\omega}^2/\alpha\beta\omega^2, \qquad \eta = 1/2\alpha$$

Hence Equation [43] reduces to

$$\gamma \frac{J_0(\gamma)}{J_1(\gamma)} = 2\beta = \frac{\lambda + 2G}{2(\lambda + G)} = 1 - \nu \qquad [46]$$

and Equation [45] reduces, for large a/h, to

$$\frac{a}{h} = \frac{2\bar{\omega}c_3\gamma}{\pi\omega c_1}$$

or

$$\omega = \frac{\gamma}{a}\sqrt{\frac{E}{\rho(1-\nu^2)}} \qquad [47]$$

That is, the frequencies are given by Equation [47], in which γ are the roots of Equation [46]. This is exactly the solution of the equations of generalized plane stress. Hence, for the low frequencies of thin plates, the present equations, as they should, give the same results as the equations of generalized plane stress. Accordingly, the lower right-hand portions of Curves I and II in Fig. 1 are essentially identical. See p. 498, reference (4).

Intermediate Frequency, Small Diameter. In this case $a/h \ll 1$. Hence, from Equation [45], $\gamma \ll 1$. Now, for small x

$$\frac{xJ_0(x)}{J_1(x)} \approx 2 \qquad [48]$$

Hence Equation [43] reduces to

$$\left(\frac{\beta}{\alpha} - \frac{2\eta}{\xi^2}\right)(1 - 2\eta) = (\xi^2 - 2\eta)\left(\frac{\beta}{\alpha} - 2\eta\right) \qquad [49]$$

or

$$\left(\frac{\omega^2}{\bar{\omega}^2} - \frac{1-\beta}{\alpha - \beta}\right)(\delta_1^2 - \delta_2^2) = 0 \qquad [50]$$

This equation has only one real, positive root

$$\frac{\omega}{\bar{\omega}} = \sqrt{\frac{1-\beta}{\alpha-\beta}} \qquad [51]$$

which is the ordinate of the first mode, illustrated by Curve II in Fig. 1 at $a/h = 0$.

Equation [51] can be rewritten in the form

$$\omega = \frac{\pi}{2h}\sqrt{\frac{E}{\rho}} \qquad [52]$$

This will be recognized as the limiting value of the exact solution for the longitudinal vibration of a thin rod in its fundamental mode. It seems remarkable that a plate theory should contain such a result; that is, at least for the fundamental mode, the equations are valid for a plate so thick that, in comparison, its diameter is negligibly small. Of course this remarkable property is not held by the second mode, since only one thickness mode is contained in the theory. For $a/h \to 0$, the second mode should couple with a second thickness mode and approach thrice the frequency of Equation [52] as a limit. In the present theory the second radial mode does, however, couple with the first thickness mode. This is evidenced, in Fig. 1, by the slight reversal of curvature of Curve III in the neighborhood of $\omega/\bar{\omega} = 1$. This effect becomes more pronounced with the higher radial modes at large values of a/h. In fact, for large a/h and $\omega/\bar{\omega} > 1$, coupling occurs, in addition, with overtones of the first thickness mode; i.e., with motions characterized by multiple circular nodes as distinguished from nodal planes. These curves are not shown in Fig. 1, but they are very similar to those found for the high-frequency flexural vibrations of a disk (9).

High Frequency, Small Diameter. Although, for very small diameters, only the lowest mode is correct for intermediate frequencies, the theory again gives good results for very high frequencies. For $\omega/\bar{\omega} \gg 1$, the first and third of Equations [44] give, approximately

$$\xi^2 = \beta/\alpha, \qquad \eta = \beta/2\alpha \qquad [53]$$

Hence Equation [43] reduces to

$$\frac{\xi\gamma J_0(\xi\gamma)}{J_1(\xi\gamma)} = \frac{2\beta}{\alpha} = \frac{2G}{\lambda + 2G} = \frac{1-2\nu}{1-\nu} \qquad [54]$$

and Equations [44] yield

$$\omega = \frac{\xi\gamma}{a}\sqrt{\frac{\lambda + 2G}{\rho}} \qquad [55]$$

Accordingly, the frequencies are given by Equation [55], in which $\xi\gamma$ are the roots of Equation [54]; but this is exactly the solution of the three-dimensional equations for the radial vibrations of an infinite rod. Thus there are two ranges of frequencies of a thin rod which are contained, exactly, in the plate theory.

Intermediate Frequency, Large Diameter. In this case, for $a/h \to \infty$, Equation [45] gives $\gamma \to \infty$. Hence one solution of Equation [43] is $\xi^2 = 0$, that is, $\omega = \bar{\omega}$. Thus the solution contains, as one of the frequencies of vibration of an infinite plate, the correct frequency of the fundamental thickness mode. This, of course, is to be expected, in view of the procedure by which the value of κ was chosen.

11 Experimental Investigation

Frequency measurements were performed[7] on 15 barium-titanate disks with diameters, thicknesses, and ratios a/h shown in columns 1, 2, 3 of Table 2.

After the disks were lapped, spotted, gold-plated, and polarized, their resonant frequencies were measured by applying the output of a tunable oscillator to a frequency counter, the frequency of oscillation being controlled by the disk under test. These frequencies are recorded in columns 4 and 5 of Table 2.

The frequencies $\omega_1/2\pi$, of the fundamental mode of vibration, were obtained without difficulty for all the disks except numbers 6 and 10, both of which appeared to be mechanically defective. The frequencies of the second mode were more difficult to measure as it was necessary not only to locate resonances, but also to evaluate their relative strengths. The frequencies recorded in column 5 are thus to be considered less reliable than those in column 4.

To compare the experimental results with those predicted by the theory, the ratios $\omega/\bar{\omega}$ were computed for each disk, using for ω the experimental values and for $\bar{\omega}$ (see Equations [28], [29], [30])

$$\bar{\omega} = \frac{\pi}{2h}\left(\frac{\lambda + 2G}{\rho}\right)^{1/2} = \frac{\pi}{2h}\left[\frac{2G(1-\nu)}{\rho(1-2\nu)}\right]^{1/2}$$

with[8]

[7] At Bell Telephone Laboratories.

[8] The values of the elastic constants were supplied by W. P. Mason and the density by L. Egerton.

$$\nu = 0.3$$
$$G = 49.3 \times 10^{10} \text{ dynes/cm}^2$$
$$\rho = 5.33 \text{ gm/cm}^3$$

The results are recorded in columns 6 and 7 of Table 2 and are plotted as circles in Fig. 1.

As may be seen, the agreement between theory and experiment is good, especially for the first mode of motion. For diameter-to-thickness ratios less than 3, the theory furnishes a considerably better description than the theory of generalized plane stress. For $a/h = 0$, there is, of course, ample confirmation, by experiments with thin rods, of the predicted value of the frequency of the first mode.

BIBLIOGRAPHY

1 "On the Calculation of Three-Dimensional Corrections for the Two-Dimensional Theory of Plane Stress," by E. Reissner, Proceedings of the Fifteenth Semi-Annual Eastern Photoelasticity Conference, Addison-Wesley Press, Inc., Cambridge, Mass., 1942, pp. 23–31.

2 "Theory of Elasticity," by S. Timoshenko and J. N. Goodier, McGraw-Hill Book Company, Inc., New York, N. Y., second edition, 1951, p. 214.

3 Reference (2), p. 148.

4 "Theory of Elasticity," by A. E. H. Love, Cambridge University Press, fourth edition, Cambridge, England, 1927, p. 176.

5 "Theory of Plates and Shells," by S. Timoshenko, McGraw-Hill Book Company, Inc., New York, N. Y., 1940, p. 299.

6 Reference (5), p. 86.

7 "Vibration Problems in Engineering," by S. Timoshenko, D. Van Nostrand Company, Inc., New York, N. Y., second edition, 1937, p. 338.

8 "Influence of Rotatory Inertia and Shear on Flexural Motions of Isotropic, Elastic Plates," by R. D. Mindlin, Journal of Applied Mechanics, Trans. ASME, vol. 73, 1951, pp. 31–38.

9 "Axially Symmetric Flexural Vibrations of a Circular Disk," by H. Deresiewicz and R. D. Mindlin, Journal of Applied Mechanics, Trans. ASME, vol. 77, 1955, pp. 86–88.

Reprinted from
Journal of Applied Mechanics
23 (1956) 277-283

(*Reprinted from the Journal of Applied Mechanics for September, 1956, pp. 430–436, with permission of Copyright Owner*)

Flexural Vibrations of Rectangular Plates[1]

By R. D. MINDLIN,[2] A. SCHACKNOW,[3] and H. DERESIEWICZ,[4] NEW YORK, N. Y.

The influence of rotatory inertia and shear deformation on the flexural vibrations of isotropic, rectangular plates is investigated. Three independent families of modes are possible when the edges are simply supported. Coupling of the modes is studied for the case of one pair of parallel edges free and the other pair simply supported. The development of the coupling is traced by means of a solution for elastically supported edges. Special attention is given to the higher modes and frequencies of vibration which are beyond the range of applicability of the classical theory of thin plates.

Plate Equations

When shear deformation and rotatory inertia terms are included, the flexural vibrations of a plate are governed by three differential equations each of which contains the deflection (w) and the two components of rotation (ψ_x, ψ_y). It was shown in a previous paper (1)[5] that the equations can be uncoupled by expressing ψ_x, ψ_y, and w in terms of three potentials (w_1, w_2, H). Thus, omitting factors $\exp(ip_jt)$, $j = 1, 2, 3$, from w_1, w_2, and H, respectively

$$\left.\begin{aligned} \psi_x &= (\sigma_1 - 1)\partial w_1/\partial x + (\sigma_2 - 1)\partial w_2/\partial x + \partial H/\partial y \\ \psi_y &= (\sigma_1 - 1)\partial w_1/\partial y + (\sigma_2 - 1)\partial w_2/\partial y - \partial H/\partial x \\ w &= w_1 + w_2 \end{aligned}\right\} \ldots [1]$$

and

$$(\nabla^2 + \delta_1^2)w_1 = 0, \quad (\nabla^2 + \delta_2^2)w_2 = 0, \quad (\nabla^2 + \omega^2)H = 0 \ldots [2]$$

where

$$\left.\begin{aligned} \delta_j^2 &= 6F_j^2[1 + g - (-1)^jQ_j]h^{-2}, \quad j = 1, 2 \\ \sigma_j &= 2[1 + g - (-1)^jQ_j]^{-1}, \quad j = 1, 2 \\ Q_j &= [(1 - g)^2 + 4gF_j^{-2}]^{1/2}, \quad j = 1, 2 \\ F_j &= p_j/p_0, \quad j = 1, 2, 3, \quad p_0^2 = \pi^2G/\rho h^2 \\ \omega^2 &= \pi^2(F_3^2 - 1)h^{-2}, \quad g = \kappa^2(1 - \nu)/2 \end{aligned}\right\} \ldots [3]$$

In Equations [3], G is the shear modulus, ν is Poisson's ratio, ρ is the density, h is the thickness of the plate, and p_0 is the frequency of simple thickness-shear vibration of an infinite plate. The symbols δ_j, ω, and σ_j have the same significance as in reference (1), but here they are expressed in terms of the frequency ratios F_j. The constant κ^2, for which alternative values were given in reference (1), has the value $\pi^2/12$ here.

The bending moments, twisting moment, and shearing forces, per unit of length, are given by

$$\left.\begin{aligned} M_x &= D\left(\frac{\partial\psi_x}{\partial x} + \nu\frac{\partial\psi_y}{\partial y}\right), \quad M_y = D\left(\frac{\partial\psi_y}{\partial y} + \nu\frac{\partial\psi_x}{\partial x}\right) \\ M_{yx} &= \frac{D(1 - \nu)}{2}\left(\frac{\partial\psi_y}{\partial x} + \frac{\partial\psi_x}{\partial y}\right) \\ Q_x &= \kappa^2Gh\left(\psi_x + \frac{\partial w}{\partial x}\right), \quad Q_y = \kappa^2Gh\left(\psi_y + \frac{\partial w}{\partial y}\right) \end{aligned}\right\} \ldots [4]$$

where

$$D = Gh^3/6(1 - \nu)$$

is the plate flexure modulus.

Boundary conditions sufficient to assure a unique solution of Equations [2] are set by specifying, at each point of the edge of the plate, one member of each of the three products

$$M_\nu\partial\psi_\nu/\partial t, \quad M_{\nu s}\partial\psi_s/\partial t, \quad Q_\nu\partial w/\partial t \ldots\ldots\ldots\ldots [5]$$

where ν and s are co-ordinates normal and tangential, respectively, to the edge.

Simply Supported Rectangular Plate

The boundary conditions of a simply supported rectangular plate of length $2a$ and width $2b$ are taken to be

$$\left.\begin{aligned} M_x &= 0, \quad \partial\psi_y/\partial t = 0, \quad \partial w/\partial t = 0 \quad \text{on} \quad x = \pm a \\ M_y &= 0, \quad \partial\psi_x/\partial t = 0, \quad \partial w/\partial t = 0 \quad \text{on} \quad y = \pm b \end{aligned}\right\} \ldots [6]$$

It may be seen from Equation [5] that the conditions on ψ_x and ψ_y could be replaced by conditions on M_{yx}, but the former are chosen because they lead to simple modes analogous to those of the classical theory of thin plates.

Equations [2] have solutions

$$\left.\begin{aligned} w_1 &= A_1 \sin \xi_1x \sin \eta_1y \\ w_2 &= A_2 \sin \xi_2x \sin \eta_2y \\ H &= A_3 \cos \xi_3x \cos \eta_3y \end{aligned}\right\} \ldots\ldots\ldots\ldots\ldots [7]$$

provided that

$$\xi_1^2 + \eta_1^2 = \delta_1^2, \quad \xi_2^2 + \eta_2^2 = \delta_2^2, \quad \xi_3^2 + \eta_3^2 = \omega^2 \ldots [8]$$

Substituting Equations [7] in [4] (where necessary) and [6], we find

$$\xi_j = r_j\pi/2a, \quad \eta_j = s_j\pi/2b, \quad j = 1, 2, 3 \ldots\ldots\ldots [9]$$

where r_j and s_j are even integers. The first two of Equations [7] represent deflections odd in both x and y. Although H generates no deflection, it produces rotations with the same symmetry as those derived from w_1 and w_2. Hence all three of Equations [7] will be termed modes odd in x and y. For modes even in x, $\sin \xi_jx$ and $\cos \xi_jx$ are interchanged in all three of Equations [7]

---

[1] This investigation was supported by the Office of Naval Research and the Signal Corps Engineering Laboratories.

[2] Professor of Civil Engineering, Columbia University. Mem. ASME.

[3] Supervisor, Engineering Computing Section, Republic Aviation Corporation, Long Island, N. Y.

[4] Assistant Professor of Civil Engineering, Columbia University. Assoc. Mem. ASME.

[5] Numbers in parentheses refer to the Bibliography at the end of the paper.

Contributed by the Applied Mechanics Division and presented at the Diamond Jubilee Annual Meeting, Chicago, Ill., November 13–18, 1955, of The American Society of Mechanical Engineers.

Discussion of this paper should be addressed to the Secretary, ASME, 29 West 39th Street, New York, N. Y., and will be accepted until October 10, 1956, for publication at a later date. Discussion received after the closing date will be returned.

Note: Statements and opinions advanced in papers are to be understood as individual expressions of their authors and not those of the Society. Manuscript received by ASME Applied Mechanics Division, July 9, 1955. Paper No. 55—A-78.

and the r_j are odd integers. Similarly, for modes even in y, sin $\eta_j y$ and cos $\eta_j y$ are interchanged and the s_j are odd integers.

Substituting Equations [9] in [8] and solving for the frequency ratios F_j, we find

$$\left.\begin{aligned} 2F_j^2 &= 1 + k\psi_j^2 + (-1)^j\Omega_j, \quad j = 1, 2 \\ F_3^2 &= 1 + \psi_3^2 \end{aligned}\right\} \quad \ldots\ldots [10]$$

where

$$\left.\begin{aligned} \psi_j^2 &= (r_jh/2a)^2 + (s_jh/2b)^2, \quad j = 1, 2, 3 \\ \Omega_j &= [(1 + k\psi_j^2)^2 - 4\kappa^4\psi_j^4/g]^{1/2}, \quad j = 1, 2, 3 \\ k &= \kappa^2(1 + g)/g \end{aligned}\right\} \quad .[11]$$

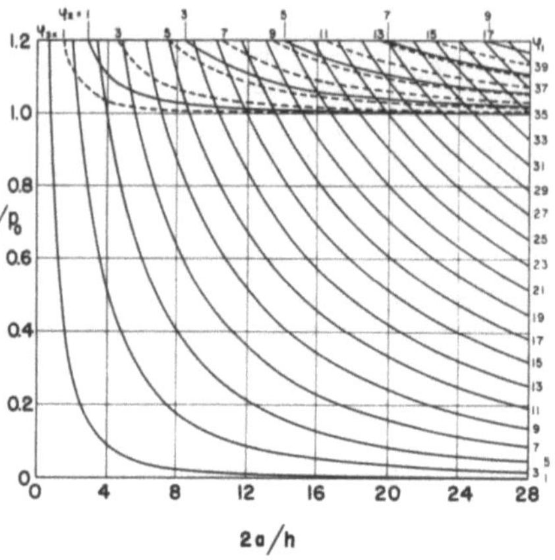

Fig. 1 Chart for Estimating Frequencies of Vibration of Simply Supported Rectangular Plates

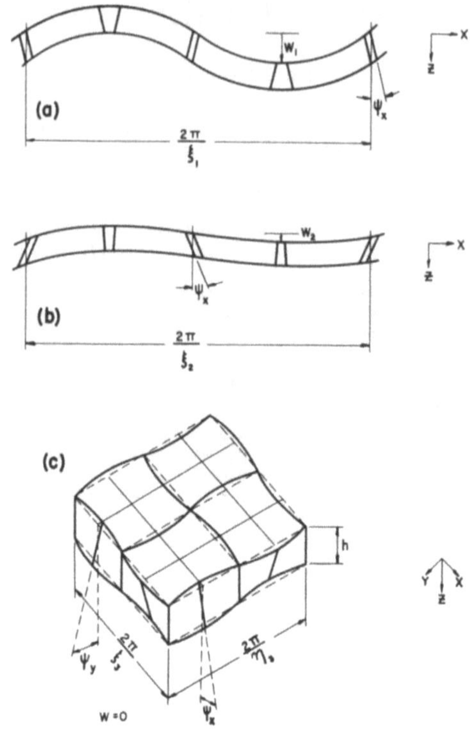

Fig. 2 The Three Types of Vibrational Modes of Simply Supported Rectangular Plates When Rotatory Inertia and Shear Deformation Are Taken Into Account
(*a*, Flexural; *b*, thickness-shear; *c*, thickness-twist.)

The three sets of frequency ratios are plotted against the length-thickness ratio, in Fig. 1, with

$$\varphi_j = (r_j^2 + s_j^2a^2/b^2)^{1/2}, \quad j = 1, 2, 3 \ldots\ldots\ldots [12]$$

as parameters and $\nu = 0.312$.

For given length-width-thickness ratios and order of mode, the frequencies of the three types of motion satisfy the inequalities $p_1 < p_3 \leqslant p_2$. The lowest of these derives from the potential w_1; its mode shape, as calculated from Equations [1], is illustrated in Fig. 2(*a*). Of the three types of motion, this is the one that corresponds most closely to that of the classical theory. In fact, for the lower modes of large, thin plates, this mode shape approaches that of classical theory, in which

$$\psi_x = -\partial w/\partial x \text{ and } \psi_y = -\partial w/\partial y$$

as illustrated in Fig. 3. Also, as p_1 diminishes, it approaches

$$p_0\psi_1^2\kappa^2/g^{1/2}$$

the frequency obtained from the classical theory. The other two frequencies are very much higher than p_1 and their mode shapes are quite different. In the case of p_2, which is associated with the potential w_2, the thickness-shear deformation predominates over the flexural deformation, as illustrated in Fig. 2(*b*). The remaining mode shape, Fig. 2(*c*), contains no deflection and the two components of rotation are so related in phase as to form a twist about the normal to the plate. For convenience of reference, the types of motion generated by w_1, w_2, and H are called flexural, thickness-shear, and thickness-twist, respectively.

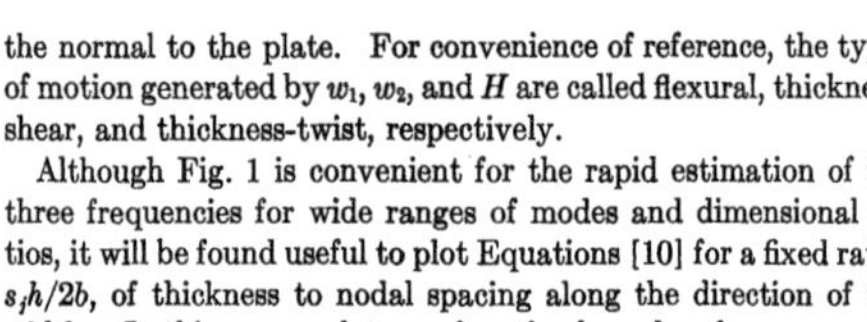

Fig. 3 Vibrational Mode of a Simply Supported Rectangular Plate According to the Classical Theory of Thin Plates

Although Fig. 1 is convenient for the rapid estimation of the three frequencies for wide ranges of modes and dimensional ratios, it will be found useful to plot Equations [10] for a fixed ratio, $s_jh/2b$, of thickness to nodal spacing along the direction of the width. In this case, each type of motion has a low-frequency cut-off given by

$$\left.\begin{aligned} 2(F_j^*)^2 &= 1 + (\psi_j^*)^2 + (-1)^j\Omega_j^*, \quad j = 1, 2 \\ (F_3^*)^2 &= 1 + (\psi_3^*)^2 \end{aligned}\right\} \quad ..[13]$$

where ψ_j^* and Ω_j^* are given by Equations [11] with $r_jh/2a = 0$. Equations [10] are plotted in Fig. 4 for $s_jh/2b = 0.2$ and $\nu = 0.312$. In this figure, $r_1 = m$, $r_2 = n$, and $r_3 = q$; i.e., the curves m, n, q = constant give the frequencies of the flexural, thickness-shear, and thickness-twist modes, respectively. Odd and even values of m, n, q identify modes even and odd, respectively, in x. All the curves apply to modes either odd or even in y. The range of F in Fig. 4 is 1.0198 to 1.20. Above 1.20 the results are not very reliable, as the theory does not contain the effects of vibrations having more than one nodal surface parallel to the faces $z = \pm h/2$. Below $F = 1.0198$ (the thickness-twist cutoff frequency for $s_jh/2b = 0.2$) the curves m = constant look much the same as those derived from the classical theory, but the frequencies are lower. The two sets of curves coalesce as F_1 diminishes to its cutoff value 0.0578.

One Pair of Parallel Edges Free

When the plate is free along $x = \pm a$ and simply supported along $y = \pm b$ the boundary conditions become

$$\left.\begin{array}{lll} M_x = 0, \quad M_{yx} = 0, \quad Q_x = 0 & \text{on} & x = \pm a \\ M_y = 0, \quad \partial\psi_x/\partial t = 0, \quad \partial w/\partial t = 0 & \text{on} & y = \pm b \end{array}\right\} \quad ..[14]$$

The solutions given by Equations [7] and [8] are applicable again, but w_1, w_2, and H cannot each satisfy all the boundary conditions, as they did for the simply supported plate. This is because, on incidence at a free edge, a wave of each of the three types gives rise, in general, to reflected waves of all three types (2).

The conditions on $y = \pm b$ are satisfied if

$$\eta_1 = \eta_2 = \eta_3 = \eta = s\pi/2b \ldots\ldots\ldots\ldots [15]$$

where s is an even or odd integer for modes odd or even, respectively, in y. Substituting Equations [7] successively in [1], [4], and [14], the boundary conditions on $x = \pm a$ are satisfied if $p_1 = p_2 = p_3$ and

$$\left.\begin{array}{l} A_1\alpha_{11} \sin \xi_1 a + A_2\alpha_{12} \sin \xi_2 a + A_3\alpha_{13} \sin \xi_3 a = 0 \\ A_1\alpha_{21} \cos \xi_1 a + A_2\alpha_{22} \cos \xi_2 a + A_3\alpha_{23} \cos \xi_3 a = 0 \\ A_1\alpha_{31} \cos \xi_1 a + A_2\alpha_{32} \cos \xi_2 a + A_3\alpha_{33} \cos \xi_3 a = 0 \end{array}\right\} \quad ..[16]$$

where, with $j = 1, 2$

$$\left.\begin{array}{ll} \alpha_{1j} = (\sigma_j - 1)(\xi_j^2 + \nu\eta^2), & \alpha_{13} = -(1 - \nu)\xi_3\eta \\ \alpha_{2j} = 2(\sigma_j - 1)\xi_j\eta, & \alpha_{23} = \xi_3^2 - \eta^2 \\ \alpha_{3j} = \sigma_j\xi_j, & \alpha_{33} = -\eta \end{array}\right\} \quad ..[17]$$

For modes even in x, sine and cosine are interchanged in Equations [16].

Equations [16] determine amplitude ratios $A_1 : A_2 : A_3$ so that, except for special cases to be considered later, all three types of motion must be present when a pair of edges is free. For a nontrivial solution the determinant of the coefficients of the A_j in Equations [16] must vanish. This leads to the equation

$$M_1 \tan \gamma - M_2 \tan \alpha\gamma + M_3 \tan \beta\gamma = 0 \ldots\ldots [18]$$

where

$$\left.\begin{array}{l} M_1 = \xi_2(\sigma_1 - 1)(\xi_1^2 + \nu\eta^2)(\sigma_2\omega^2 - 2\eta^2) \\ M_2 = \xi_1(\sigma_2 - 1)(\xi_2^2 + \nu\eta^2)(\sigma_1\omega^2 - 2\eta^2) \\ M_3 = 2(1 - \nu)\xi_1\xi_2\xi_3\eta^2(\sigma_1 - \sigma_2) \\ \gamma = \xi_1 a, \quad \alpha = \xi_2/\xi_1, \quad \beta = \xi_3/\xi_1 \end{array}\right\} \quad \ldots [19]$$

The frequency spectrum may be computed as follows: For a given Poisson ratio (which fixes g) and a given ratio, $sh/2b$, of thickness to nodal spacing along y, a choice of F fixes α, β, and M_jh^6. Equation [18] then has γ as the only variable and a set of roots may be computed by successive approximations. For the chosen frequency, to each root γ there corresponds a length-thickness ratio

$$2a/h = \gamma(h^2\delta_1^2 + h^2\eta^2)^{-1/2} \ldots\ldots\ldots\ldots [20]$$

as obtained from the first of Equations [8]. Since $h\delta_1$ is fixed by the choice of F and g, see Equations [3], and since $h\eta$ is fixed by the choice of $sh/2b$, all the quantities on the right-hand side of Equation [20] are known. Thus a set of values of $2a/h$ is obtained for the chosen F. This procedure was followed for a number of suitably spaced values of F, with $\nu = 0.312$ and $sh/2b = 0.2$, sufficient to enable plotting of the curves shown in Fig. 5. The dashed and full lines are for modes odd and even, respectively, in x.

The difference in character of the spectrum above and below each frequency F_j^* is due to the fact that ξ_1, ξ_2, and ξ_3 are real or imaginary as δ_1, δ_2, and ω, respectively, are greater or less than η, see Equations [8]. The changes occur when $\xi_j = 0$; but this is the condition used to determine the cutoff frequencies F_j^* in the case of the simply supported plate. Hence there are four frequency ranges, separated by the three F_j^*, in which the character of the frequency spectrum may be expected to be different, since one, two, or three of the transcendental functions in Equation [18] are hyperbolic, depending on the frequency range.

In the range $F < F_1^*$, all three ξ_j are imaginary and Equation [18] has no real roots. Hence there are no frequencies of vibration below F_1^*. For $sh/2b = 0.2$, $F_1^* = 0.0578$ and is not shown in Fig. 5.

In the range $F_1^* < F < F_3^*$, only ξ_1 is real and, as a result, the spectrum has the simple character shown in Fig. 5. The lower part of this range ($F \rightarrow F_1^*$) is not shown in Fig. 5 as it is given adequately by the classical theory of thin plates.

In the next higher range, $F_3^* < F < F_2^*$, ξ_1 and ξ_3 are real and ξ_2 is imaginary. This results in a more complicated spectrum but also in some simple solutions, which may be found by noting that, with ξ_1 and ξ_3 real, Equations [16] are satisfied by

$$\cos \xi_1 a = 0, \quad \cos \xi_3 a = 0, \quad A_2 = 0 \ldots\ldots\ldots [21]$$

Thus the thickness-shear type of motion is absent and Equation [18] has roots

$$\xi_1 = m\pi/2a, \quad \xi_2 = q\pi/2a, \quad m, q \text{ odd} \ldots\ldots\ldots [22]$$

These roots, and the condition $F_1 = F_3$, determine the co-ordinates of the white squares in both Figs. 4 and 5. Hence, for a given nodal spacing in the direction of the width (for modes odd or even in y) points on the frequency spectrum of a simply supported plate, where flexural and thickness-twist modes even in x have the same frequencies, are also points on the spectrum, for modes odd in x, of a plate with a pair of parallel edges free. Accordingly, the curves of F versus $2a/h$, determined by Equation [18], (the dashed curves), pass through the white squares in Fig. 5. However, the dashed curves (modes odd in x) do not pass through the black squares since these are intersections of curves m even, q even, and these points are not solutions of Equations [21]. The converse is true for modes even in x, of the plate with a pair of edges free. Thus the full curves in Fig. 5 pass through the black squares but not the white squares. Points on the curves m = constant, q = constant, Fig. 4, lying between their intersections are not points on the spectrum shown in Fig. 5, as only the intersections satisfy the condition $F_1 = F_3$ which is required for free edges.

It is interesting to observe the behavior of the lowest mode as a/h approaches zero. In this case Equation [18] becomes an

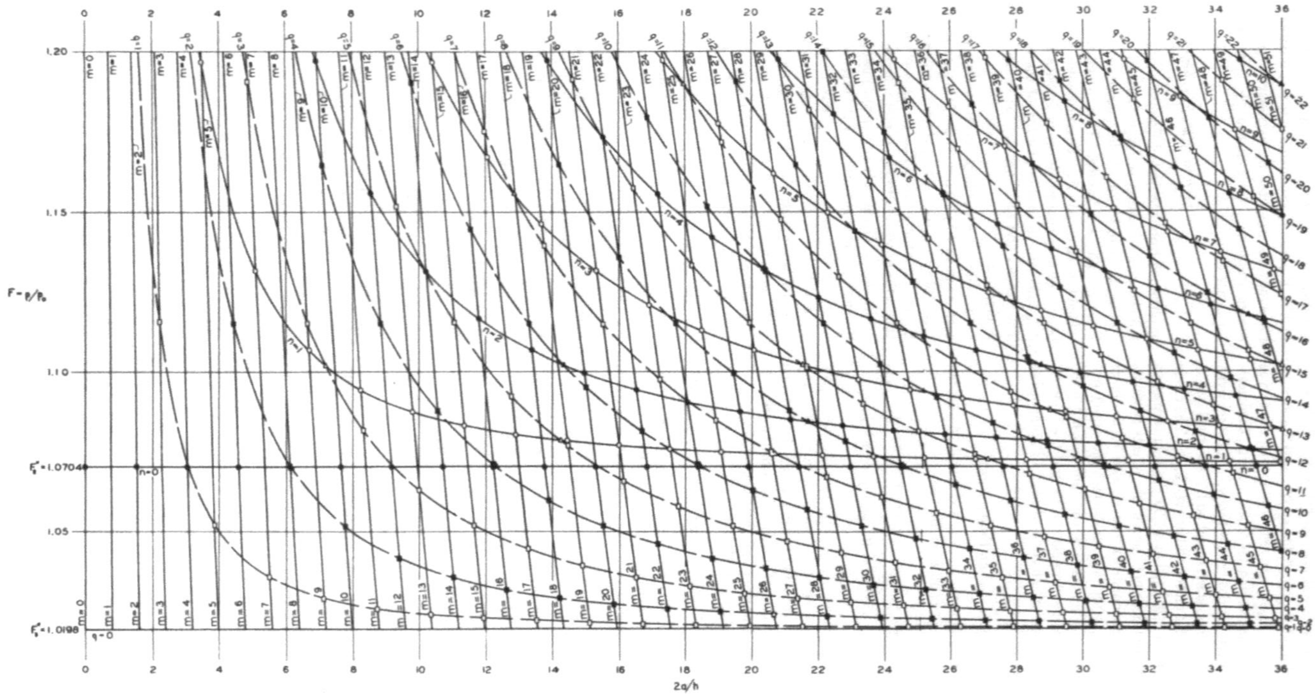

Fig. 4 Frequencies of Vibration of Simply Supported Rectangular Plates (ν = 0.312) When the Distance Between Nodes Along the Width Is Five Times the Thickness

(Curves m, n, q = constant give frequencies of flexural, thickness-shear, and thickness-twist modes, respectively. m, n, q even and odd are modes odd and even, respectively, in x. All modes may be odd or even in y.)

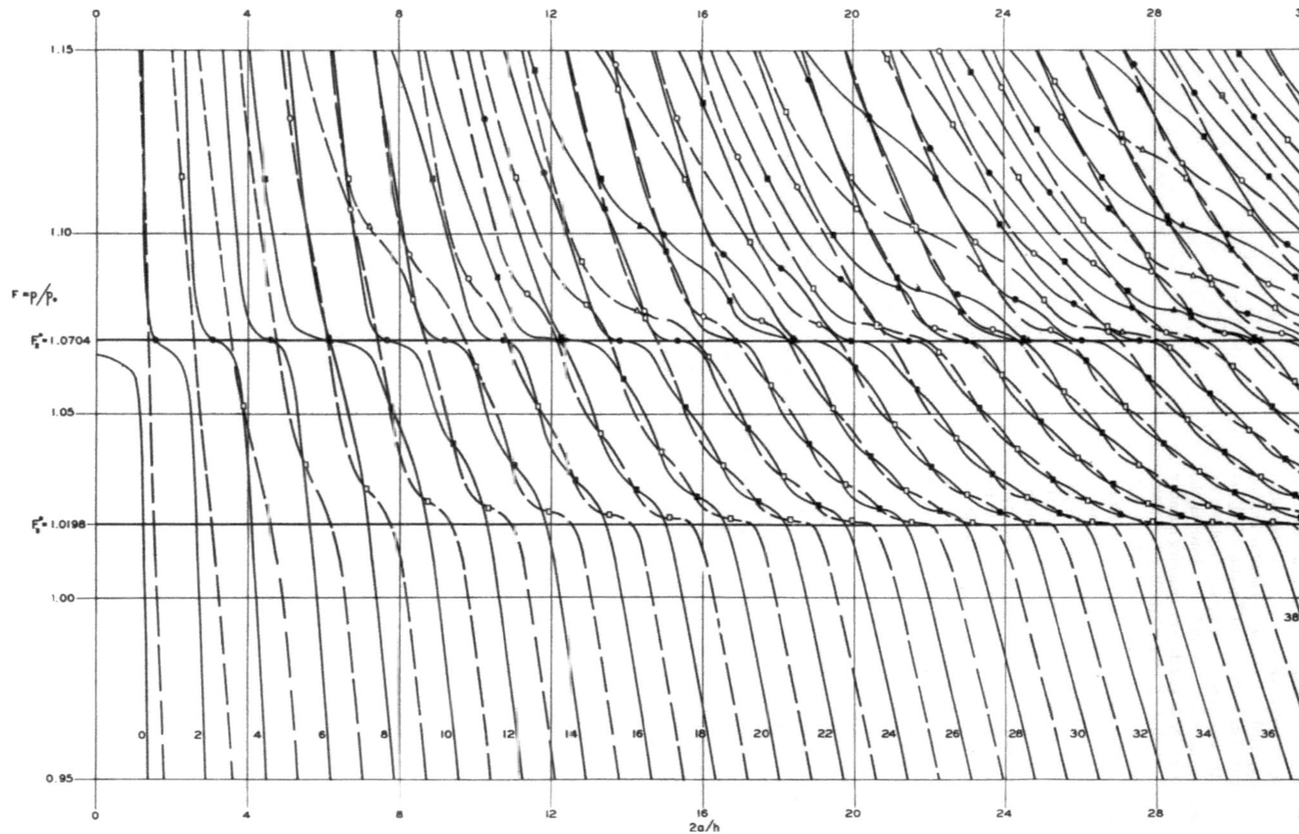

Fig. 5 Frequencies of Vibration of Rectangular Plates With One Pair of Parallel Edges Free and the Other Pair Simply Supported
(Dashed and full lines are for modes odd and even, respectively, in x.)

indeterminate form, but the limiting frequency may be found readily by recognizing that the lowest mode is even in x and that, as $a/h \to 0$, the plate degenerates to a beam, of narrow rectangular section, simply supported at $y = \pm b$. Hence the limiting frequency is given by the formula for F_2^* in Equations [13], except that $1 - \nu$, in the expression for g, must be replaced by $(1 + \nu)^{-1}$ in order to account for the difference between the modulus of elasticity E of a beam and the corresponding quantity $E/(1-\nu^2)$ for a plate. With $sh/2b = 0.2$, the result is $F = 1.0666$.

In the highest frequency range ($F > F_2^*$) all three ξ_j are real. Hence what was said of the case ξ_1, ξ_3 real and the points identified by white and black squares applies again in this range. Similar conclusions are reached for the combinations ξ_1, ξ_2 and ξ_2, ξ_3. In the former case, the curves of the coupled spectrum, for modes odd in x, pass through the intersections of m odd, n odd (white circles) but not through m even, n even (black circles) and at these intersections the thickness-twist mode is absent. In the latter case the common intersections are n odd, q odd (white triangles) and the flexural mode is absent. For the description of the coupled spectrum of modes even in x, "odd" and "even," "black" and "white" should be interchanged.

Development of Coupling

The difference between the simple spectrum of the simply supported plate, Fig. 4, and the complicated spectrum of the plate with a pair of edges free, Fig. 5, is due to the absence, in the former, and presence, in the latter, of coupling between the three types of motion. The development of the coupling, and hence the transition from one spectrum to the other, may be traced continuously by considering a case of elastically supported edges.[6] In order to simplify the discussion, the special case $\eta_j = 0$ will be considered, so that the mode shapes are independent of y and only the flexural and thickness-shear modes can couple. Then, from the first two of Equations [7] and [8]

$$w_1 = A_1 \sin \delta_1 x, \quad w_2 = A_2 \sin \delta_2 x \qquad [23]$$

and the boundary conditions are

$$M_x = 0, \quad Q_x = \mp K w \quad \text{on } x = \pm a \qquad [24]$$

where K is the spring constant of the elastic support. Substituting Equations [23] successively in Equations [4] and [24], we obtain

$$\left.\begin{aligned} &(\sigma_1 - 1)\delta_1^2 A_1 \sin \delta_1 a + (\sigma_2 - 1)\delta_2^2 A_2 \sin \delta_2 a = 0 \\ &\kappa^2 Gh(\sigma_1\delta_1 A_1 \cos \delta_1 a + \sigma_2\delta_2 A_2 \cos \delta_2 a) \\ &\qquad = -K(A_1 \sin \delta_1 a + A_2 \sin \delta_2 a) \end{aligned}\right\} \qquad [25]$$

When $K = \infty$, the edges are simply supported and Equations [25] reduce to

$$A_1 \sin \delta_1 a = 0, \quad A_2 \sin \delta_2 a = 0 \qquad [26]$$

Thus, the modes w_1 and w_2 are not coupled and the roots of Equations [26] are

$$\delta_1 = m\pi/2a, \quad \delta_2 = n\pi/2a, \quad m, n \text{ even} \qquad [27]$$

In the case of modes even in x the roots have the same form but m and n are odd. A typical portion of the frequency spectrum in the range $F > 1$ is shown in Fig. 6 by the curves m = constant, n = constant. (Note that, when $\eta = 0$, $F_2^* = F_3^* = 1$, $F_1^* = 0$.)

When $K = 0$, the edges are free and Equations [25] reduce to

$$\left.\begin{aligned} &(\sigma_1 - 1)\delta_1^2 A_1 \sin \delta_1 a + (\sigma_2 - 1)\delta_2^2 A_2 \sin \delta_2 a = 0 \\ &\sigma_1\delta_1 A_1 \cos \delta_1 a + \sigma_2\delta_2 A_2 \cos \delta_2 a = 0 \end{aligned}\right\} \qquad [28]$$

[6] This was suggested by E. Sternberg.

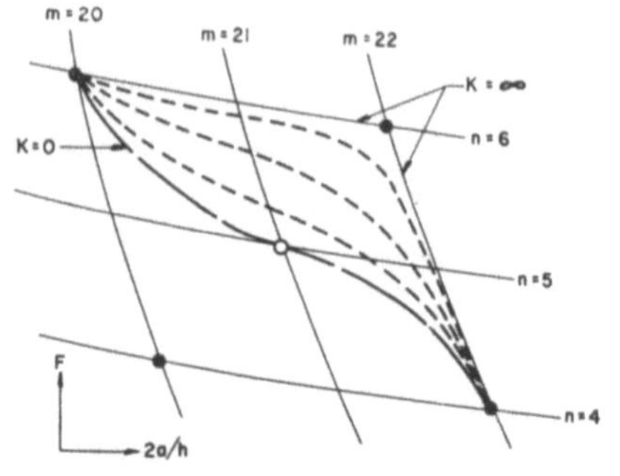

Fig. 6 Illustrating the Development of Coupling Between Flexural and Thickness-Shear Modes as the Stiffness of Elastic Support of Edges Is Diminished From $K = \infty$ (Simply Supported) to $K = 0$ (Free)

These are the same as the first and third of Equations [16] for the limiting case $\eta = 0$. It will be observed that Equations [28] have the solutions [27] for m and n both odd and even. Hence the spectral curves for modes odd in x and free edges must pass through the intersections of the spectral curves both for modes odd in x (black circles) and modes even in x (white circles) with simply supported edges. A typical curve ($K = 0$) is shown[7] in Fig. 6.

When $0 < K < \infty$, Equations [25] still have the roots determined by Equations [26], so that the curves giving the frequencies of modes odd in x pass through the intersections of m even, n even (black circles in Fig. 6). The portions of the curves between the black circles do not, however, pass through the intersections m odd, n odd (white circles) because Equations [25] are not satisfied by $\cos \delta_1 a = 0$, $\cos \delta_2 a = 0$. These portions pass above the white circles because the introduction of the elastic support is an additional constraint which serves to raise the frequencies above those for free edges, as shown by the dashed curves lying between $K = 0$ and $K = \infty$ in Fig. 6. As the support is stiffened indefinitely, these curves approach, asymptotically, the curves m even, n even. Similar results apply to modes even in x, but odd and even, black and white are interchanged.

If $\eta \neq 0$, the situation is somewhat more complicated because there are no common points such as the black circles in Fig. 6. For example, in the limiting case $K = 0$, for modes odd in x, the curves pass through the intersections m odd, n odd but not through the intersections m even, n even; and similarly, for the networks n, q and q, m, as explained in the discussion of Equations [16] and [18].

Construction of Approximate Coupled Spectrum

The co-ordinates of the points marked by circles, squares, and triangles in Figs. 4 and 5 are easy to calculate, as they are given by explicit algebraic formulas obtained by eliminating F and $2a/h$, in turn, from Equations [10], taken two at a time. Since the curves m, n, q = constant, for the simply supported plate, are monotonic, they may be drawn with little error through the

[7] More extensive sets of curves for modes both odd and even in x are given in references (3) and (4). In these, problems of crystal plates are treated; but when $\eta = 0$ the transition from crystalline to isotropic requires only a change in the definition of g. It may also be observed that Equations [23] to [28] apply equally well to beams. It is only necessary to replace $1 - \nu$ by $1/(1 + \nu)$ in the expression for g given in Equation [3]. With $\kappa^2 = \pi^2/12$ the solutions then apply to beams of narrow rectangular section. For beams of other cross sections, appropriate values of κ are given in reference (5).

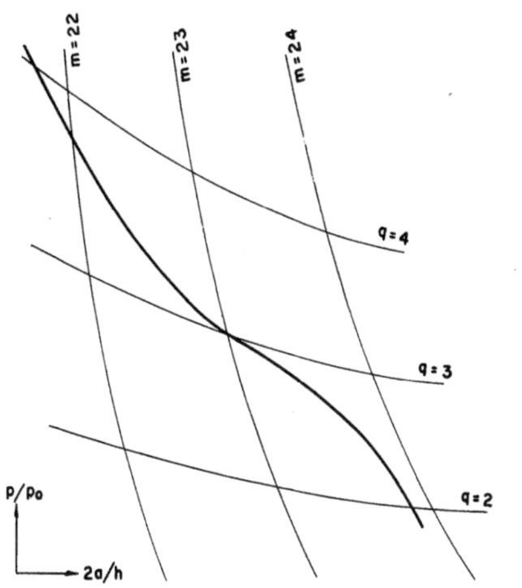

FIG. 7 CONSTRUCTION OF TYPICAL PORTION OF COUPLED SPECTRUM FOR A MODE ODD IN x. FREQUENCY RANGE $F_3^* < F < F_2^*$

plotted points, Fig. 4. On the other hand, the calculation of the complicated curves of Fig 5 is a long and tedious process, involving the determination of the roots of a transcendental equation. However, if only qualitative information is required, the spectrum may be sketched approximately, on the network of curves for the simply supported plate, in accordance with the following rules. The rules are given for modes odd in x. For modes even in x, odd and even are interchanged.

1 Starting in the low-frequency range of the spectrum, proceed to sketch a line upward. (The line should tend to the left, since, for a given mode, increasing frequencies correspond to decreasing length-thickness ratios.)

2 The line cannot cross a curve m, n, or q odd except at the intersection with another curve m, n, or q odd.

3 On approaching a curve m, n, or q odd, either from below or from the right, the line must bear upward and to the left, so that it converges, with the curve, to the nearest intersection with a second curve m, n, or q odd, at which point the crossing is made.

4 The line may cross curves m, n, or q even, but must not pass through intersections of curves m, n, or q even.

Typical applications of these rules are illustrated in Figs. 7 and 8.

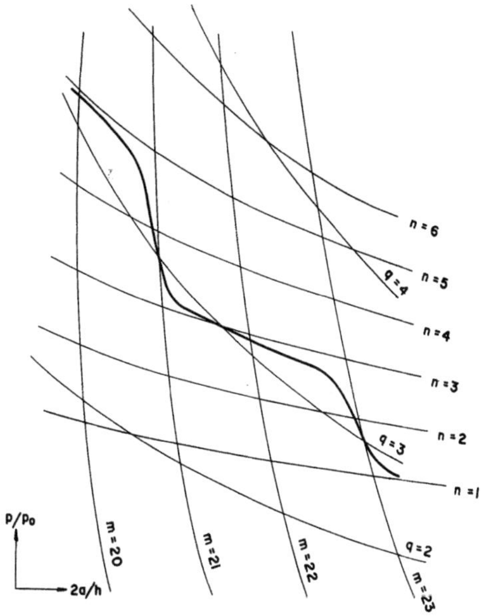

FIG. 8 CONSTRUCTION OF TYPICAL PORTION OF COUPLED SPECTRUM FOR A MODE ODD IN x. FREQUENCY RANGE $F > F_2^*$

BIBLIOGRAPHY

1 "Influence of Rotatory Inertia and Shear on Flexural Motions of Isotropic, Elastic Plates," by R. D. Mindlin, JOURNAL OF APPLIED MECHANICS, Trans. ASME, vol. 73, 1951, pp. 31–38.

2 "Reflection of Flexural Waves at the Edge of a Plate," by T. R. Kane," JOURNAL OF APPLIED MECHANICS, Trans. ASME, vol. 76, 1954, pp. 213–220.

3 "Thickness-Shear and Flexural Vibrations of Crystal Plates," by R. D. Mindlin, *Journal of Applied Physics*, vol. 22, 1951, pp. 316–323.

4 "Forced Thickness-Shear and Flexural Vibrations of Piezoelectric Crystal Plates," by R. D. Mindlin, *Journal of Applied Physics*, vol. 23, 1952, pp. 83–88.

5 "Timoshenko's Shear Coefficient for Flexural Vibrations of Beams," by R. D. Mindlin and H. Deresiewicz, Proceedings of the Second U. S. National Congress of Applied Mechanics, 1954, pp. 175–178.

Reprinted from JOURNAL OF APPLIED PHYSICS, Vol. 27, No. 12, 1462–1466, December, 1956

Simple Modes of Vibration of Crystals*

R. D. MINDLIN
Department of Civil Engineering and Engineering Mechanics, Columbia University, New York, New York
(Received June 6, 1956)

In this paper there are described solutions of the three-dimensional equations of elasticity and the equations of generalized plane stress pertaining to vibrations of crystal rectangular parallelepipeds and thin plates with traction-free faces. The modes of motion exist in crystals of certain classes for specific orientations and length-ratios of the edges. Since the solutions are exact and in closed form, they may be used to advantage in connection with the measurement of elastic constants. Special disposition of electrodes is required for piezoelectric excitation.

1. INTRODUCTION

LAMÉ[1] has described simple modes of vibration of isotropic, rectangular parallelepipeds all six of whose faces are traction-free. The modes are exact solutions of the equations and boundary conditions of the linear theory of elasticity, subject only to the restriction that one of the three dimensions of the parallelepiped must be an integral multiple of one of the remaining two. In what follows it is shown that analogous modes of vibration exist in crystals of certain classes for specific orientations of the edges of the parallelepiped with respect to the crystallographic axes. In the cases of the cubic and hexagonal systems and four classes of the tetragonal system, the solutions are exact and very similar to Lamé's. In the orthorhombic system and three classes of the trigonal system almost all of the conditions are satisfied for certain orientations of cut and certain nonintegral ratios of dimensions. The only condition left unsatisfied is the vanishing of a normal traction on one pair of faces. In view of this, the problem is recast as one in the theory of generalized plane stress. Then the solutions apply to low-frequency extensional vibrations of thin plates.

2. SOLUTION OF EQUATIONS OF MOTION

We begin by considering a stress-strain relation in which the tensor of elastic stiffness has monoclinic symmetry with x the twofold axis[2]:

$$T_p = c_{pq}S_q, \quad p,q = 1\cdots 6, \tag{1}$$

$$c_{15}=c_{16}=c_{25}=c_{26}=c_{35}=c_{36}=c_{45}=c_{46}=0. \tag{2}$$

Guided by the form of Lamé's solution, we consider the possibility of finding solutions of the crystal

* This investigation was supported by the Signal Corps Engineering Laboratories.

[1] M. G. Lamé, *Leçons sur la Théorie Mathématique de l'Élasticité des Corps Solides* (Gauthier-Villars, Paris, 1886), second edition, p. 170. These modes were employed recently by H. Ekstein and T. Schiffman, J. Appl. Phys. 27, 405 (1956), as trial functions in obtaining approximate solutions for other modes in isotropic parallelepipeds.

[2] W. P. Mason, *Piezoelectric Crystals and Their Applications to Ultrasonics* (D. Van Nostrand Company, Inc., New York, 1950), pp. 44–46.

equations of motion of the form

$$u=u(x,y,t) \quad v=v(x,y,t), \quad w=0 \tag{3}$$

$$T_4=T_5=T_6=0, \tag{4}$$

where u, v ,and w are the components of displacement and T_4, T_5, T_6 are the components of shear-stress. With these restrictions the strains reduce to

$$\left.\begin{array}{l} S_1=\partial u/\partial x, \quad S_2=\partial v/\partial y, \quad S_3=0 \\ S_4=0, \quad S_5=0, \quad S_6=\partial v/\partial x+\partial u/\partial y \end{array}\right\} \tag{5}$$

and, hence, Eqs. (1) become

$$\begin{aligned} T_1&=c_{11}\partial u/\partial x+c_{12}\partial v/\partial y \\ T_2&=c_{12}\partial u/\partial x+c_{22}\partial v/\partial y \\ T_3&=c_{13}\partial u/\partial x+c_{23}\partial v/\partial y \\ T_4&=c_{14}\partial u/\partial x+c_{24}\partial v/\partial y=0 \\ T_5&=c_{56}(\partial v/\partial x+\partial u/\partial y)=0 \\ T_6&=c_{66}(\partial v/\partial x+\partial u/\partial y)=0. \end{aligned} \tag{6}$$

Also, the stress-equations of motion reduce to

$$\partial T_1/\partial x=\rho\partial^2 u/\partial t^2, \quad \partial T_2/\partial y=\rho\partial^2 v/\partial t^2. \tag{7}$$

Substituting the first, second, and fifth of Eqs. (6) into Eqs. (7), we find the displacement-equations of motion

$$\begin{aligned} c_{11}\partial^2 u/\partial x^2-c_{12}\partial^2 u/\partial y^2&=\rho\partial^2 u/\partial t^2 \\ c_{22}\partial^2 v/\partial y^2-c_{12}\partial v^2/\partial x^2&=\rho\partial^2 v/\partial t^2. \end{aligned} \tag{8}$$

As solutions of Eqs. (8) we consider

$$\begin{aligned} u&=A\cos\xi x\sin\eta y\exp i\omega t \\ v&=B\sin\xi x\cos\eta y\exp i\omega t. \end{aligned} \tag{9}$$

This form of displacement is antisymmetric with respect to x and y. What follows applies also to the remaining three symmetries obtained by interchanging $\sin\xi x$ and $\cos\xi x$ or $\sin\eta y$ and $\cos\eta y$ or both. Substituting Eqs. (9) in (8), we find

$$\rho\omega^2=c_{11}\xi^2-c_{12}\eta^2=c_{22}\eta^2-c_{12}\xi^2. \tag{10}$$

Hence

$$\xi^2/\eta^2=(c_{22}+c_{12})/(c_{11}+c_{12}) \tag{11}$$

$$\omega=\xi(c/\rho)^{\frac{1}{2}} \tag{12}$$

$$c=(c_{11}c_{22}-c_{12}^2)/(c_{22}+c_{12}). \tag{13}$$

We have still to satisfy Eqs. (4). From the last two of these,

$$A/B=-\xi/\eta \tag{14}$$

and, from the first,

$$c_{14}A\xi+c_{24}B\eta=0 \tag{15}$$

Hence

$$\xi^2/\eta^2=c_{24}/c_{14}. \tag{16}$$

Now, Eqs. (11) and (16) require the relation

$$c_{14}(c_{22}+c_{12})=c_{24}(c_{11}+c_{12}). \tag{17}$$

If this relation is satisfied, the displacements given in Eqs. (9) are solutions of the equations of motion. The ratio of wave-numbers and the frequency are then given by Eqs. (11)–(13).

3. BOUNDARY CONDITIONS

Since $T_4=T_5=T_6=0$, the only components of traction on the coordinate planes are

$$\begin{aligned} T_1&=-(c_{11}A\xi+c_{12}B\eta)\sin\xi x\sin\eta y\exp i\omega t \\ T_2&=-(c_{12}A\xi+c_{22}B\eta)\sin\xi x\sin\eta y\exp i\omega t \\ T_3&=-(c_{13}A\xi+c_{23}B\eta)\sin\xi x\sin\eta y\exp i\omega t. \end{aligned} \tag{18}$$

Now, $T_1=0$ on $x=\pm a$ if

$$\xi=m\pi/2a, \tag{19}$$

where m is an even integer for modes antisymmetric with respect to x and an odd integer for modes symmetric with respect to x. Then, from Eqs. (19) and (12), the frequencies of vibration are

$$\omega=(m\pi/2a)(c/\rho)^{\frac{1}{2}}, \tag{20}$$

where c is given by Eq. (13), provided that Eq. (17) is satisfied.

Returning to Eqs. (18), we see that $T_2=0$ wherever

$$y=\pm n\pi/2\eta=\pm(na/m)[(c_{22}+c_{12})/(c_{11}+c_{12})]^{\frac{1}{2}}$$

in which n is an even integer for modes antisymmetric with respect to y and an odd integer for modes symmetric with respect to y.

Thus, planes $x=$constant and $y=$constant are free of traction at intervals $2a/m$ and $2b/n$, respectively, where

$$b/n=(a/m)[(c_{22}+c_{12})/(c_{11}+c_{12})]^{\frac{1}{2}}. \tag{21}$$

Accordingly, the crystal separates into prisms with traction-free faces. The cross sections of the prisms are rectangles or rectangular polygons, convex or concave, whose sides are integral multiples of $2a/m$ and $2b/n$ in the directions of x and y, respectively. The displacements of points on the traction-free faces are normal to the faces and vary sinusoidally with half wavelengths $2a/m$ and $2b/n$. These results are illustrated in Fig. 1, in which the straight lines are intersections of the traction-free faces with a plane $z=$constant. The cross section of a prism with traction-free faces may have any outline which follows the straight lines. The sinusoidal lines in Fig. 1 represent the displacements of the traction-free faces.

From Eqs. (11), (14), and the last of (18), all planes $z=$constant are traction-free if

$$c_{13}(c_{22}+c_{12})=c_{23}(c_{11}+c_{12}). \tag{22}$$

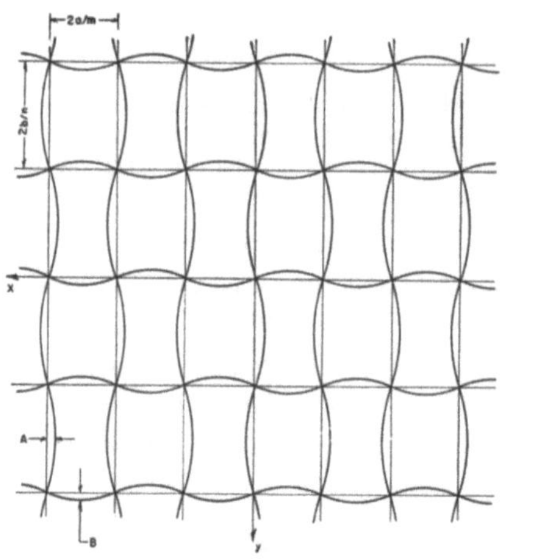

FIG. 1. Mode-shapes and cross sections of prisms with traction-free faces and plates with traction-free edges.

4. EXAMPLES

The solution of the equations of motion, given in Sec. 2, is valid if Eqs. (2) and (17) are satisfied. In this event, the results in Sec. 3 show that conditions of traction-free boundaries $x=$constant and $y=$constant are always met at intervals $2a/m$ and $2b/n$, respectively, but the condition that planes $z=$constant are traction-free is met only if Eq. (22) is satisfied. Some typical examples for various elastic symmetries follow.

Isotropic. In this case Eq. (2) is satisfied and, in addition,

$$c_{11}=c_{22}, \quad c_{13}=c_{23}, \quad c_{14}=c_{24}=0 \tag{23}$$

$$c_{11}-c_{12}=2c_{66}. \tag{24}$$

Hence, Eqs. (17) and (22) are satisfied and we have

$$a/m=b/n \tag{25}$$

$$\omega=(m\pi/2a)(2c_{66}/\rho)^{\frac{1}{2}}. \tag{26}$$

This is Lamé's result.

Cubic.—If x, y, z are the crystallographic axes,[2] Eq. (2) is satisfied and

$$c_{11}=c_{22}, \quad c_{13}=c_{23}, \quad c_{14}=c_{24}=0. \tag{27}$$

Hence Eqs. (17) and (22) are satisfied and we have

$$a/m=b/n \tag{28}$$

$$\omega=(m\pi/2a)[(c_{11}-c_{12})/\rho]^{\frac{1}{2}}. \tag{29}$$

Hexagonal.—If z is the sixfold axis and x a twofold axis, Eqs. (2), (23), and (24) again apply.[2] Hence, the results are the same as for the isotropic case, i.e., Eqs. (25) and (26).

Tetragonal.—If z is the fourfold axis and x a twofold axis (Classes 11, 12, 14, 15), Eqs. (2) and (27) are satisfied.[2] Hence, the results are the same as for the cubic system, i.e., Eqs. (28) and (29).

Orthorhombic.—If x, y, and z are the crystallographic axes,[2] Eq. (2) is satisfied and

$$c_{14}=c_{24}=0. \tag{30}$$

Hence Eq. (17) is satisfied and we have

$$a/m=(b/n)[(c_{11}+c_{12})/(c_{22}+c_{12})]^{\frac{1}{2}} \tag{31}$$

$$\omega=(m\pi/2a)[(c_{11}c_{22}-c_{12}^2)/\rho(c_{22}+c_{12})]^{\frac{1}{2}}. \tag{32}$$

However Eq. (22) is not satisfied, so that there is a normal traction (T_3) left on planes $z=$constant.

Trigonal.—In this case Eqs. (2), (17), and (22) are not satisfied in general. However, if Z is the trigonal axis, X a twofold axis (Classes 18, 20, and 21) and Y is chosen so as to form a right-handed system, we may find orientations of x, y, z with respect to X, Y, Z for which Eqs. (2) and (17) are satisfied. Let θ be the angle of rotation of the x, y, z system about the X axis (see Fig. 2). Then Eq. (2) is satisfied and the constants c_{pq}, which appear in Eq. (17), are related to the constants c_{pq}^0 referred to the X, Y, Z system according to[3]

$$c_{11}=c_{11}^0$$

$$c_{22}=c_{11}^0c^4+c_{33}^0s^4+2(2c_{44}^0+c_{13}^0)s^2c^2+4c_{14}^0sc^3$$

$$c_{12}=c_{12}^0c^2+c_{13}^0s^2-2c_{14}^0sc$$

$$c_{14}=c_{14}^0(c^2-s^2)+(c_{12}^0-c_{13}^0)sc$$

$$c_{24}=[c_{11}^0c^2-c_{33}^0s^2-2(c_{44}^0+c_{13}^0)(c^2-s^2)]sc + c_{14}^0(4s^2-1)c^2,$$

where $s=\sin\theta$ and $c=\cos\theta$. If these expressions for the c_{pq} are substituted in Eq. (17) we obtain a transcendental equation on the variable θ with coefficients depending on the constants c_{pq}^0.

Taking quartz as an example and using Mason's[4]

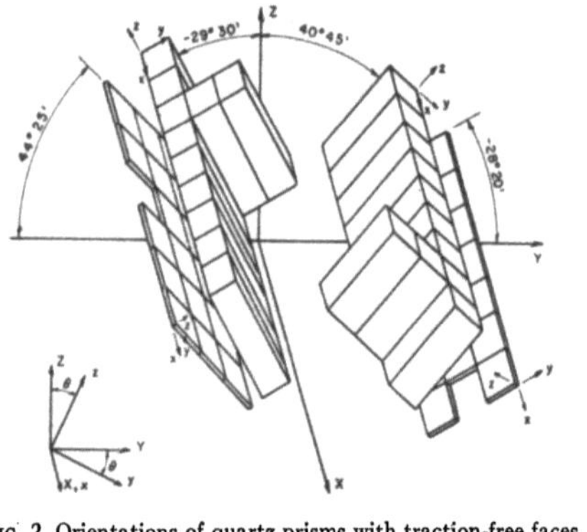

FIG. 2. Orientations of quartz prisms with traction-free faces and quartz plates with traction-free edges.

[3] R. A. Sykes, "Modes of motion in quartz crystals" in *Quartz Crystals for Electrical Circuits*, edited by R. A. Heising (D. Van Nostrand Company, Inc., New York, 1946), Chap. VI, p. 247.
[4] Reference 2, p. 84.

values (in units of 10^{10} dyne/cm^2)

$$c_{11}^0=86.05 \quad c_{14}^0=18.25$$
$$c_{12}^0=4.85 \quad c_{33}^0=107.1$$
$$c_{13}^0=10.45 \quad c_{44}^0=58.65$$

we find that Eq. (17) has the two roots

$$\theta_1=-29°32', \quad \theta_2=40°45'.$$

For these two orientations, the equations of motion are satisfied and we have, also,

$$\frac{\xi}{\eta}=-\frac{A}{B}=\frac{mb}{na}=\begin{cases}0.97, & \theta=\theta_1\\ 1.26, & \theta=\theta_2\end{cases}$$

and

$$\omega=(m\pi/2a)(c/\rho)^{\frac{1}{2}},$$

where

$$c=\begin{cases}62.64\times10^{10}\text{ dyne/cm}^2, & \theta=\theta_1\\ 92.83\times10^{10}\text{ dyne/cm}^2, & \theta=\theta_2.\end{cases}$$

Traction-free boundaries occur on planes x=constant and y=constant at intervals $2a/m$ and $2b/n$, respectively, but there remains a normal traction on all planes z=constant. Thus the result applies only to prisms of infinite length. The orientations of the prisms are illustrated in Fig. 2.

5. FACE-MODES OF THIN PLATES

In view of the fact that, in the orthorhombic and trigonal crystals, there is a normal traction left on planes z=constant, it is appropriate to resort to the approximation of generalized plane stress[5] so that, at least, we may have results applicable to low-frequency face-modes of thin plates.

We replace Eqs. (3) and (4) by

$$u=u(x,y,t), \quad v=v(x,y,t), \quad w=zf(x,y,t) \tag{33}$$

$$T_3=T_4=T_5=T_6=0 \tag{34}$$

and define components of generalized plane stress, strain and displacement:

$$(\bar{T}_p,\bar{S}_p,\bar{u},\bar{v},\bar{w})=\frac{1}{2c}\int_{-c}^{c}(T_p,S_p,u,v,w)dz, \tag{35}$$

where $2c$ is the thickness of the plate.

In this case it is convenient to express the stress-strain relation in terms of elastic compliances s_{pq} instead of elastic stiffnesses. For monoclinic symmetry, we have, in place of Eqs. (1) and (2),

$$S_{pq}=s_{pq}T_q, \quad p,q=1\cdots6, \tag{36}$$

$$s_{15}=s_{16}=s_{25}=s_{26}=s_{35}=s_{36}=s_{45}=s_{46}=0. \tag{37}$$

Then, from Eqs. (33)–(37),

$$\begin{aligned}\bar{S}_1&=s_{11}\bar{T}_1+s_{12}\bar{T}_2=\partial\bar{u}/\partial x\\ \bar{S}_2&=s_{12}\bar{T}_1+s_{22}\bar{T}_2=\partial\bar{v}/\partial y\\ \bar{S}_3&=s_{13}\bar{T}_1+s_{23}\bar{T}_2=f\\ \bar{S}_4&=s_{14}\bar{T}_1+s_{24}\bar{T}_2=0\\ \bar{S}_5&=0\\ \bar{S}_6&=\partial\bar{u}/\partial y+\partial\bar{v}/\partial x=0\end{aligned} \tag{38}$$

Also, the stress-equations of motion reduce to

$$\partial\bar{T}_1/\partial x=\rho\partial^2\bar{u}/\partial t^2, \quad \partial\bar{T}_2/\partial y=\rho\partial^2\bar{v}/\partial t^2. \tag{39}$$

To convert these to displacement-equations of motion we first solve Eqs. (38) for $\bar{T}_1$ and $\bar{T}_2$, with the result

$$\begin{aligned}\bar{T}_1&=\gamma_{11}\partial\bar{u}/\partial x+\gamma_{12}\partial\bar{v}/\partial y,\\ \bar{T}_2&=\gamma_{12}\partial\bar{u}/\partial x+\gamma_{22}\partial\bar{v}/\partial y,\end{aligned} \tag{40}$$

where

$$\begin{aligned}\gamma_{11}&=s_{22}/(s_{11}s_{22}-s_{12}^2),\\ \gamma_{22}&=s_{11}/(s_{11}s_{22}-s_{12}^2),\\ \gamma_{12}&=-s_{12}/(s_{11}s_{22}-s_{12}^2).\end{aligned} \tag{41}$$

It will be observed that Eqs. (40), (39) and the last of Eqs. (38) have the same forms as their counterparts in Sec. 2. Hence we can use Eqs. (9) for $\bar{u}$ and $\bar{v}$ and we find the same expressions for ξ/η, ω and A/B as in Eqs. (11)–(14) except that the c_{pq} are replaced by γ_{pq}. Now, from Eqs. (41),

$$\begin{aligned}(\gamma_{22}+\gamma_{12})/(\gamma_{11}+\gamma_{12})&=(s_{11}-s_{12})/(s_{22}-s_{12})\\ (\gamma_{11}\gamma_{22}-\gamma_{12}^2)/(\gamma_{22}+\gamma_{12})&=1/(s_{11}-s_{12}).\end{aligned} \tag{42}$$

Hence, we now have

$$\begin{aligned}\xi/\eta=-A/B&=[(s_{11}-s_{12})/(s_{22}-s_{12})]^{\frac{1}{2}}\\ \omega&=\xi[\rho(s_{11}-s_{12})]^{-\frac{1}{2}}.\end{aligned} \tag{43}$$

Only the condition $\bar{S}_4=0$ remains to be satisfied for a complete solution of the differential equations. Now, $\bar{T}_1$ and $\bar{T}_2$ are the same as T_1 and T_2 in Eqs. (18), with c_{pq} replaced by γ_{pq}. Using Eqs. (41) for the latter, we find $\bar{T}_1+\bar{T}_2=0$. Hence, from the fourth of Eqs. (38), we must have

$$s_{14}=s_{24} \tag{44}$$

This condition replaces Eq. (17) and there is no requirement analogous to Eq. (22).

As for boundary conditions, we have only to consider $\bar{T}_1$ and $\bar{T}_2$ since $\bar{T}_3=0$. But $\bar{T}_1$ and $\bar{T}_2$ have the same form of dependence on x and y as T_1 and T_2. Hence Eq. (19) holds again and we have traction-free planes x=constant and y=constant at intervals $2a/m$ and

[5] A. E. H. Love, *Mathematical Theory of Elasticity* (Cambridge University Press, Cambridge, 1927) fourth edition, pp. 138, 497. The term "generalized plane stress" was introduced by L. N. G. Filon, Trans. Roy. Soc. (London) Ser. A, **201** (1903), but the equations themselves were introduced earlier. For isotropic plates they were given by S. D. Poisson, Mém. Acad. Sci. (Paris) Sér. 2, **8**, 357–570 (1829) and for crystal plates (subject to the Cauchy relations) by A. L. Cauchy, *Exercices de Mathématique* (Bure Frères, Paris, 1829), Vol. 4, p. 13. The equations are also given by W. Voigt, *Lehrbuch der Kristallphysik*, (B. G. Teubner, Leipzig, 1928), second edition, p. 698

$2b/n$, respectively, but now

$$b/n=(a/m)[(s_{11}-s_{12})/(s_{22}-s_{12})]^{\frac{1}{2}} \quad (45)$$

$$\omega=(m\pi/2a)[\rho(s_{11}-s_{12})]^{-\frac{1}{2}}. \quad (46)$$

This completes the solution of the equations of generalized plane stress.

6. EXAMPLES

We have now to find crystal classes and orientations of plates which satisfy Eqs. (37) and (44).

The results obtained in Sec. 4 for the isotropic, cubic, hexagonal, and tetragonal systems remain unchanged. For the orthorhombic system the results are now given by Eqs. (45) and (46) instead of Eqs. (31) and (32). If $s_{11}=s_{22}$ the solution applies to a square plate and has been discussed by Petrzilka,[6] Bechmann,[7] and Ekstein.[8]

In the case of the trigonal system, Eqs. (37) are satisfied for any angle θ but we have to find values of θ which satisfy Eq. (44).

We have[3]

$$s_{14}=s_{14}^0(c^2-s^2)+2(s_{12}^0-s_{13}^0)sc$$

$$s_{24}=[2(s_{11}^0c^2-s_{33}^0s^2)-(s_{44}^0+2s_{13}^0)(c^2-s^2)]sc+s_{14}^0(4s^2-1)c^2.$$

Substituting these expressions into Eq. (44) and using Mason's[4] values for quartz (in units of 10^{-14} cm²/dyne)

$$s_{11}^0=127.9 \quad s_{14}^0=-44.6$$

$$s_{12}^0=-15.35 \quad s_{33}^0=95.6$$

$$s_{13}^0=-11.0 \quad s_{44}^0=197.8.$$

we find the two roots

$$\theta_1=-28°20', \quad \theta_2=44°25'.$$

Then we have, for insertion into Eqs. (45) and (46),

$$\frac{mb}{na}=\begin{cases}0.96, & \theta=\theta_1\\ 1.32, & \theta=\theta_2\end{cases}$$

$$s_{11}-s_{12}=\begin{cases}161.2\times10^{-14}\ \text{cm}^2/\text{dyne}, & \theta=\theta_1\\ 119.3\times10^{-14}\ \text{cm}^2/\text{dyne}, & \theta=\theta_2.\end{cases}$$

[6] V. Petrzilka, Z. Physik **97**, 436 (1935).
[7] R. Bechmann, Z. Physik **117**, 180 (1941).
[8] H. Ekstein, Phys. Rev. **66**, 108 (1944).

The two orientations of the quartz plates are shown in Fig. 2.

7. APPLICATIONS

Since the modes described are exact solutions, either of the three-dimensional equations or of the equations of generalized plane stress, the usual uncertainties in the relation between frequency and geometrical and physical properties of the crystal are reduced. Accordingly the solutions may be used to advantage in connection with measurement of elastic constants of crystals.

To excite the modes piezoelectrically it is necessary to dispose the electrodes properly. For example, although the *BT* and *ET* cuts of quartz are close to the cuts found in the preceding section, other modes than those described here would be excited with the usual electrode plating even if the length-width ratios were correct. This may be seen from the piezoelectric relation for quartz (referred to rotated axes):

$$\begin{aligned}
S_1&=s_{11}T_1+s_{12}T_2+s_{13}T_3+s_{14}T_4+d_{11}E_1\\
S_2&=s_{21}T_1+s_{22}T_2+s_{23}T_3+s_{24}T_4+d_{12}E_1\\
S_3&=s_{31}T_1+s_{32}T_2+s_{33}T_3+s_{34}T_4+d_{13}E_1\\
S_4&=s_{41}T_1+s_{42}T_2+s_{43}T_3+s_{44}T_4+d_{14}E_1\\
S_5&=s_{55}T_5+s_{56}T_6+d_{25}E_2+d_{35}E_3\\
S_6&=s_{65}T_5+s_{66}T_6+d_{26}E_2+d_{36}E_3.
\end{aligned} \quad (47)$$

In the coordinate system used here, the faces of the plate are at $z=\pm c$ (see Fig. 2). The customary plating on the faces then produces a field E_3 which, in turn induces strains S_5 and S_6, according to Eqs. (47). This results in the excitation of thickness-shear or face-shear modes. Now, S_5 and S_6 are not present in the modes described above. To excite these modes requires a field E_1, and this is produced by electrodes on the thin edges $x=\pm a$. In addition, of course, the length-width ratios must be those found in the preceding section. If modes with n not equal to unity are desired, the electrodes must be disposed in segments in the intervals $2b/n$, with alternating phases of the voltage applied to successive segments.

Reprinted from JOURNAL OF APPLIED PHYSICS, Vol. 28, No. 6, 669–671, June, 1957

Waves on the Surface of a Crystal*

H. DERESIEWICZ AND R. D. MINDLIN
Columbia University, New York, New York
(Received September 4, 1956; revised manuscript received March 13, 1957)

Surface waves in crystals are investigated for surface orientations and directions of propagation which preclude waves of plane strain. These waves are of importance in the study of very high frequency vibrations of certain crystal plates. It is shown that, in the AT cut of quartz, one and only one surface wave can propagate along the digonal axis. By reduction to the case of orthorhombic symmetry, it is shown that the surface wave is the result of coupling between a surface wave of plane strain and either a face-shear mode or a transverse thickness-shear mode. The former combination is associated with flexural and the latter with extensional vibrations of AT plates. Computation of the wave velocity requires the simultaneous solution of algebraic equations of the third and fifth degrees as opposed to a single cubic equation in the case of waves of plane strain.

I. INTRODUCTION

THE lowest mode of propagation of a plane extensional or flexural wave in an infinite, isotropic, elastic plate takes on the character of a surface wave as the frequency increases.[1] Accordingly, the properties of surface waves play an important role in the study of vibrations of isotropic plates at very high frequencies and the same may be expected of crystal plates.

Plane waves on a plane surface of a cubic crystal have been studied for the case in which the wave front and the surface of the crystal are parallel to a principal axis of elastic symmetry.[2] This case and the isotropic one are distinguished by the fact that the deformation constitutes a plane state of strain, i.e., there is no component of displacement normal to the sagittal plane (the plane perpendicular to both the wave front and the surface of the crystal).

When the sagittal plane is not a plane of elastic symmetry, waves of plane strain are not possible. Important examples are those of the rotated Y cuts of quartz, in which the stress-strain relation exhibits monoclinic symmetry when referred to the rotated axes. The direction of propagation of technical interest is parallel to the digonal axis and, in that case, the strains in the sagittal plane are coupled with shears transverse to it. From Ekstein's[3] solution of the problem of vibrations of a monoclinic plate, it may be concluded that, in the case of flexural vibrations, the coupled mode is one of face shear while, in the case of extensional vibrations, the coupled mode is one of transverse thickness shear. Two questions arise: (1) Are surface waves possible under these circumstances? (2) If so, is there only one limiting surface mode (and one velocity) for both flexure and extension, as in the case of isotropy? These questions are considered in the sections which follow and are answered in the affirmative. The existence of the surface wave is established and the presence of the coupled shears is exhibited in a study of the case of orthorhombic symmetry. In this instance the surface mode uncouples into a surface mode of plane strain and either a face-shear mode or a transverse thickness-shear mode. Results of computations are given for the AT cut of quartz. The computations involve the simultaneous solution of algebraic equations of the third and fifth degrees whereas only one cubic equation need be solved for the analogous problems in isotropic, cubic, and orthorhombic media.

* This investigation was supported by the Signal Corps Engineering Laboratories.
[1] Lord Rayleigh, Proc. London Math. Soc. 17, 4 (1887); 20, 225 (1889).
[2] H. Stoneley, Proc. Roy. Soc. (London) A232, 447 (1955).
[3] H. Ekstein, Phys. Rev. 68, 11 (1945).

II. SURFACE WAVES: MONOCLINIC SYMMETRY

When referred to the rotated axes, the stress-strain relation for the rotated Y cuts of quartz takes the form

$$\begin{aligned}
T_1&=c_{11}S_1+c_{12}S_2+c_{13}S_3+c_{14}S_4,\\
T_2&=c_{12}S_1+c_{22}S_2+c_{23}S_3+c_{24}S_4,\\
T_3&=c_{13}S_1+c_{23}S_2+c_{33}S_3+c_{34}S_4,\\
T_4&=c_{14}S_1+c_{24}S_2+c_{34}S_3+c_{44}S_4,\\
T_5&=c_{55}S_5+c_{56}S_6,\\
T_6&=c_{56}S_5+c_{66}S_6.
\end{aligned}\tag{1}$$

This is the stress-strain relation for monoclinic symmetry with x the digonal axis. The components of strain are expressed in terms of the components of displacement by

$$\begin{aligned}
S_1&=\partial u/\partial x, & S_4&=\partial w/\partial y+\partial v/\partial z,\\
S_2&=\partial v/\partial y, & S_5&=\partial u/\partial z+\partial w/\partial x,\\
S_3&=\partial w/\partial z, & S_6&=\partial v/\partial x+\partial u/\partial y,
\end{aligned}\tag{2}$$

and the stress equations of motion are

$$\begin{aligned}
\partial T_1/\partial x+\partial T_6/\partial y+\partial T_5/\partial z&=\rho\partial^2 u/\partial t^2,\\
\partial T_6/\partial x+\partial T_2/\partial y+\partial T_4/\partial z&=\rho\partial^2 v/\partial t^2,\\
\partial T_5/\partial x+\partial T_4/\partial y+\partial T_3/\partial z&=\rho\partial^2 w/\partial t^2.
\end{aligned}\tag{3}$$

We consider a plane wave traveling along the digonal axis in the neighborhood of the surface $y=0$. Thus,

$$\begin{aligned}
u&=Ae^{-\eta y}\sin\xi(x-ct),\\
v&=Be^{-\eta y}\cos\xi(x-ct),\\
w&=Ce^{-\eta y}\cos\xi(x-ct),
\end{aligned}\tag{4}$$

where η is real. Upon substituting (4) successively in (2), (1), and (3), we find

$$\begin{aligned}(\beta^2-g_{11}+q)A+k\beta B+h\beta C&=0,\\ k\beta A+(1-g_{22}\beta^2-q)B+(g_{56}-g_{24}\beta^2)C&=0,\\ h\beta A+(g_{56}-g_{24}\beta^2)B+(g_{55}-g_{44}\beta^2-q)C&=0,\end{aligned}\tag{5}$$

where

$$\beta=\eta/\xi,\quad q=\rho c^2/c_{66},\tag{6}$$
$$g_{rs}=c_{rs}/c_{66},\quad h=g_{14}+g_{56},\quad k=1+g_{12}.$$

The compatibility of Eqs. (5) requires that the determinant of the coefficients of A, B, and C vanish:

$$\begin{vmatrix}\beta^2-g_{11}+q & k\beta & h\beta\\ k\beta & 1-g_{22}\beta^2-q & g_{56}-g_{24}\beta^2\\ h\beta & g_{56}-g_{24}\beta^2 & g_{55}-g_{44}\beta^2-q\end{vmatrix}=0.\tag{7}$$

This is a bicubic equation relating the wave-number ratio β and the velocity c. Thus, for a given wave number ξ, there are three values of the decay constant η for each velocity; or, alternatively,

$$\beta^2=\beta_j^2(q),\quad j=1,2,3.\tag{8}$$

Amplitudes A_j, B_j, C_j may be assigned to each β_j, in which case Eqs. (5) yield the ratios

$$\begin{aligned}r&=B_j/A_j\\ &=[(\beta_j^2-g_{11}+q)(g_{55}-g_{44}\beta_j^2-q)-h^2\beta_j^2]/\Delta_j,\\ s_j&=C_j/A_j\\ &=[(\beta_j^2-g_{11}+q)(g_{24}\beta^2-g_{56})+hk\beta_j^2]/\Delta_j,\end{aligned}\tag{9}$$

where

$$\Delta_j=-\beta_j[k(g_{55}-g_{44}\beta_j^2-q)+h(g_{24}\beta_j^2-g_{56})].$$

The displacements (4) may now be written

$$\begin{aligned}u&=\textstyle\sum A_j\exp(-\beta_j\xi y)\sin\xi(x-ct),\\ v&=\textstyle\sum A_jr_j\exp(-\beta_j\xi y)\cos\xi(x-ct),\\ w&=\textstyle\sum A_js_j\exp(-\beta_j\xi y)\cos\xi(x-ct),\end{aligned}\tag{10}$$

where the summations are over $j=1, 2, 3$.

On the bounding plane the tractions produced by the displacements (10) must vanish, i.e., we require

$$T_2=T_4=T_6=0\quad\text{on}\quad y=0.\tag{11}$$

Substituting (10) successively in (2), (1), and (11) we obtain

$$\begin{aligned}d_{11}A_1+d_{12}A_2+d_{13}A_3&=0,\\ d_{21}A_1+d_{22}A_2+d_{23}A_3&=0,\\ \beta_1d_{31}A_1+\beta_2d_{32}A_2+\beta_3d_{33}A_3&=0,\end{aligned}\tag{12}$$

where, for $j=1, 2, 3$,

$$\begin{aligned}d_{1j}&=g_{12}-\beta_jr_jg_{22}-\beta_js_jg_{24},\\ d_{2j}&=g_{14}-\beta_jr_jg_{24}-\beta_js_jg_{44},\\ d_{3j}&=1+\beta_j^{-1}r_j+\beta_j^{-1}s_jg_{56}.\end{aligned}\tag{13}$$

The compatibility of Eqs. (12) requires

$$\beta_1M_1+\beta_2M_2+\beta_3M_3=0,\tag{14}$$

where

$$\begin{aligned}M_1&=d_{31}(d_{12}d_{23}-d_{13}d_{22}),\\ M_2&=d_{32}(d_{13}d_{21}-d_{11}d_{23}),\\ M_3&=d_{33}(d_{11}d_{22}-d_{12}d_{21}).\end{aligned}\tag{15}$$

It may be ascertained that Eq. (14) is the limiting form of the transcendental frequency equations which can be formed from Ekstein's[3] solution. Equation (14) is a quintic on q which must be solved simultaneously with the cubic (7) to obtain the velocities. Only those roots are admissible for which q and the associated β_j are real and positive, since only then does the solution correspond to a physically real surface wave.

III. SURFACE WAVES: ORTHORHOMBIC SYMMETRY

The complexity of the preceding solution is due to the presence of the constants c_{14}, c_{24}, and c_{56}, as a result of which the displacements u and v, in the sagittal plane, are coupled with the displacement w normal to it. Suppression of the coupling permits the identification of the relatively simple component modes.

The three coupling constants (and also c_{34}) are absent in the stress-strain relation of orthorhombic symmetry. Equations (5) then have the two solutions

$$A\neq0,\quad B\neq0,\quad C=0,\tag{16}$$

$$(\beta^2-g_{11}+q)(1-g_{22}\beta^2-q)-k^2\beta^2=0\tag{17}$$

$$r_j=B_j/A_j=-(\beta_j^2-g_{11}+q)/k\beta_j,\quad j=1,2;\tag{18}$$

and

$$A=0,\quad B=0,\quad C\neq0,\tag{19}$$

$$g_{55}-g_{44}\beta^2-q=0.\tag{20}$$

The first solution is a wave of plane strain. On applying the boundary conditions (11), we find a degenerate form of (12) which leads to

$$(g_{12}-r_1\beta_1g_{22})(\beta_2+r_2)=(g_{12}-r_2\beta_2g_{22})(\beta_1+r_1),\tag{21}$$

a degenerate form of (14). This is a quadratic on q which must be solved simultaneously with the quadratic (17) to obtain the velocity. The roots, β_1 and β_2, of the latter may be expressed as

$$\begin{aligned}\beta_1^2+\beta_2^2&=-[q-1+k^2+g_{22}(q-g_{11})]/g_{22},\\ \beta_1^2\beta_2^2&=(q-1)(q-g_{11})/g_{22},\end{aligned}\tag{22}$$

and (21) may be factored into $\beta_1-\beta_2$ and a term in which β_1 and β_2 appear only in the combinations $\beta_1^2+\beta_2^2$, $\beta_1^2\beta_2^2$, and $\beta_1\beta_2$. Upon substitution of the expressions (22) in (21), there results a quartic on q, one root of which ($q=g_{11}$) leads to zero displacement. There remains the cubic equation

$$(1-d)q^3-(1-e+2f)q^2+f(f+2)q-f^2=0,\tag{23}$$

where

$$d=c_{66}/c_{22},\quad e=c_{11}/c_{22},\quad f=(c_{11}-c_{12}^2/c_{22})/c_{66}. \tag{24}$$

In general, Eq. (23) has one real, positive root associated with real, positive β_1 and β_2. This root gives the velocity of the surface wave of plane strain. It may be observed that (23) degenerates to Rayleigh's equation for the isotropic material when the appropriate reductions of the elastic constants are made.

IV. UNCOUPLED TRANSVERSE MODES

Returning now to the second solution, given by (19) and (20), we note that β need no longer be real and positive since the displacement w is not coupled with the surface wave of plane strain. From (20), β may be either zero or imaginary and still yield real, positive q and hence a real phase velocity.

If $\beta=0$, the solution reduces to

$$u=v=0,\quad w=C\cos\xi(x-ct),\quad c=(c_{55}/\rho)^{\frac{1}{2}}. \tag{25}$$

The components of stress T_2, T_4, T_6 are zero throughout, so that the boundary conditions (11) are satisfied. All the components of strain are zero except S_5 and hence (25) is an $x-z$ face-shear mode.

Alternatively, if β is imaginary, the real part of the second solution is

$$u=v=0,\quad w=C\cos\bar{\beta}\xi y\cos\xi(x-ct), \tag{26}$$

where $\bar{\beta}=i\beta$ and the velocity is given by (20). The components of stress T_2 and T_6 are zero everywhere while

$$T_4=-Cc_{44}\bar{\beta}\xi\sin\bar{\beta}\xi y\cos\xi(x-ct). \tag{27}$$

Thus, all three boundary conditions are satisfied not only on $y=0$ but also on the parallel planes

$$y=n\pi/\bar{\beta}\xi,\quad n=1,2,3\cdots. \tag{28}$$

All the components of strain are zero except S_4 and S_5. This solution may be interpreted as a $y-z$ thickness-shear mode with phase reversals along x. The latter disappear at infinite wavelength ($\xi=0$) and the mode reduces to a transverse thickness-shear vibration. From (28) and (20), the frequency of the vibration, in a plate of thickness b, is given by the usual formula

$$\omega=(n\pi/b)(c_{44}/\rho)^{\frac{1}{2}}. \tag{29}$$

V. APPLICATION

A computation of the roots of Eqs. (14) and (7) was performed for the AT cut of quartz, using Mason's[4] values of the elastic constants referred to the principal axes. When referred to the rotated axes,[5] the constants that appear in the equations are (in units of 10^{10} dynes/cm^2):

$c_{11}=86.05$	$c_{12}=-10.49$	$c_{55}=69.84$
$c_{22}=129.9$	$c_{14}=3.453$	$c_{66}=29.41$
$c_{44}=39.06$	$c_{24}=-5.820$	$c_{56}=-12.415.$

The coupling constants c_{14}, c_{24}, and c_{56} are small in comparison with the others and hence a good first approximation is obtainable, from Eqs. (23), (17), and (20), *viz.*,

$$q=0.9054,\quad \beta_1^2=0.0210,\quad \beta_2^2=1.9202,\quad \beta_3^2=1.1115.$$

Starting with these values, a procedure of successive approximations in Eqs. (7) and (14) yielded

$$q=0.901,\quad \beta_1^2=0.022,\quad \beta_2^2=1.923,\quad \beta_3^2=1.115.$$

If the density of quartz[6] is taken as $\rho=2.65$, the velocity of the surface wave is

$$c=(qc_{66}/\rho)^{\frac{1}{2}}=3.16\times10^5\text{ cm/sec.}$$

The roots of Eqs. (7) and (14) yield one additional real velocity, given by

$$2c_{66}q=c_{55}+c_{66}-[(c_{55}-c_{66})^2+4c_{56}^2]^{\frac{1}{2}}, \tag{30}$$

but one of the β_j associated with this root is imaginary and hence the corresponding mode is not a surface wave.

Thus, one and only one surface wave can be propagated along the digonal axis in the AT cut of quartz. In an AT plate this is the limiting form of the lowest mode of both flexural and extensional vibrations as the wavelength approaches zero or as the frequency approaches infinity. As shown in the foregoing sections, the surface wave, in the case of flexure, is the result of coupling of a wave of plane strain with a face-shear mode whereas, in the case of extension, the same surface wave is the result of coupling of the same wave of plane strain with a transverse, thickness-shear mode.

---

[4] W. P. Mason, *Piezoelectric Crystals and Their Applications to Ultrasonics* (D. Van Nostrand Company, Inc., New York, 1950), p. 84.

[5] R. A. Sykes, "Modes of motion in quartz crystals" in *Quartz Crystals for Electrical Circuits*, edited by R. A. Heising (D. Van Nostrand Company, Inc., New York, 1946), Chap. VI, p. 247.

[6] See reference 4, p. 79.

Paper No.
57—APM-29

Influence of Width on Velocities of Long Waves in Plates[1]

By D. C. GAZIS[2] and R. D. MINDLIN,[3] NEW YORK, N. Y.

This paper contains a discussion of the velocities of propagation of long flexural and extensional waves in a plate or bar in the transition region between the states of generalized plane stress and plane strain; i.e., for arbitrary width-thickness ratios (h/a) of rectangular sections. The investigations are carried out mathematically on the basis of approximate equations of motion. It is found that, for flexural waves, as h/a increases from zero, the velocity remains close to that for generalized plane stress until quite large values of h/a are reached, after which the velocity rises asymptotically toward that for plane strain. For extensional waves the velocity is slightly smaller than that for generalized plane stress for all $h/a > 0$.

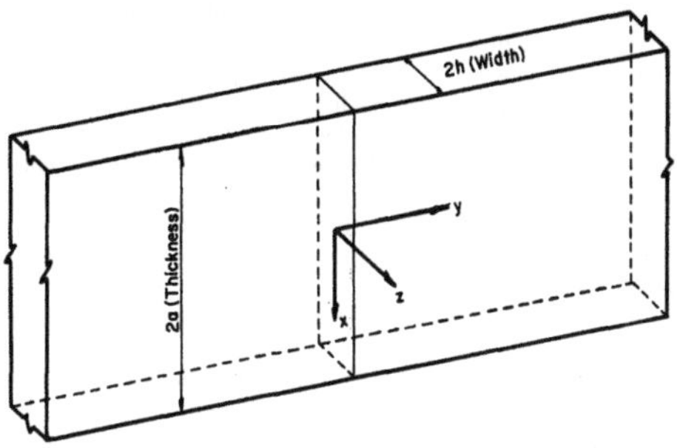

Fig. 1 Reference Co-Ordinates and Dimensions

Introduction

The velocities of flexural and extensional waves in an infinite elastic plate were deduced by Rayleigh (1)[4] and Lamb (2, 3) from the equations of elasticity. In their treatment of straight-crested waves, they employed the theory of plane strain, so that the results pertain to a plate of infinite width. By a simple change of the constants of elasticity the velocities may be converted to those for generalized plane stress, applicable to a plate of small width; i.e., a beam or bar of narrow rectangular section. It is the purpose of this paper to study the velocities in the intermediate range of widths for the case of long wave lengths of the lowest flexural and extensional modes.

Consider an elastic body bounded by traction-free planes $x = \pm a$, $z = \pm h$ and infinite in the direction y, of wave propagation, Fig. 1. For the case of plane strain (corresponding to h/a infinite) the lowest flexural mode, in the Rayleigh-Lamb solution, has as its limiting phase velocity, for waves long compared with $2a$

$$c = \eta a c_2 \sqrt{\left[\frac{2}{3(1-\nu)}\right]} = \eta a \sqrt{\left[\frac{E}{3\rho(1-\nu^2)}\right]} \qquad [1]$$

where E is Young's modulus, ν is Poisson's ratio, ρ is the density, η is the wave number, and c_2 is the velocity of shear waves in an infinite body, i.e.

$$c_2 = \sqrt{(G/\rho)}$$

where G is the shear modulus. As pointed out by Rayleigh and Lamb, the velocity given in Equation [1] is the same as that obtained from the classical (Lagrange) theory of low-frequency flexural vibrations of thin plates (4) for the case of straight-crested waves in an infinite plate. The corresponding velocity of extensional waves, from the Rayleigh-Lamb solution, is

$$c = c_2 \sqrt{\left(\frac{2}{1-\nu}\right)} = \sqrt{\left[\frac{E}{\rho(1-\nu^2)}\right]} \qquad [2]$$

and this coincides with the velocity obtained from Poisson's theory[5] of low-frequency extensional vibrations of thin plates.

At the other extreme ($h/a \rightarrow 0$) the conditions of generalized plane stress apply. To convert from plane strain it is only necessary to replace ν by ν' and E by E' where[6]

$$\nu' = \frac{\nu}{1+\nu}, \quad E' = \frac{(1+2\nu)E}{(1+\nu)^2}$$

Equations [1] and [2] then become

$$c = \eta a c_2 \sqrt{\left[\frac{2(1+\nu)}{3}\right]} = \eta a \sqrt{\left(\frac{E}{3\rho}\right)} \qquad [3]$$

for flexural waves and

$$c = c_2 \sqrt{[2(1+\nu)]} = \sqrt{\left(\frac{E}{\rho}\right)} \qquad [4]$$

for extensional waves.

At intermediate values of h/a, waves in the plane of x and y couple with waves reflecting back and forth between the boundaries $z = \pm h$ (i.e., with transverse vibrations across the width), and this alters the phase velocities.

To take into account the additional free boundaries at $z = \pm h$ within the framework of the three-dimensional theory would be a very difficult task because each of the infinity of modes in the Rayleigh-Lamb solution gives rise to an infinity of modes in the presence of the additional boundaries. If, however, attention is confined to the lowest extensional and flexural modes it is necessary to take into account only the lowest width mode inasmuch as the higher width modes will have little influence.

[1] This investigation was supported by the Office of Naval Research under a contract with Columbia University.

[2] Research Assistant, Department of Civil Engineering and Engineering Mechanics, Columbia University.

[3] Professor of Civil Engineering, Columbia University.

[4] Numbers in parentheses refer to the Bibliography at the end of the paper.

For presentation at the Applied Mechanics Division Summer Conference, Berkeley, Calif., June 13–15, 1957, of The American Society of Mechanical Engineers.

Discussion of this paper should be addressed to the Secretary, ASME, 29 West 39th Street, New York, N. Y., and will be accepted until one month after final publication of the paper itself in the Journal of Applied Mechanics.

Note: Statements and opinions advanced in papers are to be understood as individual expressions of their authors and not those of the Society. Manuscript received by ASME Applied Mechanics Division, January 19, 1957. Paper No. 57—APM-29.

[5] Reference (4), p. 497.

[6] Reference (4), p. 208.

Equations of motion already are available which take into account the first symmetric extensional mode across the width (5). These equations result from retaining the zero-order and first-order terms of an expansion of the displacements in a series of powers of the width co-ordinate (z). They are referred to in the sequel as "first-order equations" and are used here in the investigation of both the flexural and extensional velocities. From a physical point of view the first-order equations are appropriate only for small values of Poisson's ratio inasmuch as, for $\nu > 1/3$, there is at least one transverse symmetric shear mode whose frequency is less than that of the transverse extensional mode. However, if the wave length along y is large and $\nu < 1/3$, the discrepancy will not be noticeable in the final mathematical expressions for the velocities. This is because, when $\nu < 1/3$, the symmetric width mode of lowest frequency is, in fact, the extensional mode and its frequency is depressed but little by the neighboring shear mode as long as the wave length along y is large.

In the case of flexural waves, the influence of width also can be investigated within the framework of the classical theory of thin plates. This is done here and the results are compared with those obtained from the first-order equations.

The conclusions reached in all the investigations may be summarized briefly as follows: In the case of flexural waves, as the width-thickness ratio (h/a) is increased from zero, the velocity remains close to that for generalized plane stress until quite large values of h/a are reached; after which the velocity rises asymptotically toward that for plane strain. In the case of extensional waves the velocity is slightly smaller than that for generalized plane stress for all $h/a > 0$.

First-Order Equations

The plate displacements, as defined in reference (5), are expressed in terms of three potentials $\Phi_i(x, y, t)$

$$\left.\begin{aligned} v_x &= \frac{\partial\Phi_1}{\partial x} + \frac{\partial\Phi_2}{\partial x} + \frac{\partial\Phi_3}{\partial y} \\ v_y &= \frac{\partial\Phi_1}{\partial y} + \frac{\partial\Phi_2}{\partial y} - \frac{\partial\Phi_3}{\partial x} \\ v_z &= \sigma_1\Phi_1 + \sigma_2\Phi_2 \end{aligned}\right\} \quad [5]$$

where the Φ_i satisfy the equations

$$(\nabla^2 + \delta_i^2)\Phi_i = 0, \quad i = 1, 2, 3 \quad [6]$$

In Equations [6], ∇^2 is Laplace's operator in the plane of x and y and

$$\left.\begin{aligned} \delta_i^2 &= \frac{3\kappa^2}{2\beta h^2}\left[(\alpha+\beta)\frac{\omega^2}{\bar\omega^2} - 1 - (-1)^i\psi\right], \quad i = 1, 2 \\ \psi &= \left\{\left[(\alpha+\beta)\frac{\omega^2}{\bar\omega^2} - 1\right]^2 + 4\alpha\beta\frac{\omega^2}{\bar\omega^2}\left(1 - \frac{\omega^2}{\bar\omega^2}\right)\right\}^{1/2} \\ \delta_3^2 &= \frac{\omega^2}{c_2^2}, \quad \bar\omega^2 = \frac{\pi^2}{4h^2}c_1^2, \quad c_1^2 = \frac{\lambda + 2G}{\rho}, \quad c_2^2 = \frac{G}{\rho} \\ \alpha &= \frac{c_1^2}{c_3^2}, \quad \beta = \frac{c_2^2}{c_3^2}, \quad c_3^2 = \frac{4G(\lambda + G)}{\rho(\lambda + 2G)} = \frac{E}{\rho(1 - \nu^2)} \\ \kappa^2 &= \frac{\pi^2}{12}, \quad \sigma_i = \frac{h(\lambda + 2G)}{\kappa\lambda}\left(\delta_i^2 - \frac{\omega^2}{c_1^2}\right), \quad i = 1, 2 \end{aligned}\right\} \quad [7]$$

where λ, G are Lamé's constants.

It should be noted that the dimension $2h$ which was called the thickness of the plate in reference (5) is here called the width.

Let

$$\Phi_i = f_i(x)\exp[i(\eta y + \omega t)] \quad [8]$$

then Equations [6] yield

$$f_i'' = \xi_i^2 f_i, \quad i = 1, 2, 3 \quad [9]$$

where prime denotes differentiation with respect to x and

$$\xi_i^2 = \eta^2 - \delta_i^2 \quad [10]$$

The plate displacements, expressed in terms of the f_i, are

$$\left.\begin{aligned} v_x &= [f_1' + f_2' + i\eta f_3]\exp[i(\eta y + \omega t)] \\ v_y &= [i\eta(f_1 + f_2) - f_3']\exp[i(\eta y + \omega t)] \\ v_z &= [\sigma_1 f_1 + \sigma_2 f_2]\exp[i(\eta y + \omega t)] \end{aligned}\right\} \quad [11]$$

and the plate stresses, after omission of the exponential factor, are

$$\left.\begin{aligned} N_x &= 2Gh[(\eta^2 + \xi_3^2)(f_1 + f_2) + 2i\eta f_3'] \\ N_{xy} &= 2Gh[i\eta(f_1' + f_2') - (\eta^2 + \xi_3^2)f_3] \\ R_x &= \frac{2Gh^2}{3}(\sigma_1 f_1' + \sigma_2 f_2') \end{aligned}\right\} \quad [12]$$

In the problems under consideration, the boundary conditions are

$$N_x = N_{xy} = R_x = 0 \quad \text{at } x = \pm a \quad [13]$$

Flexural Vibrations. When the motion is antisymmetric with respect to the plane $x = 0$ we take as the solutions of Equation [9]

$$f_1 = A_1 \sinh \xi_1 x, \quad f_2 = A_2 \sinh \xi_2 x, \quad f_3 = A_3 \cosh \xi_3 x \quad [14]$$

The plate stresses on the faces $x = \pm a$ are now computed by means of Equations [12] and [14], and equated to zero. This results in a homogeneous system of linear equations in the A_i. The vanishing of the determinant of the coefficients of the A_i leads to the frequency equation

$$\sigma_2\xi_2 F_1 = \sigma_1\xi_1 F_2 \quad [15]$$

where

$$F_i = \frac{\tanh \xi_i a}{\tanh \xi_3 a} - \frac{4\eta^2\xi_i\xi_3}{(\eta^2 + \xi_3^2)^2}, \quad i = 1, 2 \quad [16]$$

Equation [15] may be employed for the determination of $\omega/\bar\omega$ for a given ηa; i.e., for the evaluation of the frequencies of various modes corresponding to any given wavelength. We are interested in the variation of the phase velocity c with the wave length of the lowest antisymmetric mode only, and in the region of wave lengths large in comparison with $2a$, or $\eta a \ll 1$.

It can be verified that Equation [15] has a root for vanishing $\omega/\bar\omega$ and ηa. It is assumed, further, that $c/c_2 \to 0$ as $\eta a \to 0$, which is seen to be verified by the results.

Expanding expressions [7] for δ_i and σ_i in series of powers of c/c_2 and ηa, and dropping terms above the second power, we obtain

$$\left.\begin{aligned} \delta_1^2 &= \frac{\eta^2 c^2}{c_3^2}, \quad \delta_3^2 = \frac{\eta^2 c^2}{c_2^2} \\ \delta_2^2 &= \frac{\eta^2 c^2}{c_3^2}\left(\frac{1}{\beta} + \frac{1}{\alpha} - 1\right) - \frac{\pi^2}{4h^2}\frac{c_3^2}{c_2^2} \\ \frac{\sigma_1}{\sigma_2} &= \frac{\eta^2(c^2/c_3^2 - c^2/c_1^2)}{\eta^2(1 - c^2/c_1^2) - \xi_2^2} \end{aligned}\right\} \quad [17]$$

In addition we observe that $\xi_1 a$ and $\xi_3 a$ tend to zero with ηa, hence we may write

$$\tanh \xi_i a \approx \xi_i a \left(1 - \frac{1}{3}\xi_i^2 a^2\right), \quad i = 1, 3 \ldots\ldots [18]$$

and finally

$$\left.\begin{aligned} \tanh \xi_1 a &\approx \eta a \sqrt{\left(1 - \frac{c^2}{c_3^2}\right)} \left[1 - \frac{1}{3}\eta^2 a^2 \left(1 - \frac{c^2}{c_3^2}\right)\right] \\ \tanh \xi_3 a &\approx \eta a \sqrt{\left(1 - \frac{c^2}{c_2^2}\right)} \left[1 - \frac{1}{3}\eta^2 a^2 \left(1 - \frac{c^2}{c_2^2}\right)\right] \end{aligned}\right\} \ldots [18a]$$

In virtue of Equations [17] and [18a], Equation [15] yields

$$\frac{c^2}{c_2^2} = \frac{4}{3}\eta^2 a^2 \left\{\frac{1+\nu}{2} + \frac{3\nu^2}{\pi^2}\left(\frac{h}{a}\right)^2 \left[1 - \frac{h}{\kappa a}\sqrt{\left(\frac{1-\nu}{6}\right)} \tanh\left(\frac{\kappa a}{h}\sqrt{\left(\frac{6}{1-\nu}\right)}\right)\right]\right\} \ldots\ldots [19]$$

Let

$$\gamma = \kappa \sqrt{\left(\frac{6}{1-\nu}\right)}, \quad c'^2 = \frac{E}{3\rho} = \frac{2}{3}(1+\nu)c_2^2 \ldots [20]$$

Then Equation [19] becomes

$$\frac{c}{c'} = \eta a \left[1 + \frac{6\nu^2}{\pi^2(1+\nu)}\left(\frac{h}{a}\right)^2 \left(1 - \frac{\tanh(\gamma a/h)}{\gamma a/h}\right)\right]^{1/2} \ldots\ldots [21]$$

For $\gamma a/h \geqq 1$, that is, when the width is not large in comparison with the thickness of the plate, we may write

$$\tanh(\gamma a/h) \approx 1$$

and Equation [21] yields

$$\frac{c}{c'} = \eta a \left\{1 + \frac{6\nu^2}{\pi^2(1+\nu)}\left(\frac{h}{a}\right)^2 \cdot \left[1 - \frac{2h}{\pi a}\sqrt{\left(\frac{1-\nu}{2}\right)}\right]\right\}^{1/2} \ldots\ldots [22]$$

For $\gamma a/h \ll 1$, that is for shallow plates, we have

$$\frac{\tanh(\gamma a/h)}{\gamma a/h} \approx 1 - \frac{1}{3}(\gamma a/h)^2 \left[1 - \frac{2}{5}\left(\frac{\gamma a}{h}\right)^2\right]$$

and

$$\frac{c}{c'} = \frac{\eta a}{(1-\nu^2)^{1/2}} \left[1 - \frac{\pi^2 \nu^2}{10(1-\nu)}\left(\frac{a}{h}\right)^2\right] \ldots\ldots [23]$$

The second term in the bracket of Expression [22] gives the increase of the phase velocity, over the velocity for generalized plane stress, for a relatively thin slice of a plate. Similarly, the second term in the bracket of Expression [23] gives the departure from plane strain, for a shallow plate. These equations yield the same results as Equations [1] and [3], for the limiting values of h/a.

The magnitude of the phase velocity of long flexural waves, expressed as the per cent excess over the velocity for generalized plane stress, is shown in Fig. 2, for three values of Poisson's ratio. The same curves give the per cent excess, over the case of generalized plane stress, for the value of the tangent

$$\frac{d(c/c_2)}{d(\eta a)}$$

at the origin $\eta a = c/c_2 = 0$. It is seen that as the width $2h$ increases with respect to the thickness $2a$, the velocity corresponding to plane stress is not greatly exceeded until the width becomes considerably larger than the thickness.

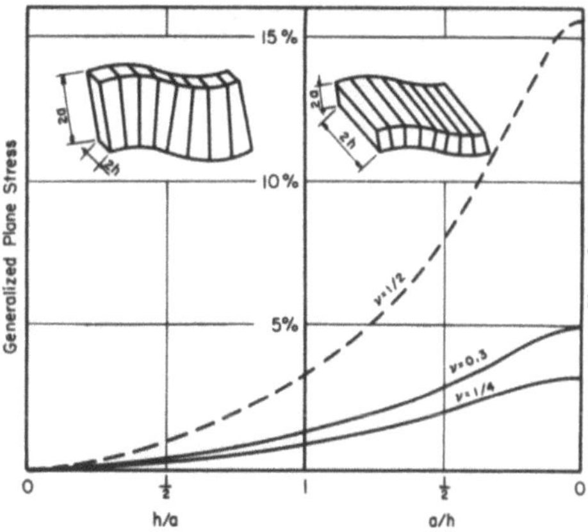

FIG. 2 VELOCITY OF LONG FLEXURAL WAVES; PER CENT EXCESS OVER GENERALIZED PLANE STRESS

Extensional Vibrations. When the motion is symmetric with respect to the plane $x = 0$ we take as the solutions of Equation [9]

$$f_1 = A_1 \cosh \xi_1 x, \quad f_2 = A_2 \cosh \xi_2 x, \quad f_3 = A_3 \sin \xi_3' x, \quad (\xi_3' = i\xi_3) \ldots\ldots [24]$$

The trigonometric, rather than the hyperbolic function is assumed for f_3, in anticipation that $\delta_3 a > \eta a$.

Proceeding as for the antisymmetric case, we derive the frequency equation

$$\sigma_2 \xi_2 G_1 \tanh \xi_2 a = \sigma_1 \xi_1 G_2 \tanh \xi_1 a \ldots\ldots\ldots\ldots [25]$$

where

$$G_i = \frac{\tanh \xi_i a}{\tan \xi_3' a} - \frac{(\eta^2 - \xi_3'^2)^2}{4\eta^2 \xi_i \xi_3'}, \quad i = 1, 2 \ldots\ldots [26]$$

For $\omega/\bar{\omega} = \eta a = 0$ Equation [25] is satisfied, thus there is a mode whose frequency approaches zero, as the wave length becomes large in comparison with the thickness. It is the behavior of this mode at long wave lengths in which we are interested.

Furthermore

$$\frac{\omega}{\bar{\omega}} = \frac{2}{\pi}\frac{c}{c_1}\eta h$$

hence, under the verifiable supposition that c/c_1 is finite and non-zero, we conclude that

$$\text{if } \frac{\omega}{\bar{\omega}} \ll 1 \quad \text{then} \quad \eta h \ll 1 \ldots\ldots\ldots\ldots\ldots [a]$$

that is, the approximations given by Equation [17] are valid for wavelengths large in comparison with both of the transverse dimensions of the plate.

We now assume $\eta a \ll 1$, expand Equation [25] and drop terms beyond the second power of ηa. It should be remarked that it follows from [a] that

$$\xi_2 a \tanh \xi_2 a \gg \eta^2 a^2$$

for all values of the ratio h/a, hence only the terms multiplied by $\xi_2 a \tanh \xi_2 a$ contribute to the final equation. We thus obtain

$$N_1 + N_2\eta^2 a^2 + N_3\eta^2 a^2 \left(\frac{h}{a}\right)^2 = 0 \qquad [27]$$

where

$$\left.\begin{aligned} N_1 &= \left(2 - \frac{c^2}{c_2^2}\right)^2 - 4\left(1 - \frac{c^2}{c_3^2}\right) \\ N_2 &= \frac{1}{3}\left[\left(\frac{c^2}{c_2^2} - 1\right)\left(2 - \frac{c^2}{c_2^2}\right)^2 - 4\left(1 - \frac{c^2}{c_3^2}\right)^2\right] \\ N_3 &= \frac{4}{\pi^2}\frac{c^2}{c_1^2}\left\{\alpha(1-\beta)\left(2 - \frac{c^2}{c_2^2}\right)^2 \right. \\ &\qquad \left. - 4[\alpha + \beta - 2\alpha\beta]\left(1 - \frac{c^2}{c_3^2}\right)\right\} \end{aligned}\right\} \qquad [28]$$

For $\eta a \to 0$, Equation [27] yields for the ratio c/c_2 the same value given by [4] for the case of generalized plane stress, or

$$c/c_2 = \sqrt{[2(1 + \nu)]}$$

Furthermore for $\eta a = 0$ and $c/c_2 = \sqrt{[2(1 + \nu)]}$

$$\frac{d(c^2/c_2^2)}{d(\eta^2 a^2)} = -\frac{4}{\pi^2}\nu^2(1 - 2\nu)\frac{h^2}{a^2} - \frac{2}{3}\nu^2(1 + \nu) \qquad [29]$$

Hence, near the origin

$$\frac{c^2}{c_2^2} = 2(1 + \nu) - \frac{2}{3}\nu^2(1 + \nu)\eta^2 a^2 - \frac{4}{\pi^2}\nu^2(1 - 2\nu)\eta^2 h^2 \qquad [30]$$

or, approximately

$$\frac{c}{c_2} = \sqrt{[2(1 + \nu)]}\left[1 - \frac{\nu^2}{6}\eta^2 a^2 - \frac{\nu^2(1 - 2\nu)}{\pi^2(1 + \nu)}\eta^2 h^2\right] \qquad [31]$$

Equation [31] describes the variation of the phase velocity with the wave length of the first extensional mode. Now, h and a are not interchangeable in this equation, inasmuch as only the first transverse extensional mode in the z-direction has been taken into account, whereas all the modes in the x-direction have been included. However Equation [31] is in agreement with the expectation that the predominant coupling effect, at long wave lengths, is associated with the larger transverse dimension; i.e., with the lower transverse frequency.

It is well known that coupling is stronger when the difference of the frequencies of the coupled modes is small. As the wave length of the extensional mode decreases from infinity, its frequency rises from zero, and approaches first the frequency of the lowest transverse mode, which corresponds to the larger transverse dimension. Hence the predominance of this dimension in Equation [31]. It should be noted that this equation is not to be used if $\nu > 1/3$, for the reason given in the "Introduction."

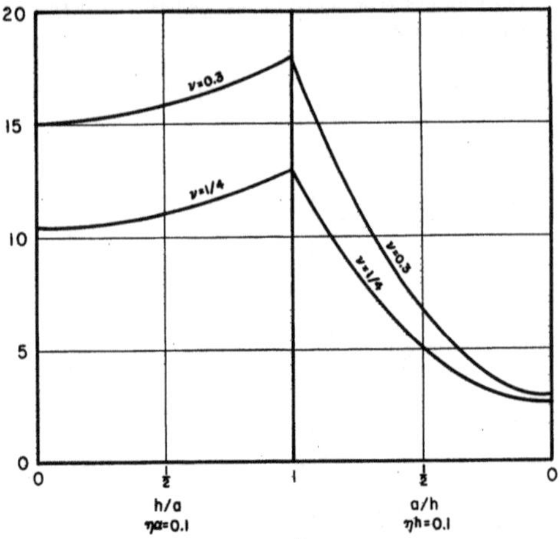

Fig. 3 Velocity of Long Extensional Waves; Decrease in 10^{-3} Per Cent From Generalized Plane Stress

The per cent decrease of the phase velocity from the limiting velocity for generalized plane stress is plotted in Fig. 3 for two values of Poisson's ratio, and for $\eta a = 0.1$ or $\eta h = 0.1$, respectively, for $a/h > 1$ or $h/a > 1$. Actually Fig. 3 should be symmetrical about the line $a/h = 1$. The asymmetry is due to the first order approximation in z, which constrains the motion in this direction, thus resulting in higher frequencies.

As is well known for the case of generalized plane stress, the phase velocity of the lowest extensional mode diminishes as the wave length diminishes. It is seen from Equation [31], and the left-hand side of Fig. 3, that this drop in velocity is accentuated as the width increases from zero to the neighborhood of the thickness.

It should be remarked that if the dimension $2h$ is increased to infinity, the first extensional mode degenerates to a zero mode, and a higher one becomes the mode of plane strain, with motion only in the plane of x and y.

Lagrange's Equation

Referring to Fig. 1, we consider a plate with the plane of y and z its middle plane and w (a function of y, z, and t) its deflection. In the classical theory of flexure, w is governed by (4)

$$D\nabla^4 w + 2\rho a \frac{\partial^2 w}{\partial t^2} = 0, \quad w = w(y, z, t) \qquad [32]$$

where

$$D = 2Ea^3/3(1 - \nu^2) \qquad [33]$$

For free edges, the boundary conditions are

$$\left.\begin{aligned} \frac{\partial^2 w}{\partial z^2} + \nu\frac{\partial^2 w}{\partial y^2} &= 0 \\ \frac{\partial^3 w}{\partial z^3} + (2 - \nu)\frac{\partial^3 w}{\partial y^2 \partial z} &= 0 \end{aligned}\right\} \text{ at } z = \pm h \qquad [34]$$

Let

$$w = Z(z)\exp[i(\eta y + \omega t)] \qquad [35]$$

Then Equation [32] yields the linear differential equation

$$Z^{\mathrm{IV}} - 2\eta^2 Z'' + \left(\eta^4 - \frac{\omega^2}{m^2}\right)Z = 0 \qquad [36]$$

where

$$m^2 = D/2\rho a \ldots\ldots\ldots\ldots\ldots\ldots [37]$$

A solution of Equation [36], symmetric in z, is

$$Z = A_1 \cosh \lambda_1 z + A_2 \cosh \lambda_2 z \ldots\ldots\ldots\ldots [38]$$

with

$$\lambda_i = \eta^2 - (-1)^i(\omega/m), \quad i = 1, 2 \ldots\ldots\ldots [39]$$

Introducing Expression [38] into Equations [34], we derive the frequency equation[7]

$$M_1 \tanh \lambda_1 h = M_2 \tanh \lambda_2 h \ldots\ldots\ldots\ldots [40]$$

where

$$M_i = \lambda_i/(\lambda_i^2 - \nu\eta^2)^2, \quad i = 1, 2 \ldots\ldots\ldots [41]$$

The quantities $\lambda_i h$ are of the order of magnitude of ηh, or smaller. Hence, for $\eta h \ll 1$, we can write

$$\tanh \lambda_i h \approx \lambda_i h \left(1 - \frac{1}{3}\lambda_i^2 h^2\right) \ldots\ldots\ldots [42]$$

Equation [40] then yields

$$\frac{\omega^2}{\eta^2}\left(1 - \frac{2\nu\eta^2 h^2}{3}\right) = \frac{1 - \nu^2}{m^2}\eta^2 - \frac{2h^2}{3}\frac{\nu(1 - \nu)}{m^2}\eta^4 \ldots [43]$$

and finally

$$\frac{c}{c'} = \eta a \left[1 + \frac{\nu^2}{3(1 + \nu)}\eta^2 h^2\right] \ldots\ldots\ldots [44]$$

where c' is given by Equation [20].

To study the case of larger ηh, we write Equation [40] in the form

$$\sqrt{(1 - r)}[(1 - \nu) + r]^2 \tanh [\eta h \sqrt{(1 - r)}] = \sqrt{(1 + r)}[(1 - \nu) - r]^2 \tanh[\eta h \sqrt{(1 + r)}] \ldots\ldots [45]$$

where r describes the departure of the phase velocity from the velocity of plane strain, and is given by

$$r = \frac{\omega m}{\eta^2} = \frac{c}{c'}\sqrt{(1 - \nu^2)} \ldots\ldots\ldots\ldots [46]$$

The value $r = 1$ corresponds to the velocity of long flexural waves in plane strain, and the value $r = (1 - \nu^2)^{1/2}$ to that in generalized plane stress.

Equation [45] constitutes a relation between ηh and r; i.e., it yields the velocity corresponding to a given ratio of the wave length to the width $2h$. Alternatively, for a constant ηa, it may be regarded as a relation between h/a and r. That is, if the ratio of the wave length to the thickness $2a$ is kept constant, Equation [45] yields the velocity corresponding to a given ratio of width to thickness.

From Equation [44] it is ascertained that, if $\eta h \to 0$

$$c/c' \to 1, \quad r \to \sqrt{(1 - \nu^2)}$$

However, for $\eta h \to \infty$, Equation [45] reduces to the algebraic equation

$$\sqrt{(1 - r)}[(1 - \nu) + r]^2 - \sqrt{(1 + r)}[(1 - \nu] - r]^2 = 0, \quad r \neq 1 \ldots\ldots [47]$$

whose only acceptable root is

$$r^2 = (1 - \nu)[(3\nu - 1) + 2\sqrt{(1 - 2\nu + 2\nu^2)}] \leqslant 1 \ldots [48]$$

Hence, the phase velocities given by Equation [45] never reach the values of plane strain, however large the ratio h/a may become. This is due to the anticlastic curvature present in the classical theory even when the width of the plate is very large. Prevention of the anticlastic curvature could be obtained, within the framework of the classical theory, by couples

$$M_z = -\nu M_y$$

along the free edges, and this would raise the phase velocity to the value for plane strain. To obtain suppression of the anticlastic curvature when the edges are free would require that finite rotations be taken into account.

[7] This frequency equation is given by Voigt (6).

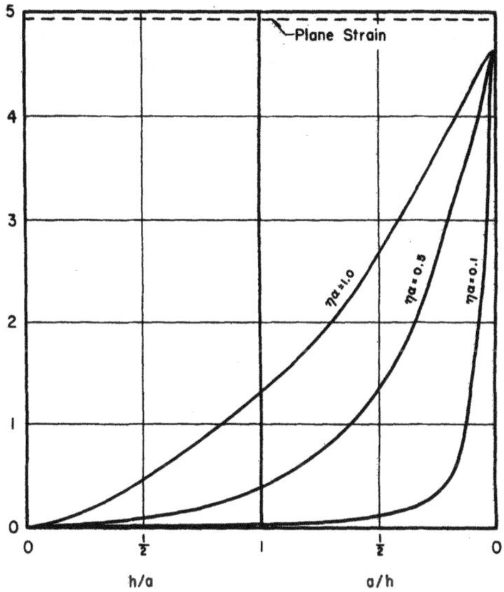

FIG. 4 VELOCITY OF LONG FLEXURAL WAVES; PER CENT EXCESS OVER GENERALIZED PLANE STRESS; SOLUTION OF LAGRANGE EQUATION FOR $\nu = 1/4$

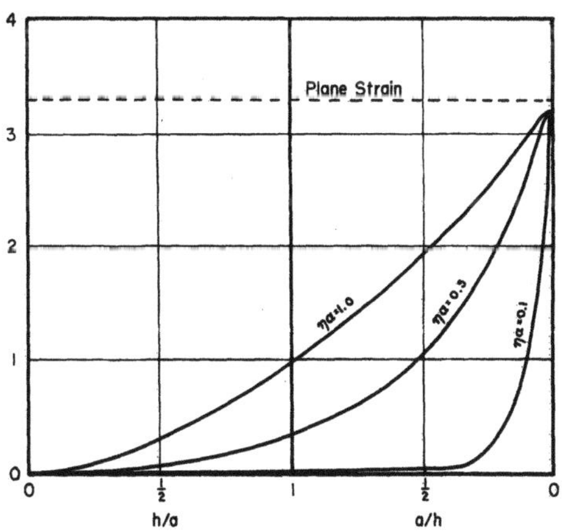

FIG. 5 VELOCITY OF LONG FLEXURAL WAVES; PER CENT EXCESS OVER GENERALIZED PLANE STRESS; SOLUTION OF LAGRANGE EQUATION FOR $\nu = 0.3$

Curves computed from the solution of Lagrange's equation are shown in Figs. 4 and 5, for $\nu = 1/4$ and $\nu = 0.3$, respectively, and for three values of ηa. These are to be compared with the results from the first-order equations shown in Fig. 2. The presence of anticlastic curvature in the Lagrange theory is the cause of the difference between the two. The first-order theory does not contain displacements which could produce anticlastic curvature and, in this respect contains an additional constraint, resulting in higher frequencies. For example, for $\nu = 0.3$ and $a = h$, Fig. 2 gives an excess of 1.3 per cent over the velocity of generalized plane stress, whereas Fig. 5 gives less than 0.05 per cent for $\eta a = 0.1$.

However, it should be noted that for $h/a \rightarrow \infty$ the solution of the first-order equations corresponds to zero amplitude of the transverse displacements v_z and hence to the correct deformation configuration of plane strain.

BIBLIOGRAPHY

1 "On the Free Vibrations of an Infinite Plate of Homogeneous Isotropic Elastic Matter," by Lord Rayleigh, Proceedings of the London Mathematical Society, vol. 20, 1889, pp. 225–234.

2 "On the Flexure of Elastic Plates," by H. Lamb, Proceedings of the London Mathematical Society, vol. 21, 1889, p. 85.

3 "On Waves in an Elastic Plate," by H. Lamb, Proceedings of the Royal Society of London, series A, vol. 93, 1917, pp. 114–128.

4 "A Treatise on the Mathematical Theory of Elasticity," by A. E. H. Love, Cambridge University Press, London, England, 1927, p. 496.

5 "High-Frequency Extensional Vibrations of Plates," by T. R. Kane and R. D. Mindlin, Journal of Applied Mechanics, Trans. ASME, vol. 78, 1956, pp. 277–283.

6 "Bemerkung zu dem Problem der transversalen Schwingungen rechteckiger Platten," by W. Voigt, Göttingen Nachrichten, 1893, p. 225.

Reprinted from
Journal of Applied Mechanics
24 (1957) 541-546

Paper No.
57—APM-39

Stress-Strain Relations and Vibrations of a Granular Medium[1]

By J. DUFFY[2] and R. D. MINDLIN[3]

A differential stress-strain relation is derived for a medium composed of a face-centered cubic array of elastic spheres in contact. The stress-strain relation is based on the theory of elastic bodies in contact, and includes the effects of both normal and tangential components of contact forces. A description is given of an experiment performed as a test of the contact theories and the differential stress-strain relation derived from them. The experiment consists of a determination of wave velocities and the accompanying rates of energy dissipation in granular bars composed of face-centered cubic arrays of spheres. Experimental results indicate a close agreement between the theoretical and experimental values of wave velocity. However, as in previous experiments with single contacts, the rate of energy dissipation is found to be proportional to the square of the maximum tangential contact force rather than to the cube, as predicted by the theory for small amplitudes.

Introduction

The relation between stress and small strain in a granular medium may be derived from a study of the behavior of the individual grains of the medium considered as elastic bodies in contact. In previous studies (1, 2, 3, 4)[4] the relation between the contact force and the displacement of the grains was based on Hertz' theory (5). In the present paper both normal and tangential components of contact force are considered. While the classical theory of Hertz is retained, the theories of Cattaneo (6), Mindlin (7), and Mindlin and Deresiewicz (8) make possible the inclusion of the tangential components of the forces at the contacts between grains. The stress-strain relation is found for the case of a medium composed of elastic spheres arranged in a face-centered cubic array.

As a result of the nonlinearity in the relation between the normal component of contact force and the displacement, the stress-strain relation is nonlinear. Furthermore, since the relation between the tangential component of contact force and displacement is inelastic, the stress-strain relation depends upon the entire loading history of the medium. Accordingly, the stress-strain relation is given as an incremental one.

The derivation itself is performed by satisfying simultaneously two independent sets of conditions. The first of these expresses the requirement that equilibrium be maintained when an arbitrary incremental load is applied to the medium. This requirement alone, however, is insufficient for a unique determination of the forces and displacements, and a second set of conditions must be imposed. This requires that relative displacements of material points in the medium be compatible. Satisfying these two sets of conditions simultaneously provides a relation giving increments in relative displacements in terms of increments in applied load.

In general, integration of the differential stress-strain relation will entail considerable difficulty. The relation is nonlinear, and, in addition, the integration must be performed over the entire loading history of the medium. However, in the investigation of problems of wave propagation, the differential relation may be used without integration, provided the stress increments accompanying the wave disturbance are small in comparison with the initial stress on the medium. Experiments were performed in which waves of small amplitude were propagated in granular bars. It was found that the wave velocities agreed closely with predicted values based on the differential stress-strain relation.

Contact theory, in addition to giving the relation between load and displacement, predicts that the frictional energy loss accompanying the wave disturbance will vary as the cube of the amplitude of the displacement when the tangential forces are small (8). In this respect the experiments of granular bars did not corroborate the theory, giving a rate of energy dissipation depending on the square of the amplitude. This result is in agreement with previous experiments (9, 10) on single contacts.

Pairs of Spheres

If two identical elastic spheres are in contact under the action of a force N, the Hertz theory gives, for the contact radius a and the displacement α

$$a = [3(1 - \nu)NR/8\mu]^{1/3} \qquad [1]$$

$$\alpha = 2[3(1 - \nu)N/8\mu R^{1/2}]^{2/3} \qquad [2]$$

where R is the radius of a sphere, and ν and μ are, respectively, Poisson's ratio and the shear modulus of the material of the spheres. From these the normal compliance

$$C = d\alpha/dN = (1 - \nu)/2\mu a \qquad [3]$$

may be obtained.

The theory was extended by Cattaneo (6) and Mindlin (7) to include a component of force T, tangential to the contact surface. According to this theory, when a monotonically increasing tangential force is added to a contact under a constant normal force, slip occurs at the contact surface, starting at the circumference of the area of contact and progressing radially inward so that the portion on which slip occurs is an annulus of outer radius a and inner radius

$$c = a(1 - T/fN)^{1/3} \qquad [4$$

---

[1] This investigation was supported by the Office of Naval Research under Contract Nonr-266(09), with Columbia University.

[2] Assistant Professor, Division of Engineering, Brown University, Providence, R. I.; formerly, Research Assistant, Department of Civil Engineering and Engineering Mechanics, Columbia University, New York, N. Y.

[3] Professor of Civil Engineering, Columbia University, New York, N. Y. Mem. ASME.

[4] Numbers in parentheses refer to the Bibliography at the end of the paper.

For presentation at the Applied Mechanics Division Summer Conference, Berkeley, Calif., June 13–15, 1957, of The American Society of Mechanical Engineers.

Discussion of this paper should be addressed to the Secretary, ASME, 29 West 39th Street, New York, N. Y., and will be accepted until one month after final publication of the paper itself in the Journal of Applied Mechanics.

 Manuscript received by ASME Applied Mechanics Division, December 3, 1956. Paper No. 57—APM-39.

where f is the coefficient of friction. The relative tangential displacement of the centers of the two spheres is given by

$$\delta = 3fN(2-\nu)[1-(1-T/fN)^{2/3}]/8\mu a \qquad [5]$$

yielding

$$S = d\delta/dT = (2-\nu)/4\mu a(1-T/fN)^{1/3} \qquad [6]$$

for the tangential compliance. The Expression [5] for the tangential displacement has been confirmed experimentally by Johnson (9).

Mindlin, Mason, Osmer, and Deresiewicz (10) studied the effects produced by reversals of direction of the tangential force. This theory shows that, as a result of slip, the relation between force and displacement changes upon reversals of the direction of the tangential force. If the tangential force oscillates between $\pm T^*(<fN)$, a stable cycle is obtained after the first quarter cycle. The existence of an annulus of slip and Equation [4] for its inner radius were verified experimentally in this investigation. Experiments by Johnson (9) confirmed the relation between load and displacement as well as the existence of an annulus of slip.

Mindlin and Deresiewicz (8) further extended the theory to include the addition of a varying force of constant obliquity across a contact surface under an initial normal force, N_0. The theory provided the load-displacement relations for first loading, first unloading, and for the subsequent stabilized cycle. In the stabilized cycle, if the tangential component of force oscillates between $\pm T^*$, the tangential compliance during loading is given by

$$S_R = \frac{2-\nu}{4\mu a}\left\{\theta + (1-\theta)\left[1-(1+\theta)\frac{L^*+L}{2(1+\theta L)}\right]^{-1/3}\right\} \qquad [7]$$

where

$$L = T/fN_0, \quad L^* = T^*/fN_0$$
$$\theta = f/\beta, \quad \beta = dT/dN \geqslant f$$

For the unloading part of the stabilized cycle the compliance, S_U, is obtained by reversing the signs of θ and L.

For small L^*, the frictional energy loss per cycle, obtained from the area of the load displacement diagram, is given by

$$F = (2-\nu)(T^*)^3(1-\theta^2)/18\mu a_0 f N_0 \qquad [8]$$

for the stabilized load cycle, where a_0 is the radius of the contact circle corresponding to the normal force N_0.

Face-Centered Cubic Array of Spheres

As a result of the inelastic behavior at the contacts, the stress-strain relation for a granular medium depends upon load history. Until load history is specified it is appropriate to express the stress-strain relation in terms of differential increments of stress and strain. It is to be noted that the incremental stress-strain relation must involve the compliances which, themselves, depend on the history of the loading.

Consider a medium composed of identical spheres in a face-centered cubic array. Take as an element of the medium the cube shown in Fig. 1. This cube is chosen to contain a sufficient portion of the medium to define the arrangement. Increments of force, dP_{ij}, act on the faces of the cube, Fig. 2, and are assumed to be distributed among the spheres in proportion to the areas of their sections exposed on the faces of the cube. Incremental stress components are defined by $d\sigma_{ij} = dP_{ij}/8R^2$, where $8R^2$ is the gross area of a face. Each sphere in the medium is in contact with twelve other spheres. By the definition of a face-centered cubic array the disposition of these contacts on the surfaces of the spheres is the same for each sphere in the medium; i.e., the spheres are indistinguishable one from another as far as disposition of contacts with other spheres is concerned. At the contacts between spheres the normal and tangential components of incremental force are designated by dN and dT, respectively. Subscripts i, j, k $(i \neq j \neq k)$ are used to identify contacts. Each one of i, j, or k may be any one of x, y, or z. A contact with its normal in the i, j-plane has force increments dN_{ij}, dT_{ij}, and dT_{kk}, in which no significance is attributed to the order in which subscripts are written. It may be seen, Fig. 3, that the contact normals always lie in the co-ordinate planes. Hence, dN_{ij} lies in the i, j-plane, while the tangential components dT_{ij} and dT_{kk}, are chosen to lie in and normal to the i, j-plane, respectively. The contact normal, however, may have either of two possible orientations in a plane and a prime is used to designate forces at a contact having a normal whose components are of unlike sign.

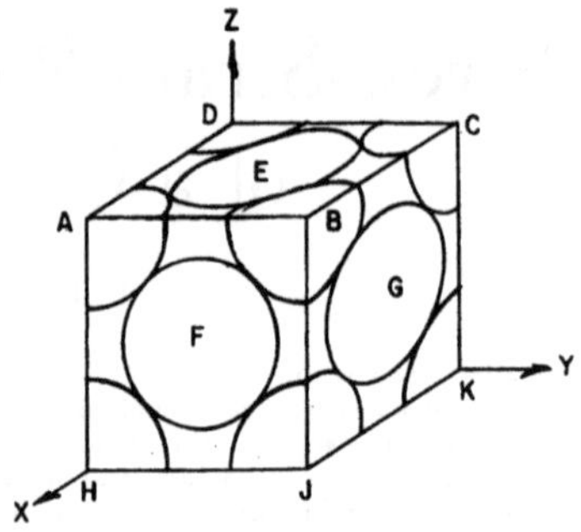

Fig. 1 Element of Volume of a Face-Centered Cubic Array of Spheres

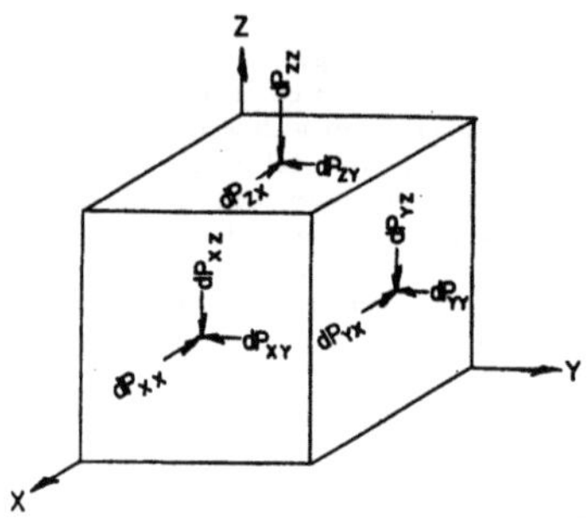

Fig. 2 Incremental Forces Acting on Faces of an Element of Volume

The first step in the derivation of the stress-strain relation for the medium is to find expressions for the increments in the forces at the contacts between the spheres in the cube (i.e., dN_{ij}, dT_{ij}, and dT_{kk}) in terms of the increments in force dP_{ij}, applied to the faces of the cube, Fig. 2. Once this has been done, then, since the contact forces dN_{ij}, dT_{ij}, and dT_{kk} and the relative displacements of the centers of the spheres are related through the compliances, the relative displacements can be obtained in terms of the unknown compliances and the increments of applied force dP_{ij}. One set of relations between the increments of contact forces, dN_{ij}, dT_{ij}, and dT_{kk} and the increments of applied force dP_{ij} is found by requiring that equilibrium be maintained when arbitrary increments in applied force dP_{ij} are added to the faces of the cube. However, equilibrium conditions alone do not pro-

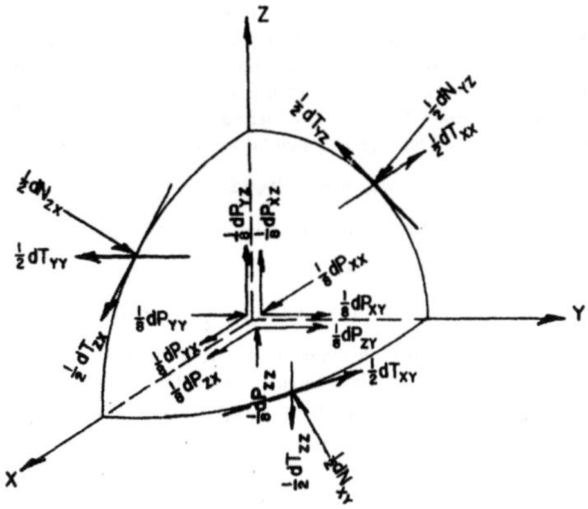

FIG. 3 FORCES ACTING ON AN OCTANT OF A SPHERE
(Octant shown here is located at origin in Fig. 1.)

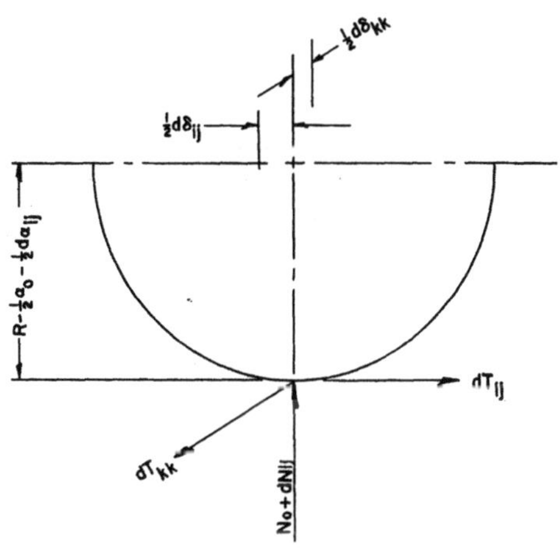

FIG. 4 COMPONENTS OF FORCE AT A CONTACT AND CORRESPONDING COMPONENTS OF DISPLACEMENT OF THE CENTER OF A SPHERE

vide sufficient equations to be able to solve for the increments in contact forces. The additional necessary conditions are provided by requiring that the relative displacements of the centers of spheres be compatible. Together, the equilibrium and the compatibility conditions are sufficient to determine the increments in the contact forces uniquely in terms of the compliances and the relative displacements of centers of spheres.

Consider the representative cube, Fig. 1, in equilibrium under an arbitrary initial stress imposed on the medium. If an arbitrary increment in stress is added to the initial stress, then the forces on the faces of the cube will be increased by corresponding increments dP_{ij}, Fig. 2. It is now desired to find the unknown increments in force at the contacts between the spheres within the cube which result from the addition of the increments in force dP_{ij} on the faces of the cube. Suppose that the stress increment in the medium is taken temporarily as homogeneous. Then, in going from one sphere to another, the contact forces will be equal at contacts having corresponding positions on the surfaces of the spheres. Since each sphere has 12 contacts, the number of unknown contact force-components will be 36. However, for each sphere the contact forces at diametrically opposed points are equal, so that only 18 of the contact forces are independent. The problem is to find these 18 increments in contact forces, dN_{ij}, dT_{ij}, and dT_{kk} for arbitrary increments in the forces dP_{ij}, applied on the faces of the cube. Not only must the cube as a whole remain in equilibrium under the action of the increments in applied forces dP_{ij}, but each sphere and each portion of a sphere, Fig. 3, within the cube also must remain in equilibrium. Since the portions of spheres in the cube are acted upon both by applied forces and by contact forces, the equations of equilibrium provide expressions relating the increments in applied forces to the increments in contact forces. By writing equilibrium equations for various portions of spheres in the cube it may be shown that there are 9 independent equations expressing equilibrium of increments in force. These are given by any three of the following equations

$$\left.\begin{aligned}
4dT_{xx} + 2\sqrt{(2)}(dN_{zx} + dN_{xy} - dT_{zx} + dT_{xy}) &\\
= dP_{xx} + dP_{xy} + dP_{zx} &\\
4dT_{xx}' + 2\sqrt{(2)}(dN_{zx}' + dN_{xy} - dT_{zx}' + dT_{xy}) &\\
= dP_{xx} + dP_{xy} - dP_{zx} &\\
4dT_{xx} - 2\sqrt{(2)}(dN_{zx}' + dN_{xy}' - dT_{zx}' + dT_{xy}') &\\
= -dP_{xx} + dP_{xy} + dP_{zx} &\\
4dT_{xx}' - 2\sqrt{(2)}(dN_{zx} + dN_{xy}' - dT_{zx} + dT_{xy}') &\\
= -dP_{xx} + dP_{xy} - dP_{zx} &
\end{aligned}\right\} \ldots [9]$$

and six others obtained by cyclical permutation of subscripts. Moment equilibrium equations are given by three of the type

$$(dT_{yz}' - dT_{yz})R - (dT_{zz}' - dT_{zz} + dT_{yy}' + dT_{yy})\sqrt{(2)}R/2 = 0 \ldots [10]$$

These equations, however, are not independent of Equations [9], so that equilibrium considerations yield only 9 equations to be solved for 18 unknown components of force. The problem is thus statically indeterminate; additional equations must be obtained from conditions of compatibility of relative displacements of centers of spheres.

Components of relative displacements of the centers of spheres are designated by $d\alpha_{ij}$, $d\delta_{ij}$, and $d\delta_{kk}$, to correspond, respectively, to the forces dN_{ij}, dT_{ij}, and dT_{kk}, Fig. 4. Consider the centers of spheres F, B, and G in the element of volume, Fig. 1, noting that spheres F and G are in contact at a point within the cube. In order that the displacements of the centers of these spheres be single valued, the vector distance around the closed path through the centers of the spheres must vanish both before and after the medium is strained. Hence the sum of the relative displacements of the centers of these spheres around the closed path must vanish. An expression of this condition for all possible paths connecting the centers of spheres in the medium is given by any three of the following equations

$$\left.\begin{aligned}
\sqrt{2}(d\delta_{zz}) &= -d\alpha_{yz}' + d\alpha_{zx} + d\delta_{yz}' + d\delta_{zx}\\
\sqrt{2}(d\delta_{zz}) &= +d\alpha_{yz} - d\alpha_{zx}' - d\delta_{yz} - d\delta_{zx}'\\
\sqrt{2}(d\delta_{zz}') &= -d\alpha_{yz} + d\alpha_{zx} + d\delta_{yz} + d\delta_{zx}\\
\sqrt{2}(d\delta_{zz}') &= +d\alpha_{yz}' - d\alpha_{zx}' - d\delta_{yz}' - d\delta_{zx}'
\end{aligned}\right\} \ldots [11]$$

and six others obtained by cyclical permutation of subscripts. Equations [11] may be expressed in terms of force increments by using

$$\left.\begin{aligned}
d\alpha_{ij} &= C_{ij}dN_{ij}, & d\alpha_{ij}' &= C_{ij}'dN_{ij}'\\
d\delta_{ij} &= S_{ij}dT_{ij}, & d\delta_{ij}' &= S_{ij}'dT_{ij}'\\
d\delta_{kk} &= S_{kk}dT_{kk}, & d\delta_{kk}' &= S_{kk}'dT_{kk}'
\end{aligned}\right\} \ldots [12]$$

where subscripts identify the contact (and no summation convention is employed). After substitution of Equations [12] into Equations [11], the latter along with Equations [9] furnish 18 independent equations containing the components of the applied forces dP_{ij}, and the 18 independent components of contact forces dN_{ij}, dT_{ij}, and dT_{kk}, which it is desired to determine. However, these 18 equations are not linear in the components of contact forces since the equations contain the compliances; and the compliances, as may be seen by Equations [3], [6], or [7], are themselves dependent upon the contact forces, either directly or through the radius of the contact circle a. Indeed, even the expressions which are to be used for compliances depend upon load history. Hence, Equations [9], [11], and [12] can be solved for the components of contact forces, dN_{ij}, dT_{ij}, and dT_{kk}, only as functions of both applied forces dP_{ij}, and compliances.

Initial Isotropic Stress

A solution was first obtained for the case of small increments in stress applied to a medium under an initial isotropic stress σ_0. If the loading history up to this initial stress is homothetic then it may be shown from considerations of symmetry that the initial contact forces are all equal and have no tangential components. Using Equations [9] and the definition of stress, we find the initial contact forces to be $N_0 = \sqrt{(2)}(R^2\sigma_0)$. In addition, since the loading history at each contact is the same, the initial compliances do not vary from contact to contact. If, to the medium under the initial stress σ_0, additional arbitrary increments in stress are applied such that the total stress never departs from the initial stress by more than a small increment, approximate values for the compliances may be found by letting $L^* \to 0$ in the expressions for compliances. Thus, using Equations [7], one obtains for the tangential compliance

$$S = S_R = S_U = (2 - \nu)/4\mu a_0 \qquad [13]$$

and the normal compliance is given by Equation [3] if a is replaced by a_0. For a given initial stress these compliances may be considered as constants; and Equations [9], [11], and [12] furnish a set of simultaneous linear equations in the increments of contact forces. The solution of these equations yields the following expressions

$$\left.\begin{aligned} dN_{xy}, dN_{xy}' &= \sqrt{(2)}[(k+1)(dP_{xx} + dP_{yy}) \\ &\quad - (k-1)dP_{zz} \pm 2k(k+3)dP_{xy}/(k+1)]/8(k+3) \\ dT_{xy} &= dT_{xy}' = \sqrt{(2)}(dP_{xx} - dP_{yy})/4(k+3) \\ dT_{zz}, dT_{zz}' &= (dP_{xz} \pm dP_{yz})/4(k+1) \end{aligned}\right\} \qquad [14]$$

where

$$k = S/C = (2-\nu)/2(1-\nu)$$

and 12 others obtained by cyclical permutation of subscripts. In these expressions the upper signs refer to unprimed contact forces and lower signs to primed contact forces.

Having determined the contact forces it is now possible to obtain expressions for relative incremental displacements of the centers of spheres in terms of applied incremental forces. For example, the x-component of the relative incremental displacement of the centers of spheres A and D, Fig. 1, is given by (see Fig. 4)

$$d\Delta_{xx} = (d\alpha_{xy} + d\delta_{xy})/\sqrt{2} + (d\alpha_{xy}' + d\delta_{xy}')/\sqrt{2}$$

or

$$d\Delta_{xx} = C[(3k+1)dP_{xx} + (k-1)(dP_{yy} + dP_{zz})]/4(k+3) \qquad [15]$$

by using Equations [12] and [14]. The same expression for $d\Delta_{xx}$ may be derived by using any other pair of neighboring spheres whose centers lie on a line parallel to the x-axis. The corresponding relations for $d\Delta_{yy}$ and $d\Delta_{zz}$ are obtained by the same method, and are expressed by a cyclical permutation of subscripts in Equation [15]. Defining incremental extensional strain as

$$d\epsilon_{ii} = d\Delta_{ii}/2\sqrt{(2)}R$$

where $2\sqrt{(2)}R$ is the length of an edge of the cube, one obtains

$$d\epsilon_{xx} = \sqrt{(2)}RC[(3k+1)d\sigma_{xx} + (k-1)(d\sigma_{yy} + d\sigma_{zz})]/2(k+3) \qquad [16]$$

for the x-component of incremental extensional strain.

Incremental shear-strain components are defined as the changes in the angles between adjacent faces of the cube. For example, the change in the angle CDA, Fig. 1, is obtained by adding to the x-component of the displacement of C relative to D the y-component of the displacement of A relative to D, and dividing by the length of a side, $2\sqrt{(2)}R$. This sum is

$$d\gamma_{xy} = (d\alpha_{xy} + d\delta_{xy} - d\alpha_{xy}' - d\delta_{xy}')/4R + (d\alpha_{xy} - d\delta_{xy} - d\alpha_{xy}' + d\delta_{xy}')/4R$$

which becomes

$$d\gamma_{xy} = [2\sqrt{(2)}RS/(k+1)]d\sigma_{xy} \qquad [17]$$

by using Equations [12], [14], and the definition of stress. The same expression for this component of incremental shear strain may be obtained by basing the derivation on the change in the angle formed by any other two edges of the cube which lie in the x, y-plane.

From differential strain-stress relations of the type of Equations [16] and [17] one obtains the differential stress-strain relations

$$\left.\begin{aligned} d\sigma_{xx} &= c_{11}d\epsilon_{xx} + c_{12}(d\epsilon_{yy} + d\epsilon_{zz}) \\ d\sigma_{yy} &= c_{11}d\epsilon_{yy} + c_{12}(d\epsilon_{zz} + d\epsilon_{xx}) \\ d\sigma_{zz} &= c_{11}d\epsilon_{zz} + c_{12}(d\epsilon_{xx} + d\epsilon_{yy}) \\ d\sigma_{yz} &= 2c_{44}d\epsilon_{yz} \\ d\sigma_{zx} &= 2c_{44}d\epsilon_{zx} \\ d\sigma_{xy} &= 2c_{44}d\epsilon_{xy} \end{aligned}\right\} \qquad [18]$$

where

$$c_{11} = 2c_{44} = \frac{1}{\sqrt{(2)}R}\left(\frac{1}{C} + \frac{1}{S}\right) = \frac{4-3\nu}{2-\nu}\left[\frac{3\mu^2\sigma_0}{2(1-\nu)^2}\right]^{1/3}$$

$$c_{12} = \frac{1}{2\sqrt{(2)}R}\left(\frac{1}{C} - \frac{1}{S}\right) = \frac{\nu}{2(2-\nu)}\left[\frac{3\mu^2\sigma_0}{2(1-\nu)^2}\right]^{1/3}$$

Initial Isotropic and Uniaxial Stress

When an isotropic stress is applied homothetically to a medium composed of identical spheres in a face-centered cubic array the forces do not vary from one contact to another. Hence, in deriving the stress-strain relation given by Equations [18], it was not necessary to distinguish between different contacts in the medium in the expressions for the compliances. As a result, integration of Equations [18] over finite increments in stress is limited to increments in which the stress remains always isotropic, because for any other stress increment the contact forces and hence the compliances will vary from contact to contact. To obtain a stress-strain relation which may be integrated over other than isotropic increments, the derivation must be performed anew, distinguishing between compliances at contacts with different loading histories.

As an example of such a derivation, consider the case in which the stress, in terms of the co-ordinate axes in Fig. 1, is given by

$$\left.\begin{aligned} \sigma_{xx} &= \sigma_{yy} = \sigma_0, \qquad \sigma_{zz} = \sigma_0 + \sigma_a \\ \sigma_{yz} &= \sigma_{zx} = \sigma_{xy} = 0 \end{aligned}\right\} \quad \ldots\ldots [19]$$

and for which the entire load history is one which always can be described by variations of σ_0 and σ_a. Then during the entire load history, the three planes through the center of the elementary cube, Fig. 1, which are parallel to the co-ordinate planes are planes of symmetry. Also the vertical planes through the diagonals of the horizontal faces are planes of symmetry. For each pair of contacts lying on opposite sides of planes of symmetry the contact forces must be mirror images of each other. Hence

$$\left.\begin{aligned} &N_{xy} = N_{xy}' \qquad N_{yz} = N_{yz}' = N_{zx} = N_{zx}' \\ &T_{xy} = T_{xy}' = 0, \qquad T_{yz} = T_{yz}' = -T_{zx} = -T_{zx}' \\ &T_{xx} = T_{xx}' = T_{yy} = T_{yy}' = T_{zz} = T_{zz}' = 0 \end{aligned}\right\} \quad \ldots [20]$$

during application of this load. Since contacts with identical load histories have equal compliances, one can define

$$\left.\begin{aligned} C_1 &= C_{xy} = C_{xy}', \qquad C_2 = C_{yz} = C_{yz}' = C_{zx} = C_{zx}' \\ S_1 &= S_{xy} = S_{xy}', \qquad S_2 = S_{yz} = S_{yz}' = S_{zx} = S_{zx}' \\ \bar{S}_1 &= S_{zz} = S_{zz}', \qquad \bar{S}_2 = S_{yy} = S_{yy}' = S_{xx} = S_{xx}' \end{aligned}\right\} \quad \ldots [21]$$

which may be inserted in Equations [12]. These compliances are still unknown. However, Equations [9], [11], and [12] may be solved simultaneously to provide expressions for the increments of contact forces in terms of the compliances and the increments of applied load. Such a solution gives

$$\left.\begin{aligned} dN_{xy}, dN_{xy}' &= B_1(C_2 + S_2)(dP_{xx} + dP_{yy}) \\ &\qquad -B_1(S_2 - C_2)dP_{zz} \pm \sqrt{(2)}B_4\bar{S}_2 dP_{xy} \\ dT_{xy} &= dT_{xy}' = 2B_3C_2S_2(dP_{xx} - dP_{yy}) \\ dT_{zz}, dT_{zz}' &= \sqrt{(2)}B_2C_2(S_2 + \bar{S}_2)(dP_{zx} \pm dP_{yz}) \\ dN_{yz}, dN_{yz}' &= B_1C_1(dP_{xx} + dP_{yy}) \\ &- B_3S_1S_2(dP_{xx} - dP_{yy}) + B_1(C_1 + S_2)dP_{zz} \pm B_5 dP_{yz} \\ dT_{yz}, dT_{yz}' &= B_1C_1(dP_{xx} + dP_{yy}) \\ &\qquad - B_3C_2S_1(dP_{xx} - dP_{yy}) - B_1(C_1 + C_2)dP_{zz} \\ &\qquad \pm B_2(\bar{S}_2 - \bar{S}_1)C_2 dP_{yz} \\ dT_{xx}, dT_{xx}' &= B_4C_1dP_{xy} \pm \sqrt{(2)}B_2C_2(S_2 + \bar{S}_1)dP_{zx} \\ dN_{zx}, dN_{zx}' &= B_1C_1(dP_{xx} + dP_{yy}) \\ &+ B_3S_1S_2(dP_{xx} - dP_{yy}) + B_1(C_1 + S_2)dP_{zz} \pm B_5 dP_{zx} \\ dT_{zx}, dT_{zx}' &= -B_1C_1(dP_{xx} + dP_{yy}) \\ &\qquad + B_3C_2S_1(dP_{yy} - dP_{xx}) + B_1(C_1 + C_2)dP_{zz} \\ &\qquad \pm B_2(\bar{S}_1 - \bar{S}_2)C_2 dP_{zx} \\ dT_{yy}, dT_{yy}' &= \pm B_4C_1dP_{xy} + \sqrt{(2)}B_2C_2(S_2 + \bar{S}_1)dP_{yz} \end{aligned}\right\} \quad \ldots [22]$$

where the upper signs refer to the unprimed contact forces and lower signs to the primed contact forces, and where

$$B_1 = 1/4\sqrt{(2)}(2C_1 + C_2 + S_2)$$
$$B_2 = 1/2\sqrt{(2)}(2C_2S_2 + 2\bar{S}_2\bar{S}_1 + S_2\bar{S}_1 + S_2\bar{S}_2 + C_2\bar{S}_1 + C_2\bar{S}_2)$$
$$B_3 = 1/4\sqrt{(2)}(2C_2S_2 + S_1C_2 + S_1S_2)$$
$$B_4 = 1/4(C_1 + \bar{S}_2)$$
$$B_5 = B_2(2\bar{S}_2\bar{S}_1 + S_2\bar{S}_2 + S_2\bar{S}_1)$$

Proceeding with the solution of the problem by the same method as that used for the initial isotropic stress, one obtains for the differential stress-strain relation

$$\left.\begin{aligned} d\sigma_{xx} &= c_{11}d\epsilon_{xx} + c_{12}d\epsilon_{yy} + c_{13}d\epsilon_{zz} \\ d\sigma_{yy} &= c_{12}d\epsilon_{xx} + c_{11}d\epsilon_{yy} + c_{13}d\epsilon_{zz} \\ d\sigma_{zz} &= c_{13}d\epsilon_{xx} + c_{13}d\epsilon_{yy} + c_{33}d\epsilon_{zz} \\ d\sigma_{yz} &= 2c_{44}d\epsilon_{yz} \\ d\sigma_{zx} &= 2c_{44}d\epsilon_{zx} \\ d\sigma_{xy} &= 2c_{66}d\epsilon_{xy} \end{aligned}\right\} \quad \ldots\ldots\ldots [23]$$

where

$$c_{11} = (C_1^{-1} + S_1^{-1} + C_2^{-1} + S_2^{-1})/2\sqrt{(2)}R$$
$$c_{12} = (C_1^{-1} - S_1^{-1})/2\sqrt{(2)}R, \qquad c_{33} = (C_2^{-1} + S_2^{-1})/\sqrt{(2)}R$$
$$c_{13} = (C_2^{-1} - S_2^{-1})/2\sqrt{(2)}R, \qquad c_{66} = (C_1^{-1} + \bar{S}_2^{-1})/2\sqrt{(2)}R$$
$$c_{44} = \frac{1}{2\sqrt{(2)}R}\left[\frac{1}{C_2} + \frac{2S_2 + \bar{S}_1 + \bar{S}_2}{2\bar{S}_1\bar{S}_2 + S_2(\bar{S}_1 + \bar{S}_2)}\right]$$

As pointed out before, the compliances in these equations depend upon stress history, and are determined for each particular case in accordance with the results given in reference (8).

To obtain the strain resulting from finite increments in stress it is necessary to integrate Equations [23] over the stress history. However, integration of these equations is limited to increments in stress describable in terms of variations of σ_0 and σ_a. To determine the compliances in Equations [23] the contact forces must be obtained by an integration of Equation [22]. Using the definition of stress, and inserting the values of stress given by Equations [19] into Equations [22], one obtains for the quantities to be integrated for the contact forces

$$\left.\begin{aligned} dN_{xy} &= B_6R^2[(3C_1 + S_2)d\sigma_0 + (C_2 - S_2)d\sigma_a] \\ dN_{yz} &= B_6R^2[(3C_1 + S_2)d\sigma_0 + (C_1 + S_2)d\sigma_a] \\ dT_{yz} &= -B_6R^2[(C_2 - C_1)d\sigma_a + (C_2 + C_1)d\sigma_a] \end{aligned}\right\} \quad \ldots [24]$$

where

$$B_6 = \sqrt{(2)}/(2C_1 + C_2 + S_2)$$

and, as a result of Equations [20]

$$dN_{xy} = dN_{xy}'$$
$$dN_{yz} = dN_{yz}' = dN_{zx} = dN_{zx}'$$
$$dT_{yz} = dT_{yz}' = -dT_{zx} = -dT_{zx}'$$

and the remaining tangential forces are zero. At each step in the loading history, the expressions for the compliances are the same in Equations [23] and [24]. If, for any particular case, Equations [24] can be integrated, the compliances can be evaluated and, hence, the coefficients in the differential stress-strain relation will be known functions of stress.

For infinitesimal increments in stress Equations [23] may be used without integration to give the relation between stress and strain. The medium must be under an initial stress whose entire history is describable in terms of σ_0 and σ_a, and the additional arbitrary increments in stress must be such that the total stress never departs from the initial stress by more than a small increment. The compliances are made to depend solely on initial stress by finding the limit as the increments in the contact forces approach zero. The values of the initial contact forces to be used to determine the compliances are obtained by an integration of Equations [24] over the past stress history up to the initial stress. Once this integration has been performed Equations [23] may be used directly for the relation between stress and strain.

Application of Differential Stress-Strain Relations to Wave Propagation in a Granular Bar

The differential stress-strain Relations [18] are applicable, without integration, to a study of the problem of wave propagation as long as the stress increments accompanying the wave disturbance are small in comparison with the initial stress on the medium. Experiments were performed in which compressional waves were propagated in granular bars. These bars were constructed of steel bearing balls arranged in a face-centered cubic array and were subjected to an initial isotropic stress σ_0. As may be seen from Equations [18] the medium is anisotropic, so that the velocity of compressional waves in a bar of the granular medium depends upon the orientation of the bar with respect to the symmetry axes of the medium. The axes of the granular bars constructed were oriented in either of two directions: bar A in the [100] direction, and bar B in the [110] direction. (These orientations were chosen to avoid a coupling with flexural waves.) If the linear dimensions of the cross section of the bar are sufficiently small in comparison with the wave length, the elementary theory of longitudinal vibrations of slender bars may be employed to find the expression for the wave velocity in terms of the material constants. In the present case, we find.

$$\left.\begin{aligned} \rho_M v_A^2 &= c_{11} - \frac{2c_{12}^2}{c_{11}+c_{12}} = \frac{2(8-7\nu)}{(8-5\nu)}\left[\frac{3\mu^2\sigma_0}{2(1-\nu)^2}\right]^{1/3} \\ \rho_M v_B^2 &= \frac{4c_{44}(c_{11}-c_{12})(c_{11}+2c_{12})}{2c_{11}c_{44}+(c_{11}-c_{12})(c_{11}+2c_{12})} \\ &= \frac{2(4-3\nu)(8-7\nu)}{(4-3\nu)^2+(8-7\nu)(2-\nu)}\left[\frac{3\mu^2\sigma_0}{2(1-\nu)^2}\right]^{1/3} \end{aligned}\right\} \quad \ldots [25]$$

where ρ_M is the mass density of the granular medium, i.e., the density of the material of the spheres multiplied by $\pi/3\sqrt{2}$, the packing of the spheres. For steel balls the velocities become

$$v_A = 1330\sigma_0^{1/6}\text{ fps}, \qquad v_B = 1350\sigma_0^{1/6}\text{ fps} \ldots\ldots [26]$$

where σ_0 is in pounds per square inch.

It is interesting to compare the velocities in Equations [25] and [26], which are based on the theories of Cattaneo (6), Mindlin (7), and Mindlin and Deresiewicz (8), with the expressions obtained when the effect of the tangential forces acting at the contacts is neglected. If only normal forces act at the contacts the elastic contants of the medium are

$$c_{11} = 2c_{12} = 2c_{44} = 1/\sqrt{2}RC$$

so that

$$\rho_M v_A^2 = \frac{2}{3}\left[\frac{3\mu^2\sigma_0}{2(1-\nu)^2}\right]^{1/3}, \qquad \rho_M v_B^2 = \left[\frac{3\mu^2\sigma_0}{2(1-\nu)^2}\right]^{1/3}$$

give the velocities of longitudinal waves for slender granular bars. These velocities are

$$v_A = 800\sigma_0^{1/6}\text{ fps}, \; v_B = 980\sigma_0^{1/6}\text{ fps}$$

for bars made of steel balls.

Description of Experiment

The granular bars were constructed of steel bearing balls arranged in a face-centered cubic array (Table 1). The balls were held in place by an externally applied hydrostatic pressure, Fig. 5. Velocities of longitudinal waves were determined by resonance tests, and the rate of energy dissipation was found from observations of the decay of free vibrations.

Table 1 Description of Granular Bars

Miller indexes for longitudinal axis of bar	Length, in.	Length to width ratio	Number of spheres in bar	Number of contacts in bar
[1, 0, 0]	4.243	7.99	1201	5796
[1, 1, 0]	5.1266	8.12	1869	9228

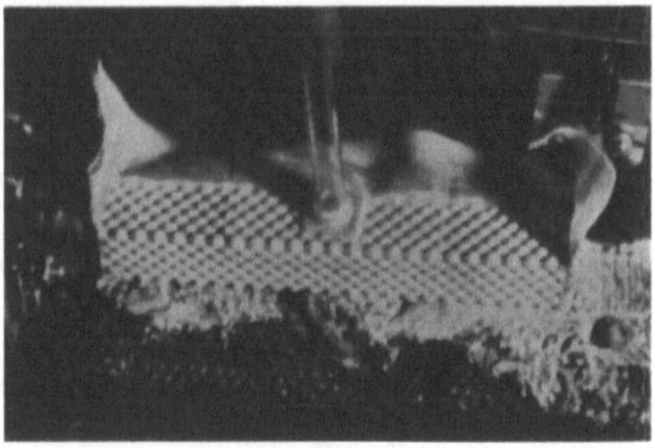

Fig. 5 Experimental Bar Under a Pressure of 1 Atm

The stainless-steel balls (AISI No. 440C) were of $^1/_8$ in. diam and of either of two dimensional tolerances: $\pm 50 \times 10^{-6}$ in. or $\pm 10 \times 10^{-6}$ in. The balls were cleaned in successive baths of carbon tetrachloride, toluol, and acetone, and carefully packed in a face-centered cubic array. During packing the arrangement of the balls was preserved by holding them in a box. The inside of the box contained indentations to prevent rolling of the balls and the box was lined with a thin rubber sheet (0.007 in. thick). After packing was completed, the rubber sheet was wrapped about the array of balls, closed, and sealed. The sealed bag was evacuated, locking the balls in place and permitting the bar to be removed from the box. Evacuation was performed through a glass tube piercing the rubber sheet at the mid-point of the bar, i.e., at a node of the odd harmonics.

During the tests the bar was supported along its entire length by a polished lucite surface lubricated by a thin coat of powdered talc. This type of support avoided bending of the bar, which would have changed the values of the compliances at the contacts between the steel balls.

Wave velocities were found from measurements of resonant frequencies of the bars. The drive was provided by an induction coil and a permanent magnet placed at one end of the bar. At the other end a crystal phonograph pick-up connected to a vacuum tube audio-voltmeter measured the response of the bar. Maximum-voltage output determined the resonant frequency, and the shape of the output was observed on an oscilloscope.

In order to measure the amplitude of displacement of the ends of the bar a short piece of a sewing-needle was inserted in a sphere at each end of the bar and made to protrude through the rubber sheet. The tip of the phonograph needle was placed against the projection. Calibration was performed at various frequencies over a range of amplitudes by comparing output voltage with known amplitudes as measured by means of a microscope.

Energy dissipation was determined by photographic observation of the decay of free vibrations. The exciting voltage was cut off by a synchronizing switch which also initiated the single sweep of the oscillograph. Observation of the steady motion immediately before cut-off was provided by using a variable time delay in the switch. Amplitude calibration was performed by measuring the voltage output of the phonograph pickup prior to cutoff. The frequency of the vibrations provided the time scale.

Results of Resonance Tests

Figs. 6 and 7 show experimental frequencies together with theoretical values. Results are in close agreement with predicted values, and closer to values based on the theory which includes both normal and tangential contact forces than to values based on the theories which neglect tangential contact forces (1, 2, 3, 4). In addition, as the dimensional tolerance of the balls is improved, the measured velocities tend to approach the theory which includes both the normal and tangential contact forces.

In the case of higher tolerance balls the dependence of wave velocity upon pressure was very nearly as predicted. Variation with the $^1/_6$th power of pressure was followed down to about 10 psi, changing slowly for lower pressures. For the [110] orientation velocities decreased more rapidly with pressure presumably because, in this orientation, the balls on the edge of the bar did not all lie within the same vertical plane, so that the packing was loosened by the weight of the balls themselves.

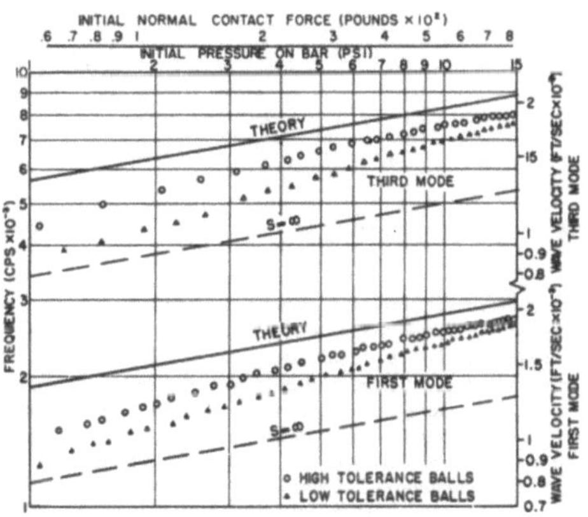

Fig. 6 Resonant Frequency of Bar as a Function of Pressure for [1, 0, 0] Orientation

In a close-packed arrangement of identical spheres under an isotropic stress the normal forces at the contacts would all be equal, as would the relative approaches of the centers of the spheres in contact. However, this is not true in practice, because of dimensional variations in the balls. A comparison of tolerances with relative approach, Table 2, shows that in this experiment there probably were contacts which did not make at all, even in the case of the better tolerance and the highest pressures used. Thus

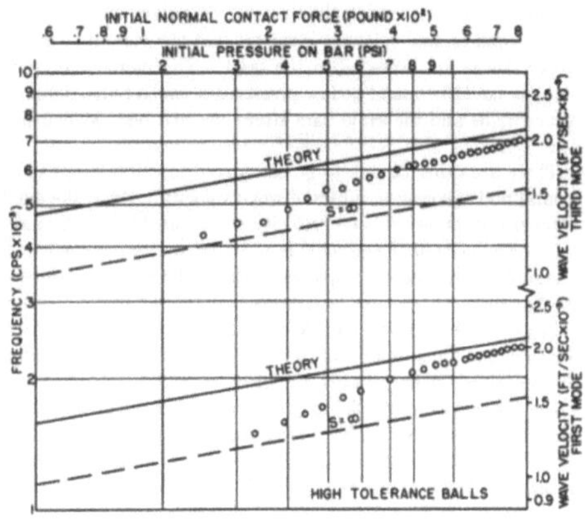

Fig. 7 Resonant Frequency of Bar as a Function of Pressure for [1, 1, 0] Orientation

the stiffness of the medium is diminished and wave velocities are reduced as compared with a bar of identical spheres. Such an effect should be more pronounced for poorer tolerances and at lower pressures. Experimental results are consistent with these conclusions.

Referring to Fig. 6, it may be seen that a change in tolerance affects the levels of the experimental curves. This may be ascribed to the fact that the difference in tolerances between the two samples of balls used is large compared to the relative approaches, so that presumably the number of contacts making is substantially larger in one bar than in the other. In addition, for the poorer tolerance balls, the change in relative approach for a pressure change of 1 atm is small as compared to the dimensional tolerance, so that the number of contacts making is probably not very different at the two pressure extremes. As indicated in Fig. 6, the change in slope with pressure for the poorer tolerance balls was much less than for the higher tolerance balls, for which, presumably, there is an appreciable change in the number of contacts making as the pressure is changed by 1 atm.

Additional improvements in the tolerance probably would produce further increases in wave velocity. However, once the dimensional tolerances become less than the relative approaches it is expected that the changes in the wave velocity would be smaller.

Since the theoretical relation between stress and strain is limited to differential increments, it was necessary in the experiment to limit stress increments to small quantities. In order to determine the effect of the magnitude of the stress increments used, measurements were made of the variation of resonant frequency with amplitude of motion. It was found that the frequency was increased by about 1 per cent when the amplitude of the end of the bar was reduced from 12×10^{-7} to 2×10^{-7} in. The frequency tests were made at the latter amplitude.

Table 2 Behavior of Spheres in Contact in a Face-Centered Cubic Medium Under an Isotropic Stress

	Hertz theory							Dimensional tolerance of balls in experiment, in. $\times 10^{-6}$	
σ_0, psi	N_0, lb	Max normal stress, psi	Max shear stress, psi	Contact radius, in. $\times 10^{-4}$	Relative approach, in. $\times 10^{-6}$	Normal compliance, $\frac{\text{in.}}{\text{lb}} \times 10^{-4}$	Tangential compliance, $\frac{\text{in.}}{\text{lb}} \times 10^{-4}$	Lower	Higher
2.00	0.01105	86300	26750	2.48	1.955	1.178	1.428	±50	±10
5.00	0.0276	117100	36300	3.36	3.60	0.869	1.053	±50	±10
10.00	0.0552	147200	45600	4.23	5.71	0.690	0.837	±50	±10
14.70	0.0812	167000	52000	4.81	7.39	0.607	0.737	±50	±10

The compliances used in the theory were derived on the basis of elastic spheres, and evidence indicates that the yield limit of the material was not exceeded in the tests. No difference in wave velocity could be found at low pressures between bars composed of new balls and the same bars after they had been loaded to 1 atm.

There are other factors to be considered in comparing experimental and theoretical values of wave velocity. First, the small difference in observed wave velocity as obtained by the first and third harmonics (0.75 per cent) is an indication that wave length was sufficiently long in comparison with the lateral dimensions of the bar to be able to use Equations [25] to give the theoretical values of wave velocity. It also should be pointed out that there were factors which were known only approximately. The effective length of the bar, that is, the distance between the planes at which the waves were reflected, is unknown, in the case of a bar composed of spheres. The length used was measured between centers of end spheres. Lastly, the values of the elastic constants of the steel balls are necessarily only approximate values.

Energy Dissipation in a Granular Bar

The logarithmic decrement δ, defined by

$$\delta = \log (A_i/A_{i+1}) \qquad [27]$$

was obtained from observation of the decay of free vibrations. In this expression A_i and A_{i+1} are successive amplitudes of the displacement of the end of the bar as determined, in this experiment, from oscillograph records of the decay.

Since the relation between stress and strain is assumed linear, the experimental values of energy loss, per cycle, of the bar may be obtained from

$$W_T = 2\delta W_E \qquad [28]$$

whenever $\delta^2 \ll 1$; where W_E is the maximum elastic energy of the bar during a cycle. Under the assumptions of the classical one-dimensional wave theory, the incremental strain energy per unit length of the bar is given by

$$\frac{1}{2} S d\sigma d\epsilon = \frac{1}{2} S \left[c_{11} - \frac{2c_{12}^2}{c_{11} + c_{12}} \right] (d\epsilon)^2$$

where S is the cross-sectional area of the bar, and where $d\sigma$ and $d\epsilon$ are, respectively, the increments in the longitudinal components of stress and strain in the bar. By taking a sinusoidal displacement distribution along the length of the bar one obtains

$$W_E = \frac{S}{l} \left(\frac{A\pi}{2} \right)^2 \left[c_{11} - \frac{2c_{12}^2}{c_{11} + c_{12}} \right] \qquad [29]$$

where A is the amplitude at the end of the bar and l is the length of the bar.

The theoretical frictional energy loss per cycle for the bar (F_T) was found by a summation of the energy dissipated at the individual contacts as given by Equation [8], and by taking a sinusoidal displacement distribution along the length of the bar. The damping tests were performed on bars oriented in the [100] direction, in which case

$$F_T = 293.9 \times 10^6 (0.816 - f^2) A^3 / f\sigma_0^{1/3} \qquad [30]$$

and, from Equations [28] and [29]

$$W_T = 4.54 \times 10^4 \sigma_0^{1/3} \delta A^2 \qquad [31]$$

for the size of bar used.

Results of Decay Tests

Values of logarithmic decrement were obtained from oscillographic recordings of the decay of free vibrations. Measurements were made of the displacement of the end of the bar as the amplitude decreased from 5×10^{-7} to 1×10^{-7} in. The results indicate that the logarithmic decrement is independent of the amplitude of displacement in this range. The energy dissipated per cycle, therefore, varies as the square of the amplitude whereas the theory of Mindlin and Deresiewicz (8) predicts variation as the cube, Fig. 8. Although the logarithmic decrement did not vary with amplitude of displacement, its magnitude depended on the initial pressure on the granular bar. For the bar composed of the better tolerance balls the values of logarithmic decrement ranged from 0.054 at a pressure of 14.7 psi to 0.129 at a pressure of 1.92 psi. For the bar composed of the poorer tolerance balls the logarithmic decrement was about 25 per cent smaller. However, the dependence of the rate of energy loss on the square of amplitude of displacement was obtained for either tolerance of balls and for the entire pressure range of the tests. The significance of these results can be better evaluated when considered in conjunction with the results of the experiments of Mindlin, Mason, Osmer, and Deresiewicz (10), and of Johnson (9) on single contacts.

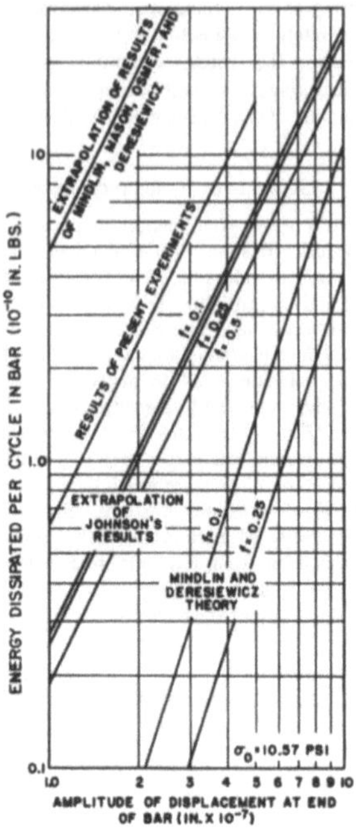

Fig. 8 Typical Variation of Energy Loss per Cycle With Amplitude

In his experiments, Johnson found, in general, close agreement between his results and predictions based on the theories of Mindlin (7) and Mindlin and Deresiewicz (8). This applies, in particular, to the tests which Johnson performed on the relation between the tangential force and the displacement for initial loading, for repeated loading, and for dynamic loading. As regards the damping tests, however, Johnson found that the energy loss was approximately of the magnitude predicted by the theory, but that the variation did not follow the predicted dependence on amplitude of displacement. In particular, he found that at very small amplitudes the damping loss followed approximately a square law and only approached the law predicted by the theory at large amplitudes. A similar dependence on amplitude of displacement had been obtained previously by Mindlin, Mason, Osmer, and Deresiewicz (10) in their experiments.

All of the measurements of energy dissipation, to date, indicate that the assumed law of Coulomb friction is a reasonably good approximation over the portion of the annulus of slip where the relative displacement of the continguous surfaces is large, but not in the neighborhood of the inner edge of the annulus where the relative displacement is small. At very small T/fN the latter region predominates and, accordingly, the theory fails to predict the correct law of energy dissipation. Since the dissipation is small, the stiffnesses are not appreciably affected by it and the theory gives good results for load versus displacement and for velocities of propagation even for small displacements of the spheres.

Fig. 8 shows the theoretical and experimental values of the rate of energy loss at the contacts. The curve marked "Extrapolation of Johnson's Results" was obtained in the following way: For small amplitudes of oscillation Johnson shows that his results tend toward a value given by the expression

$$W_D = 0.02(T^*)^2/\mu a \qquad [32]$$

for the dissipation per cycle at a contact. If this value is multiplied by a factor of

$$1 - f^2/\beta^2$$

to account for the effect of an oblique force, and if the values are summed over all contacts in the granular bar one obtains the results given in Fig. 8. A similar method was used to extrapolate the results of Mindlin, Mason, Osmer, and Deresiewicz. Although the experiments do not yield the same rate of energy loss, they all agree in that they give a dependence on the square of the amplitude of displacement whenever this amplitude is small.

With respect to dependence of energy dissipation upon pressure, the significance of a comparison of the present tests with theory is somewhat uncertain since results, in this case, varied with dimensional tolerance. The theory predicts that energy dissipation will vary with the $-1/3$ power of initial pressure, while experimental results give a dependence on approximately the $-1/6$ power.

The principal source of error in the present tests affecting measurements of energy dissipation is the degree of cleanness of the balls. Thus improper cleaning of the balls (using only acetone for instance) easily doubled the values obtained for W_T. Another factor affecting energy dissipation was the dimensional tolerance of the balls. This influences W_T mainly through the number of contacts which will or will not be made within the granular bar. It was found that the value of W_T was lower for the poorer tolerance balls indicating that in this case dissipation occurs at fewer contacts. An increase in pressure will increase the number of contacts. However, for the pressure range over which the tests were made, this increase in the number of contacts was probably small. It had been expected that the change in the dimensional tolerance would not affect the variation of W_T with amplitude of displacement. This was borne out experimentally, and thus the results showing a variation of W_T with the square of the amplitude of displacement appear reliable from this standpoint.

BIBLIOGRAPHY

1 "Theorie der akustischen Schwingungsausbreitung in gekörnten Substanzen und Experimentelle Untersuchungen an Kohlepulver," by G. Hara, *Elektrische Nachrichtentechnik*, vol. 12, 1935, pp. 191–200.

2 "Velocity of Elastic Waves in a Granular Substance," by K. Iida, *Bulletin Earthquake Research Institute*, Japan, vol. 17, 1939, pp. 783–808.

3 "On the Theory of Elastic Waves in a Granular Substance," by T. Takahashi and Y. Satô, *Bulletin Earthquake Research Institute*, Japan, vol. 27, 1949, pp. 11–16; and vol. 28, 1950, pp. 37–43.

4 "Elastic Waves Through a Packing of Spheres," by F. Gassmann, *Geophysics*, vol. 16, 1951, pp. 673–685.

5 "Theory of Elasticity," by S. Timoshenko and J. N. Goodier, McGraw-Hill Book Company, Inc., New York, N. Y., 1951, pp. 372–377.

6 "Sul contatto di due corpi elastici," by C. Cattaneo, *Accademia dei Lincei, Rendiconti*, series 6, vol. 27, 1938, pp. 342–348, 434–436, and 474–478.

7 "Compliance of Elastic Bodies in Contact," by R. D. Mindlin, JOURNAL OF APPLIED MECHANICS, Trans. ASME, vol. 71, 1949, p. A-259.

8 "Elastic Spheres in Contact Under Varying Oblique Forces," by R. D. Mindlin and H. Deresiewicz, JOURNAL OF APPLIED MECHANICS, Trans. ASME, vol. 75, 1953 pp. 327–344.

9 "Surface Interaction Between Elastically Loaded Bodies under Tangential Forces," by K. L. Johnson, Proceedings of the Royal Society, series A, vol. 230, 1955, pp. 531–548.

10 "Effects of an Oscillating Tangential Force on the Contact Surfaces of Elastic Spheres," by R. D. Mindlin, W. P. Mason, T. F. Osmer, and H. Deresiewicz, Proceedings of the First U. S. National Congress of Applied Mechanics, 1951, pp. 203–208.

Reprinted from
Journal of Applied Mechanics
24 (1957) 585-593

Reprinted from THE JOURNAL OF THE ACOUSTICAL SOCIETY OF AMERICA, Vol. 29, No. 11, 1206–1218, November, 1957

Vibrations of a Monoclinic Crystal Plate*

E. G. NEWMAN† AND R. D. MINDLIN
Department of Civil Engineering and Engineering Mechanics, Columbia University, New York, New York
(Received July 12, 1957)

There is described, in this paper, a detailed study of Ekstein's solution of the equations of elasticity for the case of the coupled vibrational modes of an infinite, monoclinic, crystal plate with traction-free faces. It is shown how the spectrum of frequencies and the character of the modes may be deduced by tracing the development of coupling from a simple solution in which the modes are uncoupled and by taking into account the asymptotic behavior at long and short wavelengths and high and low frequencies. The results of computations are given for the AT cut of quartz. The purpose of the investigation is to obtain some of the qualitative and quantitative information which is needed for the formulation of approximate equations of motion which may be solved for plates of finite dimensions.

I. INTRODUCTION

THE laws which govern the small vibrations of elastic plates are contained in the equations of the linear theory of elasticity; but general solutions of the equations cannot be expressed in terms of a finite number of elementary functions if the dimensions of the plates are finite and the bounding surfaces are free of traction. Closed general solutions are available for infinite plates with traction-free faces and these solutions reveal the existence of an infinite number of modes of vibration for each wavelength in the plane of the plate. In a finite plate with free edges, each of these modes (and each of its overtones) couples with an infinity of other modes (and all their overtones), leading to an extraordinarily complex spectrum of frequencies whose details are buried in an infinite system of transcendental equations.

The overtones of the lower modes of an infinite plate influence the high-frequency spectrum of a finite plate but the high modes of an infinite plate do not greatly affect the spectrum of a finite plate at lower frequencies. This circumstance permits the application of methods of approximation, for finite plates, in which only a finite number of the lower modes of an infinite plate are employed. However, in order to establish the approximate methods, it is necessary to have at hand quite detailed information, both qualitative and quantitative, as to the behavior of the modes of an infinite plate.

In the case of the isotropic infinite plate, the appropriate solution was given by Rayleigh,[1] in 1889, essentially in the form of a pair of transcendental equations whose roots determine the relation between the frequencies and wavelengths of the various modes. Rayleigh's frequency equations are deceptively simple in appearance. Various properties of the complicated frequency spectrum contained in them have been brought to light only gradually over a period of years[2–7] and there remain, even to this day, unexplored regions of great practical significance.

Analogous to Rayleigh's solution, there is one given by Ekstein[8] in 1945, for the monoclinic crystal, which has important applications to the rotated Y cuts of quartz. In view of the length of time it has taken to gain an understanding of Rayleigh's problem, it is not surprising that Ekstein's solution of a much more complicated problem has not yet been explored fully. Ekstein himself studied the roots corresponding to a family of high-frequency thickness-shear modes at long wavelengths. It is the purpose of this paper to describe an analysis and interpretation of the frequency spectrum covering high and low frequencies and long and short wavelengths and to give the results of detailed computations for the AT cut of quartz. The investigation is confined, for the present, to the region of real phase velocities.

At first glance, the frequency equations, which can be obtained for Ekstein's solution, appear to be hopelessly intractable. They each contain one more transcendental function than Rayleigh's, and the variable coefficients of the transcendental functions are monstrous algebraic functions of the roots of a cubic equation. Furthermore, even excluding the limiting cases of zero and infiinite wavelength and confining attention to real phase velocities, the frequency equations assume fourteen distinct forms in various portions of the spectrum. However, the equations yield readily to a technique which was used previously to study Rayleigh's frequency equations[5] and problems of high-frequency flexural vibrations of rectangular plates.[9] The technique is based on the fact that it is possible to relate such problems both mathematically and physically to much simpler problems It includes, among other devices, that employed by Holden.[3]

* This investigation was supported by the Signal Corps Engineering Laboratories.
† Now Staff Engineer, Research Laboratory of International Business Machines Corporation, Poughkeepsie, New York.
[1] Lord Rayleigh, Proc. London Math. Soc. **20**, 225 (1889).
[2] H. Lamb, Proc. Roy. Soc. (London) **93**, 114 (1917).
[3] A. N. Holden, Bell System Tech. J. **30**, 956 (1951).
[4] R. R. Aggarwal and E. A. G. Shaw, J. Acoust. Soc. Am. **26**, 341 (1952).
[5] R. D. Mindlin, "An introduction to the mathematical theory of vibrations of elastic plates," Signal Corps Engineering Laboratories, Fort Monmouth, New Jersey (1955).
[6] R. H. Lyon, J. Acoust. Soc. Am. **27**, 259 (1955).
[7] I. Tolstoy and E. Usdin, J. Acoust. Soc. Am. **29**, 37 (1957).
[8] H. Ekstein, Phys. Rev. **68**, 11 (1945).
[9] R. D. Mindlin, J. Appl. Phys. **22**, 316 (1951); **23**, 83 (1952).

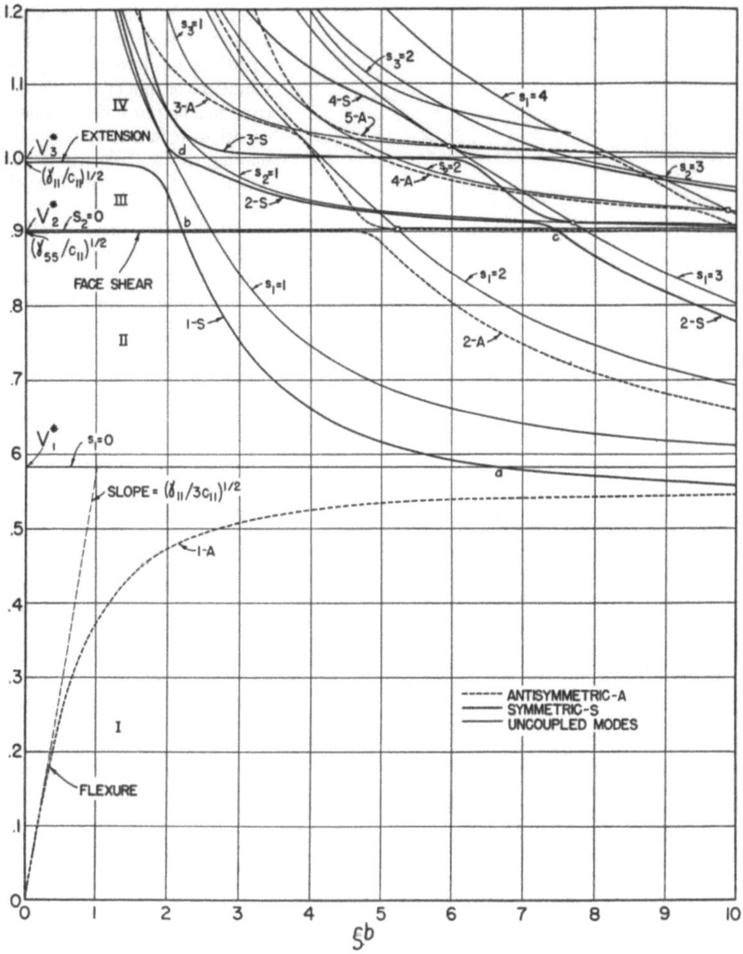

FIG. 1. Phase velocity *vs* wave number for the first five antisymmetric modes and the first four symmetric modes of an infinite, AT-cut, quartz plate.

In a crystalline medium which is infinite in all directions there are three modes of wave propagation. In the presence of a plane boundary which is free of traction, each of these waves can exist, in general, only in combination with the other two. Correspondingly, in an infinite plate with traction-free faces, each of the infinity of modes of vibration comprises three coupled waves and this coupling is what gives rise to the intricate frequency spectrum and the accompanying complicated transcendental frequency equation. There are, however, boundary conditions for which each wave can exist by itself. This is the case, for example, if, instead of requiring all three components of traction to vanish on the boundary, one of the traction conditions is replaced by the corresponding displacement condition. In a plate whose faces are under such mixed boundary conditions, each of the three waves generates only a simple family of anharmonic overtones. For a given phase velocity of an overtone, the wavelength along the plate is an integral submultiple of the wavelength of the fundamental. There are only three such families of overtones and they are not coupled. On a plot of phase velocity *versus* wave number, the spectrum is a simple grid of three families of intersecting curves (see the curves marked s_j=constant in Fig. 1). To determine these curves it is only necessary to compute one family of roots of a cubic equation. (The curves are not quite as simple as the corresponding ones for an isotropic plate because here thickness-shear and thickness-stretch displacements are coupled, through the elastic constants,

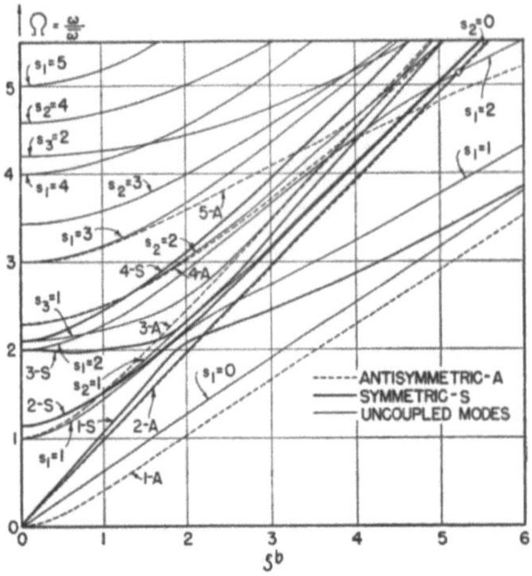

FIG. 2. Frequency *vs* wave number for the first five antisymmetric modes and the first four symmetric modes of an infinite, AT-cut, quartz plate.

irrespective of the boundary conditions. This results, for example, in the inflections in the curve $s_1=1$ in Fig. 2.)

The passage from mixed boundary conditions to free boundaries is effected through the introduction of a variable boundary condition. The condition of zero displacement is replaced by one of elastic constraint, i.e., the component of surface traction is made proportional to the corresponding component of displacement. When the coefficient of proportionality (the spring coefficient) is infinite, the boundary conditions revert to the original mixed conditions and when the spring coefficient is zero the boundaries are traction free. It is thus possible to pass continuously from the simple solution to the desired one.

An inspection of the intermediate boundary conditions shows that there are two families of roots of the transcendental equations which are always present regardless of the value of the spring coefficient. These roots are represented by certain nodes of the simple grid (the circles and squares in Fig. 1 and the intersections, in Figs. 5 and 6, of the curves $s_j=$constant). Hence it is known that the spectral curves for the free plate must pass through certain predetermined points. Also, since the elastic support constitutes a constraint, the frequencies of the mixed boundary value problem are upper bounds of the frequencies of the free plate.

Special asymptotic solutions for long and short wavelengths are either available or can be obtained. For infinite wavelength and high frequency, there are available the solutions by Koga,[10] one family of which was used by Ekstein in investigating a part of the spectrum. For long wavelength and low frequency, asymptotic solutions are readily found from Cauchy's[11] equations for low-frequency flexural and extensional vibrations of thin crystal plates. For short wavelengths corresponding to surface vibrations, the transcendental equations reduce to an algebraic equation and a single limiting phase velocity is found. An independent solution and an interpretation of this result are given in another paper.[12] For short wavelengths of the remaining modes, the phase velocities all approach that of a shear mode in accordance with the fact that the lowest family of curves in the simple grid supplies this velocity as an asymptotic upper bound and the transcendental equations supply the same velocity as a lower bound (V_1* in Fig. 1).

There are, in fact, three such velocities (V_1*, V_2*, and V_3* in Fig. 1) which play important roles. For each of them, one or more of the transcendental functions in each frequency equation vanishes, so that some of the roots can be obtained with less labor than at other phase velocities. The three velocities are those of the one compressional and two shear waves whose normals are along the digonal axis of an infinite monoclinic crystal. These velocities are also the asymptotes of the three families of curves which make up the simple grid. The asymptotes divide the spectrum into four regions which have different properties due to the absence of three, two, one, or none of the uncoupled modes with real phase velocities.

With all this advance information, it is possible to gain an understanding of the character of the frequency spectrum and the shapes of the modes without elaborate computations. In fact, the complicated spectrum of the free plate can be sketched approximately on the grid derived from the simple uncoupled modes, keeping in mind the asymptotic values at long and short wavelengths.

The final coupled modes separate into two families, one of which is essentially antisymmetric (flexural) and the other essentially symmetric (extensional) with respect to the middle plane of the plate. The two families are not coupled with each other and their frequencies are determined by separate transcendental equations; but, in contrast to the situation in the isotropic case, each family contains both antisymmetric and symmetric components.

For example, in the family of "antisymmetric" modes (whose phase velocities and frequencies are given by the dashed curves in Figs. 1 and 2) the lowest mode (1-*A*) is the usual flexural mode but the next higher mode (2-*A*), at long wavelengths, is one of face-shear, which is symmetric with respect to the middle plane. As the wavelength decreases and the frequency increases, the face-shear mode does not couple with the

[10] I. Koga, Physics 3, 70 (1932).

[11] A. L. Cauchy, *Exercises de Mathématique* (Bure Frères, Paris, 1829), Vol. 4, p. 13.

[12] H. Deresiewicz and R. D. Mindlin, J. Appl. Phys. **28**, 669 (1957).

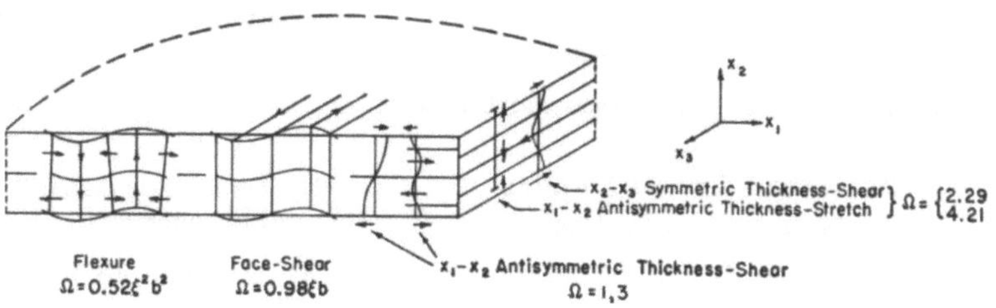

FIG. 3. Approximate shapes of the first six antisymmetric modes of the AT cut of quartz at long wavelengths.

lowest antisymmetric thickness-shear mode at the frequency of the latter (ordinate unity in Fig. 2), but it does couple, at higher frequencies (ordinate 4.75 in Fig. 2), with symmetric, transverse thickness-shear modes. The latter, in turn, are coupled through the elastic constants with antisymmetric thickness-stretch modes. Thus, certain of the thickness modes (i.e., the Koga modes) both symmetric and antisymmetric, are coupled in the antisymmetric family along with the flexure and symmetric face-shear modes. Approximate shapes, at long wavelengths, of the first six modes are shown in Fig. 3. Part of the coupling is through the elastic constants and makes its appearance at an early stage of analysis (in the cubic equation) so that it is present even in the "uncoupled" modes (s_j=constant). The remainder of the coupling is due to the boundary conditions and its development appears during the relaxation of the elastic constraint.

In the "symmetric" family, whose phase velocities and frequencies are given by the heavy, full lines in Figs. 1 and 2, the lowest mode (1-S) is the usual extensional one; but there is no mode which corresponds to the face-shear. The next higher mode (2-S), at long wavelengths, is one which is essentially antisymmetric, transverse thickness-shear and is present because of its coupling, through the elastic constants, with the lowest symmetric thickness-stretch mode (4-S). The latter has a slightly higher frequency than the lowest symmetric thickness-shear mode (3-S) and both of these symmetric modes have considerably higher frequencies than that of the antisymmetric, transverse thickness-shear mode. The extensional mode does not begin to couple with thickness modes, through the boundary conditions, until its wavelength is short enough for its frequency to reach those of the symmetric thickness-shear and symmetric thickness-stretch modes (ordinate 2 in Fig. 2) and then, through the latter, it couples with the antisymmetric, transverse thickness-shear mode (2-S). Approximate shapes of these modes, at long wavelengths, are shown in Fig. 4. Included in the family of coupled "symmetric" modes are all those thickness modes, both symmetric and antisymmetric, which are not included in the family of coupled "antisymmetric" modes.

The relative positions, in the spectrum, of the thickness-shear and thickness-stretch modes are very sensitive to the ratios of elastic constants. The positions described here are those for the AT cut of quartz. For other cuts or other materials the positions may be quite different and this may result in marked changes in the coupled spectra.[5]

After a preliminary survey along the lines indicated above, precise computations were performed, on a large digital computer, for the AT cut of quartz. In programing the machine it was a great advantage to have available the previously determined asymptotic values, the special intersection points of the simple grid and the special values of wave number at the V_j^* (through all of which spectral curves must pass) and, above all, the fact that the intersections of a line of constant phase velocity with certain curves of the simple grid define upper and lower bounds of wave number between which a root of a transcendental equation must be found.

II. EKSTEIN'S SOLUTION

We begin by rederiving Ekstein's solution in a form suitable for the analysis which follows.

The plate is referred to rectangular coordinates x_1, x_2, x_3 with its faces at $x_2=\pm b$. The material is assumed to have monoclinic symmetry, with x_1 the digonal axis. Thus the stress-strain relation is

$$T_p=c_{pq}S_q; \quad c_{pq}=c_{qp}; \quad p, q=1, 2, \cdots 6, \qquad (1)$$

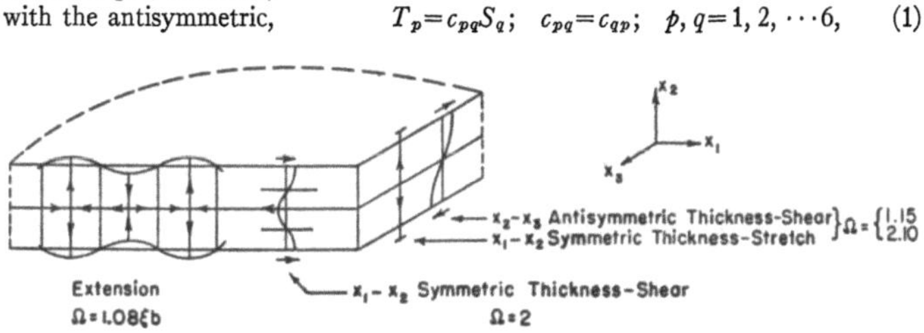

FIG. 4. Approximate shapes the first four symmetric modes of the AT cut of quartz at long wavelengths.

with

$$c_{15}=c_{16}=c_{25}=c_{26}=c_{35}=c_{36}=c_{45}=c_{56}=0. \qquad (2)$$

The strains, S_q, in terms of the displacements u_1, u_2, u_3, are

$$\begin{aligned} S_1&=\partial u_1/\partial x_1, & S_4&=\partial u_3/\partial x_2+\partial u_2/\partial x_3,\\ S_2&=\partial u_2/\partial x_2, & S_5&=\partial u_1/\partial x_3+\partial u_3/\partial x_1,\\ S_3&=\partial u_3/\partial x_3, & S_6&=\partial u_2/\partial x_1+\partial u_1/\partial x_2,\end{aligned} \qquad (3)$$

and the stresses T_p must satisfy the stress equations of motion

$$\begin{aligned}\partial T_1/\partial x_1+\partial T_6/\partial x_2+\partial T_5/\partial x_3&=-\rho\omega^2u_1,\\ \partial T_6/\partial x_1+\partial T_2/\partial x_2+\partial T_4/\partial x_3&=-\rho\omega^2u_2,\\ \partial T_5/\partial x_1+\partial T_4/\partial x_2+\partial T_3/\partial x_3&=-\rho\omega^2u_3,\end{aligned} \qquad (4)$$

where ρ is the density and ω is the (circular) frequency of vibration.

We consider only cylindrical modes of vibration, in which the components of displacement are independent of the coordinate x_3. As shown by Ekstein, there are two such solutions of the equations according as the displacement u_2 is antisymmetric or symmetric with respect to the middle plane of the plate. For the antisymmetric solution we take

$$\begin{aligned}u_1&=A_1\cos\xi x_1\sin\eta x_2,\\ u_2&=A_2\sin\xi x_1\cos\eta x_2,\\ u_3&=A_3\sin\xi x_1\cos\eta x_2,\end{aligned} \qquad (5)$$

and, for the symmetric solution,

$$\begin{aligned}u_1&=A_1\sin\xi x_1\cos\eta x_2,\\ u_2&=A_2\cos\xi x_1\sin\eta x_2,\\ u_3&=A_3\cos\xi x_1\sin\eta x_2.\end{aligned} \qquad (6)$$

Upon substituting (5) or (6) in (3), (3) in (1), and (1) in (4), we find

$$\begin{aligned}(a_1-V^2)A_1+a_6A_2+a_5A_3&=0,\\ a_6A_1+(a_2-V^2)A_2+a_4A_3&=0,\\ a_5A_1+a_4A_2+(a_3-V^2)A_3&=0,\end{aligned} \qquad (7)$$

where

$$\begin{aligned}a_1&=1+c_{66}\alpha^2/c_{11}, & a_4&=(c_{56}+c_{24}\alpha^2)/c_{11},\\ a_2&=(c_{66}+c_{22}\alpha^2)/c_{11}, & a_5&=(c_{14}+c_{56})\alpha/c_{11},\\ a_3&=(c_{55}+c_{44}\alpha^2)/c_{11}, & a_6&=(c_{12}+c_{66})\alpha/c_{11},\end{aligned} \qquad (8)$$

and

$$\alpha=\eta/\xi, \quad V^2=\rho\omega^2/c_{11}\xi^2. \qquad (9)$$

It may be observed that $V(=v/v_1)$ is the ratio of the phase velocity $(v=\omega/\xi)$ to the velocity (v_1) of the compressional wave along the digonal axis of an infinite monoclinic crystal.

The compatibility of Eqs. (7) requires that

$$\begin{vmatrix} a_1-V^2 & a_6 & a_5\\ a_6 & a_2-V^2 & a_4\\ a_5 & a_4 & a_3-V^2\end{vmatrix}=0, \qquad (10)$$

and Eqs. (7) also yield the amplitude ratios $A_1:A_2:A_3$ which may be expressed as

$$\begin{aligned}A_2/A_1&=(a_1a_4-a_5a_6-a_4V^2)/(a_2a_5-a_4a_6-a_5V^2),\\ A_3/A_1&=(a_1a_4-a_5a_6-a_4V^2)/(a_3a_6-a_4a_5-a_6V^2).\end{aligned} \qquad (11)$$

Equation (10) is a bicubic in α and V. It may be regarded as giving the three ratios of velocities corresponding to a ratio of wave numbers or as giving the three ratios of wave numbers corresponding to a velocity. Considering the latter case, for the present, we take ξ as arbitrary and thus find three wave numbers η_j, $j=1, 2, 3$, along x_2. There are then three solutions, of each symmetry, which may be summed and written in the following forms: for the antisymmetric case

$$\begin{aligned}u_1&=\cos\xi x_1\Sigma B_j\sin\alpha_j\xi x_2,\\ u_2&=\sin\xi x_1\Sigma B_jb_{2j}\cos\alpha_j\xi x_2,\\ u_3&=\sin\xi x_1\Sigma B_jb_{3j}\cos\alpha_j\xi x_2,\end{aligned} \qquad (12)$$

and, for the symmetric case,

$$\begin{aligned}u_1&=\sin\xi x_1\Sigma B_j\cos\alpha_j\xi x_2,\\ u_2&=\cos\xi x_1\Sigma B_jb_{2j}\sin\alpha_j\xi x_2,\\ u_3&=\cos\xi x_1\Sigma B_jb_{3j}\sin\alpha_j\xi x_2,\end{aligned} \qquad (13)$$

where $\alpha_j=\eta_j/\xi$ and

$$\begin{aligned}b_{2j}&=\frac{a_{1j}a_{4j}-a_{5j}a_{6j}-a_{4j}V^2}{a_{2j}a_{5j}-a_{4j}a_{6j}-a_{5j}V^2}\\ b_{3j}&=\frac{a_{1j}a_{4j}-a_{5j}a_{6j}-a_{4j}V^2}{a_{3j}a_{6j}-a_{4j}a_{5j}-a_{6j}V^2}\end{aligned} \qquad (14)$$

$$\begin{aligned}a_{1j}&=1+c_{66}\alpha_j^2/c_{11}, & a_{4j}&=(c_{56}+c_{24}\alpha_j^2)/c_{11},\\ a_{2j}&=(c_{66}+c_{22}\alpha_j^2)/c_{11}, & a_{5j}&=(c_{14}+c_{56})\alpha_j/c_{11},\\ a_{3j}&=(c_{55}+c_{44}\alpha_j^2)/c_{11}, & a_{6j}&=(c_{12}+c_{66})\alpha_j/c_{11}.\end{aligned} \qquad (15)$$

Unless otherwise specified, all summations are over $j=1, 2, 3$.

The boundary conditions for traction-free faces are

$$T_2=T_4=T_6=0 \quad \text{on } x_2=\pm b. \qquad (16)$$

When (12) and (13) are substituted in these, we obtain, for the antisymmetric solution,

$$\begin{aligned}\Sigma B_jd_{1j}\sin\alpha_j\varphi&=0,\\ \Sigma B_jd_{2j}\sin\alpha_j\varphi&=0,\\ \Sigma B_j\alpha_jd_{3j}\cos\alpha_j\varphi&=0,\end{aligned} \qquad (17)$$

and, for the symmetric solution,

$$\begin{aligned}\Sigma B_j d_{1j}\cos\alpha_j\varphi&=0,\\ \Sigma B_j d_{2j}\cos\alpha_j\varphi&=0,\\ \Sigma B_j\alpha_j d_{3j}\sin\alpha_j\varphi&=0,\end{aligned}\tag{18}$$

where $\varphi=\xi b$ and

$$\begin{aligned}d_{1j}&=c_{12}+c_{22}b_{2j}\alpha_j+c_{24}b_{3j}\alpha_j,\\ d_{2j}&=c_{14}+c_{24}b_{2j}\alpha_j+c_{44}b_{3j}\alpha_j,\\ d_{3j}&=c_{66}+c_{66}b_{2j}\alpha_j^{-1}+c_{56}b_{3j}\alpha_j^{-1}.\end{aligned}\tag{19}$$

On setting the determinant of the coefficients of the B_j equal to zero, in (17) and (18), we obtain, for the antisymmetric solution,

$$\Sigma\alpha_j M_j\cot\alpha_j\varphi=0,\tag{20}$$

and, for the symmetric solution,

$$\Sigma\alpha_j M_j\tan\alpha_j\varphi=0,\tag{21}$$

where

$$\begin{aligned}M_1&=d_{31}(d_{12}d_{23}-d_{13}d_{22}),\\ M_2&=d_{32}(d_{13}d_{21}-d_{11}d_{23}),\\ M_3&=d_{33}(d_{11}d_{22}-d_{12}d_{21}).\end{aligned}\tag{22}$$

Equations (20) and (21) are the transcendental frequency equations whose roots determine the frequency spectrum. For a given phase-velocity ratio (V), Eq. (10) determines three wave-number ratios α_j. With these, the coefficients M_j may be computed through Eqs. (15), (14), (19), and (22). The roots of (20) and (21) then yield the wave numbers, in the form of the propagation constant ξb, for the chosen V. The process is not straightforward, however, because of the fact that each of the α_j may be real, zero, or imaginary even with real ξb.

III. FOURTEEN FORMS OF THE FREQUENCY EQUATIONS

For a given, real phase-velocity ratio (V), Eq. (10) is a cubic equation in α^2 with real coefficients. In this case, the roots α_j^2 are real; but they may be positive or negative, with the result that the α_j may be real, zero, or imaginary. This affects the real forms of the transcendental equations.

The phase velocities at which the α_j change from real to imaginary may be obtained from Eq. (10) by regarding it as a cubic in V^2, setting $\alpha=0$ and solving. We find, after discarding the negative phase velocities,

$$\begin{aligned}V_{1,2}^*&=\langle\{c_{55}+c_{66}\mp[(c_{55}-c_{66})^2+4c_{56}^2]^{\frac{1}{2}}\}/2c_{11}\rangle^{\frac{1}{2}},\\ V_3^*&=1.\end{aligned}\tag{23}$$

For the AT cut of quartz (see the appendix)

$$V_1^*=0.5832,\quad V_2^*=0.9018.$$

Equations (23) give the velocities of the three plane waves in an infinite monoclinic crystal when the wave normal is along the digonal axis. The first two are shear waves and the third is compressional.

The three V_j^* divide the spectrum into four regions (see Fig. 1) with different real forms of the frequency equations and, hence, different characters of roots.

Region I ($V<V_1^*$).—Here all three α_j are imaginary. Let $\alpha_j=i\beta_j$. From (15), we see that a_{5j} and a_{6j} are imaginary; from (14), b_{2j} and b_{3j} are imaginary; from (19) the d_{ij} are real and hence, the M_j are real. Equations (20) and (21) take the real forms, respectively,

$$\beta_1M_1\coth\beta_1\varphi+\beta_2M_2\coth\beta_2\varphi+\beta_3M_3\coth\beta_3\varphi=0,\tag{24}$$

$$\beta_1M_1\tanh\beta_1\varphi+\beta_2M_2\tanh\beta_2\varphi+\beta_3M_3\tanh\beta_3\varphi=0.\tag{25}$$

Region II ($V_1^*<V<V_2^*$).—Two of the α_j are imaginary. Let $\alpha_3=i\beta_3$ and $\alpha_2=i\beta_2$. Then (20) and (21) become

$$\alpha_1M_1\cot\alpha_1\varphi+\beta_2M_2\coth\beta_2\varphi+\beta_3M_3\coth\beta_3\varphi=0,\tag{26}$$

$$\alpha_1M_1\tan\alpha_1\varphi-\beta_2M_2\tanh\beta_2\varphi-\beta_3M_3\tanh\beta_3\varphi=0.\tag{27}$$

Region III ($V_2^*<V<V_3^*$).—One of the α_j is imaginary. Let $\alpha_3=i\beta_3$. Then (20) and (21) are

$$\alpha_1M_1\cot\alpha_1\varphi+\alpha_2M_2\cot\alpha_2\varphi+\beta_3M_3\coth\beta_3\varphi=0,\tag{28}$$

$$\alpha_1M_1\tan\alpha_1\varphi+\alpha_2M_2\tan\alpha_2\varphi-\beta_3M_3\tanh\beta_3\varphi=0.\tag{29}$$

Region IV ($V>V_3^*$).—All three α_j are real and we have

$$\alpha_1M_1\cot\alpha_1\varphi+\alpha_2M_2\cot\alpha_2\varphi+\alpha_3M_3\cot\alpha_3\varphi=0,\tag{30}$$

$$\alpha_1M_1\tan\alpha_1\varphi+\alpha_2M_2\tan\alpha_2\varphi+\alpha_3M_3\tan\alpha_3\varphi=0.\tag{31}$$

In addition, there are special forms of the equations at the transition values $V=V_j^*$ where one of the α_j is zero. To obtain these, it is necessary to employ limit processes since indeterminate forms appear in the equations. We define

$$\varphi M_j^*=\lim_{\alpha_j\to 0}\alpha_jM_j\tan\alpha_j\varphi,\quad j=1,2,\tag{32}$$

$$\varphi^{-1}M_3^*=\lim_{\alpha_3\to 0}\alpha_3M_3\cot\alpha_3\varphi.\tag{33}$$

Then the results are as follows:

Transition I-II ($V=V_1^*$, $\alpha_1=0$).—There are no finite roots for the antisymmetric case. For the symmetric case

$$\varphi M_1^*-\beta_2M_2\tanh\beta_2\varphi-\beta_3M_3\tanh\beta_3\varphi=0.\tag{34}$$

Transition II-III ($V=V_2^*$, $\alpha_2=0$).—For the antisymmetric and symmetric cases, respectively,

$$\tan\alpha_1\varphi=0\tag{35}$$

$$\alpha_1M_1\tan\alpha_1\varphi+\varphi M_2^*-\beta_3M_3\tanh\beta_3\varphi=0.\tag{36}$$

Transition III-IV ($V=V_3^*$, $\alpha_3=0$).—For the antisymmetric and symmetric cases, respectively,

$$\alpha_1M_1\cot\alpha_1\varphi+\alpha_2M_2\cot\alpha_2\varphi+\varphi^{-1}M_3^*=0,\tag{37}$$

$$\alpha_1M_1\tan\alpha_1\varphi+\alpha_2M_2\tan\alpha_2\varphi=0.\tag{38}$$

In all, there are fourteen real forms of the frequency equations, for real ξb, applicable at various levels of phase velocity. As shown in the next three sections, asymptotic expressions for the frequencies can be obtained for very long and very short wavelengths.

IV. LONG WAVELENGTH, HIGH FREQUENCY

When the wavelength along the x_1 axis is infinite ($\varphi=\xi b=0$) the wavelength along the x_2 axis may be infinite ($\eta b=0$) or it may not. We consider the latter case first. The modes obtained are those found by Koga[10] but the method of solution shows which of them (symmetric and antisymmetric) are limiting forms of the antisymmetric coupled family of cylindrical modes and which are limiting forms of the symmetric coupled family.

When the u_j are independent of ξ, (5) and (6) may be written as

$$u_1=A_1\sin\eta x_2,\quad u_2=A_2\cos\eta x_2,\quad u_3=A_3\cos\eta x_2,\qquad(39)$$

$$u_1=A_1\cos\eta x_2,\quad u_2=A_2\sin\eta x_2,\quad u_3=A_3\sin\eta x_2,\qquad(40)$$

respectively. Substituting (39) or (40) in (3), (3) in (1), and (1) in (4), we obtain

$$\begin{aligned}(c_{66}-c)A_1&=0,\\(c_{22}-c)A_2+c_{24}A_3&=0,\\c_{24}A_2+(c_{44}-c)A_3&=0,\end{aligned}\qquad(41)$$

where $c=\rho\omega^2/\eta^2$. Thus, u_2 and u_3 are coupled even when $\xi=0$; but u_1 is not. For nonvanishing displacements, Eqs. (41) require

$$c=c_{66},$$

$$\begin{vmatrix}c_{22}-c & c_{24}\\ c_{24} & c_{44}-c\end{vmatrix}=0,\qquad(42)$$

$$A_2/A_3=(c-c_{44})/c_{24}.\qquad(43)$$

Hence, for each frequency there are three values of c:

$$\begin{aligned}c_1&=c_{66}\\c_{2,3}&=\{c_{22}+c_{44}\mp[(c_{22}-c_{44})^2+4c_{24}^2]^{\frac{1}{2}}\}/2,\end{aligned}\qquad(44)$$

three corresponding values of η:

$$\eta_j=\omega(\rho/c_j)^{\frac{1}{2}},\quad j=1,2,3;\qquad(45)$$

and two amplitude ratios A_2/A_3:

$$\bar{b}_j=(c_j-c_{44})/c_{24},\quad j=2,3.\qquad(46)$$

Consequently, the complete expressions for the displacements are, for the antisymmetric solution,

$$\begin{aligned}u_1&=B_1\sin\eta_1x_2,\\u_2&=\Sigma B_j\bar{b}_j\cos\eta_jx_2,\quad j=2,3,\\u_3&=\Sigma B_j\cos\eta_jx_2,\quad j=2,3,\end{aligned}\qquad(47)$$

and, for the symmetric solution,

$$\begin{aligned}u_1&=B_1\cos\eta_1x_2,\\u_2&=\Sigma B_j\bar{b}_j\sin\eta_jx_2,\quad j=2,3,\\u_3&=\Sigma B_j\sin\eta_jx_2,\quad j=2,3.\end{aligned}\qquad(48)$$

When these solutions are inserted in the boundary conditions (16), we obtain, for the antisymmetric case,

$$\begin{aligned}&\Sigma B_j(\bar{b}_jc_{22}+c_{24})\eta_j\sin\eta_jb=0,\quad j=2,3,\\&\Sigma B_j(\bar{b}_jc_{24}+c_{44})\eta_j\sin\eta_jb=0,\quad j=2,3,\\&B_1c_{66}\eta_1\cos\eta_1b=0,\end{aligned}\qquad(49)$$

and, for the symmetric case,

$$\begin{aligned}&\Sigma B_j(\bar{b}_jc_{22}+c_{24})\eta_j\cos\eta_jb=0,\quad j=2,3,\\&\Sigma B_j(\bar{b}_jc_{24}+c_{44})\eta_j\cos\eta_jb=0,\quad j=2,3,\\&B_1c_{66}\eta_1\sin\eta_1b=0.\end{aligned}\qquad(50)$$

Equations (49) and (50) have the three solutions

$$\begin{aligned}B_2=B_3&=0,\quad \eta_1b=s_1\pi/2,\\B_3=B_1&=0,\quad \eta_2b=s_2\pi/2,\\B_1=B_2&=0,\quad \eta_3b=s_3\pi/2,\end{aligned}\qquad(51)$$

where, for the antisymmetric case, s_1 is an odd integer and s_2 and s_3 are even integers while, for the symmetric case, s_1 is an even integer and s_2 and s_3 are odd.

The frequencies are obtained from (45) and are conveniently expressed as the ratio

$$\Omega=\omega/\bar{\omega},\quad\text{where}\quad\bar{\omega}=(\pi/2b)(c_{66}/\rho)^{\frac{1}{2}}.\qquad(52)$$

We have, then, for the antisymmetric case:

$$\begin{aligned}&u_1=B_1\sin(s_1\pi x_2/2b),\quad \Omega=s_1,\ s_1\text{ odd},\\&\left.\begin{cases}u_2=B_2\bar{b}_2\cos(s_2\pi x_2/2b), & \Omega=s_2(c_2/c_{66})^{\frac{1}{2}}\\u_3=B_2\cos(s_2\pi x_2/2b), & \bar{b}_2=(c_2-c_{44})/c_{24}\end{cases}\right\}s_2\text{ even},\\&\left.\begin{cases}u_2=B_3\bar{b}_3\cos(s_3\pi x_2/2b), & \Omega=s_3(c_3/c_{66})^{\frac{1}{2}}\\u_3=B_3\cos(s_3\pi x_2/2b), & \bar{b}_3=(c_3-c_{44})/c_{24}\end{cases}\right\}s_3\text{ even},\end{aligned}\qquad(53)$$

and, for the symmetric case:

$$\begin{aligned}&u_1=B_1\cos(s_1\pi x_2/2b),\quad \Omega=s_1,\ s_1\text{ even},\\&\left.\begin{cases}u_2=B_2\bar{b}_2\sin(s_2\pi x_2/2b), & \Omega=s_2(c_2/c_{66})^{\frac{1}{2}}\\u_3=B_2\sin(s_2\pi x_2/2b), & \bar{b}_2=(c_2-c_{44})/c_{24}\end{cases}\right\}s_2\text{ odd},\\&\left.\begin{cases}u_2=B_3\bar{b}_3\sin(s_3\pi x_2/2b), & \Omega=s_3(c_3/c_{66})^{\frac{1}{2}}\\u_3=B_3\sin(s_3\pi x_2/2b), & \bar{b}_3=(c_3-c_{44})/c_{24}\end{cases}\right\}s_3\text{ odd}.\end{aligned}\qquad(54)$$

From (53) we see that the antisymmetric family contains the antisymmetric x_1-x_2-thickness-shear modes and the coupled antisymmetric thickness-stretch and x_2-x_3-symmetric thickness-shear modes. From (54) the symmetric family is seen to contain the symmetric x_1-x_2-thickness-shear and the coupled symmetric thickness-stretch and x_2-x_3-antisymmetric thickness-shear modes. The order of the frequencies and the predominance of symmetric or antisymmetric amplitudes depend upon the elastic constants. For the AT cut of quartz (see the Appendix)

$$\begin{aligned}(c_2/c_{66})^{\frac{1}{2}}&=1.147,\quad (c_3/c_{66})^{\frac{1}{2}}=2.104,\\\bar{b}_2&=0.064,\quad \bar{b}_3=-15.4.\end{aligned}\qquad(55)$$

Hence, in the two coupled modes of the same order ($s_2=s_3$), the one with the lower frequency (s_2) is predominantly x_2-x_3-thickness-shear and the one with higher frequency (s_3) is predominantly thickness-stretch. In order of increasing frequency the first nine modes of the antisymmetric family are

$s_1=1,$	$\Omega=1$	x_1-x_2-antisymmetric shear
$s_2=2,$	$\Omega=2.294$	Mainly x_2-x_3-symmetric shear
$s_1=3,$	$\Omega=3$	x_1-x_2-antisymmetric shear
$s_3=2,$	$\Omega=4.209$	Mainly antisymmetric stretch
$s_2=4,$	$\Omega=4.588$	Mainly x_2-x_3-symmetric shear
$s_1=5,$	$\Omega=5$	x_1-x_2-antisymmetric shear
$s_2=6,$	$\Omega=6.882$	Mainly x_2-x_3-symmetric shear
$s_1=7,$	$\Omega=7$	x_1-x_2-antisymmetric shear
$s_3=4,$	$\Omega=8.417$	Mainly antisymmetric stretch

and the first nine modes of the symmetric family are

$s_2=1,$	$\Omega=1.147$	Mainly x_2-x_3-antisymmetric shear
$s_1=2,$	$\Omega=2$	x_1-x_2-symmetric shear
$s_3=1,$	$\Omega=2.104$	Mainly symmetric stretch
$s_2=3,$	$\Omega=3.441$	Mainly x_2-x_3-antisymmetric shear
$s_1=4,$	$\Omega=4$	x_1-x_2-symmetric shear
$s_2=5,$	$\Omega=5.735$	Mainly x_2-x_3-antisymmetric shear
$s_1=6,$	$\Omega=6$	x_1-x_2-symmetric shear
$s_3=3,$	$\Omega=6.313$	Mainly symmetric stretch
$s_1=8,$	$\Omega=8$	x_1-x_2-symmetric shear

It is apparent that the arrangement of the thickness modes and the spacing of their frequencies are very complicated. Slight changes of the elastic constants could produce marked changes of the arrangement, especially in the higher modes. Since these modes are limiting forms, as $\xi b\to 0$, of the higher cylindrical modes of the plate, a similar behavior may be expected of the latter.

The frequencies of the first few thickness modes of AT quartz are given in Fig. 2 by the ordinates of the curves $s_j=$ constant at $\xi b=0$. The characters of the modes may be identified in the lists given above. Since these modes have infinite phase velocities along x_1, they do not appear in Fig. 1.

V. LONG WAVELENGTH, LOW FREQUENCY

When both ξ and η approach zero, so do the frequencies, with the result that the solution given in Sec. II becomes highly indeterminate. A possible procedure would be to reduce the solution by means of an expansion in series, retaining only sufficient early terms. It is simpler, however, to introduce the series expansion at an earlier stage, so as to reduce the differential equations, themselves, before solving them. When this is done in an appropriate manner,[5] the equations degenerate to Cauchy's equations[11,13] for low-frequency extensional and flexural vibrations of thin crystal plates.

[13] W. Voigt, *Lehrbuch der Kristallphysik* (Edwards Brothers, Ann Arbor, Michigan, 1946), pp. 675–698.

For cylindrical vibrations of a monoclinic plate these are

$$\begin{aligned}&\gamma_{11}\partial^2u_1/\partial x_1^2+\rho\omega^2u_1=0,\\&b^2\gamma_{11}\partial^4u_2/\partial x_1^4-3\rho\omega^2u_2=0,\\&\gamma_{55}\partial^2u_3/\partial x_1^2+\rho\omega^2u_3=0,\end{aligned}\tag{56}$$

where

$$\begin{aligned}&\gamma_{11}=c_{11}-(c_{12}^2c_{44}+c_{14}^2c_{22}-2c_{12}c_{14}c_{24})/(c_{22}c_{44}-c_{24}^2),\\&\gamma_{55}=c_{55}-c_{56}^2/c_{66}.\end{aligned}\tag{57}$$

The three equations govern extensional, flexural, and face-shear motions, respectively. To determine which of these (in particular, the third) goes with the antisymmetric motions and which with the symmetric, we note that x_2 is bounded (by $\pm b$) but x_1 is not. Hence, as ξ and η become small, the displacements (5) for the antisymmetric case become

$$\begin{aligned}&u_1=0,\\&u_2=A_2\sin\xi x_1,\\&u_3=A_3\sin\xi x_1,\end{aligned}\tag{58}$$

and the displacements (6) for the symmetric case become

$$\begin{aligned}&u_1=A_1\sin\xi x_1,\\&u_2=u_3=0.\end{aligned}\tag{59}$$

Hence flexure and face-shear are included in the antisymmetric family of coupled modes and extension in the symmetric family.

Upon substituting (58) and (59) in (56) we find, for the extensional mode:

$$V=(\gamma_{11}/c_{11})^{\frac{1}{2}},\quad \Omega=2\xi b(c_{11}/c_{66})^{\frac{1}{2}}/\pi,\tag{60}$$

for the flexural mode:

$$V=\xi b(\gamma_{11}/3c_{11})^{\frac{1}{2}},\quad \Omega=2\xi^2b^2(\gamma_{11}/3c_{66})^{\frac{1}{2}}/\pi,\tag{61}$$

and, for the face-shear mode:

$$V=(\gamma_{55}/c_{11})^{\frac{1}{2}},\quad \Omega=2\xi b(\gamma_{55}/c_{66})^{\frac{1}{2}}/\pi.\tag{62}$$

These results are shown in Figs. 1 and 2, by the curves 1-*A*, 2-*A*, and 1-*S* (for small ξb) for the AT cut of quartz, for which

$$\begin{aligned}(\gamma_{11}/c_{11})^{\frac{1}{2}}&=0.9937,\\(\gamma_{11}/3c_{11})^{\frac{1}{2}}&=0.5737,\\(\gamma_{55}/c_{11})^{\frac{1}{2}}&=0.8996.\end{aligned}$$

VI. SHORT WAVELENGTH

As the wavelength approaches zero ($\xi b\to\infty$) all of the frequency equations (24) to (31) become oscillatory except Eqs. (24) and (25) for the antisymmetric and symmetric modes in the lowest region (Region I). Both of the nonoscillatory equations reduce to the same algebraic equation

$$\beta_1M_1+\beta_2M_2+\beta_3M_3=0.\tag{63}$$

This is a quintic in V^2, with parameters $\beta_j(=-i\alpha_j)$, which must be solved simultaneously with Eq. (10). There is one real, positive root, so that the antisymmetric and symmetric modes in Region I approach the same limiting phase velocity as $\xi b\to\infty$. For the AT cut of quartz the root is

$$V=0.5548. \qquad (64)$$

This is the asymptote of the two lowest curves (1-*A* and 1-*S*) in Fig. 1. An independent solution of this case, which exhibits the character of the mode and the nature of the coupling is given in another paper.[12]

The limiting phase velocity of the remaining modes is discussed in Sec. IX.

VII. MIXED BOUNDARY CONDITIONS

In order to study the interior of the spectrum we consider, first, an auxiliary problem[5] in which the displacements do not satisfy the conditions (16) for traction-free boundaries but, instead, satisfy

$$T_2=T_4=u_1=0 \quad \text{on } x_2=\pm b. \qquad (65)$$

Substituting the displacements (12) and (13) in (65) we find that we now have to satisfy, for the antisymmetric group,

$$\begin{aligned}\Sigma B_j d_{1j}\sin\alpha_j\varphi&=0,\\ \Sigma B_j d_{2j}\sin\alpha_j\varphi&=0,\\ \Sigma B_j \sin\alpha_j\varphi&=0,\end{aligned} \qquad (66)$$

and, for the symmetric group,

$$\begin{aligned}\Sigma B_j d_{1j}\cos\alpha_j\varphi&=0,\\ \Sigma B_j d_{2j}\cos\alpha_j\varphi&=0,\\ \Sigma B_j \cos\alpha_j\varphi&=0.\end{aligned} \qquad (67)$$

Comparing these with (17) and (18), we see that, whereas formerly the three B_j were coupled, they are now independent. Equations (66), for the antisymmetric modes, have the three independent families of solutions

$$\begin{aligned}B_2=B_3=0,&\quad \alpha_1\varphi=s_1\pi/2,\\ B_3=B_1=0,&\quad \alpha_2\varphi=s_2\pi/2,\\ B_1=B_2=0,&\quad \alpha_3\varphi=s_3\pi/2,\end{aligned} \qquad (68)$$

where the s_j are even integers. Equations (67), for the symmetric modes, have the same solutions with the s_j odd integers.

It is a simple matter to compute the spectrum of phase velocities of these uncoupled modes. For a chosen, real V, the one, two, or three real, positive roots, α_j^2, of (10) are computed. Then, through the relation

$$\xi b=s_j\pi/2\alpha_j, \quad s_j=1,2,3\cdots, \qquad (69)$$

obtained from (68), the abscissas ξb, corresponding to the chosen V are laid off in equal intervals. The process is then repeated, with different V, until sufficient points are obtained to draw the curves. Thus, it is only necessary to compute one family of roots of a cubic equation in order to draw the simple grid of curves marked s_j=constant in Figs. 1, 2, 5, and 6.

The cases $s_j=0$, which pertain to the antisymmetric modes, require special consideration. When any s_j is zero, so is the corresponding α_j and hence, from (12), the corresponding u_1 vanishes. Also, from (8), a_5 and a_6 are zero. Then, if $V=V_1^*$ (i.e., $s_1=0$) or if $V=V_2^*$ (i.e., $s_2=0$), the second and third of Eqs. (7) can be satisfied and hence there are possible displacements u_2 and u_3. However, if $V=V_3^*$ (i.e., $s_3=0$), the second two of (7) require $A_2=A_3=0$ and, hence, there are no displacements at all. Thus, while there are antisymmetric modes for $s_1=0$ and $s_2=0$, there is none for $s_3=0$. This accounts for the absence of lines marked $s_3=0$ in Figs. 1 and 2.

The first use of the spectrum of the uncoupled modes appears when it is noticed that certain of the roots (68) are also to be found in the conditions for free boundaries. For example, the free boundary conditions (12), for the antisymmetric modes, are satisfied by

$$\begin{aligned}&\sin\alpha_2\varphi=0,\quad \sin\alpha_3\varphi=0,\\ &B_1=0,\quad B_2/B_3=-(\alpha_3 d_{33}\cos\alpha_3\varphi)/(\alpha_2 d_{32}\cos\alpha_2\varphi).\end{aligned} \qquad (70)$$

Hence $\alpha_2\varphi=s_2\pi/2$ and $\alpha_3\varphi=s_3\pi/2$ with s_2 and s_3 even. Since both of these conditions must be satisfied simultaneously, the roots correspond to the intersections of the curves s_2 even and s_3 even. Hence, spectral curves for the antisymmetric modes of the free plate must pass through the intersections of the curves s_2 even, s_3 even. The same applies to the intersections of the curves s_3, s_1 even, and s_1, s_2 even. Since (70) are solutions of (12) only at intersection points of s_j even, the spectral curves for the antisymmetric modes of the free plate can cross the curves s_j even only at such points. For the symmetric modes, the pertinent intersections are those of the curves s_j odd and the corresponding spectral curves for the free plate can cross the curves s_j odd only at their intersection points. Typical intersection points are identified, in Fig. 1, by the two squares for s_j even and two circles for s_j odd. In Figs. 5 and 6 all the intersections are such points.

VIII. VARIABLE BOUNDARY CONDITIONS

The general behavior of the spectral curves for the free plate may be deduced from a study of a problem of an elastically restrained plate.

We consider the boundary conditions

$$T_2=T_4=0,\quad T_6=\mp ku_1,\quad \text{on } x_2=\pm b, \qquad (71)$$

where k is a variable spring coefficient which measures the restraint, on the surfaces, against displacement in the direction of x_1. As k varies from infinity to zero, the boundary conditions pass from the mixed conditions, with $u_1=0$, to the free conditions, with $T_6=0$.

Upon substitution of the displacements (12), for the antisymmetric case, in the boundary conditions (71), we obtain

$$\Sigma B_j d_{1j} \sin\alpha_j\varphi = 0,$$
$$\Sigma B_j d_{2j} \sin\alpha_j\varphi = 0, \quad (72)$$
$$\xi\Sigma B_j\alpha_j d_{3j} \cos\alpha_j\varphi = -k\Sigma B_j \sin\alpha_j\varphi.$$

It may be seen that these reduce to (17) when $k=0$ and to (66) when $k=\infty$.

At intermediate values of k, (72) are satisfied by (70). Hence the spectral curves for the antisymmetric modes pass through the intersections of the curves s_j even for all values of k. Similarly, the spectral curves for the symmetric modes pass through the intersections of the curves s_j odd for all values of k.

The transition from $k=\infty$ to $k=0$ is a relaxation of a constraint. Hence, the frequencies or phase velocities given by the simple grid are upper bounds, between intersection points, of the frequencies or phase-velocities for the free plate.

The situation at $\xi b=0$ requires special treatment since the ends of certain of the frequency curves remain fixed, as k varies, while others do not.

Considering, first, the antisymmetric modes, we substitute the displacements (47) into the boundary conditions (71) and find

$$\Sigma B_j(\bar{b}_j c_{22}+c_{24})\eta_j \sin\eta_j b = 0, \quad j=2, 3,$$
$$\Sigma B_j(\bar{b}_j c_{24}+c_{44})\eta_j \sin\eta_j b = 0, \quad j=2, 3, \quad (73)$$
$$c_{66}\eta_j \cos\eta_1 b = -k \sin\eta_1 b.$$

The first two of (73) are independent of k and are the same as the corresponding boundary conditions in (49) for the free plate. This is because these two conditions involve only the displacements u_2 and u_3, which are not elastically restrained. The solutions are those for s_2, s_3 even in Eqs. (53). Hence, the intersections of curves s_2 even or s_3 even with the vertical line $\xi b=0$ play the same role as intersection points within the body of the spectrum; i.e., the curves s_2, s_3 even between these terminal points and the next intersections with s_j even are upper bounds of spectral curves for antisymmetric modes of a free plate. For example, the curve $s_2=2$, between $\xi b=0$ and its intersection with $s_1=4$, is the upper bound of the fourth antisymmetric mode, which is known, from Sec. IV, to start out at $\Omega=2.29$ as the second thickness mode in the antisymmetric family.

As explained in Sec. VII, there is no uncoupled mode for $s_3=0$ but the curve for $s_2=0$ is present and forms the upper bound of the face-shear mode between the terminal point of the curve $s_2=0$ at $\xi b=0$ and the intersection of $s_2=0$ with $s_1=2$, after which $s_1=2$ becomes the upper bound.

The third of Eqs. (73) governs the frequencies of the thickness modes having u_1 as the sole displacement and, accordingly, dependent on k. The frequencies, from (44), (45), and (52), are given by

$$\Omega = 2\eta_1 b/\pi, \quad (74)$$

where, from the third of (73), $\eta_1 b$ are the roots of

$$\eta_1 b \cot\eta_1 b = -kb/c_{66}. \quad (75)$$

When $k=\infty$ the roots are $\eta_1 b = s_1\pi/2$, s_1 even, so that the frequencies are given by

$$\Omega = s_1, \quad s_1 \text{ even}.$$

When $k=0$ the roots are $\eta_1 b = s_1\pi/2$, s_1 odd, so that the frequencies are given by

$$\Omega = s_1, \quad s_1 \text{ odd}.$$

As k diminishes from infinity to zero, the frequency ratio of a mode drops from an even integer to the next lower odd integer. For example, (see Fig. 2) the curve for the uncoupled mode $s_1=2$ has the ordinate $\Omega=2$ at $\xi b=0$. As k diminishes from infinity to zero, the ordinate of the terminal point of the curve for the resulting coupled mode drops from 2 to 1 where it becomes the terminal point of the third antisymmetric mode of the free plate; which, at $\xi b=0$ is the first x_1-x_2-antisymmetric thickness-shear mode (see Sec. IV). Thus, the curve $s_1=2$, between $\xi b=0$ and its intersection with the curve $s_2=0$, is the upper bound of the third antisymmetric mode; which, in turn, is known from Sec. IV, to start at $\Omega=1$. The lower bound, in this interval, is the curve $s_2=0$. At the intersection point of $s_2=0$ and $s_1=2$ (the square near the upper right-hand corner of Fig. 2) the upper and lower bounds interchange, as is the case at every intersection point.

For the special case $\eta_1=0$ (i.e., $s_1=0$) the third of (73) is satisfied for all values of k. Hence the terminal point of the corresponding mode remains fixed at the origin, when k diminishes from infinity to zero, and the resulting coupled mode (the flexural mode) has the curve $s_1=0$ as its upper bound.

The behavior of the symmetric modes is very similar. In all the expressions it is only necessary to interchange sine and cosine, odd and even and reverse the sign preceding k. The terminal points of the curves s_2, s_3 odd remain fixed and play the role of intersection points. The curves s_2, s_3 odd are upper bounds. For example, the curve $s_2=1$, between $\xi b=0$ and its intersection with $s_1=3$, is the upper bound of the second symmetric mode, which is known, from Sec. IV, to start out at $\Omega=1.147$ as the lowest, essentially x_2-x_3-antisymmetric thickness-shear mode. As k diminishes from infinity to zero the terminal points of the curves s_1 odd drop to the next lower even integer. In the case of $s_1=1$, the drop is from 1 to 0 to form the lowest symmetric mode, i.e., the extensional mode which, accordingly, has $s_1=1$ as its upper bound.

IX. SKETCH OF SPECTRUM

We are now in a position to construct the spectrum approximately,[5] using the simple grid, the asymptotic values and the transition roots.

In the lowest region (Region I) of the spectrum of

phase velocities, the frequency equations (24) and (25) apply. All the transcendental functions are hyberbolic and there is at most one root of each equation for each real V. In the antisymmetric case, we know, from (61), that the curve starts at the origin with initial slope

$$dV/d(\xi b)=(\gamma_{11}/3c_{11})^{\frac{1}{2}},$$

which is 0.5737 for the AT cut of quartz. We construct a straight line through the origin at this slope (see Fig. 1). From (10), (63), and (64) we know that the asymptote of the lowest antisymmetric mode is

$$V=0.5548$$

and this is consistent with the fact that the upper bound is $s_1=0$ for which

$$V_1^*=0.5832.$$

Since the curve is monotonic, we may sketch one similar to 1-A in Fig. 1.

From Sec. VI, we know that the asymptotic value of the lowest symmetric mode, at infinite ξb, is also given by (64). However, from (60), this curve has, at $\xi b=0$, the ordinate

$$V=(\gamma_{11}/c_{11})^{\frac{1}{2}}=0.9937,$$

which is in Region III. Hence, the curve must cross ordinates V_1^* and V_2^* and we also know from Sec. VIII that its upper bound is the curve $s_1=1$. For Transition I–II, we find that (34) has only one finite root:

$$\xi b=6.62$$

which we plot at point a in Fig. 1. For Transition II–III we find the first root of (36) to be

$$\xi b=2.206$$

which we plot at point b in Fig. 1. We may now sketch curve 1-S from below up to point b. For the remainder of the curve we start at $\xi b=0$ where we already know the ordinate and also know that the initial slope is zero. We proceed to sketch a horizontal line, at this ordinate, until we approach the upper bound $s_1=1$ where the curve turns down to point b.

We know proceed to the curve for the second antisymmetric mode, whose ordinate at $\xi b=0$ is

$$V=(\gamma_{55}/c_{11})^{\frac{1}{2}}=0.8996$$

and whose slope is zero at that point. This is consistent with having as its upper bound the curve $s_2=0$ which is a horizontal line at ordinate

$$V_2^*=0.9018.$$

Hence, starting at ordinate 0.8996, we draw a horizontal line (the dashed line marked "face-shear") which crosses $s_1=1$ without turning (since the integer is odd and we are in the antisymmetric family) and continue until we approach the intersection of $s_2=0$ with $s_1=2$, which is marked with a square in Fig. 1. Here $s_1=2$ takes over as the upper bound so that the curve for mode 2-A turns down to be parallel to $s_1=2$, which approaches V_1^* asymptotically. Now, curve 2-A cannot cross V_1^* since Eq. (34) for Transition I–II has only the one root at a. Hence V_1^* is a lower bound for 2-A. Accordingly curve 2-A approaches V_1^* as its asymptote.

Since, as $\xi b\rightarrow\infty$, the curves for all the higher modes are eventually forced down to Region II by the upper bounds $s_j=$ constant, all the curves finally have $s_1=$ constant as upper bounds and V_1^* as the lower bound; but all the curves s_1 are asymptotic to V_1^* and hence the phase-velocity ratios of all the upper modes approach V_1^* as a common limit.

The three modes which have been sketched on the phase-velocity spectrum may now be transferred to the frequency spectrum (Fig. 2) by multiplying the ordinate according to

$$\Omega=\frac{2\xi b}{\pi}\left(\frac{c_{11}}{c_{66}}\right)^{\frac{1}{2}}V=1.0890\xi bV.$$

We now return to Fig. 1 and consider the next higher mode, which is the second symmetric mode (2-S) since the next higher upper bound in Region II is $s_1=3$. We compute the second root of (36), the frequency equation for the symmetric modes at Transition II–III, and find

$$\xi b=7.436$$

which is plotted as point c. Curve 2-S is started at the right-hand side of the spectrum and drawn parallel to $s_1=3$ at such a distance from it as to pass through c. Continuing past c, the curve turns left, with $s_2=1$ as its new upper bound, and passes ordinate V_3^* at point d which is the lowest root,

$$\xi b=2.167,$$

of Eq. (38), the frequency equation for symmetric modes at Transition III–IV. The curve 2-S then continues with $s_2=1$ as upper bound and $s_1=1$ as lower bound. To complete the curve at the long wavelength end, we transfer to Fig. 2 and sketch 2-S to its terminal point

$$\Omega=1.147$$

as described in Sec. IV.

The curves for successive higher modes may be sketched in a similar manner. There are no additional roots in Region I. In Region II the curves are simple ones like 2-A and 2-S. At short wavelengths (large ξb) in Regions III and IV, there are many intersections of the curves $s_j=$ constant. The typical behaviors of the coupled curves in these curves are illustrated in Figs. 5 and 6. For simplicity only one set of modes is shown. Thus, the curves $s_j=$ constant may be interpreted as the curves s_j even, in which case the coupled curves are those for the antisymmetric modes. If the curves are taken to be s_j odd, the coupled curves are the ones for the symmetric modes. The curves for the antisymmetric

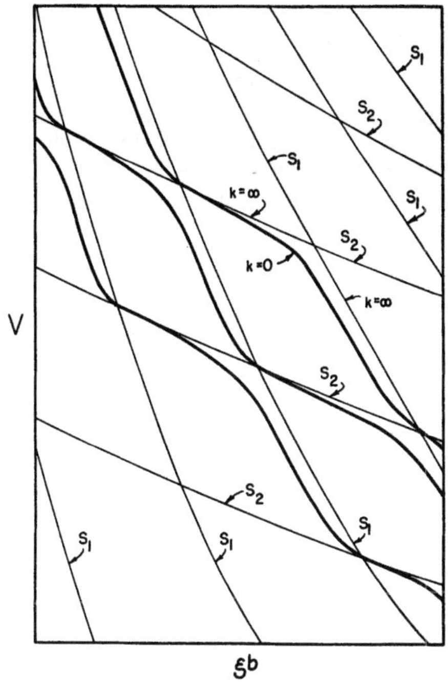

FIG. 5. Illustrating method of sketching coupled spectrum on grid of uncoupled spectrum in Region III.

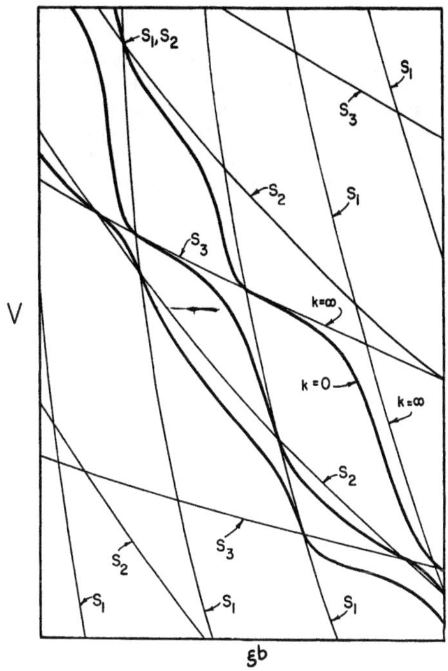

FIG. 6. Illustrating method of sketching coupled spectrum on grid of uncoupled spectrum in Region IV.

modes do not pass through the intersections of s_j odd and the curves for the symmetric modes do not pass through the intersections of s_j even. This is contrary to the situation which prevails when there are only two families of modes which couple.[9]

X. COMPUTATIONS

The curves which are shown in Figs. 1 and 2 were drawn from data obtained with a large digital computing machine. Computations were made for about 300 values of phase-velocity ratio V, between $V=300$ and $V=0$, with large increments in most of Regions I and II and small increments in the neighborhood of transitions between regions and near asymptotic values. (The roots of the transition equations, the asymptotic roots and the approximate coordinates of the intersections of the curves $s_j=$ constant were computed beforehand on a desk calculator.) For each phase-velocity ratio, the three roots of the cubic equation (7) were computed by the Newton-Raphson method and these roots were used to compute the coefficients $\alpha_j M_j$ or $\beta_j M_j$ of the transcendental equations (24) to (31).

Depending on the phase velocity, from one to seven roots of the appropriate transcendental equation were computed, amounting to about 900 roots in all. In this operation the method of "false position" was used. In general, the first approximation of a root was obtained by a second-order interpolation, using three roots obtained previously for the same mode at the next higher phase velocities. In most cases a good first approximation resulted and less than five iterations were required to compute a root to four significant figures. If the correct root was not obtained after five iterations, the method of "false position" was modified to an error halving scheme.

For each phase velocity, a pair of consecutive curves s_j even (for the antisymmetric modes) or s_j odd (for the symmetric modes) forms upper and lower bounds between which a root must be found. Throughout the computations a check was made to insure that a root was actually obtained between the proper bounds. The positions of the bounds were established by the computer for each phase velocity. When the phase velocity dropped below an intersection point, of s_j even with s_j even or s_j odd with s_j odd, a new set of bounds was computed for each root.

The computed curves supply details which do not appear in the sketching process described in the preceding section. These details include the proximity of the coupled curves to the bounds and the curvature of a coupled curve in passing from one bound to another, which is a measure of the strength of the coupling.

Of some interest is the frequency minimum which appears in the third symmetric mode (curve 3-S between $\xi b=0$ and $\xi b=1$ in Fig. 2). This is due to the coupling between the adjacent thickness-shear and thickness-stretch modes.[5] Since the group-velocity is proportional to the slope of the curves in Fig. 2, the presence of a minimum represents zero group-velocity at a finite wavelength and "negative" group velocities at longer wavelengths of the same mode. A discussion of this phenomenon, for the case of an isotropic plate, is given in a recent paper by Tolstoy and Usdin.[7] One of the implications of this circumstance, as regards finite plates, is that, in the range of wavelengths corresponding to "negative" group velocities, overtones of the anomalous mode may have lower frequencies than that of the fundamental.

APPENDIX

Mason's[14] values of the elastic constants of quartz were used in the computations. For the AT cut, the constants which appear in the equations are

$$c_{11}=86.05 \quad c_{12}=-10.49 \quad c_{55}=69.84$$
$$c_{22}=129.9 \quad c_{14}=3.453 \quad c_{66}=29.41$$
$$c_{44}=39.06 \quad c_{24}=-5.820 \quad c_{56}=-2.415.$$

[14] W. P. Mason, *Piezoelectric Crystals and Their Applications to Ultrasonics* (D. Van Nostrand Company, Inc., Princeton, 1950).

VIBRATIONS OF AN INFINITE, ELASTIC PLATE AT ITS CUT-OFF FREQUENCIES

In addition to simple thickness-vibrations, which are limiting forms of waves in a plate as the wave-length approaches infinity, there are other limiting modes of vibration at infinite wave-length. These modes have the same frequencies as those of the thickness-modes but their shapes are different. They degenerate to the thickness-modes at certain values of Poisson's ratio.

R. D. Mindlin
Columbia University

1. Plane Waves in a Plate

Rayleigh [1] and Lamb [2] have shown that, in an isotropic, elastic plate bounded by traction-free faces at $y = \pm b$, waves of plane strain are propagated parallel to the axis of x with displacements given by the real or imaginary parts (omitting a time factor $e^{i\omega t}$) of

$$u = i(\xi f + dg/dy)e^{i\xi x},\ v = (df/dy + \xi g)e^{i\xi x} \tag{1}$$

in the directions of x and y, respectively, where

$$f = A\sin\alpha y + B\cos\alpha y,\ g = C\sin\beta y + D\cos\beta y, \tag{2}$$

$$\alpha^2 = \omega^2/c_1^2 - \xi^2,\ \beta^2 = \omega^2/c_2^2 - \xi^2, \tag{3}$$

$$c_1^2 = (\lambda + 2\mu)/\rho,\ c_2^2 = \mu/\rho, \tag{4}$$

provided the conditions

$$(\xi^2 - \beta^2)f + 2\xi dg/dy = 0,\ 2\xi df/dy + (\xi^2 - \beta^2)g = 0 \tag{5}$$

are satisfied on $y = \pm b$. The functions f and g yield the dilatational and equivoluminal parts, respectively, of the deformation. The parts of f and g associated with the constants B and C give the modes symmetric with respect to the plane $y = 0$ and the remaining portions give the antisymmetric modes. The boundary conditions (5) yield the amplitude ratios and frequency equations

$$\frac{B}{C} = \frac{(\xi^2 - \beta^2)\sin\beta b}{2\xi\alpha\sin\alpha b} = \frac{2\xi\beta\cos\beta b}{(\beta^2 - \xi^2)\cos\alpha b}, \tag{6}$$

for the symmetric modes, and

$$\frac{A}{D} = \frac{(\beta^2 - \xi^2)\cos\beta b}{2\xi\alpha\cos\alpha b} = \frac{2\xi\beta\sin\beta b}{(\xi^2 - \beta^2)\sin\alpha b} \tag{7}$$

for the antisymmetric modes.

2. Simple Thickness-Modes

Lamb [2] observed that, at infinite wave-length ($\xi b = 0$), the waves in the plate degenerate to the four families of thickness-vibrations:

(i) Symmetric thickness-shear (equivoluminal):

$$u = C\beta\cos\beta y,\ v = 0,\quad \beta = q\pi/2b,\ \omega = q\pi c_2/2b,\ q = 2,4,6\ldots \tag{8}$$

(ii) Symmetric thickness-stretch (dilatational);

$$u = 0,\ v = -B\alpha\sin\alpha y,\quad \alpha = p\pi/2b,\ \omega = p\pi c_1/2b,\ p = 1,3,5\ldots \tag{9}$$

(iii) Antisymmetric thickness-shear (equivoluminal):

$$u = -D\beta\sin\beta y,\ v = 0,\quad \beta = q\pi/2b,\ \omega = q\pi c_2/2b,\ q = 1,3,5\ldots \tag{10}$$

(iv) Antisymmetric thickness-stretch (dilatational):

$$u = 0,\ v = A\alpha\cos\alpha y,\quad \alpha = p\pi/2b,\ \omega = p\pi c_1/2b,\ p = 2,4,6\ldots \tag{11}$$

These special cases may be obtained by direct substitution of $\xi b = 0$ in Equations (1)–(5), using the symmetric imaginary, symmetric real, antisymmetric imaginary and antisymmetric real parts, respectively.

3. Coupled Modes at the Cut-Off Frequencies

There are four other families of modes, at infinite wave-length, which have the same frequencies as those of the simple thickness-modes. Each of the additional modes consists of three, coupled, equivoluminal and dilatational thickness-deformations, one of which has an amplitude which varies linearly with x.

225

(i) In the symmetric, imaginary part of the Rayleigh-Lamb solution, we note that, as $\xi b \to 0$ and $w \to p\pi c_1/2b$, $p = 1,3,5\ldots$,

$$(\sin \xi x)/\xi \to x \tag{12}$$

and, from the first form of the amplitude ratio (6),

$$2B\xi/C\beta \to -k(-1)^{(p-1)/2} \sin(pk\pi/2), \tag{13}$$

where

$$k = c_1/c_2. \tag{14}$$

Hence, the limiting form of the solution is

$$u = \frac{pk\pi C}{2b}\left[-\frac{k(-1)^{(p-1)/2}}{2} \sin\frac{pk\pi}{2} \cos\frac{p\pi y}{2b} + \cos\frac{pk\pi y}{2b}\right],$$

$$v = \frac{p^2k^2\pi^2 Cx(-1)^{(p-1)/2}}{8b^2} \sin\frac{pk\pi}{2} \cos\frac{p\pi y}{2b}, \tag{15}$$

$$w = p\pi c_1/2b, \ p = 1,3,5\ldots$$

When Poisson's ratio is such that

$$k = q/p, \ q = 2,4,6\ldots, \ p = 1,3,5\ldots, \tag{16}$$

i.e., whenever there are coincidences of the frequencies of symmetric thickness-stretch and thickness-shear modes, this solution reduces to (8).

(ii) In the symmetric, real part of the Rayleigh-Lamb solution, we note, from the second form of (6), that, as $\xi b \to 0$ and $w \to q\pi c_2/2b$, $q = 2,4,6\ldots$,

$$2C\xi/B\alpha \to k(-1)^{q/2} \cos(q\pi/2k). \tag{17}$$

Hence, using (17) and (12), we find the limiting form of the solution to be

$$u = -\frac{q^2\pi^2 Bx(-1)^{q/2}}{8b^2} \cos\frac{q\pi}{2k} \cos\frac{q\pi y}{2b},$$

$$v = -\frac{q\pi B}{2kb}\left[\sin\frac{q\pi y}{2kb} - \frac{k(-1)^{q/2}}{2} \cos\frac{q\pi}{2k} \sin\frac{q\pi y}{2b}\right], \tag{18}$$

$$w = q\pi c_2/2b, \ q = 2,4,6\ldots$$

When (16) is satisfied, this solution reduces to (9).

(iii) In the antisymmetric, imaginary part of the Rayleigh-Lamb solution, we note, from the first of (7), that, as $\xi b \to 0$ and $w \to p\pi c_1/2b$, $p = 2,4,6\ldots$,

$$2A\xi/D\beta \to k(-1)^{p/2} \cos(pk\pi/2). \tag{19}$$

Hence the limiting form of the solution is

$$u = \frac{pk\pi D}{2b}\left[\frac{k(-1)^{p/2}}{2} \cos\frac{pk\pi}{2} \sin\frac{p\pi y}{2b} - \sin\frac{pk\pi y}{2b}\right],$$

$$v = \frac{p^2k^2\pi^2 Dx(-1)^{p/2}}{8b^2} \cos\frac{pk\pi}{2} \cos\frac{p\pi y}{2b}, \tag{20}$$

$$w = p\pi c_1/2b, \ p = 2,4,6\ldots$$

When Poisson's ratio is such that

$$k = q/p, \ q = 1,3,5\ldots, \ p = 2,4,6\ldots, \tag{21}$$

i.e., whenever there are coincidences of the frequencies of antisymmetric thickness-stretch and thickness-shear modes, the solution (20) reduces to (10).

(iv) In the antisymmetric, real part of the Rayleigh-Lamb solution, we note that, as $\xi b \to 0$ and $w \to q\pi c_2/2b$, $q = 1,3,5\ldots$,

$$2D\xi/A\alpha \to (-1)^{(q-1)/2} \sin(q\pi/2k). \tag{22}$$

Hence the solution has the limiting form

$$u = \frac{q^2\pi^2 Ax(-1)^{(q-1)/2}}{8b^2} \sin\frac{q\pi}{2k} \sin\frac{q\pi y}{2b},$$

$$v = \frac{q\pi A}{2kb}\left[\cos\frac{q\pi y}{2kb} + \frac{k(-1)^{(q-1)/2}}{2} \sin\frac{q\pi}{2k} \cos\frac{q\pi y}{2b}\right], \tag{23}$$

$$w = q\pi c_2/2b, \ q = 1,3,5\ldots$$

When (21) is satisfied, this solution reduces to (11).

Bibliography

1. Rayleigh, Lord, "On the free vibrations of an infinite plate of homogeneous, isotropic elastic matter," Proceedings of the London Mathematical Society, Vol. 20 (1889) pp. 225–234.

2. Lamb, H., "On waves in an elastic plate," Proceedings of the Royal Society, London, Ser. A, Vol. 93 (1917) pp. 114–128.

Reprinted from
Proceedings of the Third U.S. National Congress of Applied Mechanics (1958) 225-226

(Reprinted from the Journal of Applied Mechanics, for December, 1959, with permission of Copyright Owner)

R. D. MINDLIN
Professor of Civil Engineering, Columbia University, New York, N. Y. Mem. ASME

M. A. MEDICK[2]
Senior Staff Scientist, Avco Research and Advanced Development Division, Avco Manufacturing Company, Lawrence, Mass.

Extensional Vibrations of Elastic Plates[1]

A system of approximate, two-dimensional equations of extensional motion of isotropic, elastic plates is derived. The equations take into account the coupling between extensional, symmetric thickness-stretch and symmetric thickness-shear modes and also include two face-shear modes. The spectrum of frequencies for real, imaginary, and complex wave numbers in an infinite plate is explored in detail and compared with the corresponding solution of the three-dimensional equations.

In a previous paper [1][3] a derivation was given of approximate equations of motion that take into account the coupling between face and thickness modes of extensional vibration of elastic plates. The thickness motion considered was that of the lowest thickness-stretch mode, in which the displacement is normal to the middle plane of the plate and the middle plane is the nodal plane. The frequency of this mode depends on Poisson's ratio (ν); the frequency increasing with increasing ν. When $\nu > 1/3$, the frequency of the lowest thickness-stretch mode is higher than the frequency, independent of ν, of the lowest, symmetric thickness-shear mode. In the latter, the displacement is also unidirectional but it is parallel to the middle plane of the plate and there are two nodal planes symmetrically disposed with respect to the middle plane. In the range of Poisson's ratios commonly encountered, both thickness modes can couple with the face modes and with each other. This circumstance has a marked influence on phase and group velocities of waves and on the frequencies and shapes of high-frequency vibrational modes. In the present paper, a derivation is given of approximate equations of motion which include the effects of both thickness modes.

The approximate, two-dimensional equations are deduced from the three-dimensional equations of elasticity by a procedure based on the series-expansion methods of Poisson [2] and Cauchy [3], and the integral method of Kirchhoff [4]. A detailed exposition of the procedure, using a power series, and its application to approximations of orders zero and one are given elsewhere [5]. In approximations of the second and higher orders, awkward mathematical forms are encountered due to the lack of orthogonality of the terms of a power series. At the suggestion of W. Prager, an expansion in a series of Legendre polynomials was studied. In that case, although similar awkward forms appear as a result of the more complicated formula for the derivative, they do not occur, for the most part, until the terms of the third order are reached. Hence the expansion in a series of Legendre polynomials is a convenient one on which to base the present second-order approximation. The method of derivation and the resulting equations (with the inertia terms omitted) are closely related to E. Reissner's theory [6] of three-dimensional corrections for the equations of generalized plane stress.

The expansion in a series of Legendre polynomials of the thickness co-ordinate, followed by an integration across the thickness, converts the three-dimensional equations of elasticity into an infinite series of two-dimensional equations which then are truncated to produce the approximate equations. The full series expressions of displacement, strain, stress, energies, and equations of motion, in conjunction with an understanding of Rayleigh's [7] exact solution of the problem of vibrations of an infinite plate, are of aid in deciding what to include in various orders of approximation and in understanding the implications of what is discarded and what retained. The series expressions and Rayleigh's exact solution also supply both the motivation and the necessary data for making adjustments, of terms that are left after the truncation, in addition to an adjustment of the type made by Poisson in establishing the zero-order equations.

The adjustment analogous to Poisson's serves to uncouple the modes retained from the higher ones without seriously affecting the behavior of the lower ones. The additional adjustments are made to improve the match between the frequency spectra of an infinite plate as obtained from the approximate and exact equations. This is accomplished by the introduction of coefficients analogous to the shear coefficient in the Timoshenko beam equations [8] and the analogous equations for plates [9]. In the latter paper [9] it was shown how the shear coefficient may be chosen to effect a perfect match in one or another part of the spectrum depending upon the frequency range and mode of greatest interest in a particular application of the approximate equations. In the present case, four such coefficients are introduced and, owing to the complexity of the spectrum, several possible combinations of matching points present themselves. The choice is made here to do all the matching at zero wave number. The range of wave numbers and frequencies over which the match remains good is a measure of both the usefulness of the approximate equations and their range of applicability. In this range, solutions of the approximate equations, in the case of finite plates, may be expected to be reliable inasmuch as these solutions are composed essentially of combinations of the modes and overtones of the infinite plate.

When both the symmetric thickness-shear and thickness-stretch deformations are taken into account, important properties of the frequency spectrum contained in the exact theory are reproduced in the resulting approximate equations of the second order, whereas they do not appear in the previous approximation of the first order [1]. These properties of the exact frequency spectrum include the imaginary loop discovered by Aggarwal and Shaw [10]; the anomalous behavior of the second and third branches

[1] This investigation was supported by the Office of Naval Research and the U. S. Army Signal Engineering Laboratories.

[2] Formerly Research Assistant, Department of Civil Engineering and Engineering Mechanics, Columbia University, New York, N. Y.

[3] Numbers in brackets designate References at end of paper.

Presented at the Summer Conference of the Applied Mechanics Division, Troy, N. Y., June 18–20, 1959, of The American Society of Mechanical Engineers.

Discussion of this paper should be addressed to the Secretary, ASME, 29 West 39th Street, New York, N. Y., and will be accepted until January 10, 1960, for publication at a later date. Discussion received after the closing date will be returned.

Note: Statements and opinions advanced in papers are to be understood as individual expressions of their authors and not those of the Society. Manuscript received by ASME Applied Mechanics Division, June 25, 1958. Paper No. 59—APM-4.

with variation of Poisson's ratio [5]; the frequency minimum of the second branch [5] with its associated zero group velocity at a nonzero wave number and phase and group velocities of opposite sign at smaller wave numbers, as described by Tolstoy and Usdin [11]; and, finally, a pair of complex branches which account for edge vibrations observed in experiments.

Expansion in Infinite Series

We refer the plate to rectangular co-ordinates x_i $(i = 1, 2, 3)$ with x_1 and x_3 in the middle plane and the faces at $x_2 = \pm b$. The components of displacement u_j $(j = 1, 2, 3)$ are expressed as

$$u_j = \sum_{n=0}^{\infty} P_n(\eta) u_j^{(n)}(x_1, x_3, t), \quad (1)$$

where $\eta = x_2/b$, the $P_n(\eta)$ are the Legendre polynomials

$$P_0(\eta) = 1, \quad P_1(\eta) = \eta, \quad P_2(\eta) = (3\eta^2 - 1)/2, \ldots$$

$$\ldots P_n(\eta) = \frac{1}{2^n n!} \frac{d^n(\eta^2 - 1)^n}{d\eta^n}, \ldots$$

and the $u_j^{(n)}$, it is to be noted, are functions of the co-ordinates x_1, x_3, and the time t, only. The $u_j^{(n)}$ are the amplitudes of polynomial distributions of displacements across the thickness of the plate. For convenience, however, they will be referred to as displacements of order n or, simply, as displacements.

Stress Equations of Motion. The series expression for the displacement is substituted in the equation

$$\int_V (\tau_{ij,i} - \rho \ddot{u}_j)\delta u_j dV = 0, \quad (2)$$

which is obtained from the variational equation of motion.[4] In Equation (2) the integration is over the volume V of the plate; the τ_{ij} are the components of stress; and the summation convention for repeated indexes is employed, as are the comma notation for differentiation with respect to the co-ordinates x_i and the dot notation for differentiation with respect to time.

When the integration with respect to η, from -1 to $+1$, is performed in Equation (2), the result is

$$\int_A \sum_{n=0}^{\infty} (b\tau_{ij,i}^{(n)} - \sum_{m=1,3\ldots}^{n} D_{mn}\tau_{2j}^{(n-m)} + F_j^{(n)} - \rho b C_n \ddot{u}_j^{(n)})\delta u_j^{(n)} dA = 0, \quad (3)$$

where A is the area of the plate and

$$\tau_{ij}^{(n)} = \int_{-1}^{1} P_n(\eta)\tau_{ij}d\eta, \quad F_j^{(n)} = [P_n(\eta)\tau_{2j}]_{-1}^{1},$$
$$D_{mn} = 2(n - m) + 1, \quad C_n = 2/(2n + 1). \quad (4)$$

The $\tau_{ij}^{(n)}$ and $F_j^{(n)}$ are defined as the nth-order components of stress and face traction, respectively; while the constants D_{mn} and C_n arise from the operations

$$\frac{dP_n}{d\eta} = \sum_{m=1,3\ldots}^{n} D_{mn}P_{n-m}, \quad \int_{-1}^{1} P_m P_n d\eta = \begin{cases} 0, & m \neq n, \\ C_n, & m = n. \end{cases} \quad (5)$$

The appearance of $F_j^{(n)}$ and $\tau_{2j}^{(n-m)}$, in Equation (3), follows from an integration, by parts, of the terms in Equation (2) that contain $\partial/\partial x_2$.

Since Equation (3) must hold for all A and arbitrary $\delta u_j^{(n)}$, the quantity in parentheses must vanish and we arrive at the stress equations of motion of order n:

$$b\tau_{ij,i}^{(n)} - \sum_{m=1,3\ldots}^{n} D_{mn}\tau_{2j}^{(n-m)} + F_j^{(n)} = \rho b C_n \ddot{u}_j^{(n)}. \quad (6)$$

In the analogous equations of motion obtained from an expansion in power series,[5] an infinite series appears on the right-hand side and no series appears on the left-hand side. However, the series in Equation (6) contributes more than one term only for $n > 2$.

Strain. In the three-dimensional theory, the components of strain ϵ_{ij} are expressed in terms of the components of displacement by

$$2\epsilon_{ij} = u_{i,j} + u_{j,i}. \quad (7)$$

Inserting the series expansion from Equation (1) and using the formula for the derivative from Equation (5), we find

$$2\epsilon_{ij} = \sum_{n=0}^{\infty} \sum_{m=1,3\ldots}^{n} [(u_{i,j}^{(n)} + u_{j,i}^{(n)})P_n + (\delta_{2j}u_i^{(n)} + \delta_{2i}u_j^{(n)})b^{-1}D_{mn}P_{n-m}], \quad (8)$$

where δ_{2j} is the Kronecker symbol δ_{ij} with $i = 2$.

In order to define components of strain of order n, the summand in Equation (8) must be expressed as the product of P_n and a function independent of x_2. Considering the double sum as a triangular array and interchanging the order of summation of columns and rows, we find

$$\epsilon_{ij} = \sum_{n=0}^{\infty} P_n \epsilon_{ij}^{(n)} \quad (9)$$

where the $\epsilon_{ij}^{(n)}$, defined as the components of strain of order n, are given by

$$2\epsilon_{ij}^{(n)} = u_{i,j}^{(n)} + u_{j,i}^{(n)} + b^{-1}(2n + 1) \sum_{m=1,3\ldots}^{\infty} (\delta_{2j}u_i^{(m+n)} + \delta_{2i}u_j^{(m+n)}). \quad (10)$$

In the analogous expression obtained by a power-series expansion[6] there is no sum over m. However, the additional terms in Equation (10) do not appear until $m = 3$.

Stress-Strain Relations. The relations between the $\tau_{ij}^{(n)}$ and the $\epsilon_{ij}^{(n)}$ may be obtained by inserting the three-dimensional expressions

$$\tau_{ij} = c_{ijkl}\epsilon_{kl} = c_{ijkl}\sum_{n=0}^{\infty} P_n \epsilon_{kl}^{(n)} \quad (11)$$

in the first of Equations (4). After performing the integration, we find

$$\tau_{ij}^{(n)} = C_n c_{ijkl}\epsilon_{kl}^{(n)}. \quad (12)$$

For later use, it is convenient to have Equation (12) expressed in the reduced indicial notation, in which double indexes ranging from 1 to 3 are replaced by single indexes, ranging from 1 to 6, as follows:

$$\tau_1 = \tau_{11}, \quad \tau_4 = \tau_{23}, \quad \epsilon_1 = \epsilon_{11}, \quad \epsilon_4 = 2\epsilon_{23},$$
$$\tau_2 = \tau_{22}, \quad \tau_5 = \tau_{31}, \quad \epsilon_2 = \epsilon_{22}, \quad \epsilon_5 = 2\epsilon_{31},$$
$$\tau_3 = \tau_{33}, \quad \tau_6 = \tau_{12}; \quad \epsilon_3 = \epsilon_{33}, \quad \epsilon_6 = 2\epsilon_{12}.$$

Then Equation (12) becomes

$$\tau_p^{(n)} = C_n c_{pq}\epsilon_q^{(n)}; \quad p, q = 1, 2, \ldots 6, \; c_{pq} = c_{qp}. \quad (13)$$

[4] Reference [12], p. 167.

[5] Reference [5], p. 3.04.

[6] Ibid., p. 3.08.

Energy Densities. Using the strain energy density U given in the three-dimensional theory by

$$2U = c_{ijkl}\epsilon_{ij}\epsilon_{kl} = c_{pq}\epsilon_p\epsilon_q, \tag{14}$$

we define a plate-strain energy density

$$\bar{U} = \int_{-1}^{1} U d\eta \tag{15}$$

and find, with the aid of Equations (14), (9), and the second of (5),

$$2\bar{U} = c_{pq}\sum_{n=0}^{\infty} C_n\epsilon_p^{(n)}\epsilon_q^{(n)}. \tag{16}$$

We also note that

$$2\bar{U} = \sum_{n=0}^{\infty} \tau_p^{(n)}\epsilon_p^{(n)}, \tag{17}$$

$$\tau_p^{(n)} = \partial\bar{U}/\partial\epsilon_p^{(n)}. \tag{18}$$

Similarly, using the kinetic energy density K as given in the three-dimensional theory by

$$2K = \rho\dot{u}_j\dot{u}_j, \tag{19}$$

we define a kinetic energy density of the plate by

$$2\bar{K} = 2\int_{-1}^{1} K d\eta = \rho\sum_{n=0}^{\infty} C_n\dot{u}_j^{(n)}\dot{u}_j^{(n)}. \tag{20}$$

Extensional Vibrations of Isotropic Plates

When the plate is isotropic, motions symmetric (extensional) and antisymmetric (flexural), with respect to the middle plane, may be considered separately. In the case of the former, only those components of displacement $u_j^{(n)}$ are retained for which $j + n$ is odd. As a result, only those components of stress $\tau_{ij}^{(n)}$ and strain $\epsilon_{ij}^{(n)}$ appear for which $i + j + n$ is even.

The stress-equations of motion (6) then are

$$\begin{aligned}
&\frac{\partial\tau_1^{(0)}}{\partial x_1} + \frac{\partial\tau_5^{(0)}}{\partial x_3} + \frac{F_1^{(0)}}{b} = 2\rho\frac{\partial^2 u_1^{(0)}}{\partial t^2}\\
&\frac{\partial\tau_5^{(0)}}{\partial x_1} + \frac{\partial\tau_3^{(0)}}{\partial x_3} + \frac{F_3^{(0)}}{b} = 2\rho\frac{\partial^2 u_3^{(0)}}{\partial t^2}\\
&\frac{\partial\tau_6^{(1)}}{\partial x_1} + \frac{\partial\tau_4^{(1)}}{\partial x_3} - \frac{\tau_2^{(0)}}{b} + \frac{F_2^{(1)}}{b} = \frac{2\rho}{3}\frac{\partial^2 u_2^{(1)}}{\partial t^2}\\
&\frac{\partial\tau_1^{(2)}}{\partial x_1} + \frac{\partial\tau_5^{(2)}}{\partial x_3} - \frac{3\tau_6^{(1)}}{b} + \frac{F_1^{(2)}}{b} = \frac{2\rho}{5}\frac{\partial^2 u_1^{(2)}}{\partial t^2}\\
&\frac{\partial\tau_5^{(2)}}{\partial x_1} + \frac{\partial\tau_3^{(2)}}{\partial x_3} - \frac{3\tau_4^{(1)}}{b} + \frac{F_3^{(2)}}{b} = \frac{2\rho}{5}\frac{\partial^2 u_3^{(2)}}{\partial t^2}\\
&\frac{\partial\tau_6^{(3)}}{\partial x_1} + \frac{\partial\tau_4^{(3)}}{\partial x_3} - \frac{5\tau_2^{(2)}}{b} - \frac{\tau_2^{(0)}}{b} + \frac{F_2^{(3)}}{b} = \frac{2\rho}{7}\frac{\partial^2 u_2^{(3)}}{\partial t^2}\\
&\qquad\vdots
\end{aligned} \tag{21}$$

The components of strain that remain, in Equations (10), are

$$\begin{aligned}
\epsilon_1^{(0)} &= \partial u_1^{(0)}/\partial x_1\\
\epsilon_2^{(0)} &= (u_2^{(1)} + u_2^{(3)} + \ldots)/b\\
\epsilon_3^{(0)} &= \partial u_3^{(0)}/\partial x_3\\
\epsilon_5^{(0)} &= \partial u_3^{(0)}/\partial x_1 + \partial u_1^{(0)}/\partial x_3\\
\epsilon_4^{(1)} &= \partial u_2^{(1)}/\partial x_3 + 3(u_3^{(2)} + u_3^{(4)} + \ldots)/b\\
\epsilon_6^{(1)} &= \partial u_2^{(1)}/\partial x_1 + 3(u_1^{(2)} + u_1^{(4)} + \ldots)/b\\
\epsilon_1^{(2)} &= \partial u_1^{(2)}/\partial x_1\\
\epsilon_2^{(2)} &= 5(u_2^{(3)} + u_2^{(5)} + \ldots)/b\\
\epsilon_3^{(2)} &= \partial u_3^{(2)}/\partial x_3\\
\epsilon_5^{(2)} &= \partial u_3^{(2)}/\partial x_1 + \partial u_1^{(2)}/\partial x_3\\
\epsilon_4^{(3)} &= \partial u_2^{(3)}/\partial x_3 + 7(u_3^{(4)} + u_3^{(6)} + \ldots)/b\\
\epsilon_6^{(3)} &= \partial u_2^{(3)}/\partial x_1 + 7(u_1^{(4)} + u_1^{(6)} + \ldots)/b\\
&\;\;\vdots
\end{aligned} \tag{22}$$

The stress-strain relations (13) reduce to

$$\begin{aligned}
\tau_1^{(0)} &= 2[(\lambda + 2\mu)\epsilon_1^{(0)} + \lambda(\epsilon_2^{(0)} + \epsilon_3^{(0)})]\\
\tau_2^{(0)} &= 2[(\lambda + 2\mu)\epsilon_2^{(0)} + \lambda(\epsilon_3^{(0)} + \epsilon_1^{(0)})]\\
\tau_3^{(0)} &= 2[(\lambda + 2\mu)\epsilon_3^{(0)} + \lambda(\epsilon_1^{(0)} + \epsilon_2^{(0)})]\\
\tau_5^{(0)} &= 2\mu\epsilon_5^{(0)}\\
\tau_4^{(1)} &= 2\mu\epsilon_4^{(1)}/3\\
\tau_6^{(1)} &= 2\mu\epsilon_6^{(1)}/3\\
\tau_1^{(2)} &= 2[(\lambda + 2\mu)\epsilon_1^{(2)} + \lambda(\epsilon_2^{(2)} + \epsilon_3^{(2)})]/5\\
\tau_2^{(2)} &= 2[(\lambda + 2\mu)\epsilon_2^{(2)} + \lambda(\epsilon_3^{(2)} + \epsilon_1^{(2)})]/5\\
\tau_3^{(2)} &= 2[(\lambda + 2\mu)\epsilon_3^{(2)} + \lambda(\epsilon_1^{(2)} + \epsilon_2^{(2)})]/5\\
\tau_5^{(2)} &= 2\mu\epsilon_5^{(2)}/5\\
\tau_4^{(3)} &= 2\mu\epsilon_4^{(3)}/7\\
\tau_6^{(3)} &= 2\mu\epsilon_6^{(3)}/7\\
&\;\;\vdots
\end{aligned} \tag{23}$$

where λ and μ are Lamé's constants.

The strain energy density, in the form given in Equation (17), becomes

$$\begin{aligned}
2\bar{U} = {}& \tau_1^{(0)}\epsilon_1^{(0)} + \tau_3^{(0)}\epsilon_3^{(0)} + \tau_5^{(0)}\epsilon_5^{(0)}\\
&+ \tau_2^{(0)}\epsilon_2^{(0)} + \tau_4^{(1)}\epsilon_4^{(1)} + \tau_6^{(1)}\epsilon_6^{(1)}\\
&+ \tau_1^{(2)}\epsilon_1^{(2)} + \tau_3^{(2)}\epsilon_3^{(2)} + \tau_5^{(2)}\epsilon_5^{(2)}\\
&+ \tau_2^{(2)}\epsilon_2^{(2)} + \tau_4^{(3)}\epsilon_4^{(3)} + \tau_6^{(3)}\epsilon_6^{(3)}\\
&+ \ldots
\end{aligned} \tag{24}$$

and, finally, the kinetic energy density (20) is

$$\bar{K} = \rho(\dot{u}_1^{(0)}\dot{u}_1^{(0)} + \dot{u}_3^{(0)}\dot{u}_3^{(0)} + {}^1/_3\dot{u}_2^{(1)}\dot{u}_2^{(1)} + {}^1/_5\dot{u}_1^{(2)}\dot{u}_1^{(2)} + {}^1/_5\dot{u}_3^{(2)}\dot{u}_3^{(2)} + {}^1/_7\dot{u}_2^{(3)}\dot{u}_2^{(3)} + \ldots). \tag{25}$$

Truncation of Series

We begin by setting

$$u_1^{(n)} = 0, \quad u_3^{(n)} = 0, \quad n > 2, \qquad u_2^{(n)} = 0, \quad n > 3. \tag{26}$$

This leaves only the components $u_1^{(0)}$, $u_3^{(0)}$, $u_2^{(1)}$, $u_1^{(2)}$, $u_3^{(2)}$, and $u_2^{(3)}$, as illustrated in Fig. 1. The terms $u_1^{(0)}$ and $u_3^{(0)}$ are the amplitudes of uniform distributions that represent the thickness distributions of displacements which occur in low-frequency extensional and shear motions in the plane of the plate; $u_2^{(1)}$ is the amplitude of a linear distribution of displacement which is

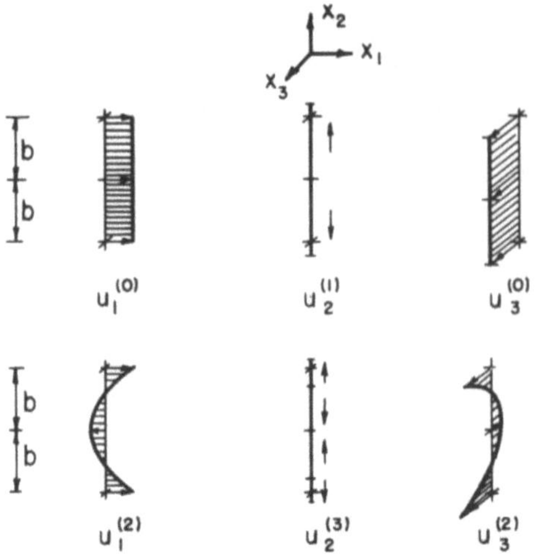

Fig. 1 Components of displacement

an approximation to the exact sinusoidal distribution in the lowest, symmetric, thickness-stretch mode; $u_1^{(2)}$ and $u_3^{(2)}$ are the amplitudes of quadratic distributions of displacements which are approximations to the sinusoidal distributions in the lowest, symmetric, thickness-shear mode, and the face-shear mode of the same order.

The last "displacement" retained ($u_2^{(3)}$) is the amplitude of a cubic distribution which is an approximation to the sinusoidal distribution of displacement in the second, symmetric, thickness-stretch mode. This is a mode of higher order than is to be included in the approximate equations of the second order which we seek. However, $u_2^{(3)}$ produces the second-order strain $\epsilon_2^{(2)}$ [see the eighth of Equations (22)] which, in turn, appears in the second-order stress-strain relations [see the seventh, eighth, and ninth of Equations (23)]. In order to permit the alternating expansion and contraction, through the thickness, which should accompany the stresses $\tau_1^{(2)}$ and $\tau_3^{(2)}$ (by coupling through Poisson's ratio) and at the same time avoid coupling with the undesired higher mode, the thickness stress $\tau_2^{(2)}$ and the velocity $\dot{u}_2^{(3)}$ are set equal to zero. The eighth of Equations (23) is then used to express $\epsilon_2^{(2)}$ in terms of $\epsilon_1^{(2)}$ and $\epsilon_3^{(2)}$:

$$\epsilon_2^{(2)} = -\lambda(\epsilon_3^{(2)} + \epsilon_1^{(2)})/(\lambda + 2\mu) \tag{27}$$

and this result is substituted in the expressions for $\tau_1^{(2)}$ and $\tau_3^{(2)}$ to obtain

$$\begin{aligned} \tau_1^{(2)} &= 2E'(\epsilon_1^{(2)} + \nu\epsilon_3^{(2)})/5, \\ \tau_3^{(2)} &= 2E'(\epsilon_3^{(2)} + \nu\epsilon_1^{(2)})/5, \end{aligned} \tag{28}$$

where $\nu = \lambda/2(\lambda + \mu)$ is Poisson's ratio and

$$E' = 4\mu(\lambda + \mu)/(\lambda + 2\mu) = E/(1 - \nu^2), \tag{29}$$

in which E is Young's modulus. In addition, the contributions of the stresses $\tau_4^{(2)}$ and $\tau_6^{(2)}$ to the strain energy [see Equation (24)] are neglected in order to destroy coupling with the unwanted higher mode when the displacements vary with x_1 and x_3.

At this stage, the stress equations of motion end with the fifth of Equations (21); the strain-displacement relations end with $\epsilon_5^{(2)}$ in Equations (22); the stress-strain relations end with $\tau_5^{(2)}$ in Equations (23), with the expressions for $\tau_1^{(2)}$, $\tau_2^{(2)}$, and $\tau_3^{(2)}$ replaced by Equations (27) and (28); the strain energy density ends with the third line in Equation (24); and the kinetic energy density ends with the fifth term in Equation (25).

Up to this point, the process of truncating the series expressions and adjusting the remaining terms is similar to the one employed by Poisson [2] to obtain the zero-order equations of extensional motions of isotropic plates,[7] the main difference being that here the process is carried on at a level two orders higher. In Poisson's equations only the zero-order displacements $u_1^{(0)}$ and $u_3^{(0)}$ survive. These are the amplitudes of uniform distributions, across the thickness of the plate, which are good approximations to the nearly uniform distributions found in the lowest extensional mode of the exact theory (at long wave lengths) and are exact for the lowest face-shear mode at all wave lengths. The additional terms $u_2^{(1)}$, $u_1^{(2)}$, and $u_3^{(2)}$, which are now included, are the amplitudes of first and second-degree polynomials and these are not good approximations to the distributions in the thickness-stretch and thickness-shear modes of the exact theory. For example, at infinite wave length along the plate, the exact distributions are sinusoidal across the thickness. It is advisable, therefore, to introduce additional adjustments to compensate, as well as possible, for the omission of the polynomials of higher degrees.

The incorrect distributions of displacements affect the frequencies mainly through the thickness strains and velocities. Accordingly, we make the substitutions

$$\begin{aligned} &\kappa_1\epsilon_2^{(0)} \text{ for } \epsilon_2^{(0)}, \qquad &&\kappa_3\dot{u}_2^{(1)} \text{ for } \dot{u}_2^{(1)}, \\ &\kappa_2\epsilon_4^{(1)} \text{ for } \epsilon_4^{(1)}, \qquad &&\kappa_4\dot{u}_1^{(2)} \text{ for } \dot{u}_1^{(2)}, \\ &\kappa_2\epsilon_6^{(1)} \text{ for } \epsilon_6^{(1)}, \qquad &&\kappa_4\dot{u}_3^{(2)} \text{ for } \dot{u}_3^{(2)} \end{aligned}$$

in the strain energy and kinetic energy densities, so that the coefficients κ_r ($r = 1, 2, 3, 4$) will be available for appropriate adjustments of the equations.

Finally, as a matter of expediency, we omit the term $u_2^{(3)}$ from the strain $\epsilon_2^{(0)}$. This term complicates the equations and may be shown to have little influence on the long wave-length end of the spectrum.

Second-Order Approximation

As a result of the truncations and adjustments described in the preceding section, the variables and equations of the second-order approximation are

Kinetic Energy Density:

$$\begin{aligned} \bar{K}^{(2)} = \rho\,[\dot{u}_1^{(0)}\dot{u}_1^{(0)} + \dot{u}_3^{(0)}\dot{u}_3^{(0)} + {}^1/_3\,\kappa_3^2\dot{u}_2^{(1)}\dot{u}_2^{(1)} \\ + {}^1/_5\,\kappa_4^2(\dot{u}_1^{(2)}\dot{u}_1^{(2)} + \dot{u}_3^{(2)}\dot{u}_3^{(2)})] \end{aligned} \tag{30}$$

Strain Energy Density:

$$\begin{aligned} \bar{U}^{(2)} = (\lambda + 2\mu)(\epsilon_1^{(0)}\epsilon_1^{(0)} + \kappa_1^2\epsilon_2^{(0)}\epsilon_2^{(0)} + \epsilon_3^{(0)}\epsilon_3^{(0)}) \\ + 2\lambda(\kappa_1\epsilon_2^{(0)}\epsilon_3^{(0)} + \epsilon_3^{(0)}\epsilon_1^{(0)} + \kappa_1\epsilon_1^{(0)}\epsilon_2^{(0)}) \\ + \mu\epsilon_5^{(0)}\epsilon_5^{(0)} + \mu\kappa_2^2(\epsilon_4^{(1)}\epsilon_4^{(1)} + \epsilon_6^{(1)}\epsilon_6^{(1)})/3 \\ + E'(\epsilon_1^{(2)}\epsilon_1^{(2)} + \epsilon_3^{(2)}\epsilon_3^{(2)} + 2\nu\epsilon_1^{(2)}\epsilon_3^{(2)})/5 + \mu\epsilon_5^{(2)}\epsilon_5^{(2)}/5 \end{aligned} \tag{31}$$

Stress-Strain Relations ($\tau_p^{(n)} = \partial\bar{U}^{(2)}/\partial\epsilon_p^{(n)}$):

$$\begin{aligned} \tau_1^{(0)} &= 2[(\lambda + 2\mu)\epsilon_1^{(0)} + \lambda(\kappa_2\epsilon_2^{(0)} + \epsilon_3^{(0)})] \\ \tau_2^{(0)} &= 2[\kappa_1^2(\lambda + 2\mu)\epsilon_2^{(0)} + \lambda\kappa_1(\epsilon_3^{(0)} + \epsilon_1^{(0)})] \\ \tau_3^{(0)} &= 2[(\lambda + 2\mu)\epsilon_3^{(0)} + \lambda(\epsilon_1^{(0)} + \kappa_1\epsilon_2^{(0)})] \end{aligned}$$

[7] Reference [12], p. 497.

$$
\begin{aligned}
\tau_5^{(0)} &= 2\mu\epsilon_5^{(0)} \\
\tau_4^{(1)} &= 2\mu\kappa_2^2\epsilon_4^{(1)}/3 \\
\tau_6^{(1)} &= 2\mu\kappa_2^2\epsilon_6^{(1)}/3 \\
\tau_1^{(2)} &= 2E'(\epsilon_1^{(2)} + \nu\epsilon_3^{(2)})/5 \\
\tau_3^{(2)} &= 2E'(\epsilon_3^{(2)} + \nu\epsilon_1^{(2)})/5 \\
\tau_5^{(2)} &= 2\mu\epsilon_5^{(2)}/5
\end{aligned} \tag{32}
$$

(The components of stress are illustrated in Fig 2.)

Strain-Displacement Relations:

$$
\begin{aligned}
\epsilon_1^{(0)} &= \partial u_1^{(0)}/\partial x_1 \\
\epsilon_2^{(0)} &= u_2^{(1)}/b \\
\epsilon_3^{(0)} &= \partial u_3^{(0)}/\partial x_3 \\
\epsilon_5^{(0)} &= \partial u_3^{(0)}/\partial x_1 + \partial u_1^{(0)}/\partial x_3 \\
\epsilon_4^{(1)} &= 3u_3^{(2)}/b + \partial u_2^{(1)}/\partial x_3 \\
\epsilon_6^{(1)} &= \partial u_2^{(1)}/\partial x_1 + 3u_1^{(2)}/b \\
\epsilon_1^{(2)} &= \partial u_1^{(2)}/\partial x_1 \\
\epsilon_3^{(2)} &= \partial u_3^{(2)}/\partial x_3 \\
\epsilon_5^{(2)} &= \partial u_3^{(2)}/\partial x_1 + \partial u_1^{(2)}/\partial x_3
\end{aligned} \tag{33}
$$

(The components of strain are illustrated in Fig. 3.)

Stress Equations of Motion:

$$
\begin{aligned}
\frac{\partial\tau_1^{(0)}}{\partial x_1} + \frac{\partial\tau_5^{(0)}}{\partial x_3} + \frac{F_1^{(0)}}{b} &= 2\rho\,\frac{\partial^2 u_1^{(0)}}{\partial t^2} \\
\frac{\partial\tau_5^{(0)}}{\partial x_1} + \frac{\partial\tau_3^{(0)}}{\partial x_3} + \frac{F_3^{(0)}}{b} &= 2\rho\,\frac{\partial^2 u_3^{(0)}}{\partial t^2} \\
\frac{\partial\tau_6^{(1)}}{\partial x_1} + \frac{\partial\tau_4^{(1)}}{\partial x_3} - \frac{\tau_2^{(0)}}{b} + \frac{F_2^{(1)}}{b} &= \frac{2\rho\kappa_3^2}{3}\,\frac{\partial^2 u_2^{(1)}}{\partial t^2} \\
\frac{\partial\tau_1^{(2)}}{\partial x_1} + \frac{\partial\tau_5^{(2)}}{\partial x_3} - \frac{3\tau_6^{(1)}}{b} + \frac{F_1^{(2)}}{b} &= \frac{2\rho\kappa_4^2}{5}\,\frac{\partial^2 u_1^{(2)}}{\partial t^2} \\
\frac{\partial\tau_5^{(2)}}{\partial x_1} + \frac{\partial\tau_3^{(2)}}{\partial x_3} - \frac{3\tau_4^{(1)}}{b} + \frac{F_3^{(2)}}{b} &= \frac{2\rho\kappa_4^2}{5}\,\frac{\partial^2 u_3^{(2)}}{\partial t^2}
\end{aligned} \tag{34}
$$

Displacement Equations of Motion:
[obtained by substituting (33) in (32) and the result in (34)]:

$$
\mu\nabla^2 u_1^{(0)} + (\lambda + \mu)\,\frac{\partial e_0}{\partial x_1} + \frac{\lambda\kappa_1}{b}\,\frac{\partial u_2^{(1)}}{\partial x_1} + \frac{F_1^{(0)}}{2b} = \rho\,\frac{\partial^2 u_1^{(0)}}{\partial t^2}
$$

$$
\mu\nabla^2 u_3^{(0)} + (\lambda + \mu)\,\frac{\partial e_0}{\partial x_3} + \frac{\lambda\kappa_1}{b}\,\frac{\partial u_2^{(1)}}{\partial x_3} + \frac{F_3^{(0)}}{2b} = \rho\,\frac{\partial^2 u_3^{(0)}}{\partial t^2}
$$

$$
\mu\kappa_2^2\nabla^2 u_2^{(1)} - \frac{3\lambda\kappa_1 e_0}{b} - \frac{3\kappa_1^2(\lambda + 2\mu)u_2^{(1)}}{b^2} + \frac{3\mu\kappa_2^2 e_2}{b} + \frac{3F_2^{(1)}}{2b} = \rho\kappa_3^2\,\frac{\partial^2 u_2^{(1)}}{\partial t^2} \tag{35}
$$

$$
\frac{E'}{2}\left[(1 - \nu)\nabla^2 u_1^{(2)} + (1 + \nu)\,\frac{\partial e_2}{\partial x_1}\right] - \frac{5\mu\kappa_2^2}{b}\left(\frac{\partial u_2^{(1)}}{\partial x_1} + \frac{3u_1^{(2)}}{b}\right) + \frac{5F_1^{(2)}}{2b} = \rho\kappa_4^2\,\frac{\partial^2 u_1^{(2)}}{\partial t^2}
$$

$$
\frac{E'}{2}\left[(1 - \nu)\nabla^2 u_3^{(2)} + (1 + \nu)\,\frac{\partial e_2}{\partial x_3}\right] - \frac{5\mu\kappa_2^2}{b}\left(\frac{\partial u_2^{(1)}}{\partial x_3} + \frac{3u_3^{(2)}}{b}\right) + \frac{5F_3^{(2)}}{2b} = \rho\kappa_4^2\,\frac{\partial^2 u_3^{(2)}}{\partial t^2}
$$

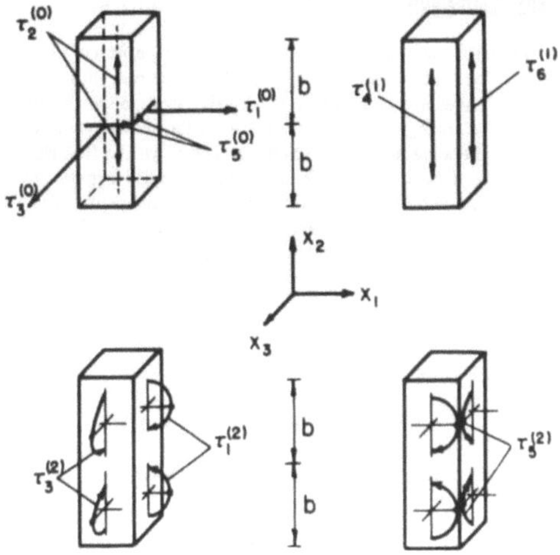

Fig. 2 Components of stress

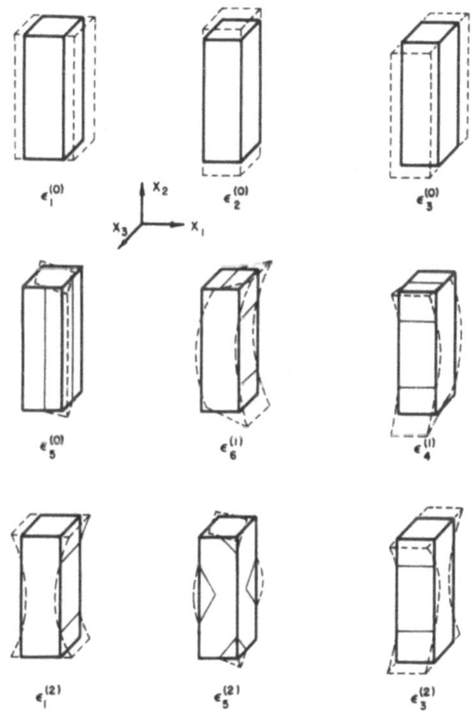

Fig. 3 Components of strain

where

$$
\nabla^2 = \frac{\partial^2}{\partial x_1^2} + \frac{\partial^2}{\partial x_3^2},\quad e_0 = \frac{\partial u_1^{(0)}}{\partial x_1} + \frac{\partial u_3^{(0)}}{\partial x_3},\quad e_2 = \frac{\partial u_1^{(2)}}{\partial x_1} + \frac{\partial u_3^{(2)}}{\partial x}.
$$

The Coefficients κ_i

The coefficients κ_i are determined from a comparison of properties of the solutions of Equations (35), for the case of straight-crested waves in an infinite plate, with the corresponding properties of the analogous solution of the three-dimensional equations. Taking the wave normal in the direction of x_1, we find that Equations (35) separate into a group of three coupled equations on $u_1^{(0)}$, $u_2^{(1)}$, and $u_1^{(2)}$, and two independent equations: one on $u_3^{(0)}$ and one on $u_3^{(2)}$. The coupled equations govern three "compressional modes" and the remaining two equations govern "face-shear" modes. To consider, first, the group of three, we set $F_j^{(n)} = 0$ and

$$u_1^{(0)} = A \sin \xi x_1 e^{i\omega t}, \quad u_3^{(0)} = 0,$$
$$u_2^{(1)} = B \cos \xi x_1 e^{i\omega t}, \tag{36}$$
$$u_1^{(2)} = C \sin \xi x_1 e^{i\omega t}, \quad u_3^{(2)} = 0$$

in Equations (35) and obtain the secular equation

$$\begin{vmatrix} a_{11} & a_{12} & 0 \\ a_{12} & a_{22} & a_{23} \\ 0 & a_{23} & a_{33} \end{vmatrix} = 0 \tag{37}$$

where
$$a_{11} = k^2 z^2 - \Omega^2$$
$$a_{22} = \kappa_2^2 z^2/3 + 4\kappa_1^2 k^2/\pi^2 - \kappa_3^2\Omega^2/3$$
$$a_{33} = E' z^2/5\mu + 12\kappa_2^2/\pi^2 - \kappa_4^2\Omega^2/5$$
$$a_{12} = 2\kappa_1(k^2 - 2)z/\pi$$
$$a_{23} = -2\kappa_2^2 z/\pi$$

and
$$z = 2\xi b/\pi, \quad \Omega = \omega/\omega_s, \quad \omega_s = \pi(\mu/\rho)^{1/2}/2b,$$
$$k^2 = (\lambda + 2\mu)/\mu = 2(1-\nu)/(1-2\nu). \tag{38}$$

We also obtain the three sets of amplitude ratios

$$A_i : B_i : C_i = 1 : a_i/z_i : a_i/b_i, \quad i = 1, 2, 3 \tag{39}$$

where

$$a_i = \pi(\Omega^2 - k^2 z_i^2)/2\kappa_1(k^2 - 2)$$
$$b_i = \frac{3}{2\pi}\left(\frac{\pi^2\kappa_4^2\Omega^2}{15\kappa_2^2} - 4\right) - \frac{2\pi(k^2-1)z_i^2}{5\kappa_2^2 k^2}$$

and the z_i are the roots of Equation (37).

It may be seen that z, which is proportional to the wave number ξ, is the ratio of the thickness of the plate to the real half wave length along the plate; k is the ratio of the velocity of dilatational waves to the velocity of equivoluminal waves in an infinite medium; and Ω is the ratio of the frequency to the frequency of the lowest antisymmetric thickness-shear mode.

The relation between Ω and z, in Equation (37), should match, as closely as possible, the corresponding relation obtained from the three-dimensional equations. The match is improved, within the framework of the present approximation, by choosing appropriate values for the coefficients κ_i; but, since the κ_i are constants, a perfect match can be made only at one value of z for each of them. Now, large enough z corresponds to frequencies high enough to enter the range of modes that have not been included in the approximate equations. In a plate vibrating at such high frequencies under practical (i.e., not mixed) edge conditions, the high modes would, in general, couple with the ones of lower order. Thus the applicability of the approximate equations to bounded plates is limited to frequencies below the lowest frequency of the lowest neglected mode. There is, then, little advantage to be gained in matching the approximate and exact solutions at short wave lengths (large z) at the expense of a good match at long wave lengths. In fact, we go to the extreme and do all of the matching in the neighborhood of $z = 0$ primarily because of the intricate behavior of the exact solution at long wave lengths and also because this choice results in a reasonably good match out to as short wave lengths as the frequency limitation permits.

When $z = 0$, Equation (37) has the three roots

$$\Omega^2 = 0, \quad 12\kappa_1^2 k^2/\pi^2\kappa_3^2, \quad 60\kappa_2^2/\pi^2\kappa_4^2 \tag{40}$$

corresponding to the amplitude ratios

$$A:B:C = 1:0:0, \quad 0:1:0, \quad 0:0:1$$

respectively. It may be seen that, at infinite wave length, the three compressional modes uncouple to form an extensional, a thickness-stretch, and a thickness-shear mode. The three limiting frequencies are to be compared with the exact values

$$\Omega^2 = 0, \quad k^2, \quad 4 \tag{41}$$

of the cut-off frequencies of the extensional, first thickness-stretch and first symmetric thickness-shear modes in Rayleigh's exact solution [7]. Hence, we set

$$\kappa_1/\kappa_3 = \pi/\sqrt{12}, \quad \kappa_2/\kappa_4 = \pi/\sqrt{15} \tag{42}$$

and these relations between pairs of κ_i are used in what follows.

We proceed to examine the roots of Equation (37) in the neighborhood of the roots given in Equations (40) or (41) (which are now the same).

First, for $z \ll 1$, $\Omega \ll 1$, we find

$$\Omega^2 = 4z^2(k^2 - 1)/k^2, \tag{43}$$

i.e., phase velocity $[E/\rho(1-\nu^2)]^{1/2}$. This is exactly the result obtained from the three-dimensional theory for the lowest mode at long wave lengths.

Next, for $z \ll 1$, $\Omega = k + \epsilon$, $\epsilon \ll 1$, Equation (37) gives

$$\frac{d\Omega}{dz} = \frac{z}{k^2 - 4}\left(\frac{\pi^2 k\kappa_2^2}{12\kappa_1^2} + \frac{(k^2-2)^2(k^2-4)}{k^3}\right) \tag{44a}$$

and in the limit, as $k \to 2$ and $z \to 0$,

$$\frac{d\Omega}{dz} = \pm\frac{\pi\kappa_2}{4\kappa_1\sqrt{3}}. \tag{45a}$$

Finally, for $z \ll 1$, $\Omega = 2 + \epsilon$, $\epsilon \ll 1$, we find

$$\frac{d\Omega}{dz} = \frac{\pi^2 z}{6(k^2 - 4)}\left(\frac{4(k^2-1)(k^2-4)}{5k^2\kappa_2^2} - \frac{\kappa_2^2}{\kappa_1^2}\right) \tag{46a}$$

and in the limit, as $k \to 2$ and $z \to 0$,

$$\frac{d\Omega}{dz} = \mp\frac{\pi\kappa_2}{4\kappa_1\sqrt{3}}. \tag{47a}$$

Equations (44a) to (47a) are to be compared with the exact values[8] given in the following correspondingly numbered equations:

$$\frac{d\Omega}{dz} = \frac{4kz}{\pi}\left(\frac{\pi}{4} + \frac{4}{k^3}\cot\frac{\pi k}{2}\right) \tag{44b}$$

$$\frac{d\Omega}{dz} = \pm\frac{2}{\pi} \tag{45b}$$

$$\frac{d\Omega}{dz} = \frac{z}{\pi}\left(\frac{\pi}{2} - \frac{4}{k}\tan\frac{\pi}{k}\right) \tag{46b}$$

[8] Reference [5], p. 2.43.

$$\frac{d\Omega}{dz} = \mp\frac{2}{\pi} \tag{47b}$$

Equating (44a) to (44b) and (46a) to (46b), we find

$$\frac{\kappa_2^2}{\kappa_1^2} = \frac{48(k^2-4)}{\pi^3 k^4}\left[\pi(k^2-1) + 4k\cot\frac{\pi k}{2}\right],$$

$$\frac{1}{\kappa_1^2} = \frac{5k^2\kappa_2^2}{4\kappa_1^2(k^2-1)(k^2-4)} \left[\frac{\kappa_2^2}{\kappa_1^2} + \frac{6(k^2-4)}{\pi^3}\left(\frac{\pi}{2} - \frac{4}{k}\tan\frac{\pi}{k}\right)\right]. \tag{48}$$

We also note, from Equations (48), that

$$\lim_{k\to 2}\frac{\kappa_2^2}{\kappa_1^2} = \frac{192}{\pi^4},$$

$$\lim_{k\to 2}\frac{1}{\kappa_1^2} = \frac{3840(\pi^2-6)}{\pi^8}. \tag{49}$$

Equations (42) and (48) give the values of the four coefficients κ_i in terms of Poisson's ratio. As a result of these relations, all of the ordinates, slopes, and curvatures of the curves Ω versus z, characterizing the frequency spectrum of the compressional modes of the approximate theory, are identical with those of the exact solution at $z = 0$. Equations (48) and, in fact, all of the equations of this second-order approximation should be used only for Poisson's ratios less than about $^7/_{16}$. Above that value, the frequency of the thickness-stretch mode is so high in the spectrum that, in the exact theory, coupling with the second symmetric thickness-shear mode becomes important and that mode is not included in the present approximation.

Turning, now, to the two face-shear modes, we set $F_j^{(n)} = 0$,

$$u_1^{(0)} = u_2^{(1)} = u_1^{(2)} = 0,$$

$$u_3^{(0)} = A_1\cos\xi x_1\, e^{i\omega t}, \tag{50}$$

$$u_3^{(2)} = A_2\cos\xi x_1\, e^{i\omega t},$$

in Equations (35), and find that $u_3^{(0)}$ and $u_3^{(2)}$ are not coupled and have frequencies given by

$$\Omega^2 = z^2,\quad 4 + z^2/\kappa_4^2 \tag{51}$$

respectively. These are to be compared with the frequencies

$$\Omega^2 = z^2,\quad 4 + z^2 \tag{52}$$

of the face-shear modes of orders zero and two obtained from the exact equations. It may be seen that the zero-order face-shear mode is reproduced exactly, in the approximate equations, for all wave lengths. The second order face-shear mode has the correct ordinate (Ω) and slope ($d\Omega/dz$) at $z = 0$ but it does not have the correct curvature because of the presence of the coefficient κ_4. By choosing $\kappa_4 = 1$, this discrepancy could be eliminated, but only at the expense of incorrect behavior of one of the coupled modes.

Frequency Spectrum

The spectrum of frequencies of the five possible modes of vibration of an infinite plate, which the approximate theory contains, is given by the five independent roots of Equations (37) and (51). We consider, first, the three compressional modes.

Equation (37) relates Ω^2 and z^2 but only real, positive values of Ω have physical significance. For real, positive Ω, the roots z^2 may be three real positive or two real positive and one real negative or one real positive and two conjugate complex. Hence the roots

$$z = x + iy$$

may be three real or two real and one imaginary or one real and two conjugate complex. In addition, the character of the spectrum is different according as Poisson's ratio is less than equal to, or greater than $^1/_3$.

We consider the case $\nu < {^1/_3}$ (i.e., $k < 2$) first. Then, when $\Omega > 2$, the three roots of Equation (37) are real ($y = 0$). They are illustrated in Fig. 4(*a*) (for $\nu = {^1/_4}$, i.e., $k^2 = 3$) by the three curves (full lines) marked ϕ_1, ϕ_2, ϕ_3 in the plane $y = 0$ and the region $\Omega > 2$. As Ω drops below $\Omega = 2$, the largest root ϕ_1 approaches zero with asymptotic behavior given by Equation (43); the intermediate root ϕ_2 approaches a minimum Ω^*, below which there are no real roots other than ϕ_1; the smallest root ϕ_3 approaches a minimum in the real plane $y = 0$ at $\Omega = 2$, $x = 0$ with asymptotic behavior given by Equation (46a) or (46b) which are the same in view of Equations (48). Continuing with the root ϕ_3, as Ω drops below $\Omega = 2$, the root is imaginary ($x = 0$), again with asymptotic behavior given by Equation (46a) or (46b) in the neighborhood of $\Omega = 2$. As Ω approaches k, from above, the root ϕ_3 forms a loop in the imaginary plane ($x = 0$) and approaches $\Omega = k$, $y = 0$ with asymptotic behavior given by Equation (44a) or (44b). As Ω continues to drop below $\Omega = k$, the root ϕ_3 becomes real, with behavior in the neighborhood of $\Omega = k$, $x = 0$ again given by (44a) but now x is negative. Upon further diminution of Ω, the root ϕ_3 approaches a minimum at $\Omega = \Omega^*$ and negative x. This portion of ϕ_3 (i.e., between $\Omega = k$ and $\Omega = \Omega^*$) is identified in Fig. 4(*a*) by $[\phi_3]$, where the brackets indicate that it is the reflection in the plane $x = 0$ that is shown. Finally, when $\Omega < \Omega^*$, the roots ϕ_2 and ϕ_3 are conjugate complex ($z = \pm x + iy$). One of them is shown, in Fig. 4(*a*), as the curve marked ϕ_2; this curve is also $[\phi_3]$, i.e., the reflection of the conjugate root ϕ_3 in the plane $x = 0$.

As Poisson's ratio approaches $^1/_3$ from below, the frequency of the thickness-stretch mode increases, approaching the frequency of the thickness-shear mode; i.e., at $z = 0$ the intercept $\Omega = k$ approaches $\Omega = 2$. Conjointly, the curvature of ϕ_3 at $\Omega = k$, $z = 0$ approaches negative infinity in the plane $y = 0$ and positive infinity in the plane $x = 0$ while both slopes remain zero; all in accordance with Equation (44a) or (44b). At the same time, the curvature of ϕ_3 at $\Omega = 2$, $z = 0$ approaches positive infinity in the plane $y = 0$ and negative infinity in the plane $x = 0$ while the slopes remain zero: all in accordance with Equation (46a) or (46b). Meanwhile, the imaginary loop shrinks toward the point $\Omega = 2$, $y = 0$.

At $\nu = {^1/_3}$, the thickness-stretch and thickness-shear modes have the same frequency and the slopes of the two branches of ϕ_3 become $2/\pi$ in accordance with Equations (45a) and (47a). This situation is illustrated in Fig. 4(*b*). Here, again, the portions marked $[\phi_3]$ are the reflections, in the plane $x = 0$ of the actual branches.

When $\nu > {^1/_3}$ the thickness-stretch mode has a frequency higher than that of the thickness-shear mode. The spectrum [illustrated in Fig. 4(*c*) for $\nu = {^3/_8}$] has now undergone an important change in that the imaginary branch loops up, from $\Omega = 2$, instead of down.

Since only powers of z^2 occur in Equation (35), there is another set of physically significant roots given by the reflections of the curves of Fig. 4 in the planes $x = 0$ and $y = 0$.

Turning, now, to the face-shear modes, the first root in Equation (51) yields the straight line marked H_0 in Fig. 4. The second face-shear mode gives the roots marked H_2 in the figures; these roots are real for $\Omega > 2$ and imaginary for $\Omega < 2$. As before, there is an additional set of roots given by the reflections of the curves H_0 and H_2 in the planes $x = 0$ and $y = 0$.

The spectra of the corresponding five modes, as computed from

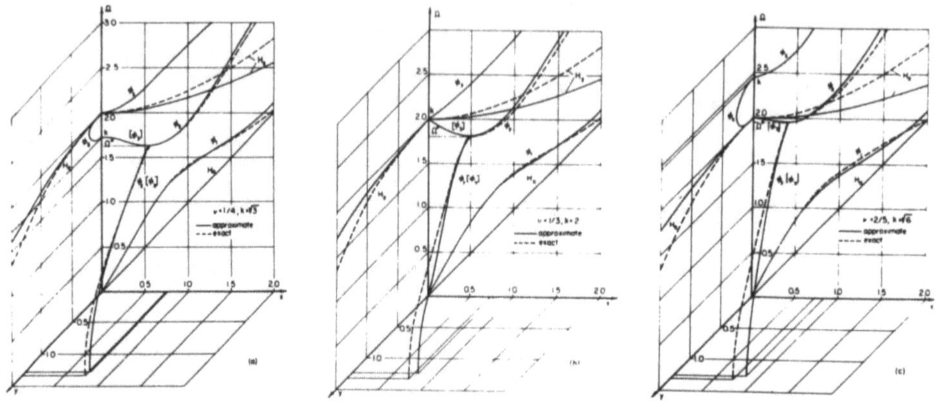

Fig. 4 Frequency spectrum of an infinite plate. Comparison of approximate and exact branches for three values of Poisson's ratio.

Rayleigh's solution of the exact equations, are also shown[9] in Figs. 4(*a*), (*b*), and (*c*) (as dashed lines). The importance of the introduction of the coefficients κ_i and their definitions, in terms of Poisson's ratio, is apparent. Without these coefficients, the extraordinarily complicated behavior of the branches of the exact frequency spectrum at long wave lengths would not be reproduced in the approximate equations. In the whole range of frequencies and wave numbers depicted, the approximate spectrum is reasonably close to the exact one; the poorest representation occurring in portions of the complex branches and the spectrum of the second face-shear mode. On the whole, fair results may be expected from solutions of the approximate equations in the case of finite plates.

The shapes of modes in the various ranges of frequency can be anticipated from the real, imaginary, or complex character of the roots. In rectangular co-ordinates, for example, real roots correspond to trigonometric mode-shapes; imaginary roots to exponential or hyperbolic mode-shapes; and complex roots correspond to modes whose shapes are given by products of trigonometric and exponential or hyperbolic functions. A striking example of the latter is to be found in the experiments with circular disks by Shaw [13].

The phase velocities (v) and group velocities (v_g) can be visualized readily from Fig. 4 inasmuch as

$$v = \frac{\omega}{\xi} = \frac{\Omega}{x}\left(\frac{\mu}{\rho}\right)^{1/2},$$
$$v_g = \frac{d\omega}{d\xi} = \frac{d\Omega}{dx}\left(\frac{\mu}{\rho}\right)^{1/2}. \qquad (53)$$

For example, the phenomenon of phase and group velocities of opposite sign, noticed by Tolstoy and Usdin [11] in Rayleigh's solution, is represented by the branch $[\phi_3]$ in the real plane. Also, the minimum at Ω^* represents zero group velocity and nonzero phase velocity. As the wave length approaches zero ($x \to \infty$) the frequencies rise beyond the range of applicability of the equations and the asymptotic behavior of the velocities of the three compressional modes are

$$v = v_g = \frac{\kappa_2}{\kappa_3}\left(\frac{\mu}{\rho}\right)^{1/2},\ \left(\frac{\lambda + 2\mu}{\rho}\right)^{1/2},\ \frac{1}{\kappa_4}\left[\frac{E}{\rho(1-\nu^2)}\right]^{1/2}. \qquad (54)$$

According to the exact theory, the first of these should be the velocity of Rayleigh surface waves and the second and third should be the velocity of equivoluminal waves.

[9] The complex roots were kindly supplied by Dr. Morio Onoe.

Additional Results

(*a*) The equations of compatibility may be obtained by eliminating the displacements from Equations (33), with the result:

$$\frac{\partial^2 \epsilon_1^{(0)}}{\partial x_3^2} + \frac{\partial^2 \epsilon_3^{(0)}}{\partial x_1^2} = \frac{\partial^2 \epsilon_5^{(0)}}{\partial x_1 \partial x_3}$$
$$\frac{\partial^2 \epsilon_2^{(0)}}{\partial x_1^2} + \frac{3\epsilon_1^{(2)}}{b^2} = \frac{1}{b}\frac{\partial \epsilon_6^{(1)}}{\partial x_1}$$
$$\frac{3\epsilon_3^{(2)}}{b^2} + \frac{\partial^2 \epsilon_2^{(0)}}{\partial x_3^2} = \frac{1}{b}\frac{\partial \epsilon_4^{(1)}}{\partial x_3}$$
$$\frac{\partial^2 \epsilon_1^{(2)}}{\partial x_3^2} + \frac{\partial^2 \epsilon_3^{(2)}}{\partial x_1^2} = \frac{\partial^2 \epsilon_5^{(2)}}{\partial x_1 \partial x_3} \qquad (55)$$
$$\frac{3}{b}\frac{\partial \epsilon_1^{(2)}}{\partial x_3} = \frac{1}{2}\frac{\partial}{\partial x_1}\left(-\frac{\partial \epsilon_4^{(1)}}{\partial x_1} + \frac{3\epsilon_5^{(2)}}{b} + \frac{\partial \epsilon_6^{(1)}}{\partial x_3}\right)$$
$$\frac{\partial^2 \epsilon_2^{(0)}}{\partial x_1 \partial x_3} = \frac{1}{2b}\left(-\frac{3\epsilon_5^{(2)}}{b} + \frac{\partial \epsilon_6^{(1)}}{\partial x_3} + \frac{\partial \epsilon_4^{(1)}}{\partial x_1}\right)$$
$$\frac{3}{b}\frac{\partial \epsilon_3^{(2)}}{\partial x_1} = \frac{1}{2}\frac{\partial}{\partial x_3}\left(-\frac{\partial \epsilon_6^{(1)}}{\partial x_3} + \frac{\partial \epsilon_4^{(1)}}{\partial x_1} + \frac{3\epsilon_5^{(2)}}{b}\right)$$

The first of (55) is the usual compatibility equation of generalized plane stress and the remaining six equations correspond to the ordinary six compatibility equations; the main differences being that here the components $\epsilon_p^{(2)}$ have a factor 3 and the operator $\partial/\partial x_2$ is replaced by $1/b$.

(*b*) It also may be shown that there are nine dislocations when the nine components of strain and their derivatives are continuous. Three of them are the two translational and one rotational dislocations of generalized plane stress. Of the remaining six, three are translational (in the displacements $u_1^{(2)}$, $u_2^{(1)}$, $u_3^{(2)}$) and three are rotational, in the component of rotation

$$\frac{1}{2}\left(\frac{3u_3^{(2)}}{b} - \frac{\partial u_2^{(1)}}{\partial x_3}\right),\ \frac{1}{2}\left(\frac{\partial u_3^{(2)}}{\partial x_1} - \frac{\partial u_1^{(2)}}{\partial x_3}\right),$$
$$\frac{1}{2}\left(\frac{\partial u_2^{(1)}}{\partial x_1} - \frac{3u_1^{(2)}}{b}\right). \qquad (56)$$

(c) A theorem of uniqueness of solutions of the approximate equations of motion may be established along the lines of Neumann's theorem.[10] This leads to the following conditions sufficient for a unique solution (in the absence of discontinuities and singularities):

(i) Initial values of $u_1^{(0)}$, $u_3^{(0)}$, $u_2^{(1)}$, $u_1^{(2)}$, and $u_3^{(2)}$ throughout the plate.

(ii) Initial values of $\dot{u}_1^{(0)}$, $\dot{u}_3^{(0)}$, $\dot{u}_2^{(1)}$, $\dot{u}_1^{(2)}$, and $\dot{u}_3^{(2)}$ throughout the plate.

(iii) One member of each of the five products $F_1^{(0)}u_1^{(0)}$, $F_3^{(0)}u_3^{(0)}$, $F_2^{(1)}u_2^{(1)}$, $F_1^{(2)}u_1^{(2)}$, $F_3^{(2)}u_3^{(2)}$ at each point of the plate; i.e., five surface conditions.

(iv) At each point on the edge of the plate (normal n tangent s) one member of each of the five products $\tau_{nn}^{(0)}u_n^{(0)}$, $\tau_{ns}^{(0)}u_s^{(0)}$, $\tau_{n2}^{(1)}u_2^{(1)}$, $\tau_{nn}^{(2)}u_n^{(2)}$, $\tau_{ns}^{(2)}u_s^{(2)}$; i.e., five edge conditions.

The requirement that the strain energy-density $\bar{U}^{(2)}$, Equation (31), be positive definite is satisfied by the addition of the requirements $\kappa_1 > 0$, $\kappa_2 > 0$ to the usual requirements $3\lambda + 2\mu > 0$, $\mu > 0$.

(d) In the case of steady vibrations, the displacements may be expressed conveniently in terms of potentials that satisfy Helmholtz equations. Omitting a factor $e^{i\omega t}$ the results are

$$u_1^{(0)} = \frac{\partial\phi_1}{\partial x_1} + \frac{\partial\phi_2}{\partial x_1} + \frac{\partial\phi_3}{\partial x_1} - \frac{\partial H_0}{\partial x_3}$$

$$u_3^{(0)} = \frac{\partial\phi_1}{\partial x_3} + \frac{\partial\phi_2}{\partial x_3} + \frac{\partial\phi_3}{\partial x_3} + \frac{\partial H_0}{\partial x_1}$$

$$u_2^{(1)} = \alpha_1\phi_1 + \alpha_2\phi_2 + \alpha_3\phi_3 \qquad (57)$$

$$u_1^{(2)} = \beta_1\frac{\partial\phi_1}{\partial x_1} + \beta_2\frac{\partial\phi_2}{\partial x_1} + \beta_3\frac{\partial\phi_3}{\partial x_1} - \frac{\partial H_2}{\partial x_3}$$

$$u_3^{(2)} = \beta_1\frac{\partial\phi_1}{\partial x_3} + \beta_2\frac{\partial\phi_2}{\partial x_3} + \beta_3\frac{\partial\phi_3}{\partial x_3} + \frac{\partial H_2}{\partial x_1}$$

$$\mu\nabla^2 H_0 + \rho\omega^2 H_0 = 0$$

$$\mu\nabla^2 H_2 + (\kappa_1^2\rho\omega^2 - 15b^{-2}\kappa_2^2\mu)H_2 = 0 \qquad (58)$$

$$\nabla^2\phi_i + \xi_i^2\phi_i = 0, \quad i = 1, 2, 3$$

where

$$\alpha_i = b[(\lambda + 2\mu)\xi_i^2 - \rho\omega^2]/\kappa_1\lambda$$

$$\beta_i = \alpha_i b\Big/\left[\frac{3}{4}(\Omega^2 - 4) - \frac{E'b^2\xi_i^2}{5\mu\kappa_2^2}\right] \qquad (59)$$

and the ξ_i^2 are, again, the roots of Equation (37). It may be seen that H_0 and H_2 are the potentials of the two face-shear modes and the ϕ_i are the potentials of the three compressional modes.

(e) The tensor, and hence invariant, characters of the quantities that appear in the second-order equations are, for the most part, apparent. For example, in the displacement equations of motion (35), if the first and second equations are regarded as the rectangular components of a vector and the fourth and fifth equations the rectangular components of another vector, the only differential operators that appear are the gradient, divergence, Laplacian, and $\partial^2/\partial t^2$: all invariants. The dependent variables are the scalar $u_2^{(1)}$ and the two vectors

$$\mathbf{u}^{(0)} = u_1^{(0)}\mathbf{k}_1 + u_3^{(0)}\mathbf{k}_3 = u_\alpha^{(0)}\mathbf{k}_\alpha + u_\gamma^{(0)}\mathbf{k}_\gamma,$$
$$\mathbf{u}^{(2)} = u_1^{(2)}\mathbf{k}_1 + u_3^{(2)}\mathbf{k}_3 = u_\alpha^{(2)}\mathbf{k}_\alpha + u_\gamma^{(2)}\mathbf{k}_\gamma, \qquad (60)$$

[10] References [11], p. 176, and [9].

while the gradient operator is

$$\nabla = \mathbf{k}_1\frac{\partial}{\partial x_1} + \mathbf{k}_3\frac{\partial}{\partial x_3} = \mathbf{k}_\alpha\frac{\partial}{\partial s_\alpha} + \mathbf{k}_\gamma\frac{\partial}{\partial s_\gamma} \qquad (61)$$

where $\mathbf{k}_1$, $\mathbf{k}_3$, and $\mathbf{k}_\alpha$, $\mathbf{k}_\gamma$ are unit vectors in the rectangular directions 1, 3, and the orthogonal curvilinear directions α, γ, respectively: all in the plane of the plate.

The appropriate strain tensors and their expression in terms of vector displacements are not quite as apparent. We define another vector displacement

$$\mathbf{u}' = \mathbf{u}^{(2)} + \mathbf{k}_2 u_2^{(1)} \qquad (62)$$

and a gradient operator

$$\nabla' = \nabla + \mathbf{k}_2/b \qquad (63)$$

where $\mathbf{k}_2$ is a unit vector normal to the plane of the plate. Then the two tensors

$$\boldsymbol{\varepsilon}^{(0)} = {}^1/_2(\nabla\mathbf{u}^{(0)} + \mathbf{u}^{(0)}\nabla),$$
$$\boldsymbol{\varepsilon}' = {}^1/_2(\nabla'\mathbf{u}' + \mathbf{u}'\nabla') \qquad (64)$$

constitute the strain. The seven equations of compatibility, for example, become

$$\nabla \times \boldsymbol{\varepsilon}^{(0)} \times \nabla = 0,$$
$$\nabla' \times \boldsymbol{\varepsilon}' \times \nabla' = 0. \qquad (65)$$

References

1 T. R. Kane and R. D. Mindlin, "High-Frequency Extensional Vibrations of Plates," Journal of Applied Mechanics, vol. 23, Trans. ASME, vol. 78, 1956, pp. A-277–283.

2 S. D. Poisson, "Mémoire sur l'Équilibre et le Mouvement des Corps Élastiques," *Mémoires de l'Académie des Sciences*, series 2, vol. 8, Paris, 1829, pp. 357–570.

3 A. L. Cauchy, "Sur l'Équilibre et le Mouvement d'une Plaque Solide," *Exercices de Mathématique*, vol. 3, 1828, pp. 328–355.

4 G. Kirchhoff, "Über das Gleichgewicht und die Bewegung einer Elastichen Scheibe," *Crelles Journal*, vol. 40, 1850, pp. 51–88.

5 R. D. Mindlin, "An Introduction to the Mathematical Theory of Vibrations of Elastic Plates," Signal Corps Engineering Laboratories, Fort Monmouth, N. J., 1955.

6 E. Reissner, "On the Calculation of Three-Dimensional Corrections for the Two-Dimensional Theory of Plane Stress," *Proceedings of the Fifteenth Semi-Annual Eastern Photoelasticity Conference*, Addison-Wesley Press, Cambridge, Mass., 1942.

7 Lord Rayleigh, "On the Free Vibrations of an Infinite Plate of Homogeneous Isotropic Elastic Matter," *Proceedings of The London Mathematical Society*, vol. 20, 1889, pp. 225–234.

8 S. Timoshenko, "On the Correction for Shear of the Differential Equation for Transverse Vibrations of Prismatic Bars," *Philosophical Magazine*, series 6, vol. 41, 1921, pp. 744–746.

9 R. D. Mindlin, "Influence of Rotatory Inertia and Shear on Flexural Vibrations of Isotropic, Elastic Plates," Journal of Applied Mechanics, vol. 18, Trans. ASME, vol. 73, 1951, pp. 31–38.

10 R. R. Aggarwal and E. A. G. Shaw, "Axially Symmetrical Vibrations of a Finite Isotropic Disk. IV," *Journal of the Acoustical Society of America*, vol. 26, 1954, pp. 341–342.

11 I. Tolstoy and E. Usdin, "Wave Propagation in Elastic Plates: Low and High Mode Dispersion," *Journal of the Acoustical Society of America*, vol. 29, 1957, pp. 37–42.

12 A. E. H. Love, "A Treatise on the Mathematical Theory of Elasticity," Cambridge University Press, London, 1927.

13 E. A. G. Shaw, "On the Resonant Vibrations of Thick Barium Titanate Disks," *Journal of the Acoustical Society of America*, vol. 28, 1956, pp. 38–50.

(*Reprinted from the Journal of Applied Mechanics for March, 1960, with permission of Copyright Owner*)

R. D. MINDLIN
Professor of Civil Engineering, Columbia University, New York, N. Y. Mem. ASME

H. D. McNIVEN
Associate Professor of Civil Engineering, University of California, Berkeley, Calif.

Axially Symmetric Waves in Elastic Rods[1]

A system of approximate, one-dimensional equations is derived for axially symmetric motions of an elastic rod of circular cross section. The equations take into account the coupling between longitudinal, axial shear, and radial modes. The spectrum of frequencies for real, imaginary, and complex wave numbers in an infinite rod is explored in detail and compared with the analogous solution of the three-dimensional equations.

1 Introduction

In a previous paper [1],[2] a derivation was given of approximate equations of motion that take into account the coupling between longitudinal and radial modes of axially symmetric vibration of elastic rods. The frequency of the radial mode depends on Poisson's ratio, ν; the frequency increasing with increasing ν. When $\nu > 0.2833$, the frequency of the radial mode is higher than the frequency, independent of ν, of the lowest axial shear mode. In the latter the displacement is parallel to the axis of the rod and has one circular node. The possibility of such a displacement was not taken into account in the previous paper. In the range of Poisson's ratios commonly encountered, both the axial shear and the radial modes can couple with the longitudinal mode and with each other. This circumstance has a marked influence on phase and group velocities of waves and on the frequencies and shapes of high-frequency vibrational modes.

In the present paper a derivation is given of approximate equations of motion which includes the effects of the additional mode. The approximation is based on expansions of the displacements in series of Jacobi polynomials in the radial co-ordinate, retaining only the earliest terms representative of axial, radial, and axial shear deformation. To offset the error introduced by omission of the terms of higher order, four compensating factors are introduced and chosen in such a way that the behavior of the first three branches of the Pochhammer frequency spectrum is reproduced at long wave lengths. Of particular interest is the approximate reproduction of the complex segments and frequency minima of the second and third branches.

2 Expansion in Infinite Series

The rod is bounded by the cylindrical surface $r = a$ and the planes $z = \pm c$ in a cylindrical co-ordinate system r, θ, z. The circumferential component of displacement u_θ is taken as zero and the radial and axial components are expressed in series of Jacobi polynomials in the radial co-ordinate:

$$u_r = \sum_{n=0}^{\infty} \mathcal{U}_n(\alpha) u_n(z, t) \tag{1}$$

$$u_z = \sum_{n=0}^{\infty} \mathcal{W}_n(\alpha) w_n(z, t) \tag{2}$$

where

$$\alpha = r/a$$

and [2][3]

$$\mathcal{U}_0(\alpha) = \alpha, \quad \mathcal{U}_1(\alpha) = \alpha - \frac{3}{2}\alpha^2, \ldots$$
$$\ldots \mathcal{U}_n(\alpha) = \alpha + \sum_{k=1}^{n} (-1)^k \binom{n}{k} \frac{(n+2)_k}{(k+1)!} \alpha^{2k+1} \tag{3}$$

$$\mathcal{W}_0(\alpha) = 1, \quad \mathcal{W}_1(\alpha) = 1 - 2\alpha^2, \ldots$$
$$\ldots \mathcal{W}_n(\alpha) = 1 + \sum_{k=1}^{n} (-1)^k \binom{n}{k} \frac{(n+1)_k}{k!} \alpha^{2k} \tag{4}$$

in which

$$\binom{n}{k} = \frac{n(n-1)(n-2)\ldots(n-k+1)}{k!} \tag{5}$$

$$(\beta)_k = \beta(\beta+1)(\beta+2)\ldots(\beta+k-1), \quad (\beta)_0 = 1 \tag{6}$$

The polynomials $\mathcal{U}_n$ and $\mathcal{W}_n$ have the properties

$$4(n+1)^2 \int_0^1 \alpha \mathcal{U}_m \mathcal{U}_n d\alpha = \begin{cases} 0, & m \neq n \\ 1, & m = n \end{cases} \tag{7}$$

$$2(2n+1) \int_0^1 \alpha \mathcal{W}_m \mathcal{W}_n d\alpha = \begin{cases} 0, & m \neq n \\ 1, & m = n \end{cases} \tag{8}$$

$$d\mathcal{W}_n/d\alpha = \sum_{k=0}^{n-1} C_{nk} \mathcal{U}_k, \quad C_{nk} = 4(k+1)^2(-1)^{n+k} \tag{9}$$

Stress Equations of Motion. The series expressions (1) and (2) are substituted in the equations

[1] This investigation was supported by the Office of Naval Research under a contract with Columbia University.

[2] Numbers in brackets designate References at end of paper.

Presented at the West Coast Conference of the Applied Mechanics Division, Stanford, Calif., September 9–11, 1959, of The American Society of Mechanical Engineers.

Discussion of this paper should be addressed to the Secretary, ASME, 29 West 39th Street, New York, N. Y., and will be accepted until April 10, 1960, for publication at a later date. Discussion received after the closing date will be returned.

Note: Statements and opinions advanced in papers are to be understood as individual expressions of their authors and not those of the Society. Manuscript received by ASME Applied Mechanics Division, November 17, 1958. Paper No. 59—APMW-2.

[3] In the notation of reference [2],

$$\mathcal{U}_n(x) = x\mathfrak{F}_n(2, 2, x^2)$$
$$\mathcal{W}_n(x) = \mathfrak{F}_n(1, 1, x^2)$$

Note, also, that

$$\mathcal{U}_n(1) = (-1)^n/n!$$
$$\mathcal{W}_n(1) = (-1)^n$$

$$\int_{-c}^{c}\int_{0}^{a}\left(\frac{\partial\sigma_r}{\partial r}+\frac{\partial\tau_{rz}}{\partial z}+\frac{\sigma_r-\sigma_\theta}{r}-\rho\frac{\partial^2 u_r}{\partial t^2}\right)\delta u_r\, rdr\,dz = 0 \quad (10)$$

$$\int_{-c}^{c}\int_{0}^{a}\left(\frac{\partial\tau_{rz}}{\partial r}+\frac{\partial\sigma_z}{\partial z}+\frac{\tau_{rz}}{r}-\rho\frac{\partial^2 u_z}{\partial t^2}\right)\delta u_z\, rdr\,dz = 0$$

which are obtained from the variational equations of motion [3]. In (10), σ_r, σ_θ, σ_z, τ_{rz} are components of stress derivable from a strain-energy-function U by differentiation with respect to components of strain ϵ_r, ϵ_θ, ϵ_z, γ_{rz}, respectively. By substituting (1) and (2) into (10), performing the integrations with respect to r and setting the coefficients of δu_n and δw_n equal to zero, we obtain stress equations of motion of order n:

$$\left.\begin{aligned} \frac{dQ_n}{dz}-\frac{P_n}{a}+\frac{(-1)^n}{n+1}\frac{R}{a} &= \frac{\rho}{4(n+1)^3}\frac{\partial^2 u_n}{\partial t^2} \\ \frac{dF_n}{dz}-\sum_{k=0}^{n-1}C_{nk}\frac{Q_k}{a}+(-1)^n\frac{Z}{a} &= \frac{\rho}{2(2n+1)}\frac{\partial^2 w_n}{\partial t^2} \end{aligned}\right\} \quad (11)$$

where

$$\left.\begin{aligned} P_n &= \int_0^1 (\sigma_\theta \mathfrak{U}_n+\sigma_r\alpha d\mathfrak{U}_n/d\alpha)d\alpha \\ Q_n &= \int_0^1 \tau_{rz}\mathfrak{U}_n\alpha d\alpha \\ F_n &= \int_0^1 \sigma_z \mathfrak{W}_n\alpha d\alpha, \end{aligned}\right\} \quad (12)$$

$$R = [\sigma_r]_{r=a},\quad Z = [\tau_{rz}]_{r=a} \quad (13)$$

Components of Strain. Upon substituting the expansions (1) and (2) in the usual strain-displacement relations, we obtain

$$\left.\begin{aligned} \epsilon_r &= \frac{\partial u_r}{\partial r} = \sum_{n=0}^{\infty}\frac{\partial \mathfrak{U}_n}{\partial\alpha}\frac{u_n}{a} \\ \epsilon_\theta &= \frac{u_r}{r} = \sum_{n=0}^{\infty}\frac{\mathfrak{U}_n}{\alpha}\frac{u_n}{a} \\ \epsilon_z &= \frac{\partial u_z}{\partial z} = \sum_{n=0}^{\infty}\mathfrak{W}_n\frac{\partial w_n}{\partial z} \\ \gamma_{rz} &= \frac{\partial u_r}{\partial z}+\frac{\partial u_z}{\partial r} = \sum_{n=0}^{\infty}\left(\mathfrak{U}_n\frac{\partial u_n}{\partial z}+\frac{\partial\mathfrak{W}_n}{\partial\alpha}\frac{w_n}{a}\right) \end{aligned}\right\} \quad (14)$$

Stress-Strain-Displacement Relations. If the strain-displacement relations (14) are substituted in the stress-strain relations

$$\left.\begin{aligned} \sigma_r &= \partial U/\partial\epsilon_r = \lambda(\epsilon_r+\epsilon_\theta+\epsilon_z)+2\mu\epsilon_r \\ \sigma_\theta &= \partial U/\partial\epsilon_\theta = \lambda(\epsilon_r+\epsilon_\theta+\epsilon_z)+2\mu\epsilon_\theta \\ \sigma_z &= \partial U/\partial\epsilon_z = \lambda(\epsilon_r+\epsilon_\theta+\epsilon_z)+2\mu\epsilon_z \\ \tau_{rz} &= \partial U/\partial\gamma_{rz} = \mu\gamma_{rz} \end{aligned}\right\} \quad (15)$$

and these, in turn, substituted in (12), we obtain, after using (14) and performing the integrations, relations between the components of stress of order n (P_n, Q_n, F_n) and the displacement amplitudes u_n, w_n of order n. The first few relations are

$$\left.\begin{aligned} P_0 &= 2(\lambda+\mu)\frac{u_0}{a}+\lambda\frac{\partial w_0}{\partial z}+\ldots \\ Q_0 &= \frac{\mu}{4}\left(\frac{\partial u_0}{\partial z}-\frac{4w_1}{a}\right)+\ldots \end{aligned}\right\} \quad (16)$$

$$\left.\begin{aligned} F_0 &= \frac{\lambda+2\mu}{2}\frac{\partial w_0}{\partial z}+\lambda\frac{u_0}{a}+\ldots \\ F_1 &= \frac{\lambda+2\mu}{6}\frac{\partial w_1}{\partial z}+\ldots \end{aligned}\right\} \quad (16)\ (cont.)$$

where the terms represented by dots contain amplitudes u_n of order higher than $n = 0$ and w_n of order higher than $n = 1$.

Energy-Densities. From the strain-energy-density of the three-dimensional theory:

$$\begin{aligned} U &= {}^1/_2[\lambda(\epsilon_r+\epsilon_\theta+\epsilon_z)^2+2\mu(\epsilon_r^2+\epsilon_\theta^2+\epsilon_z^2+{}^1/_2\gamma_{rz}^2)] \\ &= {}^1/_2(\sigma_r\epsilon_r+\sigma_\theta\epsilon_\theta+\sigma_z\epsilon_z+\tau_{rz}\gamma_{rz}) \end{aligned} \quad (17)$$

we define an energy-density

$$\begin{aligned} \bar{U} &= \int_0^1 U\alpha\, d\alpha \\ &= \frac{1}{2}\left[P_0\frac{u_0}{a}+\ldots+F_0\frac{\partial w_0}{\partial z}+F_1\frac{\partial w_1}{\partial z}+\ldots + Q_0\left(\frac{\partial u_0}{\partial z}-\frac{4w_1}{a}\right)+\ldots\right] \end{aligned} \quad (18)$$

Similarly, from the kinetic-energy-density

$$K = \frac{\rho}{2}\left[\left(\frac{\partial u_r}{\partial t}\right)^2+\left(\frac{\partial u_z}{\partial t}\right)^2\right] \quad (19)$$

we define an energy-density

$$\begin{aligned} \bar{K} &= \int_0^1 K\alpha\, d\alpha \\ &= \frac{\rho}{2}\left[{}^1/_4\left(\frac{\partial u_0}{\partial t}\right)^2+{}^1/_2\left(\frac{\partial w_0}{\partial t}\right)^2+{}^1/_6\left(\frac{\partial w_1}{\partial t}\right)^2+\ldots\right] \end{aligned} \quad (20)$$

3 Second-Order Approximation

We retain only the terms associated with displacement amplitudes u_0, w_0, and w_1. To simplify the notation, let

$$u = u_0,\quad w = w_0,\quad \psi = w_1 \quad (21)$$

$$\left.\begin{aligned} P_r &= P_0 = \int_0^1 (\sigma_r+\sigma_\theta)\alpha d\alpha \\ P_z &= P_0 = \int_0^1 \sigma_z\alpha d\alpha \\ P_\psi &= F_1 = \int_0^1 \sigma_z(1-2\alpha^2)\alpha d\alpha \\ P_{rz} &= Q_0 = \int_0^1 \tau_{rz}\alpha^2 d\alpha \end{aligned}\right\} \quad (22)$$

$$\left.\begin{aligned} \Gamma_r &= u/a & \Gamma_z &= \partial w/\partial z \\ \Gamma_\psi &= \partial\psi/\partial z & \Gamma_{rz} &= \partial u/\partial z - 4\psi/a \end{aligned}\right\} \quad (23)$$

The components of displacement, strain, stress, and surface traction are illustrated in Fig. 1. It may be observed that w, u- and ψ are the amplitudes of uniform, linear, and quadratic distributions of axial (w, ψ) and radial (u) displacements along a radius of the rod.[4]

[4] Thus the three-dimensional displacements are represented in the second-order approximation by

$$u_r \approx \alpha u(z, t),\quad u_z \approx w(z, t) + (1-2\alpha^2)\psi(z, t)$$

and the strains by

$$\epsilon_r = \epsilon_\theta \approx \Gamma_r,\quad \epsilon_z \approx \Gamma_z + (1-2\alpha^2)\Gamma_\psi,\quad \gamma_{rz} \approx \alpha\Gamma_{rz}$$

Omission of the higher-order terms limits the applicability, of what is left, to relatively long wave lengths. In this range, the uniform distribution of axial displacement, whose amplitude is w, is a good approximation to the distribution of u_z found in the lowest mode of the exact solution for waves in an infinite rod. In fact, the latter distribution approaches uniformity as the wave length approaches infinity. The linear and quadratic distributions of the other two displacements, i.e., those with amplitudes u and ψ, are not as good approximations to the Bessel-function distributions found in the second and third modes of the exact solution, even at very long wave lengths, as shown by the dotted lines in the diagrams for u and ψ in Fig. 1. It is advisable, therefore, to introduce means for compensating, as well as is possible with the terms retained, for the omission of the polynomials of higher degree. To this end, we replace Γ_r by $\kappa_1\Gamma_r$ and Γ_{rz} by $\kappa_2\Gamma_{rz}$ in the strain-energy-density and $\dot{u}$ by $\kappa_3\dot{u}$ and $\dot{\psi}$ by $\kappa_4\dot{\psi}$ in the kinetic-energy-density, where the κ_i are constants, whose values are determined in Section 5, and the dot indicates differentiation with respect to time. Then the energy-densities of the second-order approximation are

$$U_2 = {}^1/_2[2(\lambda + \mu)\kappa_1{}^2\Gamma_r{}^2 + {}^1/_2(\lambda + 2\mu)\Gamma_z{}^2 + 2\lambda\kappa_1\Gamma_r\Gamma_z + {}^1/_4\mu\kappa_2{}^2\Gamma_{rz}{}^2 + {}^1/_6(\lambda + 2\mu)\Gamma_\psi{}^2] \quad (24)$$

$$K_2 = {}^1/_2\rho({}^1/_4\kappa_3{}^2\dot{u}^2 + {}^1/_2\dot{w}^2 + {}^1/_6\kappa_4{}^2\dot{\psi}^2) \quad (25)$$

The adjusted stress-strain-displacement relations, derived from the strain-energy-density function (24), are

$$\left.\begin{aligned} P_r &= \partial U_2/\partial\Gamma_r = 2(\lambda + \mu)\kappa_1{}^2\Gamma_r + \lambda\kappa_1\Gamma_z \\ &= [2(\lambda + \mu)\kappa_1{}^2u + \lambda\kappa_1 w']a^{-1} \\ P_z &= \partial U_2/\partial\Gamma_z = {}^1/_2(\lambda + 2\mu)\Gamma_z + \lambda\kappa_1\Gamma_r \\ &= [{}^1/_2(\lambda + 2\mu)w' + \lambda\kappa_1 u]a^{-1} \\ P_{rz} &= \partial U_2/\partial\Gamma_{rz} = {}^1/_4\mu\kappa_2{}^2\Gamma_{rz} \\ &= {}^1/_4\mu\kappa_2{}^2(u' - 4\psi)a^{-1} \\ P_\psi &= \partial U_2/\partial\Gamma_\psi = {}^1/_6(\lambda + 2\mu)\Gamma_\psi \\ &= {}^1/_6(\lambda + 2\mu)\psi' a^{-1} \end{aligned}\right\} \quad (26)$$

where prime indicates differentiation with respect to z/a.

The adjusted stress-equations of motion, derived from the variational equations of motion with the constants κ_3 and κ_4 introduced into the kinetic energy, are

$$\left.\begin{aligned} P_{rz}' - P_r + R &= {}^1/_4\rho a\kappa_3{}^2\ddot{u} \\ P_z' + Z &= {}^1/_2\rho a\ddot{w} \\ P_\psi' + 4P_{rz} - Z &= {}^1/_6\rho a\kappa_4{}^2\ddot{\psi} \end{aligned}\right\} \quad (27)$$

and, from (26) and (27), the displacement equations of motion are

$$\left.\begin{aligned} \mu\kappa_2{}^2(u'' - 4\psi') - 8(\lambda + \mu)\kappa_1{}^2u \\ - 4\lambda\kappa_1 w' + 4aR &= \rho a^2\kappa_3{}^2\ddot{u} \\ (\lambda + 2\mu)w'' + 2\lambda\kappa_1 u' + 2aZ &= \rho a^2\ddot{w} \\ (\lambda + 2\mu)\psi'' + 6\mu\kappa_2{}^2(u' - 4\psi) - 6aZ &= \rho a^2\kappa_4{}^2\ddot{\psi} \end{aligned}\right\} \quad (28)$$

4 Pochhammer Equation

As shown in the next section, the solution of the equations of motion (28) for free vibrations, or waves, in an infinite bar leads to a cubic frequency equation relating the square of the frequency to the square of the wave number, with Poisson's ratio and the coefficients κ_i as parameters. The three branches of this equation should match, as closely as possible, the corresponding branches of Pochhammer's frequency equation [4]

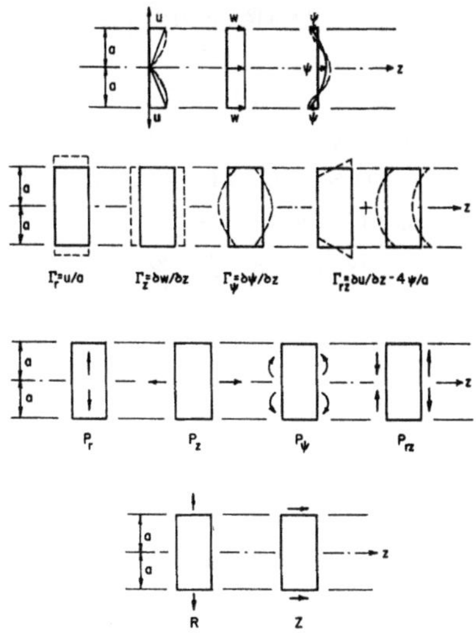

Fig. 1 Displacements, strains, stresses, and surface tractions; second-order approximation

$$(\gamma^2 - \beta^2)\mathcal{J}_1(\alpha a) + 4\gamma^2\alpha^2\mathcal{J}_1(\beta a) = 2(\gamma^2 + \beta^2)\alpha^2 \quad (29)$$

where

$$\alpha^2 = \rho\omega^2/(\lambda + 2\mu) - \gamma^2, \quad \beta^2 = \rho\omega^2/\mu - \gamma^2 \quad (30)$$

and $\mathcal{J}_1(x)$ is the Onoe function [5, 6] of the first kind and order unity:

$$\mathcal{J}_1(x) = xJ_0(x)/J_1(x) \quad (31)$$

and $J_n(x)$ is the Bessel function of the first kind. In (29) and (30), the frequency ω must be real; but the wave number along the axis of the bar γ may be real, imaginary, or complex.

The match between the three branches of the cubic equation and the analogous branches of (29) may be adjusted by means of the coefficients κ_i; but, since the κ_i are constants, a perfect match can be made, in general, only at one value of γ for each of them. Now, large enough real γ corresponds to frequencies high enough to enter the frequency range of modes that have not been included in the approximate equations. In general, the applicability of the approximate equations is limited to frequencies below the lowest frequency, for real wave numbers, of the lowest neglected mode and to the correspondingly small wave numbers. This is because, except for infinite trains of sinusoidal waves, there would be coupling of all the Pochhammer modes and their anharmonic overtones wherever they have the same frequency and wave length. There is, then, little advantage to be gained in a choice of the κ_i that makes an approximate branch match the exact one at short wave lengths (large γ) at the expense of a good match at long wave lengths. In fact, we go to the extreme and choose all of the κ_i so that the match is perfect in the neighborhood of $\gamma = 0$. This choice has the advantage of reproducing, in the approximate equations, the intricate behavior which the exact solution has at long wave lengths, as Poisson's ratio varies, while retaining a reasonably good match out to as short wave lengths as the frequency limitation permits.

When γa is small, the three branches of the Pochhammer equation, which correspond to the three branches of the approximation are given by [7]

$$\left.\begin{aligned}\Omega &= \zeta[2(1+\nu)]^{1/2}\\ \Omega &= 1 + {}^1/_2\zeta^2 C_s\\ \Omega &= k\chi/\delta + {}^1/_2\zeta^2 C_r\end{aligned}\right\} \quad (32)$$

where

$$\left.\begin{aligned}&\Omega = \omega/\omega_s,\quad \omega_s = (\delta/a)(\mu/\rho)^{1/2},\quad \zeta = \gamma a/\delta\\ &k^2 = (\lambda + 2\mu)/\mu = 2(1-\nu)/(1-2\nu)\\ &C_s = 1 - 8/[k^2\mathcal{J}_1(\delta/k) - 2]\\ &C_r = \frac{k\delta}{\chi}\left[1 + \frac{4[k^2 + 2\mathcal{J}_1(k\chi) - 4]}{k^4\chi^2 - 4(k^2-1)}\right]\end{aligned}\right\} \quad (33)$$

and δ and χ are the lowest (nonzero) roots of

$$J_1(\delta) = 0 \quad (34)$$

$$k^2\mathcal{J}_1(\chi) = 2 \quad (35)$$

respectively. Thus $\delta = 3.8317$ while χ is a function of Poisson's ratio. If Poisson's ratio is such that $k\chi = \delta$, the second and third of (32) are to be replaced by

$$\Omega = 1 \pm 2k\zeta[k^2\delta^2 - 4(k^2-1)]^{-1/2} \quad (36)$$

The first of (32) gives the frequency of the well-known longitudinal mode, which has a limiting phase velocity

$$[2(1+\nu)\mu/\rho]^{1/2} = [E/\rho]^{1/2} \quad (37)$$

where E is Young's modulus. The accompanying distribution of axial displacement u_z is uniform, in the limit as $\zeta \to 0$, and is represented in the approximate equations by w. The second of (32) has a frequency at cutoff (i.e., $\zeta = 0$) of $\omega = \omega_s$ where ω_s is given in (33), and the accompanying displacement is purely axial with one circular node. This mode, the lowest axial shear mode, is represented in the approximate equations by ψ. The mode shape associated with the third of (32), at cutoff, is characterized by a purely radial displacement which is represented in the approximate equations by u. The frequency

$$\left.\begin{aligned}\omega_r &= (\chi/a)[(\lambda+2\mu)/\rho]^{1/2}\\ \text{or}\quad \Omega_r &= \omega_r/\omega_s = k\chi/\delta\end{aligned}\right\} \quad (38)$$

of this radial mode depends on Poisson's ratio whereas ω_s does not. The two are equal when $k\chi = \delta$, i.e., when Poisson's ratio is given by the lowest root of

$$k^2\mathcal{J}_1(\delta/k) = 2,\quad \delta = 3.8317 \quad (39)$$

or, when

$$\nu = \nu_c = 0.2833 \quad (40)$$

The frequency of the radial mode is less than, equal to, or greater than the frequency of the axial shear mode according as Poisson's ratio is less than, equal to, or greater than ν_c. As Poisson's ratio increases above ν_c, the frequency ω_r approaches the frequency of the second axial shear mode, given by

$$\Omega_{2s} = \frac{\delta_2}{\delta} = \frac{7.0156}{3.8317} = 1.8309 \quad (41)$$

where δ_2 is the second root of (34). Now, the second axial shear mode has two circular nodes and such a displacement is not included in the second-order approximation. Hence the approximate equations are limited, in application, to frequencies less than that given by (41) and to the wave lengths associated with such frequencies in the first three branches. Also, the equations are limited to Poisson's ratios less than that for which ω_r is equal to the frequency of the second axial shear mode. This is obtained from the lowest root of

$$k^2\mathcal{J}_1(\delta_2/k) = 2,\quad \delta_2 = 7.0156 \quad (42)$$

from which $\nu = \nu_2 = 0.439$. Actually, the limitation on frequency and Poisson's ratio is probably slightly more severe because of possible coupling with the next higher complex branches in the case of bounded rods or transient waves.

The quantities C_s and C_r in (33) are the curvatures of the two upper branches at $\zeta = 0$. When $\nu < \nu_c$, $C_r < 0$ and $C_s > 0$ and they approach positive and negative infinity, respectively, as $\nu \to \nu_c$ from below. When $\nu_c < \nu < \nu_2$, the signs of the curvatures are reversed. From (32) it may be seen that the slopes of the two upper branches are zero at $\zeta = 0$, in the range $\nu < \nu_2$, except when $\nu = \nu_c$, in which case they can be obtained from (36).

5 The Coefficients κ_i

The intricate behavior of the second and third branches of the exact solution, at $\zeta = 0$, is to be reproduced in the corresponding branches, obtained from the approximate equations, by appropriate choice of the coefficients κ_i as functions of ν. Upon substituting

$$\left.\begin{aligned}u &= A\cos\gamma z\, e^{i\omega t}\\ w &= B\sin\gamma z\, e^{i\omega t}\\ \psi &= C\sin\gamma z\, e^{i\omega t}\\ R &= Z = 0\end{aligned}\right\} \quad (43)$$

in (28) and eliminating the constants A, B, and C, we find, for the cubic equation governing the branches of the approximate equations,

$$\begin{vmatrix} a_{11} & a_{12} & a_{13}\\ a_{12} & a_{22} & 0\\ a_{13} & 0 & a_{33}\end{vmatrix} = 0 \quad (44)$$

where

$$\left.\begin{aligned}a_{11} &= \kappa_2^2\delta^2\zeta^2 + 8(k^2-1)\kappa_1^2 - \kappa_3^2\delta^2\Omega^2\\ a_{22} &= 2\delta^2(k^2\zeta^2 - \Omega^2)\\ a_{33} &= 6(k^2\delta^2\zeta^2 + 24\kappa_2^2 - \kappa_4^2\delta^2\Omega^2)\\ a_{12} &= 4(k^2-2)\kappa_1\delta\zeta\\ a_{13} &= 12\kappa_2^2\delta\zeta\end{aligned}\right\} \quad (45)$$

When $\zeta = 0$, (44) has the three roots

$$\Omega^2 = 0,\quad \frac{24\kappa_2^2}{\delta^2\kappa_4^2},\quad \frac{8(k^2-1)\kappa_1^2}{\delta^2\kappa_3^2} \quad (46)$$

for the longitudinal (w), axial shear (ψ), and radial (u) modes, respectively. To make these frequencies the same as the cutoff frequencies obtained from (32), we set

$$\frac{\kappa_2^2}{\kappa_4^2} = \frac{\delta^2}{24},\quad \frac{\kappa_1^2}{\kappa_3^2} = \frac{k^2\chi^2}{8(k^2-1)} \quad (47)$$

Then, for $\zeta \ll 1$, (44) becomes

$$\left.\begin{aligned}\Omega &= \zeta[2(1+\nu)]^{1/2}\\ \Omega &= 1 + {}^1/_2\zeta^2\bar{C}_s\\ \Omega &= k\chi/\delta + {}^1/_2\zeta^2\bar{C}_r\end{aligned}\right\} \quad (48)$$

where

$$\left.\begin{aligned}\bar{C}_s &= \frac{k^2\delta^2}{24}\left[\frac{1}{\kappa_2^2} - \frac{3\chi^2\kappa_2^2}{\kappa_1^2(k^2-1)(k^2\chi^2-\delta^2)}\right]\\ \bar{C}_r &= \frac{k\chi\delta}{k^2-1}\left[\frac{(k^2-2)^2}{k^2\chi^2} + \frac{\kappa_i^2k^2\chi^2}{8\kappa_1^2(k^2\chi^2-\delta^2)}\right]\end{aligned}\right\} \quad (49)$$

except if $k\chi = \delta$ (i.e., $\nu = \nu_c$) in which case the second and third of (48) are

$$\Omega = 1 \pm \zeta \left[\frac{\delta^2\kappa_2^2}{32(k^2-1)\kappa_1^2}\right]^{1/2} \tag{50}$$

Comparing (48) with (32), we see that the longitudinal mode needs no adjustment, while the others will match if

$$\bar{C}_r = C_r, \quad \bar{C}_s = C_s \tag{51}$$

from which

$$\left.\begin{aligned} \frac{\kappa_2^2}{\kappa_1^2} &= \frac{8(k^2\chi^2-\delta^2)}{k^2\chi^2}\left[\frac{(k^2-1)C_r}{k\chi\delta} - \frac{(k^2-2)^2}{k^2\chi^2}\right] \\ \frac{1}{\kappa_2^2} &= \frac{24}{k^2\delta^2}\left[\frac{k^2\chi^2\delta^2\kappa_2^2}{8(k^2-1)(k^2\chi^2-\delta^2)\kappa_1^2} + C_s\right] \end{aligned}\right\} \tag{52}$$

In the case $k\chi = \delta$, we have

$$\lim_{k\chi\to\delta} \frac{\kappa_2^2}{\kappa_1^2} = \frac{128k^2(k^2-1)}{\delta^2[k^2\delta^2 - 4(k^2-1)]} \tag{53}$$

When this is substituted in (50), that equation becomes identical with (36), as required.

Equations (47) and (52) fix the four coefficients κ_i as functions of Poisson's ratio. With these values, the ordinates, slopes, and curvatures of the three branches from the approximate and exact equations are identical at $\zeta = 0$.

6 Frequency Spectrum

Equation (44) relates Ω^2 to ζ^2 but only real, positive Ω has physical significance. For real, positive Ω, the roots ζ^2 may be three real positive, two real positive, and one real negative or one real positive and two conjugate complex. Hence there are six roots

$$\zeta = \xi + i\eta \tag{54}$$

which occur in two sets of three roots each, distinguished only by reversal of the sign of ζ. In each set, the roots may be three real, or two real and one imaginary, or one real and two conjugate complex. In addition, the character of the spectrum is different according as Poisson's ratio is less than, equal to, or greater than ν_c.

We consider the case $\nu < \nu_c$ first. Then, when $\Omega > 1$, the three branches of (44), corresponding to one set of roots, are real ($\eta = 0$). They are illustrated in Fig. 2(*a*) (for $\nu = 1/4$) by the three dashed curves marked ζ_1, ζ_2, ζ_3 in the plane $\eta = 0$ and the region $\Omega > 1$. As Ω drops below $\Omega = 1$, the largest root ζ_1 approaches zero with asymptotic behavior given by the first of (48); the intermediate branch ζ_2 approaches a frequency minimum at Ω^*, below which there are no real roots other than ζ_1; the smallest root ζ_3 approaches a frequency minimum in the real plane $\eta = 0$, at $\Omega = 1$, $\xi = 0$, with asymptotic behavior given by the second of (48). Continuing with the branch ζ_3, as Ω drops below $\Omega = 1$, the root is imaginary $\xi = 0$ with asymptotic behavior, for small η, given by the second of (48). As Ω approaches Ω_r from above, the branch ζ_3 forms a loop in the imaginary plane ($\xi = 0$) and approaches $\Omega = \Omega_r$, $\eta = 0$, with asymptotic behavior given by the third of (48). As Ω drops below $\Omega = \Omega_r$, the root ζ_3 becomes real with behavior near $\Omega = \Omega_r$, $\zeta = 0$, given by the third of (48), but now ξ is negative. Upon further diminution of Ω, the branch ζ_3 approaches a frequency minimum at $\Omega = \Omega^*$ and negative ξ. The portion of ζ_3 between Ω_r and Ω^* is identified in Fig. 2(*a*) by $[\zeta_3]$, where the brackets indicate that it is the reflection of the branch in the plane $\xi = 0$ that is shown. Finally, when $\Omega < \Omega^*$, the roots ζ_2 and ζ_3 are conjugate complex ($\zeta = \pm\xi + i\eta$). One of them is shown, in Fig. 2(*a*), as the curve ζ_2 below Ω^*. This curve is also $[\zeta_3]$, i.e., the reflection of the conjugate root ζ_3 in the plane $\xi = 0$.

As Poisson's ratio approaches ν_c from below, the frequency of the radial mode increases, approaching the frequency of the axial shear mode; i.e., at $\zeta = 0$, the intercept $\Omega = \Omega_r$ approaches $\Omega = 1$. Conjointly, the curvature of the branch ζ_3 at $\Omega = \Omega_r$, $\zeta = 0$, approaches negative infinity in the plane $\eta = 0$ and positive infinity in the plane $\xi = 0$ while both slopes remain zero, all in accordance with the third of (48). At the same time, the curvature of the branch ζ_3, at $\Omega = 1$, $\zeta = 0$, approaches positive infinity in the plane $\eta = 0$ and negative infinity in the plane $\xi = 0$, while the slopes remain zero, all in accordance with the second of (48). Meanwhile, the imaginary loop shrinks toward the point $\Omega = 1$, $\eta = 0$.

At $\nu = \nu_c$, the axial shear and radial modes have the same frequency; there is no imaginary root and the slopes of the two segments of ζ_3 at $\Omega = 1$, $\zeta = 0$, are equal and are given by (50) with the positive sign. This situation is illustrated in Fig. 2(*b*). Here, again, the portions marked $[\zeta_3]$ are the reflections, in the plane $\xi = 0$, of the actual branches.

When $\nu > \nu_c$, the radial mode has a frequency higher than that of the axial shear mode. The spectrum, illustrated in Fig. 2(*c*) for $\nu = 0.31$, has now undergone an important change in that the imaginary segment of ζ_3 loops up from $\Omega = 1$ instead of down.

Since only powers of ζ^2 occur in (44), there is another set of roots given by the reflections of the curves of Fig. 2 in the planes $\xi = 0$ and $\eta = 0$. These branches correspond to modes in which the sign of the co-ordinate z is reversed.

The shape, along z, of the mode corresponding to a point on any branch can be deduced from the real, imaginary, or complex character of ζ. Real roots correspond to trigonometric shapes; imaginary roots to exponential or hyperbolic shapes; and complex roots to modes whose shapes are given by products of trigonometric and exponential or hyperbolic functions. A striking example of the latter is to be found in the experiments of Oliver [8].

The ratio of the phase velocity (v) and group velocity (v_g) to the velocity of equivoluminal waves $v_2 = (\mu/\rho)^{1/2}$ can be visualized readily from Fig. 2 inasmuch as

$$\frac{v}{v_2} = \frac{\omega}{\gamma v_2} = \frac{\Omega}{\xi} \tag{55}$$

$$\frac{v_g}{v_2} = \frac{d\omega}{v_2 d\gamma} = \frac{d\Omega}{d\xi} \tag{56}$$

Thus the slope of the straight line from the origin to a point on a branch is proportional to the phase velocity, and the slope of the branch is proportional to the group velocity. There are several anomalous velocities in Fig. 2. For example, at the frequency minimum Ω^*, the phase velocity is nonzero and finite while the group velocity is zero; between Ω^* and the lower of the two cutoff frequencies of the third branch, the phase and group velocities have opposite signs; when $\nu = \nu_c$, the point $\Omega = 1$, $\zeta = 0$ corresponds to infinite phase velocity with finite, nonzero group velocity.

The full lines in Fig. 2 are the first three branches of the Pochhammer equation. The approximate branches match the exact ones very well in the neighborhood of $\zeta = 0$; due, of course, to the choice of the κ_i. Without such a choice, the extraordinarily intricate behavior of the exact branches would not be reproduced in the approximate equations based on simple polynomial distributions of displacement. It is, in fact, remarkable that the correspondence is reasonably good for quite large values of ζ and for real, imaginary, and complex portions of the branches; especially since the simple polynomial distributions replace Bessel functions of real, imaginary, and complex arguments.

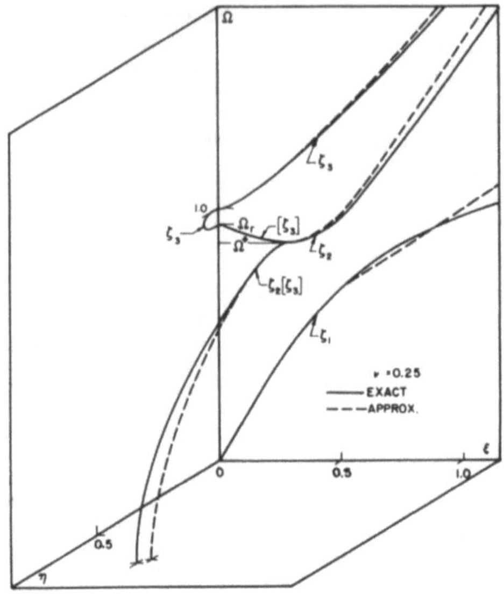

(a)

Fig. 2 Frequency spectrum

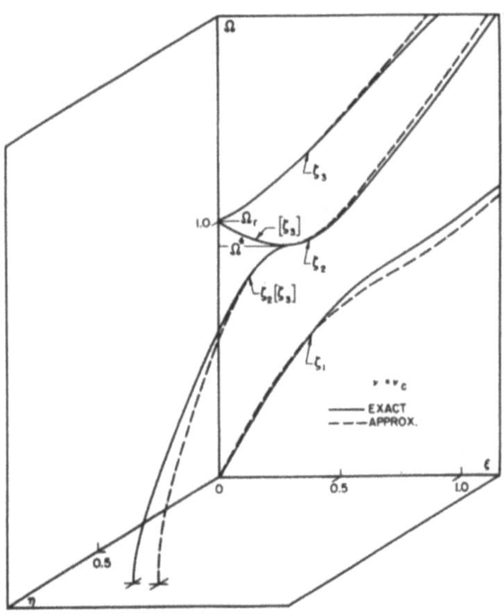

(b)

Fig. 2 Frequency spectrum

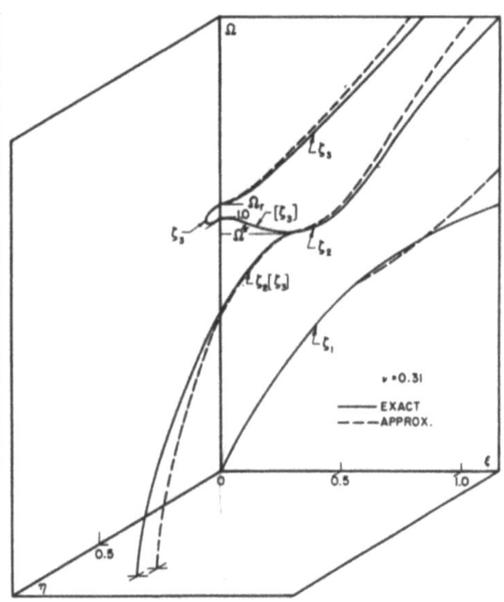

(c)

Fig. 2 Frequency spectrum

7 Uniqueness and Orthogonality

A uniqueness theorem, analogous to Neumann's [9], will establish the initial and boundary conditions appropriate to the second-order equations.

Consider two systems of displacements, strains, and stresses which satisfy the strain-displacement relations (23), the associated compatibility condition

$$\frac{\partial \Gamma_{rz}}{\partial z} + \frac{4}{a}\Gamma_\psi - a\frac{\partial^2 \Gamma_r}{\partial z^2} = 0 \tag{57}$$

the stress-strain relations (26), and the stress equations of motion (27). Let the differences between corresponding components of displacement, strain, and stress constitute a "difference system" of these quantities and let K_2 and U_2 be the kinetic and strain energy densities of the difference system. Then the sum of K_2 and U_2 in a bar of length $2c$ at time t is

$$\mathfrak{K} + \mathfrak{U} = \mathfrak{K}_0 + \mathfrak{U}_0 + \int_0^t dt \int_{-c}^{c} (\dot{K}_2 + \dot{U}_2)dz \tag{58}$$

where $\mathfrak{K}_0$ and $\mathfrak{U}_0$ are the values of $\mathfrak{K}$ and $\mathfrak{U}$ at an initial time $t = 0$.

Now $\qquad \dot{K}_2 = \rho(^1/_4\kappa_3^2\ddot{u}\dot{u} + {}^1/_2\ddot{w}\dot{w} + {}^1/_6\kappa_4^2\ddot{\psi}\dot{\psi}) \qquad (59)$

where u, w, and ψ are the displacements of the difference system; and

$$\dot{U}_2 = \frac{\partial U_2}{\partial \Gamma_r}\dot{\Gamma}_r + \frac{\partial U_2}{\partial \Gamma_z}\dot{\Gamma}_z + \frac{\partial U_2}{\partial \Gamma_\psi}\dot{\Gamma}_\psi + \frac{\partial U_2}{\partial \Gamma_{rz}}\dot{\Gamma}_{rz} \tag{60}$$

where the strains are those of the difference system. From (23) and (26), (60) becomes

$$\begin{aligned} a\dot{U}_2 &= P_r\dot{u} + P_z\dot{w}' + P_\psi\dot{\psi}' + P_{rz}(\dot{u}' - 4\dot{\psi}) \\ &= (P_r - P_{rz}')\dot{u} - P_z'\dot{w} - (P_\psi + 4P_{rz})\dot{\psi} \\ &\quad + (P_z\dot{w} + P_\psi\dot{\psi} + P_{rz}\dot{u})' \end{aligned} \tag{61}$$

where prime indicates differentiation with respect to z/a. Hence

$$\begin{aligned} a(\dot{K}_2 + \dot{U}_2) &= -(P_{rz}' - P_r - {}^1/_4\rho a\kappa_3^2\ddot{u})\dot{u} \\ &\quad - (P_z' - {}^1/_2\rho a\ddot{w})\dot{w} \\ &\quad - (P_\psi' + 4P_{rz} - {}^1/_6\rho a\kappa_4^2\ddot{\psi})\dot{\psi} \\ &\quad + (P_z\dot{w} + P_\psi\dot{\psi} + P_{rz}\dot{u})' \end{aligned} \tag{62}$$

or, using the stress equations of motion (27),

$$a(\dot{K}_2 + \dot{U}_2) = R\dot{u} + Z(\dot{w} - \dot{\psi}) + (P_z\dot{w} + P_\psi\dot{\psi} + P_{rz}\dot{u})' \tag{63}$$

Upon substituting (63) into (58), we have, finally

$$\begin{aligned} \mathfrak{K} + \mathfrak{U} &= \mathfrak{K}_0 + \mathfrak{U}_0 + a^{-1}\int_0^t dt \int_{-c}^{c} [R\dot{u} + Z(\dot{w} - \dot{\psi})]dz \\ &\quad + [P_z\dot{w} + P_\psi\dot{\psi} + P_{rz}\dot{u}]^c_{-c} \end{aligned} \tag{64}$$

Then, by the usual argument based on the positive definiteness of K_2 and U_2, the initial and boundary conditions are:

(i) Throughout the rod, the initial values of u, w, ψ and $\dot{u}$, $\dot{w}$, $\dot{\psi}$.

(ii) Throughout the rod, one member of each of the two products Ru and $Z(w - \psi)$.

(iii) At each end of the rod, one member of each of the three products P_zw, $P_\psi\psi$, and $P_{rz}u$.

By a closely related procedure, it may be shown that in two solutions

$$\begin{aligned} (u, w, \psi) &= (u_p, w_p, \psi_p) \exp i\omega_p t \\ (u, w, \psi) &= (u_q, w_q, \psi_q) \exp i\omega_q t \end{aligned} \tag{65}$$

of the homogeneous ($R = Z = 0$) stress equations of motion (27), the characteristic functions satisfy the orthogonality condition

$$\int_{-c}^{c} (3\kappa_3^2 u_p u_q + 6w_p w_q + 2\kappa_4^2\psi_p\psi_q)dz = 0 \tag{66}$$

for $\omega_p \neq \omega_q$ and homogeneous end conditions (iii).

8 Helmholtz Equations

In case of free vibrations, or waves, the displacement equations of motion (28) may be reduced to three uncoupled Helmholtz equations. Let

$$\left.\begin{aligned} u &= \varphi \exp i\omega t \\ w &= \sigma\varphi' \exp i\omega t \\ \psi &= \tau\varphi' \exp i\omega t \end{aligned}\right\} \tag{67}$$

where σ and τ are coefficients to be determined; φ is a function of z; and prime, as before, indicates differentiation with respect to z/a. Upon substituting (67) in (28), we find that the latter are satisfied if

$$\varphi'' + \delta^2\zeta^2\varphi = 0 \tag{68}$$

provided ζ is one of the three roots of (44) and

$$\left.\begin{aligned} \sigma &= a_{12}/a_{22}\delta\zeta \\ \tau &= 3a_{13}/a_{33}\delta\zeta \end{aligned}\right\} \tag{69}$$

Hence

$$\left.\begin{aligned} u &= (\varphi_1 + \varphi_2 + \varphi_3) \exp i\omega t \\ w &= (\sigma_1\varphi_1' + \sigma_2\varphi_2' + \sigma_3\varphi_3') \exp i\omega t \\ \psi &= (\tau_1\varphi_1' + \tau_2\varphi_2' + \tau_3\varphi_3') \exp i\omega t \end{aligned}\right\} \tag{70}$$

where

$$\varphi_i'' + \delta^2\zeta_i^2\varphi_i = 0, \qquad i = 1, 2, 3 \tag{71}$$

and the σ_i and τ_i are given by (69) with ζ replaced by ζ_i.

References

1 R. D. Mindlin and G. Herrmann, "A One-Dimensional Theory of Compressional Waves in an Elastic Rod," Proceedings of the First U. S. National Congress of Applied Mechanics, 1950, pp. 187–191.

2 W. Magnus and F. Oberhettinger, "Special Functions of Mathematical Physics," Chelsea Publishing Company, New York, N. Y., 1949, p. 83.

3 A. E. H. Love, "A Treatise on the Mathematical Theory of Elasticity," Cambridge University Press, London, England, 1927, p. 167.

4 L. Pochhammer, "Ueber die Fortpflanzungsgeschwindigkeiten Schwingungen in einem unbegrenzten isotropen Kreiscylinder," *Zeitschrift für Mathematik*, vol. 81, 1876, pp. 324–336.

5 M. Onoe, "Formulae and Tables, the Modified Quotients of Cylinder Functions," Report of the Institute of Industrial Science, University of Tokyo, Chiba City, Japan, vol. 4, 1955.

6 M. Onoe, "Tables of Modified Quotients of Bessel Functions of the First Kind for Real and Imaginary Arguments," Columbia University Press, New York, N. Y., 1958.

7 M. Onoe, H. D. McNiven, and R. D. Mindlin, "Dispersion of Axially Symmetric Waves in Elastic Rods," forthcoming.

8 J. Oliver, "Elastic Wave Dispersion in a Cylindrical Rod by a Wide-Band Short-Duration Pulse Technique," *Journal of The Acoustical Society of America*, vol. 29, 1957, pp. 189–194.

9 A. E. H. Love, "A Treatise on the Mathematical Theory of Elasticity," Cambridge University Press, London, England, 1927, p. 176.

(*Reprinted from the Journal of Applied Mechanics for March, 1960, pages 152–158, with permission of Copyright Owner*)

R. D. MINDLIN
Professor
of Civil Engineering,
Columbia University,
New York, N. Y.
Mem. ASME

E. A. FOX
Associate Professor
of Mechanics,
Rensselaer Polytechnic Institute,
Troy, N. Y.
Assoc. Mem. ASME

Vibrations and Waves in Elastic Bars of Rectangular Cross Section[1]

An exact solution of the equations of elasticity is found for a family of modes of vibration, or waves, in an infinite bar of rectangular cross section for certain ratios of width to depth. The solution is composed of coupled dilatational and equivoluminal waves and the four faces of the bar are free of traction.

Introduction

Solutions of the linear equations of elasticity will be described for vibrational modes, or waves, in an isotropic bar, of rectangular cross section and infinite length, with traction-free boundaries. There is a previous solution, by Lamé,[2] which yields a family of modes of equivoluminal distortion that are possible in such a bar. The present solutions involve coupled dilatational (P) and equivoluminal (S) waves.

The results for the bar can be obtained by combining two types of solution that are known for an infinite plate with traction-free faces. One of the solutions for the plate is that given by Rayleigh[3] and Lamb.[4] It is itself composed of a dilatational wave coupled with two equivoluminal waves. The other type of solution contains only equivoluminal deformation and the displacement is parallel to the surfaces of the plate. Two of these modes are required. They are special cases of Love waves[5] and correspond to what are sometimes called SH waves. A property of SH waves, in a plate with traction-free faces, is that the ratio of the thickness of the plate to the distance between nodes, measured through the thickness, is an integer. In combining the three solutions so as to satisfy additional traction-free boundary conditions at a pair of parallel sides of the plate, only those Rayleigh-Lamb modes can be used that have thickness-behavior like that of SH waves. This circumstance restricts the resulting solution to certain frequencies and ratios of width to depth of the cross section. A special property of the modes thus found is that, in addition to the four faces of the bar being traction free, they remain plane during the motion; i.e., traction-free and mixed boundary conditions are satisfied simultaneously.

For a given, permissible ratio of width to depth, the solution presented gives a set of discrete points on the branches of the frequency spectrum (frequency versus wave-number) of the bar. The complete branches are not known, but their slopes at the discrete points are determined, in Section 9, through the use of the identity of group velocity and velocity of propagation of energy.

In Section 11 it is shown that the five component waves (three in the Rayleigh-Lamb solution and two SH waves) combine into three plane waves. One of the three waves is the original dilatational wave, another is an SH wave whose displacement is parallel to one pair of parallel faces of the bar and the third is an SV wave; i.e., an equivoluminal wave whose displacement is parallel to the other pair of faces of the bar. Each of these waves, in general, progresses along the bar by oblique reflections at all four faces. At each reflection, the angles of incidence and amplitudes of the three waves are so related that the face remains free of traction.

[1] This investigation was supported, in part, by the Office of Naval Research under a contract with Columbia University.

[2] G. Lamé, "Leçons sur la théorie mathématique de l'élasticité des corps solides," Gauthier-Villars, Paris, France, second edition, 1866, p. 170.

[3] Lord Rayleigh, "On the Free Vibrations of an Infinite Plate of Homogeneous Isotropic Elastic Matter," *Proceedings of the London Mathematical Society*, England, vol. 20, 1889, pp. 225–234.

[4] H. Lamb, "On Waves in an Elastic Plate," *Proceedings of the Royal Society of London*, England, series A, vol. 93, 1917, pp. 114–128.

[5] A. E. H. Love, "Some Problems in Geodynamics," Cambridge University Press, London, England, 1926, p. 160.

Presented at the West Coast Conference of the Applied Mechanics Division, Stanford, Calif., September 9–11, 1959, of The American Society of Mechanical Engineers.

Discussion of this paper should be addressed to the Secretary, ASME, 29 West 39th Street, New York, N. Y., and will be accepted until April 10, 1960, for publication at a later date. Discussion received after the closing date will be returned.

Note: Statements and opinions advanced in papers are to be understood as individual expressions of their authors and not those of the Society. Manuscript received by ASME Applied Mechanics Division, January 5, 1959. Paper No. 59—APMW-23.

2 Equations of Elasticity

In the linear theory of elasticity, the components of stress in an isotropic body are related to the components of displacement by three equations of the type

$$\sigma_{xx} = \lambda(\partial u/\partial x + \partial v/\partial y + \partial w/\partial z) + 2\mu \partial u/\partial x \quad (1)$$

and three of the type

$$\sigma_{yz} = \mu(\partial w/\partial y + \partial v/\partial z), \quad (2)$$

where λ and μ are Lamé's constants.

Solutions of the displacement-equations of motion for the case of free vibrations, at circular frequency ω, may be expressed in the form

$$\begin{aligned} u &= \partial\varphi/\partial x + \partial H_z/\partial y - \partial H_y/\partial z, \\ v &= \partial\varphi/\partial y + \partial H_x/\partial z - \partial H_z/\partial x, \\ w &= \partial\varphi/\partial z + \partial H_y/\partial x - \partial H_x/\partial y, \end{aligned} \quad (3)$$

if

$$v_1^2\nabla^2\varphi + \omega^2\varphi = 0, \qquad v_1^2 = (\lambda + 2\mu)/\rho, \quad (4)$$

$$v_2^2\nabla^2 H_j + \omega^2 H_j = 0, \quad v_2^2 = \mu/\rho, \quad j = x, y, z, \quad (5)$$

where v_1 and v_2 are the velocities of dilatational and equivoluminal waves, respectively, in an infinite medium.

In Equations (3) to (5) and in the sequel, a time-dependent factor $e^{i\omega t}$ is omitted.

3 Rayleigh-Lamb Waves in an Infinite Plate

We consider, first, vibrations or waves in a plate bounded by traction-free surfaces $y = \pm b$. Let $H_y = 0$ and

$$\begin{aligned} \varphi &= -A\zeta \cos \xi x \cos \alpha y \cos \zeta z, \\ H_x &= B\zeta \cos \xi x \sin \beta y \sin \zeta z, \\ H_z &= -B\xi \sin \xi x \cos \beta y \cos \zeta z. \end{aligned} \quad (6)$$

1

These give modes symmetric with respect to the three co-ordinate planes. The remaining seven symmetries may be obtained by appropriate interchanges of sines and cosines of like argument, as described in Section 5. Equations (6) satisfy Equations (4) and (5) if

$$\alpha^2 + \xi^2 + \zeta^2 = \omega^2/v_1^2, \qquad \beta^2 + \xi^2 + \zeta^2 = \omega^2/v_2^2 \tag{7}$$

For the components of traction on $y = \pm b$ we have

$$\begin{aligned} \sigma_{xy} &= \mp 2\mu A\xi\alpha\zeta \sin \xi x \sin \alpha b \cos \zeta z \\ &\quad \mp \mu B\xi(\zeta^2 + \xi^2 - \beta^2) \sin \xi x \sin \beta b \cos \zeta z, \\ \sigma_{yy} &= \mu A\zeta(\beta^2 - \xi^2 - \zeta^2) \cos \xi x \cos \alpha b \cos \zeta z \\ &\quad + 2\mu B\beta(\xi^2 + \zeta^2) \cos \xi x \cos \beta b \cos \zeta z, \\ \sigma_{yz} &= \mp 2\mu A\alpha\zeta^2 \cos \xi x \sin \alpha b \sin \zeta z \\ &\quad \mp \mu B\zeta(\xi^2 + \zeta^2 - \beta^2) \cos \xi x \sin \beta b \sin \zeta z. \end{aligned} \tag{8}$$

The compatibility of the conditions obtained by setting Equations (8) equal to zero requires

$$\begin{aligned} \frac{B}{A} &= \frac{\zeta(\xi^2 + \zeta^2 - \beta^2) \cos \alpha b}{2\beta(\xi^2 + \zeta^2) \cos \beta b} \\ &= \frac{2\alpha\zeta \sin \alpha b}{(\beta^2 - \xi^2 - \zeta^2) \sin \beta b}, \end{aligned} \tag{9}$$

which are the Rayleigh-Lamb amplitude-ratio and frequency-equation.

In anticipation of a property of SH waves, mentioned in Section 1, we take only the solutions

$$\sin \alpha b = 0, \qquad \sin \beta b = 0 \tag{10}$$

of the frequency equation. Then

$$\alpha = m\pi/2b, \qquad \beta = n\pi/2b, \tag{11}$$

with m and n even integers, and

$$B/A = \zeta(\xi^2 + \zeta^2 - \beta^2)/2\beta(\xi^2 + \zeta^2)(-1)^{(n-m)/2}. \tag{12}$$

From Equations (7) and (11), the allowed frequencies are given by

$$\Omega^2 = k^2(n^2 - m^2)/(k^2 - 1) \tag{13}$$

where

$$\begin{aligned} \Omega^2 &= 4b^2\omega^2/\pi^2 v_2^2, \\ k^2 &= v_1^2/v_2^2 = (\lambda + 2\mu)/\mu = 2(1 - \nu)/(1 - 2\nu) \end{aligned} \tag{14}$$

and ν is Poisson's ratio.

The components of traction on planes $x = \pm a$ are

$$\begin{aligned} \sigma_{xx} &= \mu A\zeta(\beta^2 - 2\alpha^2 + \xi^2 - \zeta^2) \cos \xi a \cos \alpha y \cos \zeta z \\ &\quad - 2\mu B\xi^2\beta \cos \xi a \cos \beta y \cos \zeta z, \\ \sigma_{xy} &= \mp 2\mu A\xi\alpha\zeta \sin \xi a \sin \alpha y \cos \zeta z \\ &\quad \mp \mu B\xi(\xi^2 + \zeta^2 - \beta^2) \sin \xi a \sin \beta y \cos \zeta z, \\ \sigma_{xz} &= \mp 2\mu A\xi\zeta^2 \sin \xi a \cos \alpha y \sin \zeta z \\ &\quad \pm 2\mu B\xi\beta\zeta \sin \xi a \cos \beta y \sin \zeta z. \end{aligned} \tag{15}$$

4 SH Waves in an Infinite Plate

We consider, now, two SH waves, or vibrations, in a plate bounded by traction-free surfaces $y = \pm b$. Let $\varphi = H_x = H_z = 0$ and

$$\begin{aligned} H_y &= C_1\zeta \sin \eta x \cos \alpha y \sin \zeta z \\ &\quad + C_2\zeta \sin \xi x \cos \beta y \sin \zeta z. \end{aligned} \tag{16}$$

Equations (5) are satisfied if

$$\omega^2/v_2^2 = \xi^2 + \beta^2 + \zeta^2 = \eta^2 + \alpha^2 + \zeta^2 \tag{17}$$

In this solution $\sigma_{yy} = 0$ everywhere and, on $y = \pm b$, the remaining components of traction are

$$\begin{aligned} \sigma_{yz} &= \mp\mu C_1\alpha\eta\zeta \cos \eta x \sin \alpha b \sin \zeta z \\ &\quad \mp \mu C_2\beta\xi\zeta \cos \xi x \sin \beta b \sin \zeta z, \\ \sigma_{xy} &= \pm\mu C_1\alpha\zeta^2 \sin \eta x \sin \alpha b \cos \zeta z \\ &\quad \pm \mu C_2\beta\zeta^2 \sin \xi x \sin \beta b \cos \zeta z. \end{aligned} \tag{18}$$

In view of Equation (11), the surfaces $y = \pm b$ are free of traction. Also, from (17) and (11), we require

$$\eta^2 - \xi^2 = (n^2 - m^2)\pi^2/4b^2 \tag{19}$$

The components of traction on $x = \pm a$ are

$$\begin{aligned} \sigma_{xx} &= -2\mu C_1\eta\zeta^2 \cos \eta a \cos \alpha y \cos \zeta z \\ &\quad - 2\mu C_2\xi\zeta^2 \cos \xi a \cos \beta y \cos \zeta z, \\ \sigma_{xy} &= \pm\mu C_1\alpha\zeta^2 \sin \eta a \sin \alpha y \cos \zeta z \\ &\quad \pm \mu C_2\beta\zeta^2 \sin \xi a \sin \beta y \cos \zeta z, \\ \sigma_{xz} &= \pm\mu C_1\zeta(\zeta^2 - \eta^2) \sin \eta a \cos \alpha y \sin \zeta z \\ &\quad \pm \mu C_2\zeta(\zeta^2 - \xi^2) \sin \xi a \cos \beta y \sin \zeta z. \end{aligned} \tag{20}$$

5 Bar of Rectangular Cross Section

The sum of the three solutions, given in Sections 3 and 4, leaves the faces $y = \pm b$ free of traction and the faces $x = \pm a$ with the tractions given in Equations (15) and (20). The tangential components of traction on $x = \pm a$ vanish if

$$\sin \xi a = 0, \qquad \sin \eta a = 0, \tag{21}$$

i.e., if

$$\xi = p\pi/2a, \qquad \eta = q\pi/2a, \tag{22}$$

where p and q are even integers. Then the remaining component of traction vanishes if

$$\begin{aligned} C_1/A &= (\eta^2 - \alpha^2 - \zeta^2)/2\eta\zeta(-1)^{(q-p)/2}, \\ C_2/A &= \xi(\beta^2 - \xi^2 - \zeta^2)/2\zeta(\xi^2 + \zeta^2)(-1)^{(n-m)/2}. \end{aligned} \tag{23}$$

From (22) and (19) the allowed ratios of width to depth of the cross section of the bar are given by

$$a^2/b^2 = (q^2 - p^2)/(n^2 - m^2). \tag{24}$$

Thus far we have considered only solutions symmetric with respect to the three co-ordinate planes. A change of symmetry with respect to the plane of x and y is equivalent only to a shift of the origin of co-ordinates along the axis of the bar; but changes of symmetry with respect to the other two co-ordinate planes produce three additional families of modes. To change the symmetry with respect to the plane of y and z or x and z without changing formulas (12) and (23) for the amplitude ratios, it is only necessary to interchange sines and cosines of functions of x or y, change p and q or m and n from even to odd and reverse the sign of u or v. The solutions, in terms of displacements, are then

(i) symmetric-symmetric (m, n even; p, q even)

$$\begin{aligned} u &= (A\xi \sin \xi x - C_1\zeta \sin \eta x)\zeta \cos \alpha y \cos \zeta z, \\ v &= [A\alpha\zeta \sin \alpha y + B(\xi^2 + \zeta^2) \sin \beta y] \cos \xi x \cos \zeta z, \\ w &= (A\zeta \cos \xi x + C_1\eta \cos \eta x)\zeta \cos \alpha y \sin \zeta z \\ &\quad + (C_2\xi - B\beta)\zeta \cos \xi x \cos \beta y \sin \zeta z; \end{aligned} \tag{25}$$

(ii) symmetric-antisymmetric (m, n even; p, q odd)

$$\begin{aligned} u &= -(A\xi \cos \xi x - C_1\zeta \cos \eta x)\zeta \cos \alpha y \cos \zeta z, \\ v &= [A\alpha\zeta \sin \alpha y + B(\xi^2 + \zeta^2) \sin \beta y] \sin \xi x \cos \zeta z, \\ w &= (A\zeta \sin \xi x + C_1\eta \sin \eta x)\zeta \cos \alpha y \sin \zeta z \\ &\quad + (C_2\xi - B\beta)\zeta \sin \xi x \cos \beta y \sin \zeta z; \end{aligned} \tag{26}$$

2

(iii) antisymmetric-symmetric (m, n odd; p, q even)

$$
\begin{aligned}
u &= (A\xi \sin \xi x - C_1\zeta \sin \eta x)\zeta \sin \alpha y \cos \zeta z,\\
v &= -[A\alpha\zeta \cos \alpha y + B(\xi^2 + \zeta^2) \cos \beta y] \cos \xi x \cos \zeta z,\\
w &= (A\zeta \cos \xi x + C_1\eta \cos \eta x)\zeta \sin \alpha y \sin \zeta z\\
&\quad + (C_2\xi - B\beta)\zeta \cos \xi x \sin \beta y \sin \zeta z;
\end{aligned} \tag{27}
$$

(iv) antisymmetric-antisymmetric (m, n odd; p, q odd)

$$
\begin{aligned}
u &= -(A\xi \cos \xi x - C_1\zeta \cos \eta x)\zeta \sin \alpha y \cos \zeta z,\\
v &= -[A\alpha\zeta \cos \alpha y + B(\xi^2 + \zeta^2) \cos \beta y] \sin \xi x \cos \zeta z,\\
w &= (A\zeta \sin \xi x + C_1\eta \sin \eta x)\zeta \sin \alpha y \sin \zeta z\\
&\quad + (C_2\xi - B\beta)\zeta \sin \xi x \sin \beta y \sin \zeta z.
\end{aligned} \tag{28}
$$

In each of the four cases, the amplitude ratios are given by (12) and (23) and the frequency is given by (13).

6 Ratios of Width to Depth

Since Poisson's ratio is restricted to the range $-1 < \nu \leqslant 1/2$ we have, from the last of (14), $(4/3) < k^2 \leqslant \infty$; hence, from (13), we require $n > m$ and, from (24), $q > p$. Now, the difference between the squares of two integers, both even or both odd, is a multiple of four, say $4N$, where N is any integer. Since, by Equation (24), a^2/b^2 is the ratio of two such differences, it may be the ratio of any two integers. Hence, the ratio of width to depth of the bar may be the square root of any positive rational number.

For a given m and n, i.e., for a given frequency, there may be more than one pair of p and q corresponding to the same a/b. Similarly, there may be more than one pair of m and n corresponding to the same frequency. The additional number of pairs is equal to the number of unlike factors of N whose squares are not greater than N. To find all such pairs for any N we write

$$(n_i - m_i)(n_i + m_i) = 4N, \qquad N = 1, 2, 3 \ldots \tag{29}$$

Let

$$n_i = L_i + M_i, \qquad m_i = L_i - M_i, \tag{30}$$

where L_i and M_i are integers and $L_i > M_i$. If L_i and M_i are both even or both odd, n_i and m_i are both even; if one of L_i and M_i is even and the other odd, then n_i and m_i are both odd. From (29) and (30), $L_i = N/M_i$ subject to the restrictions that N/M_i is an integer and $M_i^2 \leqslant N$. Hence

$$
\left.\begin{aligned}
n_i &= N/M_i + M_i\\
m_i &= N/M_i - M_i
\end{aligned}\right\}
\begin{cases}
N = 1, 2, 3 \ldots\\
M_i = 1, 2, 3 \ldots \leqslant N^{1/2}.
\end{cases} \tag{31}
$$

The first twenty-five sets of pairs are given in Table 1. The table may be used either for n_i and m_i or q_i and p_i.

7 Frequency Spectrum

The relation between the frequency and the wave length, $\lambda' = 2\pi/\zeta$, along the axis of z, is determined by Equations (7) and (17). In view of (19), the second of (7) and both of (17) are the same and may be written as

$$\Omega^2 = s^2 + \gamma^2, \tag{32}$$

where

$$s^2 = n^2 + p^2b^2/a^2 \tag{33}$$

and

$$\gamma = 2\zeta b/\pi = 4b/\lambda', \tag{34}$$

i.e., γ is the ratio of the depth, $2b$, of the bar, to the half wave length along the axis of z. The remaining expression for the frequency, i.e., the first of (7), can be written as

$$\Omega^2 = k^2(r^2 + \gamma^2) \tag{35}$$

where

$$r^2 = m^2 + p^2b^2/a^2 \tag{36}$$

For real, positive Ω and γ, (32) and (35) are families of hyperbolas in the $\gamma - \Omega$ plane with asymptotes $\Omega = \gamma$ and $\Omega = k\gamma$ and semitransverse axes $\Omega = s$ and $\Omega = kr$, respectively. The curves are illustrated in Fig. 1 for integer-values of r and s and for $\nu = 0.31$. Since both (32) and (35) must be satisfied, permissible combinations of frequency and wave length are given by the intersections of the two families of curves. The co-ordinates of the intersection points are

$$
\begin{aligned}
\Omega &= [k^2(s^2 - r^2)/(k^2 - 1)]^{1/2},\\
\gamma &= [(s^2 - k^2r^2)/(k^2 - 1)]^{1/2}.
\end{aligned} \tag{37}
$$

The parameters r and s need not be integers, but they may not be chosen arbitrarily inasmuch as they must satisfy the relation

Table 1 Pairs of integers $n_i^2 - m_i^2 = 4N$

N	$n_i^2 - m_i^2$	n_1	m_1	n_2	m_2	n_3	m_3	n_4	m_4	n_5	m_5
1	4	2	0								
2	8	3	1								
3	12	4	2								
4	16	5	3	4	0						
5	20	6	4								
6	24	7	5	5	1						
7	28	8	6								
8	32	9	7	6	2						
9	36	10	8			6	0				
10	40	11	9	7	3						
11	44	12	10								
12	48	13	11	8	4	7	1				
13	52	14	12								
14	56	15	13	9	5						
15	60	16	14			8	2				
16	64	17	15	10	6			8	0		
17	68	18	16								
18	72	19	17	11	7	9	3				
19	76	20	18								
20	80	21	19	12	8			9	1		
21	84	22	20			10	4				
22	88	23	21	13	9						
23	92	24	22								
24	96	25	23	14	10	11	5	10	2		
25	100	26	24							10	0

$$s^2 - r^2 = n^2 - m^2 = (q^2 - p^2)b^2/a^2. \qquad (38)$$

For real, positive Ω and imaginary, positive γ, (32) are quarter-circles of radii s and (35) are quarter-ellipses of major semiaxes kr and minor semiaxes r. Imaginary values of γ do not pertain to infinite trains of traveling waves but they have significance for vibrations of bars of finite length and for transient motions. Complex values of γ, with real, positive Ω, are permissible in general, but they do not appear in the special solutions described here.

8 Phase-Velocities

The phase-velocity of the coupled waves traveling along the bar is $v = \omega/\zeta$. Hence, from (14), (34), and (37),

$$v/v_2 = [k^2(s^2 - r^2)/(s^2 - k^2r^2)]^{1/2} = \Omega/\gamma. \qquad (39)$$

Thus the slope of the straight line connecting an intersection point and the origin, in Fig. 1, gives the ratio of the phase-velocity in the bar to the velocity of equivoluminal waves in an infinite medium. It may be seen that the minimum phase-velocity, of the special waves found here, is the velocity, $v = kv_2 = v_1$, of dilatational waves in an infinite medium. This occurs when $r = 0$, i.e., when m and p are zero and, hence, only in the case of the symmetric-symmetric family of modes.

The maximum phase-velocity can be infinite ($\zeta = 0$), but the character of the solution is then different. There are two groups of modes possible for $\zeta = 0$; the modes in one group do not depend on Poisson's ratio and the modes in the second group exist only for special values of Poisson's ratio.

In the first group there are eight families of simple axial shear modes which may be obtained from equations (25) to (28) by setting $\zeta = 0$ after interchanging $\sin \zeta z$ and $\cos \zeta z$. For example, from (25) we obtain $u = v = 0$ and

$$\begin{aligned} w &= C_1'' \cos \eta x \cos \alpha y, & \Omega^2 &= q^2b^2/a^2 + m^2, \\ w &= C_2'' \cos \xi x \cos \beta y, & \Omega^2 &= p^2b^2/a^2 + n^2. \end{aligned} \qquad (40)$$

where C_1'' and C_2'' are unrelated. Similar results are obtained, for the other three symmetries, from equations (26) to (28). The modes of type (40) are those that are independent of Poisson's ratio.

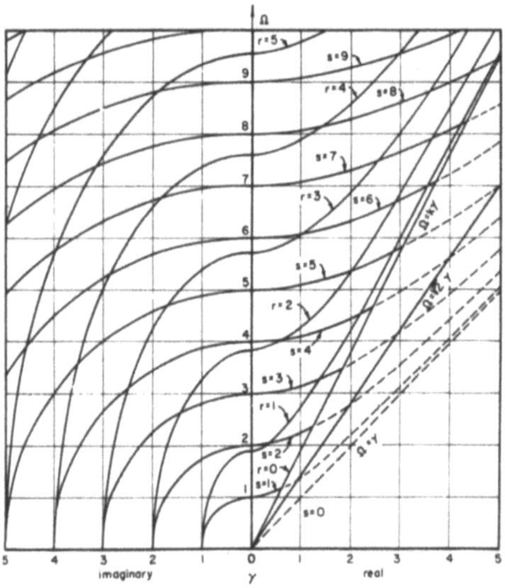

Fig. 1 Spectrum of frequencies and phase velocities

4

The second group of modes with infinite phase-velocity may be obtained from (25) to (28) by setting

$$\lim_{\zeta \to 0} A\zeta = A^*, \quad \lim_{\zeta \to 0} C_1\zeta^2 = C_1^*, \quad \lim_{\zeta \to 0} C_2\zeta^2 = C_2^*, \quad \lim_{\zeta \to 0} (\sin \zeta z)/\zeta = z. \qquad (41)$$

The solution, in the symmetric-symmetric case, then reduces to

$$\begin{aligned} u &= (A^*\xi \sin \xi x - C_1^* \sin \eta x) \cos \alpha y, \\ v &= (A^*\alpha \sin \alpha y + B\xi^2 \sin \beta y) \cos \xi x, \\ w &= z(C_1^*\eta \cos \eta x \cos \alpha y + C_2^*\xi \cos \xi x \cos \beta y), \end{aligned} \qquad (42)$$

with

$$\begin{aligned} B/A^* &= (\xi^2 - \beta^2)/2\beta\xi^2(-1)^{(n-m)/2}, \\ C_1^*/A^* &= (\beta^2 - 2\alpha^2 + \xi^2)/2\eta(-1)^{(q-p)/2}, \\ C_2^*/A^* &= (\beta^2 - \xi^2)/2\xi(-1)^{(n-m)/2}. \end{aligned} \qquad (43)$$

Corresponding solutions obtain for the remaining three symmetries. These modes exist only for special values of Poisson's ratio. From (32) and (35), with $\gamma = 0$, $\Omega = s = kr$. Hence the modes are possible only for those values of Poisson's ratio satisfying $k = s/r$. In Fig. 1, this means values of k for which there are intersections at $\gamma = 0$.

9 Propagation of Energy

In the case of an infinite train of sinusoidal waves, the velocity of propagation of energy is the group velocity $v_g = d\omega/d\zeta$. Because we have only discrete pairs of ω and ζ, the differentiation cannot be performed. However, the velocity can be obtained from considerations of work and energy.

During a period T, the average rate at which work is done on a right section of the bar is

$$W = T^{-1}\int_0^T dt \int_{-b}^{b}\int_{-a}^{a} (\sigma_{zx}\dot{u} + \sigma_{zy}\dot{v} + \sigma_{zz}\dot{w})dxdy, \qquad (44)$$

where $\dot{u}$, $\dot{v}$, and $\dot{w}$ are the components of velocity. Also, during the period T, the average total energy per unit of length of the bar (E) is twice the average kinetic energy per unit of length, i.e.,

$$E = \rho T^{-1}\int_0^T dt \int_{-b}^{b}\int_{-a}^{a} (\dot{u}^2 + \dot{v}^2 + \dot{w}^2)dxdy. \qquad (45)$$

Hence, the velocity at which energy crosses the section is

$$v_g = v_e = W/E. \qquad (46)$$

Upon performing the integrations in (44) and (45) and substituting in (46), we find

$$v_g = v_e = (v_2^2/v)(1 + P/Q), \qquad (47)$$

where v is, again, the phase-velocity and

$$\begin{aligned} P &= 4n^2(s^2 - m^2)(n^2 - m^2)\gamma^2, \\ Q &= 4n^2(s^2 - m^2)(r^2 + \gamma^2)\gamma^2 \\ &\quad + n^2(s^2 - m^2 + \gamma^2)(s^2 - 2m^2 - \gamma^2)^2 \\ &\quad + (s^2 - m^2)(n^2 + \gamma^2)(s^2 - 2n^2 + \gamma^2)^2. \end{aligned} \qquad (48)$$

A case of some interest is that for which $v = v_1$, i.e., when the phase-velocity is the velocity of dilatational waves. For example, the lowest of such modes in a square bar has

$$m, n, p, q = 0, 2, 0, 2$$

Then, for $\nu = 1/4$,

$$v_e = 5v_1/7 = 5v_2\sqrt{3}/7,$$

i.e., the velocity of propagation of energy lies between the velocities of dilatational and equivoluminal waves in an infinite medium.

10 Shapes of Modes

Inspection of Equations (25) to (28) shows that, in view of (11) and (22), $u = 0$ on $x = \pm a$ and $v = 0$ on $y = \pm b$. That is, in addition to these planes being free of traction, they remain plane. The displacement w is zero on $z = 0$ and again at every half wave length, π/ζ, along the z-axis. On these planes of constant z, although the tangential tractions σ_{zx} and σ_{zy} are zero, the normal traction is not. However, since $w = 0$, mixed boundary conditions are satisfied, so that we have solutions for the vibrations of bars of finite length with the four sides free of traction and mixed boundary conditions at the ends. However, the end conditions, although permissible mathematically, are not practical ones.

The displacement u has two wave lengths, $4a/p$ and $4a/q$, along x and one wave length, $4b/m$, along y. The displacement v has two wave lengths, $4b/m$ and $4b/n$, along y and one wave length, $4a/p$, along x. The displacement w has both wave lengths along x and both along y. The integers m and n or p and q are the ratios of the cross-sectional dimensions $2b$ or $2a$ to the half wave lengths in the corresponding directions. All the displacements have the same wave-length, $2\zeta/\pi$, along z. Examples of the mode-shapes for

$$m, n, p, q = 0, 2, 0, 2; \quad a = b, \quad \nu = 0.31, \quad \gamma = 1.23,$$
$$m, n, p, q = 1, 3, 0, 2; \quad a\sqrt{2} = b, \quad \nu = 0.31, \quad \gamma = 1.43,$$
$$m, n, p, q = 1, 3, 1, 3; \quad a = b, \quad \nu = 0.31, \quad \gamma = 1.02,$$

are illustrated in Figs. 2, 3, and 4. The first of these is one of the special cases for which the phase-velocity is the velocity of dilatational waves. In these cases a square grid on the cross section becomes a rectangular grid instead of a sinusoidal one.

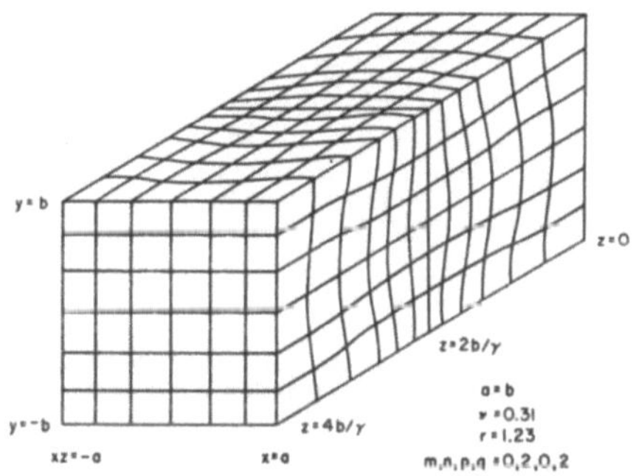

Fig. 2 Mode shape for m, n, p, q = 0, 2, 0, 2

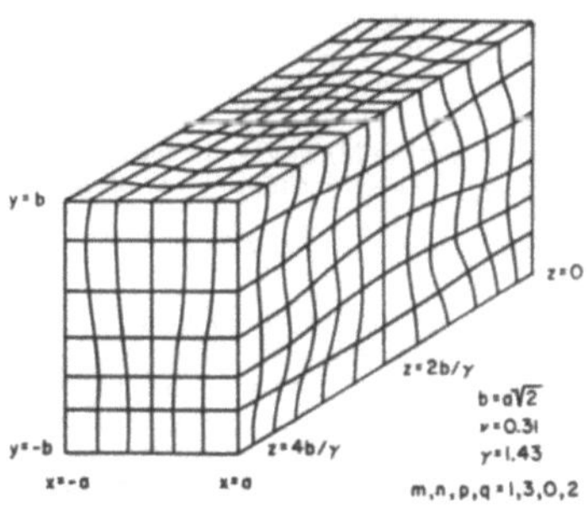

Fig. 3 Mode shape for m, n, p, q = 1, 3, 0, 2

11 Rays

The solution given in Section 5 was presented as a combination of a Rayleigh-Lamb wave and two SH waves in order to show its relation to known solutions for waves in an infinite plate. The result, however, may be expressed in simpler form. For example, Equations (25) derive from

$$\begin{aligned} \varphi &= -A' \cos \xi x \cos \alpha y \cos \zeta z, \\ H_z &= B' \cos \xi x \sin \beta y \sin \zeta z, \\ H_y &= -C' \sin \eta x \cos \alpha y \sin \zeta z, \end{aligned} \tag{49}$$

where

$$\begin{aligned} A' &= A\zeta, \\ B' &= B(\xi^2 + \zeta^2)/\zeta = A'(\xi^2 + \zeta^2 - \beta^2)/2\beta\zeta(-1)^{(n-m)/2}, \\ C' &= -C_1\zeta = A'(\alpha^2 + \zeta^2 - \eta^2)/2\eta\zeta(-1)^{(q-p)/2}. \end{aligned} \tag{50}$$

Similar forms can be written for the other three symmetries.

The three functions in (49) represent dilatational waves (φ), SV waves (H_z), and SH waves (H_y). The displacement in the SV waves is parallel to planes x = const and the displacement in the SH waves is parallel to planes y = const. It may be seen, from (49) and (50), that a rotation of 90 deg about the axis of z does not produce another family of solutions. This is not immediately apparent from the form of the solution in Section 5.

Each of the three functions in (49) may be resolved into a set of eight plane waves which progress along the bar by oblique reflections at the faces $x = \pm a$ and $y = \pm b$. The resolution may be effected by expanding the products of four trigonometric functions, for example

$$\cos \xi x \cos \beta y \cos \zeta z \cos \omega t, \tag{51}$$

that appear in Equations (25) to (28), into sums and differences of eight trigonometric functions whose arguments are the sums and differences of the original arguments, e.g.,

$$\cos (\xi x + \beta y + \zeta z - \omega t), \tag{52}$$

which is a traveling, plane wave whose normal has direction cosines proportional to ξ, β, and ζ. In the case of the SV waves, the eight sets of direction cosines are the eight combinations of

$$\pm v_2\xi/\omega, \quad \pm v_2\beta/\omega, \quad \pm v_2\zeta/\omega; \tag{53}$$

or

$$\pm pb/\Omega a, \quad \pm n/\Omega, \quad \pm \gamma/\Omega \tag{54}$$

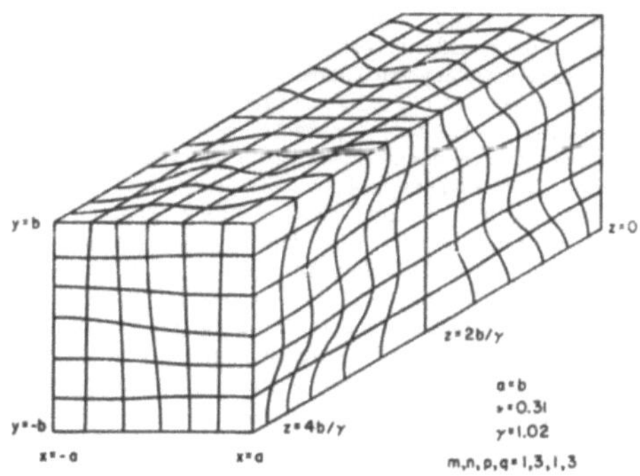

Fig. 4 Mode shape for m, n, p, q = 1, 3, 1, 3

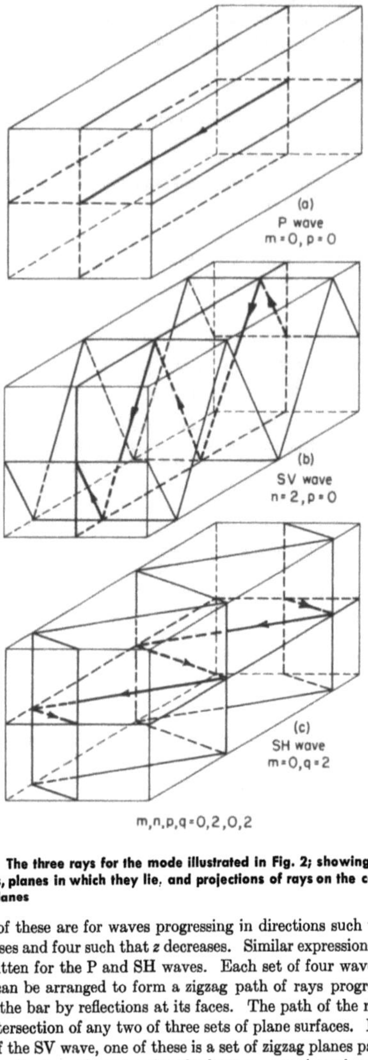

Fig. 5 The three rays for the mode illustrated in Fig. 2; showing paths of rays, planes in which they lie, and projections of rays on the co-ordinate planes

Four of these are for waves progressing in directions such that z increases and four such that z decreases. Similar expressions may be written for the P and SH waves. Each set of four wave-normals can be arranged to form a zigzag path of rays progressing along the bar by reflections at its faces. The path of the rays is the intersection of any two of three sets of plane surfaces. In the case of the SV wave, one of these is a set of zigzag planes parallel to the axis of x and intersecting the faces $x = \pm a$ in a zigzag line whose direction cosines are proportional to $\pm n$ and γ. A second is a set of zigzag planes parallel to the axis of y and intersecting the faces $y = \pm b$ in a zigzag line whose direction cosines are proportional to $\pm pb/a$ and γ. The third set of surfaces is a family of quadrangular prisms whose lateral edges are parallel to the axis of z and whose right sections are equilateral parallelograms whose sides have direction cosines $\pm pb/a$ and $\pm n$. If one of the direction cosines is zero the parallelograms degenerate to a straight line and if both direction cosines are zero the parallelograms degenerate to a point. In the case of the P waves, p and n become p and m; in the case of SH waves p and n become q and m.

The sets of intersecting planes and the paths of the rays for the modes shown in Figs. 2, 3, and 4 are shown in Figs. 5, 6, and 7.

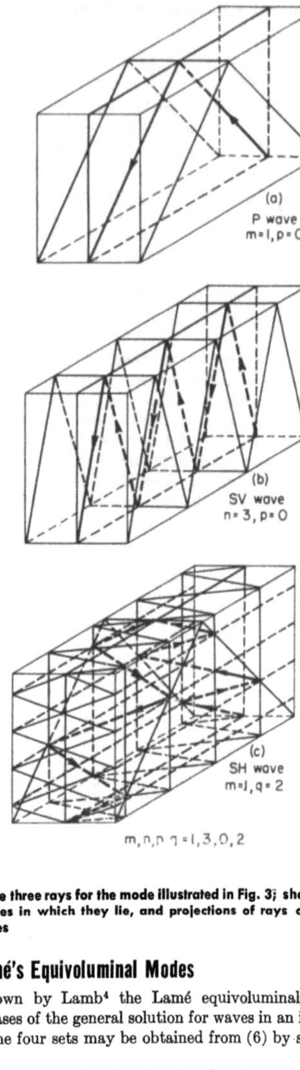

Fig. 6 The three rays for the mode illustrated in Fig. 3; showing paths of rays, planes in which they lie, and projections of rays on the co-ordinate planes

12 Lamé's Equivoluminal Modes

As shown by Lamb[4] the Lamé equivoluminal modes[2] are special cases of the general solution for waves in an infinite plate. One of the four sets may be obtained from (6) by setting

$$A = 0, \quad \xi = 0, \quad \beta = \zeta. \tag{55}$$

The boundary conditions (8) on $y = \pm b$ then reduce to a single condition, on σ_{yy}, which is satisfied by $\cos \beta b = 0$ so that

$$\beta = n\pi/2b \tag{56}$$

where n is now an odd integer instead of the even integer which it was before for the same symmetry. There are no tractions on planes normal to the axis of x so that the result is independent of the ratio a/b. The wave is solely a plane SV wave whose rays are parallel to planes $x =$ const and reflect at the surfaces $y = \pm b$ at angles of incidence and reflection of 45 deg. For modes antisymmetric with respect to the plane of x and z, n is even.

From (55), (56), and the second of (7),

$$\Omega^2 = n^2 + \gamma^2, \quad \gamma = n \tag{57}$$

so that, in Fig. 1, the frequency spectrum of these modes is given by the intersections of the line $\Omega = \sqrt{2}\,\gamma$ with the hyperbolas $s =$ const, for integer-values of s. The phase-velocity is thus $v = \sqrt{2}\, v_2$ and, by the method of Section 9, the velocity of propagation of energy is $v_e = v_2/\sqrt{2}$.

There are similar modes for SH waves whose rays are parallel to planes $y =$ const and reflect at the faces $x = \pm a$ at angles of incidence and reflection of 45 deg. This time we have

$$\Omega^2 = p^2b^2/a^2 + \gamma^2, \qquad \gamma = pb/a \tag{58}$$

and, in Fig. 1, the spectrum is given by the intersections of $\Omega = \sqrt{2}\,\gamma$ with the hyperbolas $s = pb/a$. The apparent dependence on the ratio a/b is due only to the fact that a reference length b is contained in the co-ordinates Ω and γ.

13 Relation to Complete Spectrum

The complete solution is not known for waves in an infinite bar of rectangular cross section with traction-free faces and, in fact, it cannot be expressed in terms of a finite number of elementary functions. The special solutions described in this paper give some information about the behavior of the complete spectrum in the case of bars whose cross-sectional dimensions satisfy Equation (24). For a given a/b satisfying (24), the solutions in Sections 5 and 12 yield a set of points on the branches of the complete spectrum Ω versus γ. In addition, the slopes of the branches at these points are given by the expressions for the velocity of propagation of energy, inasmuch as

$$\frac{v_e}{v_2} = \frac{v_g}{v_2} = \frac{1}{v_2}\frac{d\omega}{d\zeta} = \frac{d\Omega}{d\gamma} \tag{59}$$

How the branches behave between these points and how they behave for complex wave-numbers is not known. One family of branches must intersect $\gamma = 0$ at cut-off frequencies and mode shapes of the type given in Equation (40). These are equivoluminal cut-off modes whose displacements are parallel to the axis of the bar. The companion cut-off modes, involving coupled equivoluminal and dilatational deformation with displacements perpendicular to the axis of the bar, are not available. These, along with (*a*) the easily obtainable solutions for a bar with one pair of opposite faces traction-free and the other pair with mixed boundary conditions; and (*b*) the known behavior of the lowest branch at long wave length, could form a basis for establishing the behavior of the complete branches in greater detail. In such a study it would be important to recognize that the special solutions for the rectangular bar, given in Section 5, must be related to the unknown general solution in the same way that the special solutions for the infinite plate, given by Equations (10) to (12), are related to the general solution, Equation (9), for the plate.

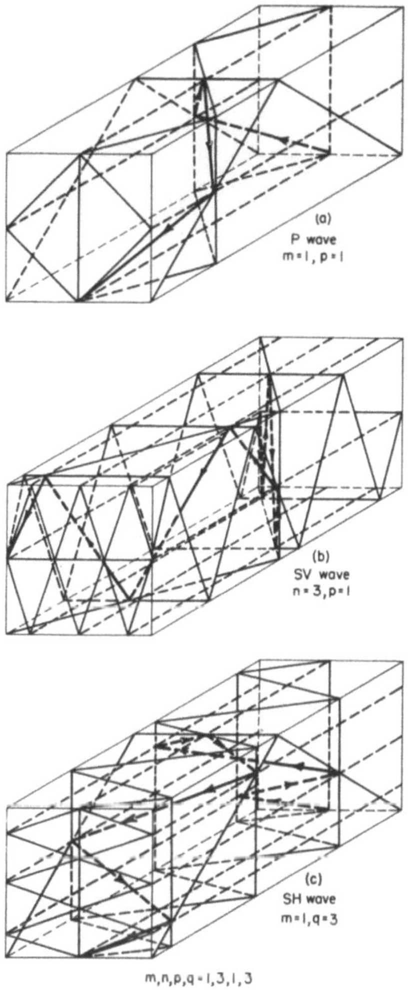

Fig. 7 The three rays for the mode illustrated in Fig. 4; showing paths of rays, planes in which they lie, and projections of rays on the co-ordinate planes

7

(Reprinted from the Journal of Basic Engineering for September, 1960, with permission of Copyright Owner)

YIH-HSING PAO
Assistant Professor of Mechanics, Cornell University, Ithaca, N. Y

R. D. MINDLIN
Professor of Civil Engineering, Columbia University, New York, N. Y.
Mem. ASME

Dispersion of Flexural Waves in an Elastic, Circular Cylinder[1]

In this paper it is shown how the branches of Pochhammer's frequency equation for flexural waves in a circular cylinder may be constructed approximately with the aid of a grid of simpler curves and asymptotic equations for long and short wave lengths. With very little computation, in comparison with that required in the direct determination of the roots of Pochhammer's equation, a qualitative view is obtained of the relations between frequency, phase velocity, group velocity, and propagation constant, for any branch, as well as some information as to the shapes of the modes.

The solutions of the linear equations of elasticity for extensional and flexural vibrations and waves in an infinitely long circular cylinder, with its surface free from traction, were given by Pochhammer [1],[2] in 1876, in the form of transcendental frequency equations which have been the subject of numerous studies in the succeeding years. The extraordinarily intricate branches of these equations give the relations between frequency, propagation constant, phase velocity, and group velocity which are required for an understanding of the physical significance of the solutions.

Recently, methods which do not require extensive numerical computation have been developed for the approximate mapping of the branches of Pochhammer's equation for extensional waves and the branches of similar equations which appear in solutions for both extensional and flexural vibrations or waves in plates [2-8]. These methods are extended, in this paper, to the case of Pochhammer's equation for flexure.

The essence of the technique of mapping has been the preliminary construction of intersecting families of simple algebraic curves which the branches cannot cross except at certain of the points of intersection. This information, along with the asymptotic behavior at long and short wave lengths, has been sufficient to permit the tracing of the branches through the grid of intersecting curves, or "bounds," in considerable detail. If the construction is performed with frequency as ordinate and propagation constant, i.e., wave number in the direction of propagation, as abscissa, the slope of the straight line connecting a point on a branch with the origin is proportional to the phase velocity and the slope of the branch itself is proportional to the group velocity. A pair of values of frequency and wave number, i.e., a point on a branch, corresponds to a mode of motion whose properties may be determined by inserting the pair into the formulas for displacement, strain, and stress. Some, and in certain cases all, of the bounds have been the branches of the frequency equation for waves in the same body with mixed, instead of traction-free, boundary conditions; one family of bounds corresponding to equivoluminal deformation, and another to rotationless dilatational deformation. Passage from mixed to traction-free boundary conditions is accompanied by coupling of the two types of deformation and leads to the complicated behavior invariably encountered with nonmixed boundary conditions. The resolution into the two types of simple deformation gives, in addition, insight into the modal composition corresponding to points on the branches both at and between intersections of bounds.

In the case of Pochhammer's equation for flexure, no such resolution is possible. Although simple bounds corresponding to equivoluminal deformation (but not to a mixed boundary-value problem) may be found, simple bounds corresponding to purely dilatational deformation are not possible. This obstacle has been surmounted, in the present study, by introducing auxiliary curves by means of which the missing bounds are constructed. Six families of curves are employed to construct the latter and the result is a second family of bounds of intricacy comparable with the branches in previous problems. Nevertheless, sufficient bounds are then available to permit the construction of the branches of Pochhammer's equation for flexure. In this paper, an analytical study of the frequency equation is summarized and an example is given of the construction of bounds and branches for Poisson's ratio 1/3 and real propagation constant.

Pochhammer's Equation

In the form given by Love [9], Pochhammer's displacements, in cylindrical co-ordinates r, θ, z, are

$$\begin{aligned} u_r &= U \cos\theta \, e^{i(\gamma z + pt)} \\ u_\theta &= V \sin\theta \, e^{i(\gamma z + pt)} \\ u_z &= W \cos\theta \, e^{i(\gamma z + pt)} \end{aligned} \tag{1}$$

where p is the circular frequency, γ is the propagation constant, and

$$\begin{aligned} U &= A \frac{\partial J_1(hr)}{\partial r} + \beta\gamma \frac{\partial J_1(\kappa r)}{\partial r} + C \frac{J_1(\kappa r)}{r} \\ V &= -A \frac{J_1(hr)}{r} - B\gamma \frac{J_1(\kappa r)}{r} - C \frac{\partial J_1(\kappa r)}{\partial r} \\ W &= iA\gamma J_1(hr) - iB\kappa^2 J_1(\kappa r) \end{aligned} \tag{2}$$

in which J_1 denotes Bessel's function of order unity and

$$\begin{aligned} h^2 &= p^2/v_1^2 - \gamma^2, \quad v_1^2 = (\lambda + 2\mu)/\rho \\ \kappa^2 &= p^2/v_2^2 - \gamma^2, \quad v_2^2 = \mu/\rho \end{aligned} \tag{3}$$

---

[1] This investigation was supported by the Office of Naval Research under a contract with Columbia University.
[2] Numbers in brackets designate References at end of paper.

Contributed by the Applied Mechanics Division and presented at the Annual Meeting, Atlantic City, N. J., November 29–December 4, 1959, of The American Society of Mechanical Engineers.
Discussion of this paper should be addressed to the Secretary, ASME, 29 West 39th Street, New York, N. Y., and will be accepted until October 10, 1960, for publication at a later date. Discussion received after the closing date will be returned.
Note: Statements and opinions advanced in papers are to be understood as individual expressions of their authors and not those of the Society. Manuscript received by ASME Applied Mechanics Division, May 13, 1959. Paper No. 59—A-84.

where λ and μ are Lamé's constants, and ρ is the density. The constant A is associated with the dilatational part of the deformation while B and C are associated with equivoluminal parts.

It is convenient to introduce dimensionless wave numbers, α, β, ζ, and frequency, Ω, in terms of the radius, a, of the cylinder and the velocity, v_2, of equivoluminal waves in an infinite medium:

$$\alpha = ha, \; \beta = \kappa a, \; \zeta = \gamma a, \; \Omega = pa/v_2 \tag{4}$$

so that (3) become

$$\alpha^2 = k^{-2}\Omega^2 - \zeta^2 \tag{5}$$

$$\beta^2 = \Omega^2 - \zeta^2 \tag{6}$$

where

$$k^2 = (\lambda + 2\mu)/\mu = 2(1 - \sigma)/(1 - 2\sigma)$$

and σ is Poisson's ratio.

Upon inserting (1) into the formulas for the components of traction across $r = a$, setting these equal to zero and eliminating the constants A, B, C, we obtain Pochhammer's frequency equation, which may be written[3]

$$J_1(\alpha)J_1{}^2(\beta)[f_1\mathfrak{J}_\beta{}^2 + f_2\mathfrak{J}_\alpha\mathfrak{J}_\beta + f_3\mathfrak{J}_\beta + f_4\mathfrak{J}_\alpha + f_5] = 0 \tag{7}$$

where

$$\begin{aligned} f_1 &= 2(\beta^2 - \zeta^2)^2, \quad f_2 = 2\beta^2(5\zeta^2 + \beta^2) \\ f_3 &= \beta^6 - 10\beta^4 - 2\beta^4\zeta^2 + 2\beta^2\zeta^2 + \beta^2\zeta^4 - 4\zeta^4 \\ f_4 &= 2\beta^2(2\beta^2\zeta^2 - \beta^2 - 9\zeta^2) \\ f_5 &= \beta^2(-\beta^4 + 8\beta^2 - 2\beta^2\zeta^2 + 8\zeta^2 - \zeta^4) \end{aligned} \tag{8}$$

and

$$\mathfrak{J}_x \equiv \mathfrak{J}_1(x) \equiv xJ_0(x)/J_1(x) \tag{9}$$

is Onoe's function [10] of the first kind and order unity: the analog of the trigonometric cotangent. The behavior of $\mathfrak{J}_1(x)$ is illustrated in Fig. 1.

In view of (5) and (6), Pochhammer's equation (7) relates the dimensionless frequency, Ω, to the dimensionless propagation constant ζ; with k, or Poisson's ratio, as a parameter. Only the squares of Ω and ζ appear in (5) and (6) so that positive and negative values are interchangeable. Negative frequencies have no physical significance but negative propagation constants are permissible and give the mirror images of the curves that are described in the sequel.

From (5) and (6) it may be seen that the plane of Ω and real positive ζ is divided into three sectors in which (7) has different behaviors according as the arguments α and β, of the transcendental functions, are real or imaginary:

Sector 1 ($0 < \Omega/\zeta < 1$): α and β imaginary
Sector 2 ($1 < \Omega/\zeta < k$): α imaginary, β real
Sector 3 ($k < \Omega/\zeta < \infty$): α and β real

In Figs. 2 and 3, the radii dividing the sectors are designated OE ($\Omega = \zeta$ or $\beta = 0$) and OD ($\Omega = k\zeta$ or $\alpha = 0$).

Calling v the phase velocity and v_g the group velocity, we have

$$v \equiv p/\gamma = v_2\Omega/\zeta, \quad v_g \equiv dp/d\gamma = v_2 d\Omega/d\zeta \tag{10}$$

so that the slope of the straight line from the origin to a point in the ζ, Ω-plane satisfying (5), (6), and (7) is proportional to the phase velocity and the slope of a branch defined by (5), (6), and (7) is proportional to the group velocity, the constant of proportionality being the velocity of equivoluminal waves in an infinite medium. A point of inflection of a branch signifies a maximum or minimum of group velocity according as the change of sign is from positive to negative or vice versa as ζ increases.

[3] In determinantal form, Pochhammer's frequency equation is

$$J_1(\alpha)J_1{}^2(\beta)\begin{vmatrix} \zeta^2 - \beta^2 + 4 - 2\mathfrak{J}_\alpha & 2\zeta(\beta^2 - 2 + \mathfrak{J}_\beta) & 2\mathfrak{J}_\beta - 4 \\ 4 - 2\mathfrak{J}_\alpha & 2\zeta(\mathfrak{J}_\beta - 2) & \beta^2 - 4 + 2\mathfrak{J}_\beta \\ 2\zeta(\mathfrak{J}_\alpha - 1) & (\beta^2 - \zeta^2)(\mathfrak{J}_\beta - 1) & \zeta \end{vmatrix} = 0$$

in which the columns derive from the constants A, B, C, respectively, and the rows from the components of surface traction in the directions of r, θ, and z, respectively.

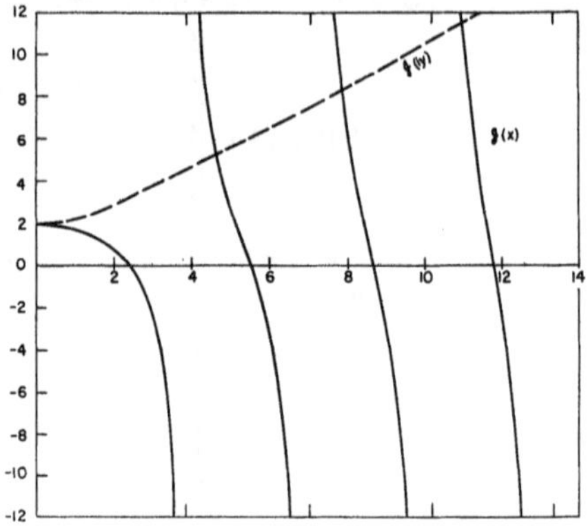

Fig. 1 Onoe's function $\mathfrak{J}_1(x) = xJ_0(x)/J_1(x)$

Asymptotic Behavior at Long Wave Length

We require the ordinates, slopes, and curvatures of the branches of (7) at $\zeta = 0$, i.e., infinite wave length along the rod.

Ordinates. When $\zeta = 0$, (7) can be expressed as a product of three functions. One of the three factors is β^2, giving zero frequency as discussed in the next section. The other two factors, set equal to zero, are

$$\mathfrak{J}(\beta_0) - 1 = 0 \tag{11}$$

$$2\mathfrak{J}(\alpha_0) + 2\mathfrak{J}(\beta_0) + \beta_0{}^2 - 8 = 0 \tag{12}$$

where, from (5) and (6), $\alpha_0 = \Omega/k$, $\beta_0 = \Omega$. The roots of (11) and (12) are determined readily with the aid of Onoe's tables and give the cut-off frequencies.

Corresponding to (11), A, C, and γ are zero, so that, from (2), the displacement is purely axial and equivoluminal. The roots of (11) are independent of Poisson's ratio and are almost uniformly spaced. The first seven frequencies are listed in the starred columns in Table 1 and are identified by circles along $\zeta = 0$ in Figs. 2 and 3. The associated modes of displacement have one diametral node and a number of circular nodes equal to one less than the order of the root of (11). These are the axial shear modes.

Corresponding to (12), $A/C = -J_1(\beta_0)/J_1(\alpha_0)$ and B and γ are zero, so that the motion is one of coupled dilatational and rotational plane strain in the plane of the cross section. This is an early indication of difficulties to be encountered subsequently. In the case of Pochhammer's solution for extensional waves in a rod or Rayleigh's solution [11] for both extensional and flexural waves in a plate, the analogous motion is purely dilatational. Equation (12) yields a nonuniform distribution of frequencies which varies markedly, but systematically, with Poisson's ratio, as discussed in the Appendix. The first ten frequencies, for $\sigma = 1/3$, i.e., $k = 2$, are given in Table 1 and are identified by crosses along $\zeta = 0$ in Figs. 2 and 3. Whether a mode of a given order is predominantly dilatational ($A/C > 1$) or equivoluminal ($A/C < 1$) depends on Poisson's ratio. The modes of resultant displacement have nodal points but no nodal diameters or circles. By the usual substitution of $\sigma/(1 + \sigma)$ for σ, this solution is converted to the analogous solution of Poisson's equations[4] of

[4] Reference [9], p. 497.

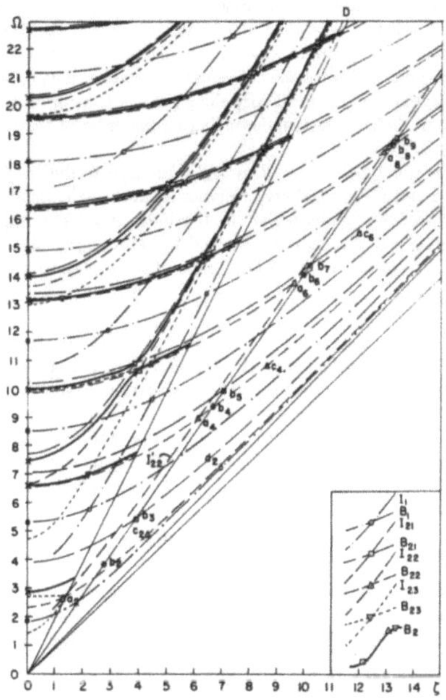

Fig. 2 Bounds, barriers, auxiliary barriers, and intersectors

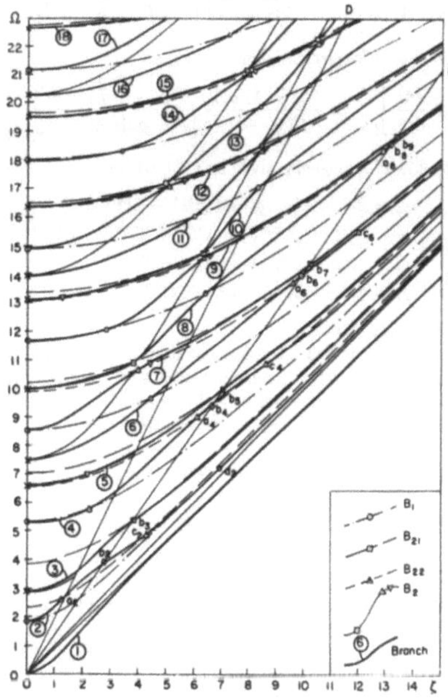

Fig. 3 Branches of Pochhammer's frequency equation traced through the grid of bounds and barriers

Table 1 Cut-off frequencies and ratios of curvatures of branches to curvatures of bounds at cut-off. Poisson's ratio = 1/3.

Branch no.	1	2*	3	4*	5	6	7*	8	9*	10	11	12*	13	14*	15	16	17*	18
Ω	0	1.84	2.86	5.33	6.55	7.48	8.53	9.97	11.70	13.14	13.91	14.86	16.35	18.02	19.50	20.26	21.16	22.67
Curv. of branch / Curv. of bound	—	3.01	1.17	0.34	0.70	0.83	1.79	1.01	0.81	0.99	0.89	1.52	1.01	0.89	1.01	0.92	1.43	1.01

* Asterisks designate axial shear modes.

extensional motion of thin plates (generalized plane stress) and then (12) is identical with the case $n = 1$ of Love's general solution for the extensional vibrations of a thin circular disk.[5]

The zero frequency in Table 1 represents the cut-off frequency of the lowest branch of (7) and also two rigid motions of the rod; one transverse translational and one rotational.

Slopes and Curvatures. As shown by Pochhammer [1], the lowest branch has the asymptotic behavior

$$\Omega = \zeta^2[(1+\sigma)/2]^{1/2} \tag{13}$$

for $\zeta \ll 1$, so that the terminal ordinate and slope are zero and the terminal curvature is

$$d^2\Omega/d\zeta^2]_{\zeta=0} = 2[(1+\sigma)/2]^{1/2} \tag{14}$$

To determine the slopes and curvatures of the upper branches at $\zeta = 0$, we note, first, that equation (7) may be expressed as a function of either α and β or Ω and ζ, where the two pairs of variables are related through (5) and (6). Then, calling the left-hand side of Pochhammer's equation P, we have, by the rule of differentiation of implicit functions,

$$\frac{d\Omega}{d\zeta} = -\frac{P_\zeta}{P_\Omega} = -\frac{P_\alpha\alpha_\zeta + P_\beta\beta_\zeta}{P_\alpha\alpha_\Omega + P_\beta\beta_\Omega} \tag{15}$$

[5] Reference [9], p. 498.

where a subscript designates the partial derivative with respect to that variable. Hence, from (5) and (6), if neither α nor β is zero,

$$\frac{d\Omega}{d\zeta} = \frac{k^2\zeta(\beta P_\alpha + \alpha P_\beta)}{\Omega(\beta P_\alpha + k^2\alpha P_\beta)} \tag{16}$$

so that the slope is zero at $\zeta = 0$ except if $P_\Omega = 0$; i.e., except if

$$[P_\alpha + kP_\beta]_{\zeta=0} = 0 \tag{17}$$

Again,

$$\frac{d^2\Omega}{d\zeta^2} = -\frac{1}{P_\Omega}\left[P_{\zeta\zeta} + 2P_{\zeta\Omega}\frac{d\Omega}{d\zeta} + P_{\Omega\Omega}\left(\frac{d\Omega}{d\zeta}\right)^2\right] \tag{18}$$

Now, at $\zeta = 0$, $d\Omega/d\zeta = 0$ if $P_\Omega \neq 0$; hence

$$d^2\Omega/d\zeta^2]_{\zeta=0} = -P_{\zeta\zeta}/P_\Omega]_{\zeta=0} \tag{19}$$

or, since $\alpha_\zeta = \beta_\zeta = 0$, $\alpha_{\zeta\zeta} = -\alpha^{-1}$, $\beta_{\zeta\zeta} = -\beta^{-1}$ at $\zeta = 0$,

$$\left.\frac{d^2\Omega}{d\zeta^2}\right]_{\zeta=0} = \left.\frac{k(kP_\alpha + P_\beta)}{\Omega(P_\alpha + kP_\beta)}\right]_{\zeta=0} \tag{20}$$

Upon insertion of (7) into (20), the curvatures of the branches at $\zeta = 0$ are found to be

$$\left.\frac{d^2\Omega}{d\zeta^2}\right]_{\zeta=0} = \left.\frac{2(5\beta^2 - 9)\mathfrak{J}_\alpha + \beta^4 - 15\beta^2 + 18}{\Omega(\beta^2 - 1)(2\mathfrak{J}_\alpha + \beta^2 - 6)}\right]_{\zeta=0} \quad (21)$$

at the cut-off frequencies given by (11) and

$$\left.\frac{d^2\Omega}{d\zeta^2}\right]_{\zeta=0} = \left.\frac{(\mathfrak{J}_\beta - 1)(g_1\mathfrak{J}_\beta^2 + g_2\mathfrak{J}_\beta + g_3) - g_4}{\Omega(\mathfrak{J}_\beta - 1)(4\mathfrak{J}_\beta^2 + g_5\mathfrak{J}_\beta + g_6)}\right]_{\zeta=0} \quad (22)$$

at the cut-off frequencies given by (12), where

$$\begin{aligned} g_1 &= 2(1 + k^2), \quad g_2 = 2(\beta^2k^2 - 6k^2 - 16) \\ g_3 &= (\beta^4k^2 - 12\beta^2k^2 - 40\beta^2 + 32k^2 + 184)/2 \\ g_4 &= 4(\beta^2 - 3)^2, \quad g_5 = 2(\beta^2 - 8) \\ g_6 &= (\beta^4 - 12\beta^2 + 4\alpha^2 + 32)/2 \end{aligned} \quad (23)$$

and $\alpha = \Omega/k$, $\beta = \Omega$.

It may be seen, from the denominators in (21) and (22), that (17) is the condition for coincidence of a root of (11) with one of (12); i.e., a coincidence of cut-off frequencies. This can occur at certain values of Poisson's ratio (Appendix) and then the curvatures are infinite and the slopes are nonzero and finite. The behavior is analogous to that in plates [5].

Asymptotic Behavior at Short Wave Length

In Sector 1, both α and β are imaginary so that the transcendental functions in Pochhammer's equation are nonoscillatory and there is one and only one root Ω for each real ζ. As $\zeta \to \infty$, so does Ω and the limiting form of (7) is

$$c^2[c^6 - 8c^4 + 8(3 - 2k^{-2})c^2 - 16(1 - k^{-2})] = 0 \quad (24)$$

where $c = \Omega/\zeta = v/v_2$. Equation (24) is Rayleigh's equation for the velocity of surface waves. It has one real, nonzero root corresponding to imaginary α and β. For $k = 2$, the root is $\Omega/\zeta = 0.932$ and this asymptote is shown as the line OR in Figs. 2 and 3. The displacement at large ζ is entirely analogous to that in Rayleigh's surface waves, being confined, in this case, to the neighborhood of the cylindrical surface. These results were found by Hudson [12].

In Sector 2, α is imaginary and β is real. Hence $\mathfrak{J}_\beta$ is oscillatory and there is an infinity of roots for each phase velocity. When β is set equal to zero, Pochhammer's equation is satisfied identically but all the amplitudes are zero. The appropriate form of the solution, for $\beta = 0$, may be obtained by returning to equations (2) and replacing the equivoluminal parts by

$$U = iB\gamma r^2 + C, \; V = iB\gamma r^2 - C, \; W = -4Br$$

It is then found that the frequency equation is

$$\zeta(\mathfrak{J}_\alpha - 2) = 0$$

Now, $\mathfrak{J}_\alpha = 2$ has no roots, for imaginary α, other than $\alpha = 0$ which is inconsistent with $\beta = 0$. Hence, except at $\zeta = 0$, no branch of (7) crosses the line $\beta = 0$ (i.e., $\Omega = \zeta$ or OE in Figs. 2 and 3) so that no mode has the phase velocity v_2.

For large ζ, in Sector 2, Pochhammer's equation reduces to

$$2\mathfrak{J}_\beta^2 + (\beta^2 - 4)\mathfrak{J}_\beta - \beta^2 = 0$$

which yields two families of roots; namely, the roots of

$$4\mathfrak{J}_\beta - 4 + \beta^2 + (16 + \beta^4)^{1/2} = 0 \quad (25a)$$

$$4\mathfrak{J}_\beta - 4 + \beta^2 - (16 + \beta^4)^{1/2} = 0 \quad (25b)$$

For a given root β, (6) is approximately

$$\Omega = \zeta + \beta^2/2\zeta \quad (26)$$

for large ζ. Thus the branches in Sector 2 are asymptotic to $\Omega = \zeta$ and their distribution is determined by the roots of (25a) and (25b). All the modes, then, approach the phase and group velocity v_2 asymptotically. Since α is imaginary and large, at large ζ near this asymptote, the dilatational part of the displacement is again a surface wave, nonoscillatory along the radius; but β is real and hence the individual components of equivoluminal displacement are oscillatory along r with a number of circular nodes in the cross section depending on the order of the root of (25a) or (25b).

In Sector 3, both α and β are real. For $\alpha = 0$ (the line OD in Figs. 2 and 3) Pochhammer's equation is satisfied identically, but the associated amplitudes A, B, C are all zero. The required nontrivial solution may be obtained by returning to equations (2) and replacing the dilatational part of the displacement by

$$U = A, \; V = -A, \; W = iA\gamma r \quad (27)$$

which are the appropriate forms for $\alpha = 0$. These represent a translation of the section in the direction of x ($= r \cos \theta$) and a rotation about y ($= r \sin \theta$). The normal component of traction on the section varies linearly with x and the tangential component is uniform and in the direction of x. Along with the same equivoluminal parts as before, (27) converts Pochhammer's equation to a quadratic in $\mathfrak{J}_\beta$. For $k = 2$, $\sigma = 1/3$, the quadratic has the solutions

$$4\mathfrak{J}_\beta = -2 - \beta^2 \pm (\beta^4 - 12\beta^2 + 100)^{1/2} \quad (28)$$

The two families of roots, β, of (28), when substituted into (5) and (6), with $\alpha = 0$, give the points at which the branches cross $\Omega = k\zeta$ (OD in Figs. 2 and 3) and so give the frequencies and wave numbers of the modes with phase velocity equal to the velocity of dilatational waves in an infinite medium.

Bounds, Barriers, and Intersectors

A set of *bounds* of the branches of a frequency equation is defined, in this paper, as a family of curves which have the following properties:

1 Their construction requires very much less computation than that required for plotting the branches of the frequency equation directly.

2 Their co-ordinates satisfy the frequency equation only at certain points which can be determined readily.

3 They pass through a family of cut-off frequencies at $\zeta = 0$.

Curves that satisfy conditions 1 and 2, but not 3, are called *barriers*. Curves that intersect the bounds or barriers at points which satisfy the frequency equation are called *intersectors*. Branches can cross bounds or barriers only at the points of intersection of the latter with intersectors.

In the present case, one set of bounds (B_1), which terminate at the cut-off frequencies of the axial shear modes, is a family of hyperbolas and the associated intersectors (I_1) are also simple curves approximated closely by hyperbolas. The second set of bounds (B_2), which terminate at the remaining cut-off frequencies, is a family of complicated curves which are themselves constructed on a grid of three pairs of families of intersecting curves designated as *auxiliary barriers*. Two of the six families of auxiliary barriers are used subsequently as barriers of the branches.

Bounds B_1 and Intersectors I_1

Consider the equation

$$B_1 \equiv \mathfrak{J}_\beta - 1 = 0 \quad (29)$$

In view of (6), (29) is the equation of a family of hyperbolas in the ζ, Ω-plane. The curves are asymptotic to $\Omega = \zeta$ (OE in Figs. 2 and 3) and pass through the cut-off frequencies of the axial shear modes (the circles at $\zeta = 0$ in Figs. 2 and 3) as may be seen by comparison with (11). Further, when (29) is substituted in the frequency equation (7), the latter is satisfied only if $\zeta = 0$, which gives the cut-off frequencies (11), or if

$$I_1 \equiv \mathcal{J}_\alpha - 1 - (\beta^2 + \zeta^2)/2\beta^2(\beta^2 - 2) = 0 \qquad (30)$$

With (5) and (29), (30) is a family of curves whose intersections with B_1 are the only points on B_1, aside from those at $\zeta = 0$, which satisfy the frequency equation. Hence B_1 is a family of bounds and I_1 is the associated family of intersectors.

Now, for real α, (30) is approximated closely by

$$I_1 \approx \mathcal{J}_\alpha - 1 = 0 \qquad (31)$$

In view of (5), (31) is a family of hyperbolas asymptotic to $\Omega = k\zeta$ (OD in Figs. 2 and 3). The intersections of the two families of hyperbolas, designated by circles in Sector 3 of Figs. 2 and 3, are, to a close approximation, points through which the branches pass, and the hyperbolas asymptotic to OE are bounds. In Sector 2, α is imaginary and (30) has only one root for each root β of (29). Thus, on each bound B_1, in Sector 2, there is one point, such as d_2 in Figs. 2 and 3, at which a branch of (7) crosses the bound. The points like d_2 on the higher bounds B_1 are far beyond the range of ζ in Figs. 2 and 3.

Bounds B_2

In previous constructions of frequency spectra by means of bounds, e.g. [3–8], the curves analogous to I_1 have also been bounds or, at least, barriers, crossed by the branches only at the intersections with B_1. Thus the branches were well determined by a grid of two intersecting families of hyperbolas. In the present case, curves such as I_1 cannot give the remaining family of cut-off frequencies (12) at $\zeta = 0$ since the cut-off modes contain coupled dilatational and equivoluminal deformation and so depend on transcendental functions of both α and β. Furthermore, two equivoluminal deformations, those associated with B and C in Equations (2), are here involved. Hence the substitution of (30) in the frequency equation (7) reduces the latter only to a quadratic in $\mathcal{J}_\beta$, instead of to (29), and the determination of the roots of the reduced equation, i.e., the determination of the points at which the branches cross I_1, involves a large amount of computation which is not worth doing since (30) does not reduce to (12) at $\zeta = 0$ in any event and so is not suitable for use as a family of bounds.

It is apparent that the additional set of bounds that is required must be a family of curves more complicated than simple hyperbolas. As a result of a study of the cofactors of Pochhammer's determinant,[3] we consider the equation

$$B_2 \equiv J_1(\alpha)J_1(\beta)[6\mathcal{J}_\alpha + 6\mathcal{J}_\beta - 24 + 3\beta^2 - \zeta^2] = 0 \qquad (32)$$

which gives the second family of cut-off frequencies (12) when $\zeta = 0$. It is shown in the following sections that the curves defined by (32) may be constructed by means of auxiliary barriers and used as the second set of bounds of Pochhammer's equation.

Auxiliary Barriers B_{21} and I_{21}. If equation (32) is expanded by multiplying through by $J_1(\alpha)$ and $J_1(\beta)$, it may be seen that if

$$B_{21} \equiv J_1(\beta) = 0 \qquad (33)$$

(32) is satisfied if and only if

$$I_{21} \equiv J_1(\alpha) = 0 \qquad (34)$$

and vice versa. In view of (5) and (6), the two families of curves (33) and (34) are hyperbolas: B_{21} asymptotic to OE and I_{21} asymptotic to OD. The only points on these two families that satisfy (32) are their intersections. Thus, B_{21} and I_{21} are auxiliary barriers of B_2. Since (34) has no roots when α is imaginary, there are no intersections in Sector 2.

Now, substitution of (33) into Pochhammer's equation reduces the latter to

$$2(\beta^2 - \zeta^2)\beta^2 J_0(\beta)J_1(\alpha) = 0 \qquad (35)$$

Since this is satisfied by $J_1(\alpha) = 0$, the intersections of I_{21} with B_{21} (except $\alpha = 0$) are roots of (7) in addition to being roots of (32). Also, from (35), (7) is satisfied at the intersections of B_{21} with

$$\beta^2 - \zeta^2 = 0 \qquad (36)$$

From (5) and (6), this is the line $\Omega = \sqrt{2}\,\zeta$ in Sector 2, marked OL in Figs. 2 and 3. Since there are no roots of (35) other than those of (34) and (36), B_{21} are not only barriers of B_2 but are also barriers of the branches of Pochhammer's equation; the branches intersecting B_{21} at and only at the intersections of B_{21} with I_{21} and OL. The former are identified by squares in Sector 3 and the latter by the squares $b_3, b_5 \ldots$ along OL in Figs. 2 and 3. From (16) it may be shown that B_{21} and the branches of (7) have the same slope

$$d\Omega/d\zeta = 1/\sqrt{2} \qquad (37)$$

at their common intersections with OL. It is shown in the last section that these intersections are points of tangency rather than crossings.

Substitution of (34) into Pochhammer's equation reduces it to

$$\alpha J_0(\alpha)J_1^2(\beta)[f_2\mathcal{J}_\beta + f_4] = 0 \qquad (38)$$

Hence, in addition to the intersections of I_{21} with B_{21}, the intersections of I_{21} with

$$f_2\mathcal{J}_\beta + f_4 = 0 \qquad (38a)$$

are also roots of (7). Equation (38*a*) plays the same role in (38) that (36) plays in (35); but (38*a*) is a family of curves whose construction requires much computation, whereas (36) is a single straight line. Hence I_{21} is not convenient for use as a family of barriers to the branches of (7) but only as a family of intersectors with B_{21}.

Bound $\zeta = \beta$. Substitution of (36) in Pochhammer's equation reduces it to

$$4\beta^4 J_1(\alpha)J_1(\beta)[1 + (1 - \mathcal{J}_\alpha)(3\mathcal{J}_\beta + \beta^2 - 5)] = 0 \qquad (39)$$

Thus, in addition to the points $b_3, b_5 \ldots$ mentioned in the preceding section, the intersections of (36) with

$$1 + (1 - \mathcal{J}_\alpha)(3\mathcal{J}_\beta + \beta^2 - 5) = 0 \qquad (40)$$

are roots of (7). These intersections are identified by the squares marked $b_2, b_4 \ldots$ along OL in Figs. 2 and 3. Since α is imaginary, (39) has no roots $J_1(\alpha) = 0$; and we have already seen that $\beta = 0$ is not a root of Pochhammer's equation. Thus line OL is a bound of (7); it is crossed by the branches at and only at the points $b_3, b_5 \ldots$ and $b_2, b_4 \ldots$.

The modes corresponding to $b_3, b_5 \ldots$ have

$$A = 0, \qquad B = -C/\zeta$$

in equations (2), so that they are equivoluminal. They are the analogs of modes found by Lamé [13] in plates and Lamb [14] in longitudinal vibrations of circular cylinders except that, in the present case, there are no transverse sections free from traction.

Corresponding to $b_2, b_4 \ldots$ there are coupled dilatational and equivoluminal modes with

$$C = 2B\zeta = 2A(1 - \mathcal{J}_\alpha)J_1(\alpha)/J_1(\beta)$$

These have no counterparts in vibrations of plates or longitudinal vibrations of cylinders.

Auxiliary Barriers B_{22} and I_{22}. Returning to (32), we see that if

$$B_{22} \equiv 3\mathcal{J}_\beta + \beta^2 - 6 = 0 \qquad (41)$$

(32) is satisfied if and only if

$$I_{22} \equiv 6\mathcal{J}_\alpha + \beta^2 - \zeta^2 - 12 = 0 \qquad (42)$$

and vice versa.[6] Hence B_{22} and I_{22} are auxiliary barriers of B_2; i.e., the curves B_2 cross B_{22} and I_{22} at and only at the intersections of B_{22} and I_{22}.

Now, if (41) is substituted in Pochhammer's equation, the resulting equation is satisfied only if $I_{22} = 0$ or if

$$I_{22}' \equiv \beta^4 - (3 + \zeta^2)\beta^2 - 3\zeta^2 = 0 \qquad (43)$$

Accordingly B_{22}, in addition to being barriers of B_2, are barriers of the branches of (7), with crossings at and only at the intersections of B_{22} with I_{22} and with I_{22}'. The intersections of B_{22} and I_{22} are shown in Figs. 2 and 3 as triangles with apex up; those in Sector 2, one to each of B_{22} since α is imaginary, are the points marked $c_2, c_4 \ldots$ The intersections of B_{22} with the single curve (43) are the points marked $a_2, a_4 \ldots$.

As in the case of auxiliary barriers I_{21}, the auxiliary barriers I_{22} are found to be not well suited for use as barriers to the branches.

Auxiliary Barriers B_{23} and I_{23}. We observe that if

$$B_{23} \equiv (7\beta^2 - \zeta^2)\mathcal{J}_\beta - 16\beta^2 - \beta^2\zeta^2 + 3\beta^4 = 0 \qquad (44)$$

(32) is satisfied if and only if

$$I_{23} \equiv 6(7\beta^2 - \zeta^2)\mathcal{J}_\alpha + 24\zeta^2 - 72\beta^2 + \zeta^4 - 4\zeta^2\beta^2 + 3\beta^4 = 0 \qquad (45)$$

and vice versa. Hence B_{23} and I_{23} are auxiliary barriers of B_2. The motivation for their choice appears in (46) and (47). Equations (44) and (45) are not as simple to plot as are the previous curves, but enough points to construct them may be obtained without difficulty by computing their intersections with the radial lines $\beta = e\zeta$ and $\alpha = e\zeta$, respectively, for a series of values of the parameter e. The intersections of B_{23} and I_{23} are shown in Figs. 2 and 3 as triangles with apex down.

It is not necessary to investigate the possible use of B_{23} or I_{23} as barriers of the branches because sufficient curves are already available. However, it can be shown that B_{23} are in fact barriers of the branches.

Construction of Curves B_2. The auxiliary barriers B_{2j} and I_{2j}, j = 1, 2, 3, are shown in Fig. 2 and the curves B_2 are traced through the appropriate intersections (marked by squares and triangles) as shown enlarged in Fig. 4. The curves B_2 are shown only in Sector 3 as they are not needed in Sector 2 and do not exist in Sector 1.

Establishment of Curves B_2 as Bounds. Subject to the conditions $J_1(\alpha) \neq 0$, $J(\beta) \neq 0$, i.e., for points not on B_{21} or I_{21}, (32) may be solved, in turn, for $\mathcal{J}_\alpha$ and $\mathcal{J}_\beta$. When these expressions are substituted in Pochhammer's equation, the latter are satisfied if and only if

$$\zeta^2 B_{22} B_{23} = 0 \qquad (46)$$

$$\zeta^2 I_{22} I_{23} = 0 \qquad (47)$$

respectively. Hence, except for points on B_{21} or I_{21}, the branches intersect B_2 where and only where B_2 crosses B_{22}, B_{23}, I_{22}, I_{23}, and $\zeta = 0$. Now, by (41) and (42), the intersections of B_2 with B_{22} and I_{22} are common and, by (44) and (45), the intersections of B_2 with B_{23} and I_{23} are common. Accordingly, the branches cross B_2 where and only where B_2 passes through $\zeta = 0$, the intersections of B_{22} with I_{22} and the intersections of B_{23} with I_{23} — except for points on B_{21} or I_{21}. But, by (33) and (34), B_2 coincides with B_{21} or I_{21} only at the intersections of the latter two and, by (33) and (35), these intersections are roots of Pochhammer's equation. Therefore the branches cross B_2 at $\zeta = 0$ and at the intersections of B_{2j} with I_{2j}, j = 1, 2, 3, and at no other points. This establishes the curves B_2 as bounds.

[6] Note that the cofactors A_{31}, A_{32}, A_{33} of the elements of the third row of Pochhammer's determinant[3] are

$$A_{31} = 2\zeta\beta^2 B_{22}$$
$$3A_{32} = \beta^2 I_{22} - 2(\zeta^2 - \beta^2)B_{22}$$
$$3A_{33} = 2\zeta[\beta^2 I_{22} + (\zeta^2 - \beta^2)B_{22}]$$

and note also that $B_2 = (I_{22} + 2B_{22})J_1(\alpha)J_1(\beta)$.

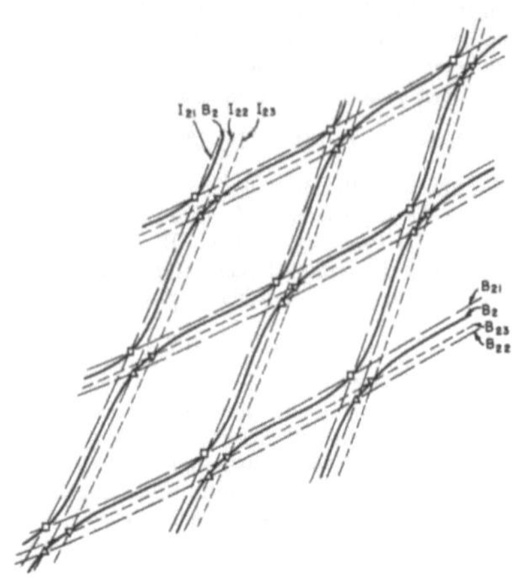

Fig. 4 Bounds B_2 traced through auxiliary barriers B_{2j} and I_{2j}

Construction of Branches

After the bounds have been drawn, as illustrated in Fig. 2 for $\sigma = 1/3$, i.e., $k = 2$, there is no further use for the auxiliary barriers B_{23}. Also, there is no further use for any of the intersectors I_1 or I_{2j}: only their intersections with their respective bounds or barriers are required for tracing the branches. Finally, the bounds B_2 are needed only in Sector 3. The remaining bounds, barriers, intersection points, and asymptotes, as well as the line OD, are shown in Fig. 3 and branches are traced through them as described later after a few preliminary remarks about asymptotic behavior.

Sequence of Branches and Bounds Near Cut-Off. Each of the branches above the first starts at a cut-off frequency, (11) or (12), where the branch coincides with a bound B_1 or B_2. The bounds have zero slopes at $\zeta = 0$, as may be ascertained by replacing P with B_1 or B_2 in (16). Whether a branch starts out above or below a bound may be determined by comparing their curvatures at $\zeta = 0$. The curvatures of the branches are given by (21) and (22) and the curvatures of the bounds may be obtained by replacing P with B_1 or B_2 in (20). Their ratios are given in Table 1 for the next seventeen modes above the first. The branch starts out above or below the bound according as the ratio is greater or less than unity.

Negative Curvature at Cut-Off. The curvatures of the bounds are always positive at $\zeta = 0$ but the curvatures of the branches may be positive or negative, depending upon Poisson's ratio and the order of the branch. As Poisson's ratio increases, the cut-off frequencies of the coupled modes increase while those of the axial shear modes remain fixed (Appendix). As the cut-off frequency of a coupled mode passes a cut-off frequency of an axial shear mode, with increasing σ, the curvatures (21) and (22) pass through infinity and change sign. The behavior is analogous to that in plates [5]. When the curvature of a branch is negative at $\zeta = 0$, the frequency drops to a minimum with increasing ζ.

In this range the phase and group velocities have opposite signs and the group velocity drops to zero at the frequency minimum [15] which is a point of intersection of a real branch with a branch corresponding to complex ζ [7]. With $\sigma = 1/3$, none of the first eighteen branches of Pochhammer's equation for flexure has a negative curvature at $\zeta = 0$ so that these phenomena do not appear in Fig. 3.

Sequence of Branches, Bounds, and Barriers at Short Wave Length. The asymptotic behavior of all the branches above the first, as $\zeta \to \infty$, is given by (25) and (26). These curves are to be compared with the barriers B_{21} and the bounds B_1, which alternate in position beginning with the root $\beta = 0$ for B_{21}, then $\beta = 1.84$ for B_1, then $\beta = 3.83$ for B_{21}, and so on. Now, the lowest root of (25) is $\beta = 0$ for (25*b*). Hence there must be a branch adjacent to $\beta = 0$; i.e., lying between OE and the lowest of bounds B_1. The next higher root of (25) and (26) is $\beta = 3.25$ for (25*a*). Hence the next branch lies between the lowest of bounds B_1 and the next barrier B_{21}. This sequence continues, with one branch between each pair of adjacent curves B_{21} and B_1, independent of the interpositions of barriers B_{22}. As β increases, the roots of (25*a*) approach the roots of $J_1(\beta) = 0$ from below and the roots of (25*b*) approach the roots of $\mathcal{J}_\beta = 1$ from below. Hence, at large ζ, the higher branches lie just below the successive curves B_{21} and B_1.

Fundamental Mode. The first branch starts at the origin as a parabola, in accordance with (13). As shown by Hudson [12], (7) has one and only one root in Sector 1 for each ζ and, as $\zeta \to \infty$, the branch approaches the phase velocity of Rayleigh surface waves (24) asymptotically. With this behavior of the branch, the phase velocity must increase monotonically, with increasing ζ, to the asymptotic velocity of surface waves and the group velocity must start at zero, increase to a maximum (at the point of inflection of the branch), and fall off asymptotically to the same velocity of surface waves. Tables of this branch have been computed by Hudson [12].

Second Branch. This starts out at the lowest cut-off frequency, $\Omega = 1.84$, of the axial shear modes, with large positive curvature ratio, as shown in Table 1, so that the branch swings sharply up above its bound B_1 and then reverses curvature in order to cross the barrier B_{22} at point a_2. The branch then continues, crossing OL at b_2, recrossing B_{22} at c_2 and, finally, crossing B_1 at d_2 so as to take its asymptotic position, with positive curvature, between B_1 and OE. In order to do this there must be a second reversal of curvature near b_2. Inspection of the resulting branch shows that, as ζ increases from zero, the phase velocity diminishes monotonically from infinity to v_2 and the group velocity starts at zero, goes through a high maximum before a_2, a shallow minimum near b_2, and then rises monotonically to v_2. Comparison can be made with computations of this branch by Abramson [16].

Third Branch. The lowest frequency, $\Omega = 2.86$, of the family of coupled cut-off modes is the starting point of the third branch. The curvature ratio is positive and slightly greater than unity so that the branch swings gradually upward away from the lowest of bounds B_2, but the branch must then reverse its curvature in order to cross OL at b_3 with the slope of the tangent to the barrier B_{21} which also crosses OL at b_3. The branch cannot cross B_{21} at b_3 because there are no points on B_{21} beyond b_3 where the branch can recross B_{21} to take its asymptotic position between B_{21} and the next lower bound B_1. Hence the branch is tangent to B_{21} at b_3 entirely from below. Since the asymptotic curvature is positive there must be another reversal of curvature near b_3. The interpretation of phase and group velocities is similar to that of the second branch, but the maximum and minimum of the group velocity are less pronounced.

Fourth Branch. Here the initial curvature ratio is positive but less than unity so that the branch hugs the underside of the second bound B_1 until it crosses B_1 at the first circled point; following which the branch turns upward toward the region where the next higher barrier B_{22} and the bound OL converge. Now, b_4 lies on OL above B_{22}, so the branch must first cross B_{22} at a_4 and then OL at b_4. Following this, the branch must recross B_{22} at c_4 and, far off to the right of the portion of the spectrum shown in Fig. 3, recross B_1 at a point d_4 so as to regain the asymptotic position between the initial bound B_1 and the next lower barrier B_{21}.

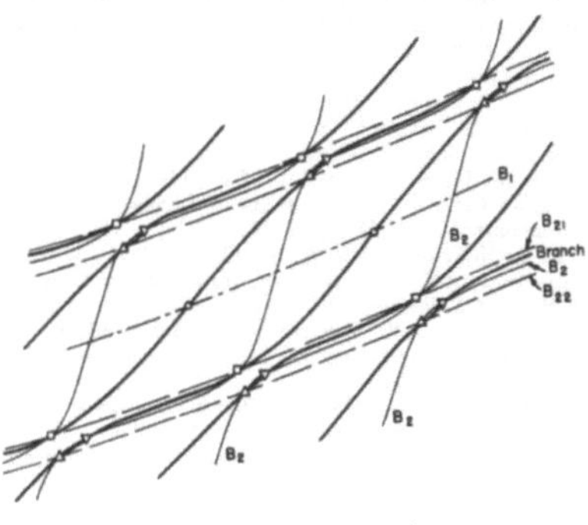

Fig. 5 Branches traced through bounds B_1 and B_2 and barriers B_{21} and B_{22} in sector 3

Higher Branches. The next two branches may be sketched in the same manner as the lower ones, but, with the seventh and eighth branches, the clusters of three intersection points of B_{2j} and I_{2j} begin to appear in Sector 3. The paths of the branches through these clusters on the barriers B_{21} and B_{22} and the bounds B_1 and B_2, at high frequencies in Sector 3, are shown in the enlarged sketch in Fig. 5. The result is the typical terrace-like structure of frequencies and phase velocities and multiple maxima and minima of group velocities of each branch. These phenomena disappear as the branch crosses OD into Sector 2. At these crossings the odd numbered branches are bounded closely above by a barrier B_{21} and below by a bound B_1. Hence the slopes of the branches are nearly the same as the slopes of B_{21}, which are all equal to $1/k$ at the intersections with OD. Thus, for the odd-numbered modes, where the phase velocity is kv_2, the velocity of dilatational waves in an infinite medium, the group velocity is approximately v_2/k which is very much less. On the other hand, the even-numbered branches behave differently. In Sector 3 adjacent to OD, the branches are squeezed against OD by the bounds B_2 immediately to the left so that the higher branches are there almost parallel to OD. However, as they cross OD, the branches must bend to the right in order to avoid the barriers B_{22} above. Hence the slopes of the branches, though high, are less than the slope of OD, i.e., for the high, even-numbered modes: where the phase velocity is the velocity of dilatational waves in an infinite medium, the group velocities are slightly less.

References

1 L. Pochhammer, "Ueber die Fortpflanzungsgeschwindigkeiten Kleiner Schwingungen in einem unbegrenzten isotropen Kreiscylinder," *Zeitschrift für Mathematik*, vol. 81, 1876, pp. 324–336.

2 A. N. Holden, "Longitudinal Modes of Elastic Waves in Isotropic Cylinders and Slabs," *Bell System Technical Journal*, vol. 30, 1951, pp. 956–969.

3 R. D. Mindlin, "Thickness-Shear and Flexural Vibrations of Crystal Plates," *Journal of Applied Physics*, vol. 22, 1951, pp. 316–323.

4 M. Onoe, "A Study of the Branches of the Velocity-Dispersion Equations of Elastic Plates and Rods," Joint Committee on Ultrasonics of the Institute of Electrical Communication Engineers and the Acoustical Society of Japan, Report, July 20, 1955.

5 R. D. Mindlin, "An Introduction to the Mathematical Theory of Vibrations of Elastic Plates," U. S. Army Signal Engineering Laboratories, Fort Monmouth, N. J., 1955.

6 R. D. Mindlin, A. Schacknow, and H. Deresiewicz, "Flexural Vibrations of Rectangular Plates," JOURNAL OF APPLIED MECHANICS, vol. 23, TRANS. ASME, vol. 78, 1956, pp. 430–436.

7 R. D. Mindlin and M. Onoe, "Mathematical Theory of Vibrations of Elastic Plates," Proceedings of the 11th Annual Symposium on Frequency Control, U. S. Army Signal Engineering Laboratories, Fort Monmouth, N. J., 1957, pp. 17–40.

8 E. G. Newman and R. D. Mindlin, "Vibrations of a Monoclinic Crystal Plate," *Journal of the Acoustical Society of America*, vol. 29, 1957, pp. 1206–1218.

9 A. E. H. Love, "A Treatise on the Mathematical Theory of Elasticity," Cambridge University Press, London, England, second edition, 1927, pp. 291–292.

10 M. Onoe, "Tables of Modified Quotients of Bessel Functions," Columbia University Press, New York, N. Y., 1958.

11 Lord Rayleigh, "On the Free Vibrations of an Infinite Plate of Homogeneous Isotropic Elastic Matter," *Proceedings of the London Mathematical Society*, vol. 20, 1888–1889, pp. 225–234.

12 G. Hudson, "Dispersion of Elastic Waves in Solid Circular Cylinders," *Physical Review*, vol. 63, 1943, pp. 46–51.

13 G. Lamé, "Lecons sur la Théorie Mathématique de l'Élasticité des Corps Solides," Gauthier-Villars, Paris, France, second edition, 1866, p. 170.

14 H. Lamb, "On Waves in an Elastic Plate," *Proceedings of the Royal Society of London*, England, series A, vol. 93, 1917, footnote on p. 122.

15 I. Tolstoy and E. Usdin, "Wave Propagation in Elastic Plates: Low and High Mode Dispersion," *Journal of the Acoustical Society of America*, vol. 29, 1957, pp. 37–42.

16 H. N. Abramson, "Flexural Waves in Elastic Beams of Circular Cross Section," *Journal of the Acoustical Society of America*, vol. 29, 1957, pp. 42–46.

APPENDIX

Variation of Cut-Off Frequencies With Poisson's Ratio

Equation (12) may be regarded as relating Ω and k and a set of barriers in the plane of k and Ω may be constructed. As suggested by the elements of Pochhammer's determinant, with $\zeta = 0$, we form the three pairs of equations

$$\begin{aligned} C_{11} &\equiv J_1(\beta_0) = 0 \\ C_{12} &\equiv J_1(\alpha_0) = 0 \end{aligned} \tag{48}$$

$$\begin{aligned} C_{21} &\equiv 2\mathcal{J}(\beta_0) + \beta_0^2 - 4 = 0 \\ C_{22} &\equiv \mathcal{J}(\alpha_0) - 2 = 0 \end{aligned} \tag{49}$$

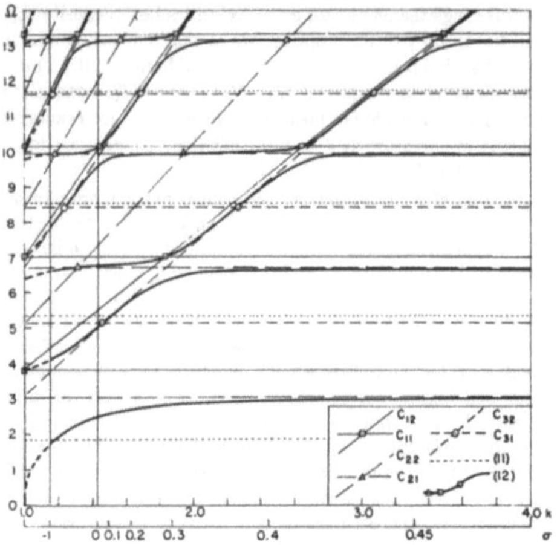

Fig. 6 Variation of cut-off frequencies with Poisson's ratio

$$\begin{aligned} C_{31} &\equiv \mathcal{J}(\beta_0) - 2 = 0 \\ C_{32} &\equiv 2\mathcal{J}(\alpha_0) + \beta_0^2 - 4 = 0 \end{aligned} \tag{50}$$

where $\beta_0 = \Omega$ and $\alpha_0 = \Omega/k$. With Ω as the ordinate and k as the abscissa, the first member of each pair is the equation of a family of horizontal lines; the second members of the first two pairs are families of inclined straight lines; and the second member of the last pair is the equation of a family of curves asymptotic to inclined straight lines. If either equation of a pair is substituted into equation (12), that equation is satisfied if and only if the other equation of the pair is satisfied. Hence the members of each pair are mutual barriers and intersectors. The branches may then be traced through the grid of lines as shown in Fig. 6. The frequencies of the axial shear modes, equation (11), are also shown in order to exhibit values of k or σ at which coincidences of cut-off frequencies occur.

(Reprinted from the Journal of Applied Mechanics for September, 1960, with permission of Copyright Owner)

D. C. GAZIS
Research Laboratories,
General Motors Corporation,
Warren, Mich.
Assoc. Mem. ASME

R. D. MINDLIN
Professor of Civil Engineering,
Columbia University, New York, N. Y.
Mem. ASME

Extensional Vibrations and Waves in a Circular Disk and a Semi-Infinite Plate[1]

A set of approximate equations of extensional motion of elastic plates is solved for the case of axially symmetric vibrations of a circular disk and the results are compared with experiments performed by E. A. G. Shaw. Special attention is given to a new mode of vibration, discovered by Shaw, in which the deformation is predominantly at the edge of the disk. It is shown how this mode arises from the complex conjugate roots of the frequency equation of an infinite plate. The properties of the edge-mode are examined in detail in a study of the reflection of straight-crested extensional waves at the edge of a semi-infinite plate.

Introduction

In a previous paper [1][2] a derivation was given of approximate, two-dimensional equations of extensional motion, of an elastic plate, which take into account the effect of coupling between extensional, symmetric thickness-stretch, and symmetric thickness-shear deformations. These equations result from retaining the zero-order, first-order, and second-order terms of an expansion of the displacements in a series of Legendre polynomials of the thickness co-ordinate and they are referred to in the sequel as "second-order equations."

As pointed out in reference [1], a number of important properties of the frequency spectrum contained in the exact theory [2, 3] of extensional motion of an infinite plate are reproduced by the second-order equations, whereas they do not appear in the zero-order approximation [4, 5] or in a first-order approximation [6]. Among these properties is a pair of segments, of the second and third branches, with complex conjugate wave-numbers, Fig. 1. Complex wave-numbers signify modes whose shapes are given by products of trigonometric and exponential functions in Cartesian co-ordinates and by functions of similar behavior in other co-ordinate systems. These mode-shapes are physically unacceptable in the case of free vibrations of an infinite plate but they are important in transients or when other boundaries, in addition to the free faces, are introduced. In this paper it is shown how the modes associated with complex wave-numbers account for edge vibrations observed experimentally by Shaw [7] in circular disks of barium titanate which were excited in axially symmetric, extensional motion.

Shaw observed coupling of the extensional modes with an unusual mode whose amplitude was large near the circular boundary of the disk and decreased rapidly toward the center. The coupling occurred at a frequency essentially independent of the radius of the plate and far below the lowest frequency associated with real wave-numbers of the second and third branches for the infinite plate. The latter circumstance implies that the usual thickness-stretch and thickness-shear modes cannot account for the observed coupling inasmuch as, for a linearly elastic material, a mode of the infinite plate has overtones in a finite plate but no subtones and hence its influence on the frequency spectrum at frequencies far below its frequency range is not appreciable. Shaw expressed the opinion that some other type of motion was in resonance with the radial extensional motion. A type of motion that has the proper frequency range and mode-shape is one corresponding to complex portions of the frequency spectrum, which were not studied until later [8, 9]. One may expect that an investigation of the extensional vibrations of circular plates within the framework of the general, linear theory of elasticity will provide a theoretical explanation of the phenomenon observed by Shaw. However, such an investigation would be encumbered by a doubly infinite series expression for each mode because each of the infinity of modes in the three-dimensional solution for an infinite plate gives rise to an infinity of modes in the presence of the additional free boundary. Now, the second-order equations contain complex branches analogous to those of the three-dimensional theory and only three modes of the infinite plate are introduced in an axially symmetric problem. Thus the approximate equations have the requisite qualitative properties

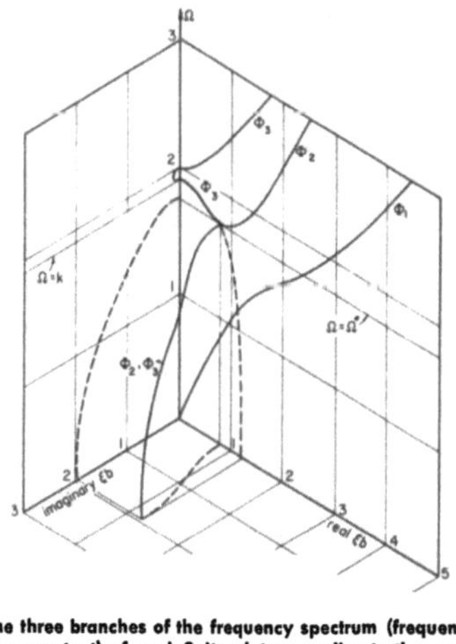

Fig. 1 The three branches of the frequency spectrum (frequency versus propagation constant) of an infinite plate according to the second-order equations. Dashed lines are projections of complex segment on co-ordinate planes.

[1] This investigation was supported, in part, by the Office of Naval Research under a contract with Columbia University.

[2] Numbers in brackets designate References at end of paper.

Presented at the Summer Conference of the Applied Mechanics Division, University Park, Pa., June 20–22, 1960, of The American Society of Mechanical Engineers.

Discussion of this paper should be addressed to the Secretary, ASME, 29 West 39th Street, New York, N. Y., and will be accepted until October 10, 1960, for publication at a later date. Discussion received after the closing date will be returned.

Note: Statements and opinions advanced in papers are to be understood as individual expressions of their authors and not those of the Society. Manuscript received by ASME Applied Mechanics Division, April 30, 1959. Paper No. 60—APM-3.

and are considerably simpler in form. Accordingly, they have been employed in this paper in an attempt to obtain an insight into the phenomena which take place at the frequency of the edge vibrations.

A characteristic equation appropriate to the circular disk was obtained within the framework of the second-order theory and evaluated numerically by means of an IBM 704 digital computer. The results of this investigation may be summarized as follows: The second-order equations include the edge vibrations but their frequencies are somewhat lower than the experimental values obtained by Shaw. The discrepancy is probably due to the fact that, although the second-order equations provide a very good approximation of the real and imaginary segments of the first three branches of the three-dimensional theory, the approximation of the complex segments is only fair. However, the theoretical results of the second-order equations are in very good qualitative agreement with the experimental results, in so far as the edge vibrations are concerned, and they are in excellent quantitative agreement over the remainder of the spectrum from zero frequency up to frequencies considerably higher than the thickness frequencies.

It is found that the edge mode, at a free edge, cannot be composed of contributions from the two complex segments alone but always involves the extensional deformation associated with the real, lowest branch. Thus the edge vibration is always linked with the usual type of extensional vibration in the interior of the plate. This situation and the detailed behavior in the neighborhood of the edge-resonance are investigated, in the last section of the paper, by means of a study of the reflection of a sinusoidal train of straight-crested extensional waves at a free, straight edge at normal incidence.

Frequency Equation for a Circular Disk

We refer the plate to cylindrical co-ordinates r, θ, and x_2 with the faces at $x_2 = \pm b$. It is required to obtain the axially symmetric, extensional modes of vibration of a plate bounded by the free faces and by a free, cylindrical surface $r = a$. The plate-displacements $u_r^{(0)}$, $u_\theta^{(0)}$, $u_2^{(1)}$, $u_r^{(2)}$, and $u_\theta^{(2)}$, as defined in reference [1],[3] are expressed in terms of three potential functions ϕ_i, $i = 1, 2, 3$, and two potential functions H_0 and H_2, all of them harmonic functions of time, according to

$$\begin{aligned}
u_r^{(0)} &= \frac{\partial}{\partial r}(\phi_1 + \phi_2 + \phi_3) - \frac{\partial H_0}{r\partial\theta} \\
u_\theta^{(0)} &= \frac{\partial}{r\partial\theta}(\phi_1 + \phi_2 + \phi_3) + \frac{\partial H_0}{\partial r} \\
u_2^{(1)} &= \alpha_1\phi_1 + \alpha_2\phi_2 + \alpha_3\phi_3 \\
u_r^{(2)} &= \Sigma_i\beta_i\frac{\partial\phi_i}{\partial r} - \frac{\partial H_2}{r\partial\theta} \\
u_\theta^{(2)} &= \Sigma_i\beta_i\frac{\partial\phi_i}{r\partial\theta} + \frac{\partial H_2}{\partial r}
\end{aligned} \tag{1}$$

where

$$\begin{aligned}
\alpha_i &= b[(\lambda + 2\mu)\xi_i^2 - \rho\omega^2]/\kappa_1\lambda \\
\beta_i &= \alpha_i b\Big/\left[\frac{3}{4}(\Omega^2 - 4) - \frac{E'b^2\xi_i^2}{5\mu\kappa_2^2}\right]
\end{aligned} \tag{2}$$

In Equations (2) λ and μ are Lamé's constants, ρ is the density, ω is the circular frequency and

$$E' = 4\mu(\lambda + \mu)/(\lambda + 2\mu), \qquad \Omega = \omega/\omega_s \tag{3}$$

[3] The second-order equations are given in invariant form in the last section of reference [1].

where

$$\omega_s = (\pi/2b)(\mu/\rho)^{1/2} \tag{4}$$

In addition, the κ_s ($s = 1$ to 4) are the correction coefficients of the second-order theory and the ξ_i ($i = 1, 2, 3$) are the three roots of the cubic equation

$$\begin{vmatrix} a_{11} & a_{12} & 0 \\ a_{12} & a_{22} & a_{23} \\ 0 & a_{23} & a_{33} \end{vmatrix} = 0 \tag{5}$$

where

$$\begin{aligned}
a_{11} &= k^2z^2 - \Omega^2 \\
a_{22} &= \kappa_2^2z^2/3 + 4\kappa_1^2k^2/\pi^2 - \kappa_3^2\Omega^2/3 \\
a_{33} &= E'z^2/5\mu + 12\kappa_2^2/\pi^2 - \kappa_4^2\Omega^2/5 \\
a_{12} &= 2\kappa_1(k^2 - 2)z/\pi \\
a_{23} &= -2\kappa_2^2z/\pi
\end{aligned}$$

and

$$z = 2\xi b/\pi, \qquad k^2 = 2(1 - \nu)/(1 - 2\nu) \tag{6}$$

with ν denoting Poisson's ratio.

For axially symmetric motion, the potentials H_0 and H_2 are set equal to zero. The remaining potentials satisfy the Helmholtz equations

$$(\nabla^2 + \xi_i^2)\phi_i = 0 \tag{7}$$

which become, in axially symmetric, cylindrical co-ordinates,

$$\left(\frac{d^2}{dr^2} + \frac{1}{r}\frac{d}{dr} + \xi_i^2\right)\phi_i = 0 \tag{8}$$

The general solution of Equation (8) is given in terms of Bessel functions by

$$\phi_i = A_iJ_0(\xi_ir) + B_iY_0(\xi_ir) \tag{9}$$

where, in the absence of singularities at the origin, the B_i are taken equal to zero.

The plate-strains, in the axially symmetric ease, are expressed in terms of the potential functions ϕ_i by means of the invariant equations of reference [1] as follows:

$$\begin{aligned}
\epsilon_{rr}^{(0)} &= \Sigma_i\phi_i'' \\
\epsilon_{\theta\theta}^{(0)} &= r^{-1}\Sigma_i\phi_i' \\
\epsilon_{22}^{(0)} &= b^{-1}\Sigma_i\alpha_i\phi_i \\
\epsilon_{r2}^{(1)} &= \frac{1}{2}\Sigma_i(\alpha_i + 3b^{-1}\beta_i)\phi_i' \\
\epsilon_{rr}^{(2)} &= \Sigma_i\beta_i\phi_i'' \\
\epsilon_{\theta\theta}^{(2)} &= r^{-1}\Sigma_i\beta_i\phi_i' \\
\epsilon_{r\theta}^{(0)} &= \epsilon_{\theta 2}^{(0)} = \epsilon_{r\theta}^{(2)} = \epsilon_{\theta 2}^{(2)} = 0
\end{aligned} \tag{10}$$

where primes denote differentiation with respect to r. Similarly, the components of plate-traction acting across the cylindrical surfaces r = const are

$$\begin{aligned}
\tau_{rr}^{(0)} &= 2\Sigma_i[(\lambda + 2\mu)\phi_i'' + \lambda(b^{-1}\kappa_1\alpha_i\phi_i + r^{-1}\phi_i')] \\
\tau_{r2}^{(1)} &= (2\mu\kappa_2^2/3)\Sigma_i(\alpha_i + 3b^{-1}\beta_i)\phi_i' \\
\tau_{rr}^{(2)} &= (2E'/5)\Sigma_i(\beta_i\phi_i'' + \nu r^{-1}\beta_i\phi_i')
\end{aligned} \tag{11}$$

Using the solution (9) for ϕ_i, one obtains, on $r = a$,

$$\tau_{rr}^{(0)} = C_1\Sigma_i A_i T_{1i}$$
$$\tau_{r2}^{(1)} = C_2\Sigma_i A_i T_{2i} \qquad (12)$$
$$\tau_{rr}^{(2)} = C_3\Sigma_i A_i T_{3i}$$

where

$$T_{1i} = \zeta_i J_1(\zeta_i) - (k^2\gamma^2/2)J_0(\zeta_i)$$
$$T_{2i} = \left[1 - \frac{15\kappa_2^2(1-\nu)R^2}{2(\zeta_i^2 - \delta^2)}\right]\zeta_i(\zeta_i^2 - \gamma^2)J_1(\zeta_i)$$
$$T_{3i} = \frac{\zeta_i^2 - \gamma^2}{\zeta_i^2 - \delta^2}\zeta_i[(1-\nu)J_1(\zeta_i) - \zeta_i J_0(\zeta_i)]$$
$$R = a/b, \quad \zeta_i = \xi_i a = R\xi_i b \qquad (13)$$
$$\gamma^2 = (\pi R\Omega/2k)^2, \quad \delta^2 = (1/8)\pi^2\kappa_4^2R^2(1-\nu)(\Omega^2 - 4)$$
$$C_1 = \frac{4\mu}{a^2}, \quad C_2 = -\frac{2\mu(\lambda + 2\mu)\kappa_2^2}{3\lambda\kappa_1 Ra^2} = \frac{C_3}{3R}$$

The problem of computing the natural frequencies of a given circular plate is now formulated in the usual manner. The boundary conditions, in the absence of external forces, are

$$\tau_{rr}^{(0)} = \tau_{r2}^{(1)} = \tau_{rr}^{(2)} = 0 \qquad (14)$$

on $r = a$ and a nontrivial solution is obtained if the determinant of the matrix of the coefficients of the A_i in Equations (12) is zero; i.e.,

$$|T_{ij}| = 0, \quad i, j = 1, 2, 3 \qquad (15)$$

The determinantal equation (15) may be evaluated as follows: A given value of Poisson's ratio ν fixes the correction coefficients κ_s ($s = 1$ to 4) as explained in reference [1]. From the cubic equation (5), the three values of $\xi_i b$ are computed for an assumed value of the frequency ratio Ω. Equation (15) is thereby reduced to a transcendental equation on R ($=a/b$). The roots R for the chosen Ω are then computed by an iteration technique as explained in the following section.

For low frequencies and long wave-lengths ($\Omega \ll 1$, $z \ll 1$) Equations (5) and (15) reduce to

$$\Omega = (2\xi_1 a/\pi R)[2/(1-\nu)]^{1/2} \qquad (16)$$
$$\xi_1 a J_0(\xi_1 a) = (1-\nu)J_1(\xi_1 a) \qquad (17)$$

respectively. These coincide with Poisson's solution [4, 10].

Numerical Solution of the Frequency Equation

A table of values of Ω and R satisfying (5) and (15) was computed by means of an IBM 704 digital computer using an "interval halving" iteration technique.

For a fixed value of Ω the determinant of Equation (15) is a function of R alone. The value of this determinant is computed at a prescribed starting point R_1 and at intervals of a specified ΔR thereafter up to and including a prescribed end point R_2. A change of sign of the determinant across an interval ΔR indicates a root in that interval. The interval is then halved and the direction of "scanning" is reversed; the process is repeated until a root is obtained within a prescribed accuracy. Additional roots are obtained by re-entering successively into the iteration subroutine, with a new starting point just beyond the last root obtained, until the entire region $R_1 \leqslant R \leqslant R_2$ is covered or until a prescribed number of roots is obtained.[4] The present computations were carried out for a maximum of eleven roots in the region $0 \leqslant \Omega \leqslant 3.0$ and $0 \leqslant R \leqslant 20$. The allowable "relative error" (i.e., the error divided by the root) was 10^{-4}.

The evaluation of the determinant in (15) is simplified by the fact that it is an even analytic function of the ζ_i. The nature of the ζ_i or, equivalently, the roots z_i of the cubic equation (5), for different ranges of Ω is seen in Fig. 1.[5] For $\Omega < \Omega^*$ one of the three z_i is real and the other two are complex conjugate; for $\Omega^* < \Omega < k$ and $\Omega > 2$ all three are real; and for $k < \Omega < 2$ two are real and one is pure imaginary. It may be ascertained that for $\Omega > \Omega^*$, where the z_i are either real or imaginary, the determinant is always real. For $\Omega < \Omega^*$, if z_3 is the complex conjugate of z_2, i.e., $z_3 = \bar{z}_2$ in the usual notation, then the terms of the third column of the determinant in (15) are the complex conjugates of the terms of the second column; i.e.,

$$T_{j3} = \bar{T}_{j2} = \mathcal{R}_{j2} - i\,\mathcal{J}_{j2} \qquad (18)$$

where $\mathcal{R}_{j2}$ and $\mathcal{J}_{j2}$ are the real and imaginary parts, respectively, of T_{j2}. Hence adding and subtracting the two columns, one obtains

$$|T_{kj}| = -4i\begin{vmatrix} T_{11} & \mathcal{R}_{12} & \mathcal{J}_{12} \\ T_{21} & \mathcal{R}_{22} & \mathcal{J}_{22} \\ T_{31} & \mathcal{R}_{32} & \mathcal{J}_{32} \end{vmatrix} \qquad (19)$$

i.e.; the characteristic determinant is pure imaginary. The determinant on the right-hand side of Equation (19) was employed, in the computations, in the region $\Omega < \Omega^*$. For very low frequencies, the roots of Equation (17), the asymptotic form of the frequency equation, may be obtained conveniently through the use of Onoe's tables [11].

The variation of the mode-shapes with frequency was also investigated numerically by computing the displacement fields of a number of modes of vibration at a few selected points of the spectrum. For a given root of the frequency equation (15), the two amplitude ratios A_2/A_1 and A_3/A_1 were determined by using two of the three compatible boundary conditions (14). These ratios were inserted in Equations (9) and (1) and the displacements $u_r^{(0)}$, $u_2^{(1)}$, and $u_r^{(2)}$ were computed for twenty points along the radius of the plate. The displacements were then normalized with respect to the maximum value of displacement-component appearing in the particular mode. It should be remarked that for $\Omega < \Omega^*$ the two amplitude ratios corresponding to the complex conjugate wave-numbers z_i are also complex conjugate. Their real and imaginary parts were computed by utilizing the reduced determinant of the right-hand side of Equation (19).

The limiting forms of the displacements corresponding to Equations (16) and (17) are

$$u_r^{(0)} = AJ_1(\xi_1 r), \quad u_r^{(2)} = 0$$
$$u_2^{(1)} = -A[\nu\xi_1 a/\kappa_1(1-\nu)R]J_0(\xi_1 r) \qquad (20)$$

These coincide with Poisson's solution except for the factor κ_1.

Discussion of the Results

The frequency ratio $\Omega = \omega/\omega_s$ is plotted against the ratio of radius to half-thickness, $R = a/b$, in Fig. 2 for the first eleven modes using Shaw's value 0.31 for Poisson's ratio. For this value of Poisson's ratio,

$$\kappa_1^2 = 0.634835$$
$$\kappa_2^2 = 1.175035$$
$$\kappa_3^2 = 0.771867$$
$$\kappa_4^2 = 1.785839$$

[4] The role of Ω and R may be reversed in the coded program to obtain roots Ω for a constant R.

[5] See also reference [1] for a detailed discussion of the frequency spectrum of the second-order theory.

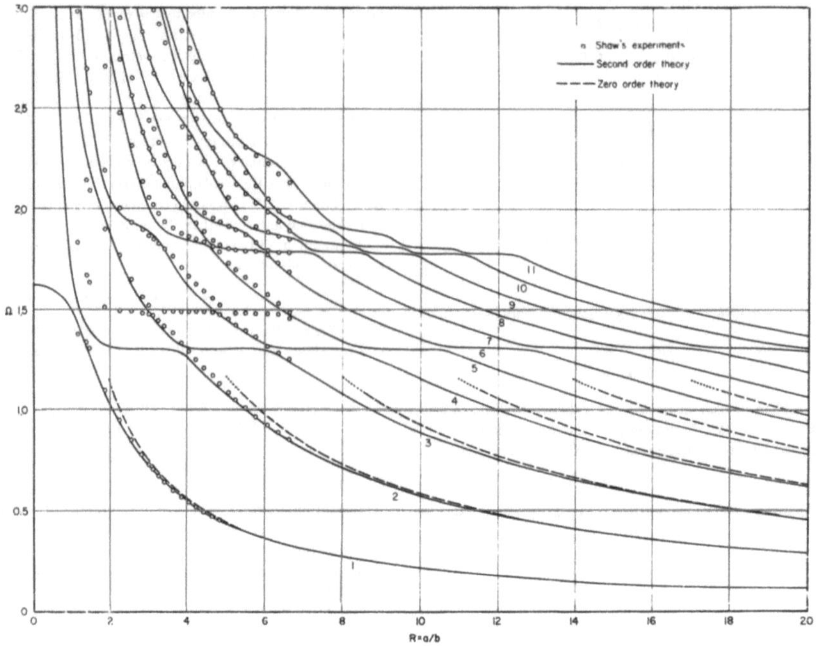

Fig. 2 Frequency spectrum of a circular disk

$$k = 1.905670$$

$$\Omega^* = 1.760744$$

In Fig. 2 are also plotted Shaw's experimental results. In order to convert Shaw's frequencies to the ordinate Ω, it is necessary to know the shear-wave velocity, $v_2 = (\mu/\rho)^{1/2}$, in barium titanate. Shaw's value of $v_2 = 2755$ m/sec, which is based on his measurements and an average of four approximate solutions by Aggarwal and Shaw, was not used. Instead, v_2 was determined from Shaw's experimental results in the neighborhood of the frequency of a family of exact solutions given by Lamb.[6] These are equivoluminal modes with circular frequency and diameter-thickness ratios

$$\omega = (\pi/b)(\mu/2\rho)^{1/2}, \quad a/b = 2\gamma_m/\pi \tag{21}$$

where the γ_m are the roots of

$$J_1'(\gamma_m) = 0, \quad \text{i.e.,} \quad \gamma_m = 1.841, 5.331, 8.536, \ldots \tag{22}$$

hence

$$a/b = 1.172, 3.394, 5.434, \ldots \tag{23}$$

If f is the frequency, Shaw's ordinate for these modes is

$$2fb = (\mu/2\rho)^{1/2} \tag{24}$$

Shaw's frequencies for the first mode near $a/b = 1.172$ are at an extreme of his range of measurements and do not appear to be reliable[7]; but the second and third roots (23) both occur at $2fb =$ 1900 m/sec. Hence

$$v_2 = (\mu/\rho)^{1/2} = 2690 \text{ m/sec} \tag{25}$$

A comparison of the experimental frequencies with the theoretical results of the second-order theory, in Fig. 2, reveals very good general agreement, except that the theoretical frequency of the edge modes ($\Omega \approx 1.31$) is lower than the experimental value ($\Omega \approx 1.47$) by about 12 per cent. An investigation was made of the possibility that the discrepancy might be due to exceptional sensitivity to Poisson's ratio. However, the result was negative. In the range $0.25 \leqslant \nu \leqslant 0.40$, the frequencies of the edge modes were found to remain in the interval $1.30 < \Omega_e < 1.35$. Computations were also made with a variety of values of the correction coefficients κ_e, obtained by matching the exact and approximate frequency spectra of an infinite plate at points other than those at infinite wave-length; but no improvement was obtained. Another source of error in the approximate theory is that the boundary conditions involve stress-resultants and stress-couples rather than tractions at all points across the thickness of the plate. However, this raises, rather than lowers, the frequency. The most likely cause of the discrepancy seems to be that, although the second-order theory provides a good approximation of the real and imaginary segments of the first three branches of the frequency spectrum of an infinite plate as given by the exact theory, it provides only a fair reproduction of the associated complex segments; and it appears that the details of the complex segments become important in the neighborhood of the frequency of the edge modes.

As the frequency diminishes, both the experimental and mathematical results approach Poisson's, as given by Equations (16) and (17) and shown by dashed lines in Fig. 2. As the frequency increases above Ω_e, the typical terrace-like structure of the high-frequency spectrum of plates begins to develop at a frequency

[6] Footnote on p. 122 of reference [3].

[7] All of the equivoluminal modes should have the same frequency and, as $R \to 0$, the frequency of the first mode of the disk should approach the frequency of the first longitudinal mode of a thin rod of length $2b$; i.e., $2fb = [(1 + \nu)\mu/2\rho]^{1/2}$.

whose asymptotic value, as $R \to \infty$ is $\Omega \approx 1.76$ in both Shaw's experiments and the second-order theory. As observed by Shaw, this is the frequency minimum Ω^* rather than the cut-off frequency of the lowest, symmetric thickness-mode which, in the present case of $\nu < 1/3$, is the thickness-stretch mode with frequency given by $\Omega = k \approx 1.9$. Accordingly, there is an important difference, in this aspect of high-frequency spectra, between extensional and flexural vibrations of plates. In the latter, the asymptotic frequency of the lowest terrace-like structure is always the cut-off frequency ($\Omega = 1$) of the lowest, antisymmetric thickness-shear mode. This is because the second branch for the infinite plate never has a real minimum other than at cut-off ($\xi b = 0$) in the case of flexure [8]. The axially symmetric flexural vibrations of a circular disk are discussed in reference [12].

Returning, now, to frequencies in the neighborhood of Ω_e, computations based on the second-order theory show that the displacements do, in fact, conform with Shaw's observations. In Fig. 3 are plotted the computed values of the displacement functions for the third mode of the disk for three frequencies: one just below the frequency Ω_e of the edge vibration, one at and one just above this frequency. Immediately above and below Ω_e the displacements are distributed across the plate in much the same manner as in Poisson's solution, Equations (20), except that the radial displacement $u_r^{(0)}$ is considerably suppressed and there is a thickness-shear displacement $u_r^{(2)}$ of about the same amplitude. As Ω approaches Ω_e, coupling with the edge mode takes place which, at $\Omega = \Omega_e$ results in large amplitudes of edge vibrations characterized by displacements which include a component large near the edge and rapidly decaying toward the center of the plate. It should be noted that, as the frequency passes from below to above Ω_e, the number of nodal circles for each mode is reduced by one. Thus each mode attains above Ω_e a mode-shape similar to the mode-shape of the next lower mode for $\Omega < \Omega_e$.

The character of the quasi-exponential decay of the displacements at $\Omega = \Omega_e$ is essentially independent of the radius at the "plateau" of edge vibrations. This is shown in Fig. 4 where the theoretical value of the normalized displacement $u_2^{(1)}$ is plotted against r'/b, where r' is the radial distance from the edge of the plate, for two values of R (3.0 and 5.0). These two values of R correspond to points on the plateaus at Ω_e on the branches of the second and third modes, respectively. It is seen that the first nodal circle from the edge occurs at a distance of r'/b approximately equal to 0.75, essentially the same in both cases. This substantiates Shaw's concluding remark that edge vibrations should be expected to occur also for circular disks with very large ratios of radius to thickness, inasmuch as the character of the edge vibrations is essentially independent of this ratio.

There are additional interesting properties of the edge mode which are exposed more readily in the problem of reflection considered in the next section.

Reflection at the Free Edge of a Semi-Infinite Plate

Consider a semi-infinite plate bounded by the free faces $x_2 = \pm b$ and a free edge at $x_1 = 0$ and extending to infinity in the direction x_1 increasing. Assume that a sinusoidal extensional wave of a normalized frequency $\Omega < \Omega^*$ is continuously generated at $x_1 = \infty$ and propagated toward $x_1 = 0$. Referring to the invariant field equations of reference [1] one may ascertain that the sinusoidal wave is associated with the dilatational potential ϕ_1 corresponding to the real root z_1 of the cubic equation (5). The potential

$$\phi_1 = A_0 \exp\,[i(\omega t + \xi_1 x_1)] \tag{26}$$

satisfies the wave equation and generates the desired type of extensional wave. The associated displacement field is

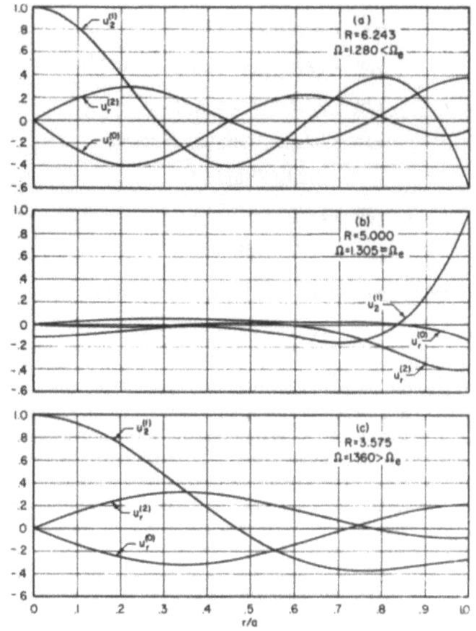

Fig. 3 Distributions of radial ($u_r^{(0)}$), thickness-stretch ($u_2^{(0)}$), and thickness-shear ($u_r^{(2)}$) displacements at frequencies in the neighborhood of frequency of edge-mode

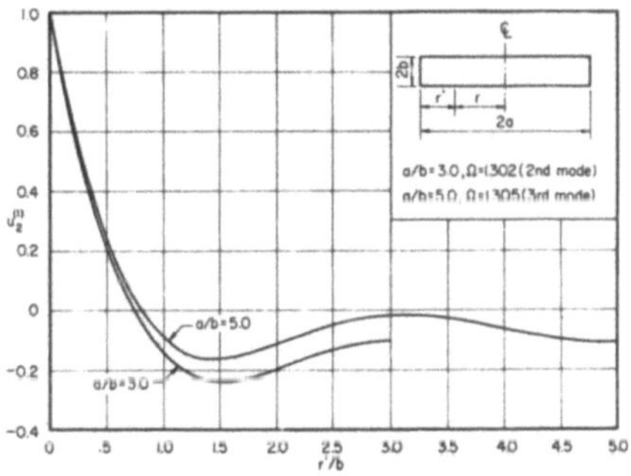

Fig. 4 Distribution of thickness-stretch displacements in two edge-modes

$$\begin{aligned} u_1^{(0)} &= A_0 \xi_1 i \exp\,[i(\omega t + \xi_1 x_1)] \\ u_2^{(1)} &= A_0 \alpha_1 \exp\,[i(\omega t + \xi_1 x_1)] \\ u_1^{(2)} &= A_0 \beta_1 \xi_1 i \exp\,[i(\omega t + \xi_1 x_1)] \\ u_2^{(0)} &= u_2^{(2)} = 0 \end{aligned} \tag{27}$$

where α_1 and β_1 are given by Equations (2). Omitting exponentials in the sequel, one obtains the plate-strains

$$\begin{aligned} \epsilon_1^{(0)} &= -A_0 \xi_1^2 \\ \epsilon_2^{(0)} &= A_0 \alpha_1 / b \\ \epsilon_1^{(2)} &= -A_0 \beta_1 \xi_1^2 \\ \epsilon_6^{(1)} &= A_0 i \xi (\alpha_1 + 3\beta_1/b) \\ \epsilon_3^{(0)} &= \epsilon_4^{(0)} = \epsilon_4^{(1)} = \epsilon_3^{(2)} = \epsilon_5^{(2)} = 0 \end{aligned} \tag{28}$$

and the plate-stresses

$$\begin{aligned}
\tau_1^{(0)} &= -2A_0\rho\omega^2 \\
\tau_2^{(0)} &= 2A_0[\kappa_1^2(\lambda + 2\mu)\alpha_1/b - \lambda\kappa_1\xi_1^2] \\
\tau_3^{(0)} &= 2A_0\lambda(\kappa_1\alpha_1/b - \xi_1^2) \\
\tau_6^{(1)} &= 2A_0(\mu\kappa_2^2/3)i\xi_1(\alpha_1 + 3\beta_1/b) \qquad (29) \\
\tau_1^{(2)} &= -2A_0(E'/5)\beta_1\xi_1^2 \\
\tau_3^{(2)} &= \nu\tau_1^{(2)} \\
\tau_5^{(0)} &= \tau_4^{(1)} = \tau_5^{(2)} = 0
\end{aligned}$$

The boundary conditions are, for a free boundary,

$$\tau_1^{(0)} = \tau_1^{(2)} = \tau_6^{(1)} = 0 \qquad (30)$$

at $x_1 = 0$.

It may be seen from Equations (29) that $\tau_1^{(0)}$ and $\tau_1^{(2)}$ are even functions of ξ_1 whereas $\tau_6^{(1)}$ is an odd function of ξ_1. Hence a simple reflection of the extensional wave, corresponding to a change of sign of ξ_1 does not make all the boundary stresses zero. It follows that all three types of waves corresponding to the three ϕ_j will be excited upon reflection of the pure extensional wave at the free boundary. The reflected waves are associated with potentials

$$\phi_j = A_j \exp\,[i(\omega t - \xi_j x_1)] \qquad (31)$$

and the corresponding displacements, plate-strains, and plate-stresses may be derived from Equations (27), (28), and (29) by substituting $-\xi_j$ for ξ_1. The boundary stresses due to the incident wave plus the stress due to the three reflected waves must satisfy the boundary conditions (30). Thus one obtains the system of three equations

$$\begin{aligned}
A_0 + A_1 + A_2 + A_3 &= 0 \\
(A_0 + A_1)T_{21} + A_2T_{22} + A_3T_{23} &= 0 \qquad (32) \\
(A_0 - A_1)T_{31} - A_2T_{32} - A_3T_{33} &= 0
\end{aligned}$$

where

$$\left.\begin{aligned}
T_{2j} &= \beta_j\xi_j^2 \\
T_{3j} &= (\alpha_j + 3\beta_j/b)\xi_j^2
\end{aligned}\right\} j = 1, 2, 3. \qquad (33)$$

From the linear system (32) one obtains the three ratios of the amplitudes of the reflected waves to the amplitude of the incident wave: A_j/A_0. It should be remarked that in selecting the proper sign for the complex wave numbers one should bear in mind that in order to be physically acceptable they should correspond to functions decreasing for increasing x_1. It is seen that the two complex wave-numbers which should be entered in Equations (31) are

$$\left.\begin{aligned}
\xi_2 &= c - id \\
\xi_3 &= -c - id
\end{aligned}\right\} c, d > 0 \qquad (34)$$

Thus in addition to a reflected pure extensional wave, one obtains two "complex-waves" corresponding to motion localized near the free edge of the plate.

An investigation of the expressions for the A_j/A_0 leads to the following conclusions: For $\Omega < \Omega^*$ the three amplitude ratios are always complex, and their moduli satisfy the relations

$$|A_1/A_0| = 1, \quad |A_2/A_0| = |A_3/A_0| = |A_e/A_0| \text{ (say)} \qquad (35)$$

The complex value of A_1/A_0 implies that the reflected pure extensional wave has a phase lag η with respect to the incident wave, given by

$$\tan \eta = -\mathcal{I}_1/\mathcal{R}_1, \qquad -\pi \leqslant \eta \leqslant \pi \qquad (36)$$

where $\mathcal{R}_1$ and $\mathcal{I}_1$ are the real and imaginary parts of the ratio A_1/A_0.

The phase of the reflected extensional wave and the magnitude of the amplitude A_e/A_0 of the complex waves were computed numerically on the IBM 704 for Ω in the neighborhood of the frequency of edge vibrations. It was found that for $\Omega \ll \Omega_e$ the amplitude $|A_e/A_0|$ is small and the phase difference between the incident and reflected extensional waves is nearly π. As Ω approaches Ω_e, $|A_e/A_0|$ increases very rapidly until, for $\Omega = \Omega_e \approx 1.314$, it reaches a maximum value of approximately 7.7, Fig. 5. At the same time the phase difference diminishes until it reaches the value 0 for $\Omega = \Omega_e$ as also shown in Fig.5. If Ω is increased beyond Ω_e, the amplitude $|A_e/A_0|$ decreases rapidly to very small values and the phase difference changes sign and increases rapidly from 0 to nearly π. In addition to $|A_e/A_0|$, Fig. 5 contains plots of $\mathcal{R}_1$ and $\mathcal{I}_1$ versus Ω.

Fig. 5 illustrates the details of the phenomena involved in the edge vibrations. This type of motion is in resonance with the pure extensional type of motion at a distinct frequency which the second-order theory gives as approximately equal to 1.314 ω_s for $\nu = 0.31$. Shaw's experiments indicate that this frequency of resonance is slightly higher, namely, in the neighborhood of 1.47 ω_s. In any case, at this frequency level a pure extensional wave gives rise to a large amplitude of edge-vibrational motion upon reflection at a free boundary, after a normal incidence. By analogy, in the case of a circular disk which has been discussed in the main part of this paper, one may visualize the edge vibrations as continuously generated, at the frequency of resonance, by reflection of extensional circular-crested waves at the free circular boundary.

An important feature, observable in Fig. 5, is that the ampli-

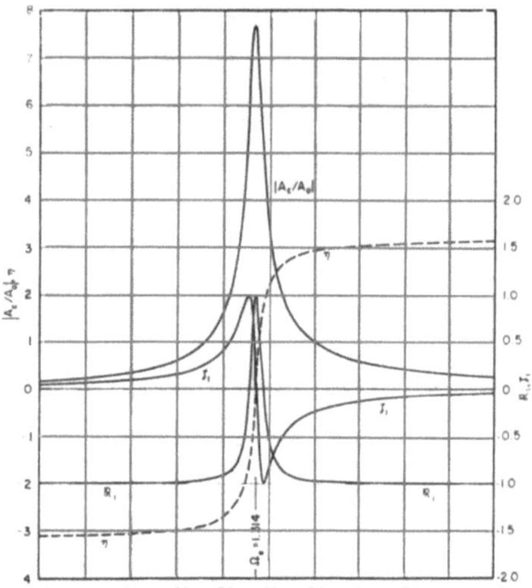

Fig. 5 Ratio ($|A_e/A_0|$) of amplitude of edge-vibration to amplitude of incident wave in the neighborhood of frequency of edge-mode. Also shown are the real ($\mathcal{R}_1$) and imaginary ($\mathcal{I}_1$) parts of the ratio (A_1/A_0) of amplitudes of incident and reflected waves and their phase difference (η).

tude ratio $|A_e/A_0|$ passes through a finite maximum and never reaches infinity. This means that the large edge-deformation is always coupled with a nondecaying extensional deformation of the usual kind. Consider, for example, a transient disturbance at the edge, initiated either by external excitation at the edge or by an incident wave from the interior of the plate. If this disturbance contains energy at frequency Ω_e, an edge-vibration will be induced. However, the edge-deformation is always coupled with extensional deformation and hence the energy that is temporarily trapped at the edge will gradually leak into the interior of the plate as an extensional wave traveling with the phase and group velocities corresponding to frequency Ω_e of the extensional mode (ϕ_1). The group velocity (v_2 times the slope of the branch in Fig. 1) at Ω_e is less than the group velocities of extensional waves at lower frequencies and hence the trailing wave generated by the edge mode will be outdistanced by the main reflected transient. These phenomena have been observed by J. Oliver [13] in experiments with rods.

References

1 R. D. Mindlin and M. A. Medick, "Extensional Vibrations of Elastic Plates," Journal of Applied Mechanics, vol. 26, Trans. ASME, vol. 81, series E, 1959, pp. 561–569.

2 Lord Rayleigh, "On the Free Vibrations of an Infinite Plate of Homogeneous Isotropic Elastic Matter," *Proceedings of the London Mathematical Society*, vol. 20, 1889, pp. 225–234.

3 H. Lamb, "On Waves in an Elastic Plate," *Proceedings of The Royal Society*, London, England, series A, vol. 93, 1917, pp. 114–128.

4 S. D. Poisson, "Mémoire sur l'équilibre et le mouvement des corps élastiques," *Mémoires de l'Académie des Sciences*, series 2, vol. 8, Paris, France, 1829, pp. 357–370.

5 A. L. Cauchy, "Sur l'équilibre et le mouvement d'une plaque solide," *Exercises de Mathématique*, vol. 3, Bure Frères, Paris, France, 1828, pp. 328–355.

6 T. R. Kane and R. D. Mindlin, "High-Frequency Extensional Vibrations of Plates," Journal of Applied Mechanics, vol. 23, Trans. ASME, vol. 78, 1956, pp. A-277-283.

7 E. A. G. Shaw, "On the Resonant Vibrations of Thick Barium Titanate Disks," *Journal of the Acoustical Society of America*, vol. 28, 1956, pp. 38–50.

8 R. D. Mindlin, M. Onoe, and M. A. Medick, "Mathematical Theory of Vibrations of Elastic Plates," Proceedings of the 11th Annual Symposium on Frequency Control, U. S. Army Signal Engineering Laboratories, Fort Monmouth, N. J., May 1957, pp. 1–38.

9 J. W. C. Sherwood, "Propagation in an Infinite Plate," *Journal of the Acoustical Society of America*, vol. 30, 1958, pp. 979–984.

10 A. E. H. Love, "A Treatise on the Mathematical Theory of Elasticity," Cambridge University Press, London, England, fourth edition, 1927, p. 498.

11 M. Onoe, "Tables of Modified Quotients of Bessel Functions of the First Kind for Real and Imaginary Arguments," Columbia University Press, New York, N. Y., 1958.

12 H. Deresiewicz and R. D. Mindlin, "Axially Symmetric Flexural Vibrations of a Circular Disk," Journal of Applied Mechanics, vol. 22, Trans. ASME, vol. 77, 1955, pp. 86–88.

13 J. Oliver, "Elastic Wave Dispersion in a Cylindrical Rod by a Wide-Band Short-Duration Pulse Technique," *Journal of the Acoustical Society of America*, vol. 29, 1957, pp. 189–194.

Reprinted from "Structural Mechanics"
PERGAMON PRESS
OXFORD - LONDON - NEW YORK - PARIS
1960

WAVES AND VIBRATIONS IN ISOTROPIC, ELASTIC PLATES

R. D. MINDLIN
Department of Civil Engineering and Engineering Mechanics
Columbia University, New York

1. INTRODUCTION

THE complexity of the theory of vibrations of plates stems from two sources: (*a*) two types of wave, dilatational and equivoluminal, are possible in an isotropic, elastic body; (*b*) neither wave can exist by itself, in general, if any part of the boundary is free of traction. The consequences of these phenomena can be understood better if the development of coupling between the two types of wave is traced from the infinite body to the half-space, then to the infinite plate and, finally, to the bounded plate. The process can be made continuous because each new plane boundary can be introduced with mixed boundary conditions, which introduce no coupling, and the subsequent passage to the free boundary can be made continuous through the introduction of a variable, elastic restraint.

At the stage of the infinite plate with free faces, the complexities become very great in the range of high frequencies, i.e. if the half-period is less than the transit time of an equivoluminal wave across the thickness of the plate. Although the formal solution of the problem was given 70 years ago, it is only in the past few years that the full implications of the frequency equation have begun to be understood fully enough so that it becomes clear how to proceed toward the solution of problems involving high frequencies in plates with boundaries.

2. EQUATIONS OF ELASTICITY

If the components of displacement are expressed as[1, 8]

$$\begin{aligned} u_1 &= \partial\phi/\partial x_1 + \partial H_3/\partial x_2 - \partial H_2/\partial x_3 \\ u_2 &= \partial\phi/\partial x_2 + \partial H_1/\partial x_3 - \partial H_3/\partial x_1 \\ u_3 &= \partial\phi/\partial x_3 + \partial H_2/\partial x_1 - \partial H_1/\partial x_2 \end{aligned} \tag{1}$$

they satisfy the equations of small motion of an isotropic, elastic medium if

$$v_1^2\nabla^2\phi = \partial^2\phi/\partial t^2 \tag{2}$$

$$v_2^2\nabla^2 H = \partial^2 H/\partial t^2,\ H = (H_1, H_2, H_3) \tag{3}$$

where

$$v_1 = [(\lambda+2\mu)/\rho]^{\frac{1}{2}},\ v_2 = [\mu/\rho]^{\frac{1}{2}} \tag{4}$$

are the velocities of dilatational and equivoluminal waves and λ and μ are Lamé's constants. The components of stress are expressed in terms of the components of displacement by the equations

$$\begin{aligned}\tau_{11}&=\lambda\Delta+2\mu\,\partial u_1/\partial x_1\\ \tau_{22}&=\lambda\Delta+2\mu\,\partial u_2/\partial x_2\\ \tau_{33}&=\lambda\Delta+2\mu\,\partial u_3/\partial x_3\\ \tau_{23}&=\mu(\partial u_3/\partial x_2+\partial u_2/\partial x_3)\\ \tau_{31}&=\mu(\partial u_1/\partial x_3+\partial u_3/\partial x_1)\\ \tau_{12}&=\mu(\partial u_2/\partial x_1+\partial u_1/\partial x_2)\end{aligned} \tag{5}$$

where

$$\Delta=\partial u_1/\partial x_1+\partial u_2/\partial x_2+\partial u_3/\partial x_3 \tag{6}$$

3. *P*- AND *SV*-WAVES OF PLANE STRAIN

Plane waves of plane strain, in the plane of x_1 and x_2 may be expressed as

$$\begin{aligned}\phi=&A_1\sin\gamma_1(x_1\sin\theta_1+x_2\cos\theta_1-v_1t)+\\ &+A_2\sin\gamma_1(x_1\sin\theta_1-x_2\cos\theta_1-v_1t)+\\ &+B_1\cos\gamma_1(x_1\sin\theta_1+x_2\cos\theta_1-v_1t)+\\ &+B_2\cos\gamma_1(x_1\sin\theta_1-x_2\cos\theta_1-v_1t)\end{aligned} \tag{7}$$

$$\begin{aligned}H_3=&C_1\cos\gamma_2(x_1\sin\theta_2+x_2\cos\theta_2-v_2t)+\\ &+C_2\cos\gamma_2(x_1\sin\theta_2-x_2\cos\theta_2-v_2t)+\\ &+D_1\sin\gamma_2(x_1\sin\theta_2+x_2\cos\theta_2-v_2t)+\\ &+D_2\sin\gamma_2(x_1\sin\theta_2-x_2\cos\theta_2-v_2t)\end{aligned} \tag{8}$$

and $H_1=H_2=0$. The four waves in ϕ are dilatational (*P*) waves whose normals make angles θ_1 with the axis of x_2 and the four waves in H_3 are equivoluminal (*SV*) waves whose normals make angles θ_2 with x_2, as illustrated in Fig. 1. In the region $x_2\leqslant 0$, the waves with amplitudes A_1, B_1, C_1, D_1

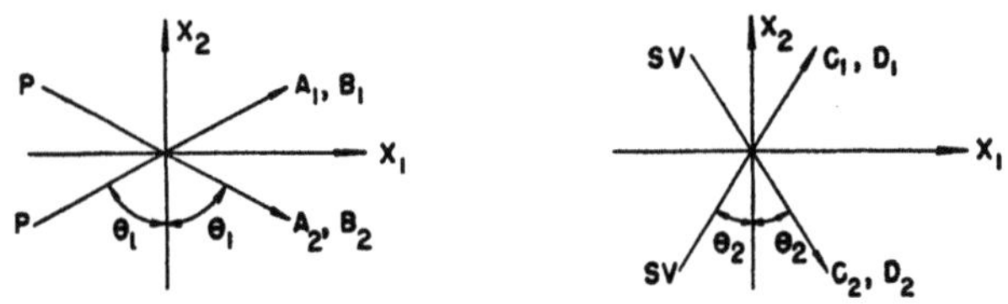

Fig. 1. *P*- and *SV*-waves.

travel toward the boundary while the remaining four waves travel away from the boundary. In all of the waves

$$u_3=\tau_{23}=\tau_{31}=0 \tag{9}$$

On $x_2=0$, the displacements are

$$\begin{aligned}u_1=&\gamma_1\sin\theta_1[(A_1+A_2)\cos\psi_1-(B_1+B_2)\sin\psi_1]-\\ &-\gamma_2\cos\theta_2[(C_1-C_2)\sin\psi_2-(D_1-D_2)\cos\psi_2]\\ u_2=&\gamma_1\cos\theta_1[(A_1-A_2)\cos\psi_1-(B_1-B_2)\sin\psi_1]+\\ &+\gamma_2\sin\theta_2[(C_1+C_2)\sin\psi_2-(D_1+D_2)\cos\psi_2]\end{aligned} \tag{10}$$

where

$$\psi_1 = \gamma_1 (x_1 \sin \theta_1 - v_1 t)$$
$$\psi_2 = \gamma_2 (x_1 \sin \theta_2 - v_2 t) \tag{11}$$

The tractions across $x_2 = 0$ are

$$\begin{aligned}\tau_{12} = &-\mu\gamma_1^2 \sin 2\theta_1 [(A_1 - A_2) \sin \psi_1 + (B_1 - B_2) \cos \psi_1] - \\ &- \mu\gamma_2^2 \cos 2\theta_2 [(C_1 + C_2) \cos \psi_2 + (D_1 + D_2) \sin \psi_2] \\ \tau_{22} = &-(\lambda + 2\mu \cos^2\theta_1) \gamma_1^2 [(A_1 + A_2) \sin \psi_1 + (B_1 + B_2) \cos \psi_1] + \\ &+ \mu\gamma_2^2 \sin 2\theta_2 [(C_1 - C_2) \cos \psi_2 + (D_1 - D_2) \sin \psi_2]\end{aligned} \tag{12}$$

4. HALF-SPACE: MIXED BOUNDARY CONDITIONS

When the boundary conditions are mixed, i.e.

$$u_2 = 0, \ \tau_{12} = 0 \text{ on } x_2 = 0 \tag{13}$$

or

$$u_1 = 0, \ \tau_{22} = 0 \text{ on } x_2 = 0 \tag{14}$$

both *P*- and *SV*-waves reflect as themselves with equal angles of incidence and reflection, as illustrated in Fig. 2. For example, if all of the amplitudes

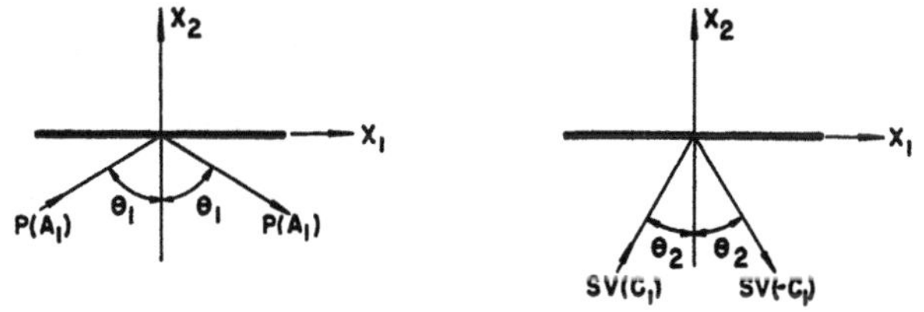

Fig. 2. Simple reflections of *P*- and *SV*-waves when the boundary conditions are mixed ($u_2 = 0$, $\tau_{12} = 0$, $\tau_{23} = 0$ on $x_2 = 0$).

are zero except A_1 and A_2, equations (13) are satisfied by (10) and (12) if $A_1 = A_2$; or equations (14) are satisfied if $A_1 = -A_2$ and similarly for the other three pairs of waves. Thus, dilatational and equivoluminal waves are not coupled when the boundary conditions are mixed.

5. HALF-SPACE: FREE BOUNDARY

Except for special cases, a plane *P*- or *SV*-wave, incident at a plane, free boundary, reflects as both *P*- and *SV*-waves[2] as illustrated in Fig. 3. To satisfy the boundary conditions

$$\tau_{12} = \tau_{22} = 0 \text{ on } x_2 = 0 \tag{15}$$

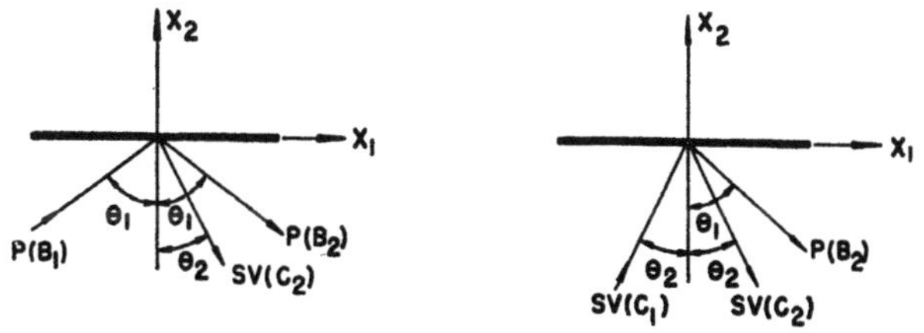

Fig. 3 General reflections of plane *P*- and *SV*-waves at a plane, traction-free boundary ($\tau_{12} = 0$, $\tau_{22} = 0$, $\tau_{32} = 0$ on $x_2 = 0$).

it is necessary that $\psi_1 = \psi_2$ in equations (11), i.e. the circular frequency,

$$\omega = \gamma_1 v_1 = \gamma_2 v_2 \tag{16}$$

and the wave-number along x_1,

$$\xi = \gamma_1 \sin\theta_1 = \gamma_2 \sin\theta_2 \tag{17}$$

must be the same for both types of wave. Then, for example, an incident P-wave

$$\phi = B_1 \cos\gamma_1 (x_1 \sin\theta_1 + x_2 \cos\theta_1 - v_1 t) \tag{18}$$

reflects as P- and SV-waves

$$\begin{aligned} \phi &= B_2 \cos\gamma_1 (x_1 \sin\theta_1 - x_2 \cos\theta_1 - v_1 t) \\ H_3 &= C_2 \cos\gamma_2 (x_1 \sin\theta_2 - x_2 \cos\theta_2 - v_2 t) \end{aligned} \tag{19}$$

with

$$\frac{B_2}{B_1} = \frac{\sin 2\theta_1 \sin 2\theta_2 - k^2 \cos^2 2\theta_2}{\sin 2\theta_1 \sin 2\theta_2 + k^2 \cos^2 2\theta_2} \tag{20}$$

$$\frac{C_2}{B_1} = -\frac{2 \sin 2\theta_1 \cos 2\theta_2}{\sin 2\theta_1 \sin 2\theta_2 + k^2 \cos^2 2\theta_2} \tag{21}$$

where

$$k^2 = v_1^2/v_2^2 = \gamma_2^2/\gamma_1^2 = (\lambda + 2\mu)/\mu = 2(1-\nu)/(1-2\nu) \tag{22}$$

and ν is Poisson's ratio. Similarly, an incident SV-wave

$$H_3 = C_1 \cos\gamma_2 (x_1 \sin\theta_2 + x_2 \cos\theta_2 - v_2 t) \tag{23}$$

reflects as P- and SV-waves

$$\begin{aligned} \phi &= B_2 \cos\gamma_1 (x_1 \sin\theta_1 - x_2 \cos\theta_1 - v_1 t) \\ H_3 &= C_2 \cos\gamma_2 (x_1 \sin\theta_2 - x_2 \cos\theta_2 - v_2 t) \end{aligned} \tag{24}$$

with

$$\frac{B_2}{C_1} = \frac{k^2 \sin 4\theta_2}{\sin 2\theta_1 \sin 2\theta_2 + k^2 \cos^2 2\theta_2} \tag{25}$$

$$\frac{C_2}{C_1} = \frac{\sin 2\theta_1 \sin 2\theta_2 - k^2 \cos^2 2\theta_2}{\sin 2\theta_1 \sin 2\theta_2 + k^2 \cos^2 2\theta_2} \tag{26}$$

Thus, in general, the presence of a free boundary serves to couple P- and SV-waves.

6. HALF-SPACE: ELASTICALLY RESTRAINED BOUNDARY

The development of the coupling between P- and SV-waves from the uncoupled state, in the case of mixed boundary conditions, to the fully coupled state, in the case of the free boundary, may be observed in the intermediate case of an elastically restrained boundary. Consider, for example, an SV-wave

$$H_3 = C_1 \cos\gamma_2 (x_1 \sin\theta_2 + x_2 \cos\theta_2 - v_2 t) \tag{27}$$

incident on the plane $x_2 = 0$ where the conditions are

$$\tau_{22} = e u_2, \quad \tau_{12} = 0 \tag{28}$$

Here e is a parameter which may vary from infinity, in which case (28) reduces to the mixed conditions (13), to zero, in which case (28) reduces to the conditions (15) for a free boundary. As is illustrated in Fig. 4, conditions (28) are satisfied if there are reflected waves

$$\begin{aligned}\phi &= A_2 \sin \gamma_1 (x_1 \sin \theta_1 - x_2 \cos \theta_1 - v_1 t) + \\ &+ B_2 \cos \gamma_1 (x_1 \sin \theta_1 - x_2 \cos \theta_1 - v_1 t) \\ H_3 &= C_2 \cos \gamma_2 (x_1 \sin \theta_2 - x_2 \cos \theta_2 - v_2 t) + \\ &+ D_2 \sin \gamma_2 (x_1 \sin \theta_2 - x_2 \cos \theta_2 - v_2 t)\end{aligned} \tag{29}$$

with

$$\begin{aligned}\frac{A_2}{C_1} &= -\frac{E}{1+E^2} \frac{k^2 \sin 4\theta_2}{\sin 2\theta_1 \sin 2\theta_2 + k^2 \cos^2 2\theta_2} \\ \frac{B_2}{C_1} &= \frac{1}{1+E^2} \frac{k^2 \sin 4\theta_2}{\sin 2\theta_1 \sin 2\theta_2 + k^2 \cos^2 2\theta_2} \\ \frac{C_2}{C_1} &= \frac{1}{1+E^2} \frac{\sin 2\theta_1 \sin 2\theta_2 - k^2 \cos^2 2\theta_2}{\sin 2\theta_1 \sin 2\theta_2 + k^2 \cos^2 2\theta_2} - \frac{E^2}{1+E^2} \\ \frac{D_2}{C_1} &= -\frac{E}{1+E^2} \frac{2 \sin 2\theta_1 \sin 2\theta_2}{\sin 2\theta_1 \sin 2\theta_2 + k^2 \cos^2 2\theta_2}\end{aligned} \tag{30}$$

where

$$E = \frac{ke \cos \theta_1}{\mu \gamma_2 (\sin 2\theta_1 \sin 2\theta_2 + k^2 \cos^2 2\theta_2)} \tag{31}$$

Thus, for a given angle of incidence and wavelength of the incident SV-wave, the amplitude of the reflected P-wave increases from zero to its value in (25), in proportion to

$$(1+E^2)^{-\frac{1}{2}} \tag{32}$$

as e drops from infinity to zero.

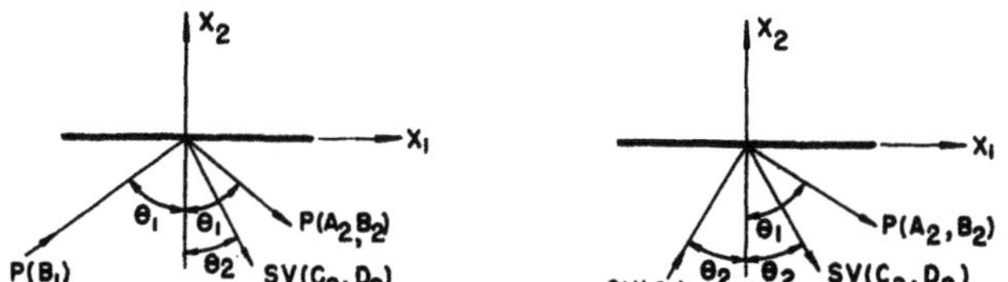

Fig. 4. Reflections of P- and SV-waves at an elastically restrained boundary ($\tau_{22}=eu_2$, $\tau_{12}=0$, $\tau_{23}=0$ on $x_2=0$).

7. HALF-SPACE: FREE BOUNDARY, SPECIAL CASES

There are special cases of reflection at a single, free boundary which are interesting because of the roles they play in vibrations or waves in plates.

Normal Incidence

When the angle of incidence, θ_1, of a P-wave is zero, (20) and (21) reduce to

$$B_2/B_1 = -1,\ C_2/B_1 = 0 \tag{33}$$

Similarly, when $\theta_2 = 0$, (25) and (26) reduce to

$$B_2/C_1 = 0,\ C_2/C_1 = -1 \tag{34}$$

Thus, at normal incidence, as illustrated in Fig. 5, P- or SV-waves reflect as themselves with a phase change of 180°.

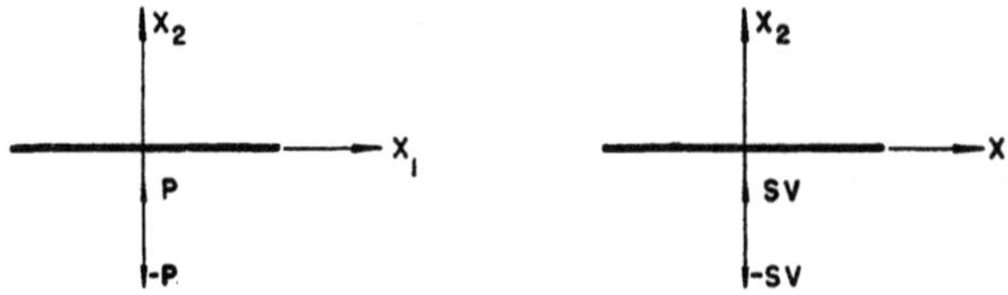

Fig. 5. Normal incidence at a traction-free boundary.

Grazing Incidence

When the incident wave is parallel to the boundary, the solutions given in Section 5 vanish; i.e. setting $\theta_1 = \pi/2$ in (18)—(21) or $\theta_2 = \pi/2$ in (23)—(26) makes the sum of the displacements zero. However, Goodier and Bishop[3] have shown, by a limiting process, that an incident P-wave

$$\phi = B_1 \cos \gamma_1 (x_1 - v_1 t) \tag{35}$$

reflects as

$$\begin{gathered} \phi = B_2' \gamma_1 x_2 \sin \gamma_1 (x_1 - v_1 t) \\ H_3 = C_2' \cos \gamma_1 \{x_1 - x_2 \sqrt{(k^2-1)} - v_1 t\} \end{gathered} \tag{36}$$

with $\sin \theta_2 = k^{-1}$ and

$$\frac{B_2'}{B_1} = -\frac{(k^2-2)^2}{4\sqrt{(k^2-1)}},\ \frac{C_2'}{B_1} = -\frac{k^2-2}{2\sqrt{(k^2-1)}} \tag{37}$$

The reflected P-wave is parallel to the boundary and its amplitude is proportional to x_2, as illustrated in Fig. 6(a). Similarly an incident SV-wave

$$H_3 = C_1 \cos \gamma_2 (x_1 - v_2 t) \tag{38}$$

reflects as

$$\begin{gathered} \phi = B_2' \exp\{\gamma_2 x_2 \sqrt{(1-k^{-2})}\} \sin \gamma_2 (x_1 - v_2 t) \\ H_3 = C_2' \gamma_2 x_2 \cos \gamma_2 (x_1 - v_2 t) \end{gathered} \tag{39}$$

with

$$\frac{B_2'}{C_1} = -\frac{k}{2\sqrt{(k^2-1)}},\ \frac{C_2'}{C_1} = \frac{k}{4\sqrt{(k^2-1)}} \tag{40}$$

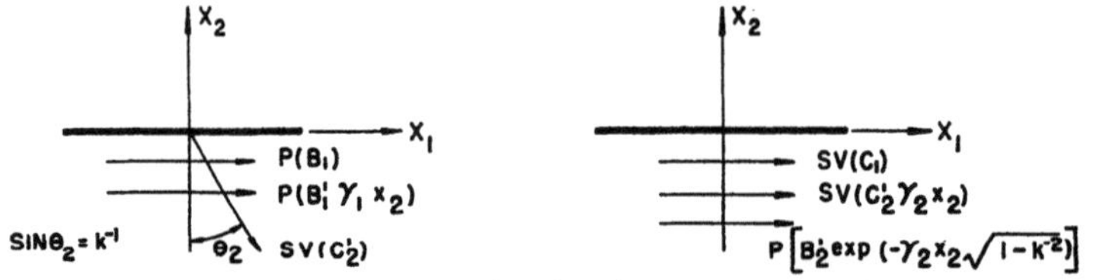

Fig. 6. Grazing incidence at a traction-free boundary (Goodier–Bishop waves).

Here both reflected waves are parallel to the boundary, as illustrated in Fig. 6(b). The amplitude of the reflected P-wave diminishes exponentially with x_2 and the amplitude of the reflected SV-wave is proportional to x_2.

Combinations of Goodier–Bishop waves yield the following three systems of waves that play important roles in the theory of plates:

$$\begin{gathered}\phi = B_1 \cos \gamma_1 (x_1 - v_1 t) \\ H_3 = C_1 \cos \gamma_1 \{x_1 + x_2 \sqrt{(k^2-1)} - v_1 t\} + \\ + C_2 \cos \gamma_1 \{x_1 - x_2 \sqrt{(k^2-1)} - v_1 t\} \\ \frac{C_2}{B_1} = -\frac{C_1}{B_1} = \frac{k^2-2}{4\sqrt{(k^2-1)}}\end{gathered} \tag{41}$$

$$\begin{gathered}\phi = A_2' \gamma_1 x_2 \cos \gamma_1 (x_1 - v_1 t) \\ H_3 = D_1 \sin \gamma_1 \{x_1 + x_2 \sqrt{(k^2-1)} - v_1 t\} + \\ + D_2 \sin \gamma_1 \{x_1 - x_2 \sqrt{(k^2-1)} - v_1 t\} \\ \frac{D_1}{A_2'} = \frac{D_2}{A_2'} = -\frac{1}{k^2-2}\end{gathered} \tag{42}$$

$$\begin{gathered}\phi = B_2' \cosh \{\gamma_2 x_2 \sqrt{(1-k^{-2})}\} \sin \gamma_2 (x_1 - v_2 t) \\ H_3 = C_2' \gamma_2 x_2 \cos \gamma_2 (x_1 - v_2 t) \\ B_2' = -2C_2'\end{gathered} \tag{43}$$

These are illustrated in Figs. 7, 8 and 9.

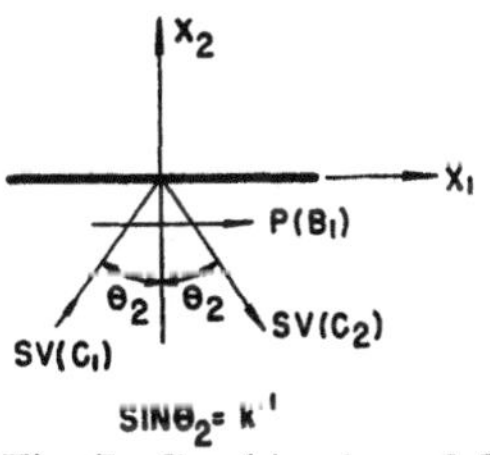

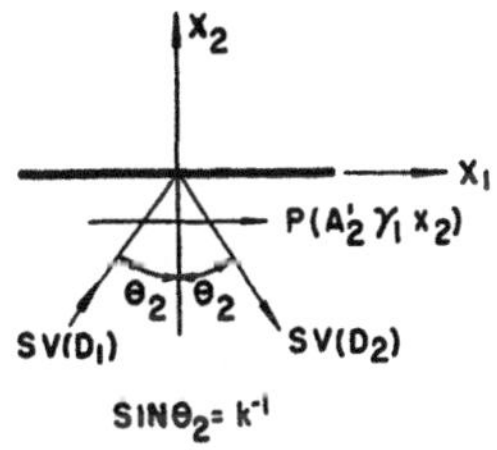

Fig. 7. Combination of Goodier–Bishop waves, equation (41).

Fig. 8. Combination of Goodier–Bishop waves, equation (42).

SV-*wave Incident at* 45°

When $\theta_2 = \pi/4$, (25) and (26) reduce to

$$B_2/C_1 = 0, \quad C_2/C_1 = 1 \tag{44}$$

Thus, if the angle of incidence is 45°, an SV-wave reflects as itself at the same angle of reflection and without change of phase, as shown in Fig. 10.

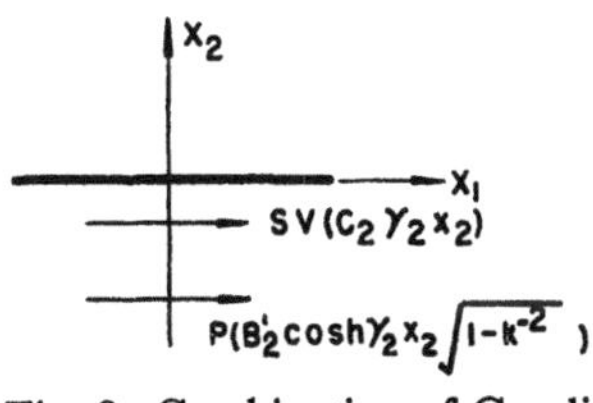

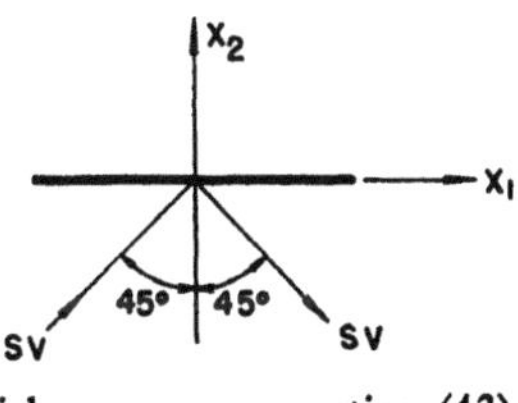

Fig. 9. Combination of Goodier–Bishop waves, equation (43).

Fig. 10. Simple reflection of an SV-wave at a traction-free boundary when the angle of incidence is 45°.

Conversion of P- *and* SV-*Waves*

When

$$\sin 2\theta_1 \sin 2\theta_2 = k^2 \cos^2 2\theta_2 \tag{45}$$

equations (20) and (21) reduce to

$$B_2/B_1 = 0, \; C_2/B_1 = -\cot 2\theta_2 \tag{46}$$

and (25) and (26) become

$$B_2/C_1 = \tan 2\theta_2, \; C_2/C_1 = 0 \tag{47}$$

Thus, for a range of combinations of angle of incidence and Poisson's ratio, given by the simultaneous solution of (45) and (17), a P-wave reflects as an SV-wave and an SV-wave reflects as a P-wave (see Fig. 11).

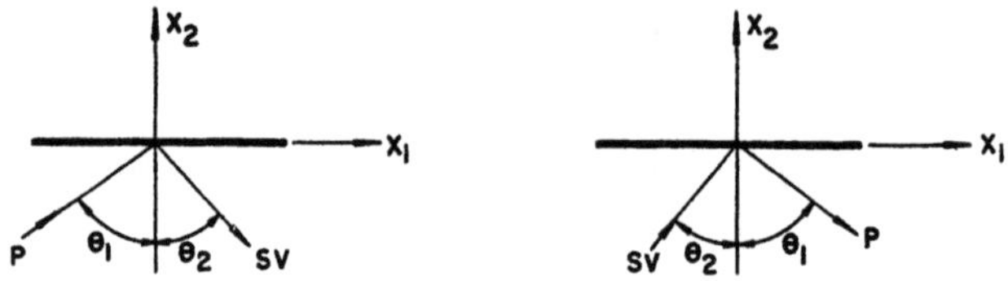

Fig. 11. Conversion of P- and SV-waves at a traction-free boundary when $\sin 2\theta_1 \cos 2\theta_2 = k^2 \cos^2 2\theta_2$, $\sin \theta_1 = k \sin \theta_2$.

Reflections of Pairs of Waves

A very general case in which P- and SV-waves reflect as themselves, at a free boundary, appears when we consider pairs of incident waves. Suppose, for example, that we take the waves with amplitudes A_1 and D_1 as the incident waves and consider the possibility of their reflection as the waves with amplitudes A_2 and D_2. Upon substituting (12) into the boundary conditions (15) and noting that

$$\lambda + 2\mu \cos^2 \theta_1 = \mu k^2 \cos 2\theta_2 \tag{48}$$

we find that the boundary conditions reduce to

$$\begin{aligned} (A_1 - A_2) \sin 2\theta_1 + (D_1 + D_2) k^2 \cos 2\theta_2 = 0 \\ (A_1 + A_2) \cos 2\theta_2 - (D_1 - D_2) \sin 2\theta_2 = 0 \end{aligned} \tag{49}$$

These are satisfied either by

$$A_1 = A_2, \; D_1 = -D_2 \tag{50}$$

$$A_1/D_1 = \tan 2\theta_2 \tag{51}$$

or by

$$A_1 = -A_2, \; D_1 = D_2 \tag{52}$$

$$A_1/D_1 = -k^2 \cos 2\theta_2 \csc 2\theta_1 \tag{53}$$

Thus, a pair of P- and SV-waves whose wave-numbers and angles of incidence satisfy (16) and (17) and whose amplitudes satisfy either (51) or (53), reflect at a free boundary as themselves with one of them undergoing a phase reversal (see Fig. 12). Similar pairs can be formed from the waves with amplitudes B_1, B_2, C_1, C_2.

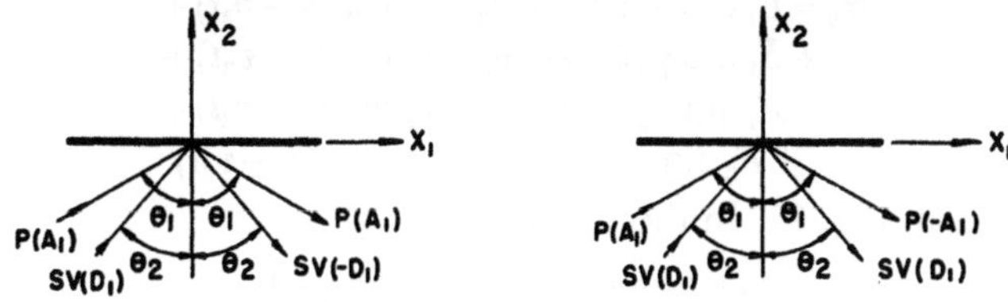

Fig. 12. Simple reflections of pairs of P- and SV-waves at a traction-free boundary.

At grazing incidence the pairs of waves reduce to the combinations of Goodier–Bishop waves given in (41)-(43). An alternative limiting process yields the Rayleigh surface-wave[4]

$$\phi = B \exp\{\xi x_2 \sqrt{(1-v^2/v_1^2)}\}\cos \xi (x_1 - vt)$$
$$H_3 = C \exp\{\xi x_2 \sqrt{(1-v^2/v_2^2)}\}\sin \xi (x_1 - vt)$$
$$\frac{C}{B} = \frac{v^2 - 2v_2^2}{2v_2\sqrt{(v_2^2 - v^2)}}$$
$$k^2\frac{v^6}{v_2^6} - 8k^2\frac{v^4}{v_2^4} + (24k^2 - 16)\frac{v^2}{v_2^2} - 16(k^2 - 1) = 0 \tag{54}$$

as illustrated in Fig. 13.

At normal incidence, direct substitution of $\theta_1 = 0$ or $\theta_2 = 0$ gives simple reflections of the type given in (33) and (34). However, a limiting process reveals other possible combinations. For example, with the pair whose amplitudes are given by (50) and (51), let

$$D_1 \tan \theta_1 = D_1' \tag{55}$$

then, in the limit,

$$\phi = A_1 \sin \gamma_1 (x_2 - v_1 t) - A_1 \sin \gamma_1 (x_2 + v_1 t)$$
$$H_3 = D_1' \gamma_1 x_1 \cos \gamma_2 (x_2 - v_2 t) - D_1' \gamma_1 x_1 \cos \gamma_2 (x_2 + v_2 t) \tag{56}$$

with

$$A_1/D_1' = 2/k \tag{57}$$

as illustrated in Fig. 14. Waves of this type differ from Goodier–Bishop waves in that the linear variation of amplitude is along the boundary rather than normal to it.

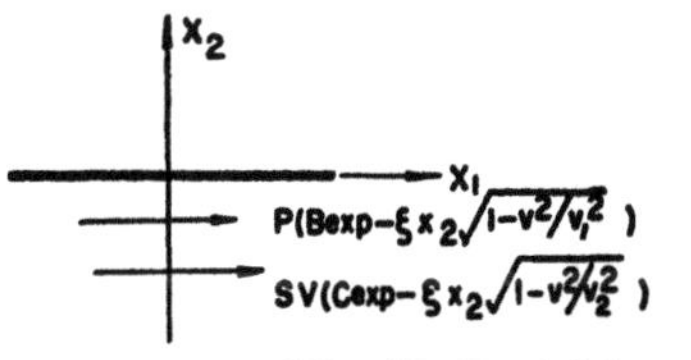

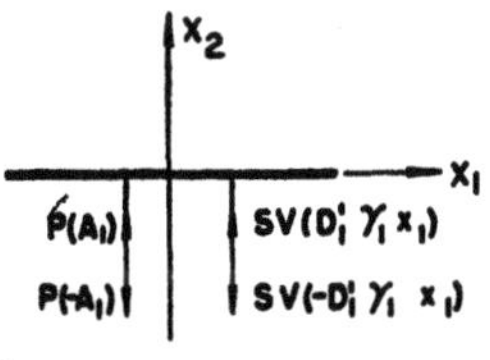

Fig. 13. Rayleigh surface waves.

Fig. 14. Example of reflection of pairs of waves at normal incidence at a traction-free boundary.

8. HALF-SPACE: *SH*-WAVES

Equivoluminal waves with wave-normal in the plane of x_1 and x_2 and displacement parallel to x_3 are given by

$$\begin{aligned} H_2 = {} & E_1 \cos \gamma_3 (x_1 \sin \theta_3 + x_2 \cos \theta_3 - v_2 t) + \\ & + E_2 \cos \gamma_3 (x_1 \sin \theta_3 - x_2 \cos \theta_3 - v_2 t) + \\ & + F_1 \sin \gamma_3 (x_1 \sin \theta_3 + x_2 \cos \theta_3 - v_2 t) + \\ & + F_2 \sin \gamma_3 (x_1 \sin \theta_3 - x_2 \cos \theta_3 - v_2 t) \end{aligned} \tag{58}$$

as illustrated in Fig. 15. In these waves

$$u_1 = u_2 = \tau_{11} = \tau_{22} = \tau_{33} = \tau_{12} = 0 \tag{59}$$

everywhere. Hence, across $x_2 = 0$, there is only one component of traction:

$$\begin{aligned} \tau_{23} = {} & \mu (E_2 - E_1) \sin \theta_3 \cos \theta_3 \cos \gamma_3 (x_1 \sin \theta_3 - v_2 t) + \\ & + \mu (F_2 - F_1) \sin \theta_3 \cos \theta_3 \sin \gamma_3 (x_1 \sin \theta_3 - v_2 t) \end{aligned} \tag{60}$$

This is zero if

$$E_1 = E_2, \; F_1 = F_2 \tag{61}$$

Thus, an *SH*-wave, incident at an appropriately oriented plane, free boundary, reflects as itself at any angle of incidence, as shown in Fig. 16.

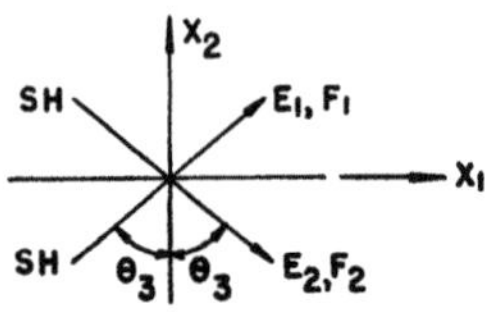

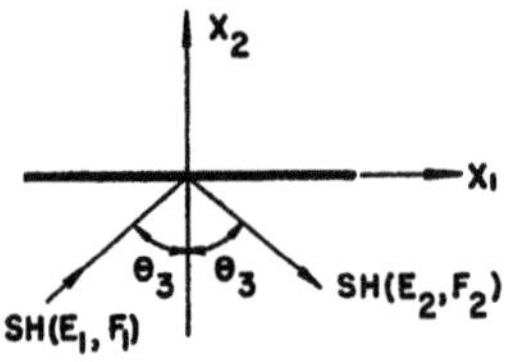

Fig. 15. *SH*-waves.

Fig. 16. Reflection of *SH*-waves at a traction-free boundary.

9. SYMMETRIC AND ANTISYMMETRIC *P*- AND *SV*-WAVES OF PLANE STRAIN

The waves given by (7) and (8) may be expressed as waves along x_1, symmetric and antisymmetric with respect to the plane $x_2 = 0$. Let

$$\begin{aligned} 2A_1 &= -2A_2 = A \\ 2B_1 &= 2B_2 = B \\ 2C_1 &= -2C_2 = C \\ 2D_1 &= 2D_2 = -D \end{aligned} \tag{62}$$

i.e. the waves are grouped in pairs as, for example, in (52). Then (7) and (8) become the real parts of

$$\phi = f(x_2) \exp i (\xi x_1 - \omega t) \tag{63}$$

$$H_3 = ih(x_2) \exp i (\xi x_1 - \omega t) \tag{64}$$

where

$$\begin{aligned} \xi &= \gamma_1 \sin \theta_1 = \gamma_2 \sin \theta_2 \\ \omega &= \gamma_1 v_1 = \gamma_2 v_2 = \xi v \end{aligned} \tag{65}$$

$$\begin{aligned} f &= A \sin \alpha x_2 + B \cos \alpha x_2 \\ h &= C \sin \beta x_2 + D \cos \beta x_2 \end{aligned} \tag{66}$$

$$\begin{aligned} \alpha &= \gamma_1 \cos \theta_1 = (\omega^2/v_1^2 - \xi^2)^{\frac{1}{2}} \\ \beta &= \gamma_2 \cos \theta_2 = (\omega^2/v_2^2 - \xi^2)^{\frac{1}{2}} \end{aligned} \tag{67}$$

The waves with amplitudes B and C are the symmetric parts of the dilatational and equivoluminal motion, respectively, the remaining parts are anti-symmetric. The waves travel in the direction x_1 with wave number ξ, phase velocity v and frequency ω; α and β are the wave numbers, along x_2, of the dilatational and equivoluminal parts, respectively. The components of displacement and stress are

$$\begin{aligned} u_1 &= i(\xi f + h') \exp i(\xi x_1 - \omega t) \\ u_2 &= (f' + \xi h) \exp i(\xi x_1 - \omega t) \end{aligned} \tag{68}$$

$$\begin{aligned} \tau_{11} &= \mu[(2\alpha^2 - \beta^2 - \xi^2)f - 2\xi h'] \exp i(\xi x_1 - \omega t) \\ \tau_{22} &= \mu[(\xi^2 - \beta^2)f + 2\xi h'] \exp i(\xi x_1 - \omega t) \\ \tau_{33} &= -\lambda(\alpha^2 + \xi^2)f \exp i(\xi x_1 - \omega t) \\ \tau_{12} &= i\mu[2\xi f' + (\xi^2 - \beta^2)h] \exp i(\xi x_1 - \omega t) \\ \tau_{23} &= \tau_{31} = u_3 = 0 \end{aligned} \tag{69}$$

where prime signifies differentiation with respect to x_2.

10. INFINITE PLATE: PLANE STRAIN, MIXED BOUNDARY CONDITIONS

We consider a plate bounded by $x_2 = \pm b$. If the boundary conditions are mixed, e.g.

$$u_2 = 0, \ \tau_{12} = 0, \text{ on } x_2 = \pm b \tag{70}$$

P- and SV-waves reflect as themselves and there is no coupling between the dilatational (f) and equivoluminal (h) parts, i.e. (70) are satisfied by (69) if

$$f'(\pm b) = 0, \ h(x_2) = 0 \tag{71}$$

or if

$$f(x_2) = 0, \ h(\pm b) = 0 \tag{72}$$

In the former case, either

$$A = 0, \ \sin \alpha b = 0; \ \alpha b = m\pi/2, \ m = 0, 2, 4 \ldots \tag{73}$$

or

$$B = 0, \ \cos \alpha b = 0; \ \alpha b = m\pi/2, \ m = 1, 3, 5 \ldots \tag{74}$$

In the latter case, either

$$D = 0, \ \sin \beta b = 0; \ \beta b = n\pi/2, \ n = 2, 4, 6 \ldots \tag{75}$$

or

$$C = 0, \ \cos \beta b = 0; \ \beta b = n\pi/2, \ n = 1, 3, 5 \ldots \tag{76}$$

The first few modes are illustrated in Fig. 17.

From (73) to (74) and (65) to (66), the phase velocities and frequencies of the dilatational part are given by

$$V^2 = v^2/v_2^2 = k^2(1 + m^2/z^2) \tag{77}$$

$$\Omega^2 = \omega^2/\bar{\omega}^2 = k^2(m^2 + z^2) \tag{78}$$

where

$$\bar{\omega} = \pi v_2/2b \tag{79}$$

$$z = x + iy = 2\xi b/\pi \tag{80}$$

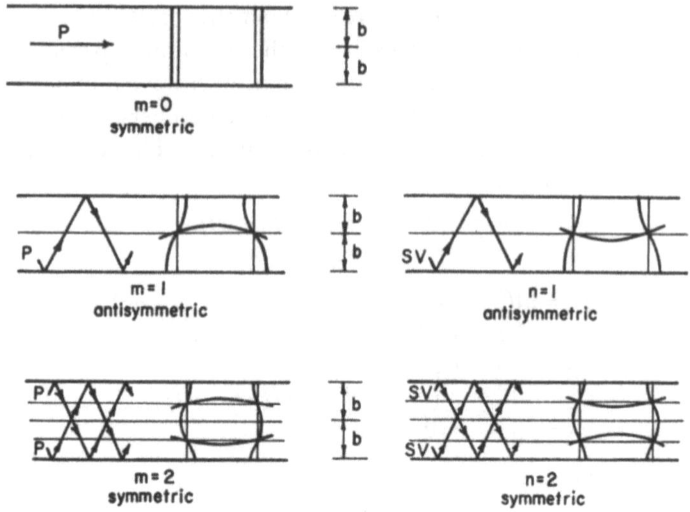

Fig. 17. Non-coupled *P*- and *SV*-waves in a plate with mixed boundary conditions $u_2=0$, $\tau_{12}=0$ on $x_2=\pm b$.

Since the frequency must be real, z, in these modes, may be real or imaginary, but not complex. Similarly, for the equivoluminal part,

$$V^2=1+n^2/z^2 \tag{81}$$

$$\Omega^2=n^2+z^2 \tag{82}$$

The phase velocities are plotted as the thin lines in Fig. 18 for real z and Poisson's ratio $\frac{1}{3}$. The frequencies are plotted as the thin lines in Fig. 19 for both real and imaginary z and Poisson's ratio 0.31. In the latter figure the vertical numerals identify the curves $m=$constant and the slanting numerals identify the curves $n=$constant.

In Figs. 18 and 19 the lines $V=k$ and $\Omega=kx$ (OD) give the phase velocity and frequencies of the fundamental dilational mode in the infinite body while the curves $m=$constant give the phase velocities and frequencies of the anharmonic overtones of the dilatational mode. Thus, the dilational mode of the infinite body has anharmonic overtones in an infinite plate with mixed boundary conditions. Similarly, the equivoluminal mode of the infinite body has anharmonic overtones in an infinite plate with mixed boundary conditions. The phase velocity of the fundamental equivoluminal mode is given by $V=1$ in Fig. 18 and its frequencies are given by $\Omega=x$ (OE) in Fig. 19. The curves $n=$constant give the phase velocities and frequencies of the anharmonic overtones of the fundamental equivoluminal mode.

11. INFINITE PLATE: PLANE STRAIN, ELASTICALLY RESTRAINED BOUNDARIES

When the boundaries of the infinite plate are elastically restrained, e.g.

$$\tau_{22}=\mp eu_2,\ \tau_{12}=0, \text{ on } x_2=\pm b \tag{83}$$

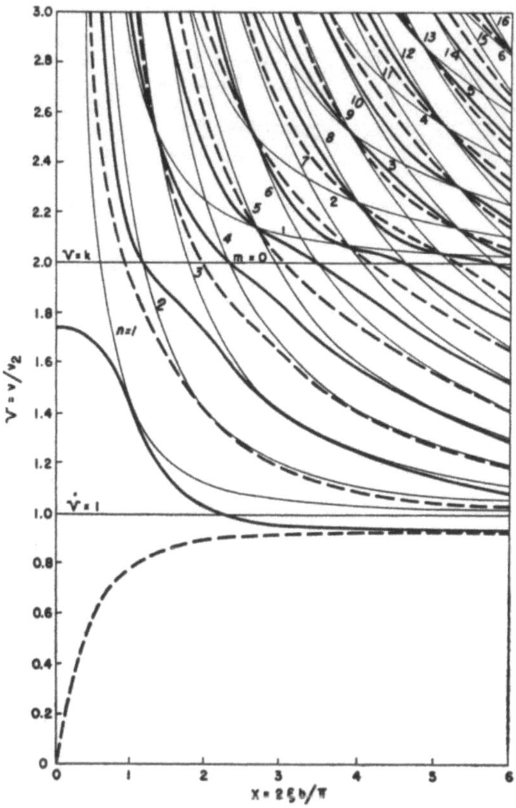

Fig. 18. Phase-velocity spectrum of an infinite plate for real wave numbers and Poisson's ratio $\frac{1}{3}$.

P- or *SV*-waves reflect as both types, so that the dilatational and equivoluminal parts, *f* and *h*, are coupled. Upon substituting (68) and (69) in (83) we obtain, for the symmetric modes:

$$\mu[B(\xi^2-\beta^2)\cos\alpha b+2C\xi\beta\cos\beta b]=e(B\alpha\sin\alpha b-C\xi\sin\beta b)$$
$$2B\xi\alpha\sin\alpha b-C(\xi^2-\beta^2)\sin\beta b=0 \tag{84}$$

and, for the antisymmetric modes,

$$\mu[A(\xi^2-\beta^2)\sin\alpha b-2D\xi\beta\sin\beta b]=-e(A\alpha\cos\alpha b+D\xi\cos\beta b)$$
$$2A\xi\alpha\cos\alpha b+D(\xi^2-\beta^2)\cos\beta b=0 \tag{85}$$

Considering the antisymmetric modes, we observe that (85) are satisfied by

$$\cos\alpha b=0,\ \cos\beta b=0$$
$$A/D=\pm 2\xi\beta/(\xi^2-\beta^2)=\mp\tan 2\theta_2 \tag{86}$$

Recalling (74) and (76), we see that (85) have some roots that are identical with some of the roots obtained for mixed boundary conditions. Since, now, ξ and ω must be the same for both the dilatational and equivoluminal parts, these roots determine pairs of wave-numbers and phase-velocities or fre-

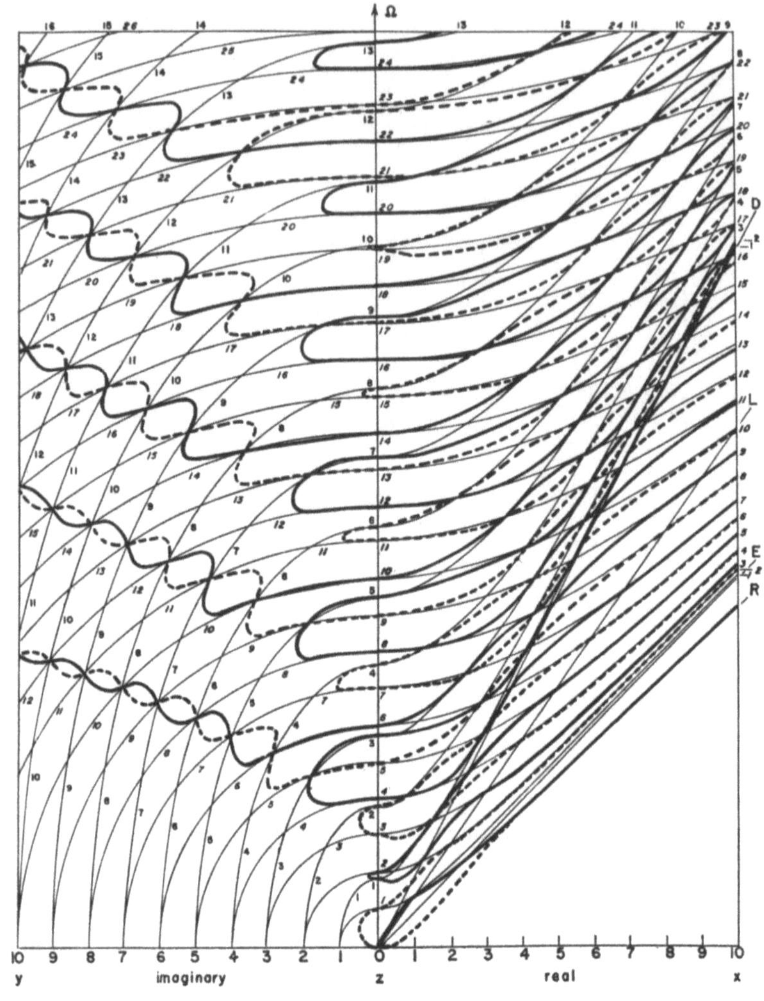

Fig. 19. Frequency spectrum of an infinite plate for real and imaginary wave-numbers ($\nu = 0.31$).

quencies which fall on the curves m = constant and n = constant, in Figs. 18 and 19, only at the intersections of curves m odd with curves n odd. If $e > 0$, $\sin \alpha b = 0$ and $\sin \beta b = 0$ are not solutions of (85) since both of (85) cannot then be satisfied by a single ratio A/D. Hence, the branches of (85) do not pass through the intersections of curves m even, n even as long as $e > 0$.

The behavior of a branch of (85), as e passes from infinity (mixed boundary conditions) to zero (traction-free boundaries) is illustrated in Fig. 20 for a portion of the spectrum. In this portion, the curves $m = 5$, $n = 13$ correspond to mixed boundary conditions ($e = \infty$). The two dashed curves marked $0 < e < \infty$ are for successive values of e, the lower curve corresponding to the

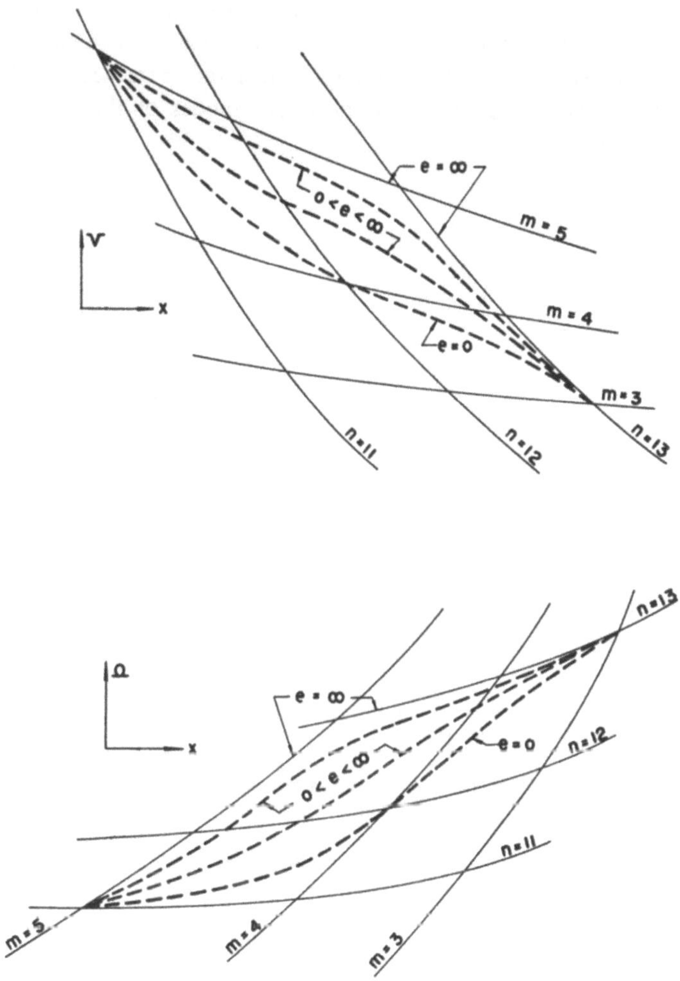

Fig. 20. Effect on the phase-velocity and frequency spectra of the antisymmetric modes of an infinite plate of the development of coupling between dilatational and equivoluminal overtones, as a result of relaxation of boundary constraint.

smaller value. When $e=0$ (traction-free boundaries), the branch of (85) passes through the intersections of m even and n even, in addition to the intersections of m odd and n odd, since

$$\sin \alpha b=0,\ \sin \beta b=0$$
$$A/D=\pm(\xi^2-\beta^2)/2\xi\alpha=\mp k^2 \cos 2\theta_2 \csc 2\theta_1 \tag{87}$$

are then solutions of (85).

The situation is very similar for the symmetric modes. It is only necessary to interchange sine and cosine, odd and even in the preceding exposition.

It may be seen that, between intersections of m even with n even and m odd with n odd, the branches of the phase-velocity (or frequency) spectrum of the infinite plate with mixed boundary conditions form upper and lower bounds of both phase-velocity (or frequency) and wave-number for the infinite plate with traction-free boundaries. This leads to the familiar terrace-like structure of the phase-velocity spectrum of the upper modes of the free plate, as illustrated in Fig. 21.

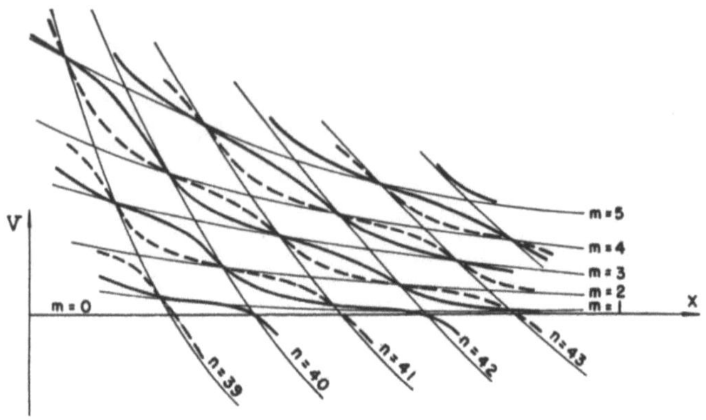

Fig. 21. Terrace-like structure of upper portion of phase-velocity spectrum of an infinite plate with traction-free faces.

12. INFINITE PLATE: PLANE STRAIN, FREE BOUNDARIES

The conditions for traction-free boundaries are obtained by setting $e=0$ in (84) and (85); and the compatibility of the resulting equations requires

$$\frac{B}{C}=\frac{2\xi\beta\cos\beta b}{(\beta^2-\xi^2)\cos\alpha b}=\frac{(\xi^2-\beta^2)\sin\beta b}{2\xi\alpha\sin\alpha b} \tag{88}$$

$$\frac{A}{D}=\frac{2\xi\beta\sin\beta b}{(\xi^2-\beta^2)\sin\alpha b}=\frac{(\beta^2-\xi^2)\cos\beta b}{2\xi\alpha\cos\alpha b} \tag{89}$$

for the symmetric and antisymmetric parts, respectively. These equations require

$$\frac{\tan\beta b}{\tan\alpha b}+\left[\frac{4\xi^2\alpha\beta}{(\xi^2-\beta^2)^2}\right]^{\pm 1}=0 \tag{90}$$

where the plus and minus signs are for the symmetric and antisymmetric parts, respectively.

Equation (90), obtained by Rayleigh[5] and Lamb[6] in 1889, is deceivingly simple in appearance and has been the subject of extensive study in the ensuing years. Examples of the branches of equation (90) are shown for real wave-numbers in Fig. 18 and for real and imaginary ξ in Fig. 19. In both figures the heavy dashed and full lines are the branches for the antisymmetric and symmetric modes, respectively.

Lamb[7], in 1917, studied the lowest symmetric and antisymmetric modes. He also identified the cut-off modes, the Lamé[8] modes (p. 170) and certain aspects of the high-frequency spectrum. Bounds α = constant, β = constant were employed by Holden[9], in 1951, to construct a portion of the spectrum of symmetric modes for real wave-numbers and this method was considerably elaborated and extended to include the antisymmetric modes and imaginary phase-velocities by Onoe[10]. Analogous bounds were employed by Mindlin[11], in 1951, to construct the branches of a similar transcendental equation, which appears in the theory of vibrations of crystal plates, and this method was applied later[12] to equation (90). The intricate behavior of the branches in the neighborhood of zero wave-number and the important role of Poisson's ratio were studied by Onoe[10], Mindlin[12] and Mindlin and Onoe[13]. The imaginary loop that connects the second and third symmetric branches, in a certain range of Poisson's ratio, had been brought to light previously in computations by Aggarwal and Shaw[14] in 1954 and a computation of the family of imaginary branches has been performed by Lyon[15]. The existence of modes at the cut-off frequencies, other than the modes identified by Lamb, was established by Mindlin[16]. The phenomenon of the occasional appearance of phase and group velocities of opposite sign was encountered and explained by Tolstoy and Usdin[17] and the criteria for the existence of this anomaly in any branch were given by Mindlin[12]. The possibility of resolving the waves in the plate into pairs of reflecting *P*- and *SV*-waves was noticed by Harrison[18]. Complex wave-numbers and phase-velocities associated with real frequencies, were found by Mindlin and Medick[19] in an approximate theory of vibrations of plates. The existence and important aspects of the behavior of the complex branches of equation (90) were established by Onoe[13] (pp. 14–17). At this writing, the understanding of the roots of equation (90) appears to be reasonably complete, except for the behavior of the higher complex branches when both the real and imaginary parts of the wave-number are large.

In Holden's procedure, the spectrum is constructed initially in α, β-space. This has an advantage of simplicity in that the bounds are intersecting families of straight lines. On the other hand, in Ω, z-space, as illustrated in Fig. 19, the bounds are intersecting arcs of circles, ellipses and hyperbolas. However, this complication is offset by the fact that the construction in the framework of these co-ordinates gives an immediate view of frequency, phase velocity and group velocity. As illustrated in Fig. 22, the ordinate gives the frequency, the slope of the straight line connecting a point on a branch with the origin gives the phase velocity and the slope of the branch gives the group velocity. In addition, by a theorem due to Onoe[13], the maxima and minima of the real and imaginary branches are the points of intersection with the complex branches.

The real and imaginary branches of equation (90) are sketched on the grid of bounds, in Fig. 19, with the aid of formulas that have been obtained for the co-ordinates and slopes of the branches at their intersections with the bounds; the ordinates, slopes and curvatures of the branches at $z=0$; and

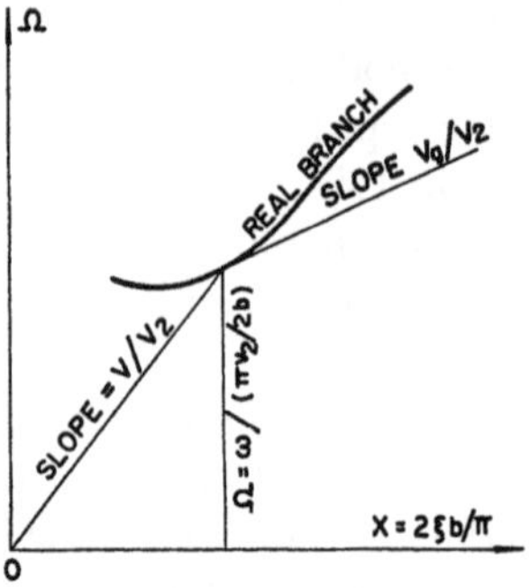

Fig. 22. Frequency, phase-velocity and group velocity.

asymptotic behavior at large wave-number. This information, supplemented by a theorem of Onoe's[10] on saddle points of the imaginary branches, suffices for the sketching of the spectrum of real and imaginary roots when Poisson's ratio is greater than 0.263. For smaller values of Poisson's ratio, i.e. when all three roots of the last of equations (54) are real, there are two real radii along which there are additional intersections of the symmetric and antisymmetric branches.

Limitation of space precludes a detailed discussion of the spectrum, but it is pertinent to the theme of this paper to point out some of the salient features, especially those that are readily related to the reflections of P- and SV-waves mentioned in the earlier sections.

From (62), (88) and (89) we see that the Rayleigh–Lamb modes are composed of pairs of P and SV-waves reflecting between $x_2 = \pm b$, as shown by Harrison[18].

At normal incidence, which, by (17) corresponds to $\xi = 0$ (i.e. $z = 0$), the two types of wave uncouple (see equations (33) and (34) and Fig. 5) and form the simple thickness–stretch and thickness–shear modes[7] whose frequencies (the cut-off frequencies) are given by

$$\Omega = kp, \ \Omega = q \tag{91}$$

respectively, where

$$p = \begin{cases} 1, 3, 5\ldots, & \text{symmetric modes} \\ 2, 4, 6\ldots, & \text{antisymmetric modes} \end{cases}$$

$$q = \begin{cases} 1, 3, 5\ldots, & \text{antisymmetric modes} \\ 2, 4, 6\ldots, & \text{symmetric modes} \end{cases}$$

In Fig. 19 the integers q are shown on the right side of the vertical centerline and the integers p are shown on the left side. It is to be noted that the cut-off frequencies $\Omega = kp$ depend on Poisson's ratio but the others do not. At the same frequencies and wave-numbers, there are other possible modes[16] composed of waves of the type given in (56) and illustrated in Fig. 14.

The slopes of the branches, at $x = 0$ and Ω given by (91), are zero (and hence the group velocities are zero) except when two cut-off modes of the same symmetry have the same frequency[12]. Such coincidences occur when Poisson's ratio satisfies the condition

$$k = q/p \tag{92}$$

The curvatures ($\partial^2\Omega/\partial x^2$) of the branches at $x=0$ may be positive or negative, according to a simple criterion[12]. Positive or negative curvature signifies phase and group velocities of the same or opposite sign, respectively. In the latter case, the branch has a real minimum (zero group velocity). Such minima are distributed sparsely among the branches and the distribution depends on Poisson's ratio. In Fig. 19, which is drawn for $\nu=0.31$, they occur in the second symmetric branch and in the third and fifteenth antisymmetric branches.

In sketching the branches on the curvilinear grid of bounds, it is convenient to start at zero wave-number. The ratio of the curvatures of branch and bound, at $z=0$, determines whether the branch starts out above or below the bound. Thereafter, a branch is confined between bounds, crossing them only at successive intersections of bounds m even with n even, and m odd with n odd. At these intersections, the corresponding modes satisfy both mixed and traction-free boundary conditions and the mode shapes across the thickness are characterized by an even number, $2m$ and $2n$, of quarter wavelengths of the dilatational and equivoluminal parts of the displacement, respectively. The wave nature of these modes is readily identified by comparing (86) and (87) with (51) and (53). The antisymmetric modes are predominantly equivoluminal ($|A/D|<|$) at the intersections m even, n even and predominantly dilatational ($|A/D|>|$) at the intersections m odd, n odd. The converse holds for the symmetric modes. Between intersections there is a continuous transition from one type of deformation to the other and from one even number of quarter wavelengths to the next.

Continuing out along the branches, with increasing real z, every branch, except the lowest symmetric and antisymmetric ones, eventually crosses the line $\Omega=kx$ (OD in Fig. 19) at which the phase-velocity is the velocity of dilatational waves in an infinite medium. At these intersections, the symmetric and antisymmetric modes are Goodier–Bishop waves of the types given in equations (41) and (42), respectively. The line OD is the bound $m=0$ and hence the symmetric branches cross it at its intersections with the bounds n even. The slopes of all the symmetric branches are the same at these points and are less than the slope of OD if $\nu\neq 0$. Hence, the group velocity is less than the velocity of dilational waves, the ratio for all the modes being

$$\frac{v_g}{v_1}=\frac{4(k^2-1)+(k^2-2)^2}{4(k^2-1)+k^2(k^2-2)^2} \tag{93}$$

The line $m=0$ is not a bound for the antisymmetric modes and the branches do not cross at its intersections with the bounds n odd. However, after the first few modes, the intersections of the bounds and branches with $m=0$ are very close together and the group velocity, from mode to mode, is nearly constant and less than v_1 for $\nu\neq 0$.

With further increase of x, all the branches, except the lowest antisymmetric one, cross the line $\Omega = x\sqrt{2}$ (OL in Fig. 19) at its intersections with the bounds $n = 1, 2, 3\ldots$ to which the branches are tangent. At these points the mode is purely equivoluminal and is composed of SV-waves reflecting at 45° angles of incidence and reflection, as shown in Fig. 10. Also, all planes $x_1 = 2b/n$, $x_2 = 2b/n$, $x_3 = $ constant are free of traction, giving rise to the Lamé[8] (p. 170) modes of vibration of rectangular parallelepipeds. The points $x = n$, $\Omega = n\sqrt{2}$ (the Lamé modes) and the points $x = 0$, $\Omega = q$ (the thickness-shear modes at cut-off) are the only points corresponding to purely equivoluminal modes for real and imaginary z. Hence they are the only points on the real and imaginary branches which remain fixed as Poisson's ratio is varied.

After crossing $\Omega = x\sqrt{2}$, all the real branches, except the lowest symmetric and antisymmetric ones, approach the asymptote $\Omega = x$, OE in Fig. 19, i.e. the limiting phase and group velocities of all the higher modes are the equivoluminal velocity. The only branch that intersects $\Omega = x$ is the lowest symmetric one. At that intersection the mode shape is a Goodier–Bishop wave of the type given in equation (43). Thereafter, the lowest symmetric branch, and also the lowest antisymmetric branch, approach the slope corresponding to the phase velocity of Rayleigh surface waves, as given by the lowest real root of Rayleigh's cubic equation (54). When $\nu < 0.263$, this equation has three real roots. The two larger roots correspond to two phase velocities, i.e. to two radial lines in Fig. 19. At intervals along these lines, there are additional intersections of the symmetric and antisymmetric branches. The waves corresponding to these points are of the type, illustrated by equations (45)–(47) and Fig. 11, in which there is complete conversion of P to SV and SV to P at each reflection.

The frequency must be real and positive, but the wave-numbers (ξ along the plate; α and β across the thickness) may be real, imaginary or complex. For real ξ, both the dilatational wave-number, α, and the equivoluminal wave-number, β, are real if $\Omega > kx$; α is imaginary and β is real if $x < \Omega < kx$ and both α and β are imaginary if $\Omega < x$. When ξ is imaginary, both α and β are real and, when ξ is complex, both α and β are complex. In rectangular co-ordinates, the mode shapes are described by trigonometric functions when the wave-number is real; hyperbolic or real exponential functions when the wave-number is imaginary; and products of trigonometric and hyperbolic or real exponential functions when the wave-numbers are complex. Imaginary wave-numbers are well known in plate theory since they appear in the elementary theory of flexure of plates; but complex wave-numbers have been encountered only recently. In an infinite plate, complex wave-numbers occur in problems of transient waves or non-uniform loading. In bounded plates, complex wave-numbers are associated with edge vibrations. Such vibrations have been observed in circular disks in experiments by Shaw[20] and have been identified mathematically with the complex wave-numbers by Gazis and Mindlin[21]. Onoe[22] has found the edge vibrations both mathematically and experimentally in rectangular plates.

Since the wave-number ratio, z, occurs in the frequency equation only to the second power, the positive or negative values of the roots are equally valid. In Fig. 19 the positive values of real z are shown except in the portions of the branches where the slope is apparently negative. The first three symmetric branches are shown in more detail in Fig. 23, exhibiting positive

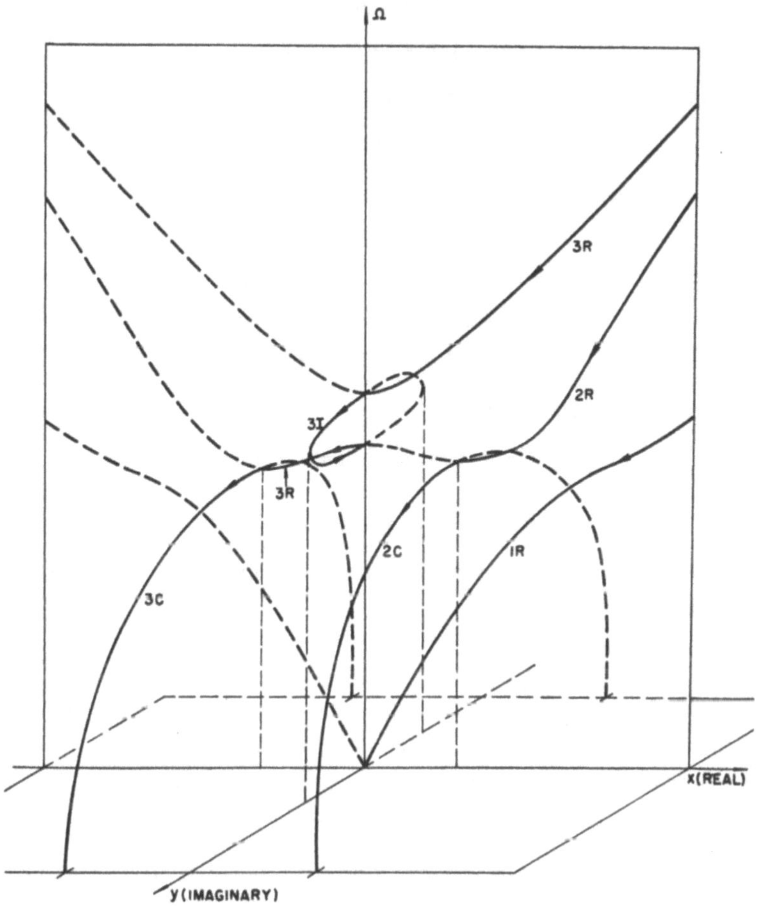

Fig. 23. First three branches (symmetric modes) showing positive (full lines) and negative (dashed lines) real (R), imaginary (I) and complex (C) parts.

and negative real, imaginary and complex roots. Here the positive roots are grouped together as full lines and the negative roots are drawn as dashed lines. It is apparent that the portion of the second branch in Fig. 19, which appears to have positive phase-velocity and negative group velocity is actually a portion of the third positive branch with positive group velocity and negative phase velocity.

13. INFINITE PLATE: *SH*-WAVES

If we set

$$2E_1 = 2E_2 = E$$
$$2F_1 = -2F_2 = F$$
$$\xi = \gamma_3 \sin \theta_3$$
$$\beta = \gamma_3 \cos \theta_3$$
$$\omega = \gamma_3 v_2 = v_2(\xi^2 + \beta^2)^{1/2} \tag{94}$$

in quation (58), we obtain the real part of

$$H_2 = (E \cos \beta x_2 + F \sin \beta x_2) \exp i\,(\xi x_1 - \omega t) \tag{95}$$

from which, omitting the exponential factor,

$$u_1 = u_2 = \tau_{11} = \tau_{22} = \tau_{33} = \tau_{12} = 0$$
$$u_3 = i\xi\,(E \cos \beta x_2 + F \sin \beta x_2)$$
$$\tau_{23} = -i\mu\xi\rho\,(E \sin \beta x_2 - F \cos \beta x_2)$$
$$\tau_{31} = -\mu\xi^2\,(E \cos \beta x_2 + F \sin \beta x_2) \tag{96}$$

The tractions are zero across $x_2 = \pm b$ if

$$F = 0,\ \beta b = n\pi/2,\ n = 0, 2, 4 \ldots \tag{97}$$

or

$$E = 0,\ \beta b = n\pi/2,\ n = 1, 3, 5 \ldots \tag{98}$$

for the symmetric and antisymmetric modes, respectively. Then, from the last of (94)

$$\Omega^2 = n^2 + z^2 \tag{99}$$

which is the same as (82) except that $n = 0$ is now permitted.

These modes are special cases of Love's[23] transverse waves in a superficial layer, i.e. they are *SH*-waves reflecting between the free faces $x_2 = \pm b$ at angles of incidence and reflection

$$\theta_3 = \tan^{-1}(z/n) \tag{100}$$

It may be seen that the frequency spectrum is given by the curves $n =$ constant, in Fig. 19, with the addition of the line $\Omega = x$, corresponding to $n = 0$.

14. BOUNDED PLATE

Rayleigh–Lamb waves in an infinite plate are composed of pairs of *P*- and *SV*-waves reflecting at the free faces of the plate. The angle of incidence and reflection (θ_1) of the *P*-waves is related to the corresponding angle (θ_2) of the *SV*-waves by

$$\sin \theta_1 = k \sin \theta_2$$

Since $k > 2/\sqrt{3}$, this requires $\theta_1 > \theta_2$. Consider now the possibility of reflection of a Rayleigh–Lamb wave at traction-free sides, say $x_1 = \pm a$, of the plate.

There the angles of incidence of the *P*- and *SV*-waves are the complements of θ_1 and θ_2, say, θ_1' and θ_2'. Hence $\theta_1' < \theta_2'$, i.e. the angle of incidence of the *P*-wave is less than that of the *SV*-wave. Accordingly, simple reflections are, in general, not possible.

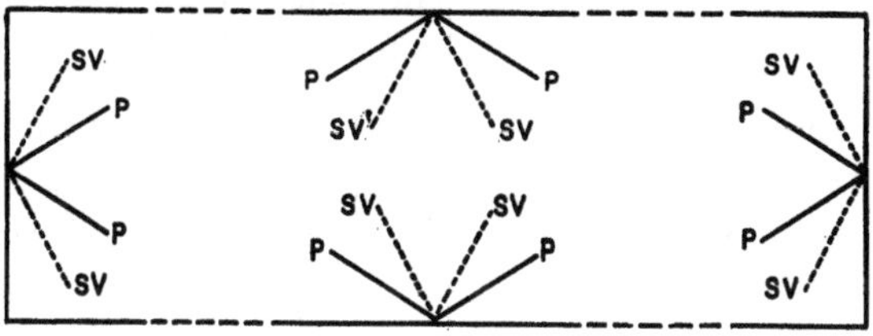

Fig. 24. Angles of incidence and reflection of pairs of *P*- and *SV*-waves at the faces and sides of a plate.

Exceptions: SV- *and* P-*Waves*

There are a few exceptional cases, involving only single incident and reflected waves, in which simple reflections are possible both at traction-free faces and traction-free sides. The simplest is that of the *SV*-wave incident at 45°, as illustrated in Fig. 10. At a side x_1 = constant the angle of incidence is again 45° so that the reflection is again a simple one. Such a wave can reflect successively at the faces of a square section to form a closed path. Two such waves, traveling in opposite directions, form Lamé's mode of vibration. The corresponding modes in a disk were found by Lamb[7] (p. 122). Another example is a special case of conversion of *P*- and *SV*-waves, see Fig. 11. Suppose, in a plate with Poisson's ratio $\frac{1}{4}$, a *P*-wave is incident at a traction-free face at an angle of incidence of 60°. In accordance with equation (45), it will be reflected as an *SV*-wave at an angle of reflection of 30°. At a side which makes an interior angle of 60° with the face, the *SV*-wave will be incident at an angle of 30° and hence will reflect as a *P*-wave at an angle of reflection of 60°. Proceeding around the sides of a 60°-120° rhombus, the wave alternates between *P* and *SV* to form a closed path. A pair of these, traveling in opposite directions, forms a vibration which is a cut-off mode of a prism of rhombic section with free sides[24]. Another such vibration has been found in a prism whose section is an equilateral hexagon[24].

Exceptions: SH-*Waves*

SH-waves, of course, reflect as themselves at any plane face parallel to the displacement. This circumstance leads, for example, to simple forms of axial-shear cut-off modes in prisms and cylinders of a variety of cross-sections[25] and to simple torsional modes in a circular disk[26].

15. THREE-DIMENSIONAL WAVES IN A BOUNDED PLATE

Another type of simple reflection at the traction-free sides of a plate involves three waves: *P*, *SV* and *SH*.

It was mentioned by Rayleigh[5] that the restriction of the solution described in Section 12 to plane strain can be removed easily. Consider, for example, the vibrations given by

$$\begin{aligned}\phi &= B\zeta \cos \xi x_1 \cos \alpha x_2 \cos \zeta x_3\, e^{i\omega t}\\ H_1 &= C\zeta \cos \xi x_1 \sin \beta x_2 \sin \zeta x_3\, e^{i\omega t}\\ H_3 &= -C\xi \sin \xi x_1 \sin \beta x_2 \cos \zeta x_3\, e^{i\omega t}\end{aligned} \tag{101}$$

These represent modes symmetric with respect to the three co-ordinate planes. The remaining seven symmetries may be obtained by appropriate interchanges of sines and cosines of like argument. The functions in (101) satisfy the equations of motion if

$$\begin{aligned}\alpha^2+\xi^2+\zeta^2 &= \omega^2/v_1^{\,2}\\ \beta^2+\xi^2+\zeta^2 &= \omega^2/v_2^{\,2}\end{aligned} \tag{102}$$

and the faces $x_2 = \pm b$ are traction free if

$$\frac{B}{C}=\frac{2(\xi^2+\zeta^2)\beta \cos \beta b}{(\beta^2-\xi^2-\zeta^2)\zeta \cos \alpha b}=\frac{(\xi^2+\zeta^2-\beta^2)\sin \beta b}{2\alpha\zeta \sin \alpha b} \tag{103}$$

whence

$$\frac{\tan \beta b}{\tan \alpha b}+\frac{4(\xi^2+\zeta^2)\alpha\beta}{(\xi^2+\zeta^2-\beta^2)^2}=0 \tag{104}$$

Comparing (104) with (90), we see that the spectra in Figs. 18 and 19 apply to the three dimensional case if, in the abscissas, ξ^2 is replaced by $\xi^2+\zeta^2$. A similar result for cylindrical waves in a plate was shown by Sezawa[27] and Goodman[28].

We may also have three-dimensional *SH*-modes

$$\begin{aligned}H_2 = {}& E'\zeta \sin \xi_1 x_1 \cos \beta_1 x_2 \sin \zeta x_3\, e^{i\omega t}+\\ &+E''\zeta \sin \xi_2 x_1 \cos \beta_2 x_2 \sin \zeta x_3\, e^{i\omega t}\end{aligned} \tag{105}$$

which satisfy the equations of motion if

$$\xi_1^{\,2}+\beta_1^{\,2}+\zeta^2=\xi_2^{\,2}+\beta_2^{\,2}+\zeta^2=\omega^2/v_2^{\,2} \tag{106}$$

and leave the faces $x_2 = \pm b$ free of traction if

$$\beta_1 = n_1\pi/2b,\ \beta_2 = n_2\pi/2b \tag{107}$$

where n_1 and n_2 are even integers, i.e. there are even numbers of quarter wavelengths through the thickness. The frequency spectra of these modes are given by the curves n even, in Fig. 19, if ξ^2 is replaced by $\xi_i^{\,2}+\zeta^2$, $i=1, 2$, in the abscissas.

Consider, now, the intersections of bounds m even, n even, in the real portion of Fig. 19. The Rayleigh–Lamb modes corresponding to these points also have even numbers of quarter wavelengths through the thickness for both the dilatational and equivoluminal parts, i.e.

$$\alpha = m\pi/2b,\ \beta = n\pi/2b \tag{108}$$

Hence the tractions on sides $x_1 = \pm a$, produced by the Rayleigh–Lamb modes, can be annihilated by the tractions produced by the two *SH*-modes if appropriate values of their wave-numbers and amplitudes are chosen. In particular:

$$n_1 = m, \; n_2 = n$$

$$\xi_2 = \xi = p\pi/2a, \; \xi_1 = q\pi/2a \tag{109}$$

$$\frac{B}{C} = \frac{2\beta(\xi^2+\zeta^2)(-1)^{(n-m)/2}}{\zeta(\beta^2-\xi^2-\zeta^2)}$$

$$\frac{E'}{B} = \frac{\zeta^2+\beta_1{}^2-\xi_1{}^2}{2\xi_1\zeta(-1)^{(q-p)/2}}$$

$$\frac{E'}{B} = \frac{\xi_2{}^2+\zeta^2-\beta_2{}^2}{2\zeta(\xi_2{}^2+\zeta^2)(-1)^{(n-m)/2}} \tag{110}$$

where p and q are even integers. In view of (102), and (106)-(109), the frequencies are given by

$$\Omega^2 = k^2(n^2-m^2)/(k^2-1) \tag{111}$$

and the solutions are valid only for

$$a^2/b^2 = (q^2-p^2)/(n^2-m^2) \tag{112}$$

These are solutions, then, for vibrations or waves in a rectangular bar, with traction-free faces, for certain ratios of width to depth[41].

16. OVERTONES OF A BOUNDED PLATE

The few cases mentioned in the two preceding sections appear to be the only ones of simple reflections at the sides of a plate that have been found. In general, a single Rayleigh–Lamb wave reflects as an infinity of waves so that exact solutions of the elasticity equations for bounded plates are generally not expressible in closed form, except for mixed boundary conditions which, unfortunately, are not readily realizable in practice. Some understanding of the complications involved in the phenomenon can be gained by examining the case of reflection of Rayleigh–Lamb waves at a pair of plane sides with mixed boundary conditions. In this case the waves reflect as themselves, so that simple modes of vibration can be found for a plate with traction-free faces $x_2 = \pm b$ and mixed boundary conditions at $x_1 = \pm a$.

From (63) and (64) we may extract, for example, the antisymmetric vibrational modes

$$\phi = A \sin \alpha x_2 \sin \xi x_1 \, e^{i\omega t}$$

$$H_3 = D \cos \beta x_2 \cos \xi x_1 \, e^{i\omega t} \tag{113}$$

The faces $x_2 = \pm b$ are traction-free if

$$\frac{A}{D} = \frac{2\xi\beta \sin \beta b}{(\xi^2-\beta^2)\sin \alpha b}$$

$$\frac{\tan \beta b}{\tan \alpha b} + \frac{(\xi^2-\beta^2)^2}{4\xi^2\alpha\beta} = 0 \tag{114}$$

and the frequency spectrum is given by the dashed lines in Fig. 19. For convenience, the frequency is plotted against the reciprocal of z in Fig. 25,

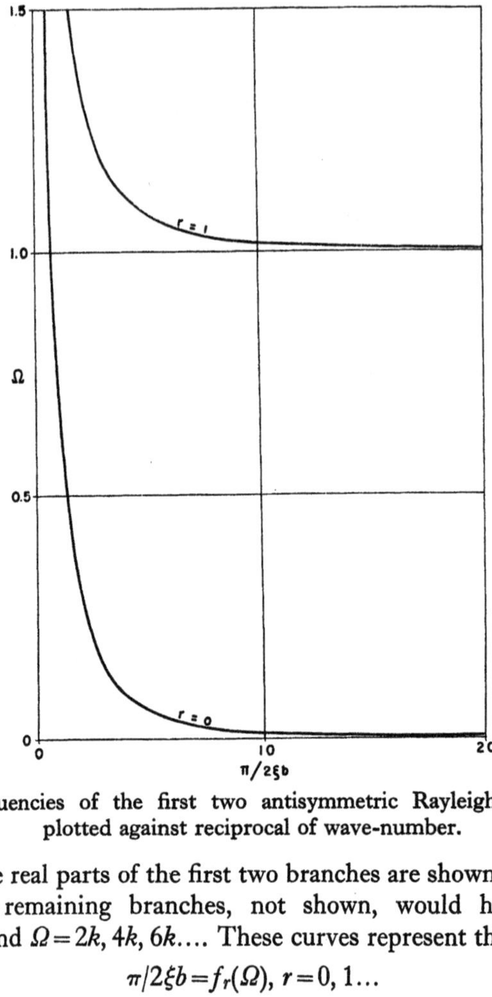

Fig. 25. Frequencies of the first two antisymmetric Rayleigh–Lamb modes plotted against reciprocal of wave-number.

where only the real parts of the first two branches are shown and are labeled $r=0, 1$. The remaining branches, not shown, would have asymptotes $\Omega=3, 5, 7\ldots$ and $\Omega=2k, 4k, 6k\ldots$. These curves represent the functions

$$\pi/2\xi b=f_r(\Omega),\ r=0, 1\ldots \tag{115}$$

The tangential component of displacement and the normal component of traction on planes $x_1=$constant are, from (113),

$$\begin{aligned} u_2&=(A\alpha \cos \alpha x_2+D\xi \cos \beta x_2) \sin \xi x_1\, e^{i\omega t} \\ \tau_{11}&=-\mu[A(\beta^2+\xi^2-2\alpha^2) \sin \alpha x_2-2D\xi\beta \sin \beta x_2] \sin \xi x_1\, e^{i\omega t} \end{aligned} \tag{116}$$

These vanish on $x_1=\pm a$ if

$$\xi=s\pi/2a,\ s=0, 2, 4\ldots \tag{117}$$

Equations (113) are for modes antisymmetric with respect to both the thickness and width. Modes symmetric with respect to the width are obtained by interchanging $\sin \xi x_1$ and $\cos \xi x_1$. Then

$$\xi=s\pi/2a,\ s=1, 3, 5\ldots \tag{118}$$

From (115), (117) and (118), the frequency spectrum for flexural vibrations of the bounded plate is given by

$$a/b = sf_r(\Omega),\ r=0, 1, 2\ldots,\ s=0, 1, 2\ldots \tag{119}$$

This is the relation between the width–thickness ratio of the plate and its frequencies of flexural vibration. Similar results for extensional vibrations are obtained by interchanging the trigonometric functions of x_2 in (113).

The curves for the fundamentals and their successive anharmonic overtones, $s=1, 2, 3\ldots$, in the plate with free faces and mixed side conditions, are obtained from the branches, $r=0, 1\ldots$ of the infinite plate with free faces by multiplying the abscissas, $\pi/2\xi b$, of each of the branches of the latter by the integers $s=1, 2, 3\ldots$. In addition, $r=1$, $s=0$ ($\Omega=1$) is the lowest antisymmetric thickness-shear mode of the infinite plate. The results are illustrated in Fig. 26. The process of obtaining the anharmonic overtones of

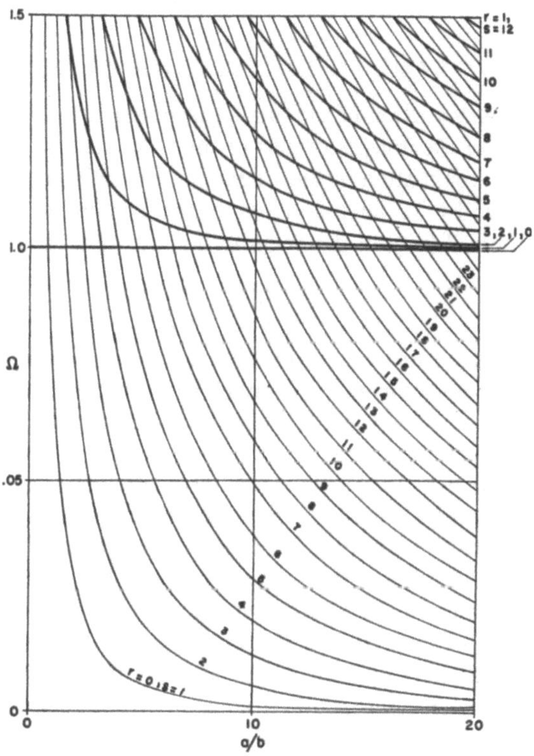

Fig. 26. Anharmonic overtones of the first two antisymmetric Rayleigh–Lamb modes in a plate with traction-free faces $x_2 = \pm b$ and mixed boundary conditions on sides $x_1 = \pm a$.

the bounded plate from the Rayleigh–Lamb spectrum is analogous to the process, in Section 10, of obtaining the anharmonic overtones of an infinite plate with mixed boundary conditions on its faces from the dilatational mode and the equivoluminal mode of an infinite body. The latter two modes are here replaced by the infinity of Rayleigh–Lamb modes, each of which has an infinity of anharmonic overtones. The frequency spectrum shown in Fig. 26 is related to the frequency spectrum of a bounded plate with traction-free faces and sides in somewhat the same way that the phase-velocity spectrum of an infinite plate with mixed conditions on its faces is related to the phase-velocity spectrum of an infinite plate with traction-free faces, as shown in Fig. 18; i.e. the frequency spectrum of the bounded plate will exhibit a terrace-like structure similar to Fig. 21. However, the curves of the type shown in Fig. 26 are not exact bounds because, first, there are more than two families of them and, secondly, they have complex parts in addition to real and imaginary parts. Nevertheless, useful approximate theories and solutions for vibrations of plates, over wider and wider frequency ranges, have been established by limiting the number of branches considered to the first one, two or more.

17. APPROXIMATE EQUATIONS OF MOTION

The low-frequency portion of the first antisymmetric branch and the low-frequency portion of the imaginary part of the second antisymmetric branch appear to good approximation in the Lagrange–Germain equation of flexure of plates. As shown by Rayleigh and Lamb, the asymptotic behavior of the antisymmetric form of (90), in the neighborhood of zero frequency, is

$$\Omega = \pi z^2 \sqrt{\left\{\frac{1}{3}\left(1-\frac{1}{k^2}\right)\right\}} \tag{120}$$

where z is either real or imaginary. This is exactly the spectrum given by the Lagrange–Germain equation and accounts for the success of that equation in the low frequency, long wavelength range, i.e. $0 < \Omega < 0.1$ in Fig. 19.

The range $0 < \Omega < 1.5$ is accommodated in equations of flexural motion by Uflyand[29] and Mindlin[30]. In the latter, the introduction of a shear correction factor results in the exact reproduction of the cut-off frequency ($\Omega = 1$) of the second antisymmetric mode and the exact reproduction of the spectrum ($\Omega^2 = 1 + z^2$) of the lowest antisymmetric *SH*-mode. These equations involve three boundary conditions at each point on the edge of the plate, as did previous equations of flexural equilibrium of plates, of the same order of approximation, by E. Reissner[31, 32]. It has been shown[30] how the suppression of the transverse shear deformation results in the reduction of the three boundary conditions to the two conditions of the Lagrange–Germain equations.

There have been numerous applications of the higher order equations of flexure. In the case of plane strain, the equations reduce to the same form as the Timoshenko[33] beam equations. Plane strain in a strip of width a and thickness b, with simply supported edges, results in the frequency spectrum shown in Fig. 26, since the simply supported edge corresponds to mixed boundary conditions and only two modes of the infinite plate are considered.

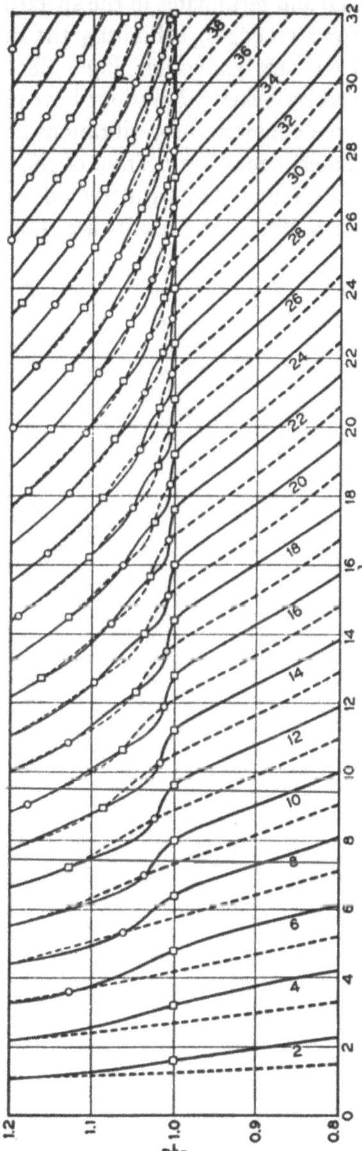

Fig. 27. Frequencies of coupled flexural modes of a strip, with free edges, vibrating in a state of plane strain[11].

As was pointed out by E. Sternberg*, the continuous passage from a simply supported edge to a free edge can be effected by introducing an elastically supported edge. As the stiffness of the elastic support diminishes from infinity to zero, the upper part of the frequency spectrum passes through the stages illustrated in Fig. 20 and terminates in the spectrum shown in Fig. 27. Parts of the spectrum for free edges, exhibiting the terrace-like structure, which occurs above $\Omega = 1$, are shown in Figs. 28 and 29, along with experimental data obtained by Sykes[34]. Solutions have also been obtained for reflection of waves at a free edge[35] and for vibrations of rectangular plates1, and circular disks[37-39]. Experimental verification of the frequencies of axially symmetric vibrations of a circular disk with free edges, in the range $\Omega <$ [36] has been obtained by Sharma[40] and the results are shown in Fig. 30.

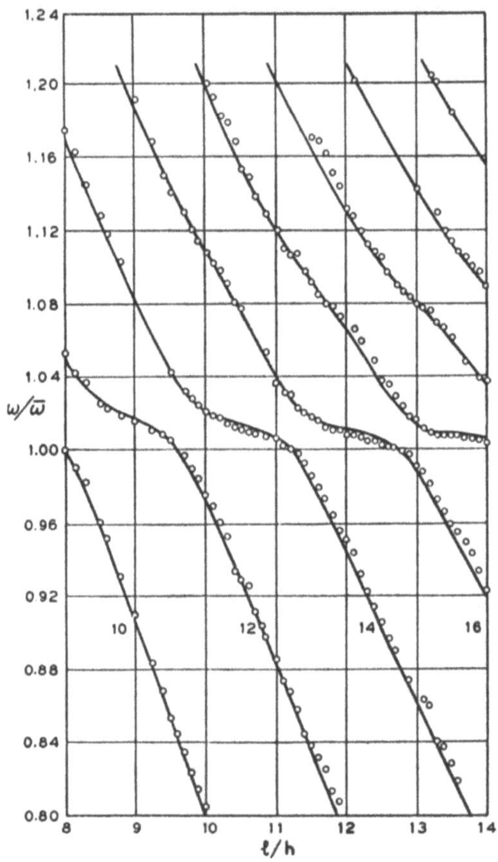

Fig. 28. Frequencies of flexural vibration of a strip with free edges. Experimental data by Sykes[34].

* Private communication.

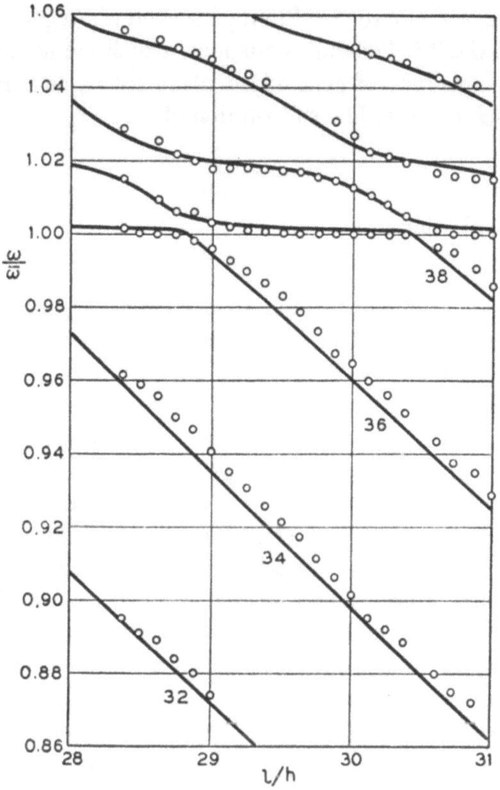

Fig. 29. Frequencies of flexural vibration of a strip with free edges. Experimental data by Sykes[34].

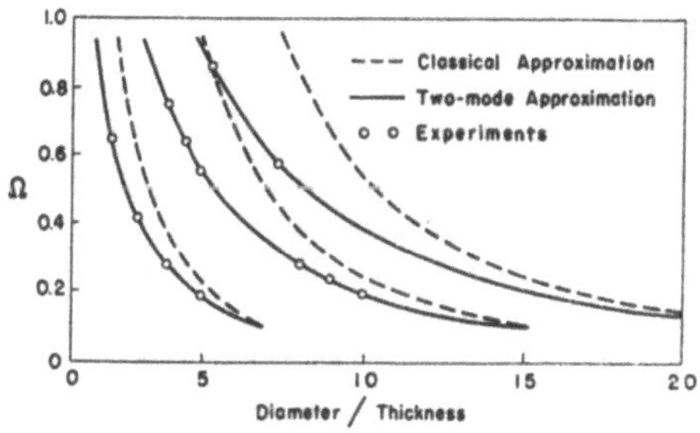

Fig. 30. Flexural vibrations of a circular disk in the range $\Omega < 1$. Comparison of experiments by Sharma[40] with results of Lagrange–Germain equation and equations taking into account first two Rayleigh–Lamb modes.

As $\Omega \to 0$, the lowest symmetric branch of equation (90) has the asymptotic behavior, as shown by Rayleigh and Lamb,

$$\Omega = 2x\sqrt{(1-k^{-2})} \tag{121}$$

This is exactly the result given by Poisson's[1] equations (pp. 357-570) of low frequency extensional vibrations of thin plates and it is a good approximation in the range $0<\Omega<0.5$. Poisson's equations, which are identical with those later called the equations of generalized plane stress, also give exactly the lowest *SH*-branch ($\Omega=x$). The substitution of

$$\lambda' = \frac{2\mu\lambda}{\lambda+2\mu} \tag{122}$$

for λ, or the substitution of

$$k' = [2/(1-\nu)]^{1/2} \tag{123}$$

for k, in the equations of plane strain, gives Poisson's equations immediately. Thus any solution of a problem of plane strain may be converted to one of generalized plane stress by making the substitutions (122) or (123). In particular the Rayleigh–Lamb spectrum is equally well the spectrum for vibrations in the plane of an infinite strip with free edges, as long as the wavelength in the plane of the strip is large in comparison with the thickness.

The next higher approximation, for the symmetric modes, has been obtained by including the symmetric branch whose cut-off frequency is $\Omega=k$. However, this has been shown to be a poor approximation[12] because of neglect of the neighboring branch whose cut-off frequency is $\Omega=2$. Equations taking both branches into account[19] yield a spectrum like that in Fig. 23 and, in addition, contain the lowest *SH*-branch exactly and the next higher symmetric *SH*-branch approximately. A recent solution[21] of these equations, for a circular disk, confirms Shaw's[20] experimental data. In particular, the complex parts of the second and third branches, along with the real first branch, generate the edge vibration which Shaw observed.

In contemplating still higher order approximations for both flexural and extensional vibrations, it does not appear to be desirable to establish higher and higher order differential equations of the type mentioned in this section. Although this method reduces the problem from three to two dimensions, each additional branch that is introduced increases the order of the differential equations by two and increases by one the degree of the algebraic equation whose roots yield the approximate branches. Furthermore, the approximations become poorer, especially for the complex branches. It would appear preferable to use numerical values for the exact branches of the Rayleigh–Lamb and *SH*-spectra and satisfy boundary conditions approximately for a finite number of them[42].

REFERENCES

1. Poisson, S. D. "Mémoire sur l'Équilibre et le Mouvement des Corps Élastiques (Addition)" *Mém. Acad. Sci. Paris* Ser. 2, Vol. 8, p. 623, 1829.
2. Knott, C. G. "Reflection and Refraction of Elastic Waves" *Phil. Mag.* Ser. 5, Vol. 48, p. 64, 1899.
3. Goodier, J. N. and Bishop, R. E. D. "A Note on Critical Reflections of Elastic Waves at Free Surfaces" *J. Appl. Phys.* Vol. 23, p. 124, 1952.
4. Rayleigh, Lord "On Waves Propagated along the Surface of an Elastic Solid" *Proc. London Math. Soc.* Vol. 17, p. 4, 1887.

5. RAYLEIGH, LORD "On the Free Vibrations of an Infinite Plate of Homogeneous Isotropic Elastic Matter" *Proc. Lond. Math. Soc.* Vol. 20, p. 225, 1888-1889.
6. LAMB, H. "On the Flexure of an Elastic Plate (Appendix)" *Proc. Lond. Math. Soc.* Vol. 21, p. 85, 1889-1890.
7. LAMB, H. "On Waves in an Elastic Plate" *Proc. Roy. Soc.* A Vol. 93, p. 114, 1917.
8. LAMÉ, G. *Leçons sur la Théorie Mathématique de l'Élasticité des Corps Solides*, 2nd Ed.; Gauthier-Villars, Paris, 1866.
9. HOLDEN, A. N. "Longitudinal Modes of Elastic Waves in Isotropic Cylinders and Slabs" *Bell System Tech. J.* Vol. 30, No. 4, Pt. 1, p. 956, 1951.
10. ONOE, M. "A Study of the Branches of the Velocity-Dispersion Equations of Elastic Plates and Rods" *Report Joint Committee on Ultrasonics of the Institute of Electrical Communication Engineers and the Acoustical Society of Japan*, 1955.
11. MINDLIN, R. D. "Thickness–Shear and Flexural Vibrations of Crystal Plates" *J. Appl. Phys.* Vol. 22, p. 316, 1951.
12. MINDLIN, R. D. *An Introduction to the Mathematical Theory of Vibrations of Elastic Plates;* U.S. Army Signal Corps Engineering Laboratories, Fort Monmouth, New Jersey, 1955.
13. MINDLIN, R. D. and ONOE, M. "Mathematical Theory of Vibrations of Elastic Plates" *Proceedings of the Eleventh Annual Symposium on Frequency Control*, pp. 17-40; U.S. Army Signal Engineering Laboratories, Fort Monmouth, New Jersey, 1957.
14. AGGARWAL, R. R. and SHAW, E. A. G. "Axially Symmetric Vibrations of a Finite Isotropic Disk. IV" *J. Acoust. Soc. Amer.* Vol. 26, p. 341, 1954.
15. LYON, R. H. "Response of an Elastic Plate to Localized Driving Forces" *J. Acoust. Soc. Amer.* Vol. 27, p. 259, 1955.
16. MINDLIN, R. D. "Vibrations of an Infinite Elastic Plate at its Cut-off Frequencies" *Proc. Third U.S. National Congress of Applied Mechanics*, p. 225, 1958.
17. TOLSTOY, I. and USDIN, E. "Wave Propagation in Elastic Plates: Low and High Mode Dispersion" *J. Acoust. Soc. Amer.* Vol. 29, p. 37, 1957.
18. HARRISON, M. "The Propagation of Elastic Waves in a Plate" *Report* No. 872; The David W. Taylor Model Basin, 1954.
19. MINDLIN, R. D. and MEDICK, M. A. "Extensional Vibrations of Elastic Plates" *J. Appl. Mech.* In press.
20. SHAW, E. A. G. "On the Resonant Frequencies of Thick Barium Titanate Disks" *J. Acoust. Soc. Amer.* Vol. 28, p. 38, 1956.
21. GAZIS, D. C. and MINDLIN, R. D. "Extensional Vibrations and Waves in a Circular Disk and a Semi-Infinite Plate" *J. Appl. Mech.* In press.
22. ONOE, M. "Contour Vibrations of Thin Rectangular Plates" *J. Acoust. Soc. Amer.* Vol. 30, p. 1159, 1958.
23. LOVE, A. E. H. *Some Problems of Geodynamics*, p. 160; Cambridge University Press, London, 1926.
24. MINDLIN, R. D. "Note on Vibrations of Prisms and Polygonal Plates" *Columbia University Report* No. CU-37-58-ONR-266(09)-CE, 1958.
25. MINDLIN, R. D. and DERESIEWICZ, H. "Timoshenko's Shear Co-efficient for Flexural Vibrations of Beams" *Proc. Second U.S. National Congress of Applied Mechanics*, p. 175, 1954.
26. ONOE, M. "Equivoluminal Vibrations of Cylinders" *Report;* Trans. Committee on Ultrasonics, Inst. Elect. Comm. Engrs., Japan, 1956.
27. SEZAWA, K. "On the Accumulation of Energy of High-frequency Vibrations of an Elastic Plate on its Surface" *Proc. Third Int. Cong. Appl. Mech., Stockholm;* Vol. 3, p. 167, 1930.
28. GOODMAN, L. E. "Circular Crested Vibrations of an Elastic Solid Bounded by Two Parallel Planes" *Proc. First U.S. Natl. Cong. of Appl. Mech.*, p. 65, 1951.

29. UFLYAND, Y. S. "The Propagation of Waves in the Transverse Vibrations of Bars and Plates" *Prikl. Mat. i Mekh.* Vol. 12, p. 287, 1948.
30. MINDLIN, R. D. "Influence of Rotatory Inertia and Shear on Flexural Vibrations of Isotropic, Elastic Plates" *J. Appl. Mech.* Vol. 18, p. 31, 1951.
31. REISSNER, E. "The Effect of Transverse Shear Deformation on the Bending of Elastic Plates" *J. Appl. Mech.* Vol. 12, p. 69, 1945.
32. REISSNER, E. "On Bending of Elastic Plates" *Quart. Appl. Math.* Vol. 5, p. 55, 1947.
33. TIMOSHENKO, S. "On the Correction for Shear of the Differential Equation for Transverse Vibrations of Prismatic Bars" *Phil. Mag.* Ser. 6, Vol. 41, p. 744, 1921.
34. SYKES, R. A. "Modes of Motion in Quartz Crystals, the Effects of Coupling and Methods of Design" in *Quartz Crystals for Electrical Circuits* (Edited by HEISING, R. A.) Chap. VI, p. 205; Van Nostrand, New York, 1946.
35. KANE, T. R. "Reflection of Flexural Waves at the Edge of a Plate" *J. Appl. Mech.* Vol. 21, p. 213, 1954.
36. MINDLIN, R. D. SCHACKNOW, A. and DERESIEWICZ, H. "Flexural Vibrations of Rectangular Plates" *J. Appl. Mech.* Vol. 23, p. 430, 1956.
37. DERESIEWICZ, H. and MINDLIN, R. D. "Axially Symmetric Flexural Vibrations of a Circular Disk" *J. Appl. Mech.* Vol. 22, p. 86, 1955.
38. DERESIEWICZ, H. "Symmetric Flexural Vibrations of a Clamped Circular Disk" *J. Appl Mech.* Vol. 23, p. 319, 1956.
39. MINDLIN, R. D. and DERESIEWICZ, H. "Thickness-Shear and Flexural Vibrations of a Circular Disk" *J. Appl. Phys.* Vol. 25, p. 1329, 1954.
40. SHARMA, R. L. "Dependence of Frequency Spectrum of a Circular Disk on Poisson's Ratio" *J. Appl. Mech.* Vol. 24, p. 642, 1957.
41. MINDLIN, R. D. and FOX, E. A. "Vibrations and Waves in Elastic Bars of Rectangular Cross-section" *J. Appl. Mech.* In press.
42. MINDLIN, R. D. "Investigations in the Mathematical Theory of Vibrations of Anisotropic Bodies: Expansion in Series of Characteristic Functions of Infinite Plates" *Columbia University Report* No. CU-4-56-SC-64687-CE, p. 16, 1956.

Reprinted from
QUARTERLY OF APPLIED MATHEMATICS
Vol. XIX, No. 1, April, 1961

HIGH FREQUENCY VIBRATIONS OF CRYSTAL PLATES*

BY
R. D. MINDLIN
Columbia University

1. Introduction. In this paper, Cauchy's [1] two-dimensional equations of coupled flexural and extensional motion of crystal plates are extended to the next higher order of approximation so as to accommodate the frequencies of the two lowest thickness-shear modes. The equations obtained are also extensions of previous equations [2] in which flexure and the same thickness-shear modes were included, but coupling with extensional modes was omitted. The new equations are deduced from the three-dimensional equations of linear elasticity by a procedure based on the series expansion methods of Cauchy [1] and Poisson [3] and the variational method of Kirchhoff [4]. Comparison of the appropriate solution of the resulting equations with Ekstein's [5] solution of the three-dimensional equations, for an infinite plate, reveals close agreement between the two frequency spectra, over the extended range of frequencies, for all five branches of the spectrum of the two-dimensional equations. This indicates that solutions of these equations, for bounded plates, will give reliable results, over the extended range of frequencies, since the modes of bounded plates, in that range, are composed essentially of coupled overtones of the first five modes of vibration of an infinite plate.

In addition to the derivation of the approximate equations, theorems of uniqueness and orthogonality are established and some general conclusions are drawn regarding solutions in rectangular coordinates and vibrations of rectangular plates.

2. Expansion in power series. The plate is referred to rectangular coordinates x_i $(i = 1, 2, 3)$ with x_1 and x_3 in the middle plane and the faces at $x_2 = \pm h$. The components of displacement u_j $(j = 1, 2, 3)$ are expanded in power series of the thickness-coordinate:

$$u_j = \sum_{n=0}^{\infty} x_2^n u_j^{(n)} = u_j^{(0)} + x_2 u_j^{(1)} + x_2^2 u_j^{(2)} \cdots , \tag{1}$$

where the $u_j^{(n)}$ are functions of x_1, x_3 and the time, t, only.

Stress-equations of motion. The series expression for u_j is substituted in

$$\int_V (T_{ij,i} - \rho u_{j,tt}) \, \delta u_j \, dV = 0 \tag{2}$$

which is obtained from the variational equation of motion, deduced from Hamilton's principle [6]. In (2), the integration is over the volume, V, of the plate; the T_{ij} are the components of stress; ρ is the density; and the summation convention for repeated indices is employed, as is the comma notation for differentiation with respect to the x_i and the time t.

When the integration with respect to x_2, from $-h$ to h, is performed in (2), the result is

*Received September 26, 1960.

$$\int_A \sum_{n=0}^{\infty} \left(T_{ij,i}^{(n)} - nT_{2j}^{(n-1)} + F_j^{(n)} - \rho \sum_{m=0}^{\infty} H_{mn} u_{j,tt}^{(m)} \right) \delta u_j^{(n)} \, dA = 0, \tag{3}$$

where A is the area of the plate and

$$T_{ij}^{(n)} \equiv \int_{-h}^{h} x_2^n T_{ij} \, dx_2 , \qquad F_j^{(n)} \equiv [x_2^n T_{2j}]_{-h}^{h}$$

$$H_{mn} \equiv \begin{cases} 2h^{m+n+1}/(m+n+1), & m+n \quad \text{even} \\ 0, \quad m+n \quad \text{odd} \end{cases} \tag{4}$$

Since (3) must hold for all A and arbitrary $\delta u_j^{(n)}$, the quantity in parentheses must vanish and we arrive at the stress-equations of motion of order n

$$T_{ij,i}^{(n)} - nT_{2j}^{(n-1)} + F_j^{(n)} = \rho \sum_{m=0}^{\infty} H_{mn} u_{j,tt}^{(m)} .$$

Strain. The three-dimensional components of strain, S_{ij} , are expressed in terms of the u_i by

$$S_{ij} = \tfrac{1}{2}(u_{i,j} + u_{j,i}). \tag{5}$$

Upon substituting (1) in (5) we obtain, after a rearrangement of terms,

$$S_{ij} = \sum_{n=0}^{\infty} x_2^n S_{ij}^{(n)}, \tag{6}$$

where

$$S_{ij}^{(n)} \equiv \tfrac{1}{2}[u_{i,j}^{(n)} + u_{j,i}^{(n)} + (n+1)(\delta_{2j} u_i^{(n+1)} + \delta_{i2} u_j^{(n+1)})]$$

in which δ_{ij} is the Kronecker symbol.

Stress-strain relations. In three dimensions,

$$T_{ij} = c_{ijkl} S_{kl} , \qquad c_{ijkl} = c_{jikl} = c_{klij} , \tag{7}$$

$$S_{ij} = s_{ijkl} T_{kl} , \qquad s_{ijkl} = s_{jikl} = s_{klij} , \tag{8}$$

where c_{ijkl} and s_{ijkl} are the elastic stiffnesses and compliances, respectively.

The expressions for the two-dimensional $T_{ij}^{(n)}$ in terms of the $S_{ij}^{(n)}$ are obtained by inserting (6) in (7) and (7) in the first of (4), with the result

$$T_{ij}^{(n)} = c_{ijkl} \sum_{m=0}^{\infty} H_{mn} S_{kl}^{(m)}. \tag{9}$$

Energy-densities. A strain-energy-density, U, and a kinetic energy-density, K, both reckoned per unit area of the plate, are defined by

$$U \equiv \tfrac{1}{2} \int_{-h}^{h} c_{ijkl} S_{ij} S_{kl} \, dx_2 = \tfrac{1}{2} c_{ijkl} \sum_{m=0}^{\infty} \sum_{n=0}^{\infty} H_{mn} S_{ij}^{(m)} S_{kl}^{(n)},$$

$$K \equiv \tfrac{1}{2} \int_{-h}^{h} \rho u_{j,t} u_{j,t} \, dx_2 = \tfrac{1}{2} \rho \sum_{m=0}^{\infty} \sum_{n=0}^{\infty} H_{mn} u_{j,t}^{(m)} u_{j,t}^{(n)} .$$

It may be noted that

$$T_{ij}^{(n)} = \partial U / \partial S_{ij}^{(n)}; \qquad \partial S_{ij}^{(n)} / \partial S_{ji}^{(n)} = 0, \qquad i \neq j.$$

3. Truncation of series. We retain only the zero and first order components of stress and strain and we write, tentatively, from (9),

$$T_{ij}^{(0)} = 2hc_{ijkl}S_{kl}^{(0)}, \tag{10}$$

$$T_{ij}^{(1)} = \tfrac{2}{3}h^3 c_{ijkl}S_{kl}^{(1)}. \tag{11}$$

The components $S_{ij}^{(0)}$ and $S_{ij}^{(1)}$ involve only the displacements $u_j^{(0)}$, $u_j^{(1)}$ and $u_j^{(2)}$ and we neglect those of higher order. Then, following Cauchy, we neglect the velocity $u_{2,t}^{(1)}$ in the kinetic energy and equations of motion and provide for the free development of the strain $S_{22}^{(0)}$ $(= u_2^{(1)})$ by setting $T_{22}^{(0)} = 0$ in (10). The condition $T_{22}^{(0)} = 0$ permits the elimination of $S_{22}^{(0)}$ from (10), with the result, again tentative,

$$T_{ij}^{(0)} = 2hg_{ijkl}S_{kl}^{(0)}, \tag{12}$$

where

$$g_{ijkl} = c_{ijkl} - c_{ij22}\,c_{22kl}/c_{2222}\,.$$

It may be observed that, because of the form of g, only five components of stress and five components of strain remain in (12).

The first order terms, $T_{ij}^{(1)}$ and $S_{ij}^{(1)}$, are treated in a similar manner except that all three velocities $u_{j,t}^{(2)}$ are neglected and free development of the three strains $S_{2j}^{(1)}$ is accommodated by setting $T_{2j}^{(1)} = 0$ in (11). Upon elimination of the $S_{2j}^{(1)}$ from (11), we arrive at

$$T_{ab}^{(1)} = \tfrac{2}{3}h^3\gamma_{abcd}S_{cd}^{(1)}; \qquad a, b, c, d = 1, 3,$$

where

$$\gamma_{abcd} = A_{abcd}/|\,s_{abcd}\,|$$

in which $|s_{abcd}|$ is the determinant

$$|\,s_{abcd}\,| = \begin{vmatrix} s_{1111} & s_{1133} & 2s_{1113} \\ s_{3311} & s_{3333} & 2s_{3313} \\ 2s_{1311} & 2s_{1333} & 4s_{1313} \end{vmatrix}$$

and A_{abcd} is the cofactor of that element of $|s_{abcd}|$ which contains s_{abcd}.

As the final step in the process of truncation, the thickness-shear strains $S_{21}^{(0)}$ and $S_{23}^{(0)}$ are replaced by $k_1S_{21}^{(0)}$ and $k_3S_{23}^{(0)}$ in the strain-energy-density, where k_1 and k_3 are correction-factors which may be used to adjust the thickness-shear frequencies to their exact values; thus compensating, in part, for the omission of terms of higher order in the series expansions. The correction-factors may be introduced conveniently by replacing g_{ijkl} with

$$g_{ijkl}^{*} \equiv k_{i+j-2}^{m}k_{k+l-2}^{n}g_{ijkl} \quad \text{(no sum)},$$

where m and n are the powers

$$m = \cos^2(ij\pi/2), \qquad n = \cos^2(kl\pi/2).$$

Thus k_{i+j-2}^{m} (or k_{k+l-2}^{n}) is equal to k_1, k_3 or 1 according as $i + j$ (or $k + l$) in g_{ijkl} is 3, 5 or neither, respectively.

4. Recapitulation. The equations remaining, after truncation of the series and adjustment of the terms retained, are [7]

*Energy-densities**

$$U = hg^*_{ijkl}S^{(0)}_{ij}S^{(0)}_{kl} + \tfrac{1}{3}h^3\gamma_{abcd}S^{(1)}_{ab}S^{(1)}_{cd}, \tag{13}$$

$$K = \rho h(u^{(0)}_{i,t}u^{(0)}_{i,t} + \tfrac{1}{3}h^2u^{(1)}_{a,t}u^{(1)}_{a,t}). \tag{14}$$

Stress-strain relations

$$T^{(0)}_{ij} = \partial U/\partial S^{(0)}_{ij} = 2hg^*_{ijkl}S^{(0)}_{kl}, \tag{15}$$

$$T^{(1)}_{ab} = \partial U/\partial S^{(1)}_{ab} = \tfrac{2}{3}h^3\gamma_{abcd}S^{(1)}_{cd}, \tag{16}$$

where $\partial S^{(0)}_{ij}/\partial S^{(0)}_{ji} = 0,\ i \neq j;\ \partial S^{(1)}_{ab}/\partial S^{(1)}_{ba} = 0,\ a \neq b.$

Stress-equations of motion

$$T^{(0)}_{ij,i} + F^{(0)}_{j} = 2h\rho u^{(0)}_{j,tt}, \tag{17}$$

$$T^{(1)}_{ab,a} - T^{(0)}_{2b} + F^{(1)}_{b} = \tfrac{2}{3}h^3\rho u^{(1)}_{b,tt}. \tag{18}$$

Strain-displacement relations

$$S^{(0)}_{ij} = \tfrac{1}{2}(u^{(0)}_{i,j} + u^{(0)}_{j,i} + \delta_{2j}u^{(1)}_i + \delta_{2i}u^{(1)}_j), \tag{19}$$

$$S^{(1)}_{ab} = \tfrac{1}{2}(u^{(1)}_{a,b} + u^{(1)}_{b,a}). \tag{20}$$

Note that, although $S^{(0)}_{22} = u^{(1)}_2$ is contained in (19), neither of the two appears in any other of the Eqs. (13)–(20) because the zero order components of strain always occur as products with g^*_{ijkl} and the latter is zero when ij or kl is 22. Hence, there are only eight components of strain and eight components of stress to be considered. The components of strain are related through the four compatibility equations

$$\begin{aligned} S^{(0)}_{11,33} + S^{(0)}_{33,11} &= 2S^{(0)}_{13,13}, & S^{(1)}_{33,1} - S^{(1)}_{13,3} &= S^{(0)}_{23,13} - S^{(0)}_{12,33}, \\ S^{(1)}_{11,33} + S^{(1)}_{33,11} &= 2S^{(1)}_{13,13}, & S^{(1)}_{11,3} - S^{(1)}_{31,1} &= S^{(0)}_{21,31} - S^{(0)}_{32,11}. \end{aligned} \tag{21}$$

Displacement-equations of motion

$$2hg^*_{ijkl}(u^{(0)}_{k,li} + \delta_{2k}u^{(1)}_{l,i}) + F^{(0)}_j = 2h\rho u^{(0)}_{j,tt}, \tag{22}$$

$$\tfrac{2}{3}h^3\gamma_{abcd}u^{(1)}_{c,da} - 2hg^*_{2bkl}(u^{(0)}_{k,l} + \delta_{2k}u^{(1)}_l) + F^{(1)}_b = \tfrac{2}{3}h^3\rho u^{(1)}_{b,tt}. \tag{23}$$

These equations are closely related to several predecessors. Thus, if the thickness-shear and flexure are eliminated by setting the transverse shear forces $T^{(0)}_{2a}$ and the couples $T^{(1)}_{ab}$ and $F^{(1)}_b$ equal to zero, the first and third of (22) reduce to the Cauchy-Voigt [1, 8] equations of low frequency extensional motion of thin plates:

$$2h\gamma_{abcd}u^{(0)}_{c,ad} + F^{(0)}_b = 2h\rho u^{(0)}_{b,tt}. \tag{24}$$

Conversely, if the extensional deformation is suppressed by setting $u^{(0)}_a = F^{(0)}_a = 0$, (22) and (23) become equations of coupled thickness-shear and flexural vibrations [2]

$$2hg^*_{a2b2}(u^{(0)}_{2,ba} + u^{(1)}_{b,a}) + F^{(0)}_2 = 2h\rho u^{(0)}_{2,tt}, \tag{25}$$

$$\tfrac{2}{3}h^3\gamma_{abcd}u^{(1)}_{c,da} - 2hg^*_{a2b2}(u^{(0)}_{2,a} + u^{(1)}_a) + F^{(1)}_b = \tfrac{2}{3}h^3\rho u^{(1)}_{b,tt}. \tag{26}$$

---

*Here and in the sequel: indices i, j, k, l range over 1, 2, 3; indices a, b, c, d range over 1, 3.

Alternatively, in the case of isotropy, (22) and (23) become

$$\mu u_{b,aa}^{(0)} + (\lambda' + \mu)u_{a,ab}^{(0)} + \tfrac{1}{2}h^{-1}F_b^{(0)} = \rho u_{b,tt}^{(0)}, \tag{27}$$

$$k^2\mu(u_{2,aa}^{(0)} + u_{a,a}^{(1)}) + \tfrac{1}{2}h^{-1}F_2^{(0)} = \rho u_{2,tt}^{(0)}, \tag{28}$$

$$\mu u_{b,aa}^{(1)} + (\lambda' + \mu)u_{a,ab}^{(1)} - 3h^{-2}k^2\mu(u_{2,b}^{(0)} + u_b^{(1)}) + \tfrac{3}{2}h^{-3}F_b^{(1)} = \rho u_{b,tt}^{(1)}, \tag{29}$$

where $\lambda' = 2\mu\lambda/(\lambda + 2\mu)$ and λ and μ are Lamé's constants. Equations (27)–(29) are also the isotropic forms of (24)–(26). Equations (27) are Poisson's equations of low frequency extensional motion of thin plates [3 and 6, p. 497] or, what is the same, the equations of motion in generalized plane stress. Equations (28) and (29) are equations of flexural motion of isotropic plates with rotatory inertia and transverse shear deformation taken into account [9]. In the one-dimensional case they have the same form as the Timoshenko beam-equations [10] and in the case of equilibrium they have the same form as Reissner's plate-equations [11]. Finally, by setting the transverse shear deformation, $S_{2a}^{(0)}$, and the rotatory inertia, $\frac{2}{3}h^3\rho u_{a,tt}^{(1)}$, equal to zero, (22) and (23) may be reduced to equations equivalent to Cauchy's [1].

5. Correction factors. The values of k_1 and k_3 are found by equating the thickness-shear frequencies obtained from (23) with the corresponding ones obtained from the three-dimensional equations.

In (23), let $F_b^{(1)} = u_j^{(0)} = 0$ and $u^{(1)} = A_b^{(1)} \exp i\omega t$, where the $A_b^{(1)}$ are constants. Then (23) become

$$g_{2b2d}^{*}A_d^{(1)} = \tfrac{1}{3}\rho h^2\omega^2 A_b^{(1)}.$$

Upon equating to zero the determinant of the coefficients of the $A_b^{(1)}$, we have

$$| g_{2b2d}^{*} - g\,\delta_{bd} | = 0, \qquad g = \tfrac{1}{3}\rho h^2\omega^2 \tag{30}$$

which gives two frequencies, say ω_1 and ω_3. These are to be equated to the two lowest roots of the frequency equation

$$| c_{2j2l} - c\,\delta_{jl} | = 0, \qquad c = 4\rho h^2\omega^2/\pi^2 \tag{31}$$

for the fundamental thickness-modes, as obtained from the three-dimensional equations [12]. The two equations yield a pair of equations which may be solved for k_1 and k_3 in terms of ratios of the c_{ijkl}.

As an example, consider the case of monoclinic symmetry with x_1 the digonal axis so that c_{2122} and c_{2123} are zero. Then the roots of (30) are

$$g_1 = k_1^2 g_{2121} = k_1^2 c_{2121}, \qquad \omega_1^2 = 3g_1/\rho h^2,$$

$$g_3 = k_3^2 g_{2323} = k_3^2(c_{2323} - c_{2223}^2/c_{2222}), \qquad \omega_3^2 = 3g_3/\rho h^2$$

and the two lowest roots of (31) are

$$c_1 = c_{2121}, \qquad \omega_1^2 = \pi^2 c_1/4\rho h^2,$$

$$c_3 = \tfrac{1}{2}\{c_{2222} + c_{2323} - [(c_{2222} - c_{2323})^2 + 4c_{2223}^2]^{1/2}\}, \qquad \omega_3^2 = \pi^2 c_3/4\rho h^2.$$

Equating corresponding frequencies, we find

$$k_1^2 = \pi^2/12, \qquad k_3^2 = \pi^2 c_3/12 g_{2323}. \tag{32}$$

6. Straight-crested waves in a monoclinic plate. The adequacy of Eqs. (13)–(23), for the prediction of frequencies of vibration of bounded crystal plates, may be judged by a comparison of the five-branched frequency spectrum of an infinite plate, as obtained from (22) and (23), with the first five branches of the spectrum obtained from the three-dimensional equations. This is because the modes of vibration of a bounded plate are composed essentially of the anharmonic overtones of the modes of an infinite plate, as described in Secs. 16 and 17 of Ref [13]. Because of its importance in technology, the frequency spectrum of straight-crested waves traveling along the digonal axis in the AT cut of quartz is chosen as the basis of comparison.

Quartz is a trigonal crystal (six elastic constants) and the AT cut is a plate which contains a digonal axis and whose normal makes an angle of 35°15′ with the trigonal axis [14]. When referred to rectangular axes in and normal to the plane of such a plate, the stress-strain relation has monoclinic symmetry (thirteen constants). With x_1 the digonal axis in the plane of the plate and x_2 the axis normal to the plate, the values of the thirteen constants, as computed from Bechmann's [15] values of the six principal constants, are, in units of 10^9 newtons per square meter,

$$\begin{array}{llll} c_{1111} = 86.74 & c_{2323} = 38.61 & c_{1122} = -8.26 & c_{2233} = -7.42 \\ c_{2222} = 129.77 & c_{3131} = 68.81 & c_{1133} = 27.15 & c_{2223} = 5.70 \\ c_{3333} = 102.84 & c_{1212} = 29.01 & c_{1123} = -3.65 & c_{3323} = 9.92 \\ & & & c_{1213} = 2.53 \end{array}$$

while

$$c_{1113} = c_{2213} = c_{3313} = c_{2313} = c_{1112} = c_{2212} = c_{3312} = c_{2312} = 0. \tag{33}$$

Then, in (22) and (23), let $F_i^{(0)} = F_b^{(1)} = 0$ and

$$u_i^{(0)} = A_i^{(0)} \exp i(\xi x_1 - \omega t), \qquad u_b^{(1)} = A_b^{(1)} \exp i(\xi x_1 - \omega t).$$

As a result, it is found that the first of (22) is coupled only with the second of (23) and the first (23) is coupled only with the second and third of (22). Thus, the general quintic, relating ω^2 and ξ^2, reduces, in this case, to a cubic and a quadratic

$$\begin{vmatrix} g^*_{1212}\xi^2 - \rho\omega^2 & g^*_{1312}\xi^2 & -ig^*_{1212}\xi \\ g^*_{1312}\xi^2 & c_{1313}\xi^2 - \rho\omega^2 & -ig^*_{1312}\xi \\ 3ih^{-2}g^*_{1212}\xi & 3h^{-2}g_{1312}\xi & \gamma_{1111}\xi^2 + 3h^{-2}g^*_{1212} - \rho\omega^2 \end{vmatrix} = 0$$

$$\begin{vmatrix} g_{1111}\xi^2 - \rho\omega^2 & -ig^*_{1123}\xi \\ 3ih^{-2}g^*_{1123}\xi & \gamma_{1313}\xi^2 + 3h^{-2}g^*_{2323} - \rho\omega^2 \end{vmatrix} = 0.$$

The roots of these equations are plotted in Fig. 1 to dimensionless coordinates

$$\Omega = \omega/(\pi/2h)(c_{1212}/\rho)^{1/2}, \qquad \phi = 2\xi h/\pi$$

The three branches marked flexure, face-shear and thickness-shear are the roots of the cubic, while the branches marked extension and thickness-twist are the roots of the quadratic.

The five branches are to be compared with the first five branches of the analogous solution, by Ekstein [5], of the three-dimensional equations. The results of detailed computations [16, 17] show correspondence to at least three significant figures over most of the range covered by Fig. 1. The asymptotic behaviors of the four branches which intersect at $\Omega = 0, \phi = 0$

$$\Omega = \phi^2(\pi/2)(\gamma_{1111}/3c_{1212})^{1/2} \quad \text{(flexure)},$$
$$\Omega = \phi(\gamma_{1313}/c_{1212})^{1/2} \quad \text{(face-shear)},$$
$$\Omega = \phi(\gamma_{1111}/c_{1212})^{1/2} \quad \text{(extension)},$$
$$\Omega = -\phi^2(\pi/2)(\gamma_{1111}/3c_{1212})^{1/2} \quad \text{(thickness-shear)}$$

are the same, for the approximate and exact equations, due to the method adopted in the derivation of the former. The remaining branch comes to zero frequency at

$$i\phi = (\gamma_{1111}c_3/\gamma_{1313}g_{1111})^{1/2} = 0.7456$$

in the approximate equations and $i\phi = 0.7467$ in the exact equations. In the high frequency range the important frequencies, in the usual applications, are near $\Omega = 1$ and the important branches are the thickness-shear and flexural branches. At $\Omega = 1$, the thickness-shear branch is exact, due to the choice $k_1^2 = \pi^2/12$; the flexural branch in the approximate equations has $\phi = 1.2483$ whereas the exact value is $\phi = 1.2417$.

The approximate equations should not be used for frequencies so high that branches higher than the fifth cannot be neglected. In the AT cut of quartz the sixth branch has a real minimum at a value of Ω slightly less than 2.0 [16]; however, due to the likelihood

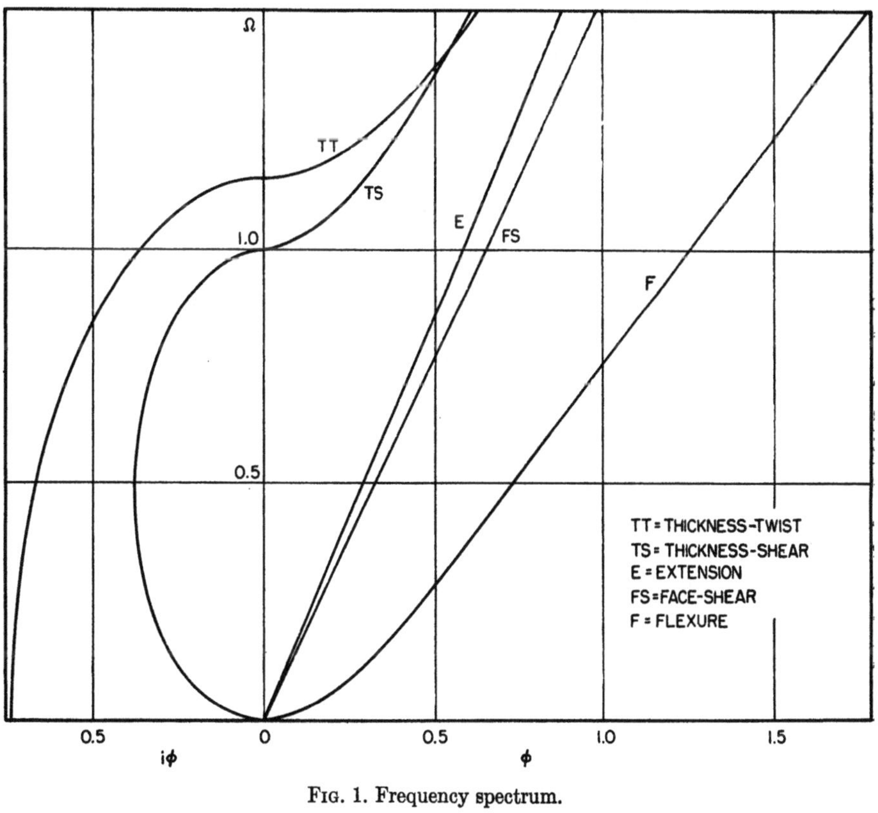

FIG. 1. Frequency spectrum.

of the contribution of an edge-mode from the complex segments of the sixth and seventh branches [17, 18], applications of the equations should be confined to $\Omega < 1.5$.

7. Uniqueness of solution. A theorem analogous to Neumann's* leads to initial and boundary conditions sufficient to assure a unique solution of equations (15)–(20).

Consider two systems of the five displacements, eight strains (omitting $S_{22}^{(0)}$) and eight stresses which satisfy the eight stress-strain equations (15)–(16), the five stress-equations of motion (17)–(18) and the eight strain-displacement equations (19)–(20). Let the differences between corresponding components of displacement, strain and stress constitute a "difference-system" of those quantities. We form the equation

$$\int_{t_0}^{t} dt \int_A [(T_{ij,i}^{(0)} + F_j^{(0)} - 2h\rho u_{j,tt}^{(0)})u_{j,t}^{(0)} + (T_{ab,a}^{(1)} - T_{2b}^{(0)} + F_b^{(1)} - \tfrac{2}{3}h^3\rho u_{b,tt}^{(1)})u_{b,t}^{(1)}]\, dA = 0 \tag{34}$$

in which all the components are those of the difference-system and t_0 is an initial time. Now

$$2h\rho \int_{t_0}^{t} dt \int_A (u_{j,tt}^{(0)}u_{j,t}^{(0)} + \tfrac{1}{3}h^2 u_{b,tt}^{(1)}u_{b,t}^{(1)})\, dA = \int_A (K - K_0)\, dA, \tag{35}$$

where K is the kinetic energy-density of the difference-system and K_0 is its value at t_0. Also

$$\begin{aligned}
T_{ij,i}^{(0)}u_{j,t}^{(0)} &+ (T_{ab,a}^{(1)} - T_{2b}^{(0)})u_{b,t}^{(1)} \\
&= (T_{ij}^{(0)}u_{j,t}^{(0)})_{,i} + (T_{ab}^{(1)}u_{b,t}^{(1)})_{,a} - T_{ij}^{(0)}u_{j,it}^{(0)} - T_{ab}^{(1)}u_{b,at}^{(1)} - T_{2b}^{(0)}u_{b,t}^{(1)}, \\
&= (T_{aj}^{(0)}u_{j,t}^{(0)} + T_{ab}^{(1)}u_{b,t}^{(1)})_{,a} - T_{ij}^{(0)}(u_{j,i}^{(0)} + \delta_{2i}u_j^{(1)})_{,t} - T_{ab}^{(1)}u_{b,at}^{(1)}, \\
&= (T_{aj}^{(0)}u_{j,t}^{(0)} + T_{ab}^{(1)}u_{b,t}^{(1)})_{,a} - \tfrac{1}{2}T_{ij}^{(0)}(u_{j,i}^{(0)} + u_{i,j}^{(0)} + \delta_{2i}u_j^{(1)} \\
&\quad + \delta_{2j}u_i^{(1)})_{,t} + \tfrac{1}{2}T_{ab}^{(1)}(u_{b,a}^{(1)} + u_{a,b}^{(1)})_{,t}, \\
&= (T_{aj}^{(0)}u_{j,t}^{(0)} + T_{ab}^{(1)}u_{b,t}^{(1)})_{,a} - (\partial U/\partial S_{ij}^{(0)})S_{ij,t}^{(0)} - (\partial U/\partial S_{ab}^{(1)})S_{ab,t}^{(1)}, \\
&= (T_{aj}^{(0)}u_{j,t}^{(0)} + T_{ab}^{(1)}u_{b,t}^{(1)})_{,a} - U_{,t},
\end{aligned}$$

where U is the strain-energy-density of the difference-system. Hence

$$\int_{t_0}^{t} dt \int_A [T_{ij,i}^{(0)}u_{j,t}^{(0)} + (T_{ab,a}^{(1)} - T_{2b}^{(0)})u_{b,t}^{(1)}]\, dA = \int_{t_0}^{t} dt \oint_C n_a(T_{aj}^{(0)}u_{j,t}^{(0)} + T_{ab}^{(1)}u_{b,t}^{(1)})\, ds - \int_A (U - U_0)\, dA \tag{36}$$

in which the line integral is around the boundary, C, of the plate and the n_a are the direction cosines of the outward normal in the plane of the plate. In the transformation from the surface integral to the line integral, it is assumed that the compatibility equations (21) are satisfied. Finally, using (35) and (36) in (34),

$$\int_A (U + K)\, dA = \int_A (U_0 + K_0)\, dA + \int_{t_0}^{t} dt \int_A (F_j^{(0)}u_{j,t}^{(0)} + F_b^{(1)}u_{b,t}^{(1)})\, dA + \int_{t_0}^{t} dt \oint_C n_a(T_{aj}^{(0)}u_{j,t}^{(0)} + T_{ab}^{(1)}u_{b,t}^{(1)})\, ds. \tag{37}$$

*Reference [6], p. 176.

Then conditions sufficient for a unique solution are established by the usual argument based on the vanishing of the right hand side of (37) and the positive definiteness of U and K. It may be seen that, in addition to the initial values of $u_j^{(0)}$, $u_a^{(1)}$, $u_{j,t}^{(0)}$ and $u_{a,t}^{(1)}$, there are five conditions to be specified at each point of the interior of the plate and five conditions at each point of the edge. In terms of orthogonal coordinates α, β, x_2, the interior conditions are one member of each of the five products

$$F_\alpha^{(0)}u_\alpha^{(0)}, \qquad F_\beta^{(0)}u_\beta^{(0)}, \qquad F_2^{(0)}u_2^{(0)}, \qquad F_\alpha^{(1)}u_\alpha^{(1)}, \qquad F_\beta^{(1)}u_\beta^{(1)}.$$

In terms of orthogonal coordinates n, s, x_2, the edge conditions are one member of each of the five products

$$T_{nn}^{(0)}u_n^{(0)}, \qquad T_{ns}^{(0)}u_s^{(0)}, \qquad T_{n2}^{(0)}u_2^{(0)}, \qquad T_{nn}^{(1)}u_n^{(1)}, \qquad T_{ns}^{(1)}u_s^{(1)}. \tag{38}$$

The first two of these products give the boundary conditions of generalized plane stress and the last three give the boundary conditions of Reissner's equations of flexure [11].

8. Orthogonal functions. A theorem analogous to Clebsch's* may also be established. Consider two solutions

$$(u_j^{(0)}, u_b^{(1)}) = (u_j^{(0)r}, u_b^{(1)r}) \exp i\omega_r t$$
$$(u_j^{(0)}, u_b^{(1)}) = (u_j^{(0)s}, u_b^{(1)s}) \exp i\omega_s t$$

of the homogeneous ($F_j^{(0)} = F_b^{(1)} = 0$) stress-equations of motion, so that the equations

$$2\rho h\omega_r^2 u_j^{(0)r} = -T_{ij,i}^{(0)r},$$
$$\tfrac{2}{3}\rho h^3\omega_r^2 u_b^{(1)r} = -T_{ab}^{(1)r} + T_{2b}^{(0)r},$$
$$-2\rho h\omega_s^2 u_j^{(0)s} = T_{ij,i}^{(0)s},$$
$$-\tfrac{2}{3}\rho h^3\omega_s^2 u_b^{(1)s} = T_{ab}^{(1)s} - T_{2b}^{(0)s}$$

are satisfied. Multiplying these equations by $u_j^{(0)s}$, $u_b^{(1)s}$, $u_j^{(0)r}$, $u_b^{(1)r}$, respectively, adding and integrating over the area of the plate, we obtain, on the left hand side,

$$2h\rho(\omega_r^2 - \omega_s^2)\int_A (u_j^{(0)r}u_j^{(0)s} + \tfrac{1}{3}h^2 u_b^{(1)r}u_b^{(1)s})\, dA$$

and, on the right hand side,

$$\int_A (T_{ij,i}^{(0)s}u_j^{(0)r} - T_{ij,i}^{(0)r}u_j^{(0)s} + T_{ab,a}^{(1)s}u_b^{(1)r} - T_{ab,a}^{(1)r}u_b^{(1)s} + T_{2b}^{(0)r}u_b^{(1)s} - T_{2b}^{(0)s}u_b^{(1)r})\, dA.$$

The latter, by a process similar to that employed in the preceding section, may be transformed to

$$2h\int_A [g_{ijkl}^*(S_{ij}^{(0)s}S_{kl}^{(0)r} - S_{ij}^{(0)r}S_{kl}^{(0)s}) + \tfrac{1}{3}h^2\gamma_{abcd}(S_{ab}^{(1)s}S_{cd}^{(1)r} - S_{ab}^{(1)r}S_{cd}^{(1)s})]\, dA$$
$$+ \oint_C n_a(T_{aj}^{(0)s}u_j^{(0)r} - T_{aj}^{(0)r}u_j^{(0)s} + T_{ab}^{(1)s}u_b^{(1)r} - T_{ab}^{(1)r}u_b^{(1)s})\, ds.$$

The integrand in the surface integral vanishes identically. The integrand in the line

*Reference [6], p. 180.

integral vanishes for any one of the thirty-two sets of edge-conditions obtained by equating to zero one member of each of the five products in (38), or for any of the associated conditions of elastic support. Then, if $w_r \neq w_s$

$$\int_A (u_j^{(0)r} u_j^{(0)s} + \tfrac{1}{3}h^2 u_b^{(1)r} u_b^{(1)s})\, dA = 0.$$

Some observations may be made regarding normal modes of rectangular plates. In the general triclinic case there are no solutions of the displacement-equations of motion, (22)–(23), with each of the five components of displacement a single product-function. However, such solutions are possible for monoclinic and higher symmetries.

Thus, in the monoclinic case, where (33) hold, (22) and (23) are satisfied by the four sets of displacements (and no others)

$$\begin{aligned} u_1^{(0)} &= A_1^{(0)} \cos(\xi x_1 + p\pi/2) \sin(\zeta x_3 + q\pi/2) \exp i\omega t, \\ u_2^{(0)} &= A_2^{(0)} \sin(\xi x_1 + p\pi/2) \cos(\zeta x_3 + q\pi/2) \exp i\omega t, \\ u_3^{(0)} &= A_3^{(0)} \sin(\xi x_1 + p\pi/2) \cos(\zeta x_3 + q\pi/2) \exp i\omega t, \\ u_1^{(1)} &= A_1^{(1)} \cos(\xi x_1 + p\pi/2) \cos(\zeta x_3 + q\pi/2) \exp i\omega t, \\ u_3^{(1)} &= A_3^{(1)} \sin(\xi x_1 + p\pi/2) \sin(\zeta x_3 + q\pi/2) \exp i\omega t, \\ &\qquad p = 0, 1; \qquad q = 0, 1 \end{aligned} \tag{39}$$

subject to a quintic equation relating ω^2, ξ^2 and ζ^2. For a plate $x_1 = \pm h_1$, $x_3 = \pm h_3$, normal modes, in which each displacement is a single product function, are possible if and only if, on $x_1 = \pm h_1$,

$$\begin{aligned} u_2^{(0)} = u_3^{(0)} = u_3^{(1)} = T_{11}^{(0)} = T_{11}^{(1)} = 0 \\ \text{or} \quad T_{12}^{(0)} = T_{13}^{(0)} = T_{13}^{(1)} = u_1^{(0)} = u_1^{(1)} = 0 \end{aligned} \tag{40}$$

and, on $x_3 = \pm h_3$,

$$\begin{aligned} u_2^{(0)} = u_3^{(0)} = u_1^{(1)} = T_{13}^{(0)} = T_{33}^{(1)} = 0 \\ \text{or} \quad T_{23}^{(0)} = T_{33}^{(0)} = T_{13}^{(1)} = u_1^{(0)} = u_3^{(1)} \quad 0, \end{aligned} \tag{41}$$

whence

$$\xi = m\pi/2h_1, \qquad \zeta = n\pi/2h_3; \qquad m, n = 0, 1, 2 \cdots \tag{42}$$

and the frequencies are given by the five roots of the quintic for each m and n.

The first set of (40) and the first set of (41) are the conditions analogous to simply supported edges in the elementary theory of flexure; but note that, in the present case, (41) are not obtained from (40) by interchange of indices 1 and 3. As in the elementary theory, solutions in closed form are not possible, in general, if all four edges are free; but closed solutions may be obtained with one pair of parallel edges free and the other pair under conditions (40) or (41). Then each of the displacements in (39) is given by a sum of five functions and one of the sets of roots (42) is replaced by the roots of a 5 × 5 transcendental, determinantal equation. There is one exceptional case of a solution in closed form, with all four edges free [19], analogous to the Lamé modes of an isotropic plate.

References

1. A. L. Cauchy, *Sur l'équilibre et le mouvement d'une plaque élastique dont l'élasticité n'est pas la même dans tous les sens*, Exercices de Mathématique **4**, 1–14 (1829)
2. R. D. Mindlin, *Thickness-shear and flexural vibrations of crystal plates*, J. Appl. Phys. **22**, 316–323 (1951)
3. S. D. Poisson, *Mémoire sur l'équilibre et le mouvement des corps élastiques*, Mém. Acad. Sci., Paris, Ser. 2, **8** 357–570 (1829)
4. G. Kirchhoff, *Über das Gleichgewicht und die Bewegung einer elastichen Scheibe*, Crelles J. **40**, 51–88 (1850)
5. H. Ekstein, *High frequency vibrations of thin crystal plates*, Phys. Rev. **68**, 11–23 (1945)
6. A. E. H. Love, *A treatise on the mathematical theory of elasticity*, Cambridge University Press, London, 1927, p. 106
7. R. D. Mindlin, *An introduction to the mathematical theory of vibrations of elastic plates*, U. S. Army Signal Corps, Fort Monmouth, N. J., Contract DA-36-039 SC-56772, pp. 6.02–6.04 (1955)
8. W. Voigt, *Lehrbuch der Krystallphysik*, B. G. Teubner, Leipzig, 2nd ed., 1928, p. 698
9. R. D. Mindlin, *Influence of rotatory inertia and shear on flexural vibrations of isotropic, elastic plates*, J. Appl. Mech. **18**, 31–38 (1951)
10. S. Timoshenko, *On the correction for shear of the differential equation for transverse vibrations of prismatic bars*, Phil. Mag., Ser. 6, **41**, 744–746 (1921)
11. E. Reissner, *On bending of elastic plates*, Quart Appl. Math. **5**, 55–68 (1947)
12. I. Koga, *Thickness vibrations of piezoelectric oscillating crystals*, Physics **3**, 70–80 (1932)
13. R. D. Mindlin, *Waves and vibrations in isotropic, elastic plates*, Proc. 1st Symp. Naval Structural Mechanics, Pergamon Press, New York, 1960, pp. 199–232
14. W. P. Mason, *Piezoelectric crystals and their application to ultrasonics*, D. Van Nostrand Co., New York, 1950, Chap. VI
15. R. Bechmann, *Elastic and piezoelectric constants of aplha-quartz*, Phys. Rev. **110**, 1060–1061 (1958)
16. E. G. Newman and R. D. Mindlin, *Vibrations of a monoclinic crystal plate*, J. Acoust. Soc. Am. **29** 1206–1218 (1957)
17. R. K. Kaul and R. D. Mindlin, *Vibrations of a monoclinic crystal plate, II*, (forthcoming)
18. D. C. Gazis and R. D. Mindlin, *Extensional vibrations and waves in a circular disk and a semi-infinite plate*, J. Appl. Mech. **27**, (1960)
19. R. D. Mindlin, *Simple modes of vibration of crystals*, J. Appl. Phys. **27**, 1462–1466 (1956)

Reprinted from
QUARTERLY OF APPLIED MATHEMATICS
Vol. XIX, No. 2, July, 1961

ON THE EQUATIONS OF EXTENSIONAL MOTION OF CRYSTAL PLATES*

BY

R. D. MINDLIN AND H. L. COOPER**
Columbia University

1. Introduction. The Cauchy-Voigt [1] equations of extensional motion of anisotropic, elastic plates may be written as

$$\begin{aligned}(L_1 + \rho\Omega^2)u + L_5 w &= 0,\\ L_5 u + (L_3 + \rho\Omega^2)w &= 0,\end{aligned} \tag{1}$$

where u and w are the components of displacement in the plane of x and z (the plane of the plate), $\Omega = i\partial/\partial t$ and the L_p, $p = 1, 3, 5$, are the quadratic operators,

$$\begin{aligned}L_1 &= \gamma_{11}\frac{\partial^2}{\partial x^2} + 2\gamma_{15}\frac{\partial^2}{\partial x\,\partial z} + \gamma_{55}\frac{\partial^2}{\partial z^2},\\ L_3 &= \gamma_{55}\frac{\partial^2}{\partial x^2} + 2\gamma_{35}\frac{\partial^2}{\partial x\,\partial z} + \gamma_{33}\frac{\partial^2}{\partial z^2},\\ L_5 &= \gamma_{15}\frac{\partial^2}{\partial x^2} + (\gamma_{13} + \gamma_{55})\frac{\partial^2}{\partial x\,\partial z} + \gamma_{35}\frac{\partial^2}{\partial z^2}.\end{aligned} \tag{2}$$

In (2), the γ_{pq} are the elastic stiffnesses relating the stress-resultants (τ_p) and the strains (ϵ_q) according to

$$\tau_p = \gamma_{pq}\epsilon_q\,; \qquad \gamma_{pq} = \gamma_{qp}\,; \qquad p, q = 1, 3, 5, \tag{3}$$

where

$$\epsilon_1 = \frac{\partial u}{\partial x}, \qquad \epsilon_3 = \frac{\partial w}{\partial z}, \qquad \epsilon_5 = \frac{\partial u}{\partial z} + \frac{\partial w}{\partial x}.$$

In the case of an isotropic plate,

$$\gamma_{11} = \gamma_{33}, \qquad 2\gamma_{55} = \gamma_{11} - \gamma_{13}, \qquad \gamma_{15} = \gamma_{35} = 0$$

and Lamé's [2] representation

$$u = \frac{\partial\varphi}{\partial x} - \frac{\partial\psi}{\partial z}, \qquad w = \frac{\partial\varphi}{\partial z} + \frac{\partial\psi}{\partial x} \tag{4}$$

reduces solutions of the displacement-equations of motion to solutions of

$$\gamma_{11}\nabla^2\varphi + \rho\Omega^2\varphi = 0, \qquad \gamma_{55}\nabla^2\psi + \rho\Omega^2\psi = 0. \tag{5}$$

In a study of the three-dimensional equations of anisotropic elasticity, G. F. Carrier [3] considered a special elastic symmetry which, in the present context, corresponds to an orthotropic plate ($\gamma_{15} = \gamma_{35} = 0$) with the additional condition

$$(\gamma_{11} - \gamma_{55})(\gamma_{33} - \gamma_{55}) = (\gamma_{13} + \gamma_{55})^2 \tag{6}$$

*Received September 12, 1960.
**Now at Queens College, New York, N. Y.

and showed that, if

$$u = (\gamma_{11} - \gamma_{55})^{1/2} \frac{\partial \varphi}{\partial x} - (\gamma_{33} - \gamma_{55})^{1/2} \frac{\partial \psi}{\partial z},$$
$$w = (\gamma_{33} - \gamma_{55})^{1/2} \frac{\partial \varphi}{\partial z} + (\gamma_{11} - \gamma_{55})^{1/2} \frac{\partial \psi}{\partial x}, \tag{7}$$

solutions of

$$\gamma_{11} \frac{\partial^2 \varphi}{\partial x^2} + \gamma_{33} \frac{\partial^2 \varphi}{\partial z^2} + \rho\Omega^2 \varphi = 0, \qquad \gamma_{55} \nabla^2 \psi + \rho\Omega^2 \psi = 0 \tag{8}$$

yield, through (7), solutions of the displacement equations of motion.

In this paper, we find the most general representation, of this type, for the Cauchy-Voigt equations; and we show, with the example of quartz, how to find orientations of crystal plates which have constants of elasticity that satisfy the conditions required for reduction of the equations of motion to the simpler forms.

2. Generalization of Carrier's representation in two dimensions. Let

$$u = \left(a_{11} \frac{\partial}{\partial x} + a_{13} \frac{\partial}{\partial z}\right)\chi \equiv M_1\chi,$$
$$w = \left(a_{31} \frac{\partial}{\partial x} + a_{33} \frac{\partial}{\partial z}\right)\chi \equiv M_3\chi, \tag{9}$$

expecting that there will be two sets of constants a_{ij} and, hence, two functions χ. Then (1) become

$$[(L_1 + \rho\Omega^2)M_1 + L_5M_3]\chi = 0, \qquad [(L_3 + \rho\Omega^2)M_3 + L_5M_1]\chi = 0. \tag{10}$$

Because of the presence of the term $\rho\Omega^2$, the only linear operator that can be a factor of the operator in brackets in the first of (10) is M_1 and in the second of (10) is M_3. Hence, if the bracketed operators in (10) are each to be expressed as a product of a linear and a quadratic operator, M_1 must be a factor of L_5M_3 and M_3 must be a factor of L_5M_1. Accordingly, we set

$$M_1M_3 = L_5\,. \tag{11}$$

Upon equating coefficients of like derivatives in (11), we find

$$a_{11}a_{31} = \gamma_{15}\,, \qquad a_{33}a_{13} = \gamma_{35}\,, \qquad a_{11}a_{33} + a_{13}a_{31} = \gamma_{13} + \gamma_{55}\,. \tag{12}$$

If (12) are satisfied, (10) become

$$M_1(L_1 + M_3^2 + \rho\Omega^2)\chi = 0, \qquad M_3(L_3 + M_1^2 + \rho\Omega^2)\chi = 0. \tag{13}$$

Now, we require the quadratic operators to be the same in both of (13). Hence, we set

$$L_1 + M_3^2 = L_3 + M_1^2 = N, \quad \text{say}, \tag{14}$$

where

$$N = b_1 \frac{\partial^2}{\partial x^2} + 2b_5 \frac{\partial^2}{\partial x\, \partial z} + b_3 \frac{\partial^2}{\partial z^2}$$

and the b_p are constants to be determined. Then solutions of

$$(N + \rho\Omega^2)\chi = 0 \tag{15}$$

will yield, through (9), solutions of the Cauchy-Voigt equations.

Upon equating coefficients in (14), we obtain

$$\begin{aligned} \gamma_{11} + a_{31}^2 &= \gamma_{55} + a_{11}^2 = b_1 , \\ \gamma_{33} + a_{13}^2 &= \gamma_{55} + a_{33}^2 = b_3 , \\ \gamma_{15} + a_{33}a_{31} &= \gamma_{35} + a_{11}a_{13} = b_5 . \end{aligned} \tag{16}$$

Equations (12) and (16) contain six relations, among the four a_{ij} and the γ_{pq}, whose consistency requires two relations among the γ_{pq}. Upon elimination of the a_{ij} from (12) and (16), these relations are found to be

$$\begin{aligned} (\gamma_{11} - \gamma_{55})(\gamma_{33} - \gamma_{55}) &= (\gamma_{13} + \gamma_{55})^2 - (\gamma_{15} + \gamma_{35})^2, \\ (\gamma_{13} + \gamma_{55})(\gamma_{15} - \gamma_{35}) &= \gamma_{35}(\gamma_{11} - \gamma_{55}) - \gamma_{15}(\gamma_{33} - \gamma_{55}). \end{aligned} \tag{A}$$

If $\gamma_{15} = \gamma_{35} = 0$, the second of A is satisfied identically and the first of A reduces to (6).

Subject to conditions A, (12) and (16) may be solved for the a_{ij} and b_p. We note, first, that the first two of (12) and the last two of (16) are satisfied by $a_{13} = a_{31}$ if $b_5 \neq 0$ and by $a_{11} = -a_{33}$ if $b_5 = 0$. Hence there are two sets of solutions:

$$\begin{aligned} a_{11} &= \pm\gamma_{15}[(\gamma_{11} - \gamma_{33})/(\gamma_{15}^2 - \gamma_{35}^2)]^{1/2} \\ a_{33} &= \pm\gamma_{35}[(\gamma_{11} - \gamma_{33})/(\gamma_{15}^2 - \gamma_{35}^2)]^{1/2} \\ a_{13} &= a_{31} = \pm[(\gamma_{15}^2 - \gamma_{35}^2)/(\gamma_{11} - \gamma_{33})]^{1/2} \\ b_1 &= \gamma_{11} + (\gamma_{15}^2 - \gamma_{35}^2)/(\gamma_{11} - \gamma_{33}) \\ b_3 &= \gamma_{33} + (\gamma_{15}^2 - \gamma_{35}^2)/(\gamma_{11} - \gamma_{33}) \\ b_5 &= \gamma_{15} + \gamma_{35} \end{aligned} \tag{17}$$

and

$$\begin{aligned} a_{11} &= -a_{33} = \pm i[(\gamma_{15}^2 - \gamma_{35}^2)/(\gamma_{11} - \gamma_{33})]^{1/2} \\ a_{13} &= \pm i\gamma_{35}[(\gamma_{11} - \gamma_{33})/(\gamma_{15}^2 - \gamma_{35}^2)]^{1/2} \\ a_{31} &= \mp i\gamma_{15}[(\gamma_{11} - \gamma_{33})/(\gamma_{15}^2 - \gamma_{35}^2)]^{1/2} \\ b_1 &= b_3 = \gamma_{55} - (\gamma_{15}^2 - \gamma_{35}^2)/(\gamma_{11} - \gamma_{33}) \\ b_5 &= 0 \end{aligned} \tag{18}$$

in which either the upper or lower signs may be taken, without loss of generality.

3. Consequences of conditions A. Conditions A take a simpler form when expressed in terms of the elastic compliances, s_{pq}, defined by

$$\epsilon_p = s_{pq}\tau_q ; \qquad s_{pq} = s_{qp} ; \qquad p, q = 1, 3, 5. \tag{19}$$

From (19) and (3),

$$\gamma_{pq} = S_{pq}/\Delta, \tag{20}$$

where S_{pq} is the cofactor of element s_{pq} in the determinant $\Delta = |s_{pq}|$. When (20) are substituted into conditions A, the latter become

$$s_{11} + s_{33} = 2s_{13} + s_{55}\,, \qquad s_{15} = s_{35} \tag{A$'$}$$

and we note that conditions A' make

$$\Delta = s_{55}(s_{11}s_{33} - s_{13}^2 - s_{15}^2).$$

Then, with $\varphi = \pm\chi/\Delta^{1/2}$ for (17) and $\psi = \pm\, i\chi/\Delta^{1/2}$ for (18), we find that (9) and (15) become

$$\begin{aligned} u &= (s_{13} - s_{33})\frac{\partial\varphi}{\partial x} + s_{15}\frac{\partial\varphi}{\partial z} + s_{15}\frac{\partial\psi}{\partial x} + (s_{13} - s_{11})\frac{\partial\psi}{\partial z}\,, \\ w &= s_{15}\frac{\partial\varphi}{\partial x} + (s_{13} - s_{11})\frac{\partial\varphi}{\partial z} - (s_{13} - s_{33})\frac{\partial\psi}{\partial x} - s_{15}\frac{\partial\psi}{\partial z} \end{aligned} \tag{21}$$

and

$$\begin{aligned} s_{33}\frac{\partial^2\varphi}{\partial x^2} - 2s_{15}\frac{\partial^2\varphi}{\partial x\,\partial z} + s_{11}\frac{\partial^2\varphi}{\partial z^2} &= \frac{\rho\,\Delta}{s_{55}}\frac{\partial^2\varphi}{\partial t^2}\,, \\ \frac{\partial^2\psi}{\partial x^2} + \frac{\partial^2\psi}{\partial z^2} &= \rho s_{55}\frac{\partial^2\psi}{\partial t^2}. \end{aligned} \tag{22}$$

We now rotate to new axes x_A, z_A according to

$$x_A = x\cos\Psi_A - z\sin\Psi_A\,, \qquad z_A = x\sin\Psi_A + z\cos\Psi_A \tag{23}$$

in order to eliminate the cross term in the first of (22). The strain-stress relations, referred to axes x_A, z_A are written as

$$\epsilon_p^A = s_{pq}^A\tau_p^A\,; \qquad p, q = 1, 3, 5,$$

where, with $s = \sin\Psi_A$, $c = \cos\Psi_A$,

$$\begin{aligned}
s_{11}^A &= s_{11}c^4 + s_{33}s^4 + (2s_{13} + s_{55})s^2c^2 - 2s_{15}sc^3 - 2s_{35}s^3c, \\
s_{33}^A &= s_{11}s^4 + s_{33}c^4 + (2s_{13} + s_{55})s^2c^2 + 2s_{15}s^3c + 2s_{35}sc^3, \\
s_{55}^A &= 4(s_{11} + s_{33} - 2s_{13})s^2c^2 + s_{55}(c^2 - s^2)^2 + 4(s_{15} - s_{35})(c^2 - s^2)sc, \\
s_{13}^A &= (s_{11} + s_{33} - s_{55})s^2c^2 + s_{13}(s^4 + c^4) + (s_{15} - s_{35})(c^2 - s^2)sc, \\
s_{35}^A &= 2s_{11}s^3c - 2s_{33}sc^3 + (2s_{13} + s_{55})(c^2 - s^2)sc + s_{15}(3s^2c^2 - s^4) + s_{35}(c^4 - 3s^2c^2), \\
s_{15}^A &= 2s_{11}sc^3 - 2s_{33}s^3c - (2s_{13} + s_{55})(c^2 - s^2)sc + s_{15}(c^4 - 3s^2c^2) + s_{35}(3s^2c^2 - s^4).
\end{aligned}$$

Now, conditions A' make

$$s_{13}^A = s_{13}\,, \qquad s_{55}^A = s_{55}\,, \qquad s_{11}^A + s_{33}^A = s_{11} + s_{33} \tag{24}$$

and if, in addition, we set

$$\tan 2\Psi_A = 2s_{15}/(s_{33} - s_{11}) \tag{25}$$

we find also

$$s_{11}^A - s_{33}^A = (s_{11} - s_{33}) \sec 2\Psi_A ,$$
$$s_{15}^A = s_{35}^A = 0. \tag{26}$$

With (21) and (24)–(26), the components of displacement, referred to axes x_A and z_A become

$$u_A = (s_{13}^A - s_{33}^A) \frac{\partial \varphi}{\partial x_A} + (s_{13}^A - s_{11}^A) \frac{\partial \psi}{\partial z_A} ,$$
$$w_A = (s_{13}^A - s_{11}^A) \frac{\partial \varphi}{\partial z_A} - (s_{13}^A - s_{33}^A) \frac{\partial \psi}{\partial x_A}. \tag{27}$$

Also, the differential equations (22) transform to

$$s_{33}^A \frac{\partial^2 \varphi}{\partial x_A^2} + s_{11}^A \frac{\partial^2 \varphi}{\partial z_A^2} = \rho(s_{11}^A s_{33}^A - s_{13}^A s_{13}^A) \frac{\partial^2 \varphi}{\partial t^2} ,$$
$$\frac{\partial^2 \psi}{\partial x_A^2} + \frac{\partial^2 \psi}{\partial z_A^2} = \rho s_{55}^A \frac{\partial^2 \psi}{\partial t^2}. \tag{28}$$

If we were to convert from constants s_{pq}^A to γ_{pq}^A by means of the inverse of (20), we should find that (28) are the same as (8) and that (27) are the same as (7), except for a constant factor. Thus, conditions A reduce the general anisotropic plate to an orthotropic plate satisfying Carrier's condition (6).

4. Slowness vector. Returning to Eqs. (1), we insert the displacements

$$u = A \exp [i\omega(\xi x + \zeta z - t)], \qquad w = C \exp [i\omega(\xi x + \zeta z - t)] \tag{29}$$

corresponding to straight-crested waves. The resulting characteristic equation, which must be satisfied by the components ξ, ζ of the slowness vector [4], is

$$(\gamma_{11}\xi^2 + 2\gamma_{15}\xi\zeta + \gamma_{55}\zeta^2 - \rho)(\gamma_{55}\xi^2 + 2\gamma_{35}\xi\zeta + \gamma_{33}\zeta^2 - \rho)$$
$$- [\gamma_{15}\xi^2 + (\gamma_{13} + \gamma_{55})\xi\zeta + \gamma_{35}\zeta^2]^2 = 0. \tag{30}$$

With ξ, ζ as real Cartesian coordinates, (30) is, in general, the equation of a pair of four-lobed curves. However, if conditions A are satisfied, the discriminant of (30) is a perfect square and the curves are a circle, corresponding to the potential ψ, and an ellipse, corresponding to φ, with principal axes oriented in accordance with (25).

The discriminant of (30) is also a perfect square if

$$(\gamma_{11} - \gamma_{55})(\gamma_{13} + \gamma_{55}) = 2\gamma_{15}(\gamma_{15} - \gamma_{35}),$$
$$(\gamma_{33} - \gamma_{55})(\gamma_{13} + \gamma_{55}) = 2\gamma_{35}(\gamma_{35} - \gamma_{15}). \tag{B}$$

In this case the locus of the tip of the slowness vector is a pair of orthogonal ellipses associated with waves whose properties are described in the following section.

5. Consequences of conditions B. In the case of conditions B, we first perform a rotation to new axes

$$x_B = x \cos \Psi_B - z \sin \Psi_B , \qquad z_B = x \sin \Psi_B + z \cos \Psi_B .$$

Referred to these axes, the stress-strain relation is

$$\tau_p^B = \gamma_{pq}^B \epsilon_q^B ,$$

where, with $s = \sin \Psi_B$, $c = \cos \Psi_B$,

$$\gamma_{11}^B = \gamma_{11}c^4 + \gamma_{33}s^4 + 2(\gamma_{13} + 2\gamma_{55})s^2c^2 - 4\gamma_{15}sc^3 - 4\gamma_{35}s^3c,$$

$$\gamma_{33}^B = \gamma_{11}s^4 + \gamma_{33}c^4 + 2(\gamma_{13} + 2\gamma_{55})s^2c^2 + 4\gamma_{15}s^3c + 4\gamma_{35}sc^3,$$

$$\gamma_{55}^B = (\gamma_{11} + \gamma_{33} - 2\gamma_{13})s^2c^2 + \gamma_{55}(c^2 - s^2)^2 + 2(\gamma_{15} - \gamma_{35})(c^2 - s^2)sc,$$

$$\gamma_{13}^B = (\gamma_{11} + \gamma_{33} - 4\gamma_{55})s^2c^2 + \gamma_{13}(s^4 + c^4) + 2(\gamma_{15} - \gamma_{35})(c^2 - s^2)sc,$$

$$\gamma_{35}^B = \gamma_{11}s^3c - \gamma_{33}sc^3 + (\gamma_{13} + 2\gamma_{55})(c^2 - s^2)sc + \gamma_{15}(3s^2c^2 - s^4) + \gamma_{35}(c^4 - 3s^2c^2),$$

$$\gamma_{15}^B = \gamma_{11}sc^3 - \gamma_{33}s^3c - (\gamma_{13} + 2\gamma_{55})(c^2 - s^2)sc + \gamma_{15}(c^4 - 3s^2c^2) + \gamma_{35}(3s^2c^2 - s^4).$$

Now, if we set $\tan 2\Psi_B = (\gamma_{13} + \gamma_{55})/(\gamma_{35} - \gamma_{15})$ and apply conditions B, we find

$$\gamma_{15}^B = \gamma_{35}^B = \gamma_{13}^B + \gamma_{55}^B = 0.$$

Hence, if u_B and w_B are the components of displacement referred to axes x_B and z_B , the equations of motion become

$$\gamma_{11}^B \frac{\partial^2 u_B}{\partial x_B^2} + \gamma_{55}^B \frac{\partial^2 u_B}{\partial z_B^2} = \rho \frac{\partial^2 u_B}{\partial t^2} , \qquad \gamma_{55}^B \frac{\partial^2 w_B}{\partial x_B^2} + \gamma_{33}^B \frac{\partial^2 w_B}{\partial z_B^2} = \rho \frac{\partial^2 w_B}{\partial t^2} \tag{31}$$

and, consequently, the introduction of displacement-potentials φ and ψ is unnecessary.

Equations (31) have the curious property that, in the waves defined by them, the directions of the respective displacements are fixed: independent of the wave-normals. The mechanism which produces this phenomenon may be identified by noting that, under conditions B,

$$\epsilon_1^B = \gamma_{33}^B[\tau_1^B + (\gamma_{55}^B/\gamma_{33}^B)\tau_3^B]/(\gamma_{11}^B\gamma_{33}^B - \gamma_{55}^B\gamma_{55}^B),$$

$$\epsilon_3^B = \gamma_{11}^B[\tau_3^B + (\gamma_{55}^B/\gamma_{11}^B)\tau_1^B]/(\gamma_{11}^B\gamma_{33}^B - \gamma_{55}^B\gamma_{55}^B).$$

Hence, since the conditions of positive-definiteness of the strain-energy-density require

$$\gamma_{11}^B > 0, \qquad \gamma_{33}^B > 0, \qquad \gamma_{55}^B > 0, \qquad \gamma_{11}^B\gamma_{33}^B - \gamma_{55}^B\gamma_{55}^B > 0,$$

both Poisson's ratios, $\nu_1^B = -\gamma_{55}^B/\gamma_{33}^B$ and $\nu_3^B = -\gamma_{55}^B/\gamma_{11}^B$, must be negative and in the ratios indicated. These are properties unlikely to be found in natural crystals.

6. Application of conditions A to quartz. The constants, say s_{pq}'', of a plate cut from a crystal, at an arbitrary orientation with respect to axes of reference fixed in the crystal, may be expressed in terms of two angles of rotation, say Θ and Φ, and the constants, say s_{mn}^0 , $m, n = 1, \cdots 6$, associated with the axes of reference. When the expressions for the s_{pq}'' , in terms of the s_{mn}^0 , Θ and Φ, are inserted in conditions A', in place of the constants s_{pq} appearing there, a pair of simultaneous, transcendental equations in Θ and Φ is obtained. If there are any real roots of the pair of equations, they give the orientations of plates whose constants s_{pq}'' satisfy conditions A'.

In the case of quartz, the compliances associated with reference axes x_0, y_0, z_0, in accordance with the I.R.E. Standards [5], are

$$s_{11}^0 = s_{22}^0 = 12.77, \qquad s_{12}^0 = -1.79,$$
$$s_{33}^0 = 9.60, \qquad s_{13}^0 = s_{23}^0 = -1.22,$$
$$s_{44}^0 = s_{55}^0 = 20.04, \qquad s_{14}^0 = -s_{24}^0 = \tfrac{1}{2}s_{56}^0 = 4.50,$$
$$s_{66}^0 = 2(s_{11}^0 - s_{12}^0),$$
$$s_{15}^0 = s_{16}^0 = s_{25}^0 = s_{26}^0 = s_{34}^0 = s_{35}^0 = s_{36}^0 = s_{45}^0 = s_{46}^0 = 0,$$

where the numerical values, in units of 10^{-12} (meter)^2/newton, are those given by Bechmann [6].

A rotation of coordinates about the axis of x_0 to axes

$$x' = x_0, \qquad y' = y_0 \cos\Theta + z_0 \sin\Theta, \qquad z' = -y_0 \sin\Theta + z_0 \cos\Theta$$

gives formulas for the thirteen non-zero compliances s'_{mn}, $m, n = 1, \cdots 6$, in terms of the s_{mn}^0 and Θ, as tabulated by Cady [7].

A further rotation, about the axis of z', to axes

$$x'' = x' \cos\Phi + y' \sin\Phi, \qquad y'' = -x' \sin\Phi + y' \cos\Phi, \qquad z'' = z'$$

gives for the six, of the twenty one, constants which we require:

$$\begin{aligned}
s''_{11} &= s'_{11} \cos^2\Phi + s'_{22} \sin^2\Phi - (s'_{11} + s'_{22} - 2s'_{12} - s'_{66}) \sin^2\Phi \cos^2\Phi,\\
s''_{33} &= s'_{33},\\
s''_{55} &= s'_{44} \sin^2\Phi + s'_{55} \cos^2\Phi,\\
s''_{13} &= s'_{13} + (s'_{23} - s'_{13}) \sin^2\Phi,\\
s''_{35} &= s'_{34} \sin\Phi,\\
s''_{15} &= s'_{24} \sin\Phi - (s'_{24} - s'_{14} - s'_{56}) \sin\Phi \cos^2\Phi,
\end{aligned} \tag{32}$$

in which the s'_{mn} are given in terms of the s_{mn}^0 and Θ by Cady's [7] formulas.

We now suppose that all of the analysis in Secs. 2 and 3 was performed in the plane of x'' and z'', so that the constants s_{pq}, appearing there, are replaced by s''_{pq}. Then, when (32) are inserted in conditions A', we find

$$[s_{14}^0(3 \sin^2\Theta - 4 \cos^4\Theta) + 2k \sin^3\Theta \cos\Theta] \sin^3\Phi$$
$$- [s_{14}^0(4 \sin^4\Theta - 3 \cos^2\Theta) + 2k \sin\Theta \cos^3\Theta] \sin\Phi = 0,$$
$$[2s_{14}^0(3 + \cos^2\Theta) \sin\Theta \cos\Theta - k \sin^4\Theta] \sin^4\Phi$$
$$+ 6[s_{14}^0(2 \cos^2\Theta - 1) \sin\Theta \cos\Theta + k \sin^2\Theta \cos^2\Theta] \sin^2\Phi$$
$$- 2s_{14}^0(3 + \sin^2\Theta) \sin\Theta \cos\Theta - k \cos^4\Theta = 0,$$

where $k = s_{11}^0 + s_{33}^0 - 2s_{13}^0 - s_{44}^0$. These equations have the four pairs of roots Θ and Φ given in the first two rows of the accompanying table. The next six rows give the values of the constants s''_{pq} computed from (32). The next row gives the values of ψ_A computed from (25), with s_{pq} replaced by s''_{pq}; and the remaining rows give the values of the constants s_{pq}^A, referred to the axes x_A and z_A defined in (23).

The first column in the table describes the well known "Z-cut" of quartz for which the Cauchy-Voigt equations reduce to isotropic form; i.e., the constants satisfy the conditions

$$s_{11}'' = s_{33}'', \qquad s_{55}'' = 2(s_{11}'' - s_{13}''), \qquad s_{15}'' = s_{35}'' = 0$$

and Eqs. (4) and (5) apply. This cut has been studied in detail by Petrzilka [8], Bechmann [9] and Ekstein [10]. Modes of vibration of the remaining three cuts may be studied by means of Eqs. (27) and (28).

As anticipated, no cut of quartz has constants of elasticity that satisfy conditions *B*.

TABLE

Orientations (Θ, Φ, Ψ) and compliances (s_{pq}, 10^{-12} meter²/newton) of the four cuts of quartz which satisfy conditions A.

	I	II	III	IV
Θ	90°	−9.65°	18.79°	18.79°
Φ	0	0	58.62°	−58.62°
s_{11}''	12.77	12.77	12.52	12.52
s_{33}''	12.77	9.52	9.77	9.77
s_{55}''	29.12	23.27	23.27	23.27
s_{13}''	−1.79	−0.49	−0.49	−0.49
s_{35}''	0	0	−0.87	0.87
s_{15}''	0	0	−0.87	0.87
Ψ_A	0	0	16.17°	73.83°
s_{11}^A	12.77	12.77	12.77	9.52
s_{33}^A	12.77	9.52	9.52	12.77
s_{55}^A	29.12	23.27	23.27	23.27
s_{13}^A	−1.79	−0.49	−0.49	−0.49
s_{35}^A	0	0	0	0
s_{15}^A	0	0	0	0

References

1. A. L. Cauchy, *Exercices de Mathématiques*, Paris, **4**, 13 (1829); W. Voigt, *Lehrbuch der Kristallphysik*, 2nd ed., Leipzig, 1928, p. 698
2. G. Lamé, *Leçons sur la théorie mathématique de l'élasticité des corps solides*, Gauthier-Villars, Paris, 1852, p. 144
3. G. F. Carrier, Quart. Appl. Math. **4**, 160 (1946)
4. J. L. Synge, J. Math. and Phys. **35**, 323 (1957)
5. Proc. Inst. Radio Eng. **37**, 1378 (1949)
6. R. Bechmann, Phys. Rev. **110**, 1060 (1958)
7. W. G. Cady, *Piezoelectricity*, McGraw-Hill Book Co., New York, 1946, p. 77
8. V. Petrzilka, Z. Physik **97**, 436 (1935)
9. R. Bechmann, Z. Physik **117**, 180 (1941)
10. H. Ekstein, Phys. Rev. **66**, 108 (1944)

Reprinted from *Problems of Continuum Mechanics*
N.I. Muskhelishvili 70th Birthday Volume (1961) 282-290

ON THE EQUATIONS OF MOTION OF PIEZOELECTRIC CRYSTALS

R. D. MINDLIN
New York, USA

§ 1. Introduction

In this paper, a systematic derivation is given of the linear equations governing small vibrations of piezoelectric crystals, including the coupling between mechanical, electrical and thermal fields. The treatment of the thermal variables is based on Biot's formulation of irreversible thermodynamics of continua [1]. Constitutive equations and equations of motion are established and a solution of the latter is exhibited, illustrating the attenuation of plane waves. In addition, a proof of a theorem of uniqueness of solution of the equations of motion is indicated and appropriate boundary conditions are enumerated.

§ 2. Cauchy's and Maxwell's equations

We take, as already established, the following definitions and equations.

a. The definition of small strain, S_{ij} $(i, j = 1, 2, 3)$, and its expression in terms of displacement:

$$2S_{ij} = u_{i,j} + u_{j,i}. \tag{2.1}$$

b. The definition of stress, T_{ij}, in terms of traction, t_i, across a surface whose unit outward normal is $\boldsymbol{\nu}$:

$$\nu_i T_{ij} = t_j. \tag{2.2}$$

c. The stress-equations of motion, omitting body forces,

$$T_{ij,i} = \rho \ddot{u}_j, \quad T_{ij} = T_{ji}. \tag{2.3}$$

d. The equations of the electromagnetic field:

$$\varepsilon_{ijk} H_{k,j} = \dot{D}_i + J_i, \quad \varepsilon_{ijk} E_{j,k} = \dot{B}_i. \tag{2.4}$$

As we shall be concerned with non-conductors at frequencies far below optical frequencies, the conduction current, J_i, and rate of change of magnetic induction, $\dot{B}_i$, may be neglected and (2.4) reduce to the equations of the quasi-stationary electric field:

$$\dot{D}_{i,i} = 0, \quad E_i = -\phi_{,i} \tag{2.5}$$

where D_i, E_i and ϕ are the electric displacement, field and potential, respectively.

§ 3. Energetic and thermodynamic considerations

We begin with the principle of conservation of energy in the form: In any portion V of a body, bounded by a surface S with unit outward normal $\boldsymbol{\nu}$, the rate of increase of energy is equal to the rate at which work is done by the tractions across S less the flux of electrical and thermal energy outward across S. Thus,

$$\int_V (\dot{K} + \dot{U})\, dV = \int_S \nu_i(T_{ij}\dot{u}_j - \phi\dot{D}_i - \dot{q}_i)\, dS, \tag{3.1}$$

where

$$K = \tfrac{1}{2}\rho\dot{u}_i\dot{u}_i \tag{3.2}$$

is the kinetic energy-density, U is the internal energy-density, $\dot{q}_i$ is the heat flux vector and $\phi\dot{D}_i$ is the form taken by the Poynting vector $\varepsilon_{ijk}E_jH_k$ in a quasi-stationary electric field.

Applying the divergence theorem to the right hand side of (3.1) and using (2.1), (2.3), (2.5) and (3.2), we have

$$\int_V \dot{U}\, dV = \int_V (T_{ij}\dot{S}_{ij} + E_i\dot{D}_i - \dot{q}_{i,i})\, dV. \tag{3.3}$$

Since this must hold for any V,

$$\dot{U} = T_{ij}\dot{S}_{ij} + E_i\dot{D}_i - \dot{q}_{i,i}, \tag{3.4}$$

which is the first law of thermodynamics for the piezoelectric medium.

By the definition of entropy density, σ, we have

$$\dot{q}_{i,i} = -\,\Theta\dot{\sigma}, \tag{3.5}$$

where Θ is the absolute temperature. Hence

$$\dot{U} = T_{ij}\dot{S}_{ij} + E_i\dot{D}_i + \Theta\dot{\sigma}. \tag{3.6}$$

Now, let

$$\Theta = \Theta_0 + \theta, \tag{3.7}$$

where Θ_0 is a constant, positive, reference temperature.

Biot's generalized free energy is defined as

$$B = U - \Theta_0 \sigma \tag{3.8}$$

instead of the usual free energy $U - \Theta\sigma$. Hence

$$\dot{B} = T_{ij}\dot{S}_{ij} + E_i\dot{D}_i + \theta\dot{\sigma}. \tag{3.9}$$

At this point it is convenient to introduce the abbreviated indicial notation in which a pair of indices, ij or kl, is replaced by a single index p or q according to the scheme

ij or kl	11	22	33	23,32	31,13	12,21
p or q	1	2	3	4	5	6

and, for the strains only,

$$S_p = \begin{cases} S_{ij},\ i = j,\ (p = 1, 2, 3), \\ 2S_{ij},\ i = j,\ (p = 4, 5, 6). \end{cases}$$

Then (3.9) becomes

$$\dot{B} = T_p\dot{S}_p + E_i\dot{D}_i + \theta\dot{\sigma}. \tag{3.10}$$

The independent variables of Biot's potential are taken to be the strain, electric displacement and entropy

$$B = B(S_p, D_i, \sigma).$$

Hence

$$\dot{B} = \frac{\partial B}{\partial S_p}\dot{S}_p + \frac{\partial B}{\partial D_i}\dot{D}_i + \frac{\partial B}{\partial \sigma}\dot{\sigma}. \tag{3.11}$$

Then, from (3.10) and (3.11),

$$T_p = \frac{\partial B}{\partial S_p}, \quad E_i = \frac{\partial B}{\partial D_i}, \quad \theta = \frac{\partial B}{\partial \sigma}. \tag{3.12}$$

The second law of thermodynamics may be written as [2]

$$\dot{\sigma} + (\dot{q}_i/\Theta)_{,i} \geq 0, \tag{3.13}$$

and this becomes, with (3.5),

$$-(\Theta_{,i}\dot{q}_i/\Theta^2) \geq 0. \tag{3.14}$$

Equation (3.14) is satisfied by the Fourier law of heat conduction

$$-\dot{q}_i = \kappa_{ij}\Theta_{,j} = \kappa_{ij}\theta_{,j}, \tag{3.15}$$

if $\kappa_{ij}\Theta_{,i}\Theta_{,j}$ is a positive definite quadratic form. By Onsager's principle, the heat conduction coefficient, κ_{ij}, is symmetric; and we then require that the three discriminants of $\kappa_{ij}\Theta_{,i}\Theta_{,j}$, obtained by omission of none, one and two of the $\Theta_{,i}$ be greater than zero, i.e.,

$$\text{(3.16)} \qquad \begin{vmatrix} \kappa_{11} & \kappa_{12} & \kappa_{13} \\ \kappa_{21} & \kappa_{22} & \kappa_{23} \\ \kappa_{31} & \kappa_{32} & \kappa_{33} \end{vmatrix} > 0, \quad \begin{vmatrix} \kappa_{11} & \kappa_{12} \\ \kappa_{21} & \kappa_{22} \end{vmatrix} > 0, \quad \kappa_{11} > 0.$$

Now, rewrite (3.5) as

$$\dot{\sigma} = -\frac{\dot{q}_{i,i}}{\Theta_0 + \theta} = -\left(1 - \frac{\theta}{\Theta_0} + \frac{\theta^2}{\Theta_0{}^2} - \cdots\right)\frac{\dot{q}_{i,i}}{\Theta_0}.$$

Hence, if the change of temperature from the reference temperature is small

$$\text{(3.17)} \qquad \theta \ll \Theta_0,$$

we have, approximately,

$$\text{(3.18)} \qquad \dot{q}_{i,i} = -\Theta_0\dot{\sigma}$$

which, with (3.15), gives

$$\text{(3.19)} \qquad \kappa_{ij}\theta_{,ij} = \Theta_0\dot{\sigma}.$$

Also, from (3.18),

$$\theta\dot{\sigma} = -\frac{\theta\dot{q}_{i,i}}{\Theta_0} = -\left(\frac{\theta\dot{q}_i}{\Theta_0}\right)_{,i} + \frac{\theta_{,i}\dot{q}_i}{\Theta_0},$$

and this becomes, with (3.15),

$$\text{(3.20)} \qquad \theta\dot{\sigma} = -\left(\frac{\theta q_i}{\Theta_0}\right)_{,i} - \kappa_{ij}{}^{-1}\Theta_0\left(\frac{\dot{q}_i\dot{q}_j}{\Theta_0{}^2}\right),$$

where $\kappa_{ij}{}^{-1}$, the reciprocal of κ_{ij}, is the heat resistance coefficient.

We now introduce Biot's definitions of entropy displacement

$$\text{(3.21)} \qquad s_i = q_i/\Theta_0$$

and dissipation function

$$\text{(3.22)} \qquad F = \tfrac{1}{2}\Theta_0\kappa_{ij}{}^{-1}\dot{s}_i\dot{s}_j,$$

so that (3.20) becomes

$$\text{(3.23)} \qquad \theta\dot{\sigma} = -(\theta\dot{s}_i)_{,i} - 2F.$$

With this result, we can rewrite the expression (3.1) of the principle of conservation of energy in a more useful form. The thermal flux term in (3.1) is

$$\int_S \nu_i \dot{q}_i \, dS = \int_V \dot{q}_{i,i} \, dV =$$

$$= - \int_V (\Theta_0 \dot{\sigma} + \theta \dot{\sigma}) \, dV =$$

$$= - \int_V [\Theta_0 \dot{\sigma} - (\theta \dot{s}_i)_{,i} - 2F] \, dV =$$

$$= \int_V (2F - \Theta_0 \dot{\sigma}) \, dV + \int_S \nu_i \theta \dot{s}_i \, dS =$$

$$= \int_V (2F + \dot{B} - \dot{U}) \, dV + \int_S \nu_i \theta \dot{s}_i \, dS.$$

Upon reinsertion of this last expression into (3.1), we have, finally,

$$\int_V (\dot{K} + \dot{B} + 2F) \, dV = \int_S \nu_i (T_{ij} \dot{u}_j - \phi \dot{D}_i - \theta \dot{s}_i) \, dS. \tag{3.24}$$

If the time derivatives were replaced by variations, (3.24) would be Biot's thermoelastic variational principle extended to include the electric field.

§ 4. Constitutive equations

The form of Biot's potential which leads to the usual linear constitutive equations is

$$B = \tfrac{1}{2} c_{ijkl}^{D\sigma} S_{ij} S_{kl} + \tfrac{1}{2} \beta_{ij}^{\sigma S} D_i D_j + \tfrac{1}{2} b^{SD} \sigma^2 - h_{kij}^{\sigma} D_k S_{ij} - q_i{}^S \sigma D_i - \gamma_{ij}{}^D \sigma S_{ij}.$$

$$c_{ijkl}^{D\sigma} = c_{jikl}^{D\sigma} = c_{ijlk}^{D\sigma} = c_{klij}^{D\sigma}, \ \beta_{ij}^{\sigma S} = \beta_{ji}^{\sigma S}, \ h_{kij}^{\sigma} = h_{kji}^{\sigma}, \ \gamma_{ij}^{D} = \gamma_{ji}^{D},$$

or

$$B = \tfrac{1}{2} c_{pq}^{D\sigma} S_p S_q + \tfrac{1}{2} \beta_{ij}^{\sigma S} D_i D_j + \tfrac{1}{2} b^{SD} \sigma^2 - - h_{ip}^{\sigma} D_i S_p - q_i{}^S \sigma D_i - \gamma_p{}^D \sigma S_p, \tag{4.1}$$

in which the symbols are those employed by Mason [3] with the exception of

$$b^{SD} = \Theta_0 / \rho C^{SD},$$

where C^{SD} is the specific heat at constant volume and constant electric displacement [1]. Then, from (3.12),

$$
\begin{aligned}
T_p &= c_{pq}^{D\sigma}\, S_q - h_{ip}^{\sigma}\, D_i - \gamma_p{}^D\, \sigma,\\
E_j &= -\, h_{jq}^{\sigma}\, S_q + \beta_{ij}^{\sigma S}\, D_i - q_j{}^S\, \sigma,\\
\theta &= -\, \gamma_q{}^D\, S_q - q_i{}^S\, D_i + b^{SD}\, \sigma.
\end{aligned}
\tag{4.2}
$$

These relations may be written in eight forms, all of which are represented on the octahedron shown in Figure 1. For example, interchanging

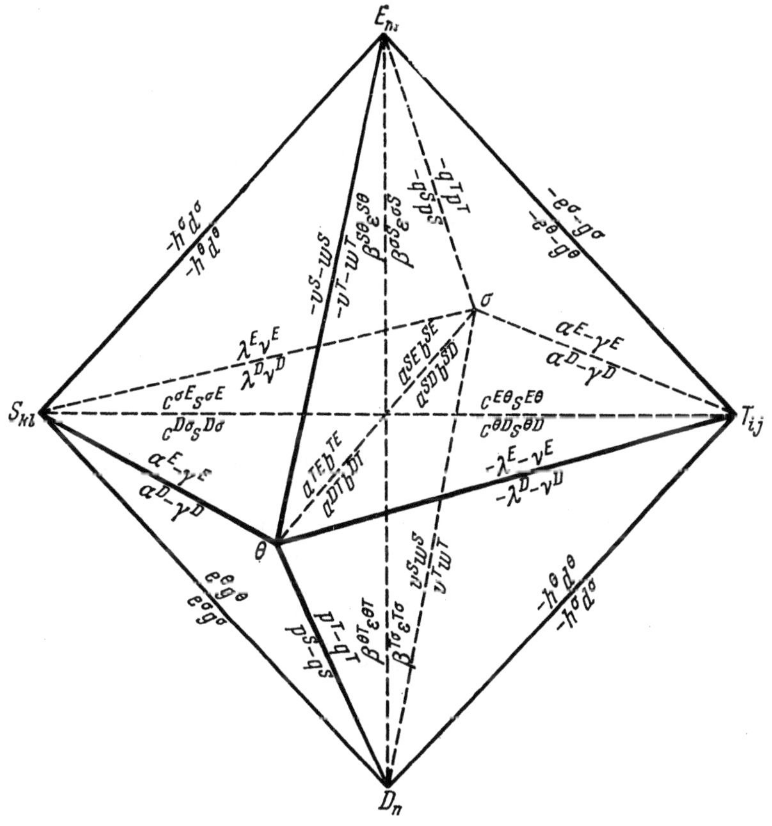

Fig. 1

the positions of θ and σ, (4.2) become

$$
\begin{aligned}
T_p &= c_{pq}^{D\theta}\, S_q - h_{ip}^{\theta}\, D_i - \lambda_p{}^D\, \theta,\\
E_j &= -\, h_{jq}^{\theta}\, S_q + \beta_{ij}^{\theta S}\, D_i - v_j{}^S\, \theta,\\
\sigma &= \lambda_q{}^D\, S_q + v_i{}^S\, D_i + a^{SD}\, \theta,
\end{aligned}
$$

[1] The symbol for the pyroelectric constant, $q_i{}^s$, should not be confused with the symbol for the heat vector q_i.

where the isothermal constants are expressed in terms of the adiabatic constants by

$$c_{pq}^{D\theta} = c_{pq}^{D\sigma} - a^{SD}\,\gamma_p{}^D\,\gamma_q{}^D, \qquad a^{SD}b^{SD} = 1,$$

$$h_{ip}^{\theta} = h_{ip}^{\sigma} + a^{SD}\,q_i{}^S\,\gamma_p{}^D, \qquad \lambda_p{}^D = a^{SD}\,\gamma_p{}^D,$$

$$\beta_{ij}^{\theta S} = \beta_{ij}^{\sigma S} - a^{SD}\,q_i{}^S\,q_j{}^S, \qquad v_i{}^S = a^{SD}\,q_i{}^S.$$

A further interchange, of E_i and D_i, gives the form convenient for the establishment of equations of motion

$$\begin{aligned} T_p &= c_{pq}^{E\theta}\,S_q - e_{ip}^{\theta}\,E_i - \lambda_p{}^E\,\theta, \\ D_j &= e_{jq}^{\theta}\,S_q + \varepsilon_{ij}^{\theta S}\,E_i + p_j{}^S\,\theta, \\ \sigma &= \lambda_q{}^E\,S_q + p_i{}^S\,E_i + a^{SE}\,\theta, \end{aligned} \tag{4.4}$$

where the new constants are given in terms of the preceding ones by

$$c_{pq}^{E\theta} = c_{pq}^{D\theta} - h_{ip}^{\theta}\,e_{iq}^{\theta}, \qquad \varepsilon_{ij}^{\theta S}\,\beta_{ij}^{\theta S} = \delta_{ij},$$

$$\lambda_p{}^E = \lambda_p{}^D + v_i{}^S\,e_{ip}^{\theta}, \qquad e_{ip}^{\theta} = \varepsilon_{ij}^{\theta S}\,h_{jp}^{\theta},$$

$$a^{SE} = a^{SD} + p_i{}^S\,v_i{}^S, \qquad p_i{}^S = \varepsilon_{ij}^{\theta S}\,v_j{}^S.$$

§ 5. Equations of motion and boundary conditions

To obtain the equations of motion, we replace the indices p and q in (4.4) by ij and kl and substitute in (2.3), (2.5) and (3.19), with the result

$$\begin{aligned} c_{ijkl}^{E\theta}\,u_{k,li} + e_{kij}^{\theta}\,\phi_{,ki} - \lambda_{ij}^{E}\,\theta_{,i} &= \rho\ddot{u}_j, \\ e_{kij}^{\theta}\,u_{i,jk} - \varepsilon_{ij}^{\theta S}\,\phi_{,ij} + p_i{}^S\,\theta_{,i} &= 0, \\ \lambda_{ij}^{E}\,\dot{u}_{i,j} - p_i{}^S\,\dot{\phi}_{,i} + a^{SE}\,\dot{\theta} &= \Theta_0{}^{-1}\kappa_{ij}\,\theta_{,ij}. \end{aligned} \tag{5.1}$$

Initial and boundary conditions sufficient for a unique solution of these equations may be established readily with the aid of the form of the principle of conservation of energy given in (3.24) in which S is now to be interpreted as the boundary of the body. The kinetic energy-density K is positive definite and, by (3.22) and (3.16), so is the dissipation function F. It is necessary, for stability, to require Biot's potential B to be positive definite and this may be done by specifying that its ten discriminants, obtained by omission of $0, 1 \cdots 9$ of its ten variables (six T_p or S_p three E_i or D_i and θ or σ), be greater than zero. Then the theorem of uniqueness and its proof may follow those given by J. Weiner [4] for the isotropic, thermoelastic case, requiring only the addition of the electrical terms. For the boundary conditions, inspection of the right hand side of (3.24) shows that at each point of the boundary S, where the unit normal is $\boldsymbol{\nu}$, any one of four mechanical conditions and either of two electrical conditions and either of two thermal conditions are sufficient as follows:

I_M : u_i (displacement),
or II_M: $\nu_i T_{ij}$ (traction),
or III_M: $\mu_i u_i$ and $\nu_j T_{ij} - \mu_i \mu_j \nu_k T_{jk}$ (the component of displacement in any direction $\boldsymbol{\mu}$ and the resultant traction in the plane perpendicular to $\boldsymbol{\mu}$),
or IV_M: $\nu_i \mu_j T_{ij}$ and $u_i - \mu_i \mu_j u_j$ (the component of traction in any direction $\boldsymbol{\mu}$ and the resultant displacement in the plane perpendicular to $\boldsymbol{\mu}$),
and I_E: ϕ (surface potential),
or II_E: $\nu_i D_i$ (surface charge)
and I_T: θ (surface temperature change),
or II_T: $\nu_i s_i$ (heat flux across surface).

In addition, conditions may be included such as the thermal radiation condition or elastic supports or the connection of the crystal to an electric circuit, if the appropriate energies are added to (3.24).

§ 6. Attenuation of plane waves

As an example of solutions of the equations of motion, we consider the case of plane waves in an infinite crystal. There is no loss of generality by taking a wave with its normal in the direction of the axis x_1

$$
\begin{aligned}
u_i &= A_i \cos \xi x_1 e^{i\omega t}, \quad i = 1, 2, 3, \\
\phi &= A_4 \cos \xi x_1 e^{i\omega t}, \\
\theta &= A_5 \sin \xi x_1 e^{i\omega t}.
\end{aligned} \tag{6.1}
$$

Substituting (6.1) into (5.1) and setting the determinant of the coefficients of the constants A equal to zero, we obtain the secular equation

$$
\begin{vmatrix}
c_{11}^{E\theta} - c & c_{16}^{E\theta} & c_{15}^{E\theta} & e_{11}^{\theta} & \lambda_1{}^E \\
c_{16}^{E\theta} & c_{66}^{E\theta} - c & c_{56}^{E\theta} & e_{16}^{\theta} & \lambda_6{}^E \\
c_{15}^{E\theta} & c_{56}^{E\theta} & c_{55}^{E\theta} - c & e_{15}^{\theta} & \lambda_5{}^E \\
e_{11}^{\theta} & e_{16}^{\theta} & e_{15}^{\theta} & -\varepsilon_{11}^{\theta S} & -p_1{}^S \\
\lambda_1{}^E & \lambda_6{}^E & \lambda_5{}^E & -p_1{}^S & -\Omega
\end{vmatrix} = 0, \tag{6.2}
$$

where

$$c = \frac{\rho\omega^2}{\xi^2}, \quad \Omega = a^{SE} + \frac{\kappa_{11}\xi^2}{i\omega\Theta_0}.$$

The roots c of (6.2) give the three complex stiffnesses.

Equation (6.2) may be condensed to the form

$$\omega\Delta^{D\theta} + i\omega_m\Delta^{D\sigma} = 0, \tag{6.3}$$

where

$$\omega_m = cC^{SE}\,\varepsilon_{11}^{\sigma S}/\kappa_{11}\,\varepsilon_{11}^{\theta S},$$

$$\Delta^{D\theta} = \begin{vmatrix} c_{11}^{D\theta} - c & c_{16}^{D\theta} & c_{15}^{D\theta} \\ c_{16}^{D\theta} & c_{66}^{D\theta} - c & c_{56}^{D\theta} \\ c_{15}^{D\theta} & c_{56}^{D\theta} & c_{55}^{D\theta} - c \end{vmatrix},$$

$$\Delta^{D\sigma} = \begin{vmatrix} c_{11}^{D\sigma} - c & c_{16}^{D\sigma} & c_{15}^{D\sigma} \\ c_{16}^{D\sigma} & c_{66}^{D\sigma} - c & c_{56}^{D\sigma} \\ c_{15}^{D\sigma} & c_{56}^{D\sigma} & c_{55}^{D\sigma} - c \end{vmatrix}.$$

Thus, $\Delta^{D\theta} = 0$ is the equation whose roots c give the three real isothermal stiffnesses, and $\Delta^{D\sigma} = 0$ is the equation whose roots give the three real adiabatic stiffnesses. Equation (6.3) shows that, as $\omega \to 0$, the stiffnesses approach their adiabatic values, and, as $\omega \to \infty$, the stiffnesses approach their isothermal values. An alternative form of (6.3) is

$$\Delta^{D\sigma} = (\Delta^{D\sigma} - \Delta^{D\theta})\,\frac{\omega^2 - i\omega\omega_m}{\omega^2 + \omega_m^2}. \tag{6.4}$$

Since the attenuation is small, the imaginary part of the quantity in parentheses may be neglected. Then the right hand side of (6.4) can be regarded as a correction on the equation for the adiabatic stiffnesses. The imaginary part gives the attenuations and has its maxima at $\omega = \omega_m$.

REFERENCES

1. Biot, M. A.: Linear thermodynamics and mechanics of solids, Proceedings of the Third U. S. National Congress of Applied Mechanics, The American Society of Mechanical Engineers, New York, 1958.
2. Eckart, C.: The thermodynamics of irreversible processes, Physical Review, 1948, *73*, No. 4, 373–382.
3. Mason, W. P.: Piezoelectric crystals and their applications to ultrasonics, D. Van Nostrand Company, New York, 1950, Chapter III.
4. Weiner, J. H.: A uniqueness theorem for the coupled thermoelastic problem, Quarterly of Applied Mathematics, 1957, *15*, No. 1, 102–105.

Reprinted from
QUARTERLY OF APPLIED MATHEMATICS
Vol. XX, No. 2, July, 1962

FORCED VIBRATIONS OF PIEZOELECTRIC CRYSTAL PLATES*

BY
H. F. TIERSTEN AND R. D. MINDLIN
Columbia University

1. Introduction. In a previous paper [1] (referred to in the sequel as I), Cauchy's two-dimensional equations of coupled flexural and extensional motion of crystal plates were extended to the next higher order of approximation so as to accommodate the two lowest thickness-shear modes. In the present paper a further extension is made to include the piezoelectric relations and the electric field equations. The equations obtained are also extensions of previous equations [2] in which the thickness-shear modes, the piezoelectric effect and the electric field equations were taken into account, but coupling with extensional modes was omitted. The new equations are deduced from the three-dimensional, linear, piezoelectric equations by a procedure based, as in I, on the series expansion methods of Cauchy and Poisson and the variational method of Kirchhoff. The theorems of uniqueness and orthogonality established in I for the approximate equations are extended to include coupling with the electric field. The solution of problems of steady, forced vibration is considered and is reduced to the solution of the associated free vibration problem plus some quadratures. The case of a plate driven by an ac-voltage applied to electrodes on its faces is investigated and a formula is given for the total surface charge. An application is made to the forced vibrations of rectangular, rotated Y-cut quartz plates.

2. Three-dimensional equations. The three-dimensional equations, from which the plate equations will be deduced, are: the variational forms of the equations of motion and electrostatics; the strain-displacement and electric field-potential relations; and the linear, piezoelectric constitutive relations. They may be written, respectively, as

$$\int_R (T_{ij,i} - \rho u_{j,tt})\, \delta u_j \, dR = 0, \qquad \int_R D_{i,i}\, \delta\varphi \, dR = 0 \tag{2.1}$$

$$S_{ij} = \tfrac{1}{2}(u_{i,j} + u_{j,i}), \qquad E_i = -\varphi_{,i}, \tag{2.2}$$

$$T_{ij} = c^E_{ijkl} S_{kl} - e_{kij} E_k , \qquad D_i = e_{ikl} S_{kl} + \epsilon^S_{ik} E_k , \tag{2.3}$$

where T_{ij} , u_j , D_i , S_{ij} , E_i are the components of stress, mechanical displacement, electric displacement, strain and electric field, respectively; ρ, R and φ are the mass density, volume and electric potential, respectively; and c^E_{ijkl} , e_{kij} , ϵ^S_{ik} are the elastic, piezoelectric and dielectric constants, respectively. We employ rectangular coordinates x_i , $i = 1, 2, 3$, and the summation convention for repeated tensor indices. A comma followed by an index or t denotes differentiation with respect to a space coordinate or time, respectively.

3. Expansion in power series. The faces of the plate are taken at $x_2 = \pm h$ and the remaining boundary is a cylindrical surface with generators perpendicular to the

*Received September 25, 1961.

faces. We expand u_j and D_i in a series of powers of the thickness-coordinate:

$$u_j = \sum_{n=0}^{g} x_2^n u_j^{(n)}, \qquad D_j = \sum_{n=0}^{g} x_2^n D_j^{(n)} + \delta_{2j} x_2^{g+1} D_2^{(g+1)}, \tag{3.1}$$

where g is a positive integer, δ_{ij} is the Kronecker delta and the variables $u_j^{(n)}$, $D_j^{(n)}$ are independent of x_2. When the series expressions are substituted in (2.1) and the integrations with respect to x_2 are performed, the result is

$$\int_A \sum_{n=0}^{g} \left(T_{ij,i}^{(n)} - nT_{2j}^{(n-1)} + F_j^{(n)} - \rho \sum_{m=0}^{g} H_{mn} u_{j,tt}^{(m)} \right) \delta u_j^{(n)} \, dA = 0, \tag{3.2}$$

$$\int_A \sum_{n=0}^{g} (D_{i,i}^{(n)} + (n+1) D_2^{(n+1)}) \, \delta\varphi^{(n)} \, dA = 0, \tag{3.3}$$

where A is the area of the plate and

$$T_{ij}^{(n)} \equiv \int_{-h}^{h} x_2^n T_{ij} \, dx_2 , \qquad F_j^{(n)} \equiv [x_2^n T_{2j}]_{-h}^{h} , \qquad \varphi^{(n)} \equiv \int_{-h}^{h} x_2^n \varphi \, dx_2 ,$$

$$H_{mn} \equiv \int_{-h}^{h} x_2^m x_2^n \, dx_2 = 2h^{m+n+1}/(m+n+1), \quad m+n \text{ even}; j = 0, m+n \text{ odd}. \tag{3.4}$$

Since the variations $\delta u_j^{(n)}$ and $\delta\varphi^{(n)}$ are arbitrary, their coefficients in (3.2) and (3.3) must vanish; and we have, as the two dimensional equations of motion and electrostatics of order n,

$$T_{ij,i}^{(n)} - nT_{2j}^{(n-1)} + F_j^{(n)} = \rho \sum_{m=0}^{g} H_{mn} u_{j,tt}^{(m)} , \qquad D_{i,i}^{(n)} + (n+1) D_2^{(n+1)} = 0. \tag{3.5}$$

If $(3.1)_2$ is substituted in the three-dimensional equation of electrostatics, $D_{i,i} = 0$, the result is

$$\sum_{n=0}^{g} x_2^n [D_{i,i}^{(n)} + (n+1) D_2^{(n+1)}] = 0,$$

which is satisfied identically by virtue of $(3.5)_2$. This was the motivation for the special form of the expansion of D_j in $(3.1)_2$.

Substituting $(3.1)_1$ in $(2.2)_1$ and rearranging terms, we obtain

$$S_{ij} = \sum_{n=0}^{g} x_2^n S_{ij}^{(n)}, \qquad S_{ij}^{(n)} \equiv \tfrac{1}{2}[u_{i,j}^{(n)} + u_{j,i}^{(n)} + (n+1)(\delta_{2j} u_i^{(n+1)} + \delta_{2i} u_j^{(n+1)})]. \tag{3.6}$$

Next, multiply $(2.2)_2$ by $x_2^n \, dx_2$ and integrate over the thickness. With $(3.4)_3$ and an integration by parts, we find

$$E_i^{(n)} \equiv \int_{-h}^{h} x_2^n E_i \, dx_2 = -\varphi_{,i}^{(n)} - \delta_{2i}(\Phi^{(n)} - n\varphi^{(n-1)}), \qquad \Phi^{(n)} \equiv [x_2^n \varphi]_{-h}^{h} . \tag{3.7}$$

At this stage we have converted (2.1) and (2.2) to two-dimensional form and there remain the constitutive relations (2.3). We substitute $(3.1)_2$ and $(3.6)_1$ in (2.3), multiply by $x_2^m \, dx_2$ and integrate over the thickness, with the result

$$T_{ij}^{(m)} = c_{ijkl}^{E} \sum_{n=0}^{g} H_{mn} S_{kl}^{(n)} - e_{kij} E_k^{(m)}, \qquad \sum_{n=0}^{g} H_{mn}(D_i^{(n)} - e_{ikl} S_{kl}^{(n)}) = \epsilon_{ik}^{S} E_k^{(m)}. \tag{3.8}$$

These may be written, in the abbreviated indicial notation, as

$$T_p^{(m)} = c_{pq}^E \sum_{n=0}^{g} H_{mn} S_q^{(n)} - e_{ip} E_i^{(m)}, \qquad \sum_{n=0}^{g} H_{mn}(D_i^{(n)} - e_{iq} S_q^{(n)}) = \epsilon_{ik}^S E_k^{(m)}, \tag{3.9}$$

where ij or kl = 11, 22, 33, 23 or 32, 31 or 13, 12 or 21 are replaced by p or q = 1, 2, 3, 4, 5, 6, respectively; and $S_p^{(n)} = S_{ij}^{(n)}$, $i = j$ and $S_p^{(n)} = 2S_{ij}^{(n)}$, $i \neq j$.

Finally, we define two-dimensional kinetic energy, internal energy and electric enthalpy [3] densities by, respectively,

$$\mathcal{K} \equiv \int_{-h}^{h} \tfrac{1}{2}\rho u_{j,t} u_{j,t}\, dx_2 = \tfrac{1}{2}\rho \sum_{m=0}^{g} \sum_{n=0}^{g} H_{mn} u_{j,t}^{(m)} u_{j,t}^{(n)}, \tag{3.10}$$

$$\mathcal{U} \equiv \int_{-h}^{h} \tfrac{1}{2}(T_{ij}S_{ij} + D_iE_i)\, dx_2 = \tfrac{1}{2} \sum_{n=0}^{g} (T_{ij}^{(n)} S_{ij}^{(n)} + E_i^{(n)} D_i^{(n)}) + \tfrac{1}{2} E_2^{(g+1)} D_2^{(g+1)}, \tag{3.11}$$

$$\mathcal{H} \equiv \int_{-h}^{h} \tfrac{1}{2}(T_{ij}S_{ij} - D_iE_i)\, dx_2 = \tfrac{1}{2} \sum_{n=0}^{g} (T_{ij}^{(n)} S_{ij}^{(n)} - E_i^{(n)} D_i^{(n)}) - \tfrac{1}{2} E_2^{(g+1)} D_2^{(g+1)}, \tag{3.12}$$

where $(3.1)_1$, $(3.1)_2$, $(3.4)_4$ and $(3.6)_1$ have been used, along with the definitions $(3.4)_1$ and $(3.7)_1$.

4. Truncation of series. In (3.9) we retain only $T_p^{(0)}$, $T_p^{(1)}$, $S_p^{(0)}$, $S_p^{(1)}$, $D_i^{(0)}$, $D_i^{(1)}$, $D_i^{(2)}$, $E_i^{(0)}$, $E_i^{(1)}$ and $E_2^{(2)}$ whence, tentatively,

$$T_p^{(0)} = 2hc_{pq}^E S_q^{(0)} - e_{ip}E_i^{(0)}, \qquad T_p^{(1)} = \tfrac{2}{3}h^3 c_{pq}^E S_q^{(1)} - e_{ip}E_i^{(1)},$$

$$\tfrac{2}{3}h^3 D_i^{(1)} = \tfrac{2}{3}h^3 e_{iq} S_q^{(1)} + \epsilon_{ik}^S E_k^{(1)}, \tag{4.1}$$

$$2h\, D_i^{(0)} + \tfrac{2}{3}h^3\, D_i^{(2)} = 2he_{iq}S_q^{(0)} + \epsilon_{ik}^S E_k^{(0)},$$

$$\tfrac{2}{3}h^3\, D_i^{(0)} + \tfrac{2}{5}h^5\, D_i^{(2)} = \tfrac{2}{3}h^3 e_{iq} S_q^{(0)} + \epsilon_{i2}^S E_2^{(2)}. \tag{4.2}$$

Eqs. (4.2) may be solved for $D_i^{(0)}$ and $D_i^{(2)}$ with the result, again tentative,

$$D_i^{(0)} = e_{iq}S_q^{(0)} + (9/8h)\epsilon_{ik}^S E_k^{(0)} - (15/8h^3)\epsilon_{i2}^S E_2^{(2)},$$

$$D_i^{(2)} = (15/8h^3)(3h^{-2}\epsilon_{i2}^S E_2^{(2)} - \epsilon_{ik}^S E_k^{(0)}). \tag{4.3}$$

The components $S_p^{(0)}$ and $S_p^{(1)}$ involve only the displacements $u_j^{(0)}$, $u_j^{(1)}$ and $u_j^{(2)}$ and we neglect those of higher order. Then, following Cauchy, as in *I*, we neglect the velocity $u_{2,t}^{(1)}$ in the kinetic energy and equations of motion and provide for the free development of the thickness-strain $S_2^{(0)}$ by setting $T_2^{(0)} = 0$ in $(4.1)_1$. The condition $T_2^{(0)} = 0$ permits the elimination of $S_2^{(0)}$ from $(4.1)_1$ and $(4.3)_1$ with the result, still tentative,

$$T_p^{(0)} = 2hc_{pq}^{*} S_q^{(0)} - e_{ip}^{*} E_i^{(0)}, \qquad D_i^{(0)} = e_{iq}^{*} S_q^{(0)} + \tfrac{1}{2}h^{-1}\epsilon_{ij}^{*} E_j^{(0)} - (15/8h^3)\epsilon_{i2}^S E_2^{(2)}, \tag{4.4}$$

where

$$c_{pq}^{*} = c_{pq}^E - c_{p2}^E c_{2q}^E / c_{22}^E, \qquad e_{ip}^{*} = e_{ip} - e_{i2} c_{p2}^E / c_{22}^E, \qquad \epsilon_{ij}^{*} = (9/4)\epsilon_{ij}^S + e_{i2}e_{j2}/c_{22}^E. \tag{4.5}$$

The first order terms are treated as in *I*: all three velocities $u_{j,t}^{(2)}$ are neglected and free development of the three strains $S_{2j}^{(1)}$ is accommodated by setting $T_{2j}^{(1)} = 0$ in $(4.1)_2$. Then, upon elimination of the $S_{2j}^{(1)}$, $(4.1)_2$ and $(4.1)_3$ become

$$T_r^{(1)} = \tfrac{2}{3}h^3 \gamma_{rs} S_s^{(1)} - \psi_{ir} E_i^{(1)}, \qquad D_i^{(1)} = \psi_{ir} S_r^{(1)} + \tfrac{3}{2}h^{-3} \zeta_{ij} E_j^{(1)}, \tag{4.6}$$

where, with $i, j = 1, 2, 3$; $r, s = 1, 3, 5$; $v, w = 2, 4, 6$:

$$\gamma_{rs} = c_{rs}^E - c_{rw}^E c_{vs}^E (c_{vw}^E)^{-1}, \qquad \psi_{ir} = e_{ir} - e_{iv} c_{rw}^E (c_{vw}^E)^{-1}, \qquad \zeta_{ij} = \epsilon_{ij}^S + e_{iv} e_{jw} (c_{vw}^E)^{-1}. \tag{4.7}$$

At this stage, the electric enthalpy is

$$\mathcal{H} = \tfrac{1}{2}(T_p^{(0)}S_p^{(0)} + T_p^{(1)}S_p^{(1)} - E_i^{(0)}D_i^{(0)} - E_i^{(1)}D_i^{(1)} - E_2^{(2)}D_2^{(2)})$$

$$= \tfrac{1}{2}[2hc_{pq}^{*}S_p^{(0)}S_q^{(0)} - 2e_{ip}^{*}S_p^{(0)}E_i^{(0)} + \tfrac{2}{3}h^3\gamma_{rs}S_r^{(1)}S_s^{(1)} - 2\psi_{ir}E_i^{(1)}S_r^{(1)} - \tfrac{1}{2}h^{-1}\epsilon_{ij}^{*}E_i^{(0)}E_j^{(0)}$$

$$- \tfrac{3}{2}h^{-3}\zeta_{ij}E_i^{(1)}E_j^{(1)} + (15/4)h^{-3}\overset{S}{\epsilon}_{2k}E_2^{(2)}E_k^{(0)} - (45/8)h^{-5}\overset{S}{\epsilon}_{22}E_2^{(2)}E_2^{(2)}], \qquad (4.8)$$

and an analogous expression may be written for the internal energy $\mathcal{U}$.

The final step in the process of truncation is to introduce the usual shear-correction factors by replacing $S_4^{(0)}$ and $S_6^{(0)}$, in $\mathcal{H}$ and $\mathcal{U}$, with $\kappa_4 S_4^{(0)}$ and $\kappa_6 S_6^{(0)}$, where the values of κ_4 and κ_6 are to be fixed so as to adjust the two thickness-shear frequencies to their exact values. The correction factors may be inserted in the appropriate positions by replacing c_{pq}^{*} and e_{ip}^{*}, in $\mathcal{H}$ and $\mathcal{U}$, by

$$c_{pq}^{**} = \kappa_p^{\alpha}\kappa_q^{\beta}c_{pq}^{*}, \qquad e_{iq}^{**} = \kappa_q^{\beta}e_{iq}^{*} \quad \text{(no sum)} \qquad (4.9)$$

where α and β are the powers $\alpha = \cos^2(p\pi/2)$, $\beta = \cos^2(q\pi/2)$. Thus, κ_p^{α} (or κ_q^{β}) is equal to κ_4, κ_6 or 1 according as p (or q) in c_{pq}^{*} and e_{ip}^{*} is 4, 6 or neither, respectively. The same definitions hold for c_{ijkl}^{**}, e_{ikl}^{**}, κ_{ij}^{α}, κ_{kl}^{β} if p and q are replaced by ij and kl.

To recapitulate, after the truncation and adjustments, we have the following*:

Electric enthalpy density: $\mathcal{H}$ is given by (4.8) with c_{pq}^{*} and e_{ip}^{*} replaced by c_{pq}^{**} and e_{ip}^{**}.

Internal energy density:

$$\mathcal{U} = \mathcal{H} + E_i^{(0)}D_i^{(0)} + E_i^{(1)}D_i^{(1)} + E_2^{(2)}D_2^{(2)} \qquad (4.10)$$

Kinetic energy density:

$$\mathcal{K} = \rho h(u_{i,t}^{(0)}u_{i,t}^{(0)} + \tfrac{1}{3}h^2u_{a,t}^{(1)}u_{a,t}^{(1)}). \qquad (4.11)$$

Constitutive relations:

$$\begin{aligned}
T_p^{(0)} &= \partial\mathcal{H}/\partial S_p^{(0)} = 2hc_{pq}^{**}S_q^{(0)} - e_{kp}^{**}E_k^{(0)},\\
T_r^{(1)} &= \partial\mathcal{H}/\partial S_r^{(1)} = \tfrac{2}{3}h^3\gamma_{rs}S_s^{(1)} - \psi_{ir}E_i^{(1)},\\
D_i^{(0)} &= -\partial\mathcal{H}/\partial E_i^{(0)} = e_{iq}^{**}S_q^{(0)} + \tfrac{1}{2}h^{-1}\epsilon_{ij}^{*}E_j^{(0)} - (15/8h^3)\overset{S}{\epsilon}_{i2}E_2^{(2)},\\
D_i^{(1)} &= -\partial\mathcal{H}/\partial E_i^{(1)} = \psi_{ir}S_r^{(1)} + \tfrac{3}{2}h^{-3}\zeta_{ij}E_j^{(1)},\\
D_2^{(2)} &= -\partial\mathcal{H}/\partial E_2^{(2)} = (15/8h^5)(3\overset{S}{\epsilon}_{22}E_2^{(2)} - h^2\overset{S}{\epsilon}_{2k}E_k^{(0)}).
\end{aligned} \qquad (4.12)$$

Equations of motion:

$$T_{ij,i}^{(0)} + F_j^{(0)} = 2\rho h u_{j,tt}^{(0)}, \qquad T_{ab,a}^{(1)} - T_{2b}^{(0)} + F_b^{(1)} = \tfrac{2}{3}\rho h^3 u_{b,tt}^{(1)}. \qquad (4.13)$$

Equations of electrostatics:

$$D_{i,i}^{(0)} + D_2^{(1)} = 0, \qquad D_{i,i}^{(1)} + 2D_2^{(2)} = 0 \qquad (4.14)$$

Strain-displacement relations:

$$S_{ij}^{(0)} = \tfrac{1}{2}(u_{i,j}^{(0)} + u_{j,i}^{(0)} + \delta_{2j}u_i^{(1)} + \delta_{2i}u_j^{(1)}), \qquad S_{ab}^{(1)} = \tfrac{1}{2}(u_{a,b}^{(1)} + u_{b,a}^{(1)}). \qquad (4.15)$$

---

*Here and in the sequel indices a, b, c, d range over 1, 3; r, s over 1, 3, 5; i, j, k, l over 1, 2, 3.

Electric field-potential relations:

$$E_i^{(0)} = -\varphi_{,i}^{(0)} - \delta_{i2}\Phi^{(0)},\ E_i^{(1)} = -\varphi_{,i}^{(1)} + \delta_{i2}(\varphi^{(0)} - \Phi^{(1)}),\ E_2^{(2)} = 2\varphi^{(1)} - \Phi^{(2)}, \tag{4.16}$$

and we note that $\Phi^{(2)} = h^2\Phi^{(0)}$.

Eqs. (4.12)–(4.16) comprise thirty-seven equations in the thirty seven variables: five each of $T_p^{(0)}$ and $S_p^{(0)}$; three each of $T_r^{(1)}$, $S_r^{(1)}$, $D_i^{(0)}$, $E_i^{(0)}$, $D_i^{(1)}$, $E_i^{(1)}$ and $u_i^{(0)}$; two $u_a^{(1)}$; and $D_2^{(2)}$, $E_2^{(2)}$, $\varphi^{(0)}$ and $\varphi^{(1)}$. The thirty-seven equations may be reduced to seven in seven variables by first substituting (4.15) and (4.16) in (4.12), with the result

$$\begin{aligned}
T_{ij}^{(0)} &= 2hc_{ijkl}^{**}(u_{k,l}^{(0)} + \delta_{2l}u_k^{(1)}) + e_{kij}^{**}\varphi_{,k}^{(0)} + e_{2ij}^{**}\Phi^{(0)},\\
T_{ab}^{(1)} &= \tfrac{2}{3}h^3\gamma_{abcd}u_{c,d}^{(1)} - \psi_{2ab}\varphi^{(0)} + \psi_{iab}\varphi_{,i}^{(1)} + \psi_{2ab}\Phi^{(1)},\\
D_i^{(0)} &= e_{ikl}^{**}(u_{k,l}^{(0)} + \delta_{2l}u_k^{(1)}) - \tfrac{1}{2}h^{-1}\epsilon_{ij}^{*}\varphi_{,j}^{(0)} - (15/4h^3)\epsilon_{i2}^{S}\varphi^{(1)} - \tfrac{1}{8}h^{-1}\epsilon'_{i2}\Phi^{(0)},\\
D_i^{(1)} &= \psi_{iab}u_{a,b}^{(1)} + \tfrac{3}{2}h^{-3}\zeta_{i2}\varphi^{(0)} - \tfrac{3}{2}h^{-3}\zeta_{ij}\varphi_{,j}^{(1)} - \tfrac{3}{2}h^{-3}\zeta_{i2}\Phi^{(1)},\\
D_2^{(2)} &= (15/8h^5)(h^2\epsilon_{2k}^{S}\varphi_{,k}^{(0)} + 6\epsilon_{22}^{S}\varphi^{(1)} - 2h^2\epsilon_{22}^{S}\Phi^{(0)}),
\end{aligned} \tag{4.17}$$

where $\epsilon'_{i2} = 4\epsilon_{i2}^{*} - 15\epsilon_{i2}^{S}$; and then substituting (4.17) into (4.13) and (4.14) to obtain

$$\begin{aligned}
&2hc_{ijkl}^{**}(u_{k,li}^{(0)} + \delta_{2k}u_{l,i}^{(1)}) + e_{kij}^{**}\varphi_{,ki}^{(0)} + F_j^{(0)} + e_{2ij}^{**}\Phi_{,i}^{(0)} = 2\rho h u_{j,tt}^{(0)},\\
&\tfrac{2}{3}h^3\gamma_{abcd}u_{c,da}^{(1)} - 2hc_{2bkl}^{**}(u_{k,l}^{(0)} + \delta_{2k}u_l^{(1)}) - e_{i2b}^{**}\varphi_{,i}^{(0)} + \psi_{iab}(\varphi_{,ia}^{(1)} - \delta_{2i}\varphi_{,a}^{(0)})\\
&\qquad\qquad + F_b^{(1)} - e_{22b}^{**}\Phi^{(0)} + \psi_{2ab}\Phi_{,a}^{(1)} = \tfrac{2}{3}\rho h^3 u_{b,tt}^{(1)},\\
&\tfrac{2}{3}h^3\psi_{2kl}u_{k,l}^{(1)} + \tfrac{2}{3}h^3 e_{ikl}^{**}(u_{k,li}^{(0)} + \delta_{2k}u_{l,i}^{(1)}) - \tfrac{1}{3}h^2\epsilon_{ij}^{**}\varphi_{,ij}^{(0)} - \zeta_{2j}(\varphi_{,j}^{(1)} - \delta_{2j}\varphi^{(0)})\\
&\qquad\qquad - 5/2\epsilon_{k2}^{S}\varphi_{,k}^{(1)} - (1/12)h^2\epsilon'_{i2}\Phi_{,i}^{(0)} - \zeta_{22}\Phi^{(1)} = 0,\\
&\tfrac{2}{3}h^3\psi_{iab}u_{a,bi}^{(1)} - \zeta_{ij}(\varphi_{,ij}^{(1)} - \delta_{2j}\varphi_{,i}^{(0)}) + 5/2\epsilon_{2k}^{S}\varphi_{,k}^{(0)} + 15h^{-2}\epsilon_{22}^{S}\varphi^{(1)}\\
&\qquad\qquad - 5\epsilon_{22}^{S}\Phi^{(0)} - \zeta_{i2}\Phi_{,i}^{(1)} = 0.
\end{aligned} \tag{4.18}$$

These are seven second order differential equations in the seven dependent variables $u_i^{(0)}$, $u_a^{(1)}$, $\varphi^{(0)}$ and $\varphi^{(1)}$ with seven non-homogeneous (forcing) terms: the five surface tractions $F_j^{(0)}$, $F_b^{(1)}$ and the two surface potentials $\Phi^{(0)}$, $\Phi^{(1)}$.

5. Uniqueness of solution. To establish a theorem of uniqueness of solution of the thirty-seven equations (4.12)–(4.16) we consider two sets of the thirty-seven variables, each set satisfying the equations, and form a system comprising the thirty-seven differences between corresponding variables in the two sets. In terms of variables in the difference-system, we form the equation

$$\int_{t_0}^{t} dt \int_A [(T_{ij,i}^{(0)} + F_j^{(0)} - 2\rho h u_{j,tt}^{(0)})u_{j,t}^{(0)} + (T_{ab,a}^{(1)} - T_{2b}^{(0)} + F_b^{(1)} - \tfrac{2}{3}\rho h^3 u_{b,tt}^{(1)})u_{b,t}^{(1)}]\, dA = 0,$$

which can be transformed, by a process analogous to that employed in *I*, to

$$\begin{aligned}
\int_A [\mathfrak{U} + \mathfrak{K}]_{t_0}^{t}\, dA = &\int_{t_0}^{t} dt \oint_C n_a(T_{aj}^{(0)}u_{j,t}^{(0)} + T_{ab}^{(1)}u_{b,t}^{(1)} - \varphi^{(0)}\, D_{a,t}^{(0)} - \varphi^{(1)}\, D_{a,t}^{(1)})\, ds\\
&+ \int_{t_0}^{t} dt \int_A [F_j^{(0)}u_{j,t}^{(0)} + F_b^{(1)}u_{b,t}^{(1)} - \Phi^{(0)}(D_{2,t}^{(0)} + h^2\, D_{2,t}^{(2)}) - \Phi^{(1)}\, D_{2,t}^{(1)}]\, dA,
\end{aligned} \tag{5.1}$$

in which the line integral is around the edge, C, of the plate and the n_a are the components of the outward normal in the plane of the plate. From (5.1), by the usual argument, there are seven conditions to be specified at each point of the interior and at each point on the edge of the plate in addition to the initial values of $u_i^{(0)}$, $u_a^{(1)}$, $\varphi^{(0)}$, $\varphi^{(1)}$, $u_{i,t}^{(0)}$ and $u_{a,t}^{(1)}$. Referred to orthogonal coordinates n, s, x_2, the edge conditions are one member of each of the seven products

$$T_{nn}^{(0)}u_n^{(0)}, \quad T_{ns}^{(0)}u_s^{(0)}, \quad T_{n2}^{(0)}u_2^{(0)}, \quad T_{nn}^{(1)}u_n^{(1)}, \quad T_{ns}^{(1)}u_s^{(1)}, \quad \varphi^{(0)}D_n^{(0)}, \quad \varphi^{(1)}D_n^{(1)}.$$

In terms of orthogonal coordinates α, β, x_2, the interior conditions are one member of each of the seven products

$$F_\alpha^{(0)}u_\alpha^{(0)}, \quad F_\beta^{(0)}u_\beta^{(0)}, \quad F_2^{(0)}u_2^{(0)}, \quad F_\alpha^{(1)}u_\alpha^{(1)}, \quad F_\beta^{(1)}u_\beta^{(1)}, \quad \Phi^{(0)}(D_2^{(0)} + h^2 D_2^{(2)}), \quad \Phi^{(1)}D_2^{(1)}.$$

The terms $D_2^{(0)} + h^2 D_2^{(2)}$ and $D_2^{(1)}$, in the interior conditions, can be interpreted by recalling that, to our approximation,

$$D_2 = D_2^{(0)} + x_2 D_2^{(1)} + x_2^2 D_2^{(2)}. \tag{5.2}$$

Hence

$$D_2^{(0)} + h^2 D_2^{(2)} = \tfrac{1}{2}[D_2(h) + D_2(-h)] \quad \text{and} \quad D_2^{(1)} = \tfrac{1}{2}h^{-1}[D_2(h) - D_2(-h)].$$

6. Orthogonal functions. Consider two solutions

$$(u_j^{(0)}, u_b^{(1)}, \varphi^{(0)}, \varphi^{(1)}) = (u_j^{(0)\mu}, u_b^{(1)\mu}, \varphi^{(0)\mu}, \varphi^{(1)\mu})e^{i\omega_\mu t},$$

$$(u_j^{(0)}, u_b^{(1)}, \varphi^{(0)}, \varphi^{(1)}) = (u_j^{(0)\nu}, u_b^{(1)\nu}, \varphi^{(0)\nu}, \varphi^{(1)\nu})e^{i\omega_\nu t},$$

of the equations of motion (4.13) and electrostatics (4.14) with $(F_j^{(0)}, F_b^{(1)}) = 0$. By a process similar to that employed in I, we find

$$\begin{aligned}
2\rho h(\omega_\mu^2 - \omega_\nu^2) &\int_A (u_j^{(0)\mu}u_j^{(0)\nu} + \tfrac{1}{3}h^2 u_b^{(1)\mu}u_b^{(1)\nu})\, dA \\
= &\int_A [\Phi^{(0)\mu}(D_2^{(0)\nu} + h^2 D_2^{(2)\nu}) - \Phi^{(0)\nu}(D_2^{(0)\mu} + h^2 D_2^{(2)\mu}) + \Phi^{(1)\mu} D_2^{(1)\nu} - \Phi^{(1)\nu} D_2^{(1)\mu}]\, dA \\
&+ \oint_C n_a(T_{aj}^{(0)\nu}u_j^{(0)\mu} - T_{aj}^{(0)\mu}u_j^{(0)\nu} + T_{ab}^{(1)\nu}u_b^{(1)\mu} - T_{ab}^{(1)\mu}u_b^{(1)\nu})\, ds \\
&+ \oint_C n_a(D_a^{(0)\nu}\varphi^{(0)\mu} - D_a^{(0)\mu}\varphi^{(0)\nu} + D_a^{(1)\nu}\varphi^{(1)\mu} - D_a^{(1)\mu}\varphi^{(1)\nu})\, ds.
\end{aligned} \tag{6.1}$$

The surface integral in (6.1) vanishes for either zero electric potential or zero electric displacement on both faces of the plate; and the line integrals vanish for homogeneous boundary conditions on the edge. Thus, for homogeneous conditions on the faces and edge, (6.1) becomes

$$\int_A (u_j^{(0)\mu}u_j^{(0)\nu} + \tfrac{1}{3}h^2 u_b^{(1)\mu}u_b^{(1)\nu})\, dA = N_{(\mu)}\,\delta_{\mu\nu}, \tag{6.2}$$

where $N_{(\mu)}$ is a normalization factor and $\delta_{\mu\nu}$ is the Kronecker delta.

7. Steady forced vibrations. A problem of technological interest is that of steady vibrations forced by an ac-voltage applied to electrodes on the faces of a plate with its edge free. This case is included in solutions of (4.18) with prescribed tractions $F_j^{(0)}$, $F_b^{(1)}$ and potentials $\Phi^{(0)}$, $\Phi^{(1)}$ on the faces and with boundary conditions

$$n_a T_{aj}^{(0)} = n_a T_{ab}^{(1)} = n_a D_a^{(0)} = n_a D_a^{(1)} = 0 \tag{7.1}$$

on the edge. Although (7.1) appear to be homogeneous boundary conditions, they are actually inhomogeneous due to the presence of the prescribed potentials $\Phi^{(0)}$ and $\Phi^{(1)}$ in (4.17). A general solution will be obtained by an adaptation of the classical device [4] of introducing an auxiliary function which transfers the boundary inhomogeneity to the differential equations, leaving a residual problem of inhomogeneous differential equations and homogeneous boundary conditions. In our case, the eqautions of motion, $(4.18)_1$ and $(4.18)_2$, and electrostatics, $(4.18)_3$ and $(4.18)_4$, are initially inhomogeneous. We shall choose the auxiliary function so that it removes the inhomogeneity not only from the boundary conditions but also from the electrostatic equations, leaving a residuai problem of inhomogeneous equations of motion, homogeneous electrostatic equations and homogeneous boundary conditions. Let

$$(u_j^{(0)}, u_b^{(1)}, \varphi^{(0)}, \varphi^{(1)}) = (u_j^{(0)A}, u_b^{(1)A}, \varphi^{(0)A}, \varphi^{(1)A})e^{i\omega t} + (u_j^{(0)R}, u_b^{(1)R}, \varphi^{(0)R}, \varphi^{(1)R})e^{i\omega t}, \quad (7.2)$$

where the superscript A identifies the auxiliary functions and R the residual solution. Inserting (7.2) in (4.17) and cancelling the factor $e^{i\omega t}$, we obtain expressions of the type

$$T_{ij}^{(0)} = T_{ij}^{(0)R} + T_{ij}^{(0)A} + e_{2ij}^{**}\Phi^{(0)}, \quad (7.3)$$

where

$$T_{ij}^{(0)R} = 2hc_{ijkl}^{**}(u_{k,l}^{(0)R} + \delta_{2l}u_k^{(1)R}) + e_{kij}^{**}\varphi_{,k}^{(0)R},$$
$$T_{ij}^{(0)A} = 2hc_{ijkl}^{**}(u_{k,l}^{(0)A} + \delta_{2l}u_k^{(1)A}) + e_{kij}^{**}\varphi_{,k}^{(0)A}.$$

With resolutions like (7.3) for all of (4.17), Eqs. (4.8) become

$$T_{ij,i}^{(0)R} + 2\rho h\omega^2 u_j^{(0)R} + T_{ij,i}^{(0)A} + 2\rho h\omega^2 u_j^{(0)A} + F_j^{(0)} + e_{2ij}^{**}\Phi_{,i}^{(0)} = 0,$$
$$T_{ab,a}^{(1)R} - T_{2b}^{(0)R} + \tfrac{2}{3}\rho h^3\omega^2 u_b^{(1)R} + T_{ab,a}^{(1)A} - T_{2b}^{(0)A}$$
$$+ \tfrac{2}{3}\rho h^3\omega^2 u_b^{(1)A} + F_b^{(1)} - e_{22b}^{**}\Phi^{(0)} + \psi_{2ab}\Phi_{,a}^{(1)} = 0,$$
$$D_{i,i}^{(0)R} + D_2^{(1)R} + D_{i,i}^{(0)A} + D_2^{(1)A} - \tfrac{1}{8}h^{-1}\epsilon'_{i2}\Phi_{,i}^{(0)} - \tfrac{3}{2}h^{-3}\zeta_{22}\Phi^{(1)} = 0,$$
$$D_{i,i}^{(1)R} + 2\,D_2^{(2)R} + D_{i,i}^{(1)A} + 2\,D_2^{(2)A} - (15/2h^3)\overset{S}{\epsilon}_{22}\Phi^{(0)} - \tfrac{3}{2}h^{-3}\zeta_{i2}\Phi_{,i}^{(1)} = 0, \quad (7.4)$$

and the boundary conditions (7.1) become

$$n_aT_{aj}^{(0)R} + n_aT_{aj}^{(0)A} + n_ae_{2aj}^{**}\Phi^{(0)} = 0,\ n_aT_{ab}^{(1)R} + n_aT_{ab}^{(1)A} + n_a\psi_{2ab}\Phi^{(1)} = 0,$$
$$n_a\,D_a^{(0)R} + n_a\,D_a^{(0)A} - \tfrac{1}{8}h^{-1}n_a\epsilon'_{a2}\Phi^{(0)} = 0,\ n_a\,D_a^{(1)R} + n_a\,D_a^{(1)A} - \tfrac{3}{2}h^{-3}n_a\zeta_{a2}\Phi^{(1)} = 0. \quad (7.5)$$

We now choose the auxiliary functions $u_i^{(0)A}$, $u_a^{(1)A}$, $\varphi^{(0)A}$, $\varphi^{(1)A}$ to be particular solutions of the differential equations

$$D_{i,i}^{(0)A} + D_2^{(1)A} = \tfrac{1}{8}h^{-1}\epsilon'_{i2}\Phi_{,i}^{(0)} + \tfrac{3}{2}h^{-3}\zeta_{22}\Phi^{(1)},$$
$$D_{i,i}^{(1)A} + 2\,D_2^{(2)A} = (15/2)h^{-3}\overset{S}{\epsilon}_{22}\Phi^{(0)} + \tfrac{3}{2}h^{-3}\zeta_{i2}\Phi_{,i}^{(1)}, \quad (7.6)$$

and to satisfy the boundary conditions

$$n_aT_{aj}^{(0)A} + n_ae_{2aj}^{**}\Phi^{(0)} = 0, \qquad n_aT_{ab}^{(1)A} + n_a\psi_{2ab}\Phi^{(1)} = 0,$$
$$n_a\,D_a^{(0)A} - \tfrac{1}{8}h^{-1}n_a\epsilon'_{a2}\Phi^{(0)} = 0, \qquad n_a\,D_{(a)}^{(1)A} - \tfrac{3}{2}h^{-3}n_a\zeta_{a2}\Phi^{(1)} = 0. \quad (7.7)$$

Then, inserting (7.6) and (7.7) in (7.4) and (7.5), respectively, we find the residual problem governed by the differential equations

$$T_{ij,i}^{(0)R} + 2\rho h\omega^2 u_j^{(0)R} + 2\rho h\omega^2 u_j^{(0)A} + G_j^{(0)} = 0, \tag{7.8}$$

$$T_{ab,a}^{(1)R} - T_{2b}^{(0)R} + \tfrac{2}{3}\rho h^3\omega^2 u_b^{(1)R} + \tfrac{2}{3}\rho h^3\omega^2 u_b^{(1)A} + \tfrac{1}{3}h^2 G_b^{(1)} = 0, \tag{7.9}$$

$$D_{i,i}^{(0)R} + D_2^{(1)R} = 0, \qquad D_{i,i}^{(1)R} + 2\, D_2^{(2)R} = 0, \tag{7.10}$$

and the boundary conditions

$$n_a T_{aj}^{(0)R} = n_a T_{ab}^{(1)R} = n_a\, D_a^{(0)R} = n_a\, D_a^{(1)R} = 0,$$

where

$$G_j^{(0)} = T_{ij,i}^{(0)A} + F_j^{(0)} + e_{2ij}^{**}\Phi_{,i}^{(0)},$$

$$G_b^{(1)} = 3h^{-2}(T_{ab,a}^{(1)A} - T_{2b}^{(0)A} + F_b^{(1)} - e_{22b}^{**}\Phi^{(0)} + \psi_{2ab}\Phi_{,a}^{(1)}).$$

The solution of the residual problem may be expressed as an infinite series

$$(u_j^{(0)R}, u_b^{(1)R}, \varphi^{(0)R}, \varphi^{(1)R}) = \sum_\mu A_\mu(u_j^{(0)\mu}, u_b^{(1)\mu}, \varphi^{(0)\mu}, \varphi^{(1)\mu}), \tag{7.11}$$

where $u_j^{(0)\mu}$, $u_b^{(1)\mu}$, $\varphi^{(0)\mu}$, $\varphi^{(1)\mu}$ are the orthogonal solutions of the homogeneous ($F_j^{(0)} = F_b^{(1)} = \Phi^{(0)} = \Phi^{(1)} = 0$) system

$$T_{ij,i}^{(0)\mu} + 2\rho h\omega_\mu^2 u_j^{(0)\mu} = 0, \qquad T_{ab,a}^{(1)\mu} - T_{2b}^{(0)\mu} + \tfrac{2}{3}\rho h^3\omega_\mu^2 u_b^{(1)\mu} = 0, \tag{7.12}$$

$$D_{i,i}^{(0)\mu} + D_2^{(1)\mu} = 0, \qquad D_{i,i}^{(1)\mu} + 2\, D_2^{(2)\mu} = 0, \tag{7.13}$$

$$n_a T_{aj}^{(0)\mu} = n_a T_{ab}^{(1)\mu} = n_a\, D_a^{(0)\mu} = n_a\, D_a^{(1)\mu} = 0. \tag{7.14}$$

Noting that, from (7.11) and (4.17), with $\Phi^{(0)} = \Phi^{(1)} = 0$,

$$(T_{ij}^{(0)R}, T_{ab}^{(1)R}, D_i^{(0)R}, D_i^{(1)R}, D_2^{(2)R}) = \sum_\mu A_\mu(T_{ij}^{(0)\mu}, T_{ab}^{(1)\mu}, D_i^{(0)\mu}, D_i^{(1)\mu}, D_2^{(2)\mu}), \tag{7.15}$$

we see, from (7.13), that the electrostatic equations (7.10) are satisfied identically by the solution (7.11). To find the A_μ which will make (7.11) satisfy the equation of motions (7.8) and (7.9), we must first express the known inhomogeneous terms, in the latter, in series form:

$$G_j^{(0)} = \sum_\mu B_\mu u_j^{(0)\mu}, \qquad u_j^{(0)A} = \sum_\mu C_\mu u_j^{(0)\mu}, \tag{7.16}$$

$$G_b^{(1)} = \sum_\mu B_\mu u_b^{(1)\mu}, \qquad u_b^{(1)A} = \sum_\mu C_\mu u_b^{(1)\mu}. \tag{7.17}$$

Multiplying (7.16) by $u_j^{(0)\nu}$, (7.17) by $\frac{1}{3}h^2 u_b^{(1)\nu}$, adding, integrating over the area of the plate and using (6.2), we find

$$B_\mu = N_{(\mu)}^{-1} \int_A (G_j^{(0)} u_j^{(0)\mu} + \tfrac{1}{3}h^2 G_b^{(1)} u_b^{(1)\mu})\, dA,$$

$$C_\mu = N_{(\mu)}^{-1} \int_A (u_j^{(0)A} u_j^{(0)\mu} + \tfrac{1}{3}h^2 u_b^{(1)A} u_b^{(1)\mu})\, dA. \tag{7.18}$$

Then, substituting (7.11), (7.15), (7.16) and (7.17) in (7.8) and (7.9), using (7.12), multiplying (7.8) by $u_j^{(0)\nu}$ and (7.9) by $u_b^{(1)\nu}$, adding, integrating over the area and then using (6.2), we find

$$A_\mu = (B_\mu + 2\rho h\omega^2 C_\mu)/2\rho h(\omega_\mu^2 - \omega^2) \tag{7.19}$$

Thus we have

$$(u_j^{(0)}, u_b^{(1)}, \varphi^{(0)}, \varphi^{(1)}) = (u_j^{(0)A}, u_b^{(1)A}, \varphi^{(0)A}, \varphi^{(1)A})e^{i\omega t} + \sum_\mu A_\mu(u_j^{(0)\mu}, u_b^{(1)\mu}, \varphi^{(0)\mu}, \varphi^{(1)\mu})e^{i\omega t} \qquad (7.20)$$

as the complete, steady state solution of (4.18) with boundary conditions (7.1).

8. Surface charge. If the ac-voltage is applied uniformly over traction-free faces, by means of a thin, perfectly conducting film, we have

$$F_j^{(0)} = F_b^{(1)} = \Phi^{(1)} = 0, \qquad \Phi^{(0)} = Ve^{i\omega t}, \qquad (8.1)$$

where V is the constant voltage drop across the thickness. The current through the plate is proportional to the integrated surface charge, Q, over the whole area of a face, where

$$Q = \int_A D_2\Big]_{x_2=-h} dA. \qquad (8.2)$$

Inserting (5.2) in (8.2) and using (4.14), the divergence theorem and boundary conditions (7.1), we find

$$Q = \int_A D_2^{(0)}\, dA. \qquad (8.3)$$

Then, substituting (7.20) in the third of (4.17) and the resulting expression for $D_2^{(0)}$ in (8.3) we have, for the total surface charge,

$$Q = V(L - \tfrac{1}{8}h^{-1}\epsilon_{22}'A + \sum_\mu A_\mu Y_\mu), \qquad (8.4)$$

where A is the area of the plate, A_μ is given by (7.19) and

$$Y_\mu = \int_A [e_{2kl}^{**}(u_{k,l}^{(0)\mu} + \delta_{2l}u_k^{(1)\mu}) - \tfrac{1}{2}h^{-1}\epsilon_{2j}^{*}\varphi_{,j}^{(1)\mu} - (15/4h^3)\overset{S}{\epsilon}_{22}\varphi^{(1)\mu}]\, dA, \qquad (8.5)$$

$$L = \int_A [e_{2kl}^{**}(u_{k,l}^{(0)A} + \delta_{2l}u_k^{(1)A}) - \tfrac{1}{2}h^{-1}\epsilon_{2j}^{*}\varphi_{,j}^{(1)A} - (15/4h^3)\overset{S}{\epsilon}_{22}\varphi^{(1)A}]\, dA, \qquad (8.6)$$

with the constant V factored out. The formula (8.4), for surface charge, takes account of the action of the crystal both as a capacitor, in the terms outside the summation sign, and as a charge generator; including, in both cases, the distortion of the field due to the finite dimensions of the plate.

9. Application to rectangular, rotated-Y-cut, quartz plates. Quartz is a trigonal crystal and a rotated-Y-cut is a plate which contains a digonal axis and whose normal makes an angle, say Θ, with the trigonal axis [3]. When referred to rectangular axes in and normal to the plane of the plate, the constitutive relations exhibit monoclinic symmetry. We take x_1 as the digonal axis in the plane of the plate. A solution [5] of the equations in I for modes independent of x_3 has given a frequency spectrum of coupled thickness-shear, flexure and face-shear modes, for rectangular AT-cut plates (Θ = 35°15′), very close to that determined experimentally by Koga and Fukuyo [6]. We shall obtain the analogous solution of the present equations for the general rotated-Y-cut and give the formula for the integrated surface charge.

With (8.1), monoclinic symmetry and independence of x_3, we may separate, from (4.18), the four equations

$$c_{66}\kappa_6^2(u_{2,11}^{(0)} + u_{1,1}^{(1)}) + c_{56}\kappa_6 u_{3,11}^{(0)} + \rho\omega^2 u_2^{(0)} = 0,$$
$$c_{56}\kappa_6(u_{2,11}^{(0)} + u_{1,1}^{(1)}) + c_{55}u_{3,11}^{(0)} + \rho\omega^2 u_3^{(0)} = 0,$$
$$\tfrac{2}{3}h^3 Y_{11}u_{1,11}^{(1)} - 2h\kappa_6^2 c_{66}(u_{2,1}^{(0)} + u_1^{(1)}) - 2h\kappa_6 c_{56}u_{3,1}^{(0)} + 2h\psi_{11}\varphi_{,11}^{(1)} + \tfrac{2}{3}h^3\rho\omega^2 u_1^{(1)} - \kappa_6 e_{26}V = 0,$$
$$\tfrac{2}{3}h^3\psi_{11}u_{1,11}^{(1)} - \zeta_{11}\varphi_{,11}^{(1)} + 15h^{-2}\epsilon_{22}^S\varphi^{(1)} - 5\epsilon_{22}^S V = 0, \tag{9.1}$$

in the four variables $u_2^{(0)}$, $u_3^{(0)}$, $u_1^{(1)}$ and $\varphi_j^{(1)}$ while $u_1^{(0)}$, $u_3^{(1)}$ and $\varphi^{(0)}$ may be taken as zero. If the piezoelectric constants e_{26} and ψ_{11} were absent, the first three of (9.1) would be the equations which were solved in reference [5] for the coupled thickness-shear, flexure and face-shear modes. The constant e_{26} couples the thickness-shear deformation with the applied voltage; and ψ_{11} couples the flexural deformation with the induced electric field.

The plate is bounded by $x_1 = \pm l$, $x_3 = \pm w$. The appropriate boundary conditions on $x_1 = \pm l$ are

$$T_{12}^{(0)} = T_{13}^{(0)} = T_{11}^{(1)} = D_1^{(1)} = 0 \tag{9.2}$$

and we ignore the boundary conditions on $x_3 = \pm w$. Substitution of (4.17) in (9.2) converts the latter to: on $x_1 = \pm l$,

$$\begin{aligned} 2h\kappa_6 c_{66}(u_{2,1}^{(0)} + u_1^{(1)}) + 2hc_{56}u_{3,1}^{(0)} + e_{26}V = 0, \quad \tfrac{2}{3}h^3\gamma_{11}u_{1,1}^{(1)} + \psi_{11}\varphi_{,1}^{(1)} = 0,\\ 2h\kappa_6 c_{56}(u_{2,1}^{(0)} + u_1^{(1)}) + 2hc_{55}u_{3,1}^{(0)} + e_{25}V = 0, \quad \tfrac{2}{3}h^3\psi_{11}u_{1,1}^{(1)} - \zeta_{11}\varphi_{,1}^{(1)} = 0. \end{aligned} \tag{9.3}$$

A set of auxiliary functions which may be used to remove the inhomogeneous terms, V, from the electrostatic equation $(9.1)_4$ and the boundary conditions (9.3), is

$$u_2^{(0)A} = \tfrac{1}{2}h^{-1}VP_2x_1, \quad u_3^{(0)A} = \tfrac{1}{2}h^{-1}VP_3x_1, \quad u_1^{(1)A} = 0, \quad \varphi^{(1)A} = \tfrac{1}{3}h^2V, \tag{9.4}$$

where

$$P_2 = (e_{25}c_{56} - e_{26}c_{55})/\kappa_6(c_{55}c_{66} - c_{56}^2), \qquad P_3 = (e_{26}c_{56} - e_{25}c_{66})/(c_{55}c_{66} - c_{56}^2)$$

Then $G_j^{(0)} = G_b^{(1)} = 0$ and the residual inhomogeneous differential equations are

$$c_{66}\kappa_6^2(u_{2,11}^{(0)R} + u_{1,1}^{(1)R}) + c_{56}\kappa_6 u_{3,11}^{(0)R} + \rho\omega^2 u_2^{(0)R} + \tfrac{1}{2}\rho\omega^2 h^{-1}VP_2x_1 = 0,$$
$$c_{56}\kappa_6(u_{2,11}^{(0)R} + u_{1,1}^{(1)R}) + c_{55}\kappa_6 u_{3,11}^{(0)R} + \rho\omega^2 u_3^{(0)R} + \tfrac{1}{2}\rho\omega^2 h^{-1}VP_3x_1 = 0,$$
$$\tfrac{2}{3}h^3\gamma_{11}u_{1,11}^{(1)R} - 2h\kappa_6^2 c_{66}(u_{2,1}^{(0)R} + u_1^{(1)R}) - 2h\kappa_6 c_{56}u_{3,1}^{(0)R} + 2h\psi_{11}\varphi_{,11}^{(1)R} + \tfrac{2}{3}\rho h^3\omega^2 u_1^{(1)R} = 0,$$
$$\tfrac{2}{3}h^3\psi_{11}u_{1,11}^{(1)R} - \zeta_{11}\varphi_{,11}^{(1)R} + 15h^{-2}\epsilon_{22}^S\varphi^{(1)R} = 0. \tag{9.5}$$

In this case, we see that $(9.1)_3$ has also become homogeneous.

The boundary conditions to be satisfied by the residual solution are

$$\begin{aligned} 2h\kappa_6 c_{66}(u_{2,1}^{(0)R} + u_1^{(1)R}) + 2hc_{56}u_{3,1}^{(0)R} = 0, \quad \tfrac{2}{3}h^3\gamma_{11}u_{1,1}^{(1)R} + \psi_{11}\varphi_{,1}^{(1)R} = 0,\\ 2h\kappa_6 c_{56}(u_{2,1}^{(0)R} + u_1^{(1)R}) + 2hc_{55}u_{3,1}^{(0)R} = 0, \quad \tfrac{2}{3}h^3\psi_{11}u_{1,1}^{(1)R} - \zeta_{11}\varphi_{,1}^{(1)R} = 0. \end{aligned} \tag{9.6}$$

We must now find the orthogonal functions for the series solution of the residual problem; i.e., we must find the solutions of (9.1) and (9.3) with $V = 0$. Consider

$$(u_2^{(0)}, u_3^{(0)}) = (A, B)\sin \xi x_1, \qquad (u_1^{(1)}, \varphi^{(1)}) = (C, D)\cos \xi x_1 . \tag{9.7}$$

These satisfy (9.1), with $V = 0$, if

$$\begin{aligned} A(\kappa_6^2 c_{66}\xi^2 - \rho\omega^2) + B\kappa_6 c_{56}\xi^2 + C\kappa_6^2 c_{66}\xi &= 0, \\ A\kappa_6 c_{56}\xi^2 + B(c_{55}\xi^2 - \rho\omega^2) + C\kappa_6 c_{56}\xi &= 0, \\ A\kappa_6^2 c_{66}\xi + B\kappa_6 c_{56}\xi + C(\kappa_6^2 c_{66} + \tfrac{1}{3}h^2\gamma_{11}\xi^2 - \tfrac{1}{3}\rho h^2\omega^2) + D\tfrac{1}{2}h^{-1}\psi_{11}\xi^2 &= 0, \\ C\psi_{11}\xi^2 - D\tfrac{3}{2}h^{-3}(15h^{-2}\epsilon_{22}^S + \zeta_{11}\xi^2) &= 0, \end{aligned} \tag{9.8}$$

For a non-trivial solution, the determinant of the coefficients of A, B, C, D must vanish, leading to a quartic equation in ξ^2 as against a cubic when the piezoelectric constant, ψ_{11}, is zero. The additional root is imaginary for all ω, i.e., it produces a non-propagating mode. Thus, there will be no additional resonances due to the piezoelectric effect. The remaining three roots are almost the same as those with $\psi_{11} = 0$, due to the fact that ψ_{11}^2 is small in comparison with $\zeta_{11}\gamma_{11}$ and $\epsilon_{22}^S\gamma_{11}$. Hence there will be little effect on the wave numbers ξ by computing all four roots from the determinant of (9.8) with $\psi_{11} = 0$. Thus we have ξ_n, $n = 1, 2, 3$, as functions of ω^2 from the first three of (9.8) and

$$\xi_4^2 = -15\epsilon_{22}^S/\zeta_{11}h^2, \tag{9.9}$$

independent of ω, from the fourth. Corresponding to ξ_n, $n = 1, 2, 3$, we may compute amplitude ratios

$$p_n = A_n/C_n, \qquad q_n = B_n/C_n, \tag{9.10}$$

from the first two of (9.8) and

$$r_n = \epsilon_{22}^S D_n/\psi_{11}C_n = \tfrac{2}{3}h^3\epsilon_{22}^S\xi_n^2/(15\epsilon_{22}^S h^{-2} + \zeta_{11}\xi_n^2), \tag{9.11}$$

from the fourth. To the approximation (9.9), $A_4 = B_4 = C_4 = 0$. Consequently, we may write (9.7), with $\xi_4' \equiv i\xi_4$, as

$$\begin{aligned} (u_2^{(0)}, u_3^{(0)}) &= \sum_{n=1}^{3} (p_n, q_n)C_n \sin \xi_n x_1, \qquad u_1^{(1)} = \sum_{n=1}^{3} C_n \cos \xi_n x_1, \\ \varphi^{(1)} &= D_4 \cosh \xi_4' x_1 + (\psi_{11}/\epsilon_{22}^S)\sum_{n=1}^{3} r_n C_n \cos \xi_n x_1 . \end{aligned} \tag{9.12}$$

We now insert (9.12) in the boundary conditions (9.3), with $V = 0$, and find

$$\begin{aligned} &\sum_{n=1}^{3} C_n[\kappa_6^2 c_{66}(p_n\xi_n + 1) + \kappa_6 c_{56} q_n\xi_n]\cos \xi_n l = 0, \\ &\sum_{n=1}^{3} C_n[\kappa_6 c_{56}(p_n\xi_n + 1) + c_{55} q_n\xi_n]\cos \xi_n l = 0, \\ &D_4\psi_{11}\epsilon_{22}^S\xi_4' \sinh \xi_4' l - \sum_{n=1}^{3} C_n(\tfrac{2}{3}h^3\gamma_{11}\epsilon_{22}^S + \psi_{11}^2 r_n)\xi_n \sin \xi_n l = 0, \\ &D_4\zeta_{11}\epsilon_{22}^S\xi_4' \sinh \xi_4' l + \psi_{11}\sum_{n=1}^{3} C_n(\tfrac{2}{3}h^3\epsilon_{22}^S - \zeta_{11}r_n)\xi_n \sin \xi_n l = 0. \end{aligned} \tag{9.13}$$

Elimination of C_n and D_4 from (9.13) results in a transcendental frequency equation relating l/h to the frequency. For a given l/h, the frequency equation has an infinity of roots, ω_μ, to each of which there correspond amplitude ratios p_n^μ, q_n^μ, r_n^μ, from (9.10) and (9.11), and amplitude ratios

$$\beta_n = C_n/C_3, \quad n = 1, 2; \quad \beta_3 = 1; \quad \beta_4 = D_4/C_3 \tag{9.14}$$

obtained from (9.13). Inserting these results in (9.12), we obtain the expressions for the orthogonal functions:

$$(u_2^{(0)\mu}, u_3^{(0)\mu}) = \sum_{n=1}^{3} (p_n^\mu, q_n^\mu)\beta_n^\mu \sin \xi_n^\mu x_1, \qquad u_1^{(1)\mu} = \sum_{n=1}^{3} \beta_n^\mu \cos \xi_n^\mu x_1,$$

$$\varphi^{(1)\mu} = \beta_4^\mu \cosh \xi_4' x_1 + (\psi_{11}/\overset{S}{\epsilon}_{22}) \sum_{n=1}^{3} r_n^\mu \beta_n^\mu \cos \xi_n^\mu x_1, \tag{9.15}$$

in which the common factor C_3 has been set equal to unity.

Little error in a computation of frequencies and amplitude ratios will result if the frequency equation is obtained from the first three of (9.13) with $D_4 = 0$. The frequency equation would then have the same form as that in reference [5] but the material constants in the coefficients of the transcendental functions would have slightly different values due to the presence of the piezoelectric constants. In such an approximation the amplitude ratios β_1^μ and β_2^μ would be obtained from the first two of (9.13) and β_4^μ from the fourth.

The normalization factor is found by inserting (9.15) in (6.2), remembering that $u_1^{(0)\mu} = u_3^{(1)\mu} = 0$, with the result

$$N_{(\mu)} = \tfrac{1}{2}A \sum_{m=1}^{3} \sum_{n=1}^{3} \beta_m^\mu \beta_n^\mu [(p_m^\mu p_n^\mu + q_m^\mu q_n^\mu + \tfrac{1}{3}h^2) S_{mn}^- - (p_m^\mu p_n^\mu + q_m^\mu q_n^\mu - \tfrac{1}{3}h^2) S_{mn}^+], \tag{9.16}$$

where $S_{mn}^{\pm} = (\xi_m^\mu l \pm \xi_n^\mu l)^{-1} \sin (\xi_m^\mu l \pm \xi_n^\mu l)$. Then, from $(7.18)_1$, we have $B_\mu = 0$ since $G_i^{(0)} = G_b^{(1)} = 0$; and, from $(7.18)_2$, noting that $u_1^{(0)A}$, $u_1^{(1)A}$ and $u_3^{(1)A}$ are also zero,

$$C_\mu = \tfrac{1}{2}AN_{(\mu)}^{-1} \sum_{n=1}^{3} (P_2 p_n^\mu + P_3 q_n^\mu)(\beta_n^\mu/\xi_n^\mu h)[(\xi_n^\mu l)^{-1} \sin \xi_n^\mu l - \cos \xi_n^\mu l]. \tag{9.17}$$

Finally, from (7.19),

$$A_\mu = C_\mu \omega^2/(\omega_\mu^2 - \omega^2). \tag{9.18}$$

When (9.18), (9.15) and (9.4) are substituted in (7.20) the solution is completed except for a formula for the correction factor κ_6. This is found by equating the frequency of free thickness-shear vibrations, from $(9.1)_3$, with Lawson's analogous solution [7] of the three-dimensional equations. The result, as obtained previously [2], is

$$\kappa_6^2 = \tfrac{1}{3}\alpha_0^2(1 + e_{26}^2/\overset{S}{\epsilon}_{22}\overset{E}{c}_{66}), \tag{9.19}$$

where α_0 is the lowest, non-zero root of

$$\tan \alpha = \alpha(1 + \overset{S}{\epsilon}_{22}\overset{E}{c}_{66}/e_{26}^2). \tag{9.20}$$

The formula (8.4) for surface charge requires expressions for Y_μ and L, in addition to A_μ. These are obtained by substituting (9.15) and (9.4) in (8.5) and (8.6), respectively, with the results:

$$Y_\mu = A \sum_{n=1}^{3} [\kappa_6 e_{26}(p_n^\mu \xi_n^\mu + 1) + e_{25} q_n^\mu \xi_n^\mu - (15/4h^3)\psi_{11} r_n^\mu](\beta_n^\mu/\xi_n^\mu l) \sin \xi_n^\mu l$$

$$- A(15/4h^3)\epsilon_{22}^S \beta_4^\mu (\xi_4' l)^{-1} \sinh \xi_4' l,$$

$$L = \tfrac{1}{2} A h^{-1}(\kappa_6 e_{26} P_2 + e_{25} P_3 - [5/2]\epsilon_{22}^S), \qquad (9.21)$$

The expression (9.18) for A_μ has, of course, the term $\omega_\mu^2 - \omega^2$ in the denominator. Hence, near a resonance only one term of the infinite series in the formula (8.4), for the surface charge Q, need be computed. We have suggested that, in computing the roots of the equations obtained by setting the determinants of the coefficients of (9.8) and (9.13) equal to zero, certain of the terms having ψ_{11} as a factor be neglected. One would thereby take into account the coupling between the applied voltage and the strain, through e_{26}, and the generation of the induced electric field by the curvature, through the amplitude ratios r_n; but a part of the small counter effect of the induced electric field on the strain would be neglected. This approximation need not be made if adequate facilities are available for computing the roots of the quartic equation obtained from (9.8) and the complete 4×4 transcendental determinantal equation which would be obtained from the equations which would replace (9.13).

References

1. R. D. Mindlin, *High frequency vibrations of crystal plates*, Quart. Appl. Math. **19**, 51–61 (1961)
2. R. D. Mindlin, *Forced thickness-shear and flexural vibrations of piezoelectric crystal plates*, J. Appl. Phys. **23**, 83–88 (1952)
3. W. P. Mason, *Piezoelectric crystals and their application to ultrasonics*, D. Van Nostrand Co., New York, 1950
4. R. Courant and D. Hilbert, *Methods of Mathematical Physics*, Interscience Publishers, New York
5. R. D. Mindlin and D. C. Gazis, *Strong resonances of rectangular AT cut quartz plates.* (Forthcoming)
6. I. Koga and H. Fukuyo, *Vibration of thin piezoelectric quartz crystal plates, especially an R_1-cut (AT-cut) rectangular plate*, J. Inst. Elec. Comm. Engrs. of Japan **36**, 59–67 (1953)
7. A. W. Lawson, *The vibration of piezoelectric plates*, Phys. Rev. **62**, 71–75 (1942)

Reprinted from THE JOURNAL OF THE ACOUSTICAL SOCIETY OF AMERICA, Vol. 34, No. 12, 1895–1901, December, 1962

Vibrations of an Infinite, Monoclinic Crystal Plate at High Frequencies and Long Wavelengths

R. K. KAUL* AND R. D. MINDLIN
Department of Civil Engineering, Columbia University, New York 27, New York
(Received December 19, 1961)

An asymptotic solution of the frequency equation of an infinite, monoclinic crystal plate is derived. Formulas are given which describe the variation of frequency with wavelength at high frequencies and long wavelengths.

INTRODUCTION

In a previous paper,[1] some progress was made on the problem of mapping the branches of the frequency equation associated with Ekstein's solution[2] of the equations of linear elasticity for the case of vibrations of an infinite, monoclinic crystal plate with traction-free faces. In the present paper, attention is focused on the region of high frequencies and long wavelengths. What is required is a set of explicit formulas for the ordinates, slopes, and curvatures of the branches at infinite wavelength. This information is helpful in extending the mapping of the branches to higher frequencies than before, and to imaginary and complex propagation constants. These extensions will be described in a subsequent paper. In addition, the formulas are useful in the establishment of approximate, two-dimensional equations of motion that can be solved for very high-frequency vibrations of bounded crystal plates. An example of the use of the ordinates, slopes, and curvatures which have been obtained[3] from Rayleigh's[4] frequency equation for the isotropic plate is given elsewhere.[5]

In the region of high frequencies and long wavelengths, there are three families of modes in which the predominant displacements are thickness shear in the direction of the digonal axis of symmetry, thickness shear at right angles to the digonal axis, and thickness stretch, respectively. The first of these was studied in some detail by Ekstein and the analysis of the remaining two is given, here, in the course of a unified treatment of all three.

I. GENERAL EQUATIONS

We choose a rectangular coordinate system x_i, $i=1, 2, 3$, with $x_2=\pm b$ as the faces of the crystal plate. An index, say j, followed by a comma designates differentiation with respect to x_j, and the summation convention for repeated indices is employed as is the dot notation for differentiation with respect to time.

The components of stress τ_{ij}, strain e_{ij}, and displacement u_i are related through

$$\tau_{ij}=c_{ijkl}e_{kl}=c_{ijkl}u_{k,l}, \tag{1}$$

where the elastic constants c_{ijkl} satisfy the properties of symmetry, $c_{ijkl}=c_{klij}=c_{jikl}$.

In the absence of body forces the three equations of motion are

$$\tau_{ij,j}=c_{ijkl}u_{k,lj}=\rho\ddot{u}_i, \tag{2}$$

where ρ is the mass density.

As solutions of (2) we take plane waves of the type

$$u_k=A_k\exp\{i(\xi_m x_m-\omega t)\}, \tag{3}$$

where the ξ_m are the propagation constants and ω is the circular frequency of vibration. In (3), ω is real while constants A_k and ξ_m are, in general, complex but the real part of u_k is understood. For the u_k to satisfy the equations of motion, Ω and the ξ_m must satisfy the characteristic equation

$$|\bar{c}_{ijkl}\bar{\xi}_j\bar{\xi}_l-\Omega^2\delta_{ik}|=0, \tag{4}$$

where

$$\bar{c}_{ijkl}=c_{ijkl}/c_{1212};\quad \bar{\xi}_i=2b\xi_i/\pi$$
$$\Omega=\omega/\bar{\omega};\qquad \bar{\omega}=(\pi/2b)(c_{1212}/\rho)^{\frac{1}{2}}$$

and δ_{ik} is the Kronecker delta.

II. FREQUENCY EQUATION

The determinantal equation (4) suggests that, for a given ratio of $\bar{\xi}_1:\bar{\xi}_3:\Omega$, there always exist three wave-numbers $\bar{\xi}_{2n}$ $(n=1,2,3)$ along the thickness direction x_2. Thus, omitting the time factor, the displacements can be written in the form

$$\begin{aligned}u_1&=\textstyle\sum_n L_{1n}\{A_nf_1\cos\xi_{2n}x_2+B_nf_2\sin\xi_{2n}x_2\},\\ u_2&=\textstyle\sum_n L_{2n}\{B_nf_1\cos\xi_{2n}x_2+A_nf_2\sin\xi_{2n}x_2\},\\ u_3&=\textstyle\sum_n L_{3n}\{B_nf_1\cos\xi_{2n}x_2+A_nf_2\sin\xi_{2n}x_2\},\end{aligned} \tag{5}$$

where

$$f_1=\sin(\xi_1x_1+\xi_3x_3);\quad f_2=\cos(\xi_1x_1+\xi_3x_3). \tag{6}$$

Here the L_{kn}, $(k=1,2,3)$ are the amplitude ratios fixed by (4) and the constants A_n and B_n are to be fixed by

*Present address: Thomas J. Watson Research Center, International Business Machines Corporation, Yorktown Heights, New York.

[1] E. G. Newman and R. D. Mindlin, J. Acoust. Soc. Am. 29, 1206 (1957).

[2] H. Ekstein, Phys. Rev. 68, 11 (1945).

[3] R. D. Mindlin, *An Introduction to the Mathematical Theory of Vibrations of Elastic Plates* (Signal Corps Engineering Laboratories, Fort Monmouth, New Jersey, 1955).

[4] Lord Rayleigh (J. W. Strutt), Proc. London Math. Soc. 20, 225 (1889).

[5] R. D. Mindlin and M. A. Medick, J. Appl. Mech. 26, 561 (1959).

the six boundary conditions on the faces $x_2=\pm b$. For tractionfree faces the boundary conditions are

$$\tau_{2j}=c_{2jkl}u_{k,l}=0 \quad \text{on} \quad x_2=\pm b. \tag{7}$$

Substituting (5) in (1) we get, for $i=2$,

$$\bar{\tau}_{2j}=\sum_n A_n\{M_{jn}f_2\cos\xi_{2n}x_2-N_{jn}f_1\sin\xi_{2n}x_2\} +\sum_n B_n\{N_{jn}f_2\cos\xi_{2n}x_2-M_{jn}f_1\sin\xi_{2n}x_2\}, \tag{8}$$

where $\bar{\tau}_{2j}=\tau_{2j}/c_{1212}$ and

$$M_{jn}=L_{1n}(\xi_1\bar{c}_{2j11}+\xi_3\bar{c}_{2j13})+\xi_{2n}(L_{2n}\bar{c}_{2j22}+L_{3n}\bar{c}_{2j23}) \tag{9}$$

$$N_{jn}=\xi_1(L_{2n}\bar{c}_{2j12}+L_{3n}\bar{c}_{2j13})+\xi_{2n}L_{1n}\bar{c}_{2j12} +\xi_3(L_{2n}\bar{c}_{2j23}+L_{3n}\bar{c}_{2j33}). \tag{10}$$

Inserting (8) in (7) we obtain a set of six homogeneous equations in A_n and B_n. For a solution other than $A_n=B_n=0$, the determinant of the coefficients must vanish. This leads to the frequency equation which, when arranged suitably, takes the form

$$\begin{vmatrix} |M_{jn}\cos\xi_{2n}b| & |N_{jn}\cos\xi_{2n}b| \\ |N_{jn}\sin\xi_{2n}b| & |M_{jn}\sin\xi_{2n}b| \end{vmatrix}=0, \tag{11}$$

where j, $n=1, 2, 3$ and every block matrix like

$$|M_{jn}\cos\xi_{2n}b|$$

represents an array of three rows for $j=1, 2, 3$ and three columns for $n=1, 2, 3$. This is a highly complicated transcendental frequency equation, the roots of which, in conjunction with the roots of (4), determine the dispersion spectrum of the crystal plate of infinite extent and finite thickness, with tractionfree faces.

III. MONOCLINIC CRYSTAL PLATE

In the case of the monoclinic system, with the digonal axis parallel to the x_1 axis of the reference system, we have $\bar{c}_{ij12}=\bar{c}_{ij13}=0$ for $ij=11, 22, 33, 23$. To avoid a multiplicity of subscripts and to shorten the notation, we now replace $\bar{c}_{ijkl}$ by its equivalent $\bar{c}_{pq}$, where p, q vary from 1 to 6 and also replace ξ_{2n} by η_n, where η represents the wavenumber in the thickness direction. Then from (9) and (10) we get

$$M_{1n}=\xi_3\bar{c}_{65}L_{1n}, \tag{12a}$$

$$M_{2n}=\xi_1\bar{c}_{21}L_{1n}+\eta_n(L_{2n}\bar{c}_{22}+L_{3n}\bar{c}_{24}), \tag{12b}$$

$$M_{3n}=\xi_1\bar{c}_{41}L_{1n}+\eta_n(L_{2n}\bar{c}_{24}+L_{3n}\bar{c}_{44}), \tag{12c}$$

$$N_{1n}=\xi_1(L_{2n}+L_{3n}\bar{c}_{65})+\eta_nL_{1n}, \tag{12d}$$

$$N_{2n}=\xi_3(L_{2n}\bar{c}_{24}+L_{3n}\bar{c}_{23}), \tag{12e}$$

$$N_{3n}=\xi_3(L_{2n}\bar{c}_{44}+L_{3n}\bar{c}_{43}). \tag{12f}$$

For straight-crested waves propagating in the x_1 direction, $\xi_3=0$ and the coefficients M_{1n}, N_{2n}, N_{3n} vanish. The determinant (11) then takes the diagonal two-block form; and the equations obtained by setting each block equal to zero are

$$\begin{vmatrix} \{\eta_nL_{1n}+\xi(L_{2n}+L_{3n}\bar{c}_{65})\}\sin\eta_nb \\ \{\bar{c}_{21}\xi L_{1n}+\eta_n(\bar{c}_{22}L_{2n}+L_{3n}\bar{c}_{24})\}\cos\eta_nb \\ \{\bar{c}_{41}\xi L_{1n}+\eta_n(\bar{c}_{24}L_{2n}+L_{3n}\bar{c}_{44})\}\cos\eta_nb \end{vmatrix}=0 \tag{13}$$

$$\begin{vmatrix} \{\eta_nL_{1n}+\xi(L_{2n}+L_{3n}\bar{c}_{65})\}\cos\eta_nb \\ \{\bar{c}_{21}\xi L_{1n}+\eta_n(\bar{c}_{22}L_{2n}+L_{3n}\bar{c}_{24})\}\sin\eta_nb \\ \{\bar{c}_{41}\xi L_{1n}+\eta_n(\bar{c}_{24}L_{2n}+L_{3n}\bar{c}_{44})\}\sin\eta_nb \end{vmatrix}=0 \tag{14}$$

in which the columns are obtained by setting $n=1, 2, 3$ and the subscript 1 has been dropped from ξ_1.

Corresponding to the frequency equation (13), the displacements are given by (5) when $B_n=0$ and, corresponding to the frequency equation (14), $A_n=0$. In the first case the displacements u_1 and u_2 are symmetric about the middle plane and u_3 is antisymmetric about this plane. We call this the essentially symmetric family. In the other case ($A_n=0$) u_1 and u_2 are antisymmetric about the middle plane and u_3 is symmetric about this plane. We call this family essentially antisymmetric about the middle plane. Because of the diagonal block form of the frequency equation (11), in the case of straight-crested waves with wave normal parallel to the digonal axis x_1, we conclude that the two families of modes are not coupled with each other and that their periods and wavelengths are governed by separate frequency equations. Unlike the isotropic case,[4] each family now contains both symmetric and antisymmetric components about the middle plane x_1-x_3.

IV. CHARACTERISTIC EQUATION FOR LONG WAVELENGTHS

The frequency equations (13) and (14) relate ξ, η_n and Ω, from which it is necessary to eliminate the η_n, through (4), in order to obtain the frequency as a function of ξ. We therefore seek to determine the roots of the characteristic equation (4), which, for the monoclinic system, takes the form

$$\begin{vmatrix} \bar{c}_{11}\bar{\xi}^2+\bar{\eta}^2-\Omega^2 & \bar{\eta}\bar{\xi}(1+\bar{c}_{12}) & \bar{\eta}\bar{\xi}(\bar{c}_{14}+\bar{c}_{56}) \\ \bar{\eta}\bar{\xi}(1+\bar{c}_{12}) & \bar{\xi}^2+\bar{c}_{22}\bar{\eta}^2-\Omega^2 & \bar{c}_{56}\bar{\xi}^2+\bar{c}_{24}\bar{\eta}^2 \\ \bar{\eta}\bar{\xi}(\bar{c}_{14}+\bar{c}_{56}) & \bar{c}_{56}\bar{\xi}^2+\bar{c}_{24}\bar{\eta}^2 & \bar{c}_{55}\bar{\xi}^2+\bar{c}_{44}\bar{\eta}^2-\Omega^2 \end{vmatrix}=0. \tag{15}$$

A general solution for arbitrary $\bar{\xi}$ is awkward, but, for $\bar{\xi}\ll 1$, the roots can easily be determined.[6]

For $\bar{\xi}=0$, (15) takes the simple form

$$\begin{vmatrix} 1-v^2 & 0 & 0 \\ 0 & \bar{c}_{22}-v^2 & \bar{c}_{24} \\ 0 & \bar{c}_{24} & \bar{c}_{44}-v^2 \end{vmatrix}=0, \tag{16}$$

where $v=\Omega/\bar{\eta}=(\omega/\eta)/(c_{66}/\rho)^{\frac{1}{2}}$. Hence the velocity ratios

[6] Lord Rayleigh (J. W. Strutt), *Theory of Sound* (Dover Publications, Inc., New York, 1945), Vol. 1, pp. 113–115.

of the three waves along the x_2 direction are

$$v_1=1, \qquad (17)$$
$$v_{2,3}=\langle\tfrac{1}{2}\{\tilde{c}_{22}+\tilde{c}_{44}\pm[(\tilde{c}_{22}-\tilde{c}_{44})^2+4\tilde{c}_{24}^2]^{\frac{1}{2}}\}\rangle^{\frac{1}{2}}.$$

Corresponding to the eigenvalues (17), the normalized amplitude ratios are

$$\bar{b}=1;\quad \bar{b}_{12}=\bar{b}_{21}=\bar{b}_{13}=\bar{b}_{31}=0, \qquad (18a)$$

$$\bar{b}_{22}=\bar{b}_{33}=(v_2^2-\tilde{c}_{44})/[(v_2^2-\tilde{c}_{44})^2+\tilde{c}_{24}^2]^{\frac{1}{2}}, \qquad (18b)$$

$$\bar{b}_{32}=-\bar{b}_{23}=\tilde{c}_{24}/[(v_2^2-\tilde{c}_{44})^2+\tilde{c}_{24}^2]^{\frac{1}{2}}, \qquad (18c)$$

where the $\bar{b}_{ij}$ satisfy the conditions of orthonormality: $\bar{b}_{ki}\bar{b}_{kj}=\delta_{ij}$, $|\bar{b}_{ij}|=1$. The normalized eigenvectors of (16) are

$$\{\bar{b}_{11}:0:0\};\quad \{0:\bar{b}_{22}:\bar{b}_{32}\};\quad \{0:\bar{b}_{23}:\bar{b}_{33}\}, \qquad (19)$$

which are the column vectors of the transformation matrix $\{\bar{b}_{ij}\}$ in the nonsingular linear transformation $A_k=\bar{b}_{kj}B_j$ which reduces (16) to the diagonal form. Here the B_j are the *normal* coordinates of the system when $\bar{\xi}=0$.

In terms of normal coordinates B_j of the system when $\bar{\xi}=0$, the original determinant (15) takes the symmetric form

$$\begin{vmatrix} v_1^2\bar{\eta}^2+\alpha_{11}\bar{\xi}^2-\Omega^2 & \alpha_{12}\bar{\xi}\bar{\eta} & \alpha_{13}\bar{\xi}\bar{\eta} \\ \alpha_{21}\bar{\xi}\bar{\eta} & v_2^2\bar{\eta}^2+\alpha_{22}\bar{\xi}^2-\Omega^2 & \alpha_{23}\bar{\xi}^2 \\ \alpha_{31}\bar{\xi}\bar{\eta} & \alpha_{32}\bar{\xi}^2 & v_3^2\bar{\eta}^2+\alpha_{33}\bar{\xi}^2-\Omega^2 \end{vmatrix}=0, \qquad (20)$$

where $\alpha_{ij}=\alpha_{ji}$ and

$$\alpha_{1j}=\epsilon^{1j}\{\bar{b}_{2j}(1+\tilde{c}_{12})+\bar{b}_{3j}(\tilde{c}_{14}+\tilde{c}_{56})\}+\tilde{c}_{11}\delta_{1j};\quad j=1,2,3$$

$$\alpha_{jj}=\epsilon^{1j}\{(\bar{b}_{2j}^2+\bar{b}_{3j}^2\tilde{c}_{55}+2\bar{b}_{2j}\bar{b}_{3j}\tilde{c}_{56})\}+\tilde{c}_{11}\delta_{1j};\quad j=1,2,3 \qquad (21)$$

$$\alpha_{ij}=\epsilon^{ij}\{\bar{b}_{ii}\bar{b}_{ij}(1-\tilde{c}_{55})+\tilde{c}_{56}(\bar{b}_{2i}^2-\bar{b}_{2j}^2)\};\quad ij=23,32$$

in which ϵ^{ij} is the two-index alternating tensor and no summation is implied.

The new form (20) is obtained from (15) by postmultiplying by the matrix $\{\bar{b}_{ij}\}$ and premultiplying by the transpose $[\bar{b}_{ji}]$. The new form reduces to the diagonal form when $\bar{\xi}=0$ and therefore, when $\bar{\xi}\ll 1$, the roots can easily be determined.

The B_j can be considered as approximately normal coordinates for $\bar{\xi}\ll 1$. When they are truly normal, a variation in any one of them, say B_i, does not entail any variation in the remaining coordinates B_j. But, in the perturbed state, $0<\bar{\xi}\ll 1$, there is a synchronous variation of the other coordinates, though the ratio $(B_j:B_i)^{(i)}$, for the ith mode, differs from zero only by small quantities of the second order.

Writing the ith equation in the form

$$(v_i^2\bar{\eta}^2+\alpha_{ii}\bar{\xi}^2-\Omega_i^2)B_i+\sum_j^{\neq i}\theta_{ij}B_j=0,$$

where

$$\theta_{12}=\theta_{21}=\alpha_{12}\bar{\xi}\bar{\eta},$$
$$\theta_{13}=\theta_{31}=\alpha_{13}\bar{\xi}\bar{\eta},$$
$$\theta_{23}=\theta_{32}=\alpha_{23}\bar{\xi}^2,$$

we find that the ith root is given by

$$\Omega_i^2=(v_i^2\bar{\eta}^2+\alpha_{ii}\bar{\xi}^2)+\sum_j^{\neq i}M_{ji}\alpha_{ij}\delta_{1i}\bar{\xi}\bar{\eta}$$
$$+M_{1i}(\alpha_{i1}\delta_{2i}+\alpha_{i1}\delta_{3i})\bar{\xi}\bar{\eta}$$
$$+(M_{2i}\alpha_{2i}\delta_{3i}+M_{3i}\alpha_{3i}\delta_{2i})\bar{\xi}^2, \qquad (22)$$

where $M_{ji}=(B_j/B_i)^{(i)}$ and no summation is implied. The ratio M_{ji} for the ith root is given by the set of the remaining two equations, and following Cramer's rule it can be shown that these correspond to the ratio of the cofactors of the coefficients of B_j and B_i in the ith row. Retaining only the lowest-order terms in $\bar{\xi}$ we find, from (20), that the ratios are

$$M_{ii}=1, \qquad i=1,2,3$$

$$M_{ij}=\frac{\alpha_{ij}}{v_j^2-v_i^2}\left(\frac{\xi}{\eta_j}\right)+\cdots, \qquad ij=12,21,13,31 \qquad (23)$$

$$M_{ij}=\frac{\alpha_{1i}\alpha_{1j}-\alpha_{ij}(1-v_j^2)}{(1-v_j^2)(v_i^2-v_j^2)}\left(\frac{\xi}{\eta_j}\right)^2+\cdots, \qquad ij=23,32$$

in which $v_i\neq v_j$; i.e., we assume nondegeneracy of the eigenvalues v_i. Substituting these values of M_{ji} in (22) and retaining only the lowest-order terms, we get the three roots as

$$\Omega_i^2=v_i^2\bar{\eta}^2+\kappa_i\bar{\xi}^2+\cdots, \qquad (24)$$

where

$$\kappa_1\doteq\{\alpha_{11}+[\alpha_{12}^2/(1-v_2^2)]+[\alpha_{13}^2/(1-v_3^2)]\},$$
$$\kappa_2\doteq\{\alpha_{22}-[\alpha_{12}^2/(1-v_2^2)]\},$$
$$\kappa_3\doteq\{\alpha_{33}-[\alpha_{13}^2/(1-v_3^2)]\}.$$

This is our approximation to the characteristic equation (15), valid for small ξ/η.

V. AMPLITUDE COEFFICIENTS

Knowing the three roots (24) of (15) and the ratios M_{ji}, we may readily find the amplitude coefficients L_{ij} since

$$L_{ij}=\left.\frac{A_i}{A_j}\right]^{(j)}\frac{\sum_n\bar{b}_{in}M_{nj}}{\sum_n\bar{b}_{jn}M_{nj}}$$

Substituting the values of M_{nj} from (23), we find

$$L_{ii}=1, \qquad i=1,2,3 \qquad (25a)$$

$$L_{i1}=\left(\frac{\bar{b}_{i2}\alpha_{21}}{1-v_2^2}+\frac{\bar{b}_{i3}\alpha_{31}}{1-v_3^2}\right)\frac{\xi}{\eta_1}+\cdots, \qquad i=2,3 \qquad (25b)$$

$$L_{1j}=\frac{\alpha_{1j}}{\bar{b}_{jj}(v_j^2-1)}\frac{\xi}{\eta_j}+\cdots, \qquad j=2,3 \qquad (25c)$$

$$L_{ij}=\frac{\bar{b}_{ij}}{\bar{b}_{jj}}+\frac{\alpha_{1i}\alpha_{1j}-\alpha_{ij}(1-v_j^2)}{\bar{b}_{jj}^2(1-v_j^2)(v_i^2-v_j^2)}\left(\frac{\xi}{\eta_j}\right)^2+\cdots. \qquad i,j=2,3 \qquad (25d)$$

By using the identities

$$\text{(i)} \qquad v_2^2+v_3^2=\bar{c}_{22}+\bar{c}_{44},$$

$$\text{(ii)} \qquad (v_2^2-\bar{c}_{22})(v_2^2-\bar{c}_{44})=\bar{c}_{24}^2,$$

$$\text{(iii)} \qquad v_2^2\bar{b}_{22}^2+v_3^2\bar{b}_{32}^2=\bar{c}_{22},$$

$$\text{(iv)} \qquad v_2^2\bar{b}_{32}^2+v_3^2\bar{b}_{22}^2=\bar{c}_{44},$$

$$\text{(v)} \qquad (v_2^2-v_3^2)\bar{b}_{22}\bar{b}_{32}=\bar{c}_{24},$$

$$\text{(vi)} \qquad (v_2^2-v_3^2)(\bar{b}_{22}^2-\bar{b}_{32}^2)=\bar{c}_{22}-\bar{c}_{44},$$

and substituting in (25) we finally obtain the following set of amplitude ratios L_{ij} when $\xi\ll1$:

$$L_{11}=L_{22}=L_{33}=1, \qquad (26a)$$

$$L_{21}=\frac{\bar{c}_{24}(\bar{c}_{14}+\bar{c}_{56})-(1+\bar{c}_{12})(\bar{c}_{44}-1)}{(v_2^2-1)(v_3^2-1)}\left(\frac{\xi}{\eta_1}\right)+\cdots, \qquad (26b)$$

$$L_{31}=\frac{\bar{c}_{24}(1+\bar{c}_{12})-(\bar{c}_{14}+\bar{c}_{56})(\bar{c}_{22}-1)}{(v_2^2-1)(v_3^2-1)}\left(\frac{\xi}{\eta_1}\right)+\cdots, \qquad (26c)$$

$$L_{12}=\frac{\bar{c}_{24}(\bar{c}_{14}+\bar{c}_{56})+(1+\bar{c}_{12})(v_2^2-\bar{c}_{44})}{(v_2^2-1)(v_2^2-\bar{c}_{44})}\left(\frac{\xi}{\eta_2}\right)+\cdots, \qquad (26d)$$

$$L_{13}=\frac{\bar{c}_{24}(1+\bar{c}_{12})+(\bar{c}_{14}+\bar{c}_{56})(v_3^2-\bar{c}_{22})}{(v_3^2-1)(v_3^2-\bar{c}_{22})}\left(\frac{\xi}{\eta_3}\right)+\cdots, \qquad (26e)$$

$$L_{32}=\frac{\bar{c}_{24}}{v_2^2-\bar{c}_{44}}+\left\{\frac{\bar{c}_{56}}{v_2^2-\bar{c}_{44}}+\frac{\bar{c}_{24}(\bar{c}_{55}-\kappa_2)}{(v_2^2-\bar{c}_{44})^2}-\frac{\bar{c}_{24}(\bar{c}_{14}+\bar{c}_{56})^2}{(v_2^2-\bar{c}_{44})^2(1-v_2^2)}-\frac{(1+\bar{c}_{12})(\bar{c}_{14}+\bar{c}_{56})}{(v_2^2-\bar{c}_{44})(1-v_2^2)}\right\}\left(\frac{\xi}{\eta_2}\right)^2+\cdots, \qquad (26f)$$

$$L_{23}=\frac{\bar{c}_{24}}{v_3^2-\bar{c}_{22}}+\left\{\frac{\bar{c}_{56}}{v_3^2-\bar{c}_{22}}+\frac{\bar{c}_{24}(1-\kappa_3)}{(v_3^2-\bar{c}_{22})^2}-\frac{\bar{c}_{24}(1+\bar{c}_{12})^2}{(v_3^2-\bar{c}_{22})^2(1-v_3^2)}-\frac{(1+\bar{c}_{12})(\bar{c}_{14}+\bar{c}_{56})}{(v_3^2-\bar{c}_{22})(1-v_3^2)}\right\}\left(\frac{\xi}{\eta_3}\right)^2+\cdots. \qquad (26g)$$

It may be noticed that the amplitude ratios in this form could be obtained directly from (15) once the values of Ω_i^2 are known. The method followed above, however, logically leads to consistent ratios without introducing indeterminacy at long wavelengths or for isotropic materials.

(a) Amplitude coefficients for thickness modes.

In the case of thickness modes, $\xi=0$, and consequently the amplitude ratios L_{ij} are

$$L_{11}=L_{22}=L_{33}=1; \quad L_{12}=L_{21}=L_{13}=L_{31}=0 \qquad (27)$$
$$L_{23}=\bar{c}_{24}/(\bar{c}_{44}-v_2^2); \quad L_{32}=\bar{c}_{24}/(v_2^2-\bar{c}_{44}).$$

(b) Amplitude coefficients for isotropic media.

In the case of isotropy, $\bar{c}_{11}=\bar{c}_{22}=(2+\lambda/\mu)$; $\bar{c}_{12}=\lambda/\mu$; $\bar{c}_{44}=\bar{c}_{55}=\bar{c}_{66}=1$; and all other $\bar{c}_{pq}$ are zero. The characteristic roots in this case are

$$\Omega^2=\bar{\eta}_1^2+\xi^2; \quad \Omega^2=(2+\lambda/\mu)(\bar{\eta}_2^2+\xi^2),$$

where λ and μ are Lamé's constants. The amplitude ratios then take the form

$$L_{11}=L_{22}=L_{33}=1; \quad L_{12}=\xi/\eta_2; \quad L_{21}=-\xi/\eta_1 \qquad (28)$$

VI. THICKNESS MODES

When the wavelength along the digonal axis is infinite, some of the amplitude ratios vanish, leading to simplification of the frequency equation. Details of the solution and of the character of these modes are well known[1,7] and therefore only the essential results are given here for use in subsequent sections.

In the case of the essentially symmetric family of modes, the displacements reduce to

$$u_1=A_1\cos\eta_1x_2,$$
$$u_2=A_2\sin\eta_2x_2+A_3L_{23}\sin\eta_3x_2, \qquad (29)$$
$$u_3=A_2L_{32}\sin\eta_2x_2+A_3\sin\eta_3x_2,$$

[7] I. Koga, Physics 3, 70 (1932).

which shows that the symmetric thickness-stretch displacement u_2 and the antisymmetric x_3-thickness-shear displacement u_3 are coupled, while the symmetric x_1-thickness-shear displacement u_1 is not. The frequency equations in this case are

$$\eta_1 \sin\eta_1 b = 0, \tag{30a}$$

$$\begin{vmatrix} (\tilde{c}_{22}L_{2n}+\tilde{c}_{24}L_{3n})\eta_n \cos\eta_n b \\ (\tilde{c}_{24}L_{2n}+\tilde{c}_{44}L_{3n})\eta_n \cos\eta_n b \end{vmatrix} = 0, \tag{30b}$$

which have the solutions $\bar{\eta}_n = s_n$; s_1 even, s_2, s_3 odd. Interchanging the sine and cosine functions in the displacements and frequency equations, we obtain the corresponding relations for the essentially antisymmetric family of modes comprising antisymmetric x_1 thickness shear and coupled antisymmetric thickness stretch and symmetric x_3 thickness shear. The solution in this case requires that s_1 be odd and s_2, s_3 be even.

The thickness frequencies, i.e., the frequencies at infinite wavelength, are then given by

$$\Omega = v_n s_n, \quad n = 1, 2, 3, \tag{31}$$

where s_1, s_2, s_3 take values even, odd, odd or odd, even, even, depending upon the type of symmetry; and v_1, v_2, v_3 are the ratios of phase velocities of the x_1 thickness shear and the coupled thickness stretch and x_3-thickness-shear modes, respectively, to the x_1-thickness-shear velocity, as given in (17).

We note that coincidence of thickness frequencies occurs if

$$v_2 = s_1/s_2$$
(x_1 thickness shear and thickness stretch), (32a)

$$v_3 = s_1/s_3$$
(x_1 and x_3 thickness shear), (32b)

$$v_2 = s_1/s_2; \quad v_3 = s_1/s_3$$
(all three types), (32c)

$$v_2/v_3 = s_3/s_2$$
(thickness stretch and x_3 thickness shear). (32d)

VII. FREQUENCY EQUATION FOR LONG WAVELENGTHS

For arbitrary wavelengths the frequency equations (13) and (14) are highly intractable. However, in the region of long wavelengths, $\xi \ll 1$, and high frequency, $\Omega \gtrless 1$, considerable simplification can be effected by using the amplitude ratios (25) or (26). Thus, writing

$$\beta_j = \xi/\eta_j; \quad L_{ij} = \sum_n L_{ij}^{(n)}\beta_j^n$$

and substituting the L_{ij} from (25) or (26) in (13) and (14), we find, after simplification, that the consecutive columns of the frequency determinant take the form

$$\{1+(L_{21}^{(1)}+\tilde{c}_{56}L_{31}^{(1)})\beta_1^2\}(\tan\eta_1 b)^{\pm 1},$$

$$\beta_1\left\{\tilde{c}_{12}+\frac{\alpha_{12}\bar{b}_{22}v_2^2}{1-v_2^2}-\frac{\alpha_{13}\bar{b}_{32}v_3^2}{1-v_3^2}\right\}, \tag{33a}$$

$$\beta_1\left\{\tilde{c}_{14}+\frac{\alpha_{12}\bar{b}_{32}v_2^2}{1-v_2^2}+\frac{\alpha_{13}\bar{b}_{22}v_3^2}{1-v_3^2}\right\};$$

$$\beta_2\left\{\frac{\bar{b}_{22}(v_2^2+\tilde{c}_{12})+\bar{b}_{32}(\tilde{c}_{14}+\tilde{c}_{56}v_2^2)}{(v_2^2-1)}+\tilde{c}_{56}\bar{b}_{22}L_{32}^{(2)}\beta_2^2\right\}$$
$$\times(\tan\eta_2 b)^{\pm 1}, \tag{33b}$$

$$v_2^2\bar{b}_{22}+\bar{b}_{22}(\tilde{c}_{24}L_{32}^{(2)}+\tilde{c}_{12}L_{12}^{(1)})\beta_2^2,$$

$$v_2^2\bar{b}_{32}+\bar{b}_{22}(\tilde{c}_{44}L_{32}^{(2)}+\tilde{c}_{14}L_{12}^{(1)})\beta_2^2;$$

$$\beta_3\left\{\frac{\bar{b}_{23}(v_3^2+\tilde{c}_{12})+\bar{b}_{22}(\tilde{c}_{14}+\tilde{c}_{56}v_3^2)}{(v_3^2-1)}+\tilde{b}_{22}L_{23}^{(2)}\beta_3^2\right\}$$
$$\times(\tan\eta_3 b)^{\pm 1}, \tag{33c}$$

$$-v_3^2\bar{b}_{32}+\bar{b}_{22}(\tilde{c}_{22}L_{23}^{(2)}+\tilde{c}_{12}L_{13}^{(1)})\beta_3^2,$$

$$v_3^2\bar{b}_{22}+\bar{b}_{22}(\tilde{c}_{24}L_{23}^{(2)}+\tilde{c}_{14}L_{13}^{(1)})\beta_3^2,$$

where positive and negative exponents stand for symmetric and antisymmetric modes, respectively. The $L_{ij}^{(n)}$ for $n=0, 1, 2\cdots$ are the coefficients of β_j^n in (25) or (26). Arrangement of these columns in determinantal form, expansion and neglect of powers of β_j^n for $n>2$ lead to the limiting form of the frequency equation:

$$1+(L_{21}^{(1)}+\tilde{c}_{56}L_{31}^{(1)})\beta_1^2\}\{\mathbf{B}+\mathbf{C}\beta_2^2+\mathbf{D}\beta_3^2+\cdots\}$$
$$\times(\tan\eta_1 b)^{\pm 1}+\beta_1\beta_2\mathbf{E}(\tan\eta_2 b)^{\pm 1}$$
$$+\beta_1\beta_3\mathbf{F}(\tan\eta_3 b)^{\pm 1}+\cdots=0, \tag{34}$$

where

$$\mathbf{B} = v_2^2 v_3^2,$$
$$\mathbf{C} = v_3^2\bar{b}_{22}\{L_{12}^{(1)}(\bar{b}_{22}\tilde{c}_{12}+\bar{b}_{32}\tilde{c}_{14})+v_2^2\bar{b}_{32}L_{32}^{(2)}\},$$
$$\mathbf{D} = v_2^2\bar{b}_{22}\{L_{13}^{(1)}(\bar{b}_{22}\tilde{c}_{14}-\bar{b}_{32}\tilde{c}_{12})-v_3^2\bar{b}_{32}L_{23}^{(2)}\}, \tag{35}$$
$$\mathbf{E} = v_3^2\{\bar{b}_{22}(v_2^2+\tilde{c}_{12})+\bar{b}_{32}(v_2^2\tilde{c}_{56}+\tilde{c}_{14})\}^2(v_2^2-1)^{-2},$$
$$\mathbf{F} = v_2^2\{\bar{b}_{32}(v_3^2+\tilde{c}_{12})-\bar{b}_{22}(v_3^2\tilde{c}_{56}+\tilde{c}_{14})\}^2(v_3^2-1)^{-2}.$$

The frequency equations for thickness modes and for the isotropic plate are contained in (33) and (34). For the thickness modes we simply set $\beta_j=0$ and obtain (31). In the case of the isotropic plate,

$$\tilde{c}_{12}=\lambda/\mu; \quad \tilde{c}_{22}=(2+\lambda/\mu); \quad \tilde{c}_{44}=1;$$
$$L_{ij}^{(1)}=\epsilon^{ij} \quad \text{for} \quad i,j=1,2;$$
$$\bar{b}_{22}=1; \quad v_2^2=(2+\lambda/\mu); \quad v_3^2=1,$$

while the remaining constants $\bar{c}_{pq}$ and the ratios $\bar{b}_{32}$, $L_{ij}^{(n)}$ vanish. Equation (34) then becomes

$$(1-\beta_1^2)\{(2+\lambda/\mu)+\lambda\beta_2^2/\mu+\cdots\}(\tan\eta_1 b)^{\pm 1}+4\beta_1\beta_2(\tan\eta_2 b)^{\pm 1}=0$$

which, in the absence of the higher-order terms represented by dots, is Rayleigh's frequency equation for an isotropic plate.[4] The dots representing higher powers in the small quantities β_j are actually all zero in the case of the isotropic plate as can be ascertained from (33).

The frequency equation (34) and the characteristic equation (24), valid at long wavelengths, give the relation between Ω and ξ which we have sought.

VIII. ASYMPTOTIC BEHAVIOR AT LONG WAVELENGTH

The replacement of (13), (14), and (15) by their asymptotic forms, (34) and (24), enables us to determine, analytically, the behavior of the branches of the frequency spectrum at long wavelengths. That is, we find explicit expressions for the ordinates, slopes, and curvatures of the branches at $\xi=0$ in the plane of Ω as ordinate and ξ as abscissa.

(a) *Ordinates.* The ordinates of the branches at $\xi=0$ are the thickness frequencies (31). It may be noticed that the frequencies of the coupled thickness stretch and x_3-thickness-shear modes are highly sensitive to small changes in elastic constants.

(b) *Slopes.* By differentiating (34) once with respect to ξ and using (24), we obtain the following expressions, at $\xi=0$, for the essentially symmetric family:

(i) Symmetric x_1 thickness shear (s_1 even)

$$d\Omega/d\xi]_0 \cos(\pi s_1/2v_2)\cos(\pi s_1/2v_3)=0.$$

Hence,

$$d\Omega/d\xi]_0=0 \quad\text{if}\quad v_2\neq s_1/s_2 \quad\text{and}\quad v_3\neq s_1/s_3. \tag{36a}$$

(ii) Symmetric thickness stretch (s_2 odd)

$$d\Omega/d\xi]_0 \sin(\tfrac{1}{2}\pi s_2 v_2)\cos(\pi s_2 v_2/2v_3)=0.$$

Hence,

$$d\Omega/d\xi]_0=0 \quad\text{if}\quad v_2\neq s_1/s_2 \quad\text{and}\quad v_2/v_3\neq s_3/s_2. \tag{36b}$$

(iii) Antisymmetric x_3 thickness shear (s_3 odd)

$$d\Omega/d\xi]_0 \sin(\tfrac{1}{2}\pi s_3 v_3)\cos(\pi s_3 v_3/2v_2)=0.$$

Hence,

$$d\Omega/d\xi]_0=0 \quad\text{if}\quad v_3\neq s_1/s_3 \quad\text{and}\quad v_3/v_2\neq s_2/s_3. \tag{36c}$$

As noted in Sec. VI, the inequalities in (36) signify noncoincidence of thickness frequencies; and in that case the slopes of all the branches are zero at $\xi=0$. If there are coincidences, the expressions for the slopes become indeterminate forms which are evaluated below.

We obtain results similar to (36) for the essentially antisymmetric family except that then s_1 is odd and s_2, s_3 are even.

(c) *Curvatures.* The formulas for the curvatures are obtained through a second differentiation of (34), again using (24), with the results:

(i) x_1 thickness shear

$$\left.\frac{d^2\Omega}{d\xi^2}\right]_0=\frac{1}{s_1}\left\{\kappa_1\mp\frac{4}{\pi s_1}\frac{v_2\mathbf{E}}{\mathbf{B}}\left[\cot\left(\frac{\pi s_1}{2v_2}\right)\right]^{\mp1}\mp\frac{4}{\pi s_1}\frac{v_3\mathbf{F}}{\mathbf{B}}\left[\cot\left(\frac{\pi s_1}{2v_3}\right)\right]^{\mp1}\right\}; \tag{37a}$$

(ii) thickness stretch

$$\left.\frac{d^2\Omega}{d\xi^2}\right]_0=\frac{1}{s_2v_2}\left\{\kappa_2\mp\frac{4}{\pi s_2}\frac{v_2\mathbf{E}}{\mathbf{B}}\left[\cot\frac{\pi}{2}(1+s_2v_2)\right]^{\mp1}\right\}; \tag{37b}$$

(iii) x_3 thickness shear

$$\left.\frac{d^2\Omega}{d\xi^2}\right]_0=\frac{1}{s_3v_3}\left\{\kappa_3\mp\frac{4}{\pi s_3}\frac{v_3\mathbf{F}}{\mathbf{B}}\left[\cot\frac{\pi}{2}(1+s_3v_3)\right]^{\mp1}\right\}; \tag{37c}$$

where negative exponent and negative sign refer to the essentially symmetric family of modes and positive exponent and positive sign refer to the essentially antisymmetric family of modes.

At coincidence of thickness frequencies, the corresponding curvatures become infinite. Upon reduction to isotropy, the formulas (37) reduce to those obtained previously for that case.[3]

The curvatures of the x_1-thickness-shear branches are contained in Eqs. (56) and (37) of Ekstein's paper.[2] These lead exactly to (37a) given above. It should be noted, however, that the first cofactor Δ_{11} in Ekstein's Eq. (51) does not contain the higher-order terms necessary to arrive at Rayleigh's frequency equation upon reduction to isotropy.

(d) *Slopes at coincidence.* The behavior of the branches at coincidence of thickness frequencies can be found by asymptotic expansion of the frequency equation (34) in the neighborhood of infinite wavelength. In this region, the frequency may be expressed as

$$\Omega^2=v_n^2s_n^2(1+\epsilon_\xi),\quad \xi\ll1,\quad \epsilon_\xi\ll1, \tag{38}$$

where ϵ_ξ is a function of ξ to be determined from the frequency equation. In the same region, Ω, ξ, and $\bar{\eta}$ are related by (24) and therefore $\bar{\eta}$, in this region, may be expressed, to the second order in ϵ_ξ and ξ, as

$$\bar{\eta}_n=(v_ms_m/v_n)\{1+\tfrac{1}{2}[\epsilon_\xi-\tfrac{1}{4}\epsilon_\xi^2-(\xi^2\kappa_n/v_m^2s_m^2)]\}. \tag{39}$$

Then, to the same order,

$$(\tan\eta_n b)^{\pm1}=\{(\pi v_ms_m/4v_n)[\epsilon_\xi-\tfrac{1}{4}\epsilon_\xi^2-(\xi^2\kappa_n/v_m^2s_m^2)]\}^{\pm1}, \tag{40}$$

if $\tan(\pi v_m s_m/2v_n)=0$, and

$$(\tan\eta_n b)^{\pm 1} = \{(\pi v_m s_m/4v_n)[\epsilon_\xi - \tfrac{1}{4}\epsilon_\xi^2 - (\xi^2\kappa_n/v_m^2 s_m^2)]\}^{\mp 1}, \quad (41)$$

if $\cot(\pi v_m s_m/2v_n)=0$.

These expansions for $\bar{\eta}_n$ and $(\tan\eta_n b)^{\pm 1}$ are now used in the frequency equation (34) to determine the function ϵ_ξ. In the process of expansion of the frequency equation only the lowest-order terms in small quantities are finally retained.

(i) Coincidence of x_1 thickness shear and thickness stretch, $v_2=s_1/s_2$. From the frequency equation we obtain

$$\epsilon_\xi = \pm \frac{4\xi}{\pi s_1 s_2}\left(\frac{\mathbf{E}}{\mathbf{B}}\right)^{\frac{1}{2}}. \quad (42)$$

Inserting this in (38) and differentiating, we find

$$\left.\frac{d\Omega}{d\xi}\right]_{\xi=0} = \pm \frac{2}{\pi s_2}\left(\frac{\mathbf{E}}{\mathbf{B}}\right)^{\frac{1}{2}}, \quad (43)$$

where s_2 is odd or even depending on the type of symmetry. For the isotropic medium, this reduces to the known result[3]

$$\left.\frac{d\Omega}{d\xi}\right]_{\xi=0} = \pm \frac{4}{\pi s_1}.$$

(ii) Coincidence of x_1 thickness shear and x_3 thickness shear, $v_3=s_1/s_3$. From the frequency equation we find

$$\epsilon_\xi = \pm \frac{4\xi}{\pi s_1 s_3}\left(\frac{\mathbf{F}}{\mathbf{B}}\right)^{\frac{1}{2}}, \quad (44)$$

so that, from (38),

$$\left.\frac{d\Omega}{d\xi}\right]_{\xi=0} = \pm \frac{2}{\pi s_3}\left(\frac{\mathbf{F}}{\mathbf{B}}\right)^{\frac{1}{2}}, \quad (45)$$

where s_3 is odd or even depending upon the type of symmetry. For the isotropic plate, the slope reduces to zero, as a consequence of uncoupling of the two modes.

(iii) Coincidence of all three branches, $v_2=s_1/s_2$; $v_3=s_1/s_3$. Using the appropriate expansions in the frequency equation, we find the three expressions

$$\epsilon_\xi = 0, \quad (46a)$$

$$\epsilon_\xi = \pm \frac{4\xi}{\pi s_1}\left[\frac{1}{\mathbf{B}}\left(\frac{\mathbf{E}}{s_2^2}+\frac{\mathbf{F}}{s_3^2}\right)\right]^{\frac{1}{2}}, \quad (46b)$$

whence, from (38),

$$\left.\frac{d\Omega}{d\xi}\right]_{\xi=0} = 0, \quad (47a)$$

$$\left.\frac{d\Omega}{d\xi}\right]_{\xi=0} = \pm \frac{2}{\pi}\left[\frac{1}{\mathbf{B}}\left(\frac{\mathbf{E}}{s_2^2}+\frac{\mathbf{F}}{s_3^2}\right)\right]^{\frac{1}{2}}. \quad (47b)$$

(iv) Coincidence of thickness stretch with x_3 thickness shear,[8] $v_2/v_3=s_3/s_2$. Expansion of the frequency equation leads to

$$\epsilon_\xi\left\{\tfrac{1}{2}\epsilon_\xi^2(\tan\tfrac{1}{2}\pi s_2 v_2 - \tfrac{1}{2}\pi s_2 v_2) - \epsilon_\xi \tan\tfrac{1}{2}\pi s_2 v_2 + \xi^2\left[\frac{\kappa_2+\kappa_3}{v_2^2 s_2^2}\tan\tfrac{1}{2}\pi s_2 v_2 + \frac{4}{\pi s_2 v_2}\left(\frac{\mathbf{E}}{s_2^2\mathbf{B}}+\frac{\mathbf{F}}{s_3^2\mathbf{B}}\right)\right]\right\} = 0. \quad (48)$$

Then (48) and (38) yield

$$d\Omega/d\xi]_{\xi=0} = 0 \quad (49)$$

for each branch,[3] as a consequence of the fact that these two thickness modes can have the same frequency only when they are not coupled.

[8] Since the v_i are nondegenerate, $s_2 \neq s_3$.

Reprinted from THE JOURNAL OF THE ACOUSTICAL SOCIETY OF AMERICA, Vol. 34, No. 12, 1902–1910, December, 1962

Frequency Spectrum of a Monoclinic Crystal Plate

R. K. KAUL* AND R. D. MINDLIN
Department of Civil Engineering, Columbia University, New York 27, New York
(Received December 19, 1961)

A sketch is constructed of the branches of Ekstein's frequency equation for an infinite, monoclinic crystal plate. The sketch is for the AT cut of quartz and covers a wide range of frequencies and propagation constants. The real, imaginary, and complex segments of the branches are included.

INTRODUCTION

THIS is the fourth in a series of studies of Ekstein's[1] solution of the equations of linear elasticity for the case of vibrations of an infinite, monoclinic crystal plate with tractionfree faces. In the first paper[2] of the series, a method was described for mapping the real segments of the branches of the frequency equation on a grid of three relatively simple families of algebraic curves which form bounds of the branches except at certain grid intersections. The method, supplemented by some computations, was used to map the real segments of the first nine branches of the spectrum of the AT cut of quartz. In the second paper,[3] a special solution was given for the phase velocity of surface waves. This supplies an asymptote for the flexural branch and the extensional branch: the higher branches having a common asymptote with that of one of the

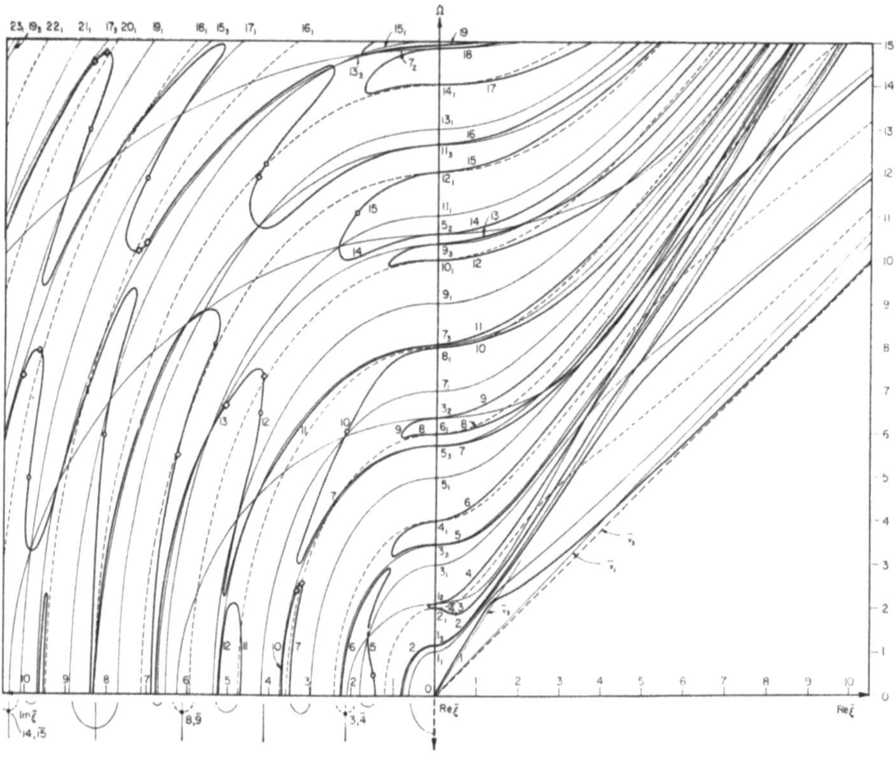

FIG. 1. Real and imaginary segments of branches of frequency equation. AT cut of quartz. Essentially symmetric family of modes.

* Present address: Thomas J. Watson Research Center, International Business Machines Corporation, Yorktown Heights, New York.
[1] H. Ekstein, Phys. Rev. 68, 11 (1945).
[2] E. G. Newman and R. D. Mindlin, J. Acoust. Soc. Am. 29, 1206 (1957).
[3] H. Deresiewicz and R. D. Mindlin, J. Appl. Phys. 28, 669 (1957).

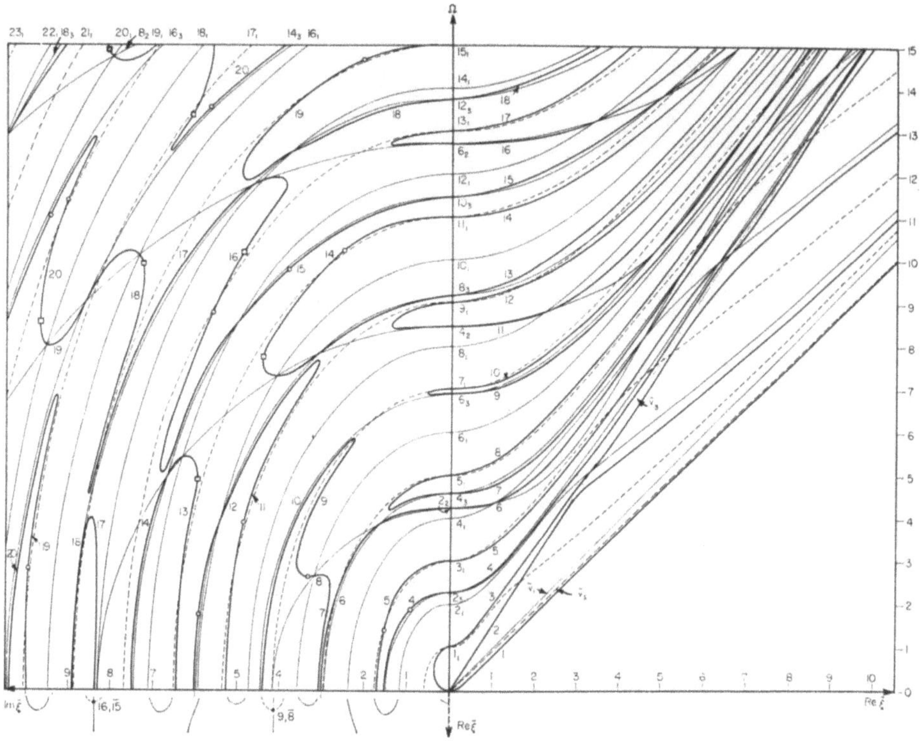

FIG. 2. Real and imaginary segments of branches of frequency equation. AT cut of quartz. Essentially antisymmetric family of modes.

families of bounds. In the third paper,[4] formulas were derived for the slopes and curvatures of the branches at infinite wavelength and frequencies greater than zero, supplementing the known formulas for infinite wavelength and zero frequency.

The bounds, asymptotes, slopes, and curvatures, along with a few roots of degenerate forms of the frequency equation on asymptotes of bounds, give enough information to sketch the real segments of the branches without detailed computation of the roots of the full frequency equation. Such a sketch is shown in Figs. 1 and 2 for about forty branches of the AT cut of quartz.

In the pure imaginary part of the spectrum, there are no asymptotes and the intersections of bounds are distributed sparsely. The bounds, slopes, curvatures, and roots at zero frequency are first used to sketch families of auxiliary bounds (Figs. 3 and 4) which are the branches of degenerate forms of the frequency equation. The bounds and auxiliary bounds form an intricate network of curves and intersections through which the imaginary segments of the branches of the full frequency equation are sketched, with results as shown in Figs. 1 and 2.

Finally, Onoe's[5] theorem on complex roots serves as part of the basis for sketching the complex segments of the branches, examples of which are shown in Fig. 5.

I. FREQUENCY EQUATION

Ekstein's frequency equation may be expressed in the form

$$|\mu_{in}|=0; \quad i, n=1, 2, 3, \tag{1}$$

where the three elements of the nth column of the determinant are

$$\begin{aligned}\mu_{1n}&=[\eta_n L_{1n}+\xi(L_{2n}+L_{3n}\bar{c}_{56})](\tan\eta_n b)^{\pm 1},\\ \mu_{2n}&=\xi\bar{c}_{12}L_{1n}+\eta_n(L_{2n}\bar{c}_{22}+L_{3n}\bar{c}_{24}),\\ \mu_{3n}&=\xi\bar{c}_{14}L_{1n}+\eta_n(L_{2n}\bar{c}_{24}+L_{3n}\bar{c}_{44}).\end{aligned} \tag{2}$$

[4] R. K. Kaul and R. D. Mindlin, J. Acoust. Soc. Am. 34, 1891 (1962).

[5] R. D. Mindlin and M. Onoe, in *Proceedings of the Eleventh Annual Symposium on Frequency Control* (U. S. Army Signal Corps Engineering Laboratories, Fort Monmouth, New Jersey, 1957), pp. 17–40.

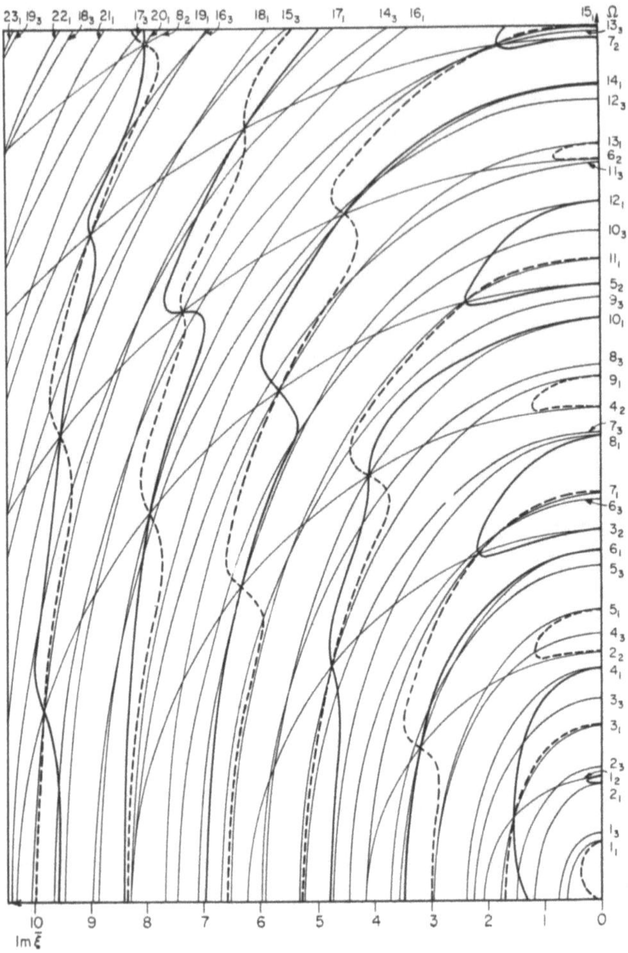

FIG. 3. First family of auxiliary bounds of imaginary segments of branches [Eqs. (22) and (28)].

The $\bar{c}_{pq}(\equiv c_{pq}/c_{66})$ are dimensionless elastic constants, b is the half-thickness of the plate, and ξ is the wavenumber (propagation constant) along the plate. The η_n are the three wavenumbers across the thickness. They are the roots of

$$|\lambda_{ij}-\delta_{ij}\Omega^2|=0,\quad \lambda_{ij}=\lambda_{ji} \tag{3}$$

in which $\Omega[\equiv\omega/(\pi/2b)(c_{66}/\rho)^{\frac{1}{2}}]$ is the dimensionless frequency, δ_{ij} is the Kronecker delta, and

$$\begin{aligned} \lambda_{11}&=\bar{\xi}^2\bar{c}_{11}+\bar{\eta}_n^2, & \lambda_{23}&=\bar{\xi}^2\bar{c}_{56}+\bar{\eta}_n^2\bar{c}_{24},\\ \lambda_{22}&=\bar{\xi}^2+\bar{\eta}_n^2\bar{c}_{22}, & \lambda_{31}&=\bar{\xi}\bar{\eta}_n(\bar{c}_{14}+\bar{c}_{56}),\\ \lambda_{33}&=\bar{\xi}^2\bar{c}_{55}+\bar{\eta}_n^2\bar{c}_{44}, & \lambda_{12}&=\bar{\xi}\bar{\eta}_n(1+\bar{c}_{12}). \end{aligned} \tag{4}$$

Here $\bar{\xi}=2\xi b/\pi$ and $\bar{\eta}_n=2\eta_n b/\pi$ are dimensionless wavenumbers. Finally,

$$L_{in}=\mathrm{cof}(\lambda_{ni}-\delta_{ni}\Omega_n^2)/\mathrm{cof}(\lambda_{nn}-\Omega_n^2). \tag{5}$$

In (2) the positive exponent pertains to the essentially symmetric family of modes and the negative exponent to the essentially antisymmetric family, as described in reference 2. The frequency equation may be written in the form

$$\textstyle\sum_n M_n(\tan\eta_n b)^{\pm 1}=0, \tag{6}$$

where $M_n=\mathrm{cof}(\mu_{1n})$.

II. THICKNESS MODES

The frequencies of the thickness modes are obtained by setting $\xi=0$ in (1) and (3), with the result

$$\Omega=v_n s_n,\quad n=1,2,3;\quad s=1,2,3,4,\cdots, \tag{7}$$

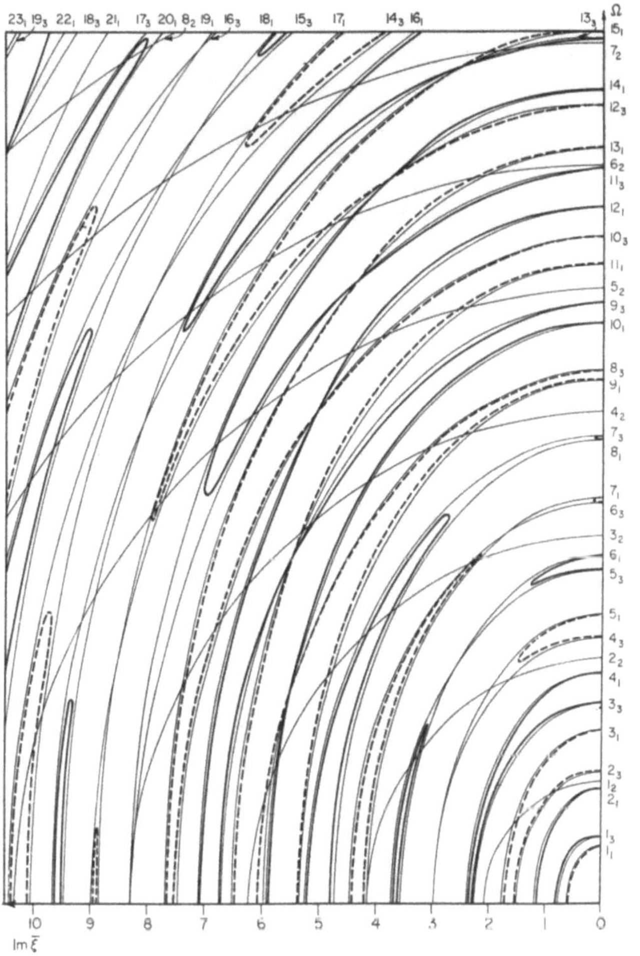

FIG. 4. Second family of auxiliary bounds of imaginary segments of branches [Eqs. (27) and (29).

where $v_1=1$ and

$$v_{2,3}=\langle\tfrac{1}{2}\{\bar{c}_{22}+\bar{c}_{44}\pm[(\bar{c}_{22}-\bar{c}_{44})^2+4\bar{c}_{24}^2]^{\frac{1}{2}}\}\rangle^{\frac{1}{2}}. \qquad (8)$$

The v_n give the velocities of waves, across the thickness of the plate, as ratios to the velocity $(c_{66}/\rho)^{\frac{1}{2}}$. The ratios v_1, v_2, v_3 are equal to $c_{66}^{-\frac{1}{2}}$ times the velocities c_1, c_3, c_2, respectively, of reference 2. With Bechmann's[6] constants for quartz (see Appendix), we find, for the AT cut,

$$v_2=2.1178, \quad v_3=1.1483. \qquad (9)$$

The integers $s_1=1, 2, \cdots$, $s_2=1, 2, \cdots$, $s_3=1, 2, \cdots$, in (7) are the same as s_1, s_3, s_2, respectively, of reference 2. Here the subscripts identify the displacement direction or the predominant displacement direction of the thickness mode. Thus the s_1 identify the orders of the thickness-shear modes whose displacements are in the direction x_1 of the digonal axis in the plane of the plate. The other two thickness modes have coupled displacements in the direction x_2, normal to the plane of the plate, and x_3, in the third orthogonal direction. The s_2 identify the orders of the modes which are predominantly thickness-stretch, and the s_3 give the orders of the thickness-shear modes with predominant displacement in the x_3 direction. In Figs. 1–5, the frequencies $\Omega=v_n s_n$ at $\xi=0$ are identified by the s_n, which appear just to the right of the center line.

[6] R. Bechmann, Phys. Rev. 110, 1061 (1958).

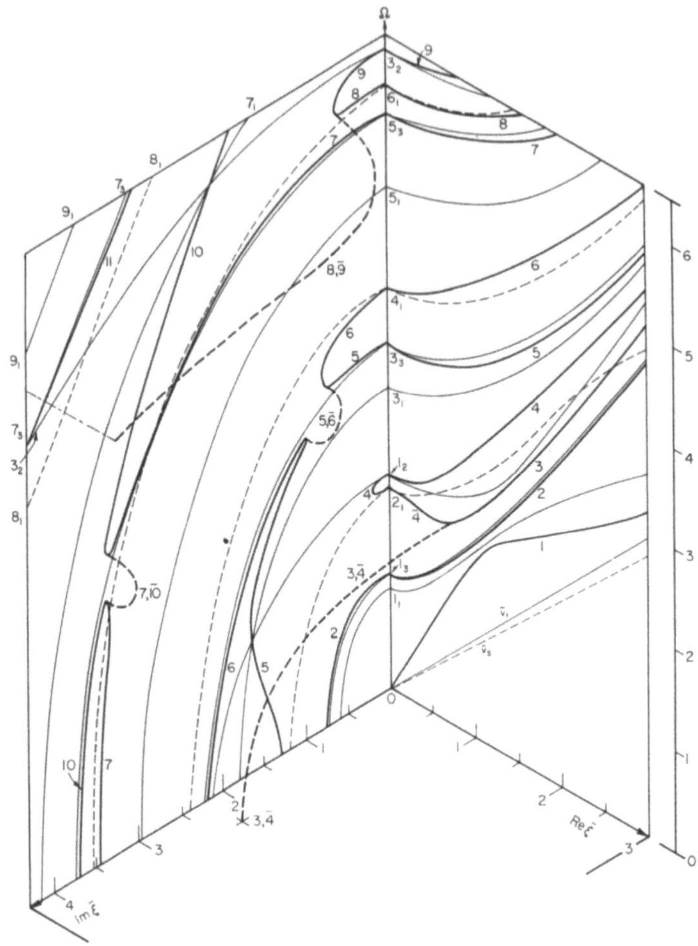

FIG. 5. Real, imaginary, and complex segments of branches. Essentially symmetric family of modes.

III. BOUNDS AND ASYMPTOTES

As described in reference 2, the equations of the bounds are given by (3) with $\bar{\eta}_n = s_n$. The bounds are drawn as the thin, full lines in Figs. 1 and 2; and they are identified by the same s_n that identify the thickness frequencies.

As the bounds pass through the thickness frequencies, at $\xi = 0$, they have the behavior given by Eq. (24) of reference 4:

$$\Omega^2 = v_n^2 s_n^2 + \kappa_n \xi^2 + \cdots. \tag{10}$$

Thus, at $\xi = 0$, the slopes of the bounds are zero and their curvatures are $\kappa_n / v_n s_n$, where the κ_n depend on the elastic constants as defined in reference 4.

The asymptotes of the bounds are obtained by setting $\bar{\eta}_n = 0$ in (3), with the result

$$\Omega = \bar{v}_n \xi, \tag{11}$$

where $\bar{v}_2 = \bar{c}_{11}^{\frac{1}{2}}$ and

$$\bar{v}_{1,3} = \langle \tfrac{1}{2}\{1 + \bar{c}_{55} \mp [(1 - \bar{c}_{55})^2 + 4\bar{c}_{56}^2]^{\frac{1}{2}}\}\rangle^{\frac{1}{2}}. \tag{12}$$

The $\bar{v}_n$ are the ratios, to $(c_{66}/\rho)^{\frac{1}{2}}$, of the velocities of waves along the digonal axis in an infinite medium; $\bar{v}_1, \bar{v}_2, \bar{v}_3$ are $(c_{11}/c_{66})^{\frac{1}{2}}$ times V_1^*, V_3^*, V_2^*, respectively, of reference 2. With Bechmann's constants, the $\bar{v}_n$ for the AT cut of quartz are

$$\bar{v}_1 = 0.99723, \quad \bar{v}_2 = 1.72910, \quad \bar{v}_3 = 1.54179. \tag{13}$$

As shown in reference 2, the asymptotes $s_1 = 0$ and $s_3 = 0$ (i.e., $\Omega = \bar{v}_1 \xi$ and $\Omega = \bar{v}_3 \xi$) are themselves bounds, but $s_2=0$ is not. The branches of the frequency equation for the essentially antisymmetric modes [negative exponent in (2)] pass through the intersections of bounds s_1 even with s_2 even, s_2 even with s_3 even, and s_3 even with s_1 even. Odd replaces even for the essentially symmetric modes. All the branches, except the lowest one of each symmetry, are asymptotic to $\Omega=\bar{v}_1\xi$. The remaining two branches are asymptotic to $\Omega=\bar{v}_s\xi$, where $\bar{v}_s$ is the ratio of the velocity of the surface wave to $(c_{66}/\rho)^{\frac{1}{2}}$.

IV. FREQUENCY EQUATION ON ASYMPTOTES

On each asymptote of the bounds, one of the η_n is zero as it changes from real to imaginary in crossing the asymptote. The number of transcendental functions in the frequency equation is then reduced, so that it is relatively easy to compute a few roots which are helpful in constructing the real segments of the branches. When an η_n is zero, indeterminate forms arise in the nth column of the frequency equation. As a consequence the elements of the column have to be redefined before taking the limit. Although the limiting forms were employed in reference 2, explicit expressions for them were not given. They are described in the following.

(a) $\eta_1=0$. The frequency equation for the antisymmetric modes has no real roots. For the symmetric modes we redefine L_{21} and L_{31}:

$$L_{21}=\frac{(\tilde{c}_{11}-v_1^2+\alpha_1^2)(\tilde{c}_{55}-v_1^2+\alpha_1^2\tilde{c}_{44})-\alpha_1^2q^2}{\alpha_1[q(\tilde{c}_{56}+\alpha_1^2\tilde{c}_{24})-p(\tilde{c}_{55}-v_1^2+\alpha_1^2\tilde{c}_{44})]},$$

$$L_{31}=\frac{(\tilde{c}_{11}-v_1^2+\alpha_1^2)(\tilde{c}_{56}+\alpha_1^2\tilde{c}_{24})-pq\alpha_1^2}{\alpha_1[p(\tilde{c}_{55}-v_1^2+\alpha_1^2\tilde{c}_{44})-q(\tilde{c}_{56}+\alpha_1^2\tilde{c}_{24})]},$$

where $p=1+\tilde{c}_{12}$, $q=\tilde{c}_{14}+\tilde{c}_{56}$ and $\alpha_1=\eta_1/\xi$. Inserting these in (2) and taking the limit as $\alpha_1\rightarrow 0$, we find

$$\mu_{11}=-\tfrac{1}{2}\pi\xi^2\bar{v}_1^2(\tilde{c}_{55}-\bar{v}_1^2)r_1,$$
$$\mu_{21}=\xi\{\tilde{c}_{12}-[\tilde{c}_{22}(\tilde{c}_{55}-\bar{v}_1^2)-\tilde{c}_{56}\tilde{c}_{24}]r_1\},$$
$$\mu_{31}=\xi\{\tilde{c}_{14}-[\tilde{c}_{24}(\tilde{c}_{55}-\bar{v}_1^2)-\tilde{c}_{56}\tilde{c}_{24}]r_1\},$$

where $r_1=(\tilde{c}_{11}-\bar{v}_1^2)/[p(\tilde{c}_{55}-\bar{v}_1^2)-q\tilde{c}_{56}]$. The remaining columns of the determinant are unchanged, but we note that η_2 and η_3 are imaginary when $\eta_1=0$. Thus the two remaining transcendental functions are hyperbolic. The frequency equation has only one real root:

$$\xi=4.3455 \tag{14}$$

for the AT cut of quartz, with Bechmann's constants.

(b) $\eta_2=0$. In this case, we define new elements $\bar{\mu}_{i2}=\mu_{i2}/L_{12}$ of the second column of (1) and redefine L_{12} and L_{32} as

$$L_{12}=\frac{(1-\tilde{c}_{11}+\alpha_2^2\tilde{c}_{22})(\tilde{c}_{55}-\tilde{c}_{11}+\alpha_2^2\tilde{c}_{44})-(\tilde{c}_{56}+\alpha_2^2\tilde{c}_{24})^2}{\alpha_2[q(\tilde{c}_{56}+\alpha_2^2\tilde{c}_{24})-p(\tilde{c}_{55}-\tilde{c}_{11}+\alpha_2^2c_{44})]},$$

$$L_{32}=\frac{p(\tilde{c}_{56}+\alpha_2^2\tilde{c}_{24})-q(1-\tilde{c}_{11}+\alpha_2^2\tilde{c}_{22})}{q(\tilde{c}_{56}+\alpha_2^2\tilde{c}_{24})-p(\tilde{c}_{55}-\tilde{c}_{11}+\alpha_2^2\tilde{c}_{44})}.$$

Then, substituting these in the revised form of the elements of the second column and taking the limit as $\alpha_2\rightarrow 0$, we find, for the symmetric modes,

$$\bar{\mu}_{12}=0,\quad \bar{\mu}_{22}=\tilde{c}_{12}\xi,\quad \bar{\mu}_{32}=\tilde{c}_{14}\xi,$$

and the same $\bar{\mu}_{22}$ and $\bar{\mu}_{32}$ for the antisymmetric modes, but

$$\bar{\mu}_{12}=\frac{2}{\pi}\left[1+\frac{(p\tilde{c}_{56}+q\tilde{c}_{11})\tilde{c}_{56}-p(\tilde{c}_{55}-\tilde{c}_{11})}{(1-\tilde{c}_{11})(\tilde{c}_{55}-\tilde{c}_{11})-c_{56}^2}\right],$$

when $\eta_2=0$, η_1, and η_3 are real; so that the remaining two transcendental functions are trigonometric and the frequency equations have an infinity of roots. These are useful in determining where the branches cross $\eta_2=0$, since such information is not given by the intersections of $\eta_2=0$ with bounds $\bar{\eta}_n=s_n$, $n=1, 3$, since, as stated above, $\eta_2=0$ is not a bound.

(c) $\eta_3=0$. This is a bound for the antisymmetric branches. Hence the roots, in that case, are given by its intersections with bounds $\bar{\eta}_1=s_1$(even). There are no intersections of $\eta_3=0$ with $\bar{\eta}_2=s_2$ since $\bar{v}_2>\bar{v}_3$. For the symmetric branches, we define new elements $\bar{\mu}_{i3}=\mu_{i3}/L_{13}$ and also

$$L_{13}=\frac{\alpha_3[p(\tilde{c}_{56}+\alpha_3^2\tilde{c}_{24})-q(1-\bar{v}_3^2+\alpha_3^2\tilde{c}_{22})}{(\tilde{c}_{11}-\bar{v}_3^2+\alpha_3^2)(1-\bar{v}_3+\alpha_3^2\tilde{c}_{22})-\alpha_3^2p^2},$$

$$L_{23}=\frac{\alpha_3^2pq-(\tilde{c}_{11}-v_3^2+\alpha_3^2)(\tilde{c}_{56}+\alpha_3^2\tilde{c}_{24})}{(\tilde{c}_{11}-v_3^2+\alpha_3^2)(1-v_3^2+\alpha_3^2\tilde{c}_{22})-\alpha_3^2p^2}.$$

Then, in the limit as $\alpha_3\rightarrow 0$,

$$\bar{\mu}_{13}=-\tfrac{1}{2}\pi\xi^2\bar{v}_3^2\tilde{c}_{56}r_3,$$
$$\bar{\mu}_{23}=\xi\{\tilde{c}_{12}+[\tilde{c}_{24}(1-\bar{v}_3^2)-\tilde{c}_{22}\tilde{c}_{56}]r_3\},$$
$$\bar{\mu}_{33}=\xi\{\tilde{c}_{14}+[\tilde{c}_{44}(1-\bar{v}_3^2)-\tilde{c}_{24}c_{56}]r_3\},$$

where $r_3=(\tilde{c}_{11}-\bar{v}_3^2)/[p\tilde{c}_{56}-q(1-\bar{v}_3^2)]$. When $\eta_3=0$, η_1 is imaginary and η_2 is real. Thus, one trigonometric and one hyperbolic function remain in the frequency equation. Of the infinity of roots, only the lowest,

$$\xi=1.4266, \tag{15}$$

is useful, as the remaining ones are very close to the intersection of the bound $\bar{\eta}_3=1$ with the bounds $\bar{\eta}_1$ odd.

The limiting form of the frequency equation on the asymptote $\Omega=\bar{v}_s\xi$, corresponding to surface waves, is given in reference 3. A recomputation of the phase velocity, with Bechmann's constants, yields

$$\bar{v}_s=0.9487 \tag{16}$$

for the AT cut of quartz.

V. REAL SEGMENTS OF BRANCHES

The sketching of the real segments of the branches on the grid of bounds may be carried out essentially as described in reference 2. However, we now have available the formulas for the curvatures of all the branches, at $\xi=0$, from reference 4. As a consequence, the sketch may be made directly in the plane of Ω and ξ and extended to as high frequencies as desired, without the necessity of computing roots of the full frequency equation over the range of frequencies. The sketch of each branch is started, at $\xi=0$, at a thickness frequency $\Omega=v_n s_n$. Knowledge of the ratio of the curvature of the branch to that of the curve $\bar{\eta}=s_n$ that passes through the same thickness frequency at $\xi=0$ indicates the path along which to start drawing the branch. If the curve $\bar{\eta}=s_n$ is a bound for the branch (s_n even for the antisymmetric branches and s_n odd for the symmetric branches), there is no uncertainty as to how to thread the branch through the intersections of the bounds. These branches are drawn first, so as to consume intersection points about which there might be some doubt. Then the remaining branches are sketched in. This procedure differs from the case of the isotropic plate,[5] for which every branch coincides with a bound at $\xi=0$.

With regard to the three real branches which pass through $\Omega=0$, their behaviors in the neighborhood of $\xi=0$ are[2]

$$\Omega=\xi\bar{\gamma}_{11}^{\frac{1}{2}},\quad \Omega=\tfrac{1}{2}\pi\xi^2(\tfrac{1}{3}\bar{\gamma}_{11})^{\frac{1}{2}},\quad \Omega=\xi\bar{\gamma}_{55}^{\frac{1}{2}} \tag{17}$$

for the extensional, flexural, and face-shear modes, respectively, where

$$\bar{\gamma}_{11}=\bar{c}_{11}-(\bar{c}_{12}^2\bar{c}_{44}+\bar{c}_{14}^2\bar{c}_{22}-2\bar{c}_{12}\bar{c}_{14}\bar{c}_{24})/(\bar{c}_{22}\bar{c}_{44}-\bar{c}_{24}^2),$$
$$\bar{\gamma}_{55}=\bar{c}_{55}-\bar{c}_{56}^2.$$

For the AT cut of quartz, with Bechmann's constants,

$$\bar{\gamma}_{11}^{\frac{1}{2}}=1.7210,\quad \bar{\gamma}_{55}^{\frac{1}{2}}=1.5375.$$

Thus, comparing the first of (17) with the third of (13), the extensional branch starts out, from the origin, above the asymptote $\Omega=\bar{v}_3\xi$ and, referring to Fig. 1, below the bound $s_1=1$. The branch cannot cross the bound $s_1=1$ and hence must cross the asymptote $\bar{v}_3$ before the latter crosses the bound. Thus, the point at which the branch crosses $\bar{v}_3$ is the one given by (15). The extensional branch is the only symmetric branch below the bound s_1. But there is the point (14) on the asymptote $\Omega=\bar{v}_1\xi$ which must be crossed by a branch. Hence the extensional branch crosses under the asymptote $\bar{v}_1$ at the point given by (14) and continues to its final asymptote $\Omega=\bar{v}_s\xi$. This procedure of sketching is typical for all the branches. The branches are the heavy curves in Figs. 1 and 2, and their orders are identified by the numbers, without subscripts, next to the curves.

VI. IMAGINARY SEGMENTS OF BRANCHES

When ξ is pure imaginary, all three $\bar{\eta}_n$ are real for all real Ω. The curvatures of the bounds and branches, at $\xi=0$, are simply the negatives of the curvatures of the real segments.

The bounds terminate at zero frequency at values of ξ which may be found by solving (3) with $\Omega=0$. We replace ξ by $i\xi'$ and η_n by $\alpha_n\xi'$ and find, for the AT cut of quartz,

$$\alpha_1=1.690\,046,\quad \alpha_2=0.482\,795,\quad \alpha_3=1.339\,796. \tag{18}$$

The integral multiples of the reciprocals of these α_n are the abscissas of the bounds at $\Omega=0$.

Certain of the branches also terminate at $\Omega=0$ and pure imaginary ξ. We write (6) in the form

$$\textstyle\sum_n M_n(\tan\alpha_n\xi' b)^{\pm 1}=0, \tag{19}$$

where, from M_n-cof(μ_{1n}), $\Omega=0$, (2) and (18),

$$M_1=3.507\,525,\quad M_2=-1,\quad M_3=-0.005\,541. \tag{20}$$

Then the roots ξ' of (19) are obtained easily. The first of them, for the symmetric modes, is $\xi'=0.74668$. This value has served as a basis for testing the corresponding quantity

$$\xi'=v_3[\bar{\gamma}_{11}/\bar{\gamma}_{55}(\bar{c}_{11}-\bar{c}_{12}^2/\bar{c}_{22})]^{\frac{1}{2}}=0.7456$$

found from a system of approximate equations of motion of plates.[7]

It may be seen, in Figs. 1 and 2, that the density of intersections of bounds is too low, when ξ is imaginary, to permit the branches to be sketched. However, the bounds may be supplemented by systems of auxiliary bounds[8] which can be constructed readily.

Consider, for example, the frequency equation for the symmetric modes in the form (6)

$$\textstyle\sum_n M_n \tan\eta_n b=0. \tag{21}$$

The curves $\bar{\eta}_3$ even are not bounds of (21), but, on $\bar{\eta}_3$ even, (21) reduces to

$$M_1 \tan\eta_1 b+M_2 \tan\eta_2 b=0. \tag{22}$$

The roots of (22) are roots of (21) when $\bar{\eta}_3$ is an even integer. Thus the intersections of the branches of (22) with the curves $\bar{\eta}_3$ even are points on the branches of (21). Furthermore points on the branches of (22) which are not on $\bar{\eta}_3$ even (nor on the single trivial curve $M_3=0$) are not roots of (21). Hence the branches of (22) are auxiliary bounds of the frequency equation (21).

Now, the auxiliary bounds are easy to construct. Their bounds are both $\bar{\eta}_1$, $\bar{\eta}_2$ even, even and $\bar{\eta}_1$, $\bar{\eta}_2$ odd, odd; so that the intersection points have a high density. The ordinates, slopes, and curvatures of the auxiliary bounds at $\xi=0$ may be obtained directly from the results in reference 4. Thus, for imaginary ξ and $\bar{\eta}_3$ even, Eq. (34) of reference 4 becomes, for the symmetric modes,

$$[1-(L_{21}^{(1)}+\bar{c}_{56}L_{31}^{(1)})\beta_1^2](\mathbf{B}-\mathbf{C}\beta_2^2-\mathbf{D}\beta_3^2)\tan\eta_1 b$$
$$-\beta_1\beta_2\mathbf{E}\tan\eta_2 b=0, \tag{23}$$

[7] R. D. Mindlin, Quart. Appl. Math. 19, 51 (1961).
[8] Y. H. Pao and R. D. Mindlin, J. Appl. Mech. 27, 513 (1960).

where $\beta_n=\bar{\xi}'/\bar{\eta}_n$ and the remaining symbols are defined in reference 4. Equation (23) is the asymptotic form of (22) for small $\bar{\xi}$. At $\bar{\xi}=0$ it has the roots

$$\Omega=v_n s_n,\quad n=1,2, \tag{24}$$

where s_1 is even and s_2 is odd. These values of Ω give the ordinates of the auxiliary bounds at $\bar{\xi}=0$. Their slopes are obtained from (36a), (36b) of reference 4:

$$\frac{d\Omega}{d\bar{\xi}'}\cos\frac{\pi s_1}{2v_2}=0,\quad \frac{d\Omega}{d\bar{\xi}'}\sin\frac{\pi s_2 v_2}{2}=0. \tag{25}$$

Thus the slopes at $\bar{\xi}=0$ are zero unless $v_2=s_1/s_2$, which is the condition for coincidence of auxiliary bounds. The curvatures are obtained from (37a) and (37b) of reference 4:

$$\frac{d^2\Omega}{d\bar{\xi}'^2}=-\frac{\kappa_1}{s_1}+\frac{4v_2\mathbf{E}}{\pi s_1^2\mathbf{B}}\tan\frac{\pi s_1}{2v_2}, \tag{26a}$$

$$\frac{d^2\Omega}{d\bar{\xi}'^2}=-\frac{\kappa_2}{s_2v_2}+\frac{4\mathbf{E}}{\pi s_2^2\mathbf{B}}\tan\tfrac{1}{2}\pi(1+s_2v_2). \tag{26b}$$

Finally, the terminal points of the auxiliary bounds at $\Omega=0$ may be found in the same way that the corresponding points of the branches were determined.

With all of the information on bounds, ordinates, slopes and curvatures at $\bar{\xi}'=0$, and terminal points, the auxiliary bounds may be sketched without difficulty; and they are shown as the thick, full lines in Fig. 3. Their intersections with the curves $\bar{\eta}_3$ even are the circles in Fig. 1.

The equation of another set of auxiliary bounds for the symmetric branches is obtained by setting $\bar{\eta}_2$ even in (21), with the result

$$M_1\tan\eta_1 b+M_3\tan\eta_3 b=0. \tag{27}$$

The branches of this equation are obtained in the same way that those of (22) were; and are shown as the thick, full lines in Fig. 4. Their intersections with $\bar{\eta}_2$ even are identified by squares in Fig. 1.

The auxiliary bounds for the branches of the essentially antisymmetric family of modes are found similarly. We first set $\bar{\eta}_3$ odd in (6), with the negative exponent, and obtain

$$M_1\cot\eta_1 b+M_2\cot\eta_2 b=0 \tag{28}$$

as an equation for auxiliary bounds; and then set $\bar{\eta}_2$ odd to obtain

$$M_1\cot\eta_1 b+M_3\cot\eta_3 b=0 \tag{29}$$

for the second equation. The branches of (28) and (29) are shown as the dashed lines in Figs. 3 and 4, respectively. The intersections of (28) with $\bar{\eta}_3$ odd and (29) with $\bar{\eta}_2$ odd are shown in Fig. 2 as circles and squares, respectively.

The bounds and auxiliary bounds now permit the sketching of the imaginary segments of the branches without difficulty. The results are shown in Figs. 1 and 2.

VII. COMPLEX SEGMENTS OF BRANCHES

When ξ is complex, say $\xi=x+iy$, $\bar{\xi}=\bar{x}+i\bar{y}$, all three $\bar{\eta}_n$ are complex for real Ω. Taking account of the fact that (3) implicitly relates the three $\bar{\eta}_n$ to $\bar{\xi}$ and Ω, the frequency equation relates one real variable, Ω, to one complex variable, $\bar{\xi}$, and hence defines a family of curves in the complex space Ω, $\bar{x}$, $i\bar{y}$. We begin by finding the intersections of these curves with the plane $\Omega=0$.

We write $\bar{\eta}_n=i\alpha_n\bar{\xi}$ in (3) and find, as before that, when $\Omega=0$, (3) has the three real roots α_n given in (18). Also, (6) becomes

$$\textstyle\sum_n M_n(\tanh\alpha_n\xi b)^{\pm1}=0 \tag{30}$$

in which the M_n are the same as in (20). However, ξ is now complex so that the left-hand side of (30) has a real and an imaginary part, each of which must be set equal to zero. Upon separating the real and imaginary parts, we find

$$\sum_n\frac{M_n(\tanh\alpha_n xb)^{\pm1}}{\cos^2\alpha_n yb[1+\tan^2\alpha_n yb(\tanh^2\alpha_n xb)^{\pm1}]}=0 \tag{31}$$

for the real part; and

$$\sum_n\frac{M_n(\tan\alpha_n yb)^{\pm1}}{\cosh^2\alpha_n xb[1+\tanh^2\alpha_n xb(\tan^2\alpha_n yb)^{\pm1}]}=0 \tag{32}$$

for the imaginary part.

On the real axis, $y=0$, (32) is satisfied identically, and (31) reduces to

$$\textstyle\sum_n M_n(\tanh\alpha_n xb)^{\pm1}=0,$$

which has no real roots. On the imaginary axis, $x=0$, (31) is satisfied identically, and (32) reduces to (19), which gives, again, the terminal points of the pure imaginary segments of the branches.

In the complex plane $\Omega=0$, the roots are the common roots of (31) and (32). Each of (31) and (32) defines a family of curves in the complex plane; and the intersections of the two families are the common roots. The two families of curves are shown on the unfolded complex plane along the lower left edges of Figs. 1 and 2. The full lines are the branches of (31), and the dashed lines are the branches of (32). The intersections are identified by the points marked 3, 4, etc., in Fig. 1 and 9, 8, etc., in Fig. 2.

A second property of the complex roots is obtained from an extension of a theorem by Onoe,[5] which was established in connection with Rayleigh's frequency equation for an isotropic plate. Each maximum and minimum of the segments of branches in the real $(\Omega,\bar{x})$

and imaginary $(\Omega,\bar{y})$ planes are a point of intersection, at normal incidence, with a complex segment. This is a general property of analytic functions of one real and one complex variable. Another property is that the space curves are continuous, with continuous derivatives of all orders.

We have now sufficient data for an approximate sketch of the complex segments; an example of which is illustrated in Fig. 5 for the essentially symmetric modes. The thick, dashed lines are the complex segments. The orders of the branches, whether the real, imaginary, or complex segments, are identified by the numbers, without subscripts, alongside the curves. The assignment of a single number to various connected real, imaginary, and complex segments, to identify a branch, is based on the rule: Each branch must extend from zero to infinite frequency, and, if n branches are considered, any plane of constant frequency must intersect the n branches in n and only n points.

Each complex segment has a conjugate which is the reflection of the segment in the imaginary plane. Further, since ξ appears in the frequency equation only as ξ^2, each complex segment and its conjugate have reflections in the real plane; each real segment has a reflection in the imaginary plane; and each imaginary segment has a reflection in the real plane. Thus, only half of the real and imaginary segments of the branches are shown in all the figures. Conjugate pairs of complex segments are represented in Fig. 5 by a single dashed segment, instead of a closed loop. Thus, the dashed segment marked 5, $\bar{6}$ represents the complex segments of both branch 5 and branch 6; the bar over 6 indicating that it is the reflection of the segment, in the imaginary plane, that is shown. Note that a real segment of branch 4, marked $\bar{4}$, is also shown as its reflection in the imaginary plane.

The complex segments are sketched only approximately as their precise properties are known only in the neighborhood of the three coordinate planes.

APPENDIX

Bechmann's[6] constants for α quartz:

$$c_{11}^{E\sigma}=86.75,\quad c_{33}^{E\sigma}=107.2,\quad c_{66}^{E\sigma}=39.88$$
$$c_{12}^{E\sigma}=6.99,\quad c_{44}^{E\sigma}=57.94,$$
$$c_{13}^{E\sigma}=11.91,\quad c_{14}^{E\sigma}=-17.91,$$

were employed in the computations. For the AT cut,[9] the elastic constants which appear in this paper are

$$\bar{c}_{11}=2.989694,\quad \bar{c}_{55}=2.371592,$$
$$\bar{c}_{12}=-0.284489,\quad \bar{c}_{56}=0.087324,$$
$$\bar{c}_{14}=-0.126136,\quad \bar{c}_{22}=4.472694,$$
$$\bar{c}_{24}=0.196479,\quad \bar{c}_{44}=1.330835,$$
$$\bar{c}_{pq}=c_{pq}/c_{66},\quad c_{66}=29.012996.$$

[9] W. P. Mason, *Piezoelectric Crystals and Their Application to Ultrasonics* (D. Van Nostrand Company, Inc., Princeton, New Jersey, 1950), p. 99.

Reprinted from THE JOURNAL OF THE ACOUSTICAL SOCIETY OF AMERICA, Vol. 34, No. 12, 1911–1918, December, 1962

Extensional Vibrations of Thin Quartz Disks

A. G. LUBOWE* AND R. D. MINDLIN
Department of Civil Engineering, Columbia University, New York 27, New York
(Received December 11, 1961)

A general solution is given for the contour modes of vibration of circular crystal disks of certain elastic symmetries. The solution is applicable to three cuts of quartz, for which the frequencies of fourteen modes are computed.

INTRODUCTION

LOW-FREQUENCY extensional (sometimes called "contour") vibrations of thin crystal plates are governed, to a close approximation, by the Cauchy–Voigt equations of motion.[1,2] In these equations, coupling between extension and flexure is neglected, and, in addition, the equations are limited in application to frequencies well below the frequency of the lowest thickness mode, due to neglect of variation of displacement across the thickness of the plate. Despite the great simplification thereby introduced, solutions of the equations, for bounded plates, still present formidable difficulties for the condition of free edges which is met in practice. Only a few exact solutions for special modes of vibration of square plates[3–5] have been found; and only one set of exact modes has been found[6] (in rectangular plates) in which coupling with flexure is absent, rather than neglected. There have been no exact solutions for circular disks and no general solutions for a plate of any shape.

The essential difficulty, in the case of the circular disk, is that, when the equations of motion are expressed in polar coordinates, they constitute a pair of coupled, second-order, partial-differential equations with coefficients which are functions of both the angular and radial coordinates. Recently, in an extension of a method introduced by Carrier,[7] Mindlin and Cooper[8] have shown how the Cauchy–Voigt equations may be reduced to a pair of uncoupled wave equations of canonical form if certain relations exist among the constants of elasticity; and they found that these relations are satisfied in four cuts of quartz. One of the cuts is the Z cut, for which it is known[8] that the Cauchy–Voigt equations reduce to isotropic form; in the other three cuts, the anisotropy remains and the constants of elasticity of all three are the same when referred to appropriate axes in the several cuts.

For these three cuts, then, it is possible to obtain solutions of the anisotropic equations of motion in the form of simple products of Bessel and trigonometric functions. However, one of the two uncoupled equations is of elementary form only in transformed space and, when its product function solutions are transformed back to real space, for insertion in the boundary conditions, infinite series result. As a consequence, the frequency equation comprises a doubly infinite array of Bessel functions.

In this paper, the frequency equation is obtained and it is shown how the roots may be computed by successive approximations. The results of computations of fourteen of the lower frequencies are given to three or four significant figures. The frequencies are closely related to those computed for the isotropic case by Onoe[9]; and, again, the mode of lowest frequency is one of circumferential order 2. However, there is a bifurcation of roots, in the anisotropic case, due to the fact that the modes are not invariant in an arbitrary rotation about an axis normal to the plane of the plate. In the isotropic case there are two families of modes of circumferential order zero: one radial and one tangential; but for each higher circumferential order there is only one family of modes. In the case of the crystal disk, there is a family of essentially radial modes of circumferential order zero and one of essentially tangential displacement of circumferential order zero. However, for each higher circumferential order, there are two families: each mode of the isotropic disk splits into two modes of slightly different frequencies and of similar shapes, but differently oriented with respect to the axes of crystal symmetry.

The three cuts of quartz have elastic symmetries in their planes, which fall in the class of the symmetry of rotated Y cuts; in fact, one of the three cuts is a rotated Y cut. Hence the results may be compared with experiments and previous theoretical conclusions, again by Onoe,[10] on such cuts. It has been thought, previously, that the mode of lowest frequency is one, observed experimentally, which has circumferential order 2 and directions of maximum displacement nearly at 45° to the axes of elastic symmetry in the plane of the plate. We also find such a mode and its frequency is 1.0% lower than an upper bound that

---

* Present address: Bell Telephone Laboratories, Inc., Whippany, New Jersey.
[1] A. L. Cauchy, *Exercices de mathématique* (Bure Frères, Paris, 1829), Vol. 4, p. 13.
[2] W. Voigt, *Lehrbuch der Kristallphysik* (B. G. Teubner, Leipzig, 1928), 2nd ed., p. 698.
[3] V. Petrzilka, Z. Physik **97**, 436 (1935).
[4] R. Bechmann, Z. Physik **117**, 180 (1941).
[5] H. Ekstein, Phys. Rev. **66**, 108 (1944).
[6] R. D. Mindlin, J. Appl. Phys. **27**, 1462 (1956).
[7] G. F. Carrier, Quart. Appl. Math. 4, 160 (1946).
[8] R. D. Mindlin and H. L. Cooper, Quart. Appl. Math. **19**, 111 (1961).
[9] M. Onoe, J. Acoust. Soc. Am. **28**, 1158 (1956).
[10] M. Onoe, J. Acoust. Soc. Am. **30**, 634 (1958).

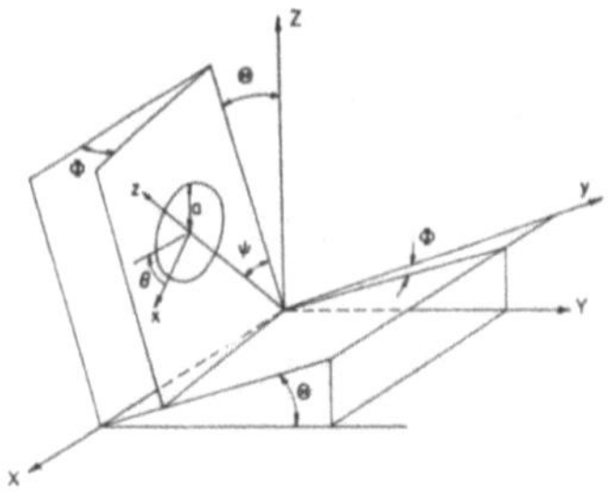

FIG. 1. Orientation of disks.

Onoe found for it by an energy method. However, this mode, in our result, is one of a bifurcated pair. The other member of the pair has a frequency 0.6% lower, and its directions of maximum displacement nearly coincide with the axes of elastic symmetry.

I. EQUATIONS OF MOTION

The Cauchy–Voigt equations may be written as

$$(L_1+\rho\omega^2)u+L_5w=0, \qquad (1)$$
$$L_5u+(L_3+\rho\omega^2)w=0,$$

where u and w are the components of displacement in the plane of x and z (the plane of the plate), ρ is the density, ω is the circular frequency, and

$$L_1=\gamma_{11}\partial^2/\partial x^2+2\gamma_{15}\partial^2/\partial x\partial z+\gamma_{55}\partial^2/\partial z^2,$$
$$L_3=\gamma_{55}\partial^2/\partial x^2+2\gamma_{35}\partial^2/\partial x\partial z+\gamma_{33}\partial^2/\partial z^2, \qquad (2)$$
$$L_5=\gamma_{15}\partial^2/\partial x^2+(\gamma_{13}+\gamma_{55})\partial^2/\partial x\partial z+\gamma_{35}\partial^2/\partial z^2.$$

In (2), the γ_{pq} are Voigt's elastic stiffnesses which relate the components of stress τ_p and the components of strain ϵ_p according to

$$\tau_p=\gamma_{pq}\epsilon_q;\quad \gamma_{pq}=\gamma_{qp};\quad p, q=1, 3, 5, \qquad (3)$$

where summation is understood over the repeated index q and

$$\epsilon_1=\partial u/\partial x,\quad \epsilon_3=\partial w/\partial z,\quad \epsilon_5=(\partial u/\partial z)+(\partial w/\partial x). \qquad (4)$$

The elastic stiffnesses γ_{pq} are related to the elastic compliances s_{pq} through

$$\gamma_{pq}=\mathrm{cof}\,s_{pq}/\det s_{pq};\quad p, q=1, 3, 5. \qquad (5)$$

Consider, now, that the axes x and z are oriented with respect to axes X, Y, Z, fixed in the crystal, in accordance with the three angles Θ, Φ, Ψ, shown in Fig. 1. The compliances s_{pq} are functions of the three angles and the compliances referred to X, Y, Z. Mindlin and Cooper showed that, if angles Θ and Φ can be found for which the relations

$$s_{11}+s_{33}=2s_{13}+s_{55},\quad s_{15}=s_{35} \qquad (6)$$

are satisfied, and if

$$\tan 2\Psi=2s_{15}/(s_{33}-s_{11}), \qquad (7)$$

then solutions of the Cauchy–Voigt equations can be reduced to Carrier's form, namely, to solutions of the equations

$$\bar{\Delta}\phi+\kappa^2\phi=0,\quad \Delta\psi+\lambda^2\psi=0, \qquad (8)$$

where

$$\bar{\Delta}=\partial^2/\partial x^2+\partial^2/\partial \bar{z}^2,\qquad \Delta=\partial^2/\partial x^2+\partial^2/\partial z^2$$
$$\bar{z}=cz,\quad c=(\gamma_{11}/\gamma_{33})^{\frac{1}{2}},\quad \kappa^2=\rho\omega^2\gamma_{11}^{-1},\quad \lambda^2=\rho\omega^2\gamma_{55}^{-1}, \qquad (9)$$

and the displacements are expressed in terms of ϕ and ψ by

$$u=\gamma_1\partial\phi/\partial x-\gamma_3\partial\psi/\partial z,\quad w=\gamma_3\partial\phi/\partial z+\gamma_1\partial\psi/\partial x, \qquad (10)$$

where

$$\gamma_1=(\gamma_{11}-\gamma_{55})^{\frac{1}{2}},\quad \gamma_3=(\gamma_{33}-\gamma_{55})^{\frac{1}{2}}. \qquad (11)$$

For such orientations of cut and orientations of axes x and z in the cut,

$$\gamma_{15}=\gamma_{35}=0,\quad \gamma_1\gamma_3=\gamma_{13}+\gamma_{55}. \qquad (12)$$

In the case of quartz, Mindlin and Cooper also found, using Bechmann's[11] values of the principal compliances, that there are four such sets of angles. With X, Y, Z taken as the IRE Standard reference axes for quartz,[12] the orientations of the axes x and z are given by

Θ	90°	−9.65°	18.79°	18.79°	
Φ	0	0	58.62°	−58.62°	(13)
Ψ	0	0	16.17°	73.83°.	

The first set is the Z cut with isótropic symmetry, while the remaining three have compliances

$$s_{11}=12.77,\quad s_{33}=9.52,\quad s_{13}=-0.49,$$
$$s_{55}=23.27,\quad s_{15}=0,\qquad s_{35}=0, \qquad (14)$$

in units of 10^{-12} m²/newton. From (5),

$$(\gamma_{11},\gamma_{33},\gamma_{13})=(s_{33},\, s_{11},\, -s_{13})/(s_{11}s_{33}-s_{13}^2),$$
$$\gamma_{55}=s_{55}^{-1}. \qquad (15)$$

II. ISOTROPIC DISK

It is convenient to have at hand a brief review of the solution for an isotropic disk.

In an isotropic medium

$$\gamma_{11}=\gamma_{33}=\gamma_{13}+2\gamma_{55},\quad \gamma_{15}=\gamma_{35}=0, \qquad (16)$$

and then $\bar{\Delta}=\Delta$ in (8). Also, in polar coordinates

$$r=(x^2+z^2)^{\frac{1}{2}},\quad \theta=\tan^{-1}(z/x), \qquad (17)$$

the radial and tangential displacements from (10) and (16) are

$$u_r=\gamma_1(\phi_r-r^{-1}\psi_\theta),\quad u_\theta=\gamma_1(r^{-1}\phi_\theta+\psi_r), \qquad (18)$$

[11] R. Bechmann, Phys. Rev. **110**, 1060 (1958).
[12] Proc. Inst. Radio Engrs. **37**, 1378 (1949).

where the subscripts r and θ on ϕ and ψ designate derivatives. The radial and tangential components of the traction across $r=$constant are

$$\tau_{rr}=\gamma_{11}\epsilon_{rr}+\gamma_{13}\epsilon_{\theta\theta}, \quad \tau_{r\theta}=\gamma_{55}\epsilon_{r\theta}, \tag{19}$$

where

$$\epsilon_{rr}=\partial u_r/\partial r, \quad \epsilon_{\theta\theta}=r^{-1}\partial u_\theta/\partial\theta+u_r/r, \quad \epsilon_{r\theta}=\partial u_\theta/\partial r-u_\theta/r+r^{-1}\partial u_r/\partial\theta. \tag{20}$$

Then, from (18) and (19),

$$\tau_{rr}=\gamma_1[2\gamma_{55}\phi_{rr}+\gamma_{13}\Delta\phi-2\gamma_{55}(r^{-1}\psi_\theta)_{,r}], \quad \tau_{r\theta}=\gamma_1\gamma_{55}[2(r^{-1}\phi_\theta)_{,r}-2\psi_{,rr}-\Delta\psi]. \tag{21}$$

Appropriate solutions of (8), of circumferential order n, are either

$$\phi=A_nJ_n(\kappa r)\cos n\theta, \quad \psi=B_nJ_n(\lambda r)\sin n\theta \tag{22}$$

or

$$\phi=C_nJ_n(\kappa r)\sin n\theta, \quad \psi=D_nJ_n(\lambda r)\cos n\theta, \tag{23}$$

where $n>0$ and J_n are Bessel's functions of the first kind and order n. Only integral orders are admissible, as the displacements must be periodic in θ of period 2π.

The frequency equation is obtained by substituting either (22) or (23) in (21), setting $\tau_{rr}=\tau_{r\theta}=0$ on $r=a$ (where a is the radius of the disk), and eliminating the constants A_n, B_n or C_n, D_n. It is apparent from the form of (21) that solutions (22) and (23) lead to the same frequency equation.[9,13] For each circumferential order n there is an infinity of roots; but the roots from (22) and (23) are the same. This is to say, a rotation of axes is immaterial in an isotropic plate.

In the case $n=0$ we have

$$\phi=A_0J_0(\kappa r), \quad \psi=0 \tag{24}$$

or

$$\phi=0, \quad \psi=D_0J_0(\lambda r), \tag{25}$$

and the frequency equations are, for (24),

$$[dJ_1(\kappa a)/da]+(\gamma_{13}/\gamma_{11}a)J_1(\kappa a)=0 \tag{26}$$

and, for (25),

$$[dJ_1(\lambda a)/da]-(1/a)J_1(\lambda a)=0. \tag{27}$$

The solution (24), (26) gives a family of modes of radial displacement, while (25) and (27) give a family of modes of tangential displacement.

Thus, in the isotropic case, there are two families of modes of circumferential order zero: one radial and one tangential; but, for each circumferential order $n>0$ there is only one family of compound modes.

III. FORMULAS FOR DISPLACEMENTS AND STRESS IN POLAR COORDINATES

If we transform (10) to polar coordinates, we obtain, for the displacements in the anisotropic plate,

$$\begin{aligned} u_r=&\tfrac{1}{2}(\gamma_1+\gamma_3)\phi_r+\tfrac{1}{2}(\gamma_1-\gamma_3)(\phi_r\cos 2\theta-r^{-1}\phi_\theta\sin 2\theta)\\ &-\tfrac{1}{2}(\gamma_1+\gamma_3)r^{-1}\psi_\theta+\tfrac{1}{2}(\gamma_1-\gamma_3)\\ &\times(\psi_r\sin 2\theta+r^{-1}\psi_\theta\cos 2\theta),\\ u_\theta=&\tfrac{1}{2}(\gamma_1+\gamma_3)r^{-1}\phi_\theta-\tfrac{1}{2}(\gamma_1-\gamma_3)(\phi_r\sin 2\theta+r^{-1}\phi_\theta\cos 2\theta)\\ &+\tfrac{1}{2}(\gamma_1+\gamma_3)\psi_r+\tfrac{1}{2}(\gamma_1-\gamma_3)\\ &\times(\psi_r\cos 2\theta-r^{-1}\psi_\theta\sin 2\theta). \end{aligned} \tag{28}$$

Also, the expressions for the components of traction are

$$\begin{aligned} \tau_{rr}=&(a_1+2a_2\cos 2\theta+a_3\cos 4\theta)\epsilon_{rr}\\ &+(a_4-a_3\cos 4\theta)\epsilon_{\theta\theta}-(a_2\sin 2\theta+a_3\sin 4\theta)\epsilon_{r\theta},\\ \tau_{r\theta}=&-(a_2\sin 2\theta+a_3\sin 4\theta)\epsilon_{rr}\\ &-(a_2\sin 2\theta-a_3\sin 4\theta)\epsilon_{\theta\theta}+(a_5-a_3\cos 4\theta)\epsilon_{r\theta}, \end{aligned} \tag{29}$$

where

$$a_1=\tfrac{1}{8}(3\gamma_{11}+3\gamma_{33}+2\gamma_{13}+4\gamma_{55}), \quad a_2=\tfrac{1}{4}(\gamma_{11}-\gamma_{33}),$$
$$a_3=\tfrac{1}{8}(\gamma_{11}+\gamma_{33}-2\gamma_{13}-4\gamma_{55}),$$
$$a_4=\tfrac{1}{8}(\gamma_{11}+\gamma_{33}+6\gamma_{13}-4\gamma_{55}),$$
$$a_5=\tfrac{1}{8}(\gamma_{11}+\gamma_{33}-2\gamma_{13}+4\gamma_{55}).$$

Then, if we substitute (28) in (20) and the result in (29), we obtain, using (11) and (12),

$$\begin{aligned} \tau_{rr}=&\tfrac{1}{2}(\gamma_1+\gamma_3)[2\gamma_{55}\phi_{rr}+\gamma_{13}\Delta\phi+\tfrac{1}{4}(\gamma_1-\gamma_3)^2(2\phi_{rr}+\Delta\phi)]\\ &+\tfrac{1}{2}(\gamma_1-\gamma_3)[(\gamma_1+\gamma_3)^2\phi_{rr}-\gamma_{13}\Delta\phi]\cos 2\theta\\ &-\tfrac{1}{2}(\gamma_1-\gamma_3)(\gamma_1+\gamma_3)^2(r^{-1}\phi_\theta)_{,r}\sin 2\theta\\ &-\tfrac{1}{8}(\gamma_1+\gamma_3)(\gamma_1-\gamma_3)^2(\Delta\phi-2\phi_{rr})\cos 4\theta\\ &-\tfrac{1}{4}(\gamma_1+\gamma_3)(\gamma_1-\gamma_3)^2(r^{-1}\phi_\theta)_{,r}\sin 4\theta\\ &-\gamma_{55}[(\gamma_1+\gamma_3)(r^{-1}\psi_\theta)_{,r}-\tfrac{1}{2}(\gamma_1-\gamma_3)\Delta\psi\sin 2\theta],\\ \tau_{r\theta}=&\tfrac{1}{4}(\gamma_1+\gamma_3)[(\gamma_1-\gamma_3)^2+4\gamma_{55}](r^{-1}\phi_\theta)_{,r}\\ &-\tfrac{1}{4}(\gamma_1-\gamma_3)[\gamma_1^2+\gamma_3^2+2\gamma_{55}]\Delta\phi\sin 2\theta\\ &-\tfrac{1}{4}(\gamma_1+\gamma_3)(\gamma_1-\gamma_3)^2(r^{-1}\phi_\theta)_{,r}\cos 4\theta\\ &+\tfrac{1}{8}(\gamma_1+\gamma_3)(\gamma_1-\gamma_3)^2(\Delta\phi-2\phi_{rr})\sin 4\theta\\ &+\tfrac{1}{2}\gamma_{55}[(\gamma_1+\gamma_3)(2\psi_{rr}-\Delta\psi)+(\gamma_1-\gamma_3)\Delta\psi\cos 2\theta]. \end{aligned} \tag{30}$$

The relative simplicity of the contribution of ψ to the stresses is due to the fact that, with (12), a radiating wave, derived from ψ, has a circular wavefront, whereas a radiating wave derived from ϕ has an elliptic wavefront. Nevertheless, even in the case of ψ the anisotropy produces terms with factors $\cos 2\theta$ and $\sin 2\theta$ in the formulas (28) and (30) for displacements and stresses, whereas terms with such factors are absent in the analogous formulas (18) and (21) for the isotropic plate.

IV. SOLUTIONS OF EQUATIONS OF MOTION

The equation of motion on ψ is the same in the anisotropic plate as in the isotropic plate, and, hence, the solutions for ψ, given in (22) and (23), may be used again. In the case of ϕ we define polar coordinates

$$\bar{r}=(x^2+\bar{z}^2)^{\frac{1}{2}}, \quad \bar{\theta}=\tan^{-1}(\bar{z}/x). \tag{31}$$

Then solutions of the first of (8) have the same form

[13] A. E. H. Love, *Theory of Elasticity* (Cambridge University Press, Cambridge, England, 1927), 4th ed., p. 498.

as those in the first of (22) and (23), but with r and θ replaced by $\bar{r}$ and $\bar{\theta}$. Again, only integral orders n are admissible, as may be deduced from the relations

$$\bar{r}\sin\bar{\theta}=cr\sin\theta,\quad \bar{r}\cos\bar{\theta}=r\cos\theta. \tag{32}$$

It is convenient to group the solutions of (8) in four categories:

Symmetry I, n even:

$$\phi=\sum_n A_nJ_n(\kappa\bar{r})\cos n\bar{\theta},\quad \psi=\sum_n B_nJ_n(\lambda r)\sin n\theta. \tag{33}$$

Symmetry I, n odd:

$$\phi=\sum_n A_nJ_n(\kappa\bar{r})\cos n\bar{\theta},\quad \psi=\sum_n B_nJ_n(\lambda r)\sin n\theta. \tag{34}$$

Symmetry II, n even:

$$\phi=\sum_n C_nJ_n(\kappa\bar{r})\sin n\bar{\theta},\quad \psi=\sum_n D_nJ_n(\lambda r)\cos n\theta. \tag{35}$$

Symmetry II, n odd:

$$\phi=\sum_n C_nJ_n(\kappa\bar{r})\sin n\bar{\theta},\quad \psi=\sum_n D_nJ_n(\lambda r)\cos n\theta. \tag{36}$$

Now the boundary conditions must be satisfied on $r=a$ for all θ. Hence it is necessary to transform the functions $\phi(\bar{r},\bar{\theta})$ to functions of r and θ. This may be done with the aid of the formula[14]

$$\frac{J_n(\bar{\omega})}{\bar{\omega}^n}=\sum_{m=0}^{\infty}\epsilon_{m+n}\frac{J_{m+n}(\zeta_1)}{\zeta_1^n}\frac{J_{m+n}(\zeta_2)}{\zeta_2^n}\frac{d^n\cos 2(m+n)\theta}{d(\cos 2\theta)^n}, \tag{37}$$

where $\bar{\omega}=(\zeta_1^2+\zeta_2^2-2\zeta_1\zeta_2\cos 2\theta)^{\frac{1}{2}}$, and ϵ_{m+n} is equal to 1 when $m+n=0$ and is equal to 2 when $m+n\neq 0$. To put (37) in the form appropriate to our needs, we set $\bar{\omega}=\kappa\bar{r}$, $\zeta_1=\kappa_1 r$, and $\zeta_2=-\kappa_2 r$, where

$$\kappa_1=\tfrac{1}{2}\kappa(1+c),\quad \kappa_2=\tfrac{1}{2}\kappa(1-c) \tag{38}$$

and obtain

$$\frac{J_n(\kappa\bar{r})}{(\kappa\bar{r})^n}=\sum_{m=0}^{\infty}(-1)^m\epsilon_{m+n}\frac{J_{m+n}(\kappa_1 r)}{(\kappa_1 r)^n}\frac{J_{m+n}(\kappa_2 r)}{(\kappa_2 r)^n}\times\frac{d^n\cos 2(m+n)\theta}{d(\cos 2\theta)^n}. \tag{39}$$

Then, with

$$F_p=(-1)^pJ_p(\kappa_1 r)J_p(\kappa_2 r) \tag{40}$$

we find, after lengthy but straightforward computations,

$$J_0(\kappa\bar{r})=F_0+2(F_1\cos 2\theta+F_2\cos 4\theta+F_3\cos 6\theta+\cdots) \tag{41}$$

$$\begin{aligned}J_1(\kappa\bar{r})\cos\bar{\theta}=-k_1(\kappa r)^{-1}[&(F_1+2F_2+3F_3+\cdots)\cos\theta\\&+(2F_2+3F_3+\cdots)\cos 3\theta\\&+(3F_3+\cdots)\cos 5\theta\\&+\cdots],\end{aligned} \tag{42}$$

where $k_1=2\kappa^2/\kappa_1\kappa_2=8/(1-c^2)$;

$$\begin{aligned}J_1(\kappa\bar{r})\sin\bar{\theta}=-k_1'(\kappa r)^{-1}[&(F_1-2F_2+3F_3-\cdots)\sin\theta\\&+(2F_2-3F_3+\cdots)\sin 3\theta\\&+(3F_3-\cdots)\sin 5\theta\\&+\cdots],\end{aligned} \tag{43}$$

where $k_1'=ck_1$;

$$\begin{aligned}J_2(\kappa\bar{r})\cos 2\bar{\theta}=k_2(\kappa r)^{-2}[&(F_2+3\bar{k}_2F_3+8F_4+\cdots)\\&+(\bar{k}_2F_2+6F_3+14\bar{k}_2F_4+\cdots)\cos 2\theta\\&+(3\bar{k}_2F_3+12F_4+\cdots)\cos 4\theta\\&+(6\bar{k}_2F_4+\cdots)\cos 6\theta\\&+\cdots],\end{aligned} \tag{44}$$

where $k_2=(1-c^2)k_1^2$ and $\bar{k}_2=(1+c^2)/(1-c^2)$;

$$\begin{aligned}J_2(\kappa\bar{r})\sin 2\bar{\theta}=k_2'(\kappa r)^{-2}[&(F_2+2F_4+\cdots)\sin 2\theta\\&+(3F_3+\cdots)\sin 4\theta\\&+(6F_4+\cdots)\sin 6\theta\\&+\cdots],\end{aligned} \tag{45}$$

[14] G. N. Watson, *Theory of Bessel Functions* (Cambridge University Press, Cambridge, England, 1944), 2nd ed., p. 362.

where $k_2'=2ck_1^2$;

$$\begin{aligned}J_3(\kappa\bar{r})\cos 3\bar{\theta}=k_3(\kappa r)^{-3}\{&[F_3+4(1+\bar{k}_3)F_4+10(\tfrac{3}{2}+\bar{k}_3)F_5+\cdots]\cos\theta\\&+[\bar{k}_3F_3+4F_4+10(1+\tfrac{3}{2}\bar{k}_3)F_5+\cdots]\cos 3\theta\\&+[4\bar{k}_3F_4+10F_5+\cdots]\cos 5\theta\\&+[10\bar{k}_3F_5+\cdots]\cos 7\theta\\&+\cdots\},\end{aligned}\tag{46}$$

where $k_3=9(1-c^2)k_1^3/2$ and $\bar{k}_3=-1-4/3(1-c^2)$;

$$\begin{aligned}J_3(\kappa\bar{r})\sin 3\bar{\theta}=k_3'(\kappa r)^{-3}\{&[F_3+4(\bar{k}_3'-1)F_4+10(\tfrac{3}{2}-\bar{k}_3')F_5+\cdots]\sin\theta\\&+[\bar{k}_3'F_3+4F_4+10(\tfrac{3}{2}\bar{k}_3'-1)F_5+\cdots]\sin 3\theta\\&+[4\bar{k}_3'F_4+10F_5+\cdots]\sin 5\theta\\&+[10\bar{k}_3'F_5+\cdots]\sin 7\theta\\&+\cdots\},\end{aligned}\tag{47}$$

where $k_3'=-ck_3$ and $\bar{k}_3'=-\bar{k}_3-4/3$. Terms of higher order, n, were not employed in the computations of frequencies.

It may be observed, in (42) to (47), that $J_n(\kappa\bar{r})\cos n\bar{\theta}$ and $J_n(\kappa\bar{r})\sin n\bar{\theta}$ produce different linear combinations of terms $F_p\cos m\theta$ and $F_p\sin m\theta$.

V. FREQUENCY EQUATIONS

When any one of the four groups of solutions (33)–(36) is substituted in (30) and the boundary conditions, $\tau_{rr}=\tau_{r\theta}=0$ on $r=a$, are set, one is confronted with doubly infinite series expressions from which to eliminate the constants A_n and B_n or C_n and D_n in order to obtain the frequency equation. The difficulty may be surmounted by forming frequency equations of successively higher orders of approximation. Each of these is itself, in general, infinite; it must be solved by successive approximations and yields frequencies of modes up to a circumferential order n. The next higher-order approximation yields a higher approximation to the roots of the previous equation and a first approximation to the roots of the next higher circumferential order (even or odd, depending on the group). Each circumferential order, of course, has an infinity of roots, as in the isotropic case.

The method of successive approximations is illustrated with the third group of solutions: symmetry II, n even. This one is chosen for the demonstration because it starts in the simplest manner: with a finite frequency equation for the zero-order approximation. Thus, in (35), we take only the first term $n=0$, i.e.,

$$\phi=0,\quad \psi=D_0J_0(\lambda r).\tag{48}$$

Upon inserting (48) in (30) and setting $r=a$, we find

$$\begin{aligned}\tau_{rr}]_{r=a}&=D_0(\qquad \bar{d}_{02}\sin 2\theta)\\ \tau_{r\theta}]_{r=a}&=D_0(d_{00}+d_{02}\cos 2\theta),\end{aligned}\tag{49}$$

where

$$\begin{aligned}d_{00}&=\tfrac{1}{2}\gamma_{55}(\gamma_1+\gamma_3)\lambda^2[(\lambda a)^{-1}J_1(\lambda a)-J_1'(\lambda a)],\\ d_{02}&=\bar{d}_{02}=-\tfrac{1}{2}\gamma_{55}(\gamma_1-\gamma_3)\lambda^2J_0(\lambda a),\end{aligned}\tag{50}$$

in which prime designates differentiation with respect to the argument.

To satisfy the boundary conditions exactly, we should have to have $d_{00}=d_{02}=0$, which is not possible if $\gamma_1\neq\gamma_3$; but we take

$$d_{00}=0\tag{51}$$

as a first approximation. Note that this is the same as (27) for the isotropic case, i.e., the family of zero-order modes of purely tangential displacement, designated, following Onoe,[9] as (T,s), where $s=0, 1, 2, \cdots$ and T indicates "tangential." Note also that, whereas (48) and (51), i.e., (25) and (27), constitute an exact solution for the isotropic disk, they are only an approximation for the crystal disk.

As the next approximation, we take the first two terms, $n=0$ and $n=2$, in (35):

$$\begin{aligned}\phi&=\qquad\qquad C_2J_2(\kappa\bar{r})\sin 2\bar{\theta},\\ \psi&=D_0J_0(\lambda r)+D_2J_2(\lambda r)\cos 2\theta.\end{aligned}\tag{52}$$

The contribution of the second term in ψ to the boundary tractions is

$$\begin{aligned}\tau_{rr}]_{r=a}&=D_2(\qquad \bar{d}_{22}\sin 2\theta+\bar{d}_{24}\sin 4\theta),\\ \tau_{r\theta}]_{r=a}&=D_2(d_{20}+d_{22}\cos 2\theta+d_{24}\cos 4\theta),\end{aligned}\tag{53}$$

where

$$\begin{aligned}d_{20}&=d_{24}=\bar{d}_{24}=-\tfrac{1}{4}\gamma_{55}(\gamma_1-\gamma_3)\lambda^2J_2(\lambda a),\\ d_{22}&=\tfrac{1}{2}\gamma_{55}(\gamma_1+\gamma_3)\lambda^2[2J_2''(\lambda a)+J_2(\lambda a)],\\ \bar{d}_{22}&=2\gamma_{55}(\gamma_1+\gamma_3)\lambda^2[(\lambda a)^{-1}J_2'(\lambda a)-(\lambda a)^{-2}J_2(\lambda a)].\end{aligned}\tag{54}$$

To find the contribution of ϕ, in (52), to the boundary tractions, we note, first, that ϕ is given by the series in (45), each term of which is of the form $(\kappa r)^{-2}F_p\sin 2m\theta$, where the F_p are defined by (40). Now, in the expressions (30) for the boundary tractions, ϕ appears only in the three forms $\Delta\phi$, ϕ_{rr}, and $(r^{-1}\phi_\theta)_r$. Hence, on $r=a$, the functions κr and $F_p(\kappa_1 r,\kappa_2 r)$ contribute to the expressions for the boundary tractions, even for the most general order n, only the three forms

$$f_{nmp}=\left[\frac{d^2}{d(\kappa a)^2}+\frac{1}{\kappa a}\frac{d}{d(\kappa a)}-\frac{m^2}{(\kappa a)^2}\right][(\kappa a)^{-n}F_p(\kappa_1 a,\kappa_2 a)],$$

$$g_{np}=[d^2/d(\kappa a)^2][(\kappa a)^{-n}F_p(\kappa_1 a,\kappa_2 a)], \tag{55}$$

$$h_{nmp}=m[d/d(\kappa a)][(\kappa a)^{-n-1}F_p(\kappa_1 a,\kappa_2 a)].$$

Further, each of the trigonometric functions $\sin 2m\theta$ in (45) is multiplied in (30) by sine or cosine of 2θ and 4θ, or by a term independent of θ. These products may be expressed as sums of terms $\cos q\theta$ or $\sin q\theta$, $q=0, 2, 4\cdots$. Collecting these results, we find that the contribution of ϕ in (52) to the boundary tractions is of the form

$$\tau_{rr}]_{r=a}=C_2\Big(\sum_{q=0}^{6}\tilde{c}_{2q2}+\sum_{q=0}^{8}\tilde{c}_{2q3}+\sum_{q=0}^{10}\tilde{c}_{2q4}+\cdots\Big)\sin q\theta,$$

$$\tau_{r\theta}]_{r=a}=C_2\Big(\sum_{q=0}^{6}c_{2q2}+\sum_{q=0}^{8}c_{2q3}+\sum_{q=0}^{10}c_{2q4}+\cdots\Big)\cos q\theta, \tag{56}$$

where $q=0, 2, 4\cdots$ and

$$\tilde{c}_{nqp}=\tilde{L}_{nqp}f_{nmp}+\tilde{M}_{nqp}g_{np}+\tilde{N}_{nqp}h_{nmp},$$

$$c_{nqp}=L_{nqp}f_{nmp}+M_{nqp}g_{np}+N_{nqp}h_{nmp}, \tag{57}$$

in which L, M, N, $\tilde{L}$, $\tilde{M}$, $\tilde{N}$ depend only on the elastic constants.

The subscripts on $\tilde{c}$ and c designate the order n of $J_n(\kappa\bar{r})$ in ϕ, the order q of the trigonometric function which $\tilde{c}$ or c multiplies in the expressions for the stresses, and the order p of the function F_p. The tilde on $\tilde{c}$ designates a product with a sine.

Upon adding (49), (53), and (56), we find on $r=a$,

$$\begin{aligned}\tau_{rr}=&[D_0\tilde{d}_{02}+D_2\tilde{d}_{22}+C_2(\tilde{c}_{222}+\tilde{c}_{223}+\tilde{c}_{224}+\cdots)]\sin 2\theta\\&+[D_2\tilde{d}_{24}+C_2(\tilde{c}_{242}+\tilde{c}_{243}+\tilde{c}_{244}+\cdots)]\sin 4\theta\\&+C_2(\tilde{c}_{262}+\tilde{c}_{263}+\tilde{c}_{264}+\cdots)\sin 6\theta\\&+\cdots.\end{aligned}$$

$$\begin{aligned}\tau_{r\theta}=&[D_0d_{00}+D_2d_{20}+C_2(c_{202}+c_{203}+c_{204}+\cdots)]\\&+[D_0d_{02}+D_2d_{22}+C_2(c_{222}+c_{223}+c_{224}+\cdots)]\cos 2\theta\\&+[D_2d_{24}+C_2(c_{242}+c_{243}+c_{244}+\cdots)]\cos 4\theta\\&+C_2(c_{262}+c_{263}+c_{264}+\cdots)\cos 6\theta\\&+\cdots.\end{aligned} \tag{58}$$

Again we cannot satisfy the boundary conditions with the whole of (58). However, as our second approximation we discard coefficients of trigonometric functions of order higher than 2 and set the remaining coefficients equal to zero:

$$\begin{aligned}&D_0d_{00}+D_2d_{20}+C_2(c_{202}+c_{203}+\cdots)=0,\\&D_0d_{02}+D_2d_{22}+C_2(c_{222}+c_{223}+\cdots)=0,\\&D_0\tilde{d}_{02}+D_2\tilde{d}_{22}+C_2(\tilde{c}_{222}+\tilde{c}_{223}+\cdots)=0,\end{aligned} \tag{59}$$

which yields the frequency equation

$$\begin{vmatrix}d_{00} & d_{20} & (c_{202}+c_{203}+\cdots)\\ d_{02} & d_{22} & (c_{222}+c_{223}+\cdots)\\ \tilde{d}_{02} & \tilde{d}_{22} & (\tilde{c}_{222}+\tilde{c}_{223}+\cdots)\end{vmatrix}=0. \tag{60}$$

This equation must be solved by successive approximations, using more and more columns of c's.

The roots of (60) yield second approximations to the roots of circumferential order zero and first approximations to the roots of circumferential order 2. Note that the shape of the mode, associated with the zero-order roots, contains terms of higher circumferential order, so that the designation "circumferential order zero" or (T,s) is only approximate. Similarly the terminology "modes of circumferential order 2" or $(2,s)_{\text{II}}$ is also not precise, but it is convenient. In the symbol $(2,s)_{\text{II}}$, an adaptation of Onoe's[9] symbol for the isotropic case, 2 designates the predominant circumferential order, $s=0, 1, 2\cdots$ designates the order of the root, and II designates symmetry II.

The next higher approximation to the roots of the group: symmetry II, n even, would be obtained from

$$\begin{aligned}\phi=&\;C_2J_2(\kappa\bar{r})\sin 2\bar{\theta}+C_4J_4(\kappa\bar{r})\sin 4\bar{\theta},\\ \psi=&\;D_0J_0(\lambda r)+D_2J_2(\lambda r)\cos 2\theta+D_4J_4(\lambda r)\cos 4\theta.\end{aligned} \tag{61}$$

Proceeding as before, and retaining terms through $\cos 4\theta$ and $\sin 4\theta$ in the stresses, we should obtain a 5×5 determinantal frequency equation. This would give a third approximation to the roots (T,s), a second approximation to the roots $(2,s)_{\text{II}}$, and a first approximation to the roots $(4,s)_{\text{II}}$. Still higher-order approximations may be obtained in a similar manner.

Turning now to symmetry I, n even, we start with the term $n=0$ in (33)

$$\phi=A_0J_0(\kappa\bar{r}),\quad \psi=0, \tag{62}$$

and proceed to the expressions for the stresses, as before. Then, dropping the coefficients of all trigonometric functions, we obtain the frequency equation

$$a_{000}+a_{001}+a_{002}+\cdots=0, \tag{63}$$

where the a_{nqp} are defined in the same way as the c_{nqp} in the paragraph following (57). In this case we see that even the approximation of lowest order leads to a frequency equation with a series of terms, which must be solved by successive approximations. The first term a_{000} is analogous to the left-hand side of (26), but is not identical with it because of the presence in a_{000} of Bessel functions of $\kappa_1 a$ and $\kappa_2 a$ rather than κa. Nevertheless, we designate the roots of (63) by the symbol (R,s), in analogy with Onoe's symbol for the radial modes in the isotropic case.

The next approximation to the frequency equation

for symmetry I, n even, is obtained with

$$\begin{aligned}\phi &= A_0J_0(\kappa\bar{r})+A_2J_2(\kappa\bar{r})\cos 2\bar{\theta},\\ \psi &= \qquad\qquad B_2J_2(\lambda r)\sin 2\theta,\end{aligned} \tag{64}$$

and leads to the equation

$$\begin{vmatrix}(a_{000}+a_{001}+\cdots) & b_{20} & (a_{202}+a_{203}+\cdots)\\ (a_{020}+a_{021}+\cdots) & b_{22} & (a_{222}+a_{223}+\cdots)\\ (\tilde{a}_{020}+\tilde{a}_{021}+\cdots) & \tilde{b}_{22} & (\tilde{a}_{222}+\tilde{a}_{223}+\cdots)\end{vmatrix}=0 \tag{65}$$

for second approximations to the roots (R,s) and first approximations to the roots $(2,s)_{\rm I}$. Higher approximations may be obtained in a similar manner. For example, the third approximation would produce a 5×5 determinant.

Passing now to modes of odd order, we consider symmetry I, with $n=1$ as the first approximation:

$$\phi=A_1J_1(\kappa\bar{r})\cos\bar{\theta},\quad \psi=B_1J_1(\lambda r)\sin\theta. \tag{66}$$

Here both ϕ and ψ contribute to the approximation of lowest order, so we are led to a 2×2 determinant for the frequency equation:

$$\begin{Vmatrix}(a_{111}+a_{112}+a_{113}+\cdots) & b_{11}\\ (\tilde{a}_{111}+\tilde{a}_{112}+\tilde{a}_{113}+\cdots) & \tilde{b}_{11}\end{Vmatrix}=0, \tag{67}$$

which gives first approximations to the roots $(1,s)_{\rm I}$. The next approximation requires the four functions

$$\begin{aligned}\phi &= A_1J_1(\kappa\bar{r})\cos\bar{\theta}+A_3J_3(\kappa\bar{r})\cos 3\bar{\theta},\\ \psi &= B_1J_1(\lambda r)\sin\theta+B_3J_3(\lambda r)\sin 3\theta,\end{aligned} \tag{68}$$

and leads to the 4×4 frequency equation

$$\begin{vmatrix}(a_{111}+a_{112}+\cdots) & b_{11} & (a_{313}+a_{314}+\cdots) & b_{31}\\ (a_{132}+a_{133}+\cdots) & b_{13} & (a_{333}+a_{334}+\cdots) & b_{33}\\ (\tilde{a}_{111}+\tilde{a}_{112}+\cdots) & \tilde{b}_{11} & (\tilde{a}_{313}+\tilde{a}_{314}+\cdots) & \tilde{b}_{31}\\ (\tilde{a}_{132}+\tilde{a}_{133}+\cdots) & \tilde{b}_{13} & (\tilde{a}_{333}+\tilde{a}_{334}+\cdots) & \tilde{b}_{33}\end{vmatrix}=0, \tag{69}$$

which gives second approximations to $(1,s)_{\rm I}$ and first approximations to $(3,s)_{\rm I}$.

Finally, in symmetry II, n odd, we begin with $n=1$:

$$\phi=C_1J_1(\kappa\bar{r})\sin\bar{\theta},\quad \psi=D_1J_1(\lambda r)\cos\theta \tag{70}$$

and obtain the first approximation

$$\begin{vmatrix}(c_{111}+c_{112}+c_{113}+\cdots) & d_{11}\\ (\tilde{c}_{111}+\tilde{c}_{112}+\tilde{c}_{113}+\cdots) & \tilde{d}_{11}\end{vmatrix}=0 \tag{71}$$

for the modes $(1,s)_{\rm II}$. For the next approximation we take

$$\begin{aligned}\phi &= C_1J_1(\kappa\bar{r})\sin\bar{\theta}+C_3J_3(\kappa\bar{r})\sin 3\bar{\theta},\\ \psi &= D_1J_1(\lambda r)\cos\theta+D_3J_3(\lambda r)\cos 3\theta,\end{aligned} \tag{72}$$

and find

$$\begin{vmatrix}(c_{111}+c_{112}+\cdots) & d_{11} & (c_{313}+c_{314}+\cdots) & d_{31}\\ (c_{132}+c_{133}+\cdots) & d_{13} & (c_{333}+c_{334}+\cdots) & d_{33}\\ (\tilde{c}_{111}+\tilde{c}_{112}+\cdots) & \tilde{d}_{11} & (\tilde{c}_{313}+\tilde{c}_{314}+\cdots) & \tilde{d}_{31}\\ (\tilde{c}_{132}+\tilde{c}_{133}+\cdots) & \tilde{d}_{13} & (\tilde{c}_{333}+\tilde{c}_{334}+\cdots) & \tilde{d}_{33}\end{vmatrix}=0 \tag{73}$$

for the second approximation to the modes $(1,s)_{\rm II}$ and the first approximation to the modes $(3,s)_{\rm II}$. Either (69) or (73) may be extended to higher orders n by adding pairs of additional terms of odd order to (68) or (72).

VI. RESULTS OF COMPUTATIONS

The lower roots of each of the eight frequency equations derived in the preceding section were computed on a desk calculator, and the results are shown in Fig. 2, along with the corresponding roots of the frequency equation for an "equivalent" isotropic disk. The latter is defined as a disk with Poisson's ratio equal to the average of the two Poisson's ratios of the anisotropic disk. Thus

$$\sigma_{\rm av}=\tfrac{1}{2}(\gamma_{13}/\gamma_{11}+\gamma_{13}/\gamma_{33})=0.0449. \tag{74}$$

All of the frequencies are given in terms of the dimensionless parameter

$$\lambda a=(\rho/\gamma_{55})^{\frac{1}{2}}\omega a=(\rho s_{55})^{\frac{1}{2}}\pi fD,$$

where f is the frequency and D is the diameter of the disk. Each frequency is a solution of the equation from which it was obtained, accurate to ± 0.001.

In Fig. 2, the frequencies of the isotropic disk are represented by circles. Each circle that corresponds to a *pair* of frequencies in the crystal disk is connected by a pair of arrows to squares which designate the frequencies in the crystal. For example, the compound mode (2,1), $\lambda a=2.338$, which is the lowest mode of

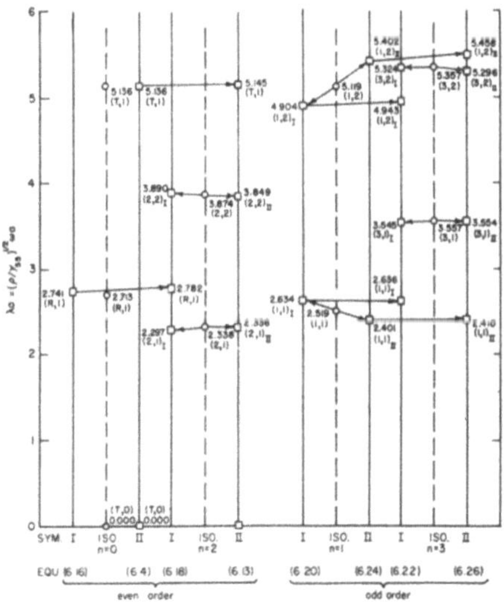

FIG. 2. Frequency spectra of quartz disks and equivalent isotropic disk.

circumferential order 2 in the isotropic disk, is connected with arrows to $(2,1)_{\mathrm{I}}$, $\lambda a=2.297$, and $(2,1)_{\mathrm{II}}$, $\lambda a=2.336$. First and second approximations to a frequency in the crystal disk are connected by an almost horizontal arrow; thus $(R,1)$, $\lambda a=2.741$ is connected with $(R,1)$, $\lambda a=2.782$. The former was computed from (63) and the latter from (65). It may be seen that the first and second approximations differ only in the third or fourth significant figure. It seems likely, therefore, that the roots obtained only as first approximations are of about the same accuracy.

The frequencies shown in Fig. 2 include all that exist in the range $0\leqslant\lambda a<6$ and circumferential order less than $n=4$. There are two pairs of additional modes of higher circumferential order at the upper end of the frequency range, as may be readily ascertained from the relatively simple frequency equation for the isotropic disk, but their frequencies were not computed for the crystal disk. Thus, the isotropic disk has $\lambda a=4.880$ for mode (4,1) and $\lambda a=5.546$ for mode (5,1).

The shapes of the modes are difficult to depict, in the interior of the plate, as there are no nodal lines. However, it may be stated that in each mode, although the displacements contain terms of all circumferential orders, the order of the equivalent isotropic disk predominates, and the amplitudes of the neighboring orders diminish rapidly. In fact, this is the reason for the rapid convergence of the frequencies.

Some idea of the shapes of the modes can be obtained from the displacements at the boundary of the disk. Of special interest are the boundary displacements of the pair of modes of lowest frequency:

$(2,1)_{\mathrm{I}}$, $\lambda a=2.297$:

$$u_r \propto 0.305+1.000\cos 2\theta+0.050\cos 4\theta+\cdots,$$
$$u_\theta \propto -0.288\sin 2\theta-0.051\sin 4\theta+\cdots.$$

$(2,1)_{\mathrm{II}}$, $\lambda a=2.336$:

$$u_r \propto 1.000\sin 2\theta+0.053\sin 4\theta+\cdots,$$
$$u_\theta \propto 0.002+0.240\cos 2\theta+0.041\cos 4\theta+\cdots.$$

Thus, the maximum displacements in $(2,1)_{\mathrm{II}}$ are nearly at 45° to the axes of elastic symmetry (x,z) of the plate. This mode is the one that is often thought to be the one of lowest frequency. However, $(2,1)_{\mathrm{I}}$ has a slightly lower frequency and has maximum displacements nearly coincident with the axes of elastic symmetry, but it produces no surface charge on the faces of a quartz disk.

An upper bound for the frequency of modes of the type $(2,1)_{\mathrm{II}}$ has been derived by Onoe,[10] using an energy method. He gives

$$fD=0.75105(1/\rho s_{55})^{\frac{1}{2}}.$$

This may be compared with our result for mode $(2,1)_{\mathrm{II}}$:

$$fD=\frac{\lambda a}{\pi}\left(\frac{1}{\rho s_{55}}\right)^{\frac{1}{2}}=0.744\left(\frac{1}{\rho s_{55}}\right)^{\frac{1}{2}}.$$

Thus, Onoe's upper bound exceeds our frequency by 1.0%.

(Reprinted from the Journal of Applied Mechanics for December, 1962, with permission of Copyright Owner)

MORIO ONOE

Associate Professor of Electrical Engineering, Institute of Industrial Science, University of Tokyo Tokyo, Japan

H. D. McNIVEN

Associate Professor of Civil Engineering, University of California, Berkeley, Calif.

R. D. MINDLIN

Professor of Civil Engineering, Columbia University, New York, N. Y. Mem. ASME

Dispersion of Axially Symmetric Waves in Elastic Rods

The relation between frequency and propagation constant, for axially symmetric waves in an infinitely long, isotropic, circular rod, as given by Pochhammer's equation, is explored. A spectrum, covering a large range of frequencies, is developed for real, imaginary, and complex propagation constants, and the influence of Poisson's ratio is described.

Axially symmetric vibrations and waves in an infinitely long, isotropic, elastic rod of circular cross section are governed by an equation formulated by Pochhammer [1].[1] Even though the equation was introduced as early as 1876, it appears that no systematic exploration of its roots was undertaken until 1941. In that year, Bancroft [2] studied the relation between the phase velocity and propagation constant of only the lowest branch of roots of the equation, although he recognized that, for a given propagation constant, there is an infinity of phase velocities, corresponding to an infinity of higher branches. He found that the shape of the lowest branch varied with changes in Poisson's ratio and he presented a spectrum that showed its influence. He showed that, as the wavelength of the vibration becomes smaller, the phase velocity approaches the velocity of Rayleigh surface waves. Further, he showed that, under the same conditions, the phase velocity of all of the higher branches approaches the velocity of equivoluminal waves in an infinite medium.

Hudson [3] later reaffirmed the asymptotic behavior at short wavelength and showed that there is one point on the lowest branch which is invariant with respect to change of Poisson's ratio.

In 1951, the analysis of the closely related transcendental frequency equations for both rods and plates was facilitated as a result of the introduction, by Holden [4] and Mindlin [5], of simple algebraic equations whose roots coincide with those of the transcendental equations at discrete points and form bounds between those points. In the paper by Holden, the spectrum of phase velocities of a rod was studied for real propagation constants. This was extended, by Onoe [6], to imaginary propagation constants.

In the present paper the analysis of Pochhammer's equation is extended over a wide range of frequencies and to complex propagation constants. In addition, the intricate behavior of the frequency spectrum with change of Poisson's ratio, which is especially pronounced for small propagation constants, is studied in detail. The methods of analysis are similar to those developed previously for plates [4–8].

[1] Numbers in brackets designate References at end of paper.

Contributed by the Applied Mechanics Division and presented at the Winter Annual Meeting, New York, N. Y., November 25–30, 1962, of The American Society of Mechanical Engineers.

Discussion of this paper should be addressed to the Editorial Department, ASME, United Engineering Center, 345 East 47th Street, New York 17, N. Y., and will be accepted until January 10, 1963. Discussion received after the closing date will be returned. Manuscript received by ASME Applied Mechanics Division, July 20, 1961. Paper No. 62—WA-60.

Pochhammer's Equation

Pochhammer's equation may be written as

$$F \equiv (2\zeta^2 - \Omega^2)^2\mathfrak{J}_1(\delta\alpha) + 4\alpha^2\zeta^2\mathfrak{J}_1(\delta\beta) - 2\alpha^2\Omega^2 = 0 \quad (1)$$

where

$$\alpha^2 = \Omega^2k^{-2} - \zeta^2; \ \beta^2 = \Omega^2 - \zeta^2 \quad (2)$$

$\Omega = \omega/\omega_s$ = dimensionless frequency
$\zeta = \gamma a/\delta$ = dimensionless propagation constant
ω = circular frequency
$\omega_s = (\delta/a)(\mu/\rho)^{1/2}$, circular frequency of first axial shear mode
γ = propagation constant (wave number along axis of rod)
a = radius of rod
δ = lowest nonzero root of $J_1(\delta_n) = 0$
k = ratio of dilatational velocity to equivoluminal velocity in an infinite, isotropic medium
$= [2(1-\nu)/(1-2\nu)]^{1/2}$, where ν is Poisson's ratio
$\mathfrak{J}_1(x) = xJ_0(x)/J_1(x)$ is the modified quotient of Bessel functions, and is tabulated for real and pure imaginary arguments [9].

It may be seen that (1) is an equation relating frequency and propagation constant.

Bounds, Slopes, and Curvatures

To construct the branches of (1), families of simple curves of the form $\Omega^2 = \zeta^2 + b_m^2$ and $\Omega^2 = k^2(\zeta^2 + c_n^2)$ are first used to form a grid on the $\zeta - \Omega$ plane. When appropriate equations are chosen governing the values of b_m and c_n, these families of curves are bounds of the branches of Pochhammer's equation between intersection points of the grid. There are four such equations and, when taken in proper pairs, each pair satisfies Pochhammer's equation at the frequencies and propagation constants of the intersections. The two families of curves that depend on one pair of the equations are drawn as full lines in Fig. 1; the other two families of curves are drawn as dashed lines. A branch is bounded by sets of full and dashed lines and passes alternately through points where full lines intersect and points where dashed lines intersect.

The sketching of the branches is made more accurate by employing explicit expressions which have been obtained for their slopes at the grid intersections and for their curvatures at zero propagation constant. At $\zeta = 0$ a branch has the same ordinate and slope (except in special cases) as a grid line. The sketch of

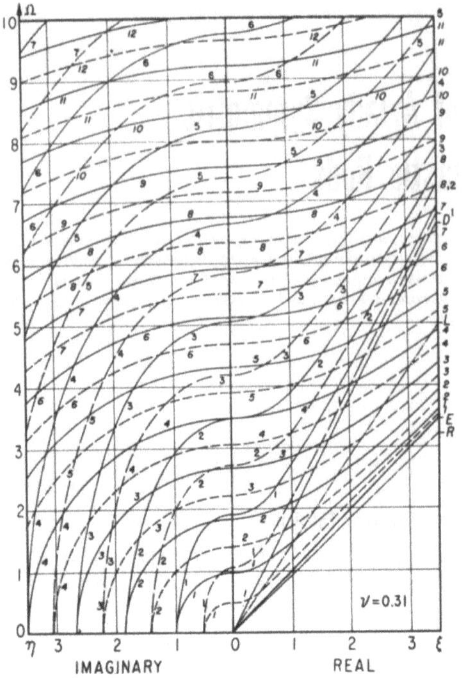

Fig. 1 Grid of bounds. These lines are bounds of the branches except at intersections of like (dashed or full) curves, where they coincide with the branches.

the branch can be started there, curving above or below the grid line depending on whether the curvature of the branch is greater or less than the curvature of its matching grid line. When the grid, slopes, and curvatures are used, there remains little flexibility in the manner in which the branches can be drawn, and the sketching is completed easily.

Equations of Bounds. The families of curves

$$\Omega^2 = \zeta^2 + \varphi_m^2 \tag{3}$$

and

$$\Omega^2 = k^2(\zeta^2 + \varphi_n^2) \tag{4}$$

are bounds of the Pochhammer branches when φ_m and φ_n satisfy the equations

$$J_1(\delta\varphi_m) = 0 \tag{5}$$

and

$$J_1(\delta\varphi_n) = 0 \tag{6}$$

respectively, and $m = n = 1$ represent the lowest (nonzero) roots. That is, as may be verified by substitution in equation (1), a pair of Ω and ζ which satisfies (3) *or* (4) does *not* satisfy Pochhammer's equation except when the pair satisfies *both* (3) and (4). The relation between frequency and propagation constant established by equations (3) and (5) is associated with equivoluminal deformation and the relation from equations (4) and (6) corresponds to dilatational deformation.

In the same way, the families of curves,

$$\Omega^2 = \zeta^2 + \sigma_m^2 \tag{7}$$

and

$$\Omega^2 = k^2(\zeta^2 + \sigma_n^2) \tag{8}$$

are bounds of the branches when σ_m and σ_n satisfy the equations

$$\mathfrak{J}_1(\delta\sigma_m) = 1 \tag{9}$$

and

$$\mathfrak{J}_1(\delta\sigma_n) = \frac{2\sigma_n^2}{\zeta^2(k^2 - 2) + k^2\sigma_n^2} \tag{10}$$

respectively, and $m = n = 1$ represent the lowest roots.

On the plane where ζ is real and equal to ξ, we have the real bounds

$$\begin{aligned}\Omega^2 &= \xi^2 + \varphi_m^2 \\ \Omega^2 &= \xi^2 + \sigma_m^2\end{aligned} \tag{11}$$

and

$$\begin{aligned}\Omega^2 &= k^2(\xi^2 + \varphi_n^2) \\ \Omega^2 &= k^2(\xi^2 + \sigma_n^2)\end{aligned} \tag{12}$$

On the plane where ζ is imaginary and equal to $i\eta$, we have the imaginary bounds

$$\begin{aligned}\Omega^2 &= \varphi_m^2 - \eta^2 \\ \Omega^2 &= \sigma_m^2 - \eta^2\end{aligned} \tag{13}$$

and

$$\begin{aligned}\Omega^2 &= k^2(\varphi_n^2 - \eta^2) \\ \Omega^2 &= k^2(\sigma_n^2 - \eta^2)\end{aligned} \tag{14}$$

The four sets of bounds are shown in Fig. 1 for positive ξ and η and Poisson's ratio 0.31.

The bounds in the figure may be identified in the following way: The φ-families are both shown as full lines, the σ-families dashed; curves φ_m = const and σ_m = const are identified by slanted numerals corresponding to the value of m; similarly, n is represented by upright numerals. For real ζ, the curves φ = const are hyperbolas with asymptote $\Omega = \xi$ for subscripts m and $\Omega = k\xi$ for subscripts n. Again, for real ζ, the curves σ = const with subscripts m are hyperbolas with asymptote $\Omega = \xi$, and with subscripts n are curves resembling hyperbolas with asymptote $\Omega = k\xi$. For imaginary ζ, the curves φ = const are quarter-circles for subscripts m and quarter-ellipses for subscripts n, while the curves σ = const with subscripts m are quarter-circles, and with subscripts n are curves closely resembling quarter-ellipses.

Ordinates of Bounds at $\zeta = 0$. The curves φ_m = constant (full, slanted) intersect the axis $\zeta = 0$ and the plane $\Omega = 0$ at $\Omega = \varphi_m$ and $\eta = \varphi_m$, respectively.

The curves φ_n = constant (full, upright) intersect the axis $\zeta = 0$ at $\Omega = k\varphi_n$ and the plane $\Omega = 0$ at $\eta = \varphi_n$.

The curves σ_m = constant (dashed, slanted) intersect the axis $\zeta = 0$ and the plane $\Omega = 0$ at $\Omega = \sigma_m$ and $\eta = \sigma_m$, respectively.

The curves σ_n = constant (dashed, upright) intersect the axis $\zeta = 0$ at $\Omega = k\sigma_n$, where σ_n are the roots of

$$\mathfrak{J}_1(\delta\sigma_n) = 2/k^2$$

and intersect the plane $\Omega = 0$ at $\eta = \sigma_n$ where σ_n are the roots of

$$\mathfrak{J}_1(\delta\sigma_n) = 1$$

It will be noted that, at the plane $\Omega = 0$, corresponding ($m = n$) full lines meet, and corresponding dashed lines meet.

The slope of all bounds at $\Omega = 0$ is infinite and at $\zeta = 0$ is zero.

Intersections of Bounds. From equations (3) and (4), the coordinates of the intersections of φ_m = const with φ_n = const are, for real ζ,

$$\xi_{mn} = [(\varphi_m^2 - k^2\varphi_n^2)(k^2 - 1)^{-1}]^{1/2}; \qquad \Omega_{mn} = k[(\varphi_m^2 - \varphi_n^2)(k^2 - 1)^{-1}]^{1/2} \tag{15}$$

and, for imaginary ζ,

$$\eta_{mn} = [(k^2\varphi_n^2 - \varphi_m^2)(k^2 - 1)^{-1}]^{1/2}; \qquad \Omega_{mn} = k[(\varphi_m^2 - \varphi_n^2)(k^2 - 1)^{-1}]^{1/2} \tag{16}$$

From equations (7) and (8), the coordinates of the intersections of σ_m = const with σ_n = const are, for real ζ,

$$\xi_{mn} = [(\sigma_m^2 - k^2\sigma_n^2)(k^2 - 1)^{-1}]^{1/2};$$
$$\Omega_{mn} = k[(\sigma_m^2 - \sigma_n^2)(k^2 - 1)^{-1}]^{1/2} \quad (17)$$

and, for imaginary ζ,

$$\eta_{mn} = [(k^2\sigma_n^2 - \sigma_m^2)(k^2 - 1)^{-1}]^{1/2};$$
$$\Omega_{mn} = k[(\sigma_m^2 - \sigma_n^2)(k^2 - 1)^{-1}]^{1/2} \quad (18)$$

It will be noted that, in equations (17) and (18), the σ_n are themselves dependent on the coordinates, so the values of the coordinates must be arrived at by an iterative procedure if greater accuracy is desired than can be obtained graphically by the intersections of the curves.

Slopes of Bounds at Intersections. All bounds have zero slope at $\zeta = 0$. At the intersection points for bounds φ = const (full lines), from (11)–(14)

$$\frac{d\Omega}{d\xi} = \frac{1}{k}\left[\frac{\varphi_m^2 - k^2\varphi_n^2}{\varphi_m^2 - \varphi_n^2}\right]^{1/2}, \qquad \text{for } \varphi_m = \text{const}$$

$$\frac{d\Omega}{d\xi} = k\left[\frac{\varphi_m^2 - k^2\varphi_n^2}{\varphi_m^2 - \varphi_n^2}\right]^{1/2}, \qquad \text{for } \varphi_n = \text{const}$$

$$\frac{d\Omega}{d\eta} = -\frac{1}{k}\left[\frac{k^2\varphi_n^2 - \varphi_m^2}{\varphi_m^2 - \varphi_n^2}\right]^{1/2}, \qquad \text{for } \varphi_m = \text{const} \quad (19)$$

$$\frac{d\Omega}{d\eta} = -k\left[\frac{k^2\varphi_n^2 - \varphi_m^2}{\varphi_m^2 - \varphi_n^2}\right]^{1/2}, \qquad \text{for } \varphi_n = \text{const}$$

The slopes of the dashed m grid lines, at their intersections with the dashed n lines, are found from

$$\frac{d\Omega}{d\xi} = \frac{1}{k}\left[\frac{\sigma_m^2 - k^2\sigma_n^2}{\sigma_m^2 - \sigma_n^2}\right]^{1/2}, \qquad \text{for } \sigma_m = \text{const}$$

$$\frac{d\Omega}{d\eta} = -\frac{1}{k}\left[\frac{k^2\sigma_n^2 - \sigma_m^2}{\sigma_m^2 - \sigma_n^2}\right]^{1/2}, \qquad \text{for } \sigma_m = \text{const} \quad (20)$$

while the slopes of the dashed n grid lines, at their intersections with the dashed m lines, are

$$\frac{d\Omega}{d\xi} = k\left[\frac{\sigma_m^2 - k^2\sigma_n^2}{\sigma_m^2 - \sigma_n^2}\right]^{1/2}\left[\frac{1 - 4\sigma_n^2D^{-1}(1 - 2k^{-2})\Omega^2}{1 - 4\sigma_n^2D^{-1}(1 - 2k^{-2})\Omega^2E}\right] \quad (21)$$

where

$$D = 4(\Omega^2k^{-2} - \xi^2)(\Omega^2 - 2\xi^2) - (\delta\sigma_n)^2(\Omega^2 - 2\xi^2)^2 - 4(\Omega^2k^{-2} - \xi^2)^2$$

$$E = k\Omega^{-1}(\sigma_m^2 - k^2\sigma_n^2)[(k^2 - 1)(\sigma_m^2 - \sigma_n^2)]^{-1/2}$$

$$\frac{d\Omega}{d\eta} = -k\left[\frac{k^2\sigma_n^2 - \sigma_m^2}{\sigma_m^2 - \sigma_n^2}\right]^{1/2}\left[\frac{1 - 4\sigma_n^2D^{-1}(1 - 2k^{-2})\Omega^2}{1 - 4\sigma_n^2D^{-1}(1 - 2k^{-2})\Omega^2E}\right] \quad (22)$$

where

$$D = 4(\Omega^2k^{-2} + \eta^2)(\Omega^2 + 2\eta^2) - (\delta\sigma_n)^2(\Omega^2 + 2\eta^2)^2 - 4(\Omega^2k^{-2} + \eta^2)^2$$

E = as above

Curvatures of Bounds at $\zeta = 0$. From (11)–(14) we have curvatures of bounds φ = const (full lines) at $\zeta = 0$

$$\frac{d^2\Omega}{d\xi^2} = \frac{1}{\varphi_m}, \qquad \varphi_m = \text{const}$$

$$\frac{d^2\Omega}{d\xi^2} = \frac{k}{\varphi_n}, \qquad \varphi_n = \text{const}$$

$$\frac{d^2\Omega}{d\eta^2} = -\frac{1}{\varphi_m}, \qquad \varphi_m = \text{const} \quad (23)$$

$$\frac{d^2\Omega}{d\eta^2} = -\frac{k}{\varphi_n}, \qquad \varphi_n = \text{const}$$

Curvatures of dashed m bounds at $\zeta = 0$ are

$$\frac{d^2\Omega}{d\xi^2} = \frac{1}{\sigma_m}, \qquad \sigma_m = \text{const}$$

$$\frac{d^2\Omega}{d\eta^2} = -\frac{1}{\sigma_m} \qquad \sigma_m = \text{const} \quad (24)$$

Curvatures of dashed n bounds at $\zeta = 0$ are

$$\frac{d^2\Omega}{d\xi^2} = \frac{k}{\sigma_n}\left[1 + \frac{4(k^2 - 2)}{(\delta\sigma_n)^2k^4 - 4(k^2 - 1)}\right]$$

$$\frac{d^2\Omega}{d\eta^2} = -\frac{k}{\sigma_n}\left[1 + \frac{4(k^2 - 2)}{(\delta\sigma_n)^2k^4 - 4(k^2 - 1)}\right] \quad (25)$$

In (25), σ_n are the roots of

$$\mathfrak{J}_1(\delta\sigma_n) = 2/k^2$$

Ordinates of Branches at $\zeta = 0$. At $\zeta = 0$, the dilatational and equivoluminal deformations are uncoupled. The uncoupled modes constitute radial motion and axial shear motion.

From (1), the radial modes have frequencies

$$\Omega = k\sigma_q \quad (26)$$

where the σ_q satisfy

$$\mathfrak{J}_1(\delta\sigma_q) = 2/k^2$$

and the axial shear modes have frequencies

$$\Omega = \varphi_p \quad (27)$$

where the φ_p satisfy

$$J_1(\delta\varphi_p) = 0$$

It can be seen that the frequencies of the radial modes change as k changes (i.e., with a change in Poisson's ratio) while frequencies of the axial shear modes are not affected. There is coincidence of frequencies when

$$\varphi_p = k\sigma_q \quad (28)$$

and two branches intersect at such points.

Curvatures of Branches at $\zeta = 0$. At $\zeta = 0$, the real segments of the branches passing through $\Omega = \varphi_p$ and $\Omega = k\sigma_q$ have the limiting curvatures:

(*a*) At $\Omega = \varphi_p$

$$\lim_{\xi\to 0}\frac{d^2\Omega}{d\xi^2} = \frac{1}{\varphi_p}\left[1 - \frac{8}{k^2\mathfrak{J}_1(\delta\varphi_p/k) - 2}\right] \quad (29)$$

(*b*) At $\Omega = k\sigma_q$

$$\lim_{\xi\to 0}\frac{d^2\Omega}{d\xi^2} = \frac{k}{\sigma_q}\left[1 + \frac{4(k^2 + 2\mathfrak{J}_1(\delta k\sigma_q) - 4)}{k^2(\delta k\sigma_q)^2 - 4(k^2 - 1)}\right] \quad (30)$$

The limiting curvatures

$$\lim_{\eta\to 0}\frac{d^2\Omega}{d\eta^2}$$

of the imaginary segments are the negatives of the curvatures given in (29) and (30). It may be observed that, when (28) is satisfied, both (29) and (30) become infinite.

Slopes of Branches at $\zeta = 0$. All branches have zero slope at ζ = 0, except when there is coincidence of frequencies, as specified by (28). In this case there are no imaginary segments at $\zeta = 0$ and the pairs of intersecting real segments of branches have the limiting slopes

$$\lim_{\xi \to 0} \frac{d\Omega}{d\xi} = \pm\, 2\varphi_p[\delta^2\varphi_p{}^4 - 4(\varphi_p{}^2 - \sigma_p{}^2)]^{-1/2} \tag{31}$$

Slopes of Branches at Intersections of Bounds. The slopes of the branches at intersections of bounds φ = const (full lines) are obtained by differentiating (1) with respect to ξ or η and setting $J_1(\delta\beta) = 0$ and $J_1(\delta\alpha) = 0$ (i.e., setting $\beta = \varphi_m$ and $\alpha = \varphi_n$) in the result. We find

(*a*) For ζ real

$$\frac{d\Omega}{d\xi} = \frac{\xi k^2}{\Omega}\left[\frac{4\xi^2\varphi_m{}^2 + (\Omega^2 - 2\xi^2)^2}{4\xi^2\varphi_m{}^2 + k^2(\Omega^2 - 2\xi^2)^2}\right] \tag{32}$$

(*b*) For ζ imaginary

$$\frac{d\Omega}{d\eta} = -\frac{\eta k^2}{\Omega}\left[\frac{4\eta^2\varphi_m{}^2 - (\Omega^2 + 2\eta^2)^2}{4\eta^2\varphi_m{}^2 - k^2(\Omega^2 + 2\eta^2)^2}\right] \tag{33}$$

The slopes of the branches at intersections of bounds σ = const (dashed lines) are obtained by differentiating (1) with respect to ξ or η and setting $\mathfrak{J}_1(\delta\beta) = 1$ and $\mathfrak{J}_1(\delta\alpha) = 2\alpha^2/[\zeta^2(k^2 - 2) + k^2\alpha^2]$ (i.e., setting $\beta = \sigma_m$ and $\alpha = \sigma_n$) in the result. We find:

(*a*) For ζ real

$$\frac{d\Omega}{d\xi} = \frac{\xi k^2}{\Omega} \left\{\frac{\delta^2\sigma_m{}^2[4\sigma_n{}^2\xi^2 + (\Omega^2 - 2\xi^2)^2] - 4\sigma_n{}^2(2\Omega^2 - \xi^2)}{\delta^2\sigma_m{}^2[4k^2\sigma_n{}^2\xi^2 + (\Omega^2 - 2\xi^2)^2] - 4\sigma_n{}^2(k^2\Omega^2 - \Omega^2 + \xi^2)}\right\} \tag{34}$$

(*b*) For ζ imaginary

$$\frac{d\Omega}{d\eta} = -\frac{\eta k^2}{\Omega} \left\{\frac{\delta^2\sigma_m{}^2\,[4\sigma_n{}^2\eta^2 - (\Omega^2 + 2\eta^2)^2] + 4\sigma_n{}^2(2\Omega^2 + \eta^2)}{\delta^2\sigma_m{}^2[4k^2\sigma_n{}^2\eta^2 - (\Omega^2 + 2\eta^2)^2] + 4\sigma_n{}^2(k^2\Omega^2 - \Omega^2 - \eta^2)}\right\} \tag{35}$$

In (32)–(35), Ω and ξ (or Ω and η) are the coordinates of the intersection under study and φ_m, σ_m, and σ_n are obtained from (11) to (14).

Special Intersections and Asymptotes

The straight line $\Omega = k\xi$ (marked *D* in Figs. 1 and 2) is the asymptote of all the real bounds described by equations (12). It can also be considered as the bound φ_n = const with $n = 0$ as given by the first of (12). The real portions of the branches, therefore, pass through intersections of this line and the bounds φ_m = const as given by the first of (11). These intersections occur at

$$\xi = \varphi_m(k^2 - 1)^{-1/2} \tag{36}$$

The line $\Omega = \xi$ (marked *E* in Figs. 1 and 2) is the asymptote of all the branches, for ζ real, except the lowest. With $\Omega = \xi$ Pochhammer's equation degenerates to

$$\mathfrak{J}_1(i\delta\alpha) = 6(1 - k^{-2}) \tag{37}$$

The lowest branch crosses the line $\Omega = \xi$ at

$$\xi = \alpha(1 - k^{-2})^{-1/2} \tag{38}$$

where α is the root of (37). After crossing $\Omega = \xi$, the lowest branch becomes asymptotic to the Rayleigh surface frequency

$$\Omega = s_1\xi \tag{39}$$

where s_1 is the real root, in the range $0 < s < 1$, of

$$s^6 - 8s^4 + 8(3 - 2k^{-2})s^2 - 16(1 - k^{-2}) = 0 \tag{40}$$

Equation (40) is obtained by substituting $\Omega = s\xi$ into the equation

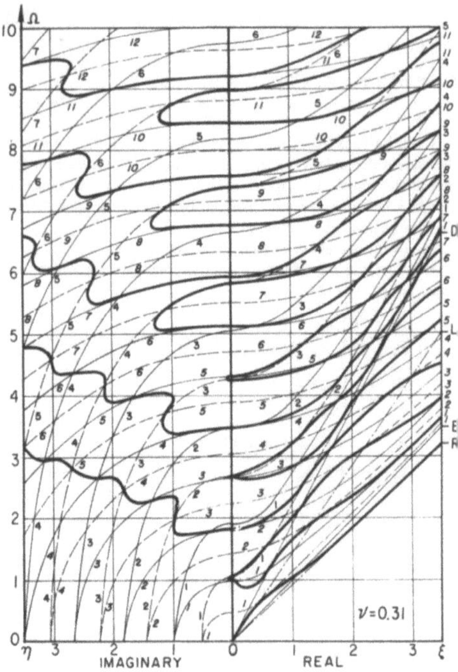

Fig. 2 Frequency spectrum showing the segments of the lowest eighteen branches for real and imaginary propagation constants

$$(\xi^2 + \beta_1{}^2)^2 = 4\xi^2\alpha_1\beta_1 \tag{41}$$

which is the limiting form of (1) for large, real ζ and imaginary α and β ($\alpha = i\alpha_1$; $\beta = i\beta_1$). Equation (41) gives the relation between frequency and propagation constant for Rayleigh surface waves. The other two roots (s_2, s_3) are real if $k^2 \leq 3.1$; i.e., if $\nu \leq 0.27$. For small ξ, the lowest branch has the asymptotic behavior

$$\Omega = [2(1 + \nu)]^{1/2}\xi \tag{42}$$

Equation (42) gives the relation between frequency and propagation constant found in the classical theory of low-frequency extensional vibrations of thin rods.

Finally, all the real branches intersect the radial line

$$\Omega = \sqrt{2}\xi \tag{43}$$

(marked *L* in Figs. 1 and 2), at $\xi = \sigma_m$. These intersections give the frequencies of Lamb's [10] equivoluminal modes of rods, which are mentioned in his famous paper on plates. The lowest of these points is the one at which Hudson found independence of Poisson's ratio in the first branch.

Saddle Points

The bounds serve to separate the branches except for occasional bounded areas, in the imaginary plane adjacent to $\eta = 0$, which are entered by two branches. When the curvature of a branch passing through $\eta = 0$, $\Omega = k\sigma_q$ is close to zero, there is some uncertainty as to whether that branch connects, in the imaginary plane, with the next lower or the next higher point $\eta = 0$, $\Omega = \varphi_p$. The situation is very sensitive to Poisson's ratio. This is illustrated in Fig. 3 which shows the projections, on the $\eta - \Omega$ plane, of contours ν = const, of the surface $F(\Omega, \eta, \nu) = 0$. It may be seen that there is a critical value of ν for which the

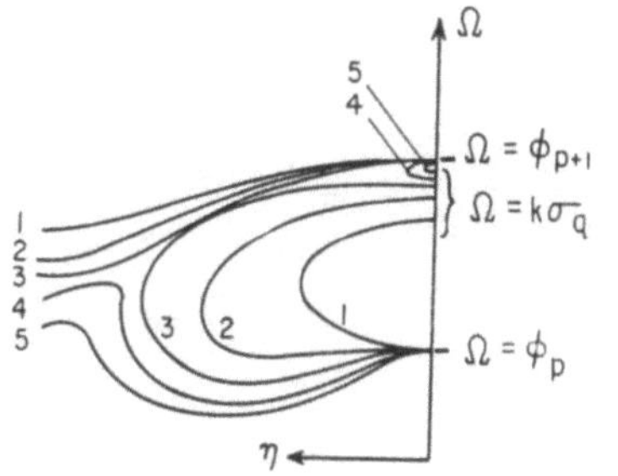

Fig. 3 Contours of constant Poisson's ratio on surface $F(\Omega, \eta, \nu) = 0$ illustrating a saddle point

surface has a saddle point at which the branches passing through the three points $(0, \varphi_p)$, $(0, k\sigma_q)$, and $(0, \varphi_{p+1})$ meet at a point in the η, Ω-plane. Below or above the critical value, the connection is below or above, respectively.

The critical values of Poisson's ratio may be found from a simultaneous solution of the three equations

$$F = 0, \quad \frac{\partial F}{\partial \Omega} = 0, \quad \frac{\partial F}{\partial \eta} = 0 \tag{44}$$

The three roots nearest $\nu = 0.3$ are given in Table 1, in which the branches are identified by the subscripts p or q of the ordinates $\Omega = \varphi_p$, $\Omega = k\sigma_q$ through which the branches pass at $\eta = 0$.

Table 1 Critical values of Poisson's ratio which give saddle points

Branches		Poisson's ratio
$p = 2, 3$	$q = 2$	0.284
$p = 4, 5$	$q = 3$	0.308
$p = 6, 7$	$q = 4$	0.317

Complex Roots

Except for the lowest branch, each branch terminates at $\Omega = 0$ with a complex segment and may have one or more additional complex segments at higher frequencies.

The points of intersection of the complex segments with the plane $\Omega = 0$ may be found as follows: Differentiate (1) with respect to Ω^2 and set

$$\lim_{\Omega \to 0} \frac{1}{\Omega} \frac{\partial \zeta}{\partial \Omega} = 0 \tag{45}$$

$$\lim_{\Omega \to 0} \alpha = \lim_{\Omega \to 0} \beta = i\zeta \tag{46}$$

The result is

$$\mathfrak{J}_1{}^2(i\delta\zeta) - (\delta\zeta)^2 = 2(1 - \nu) \tag{47}$$

the roots of which give the coordinates of the terminal points of the complex segments. The first root of (47) is given in Fig. 4 for a range of values of Poisson's ratio. For large $\delta\zeta$, $\mathfrak{J}_1(i\delta\zeta)$ may be expanded in a series of trigonometric functions and we obtain

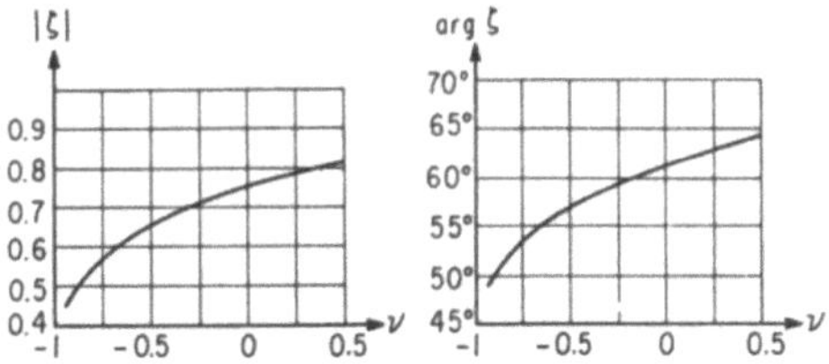

Fig. 4 Real and imaginary parts of propagation constant at zero frequency of first complex segment

the asymptotic equation

$$\sin\left(i\delta\zeta - \tfrac{1}{4}\pi\right) \approx \pm i\delta\zeta[2(1 - \nu)]^{1/2} \tag{48}$$

the roots of which, for large $\delta\zeta$, are approximately

$$\mathrm{Re}(i\delta\zeta) \approx (m + \tfrac{3}{4})\pi \tag{49}$$

$$\mathrm{Im}(i\delta\zeta) \approx \log\,[2(1 - \nu)]^{1/2}(2m + \tfrac{3}{2})\pi \tag{50}$$

Considering, now, complex segments in the neighborhood of the plane $\eta = 0$, we expand (1) in powers of η, discard powers higher than the first, equate real and imaginary parts separately to zero and find

$$F(\xi, \Omega) = 0 \tag{51}$$

$$\eta \frac{\partial F(\xi, \Omega)}{\partial \xi} = 0 \tag{52}$$

as the equations governing roots in the complex space near the real plane $\eta = 0$. The important feature of (52) is the appearance of $\partial F(\xi, \Omega)/\partial \xi$ as the coefficient of η.

Now, (51) and (52) have the usual real roots $F(\xi, \Omega) = 0$, $\eta = 0$ and also the roots

$$F(\xi, \Omega) = 0 \tag{53}$$

$$\eta \neq 0, \quad \frac{\partial F(\xi, \Omega)}{\partial \xi} = 0 \tag{54}$$

exhibiting complex segments which intersect the real segments. The character of the real segments at the intersection points is revealed if we note, from (53), that

$$\frac{\partial F(\xi, \Omega)}{\partial \Omega} \frac{\partial \Omega}{\partial \xi} + \frac{\partial F(\xi, \Omega)}{\partial \xi} = 0 \tag{55}$$

Thus, if (54) is satisfied, we must have $\partial \Omega/\partial \xi = 0$ if, as is usual, $\partial F/\partial \Omega \neq 0$. Hence, the complex segments intersect the real segments at points of zero slope of the latter. Such a point, for $\xi \neq 0$, is exemplified by the minimum in the second branch in Fig. 2. When $\xi = 0$, we have pure imaginary segments intersecting the real segments since, usually, the latter have zero slope at $\xi = 0$. An exceptional case occurs when (28) is satisfied, in which case there is coincidence of frequencies of certain radial and axial shear modes and the slopes of the intersecting real segments are not zero. In this case $\partial F/\partial \Omega = 0$ and there is no pure imaginary segment.

A similar investigation of the behavior of the roots near the plane $\xi = 0$ shows, again, that complex segments emanate from points of zero slope on the imaginary segments. In this case there is a multiplicity of maxima and minima, as may be seen in Fig. 2.

Sketch of Spectrum

With the information given in the preceding sections, it is a simple matter to sketch the real and imaginary segments of the branches with considerable precision; and the result for $\nu = 0.31$ is shown in Fig. 2. The assignment of the various segments to the appropriate branches requires consideration of the complex segments and the criterion that each branch must extend from zero to infinite frequency, with one and only one propagation constant for each frequency.

Each complex segment has a conjugate which is the reflection in the plane $\xi = 0$. Also, since ζ appears in Pochhammer's equation only as ζ^2, the complex and pure imaginary segments have mirror images in the plane $\eta = 0$ and the real segments have mirror images in the plane $\xi = 0$. In all, there are two complete sets of branches in which each branch of the set extends from zero to infinite frequency with one to one correspondence with propagation constant. In one set the slopes of the real segments are everywhere positive and in the other set negative, corresponding to positive and negative group velocity ($\partial\omega/\partial\xi$), respectively.

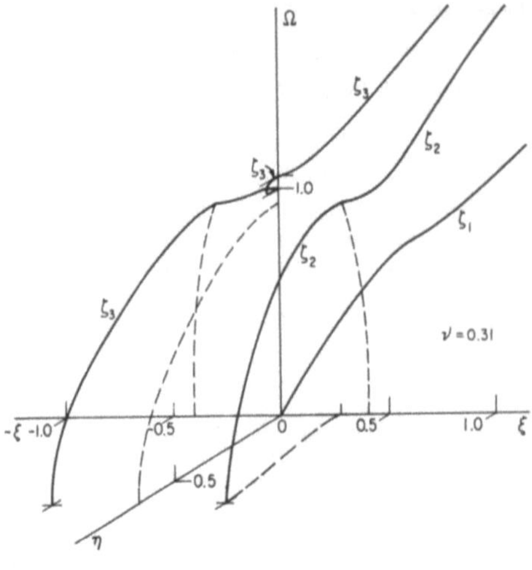

Fig. 5 Lowest three branches of frequency spectrum showing assignment of real, imaginary, and complex segments to branches

The first three branches of the positive set are shown in Fig. 5. The dashed lines in Fig. 5 are the projections of the complex segments on the coordinate planes. Comparison of the branches in the neighborhood of $\Omega = 1$ with the segments shown in Fig. 2 will reveal the appropriate assignment of segments to branches. The marked changes in these three branches, as Poisson's ratio is varied, are described in another paper [11].

The assignment of segments to the higher branches is illustrated in Fig. 6. In this figure, the dashed lines are complex segments. They are sketched in only approximately, except near the coordinate planes where their precise behavior is known from the analysis in the preceding section. The orders of the branches are given by the numbers next to the curves. A bar over a number indicates that it is the reflection of the branch in the plane $\xi = 0$ that is shown.

Near $\zeta = 0$ the spectrum is extremely sensitive to changes of Poisson's ratio. A very small change in ν may shift the branches across a saddle point and effect a drastic change in the imaginary and complex segments in that area; or the curvature of a real segment at $\xi = 0$ may change from negative to positive and thus swing a complex segment from an intersection in the plane $\eta = 0$ to one in the plane $\xi = 0$; or a frequency $\Omega = k\sigma_q$ at $\zeta = 0$ may shift from below to above $\Omega = \varphi_p$ with a resulting change in the character (essentially radial or essentially axial) of the modes of motion associated with successive orders.

As ζ increases, the changes with change of Poisson's ratio become less pronounced, though still present. In fact, the only points that do not change are those for the axial shear modes and Lamb's modes, both of which are purely equivoluminal.

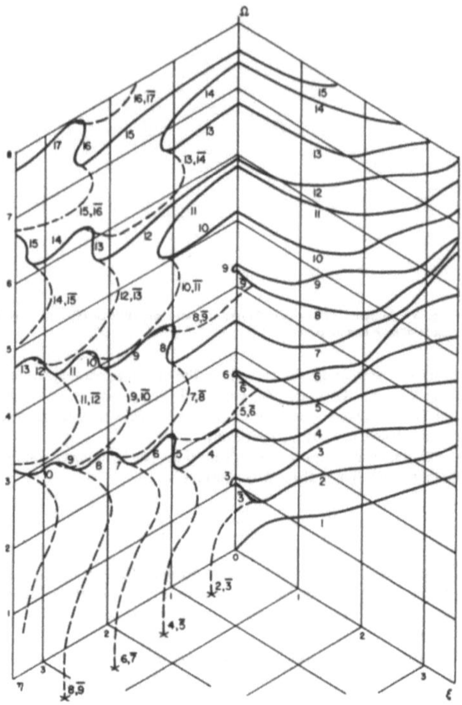

Fig. 6 Illustrating linkage of real, imaginary, and complex segments to form branches at high frequencies

References

1 L. Pochhammer, "Über die Fortpflanzungsgeschwindigkeiten Schwingungen in einem unbegrenzten isotropen Kreiscylinder," *Zeitschrift für Mathematik*, vol. 81, 1876, pp. 324–336.

2 D. Bancroft, "The Velocity of Longitudinal Waves in Cylindrical Bars," *Physical Review*, vol. 59, 1941, pp. 588–593.

3 G. E. Hudson, "Dispersion of Elastic Waves in Solid Circular Cylinders," *Physical Review*, vol. 63, 1943, pp. 46–51.

4 A. N. Holden, "Longitudinal Modes of Elastic Waves in Isotropic Cylinders and Slabs," *Bell System Technical Journal*, vol. 30, 1951, pp. 956–969.

5 R. D. Mindlin, "Thickness-Shear and Flexural Vibrations of Crystal Plates," *Journal of Applied Physics*, vol. 22, 1951, pp. 316–323.

6 Morio Onoe, "A Study of the Branches of the Velocity-Dispersion Equations of Elastic Plates and Rods," Report of Joint Committee on Ultrasonics of The Institute of Electrical Communication Engineers and the Acoustical Society of Japan, 1955.

7 R. D. Mindlin, "An Introduction to the Mathematical Theory of Vibrations of Elastic Plates," U. S. Army Signal Engineering Laboratories, Fort Monmouth, N. J., 1955.

8 R. D. Mindlin and M. Onoe, "Mathematical Theory of Vibrations of Elastic Plates," Proceedings, 11th Annual Symposium on Frequency Control, 1957, pp. 17–40, U. S. Army Signal Engineering Laboratories, Fort Monmouth, N. J.

9 Morio Onoe, "Tables of Modified Quotients of Bessel Functions of the First Kind for Real and Imginary Arguments," Columbia University Press, New York, N. Y., 1958.

10 H. Lamb, "On Waves in an Elastic Plate," *Proceedings of The Royal Society*, series A, vol. 93, 1917, footnote on p. 122.

11 R. D. Mindlin and H. D. McNiven, "Axially Symmetric Waves in Elastic Rods," Journal of Applied Mechanics, vol. 27, Trans. ASME, vol. 82, Series E, 1960, pp. 145–151.

Strong resonances of rectangular AT-cut quartz plates.

R. D. Mindlin
Columbia University
D. C. Gazis *
General Motors Research Laboratories

Introduction

In a previous paper [1], referred to in the sequel as I, there were derived two-dimensional equations of motion, of crystal plates, which take into account the coupling of flexural, extensional and face-shear deformations with each other and with the lowest thickness-orders of thickness-shear and thickness-twist deformations. The equations are intended for applications in the neighborhood of the frequency of the fundamental thickness-shear mode. Detailed measurements of the frequencies of the strong resonances, in that range, have been made by Koga and Fukuyo [2] on rectangular, AT-cut, quartz plates. In the present paper, appropriate solutions of the equations are obtained and the results of computations of the resonant frequencies are compared with the Koga-Fukuyo data.

Two-Dimensional Equations

The plate, of thickness $2h$, is referred to rectangular coordinates x_i, $i = 1, 2, 3$, with x_2 normal to the middle plane. The axis x_1 coincides with a digonal axis of the quartz and x_2 makes an angle of 35° 15′ with the trigonal axis [3]. When referred to the axes x_i, the matrix of the elastic stiffnesses c_{pq} ($p, q = 1, 2, 3, 4, 5, 6$) exhibits monoclinic symmetry, i.e.

$$c_{15} = c_{25} = c_{35} = c_{45} = c_{16} = c_{26} = c_{36} = c_{46} = 0 \qquad (1)$$

leaving thirteen non-zero constants.

*Now with International Business Machines Corporation.

The displacement-equations of motion in I have five dependent variables: three zero order components of displacement $(u_1^{(0)}, u_2^{(0)}, u_3^{(0)})$ in the coordinate directions and two components of rotation $(u_1^{(1)}, u_3^{(1)})$ of the normal to the plate. We set

$$\begin{aligned}(u_1^{(0)}, u_2^{(0)}, u_3^{(0)}) &= (\overline{u}_1, \overline{u}_2, \overline{u}_3)\, e^{i\omega t} \\ (u_1^{(1)}, u_3^{(1)}) &= (\psi_1, \psi_3)\, e^{i\omega t}\end{aligned} \qquad (2)$$

where ω is the circular frequency, $\overline{u}_i$, ψ_i are functions x_1 and x_3 only. With (1) and (2), the displacement-equations of free vibration, from I, reduce to

$$\overline{c}_{11}\overline{u}_{1,11} + \overline{c}_{13}u_{3,13} + k_3\overline{c}_{14}(\overline{u}_{2,13} + \psi_{3,1}) + c_{55}(\overline{u}_{3,13} + \overline{u}_{1,33}) + k_1c_{56}(\overline{u}_{2,13} + \psi_{1,3}) = -\rho\omega^2\overline{u}_1$$

$$k_3\overline{c}_{14}\overline{u}_{1,13} + k_3\overline{c}_{34}\overline{u}_{3,33} + k_3^2\overline{c}_{44}(\overline{u}_{2,33} + \psi_{3,3}) + k_1c_{56}(\overline{u}_{3,11} + \overline{u}_{1,13}) + k_1^2c_{66}(\overline{u}_{2,11} + \psi_{1,1}) = -\rho\omega^2\overline{u}_2$$

$$\overline{c}_{13}\overline{u}_{1,13} + \overline{c}_{33}\overline{u}_{3,33} + k_3\overline{c}_{34}(\overline{u}_{2,33} + \psi_{3,3}) + c_{55}(\overline{u}_{3,11} + \overline{u}_{1,13}) + k_1c_{56}(\overline{u}_{2,11} + \psi_{1,1}) = -\rho\omega^2\overline{u}_3 \qquad (3)$$

$$\gamma_{11}\psi_{1,11}+\gamma_{13}\psi_{3,13}+\gamma_{55}(\psi_{3,13}+\psi_{1,33}) -$$
$$3h^{-2}[k_1 c_{56}(\bar{u}_{3,1}+\bar{u}_{1,3})+k_1^2 c_{66}(\bar{u}_{2,1}+\psi_1)] = -\rho\omega^2\psi_1$$

$$\gamma_{55}(\psi_{3,11}+\psi_{1,13})+\gamma_{13}\psi_{1,13}+\gamma_{33}\psi_{3,33} -$$
$$3h^{-2}[k_3\bar{c}_{14}\bar{u}_{1,1}+k_3\bar{c}_{34}\bar{u}_{3,3}+k_3^2 c_{44}(\bar{u}_{2,3}+\psi_3)] = -\rho\omega^2\psi_3$$

where subscripts 1 and 3, following a comma, indicate differentiation with respect to x_1 and x_3 and

$$k_1^2 = \pi^2/12,$$

$$k_3^2 = \pi^2\{c_{22}+c_{44}-[(c_{22}-c_{44})^2+4c_{24}^2]^{1/2}\}/24\bar{c}_{44}$$

$$\bar{c}_{pq} = c_{pq} - c_{2p}c_{q2}/c_{22}\,; \quad p, q = 1,3,4 \tag{4}$$

$$\gamma_{pq} = \bar{c}_{pq} - \bar{c}_{4p}\bar{c}_{q4}/c_{44}\,; \quad p, q = 1,3$$

$$\gamma_{55} = c_{55} - c_{56}^2/c_{66}$$

The stress-resultants ($T_{11}^{(0)}$, $T_{33}^{(0)}$, $T_{31}^{(0)}$) in the plane of the plate, the shearing stress-resultants ($T_{23}^{(0)}$, $T_{12}^{(0)}$) normal to the plane of the plate and the bending and twisting couples ($T_{11}^{(1)}$, $T_{33}^{(1)}$, $T_{31}^{(1)}$) are written as

$$(T_{11}^{(0)}, T_{33}^{(0)}, T_{31}^{(0)}) = (N_1, N_3, N_5)\,e^{i\omega t}$$
$$(T_{21}^{(0)}, T_{23}^{(0)}) = (Q_1, Q_3)\,e^{i\omega t} \tag{5}$$
$$(T_{11}^{(1)}, T_{33}^{(1)}, T_{31}^{(1)}) = (M_1, M_3, M_5)\,e^{i\omega t}$$

Then the stress-displacement relations from I, are

$$N_1 = 2h\,[\bar{c}_{11}\bar{u}_{1,1}+\bar{c}_{13}\bar{u}_{3,3}+k_3\bar{c}_{14}(\bar{u}_{2,3}+\psi_3)]$$
$$N_3 = 2h\,[\bar{c}_{13}\bar{u}_{1,1}+\bar{c}_{33}\bar{u}_{3,3}+k_3\bar{c}_{34}(\bar{u}_{2,3}+\psi_3)]$$
$$N_5 = 2h\,[c_{55}(\bar{u}_{3,1}+\bar{u}_{1,3})+k_1 c_{56}(\bar{u}_{2,1}+\psi_1)]$$
$$Q_1 = 2hk_1\,[c_{56}(\bar{u}_{3,1}+\bar{u}_{1,3})+k_1 c_{66}(\bar{u}_{2,1}+\psi_1)] \tag{6}$$
$$Q_3 = 2hk_3\,[\bar{c}_{14}\bar{u}_{1,1}+\bar{c}_{34}\bar{u}_{3,3}+k_3\bar{c}_{44}(\bar{u}_{2,3}+\psi_3)]$$
$$M_1 = \tfrac{2}{3}h^3(\gamma_{11}\psi_{1,1}+\gamma_{13}\psi_{3,3})$$
$$M_3 = \tfrac{2}{3}h^3(\gamma_{13}\psi_{1,1}+\gamma_{33}\psi_{3,3})$$
$$M_5 = \tfrac{2}{3}h^3\gamma_{55}(\psi_{3,1}+\psi_{1,3})$$

The boundary conditions for free edges of a rectangular plate of length $2l$ and width $2w$ are

$$N_1 = N_5 = Q_1 = M_1 = M_5 = 0 \text{ on } x_1 = \pm l \tag{7}$$

$$N_3 = N_5 = Q_3 = M_3 = M_5 = 0 \text{ on } x_3 = \pm w \tag{8}$$

Selection of Solutions

The fundamental product solutions of (3), in rectangular coordinates, are

$$\bar{u}_1 = P_1 \cos(\xi x_1 + r\pi/2)\sin(\zeta x_3 + s\pi/2)$$
$$\bar{u}_2 = P_2 \sin(\xi x_1 + r\pi/2)\cos(\zeta x_3 + s\pi/2)$$
$$\bar{u}_3 = P_3 \sin(\xi x_1 + r\pi/2)\cos(\zeta x_3 + s\pi/2) \tag{9}$$
$$\psi_1 = P_4 \cos(\xi x_1 + r\pi/2)\cos(\zeta x_3 + s\pi/2)$$
$$\psi_3 = P_5 \sin(\xi x_1 + r\pi/2)\sin(\zeta x_3 + s\pi/2)$$

where $r = 0$ or 1 and $s = 0$ or 1. It would require an infinite series of functions of the type (9) to describe each mode satisfying both (7) and (8). However, simple closed solutions, satisfying either (7) or (8), may be found which adequately, though only approximately, describe the modes whose resonances were recorded by Koga and Fukuyo.

The first criterion for the choice of simple solutions from among (9) is based on the fact that only the integrated surface charge over the face of the crystal is recorded in the resonance experiments. Thus, the strong resonances will be found among those modes which depend essentially on only one of the two coordinates in the plane of the plate. Hence, with due attention to appropriate symmetry, we select, from (9), the functions

$$\bar{u}_1 = C_1 h \cos\xi x_1 + B_1 h \sin\zeta x_3$$
$$\bar{u}_2 = A_2 h \sin\xi x_1 + D_2 h \cos\zeta x_3$$
$$\bar{u}_3 = A_3 h \sin\xi x_1 + D_3 h \cos\zeta x_3 \tag{10}$$
$$\psi_1 = A_4 \cos\xi x_1 + B_4 \cos\zeta x_3$$
$$\psi_3 = C_5 \sin\xi x_1 + D_5 \sin\zeta x_3$$

Equations (3) are satisfied by (10) if

$$(\bar{\xi}^2 - 3\Omega^2)A_2 + \hat{c}_{56}\bar{\xi}^2 A_3 + \bar{\xi}A_4 = 0$$
$$\hat{c}_{56}\bar{\xi}^2 A_2 + (\hat{c}_{55}\bar{\xi}^2 - 3\Omega^2)A_3 + \hat{c}_{56}\bar{\xi}A_4 = 0$$

$$\bar{\xi}^2 A_2 + \qquad \hat{c}_{56}\bar{\xi}A_3 + (\hat{\gamma}_{11}\bar{\xi}^2 + 1 - \Omega^2)A_4 = 0 \tag{11}$$

$$(\hat{c}_{55}\bar{\zeta}^2 - 3\Omega^2)B_1 + \qquad \hat{c}_{56}\bar{\zeta}B_4 = 0$$
$$\hat{c}_{56}\bar{\zeta}B_1 + (\hat{\gamma}_{55}\bar{\zeta}^2 + 1 - \Omega^2)B_4 = 0 \tag{12}$$

$$(\hat{c}_{11}\bar{\xi}^2 - 3\Omega^2)C_1 - \qquad \hat{c}_{14}\bar{\xi}C_5 = 0$$
$$-\hat{c}_{14}\bar{\xi}C_1 + (\hat{\gamma}_{55}\bar{\xi}^2 + \hat{c}_{44} - \Omega^2)C_5 = 0 \tag{13}$$

$$(\hat{c}_{44}\bar{\zeta}^2 - 3\Omega^2)D_2 + \qquad \hat{c}_{34}\bar{\zeta}^2 D_3 - \qquad \hat{c}_{44}\bar{\zeta}D_5 = 0$$
$$\hat{c}_{34}\bar{\zeta}^2 D_2 + (\hat{c}_{33}\bar{\zeta}^2 - 3\Omega^2)D_3 - \qquad \hat{c}_{34}\bar{\zeta}D_5 = 0$$
$$-\hat{c}_{44}\bar{\zeta}D_2 - \qquad \hat{c}_{34}\bar{\zeta}D_3 - (\hat{\gamma}_{33}\bar{\zeta}^2 + \hat{c}_{44} - \Omega^2)D_5 = 0 \tag{14}$$

where

$$\bar{\xi} = \xi h, \quad \bar{\zeta} = \zeta h, \quad \Omega = \omega/\omega_s = \omega/(\pi/2h)(c_{66}/\rho)^{1/2}$$

$$\hat{c}_{11} = \bar{c}_{11}/k_1^2 c_{66}, \qquad \hat{c}_{14} = k_3\bar{c}_{14}/k_1^2 c_{66}$$
$$\hat{c}_{33} = \bar{c}_{33}/k_1^2 c_{66}, \qquad \hat{c}_{34} = k_3\bar{c}_{34}/k_1^2 c_{66}$$
$$\hat{c}_{55} = c_{55}/k_1^2 c_{66}, \qquad \hat{c}_{44} = k_3^2\bar{c}_{44}/k_1^2 c_{66}$$
$$\hat{c}_{56} = c_{56}/k_1 c_{66}, \qquad \hat{\gamma}_{pp} = \gamma_{pp}/3k_1^2 c_{66}$$

The consistency of (11)-(14) requires that the determinants of the coefficients vanish; leading to the four characteristic equations:

$$\hat{\gamma}_{11}(\hat{c}_{55} - \hat{c}_{56}^{\,2})\bar{\xi}^6 - \Omega^2[3\hat{\gamma}_{11}(1 + \hat{c}_{55}) + \hat{c}_{55} - \hat{c}_{56}^{\,2}]\bar{\xi}^4 + 3\Omega^2[\Omega^2 - \hat{c}_{55}(1 - \Omega^2) + 3\hat{\gamma}_{11}\Omega^2 + \hat{c}_{56}^{\,2}]\bar{\xi}^2 + 9\Omega^4(1 - \Omega^2) = 0 \tag{11$_1$}$$

$$(\hat{\gamma}_{55}\bar{\zeta}^2 - \Omega^2)(\hat{c}_{55}\bar{\zeta}^2 - 3\Omega^2 + 3) = 0 \tag{12$_1$}$$

$$\hat{c}_{11}\hat{\gamma}_{55}\bar{\xi}^4 + (\hat{c}_{11}\hat{c}_{44} - \hat{c}_{11}\Omega^2 - \hat{c}_{14}^2 - 3\hat{\gamma}_{55}\Omega^2)\bar{\xi}^2 - 3\Omega^2(\hat{c}_{44} - \Omega^2) = 0 \tag{13$_1$}$$

$$\hat{\gamma}_{33}(\hat{c}_{33}\hat{c}_{44} - \hat{c}_{34}^{\,2})\bar{\zeta}^6 - \Omega^2[3\hat{\gamma}_{33}(\hat{c}_{33} + \hat{c}_{44}) + \hat{c}_{33}\hat{c}_{44} - \hat{c}_{34}^{\,2}]\bar{\zeta}^4 + 3\Omega^2[\hat{c}_{44}\Omega^2 - \hat{c}_{33}(\hat{c}_{44} - \Omega^2) + 3\hat{\gamma}_{33}\Omega^2 + \hat{c}_{34}^{\,2}]\bar{\zeta}^2 + 9\Omega^4(\hat{c}_{44} - \Omega^2) = 0 \tag{14$_1$}$$

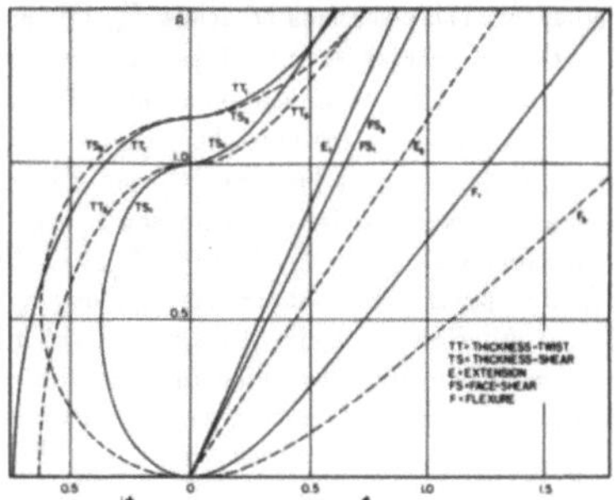

FIGURE 1. THE TEN BRANCHES OF THE CHARACTERISTIC EQUATIONS $(11)_1$–$(14)_1$.

It may be seen, from (11)-(14), that the modes separate into four groups, each containing two or three of the $\bar{u}_i$ and ψ_i, comprising ten modes in all. The relations between the frequencies and wave-numbers of the ten modes are given by the ten branches of the characteristic equations $(11)_1$-$(14)_1$. These are illustrated in Fig. 1 with dimensionless frequency ($\Omega = \omega/\omega_s$) as ordinate and dimensionless wave-number ($\phi = 2\xi h/\pi$ or $\phi = 2\zeta h/\pi$) as abscissa. Inasmuch as boundary conditions have not yet been applied, $(11)_1$-$(14)_1$ and Fig. 1 pertain to vibrations of an infinite plate.

The relative amplitudes $A_2{:}A_3{:}A_4$, $B_1{:}B_4$, $C_1{:}C_5$ and $D_2{:}D_3{:}D_5$ of the contributing displacements and rotations in each of the ten modes, as determined by (11)-(14) and $(11)_1$-$(14)_1$, vary with the frequency or wave-number; but, for convenience, the ten branches in Fig. 1 are identified by symbols (F_1, TS_1, etc.) representing the predominant deformations at long wave-lengths, as depicted in Fig. 2.

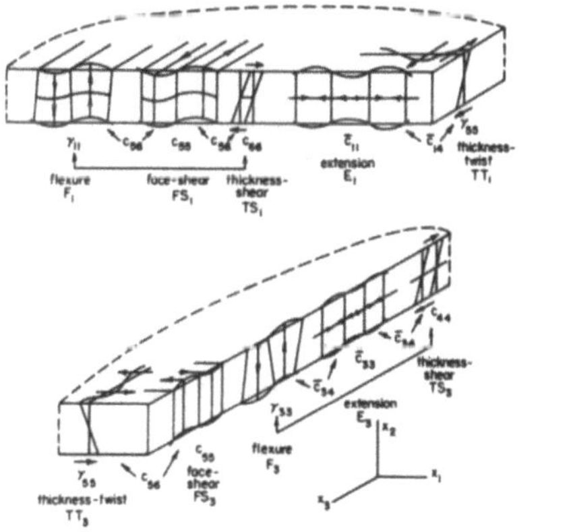

FIGURE 2. PREDOMINANT MODE-SHAPES OF SOLUTIONS A,B,C,D AT LONG WAVE-LENGTHS IN AN INFINITE PLATE.

Thus, group A, (11), comprises branches F_1, FS_1 and TS_1, i.e., flexure, face-shear and thickness-shear varying along x_1; group B, (12), comprises branches TT_3 and FS_3, i.e., thickness-twist and face-shear varying along x_3; group C, (13), comprises branches E_1 and TT_1, i.e., extension and thickness-twist varying along x_1; and group D, (14), comprises branches F_3, E_3 and TS_3, i.e., flexure, extension and thickness-shear varying along x_3.

A second criterion for choice of the modes which will be excited most strongly is based on an inspection of the piezoelectric relations to determine the strains which are forced by the application of the component (E_2) of the electric field normal to the plate. In the three-dimensional case the strains S_p are given in terms of the stresses T_p and the electric field E_i, for the AT-cut of quartz, by

$$
\begin{aligned}
S_1 &= s_{11}T_1 + s_{12}T_2 + s_{13}T_3 + s_{14}T_4 + d_{11}E_1 \\
S_2 &= s_{21}T_1 + s_{22}T_2 + s_{23}T_3 + s_{24}T_4 + d_{12}E_1 \\
S_3 &= s_{31}T_1 + s_{32}T_2 + s_{33}T_3 + s_{34}T_4 + d_{13}E_1 \\
S_4 &= s_{41}T_1 + s_{42}T_2 + s_{43}T_3 + s_{44}T_4 + d_{14}E_1 \\
S_5 &= s_{55}T_5 + s_{56}T_6 + d_{25}E_2 + d_{35}E_3 \\
S_6 &= s_{65}T_5 + s_{66}T_6 + d_{26}E_2 + d_{36}E_3
\end{aligned}
\tag{15}
$$

where s_{pq} are the elastic compliances and the d_{ip} are piezoelectric constants. It may be seen that E_2 forces only S_5 and S_6 directly. In the two-dimensional equations, S_5 is represented by $\partial\bar{u}_1/\partial x_3 + \partial\bar{u}_3/\partial x_1$ and S_6 is represented by $\psi_1 + \partial\bar{u}_2/\partial x_1$. These involve only groups A and B, as may be seen from (10). Hence, we conclude that the strong resonances are those composed of the solutions A and B.

Solutions A and B

For each Ω, $(11)_1$ has three roots $\bar{\xi}^2$ and (11) has three sets of amplitude ratios $A_2 : A_3 : A_4$. Let $\bar{A}_i$, $i = 1,2,3$, be the value of A_4 for the ith root $\bar{\xi}_i$; and let

$$A_2/A_4 = \alpha_{2i} \quad , \quad A_3/A_4 = \alpha_{3i}$$

for each root $\bar{\xi}_i$. Then, for solution A, (10) may be written as

$$
\begin{aligned}
\bar{u}_2 &= h(\bar{A}_1\alpha_{21}\sin\xi_1x_1 + \bar{A}_2\alpha_{22}\sin\xi_2x_1 + \bar{A}_3\alpha_{23}\sin\xi_3x_1) \\
\bar{u}_3 &= h(\bar{A}_1\alpha_{31}\sin\xi_1x_1 + \bar{A}_2\alpha_{32}\sin\xi_2x_1 + \bar{A}_3\alpha_{33}\sin\xi_3x_1) \\
\psi_1 &= \bar{A}_1\cos\xi_1x_1 + \bar{A}_2\cos\xi_2x_1 + \bar{A}_3\cos\xi_3x_1 \\
\bar{u}_1 &= \psi_3 = 0
\end{aligned}
\tag{16}
$$

We ignore the boundary conditions on $x_3 = \pm w$ and apply only conditions (7). Upon substituting (16) into (6) and the result into (7) we obtain

$$
\begin{aligned}
\bar{A}_1\bar{\alpha}_{11}\cos\bar{\xi}_1\bar{l} + \bar{A}_2\bar{\alpha}_{12}\cos\bar{\xi}_2\bar{l} + \bar{A}_3\bar{\alpha}_{13}\cos\bar{\xi}_3\bar{l} &= 0 \\
\bar{A}_1\bar{\alpha}_{21}\cos\bar{\xi}_1\bar{l} + \bar{A}_2\bar{\alpha}_{22}\cos\bar{\xi}_2\bar{l} + \bar{A}_3\bar{\alpha}_{23}\cos\bar{\xi}_3\bar{l} &= 0 \\
\bar{A}_1\bar{\xi}_1\sin\bar{\xi}_1\bar{l} + \bar{A}_2\bar{\xi}_2\sin\bar{\xi}_2\bar{l} + \bar{A}_3\bar{\xi}_3\sin\bar{\xi}_3\bar{l} &= 0
\end{aligned}
\tag{17}
$$

where $\bar{l} = l/h$ and

$$
\begin{aligned}
\bar{\alpha}_{1i} &= c_{55}\alpha_{3i}\bar{\xi}_i + k_1c_{56}(\alpha_{2i}\bar{\xi}_i + 1) \\
\bar{\alpha}_{2i} &= c_{56}\alpha_{3i}\bar{\xi}_i + k_1c_{66}(\alpha_{2i}\bar{\xi} + 1)
\end{aligned}
\tag{18}
$$

The frequency equation is obtained by setting the determinant of the coefficients of the $\bar{A}_i$ in (17) equal to zero, with the result

$$\tilde{A}_1\tan\bar{\xi}_1\bar{l} + \tilde{A}_2\tan\bar{\xi}_2\bar{l} + \tilde{A}_3\tan\bar{\xi}_3\bar{l} = 0 \tag{19}$$

where

$$
\begin{aligned}
\tilde{A}_1 &= \bar{\xi}_1(\bar{\alpha}_{12}\bar{\alpha}_{23} - \bar{\alpha}_{22}\bar{\alpha}_{13}) \\
\tilde{A}_2 &= \bar{\xi}_2(\bar{\alpha}_{13}\bar{\alpha}_{21} - \bar{\alpha}_{23}\bar{\alpha}_{11}) \\
\tilde{A}_3 &= \bar{\xi}_3(\bar{\alpha}_{11}\bar{\alpha}_{22} - \bar{\alpha}_{21}\bar{\alpha}_{12})
\end{aligned}
\tag{20}
$$

The roots of (19) give the relation between the dimensionless frequency, Ω, and the ratio, l/h, of length to thickness of the plate, for solution A.

For solution B, we note that $(12)_1$ has the two roots

$$\bar{\zeta}_1^2 = \Omega^2/\hat{\gamma}_{55} \quad , \quad \bar{\zeta}_2^2 = 3(\Omega^2-1)/\hat{c}_{55} \tag{21}$$

For each of these we define, from (12),

$$\beta_i = B_1/B_4 = -\hat{c}_{56}\zeta_i/(\hat{c}_{55}\bar{\zeta}_i^2 - 3\Omega^2) = -(\hat{\gamma}_{55}\bar{\zeta}_i^2 + 1 - \Omega^2)/\hat{c}_{56}$$

Then, for solution B, (10) may be written as

$$
\begin{aligned}
\bar{u}_1 &= h(\bar{B}_1\beta_1\sin\zeta_1x_3 + \bar{B}_2\beta_2\sin\zeta_2x_3) \\
\psi_1 &= \bar{B}_1\cos\zeta_1x_3 + \bar{B}_2\cos\zeta_2x_3 \\
\bar{u}_2 &= \bar{u}_3 = \psi_2 = 0
\end{aligned}
\tag{22}
$$

In this case we ignore boundary conditions (7) and apply only (8). Thus, inserting (22) in (6), the result in (8) and eliminating the constants $\bar{B}_i$, we obtain the frequency equation

$$3\bar{\zeta}_2\tan\bar{\zeta}_2\bar{w} + \hat{c}_{56}^2\,\bar{\zeta}_1^3\tan\bar{\zeta}_1\bar{w} = 0 \tag{23}$$

where $\bar{\zeta}_i = \zeta_i h$ and $\bar{w} = w/h$. From (23) and (21) we note that the former gives the relation between the dimensionless frequency, Ω, and the ratio, w/h, of width to thickness of the plate, for solution B.

Computation of Frequencies

In order to calculate the roots of (19) and (23) it is necessary to choose, first, values of the elastic stiffnesses. There are available, in the literature, numerous values differing, in many cases, by several percent. An error of one percent might be amplified to about twenty-five percent in computing the frequency of the fiftieth mode. Since, at high frequencies, neighboring modes can differ in frequency by only a fraction of a percent, no significant comparison could be made between measured frequencies and those computed on the basis of inaccurate values of stiffnesses. These inaccuracies are no doubt due, in part, to the approximate nature of the simple formulas employed in the past to convert measured dimensions, density and frequencies to stiffnesses. The most precise formulas available for this purpose are, in fact, (19) and (23) themselves. Although they do not include the piezoelectric and thermoelastic terms, they do account for mechanical coupling. Accordingly, a set of stiffnesses was computed for quartz using the Koga-Fukuyo data and (19) and (23) along with the usual formulas [4] for "rotated" stiffnesses, c_{pq}, in terms of the principal ones, $\tilde{c}_{pq}$. The results are listed in Table I. They coincide most closely with the stiffnesses given by Bechmann [5] for both constant electric field and entropy ($\tilde{c}^{E\sigma}_{pq}$) and constant electric displacement and entropy ($\tilde{c}^{D\sigma}_{pq}$), which are also listed in Table I. The values in the last column of Table I are intended only for use in the present paper. Aside from being based on only one set of experimental data, they do not include the piezoelectric and thermoelastic effects.

The measurements by Koga and Fukuyo were made on a single plate whose thickness and width were kept constant while the length was cut down in fifty-nine steps, in the range $33.79 \lesssim l/h \lesssim 42.35$, with resonances measured in the neighborhood of the fundamental thickness-shear frequency ($\Omega = 1 \pm$ about 25 percent) at each step. The dots in Fig. 3 are plotted from data kindly supplied by Dr. Koga through the courtesy of Dr. A.L. Bennett of the Georgia Institute of Technology.

TABLE I
STIFFNESSES OF QUARTZ

Bechmann		This paper
$\tilde{c}^{E\sigma}_{11} = 86.74$	$\tilde{c}^{D\sigma}_{11} = 87.49$	$\tilde{c}_{11} = 86.75$
$\tilde{c}^{E\sigma}_{33} = 107.2$	$\tilde{c}^{D\sigma}_{33} = 107.2$	$\tilde{c}_{33} = 107.2$
$\tilde{c}^{E\sigma}_{13} = 11.91$	$\tilde{c}^{D\sigma}_{13} = 11.91$	$\tilde{c}_{13} = 11.91$
$\tilde{c}^{E\sigma}_{44} = 57.94$	$\tilde{c}^{D\sigma}_{44} = 57.98$	$\tilde{c}_{44} = 57.8$
$\tilde{c}^{E\sigma}_{66} = 39.88$	$\tilde{c}^{D\sigma}_{66} = 40.63$	$\tilde{c}_{66} = 40.4$
$\tilde{c}^{E\sigma}_{14} = -17.91$	$\tilde{c}^{D\sigma}_{14} = -18.09$	$\tilde{c}_{14} = -17.8$

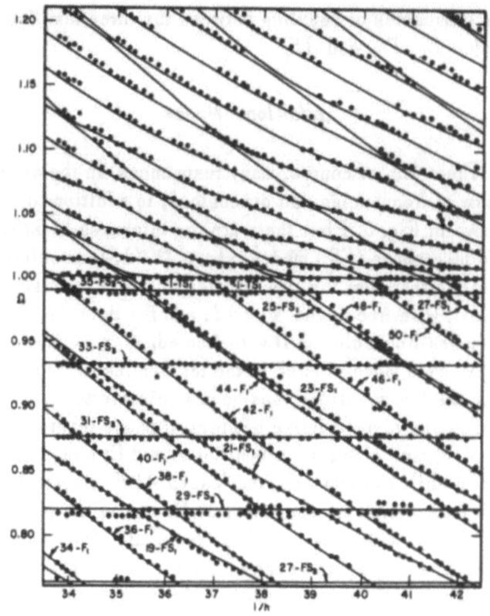

FIGURE 3: DIMENSIONLESS FREQUENCY VS. LENGTH/THICKNESS FOR RECTANGULAR, AT-CUT, QUARTZ PLATES OF CONSTANT WIDTH. THE DOTS ARE EXPERIMENTAL DATA BY KOGA AND FUKUYO: THE CURVES ARE COMPUTED FROM (19) AND (23) OF THIS PAPER.

Since the width and thickness are constant ($\bar{w} = 54.017$) and $\hat{c}_{56}$ is very small, the computation of the roots of (23) is very simple and was made on a desk calculator. In fact, except where $\bar{\zeta}^2$ is near zero (i.e., $\Omega \approx 1$) the roots of (23) are given very closely by

$$\bar{\zeta}_1 \bar{w} = m\pi/2, \quad m \text{ odd}, \quad \Omega > 0, \quad \text{for modes } FS_3$$

and

$$\bar{\zeta}_2 \bar{w} = m\pi/2, \quad m \text{ even}, \quad \Omega > 1, \quad \text{for modes } TT_3 .$$

No experimental data is given for these modes above $\Omega = 1$ and the modes TT_3 do not exist below $\Omega = 1$. Hence the only curves of this group shown in Fig. 3 are those for modes FS_3 below $\Omega = 1$ and these are horizontal lines since $\bar{w}$ is constant. The designation 29-FS_3 indicates the twenty-ninth mode of this type.

The remaining curves in Fig. 3 were computed from (19) on an IBM-704 digital computer at the General Motors Research Laboratories. It may be seen that the Koga-Fukuyo data are matched in great detail, including all the intricacies of coupling between the three families of modes F_1, FS_1 and TS_1.

Additional Modes

There are, of course, many resonances, in the same range of frequencies and dimensions, in addition to those shown in Fig. 3. First there are the thickness-twist (TT_3) and face-shear (FS_3) modes, above $\Omega = 1$, from solution B. Then there are the two families of modes from solution C, with free edges $x_1 = \pm l$; and the three families of modes from solution D with free edges $x_3 = \pm w$. Finally, each mode in each of the ten familes of modes, from the four solutions, has anharmonic overtones: corresponding to successive numbers of phase reversals of displacement along x_3 in families A and C; and along x_1 in B and D. For example, the branch 42-F_1 corresponds to a mode which is predominantly flexure in the plane of x_1 and x_2, with forty-one phase reversals of displacement u_2 along x_1 and none along x_3. This mode has overtones with the same number of reversals along x_1 and 1,2,3... reversals along x_3. Each of the overtones has only a slightly higher frequency than the preceding one. Their branches, if plotted in Fig. 3, would be a sequence of closely spaced curves paralleling the curve 42-F_1. The solutions, including these overtones, cannot be expressed in closed form. However, approximations to them can be obtained by starting with the displacements given in (9). Substitution of (9) in (3) leads to a quintic equation relating ω^2, ξ^2 and ζ^2 in place of the cubics or quadratics $(11)_1$-$(14)_1$. The displacements analogous to (16) and (22) would then be sums of five terms instead of three or two. Solutions in closed form could be obtained with free-edge conditions

$$N_1 = N_5 = Q_1 = M_1 = M_5 = 0$$

on $x_1 = \pm l$ and either

$$\bar{u}_2 = \bar{u}_3 = \psi_1 = N_5 = M_3 = 0$$

or

$$Q_3 = N_3 = M_5 = \bar{u}_1 = \psi_3 = 0$$

on $x_3 = \pm w$. Alternative solutions in closed form could be obtained with

$$\bar{u}_2 = \bar{u}_3 = \psi_3 = N_1 = M_1 = 0$$

or

$$Q_1 = N_5 = M_5 = \bar{u}_1 = \psi_1 = 0$$

on $x_1 = \pm l$ and free-edge conditions

$$N_3 = N_5 = Q_3 = M_3 = M_5 = 0$$

on $x_3 = \pm w$. For any one of the four cases, a frequency equation with five transcendental terms would be obtained, instead of the three or two of (19) and (23).

Finally, there are modes of vibration with more than one node between the faces of the plate. These have frequencies higher than those considered here. As explained in [1], applications of the equations employed in this paper are restricted to the range of frequencies shown in Fig. 1.

References

1. Mindlin, R.D., *Quart. Appl. Math.*, Vol. 19, 51 (1961).

2. Koga, I., and Fukuyo, H., *J. Inst. Elec. Comm. Engrs. of Japan*, Vol. 36, 59 (1953).

3. Mason, W.P., *Piezoelectric Crystals and Their Applications to Ultrasonics*, D. Van Nostrand Company, New York, 1950.

4. Sykes, R.A., *Quartz Crystals for Electric Circuits*, Chapter VI, Appendix C, Heising, R.A., Editor, D. Van Nostrand Company, New York, 1946.

5. Bechmann, R., *Phys. Rev.*, Vol. 100, 1060 (1958).

Reprinted from
Proceedings of the Fourth U.S. National Congress of Applied Mechanics (1962) 305-310

Offprint from "Archive for Rational Mechanics and Analysis", Volume 11, Number 5, 1962, P. 415—448

Springer-Verlag, Berlin · Göttingen · Heidelberg

Effects of Couple-stresses in Linear Elasticity

R. D. MINDLIN & H. F. TIERSTEN

Communicated by C. TRUESDELL

Contents

Introduction

In a theory of deformation of continua, originated by VOIGT [*1*] and amplified by E. & F. COSSERAT [*2*], the couple per unit area, acting across a surface within a material volume or on its boundary, was taken into account in addition to the usual force per unit area. Some typical effects of such "couple-stresses" are exhibited, in the present paper, by means of solutions of problems of wave-propagation, vibration, stress-concentration and nuclei of strain — all within the framework of a linearized form of the couple-stress theory for perfectly elastic, centrosymmetric-isotropic materials.

A modern derivation of the Cosserat equations has been given by TRUESDELL & TOUPIN [*3*]. More recently, TOUPIN [*4*] has derived the associated constitutive equations for finite deformation of perfectly elastic materials. Upon linearization, TOUPIN'S equations become identical with those which are obtained, for example by AERO & KUVSHINSKII [*5*], without first establishing constitutive equations for finite deformation. On the other hand, GRIOLI [*6*] has also obtained constitutive equations for finite deformation — equally correct, though of different form — but, upon linearization he obtains results at variance with TOUPIN'S. The source of the discrepancy is not readily apparent owing to the different procedures employed in the two treatments. In view of this situation, it seems advisable, before proceeding with solutions, to review, briefly, derivations of the equations of both authors, retaining the essentials of their reasoning but following an intermediate procedure. In this way the point of divergence is identified and it is shown how GRIOLI'S results may be brought into conformity with TOUPIN'S.

The discussion of the derivations of the finite and linearized equations occupies the first four sections of the present paper. This is followed, in Section 5, by a theorem of uniqueness of solution of the linear equations. The theorem is based on the assumption of positive definiteness of the internal energy-density. Although no proof of positive definiteness is available, some justification is given on intuitive grounds and by an example, later, of failure of uniqueness when the internal energy-density is not positive-definite. In the course of the formulation of the uniqueness theorem, it appears, at first, that six boundary conditions are required; but it is shown that five are sufficient. The reduction from six to five is analogous to the reduction from three to two conditions in the classical theory of flexure of thin plates.

In the linear theory that takes into account couple-stresses in a centro-symmetric-isotropic, elastic material, there is an additional modulus of elasticity (the ratio of couple-stress to curvature or twist, *i.e.*, a modulus of bending and twisting) with the dimensions of force. The square root of the ratio of the bending-twisting modulus to the usual shear modulus has the dimension of length. This length, l, is a material property which carries with it all of the difference between analogous equations or solutions with and without couple-stresses. The larger l may be, the greater is the difference. Presumably l is small, in comparison with bodily dimensions and wave-lengths normally encountered, as there appears to be no conclusive experimental evidence of its existence. However, even though small, its influence might become important as dimensions of a body or wave-lengths diminish to the order of the length l. The assumption of positive-definiteness of the internal energy-density requires the bending-twisting modulus to be positive. In the contrary case, l^2 would be replaced by its negative, in the equations; and the forms of solutions would be drastically different.

The propagation of plane waves is considered in Section 6. There is a non-dispersive dilatational wave, as in the usual theory, propagating at the usual velocity; but there are two rotational waves: one propagating and the other non-propagating. Both rotational waves are dispersive. The group velocity of the propagating rotational wave increases with increasing wave number and increasing l. The non-propagating wave produces a boundary layer effect.

In Section 7, the displacement equations of motion are reduced to equations corresponding to Lamé's. The equation on the potential of the dilatation is the usual one; but, in the equation on the potential of the rotation, Laplace's operator ∇^2 is replaced by $\nabla^2 - l^2\nabla^4$. In the case of steady vibration, the latter equation is separated into two Helmholtz equations.

The solution of the problem of antisymmetric, thickness-shear vibrations of an infinite plate is given in Section 8. The natural frequencies exceed the integral multiples of the fundamental which are found without couple-stresses. Since the excess increases with the order of the mode and with the magnitude of the material constant l, a laboratory experiment aimed at the measurement of l might be based on a solution of this type. The lowest mode is affected very little by the couple-stress, regardless of the value of l. However, this is due to an accident of mode shape in the case of thickness-shear vibrations. It does not occur, for example, in the solution of the problem of torsional vibrations of a cylinder, which is given in Section 9.

In preparation for solutions of problems of equilibrium involving stress concentrations and singularities, GREEN's formulas are derived, in Section 10, for functions whose governing equations contain the operators $1-l^2\nabla^2$ and $\nabla^2-l^2\nabla^4$. Following this, in Section 11, stress functions analogous to those of PAPKOVITCH and GALERKIN are introduced.

Solutions of the problems of stresses around spherical and cylindrical cavities in an infinite body under simple tension are given in Sections 12 and 14. Unlike in the classical solutions, the maximum tension at the surface depends on the radius of the cavity — appearing in the form of the ratio of the radius to the material constant l. In the limiting cases of vanishingly small cavities, the maximum tensions at the surfaces are found to be thirty to forty percent less than they are if couple-stresses are neglected.

The problem of the spherical cavity in a field of simple tension is considered again in Section 13; but this time with the bending-twisting modulus negative. Two solutions are found: exhibiting failure of uniqueness with non-positive internal energy-density.

In the last two sections, the singularities appropriate to the concentrated force, the concentrated couple and a variety of nuclei of strain are established. The center of dilatation is the same as it is without couple-stresses and the center of rotation is identical with the concentrated body-couple.

1. The Cosserat equations

Let S' be a surface which separates a portion of a material volume, V, from the remainder. At each point of S', let $\boldsymbol{n}$ be the unit vector normal to S' in a specified sense. The material on the side of S' toward which $\boldsymbol{n}$ is directed exerts loads, on the material on the other side, which are assumed to consist, at each point of S', of a force per unit area, $\boldsymbol{t_n}$, and a couple per unit area, $\boldsymbol{m_n}$. Also, at each point in V, let $\boldsymbol{f}$ be an extrinsic force per unit mass and $\boldsymbol{c}$ be an extrinsic couple per unit mass. The force-stress vector $\boldsymbol{t_n}$ and the body-force vector $\boldsymbol{f}$ are polar vectors; whereas the couple-stress vector $\boldsymbol{m_n}$ and the body-couple vector $\boldsymbol{c}$ are axial vectors. Axial vectors are taken as positive in the direction of advance of a right handed screw.

We now consider the motion of a portion V, of a material volume, bounded by a surface S with outward normal $\boldsymbol{n}$. Across S there act force-stress and couple-stress vectors, $\boldsymbol{t_n}$ and $\boldsymbol{m_n}$, and within V there act body-force and body-couple vectors $\boldsymbol{f}$ and $\boldsymbol{c}$. The equations of conservation of mass, balance of momentum and moment of momentum and conservation of mechanical energy are taken to be, respectively,

$$\frac{d}{dt}\int_V \varrho\, dV = 0, \tag{1.1}$$

$$\frac{d}{dt}\int_V \boldsymbol{v}\varrho\, dV = \int_S \boldsymbol{t_n}\, dS + \int_V \boldsymbol{f}\varrho\, dV, \tag{1.2}$$

$$\frac{d}{dt}\int_V \boldsymbol{r}\times\boldsymbol{v}\varrho\, dV = \int_S (\boldsymbol{r}\times\boldsymbol{t_n}+\boldsymbol{m_n})\, dS + \int_V (\boldsymbol{r}\times\boldsymbol{f}+\boldsymbol{c})\,\varrho\, dV, \tag{1.3}$$

$$\begin{aligned}\frac{d}{dt}\int_V \left(\tfrac{1}{2}\boldsymbol{v}\cdot\boldsymbol{v}+U\right)\varrho\, dV = \int_S \left(\boldsymbol{t_n}\cdot\boldsymbol{v}+\tfrac{1}{2}\boldsymbol{m_n}\cdot\boldsymbol{\nabla}\times\boldsymbol{v}\right) dS + \\ + \int_V \left(\boldsymbol{f}\cdot\boldsymbol{v}+\tfrac{1}{2}\boldsymbol{c}\cdot\boldsymbol{\nabla}\times\boldsymbol{v}\right)\varrho\, dV,\end{aligned} \tag{1.4}$$

where d/dt is the material time-derivative, ϱ is the mass density, $\boldsymbol{r}$ is the spatial position vector from a fixed origin, $\boldsymbol{v}$ is the material velocity $d\boldsymbol{r}/dt$, U is the internal energy per unit mass, $\boldsymbol{\nabla}$ is the spatial gradient $\partial/\partial\boldsymbol{r}$, and $\frac{1}{2}\boldsymbol{\nabla}\times\boldsymbol{v}$ is the vorticity.

Consideration of the equilibrium of forces acting on an elementary tetrahedron, as the volume of the tetrahedron shrinks to zero, leads to the definition of the usual force-stress dyadic* $\boldsymbol{\tau}$:

$$\boldsymbol{t_n} = \boldsymbol{n}\cdot\boldsymbol{\tau}. \tag{1.5}$$

An analogous treatment of the equilibrium of moments acting on the tetrahedron yields the definition of the couple-stress dyadic $\boldsymbol{\mu}$:

$$\boldsymbol{m_n} = \boldsymbol{n}\cdot\boldsymbol{\mu}. \tag{1.6}$$

Now, in (1.2),

$$\int_S \boldsymbol{t_n}\, dS = \int_S \boldsymbol{n}\cdot\boldsymbol{\tau}\, dS = \int_V \boldsymbol{\nabla}\cdot\boldsymbol{\tau}\, dV, \tag{1.7}$$

by (1.5) and the divergence theorem. With (1.7) and (1.1), (1.2) becomes

$$\int_V (\boldsymbol{\nabla}\cdot\boldsymbol{\tau}+\varrho\boldsymbol{f}-\varrho\dot{\boldsymbol{v}})\, dV = 0,$$

where $\dot{\boldsymbol{v}} = d\boldsymbol{v}/dt$. Hence

$$\boldsymbol{\nabla}\cdot\boldsymbol{\tau}+\varrho\boldsymbol{f} = \varrho\dot{\boldsymbol{v}}, \tag{1.8}$$

which is the usual force-stress equation of motion.

In (1.3),

$$\frac{d}{dt}\int_V \boldsymbol{r}\times\boldsymbol{v}\varrho\, dV = \int_V \frac{d(\boldsymbol{r}\times\boldsymbol{v})}{dt}\varrho\, dV = \int_V (\boldsymbol{v}\times\boldsymbol{v}+\boldsymbol{r}\times\dot{\boldsymbol{v}})\,\varrho\, dV = \int_V \boldsymbol{r}\times\dot{\boldsymbol{v}}\varrho\, dV,$$

$$\begin{aligned}\int_S \boldsymbol{r}\times\boldsymbol{t_n}\, dS &= \int_S \boldsymbol{r}\times(\boldsymbol{n}\cdot\boldsymbol{\tau})\, dS = -\int_S \boldsymbol{n}\cdot\boldsymbol{\tau}\times\boldsymbol{r}\, dS \\ &= -\int_V \boldsymbol{\nabla}\cdot(\boldsymbol{\tau}\times\boldsymbol{r})\, dV = \int_V [\boldsymbol{r}\times(\boldsymbol{\nabla}\cdot\boldsymbol{\tau})+\boldsymbol{\tau}\underset{\cdot}{\times}\mathbf{I}]\, dV,\end{aligned}$$

$$\int_S \boldsymbol{m_n}\, dS = \int_S \boldsymbol{n}\cdot\boldsymbol{\mu}\, dS = \int_V \boldsymbol{\nabla}\cdot\boldsymbol{\mu}\, dV,$$

* See WEATHERBURN [7] or WILSON [8] for the dyadic notation.

where $\mathbf{I}$ $(\equiv \nabla \boldsymbol{r})$ is the unit spatial dyadic. Hence (1.3) becomes

$$\int_V \boldsymbol{r} \times (\nabla \cdot \boldsymbol{\tau} + \varrho \boldsymbol{f} - \varrho \dot{\boldsymbol{v}})\, dV + \int_V (\nabla \cdot \boldsymbol{\mu} + \varrho \boldsymbol{c} + \boldsymbol{\tau} \overset{\times}{\cdot} \mathbf{I})\, dV = 0;$$

from which, with (1.8), we have the couple-stress equation of motion:

$$\nabla \cdot \boldsymbol{\mu} + \varrho \boldsymbol{c} + \boldsymbol{\tau} \overset{\times}{\cdot} \mathbf{I} = 0. \tag{1.9}$$

The antisymmetric part of $\boldsymbol{\tau}$ is $\boldsymbol{\tau}^A = -\frac{1}{2} \mathbf{I} \times (\boldsymbol{\tau} \overset{\times}{\cdot} \mathbf{I})$. Hence

$$\boldsymbol{\tau}^A = \tfrac{1}{2} \mathbf{I} \times (\nabla \cdot \boldsymbol{\mu} + \varrho \boldsymbol{c}). \tag{1.10}$$

If the couple-stress $\boldsymbol{\mu}$ and the body couple $\boldsymbol{c}$ were absent, $\boldsymbol{\tau}^A$ would vanish and $\boldsymbol{\tau}$ would be a symmetric dyadic.

Writing

$$\boldsymbol{\tau} = \boldsymbol{\tau}^S + \boldsymbol{\tau}^A, \tag{1.11}$$

where $\boldsymbol{\tau}^S$ is the symmetric part of $\boldsymbol{\tau}$, substituting (1.10) in (1.11) and the latter in (1.8), we find an alternative form of the equation of motion:

$$\nabla \cdot \boldsymbol{\tau}^S + \tfrac{1}{2} \nabla \times \nabla \cdot \boldsymbol{\mu} + \varrho \boldsymbol{f} + \tfrac{1}{2} \nabla \times \varrho \boldsymbol{c} = \varrho \dot{\boldsymbol{v}}. \tag{1.12}$$

In (1.4), with $\dot{U} = dU/dt$,

$$\frac{d}{dt} \int_V \left(\frac{1}{2} \boldsymbol{v} \cdot \boldsymbol{v} + U \right) \varrho\, dV = \int_V (\boldsymbol{v} \cdot \dot{\boldsymbol{v}} + \dot{U})\, \varrho\, dV,$$

$$\int_S \boldsymbol{t}_n \cdot \boldsymbol{v}\, dS = \int_S \boldsymbol{n} \cdot \boldsymbol{\tau} \cdot \boldsymbol{v}\, dS = \int_V \nabla \cdot (\boldsymbol{\tau} \cdot \boldsymbol{v})\, dV = \int_V [\boldsymbol{\tau} : \nabla \boldsymbol{v} + (\nabla \cdot \boldsymbol{\tau}) \cdot \boldsymbol{v}]\, dV,$$

$$\begin{aligned} \int_S \boldsymbol{m}_n \cdot \nabla \times \boldsymbol{v}\, dS &= \int_S \boldsymbol{n} \cdot \boldsymbol{\mu} \cdot \nabla \times \boldsymbol{v}\, dS = \int_V \nabla \cdot (\boldsymbol{\mu} \cdot \nabla \times \boldsymbol{v})\, dV \\ &= \int_V [\boldsymbol{\mu} : \nabla \nabla \times \boldsymbol{v} + (\nabla \cdot \boldsymbol{\mu}) \cdot \nabla \times \boldsymbol{v}]\, dV. \end{aligned}$$

Hence, (1.4) may be written as

$$\int_V \varrho \dot{U}\, dV = \int_V [(\nabla \cdot \boldsymbol{\tau} + \varrho \boldsymbol{f} - \varrho \dot{\boldsymbol{v}}) \cdot \boldsymbol{v} + \tfrac{1}{2} (\nabla \cdot \boldsymbol{\mu} + \varrho \boldsymbol{c}) \cdot \nabla \times \boldsymbol{v} + \boldsymbol{\tau} : \nabla \boldsymbol{v} + \tfrac{1}{2} \boldsymbol{\mu} : \nabla \nabla \times \boldsymbol{v}]\, dV,$$

which, by (1.8) and (1.9), becomes

$$\int_V \varrho \dot{U}\, dV = \int_V [\boldsymbol{\tau} : \nabla \boldsymbol{v} - \tfrac{1}{2} (\boldsymbol{\tau} \overset{\times}{\cdot} \mathbf{I}) \cdot \nabla \times \boldsymbol{v} + \tfrac{1}{2} \boldsymbol{\mu} : \nabla \nabla \times \boldsymbol{v}]\, dV.$$

Now, $\frac{1}{2} (\boldsymbol{\tau} \overset{\times}{\cdot} \mathbf{I}) \cdot \nabla \times \boldsymbol{v} = \boldsymbol{\tau}^A : \nabla \boldsymbol{v}$. Hence

$$\int_V \varrho \dot{U}\, dV = \int_V (\boldsymbol{\tau}^S : \nabla \boldsymbol{v} + \tfrac{1}{2} \boldsymbol{\mu} : \nabla \nabla \times \boldsymbol{v})\, dV;$$

or

$$\varrho \dot{U} = \boldsymbol{\tau}^S : \nabla \boldsymbol{v} + \tfrac{1}{2} \boldsymbol{\mu} : \nabla \nabla \times \boldsymbol{v}. \tag{1.13}$$

Equations (1.8), (1.9), (1.12) and (1.13) comprise the couple-stress theory essentially as it was left by the COSSERATS. The four equations may be identified with equations (205.2), (205.10), (205.17) and (241.4), respectively, in the article by TRUESDELL & TOUPIN [3], except for the omisssion, here, of non-mechanical terms.

28*

We note, in (1.13), that the antisymmetric part of the force-stress does not contribute to the internal energy. Also, since the scalar of $\boldsymbol{\nabla\nabla}\times\boldsymbol{v}$ is identically zero ($\mathbf{I}:\boldsymbol{\nabla\nabla}\times\boldsymbol{v}=0$), the scalar of $\boldsymbol{\mu}$ does not contribute to the internal energy. Thus, the right hand side of (1.13) is the sum of $(9-3)+(9-1)=14$ products and (1.13) may be written as

$$\varrho\dot{U}=\boldsymbol{\tau}^S:\boldsymbol{\nabla v}+\tfrac{1}{2}\boldsymbol{\mu}^D:\boldsymbol{\nabla\nabla}\times\boldsymbol{v}, \tag{1.14}$$

where $\boldsymbol{\mu}^D$ is the deviator of $\boldsymbol{\mu}$:

$$\boldsymbol{\mu}^D=\boldsymbol{\mu}-\tfrac{1}{3}\boldsymbol{\mu}:\mathbf{II}. \tag{1.15}$$

Furthermore, since $\boldsymbol{\nabla}\times\boldsymbol{\nabla}\cdot(\mathbf{II}:\boldsymbol{\mu})=0$, the scalar of $\boldsymbol{\mu}$ does not appear in the equation of motion (1.12). That is, (1.12) may be written as

$$\boldsymbol{\nabla}\cdot\boldsymbol{\tau}^S+\tfrac{1}{2}\boldsymbol{\nabla}\times\boldsymbol{\nabla}\cdot\boldsymbol{\mu}^D+\varrho\boldsymbol{f}+\tfrac{1}{2}\boldsymbol{\nabla}\times\varrho\boldsymbol{c}=\varrho\dot{\boldsymbol{v}}. \tag{1.16}$$

Thus, a peculiarity of the Cosserat equations is that the scalar of the couple-stress and, by (1.10), the antisymmetric part of the force-stress are left indeterminate.

2. Toupin's constitutive equations

In formulating constitutive equations, we shall have to employ the material position vector, $\mathring{\boldsymbol{r}}$, the material gradient, $\mathring{\boldsymbol{\nabla}}\equiv\partial/\partial\mathring{\boldsymbol{r}}$, and the material unit dyadic $\mathring{\mathbf{I}}\equiv\mathring{\boldsymbol{\nabla}}\mathring{\boldsymbol{r}}$. To convert between material and spatial gradients, $\mathring{\boldsymbol{\nabla}}$ and $\boldsymbol{\nabla}$, we note that, if $\boldsymbol{\Phi}$ is any polyadic,

$$\mathring{\boldsymbol{\nabla}}\boldsymbol{\Phi}=\mathring{\boldsymbol{\nabla}}\boldsymbol{r}\cdot\boldsymbol{\nabla\Phi},\quad \boldsymbol{\Phi}\mathring{\boldsymbol{\nabla}}=\boldsymbol{\Phi\nabla}\cdot\boldsymbol{r}\mathring{\boldsymbol{\nabla}},\quad \boldsymbol{\nabla\Phi}=\boldsymbol{\nabla}\mathring{\boldsymbol{r}}\cdot\mathring{\boldsymbol{\nabla}}\boldsymbol{\Phi},\quad \boldsymbol{\Phi\nabla}=\boldsymbol{\Phi}\mathring{\boldsymbol{\nabla}}\cdot\mathring{\boldsymbol{r}}\boldsymbol{\nabla}. \tag{2.1}$$

In particular, if $\boldsymbol{\Phi}$ is $\mathring{\boldsymbol{r}}$ or $\boldsymbol{r}$, we have

$$\mathring{\boldsymbol{\nabla}}\boldsymbol{r}\cdot\boldsymbol{\nabla}\mathring{\boldsymbol{r}}=\mathring{\boldsymbol{r}}\boldsymbol{\nabla}\cdot\boldsymbol{r}\mathring{\boldsymbol{\nabla}}=\mathring{\mathbf{I}},\qquad \boldsymbol{\nabla}\mathring{\boldsymbol{r}}\cdot\mathring{\boldsymbol{\nabla}}\boldsymbol{r}=\boldsymbol{r}\mathring{\boldsymbol{\nabla}}\cdot\mathring{\boldsymbol{r}}\boldsymbol{\nabla}=\mathbf{I}. \tag{2.2}$$

When couple-stresses are not taken into account, the specific energy of an elastic material may be expressed as a function of the material strain-dyadic

$$\mathbf{E}\equiv\tfrac{1}{2}(\mathring{\boldsymbol{\nabla}}\boldsymbol{r}\cdot\boldsymbol{r}\mathring{\boldsymbol{\nabla}}-\mathring{\mathbf{I}}),\qquad \mathbf{E}=\mathbf{E}_C, \tag{2.3}$$

where $\mathbf{E}_C$ designates the conjugate of $\mathbf{E}$. R. A. TOUPIN [4] has shown that, when couple-stresses are considered, an appropriate second variable is

$$\mathbf{K}\equiv-\mathbf{E}\times\mathring{\boldsymbol{\nabla}}=\tfrac{1}{2}\mathring{\boldsymbol{\nabla}}\mathring{\boldsymbol{\nabla}}\boldsymbol{r}\overset{\times}{\cdot}\mathring{\boldsymbol{\nabla}}\boldsymbol{r},\qquad \mathring{\mathbf{I}}:\mathbf{K}=0. \tag{2.4}$$

$\mathbf{E}$ and $\mathbf{K}$ are "appropriate" as they have the correct number, $6+8=14$, of independent components; $U(\mathbf{E},\mathbf{K})$ is invariant in a rigid rotation of the deformed body, since $\boldsymbol{r}$ appears only as $\boldsymbol{r}\cdot\boldsymbol{r}$ in (2.3) and (2.4); and the right hand side of (1.13) may be expressed as a sum of terms with $\dot{\mathbf{E}}$ and $\dot{\mathbf{K}}$ as double scalar factors, as will be shown.

Thus, we assume

$$U=U(\mathbf{E},\mathbf{K}), \tag{2.5}$$

whence

$$\varrho\dot{U}=\varrho\frac{\partial U}{\partial\mathbf{E}}:\dot{\mathbf{E}}+\varrho\frac{\partial U}{\partial\mathbf{K}}:\dot{\mathbf{K}},\qquad \mathring{\mathbf{I}}:\dot{\mathbf{K}}=0, \tag{2.6}$$

where it is understood that $\partial U/\partial\mathbf{E}=\partial U/\partial\mathbf{E}_C$, $\mathring{\mathbf{I}}:\partial U/\partial\mathbf{K}=0$.

We have now to put the energy equation

$$\varrho \dot{U} = \boldsymbol{\tau}^S : \boldsymbol{\nabla}\boldsymbol{v} + \tfrac{1}{2}\boldsymbol{\mu}^D : \boldsymbol{\nabla}\boldsymbol{\nabla}\times\boldsymbol{v}, \tag{2.7}$$

into the form $\varrho\dot{U} = \boldsymbol{\Phi}:\dot{\mathbf{E}} + \boldsymbol{\Psi}:\dot{\mathbf{K}}$, where $\boldsymbol{\Phi}$ and $\boldsymbol{\Psi}$ are dyadics to be determined, so that we may equate coefficients of $\dot{\mathbf{E}}$ and of $\dot{\mathbf{K}}$ in (2.6) and (2.7).

First we have, from (2.3) and (2.1),

$$\dot{\mathbf{E}} = \tfrac{1}{2}(\mathring{\boldsymbol{\nabla}}\boldsymbol{r}\cdot\boldsymbol{v}\mathring{\boldsymbol{\nabla}} + \mathring{\boldsymbol{\nabla}}\boldsymbol{v}\cdot\boldsymbol{r}\mathring{\boldsymbol{\nabla}}) = \tfrac{1}{2}(\mathring{\boldsymbol{\nabla}}\boldsymbol{r}\cdot\boldsymbol{v}\boldsymbol{\nabla}\cdot\boldsymbol{r}\mathring{\boldsymbol{\nabla}} + \mathring{\boldsymbol{\nabla}}\boldsymbol{r}\cdot\boldsymbol{\nabla}\boldsymbol{v}\cdot\boldsymbol{r}\mathring{\boldsymbol{\nabla}}),$$

whence

$$\dot{\mathbf{E}} = \mathring{\boldsymbol{\nabla}}\boldsymbol{r}\cdot(\boldsymbol{\nabla}\boldsymbol{v})^S\cdot\boldsymbol{r}\mathring{\boldsymbol{\nabla}}, \tag{2.8}$$

where $(\boldsymbol{\nabla}\boldsymbol{v})^S = \tfrac{1}{2}(\boldsymbol{\nabla}\boldsymbol{v} + \boldsymbol{v}\boldsymbol{\nabla})$, *i.e.*, the symmetric part of $\boldsymbol{\nabla}\boldsymbol{v}$. Then, from (2.8) and (2.2),

$$(\boldsymbol{\nabla}\boldsymbol{v})^S = \boldsymbol{\nabla}\mathring{\boldsymbol{r}}\cdot\dot{\mathbf{E}}\cdot\mathring{\boldsymbol{r}}\boldsymbol{\nabla}.$$

Hence

$$\boldsymbol{\tau}^S : \boldsymbol{\nabla}\boldsymbol{v} = \boldsymbol{\tau}^S : (\boldsymbol{\nabla}\boldsymbol{v})^S = \boldsymbol{\tau}^S : (\boldsymbol{\nabla}\mathring{\boldsymbol{r}}\cdot\dot{\mathbf{E}}\cdot\mathring{\boldsymbol{r}}\boldsymbol{\nabla}) = (\mathring{\boldsymbol{r}}\boldsymbol{\nabla}\cdot\boldsymbol{\tau}^S\cdot\boldsymbol{\nabla}\mathring{\boldsymbol{r}}):\dot{\mathbf{E}}, \tag{2.9}$$

which last has the required form.

Second, we have

$$\begin{aligned}\dot{\mathbf{K}} &= -\dot{\mathbf{E}}\times\mathring{\boldsymbol{\nabla}} = -[\mathring{\boldsymbol{\nabla}}\boldsymbol{r}\cdot(\boldsymbol{\nabla}\boldsymbol{v})^S\cdot\boldsymbol{r}\mathring{\boldsymbol{\nabla}}]\times\mathring{\boldsymbol{\nabla}},\\ &= \mathring{\boldsymbol{\nabla}}\boldsymbol{r}\cdot[(\boldsymbol{\nabla}\boldsymbol{v})^S\mathring{\boldsymbol{\nabla}}]\dot{\times}\boldsymbol{r}\mathring{\boldsymbol{\nabla}} + \mathring{\boldsymbol{\nabla}}\mathring{\boldsymbol{\nabla}}\boldsymbol{r}\underset{\cdot}{\times}[\mathring{\boldsymbol{\nabla}}\boldsymbol{r}\cdot(\boldsymbol{\nabla}\boldsymbol{v})^S],\\ &= \mathring{\boldsymbol{\nabla}}\boldsymbol{r}\cdot[(\boldsymbol{\nabla}\boldsymbol{v})^S\mathring{\boldsymbol{\nabla}}]\dot{\times}\boldsymbol{r}\mathring{\boldsymbol{\nabla}} + \mathring{\boldsymbol{\nabla}}\mathring{\boldsymbol{\nabla}}\boldsymbol{r}\underset{\cdot}{\times}[\mathring{\boldsymbol{\nabla}}\boldsymbol{r}\cdot\boldsymbol{\nabla}\mathring{\boldsymbol{r}}\cdot\dot{\mathbf{E}}\cdot\mathring{\boldsymbol{r}}\boldsymbol{\nabla}],\\ &= \mathring{\boldsymbol{\nabla}}\boldsymbol{r}\cdot[(\boldsymbol{\nabla}\boldsymbol{v})^S\mathring{\boldsymbol{\nabla}}]\dot{\times}\boldsymbol{r}\mathring{\boldsymbol{\nabla}} + \mathring{\boldsymbol{\nabla}}\mathring{\boldsymbol{\nabla}}\boldsymbol{r}\underset{\cdot}{\times}(\dot{\mathbf{E}}\cdot\mathring{\boldsymbol{r}}\boldsymbol{\nabla}).\end{aligned} \tag{2.10}$$

Then, noting that [8, p. 317]

$$\mathring{\boldsymbol{\nabla}}\boldsymbol{r} = (\mathring{\boldsymbol{r}}\boldsymbol{\nabla})_C^{-1} = \tfrac{1}{2}(\boldsymbol{\nabla}\mathring{\boldsymbol{r}})_d^{-1}\mathring{\boldsymbol{r}}\boldsymbol{\nabla}\overset{\times}{\times}\mathring{\boldsymbol{r}}\boldsymbol{\nabla}, \tag{2.11}$$

where $(\ )^{-1}$ and $(\ )_d$ denote reciprocal and determinant, we form the product, from (2.10) and (2.11),

$$\begin{aligned}\dot{\mathbf{K}}\cdot(\mathring{\boldsymbol{r}}\boldsymbol{\nabla}\mathring{\boldsymbol{r}}\overset{\times}{\times}\boldsymbol{\nabla}) = {} & \{\mathring{\boldsymbol{\nabla}}\boldsymbol{r}\cdot[(\boldsymbol{\nabla}\boldsymbol{v})^S\mathring{\boldsymbol{\nabla}}]\dot{\times}\boldsymbol{r}\mathring{\boldsymbol{\nabla}}\}\cdot(\mathring{\boldsymbol{r}}\boldsymbol{\nabla}\overset{\times}{\times}\mathring{\boldsymbol{r}}\boldsymbol{\nabla}) + \\ & + 2(\boldsymbol{\nabla}\mathring{\boldsymbol{r}})_d[\mathring{\boldsymbol{\nabla}}\mathring{\boldsymbol{\nabla}}\boldsymbol{r}\dot{\times}(\dot{\mathbf{E}}\cdot\mathring{\boldsymbol{r}}\boldsymbol{\nabla})]\cdot\mathring{\boldsymbol{\nabla}}\boldsymbol{r}.\end{aligned} \tag{2.12}$$

The first term of (2.12) contains the factor

$$\begin{aligned}\{[(\boldsymbol{\nabla}\boldsymbol{v})^S\mathring{\boldsymbol{\nabla}}]\dot{\times}\boldsymbol{r}\mathring{\boldsymbol{\nabla}}\}\cdot(\mathring{\boldsymbol{r}}\boldsymbol{\nabla}\overset{\times}{\times}\mathring{\boldsymbol{r}}\boldsymbol{\nabla}) &= 2\{[(\boldsymbol{\nabla}\boldsymbol{v})^S\mathring{\boldsymbol{\nabla}}]\cdot\mathring{\mathbf{I}}\cdot\mathring{\boldsymbol{r}}\boldsymbol{\nabla}\}\dot{\times}(\boldsymbol{r}\mathring{\boldsymbol{\nabla}}\cdot\mathring{\mathbf{I}}\cdot\mathring{\boldsymbol{r}}\boldsymbol{\nabla}),\\ &= 2(\boldsymbol{\nabla}\boldsymbol{v})^S\boldsymbol{\nabla}\dot{\times}\mathbf{I} = -2(\boldsymbol{\nabla}\boldsymbol{v})^S\times\boldsymbol{\nabla},\\ &= \boldsymbol{\nabla}\boldsymbol{\nabla}\times\boldsymbol{v}.\end{aligned}$$

Hence

$$\dot{\mathbf{K}}\cdot(\mathring{\boldsymbol{r}}\boldsymbol{\nabla}\overset{\times}{\times}\mathring{\boldsymbol{r}}\boldsymbol{\nabla}) = \mathring{\boldsymbol{\nabla}}\boldsymbol{r}\cdot\boldsymbol{\nabla}\boldsymbol{\nabla}\times\boldsymbol{v} + 2(\boldsymbol{\nabla}\mathring{\boldsymbol{r}})_d[\mathring{\boldsymbol{\nabla}}\mathring{\boldsymbol{\nabla}}\boldsymbol{r}\underset{\cdot}{\times}(\dot{\mathbf{E}}\cdot\mathring{\boldsymbol{r}}\boldsymbol{\nabla})]\cdot\mathring{\boldsymbol{\nabla}}\boldsymbol{r}, \tag{2.13}$$

from which

$$\boldsymbol{\nabla}\boldsymbol{\nabla}\times\boldsymbol{v} = 2(\boldsymbol{\nabla}\mathring{\boldsymbol{r}})_d\boldsymbol{\nabla}\mathring{\boldsymbol{r}}\cdot[\dot{\mathbf{K}} - \mathring{\boldsymbol{\nabla}}\mathring{\boldsymbol{\nabla}}\boldsymbol{r}\underset{\cdot}{\times}(\dot{\mathbf{E}}\cdot\mathring{\boldsymbol{r}}\boldsymbol{\nabla})]\cdot\mathring{\boldsymbol{\nabla}}\boldsymbol{r}. \tag{2.14}$$

Then

$$\tfrac{1}{2}\boldsymbol{\mu}^D : \boldsymbol{\nabla}\boldsymbol{\nabla}\times\boldsymbol{v} = (\boldsymbol{\nabla}\mathring{\boldsymbol{r}})_d\{(\mathring{\boldsymbol{r}}\boldsymbol{\nabla}\cdot\boldsymbol{\mu}^D\cdot\boldsymbol{r}\mathring{\boldsymbol{\nabla}}):\dot{\mathbf{K}} - [(\mathring{\boldsymbol{r}}\boldsymbol{\nabla}\cdot\boldsymbol{\mu}^D\cdot\boldsymbol{r}\mathring{\boldsymbol{\nabla}})\dot{\times}\mathring{\boldsymbol{\nabla}}\mathring{\boldsymbol{\nabla}}\boldsymbol{r}\cdot\boldsymbol{\nabla}\mathring{\boldsymbol{r}}]^S:\dot{\mathbf{E}}\}. \tag{2.15}$$

Thus, with (2.9) and (2.15), (2.7) may be written in the form

$$\varrho \dot{U} = \mathbf{\Phi} : \dot{\mathbf{E}} + \mathbf{\Psi} : \dot{\mathbf{K}}, \tag{2.16}$$

where

$$\mathbf{\Phi} = \mathring{\boldsymbol{r}}\boldsymbol{\nabla} \cdot \boldsymbol{\tau}^S \cdot \boldsymbol{\nabla}\mathring{\boldsymbol{r}} - (\boldsymbol{\nabla}\mathring{\boldsymbol{r}})_d [(\mathring{\boldsymbol{r}}\boldsymbol{\nabla} \cdot \boldsymbol{\mu}^D \cdot \boldsymbol{r}\mathring{\boldsymbol{\nabla}}) \dot{\times} \mathring{\boldsymbol{\nabla}}\mathring{\boldsymbol{\nabla}}\boldsymbol{r} \cdot \boldsymbol{\nabla}\mathring{\boldsymbol{r}}]^S, \tag{2.17}$$

$$\mathbf{\Psi} = (\boldsymbol{\nabla}\mathring{\boldsymbol{r}})_d \mathring{\boldsymbol{r}}\boldsymbol{\nabla} \cdot \boldsymbol{\mu}^D \cdot \boldsymbol{r}\mathring{\boldsymbol{\nabla}}. \tag{2.18}$$

From (2.6),

$$\varrho \dot{U} = \varrho \frac{\partial U}{\partial \mathbf{E}} : \dot{\mathbf{E}} + \varrho \frac{\partial U}{\partial \mathbf{K}} : \dot{\mathbf{K}}, \tag{2.19}$$

$$0 = A \mathring{\mathbf{I}} : \dot{\mathbf{K}}, \tag{2.20}$$

where A is an arbitrary multiplier, introduced because the nine components of $\mathbf{K}$ are not independent. Then, subtracting the sum of (2.19) and (2.20) from (2.16), we have

$$\left(\mathbf{\Phi} - \varrho \frac{\partial U}{\partial \mathbf{E}}\right) : \dot{\mathbf{E}} + \left(\mathbf{\Psi} - \varrho \frac{\partial U}{\partial \mathbf{K}} - A \mathring{\mathbf{I}}\right) : \dot{\mathbf{K}} = 0. \tag{2.21}$$

We now assume that $\boldsymbol{\tau}^S$, $\boldsymbol{\mu}^D$ and U are independent of $\dot{\mathbf{K}}$ and $\dot{\mathbf{E}}$. Hence, from (2.21),

$$\mathbf{\Phi} = \varrho \frac{\partial U}{\partial \mathbf{E}}, \tag{2.22}$$

$$\mathbf{\Psi} = \varrho \frac{\partial U}{\partial \mathbf{K}} + A \mathring{\mathbf{I}}. \tag{2.23}$$

But $\mathring{\mathbf{I}} : \mathbf{\Psi} = 0$ and $\mathring{\mathbf{I}} : \partial U / \partial \mathbf{K} = 0$; hence $A = 0$. Then solving (2.22) and (2.23) for $\boldsymbol{\tau}^S$ and $\boldsymbol{\mu}^D$, we find

$$\begin{aligned} \boldsymbol{\tau}^S &= \varrho \boldsymbol{r}\mathring{\boldsymbol{\nabla}} \cdot \frac{\partial U}{\partial \mathbf{E}} \cdot \mathring{\boldsymbol{\nabla}}\boldsymbol{r} + \varrho \left[\boldsymbol{r}\mathring{\boldsymbol{\nabla}} \cdot \left(\frac{\partial U}{\partial \mathbf{K}} \dot{\times} \mathring{\boldsymbol{\nabla}}\mathring{\boldsymbol{\nabla}}\boldsymbol{r}\right)\right]^S, \\ \boldsymbol{\mu}^D &= \varrho (\mathring{\boldsymbol{\nabla}}\boldsymbol{r})_d \boldsymbol{r}\mathring{\boldsymbol{\nabla}} \cdot \frac{\partial U}{\partial \mathbf{K}} \cdot \mathring{\boldsymbol{r}}\boldsymbol{\nabla}. \end{aligned} \tag{2.24}$$

Equations (2.24) are one form of TOUPIN's constitutive equations [4].

3. Linearization

Let us assume that the specific internal energy can be expressed as a polynomial in $\mathbf{E}$ and $\mathbf{K}$:

$$\varrho_0 U = \tfrac{1}{2} \mathbf{K} : a : \mathbf{K} + \mathbf{E} : b : \mathbf{K} + \tfrac{1}{2} \mathbf{E} : c : \mathbf{E} + \cdots, \tag{3.1}$$

where a, b, c are constant, material tetradics and the coefficients of the terms of higher degree in $\mathbf{E}$ and $\mathbf{K}$ are polyadics of higher even order. The terms linear in $\mathbf{E}$ and $\mathbf{K}$ are omitted so as to make the undeformed state ($\mathbf{E} = 0$), with mass density ϱ_0, one of zero stress. Further, with the material displacement defined as $\boldsymbol{u} = \boldsymbol{r} - \mathring{\boldsymbol{r}}$, we restrict the absolute values of the components of the displacement gradient to be small in comparison with unity:

$$|\mathring{\boldsymbol{\nabla}}\boldsymbol{u}| \ll 1. \tag{3.2}$$

Then

$$\boldsymbol{\nabla} \approx \mathring{\boldsymbol{\nabla}}, \qquad \varrho \approx \varrho_0, \qquad \mathbf{I} \approx \mathring{\mathbf{I}}, \tag{3.3}$$

$$\mathbf{E} \approx \tfrac{1}{2}(\boldsymbol{\nabla}\boldsymbol{u} + \boldsymbol{u}\boldsymbol{\nabla}) \equiv \boldsymbol{\epsilon}, \qquad \mathbf{K} \approx -\boldsymbol{\epsilon} \times \boldsymbol{\nabla} = \tfrac{1}{2}\boldsymbol{\nabla}\boldsymbol{\nabla} \times \boldsymbol{u} \equiv \boldsymbol{\varkappa}, \tag{3.4}$$

and hence

$$\varrho_0 U \equiv W \approx \tfrac{1}{2}\boldsymbol{\varkappa}:\mathsf{a}:\boldsymbol{\varkappa} + \boldsymbol{\epsilon}:\mathsf{b}:\boldsymbol{\varkappa} + \tfrac{1}{2}\boldsymbol{\epsilon}:\mathsf{c}:\boldsymbol{\epsilon} + \cdots, \tag{3.5}$$

$$\boldsymbol{\tau}^S \approx \frac{\partial W}{\partial \boldsymbol{\epsilon}} + \left(\frac{\partial W}{\partial \boldsymbol{\varkappa}} \dot{\times} \boldsymbol{\nabla}\boldsymbol{\nabla}\boldsymbol{u}\right)^S, \qquad \boldsymbol{\mu}^D \approx \frac{\partial W}{\partial \boldsymbol{\varkappa}}. \tag{3.6}$$

In (3.5), we note that $\boldsymbol{\epsilon}$ is dimensionless and $\boldsymbol{\varkappa}$ has the dimension of reciprocal of length. Hence, the dimensions of $|\mathsf{a}/\mathsf{c}|$ are the square of length and those of $|\mathsf{b}/\mathsf{c}|$ are length. Let C be one of the components of c and let L^2 be one of the positive components of a/C; *i.e.*, L is a material constant with the dimension of length. Then (3.5) may be written as

$$W \approx C\left(\tfrac{1}{2}L^2\boldsymbol{\varkappa}:\mathsf{a}^*:\boldsymbol{\varkappa} + L\boldsymbol{\epsilon}:\mathsf{b}^*:\boldsymbol{\varkappa} + \tfrac{1}{2}\boldsymbol{\epsilon}:\mathsf{c}^*:\boldsymbol{\epsilon} + \cdots\right), \tag{3.7}$$

where a^*, b^*, c^*, ... are dimensionless material constants. We now assume that a^*, b^*, c^*, ... are of order of magnitude unity and that, in addition to (3.2),

$$|L\boldsymbol{\nabla}\boldsymbol{\nabla}\boldsymbol{u}| \ll 1. \tag{3.8}$$

Then, in (3.7), terms of degree higher than the second, in $\boldsymbol{\epsilon}$ and $L\boldsymbol{\varkappa}$, are negligible in comparison with the quadratic terms; and also, in $(3.6)_1$,

$$\left|\frac{\partial W}{\partial \boldsymbol{\varkappa}} \dot{\times} \boldsymbol{\nabla}\boldsymbol{\nabla}\boldsymbol{u}\right| \ll \left|\frac{\partial W}{\partial \boldsymbol{\epsilon}}\right|.$$

Thus, to the approximations (3.2) and (3.8),

$$W = \tfrac{1}{2}\boldsymbol{\varkappa}:\mathsf{a}:\boldsymbol{\varkappa} + \boldsymbol{\epsilon}:\mathsf{b}:\boldsymbol{\varkappa} + \tfrac{1}{2}\boldsymbol{\epsilon}:\mathsf{c}:\boldsymbol{\epsilon}, \tag{3.9}$$

$$\boldsymbol{\tau}^S = \frac{\partial W}{\partial \boldsymbol{\epsilon}} = \mathsf{c}:\boldsymbol{\epsilon} + \mathsf{b}:\boldsymbol{\varkappa}, \tag{3.10}$$

$$\boldsymbol{\mu}^D = \frac{\partial W}{\partial \boldsymbol{\varkappa}} = \boldsymbol{\epsilon}:\mathsf{b} + \boldsymbol{\varkappa}:\mathsf{a}. \tag{3.11}$$

These are the linearized forms of Toupin's constitutive equations. They are identical with those of Aero & Kuvshinskii [5].

The dyadic $\boldsymbol{\epsilon}$ is the usual small-strain dyadic and $\boldsymbol{\varkappa}$ is the gradient of the small-rotation vector:

$$\boldsymbol{\varkappa} = \boldsymbol{\nabla}\boldsymbol{w}, \qquad \boldsymbol{w} = \tfrac{1}{2}\boldsymbol{\nabla}\times\boldsymbol{u}; \tag{3.12}$$

i.e., $\boldsymbol{w}$ is the vector of the small-rotation dyadic

$$\boldsymbol{\omega} = \tfrac{1}{2}(\boldsymbol{\nabla}\boldsymbol{u} - \boldsymbol{u}\boldsymbol{\nabla}). \tag{3.13}$$

Hence $\boldsymbol{\varkappa}$ may be termed a curvature-twist dyadic.

The existence of a displacement field, $\boldsymbol{u}$, requires the usual compatibility condition

$$\boldsymbol{\nabla}\times\boldsymbol{\epsilon}\times\boldsymbol{\nabla} = 0. \tag{3.14}$$

Hence we have, on $\boldsymbol{\varkappa}$, the compatibility conditions

$$\boldsymbol{\nabla}\times\boldsymbol{\varkappa} = 0, \qquad \boldsymbol{\varkappa}:\mathbf{I} = 0. \tag{3.15}$$

If $\boldsymbol{u}_0$ and $\boldsymbol{\omega}_0$ are the displacement and rotation at a point M_0, with position vector $\boldsymbol{r}_0$; $\boldsymbol{u}_1$ is the displacement at a point M_1, with position vector $\boldsymbol{r}_1$; and $\boldsymbol{r}$ is the position vector of the running point along a path connecting M_0 and

M_1, then Cesàro's theorem [9, p. 223, Eq. (2)] may be written as

$$u_1 = u_0 + (r_1 - r_0)\cdot\omega_0 + \int_{M_0}^{M_1} dr\cdot[\epsilon + \varkappa\times(r_1 - r)]. \tag{3.16}$$

Returning to (3.9), we note that, owing to the symmetry of the products $\varkappa : a : \varkappa$ and $\epsilon : c : \epsilon$,

$$\Phi : a = a : \Phi, \qquad \Phi : c = c : \Phi, \tag{3.17}$$

where Φ is an arbitrary dyadic, giving 36 relations among the 81 components of a and 36 relations among the 81 components of c. Owing to the symmetry $\epsilon = \epsilon_C$,

$$\Phi : b = \Phi_C : b, \qquad \Phi : c = \Phi_C : c, \tag{3.18}$$

giving 27 relations among the 81 components of b and 24 additional relations among the remaining components of c. Finally, owing to the property $\varkappa : \mathbf{I} = 0$, we have

$$a : \mathbf{I} = 0, \qquad b : \mathbf{I} = 0, \tag{3.19}$$

giving 9 additional relations among the components of a and 6 additional relations among the components of b. Thus, there are $81 - 36 - 9 = 36$ independent components of a; $81 - 27 - 6 = 48$ independent components of b and the usual $81 - 36 - 24 = 21$ independent components of c.

We also note that ϵ is a polar dyadic and $\varkappa$ is an axial dyadic. Since W, in (3.9), is a scalar, a and c must be polar tetradics and b must be an axial tetradic. We shall consider only centrosymmetric-isotropic materials; in which case the material constants a, b, c must be invariant in a central inversion and an arbitrary rotation of coordinate axes. Hence $b = 0$, since it is axial, while a and c can be only linear combinations of the three isotropic tetradics $\mathbf{II}$, $\mathbf{I}\times\mathbf{I}\times\mathbf{I}$ and

$$(\mathbf{I}\times\mathbf{I}) \overset{\times}{\times} (\mathbf{I}\times\mathbf{I}).$$

Thus, for a centrosymmetric-isotropic material, we may take

$$W = \sum_{i=1}^{3} a_i \varkappa : J_i : \varkappa + \sum_{i=1}^{3} c_i \epsilon : J_i : \epsilon, \tag{3.20}$$

where the a_i and c_i are scalar constants and

$$J_1 = \mathbf{II}, \qquad J_2 = (\mathbf{I}\times\mathbf{I}) \overset{\times}{\times} (\mathbf{I}\times\mathbf{I}) - J_1 - J_3, \qquad J_3 = \tfrac{1}{2}\mathbf{I}\times\mathbf{I}\times\mathbf{I}. \tag{3.21}$$

Now

$$\begin{aligned} \varkappa : J_1 : \varkappa &= \varkappa_S^2 = 0, & \epsilon : J_1 : \epsilon &= \epsilon_S^2, \\ \varkappa : J_2 : \varkappa &= \varkappa : \varkappa^S, & \epsilon : J_2 : \epsilon &= \epsilon : \epsilon, \\ \varkappa : J_3 : \varkappa &= \varkappa : \varkappa^A, & \epsilon : J_3 : \epsilon &= 0, \end{aligned}$$

where $\varkappa^S$ and $\varkappa^A$ are the symmetric and antisymmetric parts of $\varkappa$; and $\epsilon_S = \epsilon : \mathbf{I}$. Then (3.20) becomes

$$\begin{aligned} W &= a_2 \varkappa : \varkappa^S + a_3 \varkappa : \varkappa^A + c_1 \epsilon_S^2 + c_2 \epsilon : \epsilon, \\ &= \tfrac{1}{2}(a_2 + a_3)\varkappa : \varkappa + \tfrac{1}{2}(a_2 - a_3)\varkappa : \varkappa_C + c_1 \epsilon_S^2 + c_2 \epsilon : \epsilon. \end{aligned}$$

Finally, setting

$$4\eta = a_2 + a_3, \qquad 4\eta' = a_2 - a_3, \qquad \lambda = 2c_1, \qquad \mu = c_2, \tag{3.22}$$

the internal energy-density of a centrosymmetric-isotropic material becomes

$$W = 2\eta \boldsymbol{\varkappa} : \boldsymbol{\varkappa} + 2\eta' \boldsymbol{\varkappa} : \boldsymbol{\varkappa}_C + \tfrac{1}{2}\lambda \epsilon_S^2 + \mu \boldsymbol{\epsilon} : \boldsymbol{\epsilon}, \tag{3.23}$$

from which

$$\begin{aligned} \boldsymbol{\tau}^S &= \frac{\partial W}{\partial \boldsymbol{\epsilon}} = \lambda \epsilon_S \mathbf{I} + 2\mu \boldsymbol{\epsilon}, \\ \boldsymbol{\mu}^D &= \frac{\partial W}{\partial \boldsymbol{\varkappa}} = 4\eta \boldsymbol{\varkappa} + 4\eta' \boldsymbol{\varkappa}_C. \end{aligned} \tag{3.24}$$

In the analogous expressions found by GRIOLI [6], the constant η' does not appear. This is discussed in the next section.

If we replace $\boldsymbol{\epsilon}$ by $\frac{1}{2}(\nabla \boldsymbol{u} + \boldsymbol{u}\nabla)$ and $\boldsymbol{\varkappa}$ by $\frac{1}{2}\nabla\nabla \times \boldsymbol{u}$, in (3.24), the latter become

$$\begin{aligned} \boldsymbol{\tau}^S &= \lambda \nabla \cdot \boldsymbol{u} \mathbf{I} + \mu (\nabla \boldsymbol{u} + \boldsymbol{u} \nabla), \\ \boldsymbol{\mu}^D &= 2\eta \nabla\nabla \times \boldsymbol{u} + 2\eta' \nabla \times \boldsymbol{u} \nabla. \end{aligned} \tag{3.25}$$

When these are inserted in the linearized stress-equation of motion (1.16):

$$\nabla \cdot \boldsymbol{\tau}^S + \tfrac{1}{2} \nabla \times \nabla \cdot \boldsymbol{\mu}^D + \varrho \boldsymbol{f} + \tfrac{1}{2} \varrho \nabla \times \boldsymbol{c} = \varrho \ddot{\boldsymbol{u}} \tag{3.26}$$

(where $\ddot{\boldsymbol{u}}$ now designates $\partial^2 \boldsymbol{u}/\partial t^2$), we find the displacement-equation of motion

$$\mu \nabla^2 \boldsymbol{u} + (\lambda + \mu) \nabla\nabla \cdot \boldsymbol{u} + \eta \nabla^2 \nabla \times \nabla \times \boldsymbol{u} + \varrho \boldsymbol{f} + \tfrac{1}{2} \varrho \nabla \times \boldsymbol{c} = \varrho \ddot{\boldsymbol{u}}. \tag{3.27}$$

It will be observed that η' does not appear in (3.27), so that (3.27) is identical with the displacement-equation of motion that would be obtained with GRIOLI's constitutive equations.

The sign of the bending-twisting modulus, η, is of great importance as solutions of (3.27) are different in character for positive and negative η. If we require the energy density W, in (3.23), to be positive definite (implying that energy is stored, rather than produced, as a result of deformation) we find, as necessary and sufficient conditions,

$$\mu > 0, \qquad 3\lambda + 2\mu > 0, \qquad \eta > 0, \qquad -1 < \eta'/\eta < 1. \tag{3.28}$$

In this case we can establish a theorem of uniqueness of solution, of the Neumann type, as shown in Section 5. It seems reasonable to adopt $\mu > 0$, $3\lambda + 2\mu > 0$, as these are acceptable in the absence of couple-stresses. Positiveness of η conforms to the intuitive expectation that the sign of the curvature, say $\varkappa_{12}$, produced by a bending couple μ_{12}, will be the same as the sign of μ_{12}. Furthermore, an example of failure of uniqueness, for $\eta < 0$, is exhibited in Section 13. Finally, η'/η is the ratio of the transverse curvature (probably anticlastic) to the principal curvature and it seems unlikely that this ratio would lie outside the range given in $(3.28)_4$. Thus, it is plausible to assume that W is positive definite when couple stresses are considered.

With $\mu > 0$ and $\eta > 0$, we may adopt, as the material length L, in (3.8),

$$l = \sqrt{\eta/\mu}. \tag{3.29}$$

If $l=0$, there are no effects of couple-stress. Inasmuch as the ordinary theory of elasticity has been verified experimentally in great detail, l is probably very small in comparison with bodily dimensions and wave-lengths that are commonly encountered.

The indeterminacy of $\boldsymbol{\mu}:\mathbf{I}$ and $\boldsymbol{\tau}^A$, mentioned in Section 1, carries over, of course, to the linear theory. From a solution, $\boldsymbol{u}$, of (3.27), we can compute only $\boldsymbol{\tau}^S$ and $\boldsymbol{\mu}^D$ from (3.25). The remaining parts of the stress, namely $\boldsymbol{\mu}:\mathbf{I}$ and $\boldsymbol{\tau}^A$, are related through (1.10), which may be written as

$$\boldsymbol{\tau}^A=\tfrac{1}{2}\mathbf{I}\times(\boldsymbol{\nabla}\cdot\boldsymbol{\mu}+\varrho\boldsymbol{c})=\tfrac{1}{2}\mathbf{I}\times(2\eta\nabla^2\boldsymbol{\nabla}\times\boldsymbol{u}+\tfrac{1}{3}\boldsymbol{\nabla}\boldsymbol{\mu}:\mathbf{I}+\varrho\boldsymbol{c}). \tag{3.30}$$

Thus, with $\boldsymbol{u}$ and $\varrho\boldsymbol{c}$ known, there are only three scalar equations relating the scalar $\boldsymbol{\mu}:\mathbf{I}$ and the three components of $\boldsymbol{\tau}^A$.

4. Remarks on Grioli's constitutive equations

In place of (2.5), GRIOLI [6] assumed

$$U=U(\boldsymbol{\Delta},\boldsymbol{\Lambda}), \tag{4.1}$$

where

$$\boldsymbol{\Delta}=\mathring{\boldsymbol{\nabla}}\boldsymbol{r}, \tag{4.2}$$

$$\boldsymbol{\Lambda}=-\tfrac{1}{2}\boldsymbol{\Delta}\times\boldsymbol{\nabla}=-\tfrac{1}{2}(\mathring{\boldsymbol{\nabla}}\boldsymbol{r})\times\mathring{\boldsymbol{\nabla}}\cdot(\mathring{\boldsymbol{r}}\boldsymbol{\nabla})=-\tfrac{1}{2}\mathring{\boldsymbol{\nabla}}\mathring{\boldsymbol{\nabla}}\boldsymbol{r}\dot{\times}\mathring{\boldsymbol{r}}\boldsymbol{\nabla}, \tag{4.3}$$

and he noted that $\boldsymbol{\Delta}$ and $\boldsymbol{\Lambda}$ are connected by the single scalar relation

$$\boldsymbol{\Lambda}:\boldsymbol{\Delta}_C^{-1}=-\tfrac{1}{2}(\mathring{\boldsymbol{\nabla}}\mathring{\boldsymbol{\nabla}}\boldsymbol{r}\dot{\times}\mathring{\boldsymbol{r}}\boldsymbol{\nabla}):\mathring{\boldsymbol{r}}\boldsymbol{\nabla}=-\tfrac{1}{2}\mathbf{I}\times\mathbf{I}\vdots\boldsymbol{\nabla}\mathring{\boldsymbol{r}}\cdot(\mathring{\boldsymbol{\nabla}}\boldsymbol{r}\mathring{\boldsymbol{\nabla}})\cdot\mathring{\boldsymbol{r}}\boldsymbol{\nabla}=0. \tag{4.4}$$

Thus, U is a function of $9+9-1=17$ variables.

From (4.1), we have

$$\varrho\dot{U}=\varrho\frac{\partial U}{\partial\boldsymbol{\Delta}}:\dot{\boldsymbol{\Delta}}+\varrho\frac{\partial U}{\partial\boldsymbol{\Lambda}}:\dot{\boldsymbol{\Lambda}}, \tag{4.5}$$

and, from (4.4),

$$\overline{\boldsymbol{\Lambda}:\boldsymbol{\Delta}_C^{-1}}^{\,\boldsymbol{\cdot}}=\boldsymbol{\Lambda}:\dot{\boldsymbol{\Delta}}_C^{-1}+\boldsymbol{\Delta}_C^{-1}:\dot{\boldsymbol{\Lambda}}=0. \tag{4.6}$$

Now, $\boldsymbol{\Delta}_C^{-1}\cdot\boldsymbol{\Delta}_C=\mathring{\mathbf{I}}$, whence

$$\overline{\boldsymbol{\Delta}_C^{-1}\cdot\boldsymbol{\Delta}_C}^{\,\boldsymbol{\cdot}}=\boldsymbol{\Delta}_C^{-1}\cdot\dot{\boldsymbol{\Delta}}_C+\dot{\boldsymbol{\Delta}}_C^{-1}\cdot\boldsymbol{\Delta}_C=0. \tag{4.7}$$

Upon multiplying $(4.7)_2$ by $\cdot\,\boldsymbol{\Delta}_C^{-1}$, we have

$$\dot{\boldsymbol{\Delta}}_C^{-1}=-\boldsymbol{\Delta}_C^{-1}\cdot\dot{\boldsymbol{\Delta}}_C\cdot\boldsymbol{\Delta}_C^{-1}. \tag{4.8}$$

Then

$$\boldsymbol{\Lambda}:\dot{\boldsymbol{\Delta}}_C^{-1}=-\boldsymbol{\Lambda}:(\boldsymbol{\Delta}_C^{-1}\cdot\dot{\boldsymbol{\Delta}}_C\cdot\boldsymbol{\Delta}_C^{-1})=-\boldsymbol{\Delta}_C^{-1}\cdot\boldsymbol{\Lambda}_C\cdot\boldsymbol{\Delta}_C^{-1}:\dot{\boldsymbol{\Delta}}.$$

Thus, $(4.6)_2$ becomes

$$-\boldsymbol{\Delta}_C^{-1}\cdot\boldsymbol{\Lambda}_C\cdot\boldsymbol{\Delta}_C^{-1}:\dot{\boldsymbol{\Delta}}+\boldsymbol{\Delta}_C^{-1}:\dot{\boldsymbol{\Lambda}}=0, \tag{4.9}$$

which is the companion equation to (4.5), taking into account the relation $\boldsymbol{\Lambda}:\boldsymbol{\Delta}_C^{-1}=0$.

We have now to put the energy equation,

$$\varrho\dot{U}=\boldsymbol{\tau}^S:\boldsymbol{\nabla}\boldsymbol{v}+\tfrac{1}{2}\boldsymbol{\mu}^D:\boldsymbol{\nabla}\boldsymbol{\nabla}\times\boldsymbol{v}, \tag{4.10}$$

into the form $\varrho\dot{U}=\boldsymbol{\Phi}:\dot{\boldsymbol{\Delta}}+\boldsymbol{\Psi}:\dot{\boldsymbol{\Lambda}}$.

First, noting that

$$\nabla(\ )=\nabla\mathring{\boldsymbol{r}}\cdot\mathring{\nabla}(\ ),\qquad \overline{\mathring{\nabla}\boldsymbol{r}}^{\,\bullet}=\mathring{\nabla}\dot{\boldsymbol{r}}=\mathring{\nabla}\boldsymbol{v},\tag{4.11}$$

we have

$$\boldsymbol{\tau}^S:\nabla\boldsymbol{v}=\boldsymbol{\tau}^S:\nabla\dot{\boldsymbol{r}}=\boldsymbol{\tau}^S:(\nabla\mathring{\boldsymbol{r}}\cdot\mathring{\nabla}\dot{\boldsymbol{r}})=\boldsymbol{\tau}^S:(\boldsymbol{\Delta}^{-1}\cdot\dot{\boldsymbol{\Delta}})=\boldsymbol{\Delta}_C^{-1}\cdot\boldsymbol{\tau}^S:\dot{\boldsymbol{\Delta}}.\tag{4.12}$$

Next,

$$\begin{aligned}\tfrac{1}{2}\boldsymbol{\mu}^D:\nabla\nabla\times\boldsymbol{v}&=\tfrac{1}{2}\boldsymbol{\mu}^D:\nabla\mathring{\boldsymbol{r}}\cdot\mathring{\nabla}[(\nabla\mathring{\boldsymbol{r}}\cdot\mathring{\nabla})\times\boldsymbol{v}],\\&=\tfrac{1}{2}\boldsymbol{\mu}^D:\nabla\mathring{\boldsymbol{r}}\cdot[\mathring{\nabla}(\mathring{\boldsymbol{r}}\nabla)\dot{\times}\mathring{\nabla}\boldsymbol{v}-\mathring{\nabla}\mathring{\nabla}\boldsymbol{v}\dot{\times}\mathring{\boldsymbol{r}}\nabla],\\&=\tfrac{1}{2}\boldsymbol{\mu}^D:\boldsymbol{\Delta}^{-1}\cdot[\mathring{\nabla}\boldsymbol{\Delta}_C^{-1}\dot{\times}\dot{\boldsymbol{\Delta}}-\mathring{\nabla}\dot{\boldsymbol{\Delta}}\dot{\times}\boldsymbol{\Delta}_C^{-1}].\end{aligned}$$

Also,

$$\begin{aligned}\mathring{\nabla}\dot{\boldsymbol{\Delta}}\dot{\times}\boldsymbol{\Delta}_C^{-1}&=\overline{\mathring{\nabla}\boldsymbol{\Delta}\dot{\times}\boldsymbol{\Delta}_C^{-1}}^{\,\bullet}-\mathring{\nabla}\boldsymbol{\Delta}\dot{\times}\dot{\boldsymbol{\Delta}}_C^{-1}=-2\dot{\boldsymbol{\Lambda}}+\mathring{\nabla}\boldsymbol{\Delta}\dot{\times}(\boldsymbol{\Delta}_C^{-1}\cdot\dot{\boldsymbol{\Delta}}_C\cdot\boldsymbol{\Delta}_C^{-1})\\&=-2\dot{\boldsymbol{\Lambda}}+(\boldsymbol{\Delta}\nabla)\dot{\times}(\boldsymbol{\Delta}^{-1}\cdot\dot{\boldsymbol{\Delta}}),\end{aligned}$$

by (4.3) and (4.8). Hence

$$\begin{aligned}\tfrac{1}{2}\boldsymbol{\mu}^D:\nabla\nabla\times\boldsymbol{v}&=\tfrac{1}{2}\boldsymbol{\mu}^D:\boldsymbol{\Delta}^{-1}\cdot[\mathring{\nabla}\boldsymbol{\Delta}_C^{-1}\dot{\times}\dot{\boldsymbol{\Delta}}+2\dot{\boldsymbol{\Lambda}}-(\boldsymbol{\Delta}\nabla)\dot{\times}(\boldsymbol{\Delta}^{-1}\cdot\dot{\boldsymbol{\Delta}})],\\&=\tfrac{1}{2}\{[\boldsymbol{\Delta}_C^{-1}\cdot\boldsymbol{\mu}^D\dot{\times}(\mathring{\nabla}\boldsymbol{\Delta}^{-1})]_C-\boldsymbol{\Delta}_C^{-1}\cdot[\boldsymbol{\Delta}_C^{-1}\cdot\boldsymbol{\mu}^D\dot{\times}(\boldsymbol{\Delta}\nabla)]\}:\dot{\boldsymbol{\Delta}}+\boldsymbol{\Delta}_C^{-1}\cdot\boldsymbol{\mu}^D:\dot{\boldsymbol{\Lambda}}.\end{aligned}\tag{4.13}$$

Substituting (4.13) and (4.12) into (4.10), we obtain

$$\varrho\dot{U}=\boldsymbol{\Phi}:\dot{\boldsymbol{\Delta}}+\boldsymbol{\Psi}:\dot{\boldsymbol{\Lambda}},\tag{4.14}$$

where

$$\boldsymbol{\Phi}=\boldsymbol{\Delta}_C^{-1}\cdot\boldsymbol{\tau}^S+\tfrac{1}{2}[(\boldsymbol{\Delta}_C^{-1}\cdot\boldsymbol{\mu}^D)\dot{\times}(\mathring{\nabla}\boldsymbol{\Delta}^{-1})]_C-\tfrac{1}{2}\boldsymbol{\Delta}_C^{1}\cdot[(\boldsymbol{\Delta}_C^{1}\cdot\boldsymbol{\mu}^D)\dot{\times}(\boldsymbol{\Delta}\nabla)],\tag{4.15}$$

$$\boldsymbol{\Psi}=\boldsymbol{\Delta}_C^{-1}\cdot\boldsymbol{\mu}^D.\tag{4.16}$$

Then, multiplying (4.9) by an arbitrary scalar, A, and adding the result to the difference between (4.14) and (4.5), we have

$$\left(\boldsymbol{\Phi}-\varrho\frac{\partial U}{\partial\boldsymbol{\Delta}}-A\,\boldsymbol{\Delta}_C^{-1}\cdot\boldsymbol{\Lambda}_C\cdot\boldsymbol{\Delta}_C^{-1}\right):\dot{\boldsymbol{\Delta}}+\left(\boldsymbol{\Psi}-\varrho\frac{\partial U}{\partial\boldsymbol{\Lambda}}+A\,\boldsymbol{\Delta}_C^{-1}\right):\dot{\boldsymbol{\Lambda}}=0.\tag{4.17}$$

We now assume that $\boldsymbol{\tau}^S$, $\boldsymbol{\mu}^D$ and U are independent of $\dot{\boldsymbol{\Delta}}$ and $\dot{\boldsymbol{\Lambda}}$, whence

$$\boldsymbol{\Phi}-\varrho\frac{\partial U}{\partial\boldsymbol{\Delta}}-A\,\boldsymbol{\Delta}_C^{-1}\cdot\boldsymbol{\Lambda}_C\cdot\boldsymbol{\Delta}_C^{-1}=0,\tag{4.18}$$

$$\boldsymbol{\Psi}-\varrho\frac{\partial U}{\partial\boldsymbol{\Lambda}}+A\,\boldsymbol{\Delta}_C^{-1}=0.\tag{4.19}$$

Then, multiplying (4.19) by $\boldsymbol{\Delta}_C\cdot$ and using (4.16), we find

$$\boldsymbol{\mu}^D=\varrho\boldsymbol{\Delta}_C\cdot\frac{\partial U}{\partial\boldsymbol{\Lambda}}-A\,\mathbf{I};\tag{4.20}$$

and, noting that $\boldsymbol{\mu}^D:\mathbf{I}=0$,

$$A=\frac{1}{3}\varrho\,\mathbf{I}:\left(\boldsymbol{\Delta}_C\cdot\frac{\partial U}{\partial\boldsymbol{\Lambda}}\right);$$

so that

$$\boldsymbol{\mu}^D=\varrho\left(\boldsymbol{\Delta}_C\cdot\frac{\partial U}{\partial\boldsymbol{\Lambda}}\right)^D.\tag{4.21}$$

Now, substitute (4.20) into (4.15) and the latter into (4.18) to obtain

$$\begin{aligned}\boldsymbol{\Delta}_C^{-1}\cdot\boldsymbol{\tau}^S+\tfrac{1}{2}\varrho\left[\frac{\partial U}{\partial\boldsymbol{\Lambda}}\dot{\times}(\mathring{\boldsymbol{\nabla}}\boldsymbol{\Delta}^{-1})\right]_C-\tfrac{1}{2}\varrho\boldsymbol{\Delta}_C^{-1}\cdot\left[\frac{\partial U}{\partial\boldsymbol{\Lambda}}\dot{\times}(\boldsymbol{\Delta}\boldsymbol{\nabla})\right]-\varrho\frac{\partial U}{\partial\boldsymbol{\Delta}}-\\ -A\left\{\boldsymbol{\Delta}_C^{-1}\cdot\boldsymbol{\Lambda}_C\cdot\boldsymbol{\Delta}_C^{-1}+\tfrac{1}{2}[\boldsymbol{\Delta}_C^{-1}\dot{\times}(\mathring{\boldsymbol{\nabla}}\boldsymbol{\Delta}^{-1})]_C-\tfrac{1}{2}\boldsymbol{\Delta}_C^{-1}\cdot[\boldsymbol{\Delta}_C^{-1}\dot{\times}(\boldsymbol{\Delta}\boldsymbol{\nabla})]\right\}=0.\end{aligned}\tag{4.22}$$

But

$$\boldsymbol{\Delta}_C^{-1}\dot{\times}(\mathring{\boldsymbol{\nabla}}\boldsymbol{\Delta}^{-1})=\boldsymbol{\nabla}\times\boldsymbol{\Delta}^{-1}=0$$

and

$$\tfrac{1}{2}\boldsymbol{\Delta}_C^{-1}\dot{\times}(\boldsymbol{\Delta}\boldsymbol{\nabla})=-\tfrac{1}{2}[(\mathring{\boldsymbol{\nabla}}\boldsymbol{r})\times\boldsymbol{\nabla}]_C\cdot\mathring{\boldsymbol{r}}\boldsymbol{\nabla}=\boldsymbol{\Lambda}_C\cdot\boldsymbol{\Delta}_C^{-1}.$$

Hence, the coefficient of A, in (4.22), vanishes. Then, multiplying (4.22) by $\boldsymbol{\Delta}_C\,\cdot$, we have

$$\boldsymbol{\tau}^S=\varrho\boldsymbol{\Delta}_C\cdot\frac{\partial U}{\partial\boldsymbol{\Delta}}+\tfrac{1}{2}\varrho\frac{\partial U}{\partial\boldsymbol{\Lambda}}\dot{\times}(\boldsymbol{\Delta}\boldsymbol{\nabla})-\tfrac{1}{2}\varrho(\boldsymbol{\Delta}_C\boldsymbol{\nabla})\dot{\times}\frac{\partial U}{\partial\boldsymbol{\Lambda}}.\tag{4.23}$$

Since $\boldsymbol{\tau}^S$ is symmetric, so also must be the right hand side of (4.23); *i.e.*,

$$\left[\boldsymbol{\Delta}_C\cdot\frac{\partial U}{\partial\boldsymbol{\Delta}}+\frac{1}{2}\frac{\partial U}{\partial\boldsymbol{\Lambda}}\dot{\times}(\boldsymbol{\Delta}\boldsymbol{\nabla})-\frac{1}{2}(\boldsymbol{\Delta}_C\boldsymbol{\nabla})\dot{\times}\frac{\partial U}{\partial\boldsymbol{\Lambda}}\right]\dot{\times}\mathbf{I}=0.\tag{4.24}$$

The first term in (4.24) may be written as

$$\boldsymbol{\Delta}_C\cdot\frac{\partial U}{\partial\boldsymbol{\Delta}}\dot{\times}\mathbf{I}=\boldsymbol{\Delta}\dot{\times}\frac{\partial U}{\partial\boldsymbol{\Delta}}\tag{4.25}$$

and the remaining terms are

$$\frac{1}{2}\left[\frac{\partial U}{\partial\boldsymbol{\Lambda}}\dot{\times}(\boldsymbol{\Delta}\boldsymbol{\nabla})\right]\dot{\times}\mathbf{I}=\frac{1}{2}(\boldsymbol{\Delta}_C\boldsymbol{\nabla}):\frac{\partial U}{\partial\boldsymbol{\Lambda}}-\frac{1}{2}(\boldsymbol{\nabla}\cdot\boldsymbol{\Delta}_C)\cdot\frac{\partial U}{\partial\boldsymbol{\Lambda}},\tag{4.26}$$

$$-\frac{1}{2}\left[(\boldsymbol{\Delta}_C\boldsymbol{\nabla})\dot{\times}\frac{\partial U}{\partial\boldsymbol{\Lambda}}\right]\dot{\times}\mathbf{I}=\frac{1}{2}(\boldsymbol{\nabla}\cdot\boldsymbol{\Delta}_C)\cdot\frac{\partial U}{\partial\boldsymbol{\Lambda}}-\frac{1}{2}(\boldsymbol{\nabla}\boldsymbol{\Delta}):\frac{\partial U}{\partial\boldsymbol{\Lambda}}.\tag{4.27}$$

The sum of (4.26) and (4.27) is

$$\frac{1}{2}(\boldsymbol{\Delta}_C\boldsymbol{\nabla}-\boldsymbol{\nabla}\boldsymbol{\Delta}):\frac{\partial U}{\partial\boldsymbol{\Lambda}}=\frac{1}{2}\frac{\partial U}{\partial\boldsymbol{\Lambda}}\dot{\times}(\boldsymbol{\Delta}\dot{\times}\boldsymbol{\nabla})=\boldsymbol{\Lambda}\dot{\times}\frac{\partial U}{\partial\boldsymbol{\Lambda}}.\tag{4.28}$$

Hence, (4.24) may be written as

$$\boldsymbol{\Delta}\dot{\times}\frac{\partial U}{\partial\boldsymbol{\Delta}}+\boldsymbol{\Lambda}\dot{\times}\frac{\partial U}{\partial\boldsymbol{\Lambda}}=0.\tag{4.29}$$

TOUPIN [*10*, p. 885] had noted, earlier, that equations such as (4.29) are necessary and sufficient conditions for the invariance of U in a rigid rotation of the deformed material.

To summarize, GRIOLI's constitutive equations for the fourteen components of $\boldsymbol{\tau}^S$ and $\boldsymbol{\mu}^D$, in terms of a function, U, of the eighteen components of $\boldsymbol{\Delta}$ and $\boldsymbol{\Lambda}$, are [*6*, (56), (57)]

$$\begin{aligned}\boldsymbol{\tau}^S&=\varrho\boldsymbol{\Delta}_C\cdot\frac{\partial U}{\partial\boldsymbol{\Delta}}+\frac{1}{2}\varrho\frac{\partial U}{\partial\boldsymbol{\Lambda}}\dot{\times}(\boldsymbol{\Delta}\boldsymbol{\nabla})-\frac{1}{2}\varrho(\boldsymbol{\Delta}_C\boldsymbol{\nabla})\dot{\times}\frac{\partial U}{\partial\boldsymbol{\Lambda}},\\ \boldsymbol{\mu}^D&=\varrho\left(\boldsymbol{\Delta}_C\cdot\frac{\partial U}{\partial\boldsymbol{\Lambda}}\right)^D,\end{aligned}\tag{4.30}$$

subject to the four conditions [6, (54), (71)]

$$\boldsymbol{\Lambda}:\boldsymbol{\Delta}_C^{-1}=0, \tag{4.31}$$

$$\boldsymbol{\Delta}\dot{\times}\frac{\partial U}{\partial\boldsymbol{\Delta}}+\boldsymbol{\Lambda}\dot{\times}\frac{\partial U}{\partial\boldsymbol{\Lambda}}=0. \tag{4.32}$$

To obtain the linearized form of (4.30)–(4.32), we assume, first, that U can be expressed as a polynomial in $\boldsymbol{\Delta}$ and $\boldsymbol{\Lambda}$:

$$\varrho_0 U=\tfrac{1}{2}\boldsymbol{\Lambda}:\mathsf{a}:\boldsymbol{\Lambda}+\boldsymbol{\Delta}:\mathsf{b}:\boldsymbol{\Lambda}+\tfrac{1}{2}\boldsymbol{\Delta}:\mathsf{c}:\boldsymbol{\Delta}+\cdots. \tag{4.33}$$

Then, with the assumption (3.2), we have

$$\boldsymbol{\Delta}\approx\mathbf{I}+\boldsymbol{\nabla}\boldsymbol{u},\qquad \boldsymbol{\Lambda}\approx\boldsymbol{\varkappa}. \tag{4.34}$$

Hence, dropping terms linear in $\boldsymbol{\nabla}\boldsymbol{u}$ and $\boldsymbol{\varkappa}$, (4.33) becomes

$$\varrho_0 U\equiv W\approx\tfrac{1}{2}\boldsymbol{\varkappa}:\mathsf{a}:\boldsymbol{\varkappa}+\boldsymbol{\nabla}\boldsymbol{u}:\mathsf{b}:\boldsymbol{\varkappa}+\tfrac{1}{2}\boldsymbol{\nabla}\boldsymbol{u}:\mathsf{c}:\boldsymbol{\nabla}\boldsymbol{u}+\cdots. \tag{4.35}$$

Also, (4.30) reduce to

$$\boldsymbol{\tau}^S=\frac{\partial W}{\partial\boldsymbol{\nabla}\boldsymbol{u}}+\frac{1}{2}\frac{\partial W}{\partial\boldsymbol{\varkappa}}\dot{\times}(\boldsymbol{\nabla}\boldsymbol{u}\boldsymbol{\nabla})-\frac{1}{2}(\boldsymbol{u}\boldsymbol{\nabla}\boldsymbol{\nabla})\dot{\times}\frac{\partial W}{\partial\boldsymbol{\varkappa}},\qquad \boldsymbol{\mu}^D=\frac{\partial W}{\partial\boldsymbol{\varkappa}}. \tag{4.36}$$

The condition (4.31) reduces to $\boldsymbol{\varkappa}:\mathbf{I}=0$, which is satisfied identically; while (4.32) becomes

$$\mathbf{I}\dot{\times}\frac{\partial W}{\partial\boldsymbol{\nabla}\boldsymbol{u}}+\boldsymbol{\varkappa}\dot{\times}\frac{\partial W}{\partial\boldsymbol{\varkappa}}=0. \tag{4.37}$$

To complete the linearization, we must again assume (3.8). Then (4.35) to (4.37) further reduce to

$$W=\tfrac{1}{2}\boldsymbol{\varkappa}:\mathsf{a}:\boldsymbol{\varkappa}+\boldsymbol{\nabla}\boldsymbol{u}:\mathsf{b}:\boldsymbol{\varkappa}+\tfrac{1}{2}\boldsymbol{\nabla}\boldsymbol{u}:\mathsf{c}:\boldsymbol{\nabla}\boldsymbol{u}, \tag{4.38}$$

$$\boldsymbol{\tau}^S=\frac{\partial W}{\partial\boldsymbol{\nabla}\boldsymbol{u}},\qquad \boldsymbol{\mu}^D=\frac{\partial W}{\partial\boldsymbol{\varkappa}}, \tag{4.39}$$

$$\mathbf{I}\dot{\times}\frac{\partial W}{\partial\boldsymbol{\nabla}\boldsymbol{u}}=0. \tag{4.40}$$

(4.38) and (4.39) are equivalent to GRIOLI's forms; but, in effect, he retained (4.37) in place of (4.40). However, to the order of approximation employed in the passage from (4.35) and (4.36) to (4.38) and (4.39), the second vector in (4.37) is negligible in comparison with the first.

In an isotropic material,

$$W=\sum_{i=1}^{3}a_i\boldsymbol{\varkappa}:J_i:\boldsymbol{\varkappa}+\sum_{i=1}^{3}b_i\boldsymbol{\nabla}\boldsymbol{u}:J_i:\boldsymbol{\varkappa}+\sum_{i=1}^{3}c_i\boldsymbol{\nabla}\boldsymbol{u}:J_i:\boldsymbol{\nabla}\boldsymbol{u}, \tag{4.41}$$

where the J_i are again the isotropic tetradics given in (3.21). Then

$$W=a_2\boldsymbol{\varkappa}:\boldsymbol{\varkappa}^S+a_3\boldsymbol{\varkappa}:\boldsymbol{\varkappa}^A+b_2\boldsymbol{\epsilon}:\boldsymbol{\varkappa}+b_3\boldsymbol{\omega}:\boldsymbol{\varkappa}+c_1\boldsymbol{\epsilon}_S^2+c_2\boldsymbol{\epsilon}:\boldsymbol{\epsilon}+c_3\boldsymbol{\omega}:\boldsymbol{\omega}. \tag{4.42}$$

If, in addition, the material is centrosymmetric, we must have $b_2 = b_3 = 0$: since $\boldsymbol{\epsilon}$ and $\boldsymbol{\omega}$ are polar but $\boldsymbol{\varkappa}$ is axial. Then, rearranging terms for comparison with (3.23), we write (4.42) as

$$W = 2\eta\, \boldsymbol{\varkappa} : \boldsymbol{\varkappa} + 2\eta'\, \boldsymbol{\varkappa} : \boldsymbol{\varkappa}_C + \tfrac{1}{2}\lambda\, \epsilon_S^2 + \mu\, \boldsymbol{\epsilon} : \boldsymbol{\epsilon} + c_3\, \boldsymbol{\omega} : \boldsymbol{\omega}, \tag{4.43}$$

where the definitions (3.22) are employed again.

We now substitute (4.43) into the rigid rotation condition (4.40) to obtain

$$c_3\, \mathbf{I} \overset{\cdot}{\times} \boldsymbol{\omega} = 0,$$

whence $c_3 = 0$. The rôle of the condition of invariance of the energy function in a rigid rotation is, thus, to exclude dependence of W on the rotation $\boldsymbol{\omega}$.

On the other hand, at the stage analogous to (4.43), GRIOLI first eliminated dependence of W on $\boldsymbol{\omega}$, on the grounds that $\boldsymbol{\tau}^S$ is symmetric, and then redundantly applied the condition of invariance of W in a rigid rotation. However, he applied (4.37) instead of (4.40) and, with c_3 already eliminated, he obtained

$$\eta'\, \boldsymbol{\varkappa} \overset{\cdot}{\times} \boldsymbol{\varkappa}_C = 0$$

from which he concluded that $\eta' = 0$.

As remarked earlier, following (3.27), the constant η', even when properly included in the constitutive equations, does not appear in the displacement-equations of motion. Nevertheless, it may enter the displacement field through certain of the boundary conditions.

GRIOLI'S correct finite constitutive equations, (4.30)–(4.32), may be transformed to TOUPIN'S by means of a theorem by CAUCHY, previously employed by TOUPIN in another connection [*10*, p. 888], to the effect that a function of any number of vectors, which is invariant in a rigid rotation of the system of vectors, must reduce to a function of the lengths and scalar products of the vectors and the determinants of their components taken three at a time. In the case of GRIOLI'S $U(\boldsymbol{\Delta}, \boldsymbol{\Lambda})$, Cauchy's theorem requires that U must reduce to a function of the three products

$$\boldsymbol{\Delta} \cdot \boldsymbol{\Delta}_C, \quad \boldsymbol{\Lambda} \cdot \boldsymbol{\Lambda}_C, \quad \boldsymbol{\Delta} \cdot \boldsymbol{\Lambda}_C; \tag{4.44}$$

the two scalar determinants

$$(\boldsymbol{\Delta})_d = \tfrac{1}{6}\, \boldsymbol{\Delta} \overset{\times}{\times} \boldsymbol{\Delta} : \boldsymbol{\Delta}, \qquad (\boldsymbol{\Lambda})_d = \tfrac{1}{6}\, \boldsymbol{\Lambda} \overset{\times}{\times} \boldsymbol{\Lambda} : \boldsymbol{\Lambda}; \tag{4.45}$$

and the two determinants

$$\tfrac{1}{2}\, \boldsymbol{\Delta} \overset{\times}{\times} \boldsymbol{\Delta} \cdot \boldsymbol{\Lambda}_C, \qquad \tfrac{1}{2}\, \boldsymbol{\Lambda} \overset{\times}{\times} \boldsymbol{\Lambda} \cdot \boldsymbol{\Delta}_C. \tag{4.46}$$

However, all of these can be expressed in terms of the material strain dyadic $\mathbf{E}$ and TOUPIN'S variable $\mathbf{K}$. We have

$$\boldsymbol{\Delta} \cdot \boldsymbol{\Delta}_C = 2\mathbf{E} + \mathring{\mathbf{I}}, \tag{4.47}$$

$$\tfrac{1}{2}\, \boldsymbol{\Delta} \overset{\times}{\times} \boldsymbol{\Delta} \cdot \boldsymbol{\Lambda}_C = -\tfrac{1}{4}\, \mathring{\boldsymbol{\nabla}} \boldsymbol{r} \overset{\times}{\times} \mathring{\boldsymbol{\nabla}} \boldsymbol{r} \cdot (\mathring{\boldsymbol{\nabla}} \mathring{\boldsymbol{\nabla}} \boldsymbol{r} \overset{\cdot}{\times} \mathring{\boldsymbol{r}} \boldsymbol{\nabla})_C = \tfrac{1}{2} (\mathring{\boldsymbol{\nabla}} \mathring{\boldsymbol{\nabla}} \boldsymbol{r} \underset{\cdot}{\times} \mathring{\boldsymbol{\nabla}} \boldsymbol{r})_C = \mathbf{K}_C; \tag{4.48}$$

and the remaining variables in (4.44)—(4.46) can be expressed in terms of $\mathbf{\Delta}\cdot\mathbf{\Delta}_C$ and $\mathbf{K}$ as follows:

$$(\mathbf{\Delta})_d = \pm\,[(\mathbf{\Delta}\cdot\mathbf{\Delta}_C)_d]^{\frac{1}{2}}; \tag{4.49}$$

$$\begin{aligned}\mathbf{\Delta}\cdot\mathbf{\Lambda}_C &= \mathbf{\Delta}\cdot(\mathbf{\Delta}_C\cdot\mathbf{\Delta}_C^{-1})\cdot\mathbf{\Lambda}_C,\\ &= \mathbf{\Delta}\cdot\mathbf{\Delta}_C\cdot[\tfrac{1}{2}(\mathbf{\Delta})_d^{-}\,\mathbf{\Delta}\overset{\times}{\times}\mathbf{\Delta}]\cdot\mathbf{\Lambda}_C,\\ &= (\mathbf{\Delta})_d^{-1}(\mathbf{\Delta}\cdot\mathbf{\Delta}_C)\cdot\mathbf{K}_C;\end{aligned} \tag{4.50}$$

$$\begin{aligned}\mathbf{\Lambda}\cdot\mathbf{\Lambda}_C &= \mathbf{\Lambda}\cdot(\mathbf{\Delta}_C\cdot\mathbf{\Delta}_C^{-1})\cdot(\mathbf{\Delta}^{-1}\cdot\mathbf{\Delta})\cdot\mathbf{\Lambda}_C,\\ &= (\mathbf{\Lambda}\cdot\mathbf{\Delta}_C)\cdot(\mathbf{\Delta}_C^{-1}\cdot\mathbf{\Delta}^{-1})\cdot(\mathbf{\Delta}\cdot\mathbf{\Lambda}_C),\\ &= (\mathbf{\Delta}\cdot\mathbf{\Lambda}_C)_C\cdot(\mathbf{\Delta}\cdot\mathbf{\Delta}_C)^{-1}\cdot(\mathbf{\Delta}\cdot\mathbf{\Lambda}_C);\end{aligned} \tag{4.51}$$

$$(\mathbf{\Lambda})_d = \pm\,[(\mathbf{\Lambda}\cdot\mathbf{\Lambda}_C)_d]^{\frac{1}{2}}; \tag{4.52}$$

$$\begin{aligned}\tfrac{1}{2}\mathbf{\Lambda}\overset{\times}{\times}\mathbf{\Lambda}\cdot\mathbf{\Delta}_C &= (\mathbf{\Lambda})_d\,\mathbf{\Lambda}_C^{-1}\cdot\mathbf{\Delta}_C,\\ &= (\mathbf{\Lambda})_d\,\mathbf{\Lambda}_C^{-1}\cdot(\mathbf{\Lambda}^{-1}\cdot\mathbf{\Lambda})\cdot\mathbf{\Delta}_C,\\ &= (\mathbf{\Lambda})_d(\mathbf{\Lambda}\cdot\mathbf{\Lambda}_C)^{-1}\cdot(\mathbf{\Delta}\cdot\mathbf{\Lambda}_C)_C.\end{aligned} \tag{4.53}$$

Thus, taking only the positive signs in (4.49) and (4.52), each of the variables (4.44)—(4.46) can be expressed in terms of $\mathbf{E}$ and $\mathbf{K}$ and, hence, GRIOLI's $U(\mathbf{\Delta},\mathbf{\Lambda})$ can be replaced by TOUPIN's $U(\mathbf{E},\mathbf{K})$.

5. Uniqueness of solution

We consider solutions of the thirty-seven equations

{3} $$\nabla\cdot\boldsymbol{\tau}+\varrho\boldsymbol{f}=\varrho\ddot{\boldsymbol{u}}, \qquad (5.1)_1$$

{3} $$\nabla\cdot\boldsymbol{\mu}+\varrho\boldsymbol{c}+\boldsymbol{\tau}_\times\mathbf{I}=0, \qquad (5.1)_2$$

{6} $$\boldsymbol{\epsilon}=\tfrac{1}{2}(\nabla\boldsymbol{u}+\boldsymbol{u}\nabla), \qquad (5.2)_1$$

{3} $$\boldsymbol{w}=\tfrac{1}{2}\nabla\times\boldsymbol{u}, \qquad (5.2)_2$$

{8} $$\boldsymbol{\varkappa}=\nabla\boldsymbol{w}, \qquad (5.2)_3$$

{6} $$\boldsymbol{\tau}^S=\lambda\boldsymbol{\epsilon}_S\mathbf{I}+2\mu\boldsymbol{\epsilon}, \qquad (5.3)_1$$

{8} $$\boldsymbol{\mu}^D=4\eta\boldsymbol{\varkappa}+4\eta'\boldsymbol{\varkappa}_C, \qquad (5.3)_2$$

in thirty-eight dependent variables (9 components of $\boldsymbol{\tau}$, 9 of $\boldsymbol{\mu}$, 3 of $\boldsymbol{u}$, 6 of $\boldsymbol{\epsilon}$, 3 of $\boldsymbol{w}$ and 8 of $\boldsymbol{\varkappa}$).

As noted previously, the scalar of $\boldsymbol{\mu}$ cannot be determined, so that we shall have to be content with lack of uniqueness to the extent of an arbitrary value of $\boldsymbol{\mu}:\mathbf{I}$. With this reservation, we proceed in the usual manner to consider two solutions $\boldsymbol{u}'$, $\boldsymbol{\epsilon}'$, $\boldsymbol{w}'$, $\boldsymbol{\varkappa}'$, $\boldsymbol{\tau}'$, $\boldsymbol{\mu}'$ and $\boldsymbol{u}''$, $\boldsymbol{\epsilon}''$, $\boldsymbol{w}''$, $\boldsymbol{\varkappa}''$, $\boldsymbol{\tau}''$, $\boldsymbol{\mu}''$, of (5.1)—(5.3), and their differences, $\boldsymbol{u}=\boldsymbol{u}'-\boldsymbol{u}''$, *etc.*, which are also solutions, since the equations are linear. Similarly, body force and body couple differences are defined by $\boldsymbol{f}=\boldsymbol{f}'-\boldsymbol{f}''$ and $\boldsymbol{c}=\boldsymbol{c}'-\boldsymbol{c}''$. From $(5.1)_1$, we form the equation

$$\int_{t_0}^{t} dt \int_V (\nabla\cdot\boldsymbol{\tau}+\varrho\boldsymbol{f}-\varrho\ddot{\boldsymbol{u}})\cdot\dot{\boldsymbol{u}}\,dV=0, \tag{5.4}$$

on the difference variables. Now,

$$(\nabla\cdot\boldsymbol{\tau})\cdot\dot{\boldsymbol{u}}=\nabla\cdot(\boldsymbol{\tau}\cdot\dot{\boldsymbol{u}})-\boldsymbol{\tau}:\nabla\dot{\boldsymbol{u}}=\nabla\cdot(\boldsymbol{\tau}\cdot\dot{\boldsymbol{u}})-\boldsymbol{\tau}^S:\nabla\dot{\boldsymbol{u}}-\boldsymbol{\tau}^A:\nabla\dot{\boldsymbol{u}}. \tag{5.5}$$

But $\boldsymbol{\tau}^A : \nabla \dot{\boldsymbol{u}} = \frac{1}{2}(\boldsymbol{\tau} \times \mathbf{I}) \cdot \nabla \times \dot{\boldsymbol{u}}$. Hence, using $(5.1)_2$,

$$\begin{aligned}\boldsymbol{\tau}^A : \nabla \dot{\boldsymbol{u}} &= -\tfrac{1}{2}(\nabla \cdot \boldsymbol{\mu} + \varrho \boldsymbol{c}) \cdot \nabla \times \dot{\boldsymbol{u}} \\ &= -\tfrac{1}{2}[\nabla \cdot (\boldsymbol{\mu} \cdot \nabla \times \dot{\boldsymbol{u}}) - \boldsymbol{\mu} : \nabla\nabla \times \dot{\boldsymbol{u}}] - \tfrac{1}{2}\varrho \boldsymbol{c} \cdot \nabla \times \dot{\boldsymbol{u}}. \end{aligned} \tag{5.6}$$

Inserting (5.6) in $(5.5)_2$ and using $(5.2)_1$, $(5.2)_2$ and $(5.2)_3$, we have

$$(\nabla \cdot \boldsymbol{\tau}) \cdot \dot{\boldsymbol{u}} = \nabla \cdot (\boldsymbol{\tau} \cdot \dot{\boldsymbol{u}} + \boldsymbol{\mu} \cdot \dot{\boldsymbol{w}}) - \boldsymbol{\tau}^S : \dot{\boldsymbol{\epsilon}} - \boldsymbol{\mu} : \dot{\boldsymbol{\varkappa}} + \varrho \boldsymbol{c} \cdot \dot{\boldsymbol{w}}. \tag{5.7}$$

Then, noting that $\boldsymbol{\mu} : \dot{\boldsymbol{\varkappa}} = \boldsymbol{\mu}^D : \dot{\boldsymbol{\varkappa}}$, since $\boldsymbol{\varkappa} : \mathbf{I} = 0$, and using $(5.3)_1$ and $(5.3)_2$,

$$\begin{aligned}(\nabla \cdot \boldsymbol{\tau}) \cdot \dot{\boldsymbol{u}} &= \nabla \cdot (\boldsymbol{\tau} \cdot \dot{\boldsymbol{u}} + \boldsymbol{\mu} \cdot \dot{\boldsymbol{w}}) - (\lambda \epsilon_S \dot{\epsilon}_S + 2\mu \boldsymbol{\epsilon} : \dot{\boldsymbol{\epsilon}} + 4\eta \boldsymbol{\varkappa} : \dot{\boldsymbol{\varkappa}} + 4\eta' \boldsymbol{\varkappa} : \dot{\boldsymbol{\varkappa}}_C) + \varrho \boldsymbol{c} \cdot \dot{\boldsymbol{w}}, \\ &= \nabla \cdot (\boldsymbol{\tau} \cdot \dot{\boldsymbol{u}} + \boldsymbol{\mu} \cdot \dot{\boldsymbol{w}}) - \dot{W} + \varrho \boldsymbol{c} \cdot \dot{\boldsymbol{w}}, \end{aligned} \tag{5.8}$$

where W is defined as in (3.23), but in terms of the difference variables. Inserting (5.8) in (5.4) and applying the divergence theorem and an integration with respect to time, we find

$$\begin{aligned}\int_V [\tfrac{1}{2}\varrho \dot{\boldsymbol{u}} \cdot \dot{\boldsymbol{u}} + W]_{t_0}^t dV &= \int_{t_0}^t dt \int_V \varrho (\boldsymbol{f} \cdot \dot{\boldsymbol{u}} + \boldsymbol{c} \cdot \dot{\boldsymbol{w}})\, dV + \\ &\quad + \int_{t_0}^t dt \int_S (\boldsymbol{n} \cdot \boldsymbol{\tau} \cdot \dot{\boldsymbol{u}} + \boldsymbol{n} \cdot \boldsymbol{\mu} \cdot \dot{\boldsymbol{w}})\, dS. \end{aligned} \tag{5.9}$$

From the integrand of the surface integral, it would appear that the equations require six boundary conditions. However, five are sufficient. We note that

$$\begin{aligned}\boldsymbol{n} \cdot \boldsymbol{\mu} \cdot \dot{\boldsymbol{w}} &= \boldsymbol{n} \cdot \boldsymbol{\mu} \cdot \boldsymbol{n}\boldsymbol{n} \cdot \dot{\boldsymbol{w}} + \boldsymbol{n} \cdot \boldsymbol{\mu} \cdot (\mathbf{I} - \boldsymbol{n}\boldsymbol{n}) \cdot \dot{\boldsymbol{w}} \\ &= \tfrac{1}{2}\mu_{nn} \boldsymbol{n} \cdot \nabla \times \dot{\boldsymbol{u}} + \boldsymbol{n} \cdot \boldsymbol{\mu} \cdot (\mathbf{I} - \boldsymbol{n}\boldsymbol{n}) \cdot \dot{\boldsymbol{w}}, \end{aligned} \tag{5.10}$$

where $\mu_{nn} = \boldsymbol{n} \cdot \boldsymbol{\mu} \cdot \boldsymbol{n}$. Also,

$$\tfrac{1}{2}\mu_{nn} \boldsymbol{n} \cdot \nabla \times \dot{\boldsymbol{u}} = \tfrac{1}{2}\boldsymbol{n} \cdot \nabla \times (\mu_{nn} \dot{\boldsymbol{u}}) - \tfrac{1}{2}\boldsymbol{n} \times \nabla \mu_{nn} \cdot \dot{\boldsymbol{u}}. \tag{5.11}$$

By Stokes' theorem, if the surface S is smooth,

$$\int_S \boldsymbol{n} \cdot \nabla \times (\mu_{nn} \dot{\boldsymbol{u}})\, dS = 0. \tag{5.12}$$

It may be observed, from (5.11) and (5.12), that a uniform distribution of μ_{nn}, on a smooth surface, does no work.

Inserting (5.10)—(5.12) in (5.9), we find

$$\begin{aligned}\int_V [\tfrac{1}{2}\varrho \dot{\boldsymbol{u}} \cdot \dot{\boldsymbol{u}} + W]_{t_0}^t dV &= \int_{t_0}^t dt \int_V \varrho (\boldsymbol{f} \cdot \dot{\boldsymbol{u}} + \boldsymbol{c} \cdot \dot{\boldsymbol{w}})\, dV + \\ &\quad + \int_{t_0}^t dt \int_S [(\boldsymbol{n} \cdot \boldsymbol{\tau} - \tfrac{1}{2}\boldsymbol{n} \times \nabla \mu_{nn}) \cdot \dot{\boldsymbol{u}} + \boldsymbol{n} \cdot \boldsymbol{\mu} \cdot (\mathbf{I} - \boldsymbol{n}\boldsymbol{n}) \cdot \dot{\boldsymbol{w}}]\, dS. \end{aligned} \tag{5.13}$$

Thus, the normal component of the couple-stress vector on S enters only in the combination with the force-stress vector shown in the coefficient of $\dot{\boldsymbol{u}}$ in the surface integral in (5.13). Accordingly, specification of the boundary values of the three components of the force-stress vector and the three components of the couple-stress vector results in only five boundary conditions rather than six. Alternatively, the three components of displacement and only the tangential components of rotation, $(\mathbf{I} - \boldsymbol{n}\boldsymbol{n}) \cdot \boldsymbol{w}$, may be specified on the boundary.

To apply the conditions, implied in (5.13), to problems involving force-stress and couple-stress boundary values, it is necessary to express the coefficient of

$\dot{\boldsymbol{u}}$, in the surface integral, in terms of $\boldsymbol{\tau}^S$ and $\boldsymbol{\mu}^D$ since only these portions of $\boldsymbol{\tau}$ and $\boldsymbol{\mu}$ appear in the constitutive equations. To this end, we write $\boldsymbol{\tau}=\boldsymbol{\tau}^S+\boldsymbol{\tau}^A$ in (5.13). Then, from (1.10), which is an alternative form of $(5.1)_2$, we may write

$$\boldsymbol{n}\cdot\boldsymbol{\tau}^A\cdot\dot{\boldsymbol{u}}=\tfrac{1}{2}\boldsymbol{n}\cdot\mathbf{I}\times(\boldsymbol{\nabla}\cdot\boldsymbol{\mu}+\varrho\,\boldsymbol{c})\cdot\dot{\boldsymbol{u}}=\tfrac{1}{2}\boldsymbol{n}\times(\boldsymbol{\nabla}\cdot\boldsymbol{\mu})\cdot\dot{\boldsymbol{u}}+\tfrac{1}{2}\varrho\,\boldsymbol{n}\times\boldsymbol{c}\cdot\dot{\boldsymbol{u}}. \tag{5.14}$$

Noting that $\boldsymbol{\nabla}\cdot\boldsymbol{\mu}-\boldsymbol{\nabla}\mu_{nn}=\boldsymbol{\nabla}\cdot\boldsymbol{\mu}^D-\boldsymbol{\nabla}\mu_{nn}^D$, we have, then,

$$\int_S(\boldsymbol{n}\cdot\boldsymbol{\tau}-\tfrac{1}{2}\boldsymbol{n}\times\boldsymbol{\nabla}\mu_{nn})\cdot\dot{\boldsymbol{u}}\,dS=\int_S\boldsymbol{p}\cdot\dot{\boldsymbol{u}}\,dS+\tfrac{1}{2}\varrho\int_S\boldsymbol{n}\times\boldsymbol{c}\cdot\dot{\boldsymbol{u}}\,dS, \tag{5.15}$$

where

$$\boldsymbol{p}=\boldsymbol{n}\cdot\boldsymbol{\tau}^S+\tfrac{1}{2}\boldsymbol{n}\times(\boldsymbol{\nabla}\cdot\boldsymbol{\mu}^D-\boldsymbol{\nabla}\mu_{nn}^D). \tag{5.16}$$

Also, in (5.13),

$$\begin{aligned}\varrho\int_V\boldsymbol{c}\cdot\dot{\boldsymbol{w}}\,dV&=\tfrac{1}{2}\varrho\int_V\boldsymbol{\nabla}\times\boldsymbol{c}\cdot\dot{\boldsymbol{u}}\,dV-\tfrac{1}{2}\varrho\int_V\boldsymbol{\nabla}\cdot(\boldsymbol{c}\times\dot{\boldsymbol{u}})\,dV,\\&=\tfrac{1}{2}\varrho\int_V\boldsymbol{\nabla}\times\boldsymbol{c}\cdot\dot{\boldsymbol{u}}\,dV-\tfrac{1}{2}\varrho\int_S\boldsymbol{n}\times\boldsymbol{c}\cdot\dot{\boldsymbol{u}}\,dS.\end{aligned} \tag{5.17}$$

From these results, we may write (5.13) in the final form:

$$\begin{aligned}\int_V[\tfrac{1}{2}\varrho\,\dot{\boldsymbol{u}}\cdot\dot{\boldsymbol{u}}+W]_{t_0}^t\,dV&=\int_{t_0}^t dt\int_V\varrho(\boldsymbol{f}+\tfrac{1}{2}\boldsymbol{\nabla}\times\boldsymbol{c})\cdot\dot{\boldsymbol{u}}\,dV+\\&\quad+\int_{t_0}^t dt\int_S[\boldsymbol{p}\cdot\dot{\boldsymbol{u}}+\boldsymbol{n}\cdot\boldsymbol{\mu}\cdot(\mathbf{I}-\boldsymbol{n}\boldsymbol{n})\cdot\dot{\boldsymbol{w}}]\,dS.\end{aligned} \tag{5.18}$$

Consider, for example, orthogonal curvilinear or rectilinear coordinates α,β,γ and unit vectors $\boldsymbol{e}_\alpha,\boldsymbol{e}_\beta,\boldsymbol{e}_\gamma$ in the directions α,β,γ increasing. Let $\gamma=\gamma_0$ be the bounding surface S, whence $\boldsymbol{n}=\boldsymbol{e}_\gamma$. Then the integrand of the surface integral in (5.18) is

$$p_\alpha\dot{u}_\alpha+p_\beta\dot{u}_\beta+\tau_{\gamma\gamma}\dot{u}_\gamma+\mu_{\gamma\alpha}\dot{w}_\alpha+\mu_{\gamma\beta}\dot{w}_\beta, \tag{5.19}$$

where

$$p_\alpha=\tau_{\gamma\alpha}^S-\tfrac{1}{2}\boldsymbol{e}_\beta\cdot(\boldsymbol{\nabla}\cdot\boldsymbol{\mu}^D-\boldsymbol{\nabla}\mu_{\gamma\gamma}^D),\qquad p_\beta=\tau_{\gamma\beta}^S+\tfrac{1}{2}\boldsymbol{e}_\alpha\cdot(\boldsymbol{\nabla}\cdot\boldsymbol{\mu}^D-\boldsymbol{\nabla}\mu_{\gamma\gamma}^D), \tag{5.20}$$

and $\tau_{\gamma\alpha}^S=\boldsymbol{e}_\gamma\cdot\boldsymbol{\tau}^S\cdot\boldsymbol{e}_\alpha$, *etc.*; $\mu_{\gamma\alpha}=\boldsymbol{e}_\gamma\cdot\boldsymbol{\mu}\cdot\boldsymbol{e}_\alpha$, *etc.*; $w_\alpha=\boldsymbol{e}_\alpha\cdot\boldsymbol{w}$, *etc.*

If S has an edge, C, formed by the intersection of two coordinate surfaces, say $\gamma=\gamma_0$ and $\alpha=\alpha_0$, the surface integral in (5.12) is equal to $\int_C\mu_{nn}\dot{\boldsymbol{u}}\cdot d\boldsymbol{r}$, instead of zero, for each surface; and this contributes to (5.18) the additional integral

$$\int_C\{[\mu_{\gamma\gamma}]_{\gamma_0}-[\mu_{\alpha\alpha}]_{\alpha_0}\}\,\dot{\boldsymbol{u}}\cdot d\boldsymbol{r}=\int_C\{[\mu_{\gamma\gamma}^D]_{\gamma_0}-[\mu_{\alpha\alpha}^D]_{\alpha_0}\}\,\dot{\boldsymbol{u}}\cdot d\boldsymbol{r}, \tag{5.21}$$

where the integration is along the edge. The quantity in braces, in (5.21), is a force, per unit length of edge, tangent to the edge.

Conditions sufficient for a unique solution are now obtained from (5.18) in the usual manner [9, p. 176], based on the positive definiteness of W and $\dot{\boldsymbol{u}}\cdot\dot{\boldsymbol{u}}$. Returning to a single solution, $\boldsymbol{u},\boldsymbol{\epsilon},\boldsymbol{w},\boldsymbol{\varkappa},\boldsymbol{\tau},\boldsymbol{\mu}$, we see that sufficient conditions for uniqueness are

1. In V: at each point, $\boldsymbol{f}$, $\boldsymbol{\nabla}\times\boldsymbol{c}$ and initial values of $\boldsymbol{u}$ and $\dot{\boldsymbol{u}}$;

2. On S: at each point, a component of $\boldsymbol{u}$ (or $\boldsymbol{p}$), in any direction, and the resultant of $\boldsymbol{p}$ (or $\boldsymbol{u}$) in the plane at right angles; and, in the tangent plane, a component of $\boldsymbol{w}$ (or $\boldsymbol{m}_n$) and the component of $\boldsymbol{m}_n$ (or $\boldsymbol{w}$) at right angles. In orthogonal coordinates, with γ increasing outward, normal to S, these components are one member of each of the five products

$$p_\alpha u_\alpha,\ p_\beta u_\beta,\ \tau_{\gamma\gamma} u_\gamma,\ \mu_{\gamma\alpha} w_\alpha,\ \mu_{\gamma\beta} w_\beta, \tag{5.22}$$

where

$$p_\alpha = \tau^S_{\gamma\alpha} - \tfrac{1}{2} \boldsymbol{e}_\beta \cdot (\boldsymbol{\nabla} \cdot \boldsymbol{\mu}^D - \boldsymbol{\nabla} \mu^D_{\gamma\gamma}), \qquad p_\beta = \tau^S_{\gamma\beta} + \tfrac{1}{2} \boldsymbol{e}_\alpha \cdot (\boldsymbol{\nabla} \cdot \boldsymbol{\mu}^D - \boldsymbol{\nabla} \mu^D_{\gamma\gamma}); \tag{5.23}$$

3. On an edge: at each point, the force per unit length or the displacement — both tangent to the edge. If the edge is the intersection of orthogonal surfaces $\gamma = \gamma_0$ and $\alpha = \alpha_0$, the quantities are

$$[\mu^D_{\gamma\gamma}]_{\gamma_0} - [\mu^D_{\alpha\alpha}]_{\alpha_0} \quad \text{or} \quad u_\beta. \tag{5.24}$$

The reduction of the number of boundary conditions from six to five is analogous to the reduction from three to two conditions in the classical theory of flexure of thin plates.

6. Wave motion

Consider the displacement-equation of motion, (3.27), in the absence of body force and body couple:

$$\mu \nabla^2 \boldsymbol{u} + (\lambda + \mu)\, \boldsymbol{\nabla} \boldsymbol{\nabla} \cdot \boldsymbol{u} + \eta \nabla^2 \boldsymbol{\nabla} \times \boldsymbol{\nabla} \times \boldsymbol{u} = \varrho \ddot{\boldsymbol{u}}. \tag{6.1}$$

Taking the divergence and curl of (6.1), we obtain, respectively,

$$c_1^2 \nabla^2 \boldsymbol{\nabla} \cdot \boldsymbol{u} = \boldsymbol{\nabla} \cdot \ddot{\boldsymbol{u}}, \tag{6.2}$$

$$c_2^2 (1 - l^2 \nabla^2) \nabla^2 \boldsymbol{\nabla} \times \boldsymbol{u} = \boldsymbol{\nabla} \times \ddot{\boldsymbol{u}}, \tag{6.3}$$

where

$$l^2 = \eta/\mu, \qquad c_1^2 = (\lambda + 2\mu)/\varrho, \qquad c_2^2 = \mu/\varrho. \tag{6.4}$$

Thus, the dilatation is propagated non-dispersively, with velocity c_1, as it is in the absence of couple-stress; but the propagation of the rotation is influenced by couple-stress.

Consider, for example, the plane wave

$$\boldsymbol{\nabla} \times \boldsymbol{u} = \boldsymbol{d} A \exp[i\xi(\boldsymbol{n} \cdot \boldsymbol{r} - ct)] = \boldsymbol{d} A \exp[i(\xi \boldsymbol{n} \cdot \boldsymbol{r} - \omega t)], \tag{6.5}$$

where $\boldsymbol{d}$ is a unit vector, A is the amplitude, ξ is the wave number, $\boldsymbol{n}$ is the unit wave normal, c is the phase velocity and ω is the circular frequency. Upon substituting (6.5) in (6.3), we find

$$c^2 = c_2^2 (1 + l^2 \xi^2), \qquad \omega^2 = \xi^2 c_2^2 (1 + l^2 \xi^2); \tag{6.6}$$

and, solving $(6.6)_2$ for ξ^2, we find the two roots

$$\xi_1^2 = \tfrac{1}{2} l^{-2} [(1 + 4 l^2 \omega^2 / c_2^2)^{\frac{1}{2}} - 1], \qquad \xi_2^2 = -\tfrac{1}{2} l^{-2} [(1 + 4 l^2 \omega^2 / c_2^2)^{\frac{1}{2}} + 1]; \tag{6.7}$$

so that one wave number, ξ_1, is real and the other is a pure imaginary, since, by (3.28) and (3.29), l is real. Thus, there are two rotational waves: one propagating, the other non-propagating and both dispersive.

The group velocity of the real wave,

$$d\omega/d\xi_1 = c_2 (1 + 2 l^2 \xi_1^2)/(1 + l^2 \xi_1^2)^{\frac{1}{2}}, \tag{6.8}$$

increases monotonically with increasing $\xi_1 l$. This increase might produce a detectable effect of couple-stress in high frequency vibrations. If $\xi_1 l$ is small, *i.e.*, if the wave length is large in comparison with the material constant l, the group velocity is approximately the same (c_2) as without couple-stress. The phase velocity of the real wave increases monotonically with increasing $\xi_1 l$ and reaches the velocity, c_1, of the dilatational wave at

$$\xi_1 l = [(\lambda+\mu)/\mu]^{\frac{1}{2}} = (1-2\sigma)^{-\frac{1}{2}}, \tag{6.9}$$

in which

$$\sigma = \lambda/2(\lambda+\mu) \tag{6.10}$$

is Poisson's ratio.

The imaginary branch has a wave number cut-off at $\xi_2 l = \pm i$ and zero frequency. This branch is a second major source of the difference between elastic fields with and without couple-stresses. In both dynamic and static fields, it contributes local effects near boundaries and singularities.

7. Potentials of dilatation and rotation

Solutions of the equation of motion may be expressed as solutions of equations which reduce to Lamé's in the absence of couple-stress. Consider the resolution of the displacement into its lamellar and solenoidal components:

$$\boldsymbol{u} = \boldsymbol{\nabla}\varphi + \boldsymbol{\nabla}\times\boldsymbol{H}, \qquad \boldsymbol{\nabla}\cdot\boldsymbol{H} = 0. \tag{7.1}$$

Upon substituting (7.1) in (6.1), we obtain

$$\boldsymbol{\nabla}(c_1^2 \nabla^2\varphi - \ddot{\varphi}) + \boldsymbol{\nabla}\times[c_2^2(1-l^2\nabla^2)\nabla^2\boldsymbol{H} - \ddot{\boldsymbol{H}}] = 0, \tag{7.2}$$

which is satisfied by

$$c_1^2\nabla^2\varphi = \ddot{\varphi}, \qquad c_2^2(1-l^2\nabla^2)\nabla^2\boldsymbol{H} = \ddot{\boldsymbol{H}}. \tag{7.3}$$

That (7.1) and (7.3) comprise the complete solution of the equation of motion (6.1) may be proved by following exactly the steps in STERNBERG's [*11*] form of the proof for the case $l=0$. It is only necessary to replace the operator

$$\Box_2^2 = \nabla^2 - c_2^{-2}\,\partial^2/\partial t^2,$$

in STERNBERG's proof, by

$$\Box_2^2 = (1-l^2\nabla^2)\nabla^2 - c_2^{-2}\,\partial^2/\partial t^2.$$

Also, as in the case $l=0$, the condition $\boldsymbol{\nabla}\cdot\boldsymbol{H}=0$ is not necessary.

For steady vibrations ($\varphi = \overline{\varphi}e^{i\omega t}$, $\boldsymbol{H} = \overline{\boldsymbol{H}}e^{i\omega t}$), (7.3) become

$$(\nabla^2+\alpha^2)\overline{\varphi} = 0, \qquad (\nabla^2+\beta_1^2)(\nabla^2-\beta_2^2)\overline{\boldsymbol{H}} = 0, \tag{7.4}$$

where $\alpha = \omega/c_1$ and

$$\beta_1 = 2^{-\frac{1}{2}}l^{-1}[(1+4l^2\omega^2/c_2^2)^{\frac{1}{2}} - 1]^{\frac{1}{2}}, \qquad \beta_2 = 2^{-\frac{1}{2}}l^{-1}[(1+4l^2\omega^2/c_2^2)^{\frac{1}{2}} + 1]^{\frac{1}{2}}. \tag{7.5}$$

Then, with $\overline{\boldsymbol{H}} = \overline{\boldsymbol{H}}' + \overline{\boldsymbol{H}}''$, the complete solution [*11*] is

$$\overline{\boldsymbol{u}} = \boldsymbol{\nabla}\overline{\varphi} + \boldsymbol{\nabla}\times\overline{\boldsymbol{H}}' + \boldsymbol{\nabla}\times\overline{\boldsymbol{H}}'', \tag{7.6}$$

where $\overline{\varphi}$, $\overline{\boldsymbol{H}}'$ and $\overline{\boldsymbol{H}}''$ are governed by the Helmholtz equations:

$$(\nabla^2+\alpha^2)\overline{\varphi} = 0, \qquad (\nabla^2+\beta_1^2)\overline{\boldsymbol{H}}' = 0, \qquad (\nabla^2-\beta_2^2)\overline{\boldsymbol{H}}'' = 0. \tag{7.7}$$

8. Thickness-shear vibrations of a plate

We consider an infinite plate bounded by free surfaces $y=\pm b$ in a rectangular coordinate system x, y, z. From (5.22), the boundary conditions on $y=\pm b$ are

$$p_\alpha = p_\beta = \tau_{\gamma\gamma} = \mu_{\gamma\alpha} = \mu_{\gamma\beta} = 0, \tag{8.1}$$

where α, β, γ are z, x, y respectively. Thus,

$$\begin{gathered} p_\alpha = \tau^S_{yz} - \tfrac{1}{2}\mathbf{e}_x \cdot (\boldsymbol{\nabla}\cdot\boldsymbol{\mu}^D - \boldsymbol{\nabla}\mu^D_{yy}), \qquad p_\beta = \tau^S_{yx} + \tfrac{1}{2}\mathbf{e}_z \cdot (\boldsymbol{\nabla}\cdot\boldsymbol{\mu}^D - \boldsymbol{\nabla}\mu^D_{yy}), \\ \tau_{\gamma\gamma} = \tau_{yy}, \qquad \mu_{\gamma\alpha} = \mu_{yz}, \qquad \mu_{\gamma\beta} = \mu_{yx}. \end{gathered} \tag{8.2}$$

Hence, on $y=\pm b$,

$$\begin{gathered} \tau^S_{yz} - \frac{1}{2}\left(\frac{\partial \mu^D_{xx}}{\partial x} + \frac{\partial \mu_{yx}}{\partial y} + \frac{\partial \mu_{zx}}{\partial z} - \frac{\partial \mu^D_{yy}}{\partial x}\right) = 0, \\ \tau^S_{yx} + \frac{1}{2}\left(\frac{\partial \mu_{xz}}{\partial x} + \frac{\partial \mu_{yz}}{\partial y} + \frac{\partial \mu^D_{zz}}{\partial z} - \frac{\partial \mu^D_{yy}}{\partial z}\right) = 0, \\ \tau_{yy} = \mu_{yz} = \mu_{yx} = 0. \end{gathered} \tag{8.3}$$

Of main interest are the antisymmetric, thickness-shear modes of vibration of the plate. For these, we take, as solutions of (7.7),

$$\bar{\varphi} = \bar{H}_x = \bar{H}_y = 0, \qquad \bar{H}'_z = A_1 \cos\beta_1 y, \qquad \bar{H}''_z = A_2 \cosh\beta_2 y, \tag{8.4}$$

where β_1 and β_2 are defined in (7.5). Then, from (8.4) and (7.6),

$$\bar{u}_x = -A_1\beta_1 \sin\beta_1 y + A_2\beta_2 \sinh\beta_2 y, \qquad \bar{u}_y = \bar{u}_z = 0. \tag{8.5}$$

From (8.5) and (5.2), all the components of strain and curvature are zero except ε_{xy} and $\varkappa_{yz}$. Then, from (5.3), the only non-zero components of $\boldsymbol{\tau}^S$ and $\boldsymbol{\mu}^D$, across $y=$ constant, are

$$\tau^S_{yx} = -\mu(A_1\beta_1^2 \cos\beta_1 y - A_2\beta_2^2 \cosh\beta_2 y), \tag{8.6}$$

$$\mu_{yz} = -2\eta(A_1\beta_1^3 \sin\beta_1 y + A_2\beta_2^3 \sinh\beta_2 y). \tag{8.7}$$

As a result, the first, third and fifth of the boundary conditions (8.3) are satisfied identically and the remaining two reduce to

$$\tau^S_{yx} + \tfrac{1}{2}\partial\mu_{yz}/\partial y = 0, \qquad \mu_{yz} = 0, \quad \text{on } y = \pm b. \tag{8.8}$$

Substituting (8.6) and (8.7) in (8.8), we obtain

$$A_1\beta_1^2(1+l^2\beta_1^2)\cos\beta_1 b - A_2\beta_2^2(1-l^2\beta_2^2)\cosh\beta_2 b = 0, \tag{8.9}$$

$$A_1\beta_1^3 \sin\beta_1 b + A_2\beta_2^3 \sinh\beta_2 b = 0. \tag{8.10}$$

Upon eliminating A_1 and A_2 from (8.9) and (8.10), we find, after some manipulation, the frequency equation

$$\tan\gamma = f^3 \tanh f\gamma, \tag{8.11}$$

where

$$\gamma = \beta_1 b, \qquad f = [(1+g^2)^{\frac{1}{2}} + 1]/g, \qquad g = \pi\Omega l/b; \tag{8.12}$$

and

$$\Omega = \omega/(\pi c_2/2b) = \gamma\pi^{-1}\{2[(1+g^2)^{\frac{1}{2}}+1]\}^{\frac{1}{2}} = 2\gamma\pi^{-1}(1+\gamma^2 l^2/b^2)^{\frac{1}{2}}. \tag{8.13}$$

In the limit as l/b goes to zero (absence of couple stress), g goes to zero and f goes to infinity, so that the frequency equation reduces to $\cot\gamma=0$, or $\gamma=n\pi/2$, n odd; and $(8.13)_3$ reduces to $\Omega=n$. Thus, Ω is the ratio of the frequency with couple-stress to the frequency of the lowest mode without couple-stress.

To compute the frequency spectrum, Ω vs. b/l, a value of g is chosen and f is computed from $(8.12)_2$. Then the roots, γ, of (8.11) are computed for that value of f. For each root γ, Ω is computed from $(8.13)_2$ and b/l is computed from $(8.12)_3$. This process is then repeated for successive values of g. The spectrum is illustrated in Fig. 1.

As l/b or Ω becomes large, g increases and f approaches unity, in accordance with $(8.12)_3$ and $(8.12)_2$. Except for the lowest root, the roots of (8.11) are then approximately

$$\gamma=(2n-1)\pi/4,\qquad n=3,5,7,\ldots. \tag{8.14}$$

Then, from $(8.13)_3$, for large Ω and large n,

$$\Omega\approx n\left[1+(n\pi l/2b)^2\right]^{\frac{1}{2}}. \tag{8.15}$$

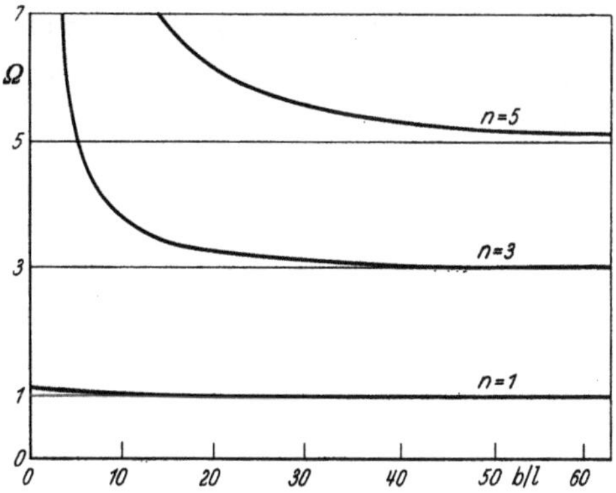

Fig. 1

The lowest root of (8.11) approaches zero as g increases. An expansion in power series yields

$$\gamma^2\approx 6/g \tag{8.16}$$

for large g. Then, from $(8.13)_2$, the limiting frequency of the lowest mode, as l/b approaches infinity, is

$$\Omega=(12)^{\frac{1}{2}}/\pi. \tag{8.17}$$

This is an increase of only about 10% over the frequency of the first mode without couple-stress. Thus, it would not be advisable to look for the effect of couple-stress by measuring the frequencies of the first mode of successively thinner plates.

On the other hand, the frequencies of higher and higher modes of the same plate eventually tend toward (8.15) with couple-stress, as opposed to $\Omega=n$ without. For detecting the effect of couple-stress, then, the most favorable experiment of this type would be made with as high modes and as thin a plate as possible. As is well known [*12*], end effects in a finite plate raise the frequencies of the lower modes above $\Omega=n$; but this effect diminishes with increasing n. Accordingly, as n increases, experimental measurements of Ω/n should first drop to a minimum, due to diminishing end effects, and then begin to increase in accordance with (8.15) if n can be made large enough in comparison with b/l. Following this, Ω/n would again drop: owing to the influence of rotatory inertia, which is not taken into account here.

The insensitivity of the lowest mode to varying b/l is due to an accident of mode shape. The displacement distribution contributed by the imaginary branch is a hyperbolic sinusoid, while that of the real branch is a sinusoid with one node. As b/l approaches zero, the displacement becomes linear, corresponding to infinite wave-length. The discrepancy between the solutions with and without couple-stress is then solely that due to the difference between a linear and a sinusoidal distribution of displacement. This is known [13] to produce a ratio $(12)^{\frac{1}{2}}/\pi$ in the frequencies. For bodies of other shape, insensitivity of the frequency of the lowest mode, to the ratio of the pertinent dimension to l, need not occur — as is illustrated in the following section.

9. Torsional vibrations of a circular cylinder

We consider an infinite, circular cylinder with free surface $r=a$ in a cylindrical coordinate system r, ϑ, z. From (5.22), the boundary conditions on $r=a$ are

$$p_\alpha = p_\beta = \tau_{\gamma\gamma} = \mu_{\gamma\alpha} = \mu_{\gamma\beta} = 0, \tag{9.1}$$

where α, β, γ are ϑ, z, r, respectively, and

$$\begin{gathered} p_\alpha = \tau^S_{r\vartheta} - \tfrac{1}{2}\mathbf{e}_z \cdot (\nabla \cdot \boldsymbol{\mu}^D - \nabla \mu^D_{rr}), \qquad p_\beta = \tau^S_{rz} + \tfrac{1}{2}\mathbf{e}_\vartheta \cdot (\nabla \cdot \boldsymbol{\mu}^D - \nabla \mu^D_{rr}), \\ \tau_{\gamma\gamma} = \tau_{rr}, \qquad \mu_{\gamma\alpha} = \mu_{r\vartheta}, \qquad \mu_{\gamma\beta} = \mu_{rz}. \end{gathered} \tag{9.2}$$

Hence, on $r=a$, the boundary conditions are

$$\begin{gathered} \tau^S_{r\vartheta} - \frac{1}{2}\left(\frac{\partial \mu_{rz}}{\partial r} + \frac{1}{r}\,\frac{\partial \mu_{\vartheta z}}{\partial \vartheta} + \frac{\partial \mu^D_{zz}}{\partial z} + \frac{\mu_{rz}}{r} - \frac{\partial \mu^D_{rr}}{\partial z}\right) = 0, \\ \tau^S_{rz} + \frac{1}{2}\left(\frac{\partial \mu_{r\vartheta}}{\partial r} + \frac{1}{r}\,\frac{\partial \mu^D_{\vartheta\vartheta}}{\partial \vartheta} + \frac{\partial \mu_{z\vartheta}}{\partial z} + \frac{\mu_{r\vartheta} + \mu_{\vartheta r}}{r} - \frac{1}{r}\,\frac{\partial \mu^D_{rr}}{\partial \vartheta}\right) = 0, \\ \tau_{rr} = \mu_{r\vartheta} = \mu_{rz} = 0. \end{gathered} \tag{9.3}$$

With the boundary conditions (9.3), a solution of (7.7) that is analogous to the solution for antisymmetric thickness-shear vibrations of an infinite plate, given in the preceding section, is one independent of ϑ and z: corresponding to torsional vibrations of the cylinder in a state of plane strain. To this end, we take, as solutions of (7.7),

$$\bar{\varphi} = \bar{H}_r = \bar{H}_\vartheta = 0, \qquad \bar{H}'_z = A_1 J_0(\beta_1 r), \qquad \bar{H}''_z = A_2 I_0(\beta_2 r), \tag{9.4}$$

where J_0 and I_0 are Bessel's function and Bessel's modified function of the first kind and order zero. Then, from (7.6),

$$\bar{u}_r = \bar{u}_z = 0, \qquad \bar{u}_\vartheta = A_1 \beta_1 J_1(\beta_1 r) - A_2 \beta_2 I_1(\beta_2 r). \tag{9.5}$$

Substituting (9.5) in (5.2), we find that all the components of strain and curvature are zero except $\varepsilon_{r\vartheta}$ and $\varkappa_{rz}$. Then, from (5.3), the only non-zero components of $\boldsymbol{\tau}^S$ and $\boldsymbol{\mu}^D$, across $r=$ constant, are

$$\tau^S_{r\vartheta} = \mu\{A_1\beta_1[\beta_1 J_0(\beta_1 r) - 2r^{-1} J_1(\beta_1 r)] - A_2\beta_2[\beta_2 I_0(\beta_2 r) - 2r^{-1} I_1(\beta_2 r)]\}, \tag{9.6}$$

$$\mu_{rz} = -2\eta[A_1 \beta_1^3 J_1(\beta_1 r) + A_2 \beta_2^3 I_1(\beta_2 r)]. \tag{9.7}$$

Hence, the boundary conditions on $r=a$ reduce to

$$\tau^S_{r\vartheta} - \tfrac{1}{2} r^{-1} \partial(r\mu_{rz})/\partial r = 0, \qquad \mu_{rz} = 0. \tag{9.8}$$

When (9.6) and (9.7) are substituted in (9.8), there result

$$A_1[\beta_1^2(1+l^2\beta_1^2)J_0(\beta_1 a)-2\beta_1 a^{-1}J_1(\beta_1 a)]- \\ -A_2[\beta_2^2(1-l^2\beta_2^2)I_0(\beta_2 a)-2\beta_2 a^{-1}I_1(\beta_2 a)]=0, \tag{9.9}$$

$$A_1\beta_1^3 J_1(\beta_1 a)+A_2\beta_2^3 I_1(\beta_2 a)=0. \tag{9.10}$$

The frequency equation is obtained by equating to zero the determinant of the coefficients of A_1 and A_2 in (9.9) and (9.10). It may be written in the form

$$\tfrac{1}{2}p\,\mathscr{J}_1(\gamma)-2-f^{-2}[\tfrac{1}{2}q\,\mathscr{I}_1(f\gamma)+2]=0, \tag{9.11}$$

where

$$\mathscr{J}_1(\gamma)=\gamma J_0(\gamma)/J_1(\gamma), \qquad \mathscr{I}_1(f\gamma)=f\gamma I_0(f\gamma)/I_1(f\gamma), \tag{9.12}$$

$$\gamma=\beta_1 a, \qquad f=[(1+g^2)^{\frac{1}{2}}+1]/g, \qquad g=2\bar{\gamma}\,\Omega l/a, \tag{9.13}$$

$$p=(1+g^2)^{\frac{1}{2}}+1, \qquad q=(1+g^2)^{\frac{1}{2}}-1, \tag{9.14}$$

$$\Omega=\omega/(\bar{\gamma}c_2/a)=(\gamma/\bar{\gamma})(p/2)^{\frac{1}{2}}=(\gamma/\bar{\gamma})(1+\gamma^2 l^2/a^2)^{\frac{1}{2}}, \tag{9.15}$$

and $\bar{\gamma}$ is the lowest root of

$$\mathscr{J}_1(\gamma)=2. \tag{9.16}$$

In the absence of couple-stress, l/a goes to zero, g goes to zero, f goes to infinity and $p=2$. Hence, the frequency equation (9.11) reduces to (9.16) and $(9.15)_2$ reduces to $\Omega=\gamma/\bar{\gamma}$. Thus Ω is the ratio of the frequency with couple-stress to the frequency of the lowest mode without couple-stress.

The frequency spectrum, Ω vs. a/l, may be computed, with the aid of ONOE's tables [*14*], by the same procedure as that described for (8.11). The results are similar to those shown in Fig. 1 but, in this case, the frequency ratios of all the modes approach infinity as a/l approaches zero; with the exception, of course, of the mode of rigid rotation, which has zero frequency for all a/l.

10. Green's formulas

In the study of problems of equilibrium, we shall make use of GREEN's formulas for a function, say ψ, for which

$$\nabla^2\psi \quad \text{or} \quad (1-l^2\nabla^2)\psi \quad \text{or} \quad (1-l^2\nabla^2)\nabla^2\psi$$

is known. The formulas may be obtained in the usual way [7, p. 31] from GREEN's identity

$$\int_S \boldsymbol{n}\cdot(\varphi\boldsymbol{\nabla}\psi-\psi\boldsymbol{\nabla}\varphi)\,dS=\int_V(\varphi\nabla^2\psi-\psi\nabla^2\varphi)\,dV. \tag{10.1}$$

When $\nabla^2\psi$ is known, we have the well known result [7, p. 32]

$$4\pi\psi=\int_S \boldsymbol{n}\cdot(r_1^{-1}\boldsymbol{\nabla}\psi-\psi\boldsymbol{\nabla}r_1^{-1})\,dS-\int_V r_1^{-1}\nabla^2\psi\,dV, \tag{10.2}$$

where

$$r_1=[(x-\xi)^2+(y-\eta)^2+(z-\zeta)^2]^{\frac{1}{2}} \tag{10.3}$$

is the distance from the field point $P(x,y,z)$ to the source point $Q(\xi,\eta,\zeta)$.

To obtain the formula applicable if $(1-l^2\nabla^2)\psi$ is known, write (10.1) in the form

$$\int_S \boldsymbol{n}\cdot(\varphi_1\nabla\psi-\psi\nabla\varphi_1)\,dS = l^{-2}\int_V[\psi(1-l^2\nabla^2)\varphi_1-\varphi_1(1-l^2\nabla^2)\psi]\,dV, \tag{10.4}$$

and let

$$\varphi_1 = -r_1^{-1}e^{-r_1/l} \tag{10.5}$$

We note that

$$(1-l^2\nabla^2)\varphi_1 = 0 \tag{10.6}$$

except at P. Then, by the usual limit process, (10.4) reduces to

$$4\pi\psi = \int_S \boldsymbol{n}\cdot(\psi\nabla\varphi_1-\varphi_1\nabla\psi)\,dS - l^{-2}\int_V\varphi_1(1-l^2\nabla^2)\psi\,dV. \tag{10.7}$$

For the case of $(1-l^2\nabla^2)\nabla^2\psi$, first replace φ and ψ, in (10.1), by

$$(1-l^2\nabla^2)\varphi_2, \qquad (1-l^2\nabla^2)\psi,$$

respectively, to obtain

$$\begin{aligned}\int_S \boldsymbol{n}\cdot[(\varphi_2-l^2\nabla^2\varphi_2)\nabla(1-l^2\nabla^2)\psi-(\psi-l^2\nabla^2\psi)\nabla(1-l^2\nabla^2)\varphi_2]\,dS \\ =\int_V[\varphi_2(1-l^2\nabla^2)\nabla^2\psi-\psi(1-l^2\nabla^2)\nabla^2\varphi_2+l^4\nabla^2\varphi_2\nabla^4\psi-l^4\nabla^2\psi\nabla^4\varphi_2]\,dV.\end{aligned} \tag{10.8}$$

Then replace φ by $l^2\nabla^2\varphi_2$ and ψ by $l^2\nabla^2\psi$ in (10.1):

$$l^4\int_S \boldsymbol{n}\cdot(\nabla^2\varphi_2\nabla\nabla^2\psi-\nabla^2\psi\nabla\nabla^2\varphi_2)\,dS = l^4\int_V(\nabla^2\varphi_2\nabla^4\psi-\nabla^2\psi\nabla^4\varphi_2)\,dV. \tag{10.9}$$

Subtracting (10.9) from (10.8), we have

$$\begin{aligned}\int_S \boldsymbol{n}\cdot[(\varphi_2-l^2\nabla^2\varphi_2)\nabla\psi-\psi\nabla(\varphi_2-l^2\nabla^2\varphi_2)+l^2\nabla^2\psi\nabla\varphi_2-l^2\varphi_2\nabla\nabla^2\psi]\,dS \\ =\int_V[\varphi_2(1-l^2\nabla^2)\nabla^2\psi-\psi(1-l^2\nabla^2)\nabla^2\varphi_2]\,dV.\end{aligned} \tag{10.10}$$

In (10.10) let

$$\varphi_2 = r_1^{-1}(1-e^{-r_1/l}) \tag{10.11}$$

and note that

$$(1-l^2\nabla^2)\nabla^2\varphi_2 = 0$$

except at P. Again, by the usual limit process,

$$\begin{aligned}4\pi\psi = \int_S \boldsymbol{n}\cdot[r_1^{-1}\nabla\psi-\psi\nabla r_1^{-1}+l^2\nabla^2\psi\nabla\varphi_2-l^2\varphi_2\nabla\nabla^2\psi]\,dS - \\ -\int_V\varphi_2(1-l^2\nabla^2)\nabla^2\psi\,dV.\end{aligned} \tag{10.12}$$

11. Stress functions

We may find a complete solution of the displacement-equation of equilibrium,

$$\mu\nabla^2\boldsymbol{u}+(\lambda+\mu)\nabla\nabla\cdot\boldsymbol{u}+\eta\nabla^2\nabla\times\nabla\times\boldsymbol{u}+\varrho\boldsymbol{f}+\tfrac{1}{2}\varrho\nabla\times\boldsymbol{c}=0, \tag{11.1}$$

in terms of stress functions which reduce to the PAPKOVITCH functions [*15*] when $l=0$. The procedure follows that in [*16*] and [*17*].

Substitute

$$\boldsymbol{u}=\nabla\varphi+\nabla\times\boldsymbol{H}, \qquad \nabla\cdot\boldsymbol{H}=0, \tag{11.2}$$

in (11.1) and obtain

$$\mu\nabla^2[\alpha\nabla\varphi+(1-l^2\nabla^2)\nabla\times\boldsymbol{H}]+\varrho\boldsymbol{f}+\tfrac{1}{2}\varrho\nabla\times\boldsymbol{c}=0, \tag{11.3}$$

where

$$\alpha=(\lambda+2\mu)/\mu. \tag{11.4}$$

Now define a vector function $\boldsymbol{B}$ by

$$4\pi l^2\boldsymbol{B}=\int_V r_1 e^{-r_1/l}[\alpha\nabla\varphi+(1-l^2\nabla^2)\nabla\times\boldsymbol{H}]\,dV. \tag{11.5}$$

Then, by (10.7),

$$(1-l^2\nabla^2)\,\boldsymbol{B}=\alpha\nabla\varphi+(1-l^2\nabla^2)\nabla\times\boldsymbol{H}; \tag{11.6}$$

and, from (11.6) and (11.3),

$$\mu(1-l^2\nabla^2)\,\nabla^2\boldsymbol{B}=-\varrho\boldsymbol{f}-\tfrac{1}{2}\varrho\nabla\times\boldsymbol{c}. \tag{11.7}$$

We must now solve (11.6) for $\nabla\varphi$ and $\nabla\times\boldsymbol{H}$. First take the divergence of (11.6) and obtain

$$\alpha\nabla^2\varphi=(1-l^2\nabla^2)\,\nabla\cdot\boldsymbol{B}, \tag{11.8}$$

the solution of which is

$$2\alpha\varphi=\boldsymbol{r}\cdot(1-l^2\nabla^2)\,\boldsymbol{B}+B_0, \tag{11.9}$$

where

$$\mu\nabla^2 B_0=\boldsymbol{r}\cdot(\varrho\boldsymbol{f}+\tfrac{1}{2}\varrho\nabla\times\boldsymbol{c}). \tag{11.10}$$

Then, substituting (11.9) in (11.6), we obtain

$$(1-l^2\nabla^2)\,\nabla\times\boldsymbol{H}=(1-l^2\nabla^2)\,\boldsymbol{B}-\tfrac{1}{2}\nabla[\boldsymbol{r}\cdot(1-l^2\nabla^2)\,\boldsymbol{B}+\boldsymbol{B}_0], \tag{11.11}$$

the solution of which is

$$\nabla\times\boldsymbol{H}=\boldsymbol{B}-l^2\nabla\nabla\cdot\boldsymbol{B}-\tfrac{1}{2}\nabla[\boldsymbol{r}\cdot(1-l^2\nabla^2)\,\boldsymbol{B}+B_0]+\boldsymbol{B}', \tag{11.12}$$

where

$$(1-l^2\nabla^2)\,\boldsymbol{B}'=0. \tag{11.13}$$

We may now substitute (11.9) and (11.12) in (11.2) to obtain

$$\boldsymbol{u}=\boldsymbol{B}-l^2\nabla\nabla\cdot\boldsymbol{B}-\alpha'\nabla[\boldsymbol{r}\cdot(1-l^2\nabla^2)\,\boldsymbol{B}+B_0]+\boldsymbol{B}', \tag{11.14}$$

where

$$\alpha'=(\lambda+\mu)/2(\lambda+2\mu)=1/4(1-\sigma). \tag{11.15}$$

Next, we insert (11.14) in (11.1) and find, with the aid of (11.7), (11.10) and (11.13),

$$\nabla\nabla\cdot\boldsymbol{B}'=0. \tag{11.16}$$

In view of (11.13), (11.16), (11.7) and the form of (11.14), $\boldsymbol{B}'$ may be absorbed in $\boldsymbol{B}$. Thus we have, finally,

$$\boldsymbol{u}=\boldsymbol{B}-l^2\nabla\nabla\cdot\boldsymbol{B}-\alpha'\nabla[\boldsymbol{r}\cdot(1-l^2\nabla^2)\,\boldsymbol{B}+B_0], \tag{11.17}$$

$$\mu(1-l^2\nabla^2)\,\nabla^2\boldsymbol{B}=-\varrho\boldsymbol{f}-\tfrac{1}{2}\varrho\nabla\times\boldsymbol{c}, \tag{11.18}$$

$$\mu\nabla^2 B_0=\boldsymbol{r}\cdot(\varrho\boldsymbol{f}+\tfrac{1}{2}\varrho\nabla\times\boldsymbol{c}). \tag{11.19}$$

The functions $\boldsymbol{B}$ and B_0 reduce to Papkovitch's functions when $l=0$. It may be observed that the property $\nabla\cdot\boldsymbol{H}=0$ was not used in the derivation.

For later use, we note that

$$\alpha\nabla\cdot\boldsymbol{u}=(1-l^2\nabla^2)\,\nabla\cdot\boldsymbol{B},\qquad \nabla\times\boldsymbol{u}=\nabla\times\boldsymbol{B},\qquad \boldsymbol{\varkappa}=\tfrac{1}{2}\nabla\nabla\times\boldsymbol{B}. \tag{11.20}$$

Further, in spherical coordinates r, ϑ, φ, if ψ is a function of r only, the complete solution of

$$(1-l^2\nabla^2)\nabla^2\psi=0 \tag{11.21}$$

is

$$\psi=a_0+b_0 r^{-1}+c_0 r^{-1}e^{-r/l}+d_0 r^{-1}e^{r/l}. \tag{11.22}$$

Also, in cylindrical coordinates r, ϑ, z, if ψ is a function or r only, the complete solution of (11.21) is

$$\psi=a_0+b_0\log r+c_0 K_0(r/l)+d_0 I_0(r/l), \tag{11.23}$$

where K_0 and I_0 are the modified Bessel functions of order zero.

The displacement may also be expressed in terms of a vector function, $\boldsymbol{G}$, which reduces to the Galerkin stress function [18] when $l=0$:

$$(\lambda+2\mu)\boldsymbol{u}=(\lambda+2\mu)\nabla^2\boldsymbol{G}-(\lambda+\mu)\nabla\nabla\cdot\boldsymbol{G}-\eta\nabla^2\nabla\nabla\cdot\boldsymbol{G}, \tag{11.24}$$

where $\boldsymbol{G}$ satisfies

$$\mu(1-l^2\nabla^2)\nabla^4\boldsymbol{G}=-\varrho\boldsymbol{f}-\tfrac{1}{2}\varrho\nabla\times\mathbf{c}. \tag{11.25}$$

The relations between $\boldsymbol{G}$ and $\boldsymbol{B}$ and B_0 are

$$\boldsymbol{B}=\nabla^2\boldsymbol{G},\qquad B_0=2(1-l^2\nabla^2)\nabla\cdot\boldsymbol{G}-\boldsymbol{r}\cdot(1-l^2\nabla^2)\nabla^2\boldsymbol{G}. \tag{11.26}$$

12. Spherical cavity in a field of simple tension

In spherical coordinates r, ϑ, φ, a stress-field of simple tension, τ_0, is given by

$$\tau_{rr}=\tfrac{1}{2}\tau_0(1+\cos 2\vartheta),\quad \tau_{\vartheta\vartheta}=\tfrac{1}{2}\tau_0(1-\cos 2\vartheta),\quad \tau_{r\vartheta}=\tau_{\vartheta r}=-\tfrac{1}{2}\tau_0\sin 2\vartheta, \tag{12.1}$$

$$\tau_{\varphi r}=\tau_{r\varphi}=\tau_{\varphi\vartheta}=\tau_{\vartheta\varphi}=\tau_{\varphi\varphi}=0,\qquad \boldsymbol{\mu}=0. \tag{12.2}$$

We wish to add a stress field $(\boldsymbol{\tau},\boldsymbol{\mu})$ which will produce a free surface at $r=a$ and vanish at infinity. From (5.22), (5.23), (12.1) and (12.2), the conditions which the additional field must satisfy on $r=a$ are

$$\begin{aligned}
p_\alpha&=\tau^S_{r\vartheta}-\tfrac{1}{2}\boldsymbol{e}_\varphi\cdot(\nabla\cdot\boldsymbol{\mu}^D-\nabla\mu^D_{rr})=\tfrac{1}{2}\tau_0\sin 2\vartheta,\\
p_\beta&=\tau^S_{r\varphi}+\tfrac{1}{2}\boldsymbol{e}_\vartheta\cdot(\nabla\cdot\boldsymbol{\mu}^D-\nabla\mu^D_{rr})=0,\\
\tau_{\gamma\gamma}&=\tau_{rr}=-\tfrac{1}{2}\tau_0(1+\cos 2\vartheta),\\
\mu_{\gamma\alpha}&=\mu_{r\vartheta}=0,\qquad \mu_{\gamma\beta}=\mu_{r\varphi}=0.
\end{aligned} \tag{12.3}$$

We take the stress functions $\boldsymbol{B}$ and B_0, for the additional field, to be of the form

$$B_x=B_y=0,\qquad B_z=B(r,\vartheta),\qquad B_0=B_0(r,\vartheta), \tag{12.4}$$

$$(1-l^2\nabla^2)\nabla^2 B=0,\qquad \nabla^2 B_0=0. \tag{12.5}$$

Then, from (11.17) and (3.25), the terms that appear in (12.3) are

$$\begin{aligned}
\tau^S_{r\vartheta}=\mu\Big\{&\cos\vartheta\frac{1}{r}\frac{\partial B}{\partial\vartheta}-2l^2\frac{\partial}{\partial r}\left[\frac{1}{r}\frac{\partial}{\partial\vartheta}\left(\cos\vartheta\frac{\partial B}{\partial r}-\sin\vartheta\frac{1}{r}\frac{\partial B}{\partial\vartheta}\right)\right]-\\
&-\sin\vartheta\frac{\partial B}{\partial r}-\frac{\lambda+\mu}{\lambda+2\mu}\frac{\partial^2}{\partial r\,\partial\vartheta}\left[\cos\vartheta(1-l^2\nabla^2)B+\frac{B_0}{r}\right]\Big\},
\end{aligned} \tag{12.6}$$

$$\tau_{rr} = \frac{\mu\lambda}{\lambda+2\mu}\left(\cos\vartheta\frac{\partial}{\partial r} - \sin\vartheta\frac{1}{r}\frac{\partial}{\partial\vartheta}\right)(1-l^2\nabla^2)\,B +$$
$$+ 2\mu\left[\cos\vartheta\frac{\partial B}{\partial r} - l^2\frac{\partial^2}{\partial r^2}\left(\cos\vartheta\frac{\partial B}{\partial r} - \sin\vartheta\frac{1}{r}\frac{\partial B}{\partial\vartheta}\right)\right] -$$
$$- \frac{\mu(\lambda+\mu)}{\lambda+2\mu}\frac{\partial^2}{\partial r^2}\left[r\cos\vartheta(1-l^2\nabla^2)\,B+B_0\right], \tag{12.7}$$

$$\mu_{r\varphi} = -2\left(\eta\frac{\partial}{\partial r} - \eta'\frac{1}{r}\right)\left(\sin\vartheta\frac{\partial B}{\partial r} + \cos\vartheta\frac{1}{r}\frac{\partial B}{\partial\vartheta}\right), \tag{12.8}$$

$$\nabla\cdot\boldsymbol{\mu}^D = 2\eta\nabla\times\boldsymbol{e}_z\nabla^2 B = -2\eta\,\boldsymbol{e}_\varphi\left(\sin\vartheta\frac{\partial}{\partial r} + \cos\vartheta\frac{1}{r}\frac{\partial}{\partial\vartheta}\right)\nabla^2 B, \tag{12.9}$$

$$\tau^S_{r\varphi} = \mu^D_{rr} = \mu_{r\vartheta} = 0. \tag{12.10}$$

For B and B_0 we take

$$B = a_1 r^{-2}\cos\vartheta + a_2 e^{-r/l}(r^{-1}l^{-1} + r^{-2})\cos\vartheta, \tag{12.11}$$

$$B_0 = a_3 r^{-1} + a_4 r^{-3}(1+3\cos 2\vartheta). \tag{12.12}$$

It may be verified that these functions satisfy (12.5) and give displacements and stresses that vanish at infinity.

Substituting (12.11) and (12.12) in (12.6)–(12.10), the result in (12.3) and equating coefficients of like functions of ϑ, we obtain a set of four equations in the constants a_1, a_2, a_3, a_4, the solution of which is

$$\begin{aligned} &a_1 = 5T, \qquad a_2 = -15 e^k k' T, \\ &a_3 = -[6-5\sigma-6k'(1-\sigma)(1+k)]\,T, \\ &a_4 = -\tfrac{1}{2}a^2\{1-4(1-\sigma)[5k^{-2} - k'(15k^{-2}+15k^{-1}+6+k)]\}\,T, \end{aligned} \tag{12.13}$$

where

$$k = a/l, \qquad k' = (3+\sigma')/[9+9k+4k^2+k^3+\sigma'(3+3k+k^2)], \tag{12.14}$$

$$\sigma' = \eta'/\eta, \qquad T = (1-\sigma)\,a^3\tau_0/\mu[7-5\sigma+18k'(1-\sigma)(1+k)]. \tag{12.15}$$

This completes the solution of the problem.

The total force-stress component $\tau_{\vartheta\vartheta}$ is

$$\tau_{\vartheta\vartheta} = \tfrac{1}{2}\tau_0(1-\cos 2\vartheta) + \frac{\mu\lambda}{\lambda+2\mu}\left(\cos\vartheta\frac{\partial}{\partial r} - \sin\vartheta\frac{1}{r}\frac{\partial}{\partial\vartheta}\right)(1-l^2\nabla^2)\,B -$$
$$- 2\mu\left[\sin\vartheta\frac{1}{r}\frac{\partial B}{\partial\vartheta} + \left(\frac{l^2}{r}\frac{\partial}{\partial r} + \frac{l^2}{r^2}\frac{\partial^2}{\partial\vartheta^2}\right)\left(\cos\vartheta\frac{\partial B}{\partial r} - \sin\vartheta\frac{1}{r}\frac{\partial B}{\partial\vartheta}\right)\right] -$$
$$- \frac{\mu(\lambda+\mu)}{\lambda+2\mu}\left(\frac{1}{r}\frac{\partial}{\partial r} + \frac{1}{r^2}\frac{\partial^2}{\partial\vartheta^2}\right)\left[r\cos\vartheta(1-l^2\nabla^2)\,B + B_0\right], \tag{12.16}$$

which yields a maximum tension at the surface:

$$[\tau_{\vartheta\vartheta}]_{r=a,\,\vartheta=\pi/2} = \frac{3[9-5\sigma+6k'(1-\sigma)(1+k)]\,\tau_0}{2[7-5\sigma+18k'(1-\sigma)(1+k)]}. \tag{12.17}$$

Thus, the maximum tension at the spherical surface depends on the ratio, k, of the radius of the cavity to the material constant l. In the absence of couple-stress, (12.17) becomes

$$\lim_{k\to\infty}[\tau_{\vartheta\vartheta}]_{r=a,\,\vartheta=\pi/2} = \frac{3(9-5\sigma)\,\tau_0}{2(7-5\sigma)}, \tag{12.18}$$

which coincides with Southwell & Gough's result [*19*, p. 361] for the case $l=0$. For a cavity with a radius vanishingly small in comparison with l, we find

$$\lim_{k\to 0}\,[\tau_{\vartheta\vartheta}]_{r=a,\,\vartheta=\pi/2}=\frac{3(11-7\sigma)\,\tau_0}{2(13-11\sigma)}. \tag{12.19}$$

In the range $0\leqq\sigma\leqq\frac{1}{2}$, (12.19) is about 30% less than (12.18).

13. Example of failure of uniqueness

In this section we take $\eta<0$ and find two solutions of the boundary value problem of the preceding section.

Let

$$\bar{\eta}=-\eta,\qquad \bar{\eta}'=-\eta',\qquad \bar{l}=i l=(\bar{\eta}/\mu)^{\frac{1}{2}}, \tag{13.1}$$

where $\bar{\eta}>0$ and $\bar{l}>0$. Then the stress functions satisfy

$$(1+\bar{l}^2\nabla^2)\nabla^2 B=0,\qquad \nabla^2 B_0=0, \tag{13.2}$$

instead of (12.5).

For B and B_0 we consider the two sets of functions

$$\begin{aligned} B&=\frac{a_1'}{r^2}\cos\vartheta+a_2'\frac{\partial}{\partial z}\left(\frac{1}{r}\cos\frac{r}{\bar{l}}\right),\\ B_0&=a_3' r^{-1}+a_4' r^{-3}(1+3\cos 2\vartheta), \end{aligned} \tag{13.3}$$

and

$$\begin{aligned} B&=\frac{a_1''}{r^2}\cos\vartheta+a_2''\frac{\partial}{\partial z}\left(\frac{1}{r}\sin\frac{r}{\bar{l}}\right),\\ B_0&=a_3'' r^{-1}+a_4'' r^{-3}(1+3\cos 2\vartheta). \end{aligned} \tag{13.4}$$

It may be verified that (13.3) and (13.4) satisfy (13.2) and yield displacements and stresses that vanish at infinity.

Upon substituting the functions (13.3) in (12.6)—(12.10), the results in the boundary conditions (12.3) and solving for the a_i', we find:

$$a_1'=\bar{T}',\qquad a_2'=-3\bar{k}'\bar{T}',$$

$$a_3'=-\frac{\bar{T}'[7\lambda+12\mu+6(\lambda+2\mu)\,\bar{k}'(\cos\bar{k}+\bar{k}\sin\bar{k})]}{10(\lambda+\mu)},$$

$$a_4'=-\frac{a^2\bar{T}'\{\lambda+\mu+2(\lambda+2\mu)\,[5\bar{k}^{-2}-3\bar{k}'(2-5\bar{k}^{-2})\cos\bar{k}-\bar{k}'(\bar{k}-15\bar{k}^{-1})\sin\bar{k}]\}}{10(\lambda+\mu)},$$

where $\bar{k}=a/\bar{l}$ and

$$\bar{T}'=\frac{5\tau_0 a^3(\lambda+2\mu)}{\mu\,[9\lambda+14\mu-18(\lambda+2\mu)\,\bar{k}'(\cos\bar{k}+\bar{k}\sin\bar{k})]},$$

$$\bar{k}'=\frac{3\bar{\eta}+\bar{\eta}'}{\bar{\eta}\,[(4\bar{k}^2-9)\cos\bar{k}+(\bar{k}^3-9\bar{k})\sin\bar{k}]-\bar{\eta}'[3\bar{k}\sin\bar{k}+(3-\bar{k}^2)\cos\bar{k}]}.$$

Similarly, for the stress functions (13.4), we are also able to satisfy the same boundary conditions with constants a_i'' which differ from the a_i' in that $\cos\bar{k}$ is replaced by $\sin\bar{k}$ and $\sin\bar{k}$ is replaced by $-\cos\bar{k}$.

Thus, with $\eta<0$, we have two solutions of the same boundary value problem — illustrating failure of uniqueness when the internal energy density is not positive definite.

14. Cylindrical cavity in a field of simple tension

The statement of the problem of the cylindrical cavity in a field of simple tension is exactly the same as in (12.1)–(12.10) except with η' set equal to zero, with φ replaced by z and with r and ϑ interpreted as the r and ϑ of cylindrical coordinates.

In place of (12.11) and (12.12), we take, as stress functions for the present problem,

$$B = a_1 r^{-1} \cos\vartheta + a_2 K_1(r/l)\cos\vartheta, \tag{14.1}$$

$$B_0 = a_3 \log r + a_4 r^{-2} \cos 2\vartheta. \tag{14.2}$$

Then, substituting in the boundary conditions and solving for the constants, we find

$$\begin{aligned} a_1 &= T', \\ a_2 &= -k'' T'/a K_1(a/l), \\ a_3 &= \tfrac{1}{2} T'[3 - 4\sigma - 2k''(1-\sigma)], \\ a_4 &= -\tfrac{1}{4} T' a^2 [1 - 2k''(1-\sigma)], \end{aligned} \tag{14.3}$$

where

$$T' = \frac{2(1-\sigma)\,\tau_0 a^2}{\mu[1+2k''(1-\sigma)]}, \qquad k'' = \frac{4}{4+k^2+2kK_0/K_1}, \tag{14.4}$$

and $k = a/l$.

The stress $\tau_{\vartheta\vartheta}$ is again given by (12.16) and we find, for the maximum tension at the surface of the cavity,

$$[\tau_{\vartheta\vartheta}]_{r=a,\,\vartheta=\pi/2} = \frac{[3+2(1-\sigma)k'']\,\tau_0}{1+2(1-\sigma)k''}. \tag{14.5}$$

Thus, the maximum tension at the surface depends on both Poisson's ratio and the ratio of the radius of the cavity to the material constant l.

We note that

$$\lim_{k\to 0} kK_0(k)/K_1(k) = 0, \qquad kK_0(k)/K_1(k) \approx k \quad \text{for} \quad k \gg 1; \tag{14.6}$$

and hence

$$\lim_{k\to 0} k'' = 1, \qquad \lim_{k\to\infty} k'' = 0. \tag{14.7}$$

Thus, in the absence of couple-stress ($k \to \infty$), the maximum tension at the surface reduces to $3\tau_0$, which coincides with the result from the Kirsch solution [19, p. 80]. For the vanishingly small cavity, we have

$$\lim_{k\to 0} [\tau_{\vartheta\vartheta}]_{r=a,\,\vartheta=\pi/2} = \frac{(5-2\sigma)\,\tau_0}{3-2\sigma}. \tag{14.8}$$

This is less than $3\tau_0$ by about thirty to forty percent over the range $0 \leqq \sigma \leqq \frac{1}{2}$.

15. Concentrated force and couple

In an infinite medium acted upon statically by body forces and body couples, we have, from (10.11), (10.12) and (11.18),

$$\boldsymbol{B} = \frac{\varrho}{4\pi\mu} \int_V \left(\frac{1}{r_1} - \frac{e^{-r_1/l}}{r_1}\right)\left(\boldsymbol{f} + \frac{1}{2}\nabla\times\boldsymbol{c}\right) dV. \tag{15.1}$$

Also, from (10.2), (10.3) and (11.19),

$$B_0 = -\frac{\varrho}{4\pi\mu}\int\limits_V \frac{\boldsymbol{r}'\cdot(\boldsymbol{f}+\frac{1}{2}\nabla\times\boldsymbol{c})\,dV}{r_1}, \tag{15.2}$$

where $\boldsymbol{r}'\cdot\boldsymbol{r}' = \xi^2+\eta^2+\zeta^2$.

For the concentrated force, we let $\boldsymbol{c}=0$ everywhere and $\boldsymbol{f}=0$ outside a region V' which contains the origin and a non-vanishing field of parallel forces $\boldsymbol{f}$. We define a concentrated force by

$$\boldsymbol{P} = \lim_{V'\to 0}\int_V \varrho\boldsymbol{f}\,dV. \tag{15.3}$$

Then, since

$$\lim_{V'\to 0} r_1 = r, \qquad \lim_{V'\to 0}\boldsymbol{r}' = 0, \tag{15.4}$$

(15.1) and (15.2) reduce to

$$\boldsymbol{B} = \frac{\boldsymbol{P}}{4\pi\mu r}(1-e^{-r/l}), \qquad B_0 = 0, \tag{15.5}$$

which are the stress functions for the concentrated force.

For the concentrated couple, we let $\boldsymbol{f}=0$ everywhere and $\boldsymbol{c}=0$ except in V', where $\boldsymbol{c}$ is a field of parallel couples. The concentrated couple is defined by

$$\boldsymbol{C} = \lim_{V'\to 0}\int_V \varrho\boldsymbol{c}\,dV. \tag{15.6}$$

Now, we note that

$$r_1^{-1}(1-e^{-r_1/l})\nabla\times\boldsymbol{c} = \nabla\times[r_1^{-1}(1-e^{-r_1/l})\boldsymbol{c}] + \boldsymbol{c}\times\nabla[r_1^{-1}(1-e^{-r_1/l})], \tag{15.7}$$

and

$$\int_V \nabla\times[r_1^{-1}(1-e^{-r_1/l})\boldsymbol{c}]\,dV = \int_S \boldsymbol{n}\times[r_1^{-1}(1-e^{-r_1/l})\boldsymbol{c}]\,dS = 0. \tag{15.8}$$

Hence, (15.1) becomes

$$\boldsymbol{B} = \frac{1}{8\pi\mu}\int\limits_V \varrho\boldsymbol{c}\times\nabla[r_1^{-1}(1-e^{-r_1/l})]\,dV. \tag{15.9}$$

Then, with (15.4) and (15.6), and noting that the gradients with respect to the coordinates of the source point and field point have opposite signs, (15.9) and (15.2) become

$$\boldsymbol{B} = -\frac{1}{8\pi\mu}\boldsymbol{C}\times\nabla\left(\frac{1}{r}-\frac{e^{-r/l}}{r}\right), \qquad B_0 = 0, \tag{15.10}$$

which are the stress functions for the concentrated couple.

16. Nuclei of strain

Because of the presence of the factor $\boldsymbol{r}$ in (11.17), it is more convenient to derive formulas for nuclei of strain in terms of the stress function $\boldsymbol{G}$ than in terms of $\boldsymbol{B}$.

For the concentrated force and couple we have, respectively,

$$\boldsymbol{G} = \boldsymbol{P}\chi/4\pi\mu, \qquad \boldsymbol{G} = -\boldsymbol{C}\times\nabla\chi/8\pi\mu, \tag{16.1}$$

where

$$\chi = \tfrac{1}{2}r + l^2 r^{-1}(1-e^{-r/l}). \tag{16.2}$$

It may be verified that (16.1) yield the stress functions $\boldsymbol{B}$ and B_0 in (15.5) and (15.10) through the relations (11.26).

The stress functions, $\boldsymbol{G}$, for various nuclei of strain may now be obtained in the usual way [*9*, p. 186]. Thus, for double forces in the directions of x, y, z, of strength p, we have, respectively,

$$\boldsymbol{G}=\boldsymbol{e}_x p\,\partial\chi/\partial x, \qquad \boldsymbol{G}=\boldsymbol{e}_y p\,\partial\chi/\partial y, \qquad \boldsymbol{G}=\boldsymbol{e}_z p\,\partial\chi/\partial z. \tag{16.3}$$

The sum of these is the stress function for a center of dilatation:

$$\boldsymbol{G}=p\boldsymbol{\nabla}\chi, \tag{16.4}$$

or, from (11.26),

$$\boldsymbol{B}=p\nabla^2\boldsymbol{\nabla}\chi=p\boldsymbol{\nabla}(r^{-1}-r^{-1}e^{-r/l}), \tag{16.5}$$

$$B_0=p[2(1-l^2\nabla^2)\nabla^2\chi-\boldsymbol{r}\cdot(1-l^2\nabla^2)\nabla^2\boldsymbol{\nabla}\chi]=3pr^{-1}. \tag{16.6}$$

By (11.17), both (16.5) and (16.6) give displacements proportional to $\boldsymbol{\nabla}r^{-1}$. Thus, the center of dilatation is the same with or without couple-stress.

For a double force in the x-direction with positive moment about the z-axis,

$$\boldsymbol{G}=q\boldsymbol{e}_x\partial\chi/\partial y, \tag{16.7}$$

where q is the strength of the double force with moment. For a double force in the y-direction with positive moment about the z-axis,

$$\boldsymbol{G}=-q\boldsymbol{e}_y\partial\chi/\partial x. \tag{16.8}$$

The sum of (16.7) and (16.8) is the stress function for a center of rotation about the z-axis:

$$\boldsymbol{G}=q(\boldsymbol{e}_x\partial\chi/\partial y-\boldsymbol{e}_y\partial\chi/\partial x). \tag{16.9}$$

This may be compared with the stress function $(16.1)_2$, specialized to a couple, of magnitude C, about the z-axis:

$$\boldsymbol{G}=-C\boldsymbol{e}_z\times\boldsymbol{\nabla}\chi/8\pi\mu=(C/8\pi\mu)(\boldsymbol{e}_x\partial\chi/\partial y-\boldsymbol{e}_y\partial\chi/\partial x). \tag{16.10}$$

Thus, with $q=C/8\pi\mu$, the center of rotation is identical with the concentrated body-couple; a conclusion which holds with or without couple-stress [*9*, p. 187].

Acknowledgements. We wish to thank Dr. R. A. Toupin for valuable discussions of the constitutive equations. The senior author also wishes to thank the Office of Naval Research for support through a contract with Columbia University.

References

[*1*] Voigt, W.: Theoretische Studien über die Elasticitätsverhältnisse der Krystalle. Abh. Ges. Wiss. Göttingen **34** (1887); Über Medien ohne innere Kräfte und eine durch sie gelieferte mechanische Deutung der Maxwell-Hertzschen Gleichungen. Gött. Abh. 72—79 (1894).

[*2*] Cosserat, E., et F.: Théorie des Corps Déformables. Paris: A. Hermann et Fils 1909.

[*3*] Truesdell, C., & R. A. Toupin: The Classical Field Theories. Encyclopedia of Physics, Vol. III/1, Secs. 200, 203, 205. Berlin-Göttingen-Heidelberg: Springer 1960.

[*4*] Toupin, R. A.: Elastic materials with couple-stresses. Arch. Rational Mech. Anal., **11**, 385—414 (1962).

[5] AERO, E. L., & E. V. KUVSHINSKII: Fundamental equations of the theory of elastic media with rotationally interacting particles. Fizika Tverdogo Tela **2**, 1399–1409 (1960); Trans.: Soviet Physics Solid State **2**, 1272–1281 (1961).
[6] GRIOLI, G.: Elasticità asimmetrica. Ann. di Mat. pura ed appl., Ser. IV, **50**, 389–417 (1960).
[7] WEATHERBURN, C. E.: Advanced Vector Analysis. London: G. Bell and Sons 1928.
[8] WILSON, E. B.: Vector Analysis. New Haven: Yale Univ. Press 1948.
[9] LOVE, A. E. H.: A Treatise on the Mathematical Theory of Elasticity, Fourth Ed. Cambridge: Cambridge Univ. Press 1927.
[10] TOUPIN, R. A.: The elastic dielectric. J. Rational Mech. and Anal. **5**, 849–914 (1956).
[11] STERNBERG, E.: On the integration of the equations of motion in the classical theory of elasticity. Arch. Rational Mech. Anal. **6**, 34–50 (1960).
[12] ATANASOFF, J. V., & P. J. HART: Dynamical determination of the elastic constants and their temperature coefficients for quartz. Phys. Rev. **59**, 85–96 (1941).
[13] MINDLIN, R. D.: Thickness-shear and flexural vibrations of crystal plates. J. Appl. Phys. **22**, 316–323 (1951).
[14] ONOE, M.: Tables of Modified Quotients of Bessel Functions of the First Kind for Real and Imaginary Arguments. New York: Columbia Univ. Press 1958.
[15] PAPKOVITCH, P. F.: The representation of the general integral of the fundamental equations of elasticity theory in terms of harmonic functions. Izv. Akad. Nauk SSSR, Phys.-Math. Ser. **10**, 1425 (1932).
[16] MINDLIN, R. D.: Note on the Galerkin and Papkovitch stress functions. Bull. Amer. Math. Soc. **42**, 373–376 (1936).
[17] MINDLIN, R. D.: Force at a point in the interior of a semi-infinite solid. Proc. First Midwestern Conference on Solid Mechanics, Urbana, Illinois, 56–59 (1953).
[18] GALERKIN, B.: Contribution à la solution générale du problème de la théorie de l'élasticité dans le cas de trois dimensions. Comptes Rendus Acad. Sci., Paris **190**, 1047 (1930).
[19] TIMOSHENKO, S. P., & J. N. GOODIER: Theory of Elasticity. New York: McGraw-Hill 1951.

Department of Civil Engineering
Columbia University
New York, New York
and
Bell Telephone Laboratories
Whippany, New Jersey

(Received July 27, 1962)

Influence of Couple-stresses on Stress Concentrations

Main features of Cosserat theory are reviewed by Lecturer and some recent solutions of the equations, for cases of stress concentration around small holes in elastic solids. are described

by R. D. Mindlin

Introduction

It is known that the apparent strengths of some materials are affected by strain gradients. For example, in alternating bending fatigue tests, the smaller the specimen size, i.e., the higher the strain gradient across the specimen, the higher the fatigue limit. Again, brittle fracture and the onset of static yielding in the presence of stress concentration occur at higher loads than might be expected on the basis of stress concentration factors calculated from the theory of elasticity. In general, increasing strain gradients appear to make some, but not all, materials stronger and to a degree that depends upon grain size.

The results of such tests are difficult to fit into a comprehensive and coherent theory partly because there have been no determinations of stress distributions by a method that takes the strain gradient into account. It cannot be expected that the interpretation of test results will be simple if a part of the phenomenon under study is excluded from the basis of interpretation. The stresses in the specimen may not be those ordinarily accepted and there may be additional forcelike quantities, accompanying strain gradients, to which a role should be assigned in a criterion of failure.

There has been available, for many years, the foundation of a theory of stress analysis in which the effect of at least a part of the strain gradient is included. In two dimensions, for example, the strain gradient has six components comprising the two components of the gradient of each of the three components of strain. The six components of the strain gradient may be arranged in six independent linear combinations: two of which are the components of the gradient of the rotation, i.e., the components of the curvature. Just as strain is associated with stress (force per unit area), so curvature is associated with couple-stress (couple per unit area) which is an additional force-like quantity with the dimensions of force per unit length. The half-century old Cosserat[1] theory takes full account of the couple-stress but the theory has been little explored until very recently.

In an elastic material, the ratio of stress to strain is a modulus of elasticity with the dimensions of force per unit area. Similarly, the ratio of couple-stress to curvature is another modulus of elasticity: one with the dimensions of force. The ratio of the modulus of curvature to the modulus of, say, shear thus has the dimensions of the square of a length. This length, l, is a material property on which the influence of couple-stresses depends strongly. If the ratio of the smallest dimension of a body to l is large, the theory has it that the effect of couple-stresses is negligible. However, when there are strain gradients and a dimension of a body approaches l, couple-stresses, according to the theory, may produce effects of appreciable magnitude.

In perfect crystals and amorphous materials like glass or plastics, l is probably submicroscopic; but might be of the order of the radius of the root of a crack. In polycrystalline metals or granular materials, l may be considerably larger. For example, in the idealized case of a simple cubic array of contiguous elastic spheres, l is about three quarters of the radius of a sphere. Thus, dimensions of structural elements and machine parts, or dimensions of holes, fillets, notches or cracks in them, may approach l for a variety of materials.

The present paper begins with a derivation of two-dimensional equations of couple-stress theory for an isotropic, elastic medium. The components of stress and couple-stress are then expressed in terms of the Airy stress function and another stress function which accounts for the couple stresses. After a transformation to polar coordinates, the stress distributions are found for a circular hole in various two-dimensional fields of stress.

R. D. Mindlin is Professor of Civil Engineering, Columbia University, New York, N. Y.
Lecture was presented at 1962 SESA Annual Meeting held in Milwaukee, Wis., on October 24–26.

Reprinted from EXPERIMENTAL MECHANICS, *January 1963*

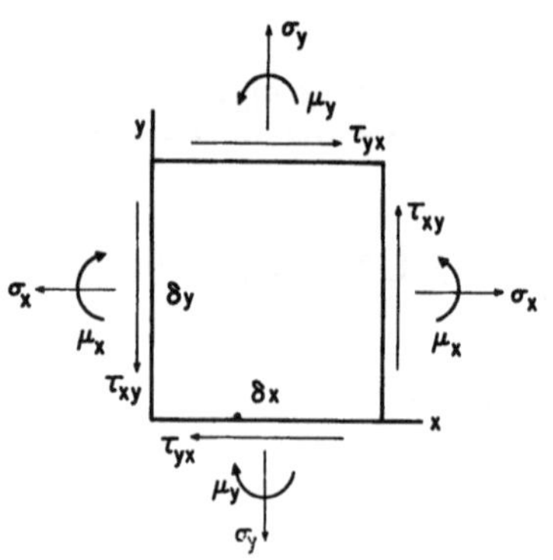

Fig. 1—Rectangular components of stress and couple-stress

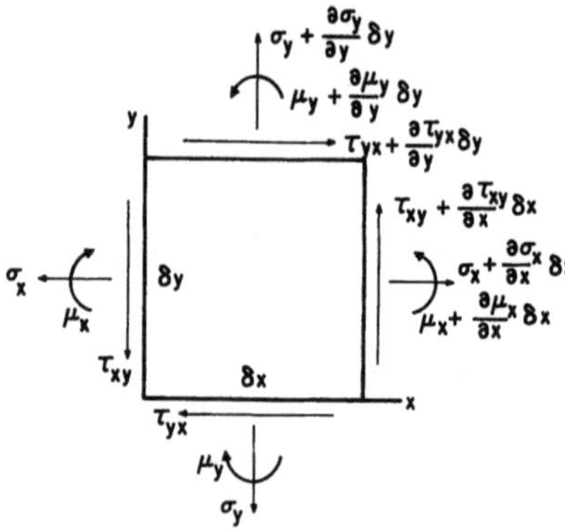

Fig. 2—Components of stress and couple-stress in a varying field

The first of the problems solved is that of the circular hole in a field of uniform tension. With couple-stresses taken into account, the stress-concentration factor, instead of having the usual value 3, depends both on Poisson's ratio and the ratio of the radius of the hole, a, to the material constant l. As a/l becomes large, the stress-concentration factor approaches 3. As a/l diminishes, so does the stress-concentration factor. For example, if $a/l = 3$, the reduction is about 12 to 20 percent, depending on Poisson's ratio. Next, the stress-concentration factors are found for the circular hole in a field of tension or compression in two perpendicular directions. For equal tensions or equal compressions, couple-stresses have no effect; but, for pure shear, the percent reduction of the usual stress-concentration factor of 4 is even larger than for simple tension: 18 to 30 percent in the case of $a/l = 3$.

In what follows, the derivation and solutions of the equations of couple-stress theory parallel those for the theory without couple-stresses in the first four chapters of *Theory of Elasticity* by Timoshenko and Goodier[2], reference to which is indicated by the letters TG and the page number. For a more general treatment and for citations of additional papers on the subject, Ref. 3 may be consulted.

Equations of Equilibrium

In Fig. 1, the rectangular components of force per unit area (σ_x, σ_y, τ_{xy}, τ_{yx}) and couple per unit area (μ_x, μ_y) are shown, in their positive senses, acting on the faces of a small rectangular element of length δx, width δy and unit thickness. If, as indicated in the figure, the stresses and couple-stresses do not vary from one side of the element to the opposite side, consideration of equilibrium of moments shows that the couples balance and, therefore, the cross shears τ_{xy} and τ_{yx} must be equal (TG, 4). If the stresses and couple-stresses vary across the element, the situation is as indicated in Fig. 2. Neglecting terms multiplied by powers and products of δx and δy in comparison with terms multiplied by δx or δy alone, the condition of equilibrium of moments now gives

$$\frac{\partial \mu_x}{\partial x} + \frac{\partial \mu_y}{\partial y} + \tau_{xy} - \tau_{yx} = 0 \tag{1}$$

so that, when varying couple-stresses are considered, the cross shears are not necessarily equal. Conversely, if the cross shears are equal, or even zero, the couple-stresses need not vanish: only the sum of the first two terms in eq (1) must then be zero.

Consideration of the equilibrium of forces in Fig. 2 leads to

$$\frac{\partial \sigma_x}{\partial x} + \frac{\partial \tau_{yx}}{\partial y} = 0, \quad \frac{\partial \tau_{xy}}{\partial x} + \frac{\partial \sigma_y}{\partial y} = 0 \tag{2}$$

which are the usual equations of equilibrium (TG, 22) except that now τ_{xy} need not be equal to τ_{yx}. Equations (1) and (2) are the Cosserat equations of equilibrium in two dimensions with body forces and body couples omitted (Ref. 1, page 137).

Deformation Produced by Stress and Couple-stress

In a state of plane strain (TG, 12) in the plane of x and y, the components of displacement, u, v, in the x and y directions, are functions of x and y only and the third component of displacement is zero. Then, in an isotropic, elastic medium, the normal components of strain, ϵ_x, ϵ_y, are related to the normal components of stress, σ_x, σ_y, by (TG, 25)

$$\begin{aligned} \epsilon_x &= \frac{1+\nu}{E}\left[\sigma_x - \nu(\sigma_x + \sigma_y)\right], \\ \epsilon_y &= \frac{1+\nu}{E}\left[\sigma_y - \nu(\sigma_x + \sigma_y)\right] \end{aligned} \tag{3}$$

where E and ν are Young's modulus and Poisson's ratio and the strains are related to the displacements by (TG, 6)

$$\epsilon_x = \frac{\partial u}{\partial x}, \quad \epsilon_y = \frac{\partial v}{\partial y} \tag{4}$$

2

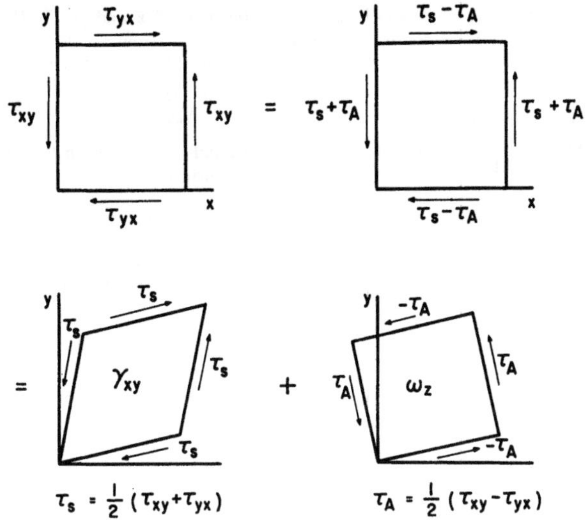

Fig. 3—Strain and unbalanced rotation produced by symmetric and anti-symmetric parts of the shear stress

As for the stress–strain relation in shear, τ_{xy} and τ_{yx} are resolved into a symmetric part τ_S and an antisymmetric part τ_A:

$$\tau_S = (\tau_{xy} + \tau_{yx})/2, \quad \tau_A = (\tau_{xy} - \tau_{yx})/2 \tag{5}$$

Then, as shown in Fig. 3, the symmetric part produces the usual shear strain (TG, 9)

$$\gamma_{xy} = \frac{1}{G}\tau_S = \frac{1+\nu}{E}(\tau_{xy} + \tau_{yx}) \tag{6}$$

where G is the modulus of shear (TG, 9)

$$G = \frac{E}{2(1+\nu)} \tag{7}$$

and γ_{xy} is related to the displacements by (TG, 6)

$$\gamma_{xy} = \frac{\partial v}{\partial x} + \frac{\partial u}{\partial y} \tag{8}$$

The antisymmetric part of the shear stress tends to produce a local rigid rotation (TG, 225)

$$\omega_z = \frac{1}{2}\left(\frac{\partial v}{\partial x} - \frac{\partial u}{\partial y}\right) \tag{9}$$

and is balanced by the couple-stresses in accordance with eq (1).

The couple-stresses produce curvatures as shown in Fig. 4. From the geometry of the figure, the curvatures κ_x and κ_y are related to the rotation ω_z by

$$\kappa_x = \frac{\partial \omega_z}{\partial x}, \quad \kappa_y = \frac{\partial \omega_z}{\partial y} \tag{10}$$

It is assumed that the curvatures are proportional to the couple-stresses:

$$\kappa_x = \frac{1}{4B}\mu_x, \quad \kappa_y = \frac{1}{4B}\mu_y \tag{11}$$

where B is a modulus of curvature or bending and the factor 4 is introduced as a convenience. Couple-stress has the dimensions of couple per unit area or force per unit length; curvature is the reciprocal of a length; hence B has the dimensions of force.

Equations of Compatibility

By eliminating the displacements between eqs (4) and (8), we obtain the usual equation of compatibility of strains (TG, 24):

$$\frac{\partial^2 \epsilon_x}{\partial y^2} + \frac{\partial^2 \epsilon_y}{\partial x^2} = \frac{\partial^2 \gamma_{xy}}{\partial x \partial y} \tag{12}$$

Similarly, eliminating the rotation from eqs (10) we find an equation of compatibility of curvatures:

$$\frac{\partial \kappa_x}{\partial y} = \frac{\partial \kappa_y}{\partial x} \tag{13}$$

Further, we see from eqs (4), (8) and (9) that

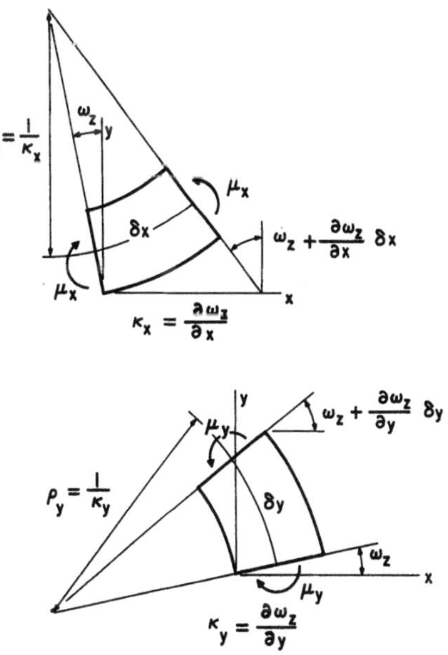

Fig. 4—Curvatures produced by couple-stresses

$$\frac{\partial \omega_z}{\partial x} = \frac{1}{2}\left(\frac{\partial^2 v}{\partial x^2} - \frac{\partial^2 u}{\partial x \partial y}\right) = \frac{1}{2}\frac{\partial \gamma_{xy}}{\partial x} - \frac{\partial \epsilon_x}{\partial y}$$

$$\frac{\partial \omega_z}{\partial y} = \frac{1}{2}\left(\frac{\partial^2 v}{\partial x \partial y} - \frac{\partial^2 u}{\partial y^2}\right) = \frac{\partial \epsilon_y}{\partial x} - \frac{1}{2}\frac{\partial \gamma_{xy}}{\partial y}$$

Thus, the components of curvature are linear combinations of components of strain gradients. Inserting these relations in eqs (10), we find two equations of compatibility of curvatures and strains:

$$\kappa_x = \frac{1}{2}\frac{\partial \gamma_{xy}}{\partial x} - \frac{\partial \epsilon_x}{\partial y}, \quad \kappa_y = \frac{\partial \epsilon_y}{\partial x} - \frac{1}{2}\frac{\partial \gamma_{xy}}{\partial y} \tag{14}$$

Note that eqs (12) and (14) imply eq (13) so that only three of the four equations of compatibility are independent.

The four equations of compatibility may be expressed in terms of the stresses and couple-stresses by using the stress-strain relations of eqs (3) and (6) and the couple-stress-curvature relations of eqs (11), with the results:

$$\frac{\partial^2 \sigma_x}{\partial y^2} + \frac{\partial^2 \sigma_y}{\partial x^2} - \nu \nabla^2(\sigma_x + \sigma_y) = \frac{\partial^2}{\partial x \partial y}(\tau_{xy} + \tau_{yx}) \tag{15}$$

$$\frac{\partial \mu_x}{\partial y} = \frac{\partial \mu_y}{\partial x} \tag{16}$$

$$\mu_x = l^2 \frac{\partial}{\partial x}(\tau_{xy} + \tau_{yx}) - 2l^2 \frac{\partial}{\partial y}[\sigma_x - \nu(\sigma_x + \sigma_y)]$$

$$\mu_y = 2l^2 \frac{\partial}{\partial x}[\sigma_y - \nu(\sigma_x + \sigma_y)] - l^2 \frac{\partial}{\partial y}(\tau_{xy} + \tau_{yx}) \tag{17}$$

where
$$\nabla^2 = \frac{\partial^2}{\partial x^2} + \frac{\partial^2}{\partial y^2} \tag{18}$$

and
$$l^2 = \frac{2(1+\nu)B}{E} = \frac{B}{G} \tag{19}$$

Thus, the material constant l, mentioned in the introduction, is a length equal to the square root of the ratio of the modulus of bending to the modulus of shear. It may be seen, from eqs (17), that, with l not zero, high stress gradients may lead to large couple-stresses. Setting $l = 0$ is equivalent to assuming that the material has no resistance to local curvature. Note, again, that only three of the four equations of compatibility are independent, as eqs (15) and (17) imply eq (16).

Stress Functions

The first of eqs (2) is a necessary and sufficient condition for the existence of a function ϕ_1, of x and y, such that

$$\sigma_x = \frac{\partial \phi_1}{\partial y}, \quad \tau_{yx} = -\frac{\partial \phi_1}{\partial x} \tag{20}$$

Similarly, from the second of eqs (2), there is another function ϕ_2 such that

$$\sigma_y = \frac{\partial \phi_2}{\partial x}, \quad \tau_{xy} = -\frac{\partial \phi_2}{\partial y} \tag{21}$$

and, finally, from eq (16),

$$\mu_x = \frac{\partial \psi}{\partial x}, \quad \mu_y = \frac{\partial \psi}{\partial y} \tag{22}$$

where ψ is a third function of x and y. If the expressions for τ_{yx}, τ_{xy}, μ_x and μ_y, from eqs (20), (21) and (22), are inserted in eq (1), the latter becomes

$$\frac{\partial}{\partial x}\left(\frac{\partial \psi}{\partial x} + \phi_1\right) + \frac{\partial}{\partial y}\left(\frac{\partial \psi}{\partial y} - \phi_2\right) = 0 \tag{23}$$

This is a necessary and sufficient condition for the existence of a function ϕ such that

$$\frac{\partial \psi}{\partial x} + \phi_1 = \frac{\partial \phi}{\partial y}, \quad \frac{\partial \psi}{\partial y} - \phi_2 = -\frac{\partial \phi}{\partial x}$$

by means of which the functions ϕ_1 and ϕ_2, in eqs (20) and (21), may be expressed in terms of ϕ and ψ. Thus we have

$$\sigma_x = \frac{\partial^2 \phi}{\partial y^2} - \frac{\partial^2 \psi}{\partial x \partial y}, \quad \sigma_y = \frac{\partial^2 \phi}{\partial x^2} + \frac{\partial^2 \psi}{\partial x \partial y}$$

$$\tau_{xy} = -\frac{\partial^2 \phi}{\partial x \partial y} - \frac{\partial^2 \psi}{\partial y^2}, \quad \tau_{yx} = -\frac{\partial^2 \phi}{\partial x \partial y} + \frac{\partial^2 \psi}{\partial x^2} \tag{24}$$

$$\mu_x = \frac{\partial \psi}{\partial x}, \quad \mu_y = \frac{\partial \psi}{\partial y}$$

whereby all the components of stress and couple-stress are expressed in terms of two stress functions ϕ and ψ. If ψ is zero, these are no couple-stresses and eqs (24) reduce to the usual expressions for the stresses in terms of the Airy stress function ϕ (TG, 26).

To derive eqs (24), all three equations of equilibrium, i.e., eqs (1) and (2), were used; but only one, eq (16), of the equations of compatibility was used. The differential equations that must be satisfied by the stress functions are obtained from the remaining equations of compatibility.

Substituting the first four of the expressions (24) into eq (15), we find

$$\nabla^4 \phi = 0 \tag{25}$$

where $\nabla^4 = \nabla^2\nabla^2$. Thus, the Airy stress function must satisfy the same differential equation (TG, 26) with or without couple-stresses. Next, substituting all of the expressions (24) into eqs (17), we find

$$\frac{\partial}{\partial x}(\psi - l^2\nabla^2\psi) = -2(1-\nu)l^2 \frac{\partial}{\partial y}\nabla^2\phi$$

$$\frac{\partial}{\partial y}(\psi - l^2\nabla^2\psi) = 2(1-\nu)l^2 \frac{\partial}{\partial x}\nabla^2\phi \tag{26}$$

i.e., $\psi - l^2\nabla^2\psi$ and $2(1-\nu)l^2\nabla^2\phi$ must be conjugate harmonic functions (TG, 182). Differentiating the first of eqs (26) with respect to x, the second with respect to y and adding, we find that ψ must satisfy the equation

$$\nabla^2\psi - l^2\nabla^4\psi = 0 \tag{27}$$

If we were to differentiate the first of eqs (26) with respect to y, the second with respect to x and subtract, we would find eq (25) again.

To summarize: in a state of plane strain without body forces or body couples, the stress functions ϕ and ψ must be solutions of eqs (25) and (27) and must be related to each other in accordance with eqs (26).

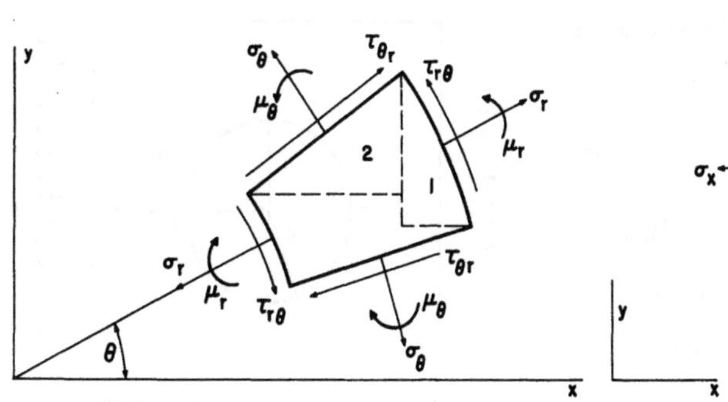

Fig. 5—Components of stress and couple-stress in polar coordinates

Polar Coordinates

In a system of polar coordinates defined by (TG, 56)

$$r = \sqrt{x^2 + y^2}, \quad \theta = \arctan(y/x) \tag{28}$$

the components of stress and couple-stress are those depicted in Fig. 5 in their positive senses. Consideration of the equilibrium of the triangular elements, extracted from Fig. 5 and shown in Fig. 6, leads to the relations (compare TG, 14)

$$\begin{aligned}
\sigma_r &= \sigma_x \cos^2\theta + \sigma_y \sin^2\theta + (\tau_{xy} + \tau_{yx}) \sin\theta\cos\theta \\
\sigma_\theta &= \sigma_x \sin^2\theta + \sigma_y \cos^2\theta - (\tau_{xy} + \tau_{yx}) \sin\theta\cos\theta \\
\tau_{r\theta} &= (\sigma_y - \sigma_x)\sin\theta\cos\theta + \tau_{xy}\cos^2\theta - \tau_{yx}\sin^2\theta \\
\tau_{\theta r} &= (\sigma_y - \sigma_x)\sin\theta\cos\theta - \tau_{xy}\sin^2\theta + \tau_{yx}\cos^2\theta \\
\mu_r &= \mu_x\cos\theta + \mu_y\sin\theta \\
\mu_\theta &= -\mu_x\sin\theta + \mu_y\cos\theta
\end{aligned} \tag{29}$$

If, now, the expressions (24) are substituted into eqs (29) and the transformations (TG, 57)

$$\frac{\partial}{\partial x} = \cos\theta\frac{\partial}{\partial r} - \frac{\sin\theta}{r}\frac{\partial}{\partial\theta}, \quad \frac{\partial}{\partial y} = \sin\theta\frac{\partial}{\partial r} + \frac{\cos\theta}{r}\frac{\partial}{\partial\theta} \tag{30}$$

are employed, eqs (29) become (compare TG, 56)

$$\begin{aligned}
\sigma_r &= \frac{1}{r}\frac{\partial\phi}{\partial r} + \frac{1}{r^2}\frac{\partial^2\phi}{\partial\theta^2} - \frac{1}{r}\frac{\partial^2\psi}{\partial r\partial\theta} + \frac{1}{r^2}\frac{\partial\psi}{\partial\theta} \\
\sigma_\theta &= \frac{\partial^2\phi}{\partial r^2} + \frac{1}{r}\frac{\partial^2\psi}{\partial r\partial\theta} - \frac{1}{r^2}\frac{\partial\psi}{\partial\theta} \\
\tau_{r\theta} &= -\frac{1}{r}\frac{\partial^2\phi}{\partial r\partial\theta} + \frac{1}{r^2}\frac{\partial\phi}{\partial\theta} - \frac{1}{r}\frac{\partial\psi}{\partial r} - \frac{1}{r^2}\frac{\partial^2\psi}{\partial\theta^2} \\
\tau_{\theta r} &= -\frac{1}{r}\frac{\partial^2\phi}{\partial r\partial\theta} + \frac{1}{r^2}\frac{\partial\phi}{\partial\theta} + \frac{\partial^2\psi}{\partial r^2} \\
\mu_r &= \frac{\partial\psi}{\partial r} \\
\mu_\theta &= \frac{1}{r}\frac{\partial\psi}{\partial\theta}
\end{aligned} \tag{31}$$

The compatibility equations (26), when transformed to polar coordinates by means of eqs (30), become

$$\begin{aligned}
\frac{\partial}{\partial r}(\psi - l^2\nabla^2\psi) &= -2(1-\nu)l^2\frac{1}{r}\frac{\partial}{\partial\theta}\nabla^2\phi \\
\frac{1}{r}\frac{\partial}{\partial\theta}(\psi - l^2\nabla^2\psi) &= 2(1-\nu)l^2\frac{\partial}{\partial r}\nabla^2\phi
\end{aligned} \tag{32}$$

where (TG, 57)

$$\nabla^2 = \frac{\partial^2}{\partial r^2} + \frac{1}{r}\frac{\partial}{\partial r} + \frac{1}{r^2}\frac{\partial^2}{\partial\theta^2} \tag{33}$$

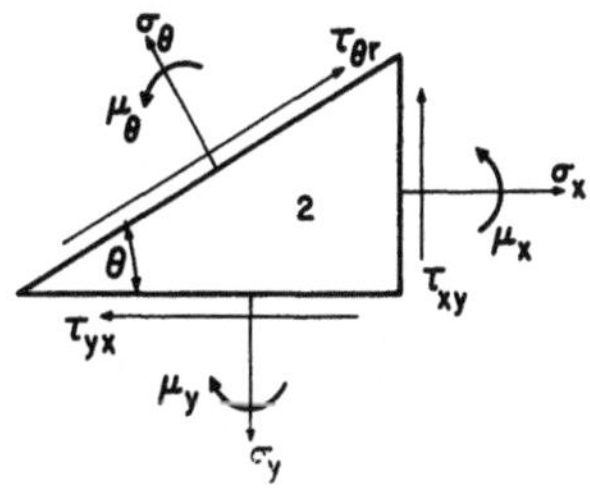

Fig. 6—Equilibrium of polar and rectangular components of stress and couple-stress

This form of the operator ∇^2 also applies to the remaining two compatibility equations (25) and (27).

Circular Hole in a Field of Uniform Tension*

Consider a cylindrical hole of radius a in a field of simple tension p, at right angles to the axis of the hole, as shown in Fig. 7. For the solution, we take (compare TG, 78)

$$\begin{aligned}
\phi &= \frac{p}{4}r^2(1 - \cos 2\theta) + A\log r + \left(\frac{B}{r^2} + C\right)\cos 2\theta \\
\psi &= \left[\frac{D}{r^2} + E\,K_2\left(\frac{r}{l}\right)\right]\sin 2\theta
\end{aligned} \tag{34}$$

where K_2 is the modified Bessel function of the second kind and second order.[4] Note that

$$K_2\left(\frac{r}{l}\right) = \frac{2l}{r}K_1\left(\frac{r}{l}\right) + K_0\left(\frac{r}{l}\right)$$

$$\frac{\partial K_0(r/l)}{\partial r} = -\frac{1}{l}K_1\left(\frac{r}{l}\right),$$

$$\frac{\partial K_1(r/l)}{\partial r} = -\frac{1}{r}K_1\left(\frac{r}{l}\right) - \frac{1}{l}K_0\left(\frac{r}{l}\right)$$

* *This problem is solved by another method in Ref. 3.*

where K_0 and K_1 are the modified Bessel functions of the second kind of orders zero and one.

The stress functions given in eqs (34) satisfy eqs (25) and (27); they also will satisfy eqs (32) if

$$D = 8(1 - \nu)l^2C \tag{35}$$

From eqs (31) and (34), we find

$$\sigma_r = \frac{p}{2}(1 + \cos 2\theta) + \frac{A}{r^2} - \left(\frac{6B}{r^4} + \frac{4C}{r^2} - \frac{6D}{r^4}\right)\cos 2\theta + \frac{2E}{lr}\left[\frac{3l}{r}K_0\left(\frac{r}{l}\right) + \left(1 + \frac{6l^2}{r^2}\right)K_1\left(\frac{r}{l}\right)\right]\cos 2\theta$$

$$\sigma_\theta = \frac{p}{2}(1 - \cos 2\theta) - \frac{A}{r^2} + \left(\frac{6B}{r^4} - \frac{6D}{r^4}\right)\cos 2\theta - \frac{2E}{lr}\left[\frac{3l}{r}K_0\left(\frac{r}{l}\right) + \left(1 + \frac{6l^2}{r^2}\right)K_1\left(\frac{r}{l}\right)\right]\cos 2\theta$$

$$\tau_{r\theta} = -\left(\frac{p}{2} + \frac{6B}{r^4} + \frac{2C}{r^2} - \frac{6D}{r^4}\right)\sin 2\theta + \frac{E}{lr}\left[\frac{6l}{r}K_0\left(\frac{r}{l}\right) + \left(1 + \frac{12l^2}{r^2}\right)K_1\left(\frac{r}{l}\right)\right]\sin 2\theta \tag{36}$$

$$\tau_{\theta r} = -\left(\frac{p}{2} + \frac{6B}{r^4} + \frac{2C}{r^2} - \frac{6D}{r^4}\right)\sin 2\theta + \frac{E}{l^2}\left[\left(1 + \frac{6l^2}{r^2}\right)K_0\left(\frac{r}{l}\right) + \left(\frac{3l}{r} + \frac{12l^3}{r^3}\right)K_1\left(\frac{r}{l}\right)\right]\sin 2\theta$$

$$\mu_r = -\frac{2D}{r^3}\sin 2\theta - \frac{E}{l}\left[\frac{2l}{r}K_0\left(\frac{r}{l}\right) + \left(1 + \frac{4l^2}{r^2}\right)K_1\left(\frac{r}{l}\right)\right]\sin 2\theta$$

$$\mu_\theta = \left\{\frac{2D}{r^3} + \frac{2E}{r}\left[K_0\left(\frac{r}{l}\right) + \frac{2l}{r}K_1\left(\frac{r}{l}\right)\right]\right\}\cos 2\theta$$

In eqs (36), the terms involving p constitute a simple tension p in the direction of x (TG, 78). The remaining terms vanish at infinity. Hence, there is a simple tension p far from the hole. On $r = a$, we must have (see Fig. 5)

$$\sigma_r = \tau_{r\theta} = \mu_r = 0 \tag{37}$$

Substituting from eqs (36) into eqs (37) we find four equations, in addition to eq (35), relating the five constants $A \ldots E$ to p. The solution of the five equations is

$$A = -\frac{pa^2}{2},\quad B = -\frac{pa^4(1 - F)}{4(1 + F)},\quad C = \frac{pa^2}{2(1 + F)}$$

$$D = \frac{4(1 - \nu)a^2l^2p}{1 + F},\quad E = \frac{palF}{(1 + F)K_1(a/l)} \tag{38}$$

where

$$F = \frac{8(1 - \nu)}{4 + \frac{a^2}{l^2} + \frac{2a}{l}\frac{K_0(a/l)}{K_1(a/l)}} \tag{39}$$

If couple-stresses are neglected, $l = 0$. Then noting that

$$\lim_{a/l \to \infty} \frac{K_0(a/l)}{K_1(a/l)} = 1$$

eqs (36) and (38) reduce to the Kirsch solution (TG, 80). With $l \neq 0$, the stress σ_θ at $r = a$, $\theta = \pm\pi/2$ (the points m in Fig. 7) is

$$\sigma_m = p\,\frac{3 + F}{1 + F} \tag{40}$$

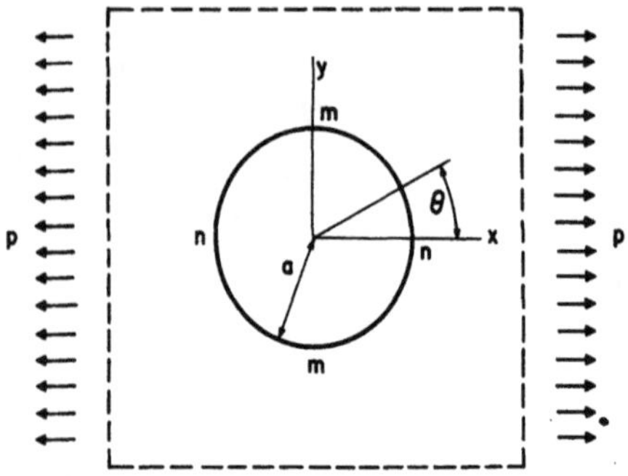

Fig. 7—Circular hole in a field of simple tension

Thus, with couple-stresses accounted for, the stress-concentration factor is seen to depend on both Poisson's ratio and the ratio of the radius of the hole to the material constant l. If couple-stresses are neglected, $F = 0$ and the stress-concentration factor reduces to the usual value 3. As a/l diminishes, so does the stress-concentration factor. For example, when $a/l = 3$,

$$\sigma_m]_{a/l=3} = \frac{3 + 0.44(1 - \nu)}{1 + 0.44(1 - \nu)}$$

so that, as ν varies from 0 to $^1/_2$, the stress-concentration factor ranges from 2.4 to 2.6. The stress-concentration factors are shown in Fig. 8 for a range of a/l. In view of the possible relation of l to material inhomogeneities, these results should probably be disregarded for a/l less than about 3. In any event, a more detailed derivation of the fundamental equations[3] brings to light an assumption that the energy of curvature is limited to the order of magnitude of the energy of strain. This, in turn, restricts the lower limit of a/l to order of magnitude unity.

Circular Hole in a Biaxial Field of Stress

In the example in the preceding section, the stress σ_θ at $r = a$ and $\theta = 0, \pi$ (the points n in Fig. 7) is

$$\sigma_n = -p\,\frac{1 - F}{1 + F} \tag{41}$$

With this result and eq (40), the stress concentration at a circular hole in a biaxial field of stress may be obtained by superposition (TG, 81). Thus, referring to Fig. 9, the stresses σ_θ at points m and n, in a field of principal stresses p and q, are

$$\sigma_m = \frac{3p - q + (p + q)F}{1 + F}$$

$$\sigma_n = \frac{3q - p + (p + q)F}{1 + F} \tag{42}$$

If $p = q$, $\sigma_m = \sigma_n = 2$. This is the same result as is obtained without couple-stresses. In fact, couple-stresses have no effect at all on the stresses around a circular hole in a field of isotropic tension or compression. The other extreme is the field of pure shear: $p = -q$. Then eqs (42) reduce to

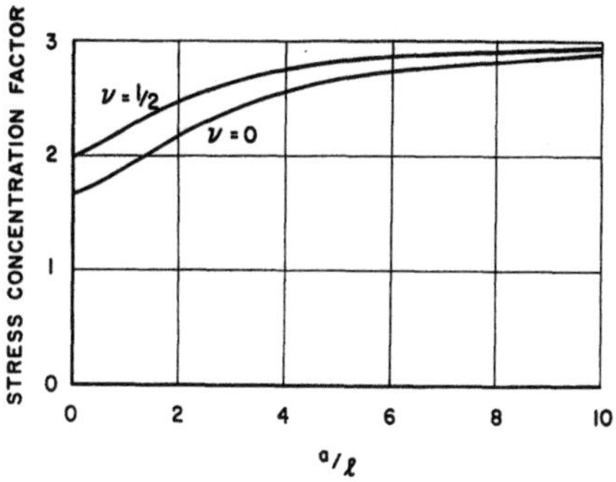

Fig. 8—Stress-concentration factor at a circular hole in a field of simple tension: variation with ratio of radius of hole to material constant *l*.

$$\sigma_m = -\sigma_n = \frac{4p}{1+F} \tag{43}$$

Here the effect of couple-stresses is even greater than in the case of simple tension. For example, for $a/l = 3$,

$$\sigma_m]_{a/l=3} = \frac{4p}{1 + 0.44(1 - \nu)}$$

so that for $\nu = 0$ the stress-concentration factor is reduced from the usual value 4 to 2.8 and for $\nu = {}^1/_2$ the reduction is from 4 to 3.2. The reduction for a range of values of a/l is shown in Fig. 10.

Concluding Remarks

A recalculation of stress-concentration factors around a circular hole, taking couple-stresses into account, results in lower values than those accepted heretofore. The fact that stress-concentration factors based on the ordinary theory of elasticity have been verified by photoelastic measurements may possibly be accounted for by very small values of the constant l in the materials used as models. The new results are in the direction which would tend to account for the apparent rise in the yield limit of polycrystalline metals and fracture strength of brittle materials when strain gradients are present. The maximum stress in the test specimen may be less than it was thought to be. In addition, the couple-stress itself may have to be included in a criterion of failure. Alternatively, in an energy criterion of failure, the energy of curvature may have to be taken into account as well as the energy of strain. At the very least, it appears that past experiments ought to be reviewed and perhaps repeated with additional measurements: the most important of which would be aimed at determining the magnitude of the material constant l. Also, it would seem to be desirable to explore the consequences of taking into account the remaining components of the strain gradient and, perhaps, second and higher gradients of the strain.

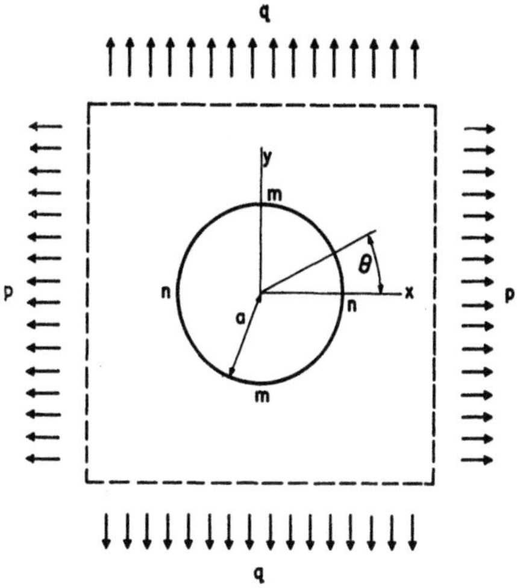

Fig. 9—Circular hole in a biaxial field of stress

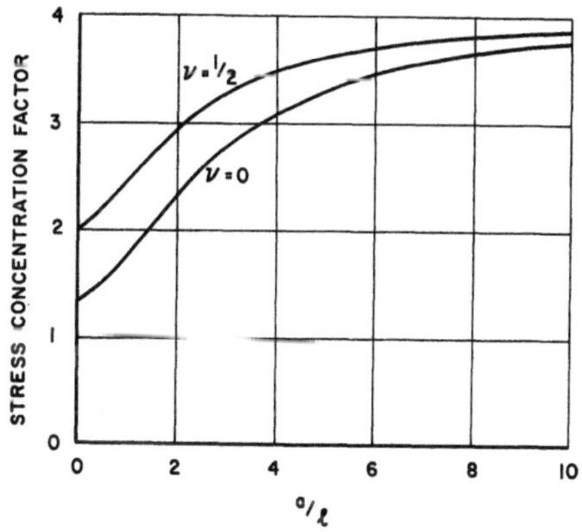

Fig. 10—Stress-concentration factor at a circular hole in a field of pure shear

Acknowledgment

The studies described in this paper were supported by the Office of Naval Research.

References

1. *Cosserat, E. and F., "Théorie des Corps Déformables," A. Hermann et Fils, Paris, 1909.*

2. *Timoshenko, S., and Goodier, J. N., "Theory of Elasticity," McGraw-Hill Book Company, New York, 2nd Edition, 1951.*

3. *Mindlin, R. D., and Tiersten, H. F., "Effects of couple-stresses in linear elasticity," Archive for Rational Mechanics and Analysis, forthcoming.*

4. *McLachlan, N. W., "Bessel Functions for Engineers," Oxford University Press, London, 1934.*

(*Reprinted from the Journal of Applied Mechanics for March, 1963, with permission of Copyright Owner.*)

J. J. McCOY
Department of Civil Engineering and Engineering Mechanics,
Columbia University,
New York, N. Y.

R. D. MINDLIN
Professor of Civil Engineering,
Columbia University,
New York, N. Y.
Mem. ASME

Extensional Waves Along the Edge of an Elastic Plate

A solution of second-order equations of extensional motion of plates is obtained for waves traveling along a free edge. The usual Rayleigh-type wave is found to become more and more confined to the edge of the plate as the frequency increases. Above a certain frequency another edge wave is found.

A WAVE of plane strain can be propagated, in an elastic half-space, as a Rayleigh surface wave[1] in which the displacements diminish exponentially into the interior of the body. As is well known, a solution of the equations of elasticity for a state of plane strain may be converted to one of generalized plane stress by replacing Poisson's ratio, ν, by $\nu/(1 + \nu)$. Rayleigh's solution then describes an extensional wave traveling along the edge of a plate. Such a wave, however, is only an approximation inasmuch as the equations of generalized plane stress take into account only the average displacements and stresses through the thickness of the plate. No account is taken of the higher modes of propagation which are possible in a plate and which must be coupled in the edge wave. Also, no indication is obtained of the depth of penetration of the wave into the plate, in terms of the plate thickness, since the thickness does not appear in the equations of generalized plane stress.

Despite the omission of the higher branches of the frequency spectrum of waves in a plate, it is reasonable to expect that the solution in generalized plane stress will yield a good approximation to the velocity of the edge wave at frequencies much lower than the frequency of the lowest thickness-mode of vibration of the plate. At such frequencies, the higher modes have only complex or pure imaginary propagation constants and experience with steady vibrations of plates indicates that their contributions are negligible.

It was the purpose of this investigation to determine the behavior of edge waves at higher frequencies and to find the decay of amplitude, with penetration into the interior of the plate, in terms of the plate thickness. It is not feasible to conduct such a study within the framework of the three-dimensional equations of elasticity; but there are available second-order equations[2] which are intermediate between those of generalized plane stress and the full linear theory and which are solvable.

Generalized plane stress takes into account only the first extensional mode and the lowest face-shear mode, whereas the second-order equations include the next three higher modes: thickness-stretch, symmetric thickness-shear, and second symmetric face-shear. At low frequencies, two of the additional modes have complex propagation constants and the third has an imaginary propagation constant. Thus the influence of such propagation constants may be investigated. Furthermore, all three additional modes have real propagation constants at higher frequencies so that there is a possibility of the appearance of new phenomena.

The results of the investigation may be summarized as follows: At very low frequencies the contributions of the complex and imaginary branches are indeed negligible and the displacement and velocity found from the equations of generalized plane stress are excellent approximations. Further, it is found that the wave is not closely confined to the edge. For example, at a frequency one twentieth of the frequency of the first antisymmetric thickness-shear mode, the amplitude of the face-shear part of the wave drops to $1/e$ of its amplitude at the edge only after 30 times the plate thickness from the edge. As the frequency increases, the contributions of the three higher branches increase and, also, the wave becomes more and more closely confined to the neighborhood of the edge. However, this has very little influence on the velocity of propagation of the wave; there is only a slight decrease of group velocity. The most interesting phenomenon occurs when the frequency is greater than that of the first symmetric thickness-shear mode. There a second edge wave makes its appearance. The additional wave is closely confined to the edge and there is very little contribution to its mode shape from the two branches of generalized plane stress. The group velocity of the additional wave is smaller than that which the first wave has at low frequency, but the two have nearly the same group velocity at corresponding frequencies.

Second-Order Equations

The plate is referred to rectangular coordinates x_i, $i = 1, 2, 3$, and is bounded by free faces at $x_2 = \pm b$ and a free edge at $x_1 = 0$. The plate displacements $u_1^{(0)}$, $u_3^{(0)}$, $u_2^{(1)}$, $u_1^{(2)}$, and $u_3^{(2)}$, as shown in Fig. 1 and defined in a previous paper,[2] are expressed in terms of three functions φ_i, $i = 1, 2, 3$ and two functions H_0 and H_2 all of them harmonic functions of time, according to

$$\begin{aligned} u_1^{(0)} &= \varphi_{1,1} + \varphi_{2,1} + \varphi_{3,1} - H_{0,3} \\ u_3^{(0)} &= \varphi_{1,3} + \varphi_{2,3} + \varphi_{3,3} + H_{0,1} \\ u_2^{(1)} &= \alpha_1\varphi_1 + \alpha_2\varphi_2 + \alpha_3\varphi_3 \\ u_1^{(2)} &= \beta_1\varphi_{1,1} + \beta_2\varphi_{2,1} + \beta_3\varphi_{3,1} - H_{2,3} \\ u_3^{(2)} &= \beta_1\varphi_{1,3} + \beta_2\varphi_{2,3} + \beta_3\varphi_{3,3} + H_{2,1} \end{aligned} \quad (1)$$

where

$$\begin{aligned} \alpha_i &= b[(\lambda + 2\mu)\xi_i^2 - \rho\omega^2]/\kappa_1\lambda \\ \beta_i &= \alpha_i b/[3(\Omega^2 - 4)/4 - E'b^2\xi_i^2/5\mu\kappa_2^2] \end{aligned} \quad (2)$$

In equations (2), λ and μ are Lamé's constants, ρ is the density, ω is the circular frequency, and

$$\begin{aligned} E' &= 4\mu(\lambda + \mu)/(\lambda + 2\mu) \\ \Omega &= \omega/\omega_s \end{aligned} \quad (3)$$

where

$$\omega_s = (\pi/2b)(\mu/\rho)^{1/2} \quad (4)$$

[1] Lord Rayleigh, "On Waves Propagated Along the Plane Surface of an Elastic Solid," *Proceedings of The London Mathematical Society*, vol. 17, 1887, pp. 4–11.

[2] R. D. Mindlin and M. A. Medick, "Extensional Vibrations of Elastic Plates," JOURNAL OF APPLIED MECHANICS, vol. 26, TRANS. ASME, vol. 81, Series E, 1959, pp. 561–569.

Contributed by the Applied Mechanics Division and presented at the Winter Annual Meeting, New York, N. Y., November 25–30, 1962, of THE AMERICAN SOCIETY OF MECHANICAL ENGINEERS.

Discussion of this paper should be addressed to the Editorial Department, ASME, United Engineering Center, 345 East 47th Street, New York 17, N. Y., and will be accepted until April 10, 1963. Discussion received after the closing date will be returned. Manuscript received by ASME Applied Mechanics Division, December 19, 1961. Paper No. 62—WA-79.

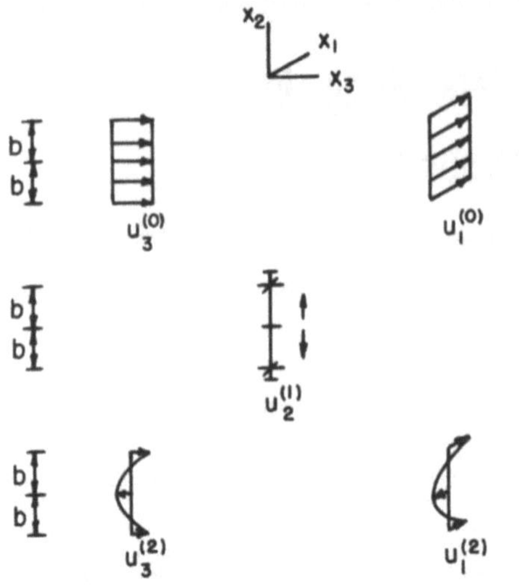

Fig. 1 Displacements taken into account in second-order equations

In addition, the κ_s (s = 1 to 4) are the correction coefficients of the second-order equations and the ξ_i (i = 1, 2, 3) are the roots of the cubic equation

$$\begin{vmatrix} a_{11} & a_{12} & 0 \\ a_{12} & a_{22} & a_{23} \\ 0 & a_{23} & a_{33} \end{vmatrix} = 0 \tag{5}$$

where

$$\begin{aligned} a_{11} &= k^2z^2 - \Omega^2 \\ a_{22} &= \kappa_2^2z^2/3 + 4\kappa^2k^2/\pi^2 - \kappa_3^2\Omega^2/3 \\ a_{33} &= E'z^2/5\mu + 12\kappa_2^2/\pi^2 - \kappa_4^2\Omega^2/5 \\ a_{12} &= 2\kappa_1(k^2 - 2)z/\pi \\ a_{23} &= -2\kappa_2^2z/\pi \end{aligned} \tag{6}$$

and

$$z = 2\xi b/\pi \qquad k^2 = 2(1 - \nu)/(1 - 2\nu) \tag{7}$$

with ν denoting Poisson's ratio.

The potentials φ_i, H_0, and H_2 satisfy Helmholtz equations:

$$\begin{aligned} \nabla^2\varphi_1 + \xi_i^2\varphi_i &= 0 \\ \mu\nabla^2H_0 + \rho\omega^2H_0 &= 0 \\ \mu\nabla^2H_2 + (\kappa_4^2\rho\omega^2 - 15b^{-2}\kappa_2^2\mu)H_2 &= 0 \end{aligned} \tag{8}$$

The five branches of the frequency spectrum of an infinite plate, associated with the five potential φ_i, H_0, and H_2 are illustrated in Fig. 2.

The plate strains (defined previously[2]) are expressed in terms of the potentials by

$$\begin{aligned} \epsilon_1^{(0)} &= \Sigma_i\varphi_{i,11} - H_{0,31} \\ \epsilon_2^{(0)} &= \Sigma_i\alpha_i\varphi_i/b \\ \epsilon_3^{(0)} &= \Sigma_i\varphi_{i,33} + H_{0,13} \\ \epsilon_5^{(0)} &= 2\Sigma_i\varphi_{i,13} + H_{0,11} - H_{0,33} \\ \epsilon_6^{(1)} &= \Sigma_i(3\beta_i/b + \alpha_i)\varphi_{i,1} - 3H_{2,3}/b \\ \epsilon_1^{(2)} &= \Sigma_i\beta_i\varphi_{i,11} - H_{2,31} \\ \epsilon_3^{(2)} &= \Sigma_i\beta_i\varphi_{i,33} + H_{2,13} \\ \epsilon_5^{(2)} &= 2\Sigma_i\beta_i\varphi_{i,13} + H_{2,11} - H_{2,33} \end{aligned} \tag{9}$$

Similarly the components of plate traction acting on the surfaces x_1 = const are

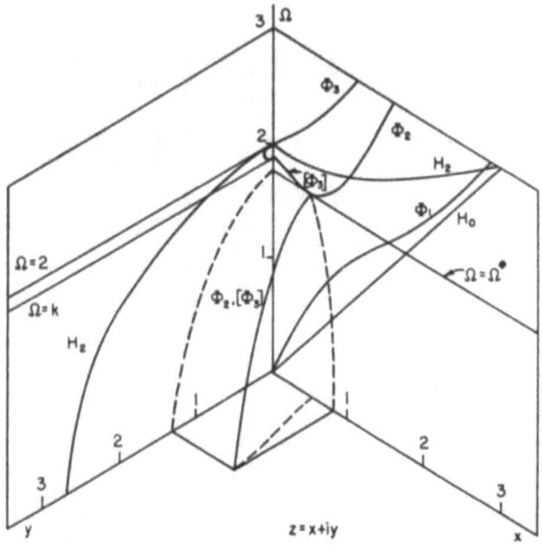

Fig. 2 Five branches of characteristic equation of second-order theory

$$\begin{aligned} \tau_1^{(0)} &= 2\{\Sigma_i[(\lambda + 2\mu)\varphi_{i,11} + (\lambda\kappa_1\alpha_i/b)\varphi_i + \lambda\varphi_{i,33}] - 2\mu H_{0,31}\} \\ \tau_5^{(0)} &= 2\mu\{2\Sigma_i\varphi_{i,13} + H_{0,11} - H_{0,33}\} \\ \tau_6^{(1)} &= (2\mu\kappa_2^2/3)\{\Sigma_i(3\beta_i/b + \alpha_i)\varphi_{i,1} - 3H_{2,3}/b\} \\ \tau_1^{(2)} &= (2E'/5)\{\Sigma_i\beta_i(\varphi_{i,11} + \nu\varphi_{i,33}) - (1 - \nu)H_{2,31}\} \\ \tau_5^{(2)} &= (2\mu/5)\{2\Sigma_i\beta_i\varphi_{i,13} + H_{2,11} - H_{2,33}\} \end{aligned} \tag{10}$$

Frequency Equation

Equations (8) will be satisfied by solutions of the form

$$\begin{aligned} \varphi_i &= A_ie^{-\zeta_ix_1}\sin\gamma x_3 \\ H_0 &= A_4e^{-\zeta_4x_1}\cos\gamma x_3 \\ H_2 &= A_5e^{-\zeta_5x_1}\cos\gamma x_3 \end{aligned} \tag{11}$$

provided that

$$\begin{aligned} \zeta_i^2 - \gamma^2 + \xi_i^2 &= 0 \\ \zeta_4^2 - \gamma^2 + \rho\omega^2/\mu &= 0 \\ \zeta_5^2 - \gamma^2 + (\kappa_4^2\rho\omega^2 - 15b^{-2}\kappa_2^2\mu)/\mu &= 0 \end{aligned} \tag{12}$$

Then, from (10) and (11), one obtains, on $x_1 = 0$,

$$\begin{aligned} \tau_1^{(0)} &= f_1\Sigma_jA_jT_{1j} \\ \tau_5^{(0)} &= f_2\Sigma_jA_jT_{2j} \\ \tau_6^{(1)} &= f_3\Sigma_jA_jT_{3j} \\ \tau_1^{(2)} &= f_4\Sigma_jA_jT_{4j} \\ \tau_5^{(2)} &= f_5\Sigma_jA_jT_{5j} \end{aligned} \tag{13}$$

where

$$\begin{aligned} T_{1j} &= 2X^2 - \Omega^2 \qquad j = 1, 2, 3 \\ T_{14} &= -2X(X^2 - \Omega^2)^{1/2} \\ T_{2j} &= 2X(X^2 - z_j^2)^{1/2} \qquad j = 1, 2, 3 \\ T_{24} &= \Omega^2 - 2X^2 \\ T_{3j} &= (\pi/2)(3\beta_j' + \alpha_j')(X^2 - z_j^2)^{1/2} \qquad j = 1, 2, 3 \\ T_{35} &= -6X/\pi \\ T_{4j} &= (\pi^2/4)\beta_j'[(1 - \nu)X^2 - z_j^2] \qquad j = 1, 2, 3 \\ T_{45} &= -(1 - \nu)X(X^2 - z_5^2)^{1/2} \\ T_{5j} &= (\pi^2/2)\beta_i'X(X^2 - z_j^2)^{1/2} \qquad j = 1, 2, 3 \\ T_{55} &= z_5^2 - 2X^2 \\ T_{15} &= T_{25} = T_{34} = T_{44} = T_{54} = 0 \end{aligned} \tag{14}$$

$$
\begin{aligned}
f_1 &= (\pi^2\mu/2b^2) \sin \gamma x_3 \\
f_2 &= (-2\mu\pi^2/4b^2) \cos \gamma x_3 \\
f_3 &= (-2\mu\kappa_2{}^2/3)(\pi^2/4b^2) \sin \gamma x_3 \\
f_4 &= (+2E'/5)(\pi^2/4b^2) \sin \gamma x_3 \\
f_5 &= (-2\mu/5)(\pi^2/4b^2) \cos \gamma x_3
\end{aligned}
\qquad (14)\ \textit{Cont.}
$$

and

$$
\begin{aligned}
\alpha_j' &= [2(1-\nu)z_j{}^2 - (1-2\nu)\Omega^2]/2\kappa_1\nu \\
\beta_j' &= \alpha_j'/[0.75(\Omega^2 - 4) - \pi^2 z_j{}^2/10(1-\nu)\kappa_2{}^2] \\
X &= 2\gamma b/\pi \\
z_5{}^2 &= (\kappa_4{}^2\Omega^2 - 60\kappa_2{}^2/\pi^2)
\end{aligned}
\qquad (15)
$$

The boundary conditions, in the absence of forces on the edge, are

$$
\tau_1{}^{(0)} = \tau_5{}^{(0)} = \tau_6{}^{(1)} = \tau_1{}^{(2)} = \tau_5{}^{(2)} = 0 \qquad (16)
$$

on $x_1 = 0$ and a nontrivial solution is obtained if the determinant of the coefficients of the A_j in equations (13) is set equal to zero; i.e.,

$$
|T_{ij}| = 0 \qquad i, j = 1, 2, 3, 4, 5 \qquad (17)
$$

The determinantal equation (17) gives the relation between dimensionless frequency, Ω, and dimensionless propagation constant, X, and may be solved in the following manner: A given value of Poisson's ratio ν fixes the correction coefficients $\kappa_s(s = 1$ to 4) as explained previously.[2] From the cubic equation (5), the three values of z_i are computed for an assumed value of the nondimensional frequency Ω. Equation (17) is thereby reduced to an algebraic equation on X_i, the dimensionless propagation constant along x_3. Since we are only interested in waves which decay with distance from the edge, the only values of X of interest are those that are greater than Ω as may be seen from the second of equations (12). For $X < \Omega$, ζ_4 is imaginary and produces a dis-

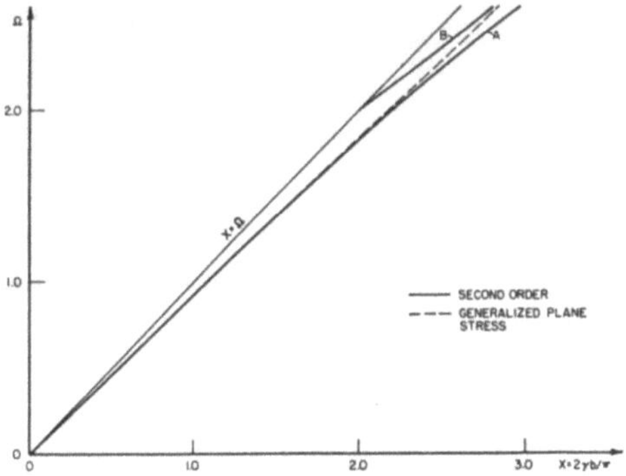

Fig. 3 Variation of frequency of edge waves with propagation constant

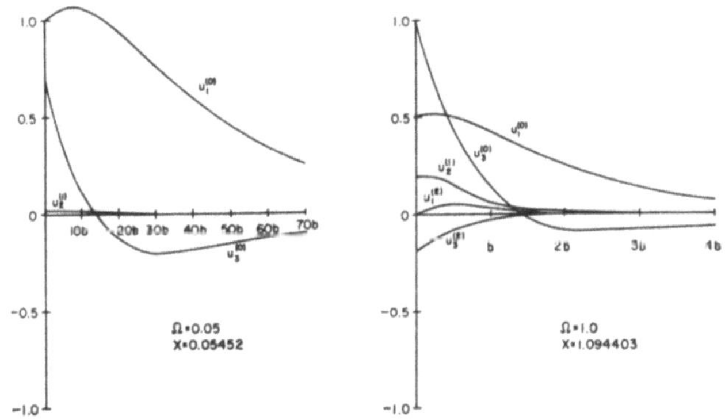

Fig. 4 Decay of amplitude with distance from edge; first edge wave; low and intermediate frequencies

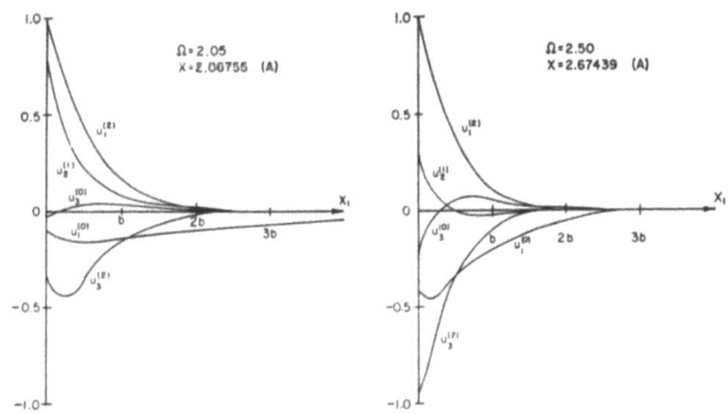

Fig. 5 Decay of amplitude with distance from edge; first edge wave; high frequency

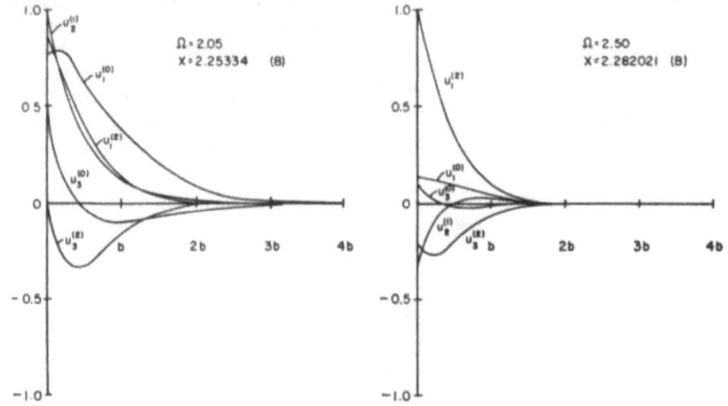

Fig. 6 Decay of amplitude with distance from edge; second edge wave

placement which does not decay into the interior of the plate.

For low frequencies and long wavelengths ($\Omega \ll 1$, $X \ll 1$) equation (17) reduces to the generalized plane-stress solution

$$(\Omega^2 - 2X^2)^2 = 4X^2[(X^2 - \Omega^2)(X^2 - z_1^2)]^{1/2}$$

where z_1 is obtained from equation (5) which reduces at low frequencies ($\Omega \ll 1$, $z \ll 1$) to

$$z_1^2 = k^2\Omega^2/4(k^2 - 1)$$

Numerical Solution of Frequency Equation

A table of values of Ω and X satisfying equations (5) and (17) was computed with the aid of an IBM 650 digital computer using an interval-halving iteration technique. Poisson's ratio was taken as 0.31.

Equation (14) shows that for $X > \Omega$ the real, imaginary, or complex character of the elements of the determinant (17) depends on the z_i. From equation (5), it may be ascertained that for $\Omega > 1.76074$ all the z_i are either real or imaginary and the elements of the determinant are all real. For $\Omega < 1.76074$, we have z_2 and z_3 as complex conjugates and therefore the elements of the second and third columns are complex conjugates. By adding and subtracting two columns, we obtain

$$T_{ij} = -4_i \begin{vmatrix} T_{11} & R_{12} & \mathcal{I}_{12} & T_{14} & T_{15} \\ T_{21} & R_{22} & \mathcal{I}_{22} & T_{24} & T_{25} \\ T_{31} & R_{32} & \mathcal{I}_{32} & T_{34} & T_{35} \\ T_{41} & T_{42} & \mathcal{I}_{42} & T_{44} & T_{45} \\ T_{51} & R_{52} & \mathcal{I}_{52} & T_{54} & T_{55} \end{vmatrix}$$

where $T_{i2} = R_{i2} + \mathcal{I}_{i2}$. All the elements are again real. This form of the determinant was employed for computations in the range $\Omega < 1.76074$.

Results

The solution of equation (17) is exhibited, graphically, on a plot with axes of nondimensional frequency (Ω), and nondimensional propagation constant (X) along the edge of the plate, Fig. 3. Two interesting results are revealed.

The first is that the velocity of the wave predicted by the elementary theory (represented by the curve A) is not essentially altered by the presence of the three additional modes. This indicates that still higher modes will also have a negligible effect on this wave at the lower frequencies.

A second result is the appearance of an additional wave (represented by the curve B) for frequencies greater than $\Omega = 2$. This wave is not predicted by the elementary theory.

It seems likely that higher orders of approximation, including more terms of the series describing the displacements, will not have a significant effect on the waves at the lower frequencies but will introduce additional waves, as the frequency increases.

Displacement-amplitude ratios were obtained at various frequencies for both of the edge waves. The results are shown graphically in Figs. 4, 5, and 6.

At the lower frequencies, where only wave A, as predicted by elementary theory, exists as an edge wave, we see that although the motion does correspond to an edge wave in that it decays with distance from the free edge, the rate of decay is extremely small.

It should be noted that, at frequencies below 1.76074, the parts of the displacements derived from the potentials φ_2 and φ_3 are not pure exponentials in x_1 but are sinusoidally decaying functions, due to the fact that ζ_2 and ζ_3 are complex. This is not revealed in the displacement curves because the amplitudes of these components are very small and their decay is very rapid.

*H*igh Frequency Vibrations of Plated, Crystal Plates

R. D. MINDLIN
Department of Civil Engineering
Columbia University
New York, New York

1. *Introduction*

In a previous paper [1], there were derived two-dimensional equations of motion, of crystal plates, in which thickness-shear, flexural, extensional, face-shear, and thickness-twist deformations were taken into account. The equations are now extended to include the effects of mass and stiffness of thin surface-platings such as are employed as electrodes on piezoelectrically-driven crystals.

The approximate equations of motion of the crystal plate, without platings, take into account the three components of displacement of the line element that is normal to the plane of the plate and the two components of rotation of the line element. Thus, the equations comprise the two equations of generalized plane stress and the three equations of flexure that include the effects of transverse shear deformation and rotatory inertia: the two sets being coupled because of the anisotropy. In practice, the platings are very thin, in comparison with the thickness of the crystal, and hence all flexural and thickness-shear stresses in them may be neglected. The motion of each plating may be described adequately by the two-equations of generalized plane stress and a simple equation which accounts for the transverse inertia of the plating.

The separate equations of the crystal and the two platings are given in sections 2 and 3 and the three sets are joined, in section 4, to produce equations of the plated crystal which are the same in number as those of the crystal plate alone and almost the same in form. The slight difference, however, results in important changes in coupling and frequencies of modes of motion. Some of these are investigated, in section 5, by means of the solution of the problem of straight-crested waves propagating in the direction of the digonal axis of a monoclinic, crystal plate with isotropic platings of different thicknesses. The additional coupling of modes of motion, resulting from the different thicknesses of the two platings, is exhibited and the circumstances under which the frequencies of various modes of motion increase or decrease are described. Formulas for the usual shear-correction factors are obtained, in section 6, from solutions of the two- and three-dimensional equations for thickness-vibrations of a plated, crystal plate. Initial and boundary conditions sufficient for a unique solution of the two dimensional equations of the plated, crystal plate are given in the final section.

2. *Equations of the Crystal Plate*

In this section there are summarized the two-dimensional, approximate equations of high frequency vibrations of crystal plates derived in an earlier paper [1]. Cartesian indicial notation is employed—with the summation convention for repeated indices and the comma notation for partial derivative. Indices i, j, k, l range over 1, 2, 3; indices a, b, c, d range over 1, 3.

The crystal plate, of thickness $2h$, is referred to rectangular coordinates x_i, with the axes of x_1, and x_3 in the middle plane. The axis of x_2 is taken as positive vertically upward so that reference may be made to "upper" and "lower" faces and platings.

The components of displacement, u_i, are approximated by the first two terms of a power series in the thickness-coordinate:

$$u_i = u_i^{(0)}(x_1, x_3, t) + x_2 u_i^{(1)}(x_1, x_3, t)$$

and the components of stress, T_{ij}, by the stress-resultants, $T_{ij}^{(0)}$, and stress-couples, $T_{ij}^{(1)}$:

$$T_{ij}^{(0)} \equiv \int_{-h}^{h} T_{ij}\, dx_2, \qquad T_{ij}^{(1)} \equiv \int_{-h}^{h} x_2 T_{ij}\, dx_2$$

The stress-equations of motion, $T_{i,ji} = \rho u_{j,tt}$, are then approximated by the five-equations

$$T^{(0)}_{ij,i} + F^{(0)}_j = 2h\rho u^{(0)}_{j,tt}, \qquad i, j = 1, 2, 3 \tag{1}$$

$$T^{(1)}_{ab,a} - T^{(0)}_{2b} + F^{(1)}_b = \tfrac{2}{3}h^3\rho u^{(1)}_{b,tt}, \qquad a, b = 1, 3 \tag{2}$$

where $F^{(0)}_j$ and $F^{(1)}_b$ are the surface loadings

$$F^{(0)}_j \equiv T_{2j}|_h - T_{2j}|_{-h}, \qquad F^{(1)}_b \equiv hT_{2b}|_h + hT_{2b}|_{-h} \tag{3}$$

and ρ is the mass density of the material of the crystal plate.

The stress-strain relations, $T_{ij} = c_{ijkl}S_{kl}$, are approximated by

$$T^{(0)}_{ij} = 2hg^*_{ijkl}S^{(0)}_{kl}, \qquad T^{(1)}_{ab} = \tfrac{2}{3}h^3\gamma_{abcd}S^{(1)}_{cd} \tag{4}$$

where

$$g^*_{ijkl} \equiv \kappa^m_{i+j-2}\kappa^n_{k+l-2}g_{ijkl} \qquad \text{(not summed)}$$

$$g_{ijkl} \equiv c_{ijkl} - c_{ij22}c_{22kl}/c_{2222}, \qquad \gamma_{abcd} \equiv c_{abcd} - c_{ab2j}c_{2kcd}(c_{2j2k})^{-1}$$

and m and n are the powers

$$m = \cos^2(ij\pi/2), \qquad n = \cos^2(kl\pi/2)$$

of the correction factors κ_1 and κ_3. The values of the κ_a are determined in section 6.

The strains in (4) are given in terms of the displacements by

$$S^{(0)}_{ij} = \tfrac{1}{2}(u^{(0)}_{i,j} + u^{(0)}_{j,i} + \delta_{2j}u^{(1)}_i + \delta_{2i}u^{(1)}_j), \qquad S^{(1)}_{ab} = \tfrac{1}{2}(u^{(1)}_{a,b} + u^{(1)}_{b,a}) \tag{5}$$

in which δ_{ij} is the Kronecker symbol.

Finally, the strain-energy and kinetic energy, per unit area of the plate, are

$$U = hg^*_{ijkl}S^{(0)}_{ij}S^{(0)}_{kl} + \tfrac{1}{3}h^3\gamma_{abcd}S^{(1)}_{ab}S^{(1)}_{cd} \tag{6}$$

$$K = \rho h(u^{(0)}_{i,t}u^{(0)}_{i,t} + \tfrac{1}{3}h^2u^{(1)}_{a,t}u^{(1)}_{a,t}) \tag{7}$$

3. *Equations of the Platings*

Quantities pertaining to the upper and lower platings will be designated by primes and double primes, respectively. The plating thicknesses, $2h'$ and $2h''$, are usually less than one thousandth of the thickness of the crystal plate [2] and the equations of the latter, that are given in section 2, are limited to frequencies not more than 50 per cent. higher than the frequency of the lowest thickness-shear mode of the plate [1]. With these limitations, the flexural stress-couples, $T^{(1)\prime}_{ab}$ and $T^{(1)\prime\prime}_{ab}$, and the thickness-shear stress-resultants,

$T_{a2}^{(0)'}$ and $T_{a2}^{(0)''}$, may be neglected in the platings. Then, all that remain of the equations analogous to (1)-(7) are, for the upper plating,

$$\delta_{jb}^0 T_{ab,a}^{(0)'} + F_j^{(0)'} = 2h'\rho' u_{j,tt}^{(0)'}, \qquad F_j^{(0)'} \equiv T'_{2j}|_{h'} - T'_{2j}|_{-h'} \tag{8}$$

$$T_{ab}^{(0)'} = 2h'\gamma'_{abcd} S_{cd}^{(0)'}, \qquad S_{ab}^{(0)'} = \tfrac{1}{2}(u_{a,b}^{(0)'} + u_{b,a}^{(0)'}) \tag{9}$$

$$U' = h'\gamma'_{abcd} S_{ab}^{(0)'} S_{cd}^{(0)'}, \qquad K' = \rho' h' u_{i,t}^{(0)'} u_{i,t}^{(0)'} \tag{10}$$

and, for the lower plating,

$$\delta_{jb}^0 T_{ab,a}^{(0)''} + F_j^{(0)''} = 2h''\rho'' u_{j,tt}^{(0)''}, \qquad F_j^{(0)''} \equiv T''_{2j}|_{h''} - T''_{2j}|_{-h''} \tag{11}$$

$$T_{ab}^{(0)''} = 2h''\gamma''_{abcd} S_{cd}^{(0)''}, \qquad S_{ab}^{(0)''} = \tfrac{1}{2}(u_{a,b}^{(0)''} + u_{b,a}^{(0)''}) \tag{12}$$

$$U'' = h''\gamma''_{abcd} S_{ab}^{(0)''} S_{cd}^{(0)''}, \qquad K'' = \rho'' h'' u_{i,t}^{(0)''} u_{i,t}^{(0)''} \tag{13}$$

where $\delta_{jb}^0 = \delta_{cb}$ when $j = c$ and $\delta_{jb}^0 = 0$ when $j = 2$.

4. *Equations of the Plated, Crystal Plate*

The condition that the platings are joined to the crystal plate are set by imposing the conditions of continuity of tractions and displacements across the interfaces:

$$T'_{2j}|_{-h'} = T_{2j}|_{h}, \qquad T''_{2j}|_{h''} = T_{2j}|_{-h} \tag{14}$$

$$u_i^{(0)'} = u_i^{(0)} + h\delta_{ia}^0 u_a^{(1)}, \qquad u_i^{(0)''} = u_i^{(0)} - h\delta_{ia}^0 u_a^{(1)} \tag{15}$$

By means of (14), the surface loadings $F_j^{(0)}$ and $F_b^{(1)}$, in (1) and (2), may be expressed in terms of the surface loadings $F_j^{(0)'}$ and $F_j^{(0)''}$ of the platings. Then the stress-equations of motion, $(8)_1$ and $(11)_1$, of the platings may be incorporated into (1) and (2) by substitution, thus forming five stress-equations of motion of the plated, crystal plate; and, in these equations, the accelerations may be expressed in terms of $u_i^{(0)}$ and $u_a^{(1)}$ by means of (15). The stress-equations of motion may then be converted to five displacement-equations of motion on the five displacements, $u_i^{(0)}$ and $u_a^{(1)}$, through the use of the three sets of stress-strain-displacement relations and (15). We proceed to carry out this process.

From $(3)_1$ and (14) and from $(3)_2$ and (14),

$$F_j^{(0)} = T'_{2j}|_{-h'} - T''_{2j}|_{h''}, \qquad F_b^{(1)} = hT'_{2b}|_{-h'} + hT''_{2b}|_{h''} \tag{16}$$

But, from the sum and difference of $(8)_2$ and $(11)_2$,

$$T'_{2j}|_{-h'} - T''_{2j}|_{h''} = \mathscr{F}_j^{(0)} - F_j^{(0)'} - F_j^{(0)''},$$

$$hT'_{2b}|_{-h'} + hT''_{2b}|_{h''} = \mathscr{F}_b^{(1)} - h(F_b^{(0)'} - F_b^{(0)''}) \tag{17}$$

where $\mathscr{F}_j^{(0)} \equiv T'_{2j}|_{h'} - T''_{2j}|_{-h''}$, $\mathscr{F}_b^{(1)} \equiv hT'_{2b}|_{h'} + hT''_{2b}|_{-h''}$

are the surface loadings on the outer faces of the platings. Hence, from (16) and (17),

$$F_j^{(0)} = \mathscr{F}_j^{(0)} - (F_j^{(0)'} + F_j^{(0)''}), \qquad F_b^{(1)} = \mathscr{F}_b^{(1)} - h(F_b^{(0)'} - F_b^{(0)''}) \tag{18}$$

Now, from the sum and difference of $(8)_1$ and $(11)_1$,

$$F_j^{(0)'} + F_j^{(0)''} = -\delta_{jb}^0(T_{ab}^{(0)'} + T_{ab}^{(0)''})_{,a} + 2h'\rho' u_{j,tt}^{(0)'} + 2h''\rho'' u_{j,tt}^{(0)''}$$

$$F_b^{(0)'} - F_b^{(0)''} = -(T_{ab}^{(0)'} - T_{ab}^{(0)''})_{,a} + 2h'\rho' u_{b,tt}^{(0)'} - 2h''\rho'' u_{j,tt}^{(0)''}$$

Hence, substituting these expressions into (18) and using (15), we have

$$F_j^{(0)} = \mathscr{F}_j^{(0)} + \delta_{jb}^0(T_{ab}^{(0)'} + T_{ab}^{(0)''})_{,a} - 2h\rho R_S u_{j,tt}^{(0)} - 2h^2\rho R_D \delta_{ja}^0 u_{a,tt}^{(1)} \tag{19}$$

$$F_b^{(1)} = \mathscr{F}_b^{(1)} + h(T_{ab}^{(0)'} - T_{ab}^{(0)''})_{,a} - 2h^3\rho R_S u_{b,tt}^{(1)} - 2h^2\rho R_D u_{b,tt}^{(0)} \tag{20}$$

where

$$R_S = (\rho' h' + \rho'' h'')/\rho h, \qquad R_D = (\rho' h' - \rho'' h'')/\rho h$$

i.e., R_S and R_D are the ratios of the sum and difference of the weights of the platings to the weight of the crystal.

Finally, substituting (19) and (20) into (1) and (2), we find the stress-equations of motion of the plated crystal:

$$\mathscr{T}_{ij,i}^{(0)} + \mathscr{F}_j^{(0)} = 2h\rho(1 + R_S)u_{j,tt}^{(0)} + 2h^2\rho R_D \delta_{ja}^0 u_{a,tt}^{(1)} \tag{21}$$

$$\mathscr{T}_{ab,a}^{(1)} - \mathscr{T}_{2b}^{(0)} + \mathscr{F}_b^{(1)} = \tfrac{2}{3}h^3\rho(1 + 3R_S)u_{b,tt}^{(1)} + 2h^2\rho R_D u_{b,tt}^{(0)} \tag{22}$$

where

$$\mathscr{T}_{ij}^{(0)} \equiv T_{ij}^{(0)} + \delta_{ia}^0\delta_{jb}^0(T_{ab}^{(0)'} + T_{ab}^{(0)''}), \qquad \mathscr{T}_{ab}^{(1)} \equiv T_{ab}^{(1)} + h(T_{ab}^{(0)'} - T_{ab}^{(0)''}) \tag{23}$$

Next, upon substituting into (23) the stress-strain relations (4), $(9)_1$, $(12)_1$ and then the strain-displacement relations (5), $(9)_2$, $(12)_2$, we find, after using (15) and then (5),

$$\mathscr{T}_{ij}^{(0)} = 2hg_{ijkl}^* S_{kl}^{(0)} + 2h\delta_{ia}^0\delta_{jb}^0(\gamma_{abcd}^S S_{cd}^{(0)} + h\gamma_{abcd}^D S_{cd}^{(1)}) \tag{24}$$

$$\mathscr{T}_{ab}^{(1)} = \tfrac{2}{3}h^3\gamma_{abcd}S_{cd}^{(1)} + 2h^3\gamma_{abcd}^S S_{cd}^{(1)} + 2h^2\gamma_{abcd}^D S_{cd}^{(0)} \tag{25}$$

where

$$\gamma_{abcd}^S \equiv (h'\gamma'_{abcd} + h''\gamma''_{abcd})/h, \qquad \gamma_{abcd}^D \equiv (h'\gamma'_{abcd} - h''\gamma''_{abcd})/h$$

The displacement-equations of motion may be obtained by substituting (5) in (24) and (25) and then (24) and (25) in (21) and (22), with the result:

$$g_{ijkl}^*(u_{k,li}^{(0)} + \delta_{2k}u_{l,i}^{(1)}) + \delta_{jb}^0\gamma_{abcd}^S u_{c,da}^{(0)} + \delta_{jb}^0\gamma_{abcd}^D u_{c,da}^{(1)} + \tfrac{1}{2}h^{-1}\mathscr{F}_j^{(0)}$$
$$= \rho(1 + R_S)u_{j,tt}^{(0)} + h\rho R_D\delta_{ja}^0 u_{a,tt}^{(1)} \tag{26}$$

$$\tfrac{1}{3}h^2\gamma_{abcd}u_{c,da}^{(1)} - g_{2bkl}^*(u_{k,l}^{(0)} + \delta_{2k}u_l^{(1)}) + h^2\gamma_{abcd}^S u_{c,da}^{(1)} + h\gamma_{abcd}^D u_{c,da}^{(0)} + \tfrac{1}{2}h^{-1}\mathscr{F}_b^{(1)}$$
$$= \tfrac{1}{3}h^2\rho(1 + 3R_S)u_{b,tt}^{(1)} + h\rho R_D u_{b,tt}^{(0)} \tag{27}$$

The strain-energy-density is found by summing (6), $(10)_1$ and $(13)_1$ and using the strain-displacement relations, (5), $(9)_2$, $(12)_2$, along with the continuity condition (15), with the result

$$\begin{aligned}\mathscr{U} &= \mathrm{U} + \mathrm{U}' + \mathrm{U}'' \\ &= hg^*_{ijkl}S^{(0)}_{ij}S^{(0)}_{kl} + \tfrac{1}{3}h^3\gamma_{abcd}S^{(1)}_{ab}S^{(1)}_{cd} + h\gamma^S_{abcd}(S^{(0)}_{ab}S^{(0)}_{cd} + h^2S^{(1)}_{ab}S^{(1)}_{cd}) \\ &\qquad + 2h^2\gamma^D_{abcd}S^{(0)}_{ab}S^{(0)}_{cd} \qquad (28)\end{aligned}$$

We note that, from (24), (25) and (28),

$$\mathscr{T}^{(0)}_{ij} = \partial\mathscr{U}/\partial S^{(0)}_{ij}, \qquad \mathscr{T}^{(1)}_{ab} = \partial\mathscr{U}/\partial S^{(1)}_{ab} \qquad (29)$$

where $\partial S^{(0)}_{ij}/\partial S^{(0)}_{ji} = 0$ if $i \neq j$; $\partial S^{(1)}_{ab}/\partial S^{(1)}_{ba} = 0$ if $a \neq b$.

The kinetic energy density is obtained by summing (7), $(10)_2$ and $(13)_2$ and using (15):

$$\begin{aligned}\mathscr{K} = K + K' + K'' &= \rho h(1 + R_S)u^{(0)}_{i,t}u^{(0)}_{i,t} + \tfrac{1}{3}\rho h^3(1 + 3R_S)u^{(1)}_{a,t}u^{(1)}_{a,t} \\ &\qquad + 2\rho h^2 R_D u^{(0)}_{a,t}u^{(1)}_{a,t} \qquad (30)\end{aligned}$$

Referring to the displacement-equations of motion, (26) and (27), we note that the effects of the platings are contained in the terms with coefficients γ^S, γ^D, R_S, R_D. In the absence of platings, these constants vanish and the equations reduce to those in [1]. If there are platings but the two are identical, only γ^D and R_D are zero. Then, if the symmetry of γ^S is a subclass of the symmetry of γ and g, the equations have the same form as those in [1], but with different constants. Various effects of like and unlike platings are examined in the following section.

5. *Straight-Crested Waves in a Monoclinic Plate with Isotropic Platings*

An important case of unlike platings is that of isotropic platings of the same material but different thicknesses. Then $\rho' = \rho''$ and

$$\begin{aligned}\gamma^S_{abcd} &= [\lambda^0\delta_{ab}\delta_{cd} + \mu(\delta_{ac}\delta_{bd} + \delta_{ad}\delta_{bc}) + \nu(\delta_{ac}\delta_{bd} - \delta_{ad}\delta_{bc})](h' + h'')/h \\ \gamma^D_{abcd} &= [\lambda^0\delta_{ab}\delta_{cd} + \mu(\delta_{ac}\delta_{bd} + \delta_{ad}\delta_{bc}) + \nu(\delta_{ac}\delta_{bd} - \delta_{ad}\delta_{bc})](h' - h'')/h\end{aligned} \qquad (31)$$

where $\lambda^0 = 2\mu\lambda/(\lambda + 2\mu)$ and λ and μ are Lamé's constants for the material of the platings. We consider the effect of such platings on the propagation of straight-crested waves along the digonal axis in a monoclinic crystal plate.

With x_1 the digonal axis, monoclinic symmetry requires

$$c_{1131} = c_{2231} = c_{3331} = c_{2331} = c_{1112} = c_{2212} = c_{3312} = c_{2312} = 0 \quad (32)$$

This is the symmetry encountered in the technologically important case of the rotated-Y-cuts of quartz[3]. There, thickness-shear vibrations, $u_1^{(1)} = A_1^{(1)} e^{i\omega t}$ are forced piezoelectrically. However, owing to the presence of boundaries $x_1 = \text{constant}$, coupling of $u_1^{(1)}$ with flexural deflection $u_2^{(0)}$ is introduced, producing waves. Then, through the constant c_{1312}, further coupling with face-shear deformation $u_3^{(0)}$ occurs. The result is that straight-crested waves of coupled flexure, thickness-shear and face-shear:

$$\begin{gathered} u_2^{(0)} = A_2^{(0)} h \sin(\xi x_1 - \omega t), \qquad u_1^{(1)} = A_1^{(1)} \cos(\xi x_1 - \omega t), \\ u_3^{(0)} = A_3^{(0)} h \sin(\xi x_1 - \omega t) \end{gathered} \quad (33)$$

are excited. This case, for the unplated crystal, was investigated in [1] and [4]. It was also shown that another type of straight-crested wave, of coupled extension and thickness-twist:

$$u_1^{(0)} = A_1^{(0)} h \cos(\xi x_1 - \omega t), \qquad u_3^{(1)} = A_3^{(1)} \sin(\xi x_1 - \omega t) \quad (34)$$

is possible in the unplated crystal; but is not coupled with (33). However, in the case of the plated crystal with unlike platings, (33) and (34) are coupled. Thus, upon substituting (31)-(34) in (26) and (27) and eliminating the amplitudes $A_1^{(0)}$ and $A_a^{(1)}$, we find the frequency equation

$$\begin{vmatrix} a_{11} & a_{12} & 0 & a_{14} & 0 \\ a_{21} & a_{22} & 0 & 0 & a_{25} \\ 0 & 0 & a_{33} & a_{34} & a_{35} \\ a_{41} & 0 & a_{43} & a_{44} & a_{45} \\ 0 & a_{52} & a_{53} & a_{54} & a_{55} \end{vmatrix} = 0 \quad (35)$$

where

$$a_{11} = [g_{1111} + (\lambda^0 + 2\mu)(h' + h'')/h]\xi^2 h^2 - \rho\omega^2 h^2(1 + R_S)$$

$$a_{22} = [\tfrac{1}{3}\gamma_{1313} + \mu(h' + h'')/h]\xi^2 h^2 + \kappa_3^2 c_{2323} - \tfrac{1}{3}\rho\omega^2(1 + 3R_S)$$

$$a_{33} = \kappa_1^2 c_{1212}\xi^2 h^2 - \rho\omega^2 h^2(1 + R_S)$$

$$a_{44} = [\tfrac{1}{3}\gamma_{1313} + (\lambda^0 + 2\mu)(h' + h'')/h]\xi^2 h^2 + \kappa_1^2 c_{1212} - \tfrac{1}{3}\rho\omega^2 h^2(1 + 3R_S)$$

$$a_{55} = [c_{1313} + \mu(h' + h'')/h]\xi^2 h^2 - \rho\omega^2 h^2(1 + R_S)$$

$$a_{12} = a_{21} = -g_{1123}^{*}\xi h, \qquad a_{35} = a_{53} = \kappa_1 c_{1312}\xi^2 h^2$$

$$a_{34} = a_{43} = \kappa_1^2 c_{1212}\xi h, \qquad a_{45} = a_{54} = \kappa_1 c_{1312}\xi h$$

$$a_{14} = a_{41} = h^{-1}(h' - h'')[(\lambda^0 + 2\mu)\xi^2 h^2 - \rho'\omega^2 h^2]$$

$$a_{25} = a_{52} = h^{-1}(h' - h'')[\mu\xi^2 h^2 - \rho'\omega^2 h^2]$$

From upper left to lower right, the elements of the main diagonal, in (35), arise from $u_1^{(0)}$, $u_3^{(1)}$, $u_2^{(0)}$, $u_1^{(1)}$ and $u_3^{(0)}$, respectively. Thus, the elements in the 3×3 block at the lower right arise from (33) while the elements in the 2×2 block at the upper left arise from (34). The two blocks are coupled through elements a_{14} and a_{25} which, as may be seen, have coefficients $h' - h''$ and hence are present only with unlike platings.

Aside from coupling due to unlike platings, the presence of platings, like or unlike, has additional influence on frequencies. The general consequences may be deduced by inspection of the elements of the main diagonal in (35). When these elements are set equal to zero, formulas are obtained for the frequencies of uncoupled modes of the five displacements. For extensional waves, we observe, from a_{11}, that the frequency is increased owing to the presence of $\lambda^0 + 2\mu$ and decreased because of $1 + R_S$. However, it may be ascertained that the difference between frequencies with and without plating is proportional to the difference between velocities of extensional waves in the plating and crystal. As a result, the frequency is increased or decreased according as the velocity of extensional waves in the plating is greater or less than the velocity of extensional waves in the crystal. Inspection of a_{55} yields the analogous result for face-shear waves. From a_{33} we see that the addition of plating always decreases the frequency of waves of deflection $u_2^{(0)}$. The effect of plating on the frequency of thickness-shear or thickness-twist waves is more complicated. From a_{44} (or a_{22}) we see that the asymptotic frequency ($\xi h = 0$) is always reduced by plating. By forming the difference between frequencies with and without plating, we find that this reduction is maintained for all wave-numbers if the velocity of extensional (or face-shear) waves in the plating is less than the corresponding velocity in the crystal. In the contrary case, the frequency with plating becomes the higher one for sufficiently large wave-numbers. The critical wave-number is that for which the ratio of the thickness of the crystal to the half-wave-length is equal to the ratio of the asymptotic velocity of thickness-shear waves $u_1^{(1)}$ (or $u_3^{(1)}$) in the crystal to the square root of the difference between the squares of the velocities of the two extensional (or face-shear) waves. All of these conclusions will, of course, be modified, somewhat, if the off-diagonal terms in (35) are taken into account.

6. *Correction Factors*

For the usual applications to a thin plate at frequencies slightly above that of the lowest thickness-shear mode of an infinite plate, reasonably good values of the correction factors κ_1 and κ_3 are found, as in [1], by equating the asymptotic thickness-shear frequencies, obtained from the two-dimensional equations, to the corresponding ones obtained from the three-dimensional equations.

In (26) and (27), let $\mathscr{F}_j^{(0)} = \mathscr{F}_b^{(0)} = u_2^{(0)} = 0$ and

$$u_a^{(0)} = A_a^{(0)} \exp i\omega t, \qquad u_a^{(1)} = A_a^{(1)} \exp i\omega t$$

Then (26) and (27) become

$$(1 + R_S)A_a^{(0)} - R_D A_a^{(1)} = 0$$

$$\rho\omega^2 h^2 R_D A_d^{(0)} - [g^*_{2b2d} - \tfrac{1}{3}\rho\omega^2 h^2(1 + 3R_S)\delta_{bd}]A_d^{(1)} = 0$$

Upon eliminating $A_a^{(0)}$ and $A_a^{(1)}$, we have

$$|g^*_{2b2d} - g\delta_{bd}| = 0, \qquad g = \tfrac{1}{3}\rho\omega^2 h^2[1 + 3R_S + 3R_D^2/(1 + R_S)] \tag{36}$$

Equation $(36)_1$ gives two roots, g_1 and g_3, which depend on the g^*_{2b2d} and, hence, on κ_1 and κ_3. Then $(36)_2$ yields two frequencies, ω_1 and ω_3, which depend on κ_1 and κ_3.

In the three-dimensional case, we take

$$u_l = (A_l \sin \eta x_2 + B_l \cos \eta x_2)\exp i\omega t \tag{37}$$

as the solution of the equations of motion $c_{ijkl}u_{l,ki} = \rho u_{j,tt}$; whence, eliminating A_l and B_l,

$$|c_{2j2l} - c\delta_{jl}| = 0, \qquad c = \rho\omega^2/\eta^2 \tag{38}$$

If $h' \ll h$ and $h'' \ll h$, we may neglect the deformation of the platings and take, as boundary conditions of the crystal:

$$\text{on} \quad x_2 = h, \quad T_{2j} = -2h'\rho' u_{j,tt}; \qquad \text{on} \quad x_2 = -h, \quad T_{2j} = 2h''\rho'' u_{j,tt} \tag{39}$$

Now, $T_{2j} = c_{2jkl}S_{kl} = c_{2jkl}u_{l,k}$. Hence (39) become

$$(c_{2j2l}\eta \cos \eta h - 2\delta_{jl}h'\rho'\omega^2 \sin \eta h)A_l - (c_{2j2l}\eta \sin \eta h + 2\delta_{jl}h'\rho'\omega^2 \cos \eta h)B_l = 0$$

$$(c_{2j2l}\eta \cos \eta h - 2\delta_{jl}h''\rho''\omega^2 \sin \eta h)A_l + (c_{2j2l}\eta \sin \eta h + 2\delta_{jl}h''\rho''\omega^2 \cos \eta h)B_l = 0$$

Upon eliminating A_l and B_l and using $(38)_2$, we find the 6×6 determinantal equation on ηh:

$$\begin{vmatrix} |(c_{2j2l}h\rho/2ch'\rho')\cot \eta h - \delta_{jl}\eta h| & |-c_{2j2l}h\rho/2ch'\rho' - \delta_{jl}\eta h \cot \eta h| \\ |(c_{2j2l}h\rho/2ch''\rho'')\cot \eta h - \delta_{jl}\eta h| & |c_{2j2l}h\rho/2ch''\rho'' + \delta_{jl}\eta h \cot \eta h| \end{vmatrix} = 0 \tag{40}$$

The required frequencies are obtained from (38) and (40) as follows. First find the two lowest roots, c_1 and c_3, of $(38)_1$; i.e., the velocities of the essentially shear waves with wave-normal in the direction of x_2. For each of c_1

and c_3, find the lowest root, $(\eta h)_1$ and $(\eta h)_3$ of (40). Then, from $(38)_2$,

$$\omega_a^2 = (\eta h)_a^2 c_a/\rho h^2, \quad \text{(not summed)} \tag{41}$$

Upon equating corresponding ω_a from (41) and $(36)_2$, we have a pair of simultaneous, algebraic equations for the κ_a.

In the case of monoclinic symmetry, c_{2122} and c_{2123} are zero. Then the roots of $(36)_1$ are

$$g_1 = g^*_{2121} = \kappa_1^2 g_{2121} = \kappa_1^2 c_{2121},$$
$$g_3 = g^*_{2323} = \kappa_3^2 g_{2323} = \kappa_3^2(c_{2323} - c^2_{2223}/c_{2222})$$

and $(36)_2$ becomes

$$\omega_a^2 = (3\kappa_a^2 g_{2a2a}/\rho h^2)[1 + 3R_S + 3R_D^2/(1 + R_S)], \quad \text{(not summed)} \tag{42}$$

Also, the two lowest roots of $(38)_1$ are

$$c_1 = c_{2121}, \qquad c_3 = \tfrac{1}{2}\{c_{2222} + c_{2323} - [(c_{2222} - c_{2323})^2 + 4c^2_{2223}]^{\frac{1}{2}}\}$$

and the 6×6 determinant in (40) resolves into the product of three 2×2 determinants: with the result that (40) may be written in the form

$$(C_j R_S \eta h \tan \eta h - 1)(C_j R_S \eta h \cot \eta h + 1) = C_j^2 R_D^2 (\eta h)^2, \quad \text{(not summed)} \tag{43}$$

where $C_j = c_j/c_{2j2j}$ (not summed). Now, if the platings were alike, R_D would be zero and (43) would reduce to

$$C_j R_S \eta h \tan \eta h - 1 = 0, \qquad C_j R_S \eta h \cot \eta h + 1 = 0 \tag{44}$$

These are the frequency equations for modes essentially antisymmetric and symmetric, respectively, with respect to the middle plane of the plate. The first of (44) would be the equation of present interest and, when $R_S \ll 1$, its lowest root is a little less than $\frac{1}{2}\pi$. Now, a typical case would be silver platings $\frac{1}{2}$ micron thick on a quartz plate one millimeter thick. Then $\rho'/\rho \approx 4$, $h'/h = 5 \times 10^{-4}$; whence $R_S \approx 4 \times 10^{-3}$ and R_D would be at most $2R_S$: for plating on one face only. Thus, returning to (43), since $R_S \ll 1$, $R_D \ll 1$, the roots of interest are

$$\eta h = \tfrac{1}{2}\pi - \varepsilon, \qquad 0 < \varepsilon \ll 1 \tag{45}$$

Inserting (45) in (43), expanding the trigonometric functions in power series of ε and dropping terms of degree higher than the third in ε, R_S and R_D, we find

$$\varepsilon = \tfrac{1}{2}\pi C_j R_S[1 - C_j R_S + \tfrac{1}{4}\pi^2 C_j^2(R_S^2 - R_D^2)], \quad \text{(not summed)} \tag{46}$$

Then, substituting (46) into (45) and the latter into (41), we have

$$\omega_a^2 = (\pi^2 c_a/4\rho h^2)\{1 - C_a R_S[1 - C_a R_S + \tfrac{1}{4}\pi^2 C_a^2(R_S^2 - R_D^2)]\}^2, \quad \text{(not summed)} \tag{47}$$

Finally, equating (47) and (42) and solving for k_a^2, we find

$$\kappa_a^2 = \frac{\pi^2}{12}\frac{c_a\{1 - C_aR_S[1 - C_aR_S + \tfrac{1}{4}\pi^2C_a^2(R_S^2 - R_D^2)]\}^2}{g_{2a2a}[1 + 3R_S + 3R_D^2/(1 + R_S)]}, \quad \text{(not summed)} \qquad (48)$$

7. *Uniqueness of Solution*

Initial and boundary conditions sufficient to assure a unique solution of the equations of the plated plate may be established by a theorem of the Neumann type, as in [1].

Consider two systems of five displacements ($u_1^{(0)}$, $u_2^{(0)}$, $u_3^{(0)}$; $u_1^{(1)}$, $u_3^{(1)}$), eight strains ($S_{11}^{(0)}$, $S_{33}^{(0)}$, $S_{23}^{(0)}$, $S_{31}^{(0)}$, $S_{12}^{(0)}$; $S_{11}^{(1)}$, $S_{33}^{(1)}$, $S_{13}^{(1)}$) and eight stresses ($\mathcal{T}_{11}^{(0)}$, $\mathcal{T}_{33}^{(0)}$, $\mathcal{T}_{23}^{(0)}$, $\mathcal{T}_{31}^{(0)}$, $\mathcal{T}_{12}^{(0)}$; $\mathcal{T}_{11}^{(1)}$, $\mathcal{T}_{33}^{(1)}$, $\mathcal{T}_{13}^{(1)}$) which satisfy the eight strain-displacement equations (5), the five stress-equations of motion (21), (22), and the eight stress-strain equations (24), (25). Let the differences between corresponding components in the two systems constitute a "difference system." We form the equation

$$\begin{aligned}\int_{t_0}^{t} dt \int_A \{&[\mathcal{T}_{ij,i}^{(0)} + \mathcal{F}_j^{(0)} - 2h\rho(1 + R_S)u_{j,tt}^{(0)} - 2h^2\rho R_D\delta_{ja}^0u_{a,tt}^{(1)}]u_{j,t}^{(0)} \\ &+ [\mathcal{T}_{ab,a}^{(1)} - \mathcal{T}_{2b}^{(0)} + \mathcal{F}_b^{(1)} - \tfrac{1}{3}h^3\rho(1 + 3R_S)u_{b,tt}^{(1)} - 2h^2\rho R_Du_{b,tt}]\, u_{b,t}^{(1)}\}\, dA \\ &= 0 \end{aligned} \qquad (49)$$

in which all the components are those of the difference-system, t_0 is an initial time and A is the area of the plate. Now, assembling products of accelerations and velocities, in (49), we have

$$\begin{aligned} 2h\rho \int_{t_0}^{t} &[(1 + R_S)u_{j,tt}^{(0)}u_{j,t}^{(0)} + \tfrac{1}{3}h^2(1 + 3R_S)u_{b,tt}^{(1)}u_{b,t}^{(1)} \\ &+ hR_D(\delta_{ja}^0u_{a,tt}^{(1)}u_{j,t}^{(0)} + u_{b,tt}^{(0)}u_{b,t}^{(1)})]\, dt \\ &= h\rho[(1 + R_S)u_{j,t}^{(0)}u_{j,t}^{(0)} + \tfrac{1}{3}h^2(1 + 3R_S)u_{b,t}^{(1)}u_{b,t}^{(1)} + 2hR_Du_{b,t}^{(0)}u_{b,t}^{(1)}]_{t_0}^{t} \\ &= \mathcal{K} - \mathcal{K}_0 \end{aligned} \qquad (50)$$

where $\mathcal{K}$ is the kinetic energy-density of the difference-system and $\mathcal{K}_0$ is its value at t_0. Further, by the same steps as in [1], but with T replaced by $\mathcal{T}$,

$$\begin{aligned}\int_{t_0}^{t} dt \int_A &[\mathcal{T}_{ij,i}^{(0)}u_{j,t}^{(0)} + (\mathcal{T}_{ab,a}^{(1)} - \mathcal{T}_{2b}^{(0)})u_{b,t}^{(1)}]\, dA \\ &= \int_{t_0}^{t} dt \oint_C n_a(\mathcal{T}_{aj}^{(0)}u_{j,t}^{(0)} + \mathcal{T}_{ab}^{(1)}u_{b,t}^{(1)})\, ds - \int_A (\mathcal{U} - \mathcal{U}_0)\, dA \end{aligned} \qquad (51)$$

in which the line integral is around the boundary, C, of the plate and the n_a are the direction cosines of the outward normal in the plane of the plate. Then, inserting (50) and (51) in (49), we have

$$\int_A (\mathscr{U} + \mathscr{K})\, dA = \int_A (\mathscr{U}_0 + \mathscr{K}_0)\, dA + \int_{t_0}^{t} dt \int_A (\mathscr{F}_j^{(0)} u_{j,t}^{(0)} + \mathscr{F}_b^{(1)} u_{b,t}^{(1)})\, dA$$
$$+ \int_{t_0}^{t} dt \oint_C n_a(\mathscr{T}_{aj}^{(0)} u_{j,t}^{(0)} + \mathscr{T}_{ab}^{(1)} u_{b,t}^{(1)})\, ds \qquad (52)$$

This has the same form as the corresponding equation in [1], but with T and F replaced by $\mathscr{T}$ and $\mathscr{F}$. Hence the initial and boundary conditions are analogous to those in [1]: initial values of $u_j^{(0)}$, $u_b^{(1)}$, $u_{j,t}^{(0)}$, $u_{b,t}^{(1)}$; five conditions at each point in the interior of the plate; five conditions at each point of the edge. In terms of orthogonal coordinates α, β, x_2, the interior conditions are: one member of each of the five products

$$\mathscr{F}_\alpha^{(0)} u_\alpha^{(0)}, \quad \mathscr{F}_\beta^{(0)} u_\beta^{(0)}, \quad \mathscr{F}_2^{(0)} u_2^{(0)}, \quad \mathscr{F}_\alpha^{(1)} u_\alpha^{(1)}, \quad \mathscr{F}_\beta^{(1)} u_\beta^{(1)}$$

In terms of orthogonal coordinates n, s, x_2, the edge conditions are: one member of each of the five products

$$\mathscr{T}_{nn}^{(0)} u_n^{(0)}, \quad \mathscr{T}_{ns}^{(0)} u_s^{(0)}, \quad \mathscr{T}_{n2}^{(0)} u_2^{(0)}, \quad \mathscr{T}_{nn}^{(1)} u_n^{(1)}, \quad \mathscr{T}_{ns}^{(1)} u_s^{(1)}$$

References

[1] Mindlin, R. D., "High Frequency Vibrations of Crystal Plates," *Quart. Appl. Math. 19*, 51–61 (1961).

[2] Heising, R. A., (ed.), *Quartz Crystals for Electrical Circuits* (D. Van Nostrand Co., New York, 1946), p. 340.

[3] Mason, W. P., *Piezoelectric Crystals and Their Application to Ultrasonics* (D. Van Nostrand Co., New York, 1950), p. 98.

[4] Mindlin, R. D., and Gazis, D. C., "Strong Resonances of Rectangular, AT-Cut, Quartz Plates," *Proc. 4th U. S. Nat'l. Cong. Appl. Mech.* (forthcoming).

Reprinted from
Progress in Applied Mechanics
The Prager Anniversary Volume
Macmillan, New York, (1963) 73-84

MIX
Papier aus verantwortungsvollen Quellen
Paper from responsible sources
FSC® C105338

If you have any concerns about our products,
you can contact us on
ProductSafety@springernature.com

In case Publisher is established outside the EU,
the EU authorized representative is:
Springer Nature Customer Service Center GmbH
Europaplatz 3, 69115 Heidelberg, Germany

Printed by Libri Plureos GmbH
in Hamburg, Germany